T4-AUB-744

THE EFFECTS OF RADIATION ON ELECTRONIC SYSTEMS

The Lichtenberg Fractal Figure. When a block of lucite or plexiglass is irradiated by a beam of high-energy electrons, a tremendous space charge is generated in this material. The electron beam-induced ionization produces ions and free electrons in the material, from which the space-charge electric field is manifest. The corresponding potential builds up to a few million volts to exceed the dielectric strength of the plastic. When the block is then suddenly grounded or rapped smartly, an electrical discharge ensues within, which is essentially a miniature version of a common lightning stroke. The self-similar fractal paths traversed by the currents during the discharge are left as a permanent record in the material, producing the arboreal-like rendition called a Lichtenberg figure. That the paths are self-similar fractals can be seen by a scanning electron microscope (SEM) examination of a path fragment (see inset), where a miniature replica of a portion of the arboreal pattern is revealed (cf. H. E. Stanley, ed., *On Growth and Form*: *Fractal and Non-Fractal Patterns in Physics*, Nijhoff Publ., 1987; *Sci. Amer.* Jan. 1987, p. 27). The figure on the cover was produced by a 3 MeV electron beam from a linear accelerator directed at normal incidence to the face of the block.

THE EFFECTS OF RADIATION ON ELECTRONIC SYSTEMS

SECOND EDITION

George C. Messenger
Consultant, Nuclear Radiation Effects
Las Vegas, Nevada

Milton S. Ash
Consultant, Radiation Effects
Santa Monica, California

VNR VAN NOSTRAND REINHOLD
New York

Library of Congress Catalog Card Number 91-18233
ISBN 0-442-23952-1

Manufactured in the United States of America

Published by Van Nostrand Reinhold
115 Fifth Avenue
New York, NY 10003

Chapman and Hall
2-6 Boundary Row
London, SE 1 8HN

Thomas Nelson Australia
102 Dodds Street
South Melbourne 3205
Victoria, Australia

Nelson Canada
1120 Birchmount Road
Scarborough, Ontario M1K 5G4, Canada

16 15 14 13 12 11 10 9 8 7 6 5 4 3 2 1

Library of Congress Cataloging-in-Publication Data
Messenger, George C.
The effects of radiation on electronic systems/George C. Messenger, Milton S. Ash. — 2nd ed.
p. cm.
Includes indes.
ISBN 0-442-23952-1
1. Electronic components and systems—Effect of radiation on. I. Ash, Milton S. II. Title.
TK7870.M4425 1991
621.381—dc20 91-18233
CIP

To our wives
Priscilla Messenger
and
Shulamite Ash
and our children.

PREFACE TO THE SECOND EDITION

This second edition of *The Effects of Radiation on Electronic Systems* constitutes a major overhaul of the first edition. Material has been added, deleted, revised, reformatted, and reinterpreted to make this edition as timely as possible, consistent with normal and natural time lags in the authoring, editing, and publishing milieu. Two additional chapters have been included, each an outgrowth of a corresponding section, as discussed below. They are on the subjects of Single Event Upset and Gallium Arsenide. Also, the number of exercises has been approximately doubled to include an average of about 14 problems per chapter for the first ten chapters. The later chapters contain no problems, as they have a more practical flavor of the kind that lends itself less to problem couching. Answers for all exercises are contained in Appendix A. A chapter by chapter detailing of this edition, featuring additional material, follows.

Chapter 1 (Electronic Properties of Semiconductors) has been enhanced to include a section on semiconductor device operating temperature, as well as an expanded semiconductor properties table.

Chapter 2 (The Field Equations) now includes a number of additions, and some expanded interpretation of surface states and surface potential.

Chapter 3 (P-N Junction Devices) contains an added discussion of diffusion capacitance, a tabulation of one-sided abrupt junction parameters, and a detailed discussion of *p-n* junction *I-V* curves for high and low carrier injection extremes.

Chapter 4 (Nuclear Environments) includes a second graph depicting the detonation double-pulse phenomenon. Further material consists of revisions, additions, and additional problems.

Chapter 5 (Neutron Damage Effects) contains an updated discussion of neutron-induced damage kinematics, including displacement cluster formation. A detailed graded base and carrier removal treatment are also included. Additional sections include bipolar annealing and displacement damage limits for silicon technologies.

Chapter 6 (Ionizing Radiation Damage Effects) has additional material consisting of the utility of the x-ray wafer probe, the latest theories of rebound in MOS devices, and updates on the effects of radiation on FETs in general. This includes the temporal aspects of threshold voltage shift, covering tunnelling and stochastic hopping. Discussions of the role of the capacitance-voltage (C-V) characteristics of FETs, how bipolar transistors degrade with ionizing dose, and how damage thresholds vary with dose rate are also included. A further addition includes a section on MOSFET Annealing.

Chapter 7 (Dose Rate Effects) includes a detailed expanded discussion of nonlinear photocurrent production, and further treatment of radiation effects on propagation delay time. Section 7.9 of the first edition on Single Event Upset (SEU) is removed and expanded to become the new Chapter 8. The discussion of transient effects in monolithically fabricated resistors (in integrated circuits) is also expanded to compare plugged and unplugged resistors.

Chapter 8 (Single Event Upset) is the new chapter mentioned above, which includes the latest funnelling and SEU modelling concepts. The South Atlantic Anomaly, proton-induced SEU, and other important SEU considerations are treated herein.

Chapter 9 (Electromagnetic Pulse) is based on the first edition's Chapter 8 on EMP, and includes expanded interpretations of the Wunsch-Bell model.

Chapter 10 (Dosimetry) is based on Chapter 9 of the first edition. Its extensions include detailed discussions of charge particle equilibrium and dose enhancement. Radiochromic dyes are also treated as an expansion of the properties of thermoluminescent materials (TLDs) from Table 9.3 of the first edition.

Chapter 11 (Gallium Arsenide) is the second new chapter. It is an expanded version of Section 10.12 of the first edition, on radiation effects in GaAs devices. It includes the latest information on gallium arsenide susceptibility to radiation environments, consisting of neutron fluence, ionizing dose, dose rate, and single event upset.

Chapter 12 (Component and Circuit Hardness Design) basically consists of Chapter 10 of the first edition. It features updates of the radiation threshold bar charts for parts and expanded part families for various incident radiation types. Also contained therein is an enlarged and updated tabulation of component-level SEU data. A section on the effects of radiation on noise is provided.

Chapter 13 (System Hardness Design) is an improved version of Chapter 11 of the first edition. A detailed exposition of the use of dose-depth curve specifications with a typical spacecraft example is included. The hardening of a strategic aircraft is surveyed. A nuclear reactor system is included, with a detailed practical discussion thereof. Ferro-electric memories are also discussed in the context of their susceptibility to radiation environments. Added sections discuss Power Transistors, A/D Converters, and Solar cell systems. The latest microprocessor radiation data is included in an updated Table 13.7, expanded from Chapter 11 of the first edition.

Chapter 14 (Statistical Aspects of Hardness Design) is Chapter 12 of the first edition. It is carried over to the second edition with the inclusion of some revisions and the addition of new references.

Chapter 15 (Hardness Assurance) is an expanded version of Chapter 13 of the first edition. Since its concepts and corresponding discussions are still valid, it is also carried over to the second edition essentially intact, except for the addition of a section on Hardness Maintenance.

Since the number of problems has been increased to a total of more than 130, it behooves the reader to seriously attempt to work them. Not doing the exercises denies the reader an important avenue for gaining a better insight and understanding of the subject material. If any difficulties arise in solving them, solutions to all problems are given in Appendix A.

In this rapidly evolving field, definitional changes are frequent. One such is "total dose." We have avoided that term herein in favor of the more accurate descriptor "ionizing dose." It is now established that ionizing dose failure levels of components are strong functions of the dose rate, so that both the ionizing dose failure threshold at the corresponding dose rate should be specified. It is strongly felt that the term "total dose" diffuses this important aspect of the definition.

It is with gratitude that we acknowledge the contributions of the many people who helped us prepare this second edition. M. Espig, M. Rose and M. Shoga, among others, graciously made available the current radiation data that we present in the appropriate chapters. Friends and colleagues, too numerous to mention, know that we thank them very sincerely with the warmest of regards for their contributions.

G. C. Messenger, Las Vegas
M. S. Ash, Santa Monica

PREFACE TO THE FIRST EDITION

Radiation effects on electronic systems as a recognizable discipline in electronic and electrical engineering has come of age in little more than the past thirty years. This discipline has become divided into two principal areas, termed Transient Radiation Effects on Electronics (TREE) and Electromagnetic Pulse (EMP). This book treats mainly TREE, but does include a chapter on EMP.

More than 85 percent of the open reference literature is represented by the principal journal, *IEEE Transactions on Nuclear Science*, NS-10, No. 5 (Nov. 1963), to date, a veritable fount of information and data on the subject; and a small number of books treating aspects of this subject or devoted to its entirety.

This book takes a modern engineering approach to TREE, while including the pillars of the subject dating from its embryonic period. It also endeavors to blend the required knowledge from solid-state and nuclear physics, as well as electronics, into a cohesive entity. Problems are included at the end of the first ten chapters as an aid for self-study or as course material. Working the problems affords the serious reader an important avenue for gaining additional insight and understanding of the subject material.

The manuscript has been used in an annual succession of senior or first-year graduate courses and concentrated short courses in Nuclear Hardening, given at the UCLA Engineering Department and Extension Division. It is suggested that about nine chapters provide enough material for a robust one quarter or one semester course. For example, a combination of Chapters 1, 3–8, 12 and 13 is one alternative. A respectable bachelor of science level in mathematics is assumed, as the applicable mathematical formulations are mainly presented with few derivational steps, whose inclusion would have easily doubled the number of pages and made the book unwieldy.

The material in this electronic engineering text should orient both the student and neophyte, as well as reacquaint and bring up to date the practitioner, enabling most to read the current literature with facility. Girded with the main ideas presented herein, one can come to grips with the problem areas in this burgeoning field. Although this book can certainly be used for reference, it is not meant to be encyclopedic. Such treatises are available, and references are cited at the end of the appropriate chapters.

A certain repetitiveness is used for emphasis, and is employed as a pedagogic device throughout the book. For example, early chapters discuss how neutrons and gamma rays damage semiconductors; later chapters then describe their translation into corresponding damage at the device, circuit, and system levels.

Chapter 1 discusses the electronic and solid-state properties of semiconductors, thereby providing an introduction to the subject matter.

Chapter 2 continues the introductory discussion, which in part consists of a refresher in electromagnetism couched to be helpful in Chapter 8 on EMP.

Chapter 3 discusses semiconductor junction devices, the most important recipients of radiation damage in the context of this book.

Chapter 4, a change of pace, describes the close-in effects of nuclear weapons, which not only provides insight into one *raison d'etre* of this subject, but contains material not usually found in books on the effects of nuclear weapons.

Chapter 5 follows the preceding important background material, and begins the central portion of the book. This chapter discusses neutron damage effects in semiconductor devices, mainly bipolars, including a derivation of the Messenger-Spratt expression for common emitter current gain degradation in bipolar transistors due to neutron damage. This is one of the few theoretical results developed in this field to date.

Chapter 6 discusses ionizing radiation damage due to x-rays and gamma rays in semiconductor devices, principally MOS transistors.

Chapter 7 describes ionizing dose rate effects on most active devices of interest. This chapter also includes a section on single event upset, which presently is enjoying a surge of interest in the technical community, and will continue to do so as the packing density of integrated circuits increases.

Chapter 8 discusses the electromagnetic pulse (EMP) output of a nuclear weapon and its effects on electronic systems. EMP can produce geographically far-reaching and dramatically deleterious effects on such systems.

Chapter 9 is on dosimetry, which is very important because of the continuing need for testing semiconductor components using nuclear radiation simulation sources to determine their radiation hardness.

Chapter 10 begins the reprise from the earlier chapters, here in the discussion of nuclear vulnerability and hardness of active devices at the component and circuit level. A set of corresponding hardening methods is also included.

Chapter 11 continues by expanding to system-hardening design, including examples from spacecraft, strategic, and tactical systems. A series of system hardening methods is also discussed.

Chapter 12 describes the statistical aspects of hardening design. This is vital when one attempts to extract the maximum amount of post-radiation information from, unfortunately, often less than a half-dozen part samples.

Chapter 13 presents the subject of hardness assurance. Hardness assurance and hardness maintenance are two very necessary methodologies utilized to ascertain that required systems-hardening indeed exists, and is maintained in the field throughout the life of the system.

Heartfelt thanks and acknowledgments are due to many people in this endeavor. Besides the generosity of the good offices of C. Kleiner, Radiation Effects Consultant, Anaheim; C. Gorton, Section Chief, Component Engineering Department, TRW; and O. Adams, Manager, Electronic Hardness Department, TRW, Redondo Beach, a number of others should be acknowledged: Drs. M. Cooper, R. McCoskey, C. Nielsen, and Ms. B. White, who patiently read the manuscript for content as well as grammar. Further, assiduous reading of the manuscript by the Electronic Vulnerability Division of the Defense Nuclear Agency (DNA) is greatly appreciated. To all others, a thousand thanks.

G. C. Messenger, Las Vegas
M. S. Ash, Santa Monica

CONTENTS

THE EFFECTS OF RADIATION ON ELECTRONIC SYSTEMS

CHAPTER 1
ELECTRONIC PROPERTIES OF SEMICONDUCTORS

1.1 INTRODUCTION

Modern electronic systems are principally composed of semiconductor devices, which are made primarily of silicon. In recent years, gallium arsenide is also being increasingly used to manufacture such devices. Gallium Arsenide is discussed in Chapter 11. This chapter will describe the pertinent electrical properties of mainly silicon from a physics viewpoint. The properties of other semiconductor materials are sufficiently similar to silicon that it can be construed as a typical semiconductor herein. This description is necessary to provide a basis for investigating the electronic behavior of semiconductor devices that are subject to nuclear radiation as a function of their basic physical structure, applied electrical bias, and signal voltages.

1.2 CRYSTAL STRUCTURE

Silicon, germanium, gallium arsenide, and other semiconductor materials usually have a crystalline structure, as opposed to an amorphous character. This crystal structure plays a central role in the energy dynamics of their operation as semiconductors. Sufficient detail of their crystalline behavior will be provided in the following sections to support a description of their function as semiconductors. Further details are beyond the scope of this book and can be found in many available books on the theory of the solid state[1] or of condensed matter.

The crystalline structure implies a periodic physical composition. That is, these materials can be characterized by a basic unit cell that is replicated ad infinitum throughout the material in all directions. There are various types of basic unit cells, such as face-centered cubic, body-centered cubic, hexagonal close-packed, diamond, etc. Silicon is representative of the diamond lattice, which is the interpenetration of two face-centered cubic lattices, as is the zincblende lattice of gallium arsenide, shown in Fig. 1.1.

Crystalline silicon, germanium, and gallium arsenide share with diamond the crystal structure called interlocking face-centered cubic. Atoms

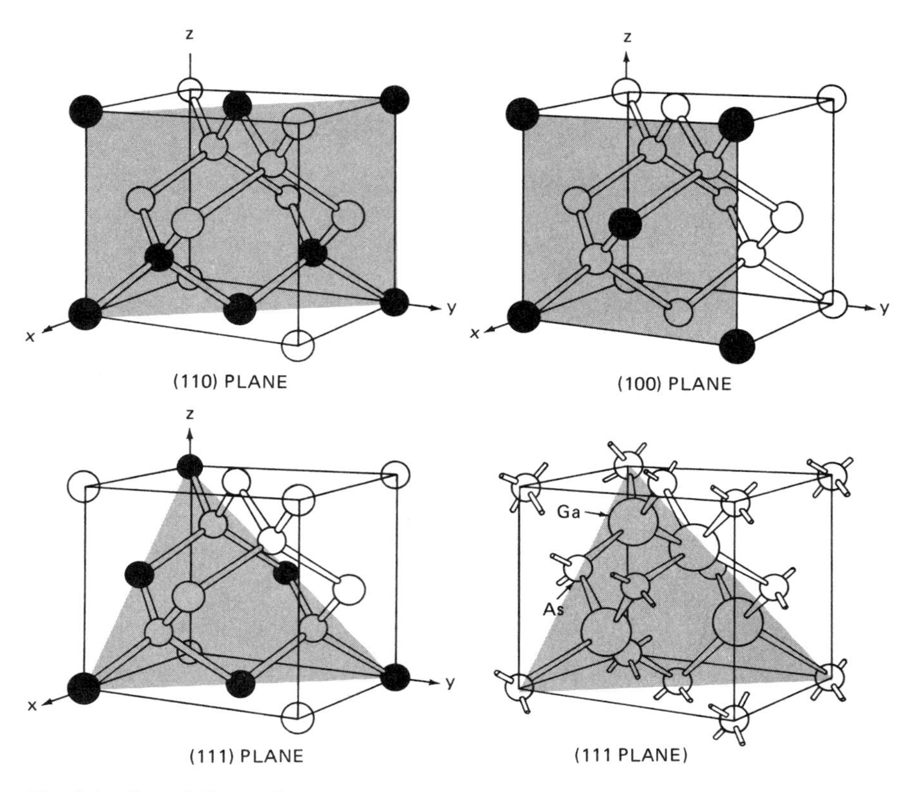

Fig. 1.1 Crystal diamond structure of silicon and germanium (top and lower left) and zincblende structure of gallium arsenide (lower right).

lie at each corner and center of each face of the unit cube. The cubes interlock, such that atoms from neighboring cubes are contained within each unit cube. The double connectors between atoms depict the tetrahedral atomic bond configuration. The unit cube axes are a rectilinear coordinate space within which directions and planes can be specified. A crystal direction is given by three coordinate numbers, called *Miller indices*, which are multiples of a unit cube length.

The connectors between the atoms making up the crystal structure, as shown in Fig. 1.1, should be construed as homopolar or covalent bonds, each consisting of a pair of valence electrons. Each atom in the crystal is seen bonded to its neighbors by four pairs of these valence electrons. For example, for germanium and silicon crystal atoms, each possesses one pair of valence electrons, as depicted in Fig. 1.6, while the other three pairs come from its neighboring atoms, as in Fig. 1.1. The details of how the electrons provide this homopolar bonding are quantum mechanical in nature. As will be seen later, an immediate form of photon-induced damage to semiconductors consists of the fracturing of these bonds, thus releasing these electrons. Their release results in changes in certain macroscopic semiconductor parameters, such as conductivity.

1.3 ENERGY LEVELS

To develop the concept of electronic energy bands and available energy states for electrons within the semiconductor of interest, it is necessary to examine electron motion in a periodic crystal structure. This can be discussed, devoid of the modern quantum mechanical apparatus, but in sufficient detail to obtain a somewhat quantitative measure of the electron kinematics and energetics involved. Nevertheless, some simple pre-quantum mechanical models of the electron are in order. Almost immediately after the birth of quantum theory, as heralded by the Planck treatment of the black body spectrum, Bohr successfully described the quantized energy levels of the electron in the hydrogen atom using a simple model, which is appropriate here. His model asserts that in a hydrogen-like atom, occupying column I in the periodic table of elements, the following rules apply, as illustrated by the hydrogen atom itself.

1. The negative electron orbits the positive proton nucleus like a satellite. Its centrifugal force is balanced, not by gravity as in a terrestrial satellite, but by its electric Coulomb attraction to the central proton nucleus, given by e^2/r^2. Thus

$$mv^2/r = e^2/r^2 \tag{1.1}$$

where e is the electron charge, r is the radius of the orbit, m is the electron mass, and v is its tangential velocity.

2. An ad hoc quantization of the angular momentum of the electron, i.e., the vector cross product $\bar{L} = \bar{r} \times \bar{p}$, where $\bar{p}$ is the linear momentum, yields for our purposes the scalar angular momentum L, which satisfies

$$L = mvr = n\hbar; \quad n = 1, 2, 3, \ldots \tag{1.2}$$

where $\hbar = h/2\pi$, and Planck's constant $h = 6.6 \cdot 10^{-27}$ erg seconds. That is, the scalar angular momentum can take on only integer multiples of $\hbar$.

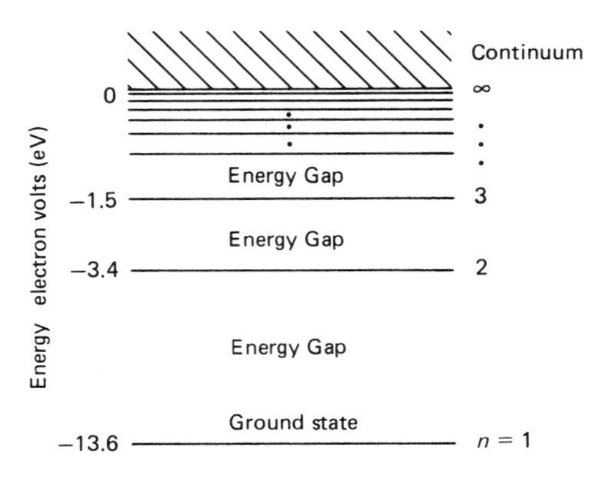

Fig. 1.2 Hierarchy of electron energy states in the hydrogen atom as predicted from the Bohr model.

3. The total energy (kinetic plus potential) of the orbiting electron is simply

$$E = \tfrac{1}{2}mv^2 - e^2/r \tag{1.3}$$

If v and r are eliminated between Eqs. (1.1)–(1.3), the resulting expression for the allowed values of the total energy is obtained to yield

$$E_n = \frac{-me^4}{2\hbar^2 n^2}; \quad n = 1, 2, 3 \ldots \tag{1.4}$$

This asserts that the total energy of the electron is quantized into allowed energy states, called *stationary states* by Bohr. This is depicted in Fig. 1.2, which is an energy level diagram and not a plot of energy versus some variable. It illustrates the allowed electron energy levels and the energy gaps between. The zero of electron energy is seen in Fig. 1.2 to be at infinite values of the quantum number n. This is the traditional manner of energy reference for bound systems, such as the electron bound by the hydrogen nucleus. Hence, the allowed energy levels of the bound electron are negative, which also simply means that an external agent must expend energy to remove the electron from this bound system. In general, for hydrogen-like atoms of atomic number Z, the preceding development gives

$$E_n = \frac{-me^4 Z^2}{2\hbar^2 n^2}; \quad n = 1, 2, \ldots \tag{1.5}$$

For positive values of energy, the electron is no longer bound and is thus free to move within the continuum of energies above the $n = \infty$ level. The Bohr model also asserts that if a bound atomic electron falls from a higher (large n) to a lower ($m < n$) energy state in this hierarchy, it gives up its energy difference $\Delta E_{nm} = h\nu_{nm}$ in the form of a photon with this energy. ν_{nm} is the frequency of the photon radiation corresponding to this energy drop. In certain instances, this energy difference ΔE_{nm} is transferred to another electron in the same atom instead of to a photon. This type of radiationless energy transfer is called an *Auger process*.

Eliminating v and E within Eqs. (1.1)–(1.4) yields the allowed Bohr radii of the orbiting electron as

$$r_n = (\hbar^2/me^2)n^2; \quad n = 1, 2, 3, \ldots \tag{1.6}$$

Correspondingly, the allowed values of the electron linear momentum are obtained similarly, using $p = mv$, to yield

$$p_n = (2mE_n)^{1/2} = me^2/\hbar n; \quad n = 1, 2 \ldots \tag{1.7}$$

Still in the pre-quantum mechanical era, de Broglie postulated that all elementary particles exhibit both wave and particle character, but not both simultaneously. This was proven by experiment, and the connection between linear momentum and particle wavelength that resulted is given by the seemingly trivial but fundamentally important expression, viz.,

$$p = h/\lambda \tag{1.8}$$

If a propagation constant describing the wavelike movement of the electron is defined by $k = 2\pi/\lambda$, then Eq. (1.8) becomes

$$p = \hbar k \tag{1.9}$$

Actually these relations are all scalar representions of the electron linear momentum vector, $\bar{p}$ and its propagation vector $\bar{k} = 2\pi\bar{n}/\lambda$, where $\bar{n}$ is a unit vector in the direction of motion of the electron. The corresponding expression for the kinetic energy then becomes

$$E_n = \frac{p^2}{2m} = \frac{\hbar^2}{2m}k_n^2 \tag{1.10}$$

Because the electron linear momentum is quantized, as given by Eq. (1.7), so is the propagation constant k_n. All this is depicted in energy-

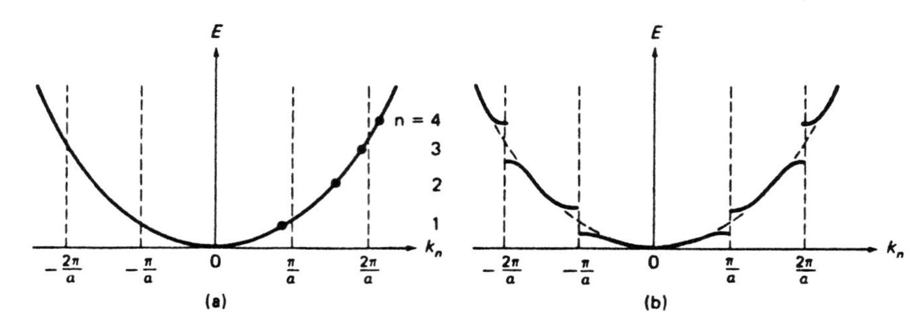

Fig. 1.3(a) Electron energy for free electrons (solid line) and bound electrons (dots).

Fig. 1.3(b) Electron energy for bound electrons in a lattice.

momentum coordinates in Fig. 1.3. In modern quantum theory, the primary particle variables are the energy E_n and the momentum represented by k_n. Corresponding velocities, radii, etc., play only minor roles in the quantum formulation. From the preceding, it is seen that free electrons or any free particle can possess any value of energy, though, for the bound electron, the energy is quantized with forbidden energy gaps between allowed energy states.

1.4 ENERGY BANDS

When many atoms combine to form a crystal lattice, the situation becomes sufficiently complicated that only modern quantum mechanics can produce the correct quantization schema, as verified by experiment. That quantization will only be sketched in what follows. The periodicity of the unit cell can be described geometrically by three basis vectors, $\bar{a}_1$, $\bar{a}_2$, $\bar{a}_3$, which are generally not mutually orthogonal. The lengths of these vectors, $|\bar{a}_1|$, $|\bar{a}_2|$, $|\bar{a}_3|$, are the interatomic dimensions of the unit cell. The volume of the unit cell is given by $V_c = |\bar{a}_1 \times \bar{a}_2 \cdot \bar{a}_3|$, which is simply the formula for the volume of a parallelepiped having these dimensions. As can be appreciated, the coordinates of every cell point in the lattice, composed of unit cells, can be represented by a ray vector, $\bar{r}_i$, from a suitable origin to the ith cell position and defined by

$$\bar{r}_i = m_i\bar{a}_1 + n_i\bar{a}_2 + p_i\bar{a}_3 \tag{1.11}$$

where m_i, n_i, and p_i are integers unique to the cell position. For reasons that will become clear, another set of ray vectors, $\bar{K}_j$, is defined through the relation

$$\bar{r}_i \cdot \bar{K}_j = 2\pi l_{ij} \tag{1.12}$$

where the l_{ij} are also integers. The vectors $\bar{K}_j$ represent all coordinate axes in what is termed the *reciprocal lattice*. If the basis vectors in the reciprocal lattice are $\bar{b}_1$, $\bar{b}_2$, $\bar{b}_3$, then, as in the real lattice

$$\bar{K}_j = g_j\bar{b}_1 + h_j\bar{b}_2 + k_j\bar{b}_3 \tag{1.13}$$

where g_j, h_j, and k_j are also integers unique to the reciprocal lattice cell point. The defining relationship between the actual and the reciprocal lattice, Eq. (1.12), implies that the scalar products between basis vectors of the two lattices satisfy

$$\bar{a}_i \cdot \bar{b}_j = 2\pi\delta_{ij} \tag{1.14}$$

where δ_{ij} is the Kronecker delta ($\delta_{ij} = 1$ for $i = j$, and 0 for $i \neq j$). A second relation between them follows, as given by

$$\bar{b}_1 = 2\pi\frac{\bar{a}_2 \times \bar{a}_3}{V_c}; \quad \bar{b}_2 = 2\pi\frac{\bar{a}_3 \times \bar{a}_1}{V_c}; \quad \bar{b}_3 = 2\pi\frac{\bar{a}_1 \times \bar{a}_2}{V_c} \tag{1.15}$$

The point of defining a reciprocal lattice is that, as can be shown from quantum theory, the reciprocal lattice corresponds to the energy-momentum lattice space. Recall the Fig. 1.3 plots, which are also in energy-momentum space for the Bohr atom. The unit cell edges in the reciprocal lattice correspond, not to interatomic distances, but to boundaries of periodic behavior of the allowed electron energy-momentum bands and gaps between bands in the reciprocal lattice. Such periodic behavior is a consequence of the quantum mechanical treatment of a unit lattice cell whose interatomic potential energy variation is periodic throughout the lattice. The energy-momentum periodicity in the reciprocal lattice is also replicated throughout it by virtue of the unit reciprocal lattice cells, called Brillouin zones, as in Fig. 1.3(b). Figure 1.4 portrays the energy-momentum description that resides in each Brillouin zone in silicon. The curves in Fig. 1.4 represent boundaries between allowed and forbidden electron states of energy within the crystal. The semiconductor action depends principally on the interaction between the lowermost nearly filled energy band of states, defined as the *valence band*, and the next, nearly empty, energy band of states immediately above the valence band. That band is defined as the *conduction band*. Between the two bands lies the forbidden energy gap, which means that it

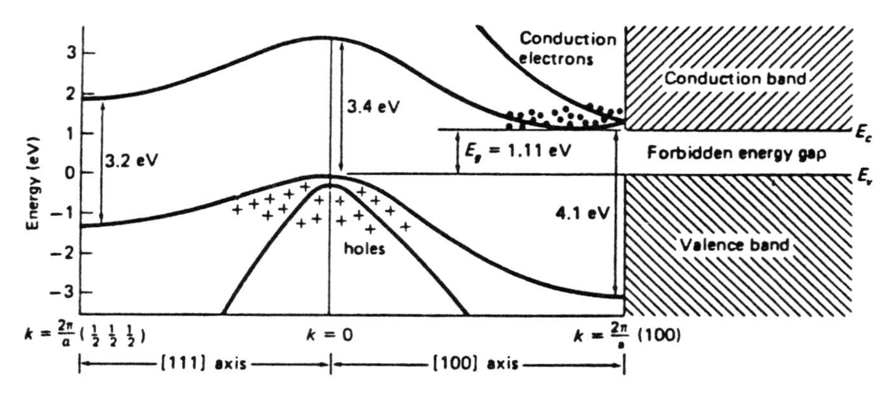

Fig. 1.4 The band structure of silicon depicting the energy gap of forbidden energy states. The triplets of numbers on the momentum abscissa show the Miller indices and axes, which are crystallographic parameters.[2] The rightmost portion depicts the stylized version of the band structure, as commonly used in discussion.

is forbidden to contain energy states, which is not universally true, as will be seen. This energy gap is about 1.11 eV wide in silicon.

The valence band simultaneously comprises the bound energy levels of the valence electrons. These electrons form the supporting bonds between neighboring atoms that make up the semiconductor crystalline structure discussed in Section 1.2. On the other hand, the conduction band consists of the continuum of energies of the unbound or free electrons in the semiconductor, which contribute to electrical conduction.

For whatever reason, if an electron is excited from the valence band, thus leaving that band to jump across the gap into the conduction band, the electron leaves a vacancy behind in the valence band where it once was energetically situated. Within the valence band another electron can jump into the vacancy. In turn, a third electron in the valence band can jump into the vacancy created by the second electron, etc. Thus, this vacancy is propagated in a direction opposite to these electrons in the valence band in domino fashion. From the quantum mechanical viewpoint, it is found that a satisfactory way of treating these vacancies is to introduce another particle to represent them. This particle is called a *hole*. It is assumed to have essentially the same mass, as well as identical other low-energy properties of the electron, except that its charge is taken as positive. This should not be confused with the positron, another electron-like, but real, particle with positive charge. It occurs only in the very relativistic high energy domain of gamma ray pair-production and pair-annihilation. The solid state energy domain under discussion corresponds to the order of but a few electron volts. Now, the semiconductor particle kinematics and energetics devolves to the motion of electrons and holes within and between the valence and conduction bands.

1.5 FERMI DISTRIBUTION, BAND-GAP ENERGY, FERMI LEVEL

To become somewhat more quantitative regarding the energetics of electrons and holes in the valence and conduction bands, it is necessary to determine (a) the number density of electron and hole states in the semiconductor, and (b) the energy state occupation probability of these electrons and holes. From the integral of the product of (a) and (b) over all pertinent energies, the actual number density of electrons and holes in the semiconductor can be obtained. Quantum theory asserts that electrons and holes obey Fermi-Dirac statistics with regard to their population dynamics. As such, the electron energy state occupation probability distribution, $f(E)$, is given by

$$f(E)\,dE = \frac{dE}{1 + \exp(E - E_f)/kT} \tag{1.16}$$

$f(E)\,dE$ is the probability that an electron occupies an energy state E within the infinitesimal energy increment dE, where T is the absolute temperature of the semiconductor, the Boltzmann constant $k = 1.38 \cdot 10^{-16}$ ergs per °K, and E_f is the Fermi energy level, also called the *chemical potential*. It is seen that $f(E_f) = \frac{1}{2}$. It will be shown that the Fermi energy level E_f lies in the forbidden energy gap. The Fermi-Dirac distribution function is shown in Fig. 1.5. $f(E)$ can also be interpreted as the fraction of levels at a given energy that are occupied when the system is in thermal equilibrium.

The number density of states, $N_e(E)$, or the distribution of available states for an electron to occupy near the bottom of the conduction band, is also well known from quantum theory and is given by

$$N_e(E) = M\sqrt{2(E - E_c)}(m_{de}^*)^{3/2}/\pi^2\hbar^3; \quad E \geqslant E_c \tag{1.17}$$

where M is the number of conduction band minima in the energy momentum reciprocal lattice space. M is often suppressed by including it in the definition of the effective mass m^*. E_c is the energy at the bottom of the conduction band, and m_{de}^* is the density of states effective mass for electrons. A digression to explain the concept of effective mass is appropriate here.[3] As can be appreciated, the detailed description of the

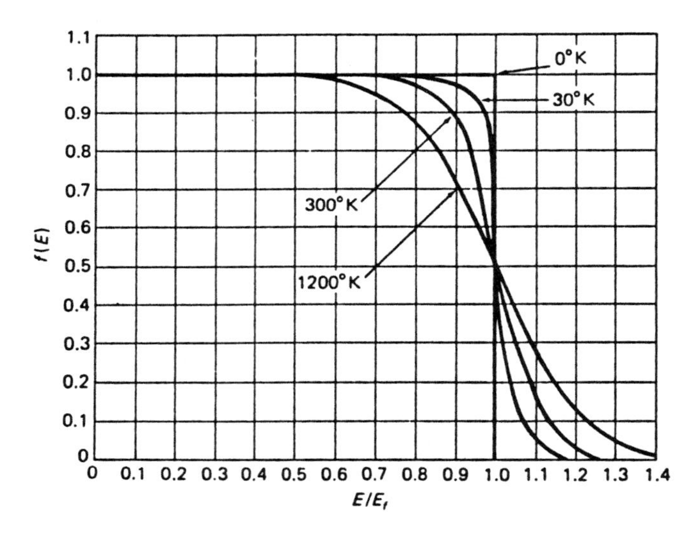

Fig. 1.5 Fermi-Dirac distribution function for various absolute temperatures.

phenomena under discussion has a quantum mechanical basis. To use many of the classical constructs, such as momentum, energy, etc., an effective mass can be defined that forms a bridge between classical and quantum mechanical descriptions for our purposes. In quantum mechanics, most particles, including the electron, can be construed as a superposition of waves. The phase of each wave is such that they add up to represent the electron position, at a given speed and time. They cancel each other everywhere else. This packet of waves representing the electron has a group velocity equal to the velocity that the electron has, and the individual waves of the packet have a distribution of velocities of their own. It is easily shown that the group velocity, v_g, is given by

$$v_g = \frac{d\omega}{dk} = \frac{dE}{\hbar\, dk} \tag{1.18}$$

where $E = \hbar\omega$; $\omega = kv(k)$, and $v(k)$ is the speed of the individual wave, each having a propagation constant k, assuming a one-dimensional space for the packet. Suppose that a force due to an impressed electric field E_x accelerates the electron. From Eq. (1.18), the acceleration is given by

$$\dot{v}_g = \frac{d^2E}{dk^2}\dot{k}/\hbar \tag{1.19}$$

where the overdot signifies differentiation with respect to time. From Newton's second law, the force $F_x = -eE_x = \dot{p}_x = \hbar\dot{k}$, so that

$$\dot{v}_g = \frac{d^2 E}{dk^2} F_x/\hbar^2 = \frac{F_x}{m^*} \tag{1.20}$$

Hence an effective mass can be defined by

$$m^* = \frac{\hbar^2}{d^2 E/dk^2} \tag{1.21}$$

In three dimensions, the above development results in an effective mass that is a tensor, e.g., with an xy component

$$(1/m^*)_{xy} = \frac{\partial^2 E}{\hbar^2 \partial k_x \partial k_y} \tag{1.22}$$

The tensor behavior is because the momentum components have different values in different directions (nonisotropic) in the reciprocal lattice. From the preceding basic definition, the various types of scalar effective masses can be derived. For example, the effective masses used in the density of state computations are, where subscripts e and h designate electron and hole respectively,

$$m_{de}^* = (m_{\parallel}^* m_{\perp}^{*2})^{1/3}; \quad m_{dh}^* = (m_{lh}^{3/2} + m_{hh}^{3/2})^{2/3} \tag{1.23}$$

and an effective mass used in conductivity computations is

$$m_{ce}^* = \frac{3m_{\parallel}^* m_{\perp}^*}{(2m_{\parallel}^* + m_{\perp}^*)}; \quad m_{ch}^* = \frac{m_{lh}^{5/2} + m_{hh}^{5/2}}{m_{lh}^{3/2} + m_{hh}^{3/2}} \tag{1.24}$$

$m_{\parallel}^*$, $m_{\perp}^*$, m_{lh}, and m_{hh} are mass components computed from the above tensor expression given in Eq. (1.22) in lattices of interest.[4]

The number density of hole states, $N_h(E)$, is analogous to $N_e(E)$ and is given by the expression

$$N_h(E) = \frac{2\sqrt{(E_v - E)}\, m_{dh}^{*3/2}}{\pi^2 \hbar^3}, \quad E \leqslant E_v \tag{1.25}$$

where E_v is the energy level at the top of the valence band, and m_{dh}^* is the effective hole mass defined in the same sense as m_{de}^*. In the conduction band then, the actual number density of electrons of a given energy E is gotten from the product of the available conduction band states, $N_e(E)$, and the probability that an electron occupies a particular energy state, namely $f(E)$. Similarly, the hole number density of energy E in the valence band is $N_h(E)(1 - f(E))$, since a hole represents the absence of an electron, and that the probability of a state not occupied by an electron (and so occupied by a hole) is $1 - f(E)$. The total number of occupied conduction band energy levels, i.e., the number density of conduction electrons, n_0, is given by

$$n_0 = \int_{E_c}^{\infty} N_e(E) f(E)\, dE \tag{1.26}$$

The upper limit of the above integral should be taken at the top of the conduction band. But the severe cutoff property of $f(E)$, as seen in Fig. 1.5, makes n_0 depend only weakly on the upper limit, so that it is taken as infinity for convenience of integration. For the usual cases where the Fermi level E_f is greater than $3kT$ below the conduction band edge in the energy gap i.e., $E_c - E_f \gg 3kT$, then using Eqs. (1.16) and (1.17) in Eq. (1.26) simplifies the integral to yield

$$n_0 = N_c \exp(E_f - E_c)/kT, \quad N_c = 2(2\pi m_{de}^* kT/h^2)^{3/2} \tag{1.27}$$

For the holes in the valence band, symmetrical results ensue when $E_f - E_v \gg 3kT$, in that for p_0 defined as the total number of occupied valence band levels, i.e., the number density of holes in the valence band, then

$$p_0 = \int_{-\infty}^{E_v} N_h(E)(1 - f(E))\, dE \tag{1.28}$$

and evaluating the integral yields

$$p_0 = N_v \exp\frac{E_v - E_f}{kT}, \quad N_v = 2(2\pi m_{dh}^* kT/h^2)^{3/2} \tag{1.29}$$

Thus far, only pure semiconductors have been considered. That is, any impurities are considered to be of negligible concentration. Such semiconductors are termed *intrinsic*. For intrinsic semiconductors, the only carriers (generic term for free electrons and holes) present are electrons that have been excited across the forbidden energy gap to the conduction band, leaving holes in the valence band. From this definition then,

$n_0 = p_0$, and their product is a constant of the material and depends on the temperature only. Or

$$n_0 p_0 = n_i^2 \tag{1.30}$$

n_i is called the *intrinsic carrier density*. This relation is very general in that under equilibrium conditions, $np = n_i^2$ holds for doped semiconductors as well, as will be seen. n_i for a particular material can be computed by inserting Eqs. (1.27) and (1.29) into Eq. (1.30) to give

$$n_i^2 = n_0 p_0 = N_c N_v \exp\frac{-E_g}{kT} = 4(2\pi kT/h^2)^3 (m_{de}^* m_{dh}^*)^{3/2} \exp\frac{-E_g}{kT} \tag{1.31}$$

where $E_g = E_c - E_v$ is the energy difference of the forbidden energy gap width. Or

$$n_i = 2\left(\frac{2\pi k}{h^2}\right)^{3/2} (m_{de}^* m_{dh}^*)^{3/4} T^{3/2} \exp\frac{-E_g}{2kT} \tag{1.32}$$

For silicon at 300°K, $n_i \cong 1.4 \cdot 10^{10}$ carriers per cm^3. For the intrinsic semiconductor, an explicit expression for the intrinsic Fermi energy level E_f can be obtained using Eqs. (1.27) and (1.29) and $p_0 = n_0$. It is

$$E_f = \tfrac{1}{2}(E_c + E_v) + (3kT/4)\ln(m_{dh}^*/m_{de}^*) \tag{1.33}$$

1.6 EXTRINSIC CARRIER DENSITIES, DONORS, ACCEPTORS

For practical semiconductor operation, very much higher carrier densities than can be obtained from intrinsic carrier generation are required. This can be accomplished by purposely introducing impurities into the pure silicon, i.e., "doping" the semiconductor crystal, thus rendering it extrinsic. As an example, if silicon is doped with phosphorous, a group V element in the periodic table, as seen in Fig. 1.6, the phosphorous atoms are diffused into the silicon lattice and substitute themselves for some of the silicon atoms in the unit crystal cells. Because the outer electron shell of phosphorous contains five electrons, four of these reform the tetravalent bond structure within the silicon lattice. The fifth electron, now being weakly bound, is easily ionized from its phosphorous atom, and so available to supplement the conduction electron carrier density within the silicon. The diffusion of impurities into the silicon results in the replace-

	Ia	IIa	Ib	IIb	III	IV	V	VI	VII
PERIOD 2	LITHIUM Li 3 2*s* 1 CORE 2	BERYLLIUM Be 4 2*s* 2 CORE 2			BORON B (ACCEPTOR) 5 2*p* 1 2*s* 2 CORE 2	CARBON C 6 2*p* 2 2*s* 2 CORE 2	NITROGEN N 7 2*p* 3 2*s* 2 CORE 2	OXYGEN O 8 2*p* 4 2*s* 2 CORE 2	FLUORINE F 9 2*p* 5 2*s* 2 CORE 2
PERIOD 3	SODIUM Na 11 3*s* 1 CORE 10	MAGNESIUM Mg 12 3*s* 2 CORE 10			ALUMINUM Al (ACCEPTOR) 13 3*p* 1 3*s* 2 CORE 10	SILICON Si 14 3*p* 2 3*s* 2 CORE 10	PHOSPHORUS P (DONOR) 15 3*p* 3 3*s* 2 CORE 10	SULFUR S 16 3*p* 4 3*s* 2 CORE 10	CHLORINE Cl 17 3*p* 5 3*s* 2 CORE 10
PERIOD 4	POTASSIUM K 19 4*s* 1 CORE 18	CALCIUM Ca 20 4*s* 2 CORE 18	COPPER Cu 29 4*s* 1 CORE 28	ZINC Zn 30 4*s* 2 CORE 28	GALLIUM Ga (ACCEPTOR) 31 4*p* 1 4*s* 2 CORE 28	GERMANIUM Ge 32 4*p* 2 4*s* 2 CORE 28	ARSENIC As (DONOR) 33 4*p* 3 4*s* 2 CORE 28	SELENIUM Se 34 4*p* 4 4*s* 2 CORE 28	BROMINE Br 35 4*p* 5 4*s* 2 CORE 28
PERIOD 5	RUBIDIUM Rb 37 5*s* 1 CORE 36	STRONTIUM Sr 38 5*s* 2 CORE 36	SILVER Ag 47 5*s* 1 CORE 46	CADMIUM Cd 48 5*s* 2 CORE 46	INDIUM In (ACCEPTOR) 49 5*p* 1 5*s* 2 CORE 46	TIN Sn 50 5*p* 2 5*s* 2 CORE 46	ANTIMONY Sb (DONOR) 51 5*p* 3 5*s* 2 CORE 46	TELLURIUM Te 52 5*p* 4 5*s* 2 CORE 46	IODINE I 53 5*p* 5 5*s* 2 CORE 46
PERIOD 6	CESIUM Cs 55 6*s* 1 CORE 54	BARIUM Ba 56 6*s* 2 CORE 54	GOLD Au 79 6*s* 1 CORE 78	MERCURY Hg 80 6*s* 2 CORE 78	THALLIUM Tl 81 6*p* 1 6*s* 2 CORE 78	LEAD Pb 82 6*p* 2 6*s* 2 CORE 78	BISMUTH Bi 83 6*p* 3 6*s* 2 CORE 78	POLONIUM Po 84 6*p* 4 6*s* 2 CORE 78	ASTATINE At 85 6*p* 5 6*s* 2 CORE 78

Fig. 1.6 Partial periodic table of the elements gives the structure of the core and valence electrons for the elements having the simplest repetitive valence properties. Partially filled energy levels for the valence electrons are shown as broken lines and filled valence electron energy levels and core-electron levels are shown as solid lines. The number of electrons in each energy level is given above each line and the symbols to the left of each line are the traditional names of the atomic orbitals or energy levels. The number in the upper right of each cell is the atomic number of the element, and the abbreviation in the upper left is the chemical symbol. Because the elements in a given column of the figure have similar valence-electron configurations, they also have similar chemical properties.

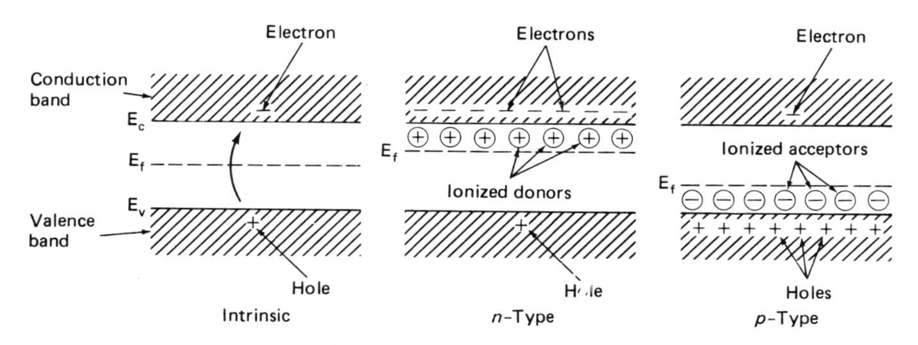

Fig. 1.7 Energy level positions of donors and acceptors and Fermi level energy shift to accommodate presence of donors or acceptors to ensure charge neutrality in the semiconductor.[4] Figure © 1973 by John Wiley & Sons.

ment of about one silicon atom in 10^5-10^{10}. As there are about $5 \cdot 10^{22}$ silicon atoms per cm^3, doping increases the concentration of free electrons from its intrinsic value of $n_i = 1.4 \cdot 10^{10}$ to about $10^{12}-10^{20}$, i.e., from about 100 to about 10 billion times! Impurity concentrations can be achieved up to the solid solubility limit, which depends on the size and valence of the impurity atom. For silicon, acceptor and donor density limits are about $10^{19}-10^{21}$ dopant atoms per cm^3. Because group V impurities add electrons to the silicon lattice, they are called *donor impurities*. The correspondingly doped semiconductor is then termed *n-type material*. Similarly, if a group III impurity atom, such as boron or aluminum, is substituted for a silicon atom in the lattice, then the resulting tetravalent bonding is reconstructed somewhat differently, as the group III impurity atom has but three electrons in its outer shell, as depicted in Fig. 1.6.

Both group III and group V impurities supply additional energy states simply by virtue of their presence in the semiconductor. By implication, this means that additional conduction electrons or additional holes are introduced, depending on whether they are group V or III atoms, respectively. These additional states for both groups lie in the forbidden gap, but very close to the gap edges, as shown in Fig. 1.7. That is, for group III impurities, the additional states lie in the gap, very close to but just above the top of the valence band. For group V impurities, the additional states also lie in the gap, very close to but just under the bottom of the conduction band. In the case of the group III energy levels, it is then very easy for a group III atom to capture or trap (Section 1.9), the needed electron from the valence band to reconstruct the corresponding tetravalent bonding in its unit cell. This then leaves a hole near the top of the valence band. With group III-type doping, the hole concentration is thereby greatly enhanced, analogous to that for group V doping. The resulting semiconductor, doped with electron acceptor atoms is called *p-type material.*

A donor state or donor trap energy level is defined as neutral if filled by an electron, and positive if vacant. An acceptor trap level, however, is defined as negative if filled by an electron and neutral if vacant. At room temperature most impurity atoms are ionized. That is, all donors have given up their electrons to the conduction band or, in the case of p-type material, all acceptors have robbed the valence band for their requisite electrons. Room temperature $E = kT$ corresponds to about 0.026 eV, which is of the same order as the dopant atom ionization energy.

Also, it turns out that the preponderance of carriers in extrinsic semiconductors are supplied by or due to the presence of the dopant atoms, with usually only an infinitesimal amount being provided by the intrinsic carrier generation.

The ionized donor or acceptor ions (positively ionized donors in n-type material, negatively ionized acceptors in p-type material) now being part of the crystal lattice are essentially physically immobile. In n-type material there are very many more electrons than holes, so that the electrons are termed the *majority carriers*, and the holes are termed the *minority carriers*. In p-type material, where the holes are the preponderant carriers, they are the majority carriers, and the electrons are the minority carriers. n and p denote electron and hole carrier densities, respectively, so that for the crystal to maintain electrical charge neutrality, charge conservation must hold, in that generally

$$n + N_A = p + N_D \tag{1.34}$$

where N_A and N_D are the acceptor and donor dopant ion densities, respectively. In light of the above charge conservation relation, the usual semiconductor contains doped material. For n-type material, $N_A \ll N_D$, or p-type material, where $N_A \gg N_D$, Eq. (1.34) becomes simplified.

For n- and p-type material, respectively, the following can be written, with Eqs. (1.27) and (1.29):

$$n \simeq N_D = N_c \cdot \exp((E_f - E_c)/kT); \quad p \simeq N_A = N_v \cdot \exp((E_v - E_f)/kT). \tag{1.35}$$

From which, for n-type material

$$E_f = E_c - kT \cdot \ln(N_c/N_D) \tag{1.36}$$

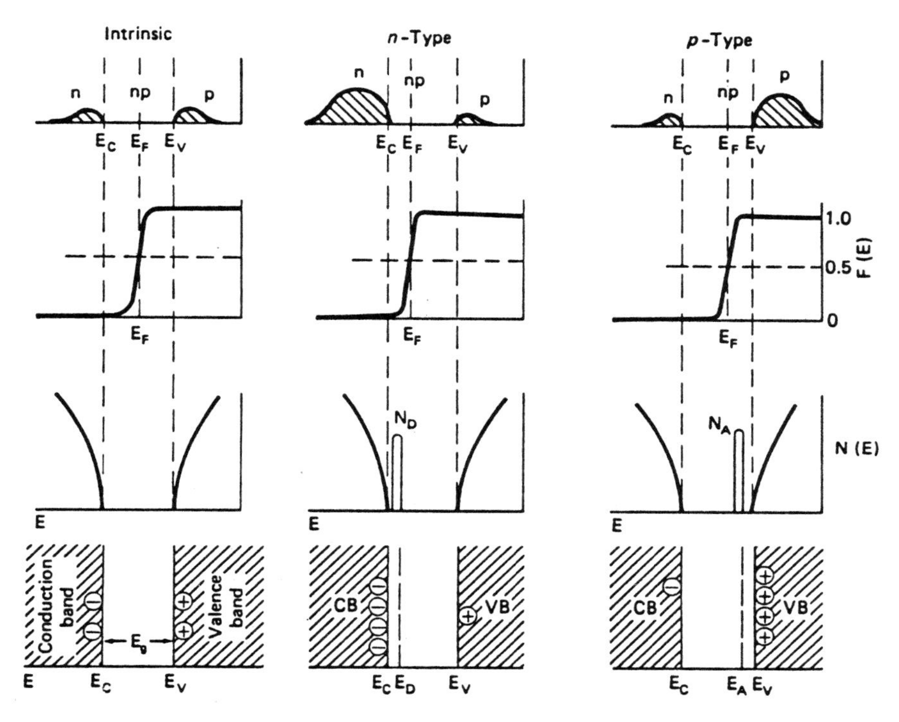

Fig. 1.8 Energy bands, number density of states, Fermi distribution, carrier densities for intrinsic, *n*-type, *p*-type materials.[21] Figure © 1969 by John Wiley & Sons.

which shows that as N_D approaches N_c from below, E_f approaches E_c, as seen in Fig. 1.7. For *p*-type material, using the appropriate inequalities from above, it is easily shown that E_f approaches E_v as N_A approaches N_v. Hence, the Fermi level E_f, which represents the electron population distribution, skews itself to accommodate dopant concentration changes in order to maintain charge neutrality within the crystal. This is also evident from Fig. 1.8, which shows the results of the integration over the product of the density of states by the Fermi distribution to yield the carrier (electron or hole) density. From the preceding, it is straightforward to show that the *np* product remains constant at its intrinsic value irrespective of reasonable dopant concentration changes. This is done by inserting Eq. (1.36) into Eq. (1.29) to obtain

$$p = \frac{N_v N_c}{N_D} \exp \frac{-E_g}{kT} \tag{1.37}$$

Comparing Eq. (1.37) with the intrinsic n_0p_0 equation, Eq. (1.31), reveals that, with the above n-type material inequalities, the hole density satisfies

$$p \cong \frac{n_i^2}{N_D} \tag{1.38}$$

Because $n \cong N_D$, it is seen that under equilibrium conditions (e.g., no applied voltage), $np = n_i^2$ holds for extrinsic materials as well. Also, using the preceding p-type material inequalities, $p \cong N_A$ yielding $n = n_i^2/N_A$, which also gives $np = n_i^2$. This is called a *mass action law* for n- and p-type materials. Implicit in the above is that thermal equilibrium holds. It is seen from these two expressions for p and n that increasing the donor dopant concentration decreases the hole density, and increasing acceptor dopant concentration decreases the electron density. This is termed *carrier suppression*. The three relations, $np = n_i^2$, $p \cong n_i^2/N_D$, and $n \cong n_i^2/N_A$, provide three simple expressions that can be used to compute carrier densities in extrinsic material.

1.7 MOBILITY, MEAN FREE TIME

Now it is well to examine carrier motion under the influence of applied electric fields, in contrast to the preceding field-free discussions. In the latter case, in silicon at 300°K, the average thermal random-walk speed of the electrons is about $100\,\text{km s}^{-1}$ with a corresponding mean free path between collisions of about $0.01\,\mu\text{m}$ and a corresponding mean free time of about 0.1 picosecond (ps). If an electric field $\bar{E}$ is applied to the semiconductor, through leads from the outside, the corresponding force on the electron is simply $\bar{F} = -e\bar{E}$. This force will accelerate the electron with a velocity $\bar{v}_n$, which, from Newton's second law, is

$$\bar{v}_n = \bar{a}\tau = -\frac{e\bar{E}\tau}{m_{ce}^*} = -\mu_n\bar{E} \tag{1.39}$$

where the negative sign indicates that the electron motion is in a direction opposite to that of the electric field. $\bar{a}$ is the acceleration, m_{ce}^* is the appropriate conductivity effective mass, and τ is the mean free time between collisions. The electron mobility, μ_n, defined as the (drift) velocity per unit electric field, is given from Eq. (1.39) as

$$\mu_n = \frac{e\tau}{m_{ce}^*} \qquad (\text{cm}^2 \text{ per volt second}) \tag{1.40}$$

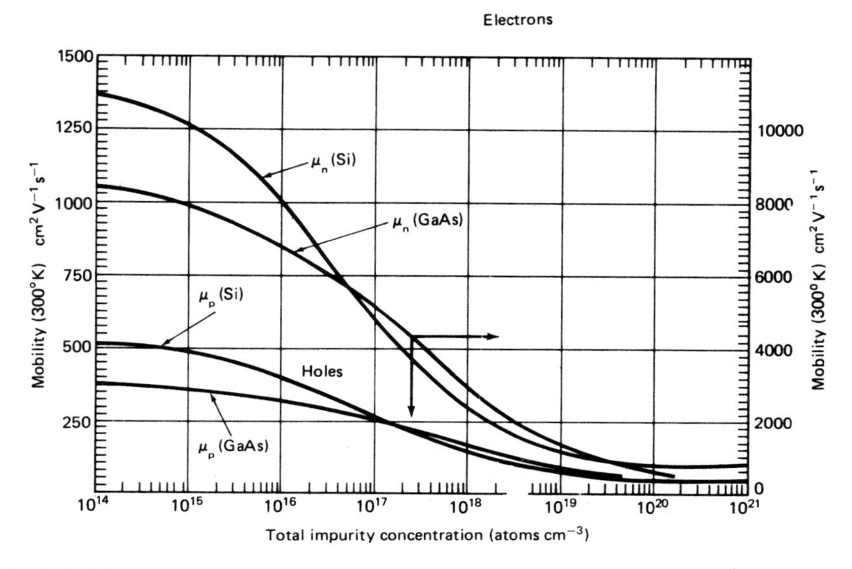

Fig. 1.9 Electron and hole mobilities in silicon and gallium arsenide at 300°K as functions of the total dopant concentration. The values plotted are the results of curve fitting both published and unpublished measurements from several sources.[22,32] Figure © 1962 by McGraw-Hill.

For holes, the analogous expression is

$$\mu_p = \frac{e\tau}{m_{ch}^*} \tag{1.41}$$

where, rewritten from Section 1.5,

$$m_{ch}^* = (m_{lh}^{5/2} + m_{hh}^{5/2})/(m_{lh}^{3/2} + m_{hh}^{3/2}) \tag{1.42}$$

$\bar{v}_p = \mu_p \bar{E}$ is the hole drift velocity in a direction opposite to that of the electrons. m_{ch}^* is the conductivity effective mass for holes.

In a pure material, the mobility depends principally on the scattering of the carriers by the lattice per se, with a minor contribution due to scattering from impurity atoms. However, in doped material, and at very low temperatures, the opposite may be true. There is also a broad temperature regime where both types of scattering contribute almost equally. Figure 1.9 depicts electron and hole mobilities in silicon and gallium arsenide at 300°C as a function of the dopant concentration.

It is instructive at this point to show how the total material mobility is composed of individual mobilities. This is done by noting that the total interaction cross section for a number of competing processes with corresponding scattering cross sections, is composed of the sum of the individual cross sections, since the individual processes are mutually exclusive. Hence, the total cross section σ_T is given by

$$\sigma_T = \sum_i \sigma_i \tag{1.43}$$

where the index i is taken over all the scattering processes in this content. The σ_i are the cross sections for the individual processes, such as impurity scattering and lattice scattering. As the corresponding mean free paths, $\lambda_i \sim 1/\sigma_i$, the total mean free path λ_T satisfies

$$\lambda_T^{-1} = \sum_i \lambda_i^{-1} \tag{1.44}$$

The mean free time between collisions $\tau_i = \lambda_i/v$, where v is the speed of the carriers. Analogously then

$$\tau^{-1} = \sum_i \tau_i^{-1} \tag{1.45}$$

From Eqs. (1.40) and (1.41), it is evident that the mobilities are proportional to mean free times; hence the total mobility μ_T is obtained as

$$\mu_T^{-1} = \sum_i \mu_i^{-1} \tag{1.46}$$

As already mentioned, two types of scattering are dominant, namely, lattice and impurity scattering. For the former, i.e., scattering by the lattice phonons, discussed in detail in Section 1.9, it is assumed that the passage of a longitudinal acoustic vibration wave through the crystal lattice, because of the influence of an incident electron or hole striking the lattice, causes alternate regions of compression and expansion therein. It is easily shown that only longitudinal components of the acoustic wave need be considered.[5] These lattice undulations result in distortion of the energy band structure, including the band gap energy width. Quantum mechanical treatment of this carrier-lattice scattering process yields a mean free time for electrons, as[6]

$$\bar{\tau} = \frac{\sqrt{8\pi}\hbar^4 c_{11}}{3(m_{ce}^* kT)^{3/2}\Xi_c^2} \tag{1.47}$$

where c_{11} is the elastic constant in the longitudinal direction. Ξ_c is the deformation potential, i.e., the change in conduction band energy per unit dilational strain. Ξ_v is the corresponding quantity for the valence band. This yields the electron mobility due to lattice scattering as

$$\mu_n = \frac{e\tau}{m_{ce}^*} = \frac{\sqrt{8\pi} e\hbar^4 c_{11}}{3m_{ce}^{*5/2}(kT)^{3/2}\Xi_c^2} \tag{1.48}$$

For hole mobility, the following ratio is obtained in easy fashion, as[7]

$$\frac{\mu_p}{\mu_n} = \left(\frac{m_{ce}^*}{m_{ch}^*}\right)^{5/2}\left(\frac{\Xi_c}{\Xi_v}\right)^2 \tag{1.49}$$

For hole or electron scattering by impurity ions, a classical treatment similar to the calculation of Coulomb scattering by charged particles is available. This is closely parallel to the Coulomb scattering of alpha particles by heavy charged nuclei as obtained by Rutherford. This yields a scattering crossection through an infinitesimal solid angle $d\Omega$, given by

$$\sigma(\theta)d\Omega = (Ze^2/2\varepsilon m^* v_0^2 \sin^2(\theta/2))^2 d\Omega; \quad \tan(\theta/2) = Ze^2/\varepsilon m^* a v_0^2 \tag{1.50}$$

Z is the atomic number of the donor or acceptor ion scatterer, m^* is the effective mass of the carrier being scattered, v_0 is the speed of the carrier, ε is the dielectric constant of the medium, e.g., silicon, a is the impact parameter, the distance of closest approach to the scatterer normal to the incoming path of the carrier, and θ is the angle through which the carrier is scattered. Using the previous development connecting the scattering crossection and the mean free time, the latter is given by

$$\tau(v_0)^{-1} = Nv_0 \int_\Omega \sigma(\theta)(1 - \cos\theta)\, d\Omega \tag{1.51}$$

where the factor $1 - \cos\theta$ is required due to the fact that τ is defined as the time it takes for the incident carrier speed to be reduced to zero.[8] N is the number of impurity ions per cm^3. The above integration results in a complicated expression for $1/\tau(v_0)$.[9] The desired quantity is $\bar{\tau}$, computed from a mean squared average that gives

$$\begin{aligned} \bar{\tau} &= \langle v_0^2 \tau(v_0)\rangle / \langle v_0^2 \rangle \\ &= 8\varepsilon^2 (kT)^{3/2}(2m^*)^{1/2}/\pi^{3/2} Z^2 e^4 N \ln\left[1 + (7\varepsilon kT/2Ze^2 N^{1/3})^2\right] \end{aligned} \tag{1.52}$$

where $\langle \cdot \rangle$ is an average over a Boltzmann distribution of carrier speeds. The Boltzmann distribution is used instead of the Fermi distribution, because the former is a good approximation to the latter at the usual ambient temperatures of interest. That is, the exponential term in the Fermi distribution denominator is very large compared with unity, so that $(\exp(E - E_f)/kT + 1)^{-1}$ is essentially $\exp - (E - E_f)/kT$, which latter is the Boltzmann distribution in the energy variable. The corresponding mobility due to impurity scattering, μ_i, is

$$\mu_i = e\bar{\tau}/m^* = 8(2/m^*)^{1/2}\varepsilon^2(kT)^{3/2}/\pi^{3/2}Z^2e^3N\ln[1 + (7\varepsilon kT/2Ze^2N^{1/3})^2] \tag{1.53}$$

This is called the *Conwell-Weisskopf mobility formula.*[10] Another treatment that is more accurate at high dopant concentrations leads to a similar expression termed the *Brooks-Herring formula* for the same quantity.[11] So that in the regime under discussion, the total mobility for electrons is given by

$$\mu_T^{-1} = \mu_i^{-1} + \mu_n^{-1} \tag{1.54}$$

1.8 CARRIER CURRENTS, DIFFUSION, CONDUCTIVITY

The carrier particle current density within the semiconductor is given simply as the expression $-n_0\mu_n\bar{E} + p_0\mu_p\bar{E}$, because the electrons and holes move in opposite directions. The corresponding total electric current density, $\bar{J}$, is then given by

$$\bar{J} = e(n_0\mu_n + p_0\mu_p)\bar{E} \tag{1.55}$$

From Maxwell's equations of electromagnetic theory, discussed in Chapter 2 in detail, the pertinent form of Ohm's law is $\bar{J} = \sigma\bar{E}$, where $\sigma = 1/\rho$ is the electrical conductivity, and ρ is the corresponding resistivity. The dependence of resistivity on dopant concentration is shown in Fig. 1.10.

Then it is seen that the conductivity as a function of the semiconductor parameters is

$$\sigma = e(n_0\mu_n + p_0\mu_p) \tag{1.56}$$

It is important now to consider perturbations to the equilibrium state of the semiconductor material. These might be due to photoexcitation from

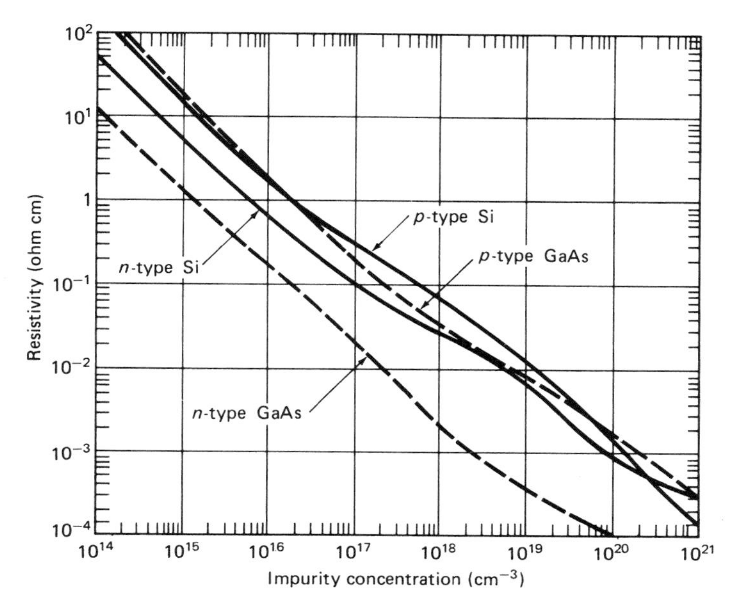

Fig. 1.10 Resistivity as a function of dopant concentration at 300°K for n- and p-type silicon[23] and gallium arsenide.[31] Figure © 1969 by John Wiley & Sons.

an external source, or direct introduction of charge producing pulses, due to ionization of the material from other than photon radiation. Such agents produce localized concentrations of charge whose transient behavior causes them to diffuse away from their source. For such charge concentrations it is assumed at the outset that Fick's law of diffusion holds. Fick's law asserts that the electron particle current density, $\bar{j}_n$, satisfies

$$\bar{j}_n = -e(-D_n \nabla n) \tag{1.57}$$

where D_n (cm^2/s) is the suitable diffusion coefficient in silicon for electrons. Correspondingly, for holes

$$\bar{j}_p = e(-D_p \nabla p) \tag{1.58}$$

and D_p is the diffusion coefficient for holes in silicon. The total electric current density due, both to diffusion and any applied field present, is given for electrons and holes, respectively, as

$$\bar{J}_n = e(n\mu_n\bar{E} + D_n\nabla n) \tag{1.59}$$

$$\bar{J}_p = e(p\mu_p\bar{E} - D_p\nabla p) \tag{1.60}$$

Charged particle transport theory asserts that the diffusion coefficients are given by, where overbars on velocities denote averages,

$$D_n = \lambda_n\bar{v}_n/3 \qquad D_p = \lambda_p\bar{v}_p/3 \tag{1.61}$$

where λ_n and $\bar{v}_n$ are the mean free path and mean thermal velocity for electrons, and λ_p and $\bar{v}_p$ are their counterparts for holes.

It is important at this juncture to show the connection between the diffusion coefficient and the mobility. From the kinetic theory of gases, it is well known that $m\bar{v}_n^2 = 3kT$ for an individual electron undergoing motion in three degrees of freedom. Inserting the corresponding mean free path $\lambda_n = \bar{v}_n\tau$, into Eq. (1.61), with the assumption that $\overline{v_n^2} \cong (\bar{v}_n)^2$, yields for electrons

$$D_n = \overline{v_n^2}\tau/3 \tag{1.62}$$

Using the definition of mobility to eliminate τ above, together with the kinetic gas relation, gives

$$\frac{D_n}{\mu_n} = \frac{D_p}{\mu_p} = kT/e \tag{1.63}$$

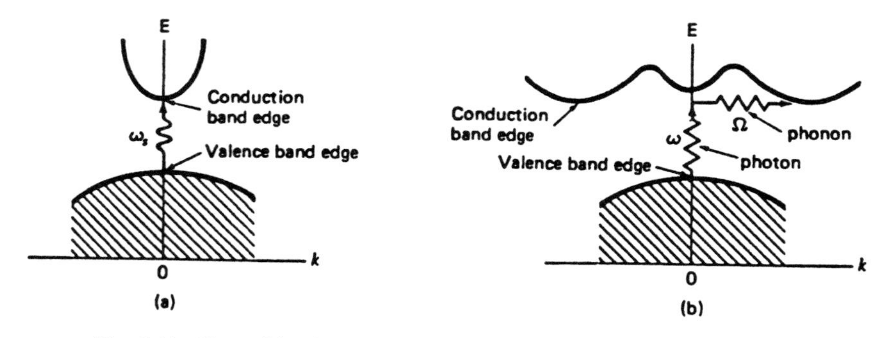

Fig. 1.11 Recombination processes in gallium arsenide (a) and silicon (b).

and kT/e is about 0.026 V at room temperature. These are called the *Einstein relations*, as they were first discovered by Einstein in his investigation of Brownian motion.

1.9 RECOMBINATION, TRAPPING

Besides the mechanism of diffusion for reducing nonequilibrium carrier concentrations, a second mechanism, called *carrier recombination*, also acts as a sink to reduce such concentrations. Under equilibrium conditions, carrier generation is balanced by carrier recombination. One immediate type is called *band-to-band recombination*, and it is shown in Fig. 1.11a and 1.12a. It is simply that an electron in the conduction band recombines with a hole in the valence band. This usually results in the radiation of a photon, whose energy is the amount that the electron gave up in descending to the lower energy valence band below the forbidden gap, which is $h\nu_g = E_g$. This energy conservation must be accompanied by a simultaneous conservation of momentum, in that $\bar{k}_n + \bar{k}_p = 0$ must also hold. Recall that $\bar{k}$ is essentially the momentum vector of the carrier. In somewhat more detail, it is seen from Fig. 1.11a that in the direct process the gap energy $E_g = \hbar\omega_g$, where $\omega_g = 2\pi\nu_g$ is the angular frequency of the photon involved. Also, there is no momentum change, for the $\bar{k}$ value of the particles involved stays constant as seen in the figure. Figure 1.11a is descriptive of gallium arsenide where direct transitions are allowed.

For silicon, however, Fig. 1.11b is appropriate. From that figure it is seen that direct energy transitions are not possible, because momentum cannot be conserved, since the valence and conduction band maxima and minima are widely separated in $\bar{k}$ space. In this case, it is also seen that the threshold energy required for the direct process is greater than the band-gap energy. Therefore a third particle must be involved in this indirect process to conserve both momentum and energy. This particle is called a *phonon.*[12] Phonons are real particles that are the quanta associated with the acoustic waves of the vibrating crystal lattice, just as photons correspond to the quanta associated with the electromagnetic field waves. Phonons travel with a velocity v_s, possess momentum $\bar{p} = \hbar\bar{K}$, angular frequency $\Omega = Kv_s$, and energy $E = \hbar\Omega$. In the above three-particle, indirect process, conservation of momentum is implied by the vector sum $\bar{k}_n + \bar{k}_p + \bar{K} = 0$. The conservation of energy relation is given by $\hbar\omega = E_g \pm \hbar\Omega$, where the positive sign corresponds to a phonon produced, and the negative sign to a phonon absorbed, as in the inverse process.

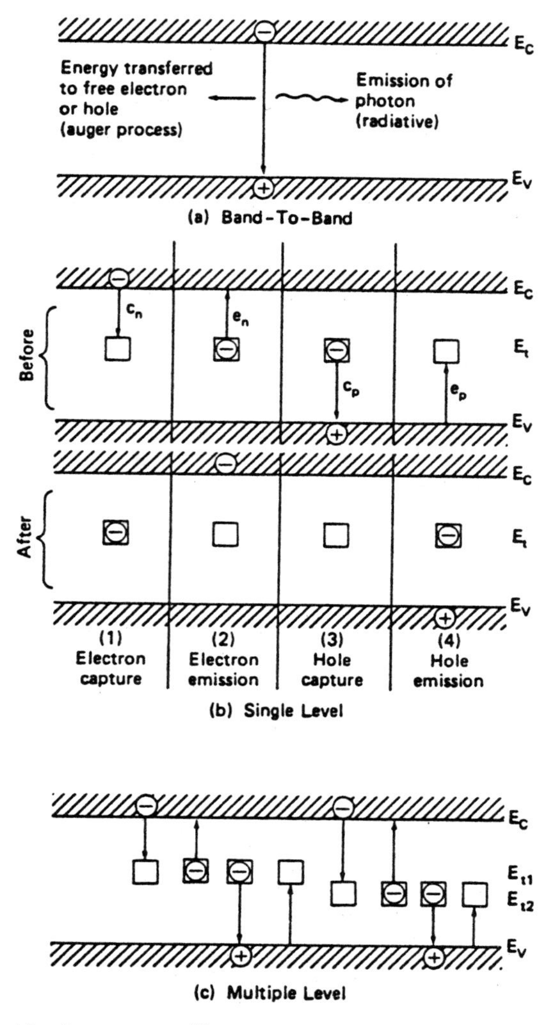

Fig. 1.12 Recombination processes.[21] Figure © 1969 by John Wiley & Sons.

In thermal equilibrium excess electrons and holes are produced in equal amounts, and they recombine at a given common rate. Also for small concentrations of excess carriers, the concomitant decay is usually exponential with a time constant τ_n or τ_p. It follows that, for both electrons and holes, where n_0 and p_0 are the equilibrium concentrations.

$$\dot{n} = -(n - n_0)/\tau_n \qquad \dot{p} = -(p - p_0)/\tau_p \tag{1.64}$$

Integrating gives, respectively

$$n = n_0 + (n(0) - n_0)e^{-t/\tau_n}; \quad p = p_0 + (p(0) - p_0)e^{-t/\tau_p} \tag{1.65}$$

$n(0)$ and $p(0)$ are the respective concentrations at the onset of the excess carrier production.

Because band-to-band recombination is not the dominant recombination mechanism in silicon, at least at low excess-carrier concentrations, it is important to examine recombination mechanisms that do characterize these processes in silicon. These are due to traps or recombination centers within the material. Certain impurities, not necessarily in columns adjacent to silicon in the periodic table, or crystal defects, by their very existence in the crystal, create allowed energy levels in the forbidden gap. Such defects or impurity atoms can thereby physically capture or trap and hold a carrier, e.g., an electron, until the oppositely charged carrier comes along and electron-hole recombination occurs, thereby annihilating the hole. This depletes the conduction band of a conduction electron, which now takes its place as a valence electron bond. This locally reconstitutes the crystalline structure of the semiconductor where a hole was located before annihilation. In Fig. 1.12(b), transitions (1) and (3), when combined, depict such a recombination.

Trap is almost synonymous with "recombination center." If the energy level corresponding to the lattice disturbances of the type discussed above has a high probability of reemitting the carrier before it recombines, it is usually termed a trap. If, however, the probability of recombination before reemission is high, then it is usually called a *recombination center*. The probability of both types of interactions occurring with a common energy is relatively high near the center of the gap as will be shown later. Thus these processes are likelier to occur for energy levels near the center of the gap, as compared to those near the gap edges. Levels near the gap center are termed *deep*, and those near the gap edges are called *shallow*. This kind of process plays an important role in the radiation damage sense, because such damage introduces deep energy levels, as will be discussed in detail later. As alluded to earlier, when the equilibrium state of the material is perturbed, in that np deviates from n_i^2, processes are manifest that restore the lattice to its equilibrium state to yield $np = n_i^2$ again. Basic recombination processes are illustrated in Fig. 1.12. Other kinds of recombination processes caused by incident damaging radiation will be discussed later.

Figure 1.12a illustrates band-to-band recombination, discussed earlier, where an electron and hole recombine with an excess of energy. The transition of an electron from the conduction to the valence band is called a *radiative transition*, if the energy excess is given up in the form of a photon. If, instead, that energy is transferred to another electron as kinetic energy, it is called an *Auger transition*. Figure 1.12b depicts a one-level recombination process, in which only one trap is involved in the gap; Fig. 1.12c depicts a multiple-level recombination process in which more than one trapping energy level is present in the gap. The single-level trap recombination processes consist of (1) conduction band electron capture by an initially neutral empty trap. (2) an electron emitted back to the conduction band by the same trap when so occupied, (3) capture of a hole from the valence band by a trap containing electrons, (4) a hole emitted from that same trap to the valence band, which is tantamount to a valence electron being captured by an initially empty trap. The single trap level recombination rate, U, ($\mathrm{cm}^{-3}\mathrm{s}^{-1}$) is given by[13,21]

$$U = \sigma_p \sigma_n v_{th}(np - n_i^2)N_t/\{\sigma_n[n + n_i \exp(E_t - E_i)/kT] + \sigma_p[p + n_i \exp - (E_t - E_i)/kT]\} \tag{1.66}$$

where σ_p and σ_n are the hole and electron capture cross sections, v_{th} is the carrier thermal velocity, N_t is the number density of traps, E_t is the trap energy level, E_i is the Fermi level of the intrinsic material, and n_i is the intrinsic carrier density. For many situations, the simplifying assumption of $\sigma_n \cong \sigma_p = \sigma$ ensues. Then from Eq. (1.66)

$$U = \sigma v_{th}(np - n_i^2)N_t/\{n + p + 2n_i \cosh(E_t - E_i)/kT\} \tag{1.67}$$

It is also seen that $U = 0$ when $np = n_i^2$, that is, under equilibrium conditions. Further, U is a maximum when $E_t \cong E_i$.[21] Hence, the deeper the trap, the more effective it is in the sense that the recombination rate is approaching its maximum value. Under low excess-carrier injection levels, i.e., δn and δp are very small compared with the majority carrier density, the recombination rate devolves to

$$U = (p - p_0)/\tau_p \tag{1.68}$$

where p_0 is the equilibrium minority carrier density, τ_p is the minority carrier lifetime, and $p = p_0 + \delta p$. Now, in n-type material, $n \cong n_0$, the equilibrium majority carrier density, and $n \gg n_i$ and p, then Eq. (1.66) yields

$$U = \sigma_p v_{th} N_t (p - p_0) \tag{1.69}$$

Comparing Eqs. (1.68) and (1.69) identifies the minority carrier lifetime relations for holes and electrons, respectively

$$\tau_p = (\sigma_p v_{th} N_t)^{-1}; \qquad \tau_n = (\sigma_n v_{th} N_t)^{-1} \tag{1.70}$$

For multiple-level traps, the recombination energetics are grossly similar to the single-trap case. The details are different, especially for high excess-carrier injection levels, i.e., where $\delta n \gg \delta p \sim n$ or p, discussed later.[14] Here, the gross minority carrier lifetime is an average, taking into account all the trapping levels, positive, negative, or neutral. Figure 1.17 shows the minority carrier lifetime, including the case of high minority carrier injection, as well as high dopant concentrations. High carrier injection results in recombination kinetics that depend mainly on phonon-aided Auger processes,[17] instead of single- or multiple-trap process models to be discussed in Section 5.9. This is reflected in small values of minority carrier lifetime, as shown in Fig. 1.17.

There is another important aspect of the trapping properties of the materials of interest. As discussed in the preceding, deep traps are more effective at capturing all carriers than shallow traps. However, it is also easily shown from Eq. (1.66) that, as the trap energy in the forbidden band gap approaches, say, the conduction band edge, electron capture probability greatly increases, in contrast to hole capture, which becomes vanishingly small. For trap energies approaching the valence band edge, the opposite is the case, as the hole trapping probability greatly increases, with electron capture becoming negligible. Recall that a state must capture both a hole and an electron to act as a recombination center.

The above properties account for the extensive use of gold as a dopant to introduce recombination centers to, for example, mitigate device latchup propensities, as discussed in Sections 7.7 and 12.4. Gold has an acceptor state at $E_c - 0.54\,\text{eV}$., which is almost at the center of the silicon band gap, making an effective recombination center in n-type silicon. Gold also has a donor state at $E_v + 0.35\,\text{eV}$. This property of gold and other materials that can exhibit a p-type behavior in n-type material, and vice versa, is called amphoteric.[34] In n-type silicon, the gold deep donor trap has a high probability of containing an electron (i.e., to be non-ionized). In p-type silicon, it has a high probability of being ionized, so that its acceptor state is effectively charge-neutral, as discussed below.

Two types of traps or trapping states have been defined. The first is the donor-like trap, which has the property of being charge-neutral when filled with an electron, and positively charged when empty. The second is the acceptor-like trap, which has the property of being negatively charged when filled with an electron, and charge-neutral when empty. Recall from Section 1.5 that the fermi-level energy E_f essentially delineates the energy region between filled ($E < E_f$) and empty ($E > E_f$) electron energy states in the band gap. Equivalently then, a donor trap is neutral when its energy state is below E_f, and positive when its energy state is above E_f. An acceptor trap is negative when its energy state is below E_f, and neutral when above E_f. For radiation-induced Si/SiO_2 interface states in the gate oxides of MOSFETs, discussed in Section 6.9, acceptor states exist mainly in the upper half of the SiO_2 band gap, while donor states exist predominantly in the lower half.[37]

With respect to the ionizing radiation damage aspects of the recombination process, there are principally two types of recombination. The first is called columnar recombination, where an incident particle, usually highly charged and heavy, produces a dense columnar ionization track of hole-electron pairs. In an insulator such as SiO_2, some fraction of the holes and electrons recombine within a few pico-seconds following the creation of the track, depending on the local electric field strength, energy, and type of the incident particle. The subsequent behavior of the unrecombined holes and electrons is discussed in detail in the context of MOSFET ionization damage, in Section 6.8. When hole-electron pairs are produced in such dense tracks or columns, recombination is a robust process, mainly because of the proximity of holes and electrons to each other. However, the electric field usually forces almost immediate separation of the unrecombined hole-electron pairs, where the single track column of charges initially separates into roughly two columns each containing charge of one sign.[36] This, of course, reduces the recombination rate, depending on the electric field strength and the material diffusional parameters. When the holes and electrons are relatively far apart, recombination becomes a much weaker process, and the second type, or geminate, recombination ensues. The latter is also the case for ionization damage tracks made by light particles such as electrons, and ^{60}Co gamma rays.[37] These tracks are rather thin and of low density. This is in contrast to the thick, high-density tracks usually made by energetic heavy particles such as alpha particles of a few MeV, or heavy cosmic ray ions like Fe^{6+}. The latter will usually cause single event upset (SEU) when they impale a dynamic RAM integrated circuit. Quantitative aspects of their SEU kinematics are discussed in Sections 8.5 and 8.7. Also, a number of incident particles of interest interact with materials in recombination processes that span both recombination types, where neither the columnar or geminate models is strictly valid.[37]

1.10 SEMICONDUCTOR TEMPERATURE

It is already evident from previous discussion that semiconductor temperature plays an important role in the physical and operational functioning of semiconductor devices. For example, military specifications for semiconductor device environments stipulate operational temperature bounds, such as from −55°C to +125°C. Many specific device properties are very sensitive to device temperature. These include fermi level energy, forbidden energy gap, mobility, bipolar common emitter gain, and others. These dependencies will be discussed in turn.

The intrinsic fermi level, E_{fi}, lies essentially at the middle of the forbidden energy gap for intrinsic semiconductors. This is evident from Eq. (1.33), where the second term is about one percent of the first term at room temperature (300°K). For the extrinsic (i.e., doped material at room temperature) case, the corresponding fermi level, E_f, can be determined from the solution of the following two sets of equations. The first set is that of the charge neutrality conservation equations, rewritten from Eq. (1.34),

$$n \cong p + N_D^+ \qquad (n \text{ material; } N_D \gg N_A) \tag{1.71}$$

where N_D^+ is the density of ionized donor atoms; a portion of the N_D donor atom density. Also,

$$n + N_A^- \cong p \qquad (p \text{ material; } N_A \gg N_D) \tag{1.72}$$

where N_A^- is the density of ionized acceptor atoms; a portion of the N_A acceptor atom density. The left side of Eq. (1.72) consists of the negative charge sources. These include the electron carrier density, n, plus the ionized portion N_A^- of the acceptor dopant atoms N_A. The N_A are neutral, but when they accept an extra electron, they become negatively charged, to yield N_A^-. The right-hand side of Eq. (1.72) consists of the positive hole density. A similar explanation applies to Eq. (1.71).

The second set of equations[21] will yield the fraction of ionized impurity atoms in terms of their respective impurity densities,

$$N_D^+ = N_D(1 + 2\exp(E_f - E_D)/kT)^{-1} \tag{1.73}$$

where E_D is the donor trap energy level. Also,

$$N_A^- = N_A(1 + 4\exp(E_A - E_f)/kT)^{-1} \tag{1.74}$$

where E_A is the acceptor trap energy level.

For n material, inserting Eq. (1.73) and the appropriate Eq. (1.35) into Eq. (1.71) yields

$$\begin{aligned} N_c \exp - (E_c - E_f)/kT = N_D(1 + 2\exp(E_f - E_D)/kT)^{-1} \\ + N_v \exp(E_v - E_f)/kT \end{aligned} \tag{1.75}$$

For p material, inserting Eq. (1.74) and the appropriate Eq. (1.35) into Eq. (1.72) yields

$$\begin{aligned} &N_c \exp - (E_c - E_f)/kT + N_A(1 + 4\exp(E_A - E_f)/kT)^{-1} \\ &= N_v \exp(E_v - E_f)/kT. \end{aligned} \tag{1.76}$$

With N_c, N_v, and all of the energy levels well known, Eq. (1.75) and (1.76) are cubic in $\exp(E_f/kT)$ for their respective materials. From the resulting computations, using equations (1.75) and (1.76) ($E_f - E_{fi}$ versus semiconductor device temperature) for n and p material impurity concentrations can be plotted. These are shown in Figs (1.13A) and (1.13B).

It is seen from Figs. (1.13A) and (1.13B) that, as the semiconductor temperature is increased, the extrinsic fermi level E_f approaches the intrinsic fermi level E_{fi} (essentially at mid-gap) for both n and p materials. Hence, the doped material is approaching an intrinsic state where $np = n_i^2$, but now n_i is very much larger for elevated temperatures, as seen from Eq. (1.32) and (3.66). At room temperature (300°K), $n_i \ll N_D$ *or* N_A is the usual case. However, as the temperature is increased, n_i, rapidly increases, e-folding for every 16°K. This means that, for very large n_i, especially when $n_i \gg N_D$ or N_A, high-temperature thermal generation of electrons from the valence band excited to the conduction band, irrespective of the dopant density or even the presence of dopants, now constitute the main source of carriers. No increase in carrier density due to ionization of dopant atoms can occur, because all dopant atoms have now been ionized. Any further increase in semiconductor temperature in this range results in carrier increase due only to thermal excitation of valence atoms. This means that all semiconductor properties manifest by dopant impurities, such as the formation of depletion layers, conductivity changes, etc., are lost at these high temperatures. Also, because of these relatively high thermally generated currents,

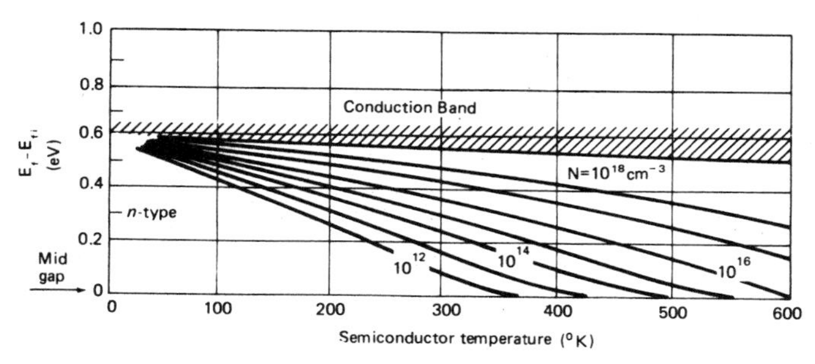

Fig. 1.13A Fermi level for n-type Si versus temperature and dopant density.[21] Figure © 1969 by John Wiley & Sons.

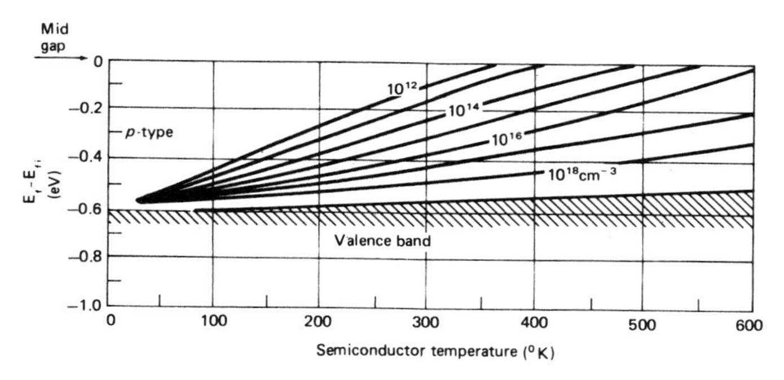

Fig. 1.13B Fermi level for p-type Si versus temperature and dopant density.[21] Figure © 1969 by John Wiley & Sons.

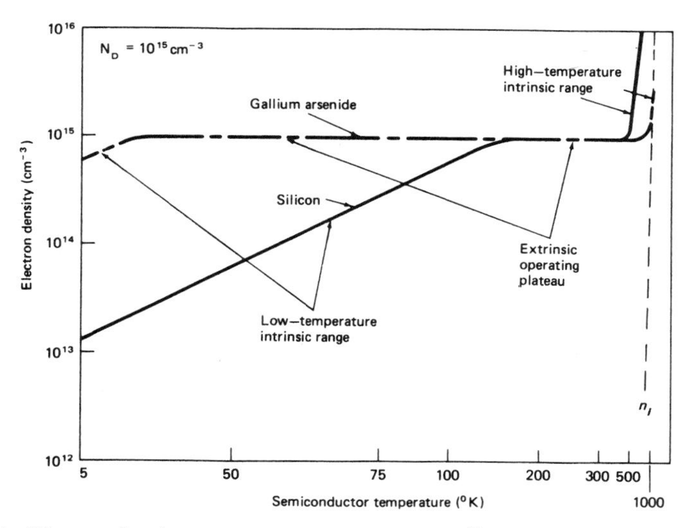

Fig. 1.14 Electron density versus temperature for silicon[21] and gallium arsenide.

this state can be construed as high injection, and is discussed in Section 1.11.

For very low temperatures, the intrinsic state also ensues, but at very low n_i levels. This is because almost all dopant atoms remain un-ionized, since the ambient $kT \ll E_D$ or E_A, so that their function in terms of doping is also minimal or non-existent. Furthermore, there are essentially no free carriers in this low temperature regime. All of this is depicted in Fig. 1.14 for silicon and gallium arsenide.

Therefore, for proper device function, temperature bounds must not be exceeded. If they are, the device cases to function as a transistor, whether bipolar or FET. However, they can usually be resuscitated by again bringing them within proper temperature bounds.

It is seen that the plateau region in Fig. 1.14, which is the extrinsic (doped) operating region for the corresponding transistors, is much elongated in the case of gallium arsenide, compared to silicon. The plateau level corresponds approximately to the (donor) dopant concentration (i.e., $N_D = 10^{15}\,\text{cm}^{-3}$) for both materials in this case. The extended plateau for GaAs means that it possesses a larger operating temperature span than does silicon. This is especially true for very low temperatures (which, for GaAs, is well into low cryogenic temperatures).

The major reason for the long plateau span in GaAs is that its dopant family of materials is different from silicon, since it is not a "column four" material. Rather, its donor dopants lie in column six in the periodic table (Fig. 1.6), such as tellurium and sulphur. The point here is that the ionization energies E_D and E_A of those dopants are at least ten times smaller than those for silicon. For example, the ionization energy of sulphur dopant is only 0.006 eV, and that of tellurium is 0.0059 eV.[21] Since the GaAs donor dopant ionization energies are so very shallow—much shallower than for silicon donors—operating temperatures in GaAs devices can be much lower than in silicon, and the corresponding dopants will still ionize to provide carriers in the performance of their dopant function. This is not the situation for silicon, as its donor dopants will not ionize for temperatures below about 200°K, as seen in Fig. 1.14.

At the very high temperature side of Fig. 1.14, the slope of the curves correspond to their respective E_g, and the extrapolated length of this portion of the curves is given from n_i, as shown in the figure.[21] In silicon, the intrinsic $n_i = 1.45 \cdot 10^{10}\,\text{cm}^{-3}$, and for GaAs, $n_i = 9 \cdot 10^{6}\,\text{cm}^{-3}$, a thousand times smaller. All of this results in a larger slope for the GaAs curve, and its displacement to the right of the silicon curve at high temperatures, as seen in Fig. 1.14.

The forbidden energy gap, E_g, plays an important role in semiconductor operaton, as can be appreciated from the previous sections. It is a determinant in the evaluation of the np product and n_i, as seen in Eq. (1.76) and associated discussion. Actually, E_g itself depends on temperature in a somewhat complex manner. The details are bound up in the quantum mechanical aspects of semiconductor physics, only the highlights of which will be discussed here. A well-known empirical relation for the variation of the gap energy with temperature for silicon is given by[28]

$$E_g \cong 1.205 - 2.8 \cdot 10^{-4} T \quad ; \quad T \geqslant 300°\text{K (eV)} \tag{1.77}$$

where the temperature is in degrees K. For very high dopant concentrations that are greater than the Mott[39] rectification potential barrier criterion ($N_A \geqslant 3 \cdot 10^{18}\,\text{cm}^{-3}$), somewhat analogous to the Schottky potential barrier criterion, at very low temperatures the band gap narrows with a decrement in Si donor material,[35]

$$|\Delta E_g| \cong 0.018 \cdot \ln(N_A/10^{17}) \tag{1.77a}$$

This should be compared with Eq. (1.77), which yields an increasing band gap with decreasing temperature for the much lesser dopant levels.

One major difficulty in the determination of the temperature dependence of E_g is the existence of the exciton. The exciton is a stable configuration of an electron-hole pair bound together by their coulomb (charge) attraction and centrifugal repulsion and existing in stable states, not unlike the hydrogen atom discussed in Section 1.3. Most electron-hole pairs, when produced by ionizing radiation or other excitation, move indepedently throughout the lattice. However, it is possible for the hole-electron pair to form an exciton, and the photon energy necessary to create this bound pair is less than E_g. The exciton itself is charge neutral, and can move freely through the crystal, transporting its energy to cause lattice excitation and subsequent annihilation by recombination of the hole with the electron. Since it is a chargeless entity, it does not contribute to the electrical conductivity.

When attempts are made to measure the variation of band gap width with temperature, the conduction band edge is partially obscured by the presence of an exciton energy state just under this band edge of energy, approximately $E_c - 0.007\,\text{eV}$. This difficulty also figures in the determination of n_i versus temperature, as well. Figure 1.15 depicts the energy gap versus temperature in silicon,[28] where an exciton energy of $E_c - 0.01\,\text{eV}$ has been used in computations yielding the figure. It is seen from Fig. 1.15 and Eq. (1.77) that, for temperatures greater than 300°K, $\Delta E_g/\Delta T$ is

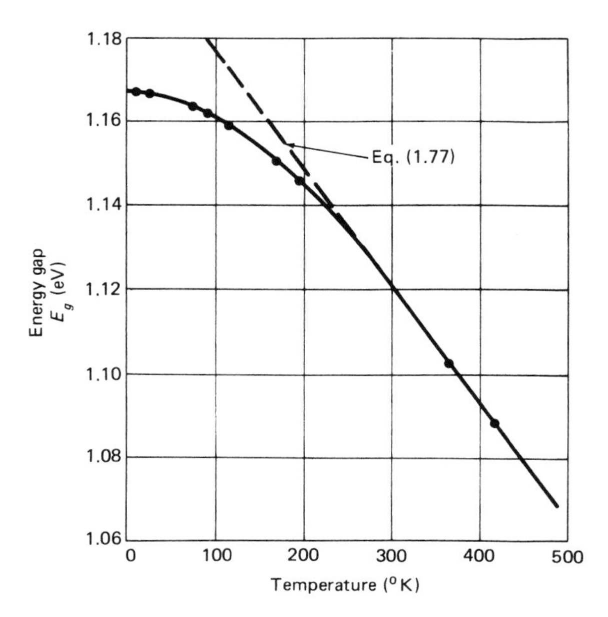

Fig. 1.15 Energy gap versus temperature for silicon.[21,28] Figure © 1983 by John Wiley & Sons.

constant with temperature. This is used to provide a voltage reference (e.g., in the manufacture and operation of band gap reference diodes).

As regards the effect of temperature on the mobility of semiconductors, the discussion in Section 1.7 provides an adequate rendition within the context and scope of this chapter. It is sufficient to state that the subtleties of mobility variation with temperature devolve to the two main types of carrier scattering mechanisms within the lattice. The first is scattering of the carriers by semiconductor impurities. This predominates at low temperatures, since phonon (lattice vibration quanta) activity is greatly diminished in this temperature range. For relatively high temperatures, the carrier scattering takes place mainly with phonons, because lattice vibrations increase rapidly with increasing temperature, and especially in intrinsic semiconductors. There is a broad intermediate temperature range where both types of scattering occur. All of this is depicted in Fig. 1.16, which shows the variation of hole mobility with temperature in *p*-type silicon, at a dopant density of $N_A = 10^{18}\,\text{cm}^{-3}$.

The effects of temperature on bipolar common emitter current gain is discussed in Section 5.10. There, the gain expression is written in terms of solid-state parameters. These parameters are investigated for their

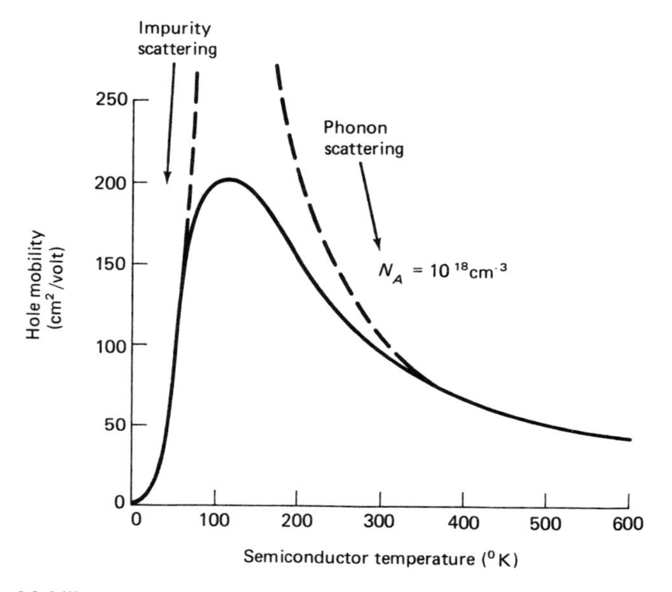

Fig. 1.16 Mobility versus temperature in p-type silicon.[28] Figure © 1983 by John Wiley & Sons.

dependence on temperature and radiation, and the results are given in Section 12.2.

The intrinsic temperature, T_i, occurs when $n_i = N_D$ or N_A, depending on the dopant. T_i is important in current filamentation occurring in second breakdown, discussed in Section 3.6.

1.11 TRANSPORT BEHAVIOR OF EXCESS CARRIERS

In a semiconductor in thermal equilibrium, the thermal generation of electron-hole pairs is balanced by their recombination, as already mentioned. The thermal generation rate is the number of electron-hole pairs generated per cubic centimeter per second, from the fracture of covalent bonds of the crystal lattice. The time between the generation of an electron or hole and its subsequent recombination is called the *mean carrier lifetime*. So that in thermal equilibrium

$$g_{On} = \frac{n_O}{\tau_{nO}} = \frac{p_O}{\tau_{pO}} = g_{Op} \tag{1.78}$$

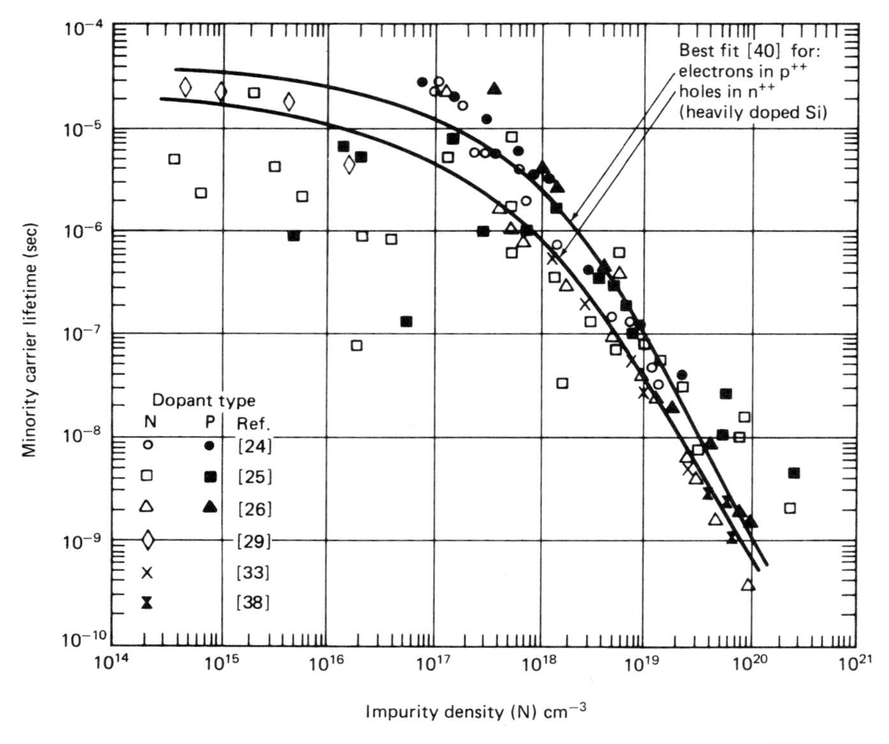

Fig. 1.17 Minority carrier lifetime versus impurity concentration in silicon.[27,28] Figure © 1991 by The IEEE.

where g_{0n} and g_{0p} are the equilibrium thermal generation rates for electrons and holes, respectively. τ_{n0} and τ_{p0} are the mean carrier lifetimes for electrons and holes. Equation (1.78) holds for all steady-state conditions, even for those not in thermal equilibrium. However, in the transient state, the generation and recombination rates for electrons and holes will not necessarily balance, so that (in general)

$$g_n = g_p; \qquad \frac{n}{\tau_n} = \frac{p}{\tau_p} \tag{1.79}$$

where g_n and g_p are the electron and hole generation rates, with n/τ_n and p/τ_p the corresponding recombination rates. The latter are also respectively given by U_n and U_p, in consonance with the previous discussions.

For a volume element in the lattice, assumed homogeneous and isotropic, an equation for continuity can be written for the carrier density, n, as

$$\frac{\partial n}{\partial t} = g_n - U_n - \frac{\nabla \cdot \bar{J}_n}{e} \tag{1.80}$$

where the right-hand side consists of the generation rate, g_n, the recombination rate U_n, for low excess carrier levels, and a loss rate due to the diffusion of carriers out of the volume element, $\nabla \cdot \bar{J}_n/e$, where $\bar{J}_n$ is the electric current density defined previously. Similarly for the hole density, the analogous continuity equation is

$$\frac{\partial p}{\partial t} = g_p - U_p - \frac{\nabla \cdot \bar{J}_p}{e} \tag{1.81}$$

The current densities are also related to their carrier concentrations through Fick's law. This relationship will be used so that ultimately equations in the carrier density variables only will ensue. Earlier, Fick's law involved carrier concentrations. Here it asserts that the electric current density is related to the corresponding carrier density by

$$\bar{J}_n = eD_n \nabla n \tag{1.82}$$

and for holes

$$\bar{J}_p = -eD_p \nabla p \tag{1.83}$$

With these equations, a pair of electron and hole current density relations can be written by summing carrier concentrations to obtain

$$\frac{\bar{J}_n}{e} = n\mu_n \bar{E} + D_n \nabla n \tag{1.84}$$

$$\frac{\bar{J}_p}{e} = p\mu_p \bar{E} - D_p \nabla p \tag{1.85}$$

The first terms on the right-hand sides are due to carrier drift under the influence of the impressed field $\bar{E}$. The second terms are the aforementioned diffusion contributions. Substituting Eqs. (1.84) and (1.85) into the continuity equations, Eqs. (1.80) and (1.81), results in a pair of transport equations in the carrier densities only. They are

$$\frac{\partial n}{\partial t} = D_n \nabla^2 n + \mu_n (\bar{E} \cdot \nabla n + n \nabla \cdot \bar{E}) + g'_n - (n/\tau_n - n_0/\tau_{n0}) \qquad (1.86)$$

$$\frac{\partial p}{\partial t} = D_p \nabla^2 p - \mu_p (\bar{E} \cdot \nabla p + p \nabla \cdot \bar{E}) + g'_p - (p/\tau_p - p_0/\tau_{p0}) \qquad (1.87)$$

where g'_n and g'_p are excess carrier generation rates over their equilibrium values. The electric field $\bar{E}$ is the sum of the applied field $\bar{E}_a$ and the internally generated bulk diffusion field $\bar{E}_i$. The latter field is created because the electrons, due to their larger diffusion coefficient, tend to outrun the holes, thus separating from them to produce $\bar{E}_i$. The direction of $\bar{E}_i$ counteracts this separation by decelerating the electrons and accelerating the holes. At high injection this leads to the electrons and holes tending to diffuse in train, with an effective diffusion coefficient whose value lies between that of the slower and faster carriers. This is called ambipolar diffusion. This ambipolar model also assumes that quasi-neutrality holds, which implies that the net mobile carrier density $p - n \cong p_0 - n_0$.

A third relationship needed to form a determinate set of equations in the three quantities n, p, and $\bar{E}$, is the appropriate Gauss's equation. This equation is discussed in detail in Chapter 2, as one of Maxwell's equations. Here, it is given by

$$\nabla \cdot \bar{E} = (4\pi e/\varepsilon)(p - n + N_D - N_A + p_a - n_d) \qquad (1.88)$$

where the right-hand side is the net sum of the associated charge densities. N_D and N_A are the ionized donor and acceptor concentrations, n_d is the density of those electrons still occupying the donor levels, i.e., the extra electrons in those donors yet to be ionized, and p_a is the hole density of un-ionized acceptor sites.

The system of equations, Eqs. (1.86)–(1.88), is approximated in the following for particular cases of interest. The approximations are (1) $\delta n \cong \delta p$, where $\delta n = n - n_0$, and $\delta p = p - p_0$. $\delta n = \delta p$ cannot strictly hold, for an internal electric field could never be generated. Nevertheless, this approximation judiciously applied, leads to fruitful results. Actually very small differences between δn and δp can generate large internal fields.[15] (2) Excess carrier generation rates are assumed equal, i.e., $g'_n = g'_p = g'$. With these two approximations, Eqs. (1.86)–(1.88) finally reduce to a single transport equation for either the electron or hole excess carrier density given by[16]

$$\frac{\partial}{\partial t}(\delta n) = D^* \nabla^2(\delta n) - \mu^* \bar{E} \cdot \nabla(\delta n) + g' - \delta n/\tau \tag{1.89}$$

where

$$D^* = \frac{(n_0 + p_0 + 2\delta p)D_n D_p}{(n_0 + \delta n)D_n + (p_0 + \delta p)D_p} \quad \text{and} \quad \mu^* = \frac{(n_0 - p_0)\mu_n \mu_p}{(n_0 + \delta n)\mu_n + (p_0 + \delta p)\mu_p} \tag{1.90}$$

and an excess carrier lifetime τ implicitly defined by

$$\delta n/\tau = (n_0 + \delta n)/\tau_n - n_0/\tau_{n0} = \delta p/\tau = (p_0 + \delta p)/\tau_p - p_0/\tau_{p0} \tag{1.91}$$

To render the ambipolar transport equation (Eq. (1.89)) tractable, δn is assumed to be much less than n_0 and p_0. Again, this is the low excess-carrier injection approximation. These approximations allow easy examination of some transport properties discussed below.

First, a digression on minority carrier lifetime and its quantitative aspects is appropriate here. Consider highly doped n-type silicon in equilibrium, where $n_0 \gg p_0$ and δp, and, for example, where the electron carrier density $n \cong 10^{16}\,\text{cm}^{-3}$, and the hole density $p \cong 10^5\,\text{cm}^{-3}$, recalling that $np \sim 10^{21}$. The equilibrium electron and hole lifetimes are still related by $n_0/\tau_{n0} = p_0/\tau_{p0}$, and $n/\tau_n = p/\tau_p$. Then assume that an excess density of electron hole pairs is injected into the material, such that $\delta n \cong \delta p \cong 10^5\,\text{cm}^{-3}$. Now, notice that the subsequent electron density has changed imperceptibly, whereas the hole density has doubled to $\sim 2 \cdot 10^5\,\text{cm}^{-3}$. As the electron density has remained essentially constant, the hole lifetime will also remain essentially constant, because the probability of hole capture by electrons has remained almost unchanged. However, the hole density has doubled, so that the electron lifetime has halved, because the electron capture probability by holes has doubled. It is seen, therefore, that the electron lifetime, i.e., the majority carrier lifetime, depends sensitively on the kinetics of the excess hole density δp. It is also seen that the hole lifetime, i.e., the minority carrier lifetime, is essentially independent of δp. Then for n-type material $n_0 \gg \delta p$, and $\tau_p = \tau_{p0}$, which asserts that the minority carrier (hole) lifetime remains essentially constant, and so it is the steady-state lifetime. From Eq. (1.91), it is then seen that the excess-carrier lifetime also yields $\tau = \tau_{p0}$. So that for low levels of carrier injection, the steady-state recombination lifetime is synonymous with the minority carrier lifetime, because both minority and majority carriers are necessarily recombining at the same rate. The two are not always the same, such as in intrinsic or extremely highly doped material. For p-type material, in the same way, the corresponding minority carrier lifetime $\tau = \tau_{n0}$.

So that for n-type material, the minority carrier lifetime is τ_{p0}, D^* and μ^* reduce to D_p and μ_p. For p-type material, $\tau = \tau_{n0}$, D^* and μ^* reduce to D_n and μ_n. Furthermore, in n-type material, the minority carrier density satisfies, from Eqs. (1.89)–(1.91).

$$\frac{\partial}{\partial t}(\delta p) = D_p\nabla^2(\delta p) - \mu_p\bar{E}\cdot\nabla(\delta p) + g' - \delta p/\tau_{p0}; \qquad n_0 \gg p_0, \delta p \quad (1.92)$$

Correspondingly, in p-type material, the minority carrier density satisfies

$$\frac{\partial}{\partial t}(\delta n) = D_n\nabla^2(\delta n) - \mu_n\cdot\bar{E}\cdot\nabla(\delta n) + g' - \delta n/\tau_{n0}; \qquad p_0 \gg n_0, \delta n \quad (1.93)$$

For the pure semiconductor, $D_i^* = 2D_nD_p/(D_n + D_p)$, $\mu_i^* = 0$, and the corresponding transport equation—for either n or p material—becomes, with no electric field, where $\delta n = \delta p \equiv \delta$,

$$\frac{\partial}{\partial t}(\delta) = D_i^*\nabla^2(\delta) + g' - \delta/\tau \quad (1.94)$$

Ambipolar conditions hold for high injection (high current), as well as for the low-injection case where conductivity modulation (Section 5.12) is a factor. These are discussed (respectively) in Section 8.7, for the early stages of single-event ionization track formation, and Section 7.10, for diffused resistors in integrated circuits.

1.12 HIGH- AND VERY-HIGH-LEVEL EXCESS-CARRIER INJECTION

Low-level carrier-injection, $\delta n/n_0$ or $\delta p/p_0 \ll 1$, processes are briefly mentioned to provide a background for comparison with high-level carrier-injection, $\delta n/n_0$ or $\delta p/p_0 \gg 1$, processes. For low-level excess-carrier densities, the carrier lifetime is limited only by simple recombination processes, such as band-to-band recombination in direct band gap lattices, such as gallium arsenide. The corresponding recombination rate is proportional to the np product, simply because each n carrier interacts with a p carrier. Thus it can be written Rnp, where R is a suitable interaction rate coefficient. Let ε_p be defined as the thermal emission rate (emissivity) for hole production. Then for an essentially intrinsic lattice, $n \cong p \cong n_i$, under equilibrium conditions the net recombination rate must vanish, i.e., $Rnp - \varepsilon_p = 0$. So that

$$\varepsilon_p \cong Rnp \cong Rn_i^2 \tag{1.95}$$

Charge neutrality requires that for low excess-carrier levels, $\delta n \cong \delta p$. The net rate of excess-carrier recombination satisfies,

$$\frac{d\delta}{dt} = R(n + \delta)(p + \delta) - Rn_i^2 = R(n\delta + p\delta + \delta^2) \tag{1.96}$$

Neglecting second-order terms, for the case of low excess-carrier density gives

$$\frac{d\delta}{dt} = R(n + p)\delta \tag{1.97}$$

For intrinsic material, the carrier lifetime τ satisfies $\tau^{-1} = \tau_n^{-1} + \tau_p^{-1}$. From Eqs. (1.95) and (1.97), the carrier lifetime can be identified as

$$\tau = \frac{1}{R(n + p)} = \frac{n_i^2}{(n + p)\varepsilon_p} \tag{1.98}$$

For strictly intrinsic material, $n = p = n_i$, and with the above definitions of carrier lifetime, the intrinsic lifetime can be written as

$$\tau = \frac{n_i}{2\varepsilon_p} \tag{1.99}$$

Now for n-type material, $\tau = \tau_p$, for the minority carriers (holes) dominate the transient behavior, as discussed earlier. Because $p \ll n$, Eq. (1.98) yields the minority carrier lifetime as $\tau_p = n_i^2/n\varepsilon_p$. For p-type material, the minority carriers (electrons) dominate the transient behavior so that, similarly, $\tau_n = n_i^2/p\varepsilon_p$. For high carrier concentrations, and/or high excess carrier injection levels, the recombination process becomes quite complicated. It now includes indirect as well as direct gap processes, together with radiative and nonradiative energy transfer. By analogy to Eq. (1.96), it can be appreciated that at high excess-carrier levels, the recombination kinetics for minority carriers satisfies, following the discarding of the constant terms,

$$-\frac{d\delta}{dt} = a_1\delta + a_2\delta^2 + a_3\delta^3 + \cdots \tag{1.100}$$

It is seen that the right-hand side of the above equation is a series in powers of the excess-carrier density, with suitably identifiable coefficients. Various terms dominate, depending on the relative concentration of carriers and the injection level. For example, the first term on the right corresponds to single-trap level recombination processes, which dominate at low excess-carrier levels, with the excess energy of recombination given off as a phonon or photon, and where $a_1^{-1} = \tau$, the minority carrier lifetime.

For higher injection levels, the second term in Eq. (1.100) dominates; because it is the square of the excess-carrier density, it corresponds to direct band-to-band recombination of two particles—an electron and a hole. The excess energy is also manifest in the form of a phonon or photon.

At very high levels of excess-carrier density, the third term in Eq. (1.100) is dominant. It corresponds to direct nonradiative recombination in indirect band gap materials like silicon. The excess energy and momentum here are taken up by a neighboring electron or hole, instead of through production of phonon or photon radiation.[18] To illustrate, for holes injected into a heavily doped n^+ region, such as into the emitter of an *npn* transistor, three particles are involved. They are the recombining electron and hole pair, with the third particle (electron) carrying off the excess energy and momentum. This can be characterized by the differential equation $d\delta/dt = -\alpha_{3p}n^2\delta$, where $\alpha_{3p} \cong 10^{-31}\,\text{cm}^6\,\text{s}^{-1}$. In a similar manner, recombination in a heavily doped p^+ region can occur involving two holes and one electron, with the hole carrying off the excess energy and momentum. This is characterized by the differential equation $d\delta/dt = -\alpha_{3n}p^2\delta$, where $\alpha_{3n} \cong 10^{-31}\,\text{cm}^6\,\text{s}^{-1}$ for silicon. These nonradiative interactions are called *Auger processes*.

1.13 APPLICATIONS OF CARRIER TRANSPORT

One particular case is of interest at this point. Assume that the electric field vanishes. This implies that with no external field the internal field is small. For one-dimensional geometry in the x direction, excess bulk generation rate $g' = 0$, and the system at steady state, Eq. (1.93), becomes, for minority carriers

$$\frac{d^2(\delta n)}{dx^2} - \frac{(\delta n)}{D_n\tau_n} = 0 \tag{1.101}$$

with an integral given by

$$n = (\delta n)_0 \exp - x/L_n \tag{1.102}$$

where a diffusion length is defined for minority n carriers as, $L_n = (D_n \tau_n)^{1/2}$, the e^{-1} attenuation length in the material. In the same way, for n-type material, the corresponding diffusion length for the minority p carriers is $L_p = (D_p \tau_p)^{1/2}$.

A set of special applications of the one-dimensional version of the transport equations, Eqs. (1.86) and (1.87), are now discussed, as this avenue leads to the investigation of surface recombination phenomena. For low excess-carrier density levels, the recombination terms in Eqs. (1.86)–(1.87) can be replaced by the term $n/\tau_n - n_0/\tau_{n0} \cong (n - n_0)/\tau_n$. Also, $\tau_n = \tau_p$, assuming that no trapping or other effects occur, while electrons and holes are being created and recombined pairwise. Under these conditions, the one-dimensional version of Eqs. (1.86) and (1.87) becomes, respectively

$$\frac{\partial n}{\partial t} = D_n \frac{\partial^2 n}{\partial x^2} + \mu_n \left(E \frac{\partial n}{\partial x} + n \frac{\partial E}{\partial x} \right) + g_n - (n - n_0)/\tau_n \tag{1.103}$$

$$\frac{\partial p}{\partial t} = D_p \frac{\partial^2 p}{\partial x^2} - \mu_p \left(E \frac{\partial p}{\partial x} + p \frac{\partial E}{\partial x} \right) + g_p - (p - p_0)/\tau_p \tag{1.104}$$

The first application is that of photoexcitation of p-type silicon. That is, the material is illuminated with light, resulting in assumed uniform electron-hole pair generation throughout the material at a rate g. It is also assumed that the electric field $\bar{E} = 0$. Equation (1.104) then yields

$$\frac{dp}{dt} = g_p - \frac{(p - p_0)}{\tau_p} \tag{1.105}$$

A similar equation holds as well for electron densities. At steady state, dp/dt vanishes, implying that $p = p_0 + g_p \tau_p$ is constant. Suppose that with an initial condition of $p(0) = p_0 + g_p \tau_p$, the illumination ceases abruptly, so that $g_p = 0$ in Eq. (1.105). The integral of the corresponding differential equation is then

$$p = p_0 + g_p \tau_p \exp - t/\tau_p \tag{1.106}$$

This expresses the principle behind the Stevenson-Keyes method of measuring carrier lifetime.[19]

The next example is that where excess carriers are injected into a semi-infinite slab of, say, n-material. Such a source of carriers might be due to incident x-rays creating electron-hole pairs at the slab surface. Again, with no field applied, and steady-state conditions, Eq. (1.104) gives, for the minority carrier density

$$D_p\left(\frac{d^2p}{dx^2}\right) - \frac{(p - p_0)}{\tau_p} = 0 \tag{1.107}$$

For the boundary conditions of constant $p = p(0)$ at the slab face, and that $\lim_{x\to\infty} p = p_0$, the integral of Eq. (1.107) is

$$p(x) = p_0 + (p(0) - p_0)\exp - x/L_p \tag{1.108}$$

Another version of the above is that where the face of the slab is being illuminated, but now excess carriers are being removed at $x = x_0$. The above condition is then changed to $p(x_0) = 0$. The integral for Eq. (1.107) is now given by

$$p(x) = p_0 + (p(0) - p_0)(\sinh(x_0 - x)/L_p)/\sinh(x_0/L_p) \tag{1.109}$$

The corresponding current density at x_0 is

$$J_p = -eD_p(dp/dx)|_{x_0} = e(p(0) - p_0)D_p/L_p \sinh(x_0/L_p) \tag{1.110}$$

A third example is that of a light pulse incident on the face of the p-type material, generating a localized source of carriers thereon. Equation (1.104), with no field applied, becomes, after the onset of the pulse

$$\partial p/\partial t = D_p(\partial^2 p/\partial x^2) - (p - p_0)/\tau_p \tag{1.111}$$

The resulting hole density is given by the solution

$$p(x,t) = p(0) + N_0[\exp - (x^2/4D_pt + t/\tau_p)]/(4\pi D_pt)^{1/2} \tag{1.112}$$

where N_0 is the areal density of holes generated by the pulse of light. This expression models the Haynes-Shockley experiment for the measurement of mobility[20] by way of the Einstein relations connecting μ_p, and D_p in Eq. 1.63.

1.14 SURFACE RECOMBINATION

A fourth example of carrier transport behavior is that of surface recombination. That is, when minority carriers reach the surface of the material from the interior, they can recombine at the surface. For p-type material, Eq. (1.104) becomes, with E = 0,

$$\frac{\partial p}{\partial t} = D_p \frac{\partial^2 p}{\partial x^2} + g_p - (p - p_0)/\tau_p \tag{1.113}$$

The boundary condition at infinity is the earlier one, i.e., $p_\infty = p_0 + g_p\tau_p$. The corresponding integral of Eq. (1.113) is

$$eD_p(\partial p/\partial x)_0 = J_{p0} = eS_p(p(0) - p_0) \tag{1.114}$$

where S_p (cm s^{-1}) is termed the *surface recombination velocity*, discussed in detail below. It is called a velocity because it has those dimensions. The boundary condition at infinity is the earlier one, i.e., $p_\infty = p_0 + g_p\tau_p$. The corresponding integral of Eq. (1.113) is

$$p(x) = p_0 + g_p\tau_p\left(1 - \frac{S_p L_p \exp - (x/L_p)}{D_p + S_p L_p}\right) \tag{1.115}$$

It is seen that in the limit of zero recombination velocity, $p(x) = p_0 + g_p\tau_p$ like the first example. For very large recombination velocities, $p(x) \sim p_0 + g_p\tau_p(1 - \exp - x/L_p)$, with the hole density approaching its equilibrium value p_0 at the surface. Further insight into the meaning of surface recombination velocity is provided by analogy to the ordinary or bulk recombination rate within the material, where the reciprocal minority carrier lifetime $\tau_p^{-1} = N_t\sigma_p v_{th}$. The surface recombination velocity is defined as $S_p = N_s\sigma_p v_{th}$, where N_s is the areal recombination center density, i.e., the number of recombination centers per square centimeter at the surface. S_p is a measure of the tendency of carriers to drift toward the surface and recombine there. If $S_p = 0$, there is no surface recombination, for there is zero current incident at the surface from the interior of the material, as seen from Eq. (1.114). If all carriers recombine at the surface upon incidence from the interior, then S_p is a maximum, and so is the current to the surface. The maximum surface recombination velocity approaches the thermal velocity v_{th} of the carriers.

The phenomenon of photoconductivity provides a further illustration of the preceding ideas. Consider a thin rectangular slab of n-type material of sides y_0 and z_0 and thickness $x_0 \ll y_0, z_0$. The slab is illuminated on both faces, with light of wavelength such that it is hardly attenuated in penetrating the material. Hence, the generation of carriers within is considered homogeneous and proportional to the light intensity. Then the excess minority carrier density varies only with the depth of penetration into the slab. Then, initially, the illumination is abruptly extinguished after having established a uniform excess-carrier density in the material. The appropriate version of Eq. (1.92) is

$$\frac{\partial}{\partial t}(\delta p) = D_p \frac{\partial^2}{\partial x^2}(\delta p) - \delta p/\tau_p; \qquad \delta p \ll n_0 \tag{1.116}$$

with surface boundary conditions on each face of the slab, where x_0 is the slab thickness

$$-D_p \frac{\partial}{\partial x}(\delta p)\bigg|_{x_0/2} = S_p \delta p(x_0/2, t); \qquad -D_p \frac{\partial}{\partial x}(\delta p)\bigg|_{-x_0/2} = S_p \delta p(-x_0/2, t) \tag{1.117}$$

and the two further conditions

$$\delta p(x, 0) = p_1; \qquad \lim_{x \to \infty} \delta p = 0 \tag{1.118}$$

The solution to Eqs. (1.116)–(1.118) is

$$\delta p(x, t) = 4p_1(\exp - t/\tau_p) \sum_{l=0}^{\infty} A_l(x) \exp - \alpha_l^2 D_p t \tag{1.119}$$

where $A_l = (\sin(\alpha_l x_0/2)\cos(\alpha_l x))/(\alpha_l x_0 + \sin(\alpha_l x_0))$, and α_l are the roots of

$$\alpha_l \tan(\alpha_l x_0/2) = S_p/D_p; \quad l = 0, 1, 2, \ldots \tag{1.120}$$

The differential conductance dC in the material, with electrodes assumed connected to the $y = 0$ and $y = y_0$ edges, is

$$dC = z_0 \sigma(x)\, dx/y_0 \tag{1.121}$$

The change in conductivity due to excess carriers present is contained within

$$\sigma(x) = \sigma_0 + \delta\sigma = \sigma_0 + e(\mu_n + \mu_p)\delta p(x) \tag{1.122}$$

Table 1.1 Properties of intrinsic germanium, silicon, gallium arsenide, silicon dioxide, silicon nitride, and aluminum oxide at 27°C unless otherwise noted. The permittivity of free space $\varepsilon_0 = 0.0886$ pF/cm.

MATERIAL	Ge	Si	GaAs	SiO_2	Si_3N_4	Al_2O_3
Type	Semiconductor	Semiconductor	Semiconductor	Insulator	Insulator	Insulator
Atomic or molecular weight	72.6	28.09	144.63	60.08	140.27	101.96
Atomic/molecular density ($10^{22}/cm^3$)	4.42	5.00	2.21	2.30	1.48	2.34
Crystal structure	Diamond-8 atoms per unit cell	Diamond-8 atoms per unit cell	Zinc-Blende-8 atoms/unit cell	Random tetrahedra	Amorphous	Hexagonal (Sapphire)
Work function (V)	4.4	4.8	4.7	–	–	–
Lattice constant (Å)	5.66	5.43	5.65	5.03	7.75	5.13
Density (g/cm^3)	5.33	2.33	5.32	2.27	3.44	3.97
Band gap energy (eV)	0.80	1.12	1.42	8.5	5.1	9.0
Effective density of states						
conduction band N_c ($10^{19}\ cm^{-3}$)	1.04	2.80	0.047	–	–	–
valence band N_v ($10^{19}\ cm^{-3}$)	0.06	1.04	0.70	–	–	–
Intrinsic carrier concentration n_i ($10^{10}\ cm^{-3}$)	2400.	1.45	0.0009	–	–	–
Diffusion coefficient						
Electrons D_n (cm^2/s)	101.	35.	223.	–	–	–
Holes D_p (cm^2/s)	49.	12.5	6.5	–	–	–
Hole-electron pair generation energy (eV)	2.8	3.6	4.8	17.	10.8	19.1
Electron mobility μ_n (cm^2/volt s)	3900.	1500.	8500.	–	–	–
Hole mobility μ_p (cm^2/volt s)	1900.	450.	400.	–	–	–
Dielectric constant ε	16.3	11.9	12.5	3.9	7.5	9.4
Breakdown field (V/μm)	8	30	35	600	900	48

Table 1.1 Continued

Melting point (°C)	937.	1415.	1238.	1650	1900	2072
Minority carrier lifetime (Max)s	10^{-3}	$2.5 \cdot 10^{-3}$	10^{-8}	–	–	–
Displ. damage thresh. energy (eV)	27.5	25.	9.9	–	–	–
Carrier generation constant (g)						
10^{13}hep/cm^3 rad(·)	11.9	4.05	6.92	0.81	–	–
Debye temperature (°K)	378	658	344	–	–	–
Vapor pressure (torr), Temp. °C	10^{-9}, 750	10^{-7}, 1050	1, 900	10^{-3}, 1450	–	10^{-6}, 725
	10^{-7}, 880	10^{-5}, 1250	100, 1050	0.1, 1750	–	10^{-4}, 1000
Specific heat, (joule/g °C) C_h	0.31	0.70	0.35	1.0	0.17	0.76
Thermal conductivity, (W/cm °C) K_{th}	0.60	1.5	0.54	0.014	0.185	0.36
Thermal diffusivity, κ (cm^2/s)	0.36	0.90	0.44	0.006	0.316	0.08
Linear coefficient of thermal						
expansion, $\Delta L/L\Delta T$ (10^{-6}/°C)	5.8	2.5	5.9	0.5	2.8	5.3
Hall effect mobility (cm^2/V sec.)						
μ_{Hn}	4500.	1300.	8500.	–	–	–
μ_{Hp}	3500.	500.	420.	–	–	–
Effective mass: m*						
electrons (m_n/m_o) $m_l(m_\parallel)$	1.64	0.97	0.068	–	–	–
(m_o = electron mass) $m_t(m_\perp)$	0.08	0.19	–	–	–	–
holes (m_p/m_o) m_{lh}	0.04	0.16	0.12	–	–	–
m_{hh}	0.33	0.50	0.50	–	–	–
Electron affinity (V) χ	4.00	4.05	4.07	1.00	–	–
Resistivity (ohm cm)	50.	$2.3 \cdot 10^5$	$7 \cdot 10^7$	$>10^{16}$	10^7	$>10^{16}$
Young's modulus (10^{11} dynes/cm^2)	10.3	13.1	8.5	8.0	–	–
Volume compressibility						
(10^{-11} cm^2/dyne)	0.13	0.10	0.13	0.12	–	–
Debye Length (μm)	0.65	24.0	2250.	0.02	–	–

where the second term is obtained from the relation between conductivity and mobility discussed in Section 1.8. So that

$$dC = z_0[\sigma_0 + e(\mu_n + \mu_p)\delta p(x)]\, dx/y_0 \tag{1.123}$$

Integrating through the slab thickness yields the conductance as

$$\begin{aligned} C = C_0 + dC &= \frac{z_0 x_0 \sigma_0}{y_0} + \frac{z_0}{y_0} e(\mu_n + \mu_p) \int_{-x_0/2}^{+x_0/2} \delta p(x,t)\, dx \\ &= C_0 + e(\mu_n + \mu_p)\delta P/y_0^2 \end{aligned} \tag{1.124}$$

where $\delta P = \delta N = \int_{-x_0/2}^{+x_0/2} \delta p \; dx$ is the total number of excess carriers in the slab, and the integrand $\delta p(x,t)$ is given by Eq. (1.119). The fractional change in the conductance is then

$$dC/C_0 = e(\mu_n + \mu_p)\delta P/\sigma_0 V \tag{1.125}$$

where V is the volume of the slab.

A method for determining the surface recombination velocity proceeds by pulsing the material with light and then waiting until the higher modes in the excess-carrier transient have damped out, so that only the $l = 0$ or fundamental mode remains. Then from Eq. (1.119), after a relatively long time compared with the damping rates of the higher modes, the single exponential variation persists, viz.

$$\delta p(x, t) \sim 4p_1 A_0(x) \exp - t/\tau_0 \tag{1.126}$$

where from Eqs. (1.119) and (1.120), the identification can be made that

$$\tau_0^{-1} = \tau_p^{-1} + \alpha_0^2 D_p; \qquad \alpha_0 \tan(\alpha_0 x_0/2) = S_p/D_p \tag{1.127}$$

If any two of the three quantities—observed lifetime τ_0, bulk lifetime τ_p, or recombination velocity S_p—are known, the third is determined from Eqs. (1.127). In the case where τ_0 and τ_p are known, the recombination velocity is obtained from

$$S_p = \sqrt{D_p(\tau_0^{-1} - \tau_p^{-1})} \tan\left(\frac{x_0}{z}\sqrt{(\tau_0^{-1} - \tau_p^{-1})/D_p}\right) \tag{1.128}$$

This method is similar to the Angstrom method for determining heat transfer parameters in materials by pulse-heating one end of a rod of this material and monitoring the thermal wave parameters within. It is also like the procedure for obtaining thermal neutron parameters by introducing a burst of neutrons into the material of interest and monitoring the asymptotic neutron density response therein.

PROBLEMS

1. (a) Eliminate the electron velocity and radius between Eqs. (1.1)–(1.3) to obtain the Bohr model energy level hierarchy, Eqs. (1.4) and (1.5). (b) Using the fact that $1\,\text{eV} = 1.6 \cdot 10^{-12}$ ergs and $h = 6.6 \cdot 10^{-27}$ ergs, compute the energy of the ground state of the Bohr atom of hydrogen to get $E_1 = -13.6\,\text{eV}$. (c) Rederive the Bohr energy levels for the hydrogenlike, weakly bound, donor orbital electrons in a medium of dielectric constant ε. Show that they are given by $E_n = -me^4Z^2/2\varepsilon^2\hbar^2n^2$. Compute the ground state donor electron energy levels in silicon where $\varepsilon \cong 12$. Is that value correct?. Check it from another source. (d) Eliminate the electron velocity and the kinetic energy within Eqs. (1.1) and (1.3) to obtain the Bohr radius of the ground state electron.
2. Discuss how the proper normalization of the Fermi probability distribution is accomplished through integration to obtain the correct equilibrium electron or hole density. What is the normalization parameter?
3. Show that the effective mass is the actual mass for a free carrier, whether electron or hole.
4. Derive the equilibrium number density of conduction electrons n_0, using Eqs. (1.16) and (1.17) and the approximation discussed to obtain its expression, as given in Eq. (1.27).
5. Obtain the expression for the Fermi level in an intrinsic material, as given in Eq. (1.33). Show that E_f lies very near the gap center.
6. Show that the number density of silicon atoms is about $5 \cdot 10^{22}$ per cm^3, using values of silicon density and atomic weight obtained from Table 1.1. Avogadro's number is $6.02 \cdot 10^{23}$ atoms per mole.
7. Using the three relations following Eq. (1.38), compute the number density of holes and electrons introduced when silicon is doped with 10^{14} phosphorous donor atoms per cm^3.
8. In pure silicon at room temperature, the mean speed of the conduction electrons is about $100\,\text{km}\,\text{s}^{-1}$, and their mean free path is about $0.01\,\mu\text{m}$. What is the corresponding mean collision rate?
9. The Fick's law expression for the carrier diffusion current given in Eq. (1.57) yields a positive electron particle current and a negative hole particle current, as in Eq. (1.58). Should the signs be reversed, and why?
10. Describe how the expression for the recombination rate U shows that the traps closest to E_i are more effective.
11. For intrinsic silicon at room temperature, compute:
    (1) The ratio of free hole-electron pairs to silicon atoms.
    (2) The intrinsic conductivity and resistivity using Eq. (1.56) and Table 1.1.

(3) Introducing dopant in the form of antimony (is it donor or acceptor?) at a concentration of 1 Sb atom per 10^7 Si atoms:
  (3a) What is the dopant concentration in atoms per cm^3?
  (3b) What is the resulting electron and hole concentration?
  (3c) By what factor is the hole concentration reduced?
  (3d) What is the resulting conductivity and resistivity?

12. Explain the role of dopants in semiconductor device operation at (a) very low temperatures, and (b) very high temperatures.

13. From a statistical viewpoint, to compare the fermi distribution with the gaussian (bell-shaped) distribution, it is realized that the latter can be associated with four statistical parameters, to include:
(a) *Mean*: the average value of a random variable x from the distribution.
(b) *Median*: the value x_{median} such that a random variable x from the distribution is equally likely to be above or below x_{median}.
(c) *Mode*: the random variable x_{mode} for which the probability distribution is a maximum.
(d) *Variance*: the deviation from the mean of the random variable, in the mean square sense.

For the fermi distribution, does the fermi level, E_f, correspond to (a), (b), (c), (d), or (e)—none of the above?

14. Both acceptor and donor atoms in silicon can be viewed as hydrogen-like atoms embedded in the silicon lattice of dielectric constant ε. This is because they have only one more or one less electron in their valence shell, with respect to silicon, as discussed in Section 1.6.
(a) From the Bohr model, calculate the ground-state energy levels of such donors and acceptors.
(b) Since germanium has a larger dielectric constant than silicon (Table 1.1), it's donor and acceptor state energy levels should be about half that of silicon. Show this.

15. It is straightforward to show that minority carrier concentration (e.g., n-type carriers in a p-type region) can be given by[35] $n_p \cong (n_i^2/p)\exp(\Delta E_g/kT)$, where ΔE_g is an incremental change in the band gap width. For a very high dopant concentration, greater than the Mott criterion[35] (as discussed in Sections 1.6 and 1.10), the majority carrier concentration can be replaced by the dopant density. Inserting the expression for ΔE_g from Eq. (1.77a) for very high dopant concentrations into the preceding for n_p, show that this implies disintegration of transistor function for $T \leqslant 210°K$.

REFERENCES

1. C. Kittel. Introduction to Solid State Physics. Wiley, New York, 1966.
2. M. L. Cohen and T. K. Bergstrasser. "Band Structure and Pseudo-potential Form Factors for Fourteen Semiconductors." Phys. Rev. 141:789, 1966.
3. W. Shockley. Electrons and Holes in Semiconductors. Van Nostrand, New York, 1950.
4. R. J. Chaffin. Microwave Semiconductor Devices. Wiley-Interscience, New York, pp. 12ff., 1973.

5. J. P. McKelvey. Solid State and Semiconductor Physics. Harper and Row, New York, pp. 50ff., 1966.
6. Ibid, p. 311.
7. McKelvey, op. cit. p. 312.
8. Ibid, p. 314.
9. Loc. cit.
10. McKelvey, op. cit. p. 316.
11. H. Brooks and C. Herring, "Impurity Scattering Mobilities," Phys. Rev. 83:879, 1951.
12. Kittel, op. cit. pp. 133ff.
13. C. T. Sah, R. N. Royce, and W. Shockley. "Carrier Generation and Recombination in p-n Junction Characteristics." Proc. IRE, 45:1228, 1957.
14. G. C. Messenger. "A Two Level Model For Lifetime Reduction Processes in Neutron Irradiated Silicon and Germanium." IEEE Trans. Nucl. Sci. NS-12, No. 6, Dec. 1967.
15. McKelvey, op. cit. pp. 326ff.
16. Ibid.
17. S. K. Ghandi. Semiconductor Power Devices. Wiley-Interscience, New York, pp. 7ff., 1977.
18. Op. cit. pp. 9ff.
19. D. T. Stevenson and R. J. Keyes. "Measurements of Carrier Lifetime in Germanium and Silicon." J. Appl. Phys. 26:190, 1955.
20. J. P. Haynes and W. Shockley. "The Mobility and Lifetime of Injected Holes in Germanium." Phys. Rev. 81:835, 1951.
21. S. M. Sze. Physics of Semiconductor Devices. Wiley-Interscience, New York, pp. 46ff., 1969; R. A. Smith. Semiconductors, 2nd ed. Cambridge Univ. Press, 1979.
22. A. B. Phillips. Transistor Engineering. McGraw-Hill, New York p. 68, 1962.
23. I. C. Irvin. "Resistivity of Bulk Silicon and Diffused Layers in Silicon." Bell System Tech. J. 41(2):387, March 1962.
24. J. D. Beck and R. Conradt. "Auger Recombination in Silicon." Solid State Comm., Vol. 13, p. 93, 1973 (German).
25. S. I. Soclof and P. A. Iles. "Grain Boundary and Impurity Effects in Low Cost Silicon Solar Cells." Proc. Photovoltaic Conf. p. 56, 1975.
26. J. Dziewior and W. Schmid. "Auger Coefficients For Highly Doped and Highly Excited Silicon." Appl. Phys. Lett. 31:346, 1977.
27. J. W. Slotboom. "Analysis of Bipolar Transistors." Ph.D. Thesis, Eindhoven Tech. Univ., October 1977.
28. R. M. Warner and B. L. Grung. Transistors: Fundamentals for the Integrated Circuit Engineer. Wiley, New York, Sec. 3-3.4, 1983.
29. C. Y. Wei and H. H. Woodbury. "Measurement of the Minority Carrier Lifetime Using an MOS Capacitor", IEEE Trans. Elect. Dev., Vol ED-32, No. 5, May 1985 pp. 957–964 (Table I).
30. H. F. Wolf. Semiconductors. Wiley-Interscience, New York, p. 223, 1971.
31. Sze, op. cit. p. 39.
32. Sze, p. 29.
33. C. Wang, K. Misiakos, and A. Neugroschel. "Minority Carrier Transport Parameters in N-Type Silicon." IEEE Trans. Elect. Dev. ED-37:1314–1322, May, 1990; EDL, Dec. 1990.
34. Warner and Grung, op. cit., Section 4-5.4.
35. K. Yano, K. Nakazato, M. Miyamoto, M. Aoki, and K. Shimohigashi. "Base-Emitter Injection in Low Temperature Pseudo-Heterojunction Bipolar Transistors." IEEE Trans. Elect. Dev. 37(10):2222–2229, Oct. 1990.

36. D. B. Brown. IEEE NSRE Conference Short Course, Reno NV, Chap. 2, 1990.
37. T. P. Ma and P. V. Dressendorfer (eds.). Ionizing Radiation Radiation Effects in MOS Devices and Circuits. Wiley-Interscience, New York, Chap. 3, 1989.
38. C. H. Wang and A. Neugroschel. "Minority Carrier Transport Parameters in Degenerate n-Type Silicon." IEEE Elect. Dev. Lett., No. 12:576–578, Dec. 1990.
39. H. K. Henisch. Rectifying Semiconductor Contacts. Oxford Press, London, Chap. 7, 1957.
40. M. E. Law, E. Solley, M. Liang, and D. E. Burke. "Self Consistent Model of Minority Carrier Lifetime, Diffusion Length, and Mobility." IEEE Elect. Dev. Lett., pp. 401–403, Aug. 1991.

CHAPTER 2
THE FIELD EQUATIONS

2.1 INTRODUCTION

As evidenced in Chapter 1, the kinematics and energetics of electrons and holes in semiconductors are influenced by applied and internally generated electromagnetic fields. In this chapter, the details of how these fields are produced, their sources, and their effects, which are important in terms of a further description of semiconductor operation, are discussed. Concurrently, a set of expressions describing field behavior will be stated, which make up an important part of the apparatus to be used in later chapters to describe radiation damage phenomenology and its effects on the semiconductor systems of interest.

2.2 MAXWELL'S EQUATIONS FOR ELECTRIC AND MAGNETIC FIELDS

Maxwell, among his many other accomplishments, presented a mathematically elegant and compact description of the behavior of electric and magnetic fields. This is embodied in his short set of equations that are now to be discussed in sequence. Using a vector formulation, these equations contain all the important, experimentally derived electromagnetic field relationships known today, in the classical, macroscopic sense. They span all the classical electromagnetic theory, as opposed to quantum electrodynamics. Also, no explicit relativistic effects are included, with respect to nonzero rest mass particles. For example, the electromagnetic theory of an ultrahigh-speed electron whose mass increases with velocity is not discussed. The Maxwell equations as given represent all the time-varying field behavior of the systems under discussion herein, in the sense that electrostatics and magnetostatics are obtained as special cases. Maxwell's equations are expressed in gaussian units.

The first Maxwell equation is the differential form of Faraday's law of induction. It contains the experiments of Faraday involving currents in circuits threaded by time-varying magnetic fields. It is

$$\nabla \times \bar{E} = -\partial \bar{B}/c\partial t \tag{2.1}$$

where c is the speed of light, and the magnetic induction is given by $\bar{B} = \mu\bar{H}$. $\bar{H}$ is the magnetic field intensity, and μ is the magnetic permeability of the medium under consideration. $\bar{E}$ is the associated electric field intensity or electric field strength. With suitable mathematical manipulation, Eq. (2.1) can be transformed into an explicit statement of Faraday's law, viz., "The induced electromotive force (emf) around a circuit embedded in a time varying magnetic field, is proportional to the time rate of change of the magnetic field linking the circuit."[1] The negative sign in Eq. (2.1) is a manifestation of Lenz's law, which states that the induced voltage is of such polarity as to oppose the magnetic flux change threading the circuit.

The second Maxwell equation is the differential form of Ampere's law, extended by Maxwell. It is

$$\nabla \times \bar{H} = 4\pi\bar{J}/c + \partial\bar{D}/c\partial t \tag{2.2}$$

where $\bar{J}$ is the current density and the electric displacement $\bar{D} = \varepsilon\bar{E}$. ε is the dielectric permittivity. The rightmost term is called the *displacement current*, due to Maxwell. Without it, these equations would not describe electromagnetic radiation.

The third Maxwell equation is the differential form of Gauss's or Coulomb's law. It is given by

$$\nabla \cdot \bar{D} = 4\pi\rho \tag{2.3}$$

where ρ is the charge density of the source from which emanates the electric field, as given from $\bar{D} = \varepsilon\bar{E}$. From this equation, with some simple integrations, the recognizable form of Gauss's law will be obtained.

The fourth Maxwell equation is simply a statement of the fact that, as far as we know in our universe there are no isolated magnetic poles, that is, unlike electric fields that can stream from a source and can extend to the infinite, all magnetic fields emanating from one pole of a magnet eventually return to end on the opposite pole. The corresponding differential form of this fact is

$$\nabla \cdot \bar{B} = 0 \tag{2.4}$$

Maxwell's equations, together with the Lorentz force on a charged particle due to the combined electric and magnetic fields, and Newton's second law of motion, provide the basis for the complete classical description of the dynamics of a charged particle in an electromagnetic field.

Macroscopically, the dynamic response of materials including semiconductors is summarized in the constitutive relations, viz., $\bar{D} = \varepsilon_s \bar{E}$, $\bar{J} = \sigma \bar{E}$, and $\bar{B} = \mu \bar{H}$. ε_s, σ, and μ, are, respectively, the dielectric permittivity, conductivity, and the permeability of the medium of interest. Frequently, in the context of radiation hardening and susceptibility, the dielectric permittivity ε_s plays the role of a "memory" function when the corresponding medium is subject to damaging radiation. That is, the electric displacement, $D(t)$, is the superposition of electric fields, $\bar{E}(\tau)$, existing for times in the past, but weighted by the memory function ε_s, i.e.,

$$\bar{D}(t) = \int_{-\infty}^{t} \varepsilon_s(t - \tau)\bar{E}(\tau)d\tau \tag{2.5}$$

where the effect on the present $\bar{D}(t)$ due to fields $\bar{E}(\tau)$, $t - \tau$ seconds ago is determined by the memory $\varepsilon_s(t - \tau)$ function to yield the above integral. Note that if the materials possess no memory, then $\varepsilon_s(t) \sim \delta(t)$ (the Dirac delta function), giving the usual $\bar{D} = \varepsilon_s \bar{E}$. As implied from Table 1.1, $\varepsilon_s = \varepsilon\varepsilon_0$, where ε is the dielectric constant and ε_0 is the permittivity of free space. Infrequently, when it is clear from the context, ε is used to mean ε_s (i.e., "dielectric constant" is sometimes meant as "dielectric permittivity").

2.3 SCALAR AND VECTOR POTENTIALS

One vector identity is that the divergence of the curl of any vector is identically zero. That is, for any vector $\bar{U}$, $\nabla \cdot \nabla \times \bar{U} \equiv 0$. Because Eq. (2.4) asserts that $\nabla \cdot \bar{B} = 0$, then a vector potential $\bar{A}$ can be defined, such that

$$\bar{B} = \nabla \times \bar{A} \tag{2.6}$$

Inserting this into the first Maxwell equation, Eq. (2.1), yields

$$\nabla \times (\bar{E} + \partial \bar{A}/c\partial t) = 0 \tag{2.7}$$

A second vector identity is that the curl of the gradient of any scalar is also identically zero. That is, for any scalar V, $\nabla \times \nabla V = 0$. So that from Eq. (2.7) a scalar potential ϕ can be defined by

$$\bar{E} = -\nabla\phi - \partial \bar{A}/c\partial t \tag{2.8}$$

Using Eqs. (2.6) and (2.8), we can eliminate the magnetic and electric fields in the second and third Maxwell equations, Eqs. (2.2) and (2.3), to reduce them to two coupled equations in the scalar and vector potentials. They are, for $\mu = \varepsilon = 1$,

$$\nabla^2\phi + \partial\nabla\cdot A/c\partial t = -4\pi\rho \tag{2.9}$$

$$\nabla^2\bar{A} - \frac{1}{c^2}\frac{\partial^2\bar{A}}{\partial t^2} - \nabla\left(\nabla\cdot\bar{A} + \frac{1}{c}\frac{\partial\phi}{\partial t}\right) = -\frac{4\pi}{c}\bar{J} \tag{2.10}$$

If Eqs. (2.9) and (2.10) can be rewritten to uncouple $\bar{A}$ and ϕ, so as to obtain separate equations for each, then they can be solved independently. The point is that then with $\bar{A}$ and ϕ known, the electric field $\bar{E}$ and the magnetic field $\bar{H}$ can be gotten in a straightforward manner from Eqs. (2.6) and (2.7). This bypasses the approach of attempting to obtain the electric and magnetic fields directly from the complicated vector relationships of their defining Maxwell equations. Now, from Eq. (2.6) it is seen that the gradient of any scalar V can be added to $\bar{A}$. This is because $\nabla \times (\bar{A} + \nabla V) = \nabla \times \bar{A} + \nabla \times \nabla V$, where $\nabla \times \nabla V \equiv 0$. Then to maintain the electric field invariant, ϕ must be able to be replaced by $\phi - \partial V/c\partial t$, as seen by substituting $\bar{A} + \nabla V$ into Eq. (2.8).

To uncouple Eqs. (2.9) and (2.10), let $\bar{A}$ be related to ϕ by

$$\nabla\cdot\bar{A} + \frac{1}{c}\frac{\partial\phi}{\partial t} = 0 \tag{2.11}$$

by choosing V appropriately. If Eq. (2.11) is inserted into Eqs. (2.9) and (2.10), $\bar{A}$ and ϕ are uncoupled, because they now satisfy

$$\nabla^2\phi - \frac{1}{c^2}\frac{\partial^2\phi}{\partial t^2} = -4\pi\rho \tag{2.12}$$

and

$$\nabla^2\bar{A} - \frac{1}{c^2}\frac{\partial^2\bar{A}}{\partial t^2} = -\frac{4\pi}{c}\bar{J} \tag{2.13}$$

Equations (2.11)–(2.13) are a set of three equations that are completely equivalent to the original Maxwell equations, Eqs. (2.1)–(2.4), and are in a useful form for applications of interest.

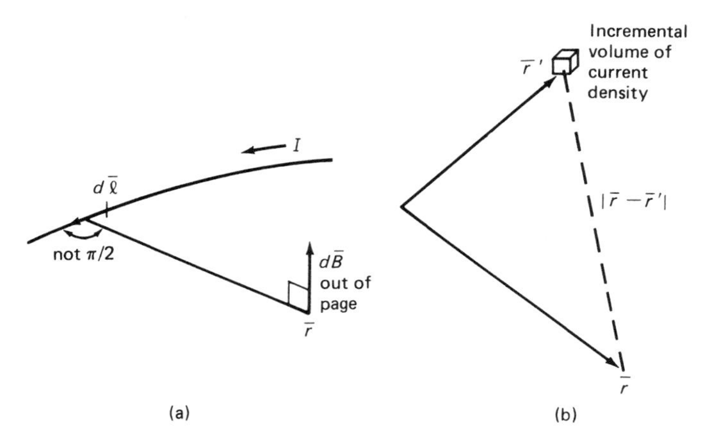

Fig. 2.1 Incremental magnetic induction due to incremental current element $Id\bar{l}$.

2.4 HALL EFFECT

An immediate application of a combined electric and magnetic field description, as discussed above, is that of the Hall effect. The Law of Biot and Savart is a consequence of Ampere's law. This second Maxwell equation, Eq. (2.2), asserts that the differential magnetic induction, $d\bar{B}$, due to a current I flowing in an incremental directed segment of a wire $d\bar{l}$ is given by

$$d\bar{B} = Id\bar{l} \times \bar{r}/cr^3 \tag{2.14}$$

as illustrated in Fig. 2.1a. In Fig. 2.1a, $\bar{r}$ is the vector distance from the incremental wire current element $Id\bar{l}$ to the field point where $d\bar{B}$ is measured, r is the corresponding distance out from a point on the wire to the field point, c is the speed of light for the charge measured in *esu*, and the magnetic induction is measured in *emu*. Instead of the current in the wire, assume a moving charge e with velocity $\bar{v}$, so that $Id\bar{l}$ can be replaced by $e\bar{v}$. Then the magnetic induction at the field point $\bar{r}$ is given by

$$\bar{B} = e(\bar{v} \times \bar{r})/cr^3 \tag{2.15}$$

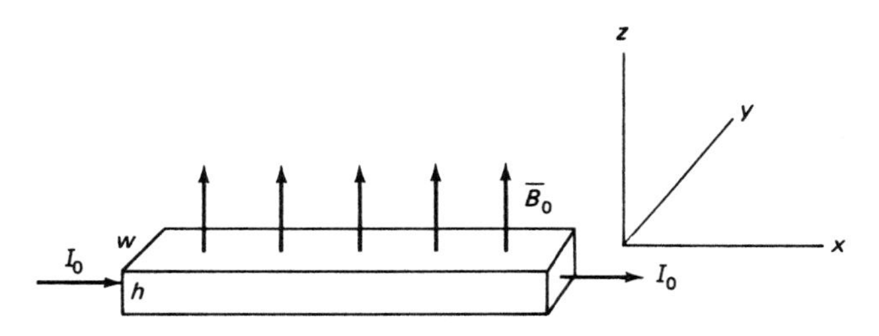

Fig. 2.2 Rectangular parallelepiped bar of semiconductor material immersed in magnetic field with current flowing as shown.

Now, insert a magnetic dipole of strength m exactly at the field point $\bar{r}$ in Fig. 2.1a. Then m is acted on by a force $\bar{F}_e$ due to the magnetic field generated by the current $e\bar{v}$ due to the moving charge e, i.e.

$$\bar{F}_e = m\bar{B} = e(\bar{v} \times m\bar{r}/r^3)/c \tag{2.16}$$

From Newton's third law, the equal and opposite reactive magnetic force $\bar{F}_e$ on the charge e, due to the magnetic induction $\bar{B}$ of the dipole m at distance r away from e, where $\bar{B} = m\bar{r}/r^3$, is

$$\bar{F}_e = e(\bar{v} \times \bar{B})/c \tag{2.17}$$

Generally, if the moving charge is subjected to both electric and magnetic fields, where the force due to the electric field is $e\bar{E}$, then the sought after total force on the charge e is

$$\bar{F} = e(\bar{E} + (\bar{v} \times \bar{B})/c) \tag{2.18}$$

In air, $\mu = 1$ so that $\bar{B} = \bar{H}$, then the above yields

$$\bar{F} = e(\bar{E} + (\bar{v} \times \bar{H})/c) \tag{2.19}$$

Now consider a rectangular parallelepiped of semiconductor material, with current flowing in the x direction, immersed in a magnetic field in the z direction, as shown in Fig. 2.2. The magnetic field produces an $e(\bar{v} \times \bar{B})/c$ force on the current in the y direction. Since I_0 flows in the x direction only, an electric (Hall) field E_y is produced to compensate this $e(\bar{v} \times \bar{B})/c$ force. The y component of the vector equation, Eq. (2.18), then must satisfy,

$$F_y = 0 = e(E_y + (v_z B_x - v_x B_z)/c) \tag{2.20}$$

The electric (Hall) field E_y, induced in the y direction due to the influence of the magnetic field B_z, is given by Eq. (2.20) to yield

$$E_y = v_x B_0/c \tag{2.21}$$

where $B_z = B_0$, with no magnetic field in the x direction. Because the current $I_0 = hwJ_x = -nev_x$, where n is the current particle number density, and J_x is the current density in the x direction, Eq. (2.21) can be rewritten as

$$E_y = RI_0B_0 \tag{2.22}$$

where the Hall coefficient R is defined for electrons as

$$R = -1/nec \tag{2.23}$$

The Hall voltage V_H that spans the width w of the bar in the y direction due to the induced field E_y, can be measured. It is given simply by the expression $V_H = wE_y$. If V_H, I_0, and B_0 are known or can be determined, then the Hall coefficient can be calculated from

$$|R| = V_H/wI_0B_0 = V_H/w^2hJ_xB_0 \tag{2.24}$$

where I_0 is the current flowing through the bar of cross-sectional area hw. Another reason for discussing the Hall effect is that its measurement provides a simple method for the determination of the carrier density n through the Hall coefficient R. Furthermore, the Hall effect also provides a simple means for obtaining the mobility. Assuming that the conductivity is known, and from Section 1.8, then

$$R\sigma = -(ne\mu)/(nec) = -\mu/c \tag{2.25}$$

The conductivity $\sigma = ne\mu$, where the majority carrier density is the main contributor to the current, so that any minority carrier current is negligible. The foregoing is somewhat oversimplified, because the carriers have a velocity distribution that should be taken into account.[2] When this is done, the magnitude of R is affected only slightly.

The Hall effect can also be used to determine the carrier concentration as well as its sign (i.e., whether the carriers are predominantly electrons or holes). The preceding development is refined to show this by using the expression for the Hall coefficient given, for a semiconductor with both n and p carriers, by[2,8]

$$R = (r/ec)(\mu_p^2 p - \mu_n^2 n)/(\mu_p p + \mu_n n)^2 \tag{2.25a}$$

$r = (\overline{\tau^2})/(\overline{\tau})^2$, where $\tau(E)$ is the mean free time between carrier collisions with the medium, which depend on energy E, and discussed in Section 1.7. Generally, $\tau = cE^{-m}$, where $m = \frac{1}{2}$ for lattice phonon scattering, and $m = -\frac{3}{2}$ for ionized impurity scattering.[2] Assuming a Boltzmann distribution function $p(E)$ for the carrier energies (velocities) (i.e., $p(E) = E^{1/2} \exp - E/kT$), then, for example,

$$\begin{aligned} \overline{\tau} &= \int_0^\infty E^{-m} p(E) dE \Big/ \int_0^\infty p(E) dE \\ &= \int_0^\infty E^{1/2-m} \exp - (E/kT) dE \Big/ \int_0^\infty E^{1/2} \exp - (E/kT) dE \end{aligned} \tag{2.25b}$$

and similarly for $(\overline{\tau^2})$. This finally results in $r = 3\pi/8 = 1.18$ for phonon scattering and $r = 315\pi/512 = 1.93$ for ionized impurity scattering.[2]

The Hall mobility, μ_H, is defined as the product of the conductivity and the Hall coefficient, for example

$$\mu_H = \sigma R \tag{2.25c}$$

and is different from the drift mobilities μ_n or μ_p. For a predominantly n-type semiconductor ($n \gg p$), from equation (2.25a),

$$R_n = -r/\text{nec} \tag{2.25d}$$

which is seen to be quite similar to the Hall coefficient in equation (2.23). For a predominantly p-type semiconductor ($p \gg n$), from equation (2.25a),

$$R_p = +r/\text{pec}. \tag{2.25e}$$

Hence, the majority carrier concentration type, whether n or p, can be identified by its sign in the case where $n \gg p$, or vice versa. This method begins to degrade as the semiconductor material approaches its intrinsic state, or is intrinsic initially, for reasons discussed in Section 1.10. For a known or measured Hall coefficient, the corresponding carrier concentration can be determined. Besides the preceding, the Hall effect can be used in certain instrumentation, such as magnetic field measurements, or where the sought for result is known to be a product of two associated Hall effect parameters.

2.5 WAVE EQUATIONS IN SEMICONDUCTORS

In Section 2.3, wave equations (2.12) and (2.13), were derived for the scalar and vector potentials. Wave equations governing the electric and magnetic fields in a homogeneous and isotropic material are derived by first taking the curl of the second Maxwell equation, Eq. (2.2), and using Eqs. (2.1)–(2.4) to give

$$\nabla \times \nabla \times \bar{H} = \nabla(\nabla \cdot \bar{H}) - \nabla^2 \bar{H} = \frac{4\pi}{c} \nabla \times \bar{J} + \frac{1}{c}\frac{\partial}{\partial t}(\nabla \times \bar{D}) \qquad (2.26)$$

which reduces to

$$\nabla^2 \bar{H} - \frac{\mu\varepsilon}{c^2}\frac{\partial^2 \bar{H}}{\partial t^2} - \frac{4\pi\mu\sigma}{c^2}\frac{\partial \bar{H}}{\partial t} = 0 \qquad (2.27)$$

In a similar manner, taking the curl of Eq. (2.1) yields

$$\nabla^2 \bar{E} - \frac{\mu\varepsilon}{c^2}\frac{\partial^2 \bar{E}}{\partial t^2} - \frac{4\pi\mu\sigma}{c^2}\frac{\partial \bar{E}}{\partial t} = 0 \qquad (2.28)$$

An important application of the above equations is the propagation of periodic plane electromagnetic waves. Such waves can be described by

$$\bar{E}(\bar{r}, t) = \bar{\varepsilon}_1 E_0 \exp(i(\bar{k} \cdot \bar{r} - \omega t)) \qquad (2.29)$$

and

$$\bar{H}(\bar{r}, t) = \bar{\varepsilon}_2 H_0 \exp(i(k \cdot \bar{r} - \omega t)) \qquad (2.30)$$

$\bar{\varepsilon}_1$ and $\bar{\varepsilon}_2$ are unit vectors that give the direction of the amplitude variations of the electric and magnetic fields, respectively, not their direction of propagation. The latter direction is given by the propagation vector $\bar{k} = 2\pi\bar{n}/\lambda$, where $\bar{n}$ is a unit vector in the direction of propagation. E_0 and H_0 are constant amplitudes, in that the actual fields are obtained by taking the real parts of Eqs. (2.29) and (2.30). $\bar{r}$ is the ray vector from a suitable origin to the plane wave front whose equation is given by $\bar{k} \cdot \bar{r} - \omega t = \text{constant}$. Inserting Eq. (2.29) into Eq. (2.28) yields the dispersion relation in the sense that $\omega = \omega(k)$, which implies that waves of different wavelength travel at different speeds. The dispersion relation is

$$c^2k^2/4\pi = \mu\varepsilon\omega^2/4\pi + i\omega\mu\sigma \tag{2.31}$$

Because the above is a complex number, let $c^2k^2/4\pi = (\bar{\alpha} + i\bar{\beta})^2$ to obtain a pair of real numbers given by

$$|\bar{\alpha}| = \frac{\omega}{2}\sqrt{(\mu\varepsilon/2\pi)}\sqrt{1 + \sqrt{1 + (4\pi\sigma/\varepsilon\omega)^2}}; \qquad |\bar{\beta}| = \omega\mu\sigma/2\alpha \tag{2.32}$$

Then the electric field plane wave can be written as

$$\bar{E}(\bar{r}, t) = \bar{\varepsilon}_1 E_0 \exp[-\bar{\beta}\cdot\bar{r} + i(\bar{\alpha}\cdot\bar{r} - \omega t)] \tag{2.33}$$

where $\bar{\alpha}$ is a vector in the same direction as $\bar{k}$. A similar relation for the magnetic field is

$$\bar{H}(\bar{r}, t) = \bar{\varepsilon}_2 H_0 \exp[-\bar{\beta}\cdot\bar{r} + i(\bar{\alpha}\cdot\bar{r} - \omega t)] \tag{2.34}$$

The velocity of the wave front gotten from $\bar{\alpha}\cdot\bar{r} - \omega t =$ constant is

$$|\bar{v}| = |d\bar{r}/dt| = \omega/\alpha = 2/\sqrt{(\mu\varepsilon/2\pi)}\sqrt{1 + \sqrt{1 + (4\pi\sigma/\varepsilon\omega)^2}} \tag{2.35}$$

Notice that $v = v(\omega)$, so that the wave will suffer dispersion as it propagates through the material as mentioned earlier. The skin depth δ is the e^{-1} attenuation distance of the wave as it penetrates the material while moving. From Eqs. (2.32) and (2.33), the skin depth is

$$\delta = 1/\beta = 2\alpha/\omega\mu\sigma \tag{2.36}$$

For most situations, $\sigma/\varepsilon\omega \gg 1$, so that from Eqs. (2.32) and (2.36), another form for the skin depth is

$$\delta = \sqrt{2/\omega\mu\sigma} = 1/\sqrt{\pi f\mu\sigma} \tag{2.37}$$

This expression shows that any incident electric field cannot penetrate a "good" conductor (i.e., one whose conductivity is extremely large). This also means that such a solid conductor cannot sustain an electric field within its interior.

For wave propagation in free space where $\sigma = 0$, and $\mu = \varepsilon = 1$, Eqs. (2.27) and (2.28) become, respectively

$$\nabla^2\bar{H} - \frac{1}{c^2}\frac{\partial^2\bar{H}}{\partial t^2} = 0 \tag{2.38}$$

and

$$\nabla^2 \bar{E} - \frac{1}{c^2}\frac{\partial^2 \bar{E}}{\partial t^2} = 0 \tag{2.39}$$

where the waves travel with the speed of light, c.

In free space with $\rho = 0$, Eqs. (2.3) and (2.4) imply that $\nabla \cdot \bar{E} = \nabla \cdot \bar{H} = 0$. Then Eqs. (2.29) and (2.30) give

$$\bar{\varepsilon}_1 \cdot \bar{k} = \bar{\varepsilon}_2 \cdot \bar{k} = 0 \tag{2.40}$$

which shows that $\bar{E}$, $\bar{H}$, and $\bar{k}$ are mutually perpendicular. This means that in free space, not only are the amplitude vibrations of the electric and magnetic fields perpendicular to each other, but both are perpendicular to the direction of propagation of the wave front. Also when Eqs. (2.29) and (2.30) are inserted into Eq. (2.1), the result is

$$\{(\bar{k} \times \bar{\varepsilon}_1)E_0 - \omega B_0 \bar{\varepsilon}_2/c\} \exp i(\bar{k} \cdot \bar{r} - \omega t)) = 0 \tag{2.41}$$

which means that

$$\bar{\varepsilon}_2 = (\bar{k} \times \bar{\varepsilon}_1)/k, \qquad B_0 = \sqrt{\mu\varepsilon} E_0 \tag{2.42}$$

again showing that $\bar{\varepsilon}_1$, $\bar{\varepsilon}_2$, and $\bar{k}$ are mutually perpendicular vectors and also that $\bar{E}$ and $\bar{B}$ are in phase, with a constant amplitude ratio. Hence, in free space $\bar{E}$ and $\bar{H}$ are transverse waves traveling in the $\bar{k}$ direction.

The power density $\bar{S}$ in these waves, that is, the energy flow rate per unit area in the $\bar{k}$ direction, can be shown to satisfy

$$\bar{S} = c(\bar{E} \times \bar{H}^*)/8\pi \tag{2.43}$$

where $\bar{H}^*$ is the complex conjugate of $\bar{H}$. Substituting Eqs. (2.29) and (2.30) into Eq. (2.43) yields

$$\bar{S} = (c/8\pi)(\varepsilon_s/\mu)^{1/2}|E_0|^2\bar{n} \tag{2.44}$$

where $\bar{n}$ is the unit vector in the direction of propagation. The time averaged power density is

$$\langle|\bar{S}|\rangle = (\varepsilon_s \bar{E} \cdot \bar{E}^* + (\bar{B} \cdot \bar{B}^*)/\mu)/16\pi = \varepsilon_s |E_0|^2/8\pi \tag{2.45}$$

The rate or speed v_s of the energy flow is

$$v_s = |\bar{S}|/\langle|\bar{S}|\rangle = c/\sqrt{\mu\varepsilon_s} \tag{2.46}$$

so that in free space $v_s = c$, since $\mu\varepsilon_s = \mu_0\varepsilon_0 = 1$.

2.6 SHORT DIPOLE ANTENNA

The radiation parameters of a simple dipole antenna are now discussed to illustrate how the vector potential $\bar{A}$ is used to determine them. Such antennas, or their complementary slot antennas, provide models for system metal protuberances, or unintentionial slits in the metal housing of electrically shielded systems. These longitudinal breeches acting as antennas can reradiate incident EMP energy from the outside into the system interior to cause malfunctions therein due to the induction of spurious voltages. This is discussed in detail in Chapter 9 on EMP.

The volume integral of Eq. (2.14), the Biot and Savart law, yields

$$\bar{B}(\bar{r}) = \int \bar{J}(\bar{r}')(\bar{r} - \bar{r}')\, d^3\bar{r}'/|\bar{r} - \bar{r}'|^3 c \tag{2.47}$$

where $\bar{r}'$ is the vector giving the location of the incremental amount of current density $\bar{J}(\bar{r}')d^3\bar{r}'$, and d^3r' is the incremental volume element, as depicted in Fig. 2.1b. $|\bar{r} - \bar{r}'|$ is the distance between the volume of the current density and the field point $\bar{r}$ where $\bar{B}$ is being calculated. Because $\bar{B} = \nabla \times \bar{A}$, Eq. (2.47) yields, with $\bar{J}(\bar{r}')(\bar{r} - \bar{r}')/|\bar{r} - \bar{r}'|^3 \equiv \nabla \times (\bar{J}(\bar{r}')/|\bar{r} - \bar{r}'|)$

$$\bar{B} = \nabla \times \bar{A} = \frac{1}{c}\int \nabla \times \left(\frac{\bar{J}(\bar{r}')}{|\bar{r} - \bar{r}'|}\right) d^3\bar{r}' = \frac{1}{c}\nabla \times \int \frac{\bar{J}(\bar{r}')\, d^3\bar{r}'}{|\bar{r} - \bar{r}'|} \tag{2.48}$$

where the rightmost expression is written using the appropriate vector identity. Hence

$$\bar{A}(\bar{r}, t) = \frac{1}{c}\int \frac{J(\bar{r}')\, d^3\bar{r}'}{|\bar{r} - \bar{r}'|} \tag{2.49}$$

Now assume that the current density $\bar{J}$ in the dipole antenna is varying sinusoidally in time, so that

$$\bar{J}(\bar{r}, t) = \bar{J}(\bar{r})\exp - i\omega\left(t - \frac{|\bar{r} - \bar{r}'|}{c}\right) \tag{2.50}$$

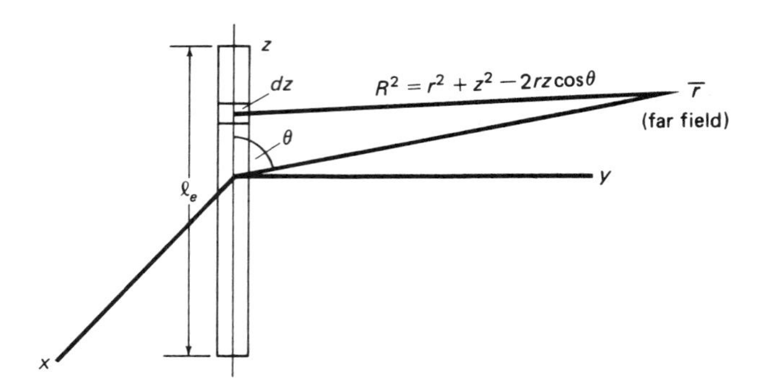

Fig. 2.3 Short dipole antenna.

Inserting this into Eq. (2.49) gives the vector potential at the field point $\bar{r}$ as

$$\bar{A}(\bar{r}, t) = \frac{1}{c} \int \frac{\bar{J}(\bar{r}')d^3\bar{r}'}{|\bar{r} - \bar{r}'|} \exp - i\omega\left(t - \frac{|\bar{r} - \bar{r}'|}{c}\right) \tag{2.51}$$

which is due to the current density $\bar{J}$ flowing in the antenna. That is, currents in the antenna at time $|\bar{r} - \bar{r}'|/c$ ago contribute to the vector potential now at time t. $|\bar{r} - \bar{r}'|/c$ is the time it takes for the effect at $\bar{r}$ to be manifest from the current source at $\bar{r}'$, as shown in Fig. 2.1b. Consider a short, center-fed dipole antenna of effective length l_e, such that $l_e/\lambda \ll 1$, as shown in Fig. 2.3. Assume that the antenna current is linear along this short antenna length, being I_0 at the center and falling to 0 at the ends. Then, the antenna current I can be written as

$$I = I_0(1 - 2|z|/l_e) \tag{2.52}$$

Inserting the above into Eq. (2.51) gives

$$A_z(r, t) = I_0 \int_{-l_e/2}^{l_e/2} dz \frac{(1 - 2|z|/l_e)}{Rc} \exp - i(\omega t - kR) \tag{2.53}$$

where R is defined in Fig. 2.3. The charge per unit length along the antenna $q_e = 2I_0/\omega l_e$ is constant on each leg of the dipole, with a corresponding dipole moment p, given as

$$|\bar{p}| = \int_{-l_e/2}^{l_e/2} zq_e(z)\,dz = I_0 l_e/2\omega \tag{2.54}$$

Because only the far field is of interest, $z/r \ll 1$. Substituting Eq. (2.54) into Eq. (2.53), with $R \cong r - z\cos\theta$, gives in a medium of permeability μ and dielectric constant ε, with $\omega = kc$ and $k = 2\pi/\lambda$

$$A_z(r,t) = \frac{-ikp\mu}{4\pi r} \cdot \exp i\omega(t - r/c) \tag{2.55}$$

As $\bar{H} = \nabla \times \bar{A}/\mu$, transformation into spherical coordinates gives the azimuthal component of the magnetic field in the far field as,

$$H_\phi = \frac{k^2}{4\pi r} p \sin\theta \cdot \exp i\omega(t - r/c). \tag{2.56}$$

The electric field is obtained from Eq. (2.1) for sinusoidally varying fields as

$$\nabla \times \bar{E} = -\frac{i\omega\bar{B}}{c} = \frac{i\omega}{c} \nabla \times \bar{A} \tag{2.57}$$

so that

$$\bar{E} = i\omega\bar{A}/c; \qquad E_\theta = -i\omega A_\theta = i\omega A_z \sin\theta \tag{2.58}$$

With Eq. (2.55) substituted into Eq. (2.58), the result yields

$$E_\theta = \frac{\omega k\mu}{4\pi r} p \sin\theta \exp i\omega(t - r/c) = \sqrt{\mu/\varepsilon} H_\phi \tag{2.59}$$

where $(\mu/\varepsilon)^{1/2}$, the impedance of free space, is 120π ohms.

Now, the angular distribution of the time averaged radiated power per unit solid angle for the dipole is, from Eqs. (2.56) and (2.59)

$$\frac{dP}{r^2 d\Omega} = \frac{1}{2} H_\phi E_\theta = \frac{1}{2}\sqrt{\mu/\varepsilon}\, H_\phi^2 = \frac{\sqrt{\mu/\varepsilon}}{32\pi^2 r^2} k^4 p^2 \sin^2\theta \tag{2.60}$$

Using Eq. (2.54), the total radiated average power is

$$P = \int_\Omega (dP/d\Omega)\, d\Omega = \sqrt{\mu/\varepsilon} I_0^2 \pi (l_e/\lambda)^2/3 = \frac{1}{2} I_0^2 R_r \tag{2.61}$$

where the corresponding radiation resistance $R_r = 80\pi^2(l_e/\lambda)^2$ ohms. For the case of the complementary antenna, which is a slot antenna of the

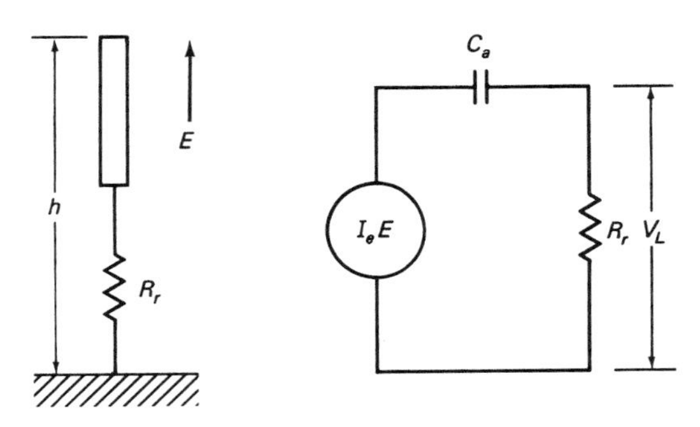

Fig. 2.4 Short monopole antenna and equivalent circuit.[7]

same length, the radiation field pattern is the same as the dipole, but with electric and magnetic field vectors interchanged, as shown from Babinet's principle.[6]

As mentioned above, the antenna structures of interest are small compared with the wavelength of the radiation of concern. Temporally, small wavelength corresponds to the case where the rise time of the wave from one end of the structure to the other, is small compared with the rise time of the incident energy pulse. In these situations, simple equivalent circuits can be used effectively to model the structure. The dipole equivalent circuit can be developed from the monopole. A short monopole model and equivalent circuit is shown in Fig. 2.4. From Thevenin's theorem, the load resistor R_r can be removed, and the output impedance of the antenna measured at that point. For a short monopole,[6] the output impedance is mainly due to the capacitance $C_a = 2h/cZ_0$, or

$$C_a = h/60c(\ln(2h/a) - 1) \tag{2.62}$$

and its effective height $l_e = h/2$. For the case of a dipole, it is seen that the required second leg is obtained from the image provided by the ground plane of the monopole in Fig. 2.4. This gives

$$C_a = h/120c(\ln(2h/a) - 1) \tag{2.63}$$

with corresponding effective height $l_e = h$.

It is easily shown that for an incident electric field E on the dipole, the voltage measured at its feedpoint is simply $V_L = l_eE$.[6] For the case where R_r is very much less than the capacitive reactance, $(\omega C_a)^{-1}$, the antenna current $I_0 = C_a l_e \dot{E}(t)$, and the maximum joule energy collected by the load is given by $J = R_r C_a (l_e E_m)^2/t_r$, where t_r is the rise time of the

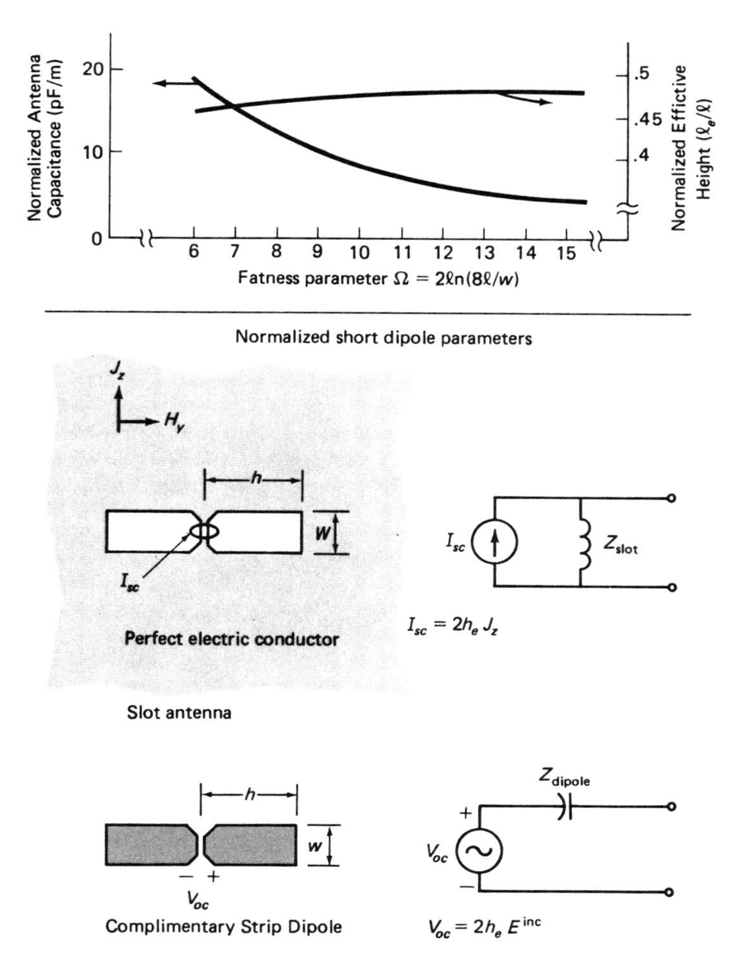

Fig. 2.5 Parameters for the short dipole, slot antenna, and its complementary strip dipole.[7]

incident electric field. E_m is the peak value of that field. On the other hand, when R_r is very much larger than the above capacitive reactance, the maximum collected energy is not dissipated by the load, but is stored in C_a and is given by $J = \frac{1}{2}C_a(l_e E_m)^2$.

These relationships hold over all frequencies of interest and are valuable in assessing experimental data or for designing electromagnetic sensors as discussed in Chapter 8.

For the complementary slot, if the antenna effective heights between the slot and the dipole are equal, i.e., $l_{e\,\text{slot}} = l_{e\,\text{dip}}$, then the input

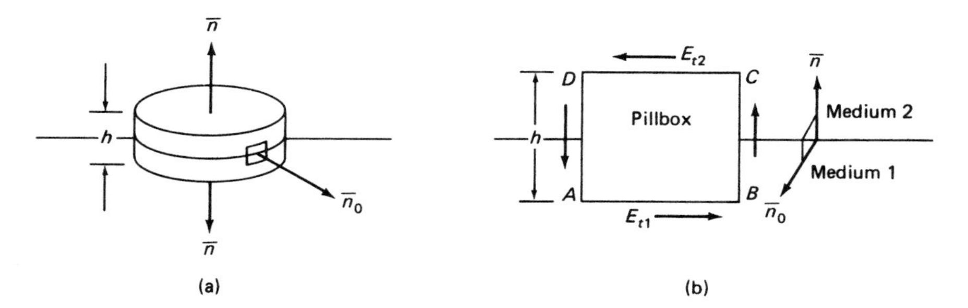

Fig. 2.6 Geometry of the interface between two distinct media showing the pillbox volume of unit area and height h.

impedance of the slot $Z_{\text{slot}} = Z_0^2/4Z_{\text{dip}} = (60\pi)^2/Z_{\text{dip}}$. The impedance and the effective height of the strip dipole differ only slightly from the cylindrical dipole, provided an antenna "fatness" parameter $\Omega = 2\ln(8l/W)$ is properly selected, as can be done from Fig. 2.5.

2.7 ELECTRIC AND MAGNETIC FIELD BOUNDARY CONDITIONS

As will be seen, it is important to investigate the behavior of electric and magnetic fields that span the boundary or interface between two distinct materials. It can be appreciated that such interfaces occur within semiconductors fabricated of complex layers of various material media.

First, the fourth Maxwell equation Eq. (2.4), asserts that $\nabla \cdot \bar{B} = 0$ throughout a given volume that can contain an interface. Figure 2.6a depicts a "pillbox" volume spanning two dissimilar media. Integrating $\nabla \cdot \bar{B}$ throughout this volume gives

$$0 = \int (\nabla \cdot \bar{B})\, d^3\bar{r} = \int (\bar{B} \cdot \bar{n})\, d^2\bar{r} \tag{2.64}$$

From the divergence theorem, this volume integral can be replaced by a surface integral of the normal component of $\bar{B}$ over the surface of the pillbox, which is the rightmost integral in Eq. (2.64). $\bar{n}$ is a unit normal vector to the surface as shown in Fig. 2.6a. Taking the limit of the surface integral in Eq. (2.64) as the height of the pillbox shrinks to zero, in such a way that the interface is always within the pillbox confines, yields

$$0 = \lim_{h \to 0} \int (\bar{B} \cdot \bar{n})\, d^2\bar{r} = (\bar{B}_2 - \bar{B}_1) \cdot \bar{n} \tag{2.65}$$

The contributions from the walls of the pillbox vanish because $\bar{B}$ and $\bar{n}$ are perpendicular to each other in the above limit, so that in this limit, $B_{1n} = B_{2n}$ across the interface. Hence, the first boundary condition developed is that the normal component of the magnetic induction is continuous across such a boundary or interface.

Second, for the tangential components of the magnetic field, integrating the second Maxwell equation, Eq. (2.2), over the surface of the pillbox gives

$$\int (\nabla \times \bar{H}) \cdot \bar{n}\, d^2\bar{r} = \frac{1}{c} \int \left(4\pi \bar{J} + \frac{\partial \bar{D}}{\partial t} \right) \cdot \bar{n}\, d^2\bar{r} \tag{2.66}$$

These surface integrals can be replaced by line integrals over closed curves bounding the pillbox surface (Stoke's theorem). Then Eq. (2.66) can be written as

$$\oint_{ABCD} \bar{H} \cdot d\bar{r} = \frac{4\pi}{c} \int \bar{J} \cdot \bar{n}_0\, d^2\bar{r} + \frac{1}{c} \int \frac{\partial \bar{D}}{\partial t} \cdot \bar{n}_0\, d^2\bar{r} \tag{2.67}$$

Again, in the limit of shrinking pillbox height h, and where $\bar{n}$ and $\bar{n}_0$ are perpendicular to each other, the above becomes

$$\bar{n} \times (\bar{H}_2 - \bar{H}_1) = \frac{1}{c} \lim_{h \to 0} \left(\frac{\partial \bar{D}}{\partial t} + 4\pi \bar{J} \right) h \tag{2.68}$$

The $\lim_{h \to 0} (\partial \bar{D}/\partial t) h = 0$, and $\lim_{h \to 0} (4\pi J h/c) = J_s$ defined as the surface current density, if any is present. Then

$$\bar{n} \times (\bar{H}_2 - \bar{H}_1) = \bar{J}_s \tag{2.69}$$

so that in the case of zero surface current density at the interface, the tangential components of the magnetic field are continuous across the boundary interface. Otherwise, the difference of the tangential components of the magnetic field yields the surface current density $\bar{J}_s$, which is normal to the boundary, as seen from Eq. (2.69). If one or both of the media are good conductors, then any surface currents will be immediately conducted into the interior of the material, so that, again, the tangential components of the magnetic field will be continuous across the boundary.

Third, analogous to the two boundary conditions involving the magnetic field, there are two boundary conditions involving the electric field. Integrating the first Maxwell equation, Eq. (2.1), over a closed surface yields

$$\int (\nabla \times \bar{E}) \cdot \bar{n}\, d^2\bar{r} = \frac{1}{c}\frac{\partial}{\partial t}\int \bar{B} \cdot \bar{n}\, d^2\bar{r} \tag{2.70}$$

Again using Stoke's theorem to convert the surface integral on the left above to a line integral along a closed path bounding that surface, as seen in Fig. 2.6b, gives

$$\oint_{ABCD} (\bar{E} \cdot \bar{n})\, d\bar{r} = \frac{1}{c}\int \frac{\partial \bar{B}}{\partial t} \cdot \bar{n}\, d^2\bar{r} \tag{2.71}$$

In the limit of shrinking height of the pillbox, h

$$\bar{n} \times (\bar{E}_{t2} - \bar{E}_{t1}) = 0 \tag{2.72}$$

Therefore, it is seen that the tangential components of the electric field are continuous across the boundary.

Fourth, and finally, integrating the third Maxwell equation, Eq. (2.3), throughout the volume of the pillbox provides

$$\int (\nabla \cdot \bar{D})\, d^3\bar{r} = 4\pi q \tag{2.73}$$

where $q = \int \rho d^3\bar{r}$, the total net charge enclosed in the pillbox. Again using the divergence theorem to transform the above volume integral to an integral over the surface bounding the pillbox volume yields

$$\int (\bar{D} \cdot \bar{n})\, d^2\bar{r} = 4\pi q \tag{2.74}$$

which is called Gauss's theorem. In the limit as the pillbox shrinks about the interface, the total charge $q = \sigma_s d^2\bar{r}$, where σ_s is the surface charge density at the interface. Then the above yields the fourth boundary condition, viz.

$$(\bar{D}_2 - \bar{D}_1) \cdot \bar{n} = 4\pi\sigma_s \tag{2.75}$$

To recapulate: for the field transition across the boundary or interface between medium 1 and medium 2, the following boundary conditions hold:

a. From Eq. (2.65) the normal components of the magnetic induction are continuous across the interface, i.e., $\bar{n} \cdot (\bar{B}_2 - \bar{B}_1) = 0$.

b. From Eq. (2.69), the tangential components of the magnetic field, are related as $\bar{n} \times (\bar{H}_2 - \bar{H}_1) = \bar{J}_s$, where $\bar{J}_s$ is the surface current density.
c. From Eq. (2.72), the tangential components of the electric field are continuous across the interface, i.e., $\bar{n} \times (\bar{E}_2 - \bar{E}_1) = 0$.
d. From Eq. (2.75), the normal components of the electric displacement are related as $\bar{n} \cdot (\bar{D}_2 - \bar{D}_1) = 4\pi\sigma_s$. There are other special boundary conditions involving the electric and magnetic field that are unique to the specific case.

2.8 SURFACE STATES AND SURFACE POTENTIAL

For our needs, another important application of Maxwell's field equations arises in the treatment of semiconductor surface considerations. This discussion will provide a basis for a treatment of the *p-n* junction, investigated in detail in Chapter 3.

The following begins with the discussion of carrier energetics in metals. The latter are examined to provide a framework for the investigation of semiconductor surfaces.

The work function of a metal or other material is the amount of energy required to overcome the surface potential barrier, in order to free carriers from the material surface. When two metals of different work functions are brought into extremely close contact, a brief current flows from the metal with the higher Fermi energy level to that with the lower. This generates an equilibrium contact potential difference between the two metals, which results in a coalescence of the Fermi levels into a common Fermi energy level. This is simply because the two metals now share common carrier populations characterized by a single Fermi level. All of the preceding is discussed in detail in Chapter 3.

In a semiconductor-metal contact, the above phenomenon also takes place, with some slight differences. Unlike a metal conductor, a semiconductor can sustain an internal electric field, so that a contact potential gradient can exist totally within the semiconductor, instead of only at the interface, as in the case of metal-metal contact. The former is depicted in Fig. 2.7 for a metal-*n*-type semiconductor interface, where the work function of the metal, ϕ_M, is larger than that of the semiconductor surface potential, ϕ_s.[5]

In Fig. 2.7, the contact potential gradient or the internal potential gradient, which is equivalent to the internal electric field, is seen to exist largely within the semiconductor. The potential energy $e\phi$ of an electron in equilibrium at the bottom of the conduction band within the semiconductor differs from that of an electron in the metal by $e(\phi_M - \phi_s)$. This

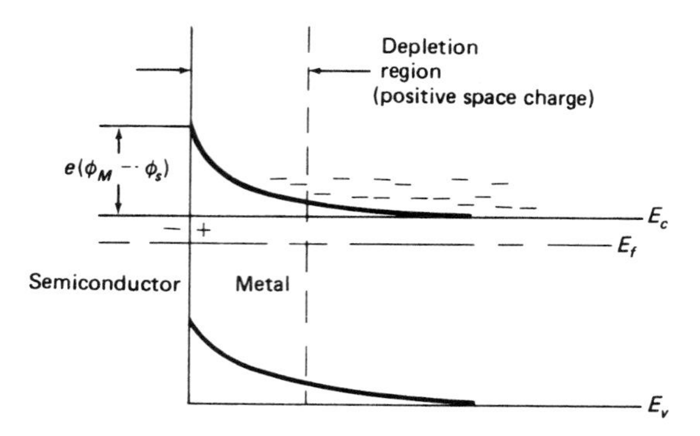

Fig. 2.7 *n*-Type semiconductor-metal interface with $\phi_M > \phi_s$.

causes the valence and conduction bands to shift with distance into the semiconductor, as seen by the curved lines in Fig. 2.7. This is discussed in detail in Chapter 3. The positive space charge in the region near the interface is the agent that produces the internal electric field that sustains the contact potential difference $\phi_M - \phi_s$ between the metal and the semiconductor. The space charge exists because of the excess concentration of ionized donor atoms over that of the electron population. If this potential difference is very large, the valence and conduction bands can be shifted with respect to the Fermi level, by such an amount that, near the interface, the valence band is closer to the Fermi level than the conduction band. Then, local to the interface, the semiconductor material is now in effect *p*-type. This subsurface is inverted in carrier population, and this regime is called an *inversion region*, as seen in Fig. 2.8a. Figure 2.8b depicts the contact interface where $\phi_M < \phi_s$. In this case, the surface is at a negative potential and the bands shift downward, as shown in the figure. It is seen that an electron accumulation region, in contrast to a depletion region, forms where the electron density is greater than that of the ionized donor atoms.

To provide a valid description of the fact that the rectification properties of silicon-metal contacts are essentially independent of the work function difference, whatever the metal,[3,4] it was postulated and experimentally verified that surface states exist. That is, there are localized electronic energy states associated with the very existence of the surface of a semiconductor, and lying in the forbidden energy gap.

Consider now the fields and potentials within the surface region of a one-dimensional, semi-infinite slab of semiconductor, when surface states are present. This is depicted in Fig. 2.9. Let the zero of the potential $\phi(x)$, where x is the principal dimension, be taken at the interior point

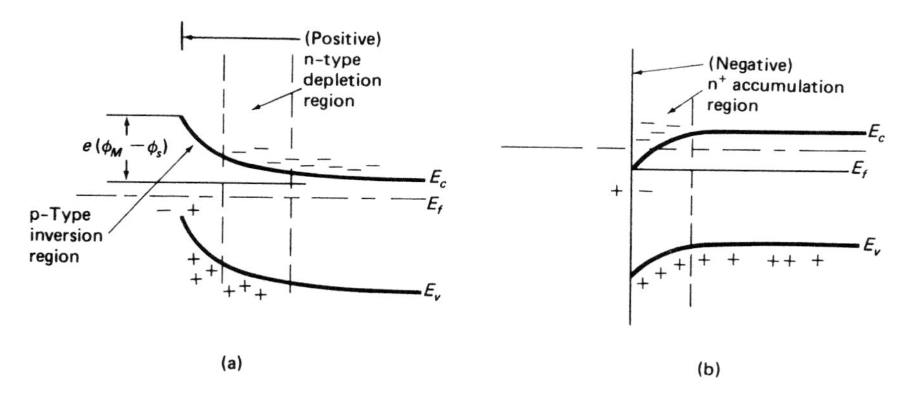

Fig. 2.8 Inversion layer at the boundary between a metal and an *n*-type semiconductor.

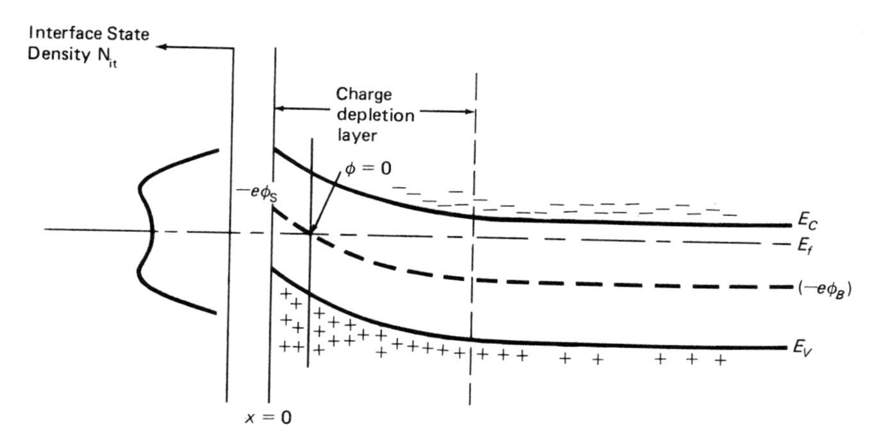

Fig. 2.9 Interaction of surface states with the *n*-type semiconductor interior to form a surface inversion layer at the interface boundary.[9]

where the material just becomes intrinsic, i.e., where $p_0 = n_0 = n_i$. The corresponding carrier densities are given from Eqs. (1.26) and (1.28), as a function of distance into the semiconductor by

$$n(x) = N_c \exp(E_f - E_c(x))/kT = N_c \exp(-(E_{ci} - E_f)/kT + e\phi(x)/kT) \quad (2.76)$$

$$p(x) = N_v \exp(E_v(x) - E_f)/kT = N_v \exp - ((E_f - E_{vi})/kT + e\phi(x)/kT) \quad (2.77)$$

where E_{ci} and V_{vi} are corresponding energies at the intrinsic point in Fig. 2.9. Inserting

$$N_c \exp(E_f - E_{ci})/kT = N_v \exp(E_{vi} - E_f)/kT = n_i \tag{2.78}$$

into Eqs. (2.76) and (2.77) yields the pair of equations

$$n(x) = n_i \exp e\phi(x)/kT; \qquad p(x) = n_i \exp - e\phi(x)/kT \tag{2.79}$$

From the third Maxwell equation, Eq. (2.3), and that here the electric field $\bar{E} = \bar{D}/\varepsilon = -\nabla\phi$, which when inserted into Eq. (2.3) yields the corresponding Poisson's equation, viz.

$$\nabla^2\phi(x) = -4\pi e(p - n + N_D - N_A)/\varepsilon \tag{2.80}$$

where the right-hand side is the net charge enclosed within the semiconductor. Using Eqs. (2.79), Eq. (2.80) becomes in one dimension

$$\phi''(x) = 4\pi e n_i\{2\sinh(e\phi(x)/kT) - (N_D - N_A)/n_i\}/\varepsilon \tag{2.81}$$

Letting $\Phi = e\phi/kT$ in Eq. (2.81) results in

$$\Phi''(x) = 2(\sinh\Phi - \sinh\Phi_B)/L_{Di}^2 \tag{2.82}$$

where $2\sinh\Phi_B \equiv (N_D - N_A)/n_i$, and L_{Di} is a plasma shielding parameter called the *intrinsic Debye length*. It is given by

$$L_{Di} = (\varepsilon kT/4\pi e^2 n_i)^{1/2} \tag{2.83}$$

Equation (2.82) is multiplied by Φ' and then integrated to give

$$x = \tfrac{1}{2}L_{Di}\int_{\Phi}^{\Phi_s}\{\cosh y - \cosh\Phi_B - (y - \Phi_B)\sinh\Phi_B\}^{-1/2}\,dy \tag{2.84}$$

where $\Phi_s = e\phi_s/kT$, and ϕ_s is the surface potential. The electric field at this surface is $E_s = -\phi_s' = -(kT/e)\Phi_s'$, obtained from Eq. (2.82) in the same way to give

$$E_s = -(2kT/eL_{Di})\{\cosh\Phi_s - \cosh\Phi_B - (\Phi_s - \Phi_B)\sinh\Phi_B\}^{1/2} \tag{2.85}$$

As seen from Eq. (2.84), the potential $\Phi(x)$ is available only implicitly as a function of distance x into the semiconductor. Eq. (2.82) cannot be integrated in closed form explicitly. However, it can for the special case of intrinsic material where $N_A = N_D$, so then $\Phi_B = 0$. This is called the *bulk intrinsic case*; then Eq. (2.82) becomes

$$\Phi'' = (2/L_{Di}^2)\sinh\Phi; \qquad \Phi(\infty) = 0, \qquad \Phi(0) = \Phi_s \tag{2.86}$$

with an integral

$$\Phi(x) = 4\tanh^{-1}\{\tanh(\Phi_s/4)\exp - (\sqrt{2}x/L_{Di})\} \tag{2.87}$$

The corresponding field at the surface is then

$$E_s = -\Phi'(0) = -(2\sqrt{2}kT/eL_{Di})\sinh(\Phi_s/2) \tag{2.88}$$

The subsurface electric field E_s, can be related to the interior space charge density σ_i through the surface charge density σ_s, using the boundary conditions discussed in the previous section. Boundary condition (d) following Eq. (2.75), between the semiconductor and the exterior (free space), where E_0 is the applied field, gives

$$(\varepsilon E_s - E_0) = 4\pi\sigma_s \tag{2.89}$$

Using Gauss's theorem, Eq. (2.74), applied to the interior plus the surface of the semiconductor, yields

$$-E_0 = 4\pi\sigma_T = 4\pi(\sigma_i + \sigma_s) \tag{2.90}$$

where the total charge $\sigma_T = \sigma_i + \sigma_s$. σ_i is the areal charge density in the interior depletion region, and σ_s is the areal charge density in the surface states. Eliminating E_0 between Eqs. (2.89) and (2.90) yields the sought-after surface electric field

$$E_s = -(4\pi/\varepsilon)\sigma_i \tag{2.91}$$

If it is desired to compute semiconductor internal parameters, such as the internal (built-in) potential, electric field, and depletion layer charge density, then surface boundary conditions must be imposed and therefore known. Assuming that the areal density and energy distribution of surface states, as well as the applied field E_0 are known, then the surface potential can be calculated. Then these quantities themselves can be used as boundary conditions for more complicated problems.

To calculate the surface potential, consider the bulk intrinsic case, where N_s is the areal density of acceptor surface states all at the same energy $\mathscr{E}_s$. From Fig. 2.10. it is seen that the energy levels of the surface states vary with the conduction and valence band edges, which in turn vary with the surface potential. That is, $\mathscr{E}_s = \mathscr{E}_{s0} - e\phi_s$, where $\mathscr{E}_{s0}$ is

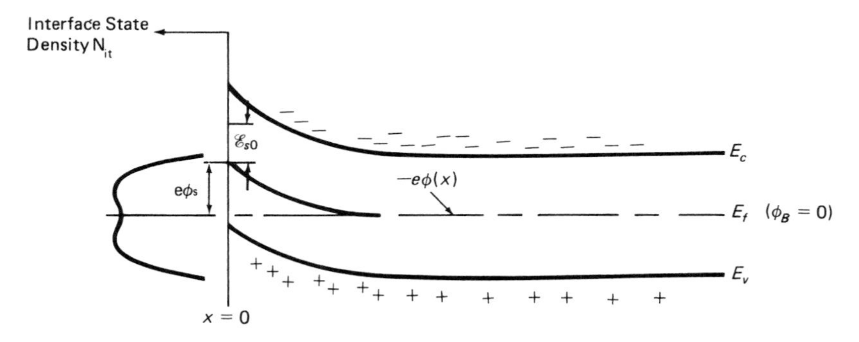

Fig. 2.10 Internal potential, conduction, and valence band edges, as a function of distance into semiconductor as measured from the interface boundary surface (intrinsic case).[9]

the energy difference between the surface level and the intrinsic point in the forbidden gap. In the bulk intrinsic case, using $E_f = 0$ as a reference, then n_s, the areal density of electrons that occupy surface states is given by Eq. (2.92), where N_s is the impurity concentration at the interface surface,

$$n_s = N_s \int_{E_v}^{E_c} \frac{\delta(E - \mathscr{E}_{s0})dE}{1 + \exp(E - e\phi_s)/kT} = \left[1 + \exp\left(\frac{\mathscr{E}_{s0}}{kT} - \Phi_s\right)\right]^{-1} N_s \tag{2.92}$$

$\delta(E - \mathscr{E}_{s0})$ is the common delta function yielding the above result. With the surface charge density $\sigma_s = -en_s$, the internal subsurface electric field E_s, using Eqs. (2.88)–(2.92) is

$$E_s = \frac{4\pi\sigma_s}{\varepsilon} + \frac{E_0}{\varepsilon} = -\frac{4\pi e N_s/\varepsilon}{1 + \exp\left(\frac{\mathscr{E}_{s0}}{kT} - \Phi_s\right)} + (E_0/\varepsilon) \tag{2.93}$$
$$= (2\sqrt{2}kT/eL_{Di})\sinh(\Phi_s/2)$$

Rewriting the two rightmost members gives

$$\frac{1}{1 + \exp\left(\frac{\mathscr{E}_{s0}}{kT} - \Phi_s\right)} = \frac{E_0}{4\pi e N_s} - \frac{2\sqrt{2}kT\varepsilon}{4\pi e^2 L_{Di} N_s}\sinh(\Phi_s/2) \tag{2.94}$$

Equation (2.94) is a quartic equation in the variable exp $-\Phi_s/2$, from which the real root can be extracted to yield the surface potential $\phi_s = kT\Phi_s/e$.

PROBLEMS

1. The gradient of any scalar, such as ∇V, can be added to the vector potential $\bar{A}$, as explained in this chapter. Show that V satisfies a free-space-type of wave equation in order that the uncoupled scalar and vector potential equations, Eqs. (2.11)–(2.13), can be written.
2. Using the Biot and Savart law, show that (a) the magnetic field at the center of a circular loop of radius r, carrying current I is $H = 2\pi I/\mu cr$, and (b) the magnetic field at the perpendicular distance r_0 from an infinitely long wire, also carrying a current I, is given by $H = 2I/\mu cr_0$. Notice that for these simple configurations it is straightforward to obtain the magnetic field directly from the appropriate Maxwell equation, instead of first computing the vector potential $\bar{A}$ and then obtaining the magnetic field from $\bar{H} = \nabla \times \bar{A}$.
3. Describe explicitly the nature of the agent by which the sign change in the Hall effect coefficient comes about for p-type material.
4. For plane periodic electromagnetic waves, show how their argument $\bar{k}\cdot\bar{r} - \omega t$ describes the propagation of a plane wave front.
5. Show that the general solution of the one-dimensional wave equation in free space for either the electric or magnetic field is $E_x = f(x - ct) + g(x + ct)$. Explain the properties of such a wave.
6. For the very short stub antenna or short stub slot in an otherwise intact missile skin, show that, if the radiation resistance R_s is defined by $P = \frac{1}{2}I_0^2R_s$, where I_0 is the antenna current and P is the radiated power, then $R_s = 20\pi^2(l_e/\lambda)^2$, where l_e is the physical length of the stub.
7. The fact that an amount of charge in a given volume is neither created or destroyed means that the charge density at any point in the volume is related to the current density in the neighborhood of the point by $\partial\rho/\partial t + \nabla\cdot\bar{J} = 0$. This is called the *equation of continuity* for charges and currents. Explain what this equation means term by term, where ρ and $\bar{J}$ are the charge density and current density, respectively.
8. The vector potential idea for solving Maxwell's equations is not new. An earlier variation, attributed to Hertz and still used, is to define a "Hertz Vector Potential" $\bar{Z}$, through $\bar{H} = (l/c)\partial(\nabla \times \bar{Z})/\partial t$. This is often used instead of letting $\bar{H} = \nabla \times \bar{A}$, because it facilitates the solution of certain electromagnetic wave propagation problems in a vacuum.

   Using $\bar{Z}$, work through the development similar to that in Section (2.3) for the vector potential $\bar{A}$, to show consistency with Maxwell's equations. In particular, show that $\bar{Z}$ satisfies a wave equation, viz., $\nabla^2\bar{Z} - (l/c^2)\partial^2\bar{Z}/\partial t^2 = 0$.
9. Consider electromagnetic spherical waves emanating from a dipole radiator, using the fact that the Hertz vector potential satisfies the wave equation, as obtained in Problem 8. Specifically:

(a) In spherical coordinates, obtain the retarded Hertz vector potential solution $\bar{Z} = (\bar{p}_0/r)\exp - i(kr - \omega t)$; $\omega t \geqslant kr$, or $t \geqslant r/c$. In the dipole moment $\bar{p} = \bar{p}_0 \exp i\omega t$, ω is the angular frequency, $k = 2\pi/\lambda$ is the wave number (where λ is the corresponding wavelength), and r is the distance from the center of the dipole to a point on the wave front.
(b) Obtain the corresponding magnetic field $\bar{H}$, and delineate the near and far electromagnetic field solutions.
(c) What is meant by the term "retarded potential"?

10. Use the temporal description of the EMP, as given in Section 9.2, and the Poynting vector $\bar{S} = \bar{E} \times \bar{H}$. The latter gives the power flux (energy flow rate per unit area) in the direction orthogonal to the $\bar{E} \times \bar{H}$ plane, for an assumed transverse EMP wave motion. Calculate for both endo- and exo-atmospheric bursts:
(a) The maximum value of the EMP power flux.
(b) The total energy fluence, W_f, in the EMP.

REFERENCES

1. J. D. Jackson. Classical Electrodynamics. Wiley, New York, 1969.
2. S. M. Sze. Physics of Semiconductor Devices. Wiley-Interscience, New York, pp. 45, 1982.
3. W. E. Mayerhoff. Phys. Rev., 71:727, 1947.
4. J. Bardeen. Phys. Rev. 71:717, 1947.
5. J. P. McKelvey. Solid State and Semiconductor Physics. Harper and Row, New York, 1967.
6. R. W. P. King. Theory of Linear Antennas. Harvard Univ. Press, Cambridge, Chap. 2, 1956.
7. Electromagnetic Pulse Handbook For Missiles and Spacecraft in Flight. Sandia Labs., Albuquerque, NM, AFWL Rept. TR73-68, Sept. 1972, Figs. 2.19 and 3.30.
8. R. A. Smith. Semiconductors, 2nd ed. Cambridge Univ. Press, London, 1979.
9. V. Nishioka, E. F. da Silva, Jr., and T. P. Ma. "Evidence For (100) Si/SiO$_2$ Interfacial Defect Transformation after Ionizing Radiation." IEEE Trans. Nucl. Sci. NS–35(6): 1227–1233, Dec. 1988.

CHAPTER 3
p-n JUNCTION DEVICES

3.1 INTRODUCTION

In the first two chapters, semiconductors were treated as though they were essentially homogeneous with respect to their material and electrical properties, including their impurity concentration. Here, the road to understanding of the operation of diodes and transistors must include nonhomogeneous semiconductors, the basic manifestation of which is the *p-n* junction. That is, semiconductors of interest are those whose dopant type changes rapidly from *n* to *p*-type across an internal boundary or interface. This thin boundary is called the *p-n* junction. It is depicted in Fig. 3.1.

As alluded to in Section 1.6 and shown below, the Fermi energy level must remain constant across a junction in thermal equilibrium, which implies that there is no net particle flow through the junction. The donors in *n*-type material provide electrons whose energies lie just above the bottom of the conduction band, as in Fig. 1.7.

The Fermi level must reflect the presence of these additional electrons, so that it must lie close to the conduction band and yet remain constant throughout both sides of the junction. Hence, the contour of the conduction band on the *n* side of the junction must conform to satisfy these two conditions simultaneously, yet maintain the band gap as shown in Fig. 3.1. Similar conditions must hold for the *p* side of the junction, in that E_f must lie close to the valence band to accommodate the increased electron concentration represented by the ionized acceptors, as in Fig. 1.7.

Under conditions of thermal equilibrium, both the applied voltage across the junction and the current through it are zero. Using the Einstein relations given by Eqs. (1.63), Eqs. (1.84) and (1.85) are then rewritten under equilibrium conditions for the *n* and *p* sides of the one-dimensional junction, respectively, as

$$\begin{aligned} J_n = J_p = 0 &= e\mu_n(n_0 E + (kT/e)\partial n_0/\partial x) \\ &= e\mu_p(p_0 E - (kT/e)\partial p_0/\partial x) \end{aligned} \tag{3.1}$$

Differentiating the Fermi level energy as a function of distance into the semiconductor, as obtained from Eqs. (1.27) and (1.29), yields

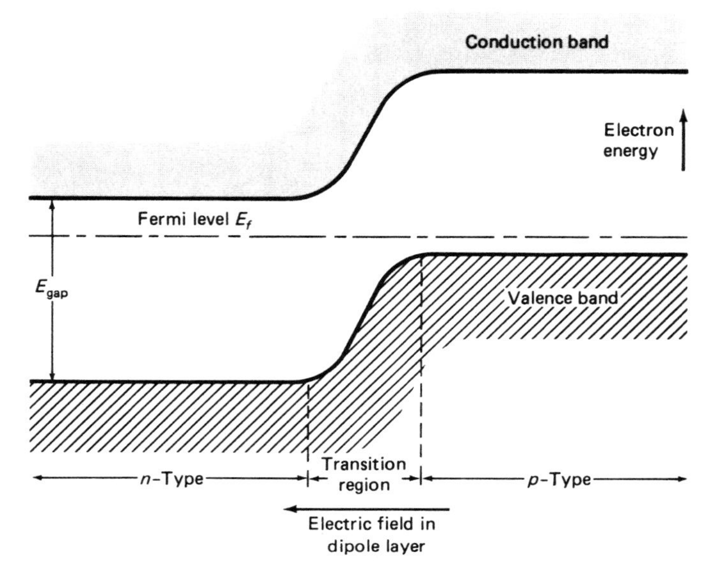

Fig. 3.1 *p-n* junction in thermal equilibrium.[1]

$$dE_f/dx = kT(\partial n_0/\partial x)/n_0 = -kT(\partial p_0/\partial x)/p_0 \tag{3.2}$$

Inserting the above into Eq. (3.1) gives

$$0 = \mu_n n_0(eE + dE_f/dx) = \mu_p p_0(eE + dE_f/dx) \tag{3.3}$$

As the applied voltage is zero, the electric field $E = 0$, so then $dE_f/dx = 0$ also. Hence, thermal equilibrium implies that the Fermi level is constant across the junction and throughout this system, as depicted in Fig. 3.1.

3.2 DEVICE FUNDAMENTALS

The *p-n* junction may consist of an essentially abrupt change in the semiconductor properties from one side of the junction to the other. Or the junction may be graded, wherein the dopant concentrations vary with distance across the junction gradually. Such junctions are shown in Fig. 3.2.

The relatively high ratio N_D/N_A makes for a junction with efficient electrical properties, as will be discussed, and is shown in Fig. 3.2.

Assume a *p-n* junction where (a) n_{n0} is the uniform concentration of

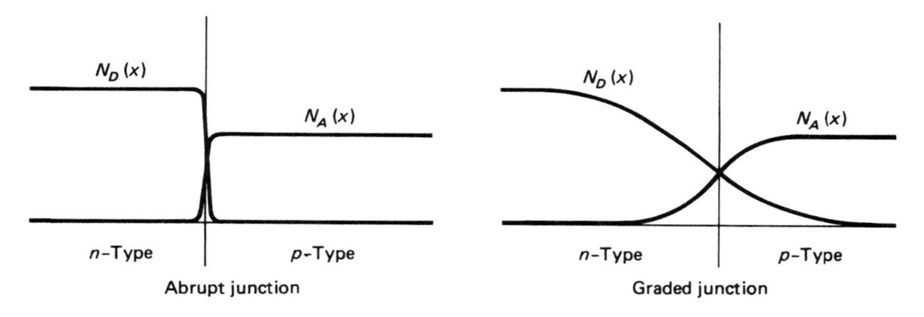

Fig. 3.2 Abrupt and graded junctions in the semiconductor.

free electrons on the n side, (b) p_{n0} is the uniform concentration of free holes on the n side, (c) p_{p0} is the concentration of free holes on the p side, and (d) n_{p0} is the concentration of free electrons on the p side of the junction. Then the mass action law at thermal equilibrium, discussed in Chapter 1, holds to yield

$$n_{n0}\,p_{n0} = p_{p0}\,n_{p0} = n_i^2 \tag{3.4}$$

3.3 DEPLETION, OR SPACE-CHARGE, REGIONS

Consider the situation at the *p-n* junction at the instant of its assumed formation. It can be appreciated that there exist tremendous concentration gradients of electrons and holes across the junction, certainly for reasonable n and p dopant concentrations on each side. Such large initial carrier concentrations immediately induce large "Fick's law" type of diffusion currents. That is, electrons from the n side diffuse across the junction into the p side, and vice versa for holes from the p side. These diffusion currents are initially of such strength that the semiconductor regions near the junction are depleted of mobile carriers. In these depletion regions, or depletion layers, on either side of the junction, the charged but immobile donor and acceptor ions now constitute regions of space charge.

Such charge separation across the junction produces an internal electrostatic field. The polarity and strength of this field are such as to oppose the above diffusion currents. A condition of dynamic equilibrium now ensues. These two depletion layers, contiguous across the junction, form an electric dipole layer straddling the junction. This is shown in Fig. 3.3.

If ϕ_n and ϕ_p are the dipole layer edge potentials most remote from the junction on the n and p sides, then from electrostatics

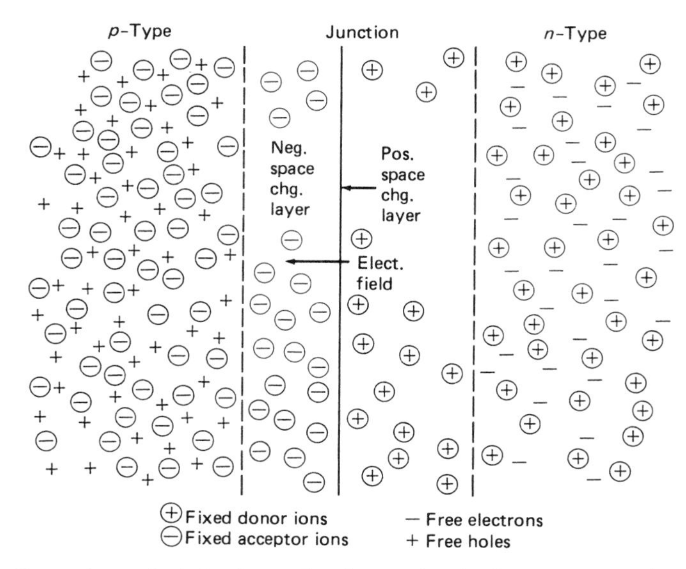

Fig. 3.3 Space charge depletion layers forming an electric dipole layer field.[4]

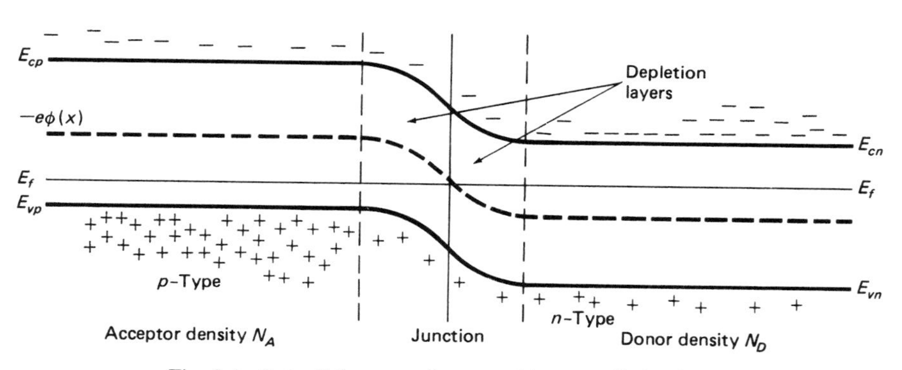

Fig. 3.4 Potential energy diagram with no applied voltage.

$$\phi_n - \phi_p = 4\pi\delta \tag{3.5}$$

where δ is the electric dipole layer strength, defined as the electric dipole moment per unit area of space charge layer. From Eq. (3.5), the potential energy of an electron in equilibrium at the bottom of the conduction band is lower on one side of the junction than the other by $4\pi e\delta$. This is shown in Fig. 3.4, where it is also seen that a contact electrostatic potential is built up between the two regions given by $\phi_0 = \phi_n - \phi_p$.

The gradient of the contact potential, $\nabla\phi_0$, represents the electric field, which prevents electrons in the conduction band on the n side of the junction from diffusing to the p side, and similarly for holes in the valence band on the p side from diffusing to the n side.

The magnitude of the contact potential ϕ_0 is simply obtained from relations involving the equilibrium carrier density on each side of the junction, but remote from the depletion layer region. From Eqs. (1.26) and (1.27), the electron density on the n side, n_{n0}, and the electron density on the p side, n_{p0}, are

$$n_{n0} = N_c \exp - (E_{cn} - E_f)/kT; \qquad N_c = 2(2\pi m_{de}^* kT/h^2)^{3/2} \tag{3.6}$$

$$n_{p0} = N_c \exp - (E_{cp} - E_f)/kT \tag{3.7}$$

where E_{cn} and E_{cp} are the conduction band edge energies on the n and p sides, as seen in Fig. 3.4. Then from Eqs. (3.4), (3.6), and (3.7), the contact potential is obtained as

$$\phi_0 = (E_{cp} - E_{cn})/e = (kT/e)\ln(n_{n0}/n_{p0}) = (kT/e)\ln(n_{n0}\,p_{p0}/n_i^2) \tag{3.8}$$

For a highly doped system, and with all donor and acceptor atoms assumed ionized, which is usually true at room temperature, so that $n_{n0} \cong N_D$, $p_{p0} \cong N_A$, Eq. (3.8) becomes

$$\phi_0 = (kT/e)\ln(N_D N_A/n_i^2) \tag{3.9}$$

It is often convenient to redefine the energy level reference to yield simplified expressions for the equilibrium concentration of the carriers. That is, multiplying Eqs. (1.27) and (1.29) yields, with $n_0 p_0 = n_i^2$

$$n_i = \sqrt{N_c N_v} \exp - (E_g/2kT) \tag{3.10}$$

Define a Fermi potential, ϕ_f, and corresponding to the center of the gap, an intrinsic potential, ψ, respectively given by

$$E_f = -e\phi_f; \qquad \tfrac{1}{2}E_g = -e\psi \tag{3.11}$$

Now, let the energy level reference be such that $E_v = 0$ and that $E_c = E_g$. Disregarding differences in the effective masses m_{de}^* and m_{dh}^*, allows $N_c \cong N_v$ in Eq. (3.10). Then dividing each of Eqs. (1.27) and (1.29) by the resulting n_i from Eq. (3.10), and using Eq. (3.11), gives for the equilibrium n and p density on each side of the junction

$$n_{n0} = n_i \exp e(\psi - \phi_f)/kT, \qquad p_{p0} = n_i \exp e(\phi_f - \psi)/kT \tag{3.12}$$

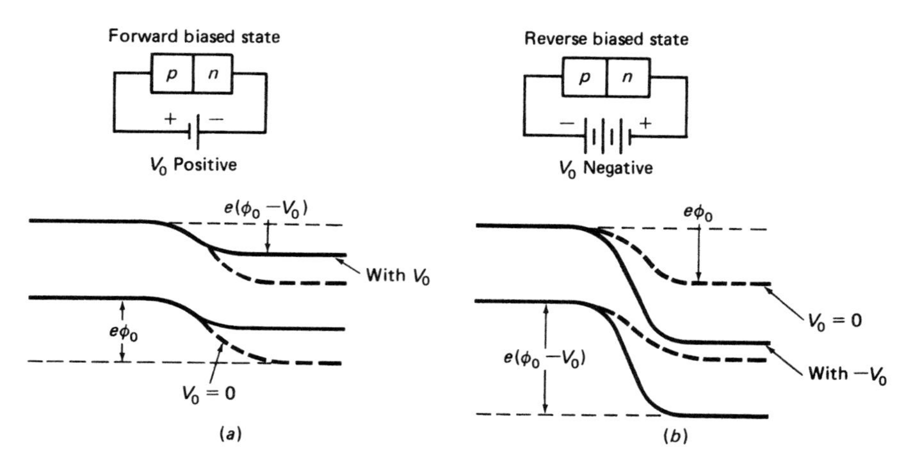

Fig. 3.5 Potential energy diagrams for forward and reverse bias.

The above, as well as slight variations, are often used for boundary conditions to obtain carrier concentrations from their respective transport equations in a straightforward manner, as will be seen in later sections of this and other chapters.

3.4 JUNCTION FIELDS AND POTENTIALS

The space charge regions or depletion layers on each side of the junction are also depleted of carriers when an external voltage V_0 is applied across the junction, as in Fig. 3.5. It is also seen that most of the applied voltage drop is across these regions, for, in their depleted state, they have a much higher resistivity than the remainder of the lattice. If the polarity of V_0 is as shown in Fig. 3.5a, then the potential barrier height is reduced as seen. The latter is now given by $e(\phi_0 - V_0)$. This external voltage polarity and corresponding state of the junction is termed the *forward bias state*. If the polarity of V_0 is reversed, Fig. 3.5b shows that the potential barrier height is increased. This is called the *reverse bias state*. Now, the potential barrier height is $e(\phi_0 - (-V_0)) = e(\phi_0 + V_0)$.

When an external voltage is applied across the junction, this system is no longer in equilibrium as currents start to flow. A unique Fermi energy level over the whole lattice is no longer the case. However, as seen in Fig. 3.6, the difference in Fermi levels on each side of the junction is $E_{fp0} - E_{fn0} = eV_0$. From Fig. 3.6 it is seen that, although the potential barrier prevents majority carriers, i.e., electrons from the n side and holes from the p side, from crossing the junction, it nevertheless induces

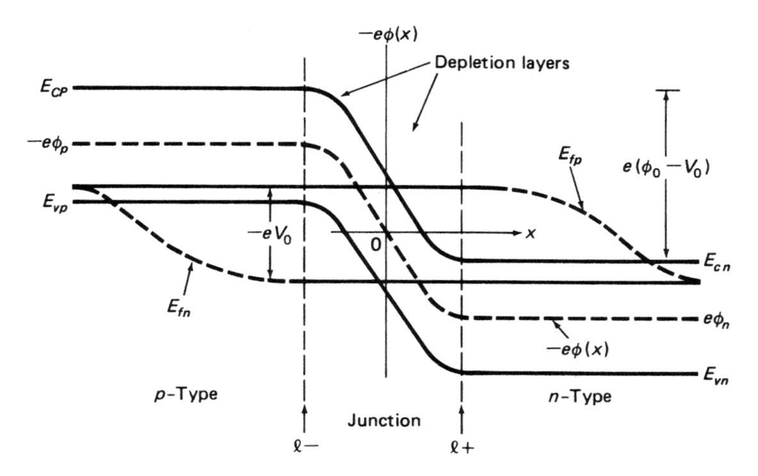

Fig. 3.6 Energy band potentials, Fermi levels, with applied voltage.

minority carriers to cross the junction to become majority carriers. In essence, the presence of an external voltage across the junction strongly affects the minority carriers, but affects the majority carriers very little. This also holds when the junction is in the forward bias state, as in Fig. 3.6. With a voltage drop V_0 across the junction, equilibrium no longer holds, so that Eqs. (3.12) can be rewritten to account now for the Fermi level difference across the junction. Define $\phi_{fn} = -E_{fn}/e$ and $\phi_{fp} = -E_{fp}/e$, with $\phi_{fp}(x) - \phi_{fn}(x) = V(x)$, to yield

$$n_{n0} = n_i \exp e(\psi - \phi_{fn})/kT; \qquad p_{p0} = n_i \exp e(\phi_{fp} - \psi)/kT \tag{3.13}$$

in the n and p regions, respectively. ϕ_{fn} and ϕ_{fp} are called *quasi-Fermi potentials*, and correspondingly E_{fn} and E_{fp} are called *quasi-Fermi energy levels*. Eliminating ψ between Eqs. (3.13), with $n_0 p_0 = n_i^2$, and $\phi_{fp}(0) - \phi_{fn}(0) = V_0$, results in expressions for the minority carrier concentrations with $n_{n0}(0) \equiv n(x_0-)$ and $p_{p0}(0) = p(x_0+)$

$$p(x_0+) = p_0 \exp eV_0/kT; \qquad n(x_0-) = n_0 \exp eV_0/kT \tag{3.14}$$

on the n and p junction boundaries. Notice that if V_0 is large and negative compared with kT/e, then both boundary carrier concentrations essentially vanish. Other variations of Eqs. (3.14) are readily evident and, again, are used as boundary conditions to obtain ready solutions to carrier transport problems.

An example of large negative V_0 boundary conditions is that of a re-

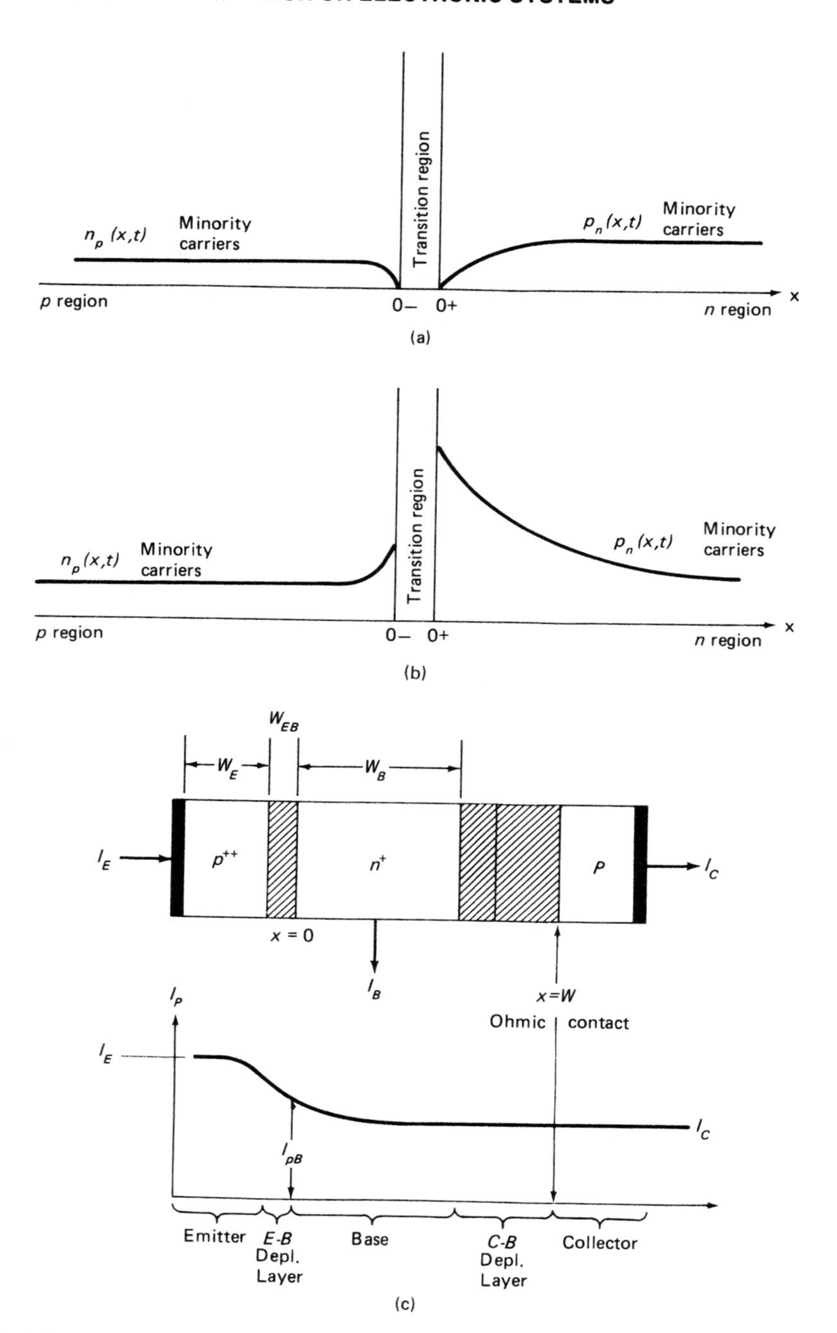

Fig. 3.6A *p-n* junction minority carrier densities for (a) reverse-biased diode (b) forward-biased diode, and (c) hole current through *n*-material side of emitter-base junction and ohmic contact of base-collector junction.

versed biased diode, as shown in Fig. 3.6A. For a forward biased diode, Eqs. (3.14) show that the boundary values of p and n on each side of the junction differ, depending on the values of n_0 and p_0 unique to the situation at hand. This is also illustrated in Fig. 3.6A.

A second example is that of the n material side of the junction ($x = 0$), and its (Fig. 3.6A) ohmic junctionless contact at $x = W$. This results in, for the minority carriers

$$p(0+) = p_0 \exp eV_0/kT; \qquad p(W) = p_0 \ll p(0+) \qquad (3.15)$$

where the left-hand equation is gotten above, and the right-hand equation implies that the minority carriers are immediately swept away from $x = W$, since this region experiences an essentially zero voltage drop V_0. These equations are used in Section 3.9 for carrier transport in a *pnp* transistor.

Referring back to the variation of potential with distance, as shown in Fig. 3.6, the potential and electric field in the depletion region can be obtained from solutions of Poisson's equation with suitable boundary conditions, as discussed in Section 2.8. In one-dimensional geometry, Poisson's equation in the potential can be written as

$$\phi''(x) = -4\pi\rho(x)/\varepsilon = -4\pi e(p - n + N_D(x) - N_A(x))/\varepsilon \qquad (3.16)$$

where $e(x)$ is the net charge density in the volume of interest. With the origin precisely at the junction, again from Eqs. (1.27) and (1.29)

$$n(x) = N_c \exp - (E_c(x) - E_{fn})/kT \qquad (3.17)$$

$$p(x) = N_v \exp - (E_{fp} - E_v(x))/kT \qquad (3.18)$$

Also, it is seen from the definition of potential, and depicted in Fig. 3.6, that

$$E_c(x) = -e\phi(x) + E_{ci} \qquad (3.19)$$

$$E_v(x) = -e\phi(x) + E_{vi} \qquad (3.20)$$

where E_{ci} and E_{vi} are the respective band-edge energies precisely at the junction origin. Now consider the case of zero applied voltage, so that a single Fermi level $E_{fn} = E_{fp} = E_f$ ensues. Also, now $N_v \exp - (E_f - E_{vi})/kT = N_c \exp - (E_{ci} - E_f)/kT$. Then Eqs. (3.16)–(3.20) can be combined to yield the basic Poisson equation in the vicinity of the junction, as

$$\phi'' - (8\pi e n_i/\varepsilon) \sinh(e\phi/kT) = (-4\pi e/\varepsilon)(N_D(x) - N_A(x)) \qquad (3.21)$$

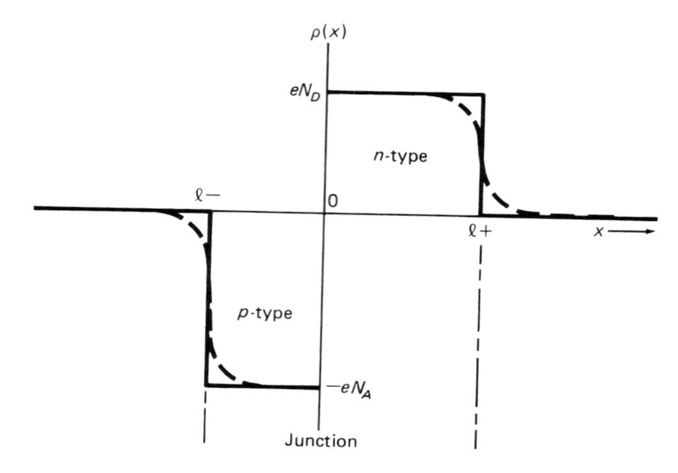

Fig. 3.7 Charge density for abrupt *p-n* junction.

For an abrupt junction, N_A is constant, and $N_D = 0$ for $x < 0$, i.e., to the left of the junction. N_D is constant, and $N_A = 0$ for $x > 0$, i.e., to the right of the junction. Then Eq. (3.21) becomes

$$\phi'' - (8\pi e n_i/\varepsilon)\sinh(e\phi/kT) = \begin{cases} (4\pi e/\varepsilon)N_A; & x < 0 \\ -(4\pi e/\varepsilon)N_D; & x > 0 \end{cases} \tag{3.22}$$

Equations (3.22) are highly nonlinear, and explicit solutions for ϕ are not obtainable in closed form, as seen from Eq. (2.82) et seq. Various approximate versions of Eq. (3.22) are tractable. One such is developed by first noting that if Eqs. (3.19) and (3.20) are inserted into Eqs. (3.17) and (3.18), the result is

$$n(x) = N_c \exp - (E_{ci} - e\phi - E_{fn})/kT \tag{3.23}$$

$$p(x) = N_v \exp - (e\phi - E_{vi} + E_{fp})/kT \tag{3.24}$$

It is seen from Eqs. (3.23) and (3.24) that the carrier densities are a sensitive function of $(E_{ci} - e\phi)/kT$ or $(e\phi - E_{vi})/kT$. Also it is noted that the charge density $\rho(x)$ quickly approaches a constant $- eN_A$ as x approaches the origin through the p side depletion layer. Similarly, $\rho(x)$ approaches a constant eN_D for positive x, as x approaches the origin through the n side depletion layer. This is depicted in Fig. 3.7, where $l_\pm$ are the depletion layer widths, and the total depletion layer width $l = l_+$

$+ l_-$. The above then implies $N_D, N_A \gg n, p$ in Eq. (3.16) or $(8\pi en_i/\varepsilon) \cdot \sinh(e\phi/kT) \ll |4\pi eN/\varepsilon|$ in Eq. (3.22), so that it becomes,

$$\phi''(x) \simeq \begin{cases} -4\pi e\, N_D/\varepsilon; & 0 < x \leqslant l_+ \\ 4\pi e\, N_A/\varepsilon; & -l_- \leqslant x < 0 \\ 0; & x < -l_-, x > l_+ \end{cases} \tag{3.25}$$

From Gauss's theorem and the definition of potential, the electric field at the junction, E_0, as a consequence of the solution below, satisfies

$$\phi'_+(0) = \phi'_-(0) = -E_+(0) = -E_-(0) = -E_0 \tag{3.26}$$

The boundary conditions are:

$$\phi'_+(l_+) = \phi'_-(-l_-) = 0 \tag{3.27}$$

The zero of potential is chosen so that

$$\phi_+(0) = \phi_-(0) = 0 \tag{3.28}$$

where the subscripts $-$ and $+$ refer to the left and right sides, respectively, of the junction in Fig. 3.7.

The solutions of Eqs. (3.25)–(3.28) are given below, where $4\pi\varepsilon_0$ is now inserted to multiply ε for conversion between CGS and SI units. $\varepsilon_0 = 8.86 \cdot 10^{-14}$ farads per cm is the permittivity of free space, and ε is the dielectric constant of the material. The solutions are

$$\phi_+ = (e/2\varepsilon\varepsilon_0)N_D(l_+^2 - (l_+ - x)^2); \qquad 0 < x < l_+ \tag{3.29}$$

$$\phi_- = (e/2\varepsilon\varepsilon_0)N_A((x + l_-)^2 - l_-^2); \qquad -l_- \leqslant x < 0 \tag{3.30}$$

To obtain the relationship between the depletion layer boundaries, $l_\pm$, in terms of the associated parameters, Eqs. (3.29) and (3.30) are inserted into the boundary conditions, to give

$$\phi_+(l_+) - \phi_-(-l_-) = \phi_0 - V_0 \tag{3.31}$$

which is simply the total drop in potential across the junction. With Eq. (3.26)

$$l_+ = (2\varepsilon\varepsilon_0 \cdot (\phi_0 - V_0)N_A/eN_D(N_A + N_D))^{1/2} \tag{3.32}$$

$$l_- = (2\varepsilon\varepsilon_0 \cdot (\phi_0 - V_0)N_D/eN_A(N_A + N_D))^{1/2} \tag{3.33}$$

The electric field at the junction, E_0, is

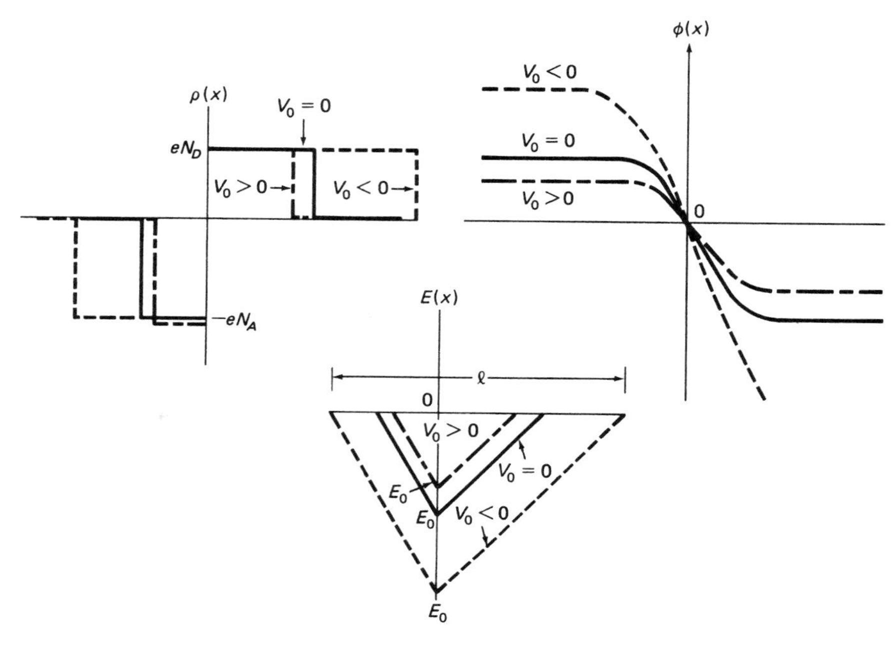

Fig. 3.8 Charge density, electric field, and potential about an abrupt junction under forward bias ($V_0 > 0$), reverse bias ($V_0 < 0$), and zero bias ($V_0 = 0$) conditions.[4]

$$E_0 = -eN_D l_+/\varepsilon\varepsilon_0 = eN_A l_-/\varepsilon\varepsilon_0 = 2(\phi_0 - V_0)/l \tag{3.34}$$

and the total depletion layer width is l, where

$$l = l_+ + l_- = [2\varepsilon\varepsilon_0(\phi_0 - V_0)(N_A^{-1} + N_D^{-1})/e]^{1/2} \tag{3.35}$$

This is depicted in Fig. 3.8.

From the above equations, it is seen that the depletion layer thickness varies inversely with the square root of the dopant concentration. When the junction is reverse biased, the space charge extends far into the material, especially if the region is lightly doped. Under forward bias conditions, large junction currents flow, resulting in appreciable voltage drops in regions outside the depletion layer.

3.5 JUNCTION CAPACITANCE

From the preceding, it is apparent that the volume within the depletion layers on each side of the junction can be perceived as a junction capacitance, due to the separation of charge across a finite region when the junction is reverse-biased. From Eqs. (3.32) and (3.33), the charge per unit area, q, is given by

$$q = eN_D l_+ = eN_A l_- = [2e\varepsilon\varepsilon_0(\phi_0 - V_0)N_D N_A/(N_D + N_A)]^{1/2} \qquad (3.36)$$

The small signal capacitance per unit junction area A_J is then obtained from Eq. (3.36) as

$$C/A_J = |dq/dV_0| = [e\varepsilon\varepsilon_0 N_D N_A/2(\phi_0 - V_0)(N_D + N_A)]^{1/2} \quad \text{F/cm}^2 \qquad (3.37)$$

Recall that for a reverse biased junction, $V_0 < 0$. Usually the dopant concentrations on each side of the junction are quite different. For example, $N_A \gg N_D$; then $N_D N_A/(N_D + N_A) \simeq N_D \equiv N$, the concentration of the lightly doped side in Eq. (3.37). In this case, the junction capacitance per unit junction area becomes

$$C/A_J = (e\varepsilon\varepsilon_0 N/2V_0)^{1/2} \qquad (3.38)$$

also for the usual situation of $|V_0| \gg \phi_0$, and depicted in Fig. 3.9.

From the expression for C/A_J, it is seen that it is an increasing function of the square root of N. For large reverse bias voltages, $C/A_J \sim V_0^{-1/2}$, characteristic of an abrupt junction. For linearly graded junctions, it will be seen that $C/A_J \sim V_0^{-1/3}$. For an abrupt junction, the easily measured junction capacitance can be used to determine the internal contact potential ϕ_0 from Eq. (3.37). For forward bias, $V_0 > 0$, but C/A_J does not become infinite as $\phi_0 - V_0$ approaches 0, because the correspondingly large current flow vitiates this electric model of the depletion layer capacitance for $V_0 \geqslant \phi_0$[14,26]. However, the forward biased junction causes a rearrangement of minority carriers outside the depletion layer, which is manifest as a diffusion capacitance.

For reverse-biased junctions, the depletion approximation in the preceding material is used to derive the depletion layer parameters, such as its width and depletion layer capacitance per unit junction area, $C_d/A_J = \varepsilon\varepsilon_0/l$, (Problem 3.2). The depletion approximation asserts that no mobile charge carriers exist in the depletion layer, in that only fixed charges (as represented by the dopant ions) are present therein. Thermal equilibrium is also assumed in the model. As implied, the depletion approximation junction capacitance model is valid for reverse-biased, and approximately valid for forward-biased, junctions[26] that satisfy the constraint $(\phi_0 - V_0) > 0$.

For forward-biased junctions where the applied voltage exceeds the

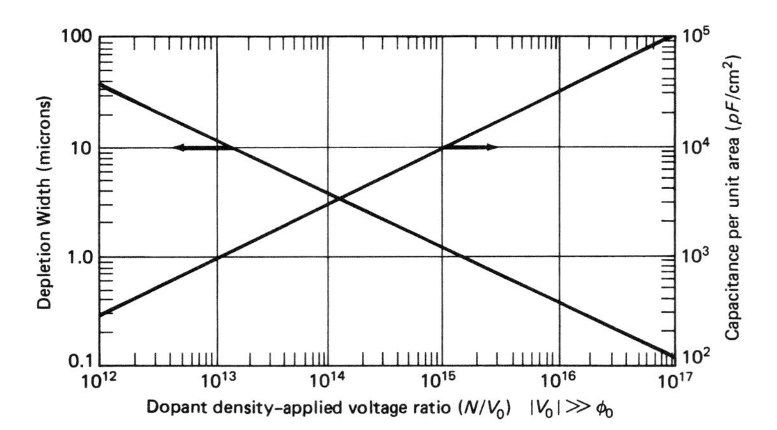

Fig. 3.9 Depletion layer width and junction capacitance per unit area versus dopant density-applied voltage ratio for abrupt junctions where N is the dopant density of the lightly doped side.[18]

built-in voltage (i.e., $(\phi_0 - V_0) < 0$), the depletion layer model is invalid. Physically, it is because there are mobile carriers in the depletion layer (i.e., there is substantial current flow through the junction), and equilibrium no longer holds. In this case, there is manifest a total junction capacitance, consisting of C_d, the remaining depletion layer capacitance, plus a diffusion capacitance C_v, which is much larger than C_d. The origin of C_v is due to the injected charge stored in the regions near the junction, exterior to the depletion region, as shown in the leftmost portion of Fig. 3.15.

The diffusion capacitance expression is different for the static (dc) and dynamic (ac) injection cases. The derivation of the static diffusion capacitance begins with the assumption that one side of the junction is so heavily doped with, say, acceptors, that the junction current I_p consists mainly of holes. The corresponding excess minority carrier charge Q will then exist almost exclusively on the n-side of the junction. Its magnitude is given by the shaded area in the n-region[17] in the leftmost portion of Fig. 3.15, multiplied by the junction area- charge product eA_J. Then, the excess hole charge given in Eq. (1.108) yields

$$\begin{aligned} Q &= \int_0^\infty eA_J(p(x) - p_0)\,dx \\ &= \int_0^\infty eA_J(p(0) - p_0)\exp - (x/L_p)\,dx \\ &= eA_J L_p(p(0) - p_0) \end{aligned} \tag{3.39}$$

Also, the hole current at $x = 0$ is available from $I_p(0) = -eD_p p'(0)A_J$ and Eq. (1.108), to give

$$I_p(0) = eA_J D_p(p(0) - p_0)/L_p \tag{3.40}$$

Combining Eq. (3.39) and (3.40), and using $L_p^2 = D_p \tau_p$, results in

$$I_p(0) = Q/\tau_p \tag{3.41}$$

This simple but important relation asserts that the junction (hole) current is proportional to the stored charge Q of the excess minority carriers. Equation (3.41) is referred to as the charge control description of a diode (Section 7.3). In this steady-state situation, the current supplies minority carriers at the same rate at which they undergo recombination.

With the above for Q, the diffusion or quasi-neutral capacitance C_v is defined as

$$C_v = dQ/dV \cong \tau_p \, dI/dV = \tau_p g \tag{3.42}$$

where the junction incremental conductance $g = dI/dV = (d/dV) \cdot (I_0 \exp(eV/\eta kT)) = (e/\eta kT) I$, so that

$$C_v = (\tau_p/\eta kT/e) I = (e\tau_p/\eta kT) I_0 \exp(eV/\eta kT) \tag{3.43}$$

where η is the junction ideality factor, usually taken as unity, here. The above C_v corresponds to hole current only. There is also an electron current diffusion capacitance as well, so that the total diffusion capacitance is their sum. Within the assumptions above, the electron current is negligible, so that its contribution to the total diffusion capacitance is neglected. For the reverse-biased junction case, $C_v = (e\tau_p/\eta kT) I_0 \exp - (e|V|/\eta kT)$, which is seen to become very small for increasing negative bias voltage V, compared to C_d. However, for a typical set of parameters (Problem 3.9) under forward bias, the ratio $C_v/C_d = 35\,500$! This has consequences for single event upset (SEU), as discussed in Chapter 8, in that it would be much more difficult to perturb a forward-biased junction information node due to spurious ionization-induced charge, than its reverse-biased counterpart, because of the corresponding very large difference in charge involved.

The dynamic diffusion capacitance corresponds to the situation for time-varying signal inputs. Letting the bias voltáge be given by $V = V_0 + V_1 \exp(i\omega t)$, $V_1 \ll V_0$, where V_0 is the dc bias voltage pedestal with a sinusoidal component $V_1 \exp(i\omega t)$ superimposed. This results in[22]

$$C_v = (e^2/kT)\,(L_p p_{n0}/\sqrt{2(1 + \sqrt{1 + (\omega\tau_p)^2})} + L_n n_{p0}/\sqrt{2(1 + \sqrt{1 + (\omega\tau_n)^2})})\cdot\exp(eV_0/kT) \quad (3.44)$$

For low frequencies, $\omega\tau_p \ll 1$, with electron current diffusion capacitance neglected,

$$C_v \cong (e^2/2kT)\,(L_p p_{n0} + L_n n_{p0})\exp(eV_0/kT) = (1/2)\tau_p g \quad (3.45)$$

For high frequencies, $\omega\tau_p \gg 1$,

$$C_v \cong (\tau_p/2\omega)^{1/2} g \quad (3.46)$$

In anticipation of the discussion of bipolar transistors in Section 3.9, it can be appreciated that the diffusion capacitance of its emitter-base junction region represents a capacitive impedance to input signals. The excess minority carrier charge Q_B stored in the base region adjacent to the depletion layer, as in the *p-n* junction above, represents the reestablishment of the steady state following injection of a current pulse, for example. This charge redistribution imposes an upper bound on the device speed. The diffusion capacitance C_{EB} corresponds to this charge rearrangement. Assuming a narrow base (i.e., $W/L_p \ll 1$, with Eq. (3.112) and $\eta = 1$), this charge is

$$Q_B = eA\int_0^W p_B(x)dx = eAp_{n0}\exp(eV_{EB}/kT)\int_0^W (1 - x/W)\,dx = (1/2)eAWp_{n0}\exp(eV_{EB}/kT) \quad (3.47)$$

Then for the transistor,

$$C_{EB} = \partial Q_B/\partial V_{EB} = ((1/2)e^2AWp_{n0}/kT)\exp(eV_{EB}/kT) = (\partial Q_B/\partial I_c)\,(\partial I_c/\partial V_{EB}) = g_m\tau_b \quad (3.48)$$

where the base transit time $\tau_b = \partial Q_B/\partial I_c$, I_c is the collector current and the transconductance $g_m = \partial I_c/\partial V_{EB}$. So that the emitter-base diffusion capacitance using Eq. (3.42) is, for a homogeneously doped base,

$$C_{EB} = g_m\tau_b = g_m W^2/2D_{pB} \quad (3.49)$$

To compute junction capacitance per unit area using Eq. (3.37), it is necessary to obtain the dopant concentrations on each side of the junction. From a practical viewpoint, the dopant concentrations for a particular device must often be gotten in the following roundabout manner, because of the manufacturer's lack of availability of this parameter for proprietary or other reasons. Parameters for which the manufacturer is relatively more forthcoming in this context include the resistivity ρ or the sheet resistance ρ_s of the emitter, base, and collector of the transistor of interest.

The sheet resistance expression is derived from the well-known formula for the bulk resistance R of a parallelepiped of length L, width W, thickness t, resistivity ρ, and cross-sectional area A, viz.

$$R = \rho L/A = \rho L/Wt \qquad \text{(ohms)} \tag{3.50}$$

with resistivity usually expressed in ohm cm. The sheet resistance is defined as $\rho_s = \rho/t$ with units of ohms, but termed ohms per square. That is, the above expression can be rewritten, with current flow in the L direction, as

$$R = \rho_s \cdot (L/W) \qquad \text{(ohms)} \tag{3.51}$$

where L/W counts the number (not necessarily an integer) of "squares" of area $W \cdot W\,\text{cm}^2$ and thickness t cm contained in series in the parallelepiped. For many reasons, sheet resistance provides easy parametrization of device processing for the manufacturer. For the present requirement, knowledge of the sheet resistance, together with the thickness of the emitter, base, and collector, allows the determination of the resistivity simply from $\rho = t\rho_s$. Then the dopant concentration is obtained from Fig. 1.10, and thence capacitance per unit junction area from Eq. (3.37) with a knowledge of the particular element voltage V_0. Pertinent junction area dimensions from forthcoming vendors or from destructive physical analyses, complete the data required to calculate this capacitance.

For graded junctions, the procedure to obtain the parameters of interest is similar to that for the abrupt junction. For an arbitrary specification of the junction gradation, Poisson's equation, Eqs. (3.21) or (3.22), must be integrated numerically. For linearly graded junctions, under boundary conditions almost identical to those of the abrupt junction, Poisson's equation becomes

$$\phi''(x) = \begin{cases} -(e/\varepsilon\varepsilon_0)m_+x & 0 < x \leqslant l_+ \\ (e/\varepsilon\varepsilon_0)m_-x & -l_- \leqslant x < 0 \\ 0 & l_+ < x < -l_- \end{cases} \tag{3.52}$$

where $m_{\pm}$ are the slopes of the linear grade of the junction, in the $n(x \geqslant 0)$ and $p(x \leqslant 0)$ regions. Again, using the boundary conditions given by Eqs. (3.26)–(3.28), as for the abrupt junction, the corresponding potentials are integrals of Eqs. (3.52) and are

$$\phi_+ = (e/2\varepsilon\varepsilon_0)m_+x(l_+^2 - x^2/3), 0 \leqslant x \leqslant l_+ \tag{3.53}$$

$$\phi_- = (e/2\varepsilon\varepsilon_0)m_-x(x^2/3 - l_-^2), -l_- \leqslant x \leqslant 0 \tag{3.54}$$

Using Eqs. (3.31) and (3.34), the depletion layer widths $l_{\pm}$ are given by

$$l_+ = [3\varepsilon\varepsilon_0(\phi_0 - V_0)/em_+(1 + M)]^{1/3}; M = (m_+/m_-)^{1/2} \tag{3.55}$$

and

$$l_- = M[3\varepsilon\varepsilon_0(\phi_0 - V_0)/em_+(1 + M)]^{1/3} \tag{3.56}$$

and the total depletion layer width l is

$$l = l_+ + l_- = [3\varepsilon\varepsilon_0(\phi_0 - V_0)(1 + M)^2/em_+]^{1/3} \tag{3.57}$$

If $m_+ = m_- = m$, then $M = 1$, so that Eq. (3.57) reduces to

$$l_m = [12\varepsilon\varepsilon_0(\phi_0 - V_0)/em]^{1/3} \tag{3.58}$$

The impurity concentration and charge density for the linearly graded junction are depicted in Fig. 3.10. The corresponding capacitance per unit area is

$$C/A_J = (me(\varepsilon\varepsilon_0)^2/12(\phi_0 - V_0))^{1/3} \tag{3.59}$$

For one-sided abrupt junctions, such as that for a metal-semiconductor called a Schottky barrier junction, the corresponding Poisson equations (Eqs. (3.25) or (3.52)) still hold, but with slightly different boundary conditions. The resulting solutions yield the parameters of interest. Also, they can be obtained by letting the dopant concentration on one side of the junction be very much larger than that on the other side, in the formulas for the sought after parameters. For a p^+n junction where $N_A \to \infty$ to represent the metal side, the preceding yields parameters analogous to those for the abrupt junction, and are given in Table 3.1.

With the advent of heterojunction devices, it is important to discuss their junction parameters, especially those that are analogous to ordinary

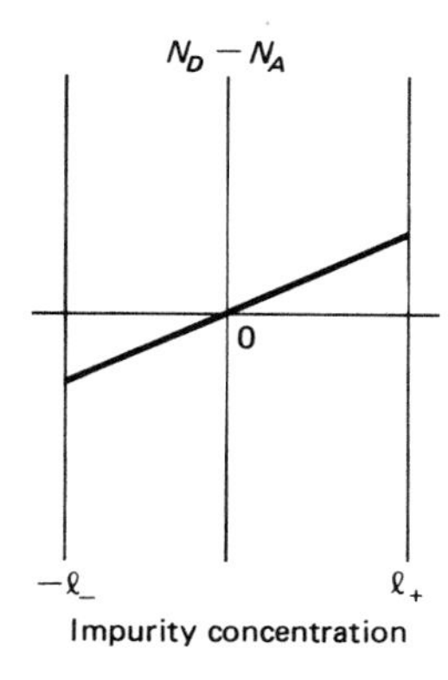

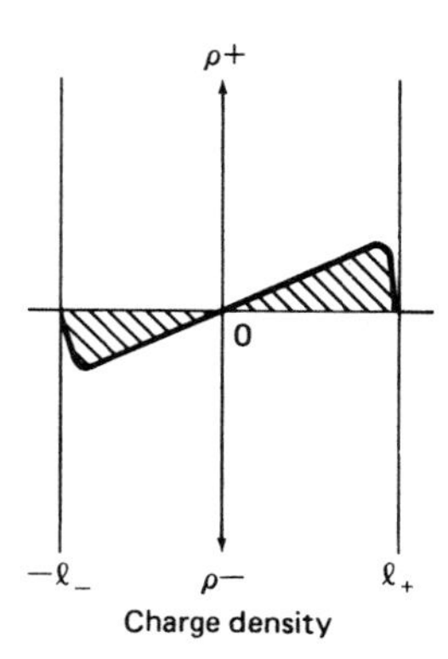

Fig. 3.10 Impurity concentration and charge density for linearly graded junctions.

Table 3.1 One-sided abrupt junction parameters.

PARAMETER	FORMULA	DESCRIPTOR
Depletion width	$l_- = (2\varepsilon\varepsilon_0(\phi_0 - \lvert V_0\rvert)/eN)^{1/2}$ $l_+ = (2\varepsilon\varepsilon_0(\phi_0 + \lvert V_0\rvert)/eN)^{1/2}$	fwd-biased metal-N_D jct fwd-biased metal-N_D jct
Maximum electric field	$E_{mo-} = 2(\phi_0 - \lvert V_0\rvert)/l$ $E_{mo+} = 2(\phi_0 + \lvert V_0\rvert)/l$	fwd-biased metal-N_D jct fwd-biased metal-N_D jct
Junction capacitance per unit area	$C/A_J = (e\varepsilon\varepsilon_0 N/2V_0)^{1/2} = \varepsilon\varepsilon_0/l$	same as Eq. (3.38)
Avalanche breakdown voltage	$V_B = \varepsilon\varepsilon_0 E_m^2/2eN$	same as Eq. (3.74)
Built-in (contact) potential	$\phi_0 = 2(kT/e)\ln(N/n_i)$	
Depletion layer space charge	$Q_{sc} = eNl = (2e\varepsilon\varepsilon_0 N(\phi_0 - \lvert V_0\rvert))^{1/2}$	

(homojunction) p-n junctions. The latter junctions are, of course, fabricated from the same substrate, such as silicon, and doped oppositely on each side. The former junctions are formed between two dissimilar semiconductors, each one doped to suit the application at hand, and not necessarily oppositely doped. When the two semiconductors have the same type of conductivity (i.e., n- or p-type, as indicated by their respective dopant contributions to conductivity, as discussed in Section 1.8), the junction is called an isotype heterojunction. When the conductivity types differ across the junction, it is called an anisotype heterojunction.[29]

Examples of heterojunctions are replete throughout the sphere of radiation effects on semiconductors. Besides heterojunctions used in optoelectronics, such as in semiconductor lasers, inadvertent heterojunctions are formed by the Si-SiO_2 interfaces in most silicon MOS devices and similar device interfaces. Also, heterojunction bipolar transistors (HBTs) such as AlGaAs/GaAs types are becoming available, with a potential use in microwave large signal applications. This is because of their speed and power advantages over GaAs MESFET technologies, even at room temperature, and especially at cryogenic temperatures.[30,31] As already mentioned, another important class of heterojunctions are metal-semiconductor junctions, called Schottky barrier diodes, which will be discussed later.[29]

In a manner closely similar to that used for homojunctions in Sections 3.4 and 3.5, the depletion layer widths and junction capacitances for heterojunctions can be obtained. This is done by solving Poisson's equation on either side of the interface of an assumed step-heterojunction. The step junction assumption is seen to be an excellent approximation for heterojunctions, since this is an actual physical step change of the properties across the junction. One modification must be made in the boundary condition for the electric field across the junction, in that Gauss's theorem (Section 2.7) now asserts that $\varepsilon_1 E_1 = \varepsilon_2 E_2$ at the heterojunction interface. This reduces to $E_1 = E_2$ for the homojunction, as in Section 3.4. $\varepsilon_1 \varepsilon_0$ and $\varepsilon_2 \varepsilon_0$ are the respective dielectric permittivities of the materials on each side of the junction, where ε_0 is the dielectric constant of free space, as in Table 1.1.

For an *n-p* anisotype heterojunction, the corresponding depletion layer widths are found to be[29]

$$l_1 = (2N_{A2}\varepsilon_2\varepsilon_1\varepsilon_0(\phi - V_0)/eN_{D1}(\varepsilon_1 N_{D1} + \varepsilon_2 N_{A2}))^{1/2} \qquad (3.60)$$

$$l_2 = (2N_{D1}\varepsilon_2\varepsilon_1\varepsilon_0(\phi - V_0)/eN_{A2}(\varepsilon_1 N_{D1} + \varepsilon_2 N_{A2}))^{1/2} \qquad (3.61)$$

with the total built-in (contact) potential $\phi = \phi_{O1} + \phi_{O2}$, where the subscripts $O1$ and $O2$ refer to each side of the heterojunction. Similarly, the (reverse bias) junction capacitance is given by

$$C_{\text{het}} = (e\varepsilon_2\varepsilon_1\varepsilon_0 N_{D1} N_{A2}/2(\varepsilon_1 N_{D1} + \varepsilon_2 N_{D2})(\phi - V_0))^{1/2} \qquad (3.62)$$

with shortened forms (as for homojunctions), when one side is lightly doped compared to the other. Further, the above expressions reduce to those for homojunctions when the materials on each side of the junction are identical.

Also, the relative voltage that can be supported on each side of the heterojunction satisfies,

$$(\phi_{01} - V_1)/(\phi_{02} - V_2) = N_{A2}\,\varepsilon_2/N_{D1}\,\varepsilon_1 \tag{3.63}$$

$V_0 = V_1 + V_2$ is the total voltage across the heterojunction, while V_1 and V_2 are the voltages across the corresponding sides of the heterojunction. For an *n-n* isotype heterojunction, the boundary conditions are complex, with more complicated expressions for the depletion layer widths and the heterojunction capacitance.[29]

The Schottky barrier junction is treated similarly. When the metal side is in close contact with the semiconductor side, the conduction and valence bands and the fermi level become uniquely related.[29] This relationship serves as a boundary condition on the solution of the corresponding Poisson equation. This then follows exactly as for the *p-n* junction in Section 3.4. Using the one-sided abrupt junction approximation results in expressions for the semiconductor depletion width[29]

$$l_{\mathrm{sch}} = ((2\varepsilon\,\varepsilon_0/eN_D)\,(\phi - V_0 - kT/e))^{1/2} \tag{3.64}$$

and junction capacitance

$$C_{\mathrm{sch}} = ((e\varepsilon\,\varepsilon_0\,N_D)/2(\phi - V_0 - kT/e))^{1/2} \tag{3.65}$$

3.6 JUNCTION BREAKDOWN

When a sufficiently high applied voltage, i.e., a strong electric field, such as that from nuclear EMP (Chapter 9) is incident on the junction, the latter breaks down and conducts current essentially like a short circuit, losing all semblance of junction behavior. The four basic effects that cause junction breakdown are: thermal instability (thermal runaway), tunnelling, avalanche multiplication (impact ionization), and second breakdown. These are discussed in turn.

Junction breakdown due to thermal instability can be prevalent in semiconductor devices with relatively small band-gap energies, such as silicon and germanium diodes or transistors, if precautions regarding safe bias voltage operation are not taken. At high values of reverse bias voltage, heat dissipation caused by the reverse leakage current increases the junction temperature. This causes a further increase in the reverse current because of an increase in thermal production of holes and electrons.[1] This is evident from Eq. (1.32), rewritten for silicon as

$$n_i = 2.23 \cdot 10^{19}(T/300)^{3/2} \exp - 0.55/kT \tag{3.66}$$

where T is the junction temperature in degrees Kelvin. From Eq. (3.66) it is apparent that an increase in temperature increases the diode reverse current, viz., $I_0 = en_i^2(D_n/p_{p0}L_n + D_p/n_{n0}L_p)$ derived in Section 3.7, because n_i increases with temperature. With the current increase, the power dissipation in the junction increases, in turn increasing the junction temperature, etc. This regenerative process, called *thermal runaway*, finally causes the junction current to increase without bound, which destroys the junction. Simple bias networks usually prevent this in actual circuitry.

The second cause of junction breakdown is due to band-to-band tunneling. That is, under conditions of very high electric field strength spanning the junction, appreciable amounts of current begin to flow through the junction from the conduction band on one side of the junction to the valence band on the other side of the junction by means of the tunnel effect. This occurs in spite of normal potential barriers impeding such flow across the junction. The full explanation of tunneling requires quantum mechanics. Quantum theoretical results yield a probability P_t of tunneling, through a rectangular, one-dimensional potential barrier, given by

$$P_t = \left[1 + \frac{U_0^2}{4E(U_0 - E)} \sinh^2 \sqrt{2m(U_0 - E)/\hbar^2}\, W\right]^{-1} \tag{3.67}$$

where U_0 is the potential barrier height, with W the barrier thickness in the x direction, and E is the energy of the incident carrier on the potential barrier. For a p-n junction, the tunnel current density, J_t, is[2]

$$J_t = \frac{E_0 V_0 e^3}{h^2} \sqrt{\frac{2m_{de}^*}{E_g}} \cdot \exp - \left(\frac{4\sqrt{2m_{de}^* E_g^3}}{3eE_0\hbar}\right) \tag{3.68}$$

where E_0 is the electric field strength at the junction, and V_0 is the applied, reverse bias voltage. For very high field strengths and dopant concentrations, the tunnel current power dissipated ultimately destroys the junction for corresponding voltage breakdown values of up to $4E_g$ (four times the gap energy). From $4E_g - 6E_g$, junction breakdown is due to a combination of tunneling and avalanche multiplication, explained below. For breakdown voltages greater than $6E_g$, breakdown is almost entirely due to avalanche multiplication. As an aside, zener diode operation is via tunneling at low voltages and avalanche multiplication breakdown at high voltages and a mixture of both at moderate applied voltage.[3] However, the band gap energies in silicon and germanium decrease with an increase in temperature,[20] in other words the tunnel breakdown voltage has a negative temperature coefficient.

So that from Eq. (3.68), for high temperatures, a given breakdown tunnel current can be attained with less reverse bias voltage. On the other hand, the avalanche multiplication mechanism possesses a positive temperature coefficient, as will be discussed.

As already alluded to, the junction breakdown effect of major importance is that due to avalanche multiplication. The avalanche breakdown voltage imposes a maximum on the safe reverse bias voltage for ordinary diodes, as well as the collector-base voltage in bipolar transistors. Avalanche multiplication is caused by *impact ionization*. Impact ionization is due to ionizing collisions occurring between electron carriers and the lattice atoms. When such ionization occurs, the resulting electron is excited across the band gap and an electron-hole pair is formed. This will result in the production of two carriers, and both holes and electrons can ionize to produce further electrons. This process then generates further pairs, and an avalanche ensues. Actually, this mechanism can be used to generate power in the microwave frequency region, besides its zener function mentioned already. Components that exploit this phenomenon are called IMPATT (impact avalanche and transit time) devices.

A model for consideration of avalanche multiplication uses the development of an expression for the corresponding ionization current that determines the breakdown of the junction. For a depletion region of width w, let I_0 be the current incident at the junction from the p side. Assume that the depletion region electric field is of sufficient magnitude to generate electron-hole pairs by impact ionization, as previously explained. The corresponding hole current I_p will thereby increase with distance through the depletion region, reaching a maximum value M_pI_0 at $x = w$. The multiplication factor is $M_p = I_p(w)/I_p(0)$. Likewise, the electron current I_n will increase in the opposite direction, from $x = w$ to $x = 0$, with the total current $I = I_p + I_n$ remaining constant. If $\alpha_n(x)$ and $\alpha_p(x)$ are the electron and hole ionization rates, then the particle current, I_p/e, satisfies

$$I_p'(x)/e = \alpha_p I_p/e + \alpha_n I_n/e \tag{3.69}$$

or

$$I'_p - (\alpha_p - \alpha_n)I_p = \alpha_n I; \qquad I_p(w) = I = M_p I_0,\ I_n(w) = 0 \tag{3.70}$$

with an integral given by

$$I_p/M_p I_0 = \left(\exp - \int_x^w (\alpha_p - \alpha_n)\,dx'\right) - \int_x^w dx' \alpha_n(x') \exp \int_{x'}^x (\alpha_p - \alpha_n)\,dx'' \tag{3.71}$$

The avalanche breakdown voltage is that which causes the multiplication to increase without bound at $x = 0$. At this voltage, M_p becomes infinite, so that the left side of Eq. (3.71) vanishes, to give

$$\int_0^w \alpha_n(x')\,dx' \exp \int_{x'}^w (\alpha_p - \alpha_n)\,dx'' = 1 \tag{3.72}$$

If $\alpha_n = \alpha_p = \alpha$, as in the case of gallium phosphide, then

$$\int_0^w \alpha(x)\,dx = 1 \tag{3.73}$$

Using empirically derived expressions[3] for the ionization rates α_n and α_p, the above avalanche breakdown criterion integrals can be evaluated for w. The solution of the one-dimensional Poisson's equation in the depletion layer, as discussed earlier, yields the electric field strength and potential therein. Then the avalanche breakdown voltage V_B is given by the area under the appropriate electric field strength curve, as in Fig. 3.8.[3] This yields:

a. one-sided abrupt junction (area under a triangular electric field curve)

$$V_B = wE_m/2 = \varepsilon E_m^2/2eN_s \tag{3.74}$$

b. linearly graded junction (area under a parabolic electric field curve)

$$V_B = 2wE_m/3 = 8/3 \cdot E_m^{3/2} \cdot (\varepsilon/2eg)^{1/2} \tag{3.75}$$

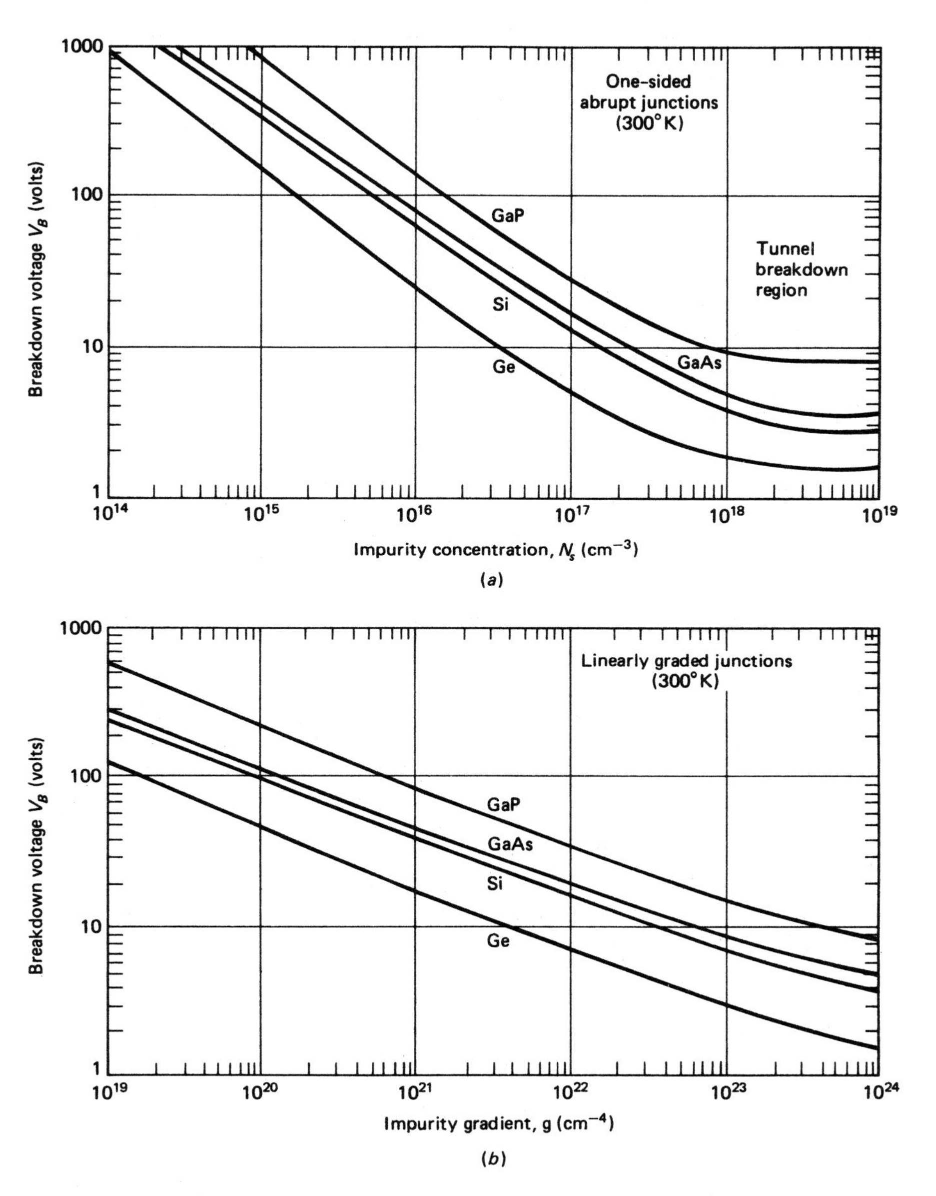

Fig. 3.11 Avalanche junction breakdown voltage versus dopant concentration for various semiconductor materials for abrupt and linearly graded junctions.[3] Figure © 1981 by John Wiley & Sons.

where E_m is the corresponding maximum electric field strength, N_s is the ionized dopant concentration of the lightly doped side of the junction, and g is the dopant density gradient.

That is, Eqs. (3.72)–(3.75) form pairwise systems of two simultaneous equations for w and E_m, depending on the semiconductor material and the junction gradation. Their solutions then provide the values of breakdown voltage V_B. Figure 3.11 depicts breakdown voltages for abrupt and linearly graded junctions for a number of materials versus N_s and g.

It should be appreciated that junction breakdown is not necessarily deleterious or destructive. For example, low-voltage zener diode breakdown, which is tunnel breakdown, occurs when both sides of the junction are heavily doped, the depletion layer is narrow, and carrier multiplication is not prevalent. The high-field region is so narrow that the colliding carriers have only a very low probability for knocking out further carriers from the lattice. This narrow-field region does not allow the carriers to accelerate very much, so they cannot gain enough energy to ionize the material to produce new carriers. However, some bound electrons in the valence band, which are not energetic enough to dissociate from their lattice atoms, now find the electric field aiding sufficiently to produce ionization via tunneling. Once this process begins, the resulting current becomes very large. This results in the zener breakdown portion of the normal diode current-voltage characteristic, as shown in Fig. 3.14. Sometimes, the avalanche breakdown component is referred to as *first breakdown*.

In diodes, bipolar and MOS transistors, and other semiconductor devices, a phenomenon called *second breakdown* occurs.[3,8] It is manifest by a sudden drop in the bipolar transistor voltage V_{CE} with almost simultaneous surge in collector current. It is felt to be due to a current localized "hot spot" in the device, caused by the negative temperature resistance coefficient of silicon that is evidenced beyond a critical temperature threshold. This temperature threshold corresponds to the state at which the thermally generated carrier concentration attains that of the dopant concentration background.[3] This phenomenon limits high-power transistor operation to a specific safe operating area (SOA) of its characteristic curves, as seen in Fig. 3.13. Chronologically, this instability occurs at the breakdown voltage at which the collector current equals I in Fig. 3.12. This is immediately followed by a sudden drop to a low-voltage region at the end of the dotted lines in Fig. 3.12. The final stage is imminent when I_c grows to cause device destruction. Associated with second breakdown are the final values of collector current, which result in the formation of a hot spot producing a localized microplasma in the device,

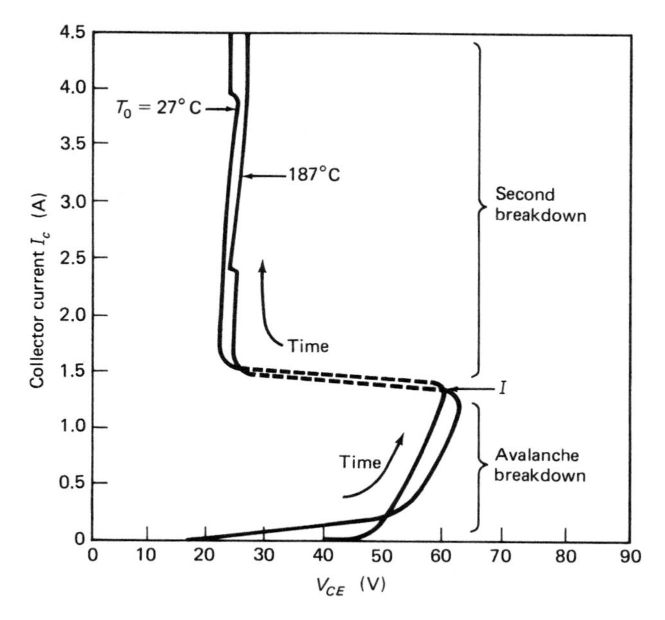

Fig. 3.12 Collector current versus collector-emitter voltage under second breakdown.[3] Figure © 1981 by John Wiley & Sons.

short-circuiting the junction. Reverse bias second breakdown is caused by an inductive load suddenly forcing a large negative polarity current through the device, resulting in a high transient breakdown reverse bias at the junction. This also causes a plasma formation and pinhole shorts at the junction. When the above current increases abruptly, it pinches or constricts, causing a current filamentation that is an attendant phenomenon of second breakdown.

When second breakdown follows avalanche breakdown, it quenches the avalanche at a particular point in the junction. This takes place usually well toward the edges of a reverse biased junction, where the fields and leakage currents tend toward a maximum. Once the avalanche is quenched there, a hot spot forms in its place. The material resistivity drops sharply, and it is heated to incandescence, producing a plasma, followed by junction destruction. When diodes in the laboratory are constructed so that the junction can be seen through a transparent cover, it is noticed that the avalanche process produces a characteristic light spectrum in the visible that can be detected. This is also noticed in high-power rf transistors that have a translucent cover. At the onset of second breakdown, the light is seen to change to a black body spectrum, indicating incandescence as the hot spot forms.[11] Hot spot temperatures are

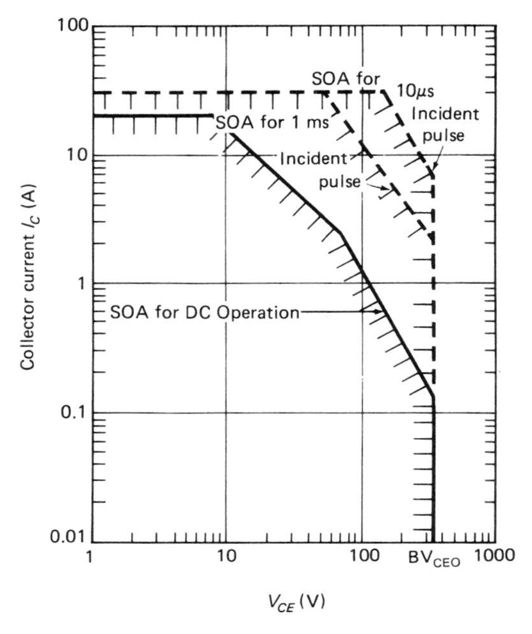

Fig. 3.13 Safe operating area (SOA) bounds for silicon transistors for peak junction temperature at 150°C.[3]

of the order of 1400°C, approaching the melting point of silicon. Another second breakdown precursor is a local temperature instability, from which second breakdown follows, but after a time delay (trigger time). The trigger time varies from about 10 μs to about 1 ms.[3]

When the transient pulse that produces second breakdown is extremely short, of the order of nanoseconds, a so-called current induced second breakdown can occur.[12,13] This is called *electrical*, as opposed to *thermal* second breakdown, the latter resulting from relatively much longer pulses. If the lifetime of the carriers of one sign differ greatly from those of the oppsite sign, a current-controlled, negative-resistance condition occurs in the device. In *p-i-n* germanium diodes, this negative resistance transforms the diode to a very low-impedance device, resulting in excessive current filamentation, presaging second breakdown.[11] However, the corresponding spectrum peaks in the infrared, indicative of a lower temperature with no plasma formation. Electrical second breakdown can be destructive for sufficiently large incident power on the junction. The prevalence of electrical second breakdown and its relationship to thermal second breakdown is still being studied.[3]

3.7 p-n JUNCTION DIODES

Using the pictorial representation of the junction as given in Fig. 3.4, at thermal equilibrium with no applied voltage, there are majority carrier electrons on the n side with sufficient energy to diffuse to the p side, the potential barrier notwithstanding. Once on the p side, the electrons as minority carriers may disappear due to electron-hole recombination. The corresponding equilibrium electron recombination current density flow is termed J_{nr}. Because at equilibrium there is zero net current flow, there must be an analogous process resulting in a generation current density J_{ng} emanating from an inverse current of electrons produced from electron-hole pairs on the p side diffusing back to the n side, so that, at equilibrium, these current densities must cancel each other, i.e., $J_{ng} + J_{nr} = 0$. The same process holds for hole current, densities so that their analogous currents also cancel for the same reasons, giving $J_{pg} + J_{pr} = 0$.

The basis for p-n junction rectifier operation is that, under reverse bias conditions, the height of the junction potential barrier increases by $|-eV_0|$ as in Fig. 3.5b. With an increased height of the potential barrier, J_{nr} and J_{pr} become very small. However, the generated current densities J_{ng} and J_{pg} are unchanged, for they depend on the unaffected rate of electron-hole pair generation. For increasingly large V_0, the junction current density approaches a saturation value $J_s = -e(J_{pg} - J_{ng})$, which depends on the semiconductor material and temperature only.

In the forward biased condition, the potential barrier is reduced, as in Fig. 3.5a, so that now J_{pr} and J_{nr} become large. Again, J_{ng} and J_{pg} are unaffected. The junction current-voltage plot is given in Fig. 3.14. In the forward biased direction, the current I consists largely of majority carriers crossing the junction to become minority carriers on the other side. For the reverse bias situation, a small reverse current flows, consisting of minority carriers from the bulk of the material anterior to the space charge regions.

It is assumed that the carrier densities are governed by Maxwell-Boltzmann statistics, i.e., the proper energy inequalities ensue, as discussed in Section 1.5, so that the probability that an electron has enough energy to vault the junction potential barrier is proportional to $\exp - e(\phi_0 - V_0)/kT$. In terms of applied voltage only, this probability is proportional to $\exp eV_0/kT$. Hence

$$J_{nr} = J_{ng} \exp(eV_0/kT); \qquad J_{pr} = J_{pg} \exp(eV_0/kT) \tag{3.76}$$

Defining a hole and electron particle current density respectively as

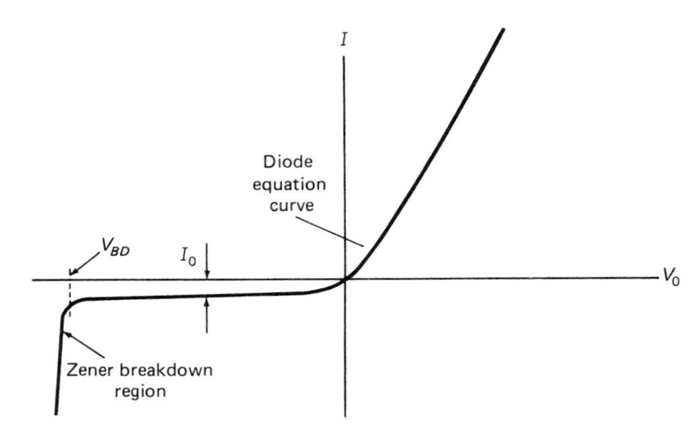

Fig. 3.14 Current-voltage characteristic for the *p-n* diode.

$J_p = J_{pr} - J_{pg}$ and $J_n = J_{ng} - J_{nr}$, then from Eqs (3.76),

$$J_n = -J_{ng}(\exp(eV_0/kT) - 1); \qquad J_p = J_{pg}(\exp(eV_0/kT) - 1) \quad (3.77)$$

Then the current $I = eA_J \cdot (J_p - J_n)$ is given by

$$I = I_0(\exp(eV_0/kT) - 1) \quad (3.78)$$

and $I_0 = eA_J(J_{pg} + J_{ng})$. I_0 is termed the saturation current and expressed below in terms of device parameters. Equation (3.78), often called the *diode equation*, is sketched in Fig. 3.14.

Actually, the real junction current is given by $I = I_0(\exp(eV_0/nkT) - 1)$, where the ideality parameter $1 \leqslant n \leqslant 2$. That n does not equal unity is due to the presence of both diffusion as well as generation-recombination currents in the depletion layer,[3] which are ignored in the abrupt junction approximation implied in Eq. (3.78), and in the preceding discussion. Note that Eq. (3.78) does not reflect diode operation in the zener breakdown region, as depicted in Fig. 3.14 and discussed in the previous section.

To obtain a more incisive picture of junction operation, recall Section 1.13, and that under equilibrium conditions for electrons in the p region, the continuity equation asserts that

$$(n_p - n_{p0})'' - (n_p - n_{p0})/L_n^2 = 0, \qquad x \geqslant -l_- \tag{3.79}$$

Similarly for holes in the n region

$$(p_n - p_{n0})'' - (p_n - p_{n0})/L_p^2 = 0, \qquad x < l_+ \tag{3.80}$$

where the diffusion lengths for the above minority carriers are $L_n = \sqrt{D_n \tau_n}$ and $L_p = \sqrt{D_p \tau_p}$. Corresponding boundary conditions for Eqs. (3.79) and (3.80) are

$$\text{a.}\ \lim_{x \to +\infty} (n_p - n_{p0}) = \lim_{x \to -\infty} (p_n - p_{n0}) = 0 \tag{3.81}$$

$$\text{b.}\ n_p(-l_-)/n_{n0} = p_n(l_+)/p_{p0} = \exp - e(\phi_0 - V_0)/kT \tag{3.82}$$

Boundary condition (b) comes about because the energy difference between the conduction band and the Fermi level on the n side is less than that on the p side by $e(\phi_0 - V_0)$, as seen in Fig. 3.5. This also implies that the obverse holds, in that $n_n(l_+)/n_p(-l_-) = p_p(-l_-)/p_n(l_+) = \exp e(\phi_0 - V_0)/kT$. The integrals of Eqs. (3.79)–(3.82) are, respectively

$$n_p = n_{p0}[1 + (\exp(eV_0/kT) - 1)\exp - (x + l_-)/L_n] \tag{3.83}$$

$$p_n = p_{n0}[1 + (\exp(eV_0/kT) - 1)\exp + (x - l_+)/L_p] \tag{3.84}$$

It is seen from the above for the reverse bias case ($V_0 < 0$) that the minority carrier electron density on the p side and the minority carrier density on the n side are less than the equilibrium value. The opposite situation holds for the forward-biased ($V_0 > 0$) junction.

Junction current densities are gotten from diffusion current densities of these minority carriers at the depletion layer boundaries, for example:

$$J_n(-l_-)/e = -D_n n_p'(-l_-) = -(n_{p0}D_n/L_n)(\exp(eV_0/kT) - 1) \tag{3.85}$$

$$J_p(l_+)/e = -D_p p_n'(l_+) = -(p_{n0}D_p/L_p)(\exp(eV_0/kT) - 1) \tag{3.86}$$

The junction current I is simply

$$I = A_J[J_p(l_+) - J_n(l_-)] = eA_J(n_{p0}D_n/L_n + p_{n0}D_p/L_p)(\exp e(V_0/kT) - 1) \tag{3.87}$$

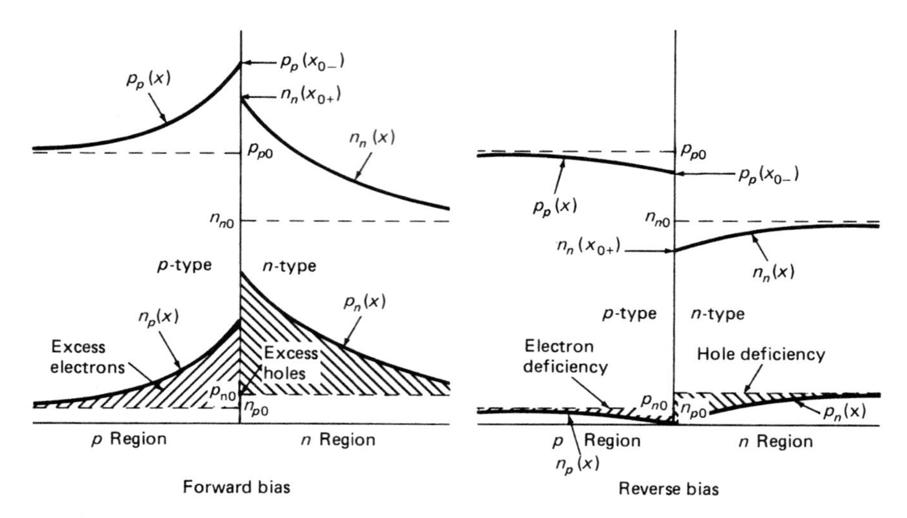

Fig. 3.15 *p-n* junction minority and majority carrier densities under conditions of forward and reverse bias. The junction region is too small to appear on this scale, so that it is within the discontinuity in the figure.

Note that this has the same form as Eq. (3.78), so that the saturation, or, more commonly called, the reverse current I_0 can be now identified as

$$I_0 = eA_J(n_{p0} D_n/L_n + p_{n0} D_p/L_p) \tag{3.88}$$

Figure 3.15 depicts the carrier concentrations under conditions of both forward and reverse bias.

By using the mass action relation $n_{n0} p_{p0} = n_i^2$, Eq. (3.88) can be rewritten as

$$\begin{aligned} I_0 &= en_i^2 A_J(D_n/p_{n0} L_n + D_p/n_{p0} L_p) \\ &= en_i^2 A_J[(D_n/\tau_n)^{1/2}/N_A + (D_p/\tau_p)^{1/2}/N_D] \end{aligned} \tag{3.89}$$

As can be noted from this expression, the saturation, or reverse, leakage current I_0 can be reduced by heavy doping, lengthening the minority carrier lifetime in both regions on each side of the junction by choosing appropriate material parameters, and maintaining the ambient device temperature as low as practicable. For graded *p-n* junctions, Eq. (3.87) can be generalized to yield[5]

$$I = en_i^2 A_J\left[D_n\left(\int_0^{l_+} N_D(x)\,dx\right)^{-1} + D_p\left(\int_0^{l_-} N_A(x)\,dx\right)^{-1}\right](\exp(eV_0/kT) - 1) \tag{3.90}$$

where l_+ and l_- are the respective depletion layer widths on the n and p sides of the junction.

As mentioned earlier, the parameter n, called the ideality factor, varies between 1 and 2 depending on whether the junction is in the very low ($n = 2$), medium ($n = 1$), or high ($n = 2$) forward bias current injection region. Equation (3.78) corresponds to the "medium" region, where diffusion current (as opposed to generation-recombination current) is predominant. Here, $n = 1$, and V_0, the forward bias, is a few tenths of a volt. For $V_0 \geqslant 0.1$ volt, the "-1" following the exponential function can be dropped in Eq. (3.78), so that in this region

$$I \simeq I_0 \exp(eV_0/nkT); \qquad n \cong 1; \qquad V_0 \geqslant 0.1 \text{ volt} \tag{3.91}$$

In detail, for forward bias where the generation-recombination processes in the depletion region are mainly capture reactions, a recombination current (as well as the diffusion current) is apparent. For forward bias voltages down to less than kT/e (1/40 volt), this total current is given to good approximation by[22]

$$I = A_J[(e\,L_p n_i^2/\tau_p N_D)(\exp(eV_0/kT) - 1) + (1/2)eW\sigma v_{th} N_t n_i(\exp(eV_0/2kT) - 1)] \tag{3.92}$$

where it is seen that the first term is the diffusion current, and the second is the recombination current. W is the junction width, N_t is the density of generation-recombination centers, and $\sigma v_{th} N_t$ is the generation-recombination rate per carrier, where v_{th} is the carrier thermal speed, and σ is the generation-recombination cross section. Inserting numerical values for the parameters, for a nominal p-n junction, Eq. (3.92) becomes[23]

$$I = 0.027(\exp(eV_0/kT) - 1) - 4.4(\exp(eV_0/2kT) - 1) \qquad (pA) \tag{3.93}$$

Hence, the second (generation-recombination) term dominates at very low forward bias levels, while the first term is predominant at larger forward bias.

For high injection levels (high current), which is the case where the injected minority carrier density is comparable to the majority carrier density, the electron and hole densities are essentially equal, due to conductivity modulation discussed in Section 5.12, both terms in Eq.

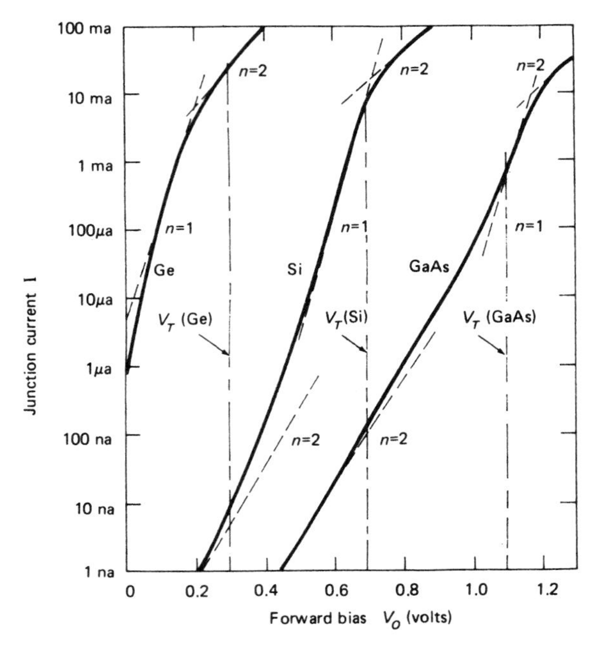

Fig. 3.16 Diode forward bias *I-V* curves showing ideality factors (n) and threshold "diode drops" for Ge, Si, and GaAs.[24] Figure © 1967 by John Wiley & Sons.

(3.92) must be considered. However, from a different avenue, high injection also implies[25]

$$np \leqslant n_i^2 \exp(eV_0/kT) \tag{3.94}$$

Since for high injection $n \simeq p$ (as stated above), and when inserted in Eq. (3.94) its square root yields

$$n \text{ or } p \cong n_i \exp(eV_0/2kT) \tag{3.95}$$

so that the high injection carrier density, and so the corresponding current, apes that of the very low injection case, where $n = 2$ also.

All of this is depicted in Fig. 3.16, which shows the forward bias *I-V* characteristics and ideality factors (n) for germanium, silicon, and gallium arsenide at room temperature.[24] Also shown in the figure are the corresponding "forward diode drops" (i.e., the assumed diode threshold turn-on forward bias voltages, V_T, used to model the diode in the "milliampere region" for circuit analysis purposes). That is, $I = 0$ for $V_0 <$

V_T. Using the forward bias transition from medium to high injection as the criterion, it is seen in Fig. 3.16 that the diode drops are 0.3 volts, 0.7 volts, and 1.1 volts for germanium, silicon, and gallium arsenide, respectively.

3.8 *p-n* JUNCTION SURFACE AND END EFFECTS

Up to this point, the material geometry usually has been considered as a one-dimensional, semi-infinite slab of semiconductor. Most previous results are applicable to finite size junction devices, as long as the n and p sides of the junction are long, compared with their respective diffusion lengths, and also that V_0 is regarded as the voltage applied across the total depletion layer region as opposed to that across the entire amount of bulk material containing the junction.

Since the major function of the *p-n* junction is ultimately rectification, such action should not take place at the connections or contacts that bring the applied voltage to the region of interest. An ohmic contact is defined as a nonrectifying metal-semiconductor contact that has a negligible resistance across it, compared to the bulk resistance of the semiconductor. The ohmic contact should not perturb device performance, but to supply corresponding current with minimum voltage drop. Ohmic contacts are usually made by soldering the wire connector to an abraded semiconductor surface, or to a heavily increased doping density at the contact surface over that in the bulk. This insures a relatively high majority carrier concentration in the contact vicinity promoting carrier equilibrium, while simultaneously keeping the corresponding minority carrier concentration low to curtail any minority carrier injection there.

For finite size devices, construed here as those whose bulk region thickness is of the order of or less than the diffusion length, the results of the previous section must be extended. For example, this must include the fact that connections or leads are contacting the device so that V_0 can be applied. In this case, Eqs. (3.79) and (3.80) are still applicable for the minority carrier densities. However, the boundary conditions are modified to take these "end effects" into account. This is done by replacing the boundary conditions given in Eqs. (3.81) and (3.82) by

$$D_n n_p'(-d_p) = S_p(n_p(-d_p) - n_{p0}) \tag{3.96}$$

$$-D_n p_n'(d_n) = S_n(p_n(d_n) - p_{n0}) \tag{3.97}$$

where d_n and d_p are shown in Fig. 3.17. S_n and S_p are the recombination velocities discussed in Section 1.13. The corresponding minority carrier

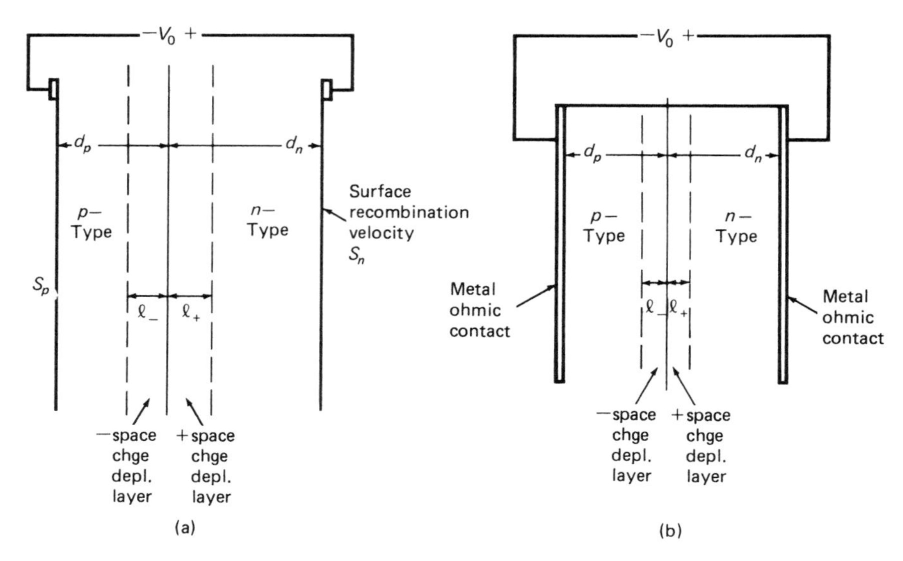

Fig. 3.17 Finite one-dimensional *p-n* junction geometry showing semiconductor surface contact and metal ohmic contact.

densities (i.e., the solutions of Eqs. (3.79) and (3.80)) with the above boundary conditions, are given by
for the p side of the junction

$$n_p = n_{p0}[1 + \cosh(x + l_-/L_n) + v_p \sinh(x + l_-/L_n)] \cdot (\exp(eV_0/kT) - 1) \tag{3.98}$$

for the n side of the junction

$$p_n = p_{n0}[1 + \cosh(x - l_+/L_p) - v_n \sinh(x - l_+/L_p)] \cdot (\exp(V_0/kT) - 1) \tag{3.99}$$

where

$$v_n = \frac{h_{np} + \tanh(d_n - l_+/L_p)}{1 + h_{np}\tanh(d_n - l_+/L_p)}; \qquad h_{np} = S_n L_p/D_p \tag{3.100}$$

$$v_p = \frac{h_{pn} + \tanh(d_p - l_-/L_n)}{1 + h_{pn}\tanh(d_p - l_-/L_n)}; \qquad h_{pn} = S_p L_n/D_n \tag{3.101}$$

and the corresponding "diode" current equation is

$$I = I_1(\exp(eV_0/kT) - 1); \qquad I_1 = e(n_{p0}D_n v_p/L_n + p_{n0}D_p v_n/L_p) \quad (3.102)$$

This equation is similar to that for the infinitely long junction, Eq. (3.88), except for the coefficients v_n and v_p. The latter coefficients include the recombination velocities, S_n and S_p, which reflect the finite aspect of the device junction. It is to be noted that if the recombination velocity has its characteristic value given by $h_{np} = 1$, then $v_n = 1$ also, and the minority hole density equation, Eq. (3.99), reduces to Eq. (3.84), corresponding to the infinite device. Also, if $h_{pn} = 1$, so does $v_p = 1$, and the above saturation current reduces to that for the infinite junction, i.e., $I_1 = I_0$.

In the limit of very large recombination velocities, i.e., $S_n \gg D_p/L_p$ and $S_p \gg D_n/L_n$, it is seen from the foregoing that

$$I_1 = e[(n_{p0}D_n/L_n)\coth(d_p - l_-)/L_n + (p_{n0}D_p/L_p)\coth(d_n - l_+)/L_p] \quad (3.103)$$

The above I_1 corresponds to the saturation current of the finite junction with leads at each end, assuming direct current (dc) to flow into and/or out of the junction in equilibrium.

This is precisely the case of high recombination velocity, so that the saturation current is I_1, given by Eq. (3.103), and the physical situation is depicted in Fig. 3.17b. For increasing values of reverse bias voltage, the depletion layer widths extend further into the bulk of the semiconductor, toward the connecting leads. If the bulk region is sufficiently thick so that $l_\pm \ll d_n, d_p$, i.e., the depletion layers never approach the ends of the bulk region, then the saturation current variation with V_0 is negligible. On the other hand, if the bulk dimensions are sufficiently narrow that the depletion layers approach or impinge upon the ends of the bulk region for a given value of V_0, then the saturation current grows without bound, as seen from the lim I_1 ($l_\pm \to d_n, d_p$) above. Then "punch-through," which is similar to junction breakdown, occurs. This is simply the situation where the depletion layers closely approach the end terminations, which act like regions of very high recombination velocity, thus increasing the reverse current. Punch-through can be minimized by, of course, maintaining enough semiconductor bulk so that the depletion layer never nears the surface device leads. However, the punch-through phenomenon can be used to provide a sudden increase in saturation current, for a given V_0, to produce a voltage regulation effect, or zener action, as depicted in Fig. 3.14.

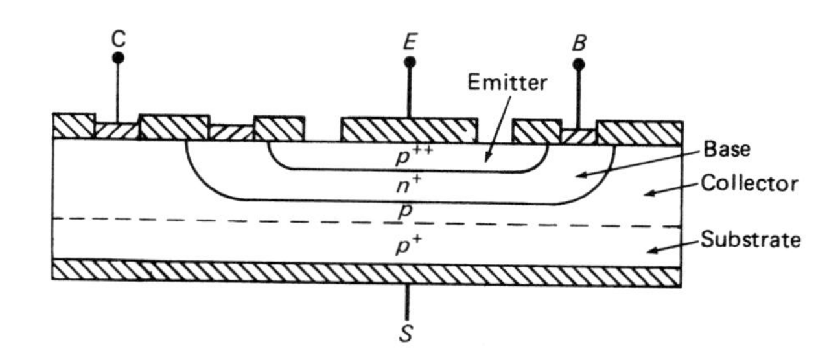

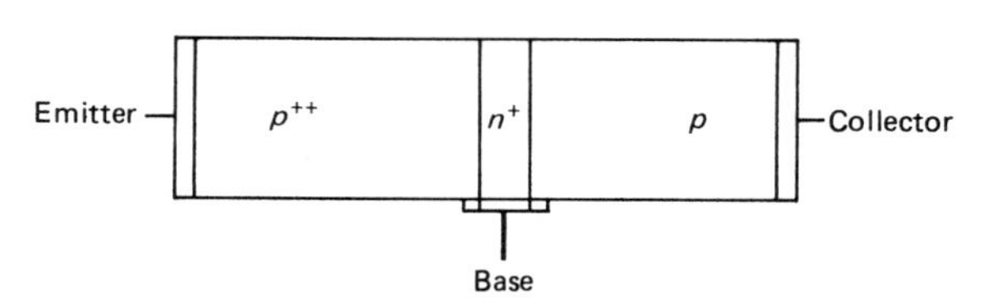

Fig. 3.18 Planar type *pnp* transistor.

3.9 THE BIPOLAR TRANSISTOR

A bipolar transistor can be defined as a semiconductor containing two junctions in close proximity. These are (a) an emitter-base junction to inject excess minority carriers into an *n*-type base, for example, and (b) a base-collector junction to recover most of them. Figure 3.18 depicts such a configuration, which is in this case, a planar *pnp* transistor. Its obverse is called an *npn* transistor. The energy band configuration of the *pnp* transistor is shown in Fig. 3.19. The upper Fig. 3.19 depicts the transistor under equilibrium conditions, as, for example, when the three transistor elements are shorted together.

The lower Fig. 3.19 shows the *pnp* transistor with the emitter-base junction forward biased, and the base-collector junction reverse biased. Because the emitter-base junction is forward biased, current flows so that large numbers of hole minority carriers are injected into the *n*-type base. Most of the holes will impinge on the base-collector junction, assuming that the two junctions are sufficiently close together, aided by the electric field, as seen in Fig. 3.19. Note that, for a reverse biased base-collector

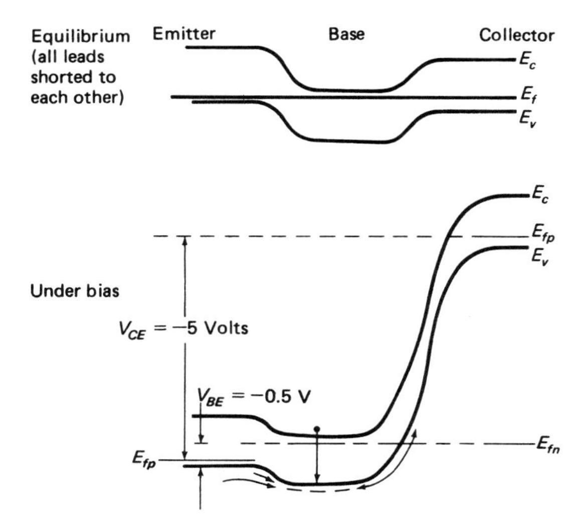

Fig. 3.19 Planar *pnp* transistor energy band configuration.[7]

junction with respect to majority carriers, a large minority carrier hole current flows across the junction to the collector. This is because, insofar as minority carriers are concerned, the base-collector junction is forward biased. This minority carrier flow provides the basis for transistor action, as will be discussed. Not all the holes reach the collector-base region. Some recombine with electrons en route in the base. Also, for the forward biased emitter-base junction some electrons are injected into the emitter together with holes injected into the base. Electrons and holes recombine in the depletion layer region contiguous to the emitter-base junction. Due to these processes, electrons flow into the base through the base lead.

The total emitter current I_E consists of a hole current reaching the collector, I_C, plus electrons that flow in through the base lead, I_B. Thus

$$I_E = I_C + I_B \tag{3.104}$$

The common base gain α, or h_{FB}, is defined as $\alpha = I_C/I_E$. The common emitter gain β, or h_{FE}, is defined as $\beta = I_C/I_B$. These are seen to be related by

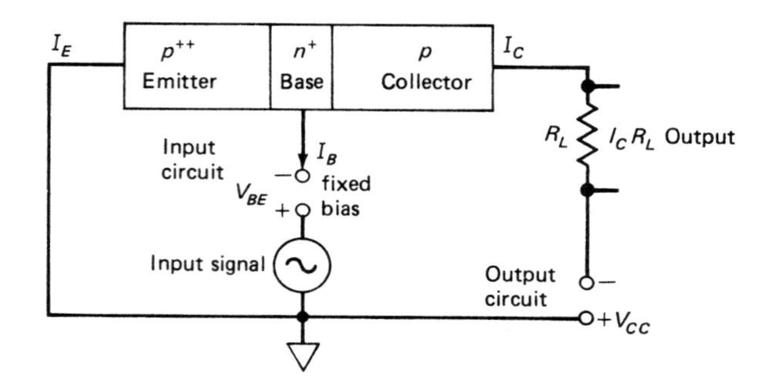

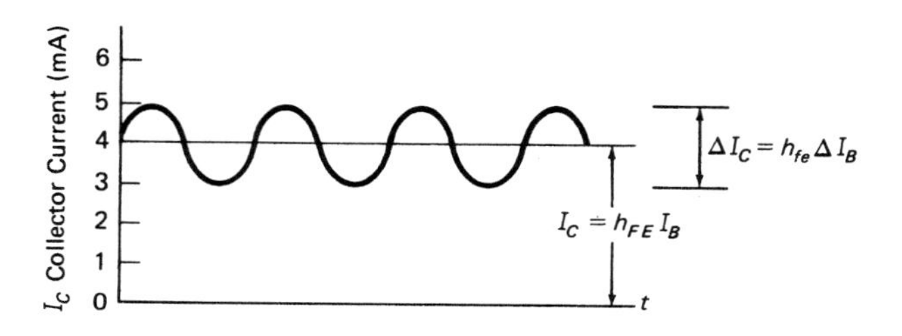

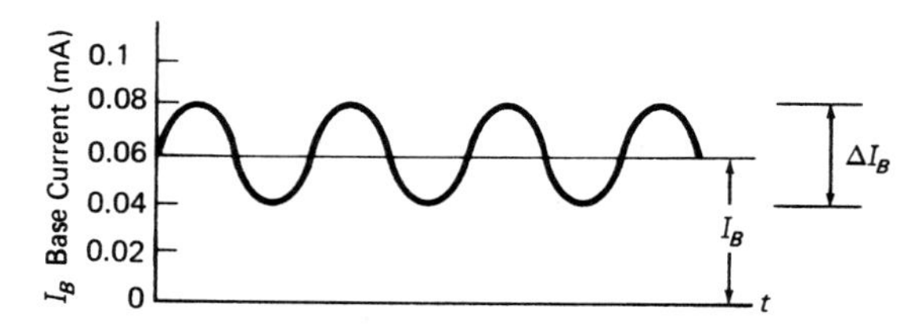

Fig. 3.20 A *pnp* transistor amplifier circuit.

$$\beta = \alpha/(1 - \alpha) \tag{3.105}$$

Under normal conditions for a high-gain transistor $\alpha \lesssim 1$, so that β will be very large. The foregoing gains are termed dc gains, as the dynamic common emitter current gain $h_{fe} = \partial I_C/\partial I_B$. Alternatively, using the definition of h_{FE} and the appropriate partial derivative chain (e.g., $\partial h_{FE}/\partial I_B = \partial h_{FE}/\partial I_C \cdot \partial I_C/\partial I_B$) yields

$$h_{fe} = h_{FE}/(1 - (I_C/h_{FE})\partial h_{FE}/\partial I_C) \tag{3.106}$$

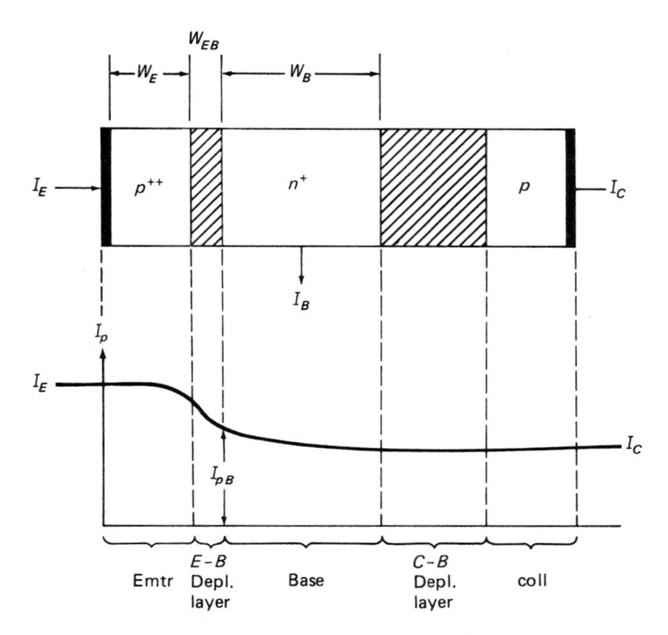

Fig. 3.21 Hole current in a *pnp* transistor.

If the current gain is independent of collector current, $\partial h_{FE}/\partial I_C = 0$, and $h_{fe} = h_{FE}$. Consider Fig. 3.20 which incorporates a *pnp* transistor in a simple common emitter amplifier circuit. For an input voltage V_{BE}, a given dc base current I_B, a collector current I_C will flow. If a small input signal is super-imposed on V_{BE}, the base current will vary with time, as shown in Fig. 3.20. This will cause a variation in the output current I_C, which is h_{FE} times larger than the input signal current. Hence, the transistor amplifies the input signal.

In Fig. 3.21, the hole current I_p in this *pnp* transistor is shown as a function of the emitter to collector distance through the base region. In the emitter, of course $I_p \cong I_E$. However, an increasing fraction of the total current is carried by electrons as the collector is approached. In Fig. 3.21, the plane through the origin is the boundary between the emitter-base depletion region and the *n*-type base material. It can be seen there that some of the emitter current, i.e., that due to the holes, is diffusing into the base. The remainder of the device current is due to electrons. Part of the latter current is injected into the *p*-type emitter, while (a) the other part is due to electrons injected into the emitter-base region, where they recombine with holes. (b) Toward the emitter region, the hole current portion decreases, because some of the current is injected into the n-type base to recombine with electrons there. (c) The remaining fraction of hole current is used up by recombination in the base. These three portions (a)–(c) together constitute the base current. The part of the total current that is due to holes will remain approximately constant beyond the base-collector depletion region interface. This is because recombination is negligible, for this depletion region is reverse biased. Analogously, the hole current in the collector field region is due to majority carriers, so that recombination does not affect it appreciably.

However, avalanche multiplication is possible. Minority carriers injected into the base are mainly influenced by diffusion at low injection levels. This diffusion in the base region is described by

$$p_n''(x) - (p_n - p_{n0})/L_p^2 = 0 \tag{3.107}$$

with boundary conditions

$$p_n(0) = p_{n0} \exp eV_B/kT; \qquad p_n(W_B) = p_{n0} \tag{3.108}$$

where W_B is the base width. The leftmost boundary condition is rewritten from Eq. (3.15) and is often called the *law of the junction.*[17] Its rederivation from another aspect is instructive and follows from the one-dimensional counterpart of the current Eq. (1.85), since the net injection current density J_p is small relative to the other terms therein. Normally, it is several orders of magnitude smaller than either the drift or diffusion component in Eq. (1.85) even when the junction is forward biased[32]. Neglecting J_p in Eq. (1.85) yields, in one dimension,

$$e\mu pE = eD_p p' \tag{3.109}$$

Using the Einstein relationship given in Section 1.8, and rewriting the above equation, where V is the corresponding potential, gives

$$E = -V' = kTp'/ep \tag{3.110}$$

Integrating Eq. (3.110) across the emitter-base junction, where the hole concentration is given by the thermal equilibrium value p_{p0} on the edge of the p-side, and correspondingly $p_n(0)$ on the n-side, yields

$$p_n(0) = p_{p0} \exp - e(V_0 - V_B)/kT \tag{3.111}$$

$V_0 - V_B$ is the junction voltage, which is the forward voltage V_0 minus

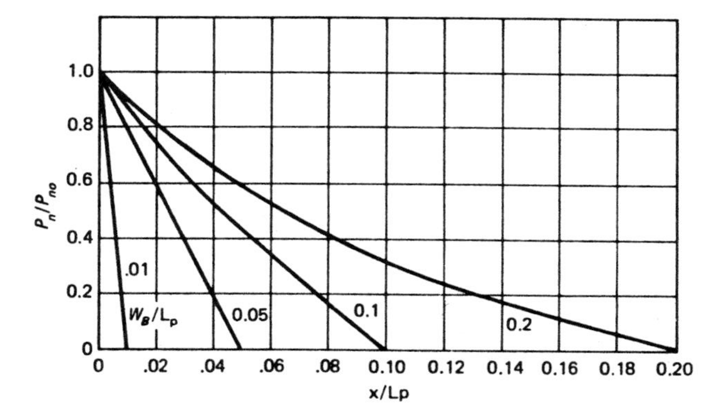

Fig. 3.22 Injected minority carrier variation with base width for small W_B/L_p ratios required for reasonable gain.

the barrier potential V_B. By using Eqs. (3.14) to write $p_{p0} = p_{n0} \exp eV_0/kT$, Eq. (3.111) which is the sought after boundary condition, can be written as

$$p_n(0) = p_{n0} \exp(eV_B/kT) \tag{3.112}$$

The rightmost boundary condition in Eq. (3.108) asserts that the base-collector junction is reverse biased. That is, the corresponding field stems majority carrier flow into the collector, but allows minority carriers to be swept into the collector, which is a key part of transistor action.

The solution to Eqs. (3.107) and (3.108) is

$$\begin{aligned} p_n(x) &= p_{n0}(1 + (\exp(eV_{EB}/kT) - 1) \cdot \operatorname{csch}(W_B/L_p) \cdot \sinh(W_B - x)/L_p) \\ &\cong p_{n0}(1 + (\exp(eV_{EB}/kT) - 1) \cdot (1 - x/W_B)); \qquad W_B/L_p \ll 1 \end{aligned} \tag{3.113}$$

and is plotted in Fig. 3.22 for the usual situation of $W_B/L_p \ll 1$ required for reasonable transistor gain. The various fractions of the emitter current are now discussed to derive an expression for the common emitter gain in terms of mainly physical parameters of the transistor. This is in contrast to these gains normally expressed in terms of electrical parameters. One fraction of the emitter current is that due to holes injected into the base, viz., I_{pB}. This is obtained from the expression for the current density j_{pB}. The latter is composed of the drift (electric field driven) component, plus the diffusion component, respectively. For example,

$$j_{pB} = e\mu_B pE - eD_B p' \tag{3.114}$$

From Eqs. (3.13) and (3.14), the mass law relation $np = n_i^2$, and that $N_v \cong N_c$, the electron density in the base, or what amounts to the dopant density $N(x)$ in the base, is given by

$$n \cong N(x) \cong n_i \exp(E_i - E_f)/kT \tag{3.115}$$

where E_i is the intrinsic Fermi level, and E_f is now the nonequilibrium Fermi level in the base, with a corresponding potential V, so that $E_f = eV$, and the electric field E in the base is given from Eq. (3.115) as

$$-E = V'(x) = -(kT/e)\, N'(x)/N(x) \tag{3.116}$$

Eliminating E between Eqs. (3.114) and (3.116), and with the Einstein relationship, the result yields

$$p'(x) + N'(x)p(x)/N(x) = -j_{pB}/eD_B \tag{3.117}$$

With the boundary condition that $p(W_B) \cong 0$, the integral of Eq. (3.117) is

$$p(x) = j_{pB} \int_x^{W_B} N(x')\, dx'/eD_B N(x) \tag{3.118}$$

Letting $N(0) = n_{B0}$, the donor concentration at the emitter-base junction, and writing the corresponding hole concentration gives

$$p_B(0) = j_{pB} \int_0^{W_B} N(x')\, dx'/eD_B n_{B0} = j_{pB} Q_B/eD_B n_{B0} = p_{B0} \exp(eV_{EB}/kT) \tag{3.119}$$

where $Q_B = \int_0^{W_B} N(x)\, dx$ is called the Gummel number, which is the per unit number of dopant atoms base area. For silicon bipolar transistors,[3] it is about $10^{12} - 10^{13}\,\text{cm}^{-2}$.

The rightmost term is the hole density at the emitter-base junction.[3] Now, assuming that the dopant density is a constant N_B, then the desired fraction of emitter current is

$$I_{pB} = j_{pB} A_J = (eD_B n_i^2/N_B W_B) A_J \exp(eV_{EB}/kT) \tag{3.120}$$

where A_J is the emitter-base junction area, and that $n_{B0} p_{B0} = n_i^2$. From a similar derivation, the second fraction of emitter current is that from electrons injected from the base into the emitter, given by

$$I_{nE} = (eD_E n_i^2/N_E W_E)A_J \exp(eV_{EB}/kT) \tag{3.121}$$

The third and last fraction of emitter current is due to electrons injected into the emitter-base depletion region where they recombine with holes. It is obtained also in similar fashion to the preceding two emitter current fractions, i.e.

$$I_{nr} = (\tfrac{1}{2}en_i W_{EB} A_J/\tau_0) \exp eV_{EB}/kT \tag{3.122}$$

where W_{EB} is the width of the emitter-base depletion region, and τ_0 is the effective carrier lifetime in a reverse biased depletion region.

An emitter in which I_{pB} and I_{nE} are small is termed *efficient*. Its corresponding emitter efficiency η is defined as

$$\eta = I_{pB}/I_E = I_{pB}/(I_{pB} + I_{nE} + I_{nr}) \tag{3.123}$$

From Eqs. (3.118)–(3.122), η can be written as

$$\eta^{-1} = 1 + \frac{N_B W_B}{D_B}\left(\frac{D_E}{N_E W_E} + \frac{W_{EB}}{2n_i\tau_0}\exp - (eV_{EB}/kT)\right) \tag{3.124}$$

or,

$$\eta^{-1} = 1 + B_B/E_E + \tfrac{1}{2}R_{EB}\sqrt{eB_B A_J/I_{pB}} \tag{3.125}$$

where the base factor $B_B = N_B W_B/D_B$ depends only on the impurities in the base region. The emitter factor $E_E = N_E W_E/D_E$ depends on only the impurities in the emitter region, and the recombination factor $R_{EB} = W_{EB}/\tau_0$ characterizes the recombination rate in the emitter-base junction depletion region.

A base transport factor α_T can be defined as the amount of minority carrier current injected into the base that reaches the collector-base depletion layer.

For a *pnp* transistor, α_T is the ratio of hole current reaching the collector to the hole current injected into the base, so that

$$\alpha_T = p_n'(W_B)/p_n'(0) \tag{3.126}$$

From Eqs. (3.113) and (3.126)

$$\alpha_T = \operatorname{sech}(W_B/L_B) \tag{3.127}$$

where L_B is the minority carrier diffusion length in the base. Usually, a well-made transistor has a W_B for which $W_B \ll L_B$, so that, from Eq. (3.127)

$$\alpha_T \cong 1 - \tfrac{1}{2}(W_B/L_B)^2 \tag{3.128}$$

The common base current gain α satisfies

$$\alpha = h_{FB} = I_C/I_E = \eta\alpha_T \tag{3.129}$$

which assumes that the leakage current of the collector-base junction, i.e., $I_{CBO} \cong 0$. When this assumption is not made, then

$$\alpha = \eta\alpha_T + I_{CBO}/I_E \tag{3.130}$$

The common emitter current gain is now given by

$$\beta = I_C/I_B = \eta\alpha_T/(1 - \eta\alpha_T) \tag{3.131}$$

Similarly, if I_{CBO} is not negligible, then

$$\beta = \left(\frac{\eta\alpha_T}{1 - \eta\alpha_T}\right)\left(1 + \frac{I_{CBO}}{\eta\alpha_T I_B}\right) \tag{3.132}$$

so that, for a high β transistor, it is desired that $\eta\alpha_T \rightarrow 1$ as closely as possible. For such a transistor, using Eqs. (3.124) and (3.128)

$$\begin{aligned} \beta^{-1} &= 1 - \eta\alpha_T \\ &= \tfrac{1}{2}(W_B/L_B)^2 + (N_B W_B/D_B)(D_E/N_E W_E) \\ &\quad + (N_B W_B/D_B)(W_{EB}/2\tau_0 n_i)\cdot \exp - (eV_{EB}/kT) \end{aligned} \tag{3.133}$$

The first term of the rightmost expression above is due to recombination in the base, the second is due to current injection into the emitter, and the third is due to recombination within the emitter-base junction depletion region. Implicit in the foregoing is β as a function of the collector current density for various recombination rates. This is shown in Fig. 3.23.

A figure that is collateral to Fig. 3.23 is Fig. 3.24, in that it depicts both DC and dynamic (signal) bipolar gain as a function of collector current. Experimental measurements of collector and base current versus base-emitter voltage for typical bipolar transistors yield, respectively,[7]

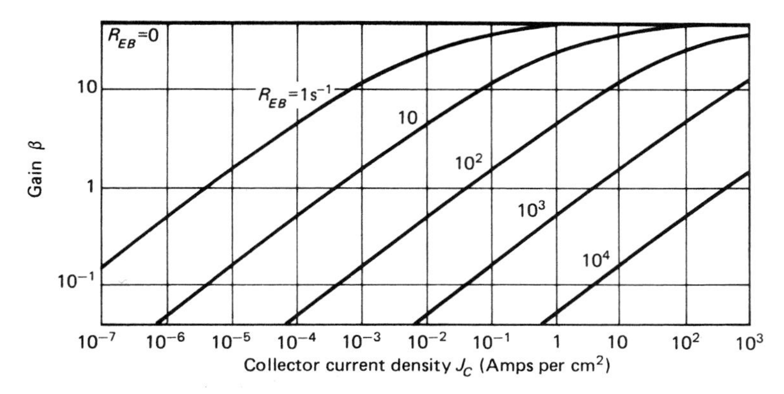

Fig. 3.23 Gain versus collector current density for various recombination rates R_{EB}.[7] Figure © 1967 by John Wiley & Sons.

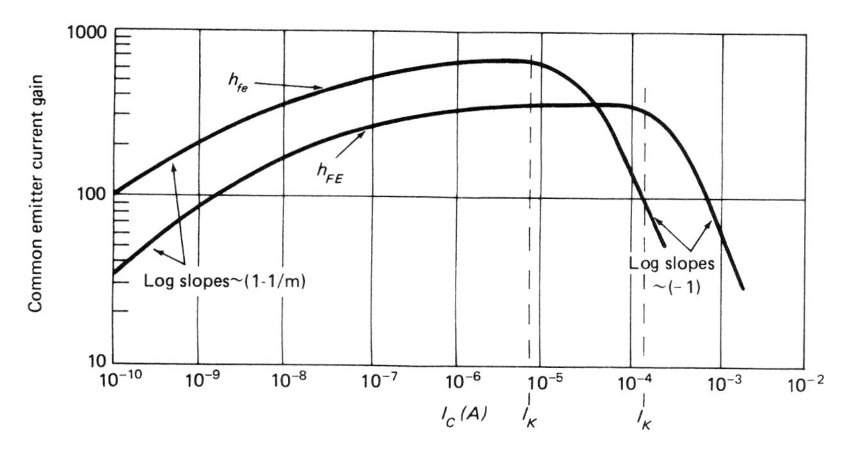

Fig. 3.24 Current gains versus collector current for a typical bipolar transistor.[7]

$$I_C \propto \exp(eV_{EB}/kT) \text{ and } I_B \propto \exp(eV_{EB}/mkT) \tag{3.134}$$

The parameter m reflects the combination of surface and emitter depletion layer recombination. m can vary from 1 to 2 depending on specific transistor design and processing parameters.[28] The parameter m is taken here as 1.7 for relatively low collector current levels. From these expressions, it is seen that the DC gain is given by

$$\beta \equiv h_{FE} = I_C/I_B \cong I_C^{1-1/m} \cong I_C^{0.41} \tag{3.135}$$

For high collector currents, the discussion in Section 3.7 results in

$$\beta \equiv h_{FE} = I_C/I_B \cong \exp(eV_{BE}/2kT)/\exp(eV_{EB}/kT) \propto I_C^{-1} \tag{3.136}$$

These variations in β for low and high current injection levels are mirrored in the low and high current slopes of the h_{FE} curves in Fig. 3.24.

Analogously, the bipolar signal gain h_{fe} can be written from Eq. (3.134) for low collector currents, where it is seen that

$$I_C \propto I_B^m, \tag{3.137}$$

as

$$h_{fe} = \partial I_C/\partial I_B \propto 1.7 I_C^{0.41} \tag{3.138}$$

From Eq. (3.136), it is also noticed that $I_C \propto I_B^{1/2}$ for high current levels, so that

$$h_{fe} = \partial I_C/\partial I_B \propto I_C^{-1} \tag{3.139}$$

Figure 3.24 shows the characteristic "knee" in the behavior of the gains with increasing collector current. The corresponding value of the knee current $I_C = I_K$ is recognized to be an important parameter in bipolar transistor curcuit models as used in SPICE, SYSCAP,[28] and other modern computer electronic circuit programs.

At very low collector current values, the recombination current at the device surface and in the emitter depletion layer, as well as the surface leakage current, can be large with respect to the base minority carrier diffusion current, so the current gain is low. At high collector currents, the base minority carriers approach the base majority carrier density, so that the former carrier density effectively increase the base doping level, which lowers the emitter efficiency. The decrease in the emitter efficiency causes the gain to drop at the high extreme of collector current,[3] as in Fig. 3.24. Also, it should be noted that this behavior can be derived analytically from detailed investigations into these types of models.[28] At very high current levels, emitter crowding (see Section 3.14) results in an even more rapid fall off of gain with current.

3.10 TRANSIT TIME/GAIN-BANDWIDTH PRODUCT

An important limitation of transistor function, and one that is affected by radiation, is that due to transit time. Previous discussions assumed instantaneous response of all parts of the transistor with respect to voltage or current changes in another portion thereof. However, carrier transit times affect this response. For example, the transit time τ_b for holes to cross the base region of *pnp* transistor is

$$\tau_b = \int_0^{W_B} dx/v(x) \tag{3.140}$$

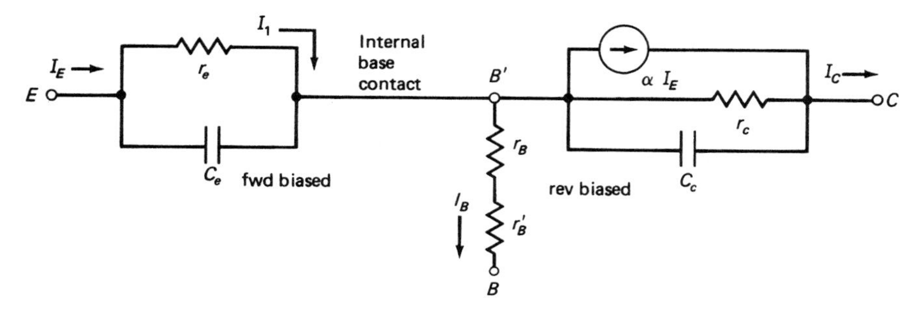

Fig. 3.25 *pnp* transistor tee equivalent circuit for the common base bipolar configuration.[16,17]

where $v(x)$ is the hole speed. Now, the base hole current I_p is simply

$$I_p = ev(x)p(x)A_J \tag{3.141}$$

Inserting Eq. (3.141) into (3.140), with Eq. (3.118), and assuming a linear hole concentration variation with distance into the base, i.e., $p(x)/p_0 = 1 - x/W_B$, results in

$$\tau_b = W_B^2/2D_B \tag{3.142}$$

The angular frequency corresponding to this time limitation can be shown to be given by[7]

$$\omega_{tr} = 1/2\pi\tau_b \tag{3.143}$$

The Tee equivalent circuit, shown in Fig. 3.25, is used to describe the gain-bandwidth or transistor cutoff frequency f_T. From Fig. 3.25, at the limit of zero frequency, i.e., dc, let $\alpha_0 = I_C/I_1$. For nonzero frequencies

$$I_E = V'_{EB}(1 + i\omega r_e C_e)/r_e = I_1(1 + i\omega r_e C_e) \tag{3.144}$$

From the foregoing, and temporarily neglecting a 180° phase shift, the following can be written

$$\alpha = I_C/I_E = \alpha_0/(1 + i\omega r_e C_e) = \alpha_0/(1 + i\omega/\omega_\alpha) \tag{3.145}$$

where $\omega_\alpha = 2\pi f_\alpha$, and $f_\alpha = 1/2\pi r_e C_e$ is called the *alpha* cutoff frequency. Then

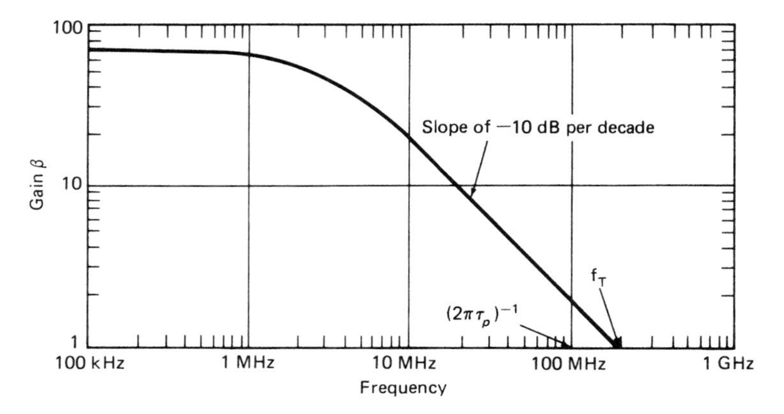

Fig. 3.26 Gain-frequency characteristic of *pnp* transistor.

$$\beta = \alpha/(1 - \alpha) = \alpha_0/(1 - \alpha_0 + i\omega/\omega_\alpha) \tag{3.146}$$

Define a frequency ω_β at which $\beta = 1/\sqrt{2}$ of its low to midfrequency value, i.e., the frequency corresponding to where the real and imaginary parts of the denominator of Eq. (3.146) are equal. That is, $\omega_\beta = \omega_\alpha/(1 + \beta_0)$ where $\beta_0 = \alpha_0/(1 - \alpha_0)$. Then, from Eq. (3.146)

$$|\beta| = \beta_0/(1 + (\omega/\omega_\beta)^2)^{1/2} \tag{3.147}$$

The cutoff frequency ω_T is that value where $\beta = 1$, i.e., the frequency intercept for unity gain of the transistor in Fig. 3.26, so that Eq. (3.147) is rewritten to give, where $\beta_0^2 \gg 1$

$$|\beta| = \beta_0/(1 + \beta_0^2(\omega/\omega_T)^2)^{1/2} \tag{3.148}$$

where is plotted in Fig. 3.26. Also, it is easily seen that $\omega_\alpha \approx \omega_T = \beta_0\omega_\beta$. From Eq. (3.148) at sufficiently large ω, $|\beta| \sim \omega_T/\omega$, and that the corresponding slope is -10 dB per decade, and shown in Fig. 3.26. Hence, the product

$$\omega|\beta| = \omega_T \tag{3.149}$$

has the constant value ω_T. $\omega|\beta|$ is called the *gain-bandwidth product* for the common emitter gain β, and it is also seen in this frequency range that only when the gain is unity does $\omega = \omega_T = 2\pi f_T$. The quantity f_T is

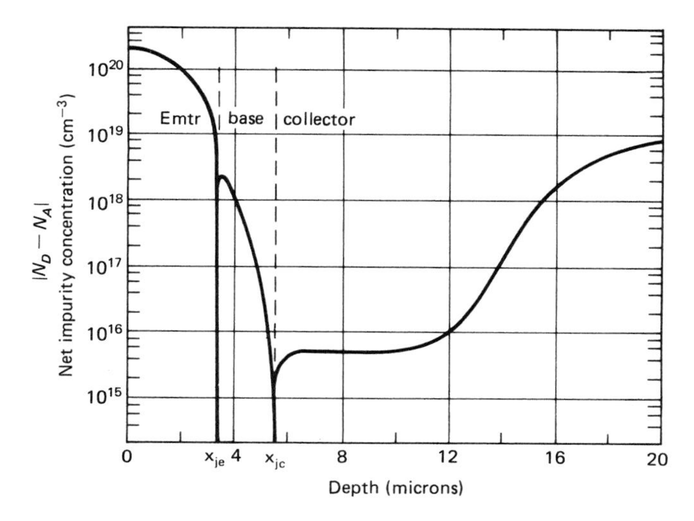

Fig. 3.27 *pnp* transistor dopant density distribution.

given by the frequency intercept for unity gain. In Fig. 3.26, it is seen that $f_T \cong 200\,\text{MHz}$.

3.11 GRADED BASE TRANSISTORS

During transistor manufacture, the double-diffusion process causes the distribution of dopant in the base to be nonuniform. An estimate of this distribution derived from diffusion parameters of the dopant in the transistor is shown in Fig. 3.27. The large change in dopant concentration throughout the base, as seen in the figure, causes a large change in the majority carrier density in the base. In equilibrium, an internal electric field must exist in the base, because no net current can flow. As discussed earlier, this field counters any diffusion currents due to majority carrier concentration gradients in the base. When minority carriers are injected into the base, they will be affected by this field. The electrons within the base diffuse toward the collector because of the dopant concentration gradient in this *pnp* transistor. Hence, the internal electric field will drive the electrons to the emitter-base junction as well as accelerate the injected holes. Now that the injected minority carrier motion is affected by both diffusion and the internal electric field, the base transit time will decrease, thus increasing f_T.

The penetration of dopant concentration c into the base region during manufacture is governed mainly by diffusion, as described by the diffusion equation in the concentration variable,[4] $c(x, t)$

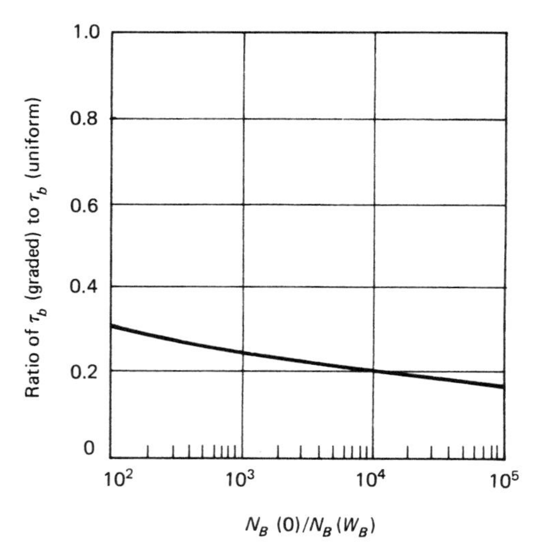

Fig. 3.28 Ratio of graded to uniform dopant impurity base transit time versus normalized dopant impurity concentration.

$$D\frac{\partial^2 c}{\partial x^2} = \frac{\partial c}{\partial t} \tag{3.150}$$

where the diffusion coefficient D is $\gg D_p$ and D_n, as this diffusion process is one of a substitutional impurity into a lattice by a vacancy mechanism.[4] The corresponding boundary conditions are given by

$$c(0, t) = c_0 \text{ (constant)}; \qquad c(\infty, t) = 0 \tag{3.151}$$

The concentration is given by the integral of Eqs. (3.150) and (3.151), which is

$$c(x, t) = c_0 erfc(x/(4Dt)^{1/2}) \tag{3.152}$$

where the complementary error function $erfc(x) = (2/\sqrt{\pi}) \int_x^\infty (\exp - y^2)\, dy$. The dopant concentration c as given in Eq. (3.152) is used in the transit time computation plotted in Fig. 3.28 as a comparison between graded and uniform base dopant concentration.[5] Further, in these computations, Eq. (3.120) is now seen to be modified to

$$I_{pB} = \frac{eD_B n_i^2 A_J \exp(eV_{EB}/kT)}{\int_0^{W_B} N_B(x)\, dx} \tag{3.153}$$

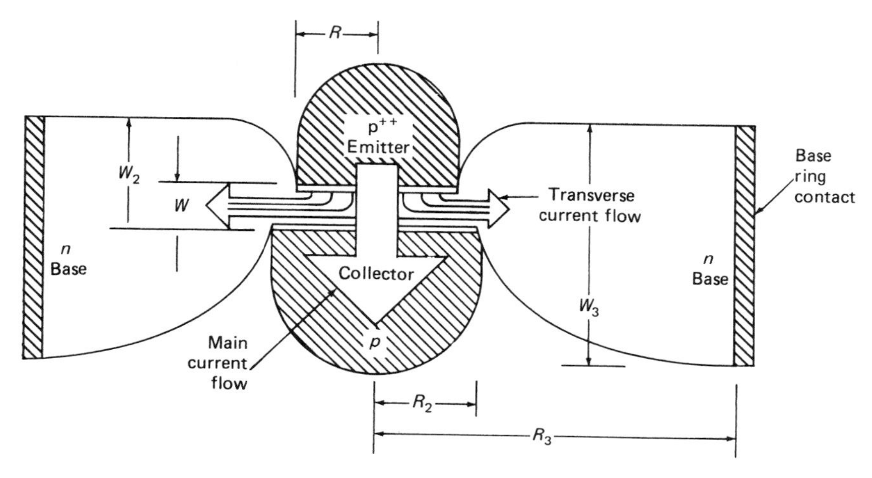

Fig. 3.29 Transverse component of base current in cylindrical geometry *pnp* transistor.[14,15]

to account for the graded base dopant concentration, where the denominator is seen to be the corresponding Gummel number.

3.12 BASE WIDTH MODULATION

In the active operating regime of the transistor, the base-emitter junction is forward biased, while the base-collector junction is reverse biased, as discussed earlier. As the magnitude of the reverse bias voltage varies, so does the width of the collector-base depletion region; hence the effective width W_B of the base region can decrease. This causes the gradient of injected minority carriers into the base to increase, so that the collector current thereby increases. This implies, in turn, an increase in current gain. However, the base current will remain unaffected, because it is mainly influenced by currents near the base-emitter junction. This modulation of the base width is called the *Early effect.*[6] It is enhanced in lightly doped base transistors.

3.13 BASE SPREADING RESISTANCE

Currents in the base of the *pnp* transistor under discussion, due to the injection of minority carriers into the emitter, recombination in the base-emitter depletion layer, and recombination in the base itself, are lumped together, as shown in Figs. 3.29 and 3.30. These currents in the base flow transversely to the remaining transistor current flow, which is marked

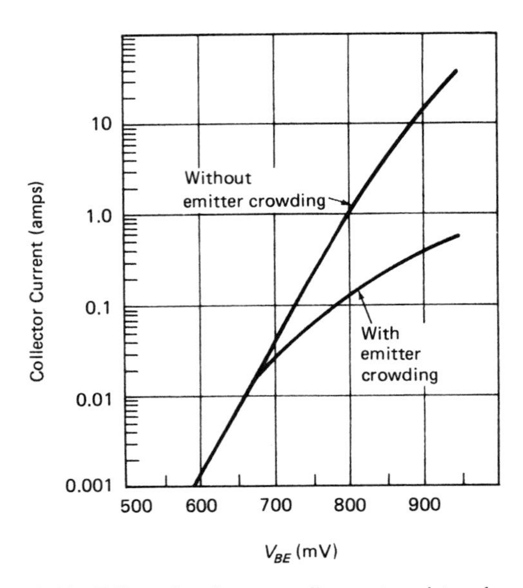

Fig. 3.30 Effect of emitter crowding on transistor characteristic.

"main current flow" in Fig. 3.29. These currents produce a transverse electric field increasing with radius, thus causing a voltage drop along the base side of the emitter-base junction. The effect of this transverse field is most pronounced in power transistors, for they operate at high current densities.

For the illustrative cylindrical transistor model as in Fig. 3.29, the above voltage drop results in the minimum voltage occurring at the emitter edge and increasing to a maximum under the center of the emitter. This voltage results in a varying emitter-base forward bias, increasing at the emitter edge over that of the inner portion of the emitter. This, in turn, causes a high current density near the emitter periphery, thus concentrating the current in that region. This is termed *emitter crowding* and is discussed in Section 3.14. These base current changes with emitter radius are reflected in what is called a *base spreading* resistance, which is different from the ohmic bulk resistance of the base.

Assume that the radial current I_r increases from zero under the emitter center to a maximum value I_1 at the emitter edge at radius R. Neglecting any voltage drop outside the collector-emitter region in Fig. 3.29, the average base transverse voltage drop is given by

$$\bar{V}_B = (1/\pi R^2) \int_0^R V_B(r) \cdot 2\pi r \, dr \tag{3.154}$$

Because the transverse electric field is increasing with radius toward the periphery, the corresponding transverse or radial current $I_r = I_1 \, (r/R)^2$, and so the radial current density J_r is given by

$$J_r = I_r/2\pi r W = I_1 r/2\pi R^2 W \tag{3.155}$$

where W is the base width.

Then the voltage drop referred to R is written using the "Ohms law" Maxwell equation of Section 2.2, viz., $J_r = \sigma_B E_r$, where the conductivity $\sigma_B = \rho_B^{-1}$

$$V_B = -\int_r^0 E_r \, dr' = \rho_B I_1 r^2/4\pi R^2 W \tag{3.156}$$

Inserting the above into Eq. (3.154) yields

$$\bar{V}_B = \rho_B I_1/8\pi W \tag{3.157}$$

Hence, the base spreading resistance for this cylindrical transistor is very nearly

$$r'_B = \rho_B/8\pi W \tag{3.158}$$

Actually, there are two other terms in this expression for r'_B, but they are usually smaller than $\rho_B/8\pi W$, depending on the transistor construction. The total expression is[15]

$$r'_B = \rho_B \left(\frac{1}{8\pi W} + \frac{\ln(R_2/R)}{2\pi W_2} + \frac{\ln(R_3/R_2)}{2\pi W_3} \right) \tag{3.159}$$

The first term is the well-known result for the resistance of a disc with current flow into one face only and out through its peripheral wall, with none out of the other face. The mean base resistivity is[7]

$$\rho_B^{-1} = \sigma_B = (e\mu_n/\pi R^2) \int_0^R (N_D(r) - N_A(r)) \cdot 2\pi r \, dr = \mu_n \bar{q}_B \tag{3.160}$$

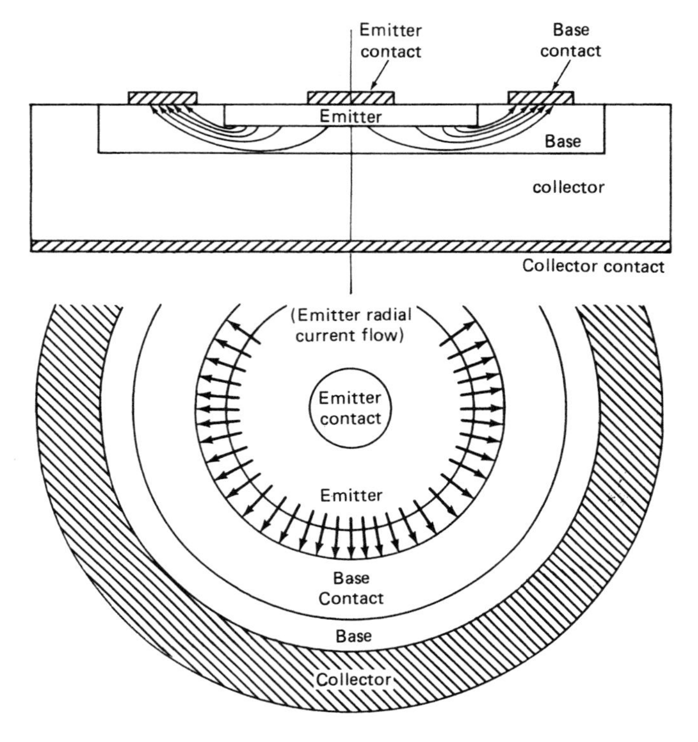

Fig. 3.31 Emitter crowded current flow.

where the mean charge density averaged over the base area is given by

$$\bar{q}_B = (e/\pi R^2) \int_0^R (N_D(r) - N_A(r)) \cdot 2\pi r \, dr \tag{3.161}$$

3.14 EMITTER CROWDING

Emitter crowding implies the concentration of emitter current flow to near the emitter periphery, instead of a homogeneous current density threading the emitter area.[9,10] Emitter crowding is produced by the component of base current that flows laterally through the base parallel to the emitter and collector junction. This is in contrast to the normal base current to the collector flowing perpendicular to the two junctions, as in Fig. 3.29. The active region in the base is that portion directly contiguous to and under the emitter. This active region is rather thin, so that its resistance is large. The corresponding current flowing through this (base spreading) resistance develops a lateral potential difference across the base. The polarity of this potential map is such as to cause a centrally lower forward emitter-base junction voltage, relative to that near the emitter rim. That is, the emitter rim is more forward biased than the emitter center. As the current flow anywhere in this region is a very sensitive function of the junction voltage, because of the exponential relationship between these two quantities, potentials of amount kT/e can cause a significant reduction in the effective emitter area.

The resistance r_c of a circular disc W cm thick, with current flowing into one face only and exiting through the edges is $r_c \cong \rho/8\pi W$, which for the radial system of Fig. 3.31 is the intrinsic or base spreading resistance r'_B as in the previous section. For a base dopant concentration N_A of $5 \cdot 10^{16}$ dopant atoms per cm^3, ρ is about 5 ohm cm. The corresponding r_c is about 100 ohms. The collector current that will cause a transverse potential of kT/e is $I_C = \beta I_B$, or $I_C = \beta kT/er_c$, which is about 12 mA for a transistor gain of 50. So the effective emitter area is reduced by a factor of e, the Naperian base, since the emitter current density ratio becomes $(\exp(-1))$, from $V = kT/e$, as inserted into $J_E/J_{E0} = \exp - (eV/kT)$. Figures 3.30 and 3.31, respectively, depict the effect of emitter crowding on the transistor characteristic and the emitter crowded current flow. The principal effect of emitter crowding is to increase the local current density at the emitter periphery and thus create a higher injection level for a given collector current. The resulting effectively smaller emitter area corresponds to a lesser overall collector current for a given V_{cc}. The deleterious effects of the high injection due to emitter crowding will be enhanced by displacement damage radiation due to neutrons, as discussed in Section 5.5. This can substantially reduce the common emitter current gain h_{FE}.

PROBLEMS

1. Discuss the approximations required to "linearize" Poisson's equation (Eq. 3.22), so that the $\sinh(e\phi/kT)$ term can be replaced by $e\phi/kT$. For an abrupt junction, solve this resulting Poisson's equation, and compare with the potentials obtained in Section 3.4.
2. For a lightly doped n or p region, the space charge extends deep into the semiconductor bulk. In this case, the maximum internal electric field E_0 becomes proportional to $(-V_0)^{1/2}$. Interpret this statement, as the latter quantity is imaginary. Show that the corresponding depletion layer capacitance per unit area is $C/A_J = \varepsilon\varepsilon_0/l$, analogous to the parallel plate capacitor formula.

3. For a linearly graded junction, show that the corresponding capacitance per unit area is proportional to $(-V_0)^{-1/3}$, and interpret this quantity in the light of problem (2).
4. High-voltage silicon zener diode action is essentially all constant equilibrium avalanche operation throughout the junction. Using Eq. (3.70), and that the sum of hole and electron currents is constant, show that in this case, $I_p = \alpha_n I/(\alpha_n - \alpha_p)$, and $I_n = -\alpha_p I/(\alpha_n - \alpha_p)$.
5. How does "punch-through" differ from junction breakdown?
6. Obtain a heuristic relationship between the transit time and the gain-bandwidth product.
7. Why is it that the built-in or contact potential cannot be measured by a voltmeter placed across the junction?
8. In Chapter 1, Problem 11, antimony was introduced as a dopant, resulting in an increased conductivity and lowered resistivity. If this material is used to make an *n*-type base 1 mil (10^{-3} inches) thick in a *pnp* bipolar transistor, what is the corresponding sheet resistance of this base?
9. The junction capacitance of an abrupt p^+-n junction, with $\tau = 10^{-6}$ seconds, measured at a reverse bias voltage of -1 volt is 5 pF. The built-in potential ϕ_0 is 0.9 volts. When the junction is forward biased at 0.5 volts, the junction current is 10 mA.
   Compute:

   (a) the depletion layer capacitance C_d when the junction is forward biased at 0.5 volts.
   (b) the corresponding diffusion capacitance C_ν.
   (c) the ratio C_ν/C_d.

10. Two silicon diodes are connected in series opposing each other, with 5 volts across all. What is the voltage across each at room temperature (i.e., for $\eta kT/e = 0.052$ volts)?
11. In Section 3.5, the sheet resistance discussion asserts that $L/W \geqslant 1$ counts the number of "squares" of area W^2 and thickness t in series. How is $L/W < 1$ to be interpreted?
12. In Section 3.9, the expression relating the bipolar transistor dynamic gain h_{fe} and the DC gain h_{FE} is given by $h_{fe} = h_{FE}/(1 - (I_C/h_{FE})(\partial h_{FE}/\partial I_C))$. Show the required derivational steps using the definitions $h_{fe} = \partial I_C/\partial I_B$ and $h_{FE} = I_C/I_B$.

REFERENCES

1. M. J. O. Strutt. Semiconductor Devices, Vol. 1, Semiconductors and Semiconductor Diodes. Academic Press, New York, Chap. 2, 1966.
2. J. L. Moll. Physics of Semiconductors. McGraw-Hill, New York, 1964.
3. S. M. Sze. Physics of Semiconductor Devices. Wiley-Interscience, New York, p. 94, 1981. S. M. Sze and G. C. Gibbons. "Avalanche Breakdown Voltages of Abrupt and

Linearly Graded Junctions in Ge, Si, GaAs, and GaP." Appl. Phys. Lett. 8:111, 1966. R. M. Warner and B. L. Grung. Transistors: Fundamentals for the Integrated Circuit Engineer. Wiley, New York, pp. 476ff, 1983.
4. J. P. McKelvey. "Solid State and Semiconductor Physics." Harper and Row, New York, p. 430, 1966.
5. J. L. Moll and I. M. Ross. "The Dependence of Transistor Parameters on the Distribution of Base Layer Resistivity." Proc. IEEE 44:72, 1956.
6. J. M. Early. "Effects of Space Charge Layer Widening in Junction Transistors." Proc. IRE 40:140, 1952.
7. A. S. Grove. Physics and Technology of Semiconductor Devices. Wiley, New York, p. 229, 1967.
8. H. A. Schafft. "Second Breakdown—A Comprehensive Review." Proc. IEEE 55:8, 1272, Aug. 1967.
9. F. Larin. Radiation Effects in Semiconductor Physies. J. Wiley, New York, pp. 97ff., 1968.
10. Hardness Assurance For Long Term Ionizing Radiation Effects on Bipolar Structures. DNA 4574F. Mission Research Corp., 1978.
11. R. A. Sunshine and M. A. Lampert. "Second Breakdown Phenomena in Avalanching SOS Diodes." IEEE Trans. Elect. Dev. ED-19(7), July, 1972.
12. H. C. Bowers and A. M. Barnett. "Filamentary Injection Currents in Semi-Insulating GaAs." IEEE Trans. ED-17(11):971–975, Nov. 1970.
13. A. L. Ward. "An Electrothermal Model of Second Breakdown." IEEE Trans. Nucl. Sci. NS-23(6):1679–1684, Dec. 1976.
14. J. Lindmeyer and C. Y. Wrigley. Fundamentals of Semiconductor Devices. Van Nostrand, Princeton, NJ, p. 92, 1965.
15. L. B. Valdes. The Physical Theory of Transistors. McGraw-Hill, New York, Chap. 15, 1961.
16. D. Casasent. Electronic Circuits. Quantum, New York, 1973.
17. J. Millman and C. C. Halkias. Integrated Electronics: Analog and Digital Circuits and Systems. McGraw-Hill, New York, 1972.
18. Larin, op. cit. Chap. 3, Fig. 3.9.
19. A. Nussbaum in R. K. Willardson and A. C. Beer (eds.). Semiconductors and Semimetals, Vol. 15, Academic Press, New York, Chap. 2, 1981.
20. R. M. Warner, Jr. and B. L. Grung. "Transistors, Fundamentals For The Integrated Circuit Engineer." J. Wiley. New York, 1983, pp. 148-ff.
21. Warner and Grung, loc. cit. pp. 428ff.
22. Sze, op. cit. p. 92.
23. Warner, Jr. and Grung, op. cit. p. 43.
24. Grove, op. cit. Fig. 6.24, p. 190.
25. H. K. Gummel. "Hole-Electron Product of P-N Junctions." Solid State Elect. *10*:209, 1967.
26. J. J. Liou and F. A. Lindholm. "Capacitance of Semiconductor P-N Junction Space Charge Layers: An Overview." Proc. IEEE. 76(11):1406–1422, Nov. 1988.
27. R. M. Burger and R. P. Donovan (eds.). Fundamentals of Silicon Integrated Device Technology, Vol. II, Bipolar and Unipolar Transistors. Prentice-Hall, Englewood Cliffs, NJ, p. 64, 1968.
28. C. T. Kleiner and G. C. Messenger. "An Improved Bipolar Junction Transistor Model For Electrical and Radiation Effects." IEEE Trans. Nucl. Sci. NS-29(6):1569–1579, Dec. 1982.
29. Sze, op. cit. Sec. 2.8ff, Sec. 5.2.2.

30. R. J. Krantz, W. L. Bloss, and M. J. O'Laughlin. "High Energy Neutron Radiation Effects in GaAs MODFETs: Threshold Voltage." IEEE Trans. Nucl. Sci. NS-35(6): 1438–1443, Dec. 1988.
31. G. A. Schrantz et al. "Neutron Irradiation Effects on AlGaAs/GaAs Heterojunction Bipolar Transistors." IEEE Trans. Nucl. Sci. NS-35(6):1657–1661, Dec. 1988.
32. R. D. Middlebrook. An Introduction to Junction Transistor Theory. J. Wiley and Sons, New York, 1957.

CHAPTER 4
NUCLEAR ENVIRONMENTS

4.1 INTRODUCTION

The thrust of this book is the investigation of the behavior of electronic systems under the influence of damaging incident radiation. It is therefore important to obtain an understanding of the sources and a somewhat detailed description of this radiation. The objective of this chapter is to supply such a description in a form that is sufficiently comprehensive to be appreciated by a rather broad spectrum of readers. There are a few references on this subject, ranging from handbooks to technical articles in the literature.[1–4] The following exposition is intended to provide a description of nuclear environments sufficient for the practioner, yet almost as comprehensive as that contained in the handbooks. Simple approximation methods and prediction techniques for nuclear weapons effects are also given. Much of the initial sections in this chapter are adapted from H.L. Brode.[4]

An appreciably detailed explanation is given for close-in and early fireball phenomena, important in tactical considerations. Emphasis is also given to indirect nuclear effects, such as shock and blast wave phenomena, as opposed to the effects of neutrons, x rays, and gamma rays. This is done to attempt to provide some balance between electronic and mechanical effects, for the direct nuclear effects are treated at length throughout this book. This is so because they are the main damage-producing agents inside electronic components and systems.

4.2 THE SOURCE

The purely scientific interest in nuclear weapon phenomenology stems from the extremely high temperatures of millions of degrees Kelvin and accompanying pressures of millions of atmospheres occurring within the nuclear detonation, not unlike the interior of the sun. Some of the novel and nonintuitive features of a nuclear detonation include:

a. The detonation gases consist of a plasma at conditions of stupendous temperature and pressure, made up principally of atomic nuclei, free electrons, and electromagnetic radiation of gamma and x rays.

Because of these extraterrestrial conditions, the hot plasma contains almost no atoms as such, but some heavy ions with most of their orbital electrons stripped off. Most of the other atoms are completely ionized.

b. The thermal radiation and visible light from such a plasma fireball detonation in the atmosphere experience two distinct intensity peaks before the fireball fades from view.
c. Certain structures, at given distances close enough to be affected by the blast, can sometimes be felled toward the detonation rather than away, as in an ordinary chemical explosion.
d. The electromagnetic pulse (EMP) resulting from a high-altitude exo-atmospheric nuclear burst can disrupt long-haul radio communications for up to a few hours, over an area on the earth the size of the continental United States, and disrupt unhardened electronic systems over somewhat shorter surface burst ranges for shorter times.

Referring to the nuclear detonation as a source, the corresponding descriptions of the fireball, thermal detonation component, and blast are relatively independent of the detailed weapon construction and its operation. This implies that, insofar as geometrical considerations are concerned, a point detonation can be assumed to a first order of approximation. Higher order geometric effects are the result of asymmetries, such as ground reflections, hydrodynamic coupling between multiple nuclear bursts, atmospheric density variations with altitude, and detailed weapon structure. The neutrons, gamma rays, and EMP outputs, however, are sensitive to the weapon kinetics. Nevertheless, useful estimates can be made of these phenomena without knowledge of specific weapon structural details.

The principal reason that the above weapon outputs are independent of the weapon construction is the tremendously high energy density of the detonation. For endoatmospheric bursts, in less than a microsecond most of the detonation energy has escaped into an air mass that, except for very low yields, is hundreds of times larger than the mass of the nuclear device itself. This then implies that most of the features of the detonation are independent of any detailed construction, because any such features are lost in the tremendous mass of engulfed air. As will be seen, this greatly facilitates the scaling of the various nuclear weapons effects and also delineates the point source characteristic of the detonation. This is in contrast to a chemical explosion, where the mass of engulfed air is comparable to the mass of the chemical explosive itself. Consequently, the shape, explosive burn mode, and other details of the chemical explosion have persistent influences on the chemical fireball and its blast and shock behavior.

4.3 DETONATION ENERGY

As mentioned, the nuclear detonation is driven by the release of stupendous amounts of energy in times of less than 1 μs. The origin of this energy is the result of extremely exothermic nuclear reactions within the bomb itself. In a thermonuclear device, these reactions are a combination of fission reactions in heavy metals like uranium and plutonium and fusion reactions in light materials like hydrogen and tritium. The latter reactions are quite similar to the nuclear reactions that drive the sun and other stars.

In contrast to the aforementioned nuclear weapons effects that are independent of the detailed bomb structure, the behavior and levels of the initial radiation, e.g., prompt neutrons and prompt gamma and x rays, are largely determined by the device construction and constituents. As such, they are independent of the external environment. The initial bomb densities and nuclear radiation characteristics are almost the same whether the detonation takes place in space, underground, underseas, or in the atmosphere, as long as the environment is not as dense as the device itself. This is highly unlikely, because the bomb materials have a much higher average density than their environment, due to the very heavy materials in their construction, such as uranium, plutonium, lead, iron, and other heavy elements. As an example of structure-dependent emanations, the fraction of bomb energy radiated as x rays depends on the extremely high temperatures, which depend on the bomb construction details and the output energy-to-mass ratio. The x-ray output can be made to vary from almost zero to about 80 percent of the total energy yield generated, depending on the construction.

It is instructive to estimate the energy density for a bomb energy yield of Y kilotons (1 KT $= 10^{12}$ calories). Assumptions implied in this estimation are that the energy is generated instantaneously and the device plasma is in equilibrium with its radiation (gamma, x, and other electromagnetic) fields. Actually, almost all of the initial energy released is in the form of hypervelocity recoiling atomic and nuclear particles. Most of these are charged particles, e.g., ions, and so have a very short range of penetration in matter, causing them to readily collide with and heat up their surrounding bomb material environment extremely quickly. Only uncharged particles, such as neutrons, gamma and x rays, and neutrinos escape from the bomb debris immediately after detonation. Even so, a good share of the gamma and x rays and neutrons are absorbed by the

Table 4.1 Hypothetical 100-lb bomb plasma parameters.[4]

YIELD (KT)	BOMB MASS (lb)	PLASMA PRESSURE (Mega-atm)	PLASMA DENSITY (g/cm^3)	PLASMA TEMP (Mega °K)	E_{rad}/E_{tot}
1	100	64	2.7	14.3	0.11
50	100	20340	205	60.5	0.93
200	100	78170	564	84.7	0.99

device materials and bomb casing upon detonation. Again, the device temperatures are in millions of degrees Kelvin, i.e., in the kilovolt range (1 eV = 12 100°K), pressures in millions of bars, and densities of the order of 10–1000 g per cm^3, as seen in Table 4.1. Yet the plasma retains the compressibility and other properties of an ideal gas, as will be discussed. In this extraterrestrial environment, the energy density of the plasma radiation is comparable to the heat energy of the plasma itself. This energy density E can be written as the sum of a material density plus a radiation density, to give

$$E = C_V T + aT^4/\rho \qquad \text{(ergs per gram)} \tag{4.1}$$

where C_V is the specific heat of the plasma, the radiation constant $a = 7.6 \cdot 10^{-13}$ ergs per cm^3 $(°K)^{-4}$, and ρ is the average plasma density. T is the plasma temperature in °K. For a perfect gas, in units of ergs per gram °K, $C_V = 3R/2\bar{\mu}$, where R is the universal gas constant, in ergs per mole °K, and $\bar{\mu}$ is the mean atomic weight, in grams per mole, of the device plasma. It is interesting to compute $\bar{\mu}$ because of the striking result. Assuming that the plasma consists of only free electrons and atomic nuclei, i.e., of completely dissociated ions, then the mean atomic weight of the plasma is given by

$$\bar{\mu} = \left(\sum_i N_i\mu_i + \sum_i N_i\mu_i Z_i/1836\right) \Big/ \left(\sum_i N_i + \sum_i N_i Z_i\right)$$

$$\cong 2 \text{ g per mole} \qquad Z_i \cong \tfrac{1}{2}\mu_i \gg 1 \tag{4.2}$$

where μ_i, N_i, and Z_i are, respectively, the atomic weight, number of moles, and the atomic number of the ith species of nuclei in the plasma. Now $Z_i \cong \frac{1}{2}\mu_i$, and $Z_i \gg 1$ is the case for essentially all of the materials that make up the plasma. Electrons are 1836 times lighter than individual nucleons, such as protons and neutrons; hence Eq. (4.2) ensues. $\bar{\mu}$ is essentially constant irrespective of the plasma constituents, as most of them are heavy elements, so that the plasma acts like a gas with an atomic weight of only 2 g per mole, characteristic of a light terrestrial gas, for which the ideal gas laws will be shown to hold. Then, for $\bar{\mu} = 2$, $C_V = 3R/4$. This results in an alternative expression for the mean energy density

$$E = 6.2 \cdot 10^7 T + 7.6 \cdot 10^{-13} T^4/\rho \qquad \text{(ergs per gram)} \tag{4.3}$$

The average energy density can also be expressed as the energy yield divided by the total mass of the bomb plasma. This allows the energy yield to be written as

$$Y = 0.78(M\theta + \tfrac{1}{2}R^3\theta^4) \qquad \text{(kilotons)} \tag{4.4}$$

where the bomb plasma mass M is measured in hundreds of pounds, θ is the mean bomb temperature in kilovolts, and R is the radius of the bomb in feet. Equation (4.4), together with the ideal gas law discussed below, is used to make up Table 4.1 for a 100-lb bomb plasma. This bomb mass is only representative and is used merely for illustration.

That is, immediately following detonation, the total bomb mass is vaporized into a 2-g-per-mole plasma, regardless of the actual specific bomb materials and detonation kinetics. Implied is essentially complete dissociation of electrons from their constituent nuclei due to the enormously high temperatures involved. This plasma has the properties of an ideal gas, as shown next. Therefore, the ideal gas thermodynamics implied in Eqs. (4.2)–(4.4) and in construction of Table 4.1 holds for the 100-lb mass of bomb plasma. This is also implied in the preceding argument, showing that the mean atomic weight of the plasma is 2 g per mole. That is, atomic identity is assumed lost within the plasma, so that it could represent almost any material or materials, as long as they are heavy in the sense of Eq. (4.2).

The ideal gas law, $P = \rho RT$ per gram of gas, holds extremely well in the interior of stellar plasmas and to a good approximation for bomb plasmas. One means to show this is through the van der Waals gas law, viz., $(P + a^2/V)(V - b) = RT$. For a plasma the van der Waals terms, a^2/V and b are meaningless. That is, the constant b accounts for the volume of the gas atoms. As there are virtually no atoms in a plasma and the nuclei and free electrons therein are millions of times smaller than atomic volumes, b is easily negligible. The constant a^2/V is a condensation pressure, manifest when a real gas begins to liquefy, which liquefaction is foreign to the plasma state, and so this quantity can be neglected also. The result of neglecting a^2/V and b then leaves the ideal gas law, viz., $PV = RT$ per mole of ideal gas.

From Table 4.1, it is seen that the pressure and density increase tremendously for increasing yield. Also the fraction of the plasma radiation density increases from about 11 percent to almost 100 percent for high yields. This is not quite true for actual nuclear detonations.

4.4 FIREBALL RADIATIVE GROWTH

The x rays that emanate from the detonation are usually at the relatively low energy of a few keV and are termed *soft* x rays. They have short mean free paths in air of about $\frac{1}{4}$–2 cm. For an endoatmospheric burst these x rays are then absorbed in the air immediately surrounding the detonation and excite this air, creating the luminous glowing fireball. The mean free path, λ, for these x rays in air is well approximated by the expression $\lambda = E^{2.78}/4(\rho/\rho_0)$ cm, where ρ is the ambient air density and ρ_0 is the sea level standard air density (0.0012393 g per cm^3). E is the x-ray energy in kilovolts. From this expression, it is seen that, as the fireball-engulfed air is heated by the early x rays to astronomically high temperatures and becomes an extremely rarefied gas, the mean free path rapidly increases. This renders the air semitransparent to the immediately following x rays, but not to visible light. Almost at the same time, the shock wave caused by the bomb debris gases quickly compresses and heats the fireball air further. This shock wave, distinct from the fireball, is almost isothermal, because as fast as the shock compresses and heats more fireball air, the radiation energy from the shock-excited air atoms is radiated out through the semitransparent fireball into the ambient air beyond. In the next few microseconds, the fireball itself reradiates energy from the x-ray-excited air atoms, so that the fireball grows by radiation diffusion during this phase. Table 4.2 depicts the evolution of the size,

Table 4.2 Early fireball evolution for uniform fireball temperatures.[4]

FIREBALL EVOLUTION TIME (μs)	FIREBALL TEMP (Mega °K)	1 KT		1 MT		100 MT	
		RADIUS (m)	ENGULFED AIR (tons)	RADIUS (m)	ENGULFED AIR (tons)	RADIUS (m)	ENGULFED AIR (tons)
0.10	7.5	0.75	.002	7.5	1.9	35	200
0.12	6	1.	.005	10	4.6	46	450
0.15	5	1.25	.009	12	8	57	860
0.17	4	1.6	0.19	16	9	74	1880
0.20	3	2.1	.043	21	43	100	4650

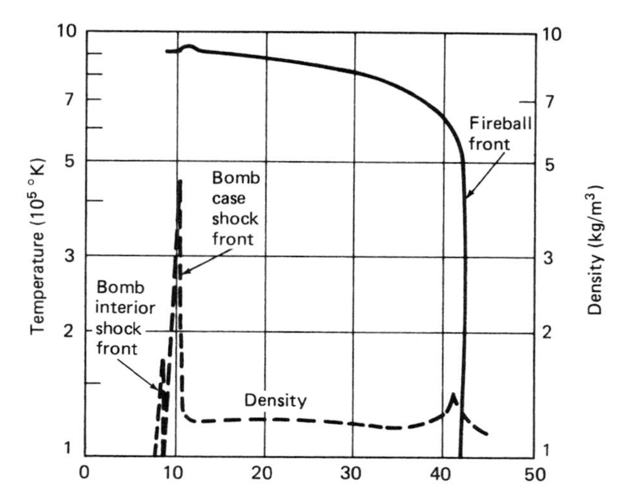

Fig. 4.1 Early fireball temperature and density profile for a 1-MT sea level burst.[4]

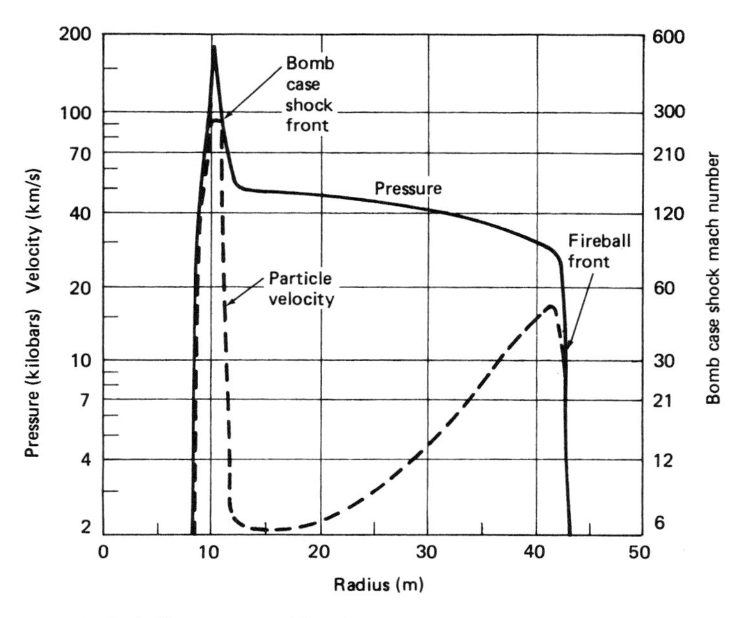

Fig. 4.2 Early fireball pressure and bomb case mach number for a 1-MT sea level burst.[4]

temperature, and air mass engulfed by very early fireballs as a function of weapon yield. Figure 4.1 depicts an early (34 μs) temperature and density profile for a 1-megaton (1-MT) sea level burst. It is seen that the fireball is roughly isothermal throughout its interior of about 30 m diameter at this time. It is also seen that a sharp density spike exists at the innermost edge of the fireball. This is the bomb case shock wave due to the very rapidly expanding material of the device. This is seen in Fig. 4.2, which shows the corresponding case shock wave pressure spike near the fireball edge. The smaller spike is the shock wave corresponding to another structural portion of the detonating bomb. In Fig. 4.2, the bomb material is shown expanding at an extremely high velocity and beginning to compress the air beyond it, which in turn is just starting to move relative to the oncoming case shock. As mentioned, the energy that the compressed fireball air acquires from being shocked is immediately radiated out through the fireball still semitransparent to x rays. In recapitulation, the two entities that make up the early detonation hydrodynamics are (a) the expanding radiation front due to the bomb x rays that create the fireball, opaque to and glowing with visible light emitted from the x-ray-excited air atoms, and (b) the case shock waves due to the bomb materials, now in the fireball interior and just beginning to expand also.

After a few seconds, the expansion of the fireball ceases suddenly, because the fireball-driving *x* rays from the nuclear reactions stop. This is because their source is blowing apart sufficiently to no longer sustain the reactions that produce this radiation. Also any energy remaining in the bomb debris is inhibited from radiating because of the increasing opacity of the bomb gases due to their rapidly falling temperature. Simultaneously, the case shock wave expansion increases as it compresses and heats more air, as the bomb debris vapors drop away from the shock wave itself. At this point, the shock wave overtakes the fireball and runs ahead through it. The fireball also now begins to form a shock wave in its transformation from radiative diffusion growth to the classic hydrodynamic shock wave. Figure 4.3 shows the typical dependence of the various shock and other fronts on radius and time. Figure 4.4 shows the fireball growth transition from a radiation front to a hydrodynamic shock front. Early on, when the fireball growth is primarily due to radiation diffusion, it is seen that the fireball interior air density is much less than ambient. In a high-yield detonation, this density can be as low as 1 percent of sea level density, as depicted in Fig. 4.5. In Fig. 4.4, it is seen that the transition from radiative to hydrodynamic behavior is a gradual one, developing over an appreciable amount of fireball thickness. Even after the hydrodynamic shock is developed, the shock front tem-

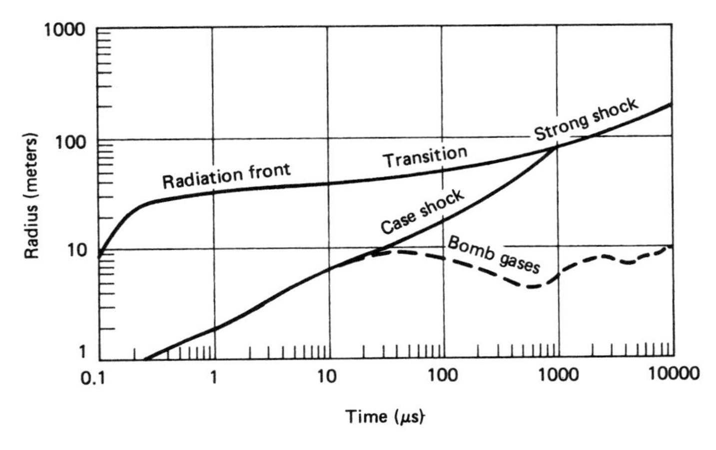

Fig. 4.3 Shock wave front radii versus time from detonation for a 1-MT sea level burst.[4]

peratures are quite high, of the order of 100 000°K. An empirical relationship for the nominal radius, R_{tr}, at which the fireball makes its transition from a radiation front to a hydrodynamic shock wave, is given by[4]

$$R_{tr} = 47Y^{0.324}/(\rho/\rho_0)^{1/2} \pm 10 \text{ percent (meters)} \tag{4.5}$$

where Y is the yield in MT, ρ is the ambient air density outside the fireball, and ρ_0 is the sea level density. This expression holds for yields from 1 KT to 4 MT and from sea level to 100 000 ft of detonation altitude.

Figure 4.6 is an illustration, rendered from photographs, of the time at which the case shock wave passes through the early fireball. As seen, there are many large blisters and hot spots on the spherical shock front. These are probably caused by high-velocity blobs of bomb debris vapor that have been flung against the inside walls of the expanding shock as it passes through the fireball. The detonation propels these blobs at high speeds as the fireball expands. When the fireball expansion is suddenly diminished, as mentioned earlier, these blobs splash against the inside of this relatively dense shock. The isothermal contours in Fig. 4.6 show the steep temperature gradients between the inside and outside of the early fireball, indicative of the expanding case shock.

4.5 SHOCK AND BLAST HYDRODYNAMICS

Fortuitously, for most hydrodynamic shock or blast wave features resulting from a chemical or nuclear detonation, there is a simple transformation

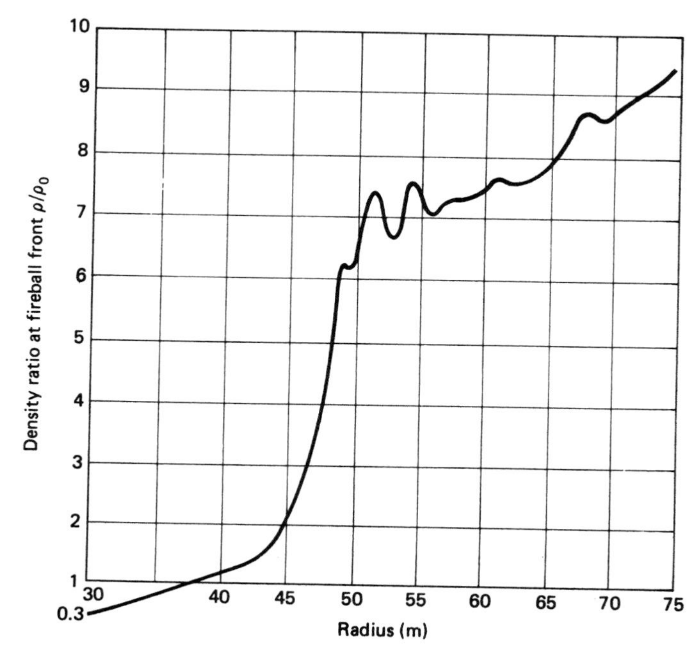

Fig. 4.4 Fireball transition from radiation diffusion phase to hydrodynamic growth phase for 1-MT sea level burst.[4]

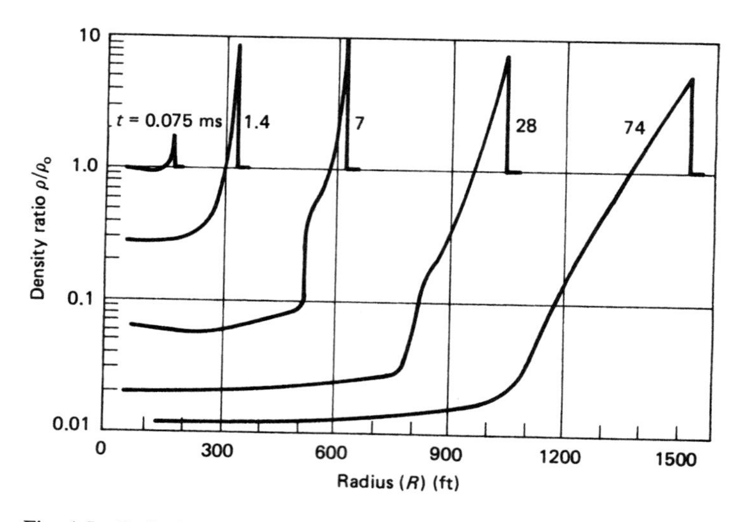

Fig. 4.5 Early fireball densities versus radius for 1-MT sea level burst.[4]

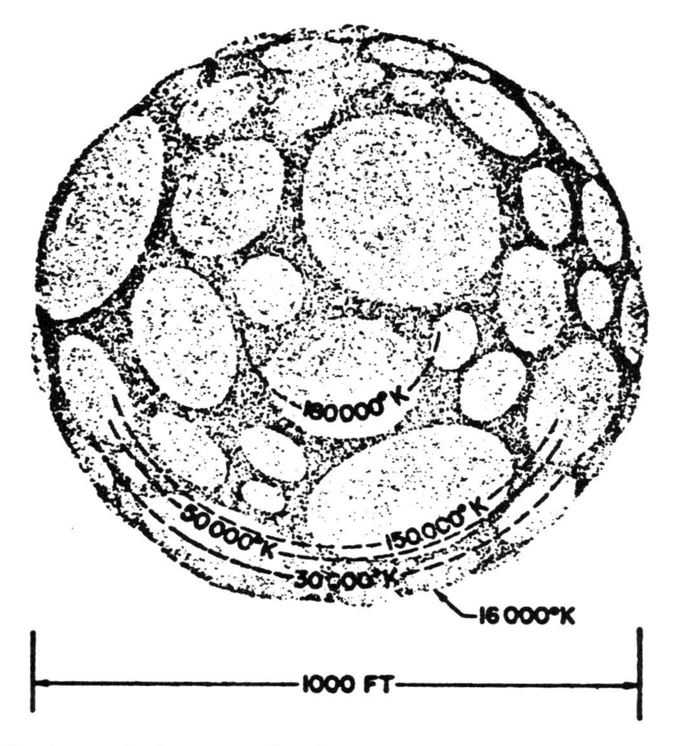

Fig. 4.6 Bomb case shock wave passing through the early fireball at 6 ms for a 1-MT sea level burst.[4]

of the hydrodynamic equation descriptions to dimensionless form. This allows the shock or blast wave parameters to be scaled from one bomb yield to another within wide limits. These limits encompass yields from a few kilotons to many megatons, as verified by experiment. This is called *Sachs scaling*, first discovered by Sachs[18] in connection with chemical explosions and adapted to nuclear detonations. When hydrodynamic parameters of interest are computed for one particular yield, this scaling provides them for any other yield within the above limits. Consider the classical hydrodynamic equations that hold for the nuclear shock or blast wave. In the following they will merely be written down and transformed to show their dimensionless character, in order to delineate the scaling relationships. No solutions will be obtained, as they are beyond the scope of this work and are not necessary.

Let ρ, R, P, E, t, and $\bar{u}$ be, respectively, the shock or blast wave density, distance from the detonation point, pressure, energy density, time, and mean particle velocity. Then the hydrodynamics equations can be written down as

$$\frac{\partial \rho}{\partial t} + \nabla \cdot (\rho \bar{u}) = 0 \text{ Equation of continuity (conservation of mass)} \quad (4.6)$$

$$\rho \frac{\partial \bar{u}}{\partial t} + \nabla P = 0 \text{ Newton's second law (conservation of momentum)} \quad (4.7)$$

$$\rho^2 \frac{\partial E}{\partial t} - P \frac{\partial \rho}{\partial t} = 0 \text{ First law of thermodynamics (conservation of energy)} \quad (4.8)$$

$$P(E, \rho) = 0 \text{ Equation of state} \quad (4.9)$$

$$u = |\bar{u}| = \frac{\partial R}{\partial t} \text{ Mean particle velocity} \quad (4.10)$$

In the spherical coordinate system of the nuclear detonation, where R is the radial dimension, the first two equations become, respectively

$$R^2 \frac{\partial \rho}{\partial t} + \frac{\partial}{\partial R}(R^2 \rho u) = 0; \qquad \rho \frac{\partial u}{\partial t} + \frac{\partial P}{\partial R} = 0 \quad (4.11)$$

Now, a displacement parameter is defined as $\alpha = (Y/P_0)^{1/3}$ (cm), where P_0 is a constant reference pressure, e.g., sea level pressure. Then a set of dimensionless variables are defined as $r = R/a$, $C_0 t/a = \tau$, $\eta = u/C_0$, $\xi = \rho/\rho_0$, $p = P/P_0$, and $\varepsilon = E/E_0$. The zero subscripts imply ambient conditions, and C_0 is the sea level speed of sound. In these variables, Eqs. (4.6)–(4.10) become, respectively

$$r^2 \frac{\partial \xi}{\partial \tau} + \frac{\partial}{\partial r}(r^2 \xi \eta) = 0 \quad (4.12)$$

$$\rho_0 C_0^2 \xi \frac{\partial \eta}{\partial \tau} + P_0 \frac{\partial P}{\partial r} = 0 \quad (4.13)$$

$$\rho_0 E_0 \xi^2 \frac{\partial \varepsilon}{\partial \tau} - P_0 p \frac{\partial \xi}{\partial \tau} = 0 \quad (4.14)$$

$$p(\varepsilon, \xi) = 0; \qquad \eta = \frac{\partial r}{\partial \tau} \quad (4.15)$$

The solutions to these equations describe the hydrodynamic behavior of the nuclear detonation shock or blast wave in dimensionless variables. Hence, these solutions can be written explicitly in terms independent of the bomb yield or ambient conditions, so that scaling between blast parameters at different yields can be accomplished. It can be shown that the boundary conditions scale in the same way. When this is strictly so, the scaling results are quite rigorous. Fortuitously, the nonscalable effects for many of the blast wave features are unimportant herein.

For scaling examples, consider the dimensionless distance from ground zero

$$r = R_1/\alpha_1 = R_2/\alpha_2 = \cdots \tag{4.16}$$

so that the ratio of two actual distances, R_1 and R_2 of two shock waves corresponding to two different device yields Y_1 and Y_2 gives

$$R_2/R_1 = \alpha_2/\alpha_1 = (Y_2/P_0)^{1/3} \cdot (P_0/Y_1)^{1/3} = (Y_2/Y_1)^{1/3} \tag{4.17}$$

Similarly, $t_2/t_1 = (Y_2/Y_1)^{1/3}$, so that these two scaling relations can be combined to give

$$R_2/R_1 = t_2/t_1 = (Y_2/Y_1)^{1/3} \tag{4.18}$$

This means that from a $Y_1 = 1$ KT nuclear burst at a point where the pressure ratio P_1/P_2 is 2, the shock wave radius is $R_1 = 238$ m, and the arrival time there is $t_1 = 0.3$ s. Then for a $Y_2 = 1$-MT detonation, the same pressure ratio occurs at $R_2 = 2380$ m, at $t_2 = 3$ s. Other scaling relationships follow in the same way.

Altitude scaling is derived similarly, but is not as universally valid, for in the upper atmosphere where the air is very rarefied, continuity of air flow and other conditions implicit in the hydrodynamic equations break down. Nevertheless, if P_1 and P_2 are pressures at (R_1, t_1) and (R_2, t_2), respectively, and P_{01} and P_{02} are the two corresponding ambient air pressures, the scaling is defined such that the pressure ratio (shock pressure to ambient) is the same. Then $P_1/P_{01} = P_2/P_{02}$, where the shock radii are related as $R_1 = R_2(P_{02}/P_{01})^{1/3}$. When this holds, it is easily seen that the times are related so that $t_1/t_2 = (P_{02}/P_{01})^{1/3} \cdot (T_{02}/T_{01})^{1/2}$, where T_{01} and T_{02} are the associated ambient temperatures. If the ambient temperatures are different, then the particle and sound speeds and times will be different by the square root of the ratio of the ambient temperatures. Hence, at the same scaled distances and times, the sound speed, absolute temperature, particle velocity, and shock wave speed U will be related for two detonations in two different atmospheres P_{01} and P_{02} by $C_{01}/C_{02} = (T_{01}/T_{02})^{1/2}$, $u_1/u_2 = C_{01}/C_{02}$, $U_1/U_2 = C_{01}/C_{02}$, and $T_1/T_2 = T_{01}/T_{02}$, where all temperatures are in degrees Kelvin.

Another example is that of the shock wave overpressure, ΔP, (pressure over atmospheric) and arrival time of the shock at one particular yield. Then these pressures for any yield can be obtained by scaling from those calculated at the reference yield. An accurate empirical expression for the overpressure (psi) and arrival time, t_s in seconds, for a shock, with Y in MT and R in thousands of feet is[4],

$$\Delta P = 3300Y/R^3 + 192(Y/R^3)^{1/2}$$

$$t_s = \begin{cases} 5.6 \cdot 10^{-5} Y^{1/3} + 5.8 \cdot 10^{-9} R^{10}/Y^3; & R/Y^3 \leqslant 144 \text{ ft} \\ \dfrac{5.8 \cdot 10^{-19} R^{10}}{Y^3 + 3.6 \cdot 10^{-14} R^{9.5} Y^{1/2}}; & 144 \text{ ft} < R/Y^3 < 522 \text{ ft} \\ 1.6 \cdot 10^{-5} R^{2.5}/Y^{1/2}; & R/Y^3 \geqslant 522 \text{ ft} \end{cases} \tag{4.19}$$

The initial decay of pressure behind the shock wave is extremely rapid, especially at high overpressures, to be followed by a much slower decay with time. The rapid early decay comes merely from the passing by of the shock wave density spike, while the later slower decay is due to the pressure drop behind the shock wave. After the shock wave passes, the pressure drops to values below sea level, creating a suction phase. This suction pressure is sufficiently strong to occasionally cause structures to be felled toward the detonation, instead of away from the detonation point, as in far weaker chemical explosions. An empirical expression for the shock wave overpressure at a particular distance and time from ground zero is given by[4]

$$\Delta P(R, t) = 62(1 + 1.6(t_s/t)^6)/(1 + 13(t/Y^{1/3})^{1.15}) \ kbar \tag{4.20}$$

where the dependence on R is implicit in the expression for the arrival time of the shock, t_s, as given in Eq. (4.19), and Y is the yield in MT. For a 1-MT detonation at sea level, after about 80 ms, the shock wave has weakened by expansion to where the shock wave temperature has fallen to about 2000°K, at which temperature the fireball no longer glows. Now the fireball becomes transparent to visible light, thus exposing the hotter interior to a remote observer. Then the apparent temperature observed at a distance begins to increase, and the corresponding bomb thermal radiation increases as well. As the shock wave expands further, the hot interior of the fireball slows up and stops expanding. By that time, the shock wave has expanded far out and is racing away from the fireball. Figures 4.12 and 4.13 indicate the temperature of the fireball and separated shock region. After the shock wave has fully separated from the fireball, and the main portion of the thermal radiation has been expelled, the fireball rises rapidly to stratospheric heights at speeds of about 300 ft per s.

The interaction of structures with air blast and shock waves is investigated by analytical techniques, some of which are empirical. As men-

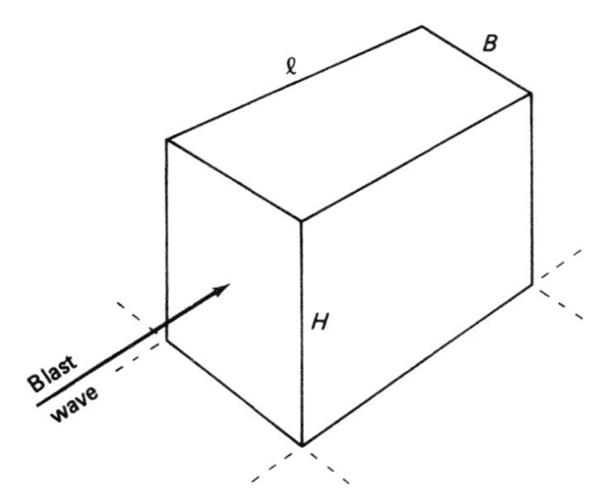

Fig. 4.7 Closed parallelepiped structure.[1]

tioned earlier, the phrases *shock wave* and *blast wave* are considered synonymous.

First, the blast or shock wave loading as a function of time is determined on the structure of interest, which is then followed by an evaluation of the structural time response to this loading. Actual structures are usually made up of complicated surface conformations that are difficult to analyze without resort to sophisticated computer programs. The following development will address idealized structures that are commonly appreciated, such as parallelepiped shapes.

The two principal parameters by which the load on structures can be represented are the overpressure, $p(t)$, and the dynamic pressure, $q(t)$.[1] The former, as already mentioned, is simply the incident pressure excess over atmospheric pressure. The major destructive effects of blast waves are due to the action of overpressure. However, in some instances, the drag forces associated with the strong transient winds behind the shock can be of greater importance. Some structures are drag-sensitive, because they are quickly enveloped by the blast wave, and any translational motion of these structures is due to these drag forces.

The dynamic pressure is defined as $q = \frac{1}{2}\rho u^2$, where ρ is the shocked air density, and u is the shock wave speed. q is sometimes called the kinetic energy density, as seen from its definition. When the blast wave strikes the front face of a parallelepiped structure, as shown in Fig. 4.7, a shock wave reflection occurs that generates a pressure from two to eight times that of the incident overpressure. As is well known, this pressure ratio approaches 2 for a weak shock wave, i.e., a sonic reflection. The incident shock wave then diffracts around the structure, exerting pressure

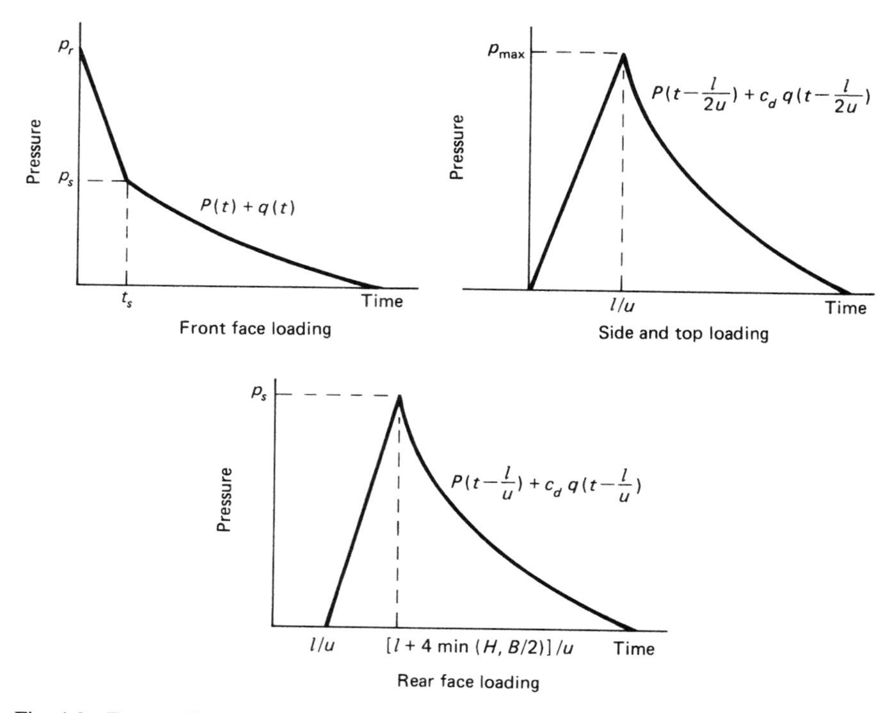

Fig. 4.8 Front, side, top, and rear face loading on parallelepiped structure.[1]

on its sides, top, and finally on its rear face. The blast wave pressure decays rapidly with time as the shock traverses the structure.

When the peak overpressure is large compared to the ambient pressure ahead of the shock wave, the peak dynamic pressure is essentially proportional to the peak overpressure.[17] Hence, the former obeys the same scaling laws as the latter. Otherwise, the dynamic pressure does not scale with yield very well, as seen in Fig. 4.10A.

With the above pressure buildup on the structure, a rolling vortex of shocked air forms on the roof, or top, of the structure, causing a low-pressure region, which follows along the top behind the advancing shock wave. After the shock vortex has passed, the resultant air inflow causes a reduction in the loading on the sides and top, which consistutes a "negative" pressure in the direction opposite to that of the advancing shock. This is the negative pressure referred to earlier in terms of objects falling toward ground zero, instead of away, after the shock has passed.

To determine the loading on the front face, the reflected pressure, p_r, is computed. This is the initial pressure on the front face. The reflected

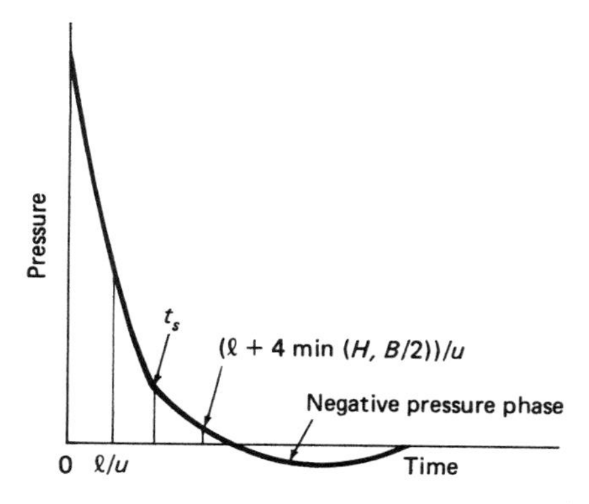

Fig. 4.9 Net horizontal loading.[1]

pressure quickly decays to a pressure, p_s, called the *stagnation pressure.* The average pressure is generally given by

$$\hat{p}(t) = p(t) - c_d q(t) \tag{4.21}$$

where the second term is called the dynamic (or *drag*) *pressure*, and c_d is a semiempirically determined drag coefficient. The value of c_d depends on the dynamic pressure magnitude, surface conformation of the front face of the structure, and its orientation with respect to the incident blast front. From experiment, t_s, the time during which the reflected pressure decays to the stagnation pressure is given by

$$t_s \cong (3/u) \min(H, B/2) \tag{4.22}$$

where min (x, y) is read as the smaller of x or y. For the front face, $c_d \cong 1$, so that at time t_s, the stagnation pressure is

$$p_s = p(t_s) - q(t_s) \tag{4.23}$$

The front face pressure as a function of time is depicted in Fig. 4.8. Although the parallelepiped loading begins to build up following the onset of the blast wave at the front face, the sides and top of the structure of Fig. 4.7 are not fully loaded until the shock wave has traveled a distance l across the structure in time $t_l = l/u$. The average pressure, p_{av}, at this time, is the sum of the overpressure plus the drag pressure, i.e.

$$p_{av} = p(t_l/2) - c_d q(t_l/2) \tag{4.24}$$

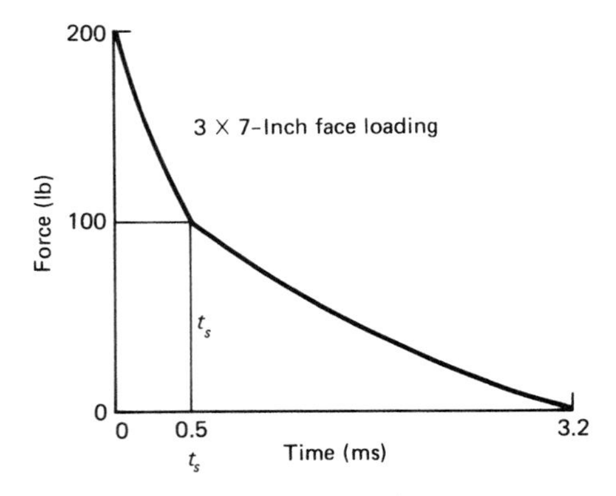

Fig. 4.10 3 × 5 × 7-Inch parallelepiped response to blast incident on 3 × 7-inch side for static overpressure of 5 psi.[1]

where the drag coefficient is given as

$$c_d = \begin{cases} 0.4, & 0 < q \leqslant 25 \text{ psi} \\ 0.3, & 25 < q \leqslant 50 \text{ psi} \\ 0.2, & 50 < q \leqslant 130 \text{ psi} \end{cases} \tag{4.25}$$

The drag coefficient c_d can have either sign, depending on the orientation of the particular face with respect to the incident blast-wave front. This pressure is shown in Fig. 4.8. The loading therefore increases up to $p_{\max}$ at time t_l; then, following t_l, it decreases, as given by

$$p(t) = p(t - t_l/2) - c_d q(t - t_l/2) \tag{4.26}$$

To compute the average loading on the rear face, it is seen that the shock front arrives at the rear face at time l/u, but requires an additional time of $(4/u)$ min $(H, B/2)$ for the pressure to increase to p_s, as seen in Figure 4.8. The corresponding pressure, p_{rear}, is

$$p_{\text{rear}} = p(t - t_l) - c_d q(t - t_l); \qquad t > [l + 4\min(H, B/2)]/u \tag{4.27}$$

The net horizontal load is the sought after quantity, being the difference between the front and rear face loads. This is shown in Fig. 4.9, where the negative pressure phase is seen, affecting objects that have barely withstood the forward pressure but that succumb to the resulting moment produced by the negative pressure, with the previously explained results. Figure 4.10 shows the results of an analysis for the case of a parallelepiped whose dimensions are $3 \times 5 \times 7$ in. The corresponding impulse I is computed to good approximation by

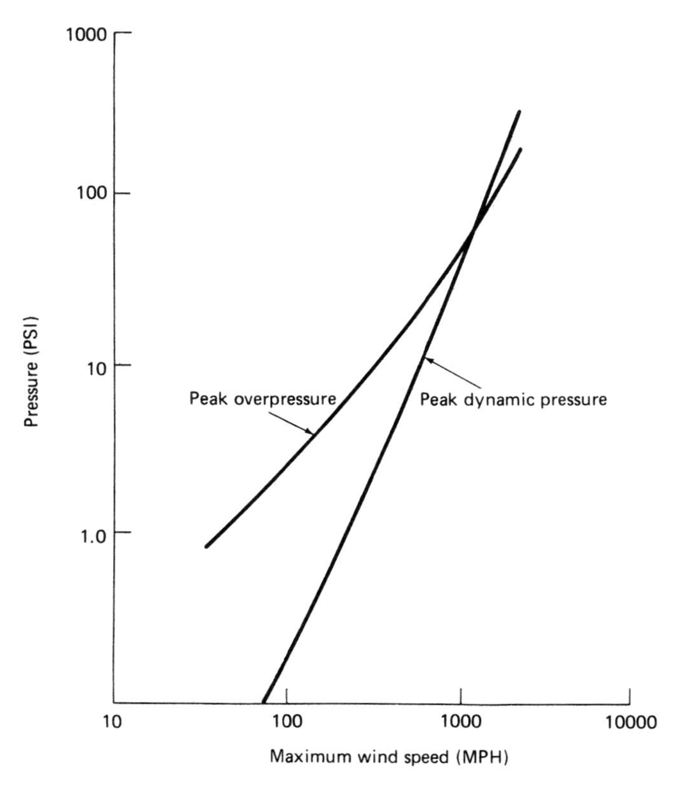

Fig. 4.10A Peak dynamic pressure and peak overpressure versus maximum wind speed at sea level for an ideal shock front.[16]

Table 4.3 Peak static and dynamic overpressures, wind velocities, and generic major damage from 1-MT surface burst from ground zero[1]

PEAK STATIC OVERPRESSURE (psi)	TYPICAL MAJOR DAMAGE TO	PEAK DYNAMIC OVERPRESSURE (psi)	MAX WIND VELOCITY (mph)	DISTANCE FROM 1-MT SURFACE BURST (mi)
200	Steel girder 2-trk RR bridge	330	2080	0.55
50	3-ft-thick brick wall bldg	40	940	0.92
10	Hvy mach tools steel-frame bldg	2	290	1.90
5	Electric tower utility pole	0.7	160	2.75
2	Bomber or commercial A/C	0.1	70	4.70

$$I = \int_0^{t_s} p(t)\,dt = 2p_m t_s \tag{4.28}$$

where p_m is the peak overpressure sustained during the passage of the shock wave. Table 4.3 shows corresponding values of hydrodynamic parameters taken from experimental data. Figure 4.10A depicts peak over-pressure and peak dynamic pressure versus shock wave-induced maximum wind speed for an ideal shock front.[16] The figure delineates the fact that the dynamic pressure is hardly evident for low wind speeds compared to the overpressure. However, at high wind speeds, it becomes the major hydrodynamic destructive mechanism. This is also evident from Table 4.3.

Now consider a circular flat plate oriented parallel to the incident shock wave direction, that is "dished-in" as a result of the shock wave passage. Its deflection δ due to impulse loading is given as

$$\delta = 0.56 I^2 a^2 / 2 m_a Y h^2 \tag{4.29}$$

where

I = impulse (dyne s/cm^2)
a = plate radius (cm)
m_a = plate areal density (g/cm^2)
Y = plate Young's modulus (dynes/cm^2)
h = plate thickness (cm)

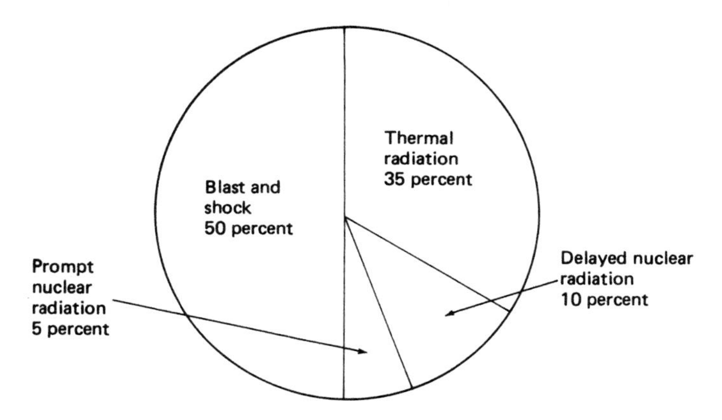

Fig. 4.11 Distribution of energy output from a fission device airburst at altitudes below 100 000 ft.[1]

For an impulse of sufficient strength, the plate will rupture due to a peak static overpressure, given by

$$p_m = 3Yh^2/2a^2 \tag{4.30}$$

For lower pressures, the plate will vibrate in flexure with an angular frequency

$$\omega_N = B[Yh^2/16a^4\rho(1 - \nu^2)] \text{ radians/s} \tag{4.31}$$

where

ρ = plate density
a = plate radius (half-side for square plate)
ν = Poisson's ratio

$$B = \begin{cases} 11.84 \text{ (totally clamped circular plate)} \\ 10.40 \text{ (totally clamped square plate)} \end{cases}$$

If ω_N is known, then the time regime of the shock loading can be separated into an impulse and static loading phase, as

$$t \begin{cases} < 1/\omega_N; & \text{Impulse Loading Phase} \\ \geqslant 1/\omega_N; & \text{Static Loading Phase} \end{cases} \tag{4.32}$$

4.6 THERMAL RADIATION

The earliest radiation from the detonation in the visible and ultraviolet wavelengths is diffuse, complex, and weak. Following this epoch,

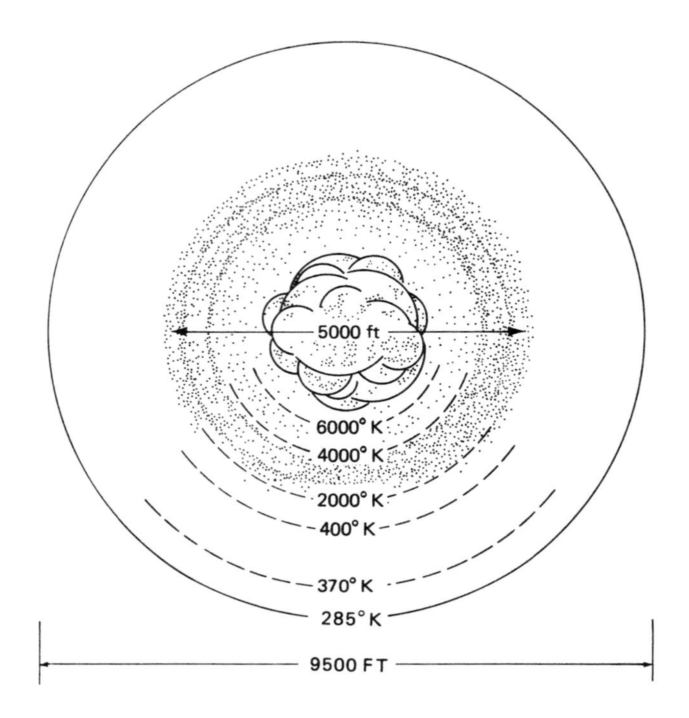

Fig. 4.12 Late fireball radii, and temperature for a 1-MT sea level burst after 1.3 s have elapsed.

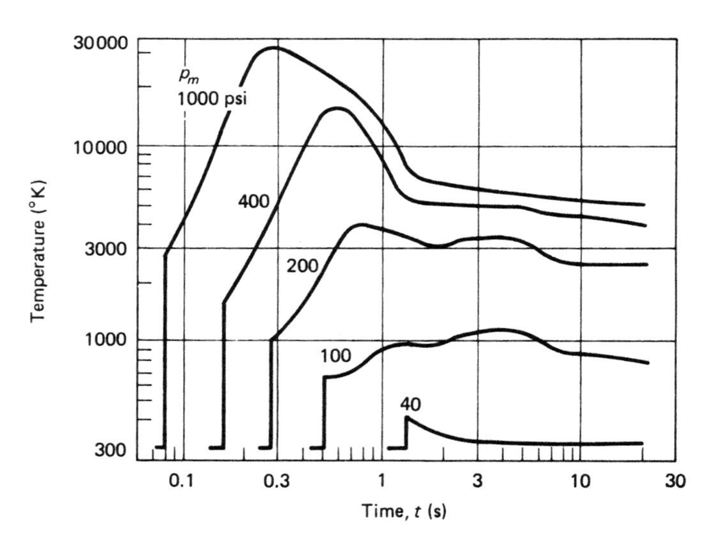

Fig. 4.13 Temperature versus time at high peak overpressures for a 1-MT surface burst.

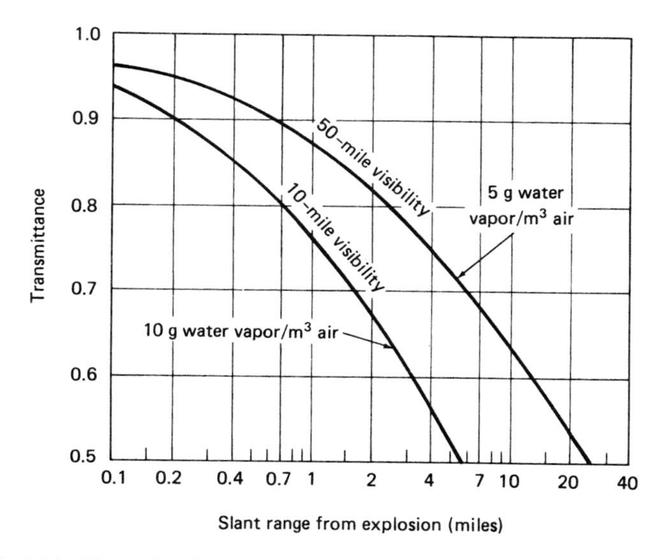

Fig. 4.14 Thermal radiation transmittance versus slant range and visibility.[1]

intense light floods out of the bomb materials as the fireball is created. Figure 4.11 depicts that better than 1/3 of the total energy output is thermal radiation from an endoatmospheric detonation, whereas the thermal radiation and the blast/shock energies are subsumed into the 80 percent of x-ray radiation output of an exoatmospheric burst, as shown in Fig. 4.16A. In certain phases of the evolution of the detonation, with respect to the radiation output in the near-visible, the black body model is suitable, i.e., the total radiated power over all wavelengths of the black body spectrum is given by

$$W = 4\pi R_s^2 \sigma T^4 \quad \text{(erg per s)} \tag{4.33}$$

where R_s is the fireball radius in cm, and $\sigma = 5.6687 \cdot 10^{-5}$ erg per cm^{2}s°K^{-4} This model is valid after the first maximum and before the minimum of light intensity emitted from the detonation. Equation (4.33) does not apply in the very early phases of fireball development, because the radiation is far from being in equilibrium, which is a requirement for the establishment of black body radiation. Until the fireball can engulf the surrounding air and excite it to luminosity, the fireball cannot be seen at a distance, i.e., the first maximum has not yet been reached. The fact that the fireball has a steep temperature gradient, and that the specific heat of

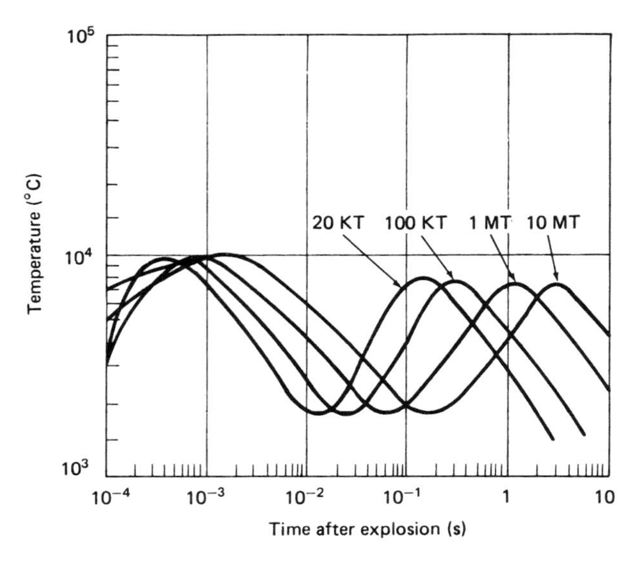

Fig. 4.15 Double pulse of thermal radiation for various air burst yields at sea level.[1]

heated air is much greater than that for cold sea level air implies that Eq. (4.33) must be modified. Such a modification must include that, at this point in time, the air temperature inside the fireball is about 20 000°K. The corresponding black body radiation at this temperature would peak in the ultraviolet. The fireball air is opaque to this radiation in its interior. Further, the luminous fireball is colder than its insides. All this can be subsumed into an expression for the interior radiation that does not emerge from the fireball. This modified black body radiation is given by

$$W_M = (1.53 \cdot 10^6 R_s^2 \phi^4)/(1 + 0.19\phi^2 + 0.06\phi^3) \text{ Watts} \tag{4.34}$$

where ϕ is the radiation temperature in electron volts (1 eV = 12 000°K).

Later, as the exposed fireball interior also begins to cool, the second thermal radiation peak is past, as shown in Figs. 4.12, 4.15, and 4.16. The corresponding temperature chronology is depicted in Fig. 4.13. Then the rising fireball begins to fade from view. Some empirical fireball thermal scaling parametric relations are given by,

$$\text{Time to first maximum} = 1.47 \cdot 10^{-4} Y^{1/3} \text{ s} \tag{4.35}$$

$$\text{Time to minimum} = 2.5 \cdot 10^{-3} Y^{1/2} \text{ s} \tag{4.36}$$

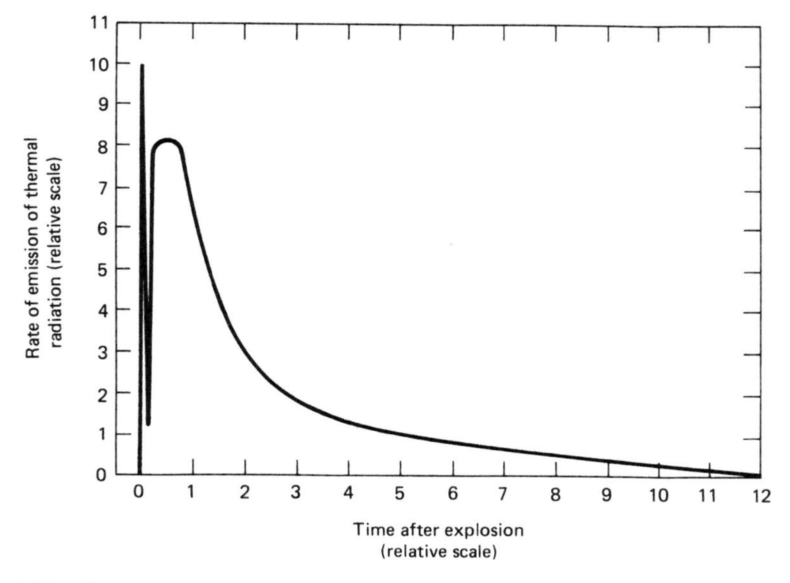

Fig. 4.16 Air burst emission of thermal radiation in two pulses.[1]

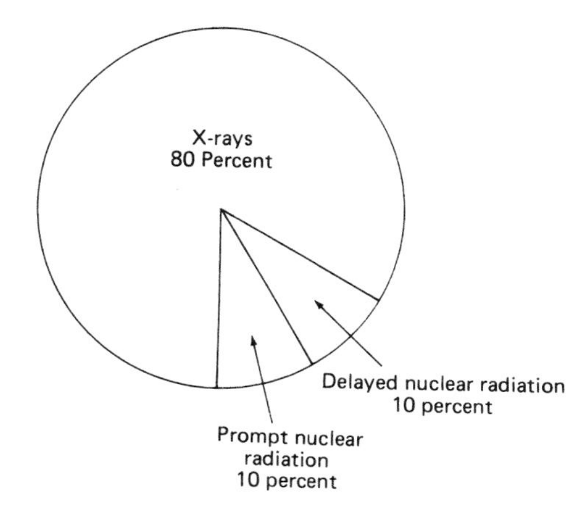

Fig. 4.16A Distribution of energy output from an exoatmospheric burst.[1]

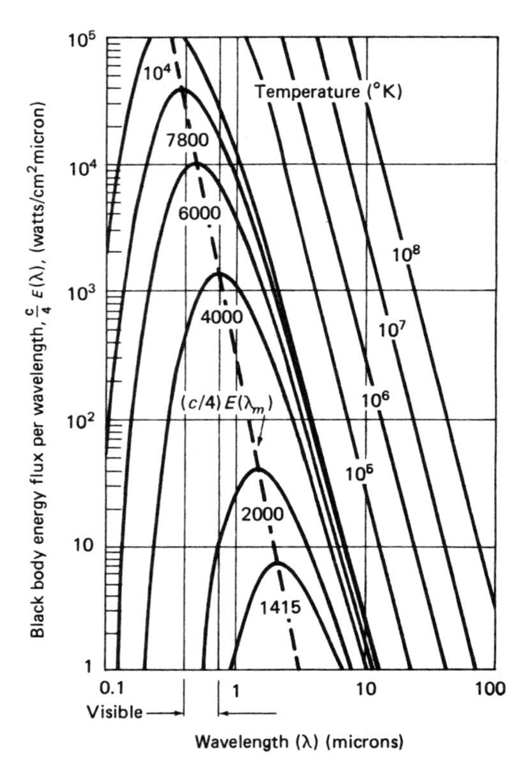

Fig. 4.17 Black body spectral radiation for various black body temperatures from 1415–100 million degrees kelvin.

Table 4.4 Visibility and condition of the atmosphere[1]

ATMOSPHERIC CONDITION	VISIBILITY	
	KILOMETERS	MILES
Exceptionally clear	280	170
Very clear	50	31
Clear	20	12
Light haze	10	6
Haze	4	2.5
Thin fog	2	1.2
Light to thick fog	1 or less	0.6 or less

$$\text{Time to second maximum} = 3.2 \cdot 10^{-2} Y^{1/2} \text{ s} \tag{4.37}$$

$$Y_{\text{thermal}}/Y_{\text{total}} = 0.40 + \frac{0.06\,Y^{1/2}}{1 + 0.5\,Y^{1/2}} \tag{4.38}$$

$$\text{Power at second maximum} = 1.2 \cdot 10^{15}\,Y^{0.6}/\eta^{0.42} \text{ Watts} \tag{4.39}$$

$$Y_{\text{visible}}/Y_{\text{total}} = 0.27 + \frac{0.0072\eta\,Y^{1/2}}{1 + Y^{1/2} + 114\eta/(1 + 8200\eta^2)} \tag{4.40}$$

where $\eta = \rho/\rho_0$ and Y is the yield in KT.

The thermal energy Q emitted as a function of distance from the detonation is

$$Q = (Y \exp - \alpha R)/R^2 \quad (\text{calories/cm}^2) \tag{4.41}$$

where Y is the yield in KT, and R is the slant range in miles from the detonation.

$T_r = \exp - \alpha R$ is the transmittance that takes into account the transport of thermal radiation through the atmosphere. A plot of transmittance is given in Fig. 4.14, with corresponding visibility factors given in Table 4.4. Visibility is defined as the horizontal distance at which a dark object on the horizon has just enough contrast with the surrounding sky to be visible in daylight.

As discussed earlier, the cooling fireball becomes transparent, so that its hot interior now shines through. The resulting thermal radiation can be represented as black body radiation, within the constraints already mentioned. Black body radiation is characterized by its emitted energy spectral density as

$$E(\lambda)\,d\lambda = (8\pi hc/\lambda^5)[\exp(hc/\lambda kT) - 1]^{-1}\,d\lambda \quad \text{erg/cm}^3 \tag{4.42}$$

where

h is Planck's constant ($6.6 \cdot 10^{-27}$ erg s)
c is the speed of light ($3 \cdot 10^{10}$ cm per s)
k is Boltzmann's constant ($1.38 \cdot 10^{-16}$ erg per °K)
T is the black body temperature (°K)
λ is the wavelength of the emitted black body radiation (cm)

The black body spectrum is depicted in Fig. 4.17. It has two important characteristics. The first is its integral, where W is the Lambertian flux, viz.

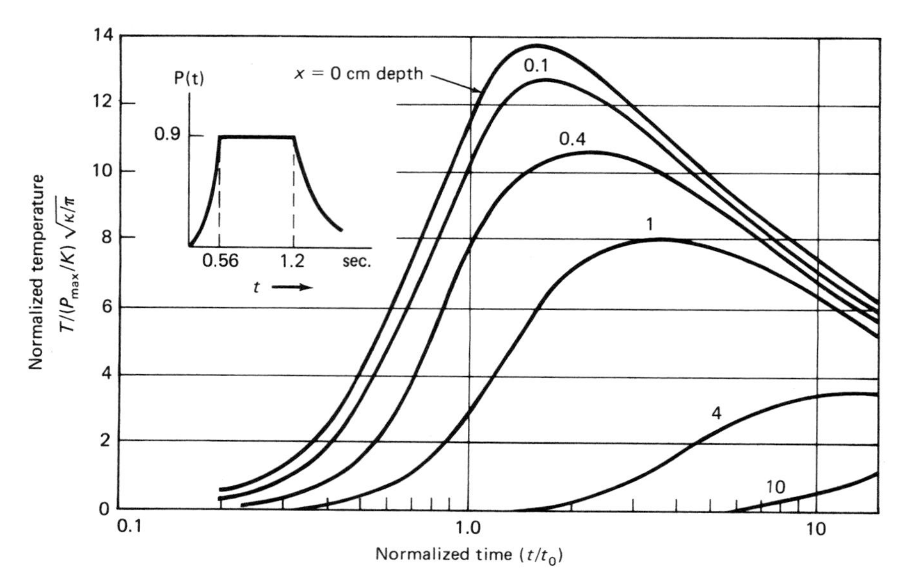

Fig. 4.18 Material temperature versus time for various semi-infinite slab depths for pulse given in Eq. (4.50)

$$\int_0^\infty E(\lambda)\, d\lambda = 8\pi^5 k^4 T^4/15c^3 h^3 \quad \text{erg/cm}^3;$$

$$W = \frac{c}{4}\int_0^\infty E(\lambda)\, d\lambda = \sigma T^4 \quad \text{erg/cm}^2\text{s} \tag{4.43}$$

This is the well-known "fourth-power" law for the total black body radiation. The second property is that equating the derivative $E'(\lambda) = 0$, yields a relationship between the black body temperature and λ_{max}, the wavelength at which $E(\lambda)$ attains its peak. It is given by

$$\lambda_{max} T = C_0 \tag{4.44}$$

This is called Wien's displacement law, where the peak of the black body spectrum is displaced toward shorter wavelengths with increasing black body temperature, and $C_0 = 2897$ micron°K. For example, for an exterior fireball temperature of about 8000°K, $\lambda_{max} = 0.37$ microns, or 3700 angstroms (Å). However, the interior black body temperature from which x rays emanate to produce the fireball by interaction with the air is of the order of a few kilovolts, or a few tens of millions of degrees Kelvin, corresponding to a λ_{max} of $\cong$ 1 angstrom.

4.7 THERMAL ABSORPTION AND HEATING

From Fig. 4.11 it is seen that more than one-third of the energy output of an atmospheric detonation is manifest in thermal radiation. Hence, it is important to determine the amount of thermal absorption and subsequent heating of materials of interest due to the incident thermal pulse. For relatively long planar structures, the one-dimensional heat conduction equation is an adequate model. It is

$$K\frac{\partial^2 T}{\partial x^2} - \rho C_h \frac{\partial T}{\partial t} = 0 \tag{4.45}$$

where

- T is the absolute temperature in the material
- x is the distance into the material
- ρ is the material density
- C_h is the material specific heat
- K is the material thermal conductivity
- κ is the thermal diffusivity of the material ($\kappa = K/\rho C_h$)
- $P(t)$ is the power density of the incident thermal pulse

The above thermal parameters are assumed constant in this rendition of thermal absorption of the incident pulse by the material. An analogous situation arises in Section 9.9, but where the material parameters depend on the temperature T. For a semi-infinite, one-dimensional slab of material of finite thickness, with boundary conditions given by

$$\text{a. } P(t) = -K\partial T(x, t)/\partial x \,\Big|_{x=0}, \quad \text{b. } T(x, 0) = 0,\ T(\infty, t) = 0; \tag{4.46}$$

the temperature is expressed as

$$T(x, t) = (\sqrt{\kappa/\pi}/K)\int_0^t P(t-\tau)\exp - (x^2/4\kappa\tau)\,d\tau/\tau^{1/2} \tag{4.47}$$

being a solution of Eqs. (4.45) and (4.46).

Assuming that the incident thermal pulse is rectangular of amplitude P_0 and pulse width t_0, then the corresponding temperatures are given from Eq. (4.47) as

Table 4.5(a) Figure of merit values for various materials heated by a thermal pulse from a nuclear detonation (CGS units).

VERY THICK SLABS $M_s = (\rho C_h K)^{1/2}$		VERY THIN SLABS $m_s = 1/\rho C_h$	
Material		Material	
Aluminum	0.52	Paint	0.030
Nickel	0.35	Quartz (SiO_2)	0.003
Iron	0.37	Sapphire (Al_2O_3)	0.007
Copper	0.82		

Table 4.5(b) Response of paint coating to incident thermal pulse of 10 cal/cm^2 on:

1. 0.1-cm Aluminum slab		
Pulse width	0.1 s	1.0 s
Front surface temp (°C)	155	155
Rear surface temp (°C)	154	154
2. 0.01-cm Paint on insulating substrate		
Pulse width	0.1 s	1.0 s
Front surface temp (°C)	1344	1088
Rear surface temp (°C)	946	1017
Mean surface temp (°C)	1145	1052
3. 0.01-cm Paint on conducting substrate		
Pulse width	0.1 s	1.0 s
Front surface temp (°C)	672	544
Rear surface temp (°C)	0	0
Mean surface temp (°C)	336	272

Table 4.6 Thermal fluences required to ignite common materials for low- and high-yield detonations.[1]

MATERIAL	LOW-YIELD (20 KT) (cal/cm^2)	HIGH-YIELD (10 MT) (cal/cm^2)
Black rayon lining	1	2
Cotton	8	13
Wool	16	35
Newspaper	3	6
White bond paper	15	30

$$T(x,t) = 2P_0/K \cdot \begin{cases} (\kappa t/\pi)^{1/2} \cdot \exp - (x^2/4\kappa t) - \frac{1}{2} x\, erfc(x/2(\kappa t)^{1/2}); \; t \leqslant t_0 \\ (\kappa t)^{1/2} ierfc(x/2\sqrt{\kappa t}) - \sqrt{\kappa(t-t_0)}\, ierfc(x/2\sqrt{\kappa(t-t_0)}); \; t > t_0 \end{cases} \tag{4.48}$$

where $erfc(x) = (2/\sqrt{\pi}) \int_x^\infty \exp - u^2 du$, and $ierfc(x) = \int_x^\infty erfc(u) du$. It is easily shown that the maximum temperature is given by

$$T_{max} = (2P_0/K)(\kappa t_0/\pi)^{1/2} \tag{4.49}$$

for the case of penetration depth into the material, such that $x \ll (\kappa t)^{1/2}$ holds. Figure 4.18 depicts the temperature in the material slab for an incident thermal pulse approximated by (Fig. 4.18, inset)

$$P(t) = \begin{cases} 0.009(t/0.056)^2; & 0 \leqslant t < 0.56 \text{ s} \\ 0.9; & 0.56 \leqslant t < 1.2 \text{ s} \\ 0.9(t/1.2)^{-3/2}; & 1.2 \leqslant t < \infty \text{ s} \end{cases} \tag{4.50}$$

For the case of thin slabs of thickness l, such as the walls of most containers of interest, the corresponding solution of Eq. (4.45), with the following boundary conditions for a rectangular pulse of amplitude P_0 and pulse width t_0.

$$T(x,0) = 0; \quad \partial T(x,t)/\partial x\Big|_{x=l} = \begin{cases} P_0/K, & t \leqslant t_0 \\ 0, & t > t_0 \end{cases}; \quad \partial T(x,t)/\partial x\Big|_{x=l} = 0 \tag{4.51}$$

is given by[10]

$$T(x,t) = \frac{P_0 t}{\rho C_h l} + \frac{P_0 l}{K}\left[\frac{3x^2 - l^2}{6l^2} - \frac{2}{\pi^2}\sum_{n=1}^{\infty}\frac{(-1)^n}{n^2}\cos\left(\frac{n\pi x}{l}\right)\exp - \left(\frac{n^2\pi^2\kappa t}{l^2}\right)\right]; \quad t \leqslant t_0 \tag{4.52}$$

The time when the maximum temperature occurs, $t_{max} \simeq t_0$, because t_0 is the allowed time for the temperature to build up, after which it starts to fall. So the maximum temperature which occurs at $x = l$, is

$$T_{max} \equiv T(l, t_0) = \frac{P_0 t_0}{\rho C_h l} + \frac{P_0 l}{K}\left(\frac{1}{3} - \frac{2\sqrt{\kappa t_0/\pi}}{l}\right) \tag{4.53}$$

Certain indices or figures of merit can be deduced to identify the thermal durability or thermal hardiness properties of materials, or material coatings, after being subject to incident thermal pulses. Table

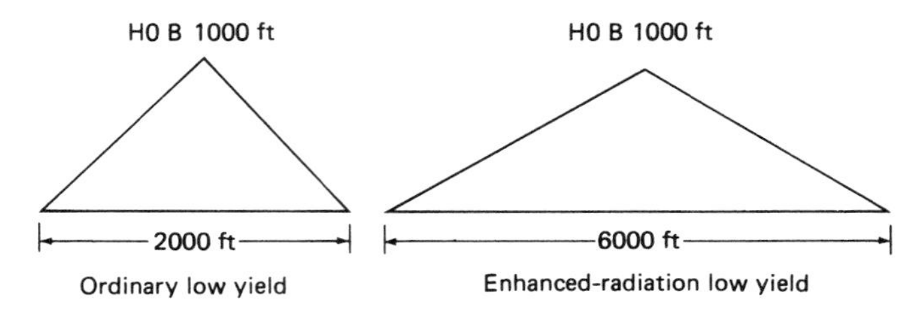

Fig. 4.19 Comparison of ordinary low-yield and enhanced-radiation devices at the 500 rad isodose radial contour; HOB is height of burst.[7]

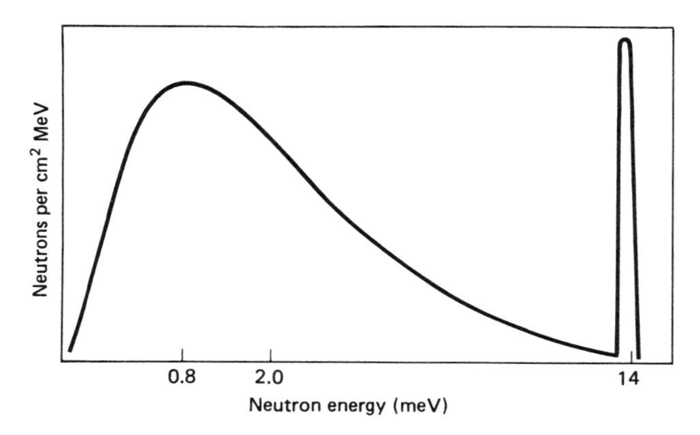

Fig. 4.20 Neutron energy spectrum from thermonuclear detonation.

4.5(a) provides such a set for figures of merit $M_s = (\rho C_h K)^{1/2}$ for thick walls, and $m_s = 1/\rho C_h$ for thin walls or such coatings as paint. Relatively large values of the figures of merit imply a material endurance to the pulse, and small values mean that the material will not stand up well under the impact of the pulse. M_s is essentially the material impulsivity per unit absolute temperature change, a parameter characterizing material erosion caused by dielectric-type breakdown.[19]

For long-time thermal response, i.e., for times long after the onset of the thermal pulse, the asymptotic forms of the temperature from Eqs. (4.47) and (4.52) yield, for a polygonal volume

$$T \sim \int_0^\infty P(t)\, dt/\rho C_h l n \tag{4.54}$$

where n is the number of sides of the many-sided volume, and l is the corresponding wall thickness of each side. As an illustration, consider a cube of six sides with $\rho = 1\,\text{g/cm}^3$, $C_h = 0.2\,\text{cal/g°K}$, $l = 3.3\,\text{mm}$, and

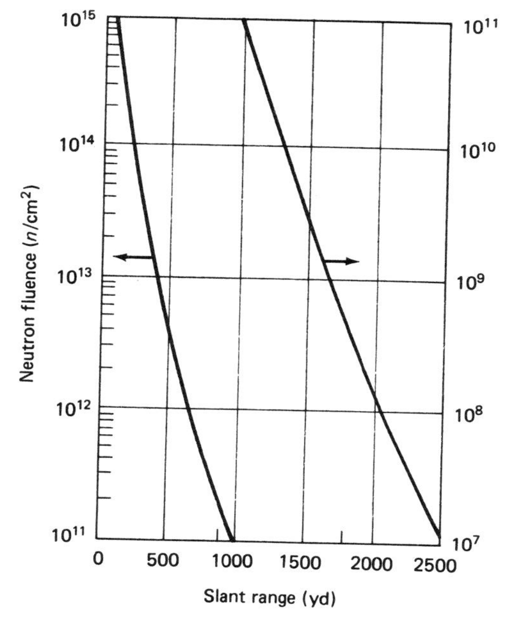

Fig. 4.21 Neutron fluence versus slant range in air of 0.9 sea level density for a 1-KT burst (fluence values are ± 15 percent).[1]

$\int_0^\infty P(t)dt = 10\,\text{cal/cm}^2$. If on one face of the cube $T_0 = 150\,C$, after a long time, Eq. (4.54) yields $T = 25°\text{C}$ attained 30 s after pulse onset, as computed from Eq. (4.48).

Table 4.6 gives the amounts of thermal fluence required to ignite common materials for low- and high-yield detonations. The high-yield detonation results in a slower rate of delivery of thermal fluence than its low-yield counterpart, as seen in Table 4.6. This accounts for the required higher ignition fluences of the high-yield bursts.

4.8 NUCLEAR RADIATION

In light of the foregoing sections on nuclear radiation output of the detonation, each type of radiation is now discussed briefly. As mentioned earlier, those of prime importance to electronic systems are elaborated on in later chapters. Nuclear radiation results from both fission and fusion reactions and is released both at the instant of the detonation and over an extended post-detonation period. Although the initial nuclear reactions take place mostly within the bomb case and are over in a few tenths of

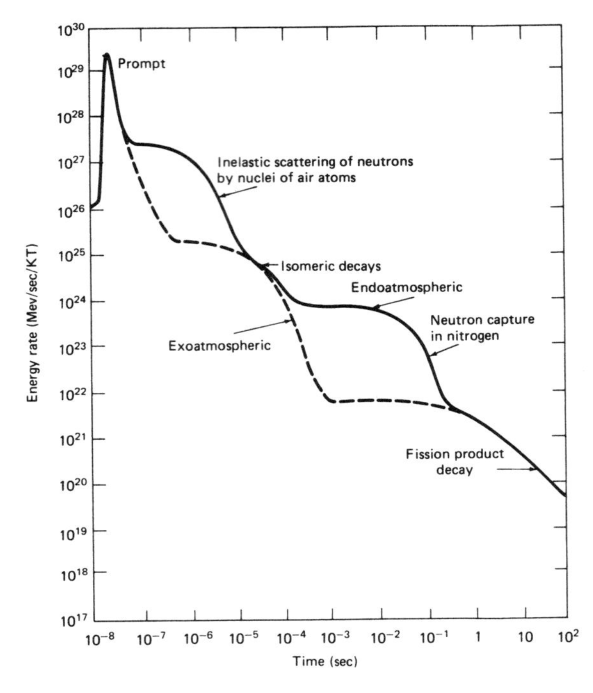

Fig. 4.22 Early time dependence of gamma ray output from a nuclear detonation.[12]

microseconds, some radiation that results persists for long periods after the burst, being radiated and reradiated from the bomb atoms and nuclei far outside as well as inside the bomb debris. Hence, radiation is divided into initial or prompt radiation, as well as residual or delayed radiation, with the latter divided into early and late. The time division between prompt and delayed radiation is not well-defined, but consensus has it at about 1 min after the detonation, with the residual radiation manifest later.

Initial or prompt radiation includes neutrons, gamma rays, x rays, alpha and beta particles, and other secondary particles. All these have their residual counterparts. For example, neutron-produced radioactive nuclei emit gamma rays with well-known half-lives, unique to the particular nuclei and much longer than the detonation epoch. Alpha particles are helium ions, i.e., helium atoms stripped of their two electrons and produced in fusion reactions in the early stages of the detonation. Being relatively heavy and charged, they have a strong propensity for interacting readily with almost all matter, so that their range in materials is very short. An ordinary piece of paper can stop all but the very highest energy alpha particles.

X rays are produced mainly from atomic reactions within the bomb. They are the principal agent that ionizes the air to create the fireball. In exoatmospheric detonations, the x rays have very long ranges measured in hundreds of kilometers; because of the very rarified atmosphere at stratospheric heights and beyond, no fireball is produced. Also, as there is no blast or shock wave, it turns out that about 80 percent of the detonation energy is in the form of x rays.

Beta particles are simply fast electrons emitted from radioactive nuclei isotopes. They are much lighter than alpha particles, but are also charged, so that their range in air is usually only a few feet. Alphas or betas thereby are not a serious biological hazard, unless they are ingested or inhaled into the body. Alphas do not play a major role in radiation damage to electronic systems, so that their discussion will be limited to special situations in ensuing chapters.

4.9 NEUTRONS

Approximately 90 percent of the neutrons generated by the fission and fusion reactions in the detonation are absorbed within the bomb case itself. However, the remaining 10 percent that escape correspond to tremendous neutron fluences in air. About 0.7 percent of the fission neutrons are released as delayed neutrons from fission decay products emitted from the detonation, from milliseconds to minutes after the detonation. Some of the initial or prompt neutrons are absorbed by the bomb materials to activate them, i.e., to make them radioactive. Neutrons are the principal particle for producing radioactive materials within and without the bomb confines by being absorbed by bomb or environment elements. They also cause ionization and atomic displacement damage in semiconductor materials, as discussed in Chapter 5.

The enhanced radiation device is a very low-yield, tactical weapon tailored to produce an enhanced neutron fluence, with a limited amount of concomitant shock and blast. Figure 4.19 compares an ordinary low-yield (0.1-KT) device with the enhanced radiation device. It is easily calculated that the 500-rad radius is at 1000 ft for the low-yield weapon and 2860 ft for the enhanced low yield radiation weapon at 1000 ft burst height. It is then appreciated that the corresponding area of the latter is about nine times larger than the former, i.e., about 0.92 mi^2, while producing roughly the same amount of blast and shock overpressure of less than 2 psi at ground zero.

The neutron energy spectrum emitted from a thermonuclear detonation

Table 4.7 Prompt dose versus distance.[4]

RANGE (mi)	AIR DENSITY (g/L)	1 KT			10 KT			100 KT			1 MT			10 MT		
		γ (GAMMA) RAYS (r)	NEUTRONS (rad)	ΔP (psi)	γ RAYS (r)	NEUTRONS (rad)	ΔP (psi)	γ RAYS (r)	NEUTRONS (rad)	ΔP (psi)	γ RAYS (r)	NEUTRONS (rad)	ΔP (psi)	γ RAYS (r)	NEUTRONS (rad)	ΔP (psi)
	1.0	430	243		4300	2400		45 000	24 000		2 900 000	240 000		390 000 000	2 400 000	
0.5	1.1	330	173	1.6	3300	1700	~6.3	35 000	17 000	~32	2 300 000	170 000	~220	310 000 000	1 700 000	~1800
	1.3	200	88		2000	880		22 000	8 800		1 400 000	88 000		200 000 000	880 000	
	1.0	9	2.1		90	21		960	210		62 000	2 100		11 000 000	21 000	
1.0	1.1	5.5	1.0	0.5	55	11	~1.8	580	110	7	38 000	1 100	~37	7 200 000	11 000	~250
	1.3	2.1	.27		21	2.7		210	27		14 000	270		3 000 000	2 700	
	1.0				3.4	.31		36	3.1		2 300	31		570 000	310	
1.5	1.1	<1	<1	0.3	1.6	.11	0.9	17	1.1	3.4	1 100	11	~15	300 000	110	~85
	1.3				<1	–		3.9	.15		250	1.5		81 000	15	
	1.0							1.7	.06		110	.59		37 000	5.9	
2.0	1.1	–	–	0.2	–	–	0.6	<1	–	2	42	.15	~8	15 000	1.5	~40
	1.3										5.8	.01		2 700	.1	
	1.0										6.1			2 700		
2.5	1.1	–	–	0.13	–	–	0.4	–	–	1.4	1.8	–	~5	910	1	~25
	1.3										<1			100		
	1.0													210		
3.0	1.1	–	–	0.10	–	–	0.3	–	–	1.0	–	–	~4	58	–	~17
	1.3													4.4		
	1.0													1.6		
4.0	1.1	–	–	0.06	–	–	0.2	–	–	0.7	–	–	~2	<1	–	~9
	1.3															

1 neutron rad (tissue) $\doteq 3 \cdot 10^8$ (1 MeV) n/cm^2
1 roentgen r (tissue) $\doteqdot$ 1 Rad (tissue)
1 electron rad (Si) $\doteq 4.35 \cdot 10^7$ (2.5 MeV) e/cm^2

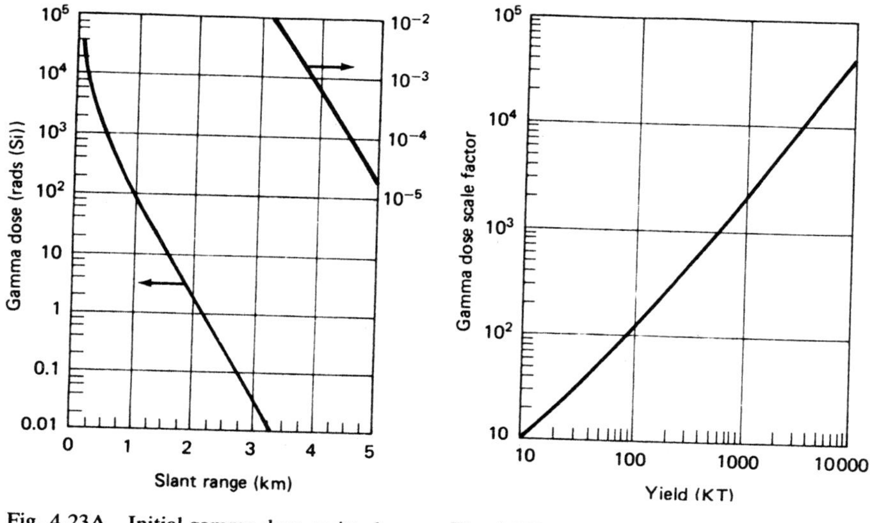

Fig. 4.23A Initial gamma dose emitted within the first minute for a 1-KT surface burst (0.9 sea level p).[1]

Fig. 4.23B Gamma dose scale factor for yields to 10 MT for use with Fig. 4.23A and Fig. 4.23C at 1-KT yield.[1]

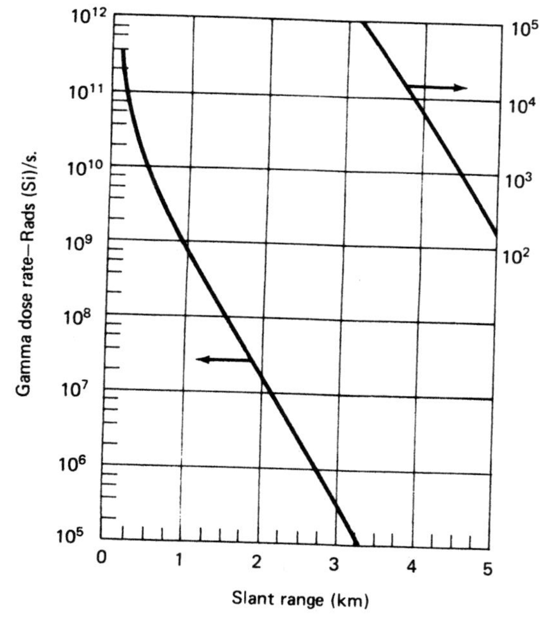

Fig. 4.23C Initial dose rate emitted within the first sec and for a 1 KT surface burst.

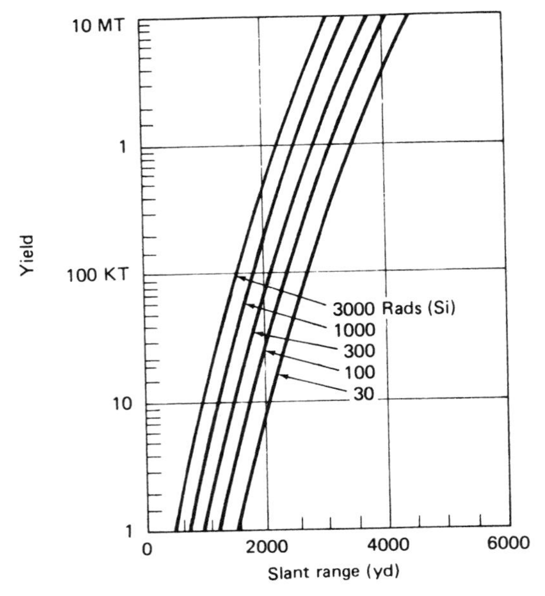

Fig. 4.24 Yield versus slant range in the atmosphere for various initial gamma-ray doses.[1]

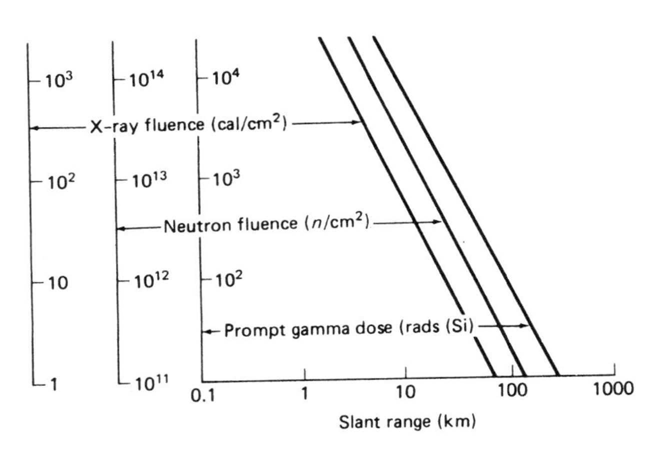

Fig. 4.25 Radiation levels for an exoatmospheric 1-MT detonation.[1]

consists of a peak at 0.8 MeV, with an average energy of about 2 MeV. In addition to this spectrum of fission-produced neutrons, as in a nuclear reactor, there is a fusion neutron peak occurring at about 14 MeV, as in Fig. 4.20.

A nuclear detonation may be viewed as a source of neutron fluence roughly proportional to the weapon yield. Actually, the neutron fluence emanating from the detonation suffers a $1/r^2$ geometrical attenuation, as well as an exponential attenuation due to neutrons absorbed and scattered in the atmosphere. This fluence is given by the empirical expression[4]

$$\Phi_n = (2 \cdot 10^{25} Y/r^2) \exp - (\rho r/2.38 \cdot 10^4) \quad \text{neutrons/cm}^2 \qquad (4.55)$$

where Y is the yield of the detonation in MT, r is the distance from the detonation in cm, and ρ is the density of the air in grams per liter ($\rho = 1.1$ for sea level air). Figure 4.21 depicts neutron fluence as a function of distance from a 1-KT detonation at 0.9 sea level density. Because the neutron fluence is shown to be proportional to the yield, as in Eq. (4.55), Fig. 4.21 can be used to scale the fluence versus distance for arbitrary yield.

4.10 GAMMA RAYS

The bomb gamma rays come from two principal sources within the detonation. They are (a) the decay of radioactive fission fragments, and (b) the result of a nucleus that has become excited by absorbing (capturing) a neutron and subsequently falling back to its ground state by emitting a gamma ray. These are called *capture*, or prompt, gamma rays. This results in a more complicated gamma spectral radiation characteristic than for the neutron fluence, for example. The fission fragment residual gamma rays are emitted with an average, overall fission species, which results in an empirical, relatively slow time relationship, given by a $t^{-1.2}$ dependence. This is compared with the capture gammas that are all generated in the first few milliseconds after the detonation.

In terms of the temporal aspects of the appearance of gamma rays and associated particles after the onset of the detonation, three categorizations can be made. They are: (a) prompt gamma rays are those that appear within one microsecond from onset. These result from the fission process directly, and (n, γ) reactions within the weapon materials[12]; (b) initial gamma rays are those that are usually emitted within the first minute following the detonation. They mostly come from the decay of excited nuclei (also called isomers), whose mean decay time for the (n, γ) reaction is up to one minute after onset; (c) delayed gamma rays are

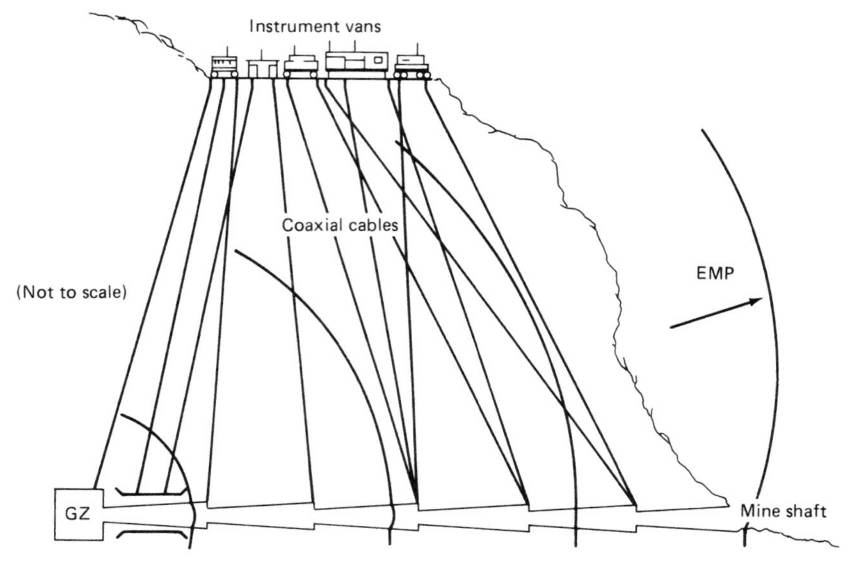

Fig. 4.26 One type of early underground test configuration.[13,14]

usually construed as those that appear after one minute from the detonation onset, corresponding to radioactive fission fragment nuclides, whose (n, γ) mean decay times are greater than one minute. Figure 4.22 depicts the dependence of the gamma ray rate of energy output per kiloton, both from endo- and exoatmospheric detonations, suggesting a succession of the temporal definitions given.

Gamma rays, even though they travel at the speed of light, are transported through the air in roughly the same fashion as neutrons, i.e., their diminution with distance is given by the product of the geometrical attenuation with the exponential attenuation. However, the relatively long mean time of gamma-ray emission allows the shock wave movement of the absorbing air to influence the gamma dose as a function of distance from the point of detonation. This hydrodynamic lowered density effect, especially for high-yield weapons, can enhance the gamma dose over that from its lower yield counterpart emanating from a point source in still air. This is depicted in Fig. 4.27, where enhanced gamma dose versus slant range is plotted. Because the shock wave is nearly spherically symmetrical about the bomb, it does not influence the spherical nature of the gamma fluence, but it does alter the behavior of the gamma absorption and scattering in air. This hydrodynamic effect is incorporated into the gamma dose expression in a way that has the same character as that of the neutron fluence, but with yield-dependent parameters. That is, the corresponding gamma dose is given by

$$D_\gamma = 7.10^{13} f Y \alpha(Y)[\exp - (\rho R/\lambda)]/R^2 \quad \text{rads (Si)} \qquad (4.56)$$

where $\alpha(Y) = (1 + 6Y^2)/(1 + 0.03Y^2 + 0.0057Y^3)$, $\lambda(\text{ft}) = 1070 + 1.5Y^2$, with Y the yield in MT, R the slant range in ft, $f \simeq \frac{1}{2}$ the fission factor, and ρ the air density in grams per liter (g/L).

The dose units are rads (Si) where 1 rad (Si) is the dose defined by the deposition of 100 ergs of energy per g of material, i.e. silicon. The rad is a CGS unit, but the international system of units (SI) defines an essentially MKS unit for absorbed dose called the *gray* (Gy). One Gy is defined as the deposition of 1 joule of radiation energy per kilogram. The equivalence between the gray and the rad is immediate, i.e. one *Gy* = 100 rads. Also 1 centigray (cGy) = 1 rad, a recently defined unit.

For tissue, a unit of dose commonly used is called the *rem* (roentgen equivalent man). It is the amount of radiation incident that produces biological damage equivalent to one rad of x radiation; 1 rem equals 1 rad of x rays multiplied by a "relative biological effectiveness."[1] The corresponding MKS (SI) unit, called the *sievert* (Sv), is to the gray what the rem is to the rad.[8] This is discussed in detail in Section 10.2.

The enhanced gamma dose, D_γ, as given by Eq. (4.56), matches most data within a factor of 2 for yields between 1 KT and 10 MT, and from 1000 ft to 20 000 ft of burst altitude. Table 4.7 gives enhanced prompt gamma-ray exposure in roentgens, neutron dose in rads (tissue), plus shock overpressure for a number of burst yields.

Figure 4.23C depicts the gamma ray dose rate $\dot{D}_\gamma$ versus the slant range from ground zero for a 1 KT yield for a prompt gamma pulse width of $t_{pw} \cong 10^{-7}$ sec. The latter value is taken from Fig. 4.22. The gamma ray dose rate is computed from $\dot{D}_\gamma \cong D_\gamma/t_{pw}$, which holds for $0 \leqslant t \leqslant t_{pw}$, where D_γ is the gamma ray dose given in Fig. 4.23A. Note that $\dot{D}_\gamma = 0$ for $t > t_{pw}$.

For yields greater than 1 KT, Fig. 4.23B provides the appropriate scale factor. The extrapolation from these dose rate sources to semiconductor damage is discussed in Chapter 7.

Figures 4.23A and B, respectively, depict the initial gamma dose for a 1-KT burst at 0.9 sea level air density, and a scale factor curve for use in obtaining other larger yields. These curves supplement those of Fig. 4.24, which depict a family of prompt gamma ray dose as a function of yield for atmospheric bursts at close-in slant ranges.

Figure 4.25 illustrates radiation levels for exoatmospheric (extraterrestrial) detonations, for comparison with endoatmospheric bursts.

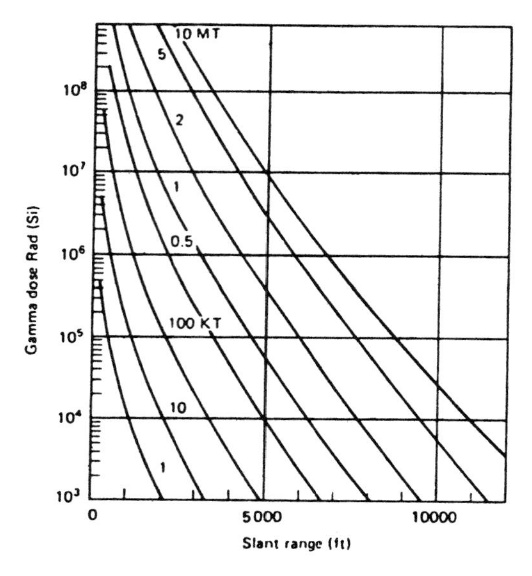

Fig. 4.27 Enhanced gamma dose due to shock rarefaction of air about ground zero versus slant range for various yields.[4]

Included are x-ray and nuetron fluence and prompt gamma-ray dose, as a function of slant range from a 1-MT detonation. In the exoatmospheric case, only geometric, or $1/4\pi R^2$, attenuation is apparent, as there is essentially no atmosphere to provide exponential attenuation. Comparing Figs. 4.24 and 4.25 for a prompt gamma dose of 1 krad (Si), it is seen that for a 1-MT burst in the atmosphere, this dose level occurs at about 2.5 km, while outside the atmosphere the same level is manifest at about 30 km, a distance factor of about an order of magnitude. For exoatmospheric x rays, the corresponding fluence is

$$\Phi_x \cong 6400Y/r^2 \text{ (calories per cm}^2\text{)} \tag{4.57}$$

where Y is the total burst yield in MT and r is in km from the detonation point. This assumes that 80 percent of the total exoatmospheric yield is x rays.

4.11 ELECTROMAGNETIC PULSE

A latter-day phenomenon in the lexicon of weapon detonation outputs is the electromagnetic pulse (EMP). Prior to circa 1960, little attention was

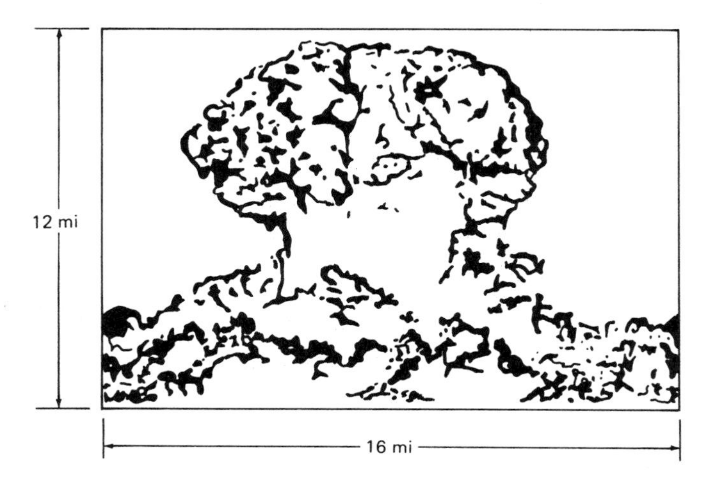

Fig. 4.28 Debris mixing with fireball and cloud following detonation of moderate yield.[9]

paid to this effect for a number of reasons. This phenomenon was first brought to the attention of test personnel when the coaxial cables used to wire their experiments would sometimes melt after a detonation a few thousand yards away.

Figure 4.26 depicts one type of early underground test configuration.[13,14] The discontinuities along the mine shaft correspond to successive annuli upon which are mounted instrumented detectors and experiments. This allows the detonation emanations to be normally incident upon them. An explosively driven pinch, shown close to ground zero (GZ), walls off the nonatomic debris following the detonation to curtail contamination of the tunnel experiments.

When a detonation occurs near the top of the sensible atmosphere, the intense fluence of prompt x rays and gamma rays emanating from the bomb collide with the air molecules, ionizing them by knocking off their outer electrons. Because these electrons are very much lighter than their parent ions, they are more mobile and thus rush outward at higher speeds than the ions. Momentarily this results in a more or less concentric, quickly expanding charge separation of tremendous size about the detonation point, consisting of an outer shell of electrons and an inner shell of ions. Such an expanding charge separation region, not unlike a concentric spherical capacitor, results in a very large transient electromagnetic field emanating from that region, due to the accelerating electron current. The mere collision or scattering of an atomic electron by a bomb gamma-ray or x-ray photon, in this case causing ionization, is called a *compton scattering event*, after Compton who first explained it theoretically. The resulting speeding electrons from the ionization encounter are called *compton electrons*. This compton electron current density, J_{cm}, can be written as[5]

$$J_{cm} = (e\lambda_e/\lambda_\gamma)\Gamma(t - R/c) \tag{4.58}$$

where λ_e and λ_γ, respectively, are the mean free paths of the electrons and gamma rays in air. $\Gamma(t - R/c)$ is the gamma ray flux (gammas/cm^2 s) from the detonation, which is related to the previously discussed gamma-ray dose, within a constant, by

$$D \sim \int_{R/c}^{t} \Gamma(t' - R/c)\,dt' \tag{4.59}$$

The argument $t' - R/c$ refers to the retardation, i.e., at a distance R from the burst, there is a delay time of R/c for the effect of a compton collision to be manifest there, and c is the speed of light.

For a near-surface burst, the situation is radically different, due to the complex interaction of the electromagnetic field with the atmosphere, as discussed in detail in Section 9.4. The result is a relatively weak air burst radiated EMP signal.

As is well known from electromagnetic theory, if the charge separation is perfectly symmetrical, i.e., spherical about the detonation point, no electromagnetic field will be radiated and therefore no EMP (Problem 4.4). For a detonation on or near the surface, enough asymmetry will exist due to the presence of the earth to ensure production of an electromagnetic pulse. In high air bursts asymmetries arising from detonating weapon structure, earth magnetic field, and atmospheric air density gradients cause the resulting charge distributions to distort sufficiently to produce an electromagnetic dipole at least, and thus corresponding radiation. For very-high-altitude bursts, the atmosphere is sufficiently rarefied to allow the electrons long mean free paths. The interaction of these electrons with the earth's magnetic field, as well as the weak but present atmospheric density gradients, supplies enough asymmetry to result in a sizeable electromagnetic pulse. These electrons can corkscrew around the earth magnetic field lines and travel back and forth to conjugate points on the earth many thousands of miles away, radiating all the while. This is called *synchrotron radiation*, because it is similar to that occurring in a synchrotron accelerator. The EMP is virtually absent from underground or undersea detonations.

Gamma rays from the detonation travel outward at the speed of light, achieving a maximal flux in a few nanoseconds, but persisting for a relatively long time, due to the addition of delayed gammas coming from the decay of fission fragments, which are part of the exploding bomb debris. Most of the compton electrons now quickly attach themselves to air atoms and molecules, at very large attachment rates. The resulting negative air ions and the original positive air ions persist for a much longer time, so that most of the charge separation is obliterated in milliseconds. Nevertheless, the large region of bomb ionization has sufficient strength to cause appreciable EMP.

The subsequent electromagnetic radiation is found to have a frequency spectrum ranging from gigahertz (GHz) to almost direct current (dc). The microwave components are attenuated in the atmosphere, but the radio frequencies are relatively unabsorbed and can travel around the world like lightning spherics. For up to a few hundred miles from the detonation, the radio frequency spectrum is continuous, with a median frequency of about 15 Hz, which frequency is related inversely to yield. At thousands of miles from the detonation, the apparent EMP is greatly influenced by atmospheric ducting properties, where the earth and the ionosphere act as a spherical waveguide. Close to the detonation, the whole frequency spectrum is present, and severe damage to electrical, electronic, and nonelectronic systems can occur. Such EMP fields are measured in hundreds of kilovolts per meter and hundreds of gauss of azimuthal magnetic field strength. EMP is discussed in much greater detail in Chapter 9.

4.12 DUST AND DEBRIS

For a surface or near-surface burst, the material from the excavated crater and the hypervelocity material bomb fragments, called *ejecta*, form the debris and dust that make up the bomb stem and the "mushroom" cloud. For a high-yield burst, even the ground beyond the crater provides large amounts of dust and smoke carried up with the rising fireball, not unlike an active volcano.

As mentioned earlier, shock wave behavior within the fireball system implies the existence of reverse winds, which can force debris back to obstruct silo slits or other shelter openings. The fireball, still with its less than ambient internal pressure, now acts like a fast-ascending balloon. This balloon, thousands of feet in diameter, rises at rates of about 300 ft per s.

The stem, filled with crater dust and debris, rises up through the clouds with top speeds of about 800 ft per s. The fireball cloud is depicted in Fig. 4.28. The corresponding wind speeds are high enough to lift a two-ton rock and carry lighter material to very high altitudes. Empirical scaling for such winds yields[5]

$$V_u \cong 500 Y^{1/4} \quad \text{(ft per s)} \tag{4.60}$$

$$M_u \cong 2.10^{-14} \eta^3 u^6 = 300 \eta^3 Y^{3/2} \quad \text{(lb)} \tag{4.61}$$

where V_u is the upward wind speed, M_u is the mass that can be lofted, Y is the yield in MT, $\eta = \rho/\rho_0$, and u is the dimensionless velocity defined in Section 4.5.

For an approximately 1-MT burst, the cloud rises to about 60 000 ft in about 5 min, depending on weather up to altitude. The dust and debris rise with the cloud, carried by the high-speed winds. Rocks will rain over a very large area for many minutes, with large boulders at ranges of the order of the stem radius (i.e., thousands of feet away from the detonation). This debris is radioactive to the extent that it can rise to thousands of rads per hour in the first scores of minutes, dropping to a few hundred rads per hour about 24 h after the burst. In this local fallout, within a few miles of the burst, the radiation levels are almost certainly fatal to man. Total radiation doses after one day could amount to about 3000 rads over a circular area of about 15-mi radius from the burst point.

The fallout parameters depend sensitively on major weapon design parameters, such as the type of bomb, yield, and whether thermonuclear or not, as well as the burst height and local soil properties. Bombs can be "salted" to enhance the production of radioactivity, but the burst position and crater soil properties are usually of overriding importance.

Shallow dirt cover over the device increases the downwind fallout levels by lofting more dust, but reduces the fallout land area coverage over that from a strictly surface detonation. On the other hand, a slight distance above the earth's surface for the detonation point reduces the downwind fallout levels. The radioactivity of the fallout varies roughly linearly with the fraction of bomb yield due to fission, in contrast to that due to fusion. The amount of local or downwind fallout is, however, determined by the scavenging action of the radioactive products by the cratered material. The amount of debris available for such scavenging action varies somewhat less than linearly with yield. This is specially true for surface bursts.

PROBLEMS

1. Develop the argument for the fact that one can stand on the ground and, due to shielding by the atmosphere, can observe with safety (wearing dark glasses) a detonation directly overhead of a 0.1-MT high-altitude burst at,

e.g., 150 000 ft. Compute the contributions from neutrons, gammas, and thermal energy[15] on the ground. Modify the exponential term in Eq. (4.55) to $\exp - \int_0^{R_0} \rho(r)\,dr/2.34 \cdot 10^4$ for neutrons. Similarly for Eq. (4.56) use $\alpha(Y) = 1$, and $\exp - \int_0^{R_0} \rho(r)\,dr/\lambda$ for gamma rays. Let R be the distance down from the detonation, R_0 be the burst height, and $\rho = \rho_0 \exp m(R - R_0)$ for the atmospheric density increase down from the burst point. m is found from the rule of thumb that air density decreases by a factor of about 10 for every 10-mi increase in altitude. Use $\rho = 1.1$ g/liter. Note that the unit of distance is different in Eqs. (4.55) and (4.56). For thermal radiation, use Eq. (4.41).

2. What is the neutron fluence level emitted from a sea level 1-MT burst at 1 mi? How many moles of neutrons are produced? Hint: Use $\lim_{s \to 0} \int \Phi_n(r) dS$ in Eq. (4.55), where S is a spherical surface around the detonation.

3. The early fallout from a burst decays with the empirical $t^{-1.2}$ law, where t is in hours after the detonation. Show that one should stay in their underground shelter for about 2 weeks after the detonation to let the ambient radiation level drop to 0.1 percent of the level existing 1 h after the burst.

4. For a spherically symmetric detonation configuration of ions and compton electrons, show that no radiation pulse, i.e., no EMP can be created. Hint: Show that in free space, a spherically symmetric electric field, given by $\bar{E}(r)$ has zero curl. Hence, no wave equ..tions can ensue, as discussed in Section 2.5.

5. The currently used CGS unit of absorbed dose for radiation is the rad; 1 rad is defined as the deposition of 100 erg per g of material. Hence, the material must be specified for completeness. In silicon, it is rads (Si). The roentgen is a somewhat obsolescent unit of radiation that persists today. It is defined as that amount of incident radiation that is required to ionize the molecules in 1 cm^3 of STP air to produce ions that, in their totality, carry off 1 esu of charge. (a) Show that 1 roentgen corresponds to the radiation exposure that deposits about 85 erg per g of air. Hint: the charge on the electron is $4.8 \cdot 10^{-10}$ esu; hence 1 esu of charge corresponds to $1/4.8 \cdot 10^{-10} = 2.08 \cdot 10^9$ ion pairs produced, and it takes about 32.5 eV to produce one ion pair in STP air. (b) Discuss whether the roentgen is a unit of exposure or a unit of dose like the rad. (c) Does 1 roentgen ionize all of the molecules in 1 cm^3 of STP air? If not, what is the fraction that it does ionize?

6. (a) As is well known, an endoatmospheric burst exhibits two distinct thermal pulse maxima soon after detonation. Why does an exoatmospheric burst exhibit only one maximum? (b) If the observed time difference between the occurrence of the double thermal pulse is 1 second, what is the corresponding yield in KT?

7. Ill-informed professional nay-sayers sometimes state that nuclear hardness efforts are generally futile. They argue that if the nuclear burst is close enough to be of consequence, everything will be annihilated, while if it is remote, it will have no effect of consequence. There is no middle ground, hence nothing can be done.

One of the many counterexamples is that of hardening aircraft avionics (aircraft electronics). Aircraft structures can usually withstand a static overpressure of 2 psi, without any critical structural problems. By using Eq. (4.19) (viz., $\Delta p = 3300(Y/R^3) + 192(Y/R^3)^{\frac{1}{2}}$), where Δp is the static over-pressure (psi), Y is the yield (MT), and R is the range (in 1000s of ft):

(a) Show that for $\Delta p = 2$ psi, the safe operating yield-range relationship is given by $Y \leqslant 8.08.10^{-14}R^3$($R$ in ft).
(b) What is the corresponding range to maintain $\Delta p \leqslant 2$ psi for a yield of 1 MT?
(c) At the range determined in (b) for 1 MT, what is the neutron fluence, using Eq. (4.55)?
(d) Similarly, as in (c), what is the total gamma dose, using Eq. (4.56)?
(e) Also, as in (c), what is the thermal fluence, using Eq. (4.41) and Fig. 4.14, for a 10 mile visibility, typical of an urban area on a clear day, at a range of 4 miles?
(f) What are some simple measures to harden the avionics against (c), (d), and (e)?
(g) If the yield-range equation of (a), $Y \leqslant 8.08 \cdot 10^{-14}R^3$, is used to eliminate the yield in Eq. (4.55), (4.56), and (4.41), compute the maximum safe radiation levels of neutron fluence, total gamma dose, and thermal energy, and the corresponding ranges.
(h) Explain what is meant by the maximum safe levels obtained in (g) above.

8. In freshman science, it is learned that the reflected pressure, or rate of change of momentum of mass m (e.g., for a volume of gas), of speed v, elastically reflected in a head-on collision (with a wall) is $p_r = d/dt(mv - (-mv)) = d/dt(2mv) \equiv 2p$, where p is the incident pressure, mv is its initial momentum, and $-mv$ is the reflected momentum. However, for a shock wave in air (blast wave) striking a wall head-on, the reflected pressure p_r is well known to be given by $p_r = 2p(1 + 3p/(7P_0 + p))$, where P_0 is the ambient pressure in front of the shock wave, and p is the peak pressure behind the shock wave. Show that, for a weak shock (small p), p_r approaches $2p$, as in the above freshman sound wave. For strong shocks (large p) show that $p_r \cong 8p$!

9. Cube root scaling, discussed in Section 4.5, applies to yield versus over-pressure range, among other parameters. If a 1 MT burst causes a peak overpressure change of 50 psi through the shock wave at a range of 4000 ft, what is the 50 psi range for a 1 KT burst?

10. You are the Army Program Manager in charge of developing a radio communications transceiver, and its corresponding radiation specifications. Prescribe these specifications for burst yields of 0.1–100 KT, based on:

(a) A man-pack version, where it is assumed that 3000 rads (tissue) will instantly incapacitate personnel, 500 rads (tissue) corresponds to 50 percent probability of death within 30 days, and 100 rads (tissue) will cause minimal discomfort but no interference with assigned military tasks.

(b) A truck-mounted version, where it is assumed that the truck will be destroyed by a peak static overpressure of 5 psi.

11. What are the size limitations on a "neutron" bomb?

REFERENCES

1. S. Glasstone (ed.). The Effects of Nuclear Weapons. US DOD, Defense Nuclear Agency (DNA), 1977, 730 pp. Obtained from Supt. of Documents, US Gov't. Printing Office, Washington D.C. About $15, hard bound, complete with weapons effects plastic computer in pocket inside rear cover.
2. J. J. Kalinowski and R. K. Thatcher. Transient Radiation Effects on Electronics. (TREE) Handbook, DASA 1420, 1969, nonclassified portion only quoted.
3. N. J. Rudie. Principles and Techniques of Nuclear Radiation Hardening, Vol. 1, Chap. 1, Western Periodical Co., N. Hollywood CA, 1976.
4. H. L. Brode. "Close-in Weapon Phenomena" Ann. Rev. Nucl. Sci. 8:153–202, 1969.
5. R. W. Hillendahl. Theoretical Models for Nuclear Fireballs. Lockheed Missiles and Space Div., Palo Alto, I. MSC B006750 DASA-1589, pt. A., 1965.
6. R. G. Sachs. The Dependence of Blast on Ambient Pressure and Temperature. Ballistic Research Labs. Report No. 466. May 1944.
7. Time Magazine, p. 30. August 24, 1981: L. W. McNaught, "Nuclear Weapons and Their Effects," Brassey's Defense Publishers, p. 20ff, 1983.
8. L. Ruby. "The Need for a Redefinition of the Sievert." Nucl. News, pp. 74–76, May 1982.
9. Hillendahl, op. cit.
10. H. C. Carslaw and J. C. Jaeger. Conduction of Heat in Solids. Oxford Press, p. 112, Ex. (i), 1959.
11. Glasstone (ed.), op. cit. p. 289, Table 740.
12. Glasstone (ed.), op. cit. p. 347.
13. G. W. Johnson and C. E. Violet. "Phenomenology of Contained Nuclear Explosions." U. C. Livermore Labs., UCRL-5124, Rev. I., (Uncl.), Dec. 1958.
14. J. C. Mark. "Research, Development and Production." Bull. At. Sci., pp. 45–56, Mar. 1983.
15. H. L. Mayer and F. Richey. "Eyeburn Damage Calculation For An Exoatmospheric Nuclear Event." J.O.S.A. 54(5):678–683, May 1964.
16. Glasstone (ed.), op. cit. Plotted from Table 3.07, p. 82.
17. R. E. Crawford, C. J. Higgins, and E. H. Bultmann. "The Air Force Manual For Design and Analysis of Hardened Structures." AFWL-TR-74-102, AFWL Kirtland AFB, NM 87117, Oct. 1974, p. 60.
18. R. G. Sachs. "The Dependence of Blast on Ambient Pressure and Temperature." Report No. 466, Ballistic Research Laboratories, May 1944.
19. T. G. Engel, M. Kristiansen, M. C. Baker, and L. L. Hatfield. "Surface-Discharge Switch Design: The Critical Factor." IEEE Trans. Elect. Dev., pp. 740–744, April, 1991.

CHAPTER 5
NEUTRON DAMAGE EFFECTS

5.1 INTRODUCTION

The transient effects of radiation on electronic systems (TREES) can be compartmentalized for discussion. Coming from five causes, four of which can emanate from a nuclear device, they are: (a) neutrons, (b) gamma and x-rays, (c) gamma- and x-ray rate, (d) galactic cosmic rays, and (e) electromagnetic pulse (EMP). Chapters 5 through 9 will discuss these causes and effects in the above order. First, consider neutron sources used for component simulation testing purposes, a very important requirement for hardness testing of components for system use.

5.2 FISSION NEUTRON SOURCES

To simulate the fission neutron component of the nuclear detonation, the most prevalent and readily available sources of fission neutrons are nuclear reactors. Their prevalence is due simply to the fact that, besides power reactors, there are appreciable numbers of research reactors operating throughout the United States. There are many reactor types, but those used most frequently for these purposes are the pulsed reactors, which better simulate the neutron output from a nuclear detonation than steady state reactors. Two prototype reactors much in use are the TRIGA and the Godiva types described below.

The TRIGA, a training, research, isotope-production reactor, is a light water-cooled, zirconium-hydride-moderated reactor.[1] Through suitable modifications, it can be operated in the steady state, as well as in the pulsed mode. The fuel is 20 percent ^{235}U-enriched ^{238}U (natural mined uranium), which is impregnated into the ZrH moderator. The reactor core is constructed of an array of ZrH moderator-enriched uranium fuel elements, which makes for a large safety margin of negative reactivity temperature coefficient (about -15 cents per degree centigrade). For pulsed operation, the moderator-fuel elements are replaced by graphite-uranium instead of ZrH-uranium. The peak pulse neutron flux is about $6 \cdot 10^{17}$ fast neutrons per square centimeter per second. This is comparable to that available in the Godiva reactor, described later. For a peak pulse power of $8 \cdot 10^4$ MW, or a peak pulse energy of about 800 MW corresponding to a core temperature rise of 900°F, the pulse half-width is about 8 ms. A cooling system of 50 kW capacity would allow a 4-h pulse period, corresponding to a mean reactor steady-state output power of about 50 kW. Electronic components are irradiated by placing them at various distances from the core, through openings, or breeches, in the system.

The Godiva reactor consisted of 93 percent enriched uranium, forming a bare sphere weighing about 50 kg, sandwich-sliced into three pieces. Each piece, or section, is suspended, so that the whole is juxtaposed for easy retraction of the two outer slices, toward or away from the central slice. Neutron pulses are induced by a fuel rod driven into a diametrical channel through the central slice, after the assembly is first brought together. Pulse peak powers from $10-10^4$ MW, or $10^{17}-10^{20}$ neutrons per square centimeter per second can be obtained, with pulse widths of 0.05–5.0 ms. One reason for the attractiveness of the Godiva reactor is that it is unmoderated, so that its neutron energy spectrum (relative neutron yield versus neutron energy) closely approximates the fission neutron portion of a nuclear weapon. What neutron moderation there is comes about from the external operating environment. Electronic components are irradiated by placing them near the core, as in the TRIGA. In certain of the Godiva reactors, the diametrical channel (glory hole) can be used to receive components that are not too large for maximum exposure to both neutrons and concomitant gamma rays. The neutron spectra of the TRIGA and Godiva reactors are shown in Fig. 5.1.

Besides the foregoing reactor neutron sources, there are a number of specialized neutron sources. One such is the accelerator-activated reactor, where a slightly subcritical reactor assembly is so constructed that it produces neutron pulses when its central region is bombarded by charged particles fired from an accelerator. The neutron spectrum of all these sources is essentially a fission spectrum. Such a spectrum of neutron energies increases in level from thermal energies to a peak at about 0.85 MeV, with an average energy at about 2 MeV, decreasing to 10 MeV and diminishing rapidly beyond. Figure 4.20, minus its 14 MeV peak, depicts a nominal fission spectrum.

5.3 FUSION NEUTRON SOURCES

Fusion neutron sources are characterized by an almost discrete monoenergetic neutron energy spectrum. An example is that of 14 MeV neutrons produced from the deuterium-tritium (D-T) nuclear fusion reaction. Fusion reactions are highly exothermic reactions usually between

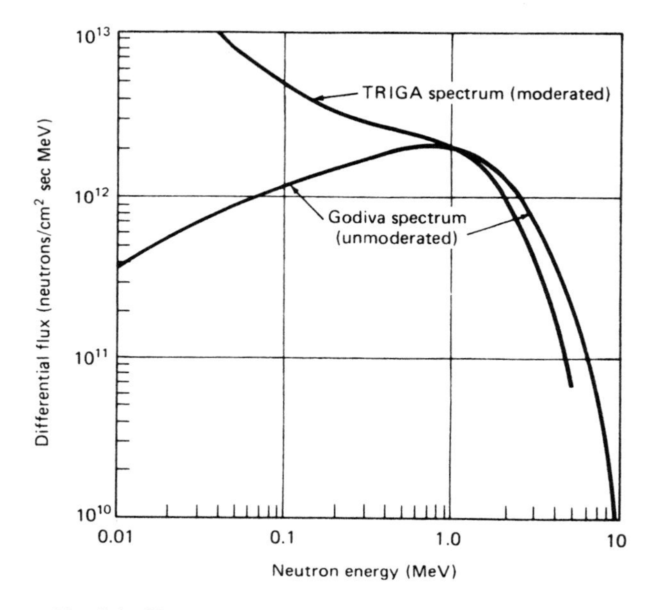

Fig. 5.1 Neutron spectra of Godiva and TRIGA reactors.

FISSION
Discovered more than 50 years ago, the fission process has revolutionized western civilization, in terms of its dual nature of producing nuclear energy through the use of nuclear reactors or nuclear detonations. The principal fissionable nuclei of interest are isotopes of uranium and plutonium. When neutrons interact with them under certain conditions, the latter are fissioned into a menagerie of energetic particles. For each fissionable nucleus, these include two fission fragment nuclei, whose masses nearly total the mass of the original fissionable nucleus. Additionally, two to five prompt neutrons are born (one of which is used to maintain the chain reaction by fissioning a neighboring fissionable nucleus), as well as gamma rays, fast electrons, delayed neutrons, neutrinos, etc. About 200 MeV is released per fission, mainly through the kinetic energy of the fission fragments. This is equivalent to about 30,000 kilowatt hours, or 25 tons of TNT, per gram of fissionable material.

light nuclei, such as the isotopes of hydrogen, viz., deutrium and tritium, as well as isotopes of helium. Readily available commercial generators of such neutrons consist mainly of D_2 gas that is ionized and thence accelerated to bombard a tritiated target to react with the tritium thereon to produce 14-MeV neutrons. Such systems can be packaged within a single container. A typical commercial generator produces pulses consisting of a peak neutron flux of about 10^6 neutrons per square centimeter per second, with a 0.1-μs half-amplitude and a triangular shape.

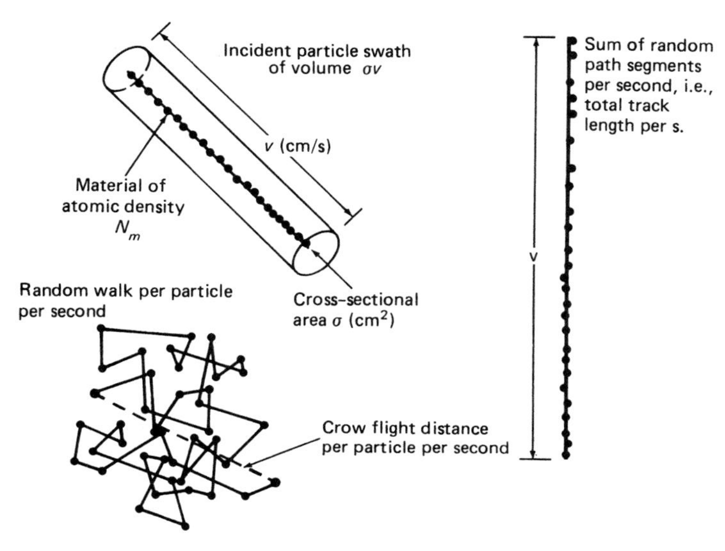

Fig. 5.2 Illustrations showing the number of collisions per second, per incident particle, in material of atomic number density N_m.

The Cockroft-Walton accelerator is another example of a neutron generator, for producing light nuclei-induced reactions. This accelerator generates the required very high beam accelerating potential through use of a series of massive diode-capacitor voltage doublers.[22] This accelerator is the early forerunner of many modern linear accelerators in use today. As with other systems, electronic components to be exposed to these neutrons are placed in the proper vicinity to interact maximally with the resulting fluxes or fluences emanating from them.

5.4 CROSS SECTIONS AND FLUENCE

At this juncture, a pedagogical digression is made to acquaint the reader with the notion of cross section and to introduce the idea of fluence and flux. An interaction cross section or merely a cross section, $\sigma(E)$, can be shown to be a dimensioned (cm^2) interaction probability, such that for an incident particle of average speed $v(E)$, with a projected area, or cross section, $\sigma(E)$, moving in a material medium with which the incident particle interacts, $\sigma(E)v(E)$ is the volume of the equivalent cylindrical swath of crossection $\sigma(E)$, and length v that the incident particle makes in 1 s, as seen in Fig. 5.2. $N_m = \rho N_0/A$ is the number density of the atoms or particles that make up the material medium with which the incident particle interacts (particles per cm^3), ρ is the density of the medium, A is the mass number (atomic or molecular weight) of the medium, and N_0 is Avogadro's number. Then $\sigma(E)v(E)N_m$ is the number of interactions or collisions of the type characterized by $\sigma(E)$ per second, per incident particle of energy E, that take place in the material medium, If $n_0(E)$ is the number density of incident particles per cm^3 per second of energy E, then $n_0(E)\sigma(E)v(E)N_m$ is the number of interactions per cm^3 per second at energy E characterized by σ. Then

$$\int n_0(E)\sigma(E)v(E)N_m t_0 dE = N_m \int \sigma(E)\varphi(E)dE \tag{5.1}$$

is the total number of the above type interactions per cm^3, where the incident spectral fluence $\varphi(E) = n_0(E)v(E)t_0$ is the number of particles per cm^2 of energy E incident on and penetrating the medium from a pulse of width t_0 seconds. Fluence, Φ, is the total number of particles penetrating a given cm^2 from all directions, and all energies. Here,

$$\Phi = \int \varphi(E)\, dE \qquad \text{(particles per cm}^2\text{)} \tag{5.2}$$

and the integration is taken over all pertinent energies, An associated parameter, called the *spectral flux*, is $\phi(E)$, where

$$\phi(E) = n_0(E)v(E) \qquad \text{(particles per cm}^2\cdot\text{s of energy } E\text{)} \tag{5.3}$$

and its integral over energy $\phi = \int \phi(E)dE$ gives the total flux. It is seen that

$$\phi = \Phi/t_0 \qquad \text{(particles per cm}^2\cdot\text{s)} \tag{5.4}$$

An archaic form of neutron fluence, from nuclear reactor theory, is $n\overline{v}t$ (neutrons/cm^2), where n is the number of neutrons per cm^3, $\overline{v}$ is their velocity(cm/s), and t(sec) is the incident neutron pulse length. $n\overline{v}$ is seen to be the corresponding neutron flux.

The mean distance that the incident particle travels between collisions is called its *mean free path* λ. It is easily shown that $\lambda = (N_m\sigma)^{-1}$(cm), and its corresponding lifetime between collisions, $t_c = (N_m\sigma v)^{-1} = \lambda/v$, where v is its average speed in the medium. The corresponding particle

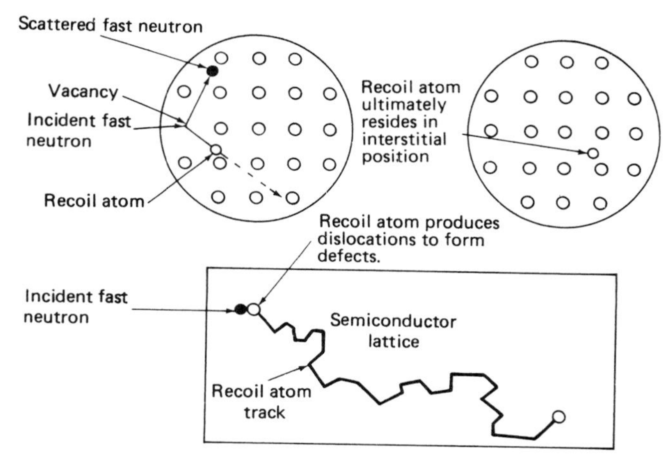

Fig. 5.3 Formation of an interstitial-vacancy pair by fast neutrons.[11]

current density is the net number of particles (in minus out) penetrating a given cm^2 of area per second, each weighted by the cosine of their angle with respect to the area normal.

5.5 NEUTRON DISPLACEMENT DAMAGE

One of the principal causes of radiation damage to electronic systems is due to neutrons. Because neutrons are relatively heavy (1840 times heavier than electrons) uncharged particles, instead of merely ionizing atoms or molecules, they collide with the lattice atoms of the semiconductor, dislodging or displacing whole atoms from their lattice sites to cause them to take up interstitial positions within the crystal. This results in disruption or distortion of the local lattice structure. The former site of the now displaced atom in the lattice is called a *vacancy*. The displaced atom is called an *interstitial*, and the interstitial-vacancy pair is called a *Frenkel defect*, as shown in Fig. 5.3. If the energy of the incident neutron is sufficiently large, it can impart enough energy to the displaced atom for it, in turn, to displace other atoms in the lattice. For highly energetic incident neutrons, this phenomenon can proceed in a cascade manner to form defects within the lattice structure.

Finally, all displaced atoms lose sufficient energy to achieve thermal equilibrium within the lattice. Some of these atoms have slipped back into isolated vacancies to reconstitute that local lattice structure. Certain of these atoms have combined with dopant or impurity atoms to produce stable defects. These defects are electrically inactive an so do not constitute recombination or trapping centers. However, mobile vacancies can combine with impurity atoms, donor atoms, or other vacancies present to produce room temperature stable defects. These defects, also called *defect complexes*, are effective recombination or trapping centers, and produce resistivity changes as discussed in Section 5.13.

Prior to their participation in defect complex formation, the vacancies are mobile and are a potent recombination agent for trapping minority carriers. For very short times following the onset of the neutron pulse, these vacancies account in large measure for greatly reducing the minority carrier lifetime, and thus the common emitter gain. This is discussed further in Section 5.14, in that this heavy loss in gain phase quickly decellerates as rapid annealing takes place within the device, and the gain rapidly assumes its asymptotic degraded value.

5.6 DISPLACEMENT KINEMATICS

Assume that an incident neutron transfers enough energy to a silicon atom to dislodge (displace) it from its lattice site. This atom is also called a *primary knock-on atom* (PKA), *primary*, *primary recoil atom*, *recoil atom*, *displaced*, or *interstitial atom*. The mean displacement energy of a lattice site is 25 eV. That is, at least 25 eV of energy must be supplied by the incident neutron to displace a silicon atom from its site in the lattice. The primary atoms can, in turn, through collisions with other lattice atoms, displace them to produce secondaries in cascade fashion. In view of the paucity of experimental demonstrations extant, a Monte Carlo program with collision kinematics corresponding to known cross section data was employed to simulate the displacement cascades and provide insight into the production of atomic displacements due to the incident neutrons.[2,3]

A major conclusion of the Monte Carlo simulation is that there is a very large kinematical variety in the cascades of displacement damage produced.[3] As will be discussed in detail, displacement damage in silicon semiconductors results in significant decreases in carrier concentration, mobility, and minority carrier lifetime.

The neutron damage depends on the total number and spatial distribution of the displaced recoil atoms in the lattice. The number of displaced atoms per cm^3, N_d, is given by[4,49]

$$N_d = N \int_{E_{th}}^{E_{max}} dE \int_{E_d}^{E_{rmax}} dE_r N_s(E_r) \sigma_D(E, E_r) \varphi_n(E) \tag{5.5}$$

where

N is the number of lattice atoms per cm^3
$\varphi_n(E)dE$ is the neutron spectral fluence in the energy increment dE (cm^{-2} eV)
$\sigma_D(E, E_r)$ is the collision cross section per recoil atom by which an incident neutron of energy E produces a recoil atom of energy E_r (cm^2)
$N_S(E_r)$ is the number of displaced atoms per unit energy interval that result per recoil atom expending all its energy E_r
E_{th} is the threshold energy required by the incident neutron to dislodge an atom from its site in the lattice
E_d is the head-on (maximum) energy that can be transferred to a recoil atom by an incident neutron of energy E_{th}
$E_{r\max}$ is the head-on (maximum) energy received by a recoil atom from an incident neutron of energy E
$E_{\max}$ is the maximum available incident neutron energy

For expository reasons, Eq. (5.5) can be rewritten as

$$N_d = \int_{E_{th}}^{E_{\max}} \varphi_n(E)\, dE \int_{E_d}^{E_{r\max}} N_s(E_r)\, dE_r/\lambda(E, E_r) \tag{5.6}$$

where the mean free path $\lambda(E, E_r) = [N\sigma_D(E, E_r)]^{-1}$. The differential $dE_r/\lambda(E, E_r)$ can be construed as the probability per centimeter of path, per unit initial energy, that an incident neutron of energy E produces a primary recoil atom of energy E_r within dE_r. Hence, $N_s(E_r)dE_r/\lambda(E, E_r)$ is the expected number of defects produced of energy E_r in dE_r, per incident neutron of energy E per cm^3. The above integration over E_r, multiplied by $\varphi_n(E)$, and the result integrated over energy E, yields the sought-after N_d. Equation (5.6), which is a double integral, can be factored into the product of two single integrals to good approximation, viz.

$$N_d \cong \int_{E_{th}}^{E_{\max}} \varphi_n(E)dE \cdot \int_{E_d}^{E_{r\max}} N_s(E_r)dE_r/\lambda, \tag{5.7}$$

since $\lambda = (N\sigma_D)^{-1}$, the neutron mean free path for colliding with lattice atoms is sufficiently slowly varying over incident neutron energies of concern. It is also factored from the integrals to give

$$N_d \cong \Phi_n \cdot \bar{n}_s/\lambda \tag{5.8}$$

with the neutron fluence $\Phi_n = \int_{E_{th}}^{E_{\max}} \varphi_n(E)dE$, and the mean number of primaries produced per primary recoil, viz., $\bar{n}_s = \int_{E_d}^{E_{r\max}} N_s(E_r)dE_r$.

Equation (5.8) merely asserts that the total number of displaced atoms per cm^3, N_d, is given by the product of the number of primaries produced per cm^3, Φ_n/λ, with the mean number of displaced atoms produced per primary atom, $\bar{n}_s$.

The collisions of the incident neutrons with the lattice atoms to form primary recoil atoms produces them homogeneously throughout the chip. This is because the neutron mean free path is very large, relative to the silicon chip dimensions. $\lambda \cong 4\,\text{cm}$ for a 1 MeV neutron in silicon, which is very much larger than any nominal chip dimension. Hence, the neutron collides with random incidence and position within the chip, thus ensuring this homogeneity.

Each primary atom then displaces secondary lattice atoms in its track, losing about 20 eV per displacement. Ultimately, its remaining kinetic energy becomes too small to cause further displacements.

It is presently generally believed that any formation of clusters of displaced atoms is minimal, and sparse along the recoil atom track. The few that do form consist of only a few displaced atoms and are positioned mainly at the ends of the recoil tracks.[51,52,61]

To estimate the level of displaced atoms being expelled from lattice positions, the microscopic total cross section for 1 MeV neutrons in silicon is $\sigma_D \simeq 5$ barns[50]. The corresponding neutron mean free path $\lambda = (N\sigma_D)^{-1} = (5 \cdot 10^{22} \times 5 \cdot 10^{-24})^{-1} = 4\,\text{cm}$. This can be interpreted as the probability that an individual incident neutron produces one primary per cm is equal to 1/4. The number of primaries then produced per cm^3 is $\Phi_n/4$. Assume that each primary yields about 500 total displacements,[52] so that $\bar{n}_s \cong 500$. Then, for a fluence of $10^{13}\,\text{cm}^{-2}$, Eq. (5.8) yields the mean number of displaced atoms as $N_d = (10^{13}/4) \cdot 500 \cong 1.25 \cdot 10^{15}$ per cm^3. The fraction of damage is then $1.25 \cdot 10^{15}/5 \cdot 10^{22}$ so that $2.5 \cdot 10^{-8}$ of the lattice atoms per cm^3 are so displaced. This seemingly small fraction is enough to cause appreciable damage, evidenced in the degradation of the macroscopic properties of the silicon. This includes increased resistivity, as discussed in Section 5.13, as well as decreased minority carrier lifetime, discussed in Section 5.9.

Contrary to earlier thought, it is now generally believed that the probability of formation of incident neutron-induced defect clusters, according to the Gossick[62,64] model, is infinitesimal. The Gossick model postulates a core of ionized defects, ostensibly composed of trapped majority carriers, and so possessing a net charge. The core is surrounded by a roughly concentric region from which the majority carriers have been removed. This region, then, has a net charge opposite to that of the core, due to the ionized dopant atoms therein. Therefore, the gross charge of the Gossick cluster is neutral, thus forming a local intrinsic region. This cluster would then act as a "giant" recombination region for minority carriers. This would cause an increase in the bulk resistivity, due to carrier removal, and the presence of a volume of intrinsic material in and around the cluster.

An argument against the formation of such clusters is derived from an examination of Fig. 5.4A, which depicts a linear variation of the damage constant ratio K_x/K_n versus nonionizing energy loss. Generically, K_n is called the damage constant in the neutron-induced common emitter gain degradation (Messenger-Spratt) equation, derived in Section 5.10. Here, a simple variant of that equation viz., $\Delta(1/\beta) = K_x\Phi_x$ will be employed, where K_x represents the damage constant for a particle type other than a neutron, such as .or electrons, protons, and photons, as shown in Fig. 5.4A. It is seen therein that the damage constant ratios for these particles correlate well over four orders of magnitude of nonionizing energy loss.[61] This means that the bipolar gain degradation per unit fluence for a given nonionizing energy loss, $K_x = \Delta(1/\beta)/\Phi_x$, is essentially the same for the interaction of each of these particles with the silicon material, regardless of the diversity of their individual interaction processes.[61] However, it is well known from physical principles that the above particles, other than the 1 MeV equivalent neutrons (Section 10.3), do not possess the requisite mass and energy to produce clusters of the Gossick type.[62] It can then be interpreted that the linear variation (dashed line) in Fig. 5.4A represents gain degradation due to isolated point defects of charged primary recoil atoms, as opposed to clusters, for all of the particles above, including the 1 MeV neutrons. Cluster models, such as the Gossick model, predict that if such clusters exist, they would be much more effective in producing gain degradation than isolated point defects. That is, a corresponding "cluster line" would appear instead of, and be displaced much higher than, the "isolated defect line" in Fig. 5.4A.

The preceding argument is strengthened by comparing the number of primary recoil atoms produced per cm^3 per unit incident neutron fluence (in silicon) with the carrier removal rate, $-\lim_{\Phi_n \to 0} n'(\Phi_n)$. The latter is the number of carriers removed per cm^3 per unit incident fluence, as discussed in Section 5.13. From the foregoing, the number of primary recoil atoms produced per cm^3 per unit fluence is $N\sigma_D \simeq 1/4$, which is of the order of unity. The silicon carrier removal rate is also well known to be of the order of unity (Section 11.2). Therefore, for each primary recoil atom produced per cm^3 per unit incident neutron fluence, only one carrier is removed per cm^3 per unit incident fluence. This implies that there are far too few excess carriers available to be captured to form a

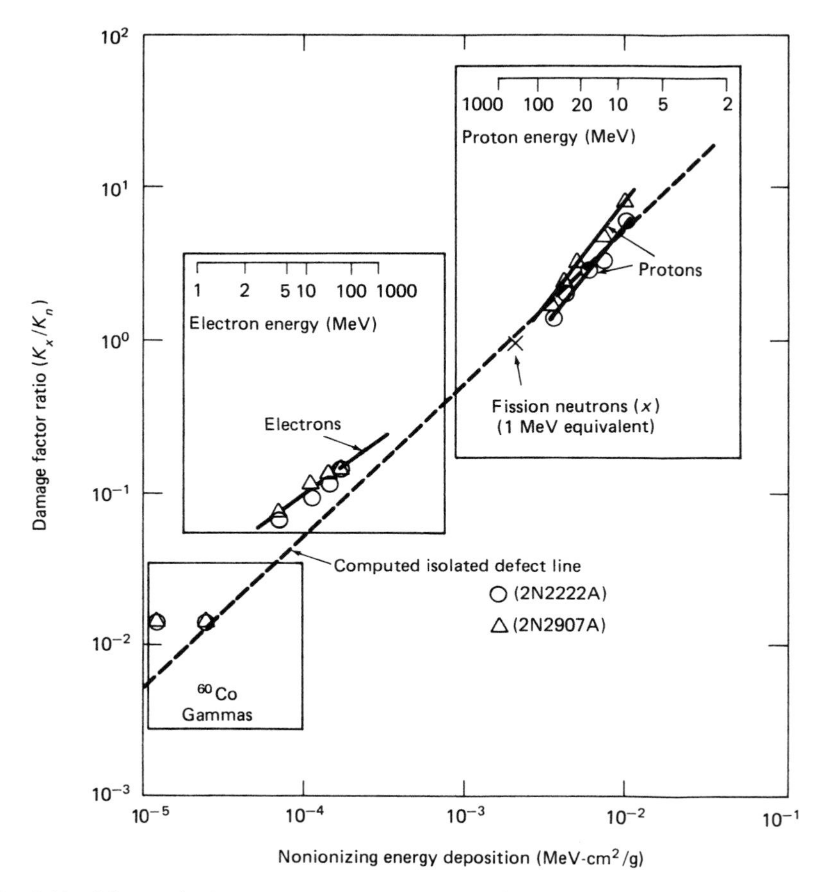

Fig. 5.4A Measured damage coefficient ratios (K_x/K_n) for electrons, protons, ^{60}Co gammas, and neutrons versus their corresponding computed nonionizing energy deposition (stopping power).[61] Figure © 1988 by The IEEE.

Gossick-type cluster. A simple geometrical argument asserts that at least four trapped carriers per cluster would be needed (Problem 5.1). Also, mutual repulsion of charge considerations would mitigate against the collection of like-charge defects to form the cluster core, subsequent to the capture of the very first majority carrier. Again, the carrier removal rate would have to be very much larger than that experimentally measured to reflect formation of these clusters. Further arguments against the formation of Gossick-type clusters can be made.[3]

Results[3] indicate that, at most, a wide variety of "spike"-type clusters can be formed near the end of the range of the primary recoil atom track, each consisting of only a few uncharged primary recoil atoms, as in Fig. 5.4B. These "spikes" do not possess the trap or recombination capabilities ascribed to the Gossick cluster model.

5.7 NEUTRON-INDUCED IONIZATION

As mentioned earlier, neutrons are capable of causing ionization in the lattice by indirect processes, even though their major influence is through damage by displacement. Because neutrons are uncharged, they cannot interact electrically with charged particles to ionize them. They nevertheless produce ionization through secondary processes, such as (a) neutron collisions that produce recoil atoms or ions, which in turn produce ionization if they are sufficiently energetic; (b) neutron collisions that excite atomic nuclei, which de-excite by emitting gamma rays that can ionize; (c) neutron collisions where the neutron is absorbed by the target atomic nucleus, which in turn emits a charged particle, such as in the (n, α) and (n, p) reactions. An alpha (α) particle is an He^{++}ion (i.e., a helium nucleus). In silicon, the (n, α) reaction corresponds to $n + Si \rightarrow {}^{25}_{12}Mg + \alpha$, while the (n, p) reaction corresponds to $n + Si \rightarrow {}^{28}_{13}Al + p$. Figure 5.4C depicts the cross sections for these reactions as a function of the incident neutron energy;[56] and (d) neutron-induced fission of trace amounts of uranium and thorium found in semiconductor chip and package materials, which produces energetic heavy fission fragment ions that cause single event upset (SEU), discussed in Section 8.9.

Such charged secondary particles have short ranges, and they interact strongly to deposit their energy within the material of origin, in contrast to exiting the material to deposit their energy elsewhere. In hydrogeneous materials, neutron collisions can dislodge protons (hydrogen nuclei) from their atomic nuclei, which protons can ionize readily. In most other materials, only very high-energy neutrons cause sufficient secondary ionization to be of any consequence.

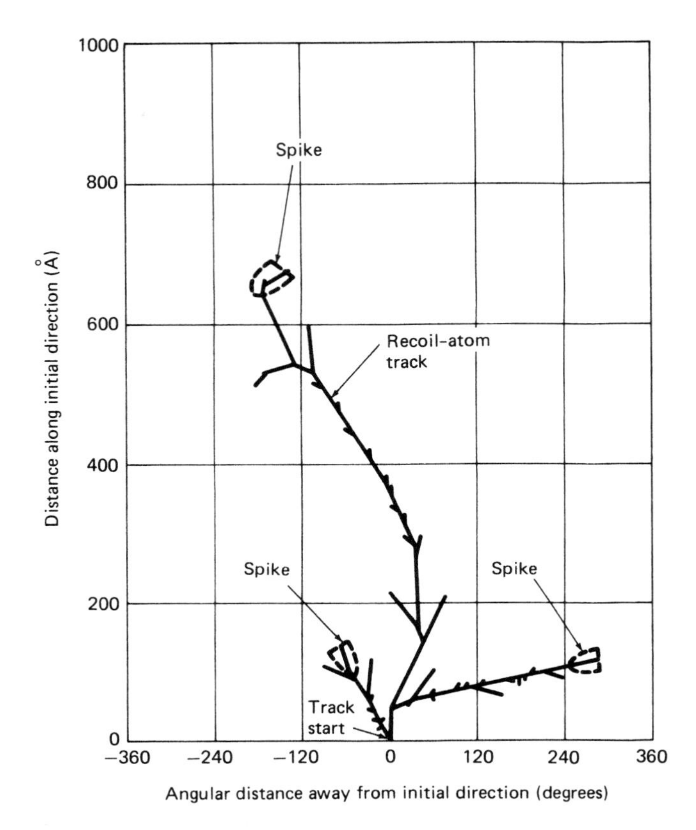

Fig. 5.4B Monte Carlo results showing recoil atom tracks terminating in "spike"-type clusters.[63]

An expression for neutron ionization stopping power, i.e., the incremental energy loss dE per incremental neutron path dx, due solely to ionization by the aforementioned processes (except fission) is given by

$$dE/dx = \sum_j (dE/dx)_j = \sum_j N \int_0^{E_{\max}} E\sigma_j(E)(1 - f_j(E))\, dE \qquad (5.9)$$

where N is the number of silicon atoms per cubic centimeter and $\sigma_j(E)$ is the cross section per neutron collision for imparting energy E to a recoil,

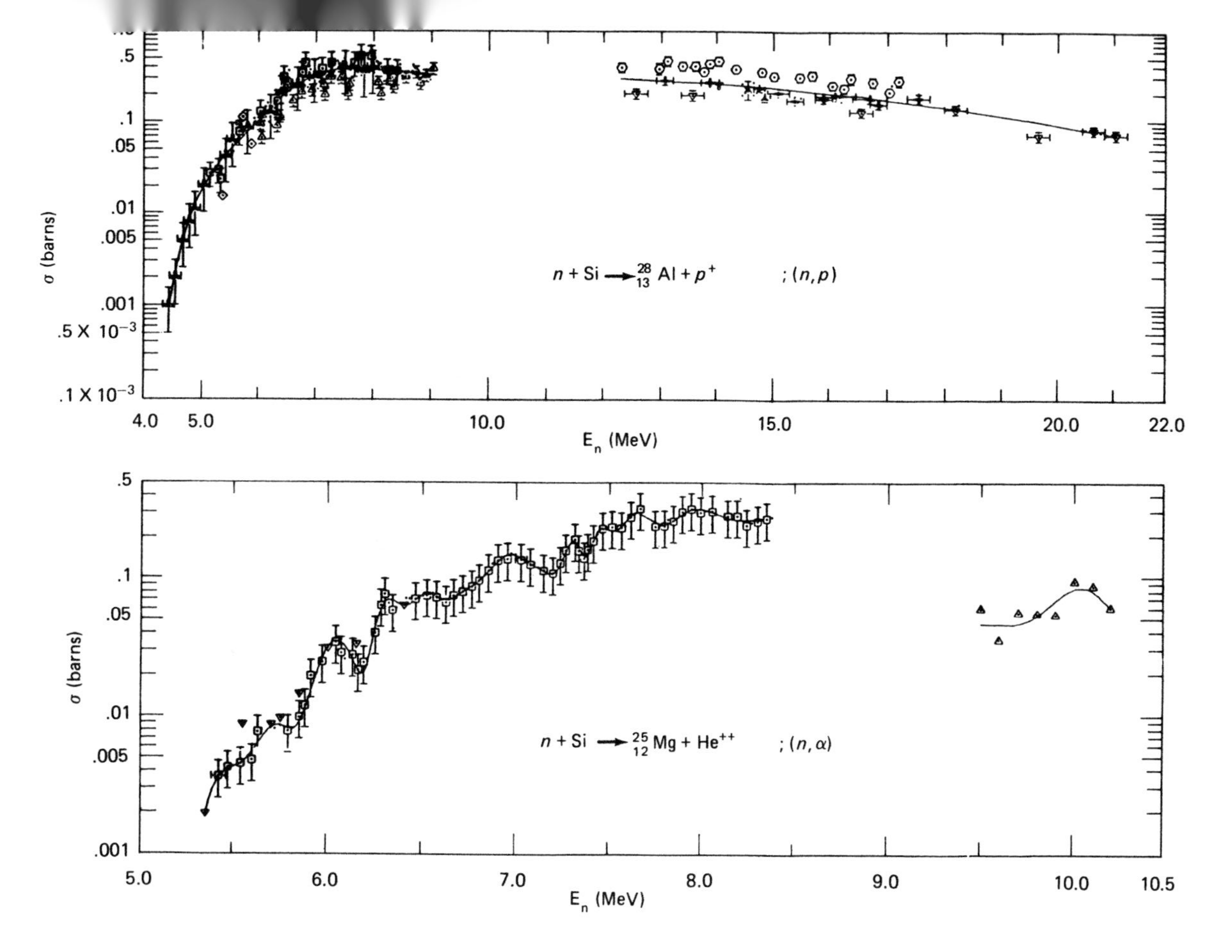

Fig. 5.4C Neutron cross sections in silicon for (n, p) and (n, α) reactions[56] (1 barn = 10^{-24} cm^2).

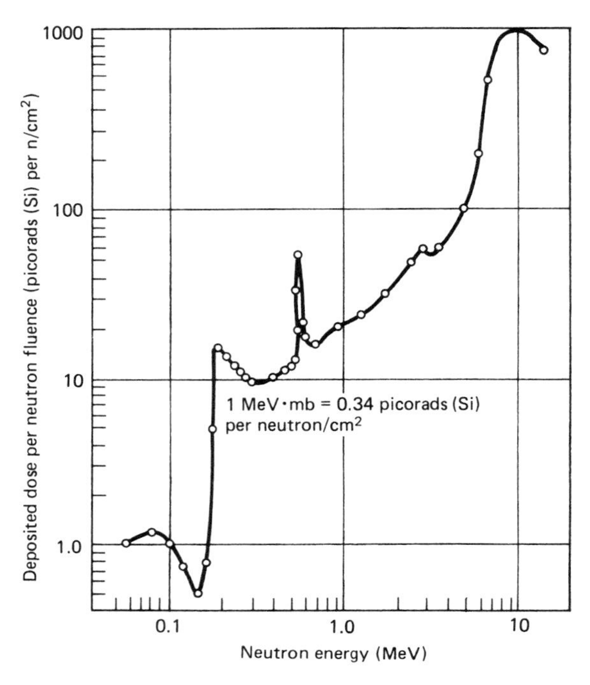

Fig. 5.5 Neutron-induced ionization in silicon.[7] Figure © 1966 by the IEEE.

or resultant, particle corresponding to the jth type ionization process. $f_j(E)$ is the fraction of particle energy consumed in jth type non-ionizing atomic displacement processes, so that $1 - f_j(E)$ is the fraction of the resultant particle energy consumed in ionization in jth type processes. The summation in Eq. (5.9) is over all processes and concomitant particle types. $f_j(E)$ for the case of recoil atoms has been calculated, and is in excellent agreement with experiment.[5,6] Also, an experimental determination of the gross neutron-induced ionization over all such processes in silicon, as a function of absorbed neutron energy, is shown in Fig. 5.5.[7] As can be seen from that figure, the neutron-induced ionization energy dependence is complicated, reflecting the complexities of the individual ionization processes.

5.8 NEUTRON SHIELDING

Because the neutron is an uncharged particle, relatively heavy, whose absorption cross section in most materials is roughly inversely proportional to its speed, shielding measures must exploit these properties. Then, in principle, the problem of shielding against neutrons devolves to (a) slowing the fast (high-energy), penetrating neutrons, (b) capturing or absorbing the thus slowed neutrons, and (c) attenuating the concomitant gamma radiation due to neutron activation of the shielding material itself that accompanies this standard approach. The latter requirement is discussed in Section 6.4, where gamma-ray shielding is examined.

The method for slowing fast, high-energy neutrons is to induce them to undergo collisions with material where their energy loss per collision is a maximum. This results in the neutrons losing most of their energy in a minimum number of collisions, and so using the minimum amount of penetration distance. Therefore, it is desired that a neutron collide with a particle whose mass is as close to that of the neutron mass as possible. This results in the maximum transfer of energy from the neutron to the material, as opposed to the other extreme of a neutron colliding with an immovable object (infinite mass). In the latter case, the neutron would merely bounce away from such a collision, retaining all the (kinetic) energy it had prior to the collision. Hence, hydrogen nuclei (protons), either free or part of a hydrogen atom, are prime candidates for maximum transfer of energy per collision, so that hydrogeneous material (solid or liquid) would be satisfactory for a practical shield. Prime examples are water, paraffin, any C_xH_y, or other hydrogen-rich hydrocarbon. Once a neutron has slowed down, i.e., thermalized to room temperature of $kT = 0.026\,\text{eV}$, then its absorption cross section is quite large for most substances. Boron, cadmium, and other common elements have very high thermal neutron absorption cross sections and so make very good thermal neutron absorbers. Note that boron (mainly ^{10}B), being in column III of the periodic table, is often used as a *p*-type dopant in many semiconductors. Because of its high absorption cross section for thermal neutrons, it can pose difficulties with regard to radiation damage if so used. Specifically, ^{10}B upon absorbing a neutron almost immediately is transformed into ^{11}B. The latter rapidly decays to produce, among other particles, ninety-four 0.478-MeV gamma rays per 100 ^{10}B neutron absorptions. The nuclear reaction is complicated, being given by

$$^{10}_{5}B + ^{1}_{0}n \rightarrow ^{11}_{5}B \rightarrow ^{7}_{3}Li^{*} + ^{4}_{2}He; \quad ^{7}_{3}Li^{*} \rightarrow ^{7}_{3}Li + \gamma\,(0.478\text{ MeV}) \tag{5.10}$$

$^{7}Li^{*}$ is a lithium nucleus in an excited state. When it emits the 0.478 MeV gamma ray, it becomes stable ^{7}Li. Hence, besides the gamma rays from the nuclear reaction, the energetic lithium and alpha particles can produce displacement as well as ionization damage in the semiconductor material.

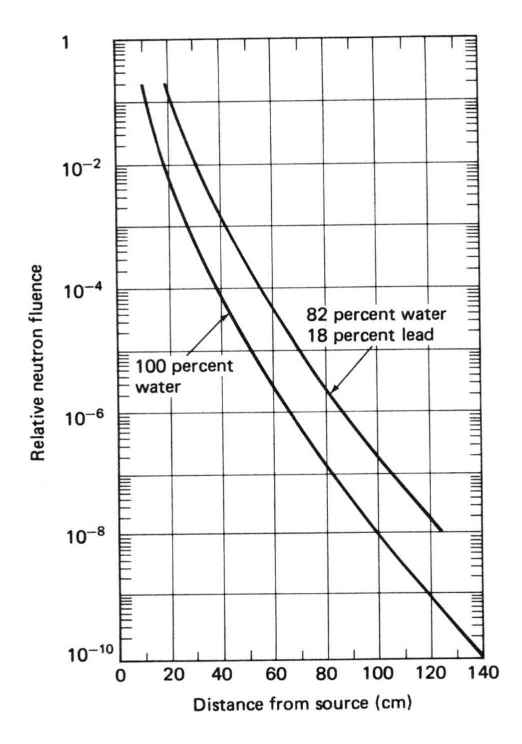

Fig. 5.6 Neutron fluence attenuation from a point fission source in water and lead shielding material.[8]

From a practical viewpoint, borates can be added to concrete, cadmium salts can be mixed with many construction materials, etc. Judicious choices of materials must be made to construct a shield to balance the three factors: adequately slowing the neutrons down, adequately absorbing them, and simultaneously producing a minimum amount of gamma rays from neutron activation of the shielding materials by neutron capture-gamma emitting (n, γ) reactions.

Figure 5.6 portrays a neutron fluence attenuation curve for a point fission source of neutrons in a shield consisting of a mixture of 82 percent water and 18 percent lead (atomic concentration), as well as one for 100 percent water. The lead, by virtue of its high atomic number, Z, i.e., relatively high concentration of electrons compared with lighter atoms, is used to shield against gamma rays by inducing them to collide with the lead electrons to cause the gamma rays to lose energy through the compton effect. This is discussed in detail in Chapter 6. As can be seen by comparing the two curves in Fig. 5.6, lead plays a minor role in interacting with and slowing the neutrons, because of its large mass, so that its displacement of water then results in a poorer neutron shield.

5.9 MINORITY-CARRIER LIFETIME/NEUTRON-DAMAGE CONSTANTS

As is well known, bipolar transistor operation can be construed as the physical and electrical behavior of majority and minority carriers, the latter from their injection into the base at the emitter-base junction, their transport through the base en route to the collector, and their retrieval, as majority carriers, by the collector, across the base-collector junction. The principal damage by incident neutrons on bipolar devices is their effect on the reduction of the minority carrier lifetime, manifesting itself principally in a reduction of the common emitter current gain. As discussed in Section 1.9, and the development given later in this section, the minority carrier lifetime is synonymous with the steady-state recombination time. Minority carrier lifetime is perhaps a misnomer in the sense that both majority and minority carriers are of necessity recombining at the same rate. A better descriptor might be carrier recombination lifetime. It has been established that nuclear radiation in general causes changes in majority and minority carrier lifetime, conductivity, and surface recombination velocity. With the exception of majority carrier lifetime, the operational severity of these parameters as they affect the usefulness of a bipolar transistor is in the order given.

Fast neutrons—those with energies between about 10 keV to approximately 10 meV, as well as gamma rays and electrons—produce changes in conductivity and minority carrier lifetime, τ, due to the same basic mechanism. All these incident particles can cause defects, which for neutrons are assumed to be vacancy-interstitial pairs, and the resulting room temperature stable defects in the lattice. The specific deleterious phenomenon is that defects introduce energy states, which correspond to actual physical recombination sites created within the lattice.

The different types of defects are known to correspond to different kinds of donor- and acceptor-like sites. Of course, these recombination sites are the agents that remove the minority carriers in their transport through the base, thus reducing the minority carrier lifetime.

This degradation of the minority carrier lifetime can be represented by a well-substantiated relation, which is a rate-balance equation written in terms of inverse lifetimes. For general time-dependent fluence $\Phi(t)$, it is

$$1/\tau = 1/\tau_i + \Phi(t)/K;\ t \leqslant t_p; \quad 1/\tau = 1/\tau_i + \Phi(t_p)/K;\ t > t_p \tag{5.11}$$

where τ_i is the unirradiated value of the minority carrier lifetime, $\Phi(t)$ is the incident fluence, t_p is the pulse duration time measured from its onset, and $K(\text{s} \cdot \text{cm}^{-2})$ is called the damage constant. For times following t_p, $\Phi(t > t_p) \equiv \Phi(t_p)$.

That is, τ does not then revert to τ_i above after the cessation of the pulse, since the corresponding radiation-induced damage has lessened τ permanently, except for its partial restoration due to annealing processes, discussed in Section 5.14 and 6.15. Usually, the epoch of concern is after the cessation of the radiation, so that the right most Eq. (5.11) is most frequently used.

Equation (5.11) is a general relation in that $\Phi(t)$ can be the proton, electron, or neutron fluence. Order of magnitude, early values of K for n- or p-type silicon[4] for 10 MeV protons are $2 \cdot 10^5$ or $5 \cdot 10^5$, and for 1 MeV electrons are 10^8 or $5 \cdot 10^9$. For 1 MeV neutrons, Fig. 5.10 provides corresponding K values versus resistivity for low and high injection levels, for both n- and p-type silicon. Refined values of K versus energy are given in Table 13.3.

The Shockley-Read model is appropriate to determine the dependence of the minority carrier lifetime on resistivity and current injection level, as discussed below. All this is accounted for in Eq. (5.11) through judicious choices of values of the damage constant K. Beyond this, K is a parameter corresponding to a particular type of damage process and dependent on material parameters. A large amount of experimental data has been amassed to determine K from populations of various semiconductor devices. As an aside, such data usually conform to a log-normal distribution of K with respect to population number.

Two important relationships for the damage constant K will be obtained involving neutron fluence, Φ_n, but valid for other fluences as well. The first is that K is a functional of the spectral neutron fluence $\varphi(E)$, which ultimately results in an expression for the common emitter current gain degradation as a function of K and Φ_n. The second is the utilization of the Shockley-Read model to derive a relation between K and the carrier concentrations, from which its dependence on resistivity and other material parameters follows.[24]

If $\sigma_c(E)$ is the cross section of neutrons of energy E for the production of recoil atoms, assumed here mainly for the creation of recombination centers in the silicon medium as discussed, then N_c is given by

$$N_c = \int_{0.01\,\text{MeV}}^{\infty} n(E)\sigma_c(E)v(E)Nt_0 dE = N \int_{0.01\,\text{MeV}}^{\infty} \sigma_c(E)\varphi(E)dE \tag{5.12}$$

which is the number of recombination centers produced per cubic centimeter by the neutrons, where:

$n(E)$ is the number density of neutrons per cm^3 of energy E
N is the number density of silicon stoms per cm^3
$v(E)$ is the average speed of neutrons of energy E
t_0 is the nominal pulse width of the incident neutron fluence
$\varphi(E) = n(E)v(E)t_0$ is the spectral neutron fluence (cm$^{-2}\cdot$MeV)
$\Phi_n = \int_{0.01\text{ MeV}}^{\infty} \varphi(E)dE$ is the neutron fluence (cm^{-2})

Let σ_m be the cross section for the absorption of minority carriers by recombination centers; then with Eq. (5.12),

$$\sigma_m v_e N_c = \sigma_m v_e N \int_{.01\text{ MeV}}^{\infty} \sigma_c(E)\varphi(E)\,dE \tag{5.13}$$

where v_e is the average speed of minority carriers, and gives the number of minority carriers captured per second by recombination centers created in the silicon due to the neutron fluence. Then the recombination rate of minority carriers, τ^{-1}, is given by the unirradiated rate, τ_i^{-1}, plus that due to the incident neutron fluence, as given by Eq. (5.11), so a rate equation can be expressed as

$$1/\tau = 1/\tau_i + \sigma_m v_e N \int_{0.01\text{ MeV}}^{\infty} \sigma_c(E)\varphi(E)dE \tag{5.14}$$

Comparing Eq. (5.11) with (5.14), it is seen that the damage constant can be given by

$$\begin{aligned} K^{-1} &= \sigma_m v_e N \int_{.01\text{ MeV}}^{\infty} \sigma_c(E)\varphi(E)\,dE/\Phi_n \\ &= \sigma_m v_e N \int_{.01\text{ MeV}}^{\infty} \sigma_c(E)\varphi(E)\,dE \Big/ \int_{.01\text{ MeV}}^{\infty} \varphi(E)\,dE \qquad (5.15) \\ &= \sigma_m v_e N \langle \sigma_c \rangle \end{aligned}$$

$\langle \sigma_c \rangle$ is the fluence averaged neutron cross section, identifiable from the right-most Eqs. (5.15), for the production of recombination centers. Equivalent cross section formulations are given in Section 10.5. Now consider a monoenergetic beam of neutrons, all at energy E_n incident on the lattice. Then the spectral neutron fluence can be written as $\varphi(E) = \Phi_0\delta(E - E_n)$, where $\delta(x)$ is the well-known delta function, and Φ_0 is a suitable amplitude. Inserting this into Eq. (5.15), or simply that $\langle \sigma_c \rangle = \sigma_c(E_n)$ for monoenergetic neutrons of energy E_n, yields

$$K^{-1} = \sigma_m v_e N \sigma_c(E_n) \tag{5.16}$$

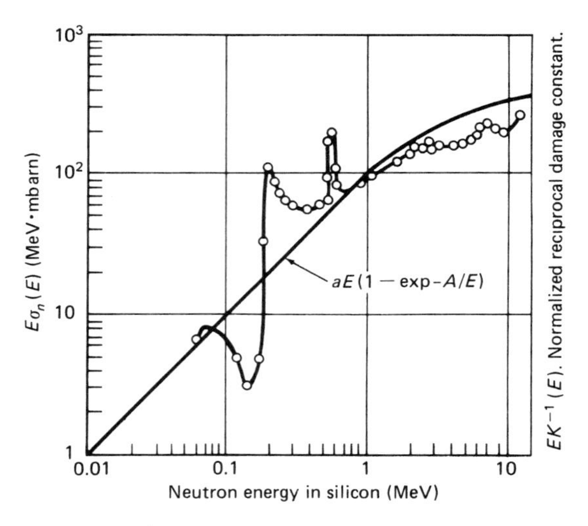

Fig. 5.7A Neutron energy (kerma) consumed by atomic displacement processes in silicon and normalized reciprocal damage constant versus incident neutron energy. Curve is normalized to 1 MeV ≐ 94 MeV · millibarns (ASTM standard). Figure © 1966 by the IEEE.

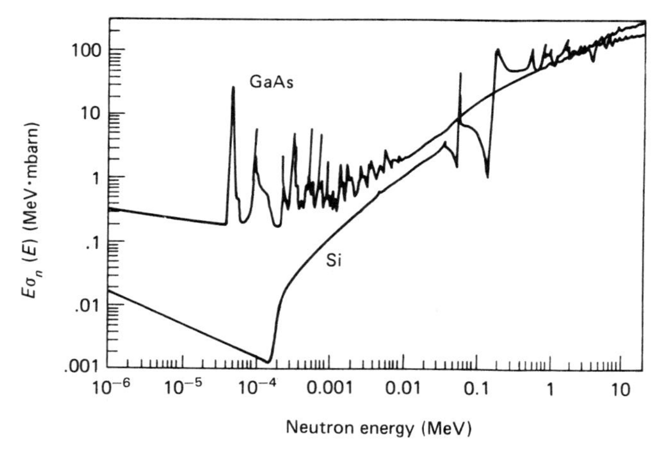

Fig. 5.7B GaAs and Si displacement kermas.[50] Figure © 1987 by the IEEE.

The cross section $\sigma_c(E_n) \equiv \sigma_n(E)$ is well known experimentally for semiconductors of interest, and is fitted by[45,46]

$$E\sigma_n(E) = a(1 - \exp - A/E)E \tag{5.17}$$

and normalized so that 94 MeV · millibarns of $E\sigma_n(E)$ kerma (Section 10.2) corresponds to 1 MeV of neutron energy, where $a \cong 102\ (\text{MeV})^{-1}$

and $A \cong 3.1$ MeV. This fit, superposed on the experimental data is shown in Fig. 5.7A. The shape of the curve of the experimental data is quite similar to that of Fig. 5.5. This cross section, $\sigma_n(E)$, corresponds to that part of the energy of the neutron that is apportioned to atomic displacement processes, as opposed to the energy that goes into neutron ionization processes.

A number of models are available to analyze the effect of defects on the minority-carrier lifetime. One of these treats the defect as a potential well, and its effects on the minority-carrier lifetime can be obtained amid a plethora of approximations.[9] Another model, used here, is the multilevel approach, where the many actual trapping levels are lumped into three representative energy levels. The resulting minority-carrier kinetics proceeds from there.[13] The damage constant K can be computed using such a three-level model, by assuming that the minority-carrier lifetime $\tau \ll \tau_i$, so that Eq. (5.11) can be rewritten as

$$1/\tau = 1/\tau_1 + 1/\tau_2 + 1/\tau_3 \cong 1/\tau_i + \Phi_n/K \tag{5.18}$$

where τ_1, τ_2 and τ_3 are the lifetimes that characterize a three-recombination-center energy level model. Obtaining expressions for τ_1 and τ_2 separately in a straightforward manner from the Shockley-Read recombination model is now shown.[14] The following is a paraphrase of the Shockley-Read model as given by Phillips.[10] This model assumes that the density of recombination centers, $N_c \ll n, p$, i.e., the steady-state, nonequilibrium concentrations of minority and majority carriers. An example of such nonequilibrium conditions occurs where an external excitation of some kind is providing carrier injection at a constant rate. Also, with respect to energy, all the recombination centers lie at a single energy level E_R in the band gap. To reiterate, this method determines τ_1, τ_2 and τ_3 one at a time, and the results are then combined. To determine the net recombination rate of minority carriers with a single recombination center energy level, the former is assumed merely to equal the capture rate, less the emission rate of carriers from a single recombination center energy level. In Section 1.5, it was shown that if the Fermi function $f(E)$ represents the probability that a state at energy E is occupied by an electron, then $1 - f(E)$ is the probability that this state is empty of an electron. Then the capture rate for electrons is given by $nC_nN_c[1 - f(E_R)]$, where C_n is the capture probability of the recombination center energy level for electrons, i.e., $C_n = \sigma_m v_e$, so that the former quantity is the expected number of electrons captured by the recombination center per cubic centimeter per second, because of the factor $1 - f(E_R)$.

The mean emission rate of electrons from recombination centers is

$e_nN_cf(E_R)$, where the emission rate for electrons is e_n. Hence the net recombination rate for electrons R_n is given by

$$R_n = nC_nN_c[1 - f(E_R)] - e_nN_cf(E_R) \tag{5.19}$$

A similar development of the net recombination rate R_p for holes can be made. Because the capture of holes is proportional to the number of empty hole states, i.e., filled electron states $f(E_R)$, and likewise the emission rate e_p of holes is proportional to the number of filled hole states, i.e., empty electron states, $1 - f(E_R)$, and C_p is the corresponding hole capture probability, then

$$R_p = pC_pN_cf(E_R) - e_pN_c[1 - f(E_R)] \tag{5.20}$$

For steady-state, nonequilibrium conditions, $R_n = R_p$, so that equating Eqs. (5.19) and (5.20) results in

$$f(E_R) = (nC_n + e_p)/[(pC_p + e_p) + (nC_n + e_n)] \tag{5.21}$$

At equilibrium, the net recombination rates of carriers vanish, i.e., $R_n - R_p = 0$.

Hence, if the Fermi level E_f were adjusted so that $E_f = E_R$, then the emission rate for electrons would equal the product of the free electron concentration n_1 by their capture probability C_n. At equilibrium then

$$e_n = n_1C_n; \qquad e_p = p_1C_p \tag{5.22}$$

In other words, when the Fermi level coincides with E_R, then $n = n_1$, and $p = p_1$, which is another statement of equilibrium conditions. Inserting Eqs. (5.22) into $R_n = R_p$, from Eqs. (5.19) and (5.20) results in

$$R_n = R_p = \frac{N_cC_nC_p(np - n_1p_1)}{(n + n_1)C_n + (p + p_1)C_p} \tag{5.23}$$

The corresponding lifetime is defined as

$$\tau = \delta n/R_n = \delta p/R_p \tag{5.24}$$

where $n = n_0 + \delta n$ and $p = p_0 + \delta p$. n_0 and p_0 are the equilibrium values of the carrier concentrations, and $\delta n = \delta p$ from charge conservation considerations. Using Eqs. (5.23) and (5.24), the minority carrier single-level lifetime can be expressed as

$$\tau = \frac{n_0 + n_1 + \delta n}{(n_0 + p_0 + \delta n)N_cC_p} + \frac{p_0 + p_1 + \delta n}{(n_0 + p_0 + \delta n)N_cC_n} \tag{5.25}$$

Now Eq. (5.25), with $N_cC_p = \tau_p^{-1}$, $N_cC_n = \tau_n^{-1}$ by definition, where τ_p and τ_n are the hole and electron mean lifetimes, respectively, can be written as

$$\tau = \frac{(n_0 + n_1 + \delta n)\tau_p + (p_0 + p_1 + \delta n)\tau_n}{n_0 + p_0 + \delta n} \tag{5.26}$$

In most cases of interest, the injection level δn and the minority carrier density will be small compared with the majority carrier density. That is, δn, $p_0 \ll n$, n_0, so that the above yields

$$\tau \simeq p_1\tau_n/n_0 + (1 + n_1/n_0)\tau_p \tag{5.27}$$

Usually, both $n_1, p_1 \ll n_0$; then Eq. (5.27) asserts that $\tau \simeq \tau_p$. This means that the "steady-state" lifetime for the capture of holes by the recombination centers N_c is the minority-carrier life lifetime. This is the reason that this recombination time is called the minority-carrier lifetime.[41] As mentioned earlier, in Section 1.11, it is a misnomer in that there is actually only one lifetime, viz., the recombination lifetime, for both minority and majority carriers necessarily recombine at the same rate. In this case, the recombination lifetime is identical to the minority-carrier lifetime. The latter is depicted in Fig. 1.13.

For the two-level portion of the model, $1/\tau = 1/\tau_1 + 1/\tau_2$, and using Eq. (5.25) yields the two-level expression for the mean lifetime, as

$$\tau^{-1} = \sum_{k=1}^{2}\left[\frac{n_0 + n_k + \delta n}{(n_0 + p_0 + \delta n)C_{pk}N_c} + \frac{p_0 + p_k + \delta n}{(n_0 + p_0 + \delta n)C_{nk}N_c}\right]^{-1} \cong \Phi_n/K \tag{5.28}$$

where C_{nk} and C_{pk} are the respective capture probabilities for energy levels E_k; $k = 1, 2$. As defined earlier, $N_c = N\langle\sigma_c\rangle\Phi_n$; then for a k level system, $R_k = N\langle\sigma_{ck}\rangle$ is the introduction rate of recombination centers per unit neutron fluence for levels for $k = 1, 2$. Rewriting Eq. (5.28) gives the sought after relation for the damage constant K as a function of carrier concentration, from which its dependence on resistivity and other material parameters of interest will be obtained. This relation is

$$K^{-1} = (2.8\cdot 10^{-8}) + \sum_{k=1}^{2}\left[\frac{n_0 + n_k + \delta n}{(n_0 + p_0 + \delta n)C_{pk}R_k} + \frac{p_0 + p_k + \delta n}{(n_0 + p_0 + \delta n)C_{nk}R_k}\right]^{-1} \tag{5.29}$$

The above corresponds to the three-level model, with two levels implied in the summation, and the additional term provides a third, mid-band, level introduced to account for the mid-band levels to provide a more accurate fit to experimental data.

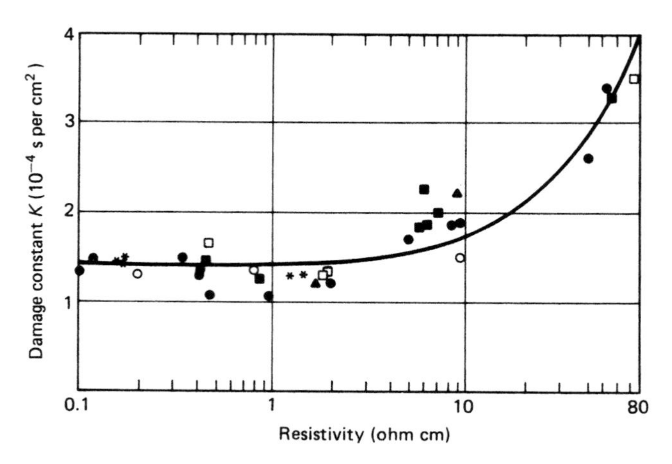

Fig. 5.8 Damage constant versus resistivity for n-type silicon.[13] Figure © 1967 by the IEEE.

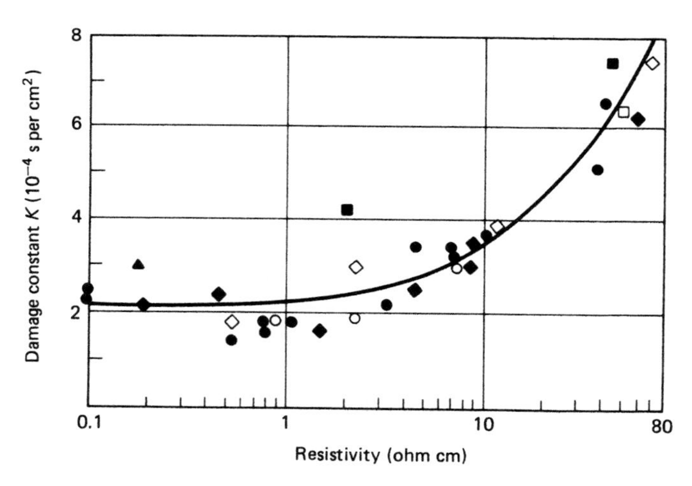

Fig. 5.9 Damage constant versus resistivity for p-type silicon.[13] Figure © 1967 by the IEEE.

As mentioned earlier, the damage constant K embodies dependencies on the resistivity of both n- and p-type semiconductors, transistor injection levels, and temperature. The damage constant as given in Eq. (5.29) is rewritten for (a) n-type and (b) p-type material at low injection levels. Let K_{ln} and K_{lp} be, respectively, values of the damage constant at low injection levels. From Eq. (5.29), in the limit small δn and p_0, and temporarily suppressing its first term,

Table 5.1 Recombination center parameters determined by least-squares fit to silicon, low-level lifetime, damage constant data.[13]

$R_1C_{p1} = 0.37 \times 10^{-6}(\text{cm}^2/\text{s})$
$R_1C_{n1} = 0.40 \times 10^{-5}$
$R_2C_{p2} = 0.68 \times 10^{-5}$
$R_2C_{n2} = 0.76 \times 10^{-6}$
$n_1 = 2.0 \times 10^{14}\,\text{cm}^{-3}$
$p_2 = 1.3 \times 10^{13}$

$$K_{ln}^{-1} = \frac{C_{p1}R_1}{1 + n_1/n_0} + \frac{C_{p2}R_2}{1 + n_2/n_0} \simeq \frac{C_{p1}R_1}{1 + n_1/n_0}; \quad C_{p2}R_2 \ll C_{p1}R_1. \tag{5.30}$$

Similarly, in the limit of small δn and n_0

$$K_{lp}^{-1} = \frac{C_{n1}R_1}{1 + p_1/p_0} + \frac{C_{n2}R_2}{1 + p_2/p_0} \simeq \frac{C_{n2}R_2}{1 + p_2/p_0}; \quad C_{n2}R_2 \gg C_{n1}R_1. \tag{5.31}$$

As an aside, for high injection levels, the limit of large δn in Eq. (5.29) yields

$$K^{-1} = \sum_{k=1}^{2} [(1/C_{pk}R_k) + (1/C_{nk}R_k)]^{-1}. \tag{5.32}$$

The Curtis experimental results for the damage constant as a function of resistivity, at low injection levels, was least-squares-fitted, using Eq. (5.30) for n-type and Eq. (5.31) for p-type silicon.[13] The results are given in Figs. 5.8 and 5.9. They show, with high confidence, that the three-level model, with one level placed in the upper half of the band gap, the second level placed in the lower half of the band gap, and the third level approximately at mid-band, adequately describes the phenomenon.

Using the parameters given in Table 5.1, which were also least-squares-fitted from the above experimental data, the damage constants K_{ln} *or* K_{lp} can be determined for any reasonable value of resistivity within experimental error. These damage constants are given by

$$K_{ln} = 10^5(1.4 + 0.086\rho + 0.0012\rho^2)/(1 + 0.038\rho) \tag{5.33}$$

$$K_{lp} = 10^5(2.1 + 0.18\rho + 0.00009\rho^2)/(1 + 0.014\rho) \tag{5.34}$$

where ρ is the material resistivity in ohm cm.

The damage constant can be shown from physical considerations to be a monotone-increasing function of resistivity, increasing rapidly for low injection levels and becoming almost independent of resistivity at very high injection levels. For low injection levels, K_{ln}^{-1} and K_{lp}^{-1} are asymptotic to

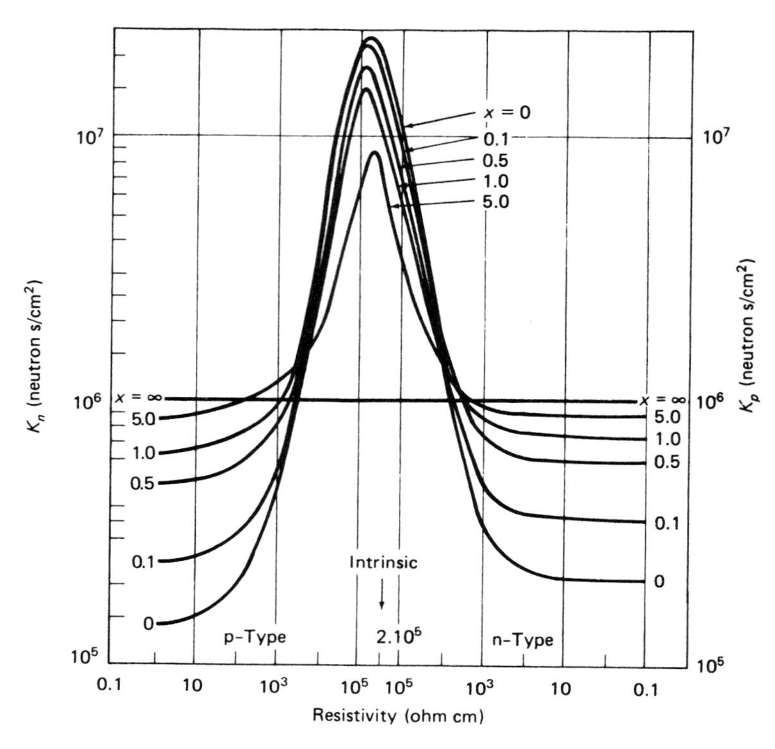

Fig. 5.10 Damage constants K_n, K_p versus resistivity for various injection ratios, $x = \delta n/n_0$, $\delta p/p_0$. Curves plotted using Table 5.1 and $\rho_n = 5 \cdot 10^{15}/n_0, \rho_p = 2.5 \cdot 10^{16}/p_0$.[14] Figure © 1967 by the IEEE.

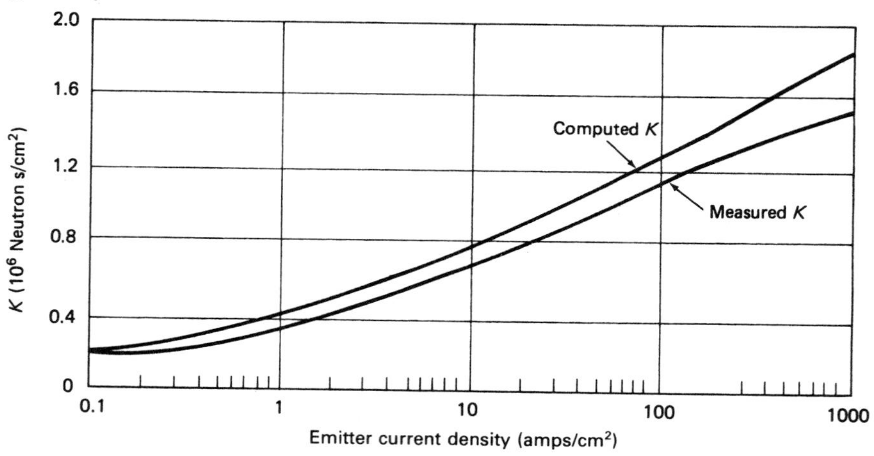

Fig. 5.11 Measured and computed K versus emitter current density.

$$\lim_{p\to 0} K_{ln}^{-1} \equiv K_{ln}^{*} \cong C_{p1}R_1 + C_{p2}R_2 \tag{5.35}$$

$$\lim_{n\to 0} K_{lp}^{-1} \equiv K_{lp}^{*} \cong C_{n1}R_1 + C_{n2}R_2 \tag{5.36}$$

In terms of injection ratios, $\delta n/n_0$, $\delta p/p_0$, the damage constant for both n- and p-type silicon as a function of resistivity comprises a one-parameter family of curves for various injection ratios and is shown in Fig. 5.10 with parameter values used from Table 5.1.

The temperature dependence of the damage constant can be introduced into Eq. (5.29) through the parameters $n_1 = AT^{3/2} \cdot \exp - (E_c - E_1)/kT$, and $p_2 = BT^{3/2} \cdot \exp - (E_v - E_2)/kT$, and the equilibrium values n_0 and p_0 are assumed temperature-independent. This is a weak restriction corresponding to the temperatures over which the dopant recombination centers are fully ionized, as most are ionized at room temperature. The resulting expressions for K_n and K_p as a function of temperature are algebraically cumbersome, but are available.[13,40]

Further descriptions of the damage constant variation with current injection, specifically with emitter current density, again follow from the multilevel model, as exemplified by Eq. (5.29). Figs. 5.12 and 5.13 include such damage constant variations.

An empirical relation can be obtained for the excess current carrier density δn as a function of the emitter current density J_E and is given by[40]

$$\delta n = (A_1 J_E^2 + A_2 J_E)/(A_3 J_E + 1) \tag{5.37}$$

with fitting coefficients A_1, A_2, and A_3. Inserting the above into Eq. (5.29) results in a quotient of two polynomials in J_E. In Fig. 5.11, the upper plot is given by

$$K = 1.23 \cdot 10^6 \left[1.68 - \frac{194 J_E^2 + 7700 J_E + 5309}{J_E^3 + 313 J_E^2 + 6280 J_E + 3410} \right] \tag{5.38}$$

The coefficients in the above expression were obtained by fitting the experimentally determined plot of K as a function of J_E, as given in Larin.[27]

For high current densities, the increase in the damage constant can be accounted for by adjusting the base resistance in the equivalent circuit. K can also change with V_{cc} and corresponding f_T, as given by

$$K/K_{VCC0} = (V_{cc}/V_{CC0})^n = f_{TVCC}/f_{TVCC0} \tag{5.39}$$

where the zero subscripted quantities refer to their values at a known nominal operating point; n is determined from experiment and usually lies between 0.2–0.5.

As discussed in Section 3.9, the common emitter gain versus collector

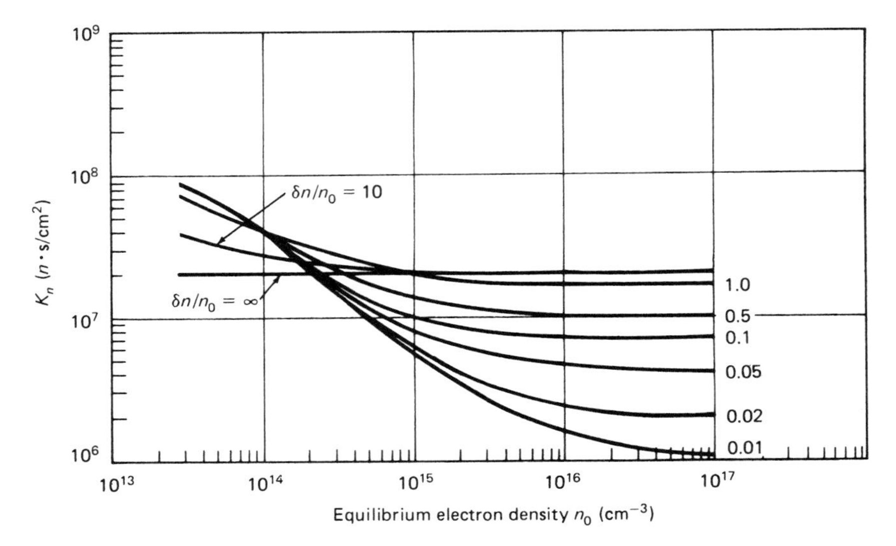

Fig. 5.12 Germanium damage constant versus equilibrium electron density for various values of injection ratio.[14] Figure © 1967 by the IEEE.

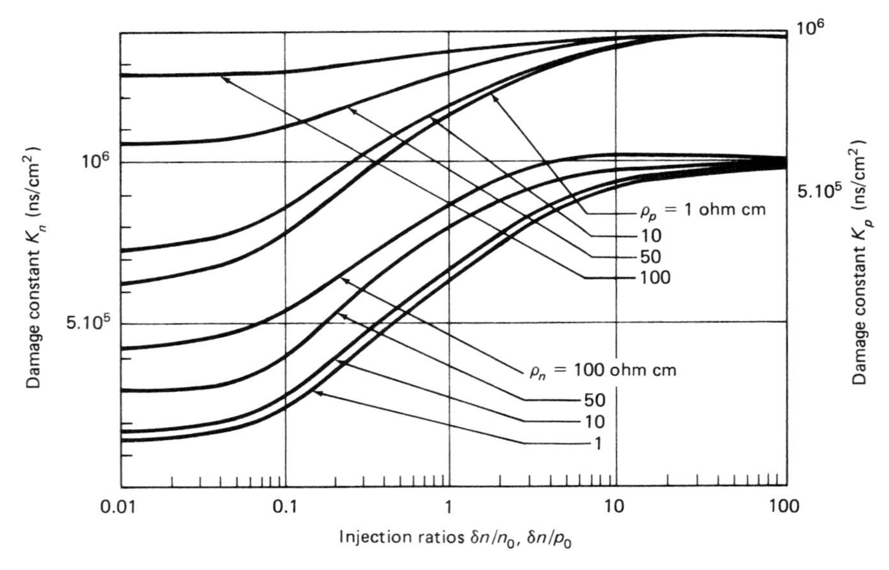

Fig. 5.13 Silicon damage constant for both n- and p-type carriers versus injection ratios δ_n/n_0, δ_n/p_0 for various resistivities.[14] Figure © 1967 by the IEEE.

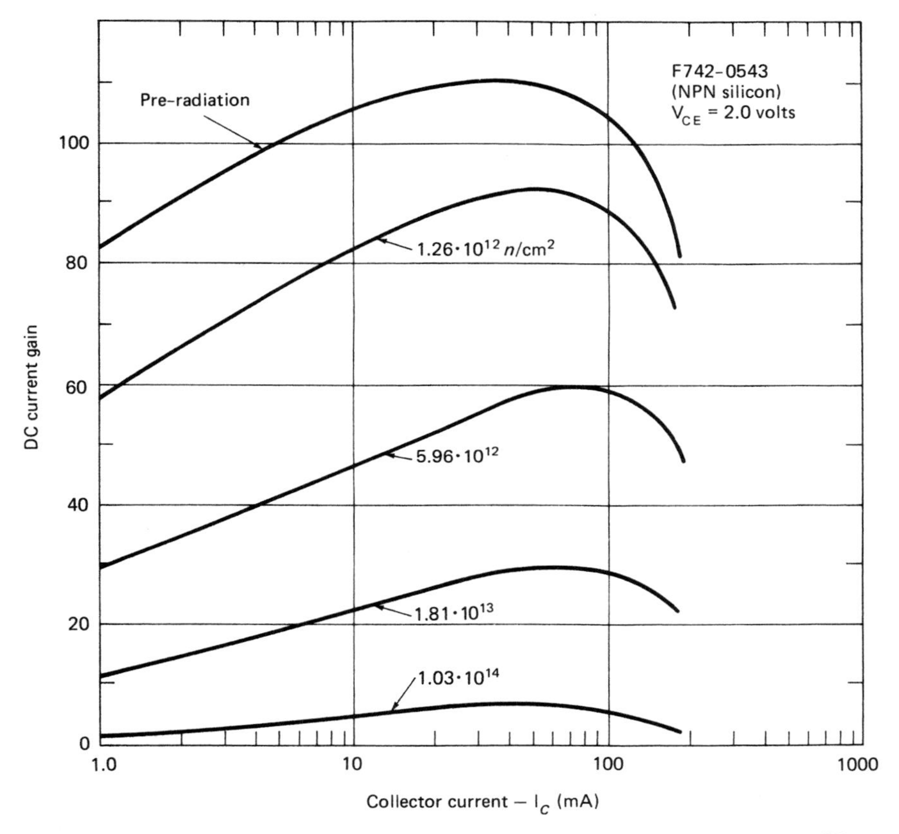

Fig. 5.13A DC current gain versus injection level as a function of neutron fluence ($E_n >$ 10KeV) for a bipolar transistor.

current attains a maximum (knee) value, as depicted in Fig. 3.24. However, this maximum is increasingly depressed with increasing neutron fluence, as shown in Fig. 5.13A.

The preceding recombination kinetics (Shockley-Read) imply low to medium excess carrier injection. For the high excess carrier injection situation, band-to-band recombination, as discussed in Section 1.10, is phonon-aided or involves a radiationless transition to emit an (Auger) electron (Section 1.9). This is because silicon is an indirect band gap semiconductor, and not direct band gap, as in gallium arsenide. The Auger recombination rate U_A is given by[58]

$$U_A = C_e(n^2p - n_o^2p_o) - C_h(p^2n - p_o^2n_o) \tag{5.39}$$

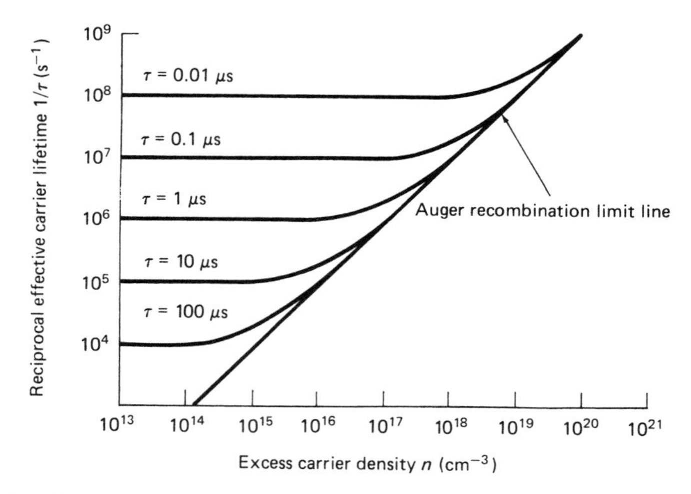

Fig. 5.13B Transition of effective carrier lifetime from low to high excess carrier densities.[59] Figure © 1979 by the IEEE.

where C_e and C_h are capture coefficients, representing the electron and hole Auger interactions, respectively. They are essentially the alpha coefficients, $\sim 10^{-31}\,\text{cm}^6/\text{s}$, in Section 1.12. Therefore, for very high injection (i.e., large n or p), the corresponding carrier recombination time, or carrier lifetime $\tau_A = n/U_A$. The preceding discussion is the antithesis of the near-equilibrium, or equilibrium carrier recombination time, as discussed in Sections 1.9 and 1.11. Typical values for these high injection lifetimes are in the neighborhood of 5 ns, whereas equilibrium lifetimes for lightly doped silicon can be up to three orders of magnitude longer.

Figure 5.13B depicts the transition of the minority carrier lifetime from low to high injection (i.e., low to high excess carrier density[59]) for the case of band-to-band recombination. The latter provides an important limitation on the deleterious effects of conductivity modulation, which would otherwise cause high dose rate-induced photocurrents to burn out monolithically diffused resistors in integrated circuits, as discussed in Section 7.10. Also, Fig. 5.13B enunciates the theoretical aspects, leading to the minority carrier lifetime behavior with impurity density shown in Fig. 1.17.

5.10 CURRENT GAIN DEGRADATION

To make the connection between the minority-carrier-lifetime degradation, Eq. (5.11), and the common emitter current gain, β, reconsider the continuity equations governing the excess minority carrier density, δn, and the corresponding current density J_n. In the low current injection approximation, from Section 1.10, the equilibrium states of these quantities are described by the simultaneous system of equations in one-dimensional geometry as[16,42]

$$D_n(x)\frac{d}{dx}\delta n(x) + \mu_n E(x)\delta n(x) = J_n(x)/e \tag{5.40}$$

$$\frac{1}{e}\frac{d}{dx}J_n(x) = \delta n(x)/\tau_n \tag{5.41}$$

The boundary conditions, from Section 3.9, are

$$\delta n(x_1) = n_p(x_1)[\exp(eV/kT) - 1], \text{ or } J_n(x_1) = J_n^{(0)} \tag{5.42}$$

$$\delta n(x_2) = 0 \tag{5.43}$$

where x_1 and x_2 are, respectively, the base-emitter and base-collector junction coordinates. For low current injection, the zero-order approximation $J_n \cong 0$ is made in Eq. (5.40). From that equation then, the internal electric field $E_i = -E$ satisfies

$$E_i = (kT/e)\, d/dx\,[\ln N_A(x)] \tag{5.44}$$

because d/dx ($\ln \delta n$) $\cong d/dx(\ln N_A)$, where N_A is the dopant density. Inserting Eq. (5.44) back into Eq. (5.40), with its right-hand side now not zero, and again using the Einstein relation between mobility and the diffusion constant gives

$$J_n(x) = (eD_n/N_A)\, d/dx\,(N_A \cdot \delta n) \tag{5.45}$$

Integrating Eqs. (5.41) and (5.45) yields the pair of coupled integral relations, i.e.

$$J_n(x) = J_n^{(0)} + e\int_{x_1}^{x} \delta n(x')\, dx'/\tau_n(x') \tag{5.46}$$

$$\delta n(x) = -\frac{1}{eN_A(x)}\int_{x}^{x_2} N_A(x')J_n(x')\, dx'/D_n(x') \tag{5.47}$$

Now, by successive substitution of δn, given by Eq. (5.47), into J_n, given by Eq. (5.46), and vice versa, the simultaneous series solution for each is obtained, that is

$$J_n(x) = J_n^{(0)} + J_n^{(1)} + \cdots \tag{5.48}$$

$$\delta n(x) = \delta n^{(0)} + \delta n^{(1)} + \cdots \tag{5.49}$$

where for the current density

$$J_n^{(0)}(x) = J_n^{(0)}; \qquad \text{constant} \tag{5.50}$$

$$J_n^{(1)} = -J_n(0)\int_{x_1}^{x} (dx'/\tau_n(x')N_A(x'))\int_{x'}^{x_2} dx'' N_A(x'')/D_n(x'') \tag{5.51}$$

and for the minority carrier density

$$\delta_n^{(0)}(x) = -(J_n^{(0)}/eN_A(x)) \int_x^{x_2} N_A(x')\,dx'/D_n(x') \tag{5.52}$$

$$\delta_n^{(1)}(x) = -(1/eN_A(x)) \int_x^{x_2} dx'(N_A(x')/D_n(x')) \int_{x_1}^{x'} \delta_n^{(0)}(x'')\,dx'' \tag{5.53}$$

From Section 3.9, the base transport factor α_T is given for arbitrary base doping distribution and narrow base width, i.e., for $W/L_n \ll 1$, by

$$\alpha_T \cong 1 - \tfrac{1}{2}(W/L_n)^2 \tag{5.54}$$

α_T generally can be written as an expansion in even powers of W/L_n as[16,42]

$$\alpha_T = 1 - u_1(W/L_n)^2 + u_2(W/L_n)^4 - \cdots \tag{5.55}$$

where the coefficients u_i are to be determined. Also, from its definition given in Section 3.9, to first order of approximation, with $J_{nr} = J_n^{(1)}$

$$\alpha_T = 1 + J_{nr}/J_n^{(0)} \cong 1 + J_n^{(1)}/J_n^{(0)} \tag{5.56}$$

Making the substitution for $J_n^{(1)}$ evaluated at the base extremities from Eq. (5.51), and also making the identification of the coefficients u_i with Eqs. (5.55) and (5.56) yields finally

$$u_1 = (D_n(x_2)\tau_n(x_2)/W^2) \int_{x_1}^{x_2} [dx'/\tau_n(x')N_A(x')] \int_{x'}^{x_2} [N_A(x'')/D_n(x'')]\,dx'' \tag{5.57}$$

$$u_2 = (D_n^2(x_2)\tau_n^2(x_2)/W^4) \int_{x_1}^{x_2} [dx'/\tau_n(x')N_A(x')] \int_{x'}^{x_2} [N_A(x'')/D_n(x'')]\,dx'' \cdot \int_{x_1}^{x''} [dy/\tau_n(y)D_n(y)] \int_{y}^{x_2} [N_A(y')/D_n(y')]\,dy' \tag{5.58}$$

Assuming that the irradiated transistor still has a reasonable amount of gain, which implies $u_2(W/L_n)^2 \ll u_1(W/L_n)^2$, then the expansion in Eq. (5.55) for α_T converges rapidly, so that the first two terms therein will suffice. Equation (5.55), with the connective between α_T and β given in Section 3.9, and that base transport is the dominant mechanism with respect to common emitter current gain, implying an emitter efficiency $\eta \simeq 1$ yields[16]

$$\beta^{-1} \simeq 1 - \alpha_T = u_1(W/L_n)^2 \tag{5.59}$$

As discussed earlier, the purview of the general minority-carrier-lifetime degradation relation

$$1/\tau = 1/\tau_i + \Phi_n/K \tag{5.60}$$

can be enlarged to include most damaging incident radiation, such as that due to protons, electrons, and gamma rays, with the proper choice of damage constant K. If Eq. (5.60) is inserted into Eq. (5.59), the result is, using $L_n^2 = D_n \tau_n$

$$1/\beta = 1/\beta_i + u_1 W^2 \Phi_n / D_n K \tag{5.61}$$

where $\beta_i^{-1} = u_1 W^2 / D_n \tau_i$.

For nonsteady-state conditions, Eq. (5.41) is modified so that the current density and the excess minority carrier density depend on time as well, i.e., $\delta n \equiv \delta n(x, t)$, $J_n \equiv J_n(x, t)$, to give

$$\frac{\partial}{\partial t} \delta n(x, t) = \frac{\partial}{\partial x} J_n(x, t)/e - \delta n(x, t)/\tau_n \tag{5.62}$$

To investigate this equation in the frequency domain for reasons that will become apparent, its Fourier transform with respect to time is taken, to yield

$$\delta \bar{n}(x, \omega)(1 + i\omega\tau_n)/\tau_n = \frac{d}{dx} \bar{J}_n(x, \omega)/e \tag{5.63}$$

where the overbar signifies the corresponding transform variable. Note that Eq. (5.63) is formally identical to Eq. (5.41) but where τ_n is replaced by $\tau_n/(1 + i\omega\tau_n)$. Similarly, because $L_n^2 = D_n \tau_n$, L_n can be replaced by $L_n/\sqrt{1 + i\omega\tau_n}$. Instead of Eq. (5.55), the expansion of the base transport factor can now be written as

$$\alpha_T = 1 - u_1 (W/L_n)^2 [1 + i\omega\tau_n] + u_2 (W/L_n)^4 [1 + i\omega\tau_n]^2 \ldots \tag{5.64}$$

Then, to within a single time constant approximation

$$\alpha_T \cong 1 + i\omega u_1 (W/L_n)^2 \tau_n = 1 + i\omega/\omega_T \tag{5.65}$$

where $\omega_T = 2\pi f_T = D_n/u_1 W^2$ is the unity gain corner frequency, tantamount to the gain-bandwidth product angular frequency of the transistor, i.e., $u_1 = D_n/\omega_T W^2$. Then inserting this u_1 into Eq. (5.61) yields the Messenger-Spratt common emitter gain degradation relation as[16,42]

$$\Delta(1/\beta) \equiv 1/\beta - 1/\beta_i = \Phi_n/K\omega_T \tag{5.66}$$

This relation holds for arbitrary base doping distribution and with both τ and D being functions of distance within the base. However, the irradiated transistor must still retain appreciable gain, as discussed.

Ramsey and Vail have shown that the emitter efficiency contribution resulting from recombination in the emitter-base field region can be related to the emitter time constant R_eC_e.[15] The angular frequency ω_T includes only effects due to the transport of the minority carriers in the base. It is easily shown that ω_T can be extended to include the emitter time constant as well.[15] As discussed in Section 3.10, the thus augmented angular frequency ω'_T from Eq. (5.65) satisfies

$$1/\omega'_T = 1/\omega_T + R_eC_e \tag{5.67}$$

The above equations hold for bipolars. For other devices, such as solar cells, their parameters, including the diffusion length and storage time, must be expressed as a function of minority carrier lifetime. Then in those cases, $\Delta(1/\tau) \equiv 1/\tau - 1/\tau_i = \Phi_n/K$ is used, and Eq. (5.29) is employed for the dependence of K on neutron fluence with regard to injection level and resistivity.

Another approach to the determination of common emitter gain degradation is to express this gain solely in terms of physical transistor parameters and then correlate the damaging effects of the incident neutron fluence, Φ_n, with one or more of these specific parameters. For illustration, consider a *pnp* transistor with the ensuing development holding for both *pnp* and *npn* transistors. The *pnp* transistor emitter current density is composed of hole minority carriers injected into the base, i_{EP}, plus electrons extracted from the base i_{EN}. Some of the hole current recombines, en route through the base, with electrons that enter through the base lead. This recombination rate can be thought of as a separate current density i_R. Now, the collector current density is made up of the unrecombined minority carrier holes and a current density i_{CO}, consisting of thermally produced holes in the base and collector. An excellent transistor would have i_{CO}, i_{EN}, and i_R all small with respect to i_{EP}. The symbolic representation of the foregoing asserts that the base current density i_B is given by

$$i_B = i_R + i_{EN} - i_{CO} \tag{5.68}$$

and emitter current density i_E is given by

$$i_E = i_{EP} + i_{EN} \cong i_{EP} \tag{5.69}$$

From Eq. (5.68)

$$\partial i_B/\partial i_{EP} = \partial i_R/\partial i_{EP} + \partial i_{EN}/\partial i_{EP} - \partial i_{CO}/\partial i_{EP} \tag{5.70}$$

The rightmost term of Eq. (5.70) vanishes because i_{CO} is independent of

i_{EP} from physical considerations. Now, the common emitter current gain, β, is to good approximation

$$\beta \cong \partial i_C/\partial i_B \cong \partial i_{EP}/\partial i_B \tag{5.71}$$

because the base contribution to the emitter current is negligible. From Eqs. (5.70) and (5.71) then

$$1/\beta \cong \partial i_R/\partial i_{EP} + \partial i_{EN}/\partial i_{EP} = \partial i_{SR}/\partial i_{EP} + \partial i_{VR}/\partial i_{EP} + \partial i_{EN}/\partial i_{EP} \tag{5.72}$$

where i_R has been apportioned into its corresponding surface current density i_{SR} and volume recombination current density i_{VR}.

For these cases of assumed low-level injection, diffusion is the principal phenomenon occurring, in contrast to any drift contributions due to internal electric fields, and is described by the steady-state, one-dimensional diffusion equation, viz., $(p_n - p_{n0})'' - (p_n - p_{n0})/L_p = 0$. Solutions comprise a family of hyperbolic functions, as discussed in Section 3.9. Such solutions for small W/L_p, with $i_{VR} = i_{EP} - i_{EN}$, are given where W is the base width, by[17,23]

$$\partial i_{VR}/\partial i_{EP} \cong i_{VR}/i_{EP} = 1 - \operatorname{sech}(W/L_b) \cong \tfrac{1}{2}(W/L_b)^2 \tag{5.73}$$

$$\partial i_{EN}/\partial i_{EP} \cong i_{EN}/i_{EP} \simeq \sigma_b W/\sigma_e L_e \tag{5.74}$$

where subscripts b and e refer to the base and emitter, respectively. The following approach is taken for the surface recombination term, $\partial i_{SR}/\partial i_{EP}$. The surface recombination hole current can be written as

$$I_{SR}/e = SA_s p \cong SA_s p_e \tag{5.75}$$

where S is the recombination velocity, A_s is the surface area for surface recombination, p is the hole density near the surface A_s, and p_e is the hole density at the emitter junction. Fick's law asserts that

$$J_p = -eD_p \nabla p_e = I_{EP}/A_e \tag{5.76}$$

where J_p is the hole current density, and A_e is the cross-sectional area of the conduction path, which is roughly the same as the emitter junction area. For one-dimensional plane geometry, integrating Eq. (5.76) over the base width yields

$$p_e = I_{EP} W/eA_e D_p \tag{5.77}$$

From Eqs. (5.72)–(5.75)

$$\partial i_{SR}/\partial i_{EP} \cong i_{SR}/i_{EP} = SA_s W/D_p A_e \tag{5.78}$$

Combining Eqs. (5.73), (5.74), and (5.78) yields the basic gain formula solely in terms of the transistor physical parameters. It is

$$1/\beta = SA_s W/D_b A_e + \sigma_b W/\sigma_e L_e + \tfrac{1}{2}(W/L_b)^2 \tag{5.79}$$

which is derived under the implication that the transistor base is homogeneously doped. The effects of a graded base transistor are discussed in Section 5.11. The first term on the right is regarded as the surface recombination term, the second is the emitter efficiency term, and the third is the volume recombination term. Notice that Eq. (5.79) is similar to Eq. (3.111), except that surface parameters are involved in the former. That is, Eq. (5.79) is an equivalent second form of the gain expression involving physical parameters only, with Eq. (3.111) being the first.

To illustrate the design applicability of Eq. (5.79), a brief investigation of how β degrades with both incident neutron radiation and case temperature of the device under consideration will be made.

For most well-designed modern transistors, the second term dominates the pre-irradiated or initial gain. In ionizing radiation environments, the first term becomes dominant with increasing ionizing radiation fluence. In the neutron environment, the third term dominates for increasing neutron fluence. For protons (e.g., the Van Allen Belts), the first and third terms are the important ones as the proton fluence increases. In an electron environment (e.g., the Van Allen Belts), like ionizing radiation, the first term is the important one with increasing electron fluence. At reasonable operating temperatures, the emitter efficiency term $\sigma_b W/\sigma_e L_e$ dominates the variation with temperature. This assumes that conductivity modulation, emitter crowding, and other high current density effects are ignored. The denominator of the emitter efficiency term is essentially independent of temperature, due to the assumed high emitter dopant density, and the numerator varies approximately inversely as the square of the temperature reflecting the corresponding mobility μ_b variation over the normal range of base conductivity σ_b for most silicon transistors. From Eq. (5.79), neglecting the first and third terms in this context, and where the temperature is in degrees K, gives

$$1/\beta(T) \simeq (T_0/T)^2/\beta_0 \tag{5.80}$$

and that $1/\beta_0 = \sigma_b W/\sigma_e L_e$. Then in terms of device case temperature changes only, Eq. (5.80) gives

$$\Delta(1/\beta)_{temp} = -\beta_0^{-1}[1 - 1/(1 + \Delta T/T_0)^2] \tag{5.81}$$

with $\Delta T = T - T_0$, where T_0 is the initial ambient temperature. From Eq. (5.66) the gain degradation for radiation is added to Eq. (5.81) to yield a total gain change due both to temperature and radiation effects,

$$\Delta(1/\beta) = \Phi_n/\omega_T K - \beta_0^{-1}[1 - 1/(1 + \Delta T/T_0)^2] \tag{5.82}$$

Note that both the effects of case temperature and radiation are summed and not differenced with respect to beta degradation, even though it is not evident from Eq. (5.82) at first glance.

For high values of fluence, the third term of Eq. (5.79), or volume recombination term, exhibits large changes with fluence, and it becomes predominant. The temperature dependence of $\Delta(1/\beta)$ then becomes rather flat, and linear superposition of fluence and temperature, as in Eq. (5.82) for modest fluence values, no longer holds. These and other ramifications are discussed in detail in Section 13.6.

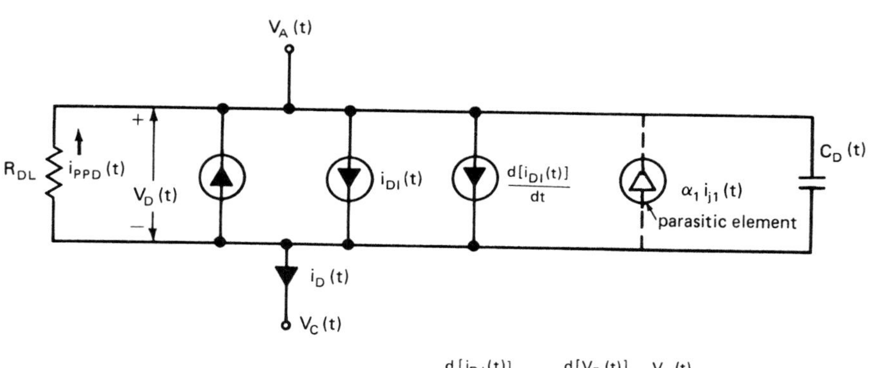

Diode node equation: $i_D(t) = i_{DI}(t) + T_D \frac{d[i_{DI}(t)]}{dt} + C_D \frac{d[V_D(t)]}{dt} + \frac{V_D(t)}{R_{DL}} - i_{PPD}(t)$

Associated equations: $i_{DI}(t) = I_S[(\exp[V_D(t)/M_D\theta]) - 1]$

$C_D(t) = C_{DO}/(1 - V_D(t)/V_{DBI})^{1/2}$

$\theta = kT/e$ (0.026 volts at 300°K)

Refining parameters:

C_{DO} = Diode zero voltage capacitor
V_{DBI} = Diode built in voltage
T_D = Diode time constant
α_1 = Current transfer coefficient
I_S = Diode reverse saturation current
M_D = Diode constant (usually $1 < M_D < 2$)
R_{DL} = Diode leakage resistance
$i_{PPD}(t)$ = Diode primary photocurrent (see Sec. 7.2)

Fig. 5.14 Radiation-inclusive Ebers-Moll large-signal diode model.

5.11 RADIATION-INCLUSIVE EBERS-MOLL MODELS

The description of how the incident neutron fluence affects bipolar devices from the electronic circuit and system viewpoint begins with the Ebers-Moll equations for the pertinent device element currents.[25,39] Figure 5.14 shows the radiation-inclusive Ebers-Moll equivalent circuit for large signals for the diode, where radiation is introduced through photocurrent production, as discussed in Section 7.2. Figure 5.15 is that for the field effect transistor, and Fig. 5.16 models the bipolar transistor. The additional loop (dotted line) on the far right of Fig. 5.14 corresponds to the parasitic junction that sometimes exists within such systems by virtue of the fabrication process. The subscripts on the definitions below are *s* for source, *g* for gate, *d* for drain, and *b* for body/substrate:

R_{gs}, R_{gd}—surface leakage resistance between the metallizations
R_s, R_d—dynamic resistance simulating dc channel characteristics
C_{gd}, C_{gs}—gate to drain and gate to source capacitance due to the overlap of gate metal and source-drain diffusions
C_g—channel capacitance. C_B, R_B, and C_c are adjusted to give C_g
D_{sb}, D_{db}—source/substrate and drain/substrate built-in junctions
I_s, I_D—source and drain radiation-induced current generators

Figure 5.15 depicts the MOSFET transistor model, which can be applied to an MOS enhancement or depletion mode, *n* or *p* channel device. The values of some circuit component and device parameters in the MOSFET equivalent circuit of Fig. 5.15 depend on the incident neutron fluence. One of the MOSFET, especially JFET, parameters most sensitive to neutron fluence is the transconductance $g_m = g_{m0} \exp - \Phi_n/K$, where g_{m0} is the preirradiation value of transconductance. *K* is a function of resistivity and is obtained from the equations and parameters discussed in Section 5.13 on carrier removal.

The corresponding model for the radiation-inclusive JFET equivalent circuit can be obtained from the 1969 *TREE Handbook*.[18] The following is a list of circuit parameters for the bipolar model:

R_C, R_B, R_E—bulk collector, base, emitter resistance
C_C, C_E—base-collector, base-emitter depletion capacitances
R_{CL}, R_{EL}—base-collector, base-emitter junction leakage resistance
M_{AC}, M_{AE}—base-collector, base-emitter junction breakdown current multipliers
I_{DC}, I_{DE}—base-collector, base-emitter junction breakdown currents
α_N, α_I—normal and inverse gain, respectively, $\alpha_N = \beta_n/(1 + \beta_0)$

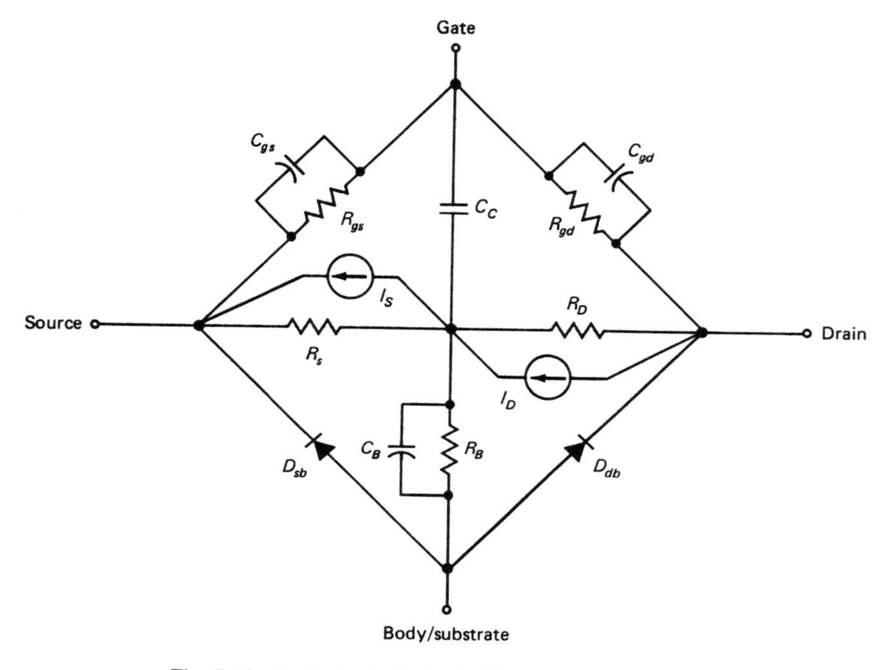

Fig. 5.15 Radiation-inclusive field effect transistor model.

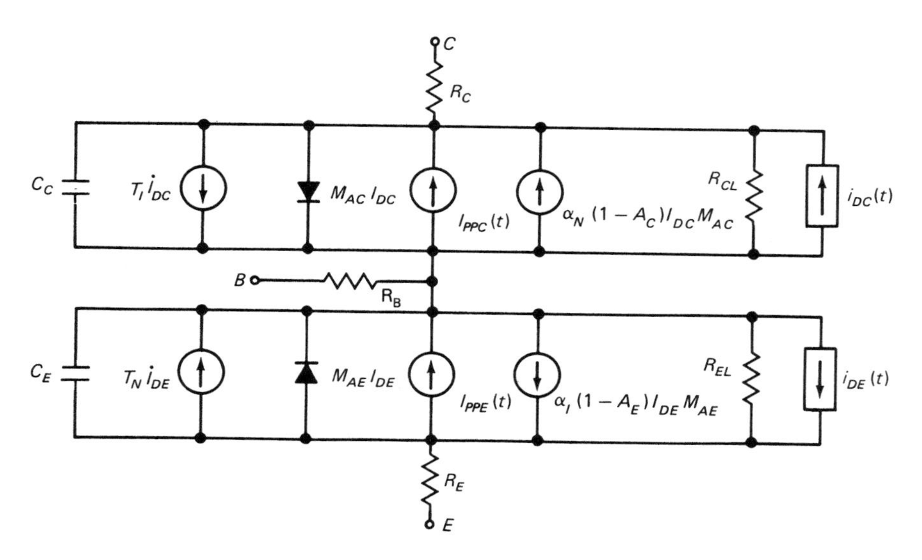

Fig. 5.16 Radiation-inclusive bipolar transistor model.

A_C, A_E—collector, emitter junction deep storage transport factors[43,44]
I_{PPC}, I_{PPE}—collector, emitter primary photocurrent
i_{DC}, i_{DE}—collector, emitter junction currents
T_I, T_N—base-collector, base-emitter junction deep storage time constants[43,44]

The value of certain elements in the bipolar transistor equivalent circuit of Fig. 5.16 depend on the neutron fluence. These include the normal gain α_N, where $\Delta(1/\alpha_N) = \Phi_n/\omega_N K$, and the inverse gain α_I, where $\Delta(1/\alpha_I) = \Phi_n/\omega_I K$.

Minority carrier damage effects are manifest at relatively low fluence levels of $10^{11} - 10^{12}$ neutrons per cm^2. Such fluences are orders of magnitude below the high fluences of $10^{14} - 10^{15}$ that deleteriously affect resistivity, carrier removal, mobility, and the diffusion parameters. These, in turn, cause changes in the values of almost every element in the above equivalent circuit.

Without any explicit statement about the internally generated electric field, other than that it is a function of position within the bipolar transistor, and that the transistor can be described as a two-port network,[19] the Ebers-Moll equations for the bipolar transistor element currents arising from both minority carrier diffusion and drift due to internal fields can be written as[19,21]

For the emitter current

$$I_E(x,t) = a_{11}[\exp(eV_{BE}/kT) - 1] + a_{12}[\exp(eV_{BC}/kT) - 1] \qquad (5.83)$$

For the collector current

$$I_C(x,t) = a_{21}[\exp(eV_{BE}/kT) - 1] + a_{22}[\exp(eV_{BC}/kT) - 1] \qquad (5.84)$$

where the coefficients a_{ij} are to be determined, and $I_B + I_C + I_E = 0$. $V_{BE}(x, t)$ and $V_{BC}(x, t)$ are the corresponding junction potentials. For notational convenience, let $\psi_{BE}(x, t) = \exp(eV_{BE}/kT) - 1$ and similarly for ψ_{BC}. A central assumption of this characterization is that the two-port network model of the transistor is quasi-linear with respect to time. This is meant in the sense that the Laplace transform counterparts of Eqs. (5.83) and (5.84) can be written, where the overbars indicate the Laplace transformed variables with respect to time, with s being the Laplace transform variable, as

$$\bar{I}_E(x,s) = \bar{a}_{11}(x,s)\bar{\psi}_{BE}(x,s) + \bar{a}_{12}(x,s)\bar{\psi}_{BC}(x,s) \qquad (5.85)$$

$$\bar{I}_C(x,s) = \bar{a}_{21}(x,s)\bar{\psi}_{BE}(x,s) + \bar{a}_{22}(x,s)\bar{\psi}_{BC}(x,s) \qquad (5.86)$$

This strictly implies that $a_{ij}(x, t)$ has a delta function behavior in time (e.g., $a_{ij}(x, t) = a_{ij}(x) \cdot \delta(t)$). Often, a_{ij} does not depend on time explicitly, but that the a_{ij} are only parameter-dependent, as in Eqs. (5.93)–(5.96). However, the a_{ij} are dependent on neutron fluence, as will be discussed.

It should be noted that, for low injection levels, where V_{BE} and V_{CE} are small, i.e., V_{BE}, $V_{CE} \ll kT/e$, then Eqs. (5.85) and (5.86) describe a linear system with an admittance matrix $\{\bar{a}_{ij}(x, s)\}$, so that within a constant factor

$$\bar{I}_E(x,s) = \bar{a}_{11}(x,s)\bar{V}_{BE}(x,s) + \bar{a}_{12}(x,s)\bar{V}_{BC}(x,s) \tag{5.87}$$

$$\bar{I}_C(x,s) = \bar{a}_{21}(x,s)\bar{V}_{BE}(x,s) + \bar{a}_{22}(x,s)\bar{V}_{BC}(x,s) \tag{5.88}$$

The above system of equations provides a good approximation for the low-injection case for minority carriers into, for example, the p-type region of an npn transistor. The method for obtaining the coefficients a_{ij} is through a corresponding one-dimensional description of minority carrier transport[19] gotten by the usual three continuity equations in the minority carrier concentration $n(x, t)$, hole concentration $p(x, t)$, and electron current density of minority carriers $J_n(x, t)$, given in Section 1.10. These equations are then combined by converting them into integral equations, which are solved simultaneously,[19,20] in a fashion similar to Eqs. (5.46)–(5.53), to obtain the excess minority carrier density $\delta n(x, t) = n_0(x, t) - n(x, 0)$. The resulting asymptotic excess minority carrier densities $\delta n(x, \infty)/n_0$ for a number of realizable base dopant distribution profiles are depicted in Figs. 5.17 and 5.18, where W is the base width.

The current densities, as found from these same equations, are multiplied by the base cross-sectional area A to obtain the currents themselves. Then taking the proper linear combinations of the currents for the particular problem at hand allows the determination of the coefficients a_{ij}, and hence the reconstitution of the Ebers-Moll equations (5.83) and (5.84).

For example, consider the exponentially graded dopant distribution, and with the Laplace transform variable $s \to i\omega$, the corresponding Fourier transforms are extracted, so that the emitter and collector currents from the aforementioned solutions at $x = W$ are given by[19]

$$\begin{aligned}\bar{I}_E(W,\omega) = {} & \frac{AD_n e\delta n_{p0}}{W}(\sigma W \coth(\sigma W) + \eta/2)\bar{\psi}_E(\omega) \\ & - \frac{AD_n e\delta n_{pW}}{W\exp(\eta/2)}(\sigma W \operatorname{csch}(\sigma W))\bar{\psi}_C(\omega)\end{aligned} \tag{5.89}$$

$$\begin{aligned}\bar{I}_C(W,\omega) = {} & -\frac{AD_n e\delta n_{p0}}{W}(\sigma W \operatorname{csch}(\sigma W))\,\bar{\psi}_E(\omega) \\ & + \frac{AD_n e\delta n_{pW}}{W}(\sigma W \coth(\sigma W) - \eta/2)\,\bar{\psi}_C(\omega)\end{aligned} \tag{5.90}$$

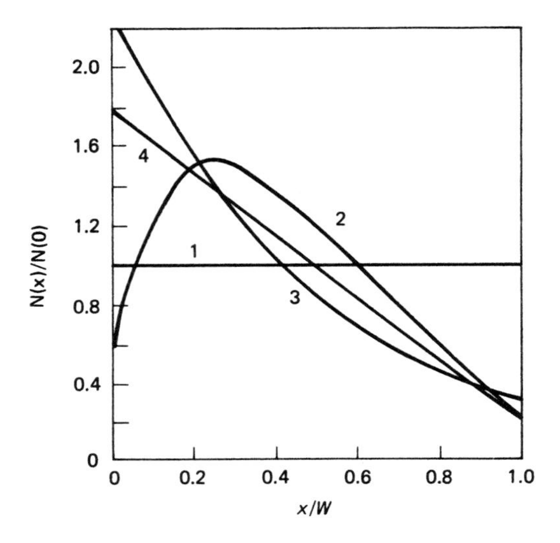

Fig. 5.17 Various base dopant diffusion profiles: 1, constant dopant distribution; 2, double-diffused distribution; 3, exponential distribution; 4, linear dopant distribution.[19]

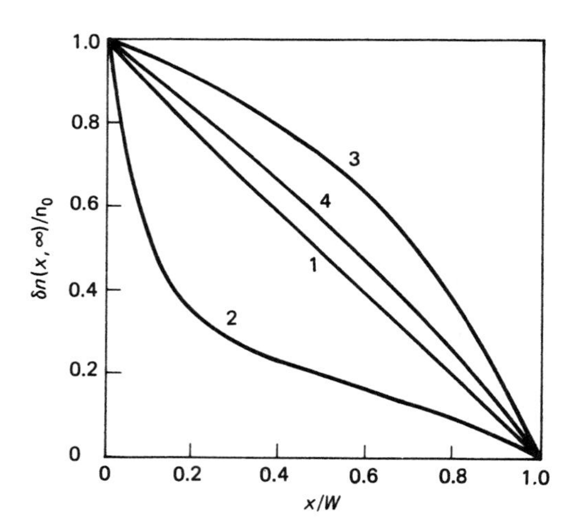

Fig. 5.18 Excess minority carrier densities $\delta n(x, \infty)$ for various base dopant diffusion profiles: 1, constant dopant distribution; 2, double-diffusion distribution; 3, exponential distribution; 4, linear dopant distribution.[19]

where $\delta\bar{n}(0,\omega) = \delta n_{p0}\bar{\psi}_E(\omega)$, $\delta\bar{n}(W,\omega) = \delta n_{pW}\bar{\psi}_C(\omega)$, $\psi_E(t) = \exp(e\phi_E/kT) - 1$ and similarly for $\psi_C(t)$. δn_{p0} and δn_{pw}, are the equilibrium counterparts. Also $\sigma = (\eta/2W)\sqrt{1 + (4W^2/\eta^2L_n^2)(1 + i\omega\tau_n)}$, and $\eta = \ln(N_{AO}/N_{AW})$ where N_{AO} and N_{AW} are the acceptor dopant concentrations at the base extremities. If the collector is not reverse biased, then the second term in Eq. (5.90) contains a correction term.[19] The limit of zero ω, i.e., the dc emitter and collector currents $\bar{I}_E(x, 0)$ and $\bar{I}_C(x, 0)$, yields the dc Ebers-Moll equations.[21] Another simplification can be made, which is to approximate the coefficients in Eqs. (5.89) and (5.90) by simple single-pole or single-zero functions in ω.[21] This gives, finally, for small σW,

$$\bar{I}_E(W, \omega) = a_{11}(1 + i\omega\tau_{11})\bar{\psi}_E(\omega) - a_{12}\bar{\psi}_C(\omega)/(1 + i\omega\tau_{12}) \tag{5.91}$$

$$\bar{I}_C(W, \omega) = -a_{21}\bar{\psi}_E(\omega)/(1 + i\omega\tau_{21}) + a_{22}(1 + i\omega\tau_{22})\bar{\psi}_C(\omega) \tag{5.92}$$

where

$$a_{11} = \frac{AD_n e\delta n_{p0}}{W}\left[1 + \frac{1}{3}(W/L_n)^2 + \eta^2/12 + \eta/2\right] \tag{5.93}$$

$$a_{22} = \frac{AD_n e\delta n_{pW}}{W}\left[1 + \frac{1}{3}(W/L_n)^2 + \eta^2/12 - \eta/2\right] \tag{5.94}$$

$$a_{12} = \frac{AD_n e\delta n_{pW}}{W\exp(\eta/2)}\left[1 + \frac{1}{6}(W/L_n)^2 + \eta^2/24\right]^{-1} \tag{5.95}$$

$$a_{21} = a_{12}\exp\eta \tag{5.96}$$

with time constant τ_{ij}.[25] The forward and inverse common base current gains are, respectively[26,53]

$$\alpha = (a_{21}/a_{11})/(1 + i\omega\tau_{11})(1 + i\omega\tau_{21}) \tag{5.97}$$

$$\alpha_i = (a_{12}/a_{22})/(1 + i\omega\tau_{22})(1 + i\omega\tau_{12}) \tag{5.98}$$

which are the corresponding two-pole approximations.

Utilizing the preceding development, minority carrier densities for various base dopant diffusion profiles, as shown in Figure 5.18, are obtained. From the foregoing expression, the corresponding common emitter gain β can also be gotten for transistors whose base is either doped in graded fashion or homogeneously doped. For these dopant profiles, the gain degradation due to displacement damage (e.g., from incident neutron fluence) follows in a straightforward manner.

At low signal frequencies (the limit of small ω), the *pnp* transistor emitter and collector currents can be described to good approximation by the Ebers-Moll equations (Eqs. (5.83) and (5.84)). The coefficients a_{ij}, $i, j = 1, 2$, in those equations are given by Eq. (5.93)–(5.96). The above equations, as written, correspond to *npn* transistors. By interchanging donor and acceptor parameters, they are easily converted to correspond to *pnp* transistors, discussed in the following paragraphs.

The common base dc current gain, from Eq. (5.97), is $\alpha = a_{21}/a_{11}$. Since $\beta = \alpha/(1 - \alpha)$, using the parameters given in Eq. (5.93)–(5.96) through quadratic terms, it can be written as[54] shown in Eq. (5.99), where now $\eta = \ln(N_{DE}/N_{DC})$, N_{DE}, and N_{DC} are the dopant concentrations at the base extremities,

$$\beta^{-1} = (\exp(\eta/2))(1 + \eta^2/8 + 1/2(W_b/L_p)^2 - (\eta/2)(1 + \eta^2/24 + (1/6)(W_b/L_p)^2)) - 1 \tag{5.99}$$

and W_b is the base width. Again, this expression applies to homogeneous dopant base transistors, where $\eta = 0$, as well as graded base transistors.

As discussed in Section 3.4, the relation between the internal field E and the internal potential ψ is $E = -\psi'(x)$. As also discussed in that section, assuming a base donor dopant concentration $N_D(x)$,

$$n \approx N_D = n_i \exp(e(\psi - \phi_f)/kT) \tag{5.100}$$

where the Fermi potential ϕ_f is defined from $E_f = e\phi_f$. Differentiating Eq. (5.100) to obtain ψ', and so the internal field, yields

$$E(x) = -(kT/e)N'_D/N_D \tag{5.101}$$

The base minority carrier (hole) current density satisfies, from Section 1.10,

$$J_p = e\mu_p pE - eD_p p' \tag{5.102}$$

Substituting for E in the above equation, from Eq. (5.101), with the Einstein relation, yields

$$p' + (N'_D/N_D)p = -J_p/eD_p; \qquad p(W_b) = 0 \text{ (Law of the Junction)} \tag{5.103}$$

The integral is given by

$$p(x) = (J_p/eD_pN_D) \int_x^{W_b} N_D(x')\, dx' \tag{5.104}$$

The transit time for holes in the n-type base, which is essentially the minority carrier lifetime, is

$$\tau_b = \int_0^{W_b} dx/v \tag{5.105}$$

Eliminating v in the above integral from $J_p = pev$, and using p from Eq. (5.104) yields

$$\tau_b = (1/D_p) \int_0^{W_b} (dx/N_D(x)) \int_x^{W_b} N_D(x')\, dx' \tag{5.106}$$

For a descending exponentially graded base dopant concentration, assume[55] that $N_D(x) = N_0(\exp - (x/L)) - N_1$, where N_0 and N_1 are the dopant densities at the emitter and collector extremities of the base, respectively. The collector junction is at $x = W_b$, where $N_D(W_b) = 0$. Therefore

$$W_b/L = \ln(N_0/N_1) = \ln(N_{DE}/N_{DC}) = \eta \tag{5.107}$$

Inserting the exponential dependence for N_D into Eq. (5.106) yields[55]

$$\tau_b = (W_b^2/D_p)[\eta - \int_1^{\exp-\eta} (\ln x dx/(1 - x))]/\eta^2 \tag{5.108}$$

or, suitably approximating the above integral finally yields

$$\tau_b \cong (W_b^2/D_p)(\eta + (\exp - \eta) - 1)/\eta^2 \tag{5.109}$$

As can be seen in the limit of zero η, the expression for the homogeneously graded base, $\tau_b \cong W_b^2/2D_p$ must ensue from Eq. (5.109), as discussed in Section 3.10.

The same expression for τ_b holds for an *npn* transistor, except for the diffusion constant, giving

$$\tau_b \cong (W_b^2/D_n)(\eta + (\exp - \eta) - 1)/\eta^2 \tag{5.110}$$

Since the gain-bandwidth product f_T is related to τ_b by $\tau_b = 1/\omega_T$, then from Eq. (5.110),

$$f_T = (1/2\pi)((W_b^2/D_n)(\eta - 1 + \exp - \eta)/\eta^2)^{-1} \tag{5.111}$$

For a bipolar transistor whose common emitter current gain has been degraded by neutron-induced displacement damage, Eq. (5.66) can be rewritten, using Eq. (5.111), as

$$\Delta(1/\beta)/\Phi_n = (W_b^2/D_n K)(\eta - 1 + \exp - \eta)/\eta^2 \tag{5.112}$$

Usually, W_b^2/D_n is computed using a measured value of f_T. Therefore, from Eq. (5.111),

$$W_b^2/D_n = (1/2\pi f_T)[\eta^2/(\eta - 1 + \exp - \eta)] \tag{5.113}$$

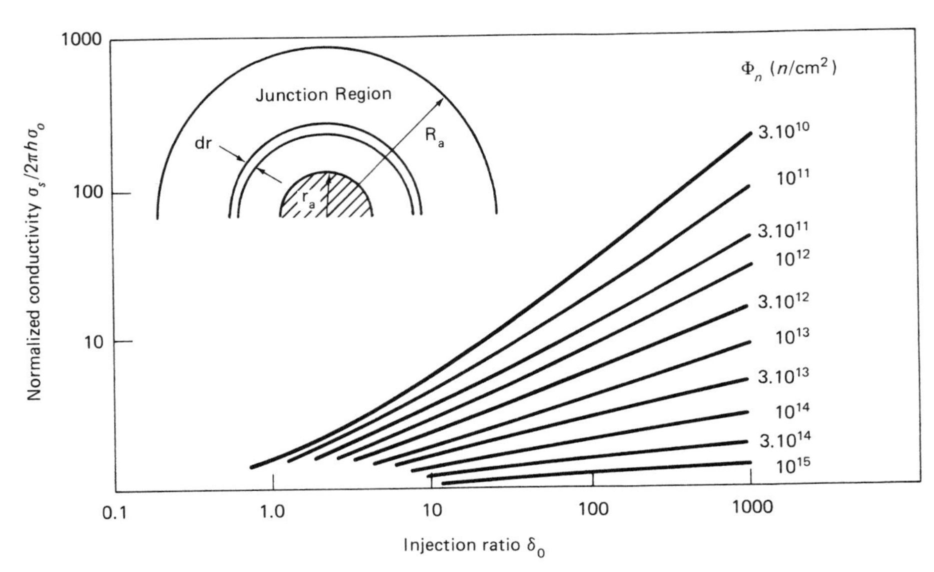

Fig. 5.19 Normalized conductivity in a cylindrical *p-n* junction versus the injection ratio for various levels of incident neutron fluence.

Or, from Eq. (5.66), the damage constant K, which is identical for graded or homogeneous bases, is calculated from

$$K = \Phi_n/\omega_T \cdot \Delta(1/\beta) \tag{5.114}$$

where $\omega_T = 2\pi f_T$, and f_T is obtained from Eq. (5.111). Φ_n and the corresponding $\Delta(1/\beta)$ are obtained from measurements of the family of transistors of interest.

5.12 CONDUCTIVITY MODULATION

For very high currents the injected carrier density can greatly exceed the dopant concentration in the pertinent region of a bipolar transistor. Hence, for charge neutrality to be maintained, with the stationary charge represented by the dopant concentration now negligible by comparison, the electron and hole densities must become virtually equal therein. Therefore, the majority carrier density changes will stay in phase with the minority carrier changes. This is called *conductivity modulation*, because the (now increased) conductivity depends on the carrier concentration, as discussed in Section 1.8. This results in a decrease in the rate of injection with increasing forward bias, so that the collector current is no longer proportional to $\exp(eV_{EB}/kT)$, but becomes more closely proportional to $\exp(eV_{EB}/2kT)$.[27] This results in a lesser slope and increased flattening of the collector characteristic curve for high current levels, as discussed in Section 3.7.

However, high incident radiation fluences can reduce conductivity modulation. For example, neutron-induced minority carrier lifetime degradation (reduction), and accompanying carrier removal, both contribute to lowering the effective injection level, resulting in a diminished conductivity modulation. For an illustration of how conductivity modulation is inhibited by neutrons, consider the case of a neutron fluence pulse, incident on a cylindrical *p-n* junction, whose half-section is shown in the inset of Fig. 5.19. For radial current flow across the junction, its cylindrical surface area is A and its height is h. The incremental change in the n-type semiconductor material resistance dR_s, due to the introduction of excess carriers, is given by

$$dR_s = dr/A\sigma = dr/2\pi rhe[(n_0 + \delta n)\mu_n + \mu_p\delta_p] \tag{5.115}$$

where the conductivity σ is defined in Section 1.8, and n_0 is the equilibrium concentration of carriers in this material. The appropriate steady-state solution of the diffusion equations, given in Section 1.11 in cylindrical (radial) coordinates, is

$$\delta_n/n_0 \cong \delta p/n_0 = (\delta p_0/n_0)\,K_0(r/L_p); \qquad r_a \leqslant r \leqslant R_a \tag{5.116}$$

$\delta n \cong \delta p$ because of hole-electron pairwise existence at high carrier injection, and $K_0(x)$ is the modified Bessel function. The diffusion length $L_p = (D_p\tau_p)^{1/2}$ is related to the neutron fluence, using Eq. (5.11), as

$$1/L^2 = 1/L_{p0}^2 + K_L\Phi_n(t),\ t < t_p;\ 1/L^2 = 1/L_{p0}^2 + K_L\Phi_n(t_p),\ t \geqslant t_p \tag{5.117}$$

K_L is the diffusion length damage constant, L_{p0} is the unirradiated diffusion length, and t_p is the pulse duration time. $\Phi_n \equiv \Phi_n(t_p)$ below, as the epoch of interest, occurs after the neutron radiation has subsided. From the conductivity expression (Section 1.8),

$$\delta_p(1 + \mu_p/\mu_n)/n_0 \cong \delta_0 K_0(r/L_p) \tag{5.118}$$

where δ_0 is essentially the injection ratio. Inserting Eq. (5.116) and (5.118) into Eq. (5.115), and integrating the result across the junction region, yields

$$\sigma_s^{-1} = R_s(\Phi_n) = (1/2\pi\sigma_{s0}) \int_{r_a}^{R_a} dr/[1 + \delta_0 K_0(r/L_p(\Phi_n))]\, r \quad (5.119)$$

where $\sigma_{s0} = e\mu n_0 h = \sigma_0 h$ is the pre-irradiation sheet conductance.

For $r_a = 5\mu m$, $R_a = 50\mu m$, $K_L^{-1} = 2.5 \cdot 10^6$, and $L_{p0} = 100\mu m$, the integral in Eq. (5.119) is evaluated numerically. The results are plotted in Fig. 5.19 for various levels of incident neutron fluence. The figure shows, for this nominally dimensioned junction, that the conductivity is a sensitive function of injection ratio for a lightly irradiated device, and that this sensitivity decreases for increased levels of neutron fluence. That is, conductivity modulation is suppressed by high levels of neutron fluence. Similar diminution of conductivity modulation is seen in diffused resistors exposed to high levels of prompt ionizing dose rates, discussed in Section 7.10.

5.13 RESISTIVITY EFFECTS

To investigate the effects of incident neutrons on the resistivity of semiconductors for the usual situation of nominally low injection ratios, it is necessary to describe the types of trapping mechanisms and how they contribute to resistivity changes. The displacement damage wrought by neutrons, discussed earlier, produces a number of different trapping defect complexes in the semiconductor material. Those important with respect to resistivity are now enumerated and discussed for n-type material. The same descriptions will hold for p-type material where the word *donor* is replaced by *acceptor*, and n is replaced by p in the appropriate phraseology. First it is important to realize that neutron-induced displacement of lattice donor atoms per se is considered negligible due to the low relative concentration of impurity atoms. The trapping mechanisms follow:

a. If a vacancy, from a vacancy-interstitial pair, comes to reside in a lattice position adjacent to a donor atom, this pair then forms a donor-vacancy defect trapping complex. This complex acts as a discrete trapping energy level, E_{DV}, in the forbidden gap. By virtue of the donor involved, the charge polarity of this complex is initially positive. This complex can capture, i.e., temporarily physically immobilize an electron from the conduction band, to change the defect complex charge from positive to neutral. If the Fermi energy level, E_f, is greater than E_{DV}, then this complex can capture an additional electron, thus changing its charge polarity to negative. Whether or not E_{DV} is greater than E_f depends on the concentration of un-ionized donors, as discussed in detail in Section 1.9, which in turn depends on the incident fluence level.

b. When two vacancies come to exist side by side in the lattice in adjacent positions, the resulting defect complex formed is called a *divacancy complex*. Besides trapping electrons, such a complex can cause homopolar bond stress leading to bond rupture. There are five different charge polarity manifestations, or charge states, of the divacancy complex. These consist of double minus, single minus, neutral, plus, and double plus. The initial divacancy charge state is neutral. However, it can capture either one or two electrons from the conduction band, corresponding to the negative charge states. Or, homopolar bond stress and subsequent bond rupture can cause the release of one or two electrons from the defect complex to the conduction band, resulting in net positive charge states.

c. When the semiconductor is formed during manufacture, atomic oxygen can be introduced as an inadvertent impurity. An oxygen atom can occupy the lattice site of a silicon vacancy, thus annihilating the latter. If this oxygen atom finds itself adjacent to another vacancy, an oxygen-vacancy defect complex can form, similar in effect to a donor-vacancy defect complex. As in the case of the donor-vacancy complex, if the Fermi level is greater than E_{OV}, the energy state of the oxygen-vacancy defect complex, it can also capture an electron from the conduction band to change its charge polarity to negative.

The increase in resistivity produced in n-type silicon is explained principally through the introduction of divacancy and donor-vacancy defect complexes, which are caused to form by incident neutrons. These defect complexes deplete the conduction band of its electrons (majority carriers) by capturing them. This is called *carrier removal*. Carrier removal thus decreases the conductivity of the semiconductor, and so increases its resistivity. Contributions due to oxygen-vacancy defect complexes are usually assumed negligible because of the relatively low concentration of oxygen within the lattice.

The important trapping energy states for n-type material as determined experimentally by electron paramagnetic resonance (EPR) methods, are at $E_c - 0.55\,\text{eV}$, corresponding to the divacancy defect complex with a single electron charge, $E_c - 0.40\,\text{eV}$, for the divacancy defect complex with a double electron charge, and $E_c - 0.40\,\text{eV}$, for the single negatively charged donor-vacancy defect complex, where E_c is, of course, the energy level at the bottom of the conduction band.

As alluded to earlier, if incident neutron radiation produces acceptor type defects, with energy levels below the Fermi level in n-type material, these defects will capture majority carriers, electrons in this case, from the conduction band, resulting in carrier removal and so decrease the carrier concentration. In the discussion in the latter part of this section, the resistivity ρ is obtained empirically as

$$\rho_{n,p} = \rho_i \exp(\Phi_n/k_{n,p}) \tag{5.120}$$

where ρ_i is the pre-radiation resistivity, and k_n and k_p are empirically determined[28] constants for n and p silicon. For an incident neutron fluence, Φ_n/k_n can be construed as the fractional density of trapped carriers normalized to unit fluence. Therefore, $1 - \Phi n/k_n$ is the fractional density of the so normalized untrapped (free) carriers. Then the carrier density, n, can be written as proportional to the ratio of untrapped fractional carrier density to that of the increase in trapped fractional carrier density, i.e.,

$$n \cong n_{0i}(1 - \Phi_n/k_n)/(1 + \Phi_n/k_n) \approx n_{0i} \exp - 2\Phi_n/k_n \tag{5.121}$$

The initial carrier removal rate is defined by $\lim_{\Phi_n \to 0} -n'(\Phi_n)$, which is the negative of the coefficient of the second term in the expansion of $n(\Phi_n)$ about $\Phi_n = 0$. Now, for not too large values of neutron fluence, the definition of carrier removal rate, and Eq. (5.121),

$$n \cong n_{0i} + \Phi_n \lim_{\Phi_n \to 0} (dn/d\Phi_n) \cong n_{0i}(1 - 2\Phi_n/k_n) \tag{5.122}$$

The carrier removal rate which represents their capture, here by neutron produced defect complex trapping centers, discussed in Section 5.5, is given by,

$$-\lim_{\Phi_n \to 0} dn/d\Phi_n = 2n_{0i}/k_n \tag{5.123}$$

which depicts how the carrier concentration decreases due to neutron fluence in this context. The same can be gotten from the derivative of Eq. (5.121) with respect to Φ_n in the limit of zero fluence. For 1 MeV neutrons, 1 MeV electrons, and 10 MeV protons in n and p silicon, the order of magnitude values for $n_{0i}/k_{n,p}$ (cm^{-1}) are (1, 5), (0.2, .005), and (100, 100), respectively.[4]

For large values of neutron fluence, the mobility becomes dependent on Φ_n, in the expression for resistivity $\rho = (e\mu_n n)^{-1}$. To obtain the corresponding fluence-dependent expression for mobility, the Conwell-Weisskopf relation, Eq. (1.53) is used as a model[48] for both n and p material, respectively, to yield the following empiricisms.

$$\begin{aligned} \mu_n &= 65 + 1265(1 + 6.47 \cdot 10^{-13} N_D{}^{0.72})^{-1}, \\ \mu_p &= 47.7 + 447(1 + 1.71 \cdot 10^{-13} N_A{}^{0.76})^{-1} \end{aligned} \tag{5.124}$$

The dependence on neutron fluence is evidenced through the donor concentration, N_D, for n-type material. It is related to the neutron fluence by $N_D = n + 1.17\Phi_n$, which is an approximation to a corresponding charge conservation relation between the particles involved.[48] Similar relations hold for p-type material. Also, a semi-empirical mobility expression of familiar form can be written as $\mu^{-1} = \mu_o^{-1} + K_\mu \Phi_n$, where for a 2 ohm-cm resistivity in n- and p-material, with a nuclear reactor neutron spectrum, $K_\mu = 3 \cdot 10^{-19}$ volt · sec · per neutron.[47] A more detailed description of mobility degradation, including electron and photon radiation effects, is available.[60]

To obtain the resistivity as a function of neutron fluence from fundamental considerations, in contrast to the above empiricisms, the conduction band bulk electron concentration must be known, in order to insert it into $\rho_n^{-1} = \sigma_n = ne\mu_n$. The conduction band electron concentration is a complicated function of the defect concentrations, which in turn are functions of the incident fluence. To obtain the defect complex trap concentrations as functions of fluence, a defect complex trap kinetics description must be given. An asymptotic version of such a kinetics model, suitable for this task, is described that corresponds to the situation in the semiconductor materials relatively long after the incident neutron pulse has been introduced and concentration equilibrium has been attained.

For n-type material, with the proviso that a similar model holds for p-type material as well, assume that a stoichiometric mass action law holds between the interacting concentrations to produce the various complexes. The particular coupling, or rate reaction, coefficients K_{IJ} are assumed to be known.

N_{DV}, N_D, and N_V are, respectively, the concentrations of donor vacancies, donors, and vacancies. Then the mass action type of relationship ensues, giving the algebraic expression

$$N_{DV} = K_{DV} N_D N_V \tag{5.125}$$

and K_{DV} is the rate coefficient for the formation of donor-vacancy complexes. Now, by simple conservation of donor atoms in the material

$$N_D = N_{Di} - N_{DV} \tag{5.126}$$

N_{Di} is the initial donor concentration, i.e., prior to the introduction of an incident neutron pulse. Equation (5.126) simply asserts that the remaining donor concentration after the pulse is the difference between the initial concentration and those donors used up in the production of donor-vacancy complexes.

Similarly for oxygen-vacancy complexes

$$N_{OV} = K_{OV} N_O N_V \tag{5.127}$$

where N_{OV} and N_O are, respectively, the concentration of oxygen-vacancy complexes and oxygen atom impurities. Likewise, the conservation of oxygen atoms asserts that

$$N_O = N_{Oi} - N_{OV} \tag{5.128}$$

where N_{Oi} is the initial concentration of oxygen atoms. For divacancies, with N_{VV} as their concentration

$$N_{VV} = \tfrac{1}{2} K_{VV} N_V^2 \tag{5.129}$$

Conservation of vacancies will prescribe the corresponding equation for the vacancy concentration. This is gotten by an audit of vacancy production, assumed proportional to neutron fluence level, less those vacancies now residing in their respective complexes. This yields

$$N_V = K_V \Phi_n - N_{DV} - N_{OV} - 2N_{VV} - N_{IV} \tag{5.130}$$

where N_{IV} is an interstitial-vacancy complex. It repairs the lattice locally and does not produce a trapping defect complex energy state, but does participate in this kinetics description, since it is a sink for vacancies. It satisfies a rate equation

$$N_{IV} = K_{IV} N_I N_V \tag{5.131}$$

where N_I is the interstitial concentration. The companion conservation equation is

$$N_I = K_V \Phi_n - N_{IV} \tag{5.132}$$

Equations (5.125)–(5.132) comprise a determinate set of eight simultaneous algebraic equations in the eight defect complex variables, each of which is ultimately a function of fluence. With the above defect complex concentrations at hand, the electron density, now $n(\Phi_n)$, can be obtained although implicitly from its charge conservation equation,[29,30] where $p = n_i^2/n$, as

$$n(\Phi_n) = n_i^2/n(\Phi_n) + N_{Di} - \left[\frac{2N_{DV}(\Phi_n)}{1 + n_i A_{DV}/2n(\Phi_n)} + \frac{N_{OV}(\Phi_n)}{1 + n_i A_{OV}/2n(\Phi_n)}\right.$$

$$\left. + \frac{N_{VV}(\Phi_n)}{1 + n_i A_{VV1}/2n(\Phi_n)} + \frac{N_{VV}(\Phi_n)}{1 + n_i A_{VV2}/2n(\Phi_n)}\right] \quad (5.133)$$

where the coefficient $A_{DV} = \exp - (E_i - E_{DV})/kT$ and similarly for the other A_{iV} coefficients. E_i is the intrinsic Fermi energy level, A_{VV1} and A_{VV2} correspond to the two divacancy defect complex energy states at $E_c - 0.40$ and $E_c - 0.55\,\text{eV}$ in silicon. The degeneracy factor is taken as $\frac{1}{2}$. The initial carrier removal rate $\lim_{\Phi_n \to 0} -n'(\Phi_n)$ can be obtained in principle from Eq. (5.133) by differentiation.

An abbreviated kinetics model of the above type for low fluence levels in n-type material includes only donor and oxygen vacancies. Therefore, using only Eqs. (5.125)–(5.128), and that for low fluence levels $N_V \cong K_V \Phi_n$, since this production term dwarfs any vacancy-complex concentrations, as opposed to Eq. (5.130), yields the approximate charge conservation equation $n = N_D - N_{DV} - N_{OV}$ and

$$\begin{aligned} N_{DV} &= [K_V K_{DV} N_{Di} \Phi_n/(1 + K_V K_{DV} \Phi_n)], \\ N_{OV} &= [K_V K_{OV} N_{Oi} \Phi n/(1 + K_V K_{OV} \Phi_n)] \end{aligned} \quad (5.134)$$

Then, eliminating N_{DV} and N_{OV} in that equation gives

$$n = N_{Di}(1 - K_V K_{DV} \Phi_n)/(1 + K_V K_{DV} \Phi_n) - N_{Oi} K_V K_{OV} \Phi_n/(1 + K_V K_{OV} \Phi_n) \quad (5.135)$$

The corresponding initial carrier removal rate is obtained from Eq. (5.135) as

$$-\lim_{\Phi_n \to 0} n'(\Phi_n) = K_V(2K_{DV} N_{Di} + K_{OV} N_{Oi}) \quad (5.136)$$

The resistivity is given as

$$\rho = (e\mu_n n)^{-1} = \rho_i n_0/n(\Phi_n) \quad (5.137)$$

where ρ_i is the unirradiated resistivity. Using Eq. (5.135) in (5.137) yields the resistivity as a function of fluence, viz.

$$\rho = \rho_i \frac{1 + (K_{DV} + K_{OV}) K_V \Phi_n}{1 - (K_{DV} - K_{OV}(1 - N_{Oi}/N_{Di})) K_V \Phi_n} \approx \rho_i \exp(2K_{DV} K_V \Phi_n) \quad (5.138)$$

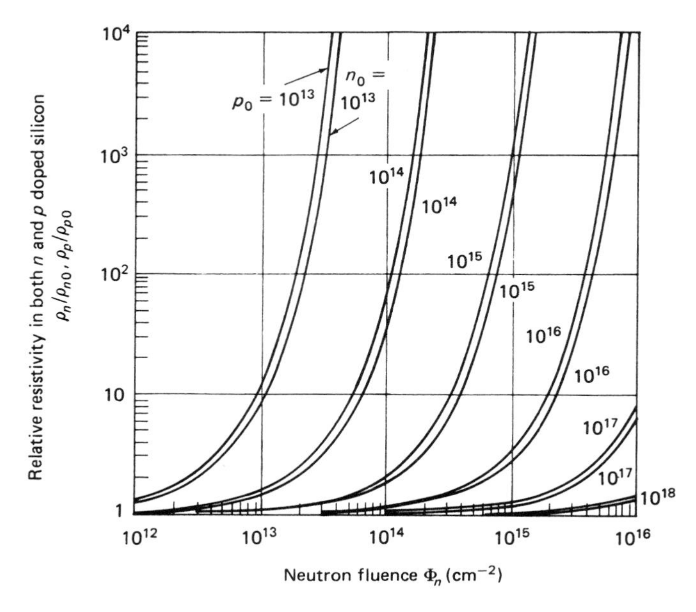

Fig. 5.20 Resistivity relative to its pre-irradiation level versus neutron fluence for various dopant densities in silicon from Buehler's resistivity model.[28] Figure © 1968 by the IEEE.

where the initial ratio of oxygen atoms to donor atoms, N_{Oi}/N_{Di} is assumed very small and μ_n is assumed to be independent of Φ_n. This exponential behavior of resistivity with neutron fluence correlates well with the Buehler data.[28] His expression for resistivity are $\rho_{n,p} = \rho_{n0,p0} \exp(\Phi_n/k_{n,p})$, with $k_n = 444n_0^{0.77}$, and $k_p = 387p_0^{0.77}$ for n- and p-type silicon, respectively, as shown in Figs. 5.20–5.21. An approximate expression for resistivity at low fluence levels is

$$\rho = \frac{\rho_i}{1 - K_1\Phi_n/N_{Di} - 2K_{DV}\Phi_n/(1 + K_{DV}\Phi_n)}; \qquad K_1 = 0.7\ \text{cm}^{-1},\ K_{DV} = 1.2 \cdot 10^{-17}\ \text{cm}^2 \tag{5.139}$$

With respect to initial carrier removal rate, Fig. 5.22 compares carrier removal data of Stein and Gereth with that predicted using the preceding development with the same values of the constants K_1 and K_{DV}.[29] It turns out that a better fit to the experimental data is obtained for $K_1 = 0.3\ \text{cm}^{-1}$.

With respect to the preceding, the variation of mobility can now be explained more fully. The predominant damaging effect is seen to be the trapping of charged carriers by the divacancy and donor-vacancy levels, which increases the density of coulomb scattering centers. Referring to

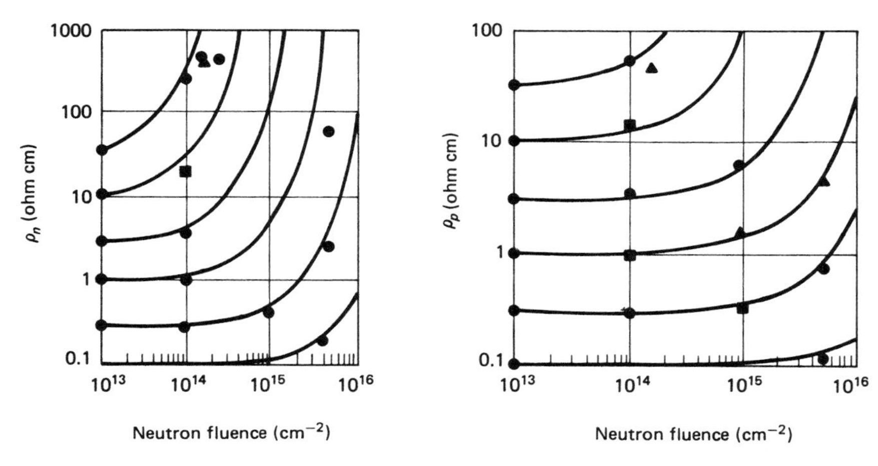

Fig. 5.21 Exponential resistivity-fluence dependence in silicon normalized to its preirradiation level assumed to be 1E13.[28] Figure © 1968 by the IEEE.

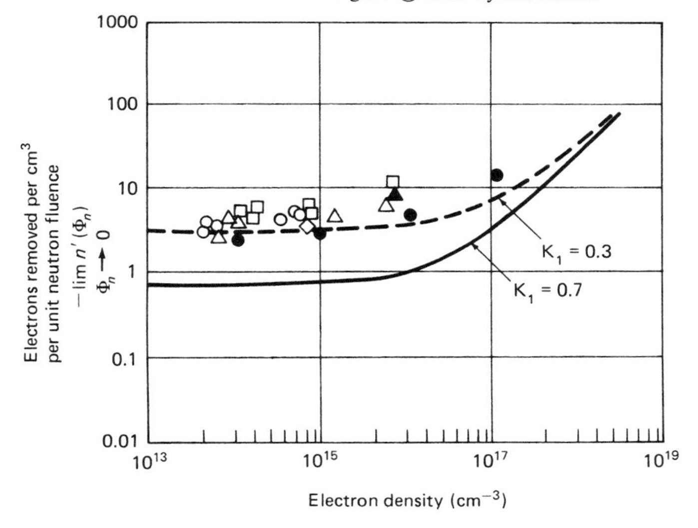

Fig. 5.22 Initial carrier removal rate versus carrier density—comparison of experiment and theory.[29] Figure © 1968 by the IEEE.

the earlier discussion of mobility in Section 1.7, scattering by ionized particle centers comprises one of its components. In many cases of interest, the contribution of the scattering mechanism to mobility is relatively small, while the preponderant neutron-induced changes are in carrier density and resistivity. However, for cases where the mobility degradation is important, that is, when it is written as

$$\mu^{-1} = \mu_l^{-1} + \mu_i^{-1} \tag{5.139a}$$

where μ_l is the mobility component due to lattice vibrations (i.e., unaffected by neutron fluence), and μ_i is the mobility component due to the presence of ionized impurity centers, the density of ionized impurity centers is equal to the initial density of ionized donors plus the carrier removal density. Thus, the influence of the ionized impurity centers increases with neutron fluence, and can be determined using Eq. (5.135). That is, a more accurate expression for mobility change is found by rewriting Eq. (1.53) as

$$\mu_i^{-1} \cong K_1 N(\Phi_n) \ln[1 + K_2 N(\Phi_n)^{-2/3}] \tag{5.139b}$$

rewriting Eq. (5.135) as

$$n \approx N(\Phi_n) = N_{Di} + K_3\Phi_n/(1 + K_4\Phi_n) - K_5\Phi_n/(1 + K_6\Phi_n) \tag{5.139c}$$

and inserting $N(\Phi_n)$ from Eq. (5.139c) into (5.139b), where the constants $K_1 \ldots K_6$ are identified in Eq. (1.53) and (5.135). However, for most cases of interest, μ^{-1} is linear in Φ_n, and so can be written as

$$\mu^{-1} = \mu_0^{-1} + b\Phi_n \tag{5.139d}$$

Also, both μ_l and μ_i are strongly temperature dependent, so that mobility degradation becomes increasingly important as the ambient temperature is reduced.

5.14 BIPOLAR ANNEALING

The words *anneal* and *annealing* refer to the partial or total self-healing of an electronic component or system after exposure to damaging nuclear radiation. In the case of damaging neutron radiation, most of the displaced atom defects resulting from incident neutrons are not stable at room temperature. This implies that as a result of thermal motion many of these defects will anneal, that is disappear due to vacancy-interstitial recombination. However, and no less important, these defect atoms can also form associations with impurities that exist in the lattice, which are stable at room temperature. Hence, annealing can cause improvement in the postradiation integrity of the lattice through recombination, as already mentioned, but also the formation of stable impurity defects and divacancies can cause further degradation.

It should be realized that there exists little in the way of an annealing theory that is comprehensive. Mainly empirical approaches that rely on experimental data for specific families of parts subjected to specific types of radiation, such as gammas or neutrons are available. For example, reliable annealing information in the form of curves, etc., is presented herewith mainly for bipolar devices irradiated by neutrons. Annealing processes in MOS devices are discussed in Section 6.15.

Transient or rapid annealing of semiconductor device performance degradation following pulsed neutron exposure is predominantly a bulk effect, in contrast to a surface effect. Annealing occurs in neutron-irradiated transistors over a whole range of temperatures above and below room ambient. Annealing after neutron irradiation at room temperature occurs partly because most of the neutron-induced, displaced atom defects are not stable at room temperature, as already mentioned.

The preceding can be couched in a time-dependent annealing factor $F(t)$. For bipolar transistors, the annealing factor can be expressed as[30]

$$F(t) = \frac{h_{FE}^{-1}(t) - h_{FE}^{-1}(0)}{h_{FE}^{-1}(\infty) - h_{FE}^{-1}(0)}; \qquad \dot{F}(t) \leqslant 0,\ F(0+) \geqslant 1,\ F(\infty) = 1 \tag{5.140}$$

where $h_{FE}(t)$ is the common emitter current gain of the transistor at an elapsed time t after neutron irradiation, and $h_{FE}(0)$ is the preirradiation gain. For $t \geqslant t_p$, where t_p is the neutron pulse duration, the above can be rewritten to obtain the gain, following t_p, as

$$h_{FE}(t) = \frac{h_{FE}(0)}{1 + F(t)[h_{FE}(0)/h_{FE}(\infty) - 1]}; \qquad 0+ \leqslant t < \infty \tag{5.141}$$

where $F(t)$ jumps discontinuously at zero time, as in Eq. (5.140) and shown in Fig. 5.23. Also, the annealing factor can be defined implicitly in terms of an effective neutron fluence, as

$$\Phi_{\text{neff}} = F(t)\Phi_n \tag{5.142}$$

The effective fluence, $\Phi_{\text{neff}} > \Phi_n$, is that fluence required to produce the same transistor degradation as that sustained at time t following exposure to the actual lesser fluence Φ_n. At very long times after exposure, Φ_{neff} approaches Φ_n as the annealing proceeds, so that F approaches unity.

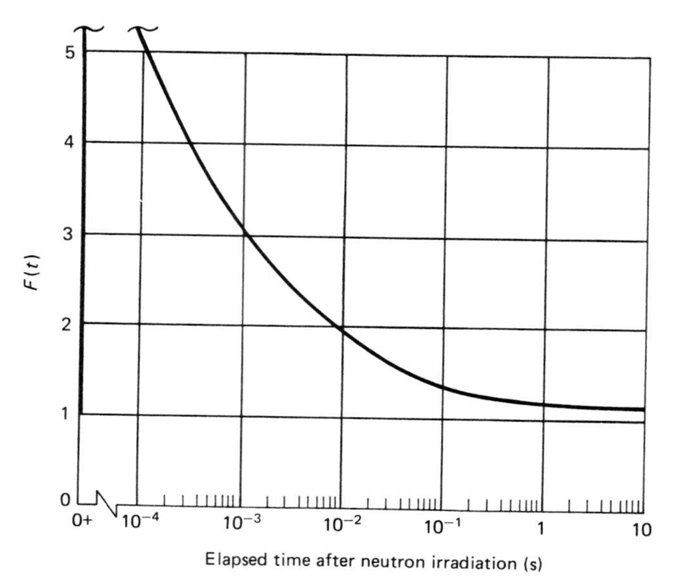

Fig. 5.23 Typical annealing factor for an integrated circuit subject to a neutron fluence pulse.[30] Figure © 1966 by the IEEE.

A typical annealing factor variation with time for an integrated circuit subject to a neutron pulse is shown in Fig. 5.23.[30]

Annealing that immediately follows pulsed neutron radiation on bulk silicon, silicon transistors, and silicon solar cells, has been studied in detail.[31,32,33] It has been found that a principal parameter affecting the rate of annealing, in the post-neutron-radiation phase, is the carrier density level within the active region, inside that portion of the device that determines its postradiation characteristics. For example, in *p*-type silicon, it is known that the annealing factor $F(t)$ in the first 100 ms after the incidence of the neutron burst can be reduced by a factor of 5 by varying the minority carrier (electrons) injection level from 10^{-5} to 0.1. Figure 5.24 depicts the variation in the rapid annealing factor with time following the succession of neutron bursts at repetition rates given in the figure for various injection levels of collector current, I_C, at constant temperature. This is called *isothermal annealing*, as opposed to isochronal annealing. An example of (slow) isochronal annealing is shown in Fig. 5.25, and Fig. 5.26 illustrates slow isothermal annealing.

Figure 5.27 depicts the rapid annealing factor versus the elapsed time after a neutron pulse for bipolar transistors in various bias states, as expressed by their base-emitter voltage V_{BE}. Notice from the figure that the worst case with respect to gain degradation is when the transistor is biased off ($V_{BE} = 0$) after the onset of the neutron pulse, because

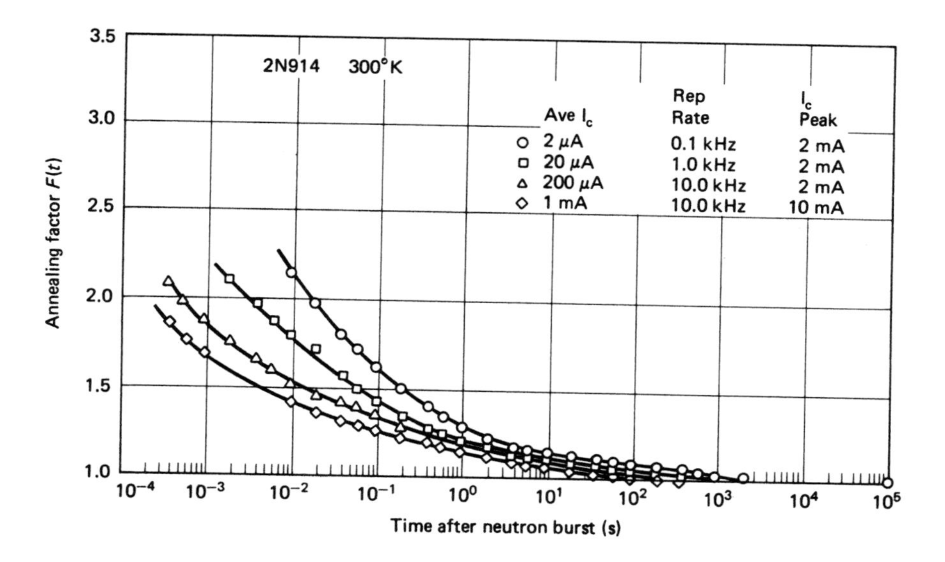

Fig. 5.24 Annealing factor versus post neutron irradiation time for various injection levels and repetition rates.[34] Figure © 1970 by the IEEE.

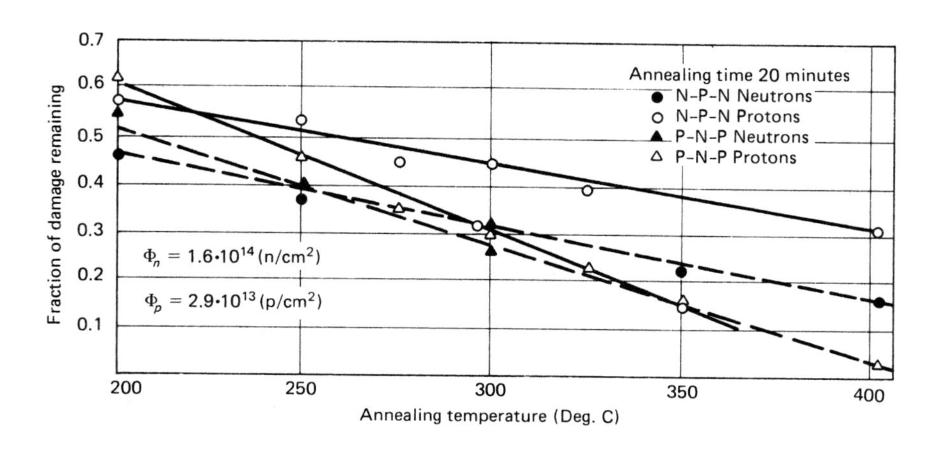

Fig. 5.25 Isochronal slow annealing characteristics for fast neutrons and 16.8 MeV protons for typical bipolar transistors.[34] Figure © 1970 by the IEEE.

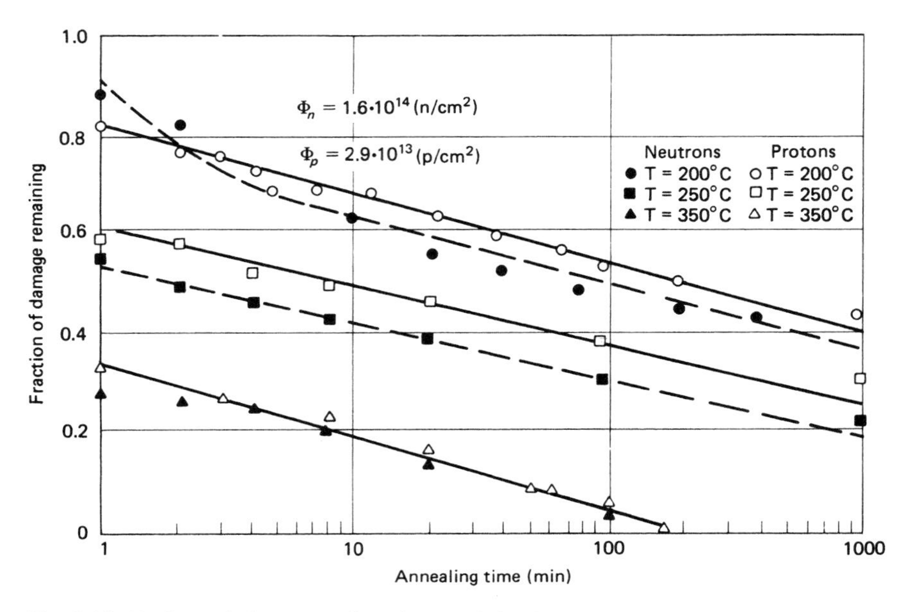

Fig. 5.26 Isothermal slow annealing characteristics for typical bipolar transistors for fast neutrons and 16-MeV protons.[34] Figure © 1970 by the IEEE.

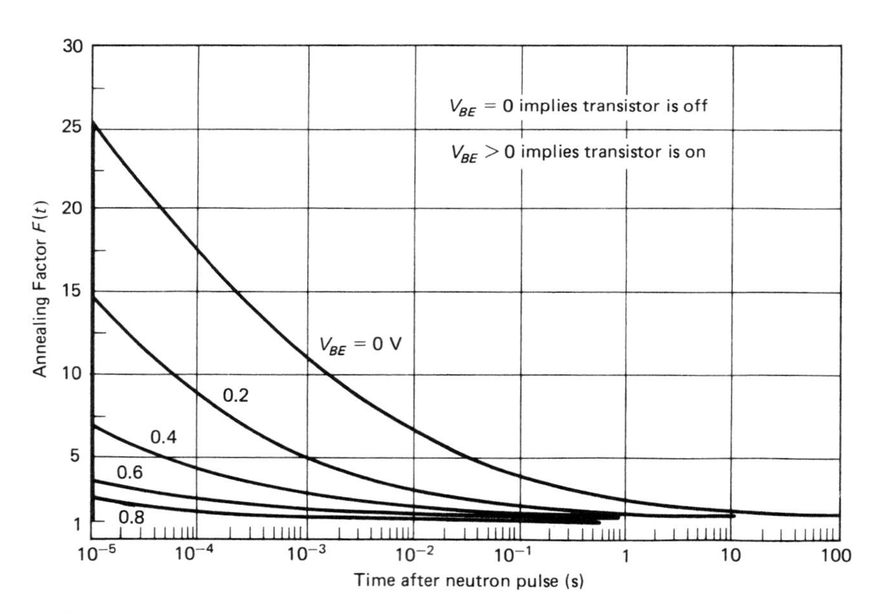

Fig. 5.27 Annealing factor versus time after neutron pulse for bipolar transistors for various V_{BE} at room temperature.[35]

the corresponding $F(0+)$ is about 26, which is its maximum value. This means that roughly the gain of this transistor immediately after the neutron pulse is instantaneously only 4 percent of its value following the annealing phase. This is seen from Eq. (5.141), which asserts that $h_{FE} \simeq h_{FE}(\infty)/F(t)$ when $h_{FE}(0)/h_{FE}(\infty) \gg 1$. This is partially attributed to the nascent vacancies produced immediately following the neutron pulse. Prior to their formation of defect complexes, they act as recombination centers as discussed in Section 5.5. The usual expressions for bipolar current gain degradation as a function of neutron fluence imply asymptotic gain conditions, i.e., relatively long after rapid annealing takes place. As seen from Fig. 5.27, the annealing factor drops precipitously even for very small values of bias, as expressed by the nonzero values of V_{BE}. Hence electronic systems anticipated to experience a very short neutron pulse should be biased on, even by special circuitry if necessary to keep F small for minimum performance degradation. The rapid annealing process occurring in transistors due to neutrons, as in the foregoing discussion, can be modeled using an approach due to McMurray and Messenger.[36]

One consequence of the fact that rapid annealing depends on the electron density in the device is that the rapid annealing factor is diminished for increasing device hardness.[36] This is because the indirect ionization electrons produced by neutrons are accelerating the rapid annealing process. This means that a hard device requiring a relatively large neutron fluence to produce significant damage will also be provided with a high density of electrons to enhance rapid annealing, whereas a "soft" device requiring a relatively smaller neutron fluence to produce damage will only be provided with a correspondingly small electron density to promote annealing. For very hard devices (e.g., hard to 10^{15} neutrons per cm^2, the magnitude of the worst case (highest fluence) annealing factor is approximately 2–3, as in Fig. 5.28. For very soft devices (hard to less than 10^{12} neutrons per cm^2) the annealing factor can be as high as 25, as in Fig. 5.27.

To investigate the temperature dependence of transient or rapid annealing, experiments have been performed on a variety of semiconductor types, ranging from bipolar transistors to solar cells. Solar cell annealing is discussed in Section 13.14. Figure 5.29 shows typical experimental results for current gain degradation annealing factor in typical *npn* bipolars for a range of temperatures.[37] To separate the temperature and injection level effects, the experiments were performed by pulsing the collector current. This allows the mean injection level to be varied through the "duty cycle," which aids in maintaining equal junction and case temperatures. For example, at 75°C (348°K) $F(t) = 1.25$ at 0.1 s, and at −60°C (213°K), at the same time, $F(t) = 2.25$. This indicates that a

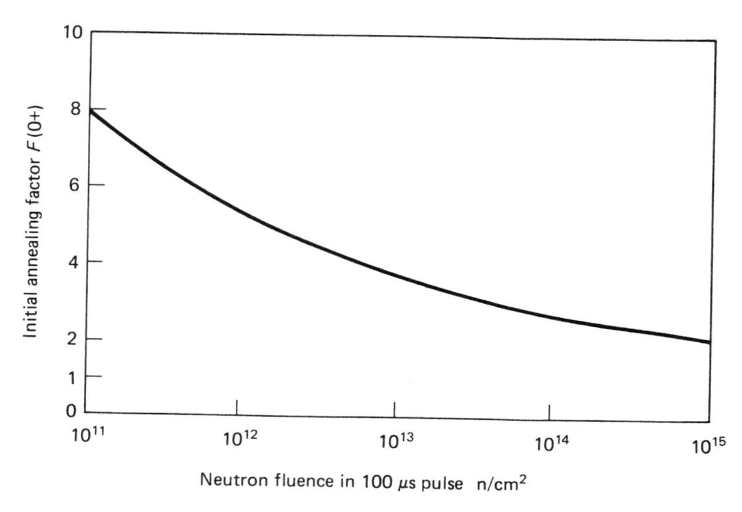

Fig. 5.28 Initial annealing factor versus incident neutron pulse fluence for relatively hard bipolar transistor.

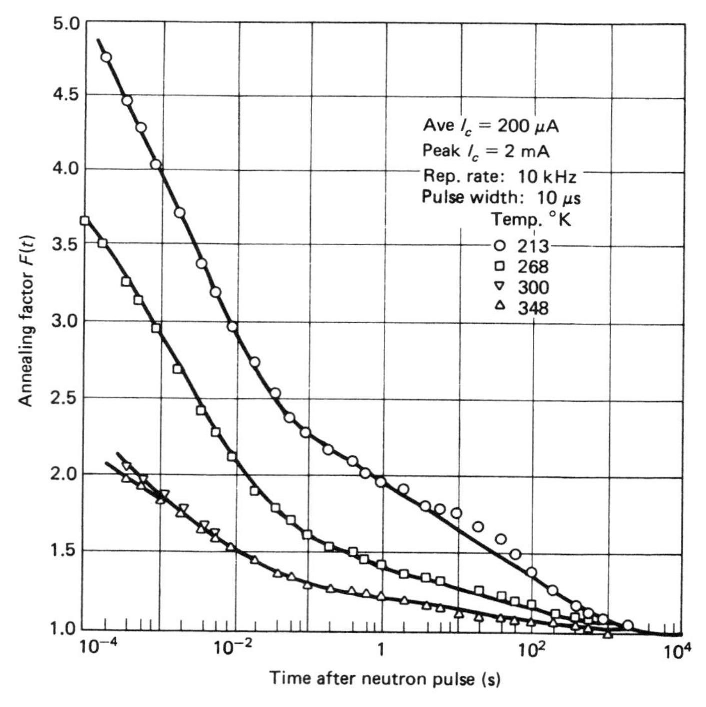

Fig. 5.29 Rapid annealing factor for bipolar transistors at various temperatures.[34] Figure © 1970 by the IEEE.

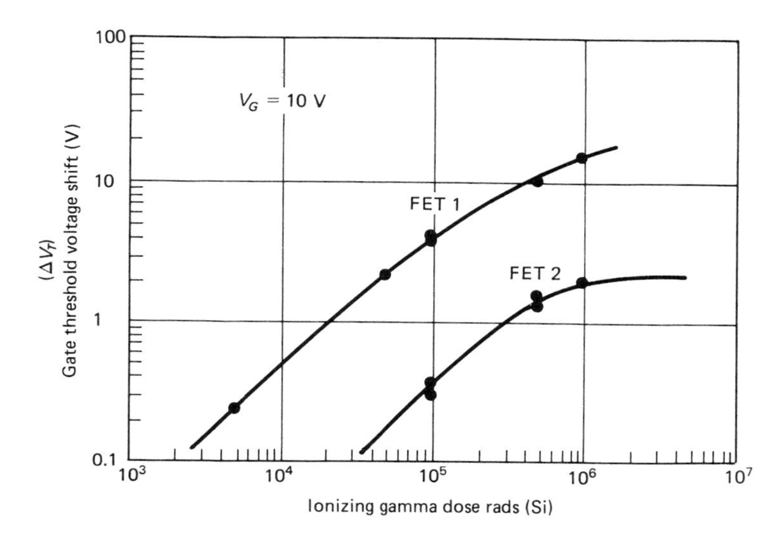

Fig. 5.30 Gate threshold voltage shifts versus ionizing gamma dose for two *n*-channel FETs.[38] Figure © 1977 by the IEEE.

system designed to operate at lower temperatures must have a design margin that accommodates a larger rapid annealing factor (much greater damage) than at room temperature.

In certain applications, such as electronic systems in spacecraft, components are subjected to long-term, relatively low rates of ionizing radiation, mainly from the Van Allen belts. These electronic components are being irradiated and are undergoing annealing simultaneously. Where applicable, one method for handling part degradation due to such radiation is that of linear superposition, borrowed from the transfer function schema of modern linear control theory.[38] Its central idea is that if a time-dependent cause (incident radiation) and a time-dependent effect (parameter degradation) are connected by a linear operator, such as through a linear differential equation with constant coefficients, or a linear integral equation, then the power of this method can be exploited.

As an illustration, consider radiation-induced *n*-channel FET gate threshold voltage shifts.[38,57] This type of parameter degradation is discussed in detail in Chapter 6, but its knowledge is not necessary here with respect to annealing dynamics. Figure 5.30 depicts the gate threshold voltage shift, $-\Delta V_T$, for two *n*-channel FETs versus ionizing dose γ_T, rad (Si). It is seen from Fig. 5.30 that for either transistor, $-\Delta V_T$ is algebraically linear with γ_T up to well beyond 100 krad (Si) and is approximately so for doses to 1 mrad (Si). This suggests that $-\Delta V_T$ and γ_T could be related in the sense of a linear operator dependence between them.

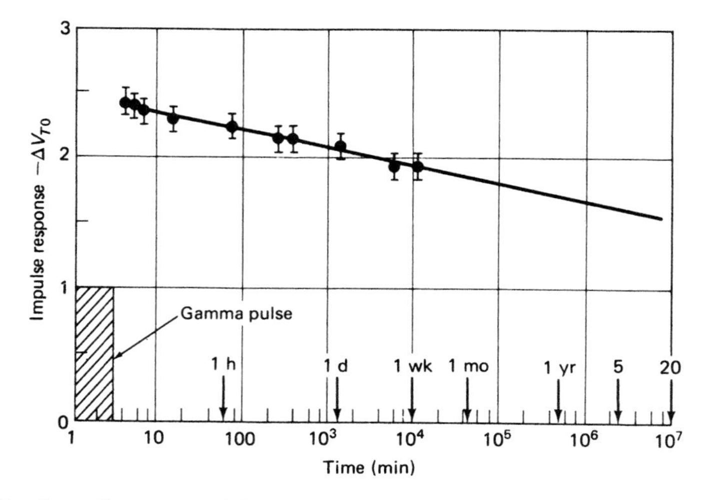

Fig. 5.31 Annealing curve of impulse response for *n*-channel field effect transistor.[38] Figure © 1977 by the IEEE.

Figure 5.31 is a slow annealing curve, as it represents the postradiation ΔV_{T0} recovery threshold voltage shift, as a function of times long after exposure to the relatively short ionizing radiation pulse. Hence, this incident pulse of radiation can be construed as an impulse (delta function), and the curve of ΔV_{T0} in Fig. 5.31 as the corresponding impulse response. Assuming that the above linearity holds, then a central theorem from linear control theory asserts that a convolution integral can be written as[39]

$$\Delta V_T(t) = \int_0^t \dot{\gamma}(\tau)\Delta V_{TO}(t-\tau)\,d\tau \quad \text{(volts)} \tag{5.143}$$

where ΔV_{T0} is the impulse response, in volts per dose, and $\dot{\gamma}(t)$ describes the ionizing radiation environment in dose per second. Of course, the foregoing is not confined to ionizing radiation, but can be employed with other types of radiation as well. The Laplace transform of the above yields

$$\Delta \bar{V}_T(s) = s\bar{\gamma}(s)\Delta \bar{V}_{TO}(s) \tag{5.144}$$

where the overbar signifies the corresponding Laplace transform, and s is the Laplace transform variable. $\overline{\Delta V_T}(s)$, is given by the product of the "input" $s\bar{\gamma}(s)$ by the "transfer function" $\overline{\Delta V_{T0}}(s)$, the transform of the impulse response. The impulse response given in Fig. 5.31 is well approximated by

$$-\Delta V_{TO} = (-c_1 \ln(t/t_A) + c_2)/\gamma_0; \qquad \gamma_0 = \int_0^{\infty} \dot{\gamma}(t)\,dt \tag{5.145}$$

where $c_1/\gamma_0 = 0.06$, and $c_2/\gamma_0 = 2.5$, $\gamma_0 = 50\,\text{krad (Si)}$, giving

$$\Delta V_{T0} = 0.06 \cdot \ln(t/t_A) - 2.5 \tag{5.146}$$

where the time after exposure is measured in minutes, and $t_A = 1\,\text{min}$.

As an example of the foregoing, consider the case of two n-channel FETs from different date codes (lots): (a) Device A is irradiated by a $\Delta_0 = 50\,\mu\text{s}$ uniform dose rate exposure, assumed sufficiently short so that the resulting curve of threshold voltage shift can be construed as its impulse response ΔV_{T0}. (b) Device B is irradiated for $t_0 = 71\,\text{s}$ also at a uniform dose rate, i.e., a rectangular $\dot{\gamma}$ pulse of constant amplitude, and width 71 s. Using Eq. (5.143) to compute this response gives (with $t_A = 1$ minute)

$$\Delta V_T(t) = \int_0^{t_0} \dot{\gamma}(\tau)\Delta V_{TO}(t - \tau)\,d\tau \tag{5.147}$$

so that for constant $\dot{\gamma} = \Gamma$, and $-\Delta V_{T0} = (-c_1 \ln(t/t_A) + c_2)/\gamma_0$, substituting into the above yields

$$\Delta V_T(t) = \begin{cases} (\Gamma t_0/\gamma_0)[-c_1(u \ln u + u \ln t_0) + (c_1 + c_2)u]; & u_1 \leqslant u \leqslant 1 \quad (5.148) \\ (\Gamma t_0/\gamma_0)\left[-c_1\left(u \ln \dfrac{u}{u-1} + \ln(u-1) + \ln(t_0/t_A)\right) + c_1 + c_2\right]; & u > 1 \quad (5.149) \end{cases}$$

where $u = t/t_0$. t_0 is the pulse width of the uniform $\dot{\gamma}$ exposure, and $u_1 = \Delta_0/t_0 = 7 \cdot 10^{-7}$. Figure 5.32 shows the results of using Eqs. (5.148) and (5.149) for the above uniform dose rate Γ for t_0 of 71 s. c_1 and c_2, and γ_0 are fitted to the corresponding impulse response given in Fig. 5.32. It is seen from the figure that the convolution integral expression agrees well with the measured values of ΔV_T.

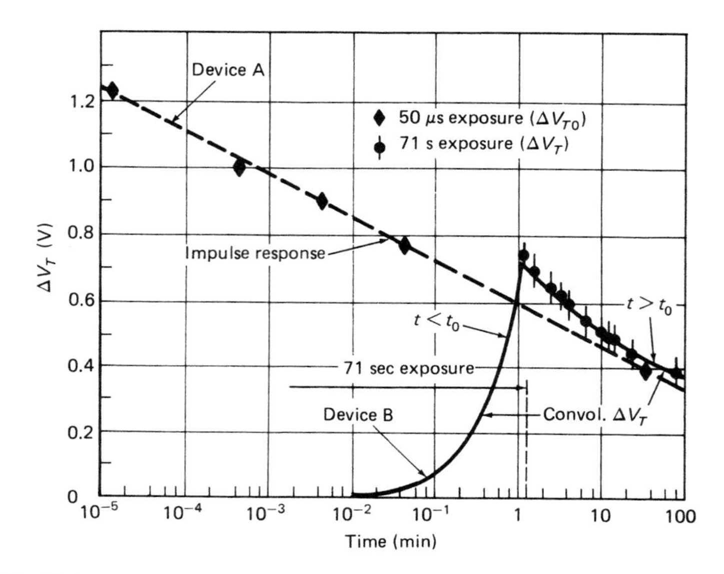

Fig. 5.32 *N*-channel FET threshold voltage shift impulse response annealing curve with corresponding threshold voltage shift.[38] Figure © 1977 by the IEEE.

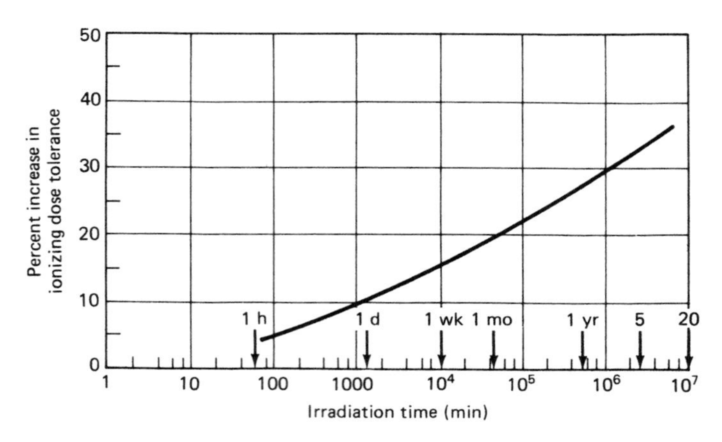

Fig. 5.33 Increase in ionizing dose tolerance for low dose rate exposures.[38] Figure © 1977 by the IEEE.

Inserting Eq. (5.145) into (5.143) for a constant dose rate yields

$$-\Delta V_T = (\Gamma t/\gamma_0)(-c_1 \ln(t/t_A) + (c_1 + c_2)) \tag{5.150}$$

which asserts that the gate threshold voltage shift is smaller if the exposure for a given dose occurs over a longer duration. In general then, it is desired to determine the increase in ionizing dose tolerance that results from very low rate exposures, due to allowing the annealing process to proceed over a longer period. This can be obtained from a comparison of the same gate threshold voltage shifts, from two different dose rates. For convenience, the high dose rate is taken as that following a short (1-min) exposure, which value of ionizing dose is obtained from Eq. (5.145) at 1 min to give $-\Delta V_{T0}(1) = c_2/\gamma_0$. For the ΔV_T corresponding to low dose rate exposure, Eq. (5.150) is used. Equating the two threshold voltage shifts provides a measure of ionizing dose tolerance $\xi(t)$ where

$$\xi(t) \equiv \Gamma t = \frac{c_2}{c_1 + c_2 - c_1 \ln(t/t_A)} \tag{5.151}$$

Defining the change in ionizing dose tolerance as $(\xi(t) - \xi(t_A))/\xi(t_A)$, where ξ is given by Eq. (5.151), then

$$\frac{\xi(t) - \xi(t_A)}{\xi(t_A)} = \frac{\ln(t/t_A)}{1 + c_2/c_1 - \ln(t/t_A)} \tag{5.152}$$

Figure 5.33 is plotted for the same n-channel FET under consideration, with the constants c_1 and c_2 given following Eq. (5.145). The figure shows an appreciable increase in ionizing dose tolerance for every low dose rate conditions. For devices that have a more rapidly increasing transient annealing curve than these devices, as depicted in Fig. 5.31, the corresponding increases in ionizing dose tolerance would be notable. It is well known that device transient annealing rates can vary greatly for different fabrication processes. Also, p-channel FETs anneal at a faster rate than their n-channel counterparts. Hence, gains in ionizing dose tolerance in low dose environments would probably be more dramatic for p-channel FETs.

5.15 OUTPUT LEAKAGE CURRENT EFFECTS

To investigate the effect of incident neutron radiation on the leakage current of junction devices, such as all types of diodes and bipolar transistors, rewrite Eq. (1.60) yielding the equilibrium hole current density in the depletion layer of the semiconductor as

$$J_p = e\mu_p pE - eD_p\, dp(x)/dx \tag{5.153}$$

assuming the validity of the one-dimensional approximation, where x is the distance into the material normal to the junction. In the depletion layer, the net hole current density J_p is several orders of magnitude smaller than either the drift current component (first rightmost term above) or the diffusion component (second rightmost term above), even when the junction is forward biased. Hence, Eq. (5.153), where the left side is considered negligible, yields

$$eD_p\, dp(x)/dx \simeq e\mu_p pE \tag{5.154}$$

Using the Einstein relation to eliminate the mobility, and integrating the internally generated field E in Eq. (5.154) across the junction yields

$$p/p_p = \exp(e(\phi + \phi_0)/kT) \tag{5.155}$$

where ϕ_0 is the junction potential, and ϕ is the applied potential. p_p is the thermal equilibrium hole concentration in the p-material. Using p_p above implies that it is essentially equal to the majority carrier density on the p side of the junction. With zero applied potential, $\phi = 0$, so that

$$p_n/p_p = \exp -e|\phi_0|/kT \tag{5.156}$$

where p_n is the thermal equilibrium hole concentration on the n-side of the junction. Combining Eqs. (5.155) and (5.156) gives the "Law of the Junction," viz.

$$p = p_n \exp(e\phi/kT) \tag{5.157}$$

Likewise for n-type material

$$n = n_p \exp(e\phi/kT) \tag{5.158}$$

Notice the sensitivity of the particle densities, n and p, as a function of the applied potential ϕ. Because kT/e is about 1/40 eV at room temperature, if the junction is forward biased, $\phi > 0$, by only a few tenths of a volt, then large minority carrier currents result at the edge of the depletion region, and also large diffusion current flow. If the junction is reverse biased, $\phi < 0$, by a few tenths of a volt, the corresponding minority carrier current is essentially zero. Then the diode equation (3.78), becomes

$$I = I_1[\exp(eV_0/kT) - 1] \equiv I_1\psi(V_0) \tag{5.159}$$

where now V_0 is the applied junction potential, and I_1 is the saturation current given from Eq. (3.78), so that

$$I = Aen_i^2[(D_n/L_nN_A)\coth(W_p/L_n) + (D_p/L_pN_D)\coth(W_n/L_p)]\psi(V_0) \tag{5.160}$$

To obtain the leakage current, which is the saturation current when the junction is reverse biased, so that $\psi(-V_0) = -1$, for even small $|V_0|$ for reasons already discussed, Eq. (5.160) now yields

$$-I_1 = Aen_i^2[(D_n/L_nN_A)\coth(W_p/L_n) + (D_p/L_pN_D)\coth(W_n/L_p)] \tag{5.161}$$

For a narrow base width, i.e., $W_{n,p} \ll L_{n,p}$, Eq. (5.161) yields the leakage current as

$$-I_{1n} \cong Aen_i^2[(D_n/W_pN_A) + (D_p/W_nN_D)] \tag{5.162}$$

while for a wide base

$$-I_{1w} \cong Aen_i^2\ [\sqrt{D_n/\tau_n}/N_A + \sqrt{D_p/\tau_p}/N_D] \tag{5.163}$$

The impact of radiation on the leakage current is introduced through its effect on the minority carrier lifetime, from $\tau^{-1} = \tau_i^{-1} + \Phi_n/K$ from Eq. (5.11). Using $L_{n,p}^2 = D_{n,p}\tau_{n,p}$ in Eq. (5.163) inserts the influence of radiation thereby. For example, from Eq. (5.163) in a dominantly n-region, with $\tau_n^{-1} \approx \Phi_n/K$

$$-I_{1w} \cong Aen_i^2\sqrt{D_n\Phi_n/K}/N_A \tag{5.164}$$

Similarly, for a graded base, using Eqs. (5.89) and (5.90),

$$-I_1 \cong Aen_i^2[(\eta_pD_n/W_pN_A) + (\eta_nD_p/W_nN_D)] \tag{5.165}$$

However, with no explicit dependence on minority carrier lifetime, the leakage current will manifest only second-order effects due to neutron fluence in this case.

5.16 DISPLACEMENT DAMAGE LIMITS FOR SILICON TECHNOLOGIES

Modern hardening goals are driving semiconductor technologies to their technological and fabrication limits. In the past, displacement damage has been perceived as a problem for bipolar silicon technologies, and the dominant physical effect has been minority carrier lifetime reduction, leading to decreases in the current gain of active devices. Displacement damage effects have been summarily dismissed for all MOS technologies, for reasons discussed in Section 6.8. Their survival fluence is usually estimated at about 10^{15} neutrons per cm^2, which is more than sufficient for most applications in reasonably conceivable radiation environment scenarios.

Now that a few requirements are emerging for fluence levels of 10^{15} neutrons per cm^2 and beyond, it is appropriate to explore the theoretical and practical limitations for silicon. The basic damage effects are: (1) minority carrier lifetime reduction, and (2) resistivity increase, which can be described in terms of carrier removal and mobility reduction.

Minority carrier lifetime reduction is the principal limitation for low-frequency devices. However, with the new generation of bipolar "Very High Speed Integrated Circuit" (VHSIC) devices, the gain-bandwidth product can be much greater than one GHz, and minority carrier lifetime is no longer a dominant factor, since the corresponding lifetimes are extremely short. Instead, the major radiation damage problem is that of resistivity increase. For all majority carrier devices (i.e., MOS and JFET devices), resistivity increase has always been the dominant displacement damage mechanism. Of course, preponderant attention has been given to the ionizing dose sensitivities of MOS devices. However, the state-of-art has now reached a point where resistivity increase is the final limiting displacement damage mechanism for all silicon semiconductor devices. Resistivity is such a basic parameter that the corresponding failure analysis can be greatly simplified by relating failure directly to a specific percentage change in resistivity. For hardened programs with high confidence and high reliability requirements, a resistivity change limit of 10 percent is practically reasonable. For marginal applications, where one is forced to extract the maximum possible in terms of hardness, a resistivity change of 100 percent might be a reconcilable limit. However, a resistivity change of 1000 percent would be regarded as catastrophic and represents an absolute upper limit.

To employ this concept, all that is needed is a credible quantitative relationship between resistivity and neutron fluence. Such a semi-empirical relationship is given, based on resistivity discussions, in Section 5.13 for

Table 5.2 Silicon semiconductor displacement damage resistivity limits as a function of maximum breakdown voltage for 1 MeV equivalent neutron fluence levels.

SEMICONDUCTOR PARAMETERS			Φ_n(1 MEV EQUIV.) FLUENCE FOR $\Delta\rho/\rho_i$ OF:		
Breakdown voltage (V)	N_{Di} (cm^{-3})	Resistivity (Ohm-cm)	10 Percent	100 Percent	1000 Percent
1000	$2 \cdot 10^{14}$	25.0	$3 \cdot 10^{12}$	$2 \cdot 10^{13}$	$5 \cdot 10^{13}$
300	10^{15}	5.0	$1.7 \cdot 10^{13}$	$8.5 \cdot 10^{13}$	$2.1 \cdot 10^{14}$
100	$5 \cdot 10^{15}$	1.0	$8.3 \cdot 10^{13}$	$4.3 \cdot 10^{14}$	$1.1 \cdot 10^{14}$
60	10^{16}	0.60	$1.7 \cdot 10^{14}$	$8.5 \cdot 10^{14}$	$2.1 \cdot 10^{15}$
30	$2.5 \cdot 10^{16}$	0.25	$4.2 \cdot 10^{14}$	$2.1 \cdot 10^{15}$	$5 \cdot 10^{15}$
15	$8 \cdot 10^{16}$	0.10	$1.3 \cdot 10^{15}$	$6.8 \cdot 10^{15}$	$1.7 \cdot 10^{16}$
5	$2 \cdot 10^{17}$	0.05	$3 \cdot 10^{15}$	$1.7 \cdot 10^{16}$	$4.2 \cdot 10^{16}$

n-type silicon, and a similar expression is available for p-type silicon. The former is

$$\rho = \rho_i[(1 + (3/N_{Di})\,\Phi_n)/\{1 - (3/N_{Di})\,\Phi_n - (2.4 \cdot 10^{-17}\Phi_n)/(1 + 2 \cdot 10^{-17}\Phi_n)\}] \tag{5.166}$$

where N_{Di} is the initial (pre-radiation) donor concentration. It is obvious from Eq. (5.166) that increasing N_{Di} or decreasing the initial resistivity, ρ_i, reduces the displacement damage effect. Therefore, one should use the lowest resistivity material that is practical. The most important device parameter that is defined by the initial resistivity is the breakdown voltage. This produces another useful simplification, which is that the limitation for hardened silicon semiconductor components is reduced to its breakdown voltage considerations. The relationship between doping density and breakdown voltage is shown in Fig. 3.11, while that between doping density and resistivity is depicted in Fig. 1.10. One can then select a value of breakdown voltage from component and circuit considerations, with the corresponding N_{Di}, and ρ from the figures, as described. Insert the values of these quantities into Eq. (5.166), which can then be solved for the fluence levels at which 10, 100, and 1000 percent change in breakdown voltage will occur. This has been done and is tabulated in Table 5.2. For example, if one considers 30V as the minimum useful breakdown voltage range for semiconductor devices in a system, the neutron hardness level should be specified at less than $4 \cdot 10^{14}$ neutrons per cm^2, and could perhaps be extended to 10^{15} n/cm^2 for a few special devices. To be comfortable at 10^{15} neutrons per cm^2, device voltage breakdown limits would have to be less than 15 volts.

PROBLEMS

1. In the expression for the base transport factor α_T, why is ω_T the unity gain frequency, i.e., the gain-bandwidth product angular frequency?
2. How does the Ebers-Moll model devolve to a strictly linear system in the limit of low transistor element voltages?
3. In the exponentially graded base version of the Ebers-Moll equations, (5.89) and (5.90), insert the impact of the incident neutron fluence through the minority carrier lifetime, i.e., $\tau^{-1} \cong \Phi_n/K$. Use the connecting relation between diffusion length L, diffusion constant D, and the minority carrier lifetime τ, and rewrite the resulting Ebers-Moll equations, delineating the neutron fluence Φ_n explicitly.
4. Modify the standard diode current equation, (3.78), to include the effect of the increased leakage current due to the incident neutron fluence.
5. Show that the effects of the neutron fluence and case temperature are indeed additive by obtaining an alternative expression for the inverse gain, β^{-1}, from Eq. (5.82).
6. From the eight equations comprising the asymptotic resistivity kinetics model, obtain the fifth-degree algebraic equation satisfied by the vacancy density N_v. Discuss how to reduce it to a quadratic or cubic equation, in terms of ignoring the effects of certain of the defect complexes.
7. Obtain the graded base expression for the leakage current in Eq. (5.165), and insert the effect of neutron fluence degradation using the methods of Problem 4.
8. Using the Messenger-Spratt equation, Eq. (5.66), show, for a Darlington configuration of two identical bipolar transistors, that $\Delta(1/\beta_D) = (2/\beta_0 + K_1\Phi_n) K_1\Phi_n$. β_D is the post radiation Darlington gain, β_0 is the pre radiation gain of the individual transistors making up the Darlington circuit, and $K_1 = (\omega_T K)^{-1}$ is the corresponding damage constant of the individual transistor. Recall that $\beta_D = \beta^2$, where β is the post radiation gain of each transistor.
9. If one fast neutron collides with a silicon nucleus, the result might be a cascade of about 1000 silicon recoil atoms per $10^{-16}\,cm^3$.[49] What is the fluence level that would maintain the same cascade intensity in one gram of silicon?
10. To ameliorate the deleterious effects of single event upset (SEU), discussed in Section 7.9, series resistors can be inserted (monolithically) in the feedback paths of flip-flop integrated circuits. This slows their response, so that they are relatively insensitive to the incident cosmic ray SEU-inducing pulses, which are measured in fractions of pico seconds. How will such devices fare in an appreciably heavy concurrent neutron environment?
11. A single 1 MeV (fast) neutron, whose mean free path $\lambda = 4\,cm$ in silicon, is assumed to produce a sufficient number of primary displacement atoms to yield a total of about 10^{19} recoil atom displacements per cm^3 in silicon. What

is the number of displacements produced per primary recoil, if the fluence is the same as that computed in Problem 9?

12. Consider a graded base transistor, as discussed in Sections 3.11 and 5.11. Assume that the graded base dopant concentration can be approximated by a descending exponential with a distance from the base-emitter boundary X_{jE} to the emitter-collector boundary X_{jC}. Referring to Fig. 3.25, the dopant density at X_{jE}, $|N_D - N_A|_E \cong 2 \cdot 10^{18}\,\text{cm}^{-3}$, and at X_{jC}, $|N_D - N_A|_C \cong 5 \cdot 10^{15}\,\text{cm}^{-3}$. Also, the base width is essentially $|X_{jE} - X_{jC}| = 2 \cdot 10^{-4}\,\text{cm}$. Calculate:

(a) The internal (built-in) electrostatic field E_0 in the base, corresponding to the contact (built-in) potential ϕ_0.
(b) The "enhancement" factor η.

13. Make the argument for the minimum number of carriers required to delineate a cluster.

14. The ratio of the gain-bandwidth product in the normal direction (emitter-base-collector), f_{TN}, to that in the reverse direction (collector-base-emitter) f_{TI} for an exponentially graded base 2N2222A transistor, and whose $f_{TN} = 300\,\text{MHz}$, is about 100. Determine:

(a) The value of η.
(b) The internal electrostatic field E_0 in the base.

15. For a descending exponentially graded base, it is seen from Eq. (5.101) that the internal field E is constant (e.g., for $N_D(x) \sim \exp - x/L$, $N_D'/N_D = -1/L$) therein. From Problem 14, $E_0 = (kT/e)(\eta/W_b)$. If this E_0 is inserted into Eq. (5.102) initially, instead of the field given explicitly in terms of N_D'/N_D as in Eq. (5.101), then the development proceeds as in Section 5.11, resulting in a simple differential equation. The integral of that equation yields Eq. (5.109) directly. Show this.

16. Equation (5.8) shows that, to good approximation, a linear relation exists between the total number density of neutron-induced defects, N_D, produced in silicon, and the corresponding incident fluence Φ_n. As discussed in Section 5.14 in an analogous situation, such quantities can then be connected by a convolution integral, as in Eq. (5.143). That is, $N_D(t) = \int_0^t \Phi_n'(\tau) N_u(t - \tau)\, d\tau$, where $N_u(t)$ is the defect density due to a unit impulse of fluence $\delta\Phi_n = \dot{\Phi}_n \delta(t)$, where $\delta(t)$ is the delta function. For an $N_u(t)$ found to be given by[36] $N_u \cong N_{u0} \exp - \alpha_0 t$, express N_D for a fluence pulse $\Phi_n(t) = \dot{\Phi}_{n0} t$ of duration t_0. N_{u0} and Φ_{n0} are the respective constants.

REFERENCES

1. M. S. Ash., Nuclear Reactor Kinetics, 2nd ed. McGraw-Hill, New York, Chap. 9, 1979.
2. A. M. Mazzone. "Monte Carlo Methods in Defect Migration." IEEE ED-32(10):1925–1929. Oct. 1985.

3. G. P. Mueller and C. S. Guenzer. "Simulations of Cascade Damage in Silicon." IEEE Trans. Nucl. Sci. NS-27(6):1474, Dec. 1980.
4. V. A. J. Van Lint. Mechanisms of Transient Radiation Effects. Gulf Radiation Technology Doc. no. GA 8810, Aug. 1968.
5. J. Linhard, M. Scharff, and H. E. Schiott. Kgl. Vidskb. Slskbd., Fys. Medd. 133(14), 1963.
6. A. R. Sattler. Phys. Rev. 138:A1815, 1965.
7. E. C. Smith et al. "Theoretical and Experimental Determination of Neutron Energy Deposition in Silicon." IEEE Trans. Nucl. Sci. NS-13(6), Dec. 1966.
8. E. P. Blizard and R. Clifford. Report No. ORNL 51-10-70, Oak Ridge National Labs., Oak Ridge, TN, March 7, 1952.
9. H. J. Stein. "Comparison of Neutron and Gamma Ray Damage in *n*-Type Silicon." J. Appl. Phys. 37:3382, 1966.
10. A. B. Phillips. Transistor Engineering. McGraw-Hill, New York, 1962.
11. L. W. Ricketts. Fundamentals of Nuclear Hardening of Electronic Equipment. Wiley-Interscience, New York, pp. 59, 185, 1972.
12. O. L. Curtis, Jr. "Effects of Oxygen and Dopant on Lifetime in Neutron Irradiated Silicon." IEEE Trans. Nucl. Sci. NS-13(6):33, Dec. 1966.
13. G. C. Messenger. "A Two Level Model For Lifetime Reduction Processes in Neutron Irradiated Silicon and Germanium." IEEE Trans. Nucl. Sci. NS-14(6), Dec. 1967.
14. G. C. Messenger. Basic Neutron Displacement Damage Effects in Silicon Semiconductors. Intl. Conf. Rad. Phys. Semiconductors and Related Mat'ls., Tiflis, USSR., p. 6, Sept. 13–19, 1979.
15. C. Ramsey and P. Vail. "Current Dependence of the Neutron Damage Factor." IEEE Trans. Nucl. Sci. NS-17(6):310–316, Dec. 1970.
16. G. C. Messenger. "A General Proof of the Beta Degradation Equation For Bulk Displacement Damage." IEEE Trans. Nucl. Sci. NS-20, Feb. 1973.
17. J. J. Loferski. "Analysis of the Effect of Nuclear Radiation on Transistors." J. Appl. Phys. 29(1), Jan. 1958.
18. R. K. Thatcher and J. J. Kalinowski (eds.). 1969 TREE Handbook, p. F-42.
19. R. M. Burger and R. P. Donovan (eds.). Fundamentals of Silicon Integrated Device Technology, vol. 2, Bipolar and Unipolar Transistors. Prentice-Hall, Englewood Cliffs, N.J. 1963.
20. Burger and Donovan (eds.), loc. cit.
21. C. S. Meyer, D. K. Lynn, and D. J. Hamilton (eds.). Analysis and Design of Integrated Circuits. Motorola Semiconductor Engineering Staff, Phoenix, AZ. McGraw-Hill, New York, p. 40, 1968.
22. R. E. Lapp and H. L. Andrews. Nuclear Radiation Physics. Prentice-Hall, Englewood Cliffs, NJ, Chap. 12, 1948.
23. W. M. Webster. "On the Variation of Junction Transistor Current Amplification Factor With Emitter Current." Proc. IRE 42:914–920, June 1954.
24. W. Schockley and W. T. Read, Jr. "Statistics of the Recombination of Holes and Electrons." Phys. Rev. 87(5), Sept. 1952.
25. J. J. Ebers and J. L. Moll. "Large Signal Behavior of Junction Transistors." Proc. IRE 42:1761–1772, Dec. 1954.
26. Meyer et al., op. cit. p. 71.
27. F. Larin. Radiation Effects in Semiconductor Devices. Wiley, New York, pp. 163, 185, 1968.
28. M. G. Buehler. "Design Curves For Predicting Fast Neutron Induced Resistivity Changes in Silicon." Proc. IEEE 56(10):1741–1743, Oct. 1968.

29. H. J. Stein and R. Gereth. "Introduction Rates of Electrically Driven Active Defects in *n* and *p*-Type Silicon by Electron and Neutron Irradiation." J. Appl. Phys. 39(6):2890–2904, May 1968.
30. H. H. Sander and B. L. Gregory. IEEE Trans. Nucl. Sci. NS-13(6), Dec. 1966.
31. J. W. Harrity and C. E. Mallon. Short Term Annealing in Semiconductor Materials and Devices. A. F. Weap. Lab. AFWL-TR-67-45, 1967.
32. J. W. Harrity, V. A. J. Van Lint, and R. A. Poll. Short Term Annealing of Beta Degradation in *pnp* Transistors Exposed to Pulsed Neutron Fluence. General Dynamics, GA-6656, 1965.
33. Harrity and Mallon, op. cit.
34. B. L. Gregory and H. H. Sander. "Transient Annealing of Defects in Irradiated Silicon Devices." Proc. IEEE 58(9):1328–1341, Sept. 1970.
35. H. H. Sander and B. L. Gregory. Circuit Applications of Transient Annealing. Sandia Report no. SCR-72-2703, Sandia Corp. Albuquerque, NM. Sept. 1971. Figure plotted from nomogram.
36. L. A. McMurray and G. C. Messenger. "Rapid Annealing Factor For Bipolar Silicon Devices Irradiated by a Fast Neutron Pulse." IEEE Trans. Nucl. Sci. NS-28(6):4392–4396, Dec. 1981.
37. G. D. Watkins. Radiation Effects in Semiconductors. Plenum Press, New York, pp. 67–81, 1968.
38. C. F. Derbenwick and H. H. Sander. "CMOS Hardness Prediction For Low Dose Rate Environments." IEEE Trans. Nucl. Sci. NS-24(6):2244–2247, Dec. 1977.
39. M. E. Van Valkenberg. Network Analysis. Prentice-Hall, Englewood Cliffs, N.J., Chap. 7, 1965.
40. D. H. Alexander, J. E. Cooper, H. C. Heaton, R. N. Land, G. C. Messenger, and C. R. Viswanathan. "Neutron Damage Constants For Bipolar Transistors." Rockwell Int'l. Autonetics Div., Report no. X9-1283/601, pp. 20 ff., July 1969.
41. V. A. J. Van Lint, T. M. Flanagan, R. E. Leadon, J. A. Naber, and V. C. Rogers. Mechanisms of Radiation Effects in Electronic Materials. J. Wiley, New York, Vol. 1, pp. 142 ff., 1980.
42. G. C. Messenger and J. P. Spratt. "The Effects of Neutron Irradiation on Germanium and Silicon." Proc. IRE 46:1038–1044, June 1958.
43. J. Gover, A. Grinberg, and A. Seidman. "Computation of Bipolar Transistor Base Parameters For General Distribution of Impurities in Base." IEEE Trans. Elect. Dev. ED-19(8), 967–975, Aug. 1972.
44. C. D. Maldonado and C. T. Kleiner. "Modification to the TRAC Ebers-Moll Model For Improved High Frequency and Collector Storage Time Predictions." IEEE Trans. Nucl. Sci. NS-18(2), 1971.
45. G. C. Messenger. "Radiation Effects in Microcircuits." IEEE Trans. Nucl. Sci. NS-13(6):141–159, Dec. 1966.
46. G. C. Messenger. "Displacement Damage in Silicon and Germanium Transistors." Trans. Nucl. Sci. NS-12(2), April 1965.
47. J. E. Gover and T. A. Fisher. "Radiation Hardened Microelectronics For Accelerators." IEEE Trans. Nucl. Sci. NS-35(1):160–165, Feb. 1988.
48. M. G. Buehler. Texas Instruments Corp. Report No. 03-71-01. Dallas, TX, 75222, 1971.
49. Van Lint, op. cit. p. 303.
50. D. J. Garber and R. R. Kinsey. "Neutron Cross sections, Vol. II, Curves." Brookhaven Nat'l Labs. BNL-325 (TID 4500), Third Edition, p. 69, Jan. 1976; T. F. Luera, J. G. Kelly, H. J. Stein, M. S. Lazo, C. E. Lee, L. R. Dawson. "Neutron Damage Equi-

valence for Silicon, Silicon Dioxide and Gallium Arsenide." IEEE Trans. Nucl. Sci. NS-34(6):1557–1563, Dec. 1987.

51. T. J. Magee. "Direct Observation of Small Distributed Clusters in 14 MeV Neutron Irradiation Silicon." DNA TREE Program Conf., White Oak, MD, pp. 27–30, April 1982.
52. G. P. Mueller, N. D. Wilsey, and M. Rosen, "The Structure of Displacement Cascades in Silicon." IEEE Trans. Nucl. Sci. NS-29(6):1493–1497, Dec. 1982.
53. Burger and Donovan (eds.), op. cit. p. 61.
54. G. C. Messenger. "Current Gain Degradation Due to Displacement Damage For Graded Base Transistors." Proc. IEEE, pp. 413–414, March 1967.
55. J. L. Moll and I. M. Ross. "The Dependence of Transistor Parameters on the Distribution of Base Layer Resistivity." Proc. I.R.E. 44:72–78, Jan. 1956.
56. Garber and Kinsey, op. cit. p. 72.
57. F. B. McLean and H. E. Boesch, Jr. "Time-Dependent Degradation of MOSFET Channel Mobility Following Pulsed Irradiation." HDL-PP-NWR-89-1, Harry Diamond Labs, Adelphi, MD, Sept. 1989; IEEE Trans. Nucl. Sci. NS-36(6):1772–1783, Dec. 1989.
58. T. P. Ma and P. V. Dressendorfer (eds.). Ionizing Radiation Effects in MOS Devices and Circuits. J. Wiley-Interscience, Chap. 9, 1989.
59. G. C. Messenger. "Conductivity Modulation Effects in Diffused Resistors at Very High Dose Rate Levels." IEEE Trans. Nucl. Sci. NS-26(6):4725–4729, Dec. 1979.
60. R. J. Chaffin. Microwave Semiconductor Devices: Fundamentals and Radiation Effects. J. Wiley-Interscience, New York, 1973.
61. C. J. Dale, P. W. Marshall, E. A. Burke, G. P. Summers, and E. A. Wolicki. "High Energy Electron Induced Displacement Damage in Silicon." IEEE Trans. Nucl. Sci. NS-35(6):1208–1214, Dec. 1988.
62. B. R. Gossick. "Disordered Regions in Semiconductors Bombarded by Fast Neutrons." J. Appl. Phys. 30:1214, 1959.
63. J. J. Kalinowski and R. K. Thatcher. TREE Handbook. Battelle Mem. Inst. 2nd Ed., Sec.E-6, 1969.
64. J. R. Srour. "Basic Mechanisms of Radiation Effects on Electronic Materials, Devices and Integrated Circuits." DNA Report, Washington, DC, Aug. 1982.

CHAPTER 6
IONIZING RADIATION DAMAGE EFFECTS

6.1 INTRODUCTION

This chapter addresses the effects of ionizing dose, primarily from gamma rays and x-rays, on electronic systems. The corresponding absorption of these rays in materials of interest is referred to as the ionizing dose, as opposed to the gamma- and x-ray dose rate discussed in the next chapter. However, the damage from the deposition of ionizing dose is a function of both the dose and the dose rate. Gamma rays are members of the family of photons that are quantized manifestations of electromagnetic energy. Other members are x, ultraviolet, visible, and infrared rays, etc., across the energy spectrum to include radio waves. These rays, considered as waves of electromagnetic energy, are also characterized by their wavelength. High-energy gamma rays possess the shortest wavelengths, of the order of fractions of an angstrom (10^{-8} cm) to a few angstroms for x-rays. These wavelengths increase to very long radio waves, whose wavelength can be up to a few miles. Gamma and x-rays are discussed because they are the substantive portion of the photon output of a nuclear detonation. Their nature and behavior are in marked contrast to those of neutrons discussed in the previous chapter. These fleeting, almost evanescent particles, in common with all photons, have zero rest mass. This means that their existence ceases by annihilation if and when they are brought to rest, as when absorbed by matter. All photons travel with the speed of light, are uncharged, and interact mainly with free electrons or electrons bound to an atomic system. They can also interact with atomic nuclei. Their energy $E = h\nu = hc/\lambda$.

Depending on their energy, x-rays and gamma rays interact with matter in three principal ways: (a) At the low-energy extreme for x-rays, of the order of a few KeV, their interactions are mainly through the photoelectric effect. This is described briefly as when an x-ray penetrates the usually innermost electron shell structure of an atom and gives up all its momentum and energy, thereby being annihilated. This energy excites the atom, causing it to expel one of its innermost shell electrons, thereby ionizing the atom. The expelled, swiftly moving electron carries off part of the energy supplied by the annihilated x-ray as kinetic energy. Another electron within the atom now essentially deexcites the atom by dropping into the energy vacancy in the electron shell previously occupied by the expelled electron. The energy difference of this latter electron between its old and new state is now expelled from the atom in the form of a photon. This photon has less energy than the initial one, so that its wavelength is longer, usually in the ultraviolet or visible region, depending on the material. This radiated photon is called *fluorescence radiation*, which when emanating from the air atoms surrounding a nuclear detonation forms the glowing fireball. Infrequently, the energy difference is instead transferred to another (Auger) electron, which is then also expelled. In either the direct photon or Auger process, this cascade continues until the weakly bound outer electrons are participating in only low-energy optical photon transitions. (b) For higher energy photons, namely in the preponderant energy regime of gamma rays emanating from a nuclear burst, the main interaction with matter is through the compton effect. This is simply a collision between an incident photon and an electron that is free or relatively weakly bound to an atom. In the compton effect, only part of the photon energy is transferred to the electron, which even if weakly bound can be propelled out of the atom thus ionizing it. In any event, the x-ray photon, as a result of the scattering (collision) careens off in a new direction, but with less energy and a longer wavelength than it had prior to the encounter. (c) For very high energy photons, in the regime of high-energy gamma rays, a third interaction occurs called *pair production*, or *pair creation*. If a sufficiently energetic photon finds itself near an atomic nucleus, it can be spontaneously annihilated. In its place instantly appears a fast-moving electron, plus a fast-moving positron. The positron is a particle with all the properties of an electron, except that its charge is positive. Conservation of energy is assured, for the kinetic energies of the moving electron and positron, plus their mass-energy equivalent, $E = mc^2$, add up to the energy supplied by the initiating gamma photon. The atomic nucleus is necessary in that it is the third participating particle in the interaction, so that the momenta of the electron, positron, and nucleus add vectorially to the momentum of the initiating gamma ray for momentum conservation. The positron is an actual, experimentally observed particle. It should not be confused with the hole, which is basically a construct used to facilitate description of the theory of the solid state.

Fissions occurring in the nuclear detonation provide prompt gamma rays and x-rays that are expelled from the burst as part of the menagerie of particles and radiation energy from this process. On the other hand, delayed gamma and x-rays are emitted from the radioactive fission fragment components of the bomb fuel debris, as well as from the neutron-activated atomic nuclei of the weapon structure. The delayed gammas are emitted from their radioactive nuclei with characteristic half-lives ranging from microseconds to days. The details of the foregoing will be discussed. Now a description of gamma- and x-ray sources used to simulate this component of the nuclear weapon output is given.

6.2 X-RAY AND GAMMA-RAY SOURCES

Because the garden variety of x-ray machine as used in medicine can be construed as a prototype x-ray producer for simulation purposes, a brief description of its operation is given. It is a fact of classical electrodynamics that if charged particles are accelerated, then they must radiate electromagnetic energy, i.e., photons. Hence, the medical x-ray machine simply consists of an electron gun, not unlike that in a television picture tube, which accelerates electrons onto a metal target, usually of a high atomic number (high Z) to enhance the interaction probability and thus the x-ray output. These electron collisions with the target are almost always near-misses, as is so with most atomic and nuclear collisions. Nevertheless, these result in a change in direction of the electron as it approaches and is deflected from the target atom. This direction change about the target atom as an instantaneous center corresponds to acceleration of the incoming electron, causing it to radiate electromagnetic energy. The collision parameters are such that the wavelengths of the radiated photons lie in the x-ray region. This type of radiation production is called *bremsstrahlung* radiation, which means braking radiation.

6.3 SPECIFIC SOURCE TYPES

The flash x-ray machine (FXR) corresponds to those types of machines that generate a highly intense pulse of *bremsstrahlung* radiation (x-rays) for simulation purposes. The electrons that are accelerated to produce these x-rays are stored as charge in a pulse-forming coaxial-type short transmission line. This pulse former line is then switched to a field emission diode, where the electrons are accelerated toward, and deflected from, a target of high Z material such as tungsten. The electrons can be accelerated to well beyond 10 MeV, with corresponding currents of a few milliamperes per pulse. Such machines deliver *bremsstrahlung* pulses of very high dose rate over a relatively large solid angle, so that sizeable systems can be irradiated. FXR machine sizes span a broad selection of radiation intensities, energies, and beam spatial distributions. FXR parameters include pulse-forming network geometry, accelerating voltage, associated electronic circuitry, and the amount of stored charge capability. Limitations exist; for example, the pulse width is fixed for a given machine. The physical size of the machine usually correlates with the amount of pulse energy it can deliver. For example, at the low energy end of the scale, the *Febetron* class of machine sizes is about that of a large home hot water boiler, horizontally mounted, providing pulse energy of a few hundred joules. The Aurora machine, one of the largest known, whose Blumlein pulse generators alone are the size of a square array of four electrical blast furnaces, delivers megajoules of energy per pulse.

Another type of x-ray producer for simulation is the linear accelerator, or LINAC. It generates a resonant radio frequency electromagnetic field that bunches electrons emanating from a source into a tightly packed pulse. These electrons are then accelerated toward a target, usually made of tungsten, to yield *bremsstrahlung* radiation, not unlike the FXR. Or the target can be made of a low or moderate Z material, such as aluminum, allowing most electrons to pass through on their way toward the system under test. Hence, the LINAC can provide either photons or electrons, depending on the type of target used. Although the LINAC delivers much less fluence that the FXR, its pulse parameters, including pulse width, are readily variable, in contrast to the fixed pulse width of the FXR. Also, the shot repetition rate of the LINAC is about the same as the FXR, measured in tens of pulses per hour. As with all these machines, the device under test (DUT) can be placed at, in, or near, the target, depending on the type of machine and the device to be irradiated.

For gamma-ray sources for simulation purposes, radioactive isotopes provide substantive amounts of radiation exposure. A popular isotope is ^{60}Co, with a half-life of 5.3 years, which emits two characteristic gamma rays of 1.17 and 1.33 MeV. ^{60}Co is produced by neutron irradiation of stable ^{59}Co, ultimately decaying to stable ^{60}Ni, emitting the above two gamma rays in the process. For a source strength of 100 kCi, the gamma ray flux is of the order of 10^{11} photons per $\mathrm{cm}^2 \cdot \mathrm{s}$ at about one foot from the source. One curie (Ci) corresponds to $3.7 \cdot 10^{10}$ disintegrations of the source isotope per second, each disintegration resulting in the emission of both of the above gamma rays in its gamma ray decay scheme. In terms of the corresponding dose rate of $4 \cdot 10^{-3}$ rad (air) per second per curie, at about one foot from the above ^{60}Co source, dose rates of 500 to 1000 rad (air) per second can be readily obtained. Even though these source strengths are orders of magnitude weaker than other types of gamma sources, they are adequate for irradiating semiconductor electronics, such as transistors and integrated circuits, as well as for the examination of point defect dislocation phenomena occurring in metals of high purity.

Another important ionizing dose simulation source is the x-ray wafer level probe. Recalling the integrated circuit manufacturing process, from the crystalline boule to wafer slice to chip lithography and packaging, it is realized that wafer chip lot quality is a prime consideration. This can prove inordinately expensive from the radiation hardness viewpoint, in that a poor-quality lot yield will be discovered only after radiation testing of each chip secured in its package has taken place. Radiation testing and screening of chips at the wafer level after the lithographic process but prior to dividing the wafer and packaging each chip, would be extremely expedient in terms of value added time and money saved. Apparatus is available to accomplish this task (e.g., ARACOR). A wafer chuck holds the wafer, while the probe system electrically connects each chip on the wafer in succession to appropriate test and monitoring equipment. A highly collimated x-ray beam (e.g., a 10 KeV tungsten target tube) is incident on the individual chip, after which the chip is viewed microscopically and marked accordingly, depending on the x-ray ionization dose test results. Also, with a laser beam directed at each chip, nondestructive dose-rate and latchup tests can be conducted. Initial x-ray and ^{60}Co correlation results for polysilicon and aluminum gate MOS devices are claimed to be within 30 percent.[53] The general correlation of the x-ray wafer probe with standard ionizing radiation source results shows much promise.

Other sources of ionizing radiation include operating nuclear reactors, which provide fission product gamma rays and fast electrons, as well as fission spectrum neutrons. Also, spent nuclear reactor fuel elements are highly radioactive, yielding gamma rays of a strength comparable to other radioactive sources. The useful life of a spent reactor fuel element for these purposes is about 1 month following its discharge from the reactor. The average energy of fuel element gamma rays is about 0.7 MeV. Still another radioactive isotope source obtained from fission products whose parameters are suitable in this context is ^{137}Cs. The energy of gamma rays from cesium is about 0.66 MeV, with a corresponding half-life of cesium of about 30 yr.

6.4 GAMMA-RAY AND X-RAY SHIELDING

To address the problems of coping with gamma and x rays incident on electronic systems of interest, as well as the difficulties of handling radioactive, high-energy, gamma-ray sources, it is well to examine the fundamentals of the attenuation of these photons in materials of interest.

In their passage through matter, gamma and x rays are removed as already discussed. This results in a decrease in their intensity or fluence as a function of distance of penetration into the material. The fact that the extent of loss in an incremental thickness, dx, anywhere in the material is proportional to the intensity at that point and to the depth x that the radiation has traversed leads to

$$dI(x) = -\mu I(x)\,dx \tag{6.1}$$

where I is the radiation intensity expressed as photons per cm^2, or as energy units per cm^2 for monoenergetic photons, and the factor μ, in units of cm^{-1}, is called the *linear attenuation coefficient* of the material. The integral of Eq. (6.1) yields the well-known expression describing the exponential attenuation behavior for gamma and x-ray as

$$I(x) = I(0)\exp - \mu x \tag{6.2}$$

where $I(0)$ is the initial intensity. This equation, in reflecting the proportionality relationship in Eq. (6.1), implies that μ is the attenuation coefficient for absorption only. That is, if the compton scattering collisions are to be included, Eq. (6.1) does not hold in general. This is because multiple collisions can take place that scatter the photon, first out of the direction of penetration, then back into that direction, and so the photon can scatter back into the initial direction of the main photon stream. To include this multiple scattering transforms this simple analysis into the domain of photon transport theory. Certain approximations, which result in scattering buildup factors, can sometimes be used in lieu of the full-blown photon transport theory. Nevertheless, the above exponential attenuation behavior describes the situation with sufficiently good approximation so that it can be still used effectively. For example, the fact that the attenuation is exponential implies that the percentage, or fraction, of photons removed from the photon stream is constant for constant μ, as seen from Eq. (6.1). Hence, this fraction is independent of the initial intensity $I(0)$. Therefore, it requires the same thickness of material to decrease the intensity from its initial value of 100 percent to 50 percent, as it does to reduce it from 10 percent to 5 percent.

It is often advantageous to rewrite Eq. (6.1) as $dI = -(\mu/\rho)\,I\rho dx$, so that for a given energy

$$I(x) = I(0)\exp - \kappa\rho x \tag{6.3}$$

since κ is almost constant over many materials (Table 6.1), where the mass absorption coefficient $\kappa = \mu/\rho$, (cm^2 per g), and the incremental areal density of penetration becomes $\rho\,dx$ (g per cm^2). The linear attenuation coefficient $\mu = \lambda^{-1}$ where λ is the mean free path. This is easily shown by computing the mean free path as $\lambda = \int_0^\infty xI(x)\,dx / \int_0^\infty I(x)\,dx$, where I is given from Eq. (6.2). Also, it can be shown that

Table 6.1 Gamma-ray attenuation for air and lead versus gamma energy.

E Mev	AIR $\mu/\rho\,(\mathrm{cm}^2/\mathrm{g})$	AIR $\mu\,(\mathrm{cm}^{-1})$	LEAD $\mu/\rho\,(\mathrm{cm}^2/\mathrm{g})$	LEAD $\mu\,(\mathrm{cm}^{-1})$
0.5	0.10	1.1×10^{-4}	0.15	1.7
1	0.07	0.81×10^{-4}	0.07	0.8
2	0.05	0.57×10^{-4}	0.05	0.52
5	0.03	0.35×10^{-4}	0.04	0.50
10	0.02	0.26×10^{-4}	0.05	0.61

$\mu = \lambda^{-1} = N\sigma Z$, where NZ is the number of material electrons per cm^3, since N is the number of target atoms per cm^3, and σ is the microscopic interaction cross section (i.e., the interaction probability per photon for all photon processes in the material). Because the incident photons lose energy mainly by ionization, as discussed, these losses per centimeter of photon path are proportional to the average concentration of electrons in the material, namely NZ, where Z is the atomic number. So that the incremental path energy loss of a mono-energetic photon fluence, called stopping power (defined as $-dE/dx$), is given in terms of the material parameters by

$$-dE/dx \sim NZ \tag{6.4}$$

or

$$-dE/\rho\,dx \sim NZ/\rho \tag{6.5}$$

The negative sign merely signifies energy lost to the material. Now $N = \rho N_0/A$, where N_0 is Avogadro's number, and A is the mass number (atomic weight) of the material. Hence, $NZ/\rho = N_0 Z/A$. $Z/A \simeq \frac{1}{2}$ is relatively constant for most elements or materials, so that the stopping power is roughly constant for a broad segment of pertinent materials when the incremental areal density or incremental path length is expressed as $\rho\,dx$, or in units of grams per cm^2 of material, as in Eq. (6.5). Therefore, the areal density $\int_0^x \rho\,dx'$ (grams per cm^2), or loosely the material mass, is often used as the unit of penetration instead of simply the penetration distance in cm. Table 6.1 yields μ and μ/ρ for gamma rays being attenuated in air and in lead as a function of the energy of the incident photon. Table 6.2 gives some tenth value thicknesses (amount of material thickness needed to attenuate to 10 percent) for gamma rays emitted from fission products for various construction materials. As can

Table 6.2 Tenth value attenuation thicknesses for shielding construction materials for fission product gamma rays.

MATERIAL	DENSITY (lb per ft^3)	TENTH VALUE (in.)
Steel	490	3.7
Concrete	144	12
Earth	100	18
Water	62.4	26
Wood	34	50

be appreciated, effective practical gamma-ray shielding uses materials with as high an atomic number, Z, as practicable consistent with other requirements. Iron, as carbon steel or stainless steel, is used extensively for shielding in nuclear reactors. Iron is plentiful, machine-workable, relatively inexpensive, and can endure reasonably high temperatures. This is in contrast to lead, which is heavier than iron (higher Z), but has a relatively low melting point and thus is unable to stand high ambient temperatures. For gamma rays of about 2 MeV, iron and lead are nearly equal as shields. At low energies, such as in the x-ray regime, and for gamma rays of energies greater than 2 MeV, the increased mass attenuation efficiency of lead makes it a better shield than iron, as seen in Fig. 6.1. Figure 6.2 depicts similar curves for semiconductors of interest.

Tantalum and tungsten are both high-Z materials, and hence make good shielding materials for photons. However, they are both expensive, and are therefore restricted to special-purpose shielding applications, such as use at very high temperatures, or in missiles and spacecraft. Another point mitigating against their use, for some applications, is that when neutrons are also present in the incident fluence, they can activate both materials (i.e., make them radioactive), with a corresponding emission of gamma rays of up to 6 MeV and 7.4 MeV for tantalum and tungsten, respectively.

6.5 IONIZATION DAMAGE

The subject of ionization was discussed quantitatively in the introductory section insofar as it is caused by gamma and x-rays. Here, investigations will be made into the details of ionization in the context of photon damage to semiconductor devices, specifically with respect to degradation of electronic parameters of components and systems. All the photon inter-

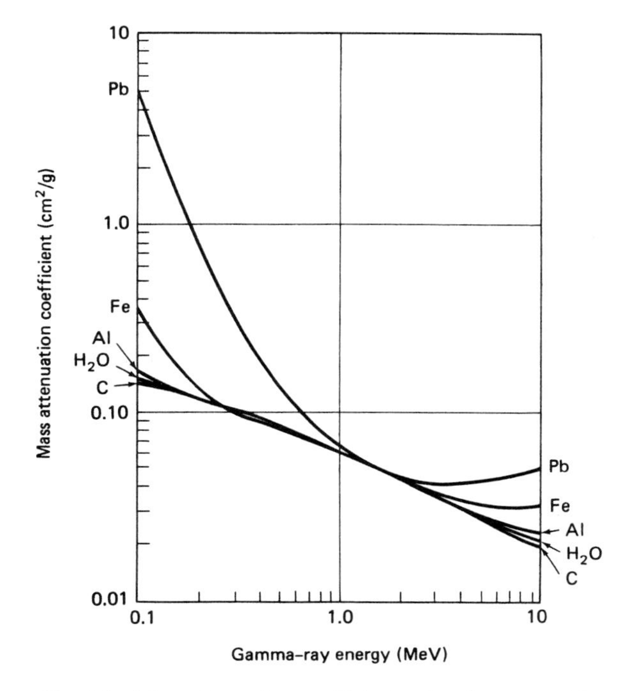

Fig. 6.1 Mass attenuation coefficients of various materials.

actions already discussed, i.e., the photoelectric effect, compton effect, and pair production, are ionization-causing processes in that free electrons, and also hole-electron pairs, are produced. Also, electrons originating from an electron-ion pair, i.e., from ionization, can themselves ionize atoms. This is done simply through electron-atom collisions that transfer sufficient energy to knock an electron out of an atom to produce further ion-electron pairs.

The principal ionization-induced changes in bulk material are conductivity increases through production of excess charged carriers (electrons and holes), trapped charges mainly in insulators, production of electric and magnetic fields, and chemical effects. Upon being released from an ionized atom, free electrons with enough energy are excited from the valence band and span the forbidden gap to easily reach the conduction band energy level states, thus creating hole-electron pairs. The energy difference that the electron has, over that used to span the gap, is converted into creating secondary hole-electron pairs, as already mentioned, or is transferred to the lattice as thermal energy. The electron, after

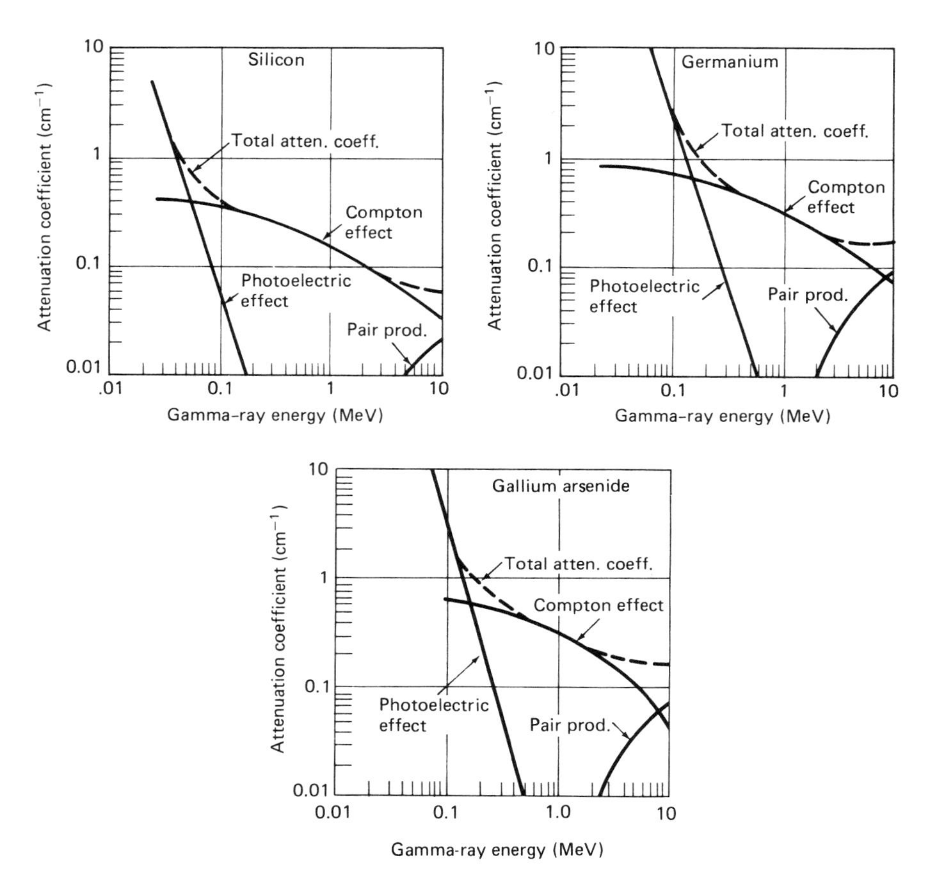

Fig. 6.2 Mean attenuation coefficients of gamma radiation in silicon, germanium, and gallium arsenide.

dissipating this energy, now has enough energy to reside in an energy state just above the lower edge of the conduction band. Similarly for the hole, but enough energy to reside just below the upper edge of the valence band. In silicon, experiments reveal that about 3.6 eV is expended to create an hole-electron pair. This is greater than three times the band gap energy of 1.1 eV for silicon. For semiconductors and insulators, the hole-electron pair creation energy is 2–3 times that of the corresponding band gap energy (Fig. 6.2A).

The specific ionization density, i.e., the number of hole-electron pairs created per incident rad absorbed in the material is constant and independent of temperature. For silicon, this is about $4.05 \cdot 10^{13}$ hole-electron pairs per $cm^3 \cdot$ rad (Si).

As already mentioned, a significant effect of ionizing photon radiation

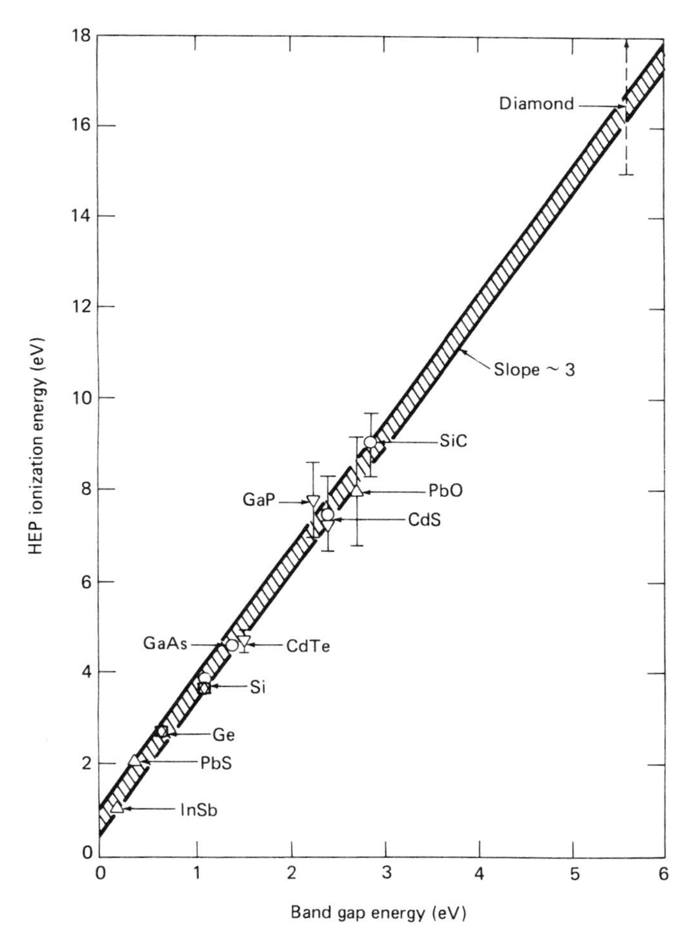

Fig. 6.2A Mean energy required to create a hole-electron pair versus band gap energy for several semiconductors.[79]

is the increase in the bulk conductivity of the material. From Eq. (1.56) it is apparent that the increase in conductivity due to the excess carriers thus formed is given by

$$\delta\sigma = e(\mu_n \delta n + \mu_p \delta p) \tag{6.6}$$

where $\delta\sigma$ is the difference between the post- and preradiation values of σ. δn and δp are the excess carrier concentrations. The mobilities involved can be assumed to be constant; however, they often actually diminish a little, due simply to electron and hole coulomb scattering with the ionized atoms.

Some of the electrons from hole-electron pairs merely escape by leaking from the surface of the material. Hence, the material now has a net positive charge due to the positive ion excess. If those electrons are captured in a contiguous material, the latter then will have an excess negative charge, causing an electric field to be generated due to the charge separation between the negative electrons in the adjacent material and the positive ions in the parent material. This field gives rise to a potential difference across the interface between the two materials. Current will start to flow across the interface, whose intensity is a function of the potential difference and the material conductivity, satisfying Ohm's law.

In insulators, excess carriers created by ionizing radiation hole-electron pairs principally from the fracture of homopolar (e.g. $Si-O_2$) bonds will be propelled (drift) due to internal electric fields produced by the radiation-induced charge separation. The effect of ionizing radiation increases the conductivity, but not enough to significantly affect the rate of charge recombination, so that charges will become trapped in the insulator material. Recall that the conductivity in semi-conductors irradiated by neutrons is decreased, due to neutron induced carrier removal. (Sec. 5.13) In high-quality insulators (low loss tangent) the recombination process can last for days after irradiation. Carriers therein can stay trapped at defect-complex sites. This can lead to permanent degradation of the dielectric properties of the insulator, even though it manifests charge neutrality. As already discussed, certain types of defect complexes can also rupture homopolar chemical bonds. This can result in permanent changes in the chemical composition of the material. An example in ice crystals is provided as one of the many radiation-induced chemical reaction sequences in water, where $h\nu$ represents the energy of an incident photon i.e.,

$$h\nu + H_2O \rightarrow H_2O^+ + e^-$$

$$e^- + H_2O^+ \rightarrow H + OH$$

$$H + H \rightarrow H_2$$

$$OH + OH \rightarrow H_2O_2$$

The formation of the free radicals H^+ and OH^- and hydrogen peroxide H_2O_2 are the results of the incident radiation. In the context of transient effects of radiation on electronics, for reasonably high dose rates, the number of ruptured homopolar bonds is negligible. However, for super-high radiation rates, it becomes appreciable. Ionization energy losses in materials result in an increase of internal energy in the material, i.e., heating. If the duration of the incident radiation pulse is shorter than the characteristic heat conduction time in the material, and the energy of the incident radiation pulse is large enough, then the heat cannot escape fast enough to prevent the material from melting or vaporizing. Likewise, if the duration of the pulse is short with respect to the speed of sound in the material, then large shock wave pressures can be generated leading to material spallation and fracture.

6.6 IONIZATION EFFECTS ON SURFACE STATES

As is apparent, lattice defects, ionizing-radiation-induced defect complexes, lattice dislocations, or nearly any foreign element introduced into the semiconductor lattice results in the appearance of energy-trapping states in the forbidden energy gap. It is therefore not too difficult to appreciate the introduction of such energy states by the lattice discontinuities due to the mere existence of the boundary surface, i.e., the existence of the end of the lattice. Such energy levels are called *surface energy states*. If they correspond to deep energy levels in the material energy gap, they should act as efficient surface recombination centers. Theoretical estimates of the numbers of surface states are on the order of 10^{15} per cm^2 of surface, or roughly the same as the surface atomic density. Such densities have been observed in the laboratory in vacuum. With exposure to air, their number drops radically to about 10^{11} per cm^2 of surface. This is due to the oxide layer that forms on the surface when exposed to ambient conditions. Surface recombination velocities, discussed in Section 1.14, are on the order of 100 cm per s, but only 1–10 cm per s on thermally oxidized silicon. When silicon is irradiated with photons, both the surface recombination velocity and the density of surface states increase. It is an experimental fact that the surface recombination velocity is proportional to the density of surface states. Consider the situation where there is an increased concentration of recombination centers in a thin layer just under the surface. It is anticipated that the recombination rate, U, will be increased near the surface, because of the increase in the number of surface states due to the radiation, so that the excess carrier density should be diminished in the neighborhood of the surface. To restore the concentration of excess carriers near the surface, carriers from the bulk of the material will begin to diffuse toward the surface due to the concentration gradient, in accordance with Fick's law. The carriers that diffuse into the subsurface region of increased recombination rate are the equivalent source of all those carriers that recombine there. The recombination rate of, e.g., p-type carriers in the subsurface is

$$U_s = \sigma_p v_{th} N_c l_s (p_n - p_{n0}) \tag{6.7}$$

where p_n is the average minority carrier concentration in the subsurface layer, N_c is the concentration of subsurface recombination centers, σ_p is the recombination probability (cross section) per carrier, and l_s is the thickness of the subsurface layer. Because the fluence of minority carriers into the subsurface region must equal U_s, then

$$D_p(\partial p_n/\partial x)|_{x=0} = \sigma_p v_{th} N_c l_s (p_n - p_{n0}) \tag{6.8}$$

Comparing Eq. (6.8) with Eq. (1.114) yields the expression for the surface recombination velocity, S_p, for holes, as

$$S_p = \sigma_p v_{th} N_s \tag{6.9}$$

where $N_s = N_c l_s$ is the areal number of recombination centers within the subsurface. It is seen that surface recombination can be construed as bulk recombination for a high density of recombination centers within a very thin subsurface volume. However, if negatively charged ions are present on the free surface of, say, an n-type semiconductor, they will perturb the carrier distribution within. Electrons will be repelled from the inside surface, while holes will be attracted there. Consequently, space charge neutrality will no longer hold in a thin subsurface region. This region is called the *surface space charge region*. When irradiated with photons, hole-electron pairs will be generated throughout its interior, and carriers will diffuse toward the surface to recombine as already described. By analogy with Eq. (1.66), the rate of carriers recombining on a unit surface area is

$$U_s = \frac{\sigma_p \sigma_n v_{th} N_s (n_s p_s - n_i^2)}{\sigma_n(n_s + n_i \exp(E_t - E_i)/kT) + \sigma_p(p_s + n_i \exp - (E_t - E_i)/kT)} \tag{6.10}$$

where n_s and p_s are the electron and hole concentrations at the surface. For convenience, assume that most of the recombination centers are those whose energies are such that $E_t \cong E_i$, and that the crossections for recombination of holes and electrons are equal, i.e., $\sigma_n = \sigma_p = \sigma$; then from Eq. (6.10)

$$U_s = S_{p0}(n_s p_s - n_i^2)/(n_s + p_s + 2n_i); \qquad S_{p0} = \sigma v_{th} N_s \tag{6.11}$$

where S_{p0} is the surface recombination velocity in the absence of a surface space charge layer. Now, the fluence of minority carriers reaching

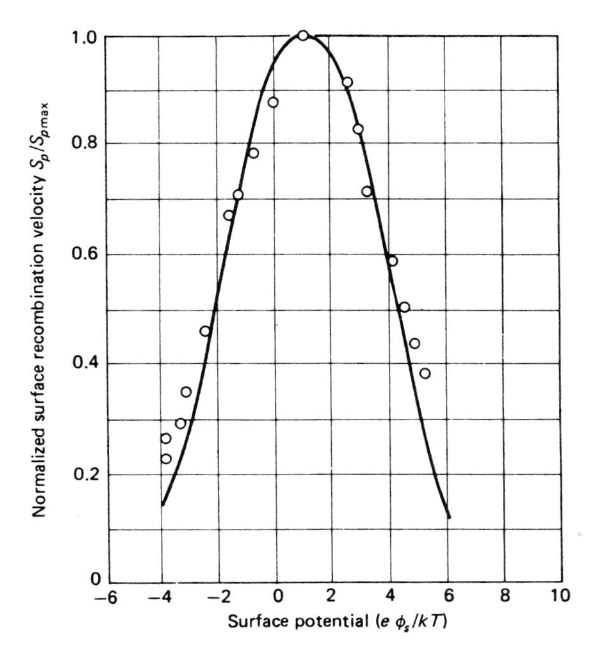

Fig. 6.3 Surface recombination velocity versus surface potential for Germanium.[17]

the surface must be given by U_s. For moderate recombination rates, so that the fluence of carriers can be approximated by the fluence reaching the surface space charge layer, then

$$D_p(\partial p_n/\partial x)|_{x=l_s} = S_{p0}(n_s p_s - n_i^2)/(n_s + p_s + 2n_i) \tag{6.12}$$

Now assume $n_s p_s$ is constant throughout the surface space charge layer, even though equilibrium does not exist within. Then

$$n_s p_s = p_n(l_s) n_s(l_s) \cong p_n(l_s) N_D \tag{6.13}$$

where N_D is the donor concentration, and that $p_{n0} N_D = n_i^2$. Then, from Eq. (6.12)

$$\begin{aligned} D_p(\partial p_n/\partial x)|_{x=1_s} &= S_{p0} N_D (p_n(l_s) - p_{n0})/(n_s + p_s + 2n_i) \\ &= S_p(p_n(l_s) - p_{n0}) \end{aligned} \tag{6.14}$$

where the actual surface recombination velocity is

$$S_p = S_{p0} N_D/(n_s + p_s + 2n_i) \qquad (6.15)$$

As the surface charge varies, n_s, p_s, and S_p will also vary. Note that $\min(n_s + p_s) = 2n_i$. That this is the case is seen by the fact that $\min(p_s + n_s)$ implies $d(n_s + p_s) \equiv dn_s + dp_s = 0$. Since $n_s p_s = n_i^2$, $n_s dp_s + p_s dn_s = 0$. Substituting yields $n_s dp_s + p_s(-dp_s) = 0$, giving $n_s = p_s = n_i$. Hence, $\min(n_s + p_s + 2n_i) = 4n_i$. The surface recombination velocity then achieves its maximum value given by

$$S_{p\max} = S_{p0} N_D/4n_i \qquad (6.16)$$

This corresponds to the situation when the surface fields (i.e., surface potentials) due to each charge polarity nearly cancel,[17] as seen in Fig. 6.3.

6.7 CHARGE-TRAPPING

Charge-trapping means the witholding of individual charge motion at trapping sites in the material. It occurs most readily in insulators, so that a brief discussion of the solid state aspects of insulator function is presented. For insulators that are made of crystalline substances, the forbidden energy gap theory and all its ramifications hold. The gap width of insulators is about triple that of semiconductors, more than 8 eV in silicon dioxide. Insulator mobilities are very much less than their values in semiconductors.

Corresponding mobilities are of the order of 1–50 cm^2 per volt second in silicon dioxide and are much less for amorphous, or noncrystalline, materials. As the insulator band gap is relatively wide, this implies that a very high impurity concentration of about 0.1 percent of the atomic concentration of the material itself can be sustained, and with the material nevertheless exhibiting very high resistivity.

For insulators composed of amorphous materials, including noncrystalline substances, there is a semblance of a gap apparent, but their properties in this context are considered those of a bulk material with low conductivity.

If radiation such as gamma rays, x-rays, or electrons, is incident on an insulator, the injected electrons, or those produced within the material through ionization result in the buildup of the trapped component of these charges therein. These charges and their neutralizing charge counterparts, which are manifest on nearby conductors, result in the generation of an electric field. As the electric field strength increases with the production of these charges, the increased conductivity can sometimes allow sufficient countercurrent back to the surface to establish an equilibrium current flow within the material. However, if the conductivity increase is not sufficient, and the electric field continues to increase, then electrical breakdown occurs, and the charge returns to the surface. This is the manner by which the Lichtenberg arboreal figures are produced in some transparent materials (e.g., Lucite) that have been subjected to ionizing radiation. These figures appear when the electrically "supercharged" Lucite is discharged by rapping it against a hard surface or by quickly bringing a grounded lead near the thus-charged material. A Lichtenberg (fractal) figure is the basis of the design of the jacket of this book, and shown in the frontispiece.

In insulators, the conductivity may be so very low that the radiation-induced space charge can persist for many days, as mentioned. Also, both electron and hole carriers that are trapped cannot recombine with their opposite numbêr, which are also trapped elsewhere in the material. Further, the properties of the material can undergo changes, even though there is no net macroscopic charge evidenced, i.e., the insulator as a whole remains electrically neutral.

Trapped carriers can also change the optical properties of materials. For example, the creation of *F* centers in alkali halides, such as potassium chloride, or in various glasses are caused by certain traps. Specifically, when an electron from an ion-electron pair is trapped in a negative ion-vacancy type of defect complex, the result is called an *F center*. These centers can now absorb light, as seen by the discoloration of the glass. In potassium chloride, light whose wavelength is 546 nm is strongly absorbed. The strength of the optical absorption can be correlated with the number density of radiation-induced ionization electrons from an incident pulse of radiation.[1] As seen in the next section, charge-trapping due to ionizing radiation occurs in the dielectric portions of field effect devices (FETs). This plays a role in the operation of these devices, including bending of the energy bands, as discussed therein.

6.8 FIELD EFFECT TRANSISTORS

In keeping with the tenor of the early chapters, with respect to the discussion of bipolar transistors, a description of the field effect transistor (FET) is given. Chapters 1 and 3 are concerned with bipolars whose majority and minority carriers are *n* or *p*, depending on whether the transistor is a *pnp* or *npn* type. The field effect transistor, by contrast, is a single (majority) carrier device, so that minority carrier lifetime degradation is essentially unimportant to its operation. A basic FET is the junction or JFET. The insulated gate FET (IGFET) is more commonly referred to as the *metal-oxide semiconductor field effect transistor* (MOSFET). Their advantages over bipolars include much less chip space

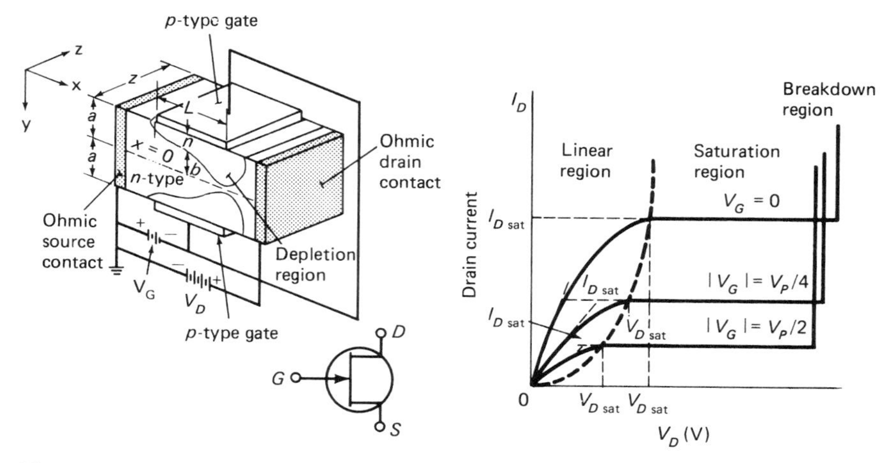

Fig. 6.4 *N*-channel JFET depicting its structure, symbol, and characteristic curves.[34]

per transistor as part of an integrated circuit, fewer processing steps, high input impedance, zero offset voltage, better temperature stability, and less noise. Their primary electrical disadvantage is relatively poor high frequency performance compared to bipolar transistors.

Physically, an *n*-channel JFET, Fig. 6.4, consists electrically of a channel of *n*-type semiconductor with a source *S* at one end and a drain *D* at the other end. The drain to source current, I_D, flows through the channel. The *n*-channel lies between two *p*-type regions whose common terminal is called the gate, *G*. These two *p*-type regions form two *p-n* junctions, one on each side of the channel, as in Fig. 6.4. For a common source configuration, the two gate-channel *p-n* junctions are reverse biased. For a positive source-drain voltage, a source to drain current flows. This current consists only of majority carriers, which are electrons for this *n*-channel JFET, or holes for a *p*-channel JFET. The current flows from source to drain in either case. The device obtains its name from the fact that its conductance and current flow are controlled by the electric field across the gate-channel interface. In the drain current-drain-source voltage characteristic curves, when the gate-source voltage is equal to $-V_p$, the pinchoff voltage, then the drain current becomes independent of V_{DS}, but not zero. The pinch-off state does not mean that the drain current is cut off, it merely means that the electric field has pinched off the channel. That is, the depletion regions of the two-gate channel junctions have consumed all the channel, and any subsequent current flow is in accordance with the bulk resistance of the depletion and channel regions, so that the system acts like a passive device. This

Table 6.3 Properties of the four MOSFET types.

MOSFET PROPERTY	(1) ENHANCEMENT MODE	(2)	(3) DEPLETION MODE	(4)
Channel carrier type	n	p	n	p
Substrate material type	p	n	p	n
Energy bands bend at channel-gate interface	down	up	up	down
Built-in channel	no	no	yes	yes
Drain current at $V_g = 0$	off	off	on*	on*
Gate voltage (V_g) polarity	+	−	−	+
Drain voltage (V_D) polarity	+	−	+	−
Source voltage (V_s) polarity	−	+	−	+
Pinchoff voltage (V_p) polarity	+	−	−	+
Threshold voltage (V_T) polarity	+	−	−	+

* Depletion mode devices normally have doped built-in channels, so they have an initial inversion layer and so are normally on with zero V_G. However, sufficient V_G of negative polarity can turn them off.

corresponds to the saturation behavior as depicted in the rightmost extremities of the characteristic curves in Fig. 6.4. The JFET also will be discussed further in Chapter 7, insofar as its performance degradation with respect to transient photon pulses is concerned. The JFET is very resistant to ionizing dose radiation, mainly because its gate contains no oxide layer, unlike the MOSFET gate oxide insulator device.

There are two main types of insulated gate FETs, as seen in Table 6.3. They are the depletion mode and enhancement mode MOSFET, and each can possess a p-channel or n-channel. A p-channel enhancement mode device means that the normally n-material substrate carrier population, contiguous to the gate, is inverted to form a p-channel there, and vice-versa for the n-channel device. Importantly, in the depletion mode device, the built-in channel is normally on, i.e., source-drain current flows with zero gate bias. The p-channel is built-in by doping the n-substrate in that neighborhood with acceptor atoms, and donor atoms for the n-channel device. Also, heavily doped p regions at each end of the channel then connect to the source and drain. Of central importance is the silicon dioxide insulator layer parallel to and astride the length of the channel called the gate oxide, upon which is deposited a metal contact, as in Figs. 6.5, 6.6, and 6.9 for a p-channel device. This is called the gate terminal and the gate system is referred to as a "metal gate". A "polysilicon gate" uses heavily doped polysilicon as the gate terminal instead of metal. When a positive bias is applied to the gate, it induces an electric field in the oxide layer of such polarity that it repels (depletes)

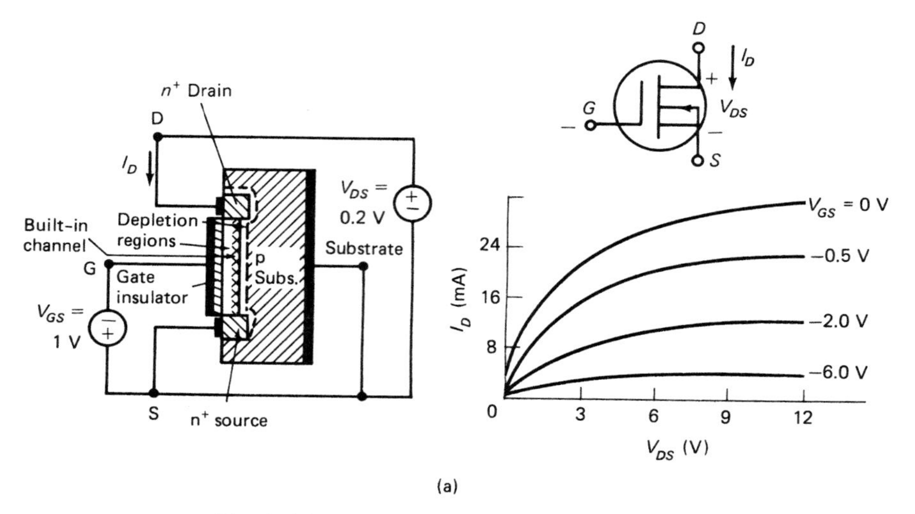

Fig. 6.5(a) *n*-channel depletion mode MOSFET.

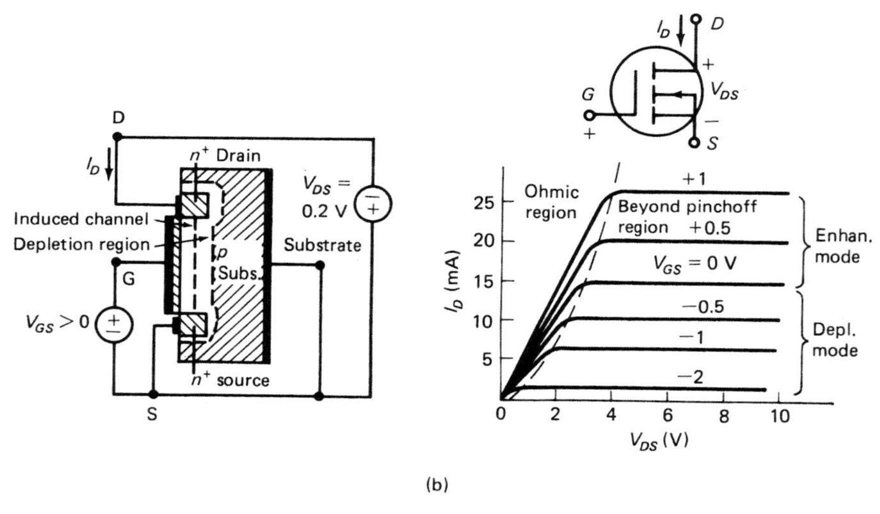

Fig. 6.5(b) *n*-channel enhancement mode MOSFET.

the holes in the normally-on channel below. This decreases the effective conductivity of the channel. If the gate bias is increased to V_p, the pinchoff voltage, the drain-source current becomes very small and essentially independent of the drain-source voltage, as in the operation of the JFET described previously.

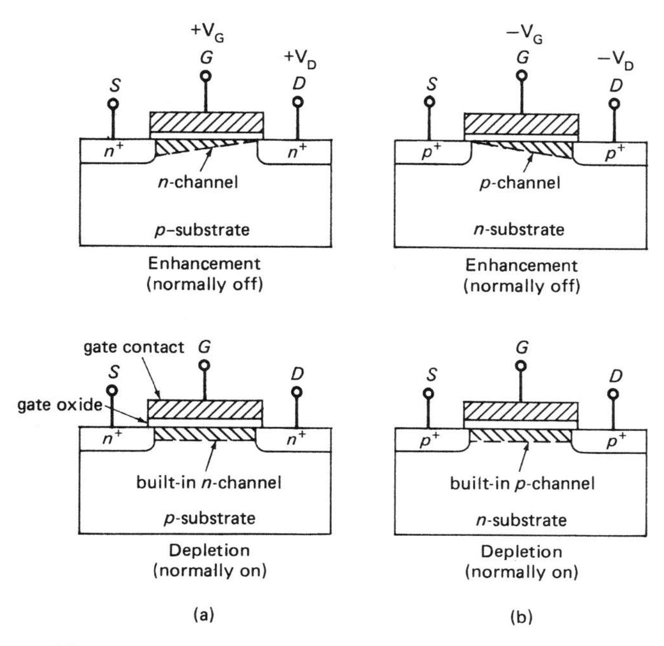

Fig. 6.6 Basic MOSFET types: (a) n-channel; (b) p-channel.

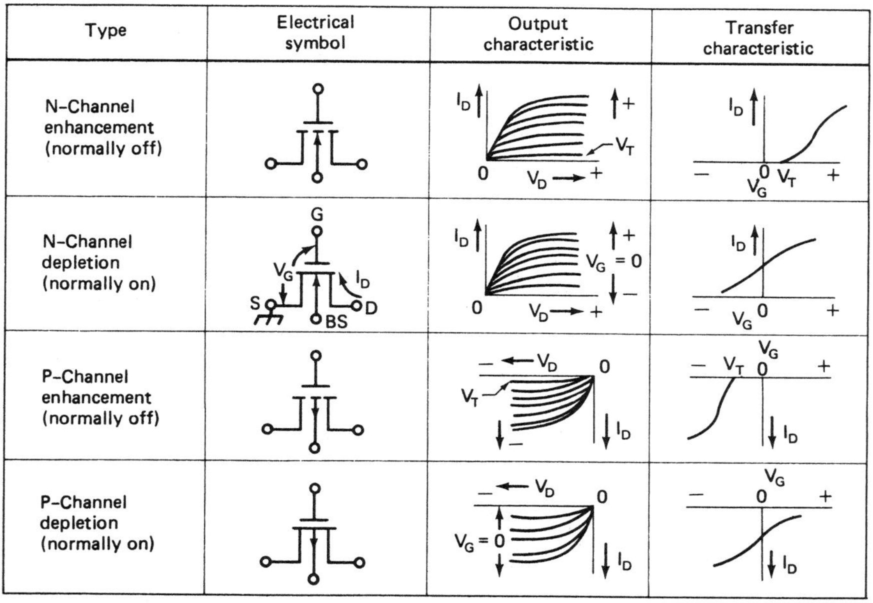

Type	Electrical symbol	Output characteristic	Transfer characteristic
N-Channel enhancement (normally off)		I_D ↑ +, V_T, 0, V_D → +	I_D, −, 0, V_G, V_T, +
N-Channel depletion (normally on)	G, V_G, I_D, S, D, BS	I_D ↑ +, $V_G = 0$, ↓ −, 0, V_D → +	I_D, −, V_G, 0, +
P-Channel enhancement (normally off)		− ← V_D, 0, V_T, ↓ −, ↓ I_D	V_G, −, V_T, 0, +, ↓ I_D
P-Channel depletion (normally on)		− ← V_D, 0, $V_G = 0$, ↓ I_D	V_G, −, 0, +, ↓ I_D

Fig. 6.7 Characteristics and symbols of the four types of MOSFETs.[26]

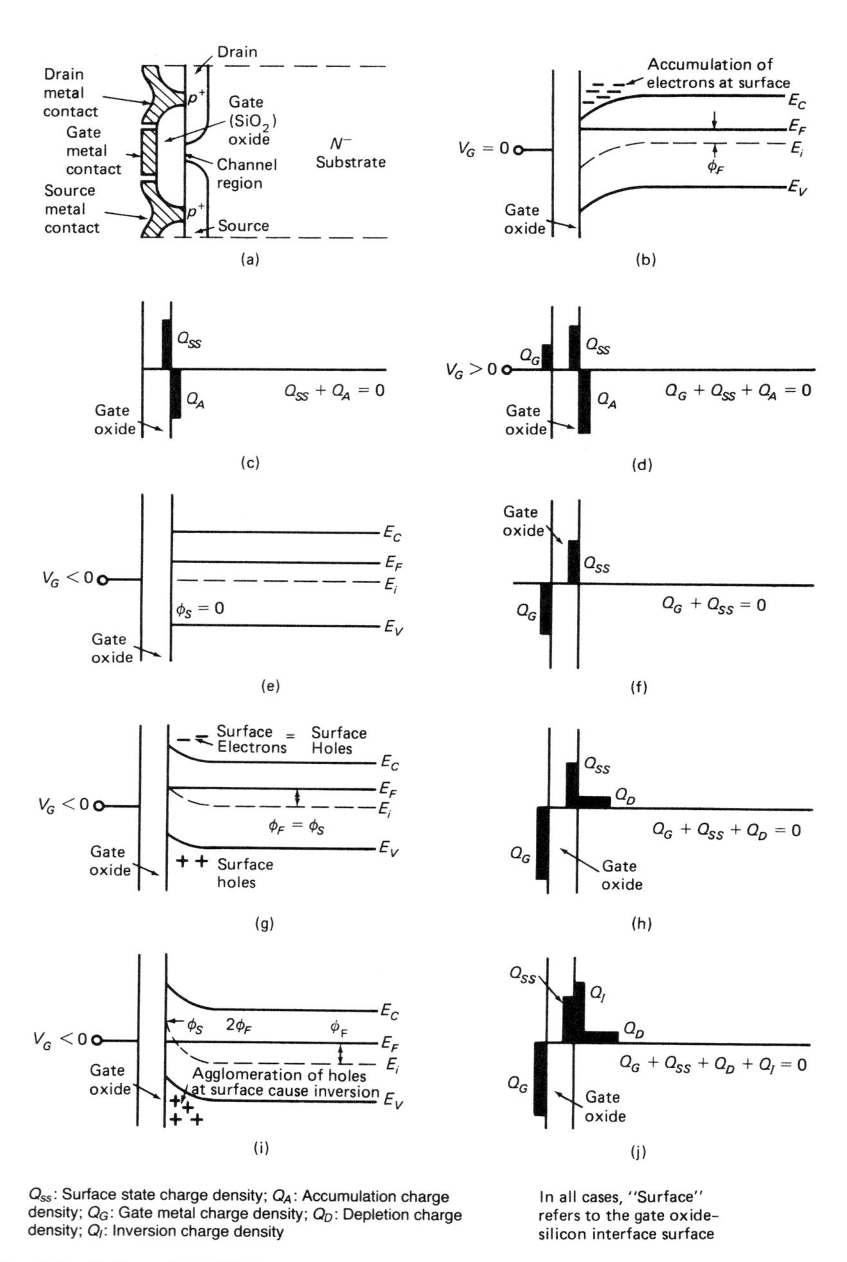

Fig. 6.8 *P*-channel MOSFET energy band and charge density distribution diagrams under conditions of accumulation, flat band, depletion, and inversion.[2] Legend: (a) device structure; (b) accumulation condition energy bands due to surface states; (c) charge density distribution due to surface states; (d) charge density distribution due to surface states and positive gate voltage V_G; (e) energy bands for flat band case; (f) charge density distribution for flat band case; (g) energy bands for depletion case; (h) charge density distribution for depletion case; (i) energy bands for inversion case; (j) charge density distribution for inversion case.

The enhancement mode MOSFET differs from the depletion mode MOSFET in that it does not have a built-in channel and is normally off with zero gate bias. Instead, the gate electric field ($+V_G$ for n-channel, $-V_G$ for p-channel) enhances the conductivity of the substrate directly under the oxide layer, to induce a channel there by attracting corresponding mobile carriers from the substrate, and so invert the carrier concentration in this channel to match both source and drain impurity type. Since there is a potential variation along the channel because of its finite resistance, the corresponding voltage drop results in a variation of charge per unit area (i.e., a change in channel thickness along its length). This is seen in the upper part of Fig. 6.6.

It is important at this juncture to further appreciate some of the other properties and characteristics of MOSFETS in order to understand better how they degrade due to incident radiation. The following refers to p-channel devices, either enhancement or depletion mode, on an n-type substrate. They correspond to columns (2) and (4) in Table 6.3. The band energy diagram of the p-channel depletion mode device, in accumulation, is shown in Fig. 6.8(b). In that figure, the drain-source voltage is assumed small and the intrinsic Fermi level energy E_i lies approximately midway between the conduction and valence band energies. With a small positive gate voltage, the corresponding negative surface charge density Q_A, made up of electrons from within the substrate, is attracted to and accumulate at the channel-oxide layer interface ($x = 0$). This accumulation of charge results in a downward bending of the conduction and valence bands.

Figure 6.8(c) shows the corresponding positive charge density Q_{ss}, which is the SiO_2 gate-channel interface surface charge density. Q_{ss} must be balanced by the negative charge Q_A accumulated near the gate interface due to the above electrons. In the figures the charge distributions are represented as sheet charge distributions.

For increasing positive gate bias, additional energy band bending and charge accumulation result. Charge neutrality must still hold so that $Q_G + Q_{ss} + Q_A = 0$, where Q_G is the gate charge due to the applied gate bias, as in Fig. 6.8(d).

If the gate bias is zero, the effect of Q_{ss} is countered by a negative gate charge, and no bending of the bands occurs, which is known as the *flat band condition*, depicted in Figs. 6.8(e) and (f), with the corresponding surface potential $\psi_s = 0$.

For negative gate bias, the electrons in the channel are repelled, which causes a depletion region to form with a concomitant positive space charge Q_D. This is shown in Figs. 6.8(g) and (h). When the intrinsic Fermi level is bent enough to intersect the actual Fermi level E_f at $x = 0$, the surface has been transformed from its initial n-type to intrinsic ($n = p = n_i$), and the surface and Fermi potentials are approximately equal, i.e., $\psi_s \cong \psi_f$. Further negative gate bias now induces positive mobile holes at the gate-channel interface, in contrast to extending the depletion region.

Note that the depletion region charge and mobile inversion surface charge have the same polarity. Hence, these charges must balance the charge Q_s in the SiO_2 insulator plus the gate charge to maintain electrical neutrality. Further increases in the negative gate bias result in a larger and larger amount of charge within the semiconductor to be contributed by the holes, as seen in Figs. 6.8(i) and (j).

Beyond the point where $E_i > E_f$, electrons are suppressed below the intrinsic energy level, while holes in the channel are raised above this energy level. This is construed as an inversion layer of holes, because the hole population concentration is greater than their normal hole concentration in the n-channel material. They have become the majority carriers in the channel and semiconductor substrate bulk. This transistor now starts to conduct between the p-type source and drain by holes.

The gate threshold voltage V_T is defined as that voltage where the surface potential $\psi_s = |2\psi_f|$, which is taken as the state in which the channel starts to conduct. V_T corresponds to a gate charge Q_G which just counterbalances the charge in the surface states Q_s plus the charge in the depletion region that supports a voltage of $2\psi_f$. An alternate definition of V_T is the threshold voltage required to invert the channel surface and allow conduction between the source and drain. For a p-channel device, a negative gate potential is required for inversion, while a positive gate potential is required for an n-channel device.

To appreciate the preceding in a quantitative manner, consider again a p-channel MOSFET where p-source and drain regions are used, as seen in Fig. 6.9. Figure 6.10 should be studied together with Figs. 6.7 and 6.8

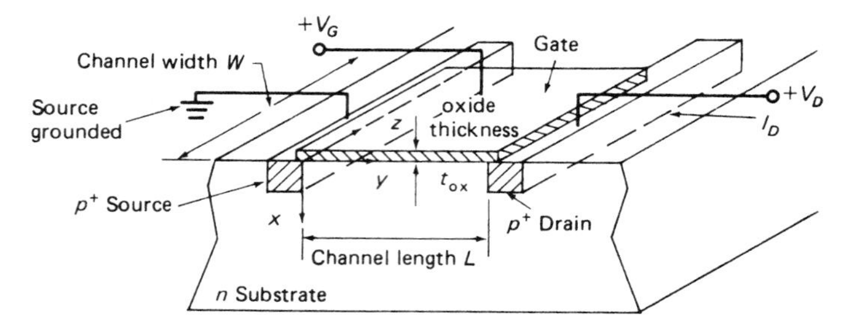

Fig. 6.9 P-channel MOSFET showing spatial coordinates and dimension labels.

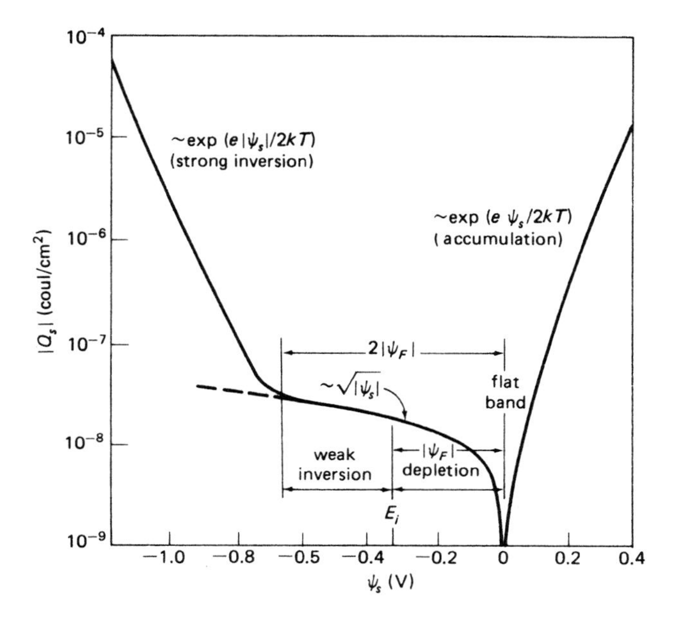

Fig. 6.10 *P*-channel gate oxide channel interface surface charge Q_s versus surface potential ψ_s.[34]

Table 6.4 *P*-channel MOSFET surface charge and surface potential regions

ψ_s	$\lvert Q_s \rvert$	SURFACE CARRIER STATE	FIG. 6.8
Positive	$\exp(e\psi_s/2\,kT)$	Accumulation	(b)
0	0	Flat band	(e)
Small negative $\lvert\psi_s\rvert < \lvert\psi_F\rvert$	$\sqrt{\lvert\psi_s\rvert}$	Depletion	(g)
Negative $\lvert\psi_s\rvert > \lvert\psi_E\rvert$	$\sqrt{\lvert\psi_s\rvert}$	Weak inversion	$\frac{1}{2}$[(g) + (i)]
Large negative $\lvert\psi_s\rvert \gg 2\lvert\psi_F\rvert$	$\exp(e\lvert\psi_s\rvert/2\,kT)$	Strong inversion	(i)

and Table 6.4. Figure 6.10 depicts net surface charge $|Q_s|$ at the oxide-channel interface for *n*-type silicon versus the surface potential $\psi_s = E_s/e$. Generally, the potential ψ is defined to be zero in the silicon bulk and referred to the intrinsic Fermi level, as in Fig. 6.8(i). At the silicon surface $\psi = \psi_s$; for positive ψ_s, $Q_s \sim \exp(e\psi_s/2kT)$, as in Fig. 6.10, and is called the *accumulation region*, shown in Fig. 6.8(b). For the flat band condition, Fig. 6.8(e), both $|Q_s| = \psi_s = 0$. The interval $0 < |\psi_s| < |\psi_F|$, where $\psi_F = E_F/e$ is the Fermi potential as depicted in Fig. 6.8(g), and $|Q_s| \sim |\psi_s|^{1/2}$. When $|\psi_s| \gg |\psi_F|$, $|Q_s| \sim \exp(e|\psi_s|/2kT)$, which is the strong inversion regime,[34] whose lower bound is usually given by $|\psi_s| = 2|\psi_F|$. The preceding is delineated in Table 6.4.

From Fig. 6.9 it is seen that the channel current I_c can be written as the integral over the cross section of the channel of the current density $J_c(x, y)$, viz.

$$I_c = \int_{\text{chan}} J_c(x, y) W \, dx \tag{6.17}$$

where W is the channel width in the z direction and is perpendicular to the channel current flow. Now

$$J_c(x, y) = \sigma(x) E_y = e\mu_p p(x) E_y \tag{6.18}$$

$\sigma(x)$ is the conductivity in the channel, $p(x)$ is the channel carrier density, μ_p is the hole mobility, and E_y is the y component of the applied electric field in the direction of current flow. Using Eq. (6.18) in (6.17) yields, with $E_y = -dV/dy$ and V as the corresponding potential

$$-I_c = W\mu_p(dV/dy) \int_{\text{chan}} ep(x) \, dx \tag{6.19}$$

The integral factor represents the mobile charge per unit area in the channel. As the total system must be electrically neutral with attention to charge sign, then

$$Q_G + Q_{ss} + Q_c + Q_D = 0 \tag{6.20}$$

because $Q_G + Q_{SS}$ is all the charge outside the active region, and $Q_c + Q_D$ is all the charge within the active region, where Q_C is the channel charge corresponding to I_c.

From Gauss's theorem, as discussed in Section 2.7, the charge induced by the gate bias satisfies, with $D = \varepsilon E$

$$\int \varepsilon_{\text{ox}} E_{\text{ox}} \, dS = Q_G \tag{6.21}$$

Here, this is interpreted as the surface integral of the electric field E_{ox} over the gate incremental area dS, in Fig. 6.9. Or, because $\int_{chan} dS = W\,dy$, $E_{ox} = -dV_{ox}/dx$, and L is the channel or gate length, then

$$\varepsilon_{ox} E_{ox} W\,dy = -\varepsilon_0 (dV_{ox}/dx) W = Q_G \tag{6.22}$$

Now, $-dV_{ox}/dx \simeq (V_G - V(y))/t_{ox}$, where t_{ox} is the thickness of the gate oxide insulator. With the capacitance per unit area of the oxide layer, $C = \varepsilon_{ox}/t_{ox}$, the voltages in the channel satisfy

$$(V_G - V(y))C = Q_G \tag{6.23}$$

Inserting Eq. (6.23) in (6.20) gives

$$Q_c = -[V_G - V(y)]C - [Q_{ss} + Q_D] \tag{6.24}$$

Because $Q_c = e \int p(x)\,dx$ in Eq. (6.19), using Eq. (6.24) yields

$$-I_c\,dy = W\mu_p(dV/dy)\{-[V_G - V(y)]C - [Q_{ss} + Q_D]\} \tag{6.25}$$

Integrating both sides of Eq. (6.25) over the length of the channel, as well as up to V_D, gives the sought after channel current, with $V_T = -(Q_{ss} + Q_D)/C$, as

$$I_c = -\nu[-(V_G - V_T)V_D + \tfrac{1}{2}V_D^2] \tag{6.26}$$

where V_D is the drain voltage, $C_{ox} = C \cdot W \cdot L$ is the oxide layer capacitance, and $\nu = W\varepsilon_{ox}\mu_p/Lt_{ox}$. As $I_c + I_D = 0$, where I_D is the drain current

$$I_D = -\nu[(V_G - V_T)V_D - \tfrac{1}{2}V_D^2]; \qquad |V_D| \ll |V_G - V_T| \tag{6.27}$$

Equation (6.27) assumes zero drain or source resistance. It can be modified in a straightforward manner to include them to give[24]

$$I_D = -\nu(V_G - V_p)^2/\{1 - \nu R_s(V_G - V_P) + \sqrt{1 - 2\nu R_s(V_G - V_P)}\} \tag{6.28}$$

where, again, V_p is the pinchoff voltage.

The threshold voltage V_T is a complicated function of gate oxide layer thickness t_{ox}, channel concentration, surface state densities N_{ss}, and substrate voltage V_{sub}. It can be written as[25]

a. For MOSFETs fabricated with n^+ polysilicon gate terminals

$$V_T = -(eN_{ss}/C_{ox})^{+}_{-} V_f - 0.45 + C_{ox}^{-1}\sqrt{2e\varepsilon_0\varepsilon_{Si}N(|V_{sub}| + 2V_f)} \quad (6.29)$$

b. For MOSFETs fabricated with p^+ polysilicon gate terminals

$$V_T = -(eN_{ss}/C_{ox})^{+}_{-} V_f + 0.65 - C_{ox}^{-1}\sqrt{2e\varepsilon_0\varepsilon_{Si}N(|V_{sub}| + 2V_f)} \quad (6.30)$$

where $\varepsilon_0\varepsilon_{Si}$ is the dielectric constant of silicon, $C_{ox} = \varepsilon_0\varepsilon_{SiO_2}/t_{ox}$ is the gate oxide capacitance per unit area, V_f is the Fermi voltage, i.e., $V_f = (kT/e)[\ln(N + N_i)]/N_i$, N is the atomic concentration of silicon, and N_i is the intrinsic carrier concentration of silicon. Note that Eq. (6.27) holds only if $|V_D| \ll |V_G - V_T|$. When $|V_D| \geqslant |V_G - V_T|$, the device is in the saturation or pinchoff region, and Eq. (6.27) is no longer valid. Here, the channel or inversion layer is not fully formed, and the gate voltage no longer controls the channel conductance, as implied from Eq. (6.27), which defines pinchoff behavior. Once saturation has been reached, the voltage drop across the inverted portion of the channel tends to remain at $V_G - V_p$, while V_D still changes. Inserting this into Eq. (6.27) gives the drain current in the saturation, or pinchoff, condition, which is

$$I_D = -\tfrac{1}{2}\nu(V_G - V_P)^2 \quad (6.31)$$

The term *pinchoff* is usually associated with depletion mode devices, including JFETs, where pinchoff occurs at the gate voltage corresponding to drain current I_{DSS}, as in Fig. 6.4.

In enhancement mode devices, the term *threshold voltage* has come into common usage to mean that gate voltage necessary to initiate conduction. Once saturation has been reached, the drain current I_{DSS} is constant, as given by Eq. (6.31) and shown in Fig. 6.4 for the JFET as well. In the saturation regime, the transconductance is given by

$$g_m = \partial I_D/\partial V_G = -\nu(V_G - V_P)|_{V_{D=const}} \quad (6.32)$$

The principal ionizing radiation damage mechanism in MOS devices results from the creation of electron-hole pairs from the breaking of silicon-oxygen bonds in the SiO_2 insulator gate.[12] This produces the build up of trapped positive charge (mainly holes) in the insulator, and trapped negative charge concentrated at the insulator-channel interface.[15,39,40] Besides the hole-electron pairs that recombine following the onset of an ionizing radiation pulse, the applied gate voltage rapidly sweeps the electrons out of the oxide insulator, because of their very large mobility, compared with that of the corresponding holes. The relatively immobile holes become trapped in the SiO_2 gate insulator near the silicon channel

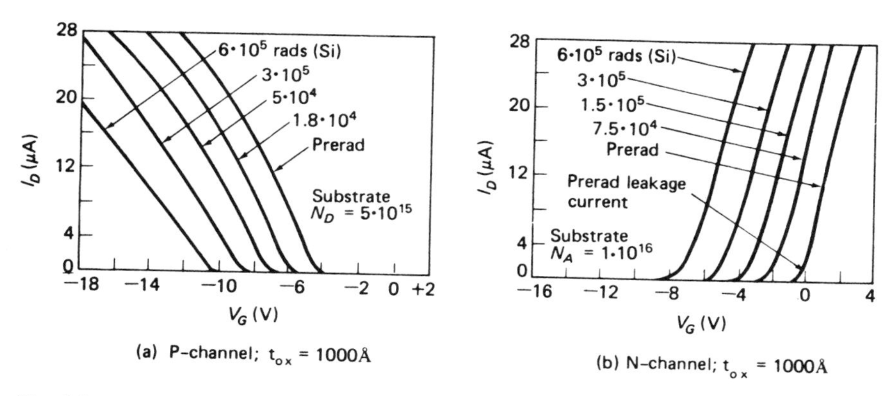

Fig. 6.11 Radiation-induced threshold voltage shift in enhancement MOSFET $I_D - V_G$ characteristics.[22]

interface for positive gate voltages, or near the SiO_2 gate metal interface for negative gate voltages. These trapped positive charges are the cause of the negative shift in the $I_D - V_G$ characteristic curves in Fig. 6.11. For both *n*- and *p*-channel enhancement mode MOS transistors, as in Figs. 6.11(a) and 6.14A, a negative gate voltage is required to accumulate positive charge (holes) at the semiconductor-insulator ($Si-SiO_2$) interface, which is necessary to invert the surface population to allow majority carrier hole conduction in the channel between the *p*-type drain and source. Ionizing radiation-induced positive charge (holes) in the insulator will then require a greater negative gate voltage to compensate the positive charge to achieve surface inversion, and thus transistor turn-on. The increase in turn-on voltage, or threshold voltage, is seen in Fig. 6.11.

As discussed earlier, the amount of radiation induced threshold voltage shift in MOS devices is a complicated function of (a) the gate bias during irradiation, (b) the gate insulator material and its thickness, (c) processing and doping methods used in securing the gate insulator onto the silicon surface, (d) to a relatively lesser extent, radiation-induced mobility changes (e) the dose rate at which the ionizing dose is accumulated, and (f) the temperature of the device during irradiation. Interestingly, at high dose levels, an *n*-channel enhancement mode MOS device gate threshold voltage shift can become positive from being initially negative at lower dose levels, due to interaction between the interface states and gate oxide trapped hole contributions, as shown in Fig. 6.14A.[15,57] Following exposure to radiation, the *n*-channel threshold voltage approaches earlier positive values with continued bias. This bias annealing with time returns the *n*-channel threshold voltage to a value more positive

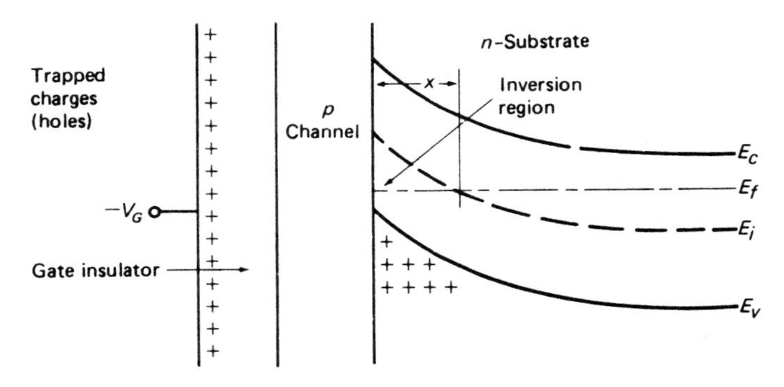

Fig. 6.12 Ionizing radiation effects in enhancement mode P-MOS transistor.

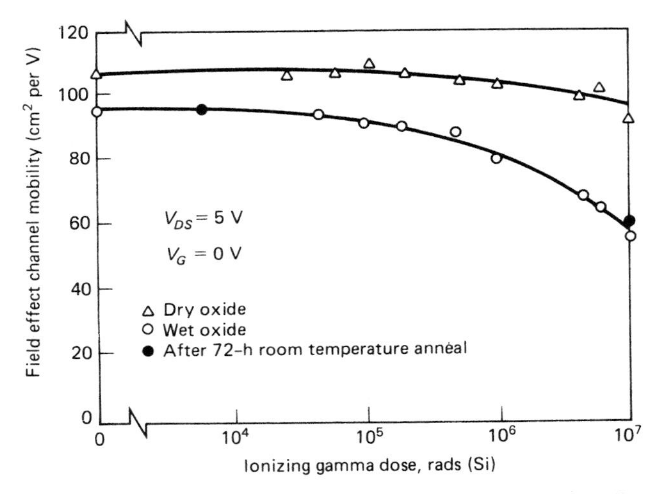

Fig. 6.13 Channel mobility degradation in MOS transistors.[42]

than its initial value. This is called rebound, or super recovery, and has been observed in irradiated microprocessors whose function was interrupted, but recovered following cessation of radiation exposure.[54]

Rebound occurs because of the differing temporal behavior of four competing mechanisms. The first is the trapping of holes in the gate oxide following the onset of ionizing radiation. This produces an almost immediate negative shift of the gate threshold voltage, V_T, in n-channel devices. The second mechanism is the much slower positive shift of V_T due to the accumulation of electrons in the SiO_2 gate-Si substrate interface states. The third is the "annealing out" of the oxide states, while the

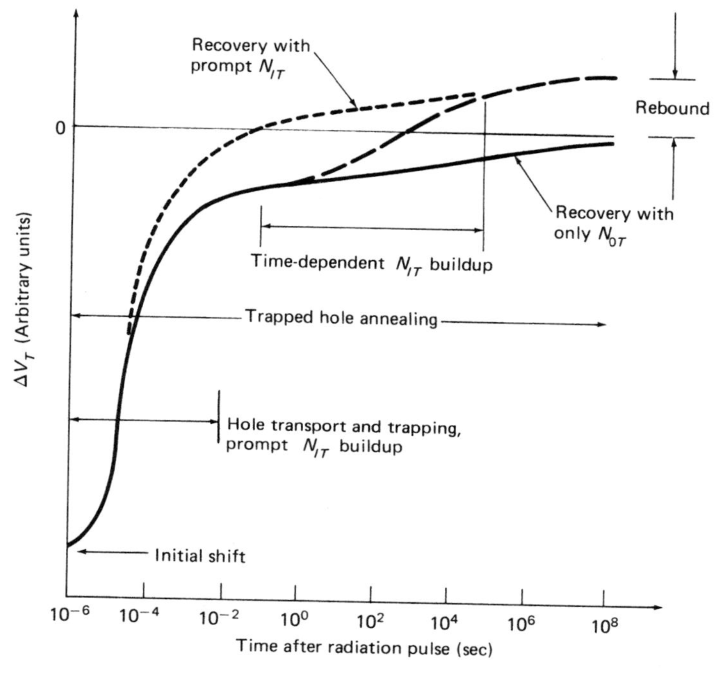

Fig. 6.14 Time-dependent threshold voltage shift depicting super-recovery (rebound) of N-channel MOSFET following pulsed ionizing radiation effects of bulk oxide (hole) trap density, N_{OT}, and interface (electron) trap density, N_{it}.[55] Figure © 1987 by the IEEE.

fourth is the "annealing out" of the interface states. If the magnitude of the changes produced by these mechanisms is comparable, then the ultimate sign of ΔV_T will depend on dose rate levels, resulting in a complex behavior of the devices with respect to the incident ionizing radiation parameters. In essence, the effects of ionizing radiation are due to dose and dose rate, and are time dependent.[56,64]

To reiterate in detail,[55] the ionizing radiation produces hole-electron pairs in the gate oxide, which become available mainly as a result of the radiation induced bond fractures therein. Recall Fig. 1.1, where the double connectors between the atoms depict the covalent valence electron pair bond structure between nearest atom neighbors.

Being much more mobile than the holes, most of the electrons produced (that do not recombine with the holes present) are rapidly swept out of the gate oxide and collected at the gate electrode. The unrecombined relatively immobile holes cause an initial negative shift in V_T to counteract their positive charge. However, from an epoch beginning at roughly 100 nanoseconds and extending to about 1 second (Fig. 6.14) at ambient temperatures, the hole migration is toward the silicon substrate (lower portion of Fig. 6.18) for a positive gate bias condition. It consists of a stochastic hopping transport[61] through the gate oxide aided by the electric field present. The preceding depends on a number of parameters, including gate oxide thickness, applied field magnitude, temperature, oxide processing methods, etc. The relatively long-lived negative ΔV_T shift is the commonly observed form of ionizing radiation damage in MOS devices.

As mentioned above, the incident radiation induces SiO_2/Si interface traps, which usually capture electrons. These trapping states correspond to energy levels within the silicon band gap. Their occupancy is determined by the position of the fermi level at the interface,[74] as discussed in Sections 1.6 and 1.9. Generally, there can be both prompt interface traps, immediately manifest upon the onset of the ionizing radiation, as well as a delayed interface trap build-up that continues for long periods of time at ambient temperature. Both these "fast" and "slow" trap states are usually occupied by electrons, resulting in a positive ΔV_T.[55] As seen below, the negatively charged interface states result in a positive ΔV_T component, compensating the positive oxide hole states that yield a negative ΔV_T.

The threshold voltage shift ΔV_T can be divided into three components[55]

$$\Delta V_T = -\Delta V_{st} - \Delta V_{0t} + \Delta V_{it} \tag{6.33}$$

where;

$|\Delta V_{st}| = (e/C_{0x}t_{ox}) \int_0^{t_{ox}} x p_h(x,t)\,dx$; short-term contribution due to early trapping of holes.

$|\Delta V_{0t}| = (e/C_{ox})\,\Delta P_{ot}(t)$; long-term contribution from deep trapping of holes near the SiO_2 gate-Si substrate interface.

$|\Delta V_{it}| = \Delta Q_{it}(t)/C_{ox}$; long-term contribution of electron trapping at SiO_2/Si interface.

$\Delta P_{ot}(t)$ is the areal density of deep trapped holes near the SiO_2/Si interface, which undergo very long term annealing, as discussed. $p_h(x, t)$ is the density of free holes in the oxide. ΔQ_{it} is the trapped interface charge density. Actually, it can contribute either a net positive or negative charge, depending on the position of the fermi level at the interface,[74] as explained in Section 1.9. Usually, ΔQ_{it} is negative for n-channel devices, and positive for p-channel devices, when the gate bias level is at the threshold voltage.

Second-order effects include long-term annealing of the deep trapped

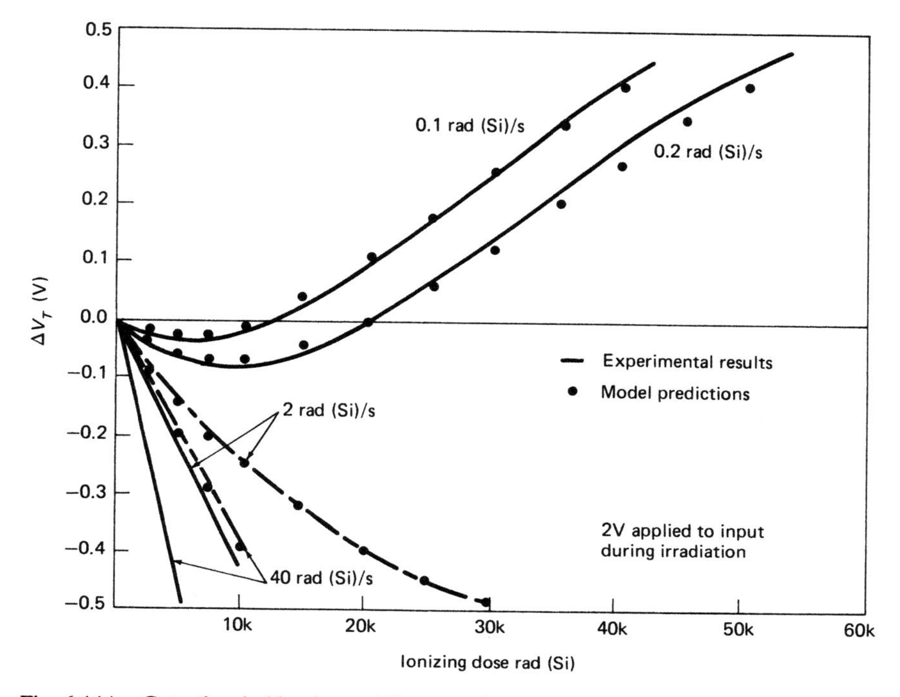

Fig. 6.14A Gate threshold voltage shift versus ionizing dose for various dose rates for a Z-80 NMOS microprocessor.[57]

holes due to tunnelling, as discussed in Section 6.9. Also, ionizing radiation-induced interface traps can exist immediately after the onset of radiation, as well as their accrual over relatively long times.[55]

A temporal model for super recovery is developed[57] using the assumption that the relation between threshold voltage shift and ionizing radiation dose rate excitation can be represented by a linear operator, as discussed in Section 5.14. The impulse response, ΔV_{T0}, here consists of an empirical fit to data for the Z 80 microprocessor,[57] employed as an illustration. Using the notation of Section 5.14, it is

$$\Delta V_{T0} = \gamma_D(\lambda_2 C_2 t/(1 + \lambda_2 t) - (|C_0| - C_1 \ln(1 + \lambda_1 t)) \exp - \lambda_2 t) \quad (6.34)$$

where the corresponding parameter values are (γ_D is the ionizing dose)

$$\begin{aligned} C_0 &= -84.3\,\mu\text{V/rad(Si)} \\ C_1 &= 7.9\,\mu\text{V/rad(Si)}; \qquad \lambda_1 = 2.5 \cdot 10^{-3}\,\text{sec}^{-1} \\ C_2 &= 20.0\,\mu\text{V/rad(Si)}; \qquad \lambda_2 = 2.5 \cdot 10^{-6}\,\text{sec}^{-1} \end{aligned}$$

The threshold voltage shift $\Delta V_T(t;\, \gamma_D,\, \dot{\gamma})$ is then given by the convolution integral, Eq. (5.143), as

$$\Delta V_T(t;\, \gamma_D,\, \dot{\gamma}) = \int_0^t \dot{\gamma}(\tau)\, \Delta V_{T0}(t - \tau;\, \gamma_D,\, \dot{\gamma})\, d\tau \tag{6.35}$$

For a constant dose rate level, $\dot{\gamma}_c$, of duration t_0, Eq. (6.35) yields

$$\Delta V_T(t;\, \gamma_D,\, \dot{\gamma}_c) = \begin{cases} \dot{\gamma}_c \displaystyle\int_0^t \Delta V_{T0}(\tau;\, \gamma_D,\, \dot{\gamma}_c)\, d\tau;\ t \leqslant t_0 & \text{(6.36A)} \\ \dot{\gamma}_c \displaystyle\int_0^{t_0} \Delta V_{T0}(t - \tau;\, \gamma_D,\, \dot{\gamma}_c)\, d\tau;\ t > t_0 & \text{(6.36B)} \end{cases}$$

ΔV_T is obtained by inserting ΔV_{T0} from Eq. (6.34) into either Eq. (6.36A) or (6.36B). Equation (6.36A) is employed herein since the ionizing radiation duration is comparable to the annealing times of interest, and its ΔV_T is plotted in Fig. 6.14A. This figure depicts ΔV_T versus ionizing dose for various $\dot{\gamma}_c$, with experimentally determined points superimposed. Also seen in the figure is the super-recovery behavior, where ΔV_T, having early negative values, ultimately returns to and passes through zero shift to rebound or overshott to positive ΔV_T (also shown in Fig. 6.17).

The preceding ΔV_T behavior cloaks a very important and far-reaching result. This is that ionizing dose failure levels, certainly for MOS devices, are strong functions of the dose rate of the incident ionizing radiation. Equivalently, there is no unique "total dose" response for semiconductor devices. This is seen by rewriting the threshold voltage shift as,

$$\Delta V_T = F(t;\, \gamma_D,\, \dot{\gamma}_c) \tag{6.37}$$

where $F(t;\, \gamma_D,\, \dot{\gamma}_c)$ is the right-hand side of Eq. (6.36), as appropriate. Failure of the device can be construed as exceeding a ΔV_T failure threshold, called ΔV_{TF}. That is, $\Delta V_T \geqslant \Delta V_{TF}$ yields an untenable distortion of the device electrical characteristics, thus constituting device failure. Then, the corresponding ionizing dose to failure γ_{DF} is obtained by inverting Eq. (6.37), to yield

$$\gamma_{DF} = G(\gamma_D/\dot{\gamma}_c;\, \Delta V_{TF},\, \dot{\gamma}_c) \tag{6.38}$$

$\gamma_D/\dot{\gamma}_c = t$ is simply a parameterization of the time variable, and G is the function obtained from the inversion. The failure dose, γ_{DF}, versus the

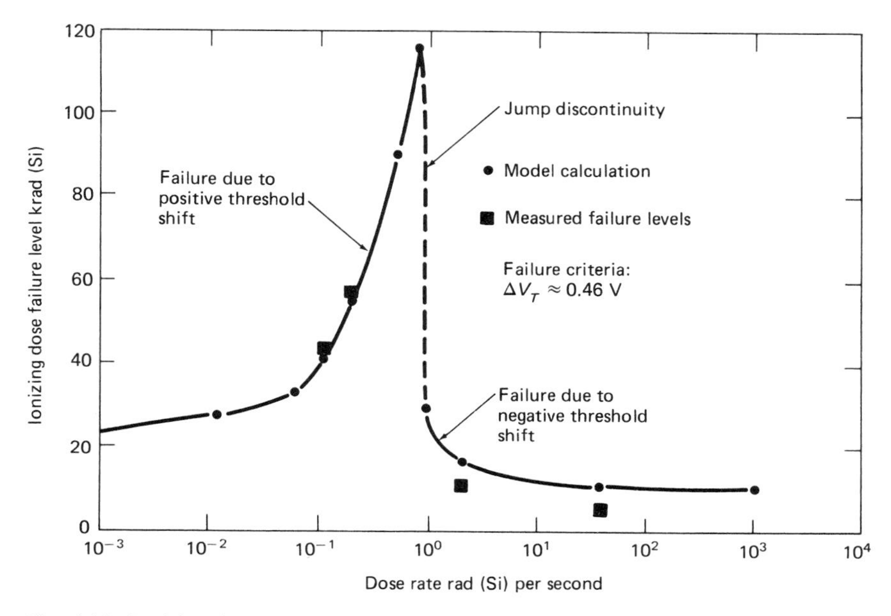

Fig. 6.15 Ionizing dose failure level versus corresponding dose rate for a given threshold voltage shift ΔV_T failure criterion.[57] Figure © 1984 by the IEEE.

dose rate, $\dot{\gamma}_c$, in Eq. (6.38) is plotted for a given ΔV_{TF} in Fig. 6.15. The striking result, as seen in the figure, is that the failure ionizing dose is a strong function of the dose rate $\dot{\gamma}_c$, and not nominally independent of dose rate, as frequently assumed heretofore. Experimental results assert that this behavior holds in general without the constraint of the threshold voltage shift and dose rate being related through a linear operator, as above. From the previous discussion, it is seen in the figure that the low dose-rate regime can be identified with failure due mainly to positive ΔV_T, since the time to fail $t_f = \gamma_{DF}/\dot{\gamma}_c$ is large for small $\dot{\gamma}_c$ for a given γ_{DF}. Recall that the positive ΔV_T is due to electrons occupying "slow" trapping states (at the SiO_2/Si interface) that anneal at very high temperature only after a relatively long time. Likewise, a large $\dot{\gamma}_c$ corresponds to small times during which oxide hole trapping causes a negative ΔV_T shift, as discussed.

Hence, the preceding delineates the fact that there is no unique or nominal ionizing dose device failure level, but that the latter depends on the incident dose rate. This implies that the response to ionizing radiation is a complex function of ionizing dose, dose rate, bias, temperature, and time after irradiation. So that, for modern microcircuits and like semiconductor devices, it would be difficult to specify only a meaningful ionizing dose requirement as part of a system radiation specification. One approach taken is to specify the pair (γ_{DF}, $\dot{\gamma}_c$), where $\dot{\gamma}_c$ would correspond to the ambient environment of the system that employs this device. For an incident dose rate pulse width of duration t_0, which is much less than the annealing times of interest, the analysis becomes somewhat more complicated, but the same general result of γ_{DF} being a function of the dose rate still holds. Also, successfully extrapolating a device failure dose at a given dose rate from an ionizing dose test source (such as ^{60}Co), to a use environment dose rate, is still under investigation.

It is now well established that ionizing dose effects are both dose-rate and time dependent.[56,64,81] It is known that they are due to the temporal behavior of the corresponding growth and annealing phenomena.[64] It is apparent that they can result in difficulties in hardness assurance testing. All of this implies that, in situations where such time-dependent effects play a role, device radiation hardness will depend on its mission ionizing dose and dose rate radiation specifications, as well as on the device parameters per se.

Fortunately, for many important use situations, such as threshold voltage shift, these effects, as expressed by Eqs. (6.35)–(6.38), can be

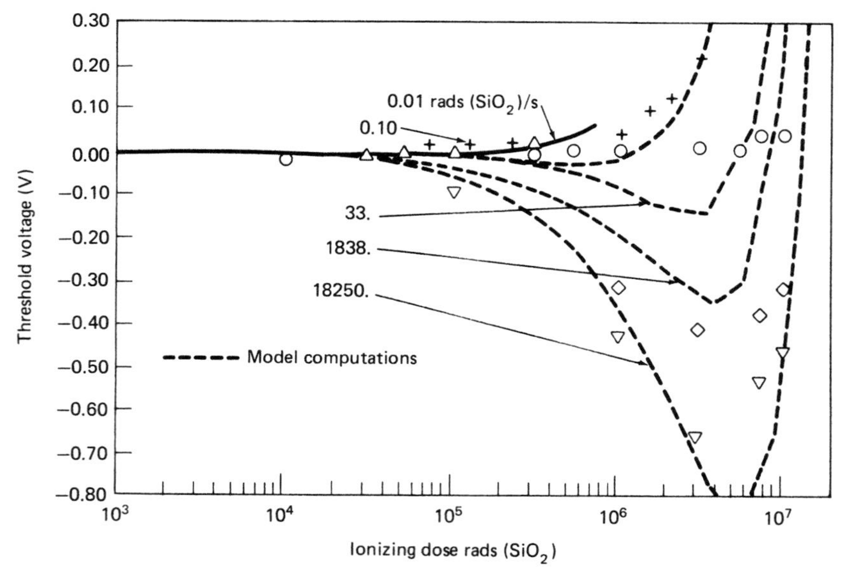

Fig. 6.16 Threshold voltage for radiation-hardened n-channel CMOS process (t_{ox} = 350Å) versus ionizing dose for various dose rates.[64] Figure © 1989 by the IEEE.

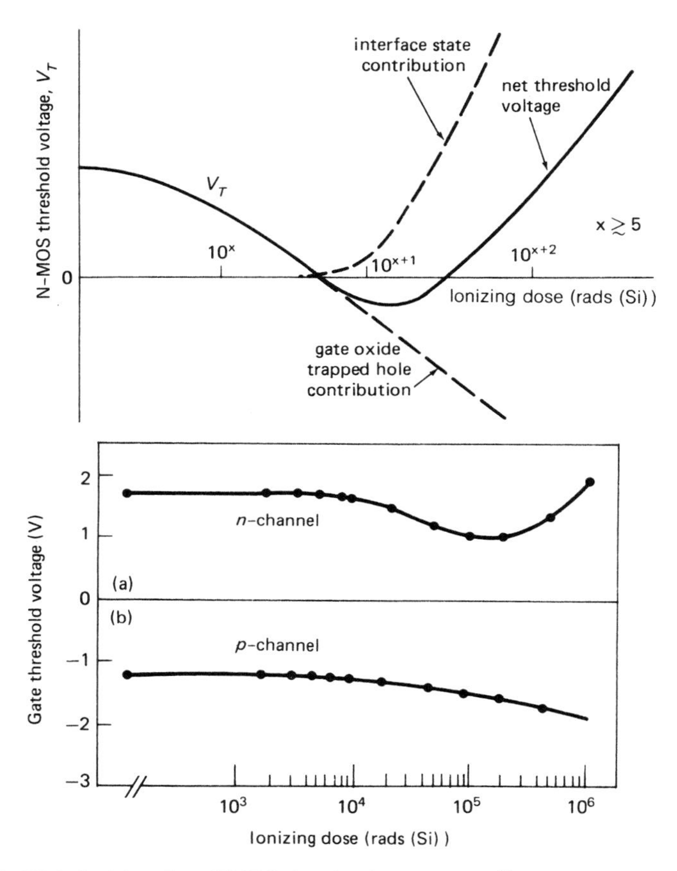

Fig. 6.17 High ionizing dose NMOS threshold voltage shift[15] (upper). *N*- and *p*-channel threshold voltage shift of inverter gates in 1802 CMOS microprocessor[68] (lower). Figure © 1977 by the IEEE.

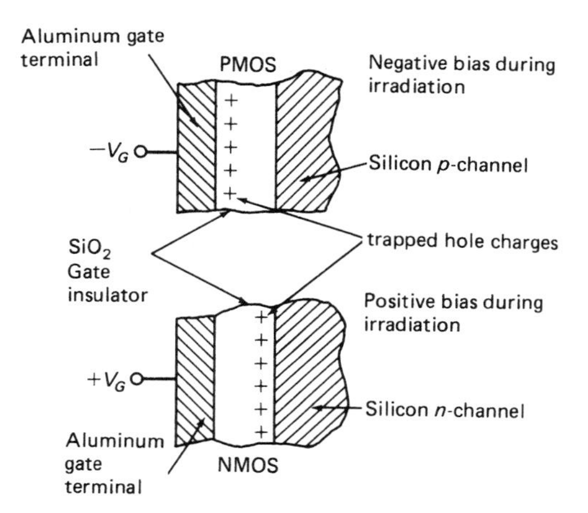

Fig. 6.18 Hole charge-trapping in gate insulators in enhancement mode MOS transistors.

reduced to straight-forward algebraic expressions.[64] Figure 6.16 depicts the gate threshold voltage shift of a radiation-hard CMOS process versus ionizing dose induced under various dose rates from 0.01–18250 rad (SiO_2) per second.[64] It is seen in the figure that, for this device, the net threshold voltage shift is initially negative for dose rates greater than about 33 rads per second, reaching minima whose depth increases with dose rate. Also, it is to be noticed that all of the curves rebound, to a greater or lesser extent, ultimately to positive threshold voltage shifts. Further, it can be seen from the figure that extrapolation from about 33 rads (SiO_2) per second, which is a nominal value for a ^{60}Co facility, to 0.01 rad (SiO_2) per second for a 10-year spacecraft mission is possible, although difficult.[64]

Radiation sensitivities of MOS transistors and integrated circuits vary over a wide range. Radiation damage has been observed at ionizing dose levels as low as 1000 rad (Si), while corresponding hardened MOS devices, discussed in Chapter 12, have a measured tolerable threshold voltage shift after enduring as much as 10^7 rad (Si). Needless to say, radiation-induced threshold voltage shift is of prime concern in the operational application of MOS devices in a radiation environment. The surface states that induce threshold voltage shift also cause a decrease in channel mobility, for they are close to the channel and provide additional scattering centers for channel carriers. Figure 6.13 depicts channel mobility degradation versus ionizing dose level.[42]

6.9 EFFECTS ON MOS AND CMOS STRUCTURES

The complementary MOS (CMOS) structure combines a *p* and *n* channel (enhancement mode) transistor pair on the same substrate. This arrangement as used in high-speed circuitry, dissipates significant power only during its change of state. Only one power supply is required, in contrast to other circuits that need two supplies, one for each polarity. CMOS is characterized by low standby power dissipation (nanowatts), and high-speed operation in the range of 20–100 MHz, which is comparable to many bipolar families, and it provides an optimum speed-power product for large-scale integrated circuits. CMOS systems also have high noise immunity, where noise voltages can be tolerated up to 40 percent of the supply voltage. CMOS also can be made compatible with transistor-transistor logic (TTL) input and output levels using associated current boost circuits. For linear analog devices and circuits, CMOS physical and electrical topology provides a unique complementary symmetry that has many applications, and for which no analog existed during the vacuum tube era.

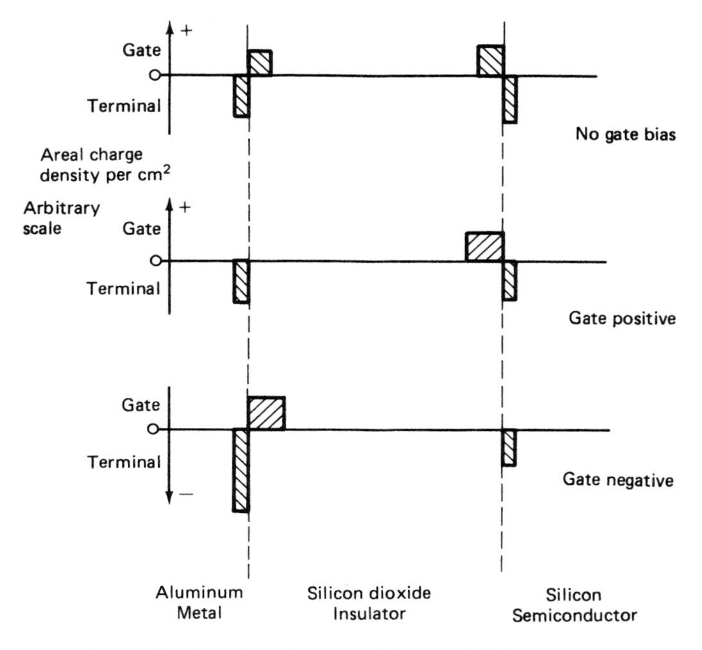

Fig. 6.19 Interface charge buildup in PMOS transistor.

Other parameters that characterize CMOS devices, especially in terms of CMOS LSI, such as very small chip area per unit cell, very low power consumption, and very low density of defects during manufacture, coincide with those that aid in optimizing their radiation hardness. This attribute is shared with most MOS structures. Being majority carrier devices, damage from neutron radiation, through its influence on minority carrier lifetime, is almost nonexistent and for other types of damage is only minimal and indirect, such as neutron secondary ionization processes and carrier removal.

As discussed previously, incident photons produce hole-electron pairs in the MOS SiO_2 gate insulator. A fraction of the holes become trapped therein, as the various electrode bias voltage induced fields sweep out the electrons. The trapped positive (hole) charge can partially anneal out in time. This positive charge buildup in the silicon dioxide gate layer adjacent to the channel causes the threshold voltage of *n*-channel transistors to shift toward negative gate voltages and that of the *p*-channel transistors to also shift toward negative gate voltages, as in Fig. 6.11. This effect is accentuated by the fact that, in the NMOS transistor, the

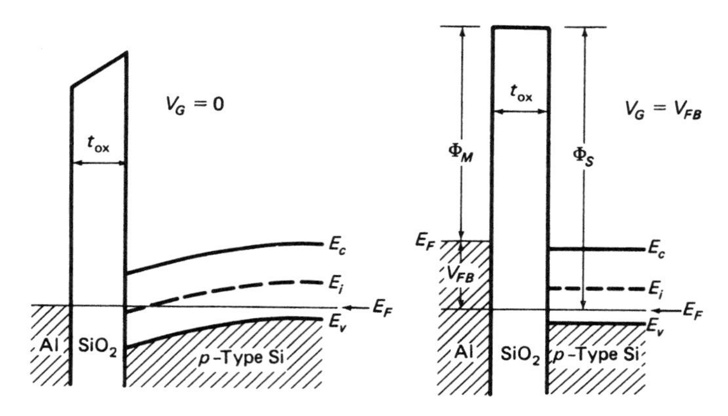

Fig. 6.20 Effect of work function difference in MOS structures.[12]

trapped positive charges migrate toward the silicon-silicon dioxide interface (positive gate bias), but in the PMOS transistor, they migrate toward the metal gate electrode (negative gate bias), as seen in Figs. 6.18 and 6.19. The smaller the span of distance between these charges and the gate terminal, the less effective the spurious field becomes, and so the less the threshold voltage shift becomes. Hence, the PMOS transistor is about an order of magnitude more radiation-resistant than the NMOS for this reason. Also, that these charges actually exist approximately in sheets, as in the figures, is well known and can be verified in straightforward fashion.[12] As mentioned, the trapped charges are a factor in causing the energy band structure to bend, as in Fig. 6.12. In equilibrium, the Fermi levels in the silicon and insulator are equal. Therefore, there will be an electrostatic potential variation from one region to another, as depicted in Fig. 6.12. The corresponding energy level differences can be couched in terms of work function energy differences between the silicon semiconductor and the SiO_2 insulator. The work function is defined as the amount of energy needed to extract an electron out of the material from its Fermi level within.[88] The gate bias voltage required to counterbalance the work function energy difference and so restore the energy bands to their "flatband" condition is called the *flatband* voltage, V_{FB}. Then $V_{FB} = \varphi_M - \varphi_S \equiv \varphi_{MS}$, where φ_M and φ_S are, respectively, the work function in the silicon substrate body and the silicon dioxide insulator layer, as in Fig. 6.20. Experimental results reveal how V_{FB} shifts as a function of gate bias for an irradiated device, as shown in Fig. 6.21.[19] Threshold voltage shifts ΔV_T, and flatband voltage shifts ΔV_{FB} vary roughly as the cube of the silicon dioxide insulator thickness.[19] Hence,

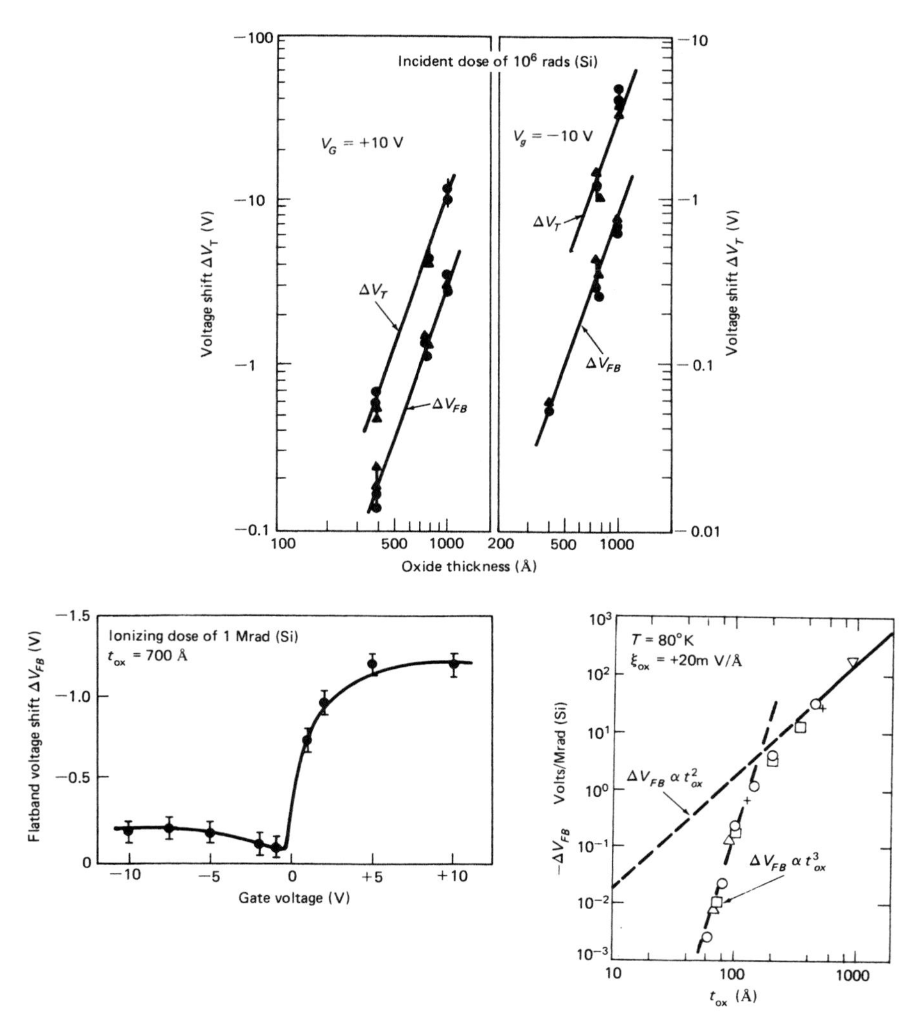

Fig. 6.21 Radiation-induced flatband and threshold voltage shifts versus gate oxide thickness at 1 Mrad (Si).[19,66,67] Figure © 1976 by the IEEE.

the thinner the oxide, the harder is the device since a minimum voltage shift is being approached. Investigations[43] point toward tunnelling of electrons into the gate oxide from the contiguous structures, that recombine to annihilate the holes therein remaining from the ionizing radiation. For thin oxides, the recombination is more complete thus reducing ΔV_T. This is depicted in Fig. 6.22.

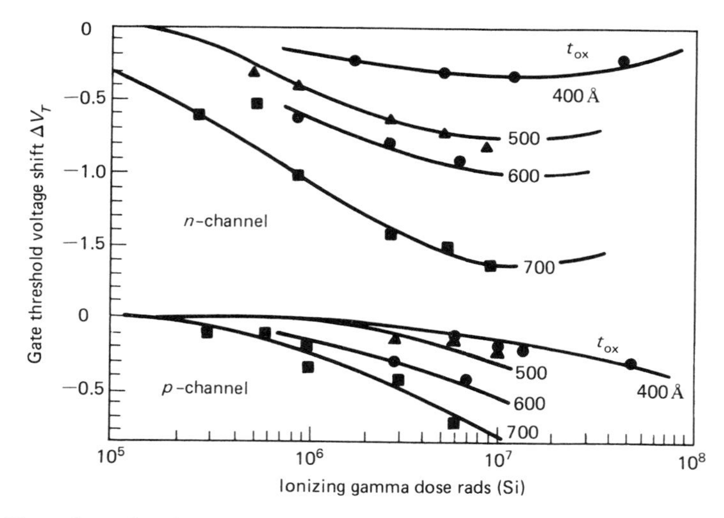

Fig. 6.22 *n*-channel and *p*-channel MOS gate threshold voltage shift versus ionizing dose for various gate oxide thicknesses.[19] Figure © 1984 by the IEEE.

Figure 6.22 (upper) also depicts the apparent tendency for the *n*-channel threshold voltage V_T curves to begin to shift upward for very high ionizing dose levels. This produces the device rebound feature shown in Fig. 6.17 and discussed in Section 6.8. This is in contrast to Fig. 6.22 (lower), the *p*-channel curves, in which V_T decreases to more negative values with very high dose levels, exhibiting no rebound. For the *n*-channel devices, at a given dose level V_T begins to increase toward positive values (i.e., rebounding because of the compensating contribution from negative charge interface states. However, for the *p*-channel device (negligible electron acceptor states), positive oxide-trapped charges, plus the above positively charged gate-oxide donor interface traps, together reinforce the monotone decrease in V_T to more negative values, as shown.

To introduce the analytical dependence of ΔV_T on gate oxide thickness and other device parameters, the dependence of ΔV_{FB} on the same parameters is examined first. A simple heuristic model is given below to show the variation of flatband voltage shift as a function of oxide layer thickness.[37] This model considers charge buildup due to hole-trapping in the gate insulator oxide, caused by ionizing radiation. The corresponding shift in the flatband voltage, ΔV_{FB}, is assumed to be caused by the dominant hole-trapping process. Fast interface surface states are also generated at the oxide-silicon interface, due to the ionizing radiation. As alluded to in Section 2.8, some fast surface states are those produced at the oxide insulator-silicon interface. These states give rise to the sheet charge Q_{ss} at the interface, as discussed in Section 6.8. Such states are called *fast*, because their charge can be readily exchanged with the silicon material.[38] This means that electrons or holes in the conduction or valence bands have relatively high transition rates into or out of the corresponding interface trap states. *Slow* trap states possess relatively low trap transition rates. The surface states apparently contribute a net negative sheet charge under flatband conditions. This tends to partially compensate for the effect of the hole-trapped charge in the oxide bulk. For the case of ionizing radiation under positive gate bias, an increased hole trap density near the oxide-silicon interface ensues. This would contribute a net positive sheet charge. Hence, these are the two cases of positive and negative gate bias that this one-dimensional model addresses.

Assume an origin ($x = 0$) at the metal gatesilicon dioxide interface, and that $x = t_{ox}$ is at the silicon dioxide-silicon interface, for an oxide layer of thickness t_{ox}. The kinetics model[37] is represented by equations in the trapped and free hole densities $p_T(x, t)$ and $p(x, t)$ respectively. They are

$$\partial p_T/\partial t = \sigma v p(N_T - p_T) = \dot{\gamma}(t) - \partial(pv)/\partial x; \tag{6.39}$$

$$p(x, 0-) = p_T(x, 0) = 0; \qquad p_T(x, \infty) = N_T \tag{6.40}$$

where:

N_T is the density of hole traps in the gate oxide.
σ is the capture cross section of the trap sites in the forbidden gap.
v is the speed of the free holes.
$\dot{\gamma} = \gamma_0\delta(t)$ is the generation rate of γ_0 hole-electron pairs per cm^3 per second due to the radiation pulse approximated by a delta function.

The leftmost Eq. (6.39) asserts that the net rate of change of trapped holes per cm^3 is given by their net trapping rate, because $N_T - p_T$ is the number density of trapping states still unoccupied. The outer Eq. (6.39) is also a balance equation for the rate of change of trapped holes per cm^3. They are given as the difference between the production rate of free holes per cm^3 due to radiation, less their diffusion per cm^3 out of an incremental volume of interest in the gate oxide. The equilibrium solution of Eqs. (6.39) and (6.40), for p_T, is given by

$$p_T \cong \begin{cases} N_T(1 - \exp - \sigma N_T x); & \text{for small } t \ (p_T \ll N_T) \\ N_T(1 - \exp - \sigma \gamma_0 x); & \text{for large } t \ (p_T \simeq N_T) \end{cases} \tag{6.41}$$

where the general form is

$$p_T = N(1 - \exp - (x/\lambda)) \tag{6.42}$$

and $\lambda^{-1} = N_T\sigma$ or $\sigma\gamma_0$ for small or large times, respectively. Now, the flatband voltage shift satisfies a one-dimensional Poisson equation in the oxide, viz.

$$(\Delta V_{FB})'' = -ep_T/\varepsilon_{ox} \tag{6.43}$$

Integrating the above equation twice from one oxide interface to the other yields, for positive gate bias

$$V_{FB} = -e\int_0^{t_{ox}} p_T(x)x\,dx/\varepsilon_{ox} \tag{6.44}$$

Inserting p_T from Eq. (6.42) into Eq. (6.44) yields

$$\Delta V_{FB} = -(eN_T/\varepsilon_{ox})\left[t_{ox}^2/2 - \lambda^2\int_0^{t_{ox}/\lambda}(\exp - x)\,x\,dx\right] \tag{6.45}$$

For small t_{ox}/λ which corresponds to a thin oxide layer,

$$\Delta V_{FB} \cong -(eN_T/3\varepsilon_{ox})(t_{ox}^3/\lambda) \tag{6.46}$$

For large t_{ox}/λ corresponding to a relatively thick oxide layer,

$$\Delta V_{FB} \sim -(eN_T/\varepsilon_{ox})(t_{ox}^2/2). \tag{6.47}$$

which is depicted in Fig. 6.21 (lower right).

For irradiation under negative bias, the resulting ΔV_{FB} is the same function of oxide thickness.[37]

Recent experimental work has shown that the gate threshold voltage shift for MOS devices can be given by the empiricism[29]

$$\Delta V_T \sim \gamma_0 N_T t_{ox}^n \tag{6.48}$$

where $1 \leqslant n \leqslant 3$, and γ_0 is the ionizing dose absorbed in the device.

A somewhat more elaborate one-dimensional model[46] for the gate threshold voltage shift ΔV_T, which includes tunneling, is discussed using the same coordinates as the flat band voltage shift model in the foregoing. Assuming a positive gate bias voltage (n-channel enhancement mode or p-channel depletion mode MOSFET) the ionizing radiation-induced hole-electron pairs generated in the gate oxide will drift under the influence of the corresponding electric field. The free electrons will be swept toward the gate electrode-gate oxide interface ($x = 0$), making the electron concentration essentially zero at the gate oxide-silicon channel opposing interface ($x = t_{ox}$). This is in contrast to that for the positively charged holes, which are swept toward $x = t_{ox}$, making their concentration essentially zero at the $x = 0$ interface.

Since, in this one-dimensional model, the electrons are virtually all traveling to the left toward the gate electrode-silicon gate oxide interface, and holes to the right toward the silicon gate oxide-channel interface, the corresponding flux and current magnitudes are equal for each. This is apparent from the usual definition of these quantities, as discussed in Section 5.4.

The corresponding one-dimensional transport equation can be written[46] using those in Section 1.4 as guidelines. For the conduction band (free) electrons,

$$\partial n(x, t)/\partial t = \partial j_n(x, t)/\partial x - \sigma_n p_T(x, t) j_n(x, t) + eg_0 \dot{\gamma}_T(t)\pi; \quad n(t_{ox}, t) = 0 \tag{6.49}$$

where:

$n(x, t)$ = conduction band (free) electron concentration.
$j_n(x, t)$ = conduction band electron current density.
$p_T(x, t)$ = trapped hole concentration.
$\dot{\gamma}_T(t)$ = incident ionizing dose rate.
σ_n = electron capture cross section (for positive coulombic trapping).
eg_0 = hep (hole-electron pair) generation constant, (Eq. (7.26)).
π = hep escape probability (probability of hole-electron pairs escaping the trapping processes).

Equation (6.49) is simply an audit of the source and sink rates that contribute to the electron concentration rate. The first term (sink) on the right-hand side of Eq. (6.49) represents the rate of electrons diffusing out of the infinitesimal one-dimensional volume element in the gate oxide. The second term (sink) is the electron trapping rate, since $\sigma_n p_T$ is the amount of geminate trapping per unit electron current density. Only geminate (hole-electron pairwise) trapping is assumed herein,[46] which is the usual assumption for this type of model. So $\sigma_n p_T j_n$ can then represent the electron trapping rate in terms of the trapped hole concentration. The third term (source) is the rate of production of hole electron pairs in the gate oxide due to the incident ionizing radiation.

Similarly, for the valence band (free) holes,

$$\partial p(x, t)/\partial t = -\partial j_p(x, t)/\partial x - \sigma_p(N_T - p_T(x, t))j_p(x, t) + eg_0\dot{\gamma}_T\pi; \qquad p(0, t) = 0. \tag{6.50}$$

where:

$p(x, t)$ = valence band (free) hole concentration.
$j_p(x, t)$ = valence band hole current density.
N_T = concentration of gate oxide hole traps.
σ_p = hole capture cross section (for neutral hole traps).

Equation (6.50) is a hole concentration rate audit, immediately analogous to Eq. (6.49). The term $N_T - p_T(x, t)$ is the concentration of gate trapping states still unoccupied (i.e., the concentration of available hole traps), so that $\sigma_p(N_T - p_T)$ is the amount of hole trapping per unit hole current density.

A similar equation exists for the gate oxide trapped hole concentration, p_T,

$$\partial p_T(x, t)/\partial t = \sigma_p(N_T - p_T(x, t))j_p(x, t) - \sigma_n p_T(x, t)j_n(x, t); \qquad p_T(x, 0) = 0 \tag{6.51}$$

The terms on the right-hand side of Eq. (6.51) are the geminate trapping rate, less the second term, which is the electron concentration trapping rate. This difference amounts to the hole trapping rate. The initial condition, $p_T(x, 0) = 0$, simply asserts that no holes are initially trapped.

To integrate Eqs. (6.49)–(6.51), the temporal setting is such that, following the onset of ionizing radiation, the electron and hole concentrations reach equilibrium relatively quickly (electrons in 10^{-13} sec, holes in 10^{-7} sec) compared to the initial dose rate ($\dot{\gamma}_T$ = 40 rads/sec herein). This then allows $\partial p/\partial t \cong \partial n/\partial t \cong 0$ in Eqs. (6.49) and (6.50). Also, the capture cross sections, σ_n and σ_p, are sufficiently small so that in this epoch the ionizing radiation-induced source term dominates the sink terms. Then Eq. (6.49) can be written as

$$0 \cong \partial j_n/\partial x + eg_0\dot{\gamma}_T\pi; \qquad j_n(t_{ox}, t) = 0 \tag{6.52}$$

with an integral given by

$$j_n(x, t) = eg_0\dot{\gamma}_T\pi(t_{ox} - x) \tag{6.53}$$

Similarly, Eq. (6.50) becomes

$$0 \cong -\partial j_p/\partial x + eg_0\dot{\gamma}_T\pi; \qquad j_p(0, t) = 0 \tag{6.54}$$

with an integral given by

$$j_p(x, t) = eg_0\dot{\gamma}_T\pi x \tag{6.55}$$

Inserting the current densities from Eqs. (6.53) and (6.55) into Eq. (6.51), for $\dot{\gamma}_T$ constant, yields

$$(1/eg_0\dot{\gamma}_T\pi)\,\partial p_T/\partial t + (\sigma_n t_{ox} + (\sigma_p - \sigma_n)x)p_T = \sigma_p N_T x; \qquad p_T(x, 0) = 0 \tag{6.56}$$

with an integral given by

$$p_T = (N_T x/(rt_{ox} + (1 - r)x)) \cdot \{1 - \exp - [eg_0\dot{\gamma}_T\pi\sigma_p\{rt_{ox} + (1 - r)x\}\,t]\} \tag{6.57}$$

where $r = \sigma_n/\sigma_p$.

It is well established[47] that the gate oxide midgap voltage shift $\Delta V_{mg} \cong \Delta V_{ot}$ can be expressed as the first moment integral of the trapped hole bulk charge, as is the flat band voltage shift in Eq. (6.44), as in

$$\Delta V \equiv \Delta V_{ot} \cong \Delta V_{mg}(t) = \int_0^{t_{ox}} ep_T(x, t)\,x\,dx/\varepsilon_{ox}, \tag{6.58}$$

where ε_{ox} is the dielectric permittivity of the gate oxide.

The first moment integral (centroid) of the charge, as in Eq. (6.58), is a characteristic of functions that satisfy a Poisson equation. The midgap voltage shift is essentially the same as the bulk oxide charge voltage shift, ΔV_{ot}, since the former does not include contributions to the voltage shift from the existence of interface states. As discussed in Sections 1.6 and 1.9, the interface donor states exist in the lower half of the gate oxide band gap, and are neutral when filled. The interface acceptor states exist in the upper half of the oxide band gap, and are neutral when empty. A gate voltage such that the fermi level E_f at the interface surface equals the midgap energy ($E_f = E_{mg}$) is called the midgap voltage[74], i.e., $V_G = V_{mg}$ there. Then V_{mg} is a measure of the situation where the interface states have zero influence on voltage shifts, such as ΔV_T. This is because the net charge of the interface traps is neutral, since all of the donor traps (below midgap) are filled, and likewise for all of the empty acceptor traps (above midgap). ΔV_{mg} can be determined using subthreshold C-V techniques[85] (Section 6.10). Subthreshold implies channel conduction or leakage current when the device is off.

It is presently generally hypothesized that the actual atomic structure singly comprising both the donor and acceptor interface trap in the gate oxide is a P_b center.[74] The center consists of a trivalent silicon atom bonded to three other Si atoms lying in the Si/SiO_2 interface, with the remaining Si dangling bond electron protruding into the oxide. Energetically, this center consists of an amphoteric defect having ground states that can capture 0, 1, or 2 electrons, with 1- or 2- electron transitions, all centered about the midgap energy. The oxide hole trap is hypothesized as an E' center,[74] consisting of a trivalent silicon atom bonded to three oxygen impurity atoms in the oxide, with a remaining dangling bond electron constituting the trap, per se.

For the interface surface fermi level between the valence band and midgap i.e., $E_v < E_{fs} < E_{mg}$ (lower half of the bandgap), the positively charged empty P_b center can trap an electron, becoming neutral (donor trap). For $E_{mg} < E_{fs} < E_c$ (upper half of the bandgap), a now neutral "empty" P_b center can trap a second electron, becoming negatively charged (acceptor trap).[74] When $E_{fs} = E_{mg}$, the P_b centers are neutral, in that these interface traps possess zero net charge, as already discussed. Also, computations of P_b center interface state density with energy result in a dual-peaked structure of the corresponding density distribution function, as shown in Figs. 2.9 and 2.10.[74,89]

Substituting p_T from Eq. (6.57) into Eq. (6.58) yields

$$\Delta V = (eN_T/\varepsilon_{\text{ox}}) \int_0^{t_{\text{ox}}} (x^2 dx/(rt_{\text{ox}} + (1-r)x) \cdot \{1 - \exp - [eg_0\dot{\gamma}_T\pi\sigma_p(rt_{\text{ox}} + (1-r)x)t]\} \tag{6.59}$$

The initial midgap voltage shift, ΔV_i, is given by integrating Eq. (6.59) and taking the limit of the result for small time to obtain, where $\gamma_T = \dot{\gamma}_T t$,

$$\Delta V_i/\gamma_T = \lim_{t \to 0} \Delta V/\gamma_T = N_T e g_0 \pi\sigma_p t_{\text{ox}}^3/3\varepsilon_{\text{ox}} \tag{6.60}$$

The maximum voltage shift is seen to be obtained by

$$\Delta V_{\max} = \lim_{t \to \infty} \Delta V = eN_{\text{eff}} t_{\text{ox}}^2/\varepsilon_{\text{ox}} \tag{6.61}$$

where $N_{\text{eff}} = (N_T/2[r-1]^3)(r^2[2(lnr) - 3] + 4r - 1)$.

Hence, as asserted earlier, it is seen that $\Delta V_T \sim t_{\text{ox}}^n$, where $n \cong 3$

for early times following the $\dot{\gamma}$ pulse onset. For later times, as ΔV_T approaches its maximum, $n \cong 2$. The electric field dependence of $\Delta V_i/\gamma_T$ emanates from σ_p and π, since all the other quantities are field independent. This holds for ΔV_{max} as well. As seen in the limit of zero electron capture cross section (i.e., lim $r \to 0$), $V_{max} = \frac{1}{2}(e/\varepsilon_{ox}) N_T t_{ox}^2$ for no electric field from Eq. (6.59).

It is also well known that tunneling of electrons into, or hole tunneling out of, the gate oxide, at the gate oxide-channel interface, heavily influences ΔV. To include tunneling, an additional term, $-p_T g(x, E)$, is added to the right-hand side of Eq. (6.51), where $g(x, E)$ is the equilibrium tunneling rate. The tunneling rate is given by[46] (where E_t is the trapping energy level)

$$g(x, E) = (h/16m^* \sigma_p) \exp - 2[(2m^* E_t)^{1/2} (t_{ox} - x)/\hbar] \qquad (6.62)$$

where m^* is the effective mass discussed in Section 1.5, h is Planck's constant, and $\hbar = h/2\pi$. To delineate the sharp cutoff behavior of g, near the gate oxide-channel interface, it is instructive to substitute nominal parameter values into Eq. (6.62). For $m^* = 0.42 m_e = 0.42 \cdot 10^{-27}$ gm, $h = 6.6 \cdot 10^{-27}$ erg sec, $E_t = 3.1$ eV, $\sigma_p = 1.4 \cdot 10^{-14}$ cm^2, and $t_{ox} = 250$Å,

$$g(x, 3.1) = g_T \exp - 307(1 - x/t_{ox}); \qquad x \leqslant t_{ox};\ g_T = 7 \cdot 10^{13}\ \text{sec}^{-1} \qquad (6.63)$$

hence it is seen that g is essentially zero except for $x = t_{ox}$, where $g = g_T$.

The corresponding $N_{eff\,t}$, which includes the effect of tunneling,[46] for $\Delta V_{max} = (e/\varepsilon_{ox}) N_{eff} t_{ox}^2$ is

$$\begin{aligned} N_{eff\,t} = {} & (N_T/(1 - r)^3)\{r^2 \cdot \ln[(r^{-1} - 1)(1 - \Delta x/t_{ox}) + 1] \\ & + (1/2)(1 - r)^2 (1 - \Delta x/t_{ox})^2 - r(1 - r)(1 - \Delta x/t_{ox})\} \end{aligned} \qquad (6.64)$$

with the thickness of the charge-depleted layer, Δx, due to the tunneling drain of carriers near the gate oxide-channel interface, given by

$$\Delta x = (\hbar/2(2m^* E_t)^{1/2}) \ln[g_T/e g_0 \dot{\gamma}_T \pi t_{ox} \sigma_p (1 + 0.2r)] \qquad (6.65)$$

From Eq. (6.58) it is seen that

$$\Delta V_{max} = \int_0^{t_{ox}} p_T(x, \infty)\, x dx/\varepsilon_{ox} \qquad (6.66)$$

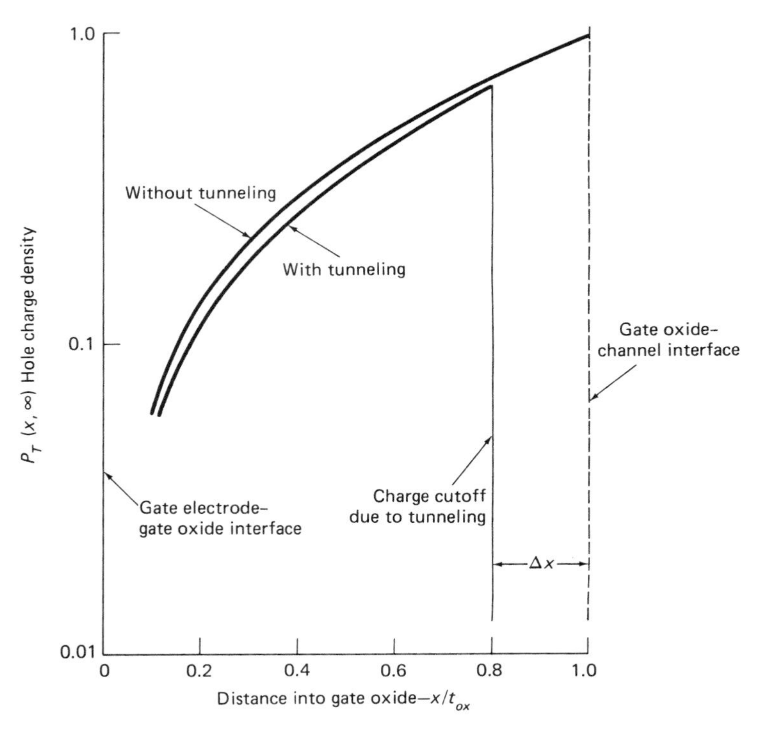

Fig. 6.23 Gate oxide hole charge versus distance in oxide ($t_{ox} \sim 225$Å) between interfaces with and without tunneling.[46] Figure © 1987 by the IEEE.

The integral yields the first moment of charge, here—the trapped hole positive charge. Figure 6.23 depicts the hole charge, $p_T(x, \infty)$, versus oxide distance from the gate electrode interface, with and without tunneling. It is seen from the figure that tunneling depletes a hole-charge layer of thickness Δx on the gate-oxide side of the gate oxide-channel interface. So that, with tunneling,

$$\Delta V_{\max t} = \int_0^{t_{ox}-\Delta x} p_T(x, \infty) x dx / \varepsilon_{ox} \qquad (6.67)$$

The preceding development provides midgap voltage shift versus gate oxide thickness t_{ox}. Again, midgap voltage shift is essentially the same as oxide-trapped hole voltage shift, as opposed to any voltage shift due to the presence of interface states.

Generally, the interface state contribution to the voltage shift corresponds to carrier trapping that provides a positive voltage shift, such as that discussed in Section 6.8 in connection with rebound. Hence, midgap voltage shift usually overestimates the total shift, since it neglects the effect of interface states.[74] The above theoretical development has been validated by comparison with experimental results obtained on t_{ox} = 225Å thermal oxide on p-type silicon test capacitors irradiated under bias at room temperature.[46]

Increases in leakage current are also caused by buildup of trapped charge. Leakage currents can modulate the channel conductance in a random fashion, inducing extraneous currents in the circuitry. In linear CMOS devices, such leakage currents must be considered with respect to their impairment of the high input impedance normally enjoyed by FETs. Drain to source leakage photocurrents in garden variety MOS transistors are of the order of $10\,pA/cm^2$ per rad (Si)/sec.

MOS processing trends are toward heavily doped polysilicon (semi-amorphous silicon) gate metallization, supplanting metal (Al) gate metallization. This is due to the easier mask alignment problems in the manufacture, especially for small feature sizes of the order of one micron and less. However, the higher temperatures involved in the polysilicon gate metallization processing compared with the aluminum gate metallization process makes for a degradation in the polysilicon gate device ionizing dose hardness. This is due probably to the surface effects manifest by the higher temperatures in the thermal cycling associated with polysilicon gate metallization processing.[44]

For cryogenic temperature applications, it is important to know how ΔV_T varies with ionizing dose. Generally, ΔV_T increases at cryogenic temperatures more rapidly and over a greater range than that at ambient (room) temperature. This is shown in Figs. 6.23A, 6.24, and Table 6.5. In Table 6.5, it is seen that the above usually holds true, in that the device damage at cryogenic temperatures is generally greater than or equal to that at room ambient. The phenomonological explanation of greater ΔV_T experienced at cryogenic than at room temperatures is straight forward. It is a variation of that discussed earlier, included in Section 6.8. At very low temperatures, after the electrons have been swept out of the gate oxide by the electric field, the holes generated therein are quickly trapped. This freezes-in the damage at this juncture, unlike that for room temperature holes, which begin to stochastically hop away from their initial trapping sites, and eventually to anneal.

At cryogenic temperatures, HgCdTe infrared focal plane arrays use silicon CMOS/SOS readout circuitry, mainly because sapphire and HgCdTe thermal expansion coefficients match closely.[86] Ionizing radiation and cryogenic temperature-induced reduction in avalanche break-

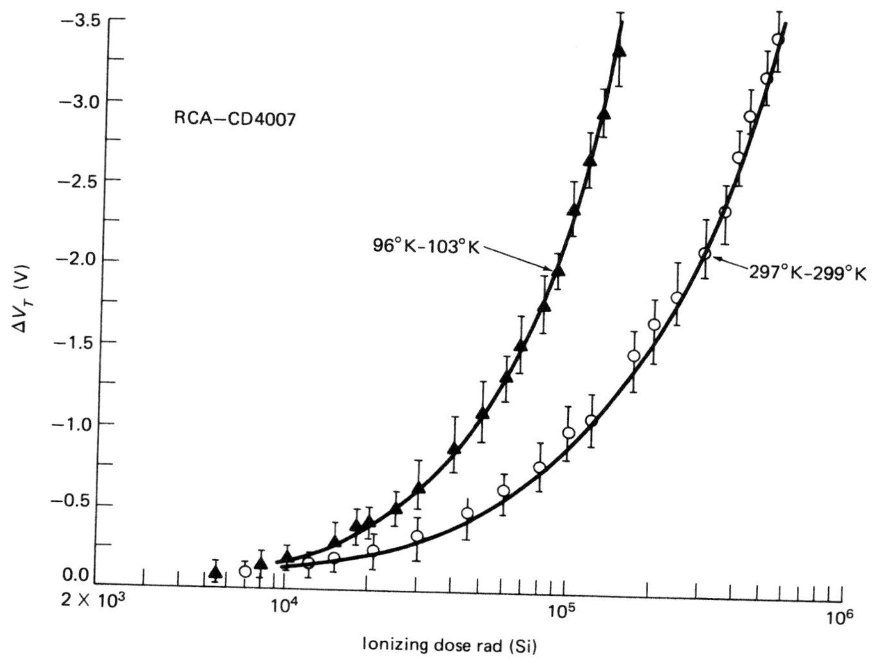

Fig. 6.23A PMOS gate V_T shift versus 10 MeV proton dose for an applied V_{GS} of −5 V at cold and room temperatures.[80]

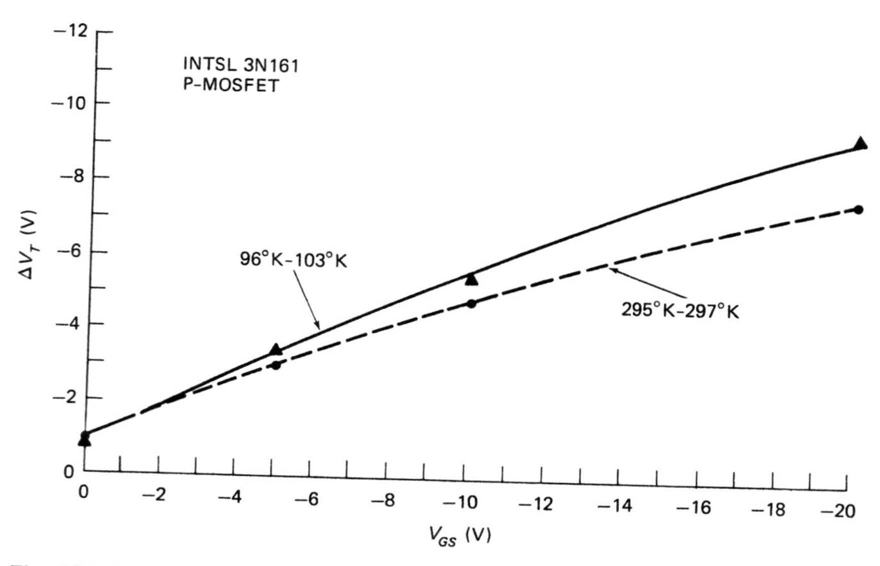

Fig. 6.24 PMOS gate V_{th} shift versus V_{GS} for a 10 MeV proton dose of 50 Krad (Si) at cold and room temperatures.[80]

down voltage can result in avalanche-generated holes. They can forward bias the source-body junction, causing a large increase in drain current. This is called the kink effect,[86,87] and is detrimental to the operation of these arrays. It can be reduced significantly by implanting a p^+ doped conducting path beneath the SOS-MOSFET field oxide.[87]

6.10 MOSFET CAPACITOR

MOSFETs depend for their operation on their surfaces and interfaces, as discussed in this chapter. With the advent of the MOSFET planar technology, where the oxidized silicon surface and gate-substrate interface play a major role, the use of the gate voltage variable MOS capacitor is a very powerful investigative tool. This is because the MOS capacitor is a simple structure, easy to fabricate and instrument, while providing a close simulation of the MOSFET transistor. For most investigations, the MOS capacitor admittance, $i\omega C$ with C defined below, as a function of gate bias and frequency, is the operative parameter from which the properties of interest are derived. Some important properties of a MOS transistor that are affected by ionizing radiation, and can be measured using a MOS capacitor, include: interface trap density and capture probability as a function of energy in the band gap, band-to-band tunneling in the silicon and SiO_2 gate insulator, properties of electron and hole traps in the SiO_2 gate insulator, SiO_2 thickness and breakdown field, lifetime in the bulk silicon, and surface recombination velocity.[48]

A MOSFET capacitor can be construed as consisting of a MOSFET transistor, where bias is applied between the metal-SiO_2 gate and the silicon substrate or bulk (ground), usually with no source or drain depositions. This can be visualized in Figs. 6.5a or 6.5b, but with the drain and source floating. The MOS total capacitance C is the series sum of the various planar element capacitances between the gate metal ohmic contact and the substrate bulk ground contact.

For illustration, a simple MIS (metal-insulator-silicon) or MOS (metal-oxide-silicon) capacitor will be discussed.

The total capacitor merely consists of a slice of n or p silicon bulk material, with a thin contiguous layer of SiO_2, followed by a metal electrode on top of the SiO_2. The opposite electrode is the ohmic contact on the far side of the silicon bulk. The capacitance and conductance versus bias characteristics, for various ionizing radiation levels, contain information about oxide and interface states, among other parameters.[49]

The small signal differential admittance of the ideal (lossless) capacitor, with no surface states, consists of the SiO_2 gate insulator capacitance per unit area, C_{ox}, between the metal electrode and the SiO_2-Si interface.

Table 6.5. Device threshold voltage shift per unit ionizing dose, $\Delta V_T/\gamma_T$, compared at cryogenic (103°K–96°K) and room (293°K–300°K) temperatures.[80]

SERIAL	DEVICE TYPE	ENHANCEMENT MODE CHANNEL	IN SITU V_{GS}(V)	DEVICE γ_T HARDNESS	DAMAGE PER IONIZING DOSE* COMPARED AT CRYOGENIC AND AMBIENT TEMPERATURES		
2	RCA Z-CD4007	p	0.0	100 Krad (Si)	$\Delta V_T/\gamma_T$ at Cryogenic Temperature	=	$\Delta V_T/\gamma_T$ at Room Temperature
3	"	p	−5.0	"	"	>	"
4	"	n	0.0	"	"	=	"
5	"	n	+5.0	"	"	=	"
6	FCH-4007UB	p	0.0	Unhardened	"	>	"
7	"	p	−5.0	"	"	>	"
8	"	n	0.0	"	"	=	"
9	"	n	+5.0	"	"	<	"
10	INSL 3N-161 PFET	p	$V_{GS} < 0$	"	"	>	"
11	INSL 3N-171 NFET	n	$V_{GS} < 0$	"	"	>	"
12	"	n	$V_{GS} > 0$	"	"	<	"

* Absorbed at the same rate.

This is in series with the p-type silicon depletion layer capacitance per unit area, $C_s = \varepsilon_s \varepsilon_0 / l$, where l is the depletion layer thickness of the SiO_2-Si junction.

The total capacitance satisfies $C^{-1} = C_{ox}^{-1} + C_s^{-1}$. Using Eq. (3.35) for the depletion layer thickness, assuming p-type silicon ($N_A \gg N_D$), then it is straightforward to show (Section 3.5) that

$$C/C_{ox} = [1 + (2\varepsilon_0 \varepsilon_{ox}^2 |V_G| / eN_D \varepsilon_s t_{ox}^2)^{1/2}]^{-1} \tag{6.68}$$

where V_G is the gate voltage, and ε_{ox}, ε_s, and ε_0 are the dielectric constants of the oxide insulator, silicon bulk, and the free space permittivity, respectively. t_{ox} is the SiO_2 insulator thickness. Note that Eq. (6.68) holds only when the oxide insulator is depleted (reverse biased). For $V_G > 0$, there is no depletion region, as discussed in Section 3.5. However, for $V_G > 0$, the insulator is accumulated (Figs. 6.8 and 6.10), so that the silicon bulk acts as a resistor in series with $\mathbf{C}_{ox}$; thence $C \cong C_{ox}$ only, in this state. On the other hand, for strong inversion and when the surface potential $\psi_s = 2\psi_F$ (Fig. 6.10), the width of the depletion region will not increase with further increase in V_G. This corresponds to the onset of strong inversion, which is equivalently the onset of MOS transistor "turn-on", so that $V_G = V_T$, the threshold turn-on voltage. Then C/C_{ox} becomes constant at the voltage given by $V_G = V_T$. This is shown in Fig (6.25).

Equation (6.68) is plotted in the right half of Fig. 6.25 for $0 \leqslant V_G \leqslant V_T$.

The SiO_2 insulator capacitance C_{ox} is essentially independent of the frequency of the gate signal voltage. The depletion layer capacitance C_s is also frequency independent in the accumulation and depletion regimes. However, C_s is a function of frequency in the inversion regime, depending on how well the carriers stay in phase with the signal frequency. C_{ss}, a surface state capacitance, can be added to this model in parallel with C_s to represent their presence. For sufficiently high signal frequency, the surface states in their exchange of charge (transition rate of electrons or holes to and from-surface charge-trapping states) with the SiO_2 cannot stay in phase. Then $C_{ss} \simeq 0$, and the system reverts to its ideal state, where $C^{-1} = C_{ox}^{-1} + C_s^{-1}$. By measuring ΔV, the voltage difference between the measured and ideal characteristic, the oxide state density N_{ss} can be obtained merely from

$$eN_{ss} = C_{ox} \Delta V \tag{6.69}$$

where eN_{ss} is the total charge trapped in the oxide states. Figure 6.26 depicts both the actual and ideal C-V curves.[49]

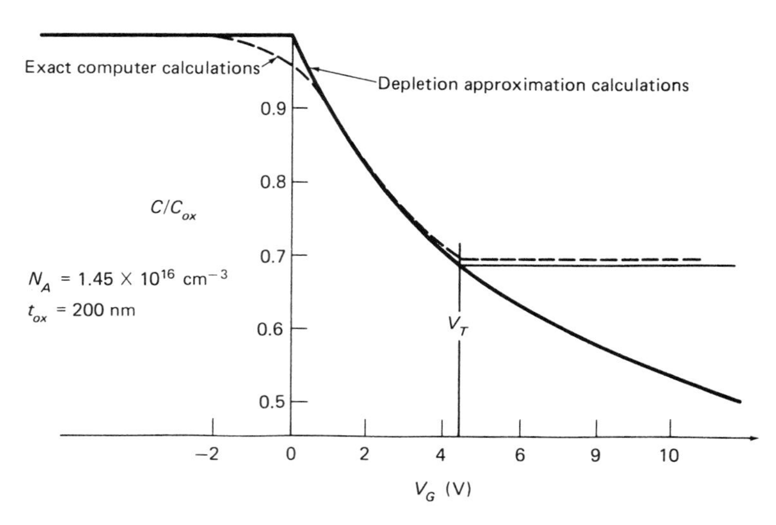

Fig. 6.25 The capacitance-voltage characteristics of a MOS structure.[38] Figure © 1967 by John Wiley & Sons.

The surface density of interface trap states at the SiO_2—Si interface, $D_{it}(E)$, (number density of interface traps per $cm^2 \cdot eV$ of band gap energy) is an important parameter, in terms of the proper function of the MOS transistor. This can be appreciated in one sense by the fact that, for high D_{it}, the charges in these traps can form an electrostatic shield across the above model, as well as in the actual MOSFET. This can essentially render the device inoperable. An associated parameter, N_{it}, the surface charge density of interface traps per cm^2, is given by the following integrals: $\int_{E_f}^{E_c+e\phi_s} D_{it}(E)\,dE$ for donor-like traps, and $\int_{E_v+e\phi_s}^{E_f} D_{it}(E)\,dE$ for acceptor-like traps. $e\phi_s$ must be added to the bulk band edges E_c and E_v of the integration limits to yield the band-edge energy at the silicon interface.

One way to obtain D_{it} is through a conductance measurement of the SiO_2 insulator—Si bulk channel, which is induced by (for instance) $V_G > 0$ for p-type silicon. This conductance is given by [38] $g = (W/L)\,\mu_n C_{ox}(V_G - V_T)$, where W and L are the channel dimensions shown in Fig. 6.9. A comparison of the conductance versus V_G with the corresponding C-V plot is given in Fig. 6.27. Figure 6.29 shows undistorted C-V curves indicative of bulk oxide states only. It has been established that the stretching distortion in the C-V curves, as seen in Fig. 6.28, is due to the existence of interface states.[48] Lateral charge non-uniformities can also

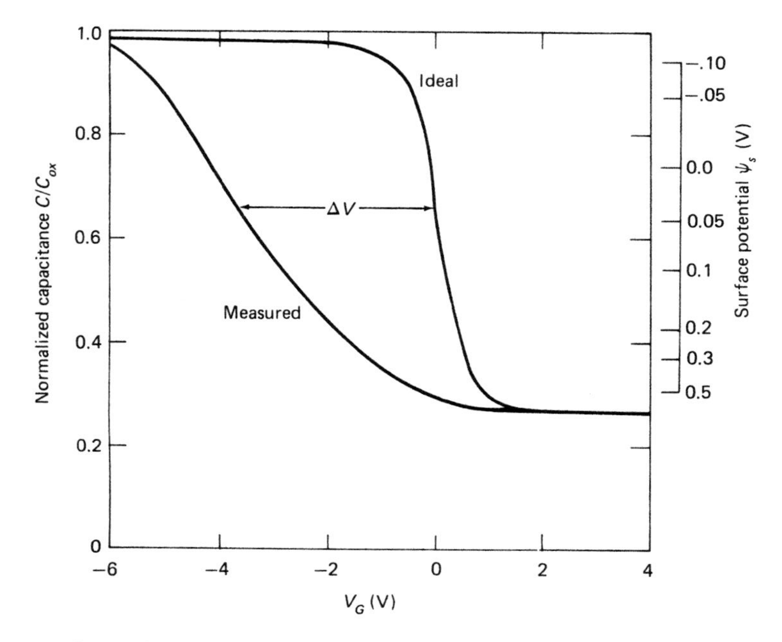

Fig. 6.26 Comparison between ideal and measured *C-V* curves.[49] Figure © 1966 by the IEEE.

cause a similar distortion. One method to resolve this ambiguity is to measure the surface recombination velocity, S_p.

Assuming interface state traps are produced uniformly across the interface, it is easily shown that S_p is proportional to D_{it}.[48] The conduction method for measuring these parameters independently delineates the interface traps produced by ionizing radiation.[48] This results in the determination of D_{it} from a measured $S_p(\gamma)$, where γ is the ionizing dose, given by[48]

$$S_p = (1/2)(\sigma_n \sigma_p)^{1/2} v_{th} \pi k T D_{it} \tag{6.70}$$

where σ_n and σ_p are the electron and hole trap capture cross sections, respectively, and v_{th} is the thermal velocity.

For SOS or SOI MOSFETs, *C-V* measurements for determining D_{it} are complex in their interpretation. The charge pumping method helps to overcome this difficulty, and compares well quantitatively with the *C-V* method.[50] In this technique, both source and drain are reverse

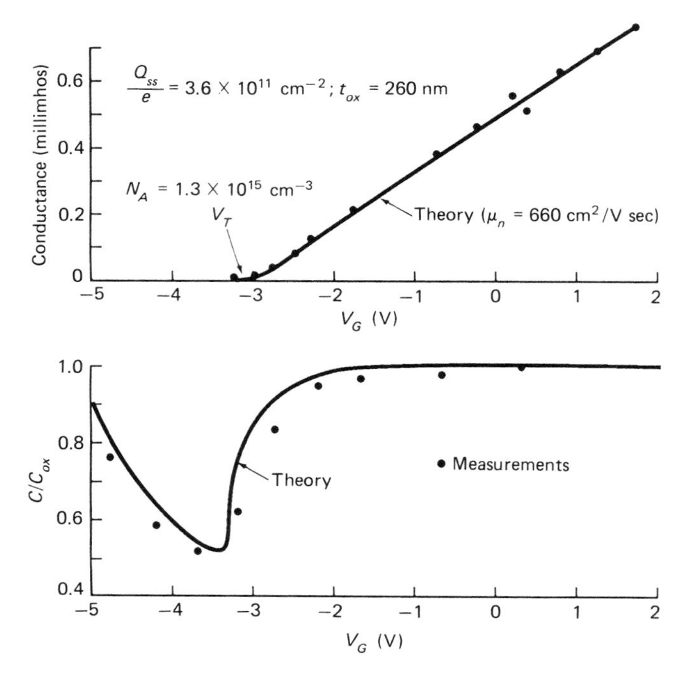

Fig. 6.27 Channel conductance and gate capacitance of a MOS transistor. Figure © 1967 by John Wiley & Sons.

biased, and the channel region is alternately biased into accumulation and inversion, by positive and negative square wave switching of the gate bias. Charge captured by the interface traps during inversion recombines in the silicon bulk during accumulation. This repetitive inversion and accumulation in the channel region results in a charge pumping current proportional to the interface state charge density, D_{it}, so that measuring the pumping current magnitude yields D_{it}.[50]

6.11 SNOS/MNOS DEVICES

The MNOS (metal-nitride-oxide-silicon) or SNOS (silicon-nitride-oxide-silicon) transistor bears a close similarity to the MOS transistor. MNOS implies a metal gate contact, while SNOS implies a polysilicon gate contact. In the MOS transistor, the gate insulator is a single layer of SiO_2 but in the SNOS or MNOS device, the gate insulator is a double layer, consisting of a thick slab of silicon nitride (Si_3N_4) contiguous with the relatively thin slab of SiO_2. The nitride layer is closer to the gate contact,

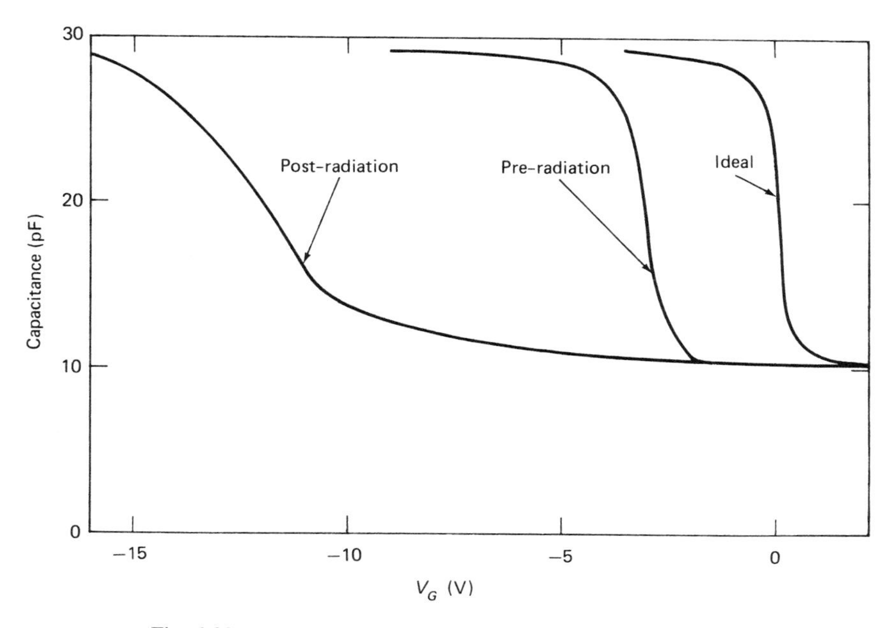

Fig. 6.28 *C-V* curves for electron fluence with interface states.[49]

and the silicon dioxide layer is closer to the silicon substrate. The channel is still located near the latter interface. A major electrical difference between the MOS and "NOS" transistors is that, whereas the threshold voltage V_T in the MOS transistor is fixed by physical construction of the device, V_T in the NOS transistor can be set and reset electrically by the gate bias. Furthermore, it will remain set for an extended period even after power is turned off, because V_T is determined by the trapped charge at the Si_3N_4-SiO_2 interface, as seen in the following. As discussed previously, the threshold voltage shift in MOS devices depends on the accumulation or removal of trapped positive charge, mainly at the Si/SiO_2 interface. In the NOS device, the two regions correspond to the oxide and nitride layers that form the gate insulator, with stored charge trapped between. The amount of stored charge depends on the amplitude, duration, and polarity of the gate signal input. This charge controls the drain current-gate voltage characteristic, so that the NOS system can serve as a memory device. To illustrate, an *n*-channel NOS can be switched to a high threshold voltage state by applying a positive voltage pulse to the gate. This will drive electrons from the silicon into the SiO_2 layer, which will reach the nitride layer after tunneling through the potential barrier

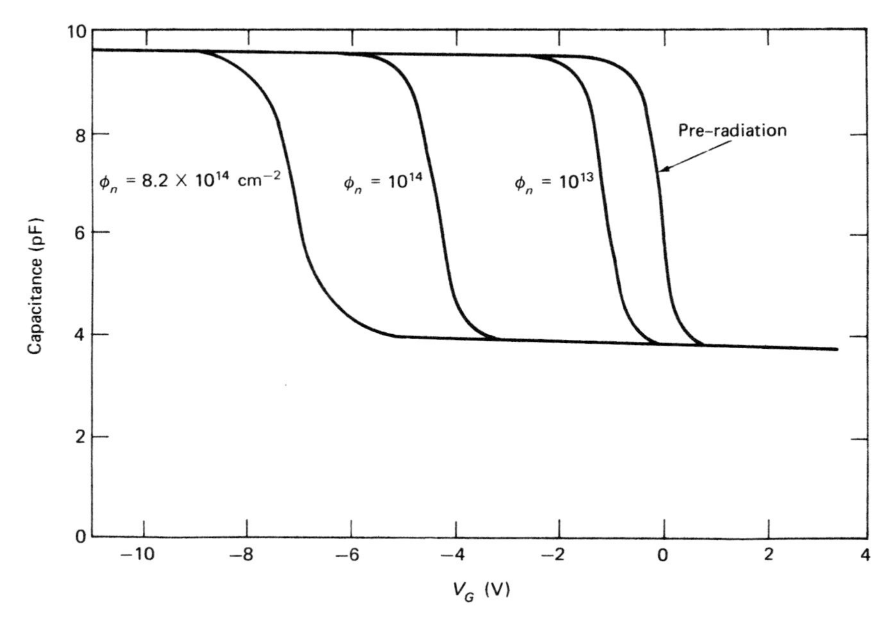

Fig. 6.29 *C-V* curves for electron fluence with oxide bulk states only.[49] Figure © 1966 by the IEEE.

represented by the thin oxide layer. Most of this charge will reside near the oxide-nitride interface. To switch to a low threshold voltage state, a large negative gate voltage pulse is applied, driving the electrons back into the silicon substrate from the gate. The *p*-NOS device operates in a complementary manner. The two different threshold levels can represent two memory states, which can be arbitrarily designated.[36] In the context of memory devices, a burst of ionizing radiation can wreak havoc with unhardened semiconductor memories through production of hole-electron pairs therein. All bias voltage balances and charges stored dynamically on junctions or nodes are thus lost in MOS devices. The NOS transistor is nonvolatile in this respect, for its information-storage capability is not dependent on the maintenance of an external power source and bias configuration.

Ionizing radiation damage commonalities in MOS and NOS devices include parasitic *p-n* junction photocurrents, channel conductance changes, and Si-SiO_2 interface surface state manifestations. However, the NOS is quite radiation-resistant to ionizing dose, dose rate, and neutron fluence.[35] Correspondingly, three typical failure threshold levels in unhardened devices are, respectively: $\sim 10^3$ rads (Si) ionizing dose, $\sim 10^9$ rads (Si) per second dose rate (due to the upset of associated read/write circuitry, and not the memory cells per se), and $\sim 10^{15}$ neutrons per cm^2.[35]

6.12 PMOS/NMOS DEVICES

P-channel MOS devices are affected by ionizing radiation in the same manner as the other members of the MOS family. That is, the principal effects of ionizing radiation are changes in the threshold voltage, flatband voltage, and channel mobility due to trapped charge accumulation in the gate insulator, as well as interface state buildup. Techniques for minimizing the effects of such radiation on *p*-channel devices using an optimal thermal SiO_2 gate insulator have been employed. When such gates are used, the corresponding *p*-channel devices can be subjected to better than 1 Mrad (Si) without apparent degradation. Optimization consists of minimizing the gate threshold voltage shift. The techniques that accomplish this include the use of ⟨100⟩ Miller index silicon structure, oxidation of the gate in a dry oxygen atmosphere and metal aluminum deposit used for the gate terminal that has been especially evaporated and sintered.[21] After irradiation with 1 Mrad (Si), these devices exhibit threshold voltage shifts of less than 1 V over a 30-V range of negative gate bias.

NMOS is the mainstay of large-scale commercial integrated circuits (LSI) and very large-scale (VLSI) integrated circuit technology today. This is because NMOS devices can be made for low threshold voltage operation, resulting in compatibility with a single 5-Volt supply. Since electron mobility is much greater than hole mobility, NMOS devices are much faster than their PMOS counterparts. Some NMOS devices fail at ionizing doses as low as 1 krad (Si), and the range of 1–3 krad (Si) seems to be the safe upper limit range for ordinary NMOS. The PMOS transistor becomes harder to turn on, as in Fig. 6.11(a), but modified circuitry can easily obviate this difficulty. However, the NMOS device could be made to turn on permanently, as in Fig. 6.11(b). Also, the NMOS positive gate voltage sweeps these holes away from the gate toward the gate oxide-channel interface, adding to the positive charge states in that vicinity, to cause additional threshold voltage shift. Further, such positive charges produced in the NMOS field oxide, i.e., in the oxides other than gate, can turn on normally inverted silicon beneath them to produce leakage currents. However, negative gate PMOS would attract gate oxide holes toward the gate terminal, away from the gate-channel interface, which is a less sensitive position insofar as disrupting MOSFET operation.

The field oxide is very sensitive to ionizing radiation induced leakage currents, and is usually the primary cause of failure of MOS devices. If the field oxide difficulty is not surmounted, LSI systems can fail at doses of less than 10 krads (Si), even though megarad gate insulator oxides have been incorporated into their manufacture. On the one hand, threshold voltage shifts and leakage currents can often be circumscribed by ingenious circuit design, while on the other hand, some dynamic RAMS can tolerate only a 200-mV shift in threshold voltage.

To minimize the ionizing radiation sensitivity of NMOS devices, a number of manufacturing steps are taken that are similar to those used for PMOS.[22] These also include the use of aluminum instead of a polysilicon gate terminal, very heavy impurity diffusion using ion implantation techniques to essentially double the field inversion potential, and fixed bias on the substrate, as opposed to allowing the substrate bias voltage to follow that of the internal bias generator (in RAMs). All these techniques together can increase the radiation susceptibility level of NMOS to greater than 1 Mrad (Si).

With the trend toward submicron feature size devices, the short-channel NMOS has long been a candidate for increasing chip packing density. The short-channel device implies a channel length comparable with the depletion layer width of the source and drain junctions. The present long-channel NMOS has a channel length much longer than the depletion layer widths. This allows a one-dimensional approach to

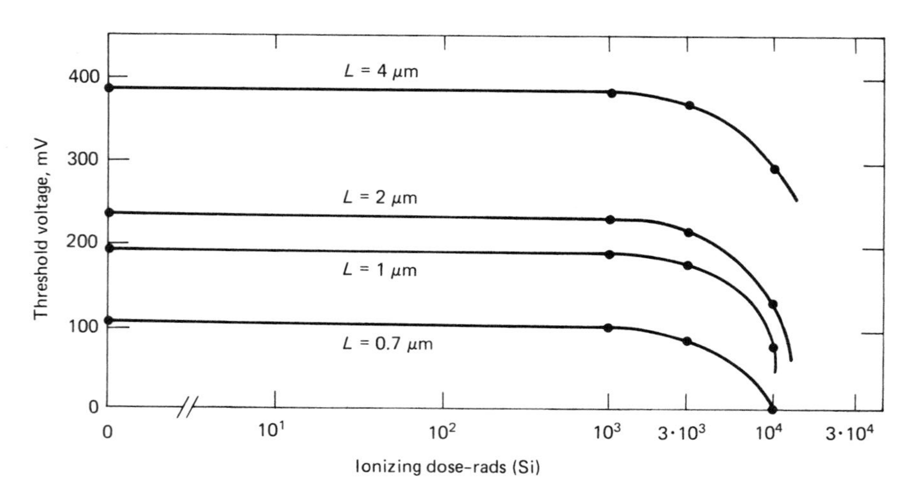

Fig. 6.30 Dependence of NMOS threshold voltage V_T on ionizing dose for various device channel lengths L.[84] Figure © 1981 by the IEEE.

modeling the potential distribution in the channel.[82] However, as the channel length is reduced, the depletion layer widths become comparable to the channel length, and the channel potential becomes dependent on both the (transverse) gate-channel field and the (longitudinal) drain-source field. This then requires a two-dimensional potential model.[82] Among other factors, the shorter channel results in a reduction in the gate threshold voltage V_T, because the fields from the source and drain now extend into the channel region, causing charge sharing between the channel and source-drain regions. This results in reduced charge drift toward the gate for a given surface potential, thus reducing V_T.[83] Parenthetically, the subthreshold current for a short channel device depends strongly on V_D. Even though the threshold voltage level V_T depends on the channel length, it and its shift ΔV_T, are relatively independent of the ionizing dose up to about 3 Krad (Si), irrespective of whether the device has a long or short channel. For ionizing dose levels greater than 3 Krad (Si), the V_T level drops precipitously, while $|\Delta V_T|$ increases for these devices,[84] as seen in Fig. 6.30.

6.13 CMOS/SOS/SOI/DEVICES

CMOS/SOS (silicon-on sapphire) devices are distinguished by a sapphire (Al_2O_3) substrate, instead of bulk silicon substrate. They are attractive because of the greatly reduced junction leakage photocurrent production over that in bulk silicon, as discussed below, as well as being impervious to latchup, as discussed in Chapter 12. Figure 6.31 compares four CMOS inverters, two formed on a bulk silicon substrate, and two formed on an insulator substrate. Figure 6.31 (a) and (b) depict CMOS and CMOS-Epi on bulk silicon inverters showing necessary structural adjuncts that include silicon channel stoppers, field oxides, and *p*-wells. In contrast, Figs. 6.31(c) and (d) show CMOS inverters formed on sapphire (CMOS/SOS) and silicon dioxide (SOI) substrates, respectively. It is seen there that the corresponding device structural changes result in the removal of the channel stoppers, field oxides, and *p*-wells. This reduction in the amount of silicon material available for production of hole-electron pairs and photocurrent, discussed in Chapter 7, plus the dielectric isolation afforded by the insulating substrate results in a greatly increased dose rate upset threshold for these devices. CMOS/SOS photocurrents are on the order of 0.1 pA per cm^2 per rad (Si)/sec., hardly 1 percent that of ordinary MOS devices.[18] After 1 Mrad (Si) irradiation, CMOS/SOS gate threshold voltage shifts are less than 1.2 V for the *n*-channel portion and less than 2.7 V for the *p*-channel portion of the device, under the most adverse bias levels.[23] The photocurrent leakage in the back channel

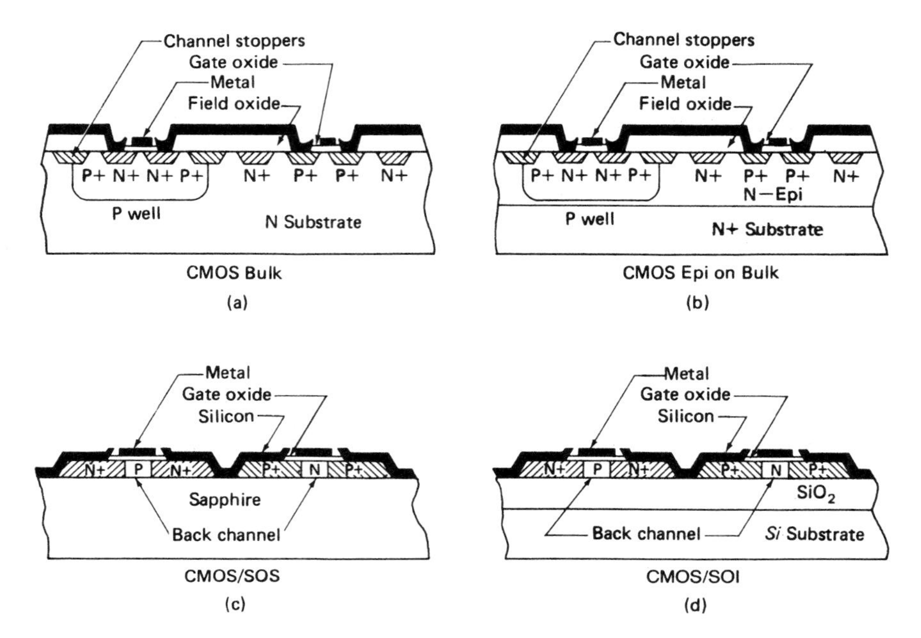

Fig. 6.31 Comparison of CMOS Bulk, CMOS Epi-on-Bulk, CMOS/SOS and silicon on insulator (SOI) structures.[18]

(Figs. 6.31(c) and (d)), i.e., the channel that can form opposite the gate channel between the channel diffusion (*n* or *p*) and the underlying sapphire (CMOS/SOS) or silicon dioxide (SOI),[61] is of the order of only 50–500 nA per cm^2 per mil of channel width. This back channel forms a parasitic MOS transistor, as shown in Fig. 6.31(c). Due to their physical construction CMOS/SOS and SOI devices comprise less volume than bulk CMOS so that their packing density can be increased, which makes for increased speed over that of bulk CMOS. Further topological improvements include capacitance reduction from metal crossover and gate overlap regions.[23]

6.14 BIPOLAR DEVICES

In semiconductor devices, as discussed in previous sections, ionizing radiation damage is caused mainly by charges trapped on or near the surfaces of their insulating layers and interfaces. It was also shown that these charges cause first-order changes in MOS gate threshold voltage. Damage to bipolars from ionizing radiation is generally much less than that to MOS devices, for reasons discussed below. In junction devices, such as bipolar and JFET transistors, trapped charges in their surface layers produce inversion layers that expand the effective surface area. This results in increased surface state generation-recombination currents that decrease the minority carrier lifetime, and so the current gain in bipolars (Section 5.10). These spurious currents especially impact bipolar gain at low operating currents. Gain degradation dispersions are larger for low current levels, as in Fig. 6.37.

Device susceptibility to radiation-induced charge production for a given dose depends strongly on its oxide layer quality. Since the quality varys widely between vendors, as well as from batch to batch for a particular vendor, vendors evidently pay little attention to radiation hardening considerations with respect to oxide layer quality control for commercial devices.[32] For bipolar transistors, decreased gain and increased leakage current are the two most important parameters degraded by ionizing radiation and neutron radiation. To reiterate, in bipolars as in MOS devices, the principal factors that degrade performance due to secularly occurring ionizing radiation, such as in a spacecraft enduring Van Allen belt charged particles, are trapped positive charge build-up in their oxides near silicon surfaces, build-up of negative charge at the Si/SiO_2 interface, and the corresponding creation of surface states at these interfaces.

For most bipolars, the occurrence and importance of such surfaces are very much less than in MOS devices, as the former depend on junctions for operation, and the latter depend on these surfaces and corresponding interfaces for their operation. Furthermore, the surface doping concentrations for bipolars are up to three orders of magnitude larger than for MOS, so that bipolars are generally more tolerant than MOS devices to ionizing radiation.

Different types of radiation, and different temperature levels, affect bipolar transistor common emitter gain β in different ways. Recall from Section 5.10 that β^{-1} is given by the sum of three terms. The first term, with subscripts modified for this context, SA_sW/D_bA_e, is called the surface recombination term, the second, σ_bW/σ_eL_e, is called the emitter efficiency term, and the third, $\frac{1}{2}(W/L_b)^2$, is called the volume recombination term. Corresponding to each radiation type and the ambient temperature, one of these terms is predominant. As discussed in Section 5.10, where the parameters are defined, the emitter efficiency term is the one that is most sensitive to temperature, so that the development of β^{-1} as a function of device temperature stems from that term. Under neutron irradiation, it is the minority carrier lifetime, τ, that is the very sensitive parameter, as discussed in the previous chapter. Hence, the above volume

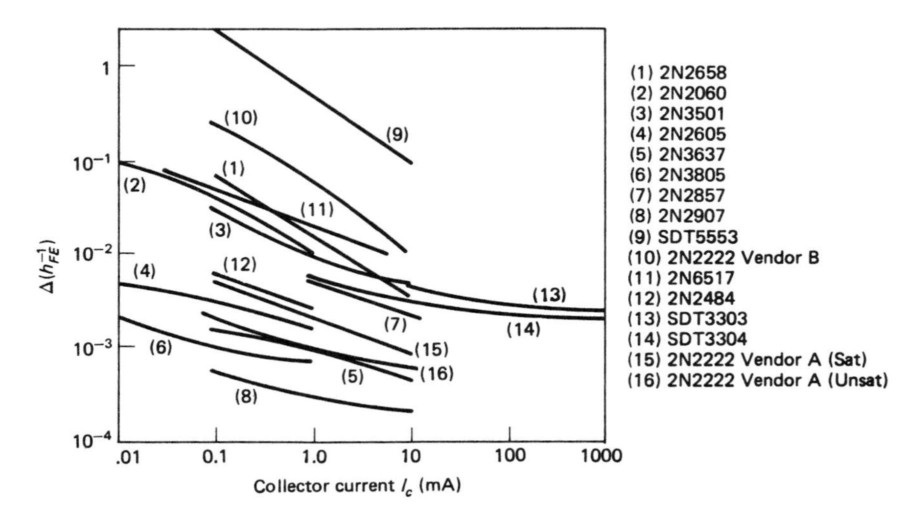

Fig. 6.32 Mean Δh_{FE}^{-1} versus I_c for different bipolar transistors subject to 125 Krad (Si) in situ ionizing radiation dose.[29]

recombination term, rewritten as $\frac{1}{2}W^2/D_b\tau_b$ is the dominant term, and from which the Messenger-Spratt expression for gain degradation in terms of neutron fluence ensues. Here, it is the surface recombination term that is predominantly sensitive to ionizing dose. Therefore, the remaining two terms in Eq. (5.79) are neglected, since they will subtract away when the expression for $\Delta(1/\beta_s) \equiv 1/\beta_s - 1/\beta_{so}$ is formed, where β_s is the corresponding common emitter gain component due to surface recombination. Then, to good approximation

$$1/\beta_s \cong SA_sW/D_bA_e = I_s/I_e \qquad (6.71)$$

with the current ratio I_s/I_e, which is seen to emerge from Eq. (5.78) and its preceding development. I_s is the surface recombination current, and I_e is the corresponding emitter current, as discussed in Section 5.10.

With no loss of generality, a circular bipolar transistor geometry is used for discussion, as depicted in Fig. 6.33. Then, for the corresponding circular emitter-base and base-collector junctions, Eq. (6.71) becomes

$$1/\beta_s = S \cdot 2\pi R_e W^2/D_b\pi R_e^2 \cong 2S/(D_b/W^2)\, R_e \cong S(D_\gamma)/\pi f_T R_e \qquad (6.72)$$

Hence the rightmost expression is due to the fact that $D_b/W^2 \simeq \omega_T \equiv 2\pi f_T$, the gain-bandwidth product angular frequency as discussed in Section 5.10, with

R_e = emitter radius
W = base width
D_γ = ionizing dose level
$S(D_\gamma)$ = surface recombination velocity
D_b = base diffusion constant
$A_e = \pi R_e^2$, emitter area

Not all of the minority carrier current leaving the emitter is sunk by the collector. Besides the normal base current, some of these carriers follow strongly curvilinear trajectories that end up at the base surface periphery, as in Fig. 6.33. The amount of this base surface recombination current can be represented by its threading an effective base surface recombination area A_s (Section 5.10), assumed to be the ringed portion of the base of width W circumscribing the emitter, as in the figure. Therefore

$$A_s = \pi((R_e + W)^2 - R_e^2) = 2\pi R_e W(1 + W/2R_e) \cong 2\pi R_e W \tag{6.73}$$

since $W/2R_e \ll 1$. The surface recombination velocity $S(D_\gamma)$ is an increasing function of the ionizing dose level,[63] D_γ, as illustrated in Fig. 6.34. Also, the detailed dependence of $S(D_\gamma)$ is given from Eq. (6.70) by

$$S(D_\gamma) \cong (1/2)(\sigma_{sn}\sigma_{sp})^{1/2}\, v_{th}\pi kTD_{it}(D_\gamma) \tag{6.74}$$

where σ_{sn}, σ_{sp} are the capture cross sections of the fast surface states for electrons and holes, respectively. v_{th} is the thermal velocity of the carriers, as discussed in Section 1.13. From Eq. (6.72) it follows that

$$\Delta(1/\beta_s) \equiv 1/\beta_s - 1/\beta_{so} = \Delta S(D_\gamma)/\pi f_T R_e; \qquad \Delta S(D_\gamma) \equiv S(D_\gamma) - S(0) \tag{6.75}$$

since the parameter most susceptible to ionizing dose is $S(D_\gamma)$. The degradation of β_s due to ionizing dose can be reduced by increasing R_e and f_T, and reducing $S(D_\gamma)$. Further, it is seen that, for a fixed base width W, the emitter perimeter-to-area ratio $2\pi R_e/\pi R_e^2 = 2/R_e$ should be minimized, to achieve the manufacture of the hardest devices. This fact has been confirmed experimentally.[62] However, bipolar power transistors are intentionally designed with maximally large emitter perimeter-to-area ratios, in order to reduce the base spreading resistance and consequently emitter crowding. This is often accomplished by using an interdigitated base-emitter structure, as shown in Fig. 6.35.

Since this type of device fabrication conflicts with the preceding minimization of emitter perimeter-to-area ratio, such transistors are extremely

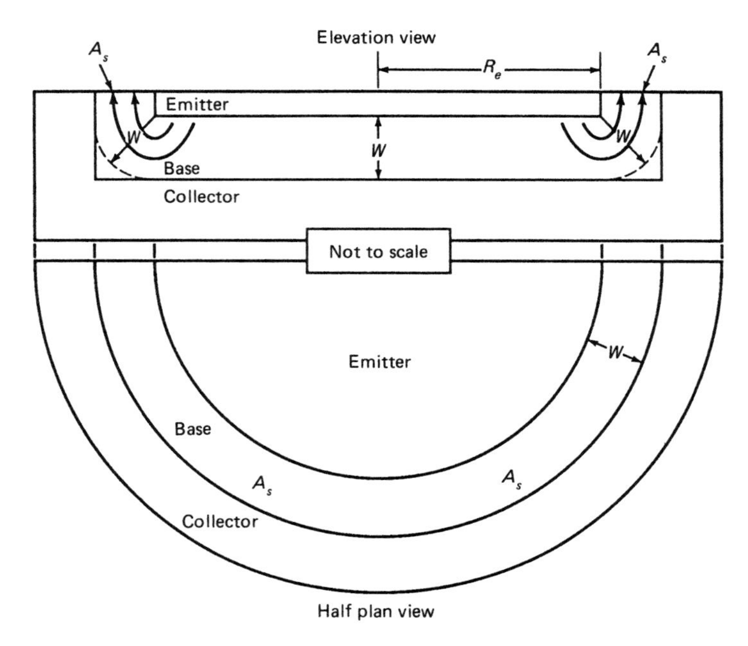

Fig. 6.33 Geometry of a cylindrical bipolar transistor.

susceptible to common emitter gain degradation from ionizing radiation dose D_γ. This gain degradation is especially noticeable at very low collector current levels. This can be seen by noting that the surface recombination current, I_s, is proportional to $\exp(eV_{EB}/2kT)$ for the forward bias condition, as discussed in Sections 3.7 and 3.14. Hence, from Eq. (6.71),

$$1/\beta_s = I_s/I_e \propto I_e^{-1/2} \tag{6.76}$$

since $I_e \propto \exp(eV_{EB}/kT)$. Figures 6.32 and 6.36 provide examples of this variation of inverse gain with $I_c \cong I_e$. Thus,

$$\Delta(1/\beta_s) \equiv (1/\beta_s - 1/\beta_{so}) \propto I_e^{-1/2} \tag{6.77}$$

which for $\beta_s \ll \beta_{so}$ implies large $\Delta(1/\beta_s)$ for very low currents. Also, this gain degradation sensitivity is manifest at high bias current levels as well, since emitter crowding further reduces the effective emitter area.

The ionizing dose level dependence of the surface recombination velocity $S(D_\gamma)$ can usually be approximated by a power law function of

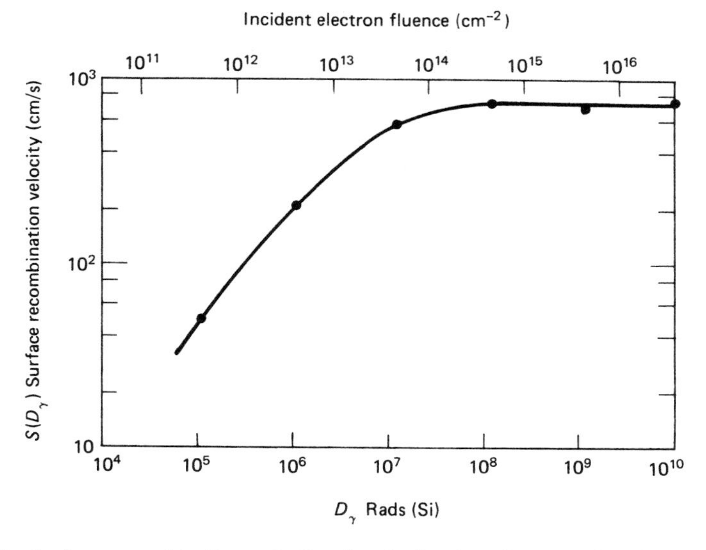

Fig. 6.34 Surface recombination velocity of a depleted surface as a function of ionizing dose. The preradiation value $S(0) = 5\,\text{cm/s}$.[63] Figure © 1967 by the IEEE.

D_γ. Also, bipolar transistors can operate under conditions where S approaches its saturation (asymptotic maximum) value, as discussed in Section 6.6. This implies that, since ($\beta_s \propto 1/S(D_\gamma)$ from Eq. (6.72), $\Delta(1/\beta_s)$ can be modelled by a "saturation" type function, viz.,

$$\Delta(1/\beta_s) \equiv 1/\beta_s - 1/\beta_{so} = (1/\beta_{ss})(1 - \exp - \alpha D_\gamma^N) \tag{6.78}$$

where N, α, and the saturation value of the gain, β_{ss}, are empirically determined constants to fit experimental data for the particular transistor of interest. This can be seen by noting the saturation behavior of the $S(D_\gamma)$ curve in Fig. 6.34. Frequently, $N = 1$, so that the sought after relation between common emitter gain degradation and ionizing dose is given by

$$\Delta(1/\beta_s) = (1/\beta_{ss})(1 - \exp - \alpha D_\gamma) \cong \alpha' D_\gamma \tag{6.79}$$

where $\alpha' = \alpha/\beta_{ss}$ for values of αD_γ sufficiently small that the exponential above can be approximated by the first two terms in its series expansion. That this is the case physically, can be appreciated by the fact that bipolar transistors can maintain reasonably large values of gain for very high D_γ, as discussed below. It is seen that this expression for inverse gain degra-

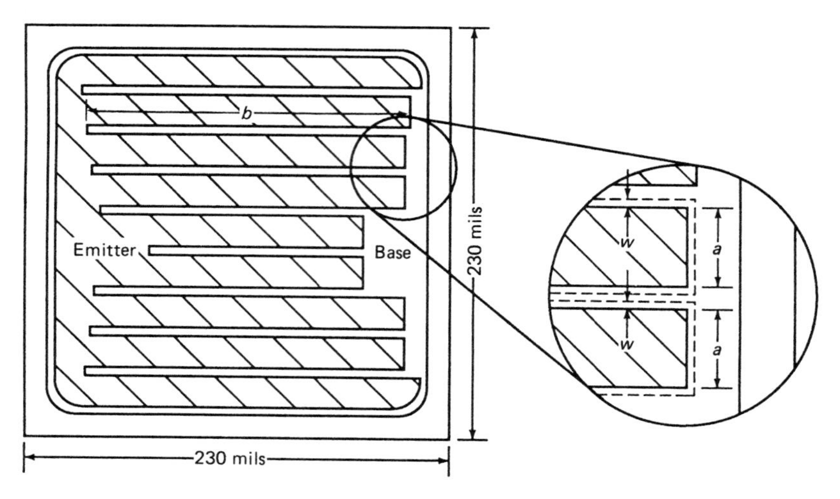

Fig. 6.35 Interdigitated chip geometry of the Motorola 2C6547 *NPN* power bipolar transistor. The C in the designator implies chip per se availability.

dation, $\Delta(1/\beta_s)$, resembles the Messenger-Spratt form for $\Delta(1/\beta)$ in a neutron environment, as given in Section 5.10.

As mentioned, most bipolar transistors usually maintain significant gain up to very high levels of ionizing dose (e.g., up to 10 Mrads (Si)). An estimate of the saturation value of the common emitter gain, β_{ss}, can be made on purely geometric grounds. Since the collector can be construed as an infinite sink for minority carriers, those that are drawn toward the collector in the base region directly under the emitter, but within a radius $R_e - W$, are assumed to be sunk by the collector (Fig. 6.33). Whereas those that emanate from the emitter between the radii $R_e - W$ and R_e are assumed to drift to the ring surface area to recombine there, and so are lost to the collector current. Using this admittedly rough approximation for the current flows, it is seen that the transistor-saturated common base gain, alpha, can be written as a ratio of corresponding areas

$$\alpha_{ss} = I_c/I_e \cong \pi(R_e - W)^2/\pi R_e^2 \cong 1 - 2W/R_e \tag{6.80}$$

where $(W/R_e)^2 \ll 1$. Then, the saturation common emitter gain is given by

$$\beta_{ss} = \alpha_{ss}/(1 - \alpha_{ss}) \cong R_e/2W \tag{6.81}$$

with the proviso that W at the edge of the emitter is the correct value to be used. That value can often be up to twice its effective value at the center of the emitter. For interdigitated power transistor structures, a comparable argument easily shows that β_{ss} is extremely small (Problem 6.13). Vendor device catalogs reveal that corresponding saturation gains for power bipolar transistors are about 5–10, or even lower in some cases.

With respect to ionizing radiation, the general question of how devices fare during a spacecraft mission life of a decade or more, simultaneously enduring radiation damage and annealing, as discussed in Section 5.14, is still under study. Other aspects of ionizing radiation damage in bipolar transistors result in upset, latchup, and burnout, as discussed in Chapter 7.

Bipolar transistor gain degrades as a result of the effects of ionizing radiation due to an increase in base current ΔI_B, for fixed collector current I_c, since $\beta = I_c/I_B$. Quantitatively, it is well known that $\Delta(1/\beta) \sim \Delta I_B$ for not too large ΔI_B. To illustrate, Fig. 6.32 shows how the mean value of $\Delta(1/\beta)$ decreases for increasing in situ collector current, for a representative family of bipolar types. In situ collector current implies that the device is biased, and collector current is measured during irradiation. From Fig. 6.32 it is seen that $1/\beta \sim (I_c)^{-1/2}$, with an approximate upper bound given by $1/\beta \lesssim 0.01/\sqrt{I_c(mA)}$.

$\Delta(1/\beta)$ often manifests an exponential dependence on ionizing dose, D, namely D^n below 1 Mrad (Si), as in Fig. 6.36, where $0.5 \leqslant n \leqslant 1$ depending on the particular transistor.[27] Because ionization damage ultimately results in complex surface effects, and because device structures are topologically very diverse, there is at present no theoretically rooted gain degradation expression as a function of ionizing radiation fluence or absorbed dose, as there is for bipolar gain degradation due to neutron fluence. Figure 6.37 depicts the relative gain degradation dispersion as a function of ionizing dose level for 43 bipolar transistor types, measured under normally utilized collector currents.[18] Little correlation seems to exist between measured relative gain degradation levels and bipolar parameters, such as transistor function (switching, small signal, etc.), power level, or f_T. Generally the gain dispersion is much larger than that shown in Fig. 6.37 for bipolars with very small collector currents.[18] The preceding is an indication of some of the difficulties involved in attempting to predict bipolar gain degradation with ionizing radiation dose.

Degradation of gain due to ionizing radiation is a minimum at the collector current level corresponding to the preradiation maximum gain operating point.[32] For higher currents, the sensitivity to ionizing radiation

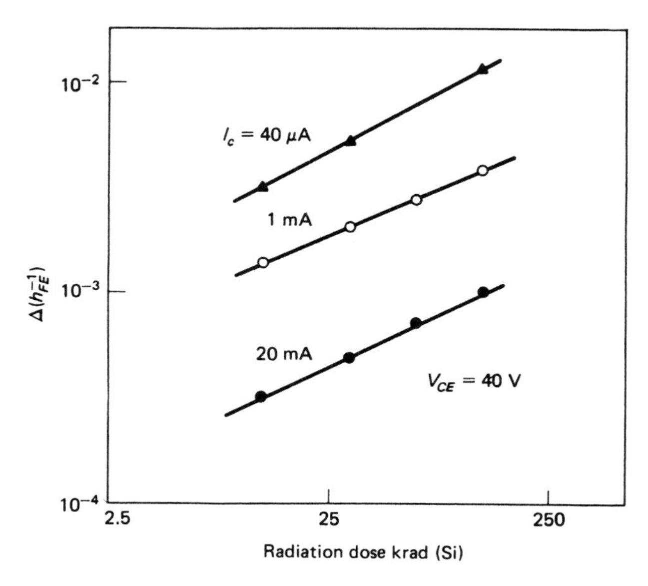

Fig. 6.36 Mean reciprocal gain change versus ionizing radiation dose for 2N2222 transistors operating in unsaturated mode.[32]

increases. It has been shown that for bipolar, high-operating-point current density levels ($\sim 1000\ A/cm^2$), the increase in base current required to maintain constant collector current during such irradiation leads to an effective increase in base width because of the Kirk effect,[30] as first described by Messenger.[45] That is, the high-injection-level, minority carrier current into the base side of the base-collector junction depletion region results in the requirement of less exposed dopant ion space charge from the base to maintain the space charge electric field, as discussed in Section 3.3. On the collector side, more exposed space charge is required to maintain the corresponding electric field, for the latter is being partially cancelled by the incoming minority carrier charge. Hence, the result is a direct expansion of the base-collector depletion region from the base into the collector, thus increasing the effective base width. This increased base width results in an increase of both Gummel Number and base transit time, and so reduces f_T, as discussed in Section 3.10. More important perhaps is that the common emitter current gain, β, is also reduced, as seen from the expression for gain in terms of the bipolar physical parameters discussed in Sections 3.9 and 5.10.

Leakage currents resulting from radiation-induced surface ionization vary over many orders of magnitude, depending on the structure, surface

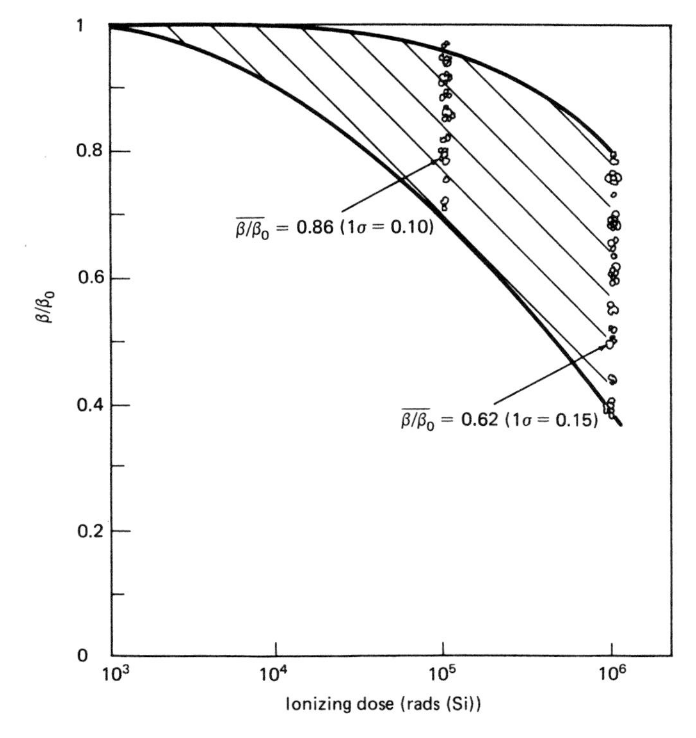

Fig. 6.37 Relative gain degradation versus ionizing radiation dose for 43 different bipolar transistors, from low-power signal to high-power, low-frequency types ($\beta/\beta_0 = 1$ assumed at 1 krad (Si)).[18] Figure © 1972 by the IEEE.

conditions, and bias for almost all families of semiconductor devices. As mentioned, leakage currents generated in passivated silicon surfaces are strongly dependent on the type and concentration of impurities (both advertent and inadvertent) introduced into the silicon dioxide layer during the manufacturing process. Generally, surface passivation and low bias operation aid in reducing radiation-induced leakage currents. Also, the leakage current levels drop off rapidly in the postradiation epoch, especially after the removal of bias voltages. However, when bias is restored, these currents return to nearly their former levels, indicating that permanent damage has been incurred with respect to increased current leakage. It is noteworthy that often the radiation-induced leakage current is described as an I_{CBO}, whereas its counterpart, I_{CEO}, can be much greater, since it is enhanced by the gain, i.e., $I_{CEO} \simeq \beta I_{CBO}$.[31] Also, the radiation-induced leakage current across the base-emitter junc-

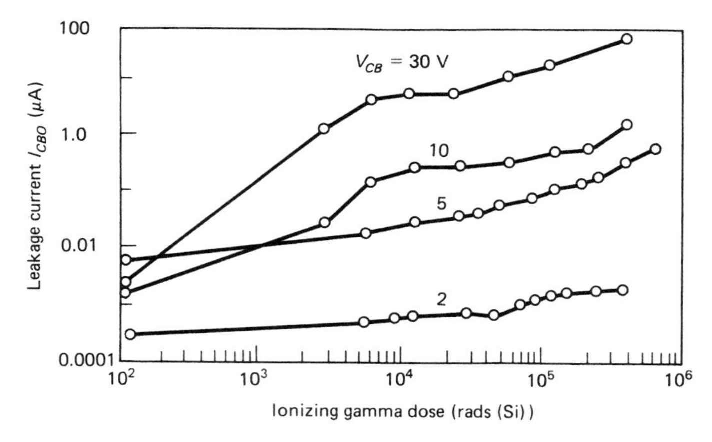

Fig. 6.38 Effects of ionizing gamma dose on leakage current for a typical bipolar transistor.[41] Figure © 1980 by the IEEE.

tion is usually negligible. Figure 6.38 depicts ionizing dose induced I_{CBO} leakage current for a typical bipolar transistor.[41]

Bipolar transistors subjected to ionizing dose levels greater than 10 krads (Si) manifest increasing levels of $1/f$ (pink) noise. The noise levels in *npn* transistors vary more slowly with frequency as a function of ionizing dose than their *pnp* counterparts,[30] but approach a common level for dose levels of 1 Mrad (Si) or greater.

Junction field effect transistors (JFETs) suffer less from ionizing radiation bulk damage than bipolars. An exception is the *n*-channel JFET that is lightly doped in the gate region, which is particularly susceptible to radiation-induced spurious carrier concentration inversion.[32] That is, the positive space charge generated in the oxide insulator induces an *n*-type inversion layer on the *p*-type gate surface region, thus generating leakage current when the junction is back-biased.[33] For example, for the gate junction reverse-biased under irradiation, the *n*-channel JFET gate current, I_{GS}, can increase above 1 μA after being subjected to 200 krad (Si), with similar levels for the drain current, I_{DS}. However, I_{GS} for most *p*-channel JFETs is less than 3 nA after enduring 2.5 Mrad (Si).

JFET $1/f$ noise at 10 Hz increases considerably under ionizing radiation, though the high-frequency noise figure increases only slightly.[65] $1/f$ noise for low frequencies (flicker noise) is related to radiation-induced surface condition changes.[32] JFETs also possess a high radiation tolerance to bulk damage, as compared with bipolars again because the former are majority carrier devices. For *n*-channel MOS devices, pre-irradiation $1/f$

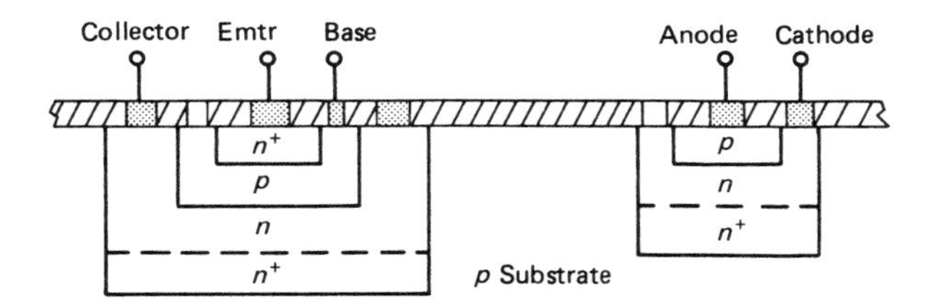

(a) Construction of *npn* bipolar and diode on common substrate

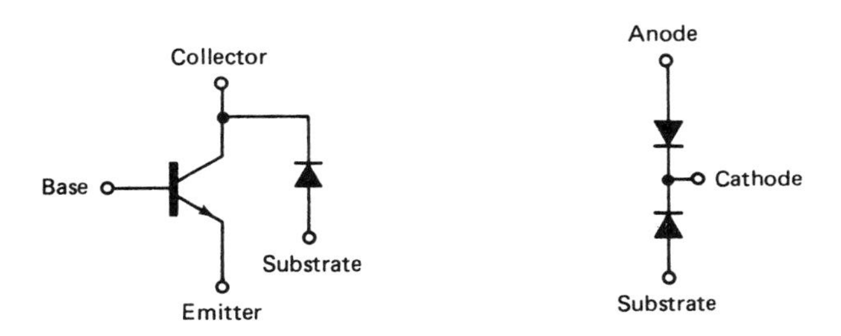

(b) Above transistor, diode, and parasitic diode equivalent circuits

Fig. 6.39 Junction-isolated integrated circuit fragment consisting of an *npn* transistor and diode on common substrate, showing the formation of a parasitic substrate diode.[18] Figure © 1972 by the IEEE.

noise correlates strongly with the gate threshold component of voltage shift (ΔV_{ot}) due to ionizing radiation-induced oxide trap states.[65] Noise is discussed in detail in Section 12.14.

Present-day integrated circuits composed of bipolar transistors are mainly built in what is termed a *junction-isolated* (JI) manner. Figure 6.39 is a sectional view of one such *npn* transistor and an adjacent diode, both embedded in a common *p* substrate.[11] Such transistors normally contain at least four regions of diffused material in their construction, e.g., $n+$, p, n, and $n+$, as in Fig. 6.39(a). The plus signs indicate greater than normal dopant concentrations to, for example, improve emitter efficiency. One reason for including the submerged $n+$ layer is that it provides a relatively high conductance path for collector current, to achieve a sharp knee on its characteristic curve, thus reducing V_{CE} (sat). Figure 6.39 (b) includes the equivalent circuit of the *npn* transistor, where a parasitic substrate diode is shown formed between the *p* substrate and the $n+$ or n collector. Because the substrate material is usually held at

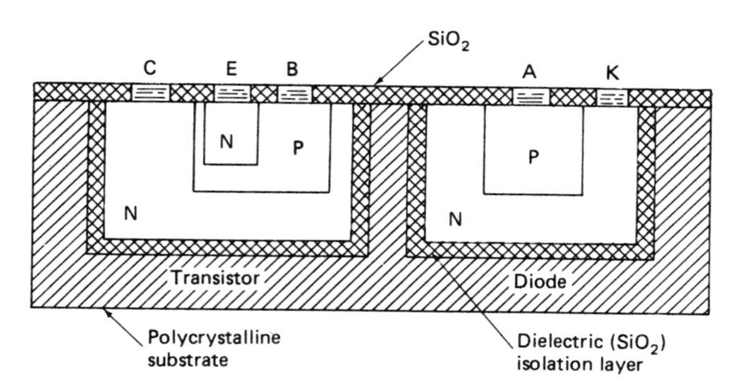

Fig. 6.40 Dielectrically isolated IC fabrication.[18] Figure © 1972 by the IEEE.

the lowest bias voltage, this fact effectively back-biases the substrate diode and thus isolates the transistor from other active devices, such as the adjacent diode. This parasitic, or substrate, diode is a junction isolation diode, from which is derived the term *junction-isolated* integrated circuit for this type of IC configuration. The junction shunting capacitance of this substrate diode reduces the gain bandwidth product, f_T, through reducing the high-frequency characteristic of the transistor. Also, a *pnp* parasitic transistor is formed by the base-collector and *p* substrate. Its operation is, however, suppressed because the minority carrier lifetime of the parasitic transistor is kept relatively short by the high dopant density of the submerged $n+$ layer.

There is another method used to isolate individual transistors in MSI to VLSI systems. It is called *dielectric isolation*, and corresponding dielectrically isolated integrated circuits (DIIC) are constructed as shown in Fig. 6.40. Each unit transistor of the IC is electronically decoupled from its neighbor by submerging each in a dielectric lined tub, as in the figure.

To integrate an *npn* and *pnp* transistor on the same chip, as in a linear device, the *pnp* is built as a lateral transistor with emitter, base, and collector side by side. This is an contrast to the respective nesting of these elements in the usual vertical bipolar transistor. The former suffer somewhat reduced gain, but are important in *dc* biasing applications.

Linear monolithic integrated circuits composed of bipolar transistors are sensitive to ionizing dose, depending on device design and construction.[42] Also, they are among the microcircuits (ICs) most sensitive to dose rate (Section 7.3). Linears are integrated circuits (microcircuits) whose transistors operate over the linear, or active, region of their characteristic curve, such as an operational amplifier or comparator. Lateral

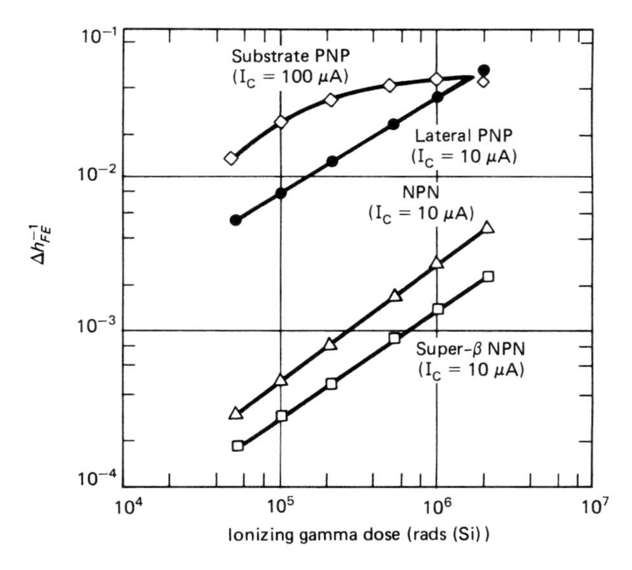

Fig. 6.41 Gain reduction as a function of ionizing dose damage of individual transistors comprising a linear integrated circuit.[42] Figure © 1989 by the IEEE.

and substrate *pnp* transistors usually are more sensitive to ionizing dose radiation than vertical or other transistors within the linear *IC*. Figure 6.41 shows the ionizing dose gain degradation of four bipolar types comprising an operational amplifier *IC*. Note that gain in the substrate *pnp* transistor marginally diminishes even at moderate ionizing dose levels. Further details on ionizing-dose damage to linear devices is given in Section 12.3. Generally, the input stages of linears are often the most sensitive to ionizing dose damage. In addition, some linear devices employ both lateral and substrate *pnp* transistors. In these cases, some linear devices can fail at ionizing dose levels as low as 10 krads (Si). Usually, oxides in bipolar devices serve only as passivation layers, so that the corresponding surface properties are usually of lesser importance compared to their electrical properties. As discussed earlier, however, transistors operated at very low currents would promote surface considerations to first-order importance, because low currents are usually surface currents. Thus, any imbalance in degradation would lead to failure in many of these linear devices. Low current operation should therefore be avoided, together with maintaining surface doping concentrations of any *p*-type silicon sufficiently low to minimize spurious inversion, and thus undesired channel production.

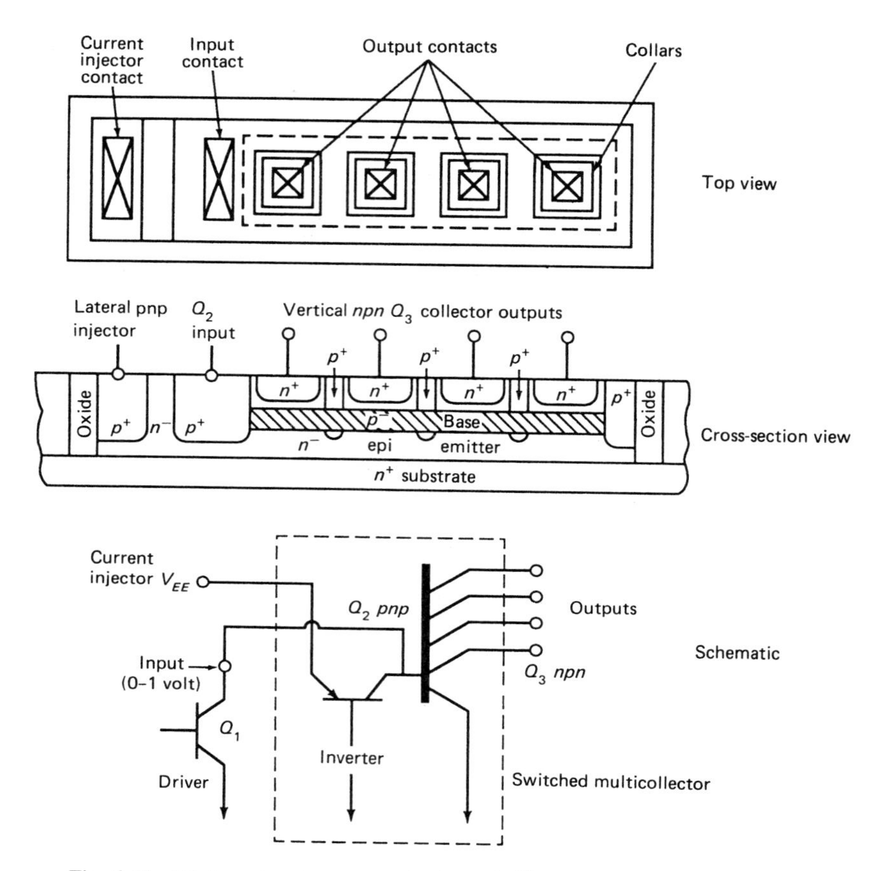

Fig. 6.42 IIL inverter structure and schematic.[58] Figure © 1975 by the IEEE.

Another integrated circuit type and technology utilizes a lateral *pnp* transistor coupled to vertical *npn* transistor(s). It is called bipolar current injection logic or integrated injection logic (IIL or I^2L) devices. As seen in Fig. 6.42, the basic I^2L array consists of a single *pnp* current injection lateral transistor, direct coupled to a multicollector *npn* vertical transistor inverter configuration. IIL technology is attractive because it offers high packing density, high reliability, low speed-power product (0.2–2.0 pJ) and low-cost devices for low-power medium-speed applications.[58] It is a relatively hard technology to ionizing dose and dose rate radiation, as shown in Table 6.6 and Sections 12.2–12.4. Also, it is virtually latchup free because of its low V_{CC} (~1 volt), as discussed in Section 12.4.

As depicted in Fig. 6.42, this device consists of a lateral *pnp* transistor Q_2 current source, supplying bias current to the switched multicollector vertical *npn* transistor Q_3. When the input driving gate transistor Q_1 is ON, so is Q_2, since its collector current is being sunk by Q_1, hence tieing the base of Q_3 to the V_{SAT} LOW of Q_1. This keeps the multicollector transistor Q_3 OFF. Contrariwise, when Q_1 is OFF, the *pnp* transistor Q_2 bias current is now diverted to the base of Q_3, turning it ON. When Q_3 is ON, it can sink the currents of any or all of the Q_3 collector inverting outputs that are connected.[58] IIL has a number of interesting applications including IIL microprocessors, as discussed in Section 13.11.

With competition from the MOS technologies with respect to device performance and integration level, i.e., MSI, LSI, VLSI, and cost, the bipolar technology has countered with certain measures. These include expanding the TTL technology to its limits of power reduction and

Table 6.6 Comparison of IIL with other modern digital integrated circuit technologies.[59] Table © 1977 by the IEEE.

	STTL	ECL	IIL	NMOS	CMOS	CMOS/SOS
Cell Density	C	C−	B	A	C	B
Switching Speed	B	A	C	C	C−	B
Static Power Dissipation	C−	D	B	C−	A	A
Dynamic Power Dissipation	B	B	A	B	B	B
Speed-Power Product	C	B	A	B	B	B
Output Drive Capability	B	B	C	C	C−	D
Noise Immunity	B	C	D	C	A	A
Temperature Range	B	B	B	C−	C	C−
Neutron Damage (n)	B	A	C	A	A	A
Long-Term Ionization Damage (γ_T)	B	A	B	D	C	C
Transient Logic Upset Level ($\dot{\gamma}$)	C	C	B	C	B	A

A; Superior, B: Good, C: Average, C−: Below Average, D: Weak

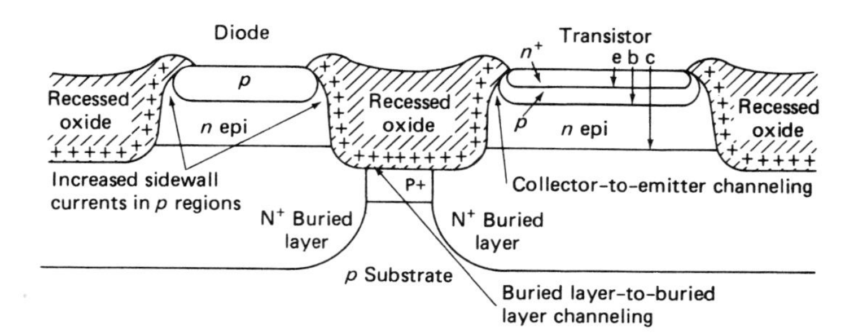

Fig. 6.43 Vertical section of recessed field oxide bipolar integrated circuit with walled emitters.[42] Figure © 1989 by the IEEE.

increased packing density, as well as new concepts, e.g., integrated injection logic (I^2L), integrated Schottky logic, (ISL), and Schottky transistor logic (STTL).

The introduction of recessed oxides (ROX) in the early 1970s has aided greatly in making these concepts a reality, but at some sacrifice in increased sensitivity to ionizing radiation effects. The recessed oxide is a field isolation oxide, to distinguish it from a gate oxide, which compartmentalizes each transistor or other active device in an integrated circuit through dielectric isolation by extending the oxide down from the surface as deep as all active elements, as in Fig. 6.43. While providing lateral dielectric isolation, it also acts as a diffusion stop and minimizes junction capacitances. It is anticipated that recessed field oxides will be accepted by the MOS industry as well. Figure 6.43 shows such a transistor, with the recessed oxide participating to form a "walled emitter" *npn* transistor and an adjacent diode.

As discussed, the two ionizing dose effects, viz., hole-trapping in the gate oxide near the Si-SiO_2 interface and the generation of fast states thereon, produce three deleterious effects in recessed oxide devices. They include (a) inversion of the $p+$ region at the bottom of the recessed oxide, to cause channeling of carriers between adjacent buried layers, as depicted in Fig. 6.43; (b) an inversion of the *p*-type base region along the recessed oxide sidewall, also causing channeling of current between emitter and collector of the *npn* transistor in Fig. 6.43; and (c) the surface recombination velocity is increased along the sidewall due to fast surface-state traps, thus increasing the *p*-region sidewall spurious current, also seen in Fig. 6.43. All these radiation-induced phenomena produce, among other things, gain reduction, and changes in the current-voltage characteristic of the *p-n* diode. It has been reported that these failure mechanisms can occur at ionizing dose levels as low as 10 krads (Si).[42]

Where channeling is the primary failure mechanism, it is known that the corresponding functional degradation diminishes in seconds to minutes after irradiation. This appears to be due to transient annealing of trapped charges in the oxides as they leak away. Further, channel-induced functional failures occur at certain low values of ionizing dose, and disappear at higher dose levels. This seems to be caused by the saturation of trapped oxide charge damage as a function of dose level, followed by annealing with an apparent corresponding marginal increase in the interface states.[42] Additionally, as with MOS oxides, the amount and distribution of trapped oxide charge depends very sensitively on bias conditions. Because of these ionizing dose-induced vagaries, the failure levels of the above sidewall-isolated devices can span three orders of magnitude, from one date code (dated part batch) to the next. Until radiation hard designs and processes are fully implemented, any bipolar integrated circuit using the recessed oxide process should be carefully evaluated before use in a hardened system.

6.15 MOSFET ANNEALING

This section discusses annealing mainly for MOS devices, as opposed to Section 5.14, which accomplishes the same for bipolar devices. First, there is nothing fundamentally different between the two device types with respect to their annealing properties at the atomic level, as they are both manufactured from silicon and SiO_2. The annealing differences emerge from the different device technologies and processing methods unique to bipolar and MOS devices. For MOS devices, the level of device response due to annealing depends on a number of factors, including temperature, bias, types of radiation environments (e.g., neutron fluence, ionizing dose, dose rate, etc.), and whether or not the device is being irradiated during the annealing epoch. Strictly, all operating devices are annealing while being irradiated. Figure 6.44 depicts the gate threshold voltage of an *n*-channel MOS transistor versus irradiation and annealing following the radiation phase.

An important annealing mechanism in MOSFETs results from deeply trapped positive charge (holes) in the SiO_2 gate. The holes are not permanently trapped, but "anneal out" over times from milliseconds to years, as mentioned in Section 6.8. This annealing of hole traps is often

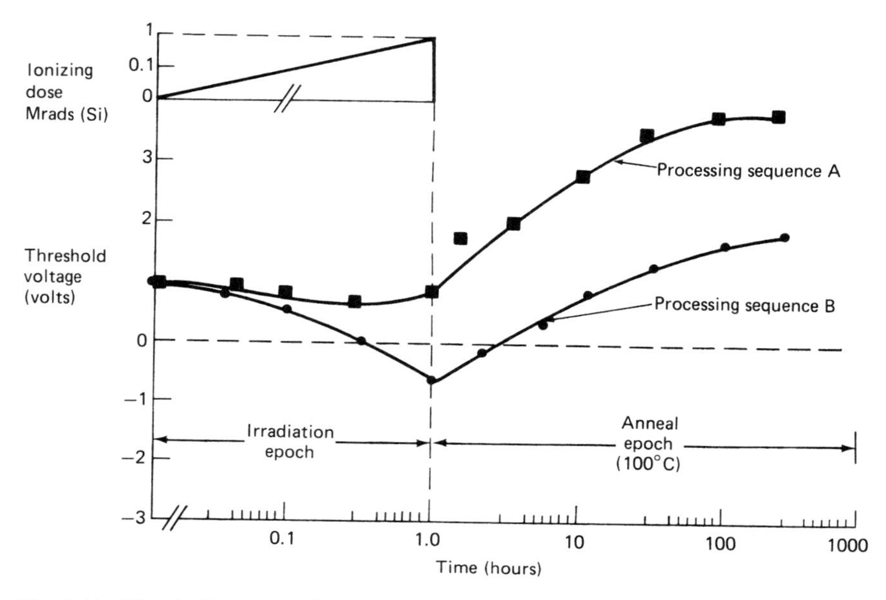

Fig. 6.44 Threshold voltage of an n-channel MOSFET during irradiation to 1 MRad (Si) and in the following annealing epoch.[69] Figure © 1989 by John Wiley & Sons.

the main signature of the long-term threshold voltage shift in MOS devices. Annealing is analytically represented in so-called tunneling and thermal annealing models, respectively.[69]

Figure 6.45 depicts the annealing of ΔV_{ot}, the oxide-trapped charge component of ΔV_T, discussed in Section 6.8, in a biased irradiation–post-radiation sequence. That is, the figure shows an initial period or irradiation of the device under bias and at elevated temperature (100°C), followed by a cessation of radiation, but maintaining one bias level for the first portion of the post-radiation phase.[70] This is followed by a switch in bias voltage, which begins a second post-radiation phase. The preceding implies that some positive charge does not anneal and that access to the annealing of the remainder of the trapped charge is bias dependent.

As is the situation with bipolar transistors, discussed in Section 5.14, isochronal and isothermal annealing tests have been performed in MOSFETs as well. Figures 6.46 and 6.47 depict isochronal and isothermal annealing, respectively. In Fig. 6.46, the devices were irradiated with 1.5 MeV electrons, followed by post-radiation annealing at zero gate bias.[71] In the figures, the recovery below "zero" damage is inter-

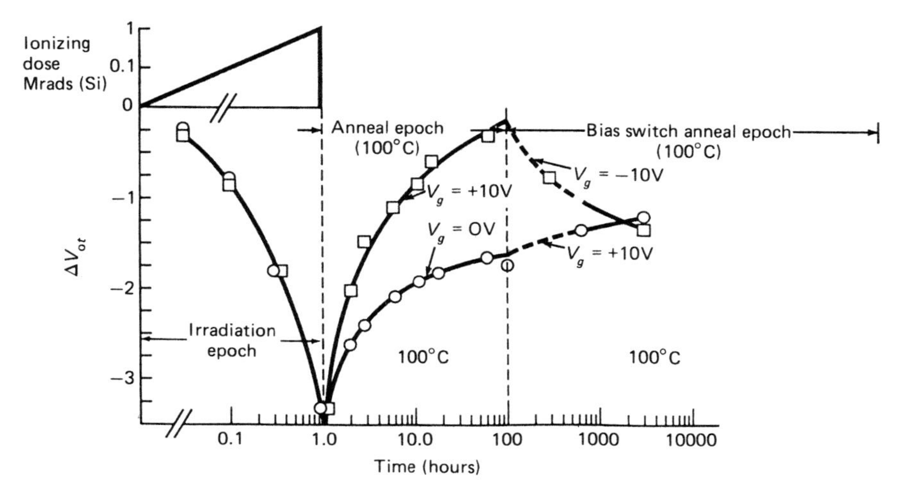

Fig. 6.45 Annealing of oxide trapped charge in an *n*-channel MOSFET during irradiation to 1 Mrad (Si) followed by an annealing epoch with bias (V_g) switches.[70]

preted with respect to the annealing of damage that is process-induced, probably from interface traps,[71] i.e., an indicator of ΔV_T rebound.

Figure 6.47 depicts isothermal annealing data on devices similar to those from which isochronal data is obtained. Isothermal data of ΔV_T versus time is taken from previously irradiated devices that were held at various constant temperatures. On the other hand, isochronal annealing data is obtained by taking readings over a constant time interval (e.g., 10 minutes) for the device at various temperatures. The isochronal data curve (Fig. 6.46) can be construed as a composite of the various isothermal curves from Fig. 6.47 for fixed time intervals and both curves are mutually consistent.

It is important to note that annealing of radiation-induced interface traps in MOS devices is not evident at ambient temperatures, except for devices with a very high surface density of traps ($\geqslant 10^{12}$ per cm^2).[73] Only for elevated temperatures ($\geqslant$100 °C) is there significant annealing.[74]

The gate bias dependence during the annealing epoch following irradiation is also illustrated in Fig. 6.48. The principal reason for this bias dependence seems to be that annealing of the oxide-trapped charge component ΔV_{ot} depends on field-induced motion and thermal excitation. This results in charge compensation.[76] Charge compensation means that electrons from the bulk tunnel into the traps themselves, and re-

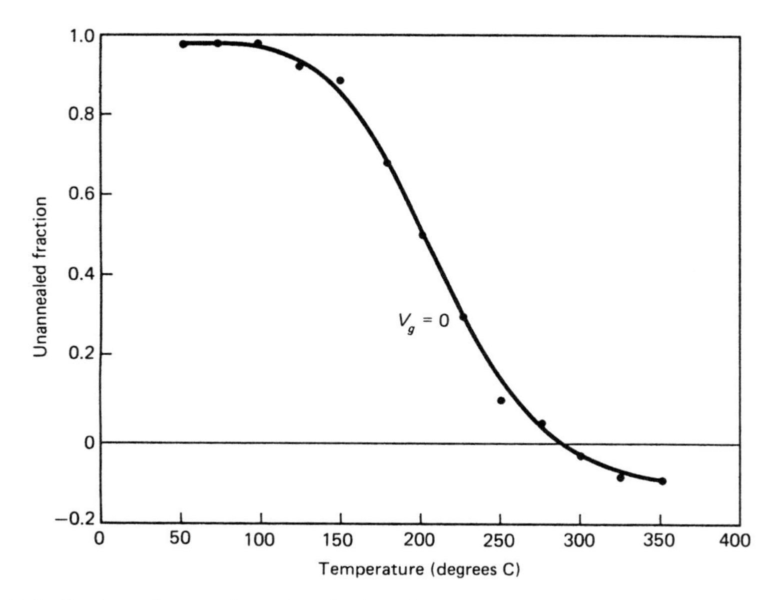

Fig. 6.46 Isochronal annealing of radiation damage from ΔV_T in MOSFETs over 10 minute (5 sample mean) step annealing epochs.[71]

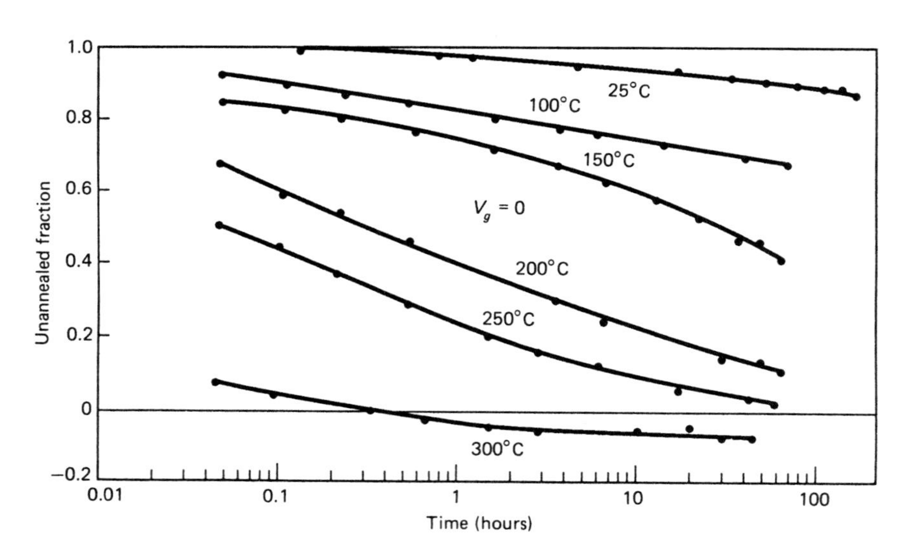

Fig. 6.47 Isothermal annealing of radiation damage from ΔV_T in MOSFETs for constant temperatures in the annealing epoch.[72]

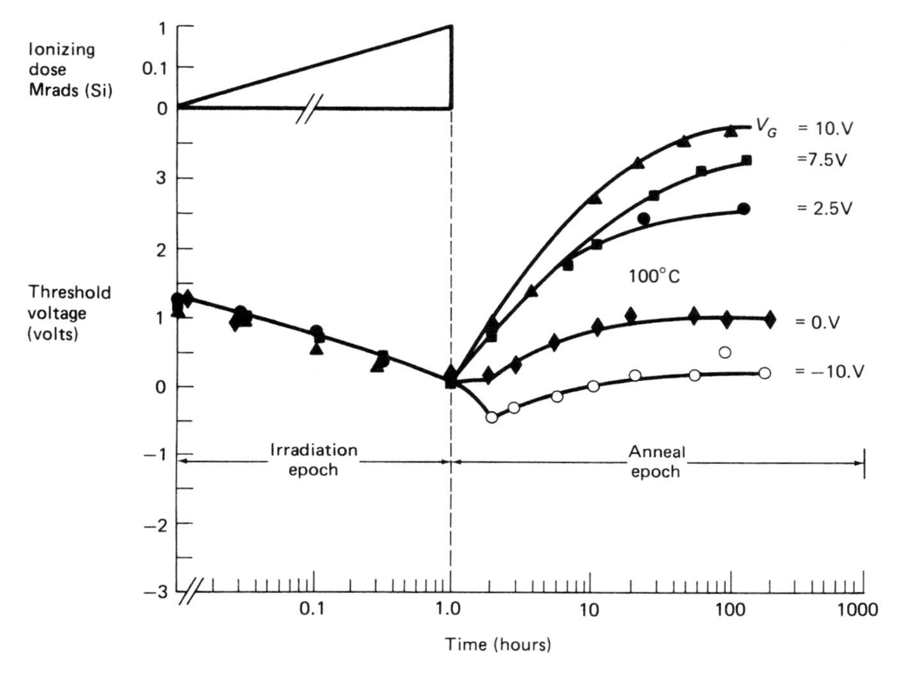

Fig. 6.48 Threshold voltage shift of *n*-channel MOSFETs during irradiation to 1 MRad (Si) followed by an annealing epoch under various gate biases.[75] Figure © 1984 by the IEEE.

combine with the holes to annihilate them. This is also called tunnel annealing, and it is very bias dependent, as in Fig. 6.48, and corresponds well to a tunneling model.[77] Compensation here also implies that, following the radiation epoch, the device will stabilize with either time or the application of gate bias.[78]

In summary, it is seen that the annealing phenomenon is a very complicated process in MOSFETs. Other parameters, such as manufacturing processes, inadvertent impurities, and high process temperatures, will influence the annealing properties of MOS devices. Many details of how these and the previously mentioned parameters interact can be found in the references, but the fact is that it is a complex aspect of MOS devices, and is still under study.

PROBLEMS

1. The expression $E^2 = p^2c^2 + m_0^2c^4$ is a relation between the total energy, E, of any elementary particle, its momentum, p, and its rest mass, m_0, where c is the speed of light. Gamma rays and x rays have zero rest mass, and their energy is given by $E = h\nu$. Show that their momentum is given by $p = E/c = h/\lambda$, where λ is their wavelength. Is this imparted momentum enough to structurally damage a 1-gram transistor if it absorbs a radiation dose of 100 Mrad (Si) in 1 μs? Its military specifications, MIL-STD-883C/Method 2002.3 Test Condition B, state that it must endure an impulse corresponding to a force of 1500 g acceleration for 0.5 ms. Will the thermal damage melt the transistor? Hint: $Q = mC_h\Delta T$.
2. A ^{60}Co gamma ray source has a half-life of 5.3 yr. How long can this source be used before its activity falls to 1.6 percent of its initial activity?
3. (a) For exponential gamma-ray attenuation $I(x) = I_0 \exp - \mu x$, where μ is the linear attenuation coefficient, show that the mean free path $\lambda_0 = \mu^{-1}$ by evaluating $\lambda_0 = \int_0^\infty xI(x)\,dx / \int_0^\infty I(x)\,dx$. (b) In a thin slab of material, the collision probability of a gamma ray is given by $\lim_{\Delta x \to 0} N\sigma\Delta x$, where N is the number of target atoms per cm^3 in the slab, σ is the total crossection of the individual gamma ray, and Δx is the slab thickness. If $I(x)$ is the gamma-ray intensity incident on the front face of the slab, and $I(x + \Delta x)$ is the intensity leaving the back face, show that $I'/I = -N\sigma$, and that $I(x) = I_0 \exp - N\sigma x$, so that $\lambda_0 = (N\sigma)^{-1}$.
4. Why is the areal density of penetration ρx (grams per cm^2) often used in describing the penetration properties of materials, instead of the simple penetration distance?
5. In pair production, what is the minimum amount of energy that the initiating gamma ray must have, where the mass-energy equivalent of the electron is 0.511 MeV?
6. Specifically, why are MOS devices less sensitive to neutron radiation than bipolar devices?
7. What is the primary damage mechanism in MOSFETs from incident ionizing radiation?
8. What is the difference between the flat band voltage, V_{FB}, and the threshold voltage, V_T, in CMOS devices, and how are they affected by ionizing radiation?
9. Both flatband and threshold voltage shift vary as t_{ox}^n, where t_{ox} is the thickness of the gate oxide insulator. Determine n from Fig. 6.21.
10. (a) Why is an MNOS or SNOS device inherently harder than an ordinary MOS transistor?
    (b) Same for PMOS over NMOS.
11. What is the attractive feature of CMOS/SOS and CMOS/SOI over bulk CMOS and why?
12. If interface state trapping contributes to positive gate voltage shift, with positive gate bias, what type of carrier is being trapped and where?

13. In Section 6.14, it is mentioned that a purely geometric argument can be made to show that bipolar power transistors having an interdigitated base-emitter structure possess very small saturated common emitter gains. Develop this argument. Hint: To obtain the expression for the corresponding gain, use the infinite sink collector assumption employed for the circular bipolar transistor. Obtain $\beta_{ss} = (a - 2w)/2w$, where a is the individual digit width and w is the base dimension beyond the emitter as in Fig. 6.35, and complete the argumentation.

14. The general problem of interpretation and extrapolation of the results of device ionizing dose or other radiation test data is clearly important, certainly from the following viewpoint: A commonly accepted maximum duration of any radiation, burn-in, or life test of a device lot is about 700 hours (~4 weeks). The majority of radiation tests last far less than 150 hours (~1 week), because of the press of test time availability, supplier parts delivery, scheduling test equipment maintenance, etc. How is such test data to be extrapolated for device use that amounts to 100 000 hours (~10 years)? 100 000 hours is a round number representing an average spacecraft or other unmonitored system life, such as a submarine cable repeater. Discuss the problem in terms of obtaining actual field data, and reliability prediction methodologies.

15. Show that a kerma level $E\sigma(E)$ of 1 MeV · millibarn(mb) per neutron per cm^2 converts to $3.4 \cdot 10^{-13}$ rads (Si) deposited per neutron per cm^2.

REFERENCES

1. J. D. Jackson. Classical Electrodynamics. John Wiley, New York Chap. 15.
2. R. H. Crawford. MOSFET in Circuit Design. McGraw-Hill, New York, p. 23, Fig. 2.1, 1967.
3. R. W. Klingensmith, D. J. Hamman, R. K. Thatcher, and M. L. Green (eds.). Transient Radiation Effects on Electronics (TREE) Simulation Facilities, 1st ed., Battelle Columbus Labs., Columbus, OH.
4. Klingensmith et al., op. cit. Chap. 4.
5. R. K. Thatcher and M. L. Green (eds.). TREE Preferred Procedures, 2nd ed. Chap. 4, Battelle Columbus Labs., Columbus OH, June 1972.
6. S. Glasstone, A. Sesonske. Nuclear Reactor Engineering. Van Nostrand, New York, Chap. 2, 1963.
7. R. D. Evans. The Atomic Nucleus. McGraw-Hill, New York, Chap. 22, 1955.
8. Glasstone et al., op. cit. Fig. 2.11.
9. V. S. Vavilov and N. A. Ukhin. Radiation Effects in Semiconductors and Semiconductor Devices. Translated from Russian by Freund Publishing House, Tel Aviv, Part 2, Chap. 3, 1977.
10. V. A. J. Van Lint, T. M. Flanagan, R. E. Leadon, J. A. Naber, and V. C. Rogers. Mechanisms of Radiation Effects in Electronic Materials, Vol. 1. Wiley-Interscience, New York, Chap. 4, 1980.
11. Van Lint et al., loc. cit.

12. A. S. Grove. Physics and Technology of Semiconductor Devices. Wiley, New York, Chap. 5, 1967.
13. Van Lint et al., ibid. Chap. 3.
14. Van Lint et al., loc. cit.
15. J. P. Raymond. IEEE Nuclear and Space Radiation Effects (NSRE) Short Course Notes, Monterey CA, Fig. 11, p. 1.26, July 1985.
16. J. P. Mitchell and D. E. Wilson. "Surface Effects of Radiation on Semiconductor Devices." Bell Syst. Tech. J. 46(1), 1–80, Jan. 1967.
17. A. Many and D. Gerlich. Phys. Rev. 107:404, 1957.
18. K. M. Schleiser and P. E. Norris. "CMOS Hardening Techniques." IEEE Trans. Nucl. Sci. NS-19(6):275–281, Dec. 1972.
19. G. F. Derbenwick and B. L. Gregory. "Process Optimization of Radiation Hardened CMOS Integrated Circuits." IEEE Trans. Nucl. Sci. NS-22(6), Dec. 1975; IEEE Nuclear Science Radiation Effects (NSRE) 1984 Conference Short Course, Colo. Sprgs., sect. 1.
20. H. A. R. Wegener, M. B. Doig, P. Marraffino, and B. Robinson. "Radiation Resistant MNOS Memories." IEEE Trans. Nucl. Sci. NS-19(6):29, Dec. 1972.
21. K. G. Aubuchon. "Radiation Hardening of PMOS Devices by Optimization of the Thermal SiO_2 Gate Insulator." IEEE Trans. Nucl. Sci. NS-18(6), Dec. 1971.
22. R. E. King. "Radiation Hardening of Static NMOS RAMs." IEEE Trans. Nucl. Sci. NS-26(6):5060, Dec. 1979.
23. R. A. Kjar, S. N. Lee, R. K. Pancholy, and J. L. Peel. "Self Aligned Radiation Hard CMOS/SOS." IEEE Trans. Nucl. Sci. NS-23(6):1610, Dec. 1976.
24. P. Richman. MOS Field Effect Transistors and Integrated Circuits. Wiley-Interscience, New York, 1973.
25. Crawford, op. cit. p. 29.
26. R. C. Gallagher and W. S. Corak. "A Metal Oxide Semiconductor (MOS) Hall Element." Solid State Electronics 9:571, 1966.
27. TRW Systems Group. Spacecraft Hardening Design Guidelines Handbook. Redondo Beach, CA, Sec. 6.2, Ionization Damage, Dec. 1974.
28. TRW VHSIC Program. Chap. 4, "CMOS Technology." pp. 4–12, Spring 1982.
29. R. K. Thatcher and J. J. Kalinowski. Transient Radiation Effects on Electronics. (TREE) Handbook, Battelle Memorial Institute Columbus, OH, Sept. 1969.
30. T. Kirk. "A Theory of Transistor Cutoff Frequency (f_r) Fall-off at High Current Densities. IEEE Trans. Elect. Devices. ED-9(2):164, March 1962.
31. A. G. Stanley. Effect of Electron Irradiation on Electronic Devices. MIT Lincoln Labs., Tech. Rept. no. 403, Nov. 3, 1965.
32. W. E. Price, K. E. Martin, and M. K. Gauthier. "Total Dose Hardness Assurance Guidelines For Semiconductor Devices." Military Handbook 279, 29 Jan., 1985.
33. A. G. Stanley. "Effect of Space Radiation Environment on Micropower Circuits Using Bipolar, Junction-Gate, and Insulated Gate Field Effect Transistors." IEEE MEREM Record, 1966.
34. S. M. Sze. Physics of Semiconductor Devices. Wiley, New York, Chap. 7, 1981.
35. G. J. Brucker. "Interaction of Nuclear Environment With MNOS Memory Device." IEEE Trans. Nucl. Sci. NS-21(6):186–192, Dec. 1974.
36. J. J. Chang. "Non-Volatile Semiconductor Memory Devices." Proc. IEEE, 64(7): 1039–1059, July 1976.
37. C. R. Viswanathan and J. Maserjian. "Model For Thickness Dependence of Radiation Charging in MOS Structures." IEEE Trans. Nucl. Sci. NS-23(6):1540–1545, Dec. 1976.
38. Grove, op. cit. Sect. 9.3.

39. D. F. Barbe (ed.). Charge Coupled Devices. Topics in Applied Physics, Vol. 38. Springer-Verlag, Berlin, Chap. 6, 1980.
40. J. R. Srour et al. "Basic Mechanisms of Radiation Effects on Electronic Materials, Devices, and Integrated Circuits." DNA-TR-82-20, sec. 4, Aug. 1, 1982.
41. A. H. Johnston. IEEE NSRE Conference Short Course, Cornell University, Ithaca, N.Y., July 1980.
42. R. L. Pease, R. M. Turfler, D. Platteter, D. Emily, and R. Blice. "Total Dose Effects in Recessed Oxide Digital Bipolar Microcircuits." IEEE Trans. Nucl. Sci. NS-30(6): 4216–4223, Dec. 1989.
43. J. M. Benedetto et al. "Hole Removal in Thin Gate MOSFETs by Tunnelling." IEEE Trans. Nucl. Sci. NS-32(6):3916–3920, Dec. 1985.
44. W. R. Dawes, Jr. IEEE NSRE Conference Short Course, Monterey, CA, July 1985.
45. G. C. Messenger. "An Analysis of Switching Effects in High Power Diffused Base Silicon Transistors." I.R.E. Electron Devices Conference, Washington D.C., Oct. 30, 1959.
46. R. J. Krantz, L. W. Aukerman, and T. C. Zeitlow. "Applied Field and Total Dose Dependence of Trapped Charge Buildup in MOS Devices." IEEE Trans. Nucl. Sci. NS-34(6):1196–1201, Dec. 1987.
47. E. H. Snow, A. S. Grove, B. E. Deal, and C. T. Sah. "Ion Transport Phenomena in Insulating Films." J. Appl. Phys. 36(5):1664–1673, May 1965.
48. E. H. Nicollian and J. R. Brews. MOS (Metal Oxide Semiconductor) Physics and Technology. Wiley-Interscience, New York, Sec. 1.3.3, 1982.
49. K. H. Zaininger. "Irradiation of MIS Capacitors With High Energy Electrons." IEEE Trans. Nucl. Sci. NS-13(6):237–247, Dec. 1966.
50. T. J. Russell, H. S. Bennett, M. Geitan, J. S. Suehle, and P. Roitman." Correlation Between CMOS Transistor and Capacitor Measurements of Interface Trap Spectra." IEEE Trans. Nucl. Sci. NS-33(6):1228–1233, Dec. 1986.
51. R. M. Warner and B. L. Grung. Transistors: Fundamentals For The Integrated Circuit Engineer. Wiley-Interscience, New York, 1983, p. 725.
52. W. E. Spicer, et al. "Unified Mechanism For Schottky Barrier Formation in III-V Oxide Interface States." Phys. Rev. Lett., Vol. 44, No. 6, Feb. 11, 1980.
53. C. M. Dozier, D. M. Fleetwood, D. B. Brown, and P. S. Winokur. "An Evaluation of Low Energy X-Ray and ^{60}Co Irradiation of MOS Transistors." IEEE Trans. Nucl. Sci. NS-34(6):1535–1539, Dec. 1987.
54. J. S. Browning, J. R. Schwank, C. L. Freshman, M. Conners, and G. A. Finney. "Total Dose Characterization of CMOS Technology at High Dose Rates and Temperatures." IEEE Trans. Nucl. Sci. NS-35(6):1557–1562, Dec. 1988.
55. F. B. McLean and T. R. Oldham. "Basic Mechanisms of Radiation Effects in Electronic Materials and Devices." IEEE NSRE Conf. Short Course, Snowmass CO, July 27, 1987, Sec. 1.
56. A. H. Johnston and S. B. Roeske. "Total Dose Effects at Low Dose Rate." IEEE Trans. Nucl. Sci. NS-33(6):1487–1492, Dec. 1986.
57. A. H. Johnston. "Super-Recovery of Total Dose Damage in MOS Devices." IEEE Trans. Nucl. Sci. NS-31(6):1427–1433, Dec. 1984.
58. J. P. Raymond, T. Y. Wong, and K. K. Schuegraf. "Radiation Effects on Bipolar Integrated Injection Logic." IEEE Trans. Nucl. Sci. NS-22(6):2605–2610, Dec. 1975.
59. J. P. Raymond and R. L. Pease. "A Comparative Evaluation of Integrated Injection Logic." IEEE Trans. Nucl. Sci. NS-24(6):2327–2335, Dec. 1977.
60. J. H. Srour and J. M. McGarrity. "Radiation Effects on Microelectronics in Space." Proc. IEEE 76(11): 1443–1469, Nov. 1988.

61. T. P. Ma and P. V. Dressendorfer (eds.). Ionizing Radiation Effects in MOS Devices and Circuits. J. Wiley, New York, 1989.
62. R. L. Pease, F. N. Coppage, and E. D. Graham. "Dependence of Ionizing Radiation Induced h_{FE} Degradation on Emitter Periphery." IEEE Trans. Nucl. Sci. NS-21(2): 41–42, Apr. 1974.
63. E. H. Snow, A. S. Grove, and D.J. Fitzgerald. "Effects of Ionizing Radiation on Oxidized Silicon Surfaces and Planar Devices." Proc. IEEE, 55(7):1168–1185, July 1967.
64. D. B. Brown, W. C. Jenkins, and A. H. Johnston. "Application of a Model For Treatment of Time Dependent Effects on Irradiation of Microelectronic Devices." IEEE Trans. Nucl. Sci. NS-36(6):1954–1962, Dec. 1989.
65. J. H. Scofield, T. P. Doerr, and D. M. Fleetwood. "Correlation Between Pre-Irradiation 1/*f* Noise and Post Radiation Oxide Trapped Charge in MOS Transistors." IEEE Trans. Nucl. Sci. NS-36(6):1946–1953, Dec. 1989.
66. H. E. Boesch, Jr. and J. M. McGarrity. "Charge Yield and Dose Effects in MOS Capacitors at 80°K." IEEE Trans. Nucl. Sci. NS-23(6):1520–1525, Dec. 1976.
67. N. S. Saks, M. G. Ancona, and J. A. Modolo. "Radiation Effects in MOS Capacitors With Very Thin Oxides at 80°K." IEEE Trans. Nucl. Sci. NS-31(6):1249–1255, Dec. 1984.
68. E. E. King and R. L. Martin. "Effects of Total Dose Ionizing Radiation on the 1802 Microprocessor." IEEE Trans. Nucl. Sci. NS-24(6):2172–2176, Dec. 1977.
69. T. P. Ma and P. V. Dressendorfer (eds.), op. cit. Chap. 1.
70. J. R. Schwank, P. S. Winokur, P. J. McWhorter, F. W. Sexton, P. V. Dressendorfer, and D. C. Turpin. "Physical Mechanisms Contributing to Device Rebound." IEEE Trans. Nucl. Sci. NS-31(6):1434–1438, Dec. 1984.
71. V. Danchenko and V. D. Desai." Characteristics of Thermal Annealing of Radiation Damage in MOSFETs." J. Appl. Phys., Vol. 39, p. 2417, 1965.
72. V. Danchenko and V. D. Desai, op. cit.
73. Y. Nishioka, E. F. daSilva, and T. P. Ma. "Radiation Induced Interface Traps in $Mo/SiO_2/Si$ Capacitors." IEEE Trans. Nucl. Sci. NS-34(6):1166–1177, Dec. 1987.
74. T. P. Ma and P. V. Dressendorfer (eds.), op. cit. Chap. 4.
75. Schwank et al., op. cit.
76. T. P. Ma and P. V. Dressendorfer (eds.), op. cit. Chap. 5.
77. Schwank et al., op. cit.
78. Ma and Dressendorfer (eds.), op. cit. Chap. 5.
79. C. A. Klein. "Bandgap Dependence and Related Features of Radiation Ionization Energies in Semiconductors." J. Appl. Phys., Vol. 79, p. 2029, 1968.
80. R. W. Tallon, A. H. Hoffland, W. T. Kemp, and J. W. Brouse. "Proton Responses on MOS Electronics Operating at Cryogenic Temperatures." Report No. WL-TR-89-80, Kirtland AFB, NM. Jan. 1990.
81. J. R. Coss, C. A. Goben, and W. E. Price. "Comparison of the Effects of Ionizing Radiation at Twelve Dose Rates From 0.0015 to 100 Rads (Si)/Sec. "J. P. L. Rept. 15, Dec. 1989—J. Elect. Mat. Jul., 1990.
82. Sze, op. cit. Chap. 8.
83. Ma and Dressendorfer (eds.), op. cit. Chap. 1.
84. J. Y. Chen, R. Martin, and D. O. Patterson. "Radiation Hardness of Submicron NMOS." IEEE Trans. Nucl. Sci. NS-28(6): 4314–4316, Dec. 1981.
85. P. S. Winokur, J. R. Schwank, P. J. McWhorter, P. V. Dressendorfer, and D. C. Turpin. "Correlating the Radiation Response of MOS Capacitors and Transistors." IEEE Trans. Nucl. Sci. NS-31(6):1453–1460, Dec. 1984.

86. F. J. Kub, C. T. Yao, and J. R. Waterman. "Radiation Hardened SOS MOSFET Technology for Infrared Focal Plane Readouts." IEEE Trans. Nucl. Sci., NS-37(6): 2020–2026, Dec. 1990.
87. I. M. Hafez, G. Ghibaudo, and F. Balestra. "Analysis of the Kink Effect in MOS Transistors." IEEE Trans. Elec. Devices, ED-37:818–824, March 1990.
88. H. F. Wolf. Semiconductors. Wiley-Interscience, New York, Sec. 5-1, 1971.
89. Y. Nishioka, E. F. da Silva, Jr., and T. P. Ma. "Evidence For (100) Si//SiO_2 Interfacial Defect Transformation After Ionizing Radiation." IEEE Trans. Nucl. Sci. NS-35(6): 1227–1233, Dec. 1988.

CHAPTER 7
DOSE-RATE EFFECTS

7.1 INTRODUCTION

This chapter deals mainly with gamma-ray and x-ray dose-rate effects, that is, damage effects in electronic systems due to the rapid time variation of photon radiation as from nuclear detonations and other transient sources. This transient radiation is referred to as ionizing dose rate, or merely dose rate.

7.2 DIODE PHOTOCURRENT

The principal effect due to gamma- and x-ray photons is that of ionization of the materials, such as semiconductors, upon which this radiation falls. The types of damaging radiation and their sources are discussed in Sections 6.2 and 6.3. The following detailed description of diode photocurrent production is adapted from Wirth and Rogers.[1]

When diodes, transistors, and integrated circuits are exposed to ionizing radiation, the ionization produces hole-electron pairs. Assuming that the transient radiation penetrates the device in a homogeneous and isotropic manner, hole-electron pairs are uniformly generated throughout the semiconductor. Some of these charge carriers produced in the vicinity of a junction will cross the junction, in turn producing transient currents that will appear at the diode leads. At rather high incident pulse radiation intensities, which are those implied here, these photocurrents can be comparable to and even greater than normal signal levels.

Ionization in bulk material results in generation of hole-electron pairs at a rate proportional to the incident radiation flux. However, the number of hole-electron pairs produced depends on the total energy absorbed by the material of interest.

When gamma rays penetrate material, they transfer their energy into the material specifically through three processes, depending on their initial energy. As discussed in detail in Section 6.1, they include, in order of increasing incident radiation energy: photoelectric effect, compton effect, and pair production. In all of these processes, the photon energy that is absorbed mainly causes ionization and excitation of the resulting electrons, with corresponding generation of hole-electron pairs in the material.

Experiments have determined that the average energy supplied by the incident photon required to create a hole-electron pair is 3.6 eV in silicon, 2.8 eV in germanium, and 4.8 eV in gallium arsenide. Energy absorbed by the material from the radiation environment is usually given in units of absorbed dose called rads. One rad is defined as the absorption of 100 ergs of radiation energy per gram of material. Because a given photon fluence is absorbed to a greater or lesser degree in different materials, the corresponding material must be specified, such as 100 rads (Si). Because 3.6 eV is required to generate a hole-electron pair in silicon, the carrier generation constant, g, for silicon is

$$g = 4.05 \cdot 10^{13} \text{ hole-electron pairs per cm}^3 \cdot \text{rad (Si)} \qquad (7.1)$$

For germanium and gallium arsenide, it is

$$\begin{aligned} g_{\mathrm{Ge}} &= 1.2 \cdot 10^{14}\,\text{hep/cm}^3 \cdot \text{rad(Ge)} \\ g_{\mathrm{GaAs}} &= 6.9 \cdot 10^{13}\,\text{hep/cm}^3 \cdot \text{rad(GaAs)} \end{aligned} \qquad (7.2)$$

The photocurrent produced in response to the incident radiation absorbed in diodes and transistors will be discussed in turn. The response of the *p-n* diode can be described by investigating the changes in the concentration of minority carrier densities due to the ionization. Figure 7.1 depicts minority carrier densities in a *p-n* diode, both forward and reverse biased, before and immediately after an incident radiation pulse. Consider first the reverse biased diode, Fig. 7.1(a) irradiated by a very short pulse of ionizing radiation, assumed to create a uniform concentration of hole-electron pairs throughout the device volume. The carriers thus generated within the junction will be carried across the junction, and collected in nanoseconds due to the very strong depletion region electric field. These particular hole and electron carriers constitute a current that can be considered as being in phase with the time behavior of the incident transient radiation pulse. This prompt photocurrent, i_p, can be considered to flow from the n to the p region, because holes are swept into the p-region and electrons into the n-region.

The electrons and holes that are generated beyond the depletion regions of the *p-n* junction also produce a transient increase in minority carrier density. However, on the average, only carriers within one diffusion length, i.e., $L_n = (D_n \tau_n)^{1/2}$, on each side of the junction are collected as above. Carriers beyond one diffusion length have a high probability of recombining before reaching the junction, and so con-

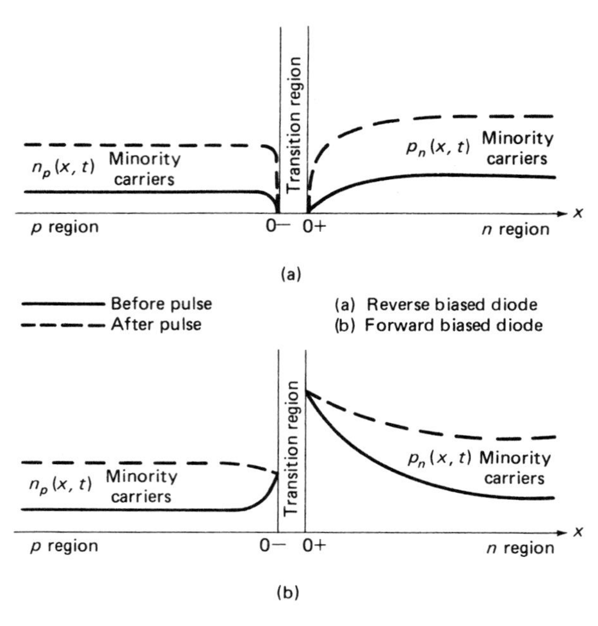

Fig. 7.1 *p-n* junction minority carrier densities prior to and immediately after an incident pulse of ionizing radiation (a) reverse-biased diode; (b) forward-biased diode.[1]

tribute little to the photocurrent. But, because of the increase in minority carrier density gradients at the junction due to the onset of the radiation pulse, the corresponding diffusion gradients increase, thereby enhancing the diffusion of carriers toward the junction. Thus a portion of the carriers remote from the junction for this reason diffuse toward and reach the junction with a corresponding delay time. This current is often called the *delayed component of photocurrent*, i_d. Both i_p and i_d are transient currents superimposed on the usual leakage currents.

In the case of the forward biased junction, the applied bias does not completely cancel the depletion region field potential. Hence, an electric field is still maintained in the junction neighborhood, so that a prompt photocurrent will flow from the n to the p region. As seen in Fig. 7.1(b), the carriers produced in the bulk material decrease the carrier density gradient at the junction, thus resulting in a decrease in the forward conduction current. The total current can now be construed as the normal conduction current plus the prompt and delayed photocurrent that flows from the n to the p region.

Now expressions for the prompt and delayed photocurrent in the *p-n* diode will be obtained. The expression for the primary photocurrent flowing across the junction can be gotten from the zero electric field approximation, a one-dimensional version of the continuity and diffusion equations for minority carriers discussed in Section 1.11. $p_n(x, t)$, the minority (hole) carrier density in the n-region, which is to the right of the junction in Fig. 7.1, satisfies

$$\partial p_n/\partial t = D_p \partial^2 p_n/\partial x^2 - (p_n - p_{n0})/\tau_p + G(t); \; J_p = eD_p \partial p_n/\partial x; \; G = g\dot{\gamma}(t) \quad (7.3)$$

where p_{n0} is the equilibrium counterpart of p_n, $G(t)$ is the carrier generation rate due to the ionizing radiation, and J_p is the corresponding photocurrent density. Similarly, for $n_p(x, t)$, the minority carrier density in the p region, which is to the left of the junction in Fig. 7.1, and satisfies

$$\partial n_p/\partial t = D_n \partial^2 n_p/\partial x^2 - (n_p - n_{p0})/\tau_n + G(t); \qquad J_n = eD_n \partial n_p/\partial x \quad (7.4)$$

where n_{p0} is the equilibrium part of n_p.

As discussed, the photocurrent is composed of two components. The first is the prompt photocurrent, i_p, produced by ionization within the depletion region, i.e.

$$i_p = eAW_t G(t) \quad (7.5)$$

where A is the area of the junction, and W_t is the depletion region width. The second is the delayed photocurrent, i_d, which is the photocurrent resulting from the diffusion of the photoproduction of carriers in the neighborhood of the junction outside the depletion region, with a corresponding delay time. It is given by the difference of the boundary conditions on each side of the junction, as

$$i_d = eA(D_p \partial p_n(0+, t)/\partial x - D_n \partial n_p(0-, t)/\partial x) \quad (7.6)$$

where $0-$ is the coordinate corresponding to the immediate left of the junction, in the p region, and $0+$ is that to the immediate right of the junction in the n region. The values of the carrier densities to the immediate right and left of the junction are, respectively, as discussed Section 3.3

$$p_n(0+, t) \equiv p_{n0} \exp(eV_0/kT); \qquad n_p(0-, t) \equiv n_{p0} \exp(eV_0/kT) \quad (7.7)$$

where V_0 is the applied voltage across that junction. The fact that the carrier densities remain finite is written as

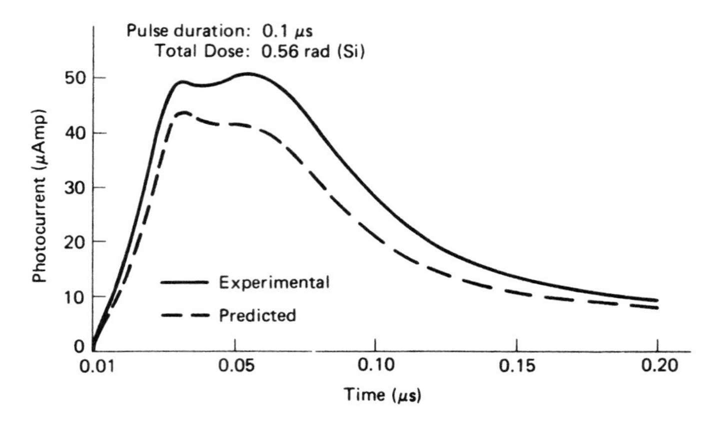

Fig. 7.2 Primary photocurrent for the 2N1051 collector-base junction.[1] Figure © 1964 by the IEEE.

$$\lim_{x \to \infty} |p_n(x, t)| < \infty; \qquad \lim_{x \to -\infty} |n_p(x, t)| < \infty \tag{7.8}$$

For a constant applied bias V_0 to the junction, the solution of Eqs. (7.3)–(7.8) yields the total primary photocurrent, $I_{pp} = i_p + i_d$ as

$$I_{pp}(t) = eA[W_t G + \frac{1}{\sqrt{\pi}} \int_0^t G(t-u)(\sqrt{D_n} e^{-u/\tau_n} + \sqrt{D_p} e^{-u/\tau_p}) \frac{du}{\sqrt{u}}] \tag{7.9}$$

For the case of present interest, which is that of a rectangular ionizing radiation pulse of amplitude $G = g_0 \dot{\gamma}$ and width t_p, Eq. (7.9) reduces to

$$I_{pp} = eAG \begin{cases} W_t + L_n erf(\sqrt{t/\tau_n}) + L_p erf(\sqrt{t/\tau_p}); & 0 \leqslant t \leqslant t_p \\ L_n[erf(\sqrt{t/\tau_n}) - erf(\sqrt{(t-t_p)/\tau_n})] & \\ \quad + L_p[erf(\sqrt{t/\tau_p}) - erf(\sqrt{(t-t_p)/\tau_p})]; & t > t_p \end{cases} \tag{7.10}$$

where $erf(x) = 2 \int_0^x \exp(-y^2) dy / \pi^{1/2}$, and L_n, L_p are the corresponding diffusion lengths in the n and p region. Figure 7.2 depicts the primary photocurrent obtained from Eq. (7.10) and compared with that obtained from experimental flash x-ray data.[1]

For very short radiation pulse widths, that is, t_p is very small compared with the minority carrier lifetimes τ_p and τ_n, then from Eqs. (7.10), the photocurrent is given by

$$I_{pp}(t) = \begin{cases} i_p(t) = eAW_t G(t); & 0 \leqslant t \leqslant t_p \\ i_d(\mathrm{t}) = \dfrac{eAGt_p}{\sqrt{\pi t}} \left(\sqrt{D_n} e^{-t/\tau_n} + \sqrt{D_p} e^{-t/\tau_p} \right); & t > t_p \end{cases} \tag{7.11}$$

The product Gt_p is the total carrier density generated during the incident radiation pulse and, of course, is proportional to the total incident dose. However, in many devices τ_n and τ_p are small, and corresponding terms in Eqs. (7.11) can be neglected. For steady-state photocurrent, such as from nuclear power reactor gamma rays, it is appreciated from the previous discussion that the steady-state photocurrent I_{ss} is given from the upper Eq. (7.10), for unbounded time and pulse width, as

$$I_{ss} = eAG_{ss}(W_t + L_p + L_n); \qquad G_{ss} = g\dot{\gamma}_{ss} \tag{7.12}$$

7.3 TRANSISTOR PHOTOCURRENT

For a bipolar transistor connected in a grounded emitter configuration, and biased to operate in the active region, the hole-electron pairs due to the radiation will both diffuse and drift across each junction. The diffusion ensues in a manner similar to the diode already discussed. There will be prompt photocurrents in each junction and delayed photocurrents caused by the diffusion of minority carriers from one region to another. For an *npn* transistor, electrons diffuse out of the base, and holes are injected into the base from the adjoining regions. The corresponding primary photocurrents are given by expressions already developed. Now, the hole photocurrent entering the base is amplified by the current gain, h_{FE}, of the transistor. This results in a collector current transient called *secondary photocurrent*. This is in addition to the collector component of the primary photocurrent. Expressions for the former are now derived using the charge control transistor model.[2] In this model, the excess base minority carrier charge $Q(t)$ crossing the base-emitter junction is given by

$$i_b(t) = \dot{Q}(t) = -Q/\tau_b \tag{7.13}$$

where i_b is the majority carrier current entering the base, and τ_b is the minority carrier lifetime in the base. The corresponding collector current is

$$i_c(t) = h_{FE}Q/\tau_b \tag{7.14}$$

Equations (7.13) and (7.14) are good approximations to the currents that would be obtained from solutions of the diffusion equations of the previous section. The above expressions are valid under the following conditions:

a. The transistor is biased to operate in the active region.

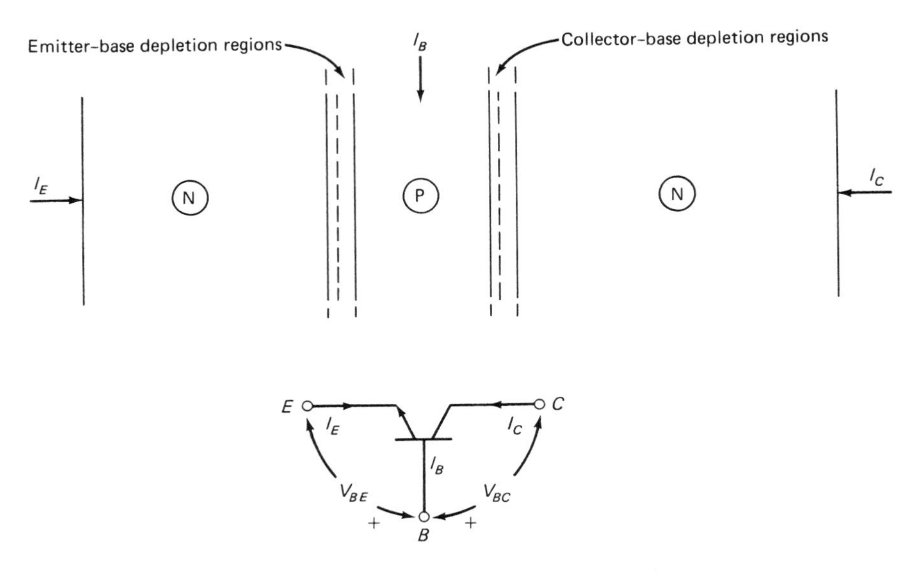

Fig. 7.3 One-dimensional transistor model.

b. During one base transit time, changes in the base current, collector current, and incident radiation pulse are relatively small.

c. The one-dimensional geometry model of the transistor is adequate, as in Fig. 7.3.

d. The base region is field-free or it has an impurity grading such that the built-in field satisfies

$$E(x) = (D/\mu)\, d/dx(\ln F(x)) = (kT/e)\, d/dx(\ln F(x)) \tag{7.15}$$

where $F(x)$ is related to the impurity grading through the corresponding potential. D and μ are the minority carrier diffusion constant and mobility, respectively. For typical values of F, if $E(x) = -dV_{BE}/dx$, then $F(x) = \exp - (eV_{BE}(x)/kT)$. For a zero impurity grading, it is seen that V_{BE} is constant, so that the built-in electric field vanishes. This development assumes that the collector voltage and base current are constants, and therefore the collector depletion layer capacitance, as well as external circuit parameters, need not be considered. A further assumption is that the emitter depletion layer capacitance is also ignored, in order to obtain a linear model yielding closed-form expressions for the desired currents. This is valid for many transistors operating in the active region, because the fractional change in charge accumulated in the emitter depletion capacitance during the radiation pulse is small. For transistors operating near cutoff, such as in digital systems, or for very-high-frequency transistors, the above approximations begin to lose their validity.

In an ionizing radiation environment, the base majority carrier current can be construed as having six components. These are discussed in turn.

The first component is a diffusion current, i_1, from the collector obtained as the appropriate term from Eq. (7.9)

$$i_1(t) = eA_c(D_p/\pi)^{1/2} \int_0^t G(t-\tau)\exp-(\tau/\tau_c)\,d\tau/\tau^{1/2} \tag{7.16}$$

where A_c is the collector area, and τ_c is the minority carrier lifetime in the collector.

The second component is the collector junction depletion region current, i_2, given from the upper Eq. (7.11) as

$$i_2(t) = eAW_{sc}G(t) \tag{7.17}$$

where W_{sc} is the thickness of the collector junction depletion region.

The third component is the emitter junction depletion region current i_3, also obtained from the upper Eq. (7.11) as

$$i_3(t) = eAW_{se}G(t) \tag{7.18}$$

where W_{se} is the thickness of the emitter junction depletion region.

The fourth component i_4 results from carrier generation within the base region. Because all minority carriers generated in the base diffuse to the collector very quickly

$$i_4 = eAW_bG(t) \tag{7.19}$$

where W_b is the base width.

The fifth component, i_5, is the diffusion current from the emitter. Because the voltage across the emitter junction varies during the epoch of the radiation pulse, the corresponding boundary conditions are time-varying. They are given in Eq. 7.7 and rewritten here for the base-emitter junction as

$$p_{ne}(0,t)/p_{n0} = n_{pb}(0,t)/n_{p0} = \exp(eV/kT) \tag{7.20}$$

where e and b in the subscripts refer to emitter and base, respectively. Because the base minority carrier density, $n_{pb}(0,\ t)$ decreases linearly from a maximum at the emitter junction to zero at the collector junction, then $n_{pb}(0,\ t) = \dot{Q}A = Q/(W_bA/2)$, so that, from Eq. (7.20)

$$p_{ne}(0,t) = (p_{n0}/n_{p0})n_{pb}(0,t) = (p_{n0}/n_{p0})(2Q(t)/AW_b) \tag{7.21}$$

Using Eqs. (7.13)–(7.21), i_5 can be obtained from

$$\dot{Q} = -Q/\tau_b + i_1 + i_2 + i_3 + i_4 + i_5 \tag{7.22}$$

The sixth component i_6 is the normal base current.

For the case of emitter efficiency assumed to be unity, and for a constant (step function) radiation level corresponding to a constant carrier generation rate G, Eq. (7.22) yields the collector current, given by

$$\begin{aligned} i_c(t) = eAG\{W_b + W_{sc} + L_c + h_{FE}[(W_b + W_{sc} + W_{se})(1 - e^{-t/\tau_b}) \\ + (L_e - L_{eb}) + (L_c - L_{cb})]\} \end{aligned} \tag{7.23}$$

where

$$\begin{aligned} L_e &= \sqrt{D_p\tau_e}\, erf\sqrt{t/\tau_e};\ L_{eb} = \sqrt{D_p(\tau_e^{-1} - \tau_b^{-1})e^{-t/\tau_b}} \cdot erf\sqrt{t(\tau_e^{-1} - \tau_b^{-1})} \\ L_c &= \sqrt{D_p\tau_c}\, erf\sqrt{t/\tau_c};\ L_{cb} = \sqrt{D_p(\tau_c^{-1} - \tau_b^{-1})e^{-t/\tau_b}} \cdot erf\sqrt{t(\tau_c^{-1} - \tau_b^{-1})} \end{aligned} \tag{7.24}$$

From Eq. (7.23), the primary and secondary photocurrents are delineated, where it is seen that the latter is made up from the term that is multiplied by h_{FE}. The steady-state counterpart, I_c, of i_c in Eq. (7.23) is obtained by allowing the time to get very large to give in the limit

$$I_c = eAG[W_b + W_{sc} + L_{c\infty} + h_{FE}(W_b + W_{sc} + W_{se} + L_{c\infty} + L_{e\infty})] \tag{7.25}$$

where

$$L_{c\infty} = \sqrt{D_p\tau_c}, \qquad L_{e\infty} = \sqrt{D_p\tau_e}, \qquad \text{and} \qquad L_{eb\infty} = L_{cb\infty} = 0$$

An alternative model for secondary photocurrent is to consider the common emitter photocurrent response as an ideal bipolar transistor with diode primary photocurrent generators in parallel with the collector-base and emitter-base junctions, as shown in Fig. 7.4. The secondary photocurrent can be perceived by considering the enhancement of the primary photocurrent through the common-emitter transient collector photocurrent, as a function of bias. If the input impedance vanishes, which is the case when the emitter is shorted to the base, the transient collector current will simply be the collector primary photocurrent, I_{ppc}. For nonzero input impedance, the principal photocurrents are effectively the

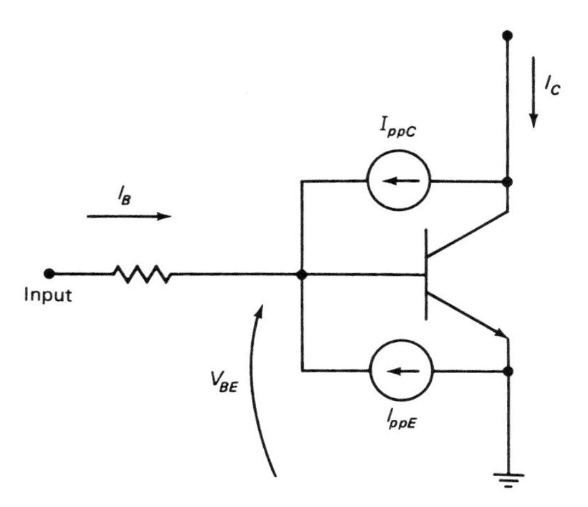

Fig. 7.4 Circuit model of transistor photocurrent response.[3] Figure © 1982 by the IEEE.

input base current multiplied by the common-emitter current gain, yielding the transient collector secondary photocurrent, or $h_{FE}(I_{ppc} + I_{ppE})$. This is equivalent to the secondary photocurrent term in Eq. (7.23). Usually, but depending on the collector current and base-emitter impedance, the emitter primary photocurrent is small compared with the secondary photocurrent, which dominates the common-emitter transistor photo response. This variation of both primary and secondary collector current with base-emitter resistance is shown[3] in Fig. 7.5. The Ebers-Moll model for a bipolar transistor including photocurrent sources is depicted in Fig. 7.14.

Alternatively, an approximate expression for primary plus secondary (amplified) photocurrent is extrapolated from Eq. (7.10) to yield

$$i_c(t) = eAG(D\tau)^{\frac{1}{2}}(erf(t/\tau)^{\frac{1}{2}})(1 + \beta(t)) \tag{7.25A}$$

where $\beta(t)$ is the inverse fourier transform of the transistor gain-bandwidth characteristic, $\overline{\beta}(\omega)$, as discussed in Section 3.10 and whose modulus $|\overline{\beta}(\omega)|$ is depicted in Fig. 3.26, i.e.

$$\beta(t) = (1/2\pi)\int_{-\infty}^{+\infty} \overline{\beta}(\omega)\, \exp i\omega t\, d\omega. \tag{7.25B}$$

$\beta(t)$ reflects both its early time behavior following the incident radiation, as discussed in Section 5.14, as well as its late time asymptotic levels, given by Eq. (5.66).

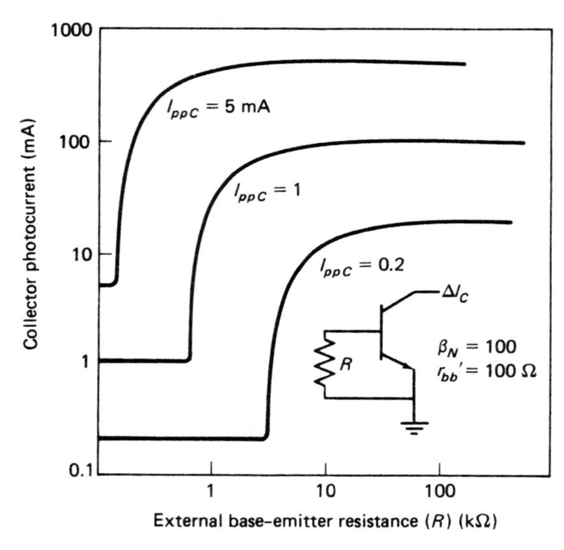

Fig. 7.5 Collector photocurrent versus base-emitter resistance R, where I_{ppc} is the corresponding primary photocurrent.[3] Figure © 1982 by the IEEE.

For a pulse width, t_p, that is short with respect to the minority carrier lifetime ($t_p \ll \tau$), Eq. (7.25A) can be written as

$$i_c(t) \cong eAG(Dt_p)^{\frac{1}{2}}(1 + \beta(t)) \tag{7.25C}$$

while, for a pulse width long with respect to the minority carrier lifetime ($t_p \gg \tau$), the corresponding photocurrent is

$$i_c(t) \cong eAG(D\tau)^{\frac{1}{2}}(1 + \beta(t)) \tag{7.25D}$$

7.4 NONLINEAR PHOTOCURRENT

It is seen from Eq. (7.5) that the photocurrent is linear with respect to dose rate. That is, the primary photocurrent is

$$i_{ppc} = eAW_t g_0 \dot{\gamma} \text{ (amps)} \tag{7.26}$$

since $eg_0 = (1.6 \cdot 10^{-19}$ coulombs/carrier$) \cdot (4.05 \cdot 10^{13}$ carriers/cm^3 rad (Si)) or 6.5 μA per cm^3-rad/s, where the dose rate is given in rads per second.

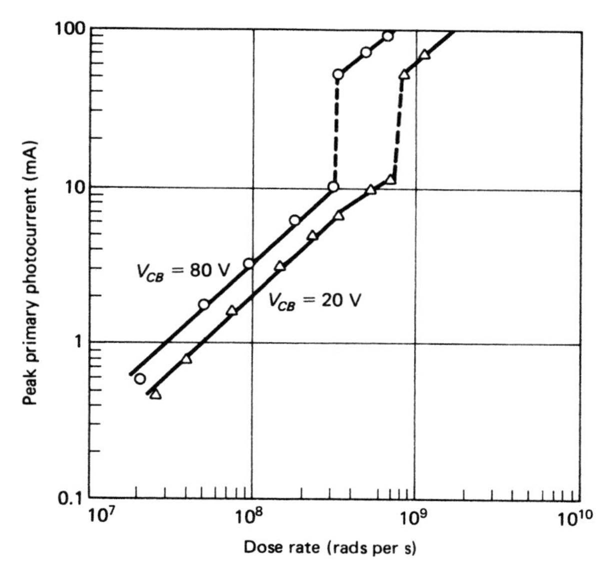

Fig. 7.6 Transistor nonlinear photocurrent behavior with dose rate for two values of collector-base voltage.[4] Figure © 1966 by the IEEE.

AW_t is the effective volume for photocurrent production in and near the junction.

However, for high dose rates of $\sim 10^{10}$ rads per s and higher, the preceding linearity begins to no longer hold. For such dose rates, the photocurrent response takes on an anomolous, nonlinear behavior, as seen in Fig. 7.6. From the figure, it is seen that for relatively low dose rates, the linear proportionality between photocurrent and dose rate holds. However, at a certain level of dose rate, there is a discontinuous jump increase in the photocurrent. For greater values of dose rate, the linear behavior again ensues, until device saturation currents are reached.

As already mentioned, when the incident radiation is sufficiently penetrating, hole-electron pairs are produced homogeneously throughout the device. The holes collected in the active region of the base, as well as those that diffuse into the region from the collector and emitter, flow laterally outward toward the base contact, as in Fig. 7.7. This current produces a lateral voltage drop outward parallel to the base-emitter and base-collector junctions, as in the figure. The center of the emitter junction is more forward biased than the periphery, and the emitter potential and that of the central portion of the base stay in phase.

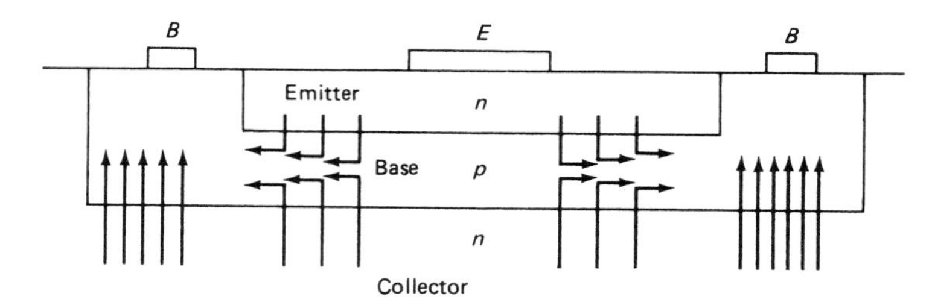

Fig. 7.7 Lateral photocurrent flow within a cylindrical transistor.[4] Figure © 1966 by the IEEE.

If the corresponding lateral currents are large enough, as in the case of a high-dose-rate incident pulse, this base-emitter potential drop can exceed the breakdown voltage BV_{EBO}. If this happens, the emitter and base are suddenly connected through this low-impedance breakdown path. The device is now in a common emitter configuration insofar as the base photocurrent drive is concerned. This photocurrent is thereby amplified to produce a very large collector current, which results in the anomolous photocurrent under discussion. Under certain conditions, the lateral photocurrent distribution is the reverse of the preceding[4] The photocurrent density can become electrically unstable, producing enhanced photocurrents near the center of the emitter.[4] Theoretical results can predict approximately the dose-rate level at which the anomolous behavior occurs in many cases.[5]

A nonlinear photocurrent model is now introduced in the following. In a manner similar to the generation of secondary photocurrent as discussed in the previous section, a secondary photocurrent is suddenly produced, at a given dose rate level, immediately following the onset of V_{CEO} breakdown, as discussed above. This results in a jump discontinuity of photocurrent at that level, as depicted in Fig. 7.6. Referring to Fig. 7.8, let the photocurrent density, $j_p(r, t)$ be given by

$$j_p(r, t) = eW_tG(t)(1 + C_1 j_e(r, t)) \tag{7.27}$$

where $G(t) = g\dot{\gamma}(t)$, the dose rate induced carrier generation rate

W_t = sum of depletion region widths about the base-collector junction

$j_e(r, t)$ = emitter current density

C_1 = experimentally determined constant specific to the device of interest

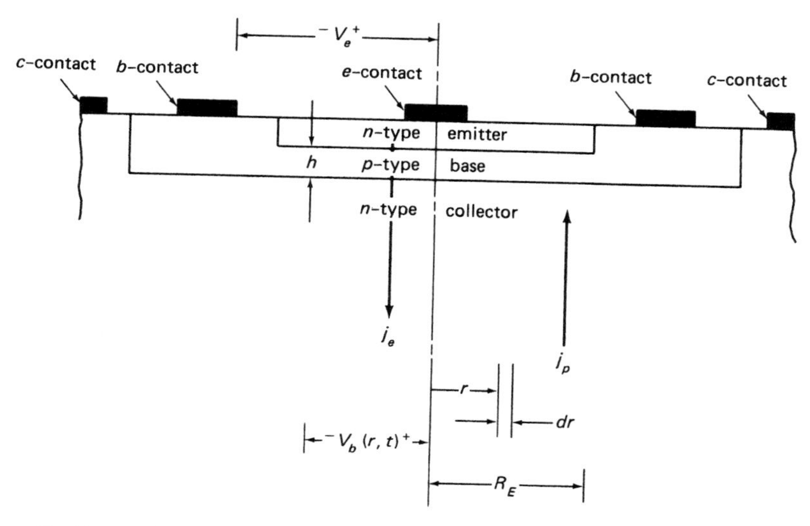

Fig. 7.8 Cylindrical *npn* transistor cross section.[8] Figure © 1988 by the IEEE.

The second term on the right hand side of Eq. (7.27) represents the collector multiplication (secondary photocurrent) contribution from emitter current injected into the collector.

The transverse voltage $V_b(r, t)$ in the base region, as depicted in Fig. 7.8 can be related to $j_p(r, t)$ using Ohm's Law. That is, simply, at a given instant

$$dV_b(r, t) \cong (\partial V_b/\partial r) \cdot dr = -dR \int_0^r j_p(r, t) dA$$

$$= -(\rho_s dr/2\pi r) \int_0^r j_p(r', t) 2\pi r' \, dr'; \; V_b(R_E, t) = 0 \quad (7.28)$$

where dR is the incremental base resistance, ρ_s is the corresponding sheet resistance (discussed in Section 3.5), and R_E is the emitter radius, as in Fig. 7.8. That is, the V_b zero reference is taken at the (radial) position of the emitter-base junction interface. It should be noted that j_p is the transverse photocurrent density in the base, under the emitter, injected from the collector, where it was flowing normal to the base, as in Fig. 7.8.

Hence, $dA = 2\pi r dr$ and $dR(r) = \rho dr/2\pi rh = \rho_s dr/2\pi r$, where h is the base thickness, and ρ_s is the base sheet resistance ρ/h.

Under transient conditions caused by the incidence of the dose rate-induced photocurrent pulse, the emitter-base depletion region capacitance charging current $I_e(t)$ is evidenced through

$$\dot{V}_e(t) = (I_e/(t) - I_{ep}(t))/C; \; V_e(0) = 0. \tag{7.29}$$

where; $V_e(t)$ = emitter-base voltage

$$I_{ep}(t) = \int_0^{R_E} j_p(r, t) 2\pi r dr, \text{ the emitter-base photocurrent}$$

$$I_e(t) = \int_0^{R_E} j_e(r, t) 2\pi r dr, \text{ the emitter current}$$

C = emitter-base depletion region capacitance

$$j_e(r, t) = J_{e0} \exp[(e/kT)\{V_b(r, t) - V_e(t)\}] \tag{7.30}$$

J_{e0} is the emitter junction saturation current density, as discussed in Section 3.7.

Inserting Eqs. (7.27) and (7.30) into Eq. (7.28) yields

$$r\partial V_b(r, t)/\partial r = -e\rho_s W_t G(t) \int_0^r \{l + C_l J_{e0} \exp[(e/kT)\{V_b(r', t) - V_e(t)\}]\} r' dr' \tag{7.31}$$

Similarly, inserting Eq. (7.30) into Eq. (7.29) yields

$$C\dot{V}_e(t) = 2\pi J_{e0} \int_0^{R_E} \exp[(e/kT)\{V_b(r, t) - V_e(t)\}] r dr - I_{ep}(t) \tag{7.33}$$

Equations (7.31) and (7.32) are the simultaneous coupled pair representing this model of transverse junction breakdown.[4]

For relatively small values of dose rate, the collector multiplication (second term in the integrand of Eq. (7.31)) is negligible. With V_b^s defined below, the resulting Eq. (7.31) is given by

$$\partial V_b^s/\partial r = -e\rho_s W_t G(t) r/2; \; V_b^s(R_E, t) = 0 \tag{7.33}$$

with an integral

$$V_b^s(r, t) = e\rho_s W_t G(t)(R_E^2 - r^2)/4 \tag{7.34}$$

which is seen to be linear in the dose rate, since $G(t) = g\dot{\gamma}$. V_b^s is the transverse voltage corresponding to the linear dose rate region, as shown in Fig. 7.6 for relatively low dose rates.

Integrating Eq. (7.31) yields

$$V_b(r, t) = [e\rho_s W_t G(t)(R_E^2 - r^2)/4] + e\rho_s W_t C_1 J_{e0} G(t) \int_r^{R_E} (dr'/r') \cdot \int_0^{r'} [\exp(e/kT)\{V_b(r'', t) - V_e(t)\}]\, r''dr'' \tag{7.35}$$

where it is seen that the first term is the contribution without collector multiplication, V_b^s, which obtains for relatively low dose rate. At room temperature, $1/kT \cong 40(eV)^{-1}$ in the exponential of the above integral. Hence, it can be appreciated that a small increase in $V_b - V_e$ can result in a very large increase in the transverse voltage $V_b(r, t)$ on the left-hand side of Eq. (7.35). A first-order approximation to V_b is obtained by inserting V_b^s for V_b in the integrand of Eq. (7.35). This results in

$$V_b(r, t) \cong V_b^s + (2kT/e)\, C_1 J_{e0}\{\ln(R_E/r) - \int_r^{R_E} \exp - [(e^2/4kT)\, \rho_s W_t G(t)\, r'^2]\, dr'/r'\} \exp(e/kT)(V_b^s(0, t) - V_e) \tag{7.36}$$

From Eq. (7.36) it is again seen that, for $V_b \ll V_e$, and hence small $G(t)$, the second term above is very small. This corresponds to low values of dose rate and so leaves only the first term in Eq. (7.36), as expected. As $G(t)$ (i.e., $\dot{\gamma}$) increases, there is a threshold value of $\dot{\gamma}$ embodied in $G(t)$ that will cause V_b to exceed V_e, causing a large sharp transition to a very large level of photocurrent, as perceived from Eq. (7.36).

Again, the sharp transition of V_b at the dose rate threshold causes an emitter-base junction breakdown, which is tantamount to a low impedance path between the emitter and base. This causes the device to assume a common emitter configuration, whereby the primary photocurrent is now amplified, causing the anomalous photocurrent behavior, as depicted in Fig. 7.6.

Consideration of another photocurrent model, also somewhat more complicated than that of Wirth-Rogers, leads to a photocurrent superlinearity resembling a discontinuity in $I_{pp}/\dot{\gamma}$ at a given dose rate threshold.[6] This model includes the effects of both time-dependent internal electric field and minority carrier lifetime. It reduces to the Wirth-Rogers model if (a) the electric field in the quasineutral region (beyond the depletion layer) is assumed to vanish; (b) high injection

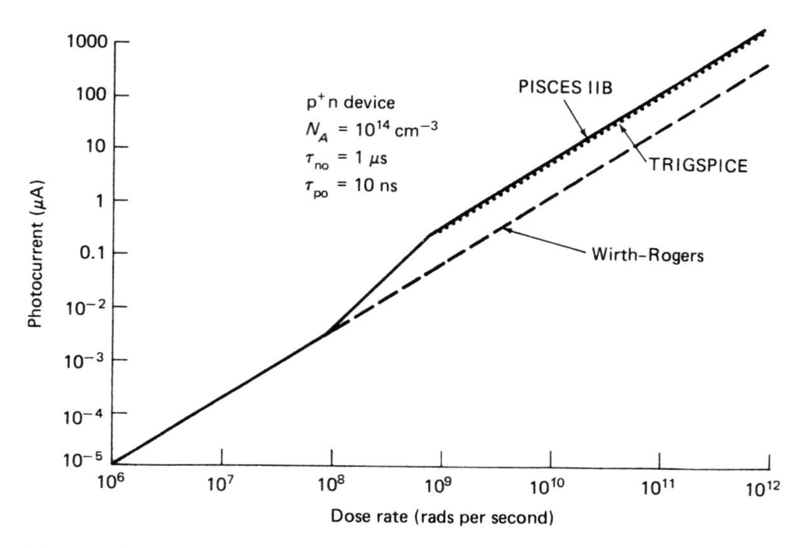

Fig. 7.9 Nonlinear photocurrent models compared to low photocurrent linear model.[8] Figure © 1988 by the IEEE.

effects on minority carrier lifetime are ignored; and (c) the junction neighborhood is allowed to be represented by an infinite one-dimensional half space.

The inclusion of the electric field and the high injection minority carrier lifetime effects increase the effective photocurrent collection volume, since increases in minority carrier lifetime increase the diffusion length[16] ($L^2 = D\tau$). Also, the electric field in the bulk is in the direction that drives the minority carriers toward the junction, which enhances the increase in effective collection volume. All of this results in a collection volume increase that is dose-rate and impurity-concentration dependent. The Wirth-Rogers model maintains a constant collection volume with dose rate, resulting in a constant $I_{pp}/\dot{\gamma}$. In the nonlinear models, when the minority carrier lifetime increase saturates, the photocurrent again becomes linear with dose rate, as shown in Figs. 7.6 and 7.9.

It should be realized that the preceding comprises approximations to the full solutions of the ambipolar transport equations, as discussed in Section 1.10. These equations encompass the appropriate approximations to use for high injection, where the relationship between photocurrent and the incident transient radiation is nonlinear. The ensuing very large photocurrents are induced by correspondingly high levels of incident transient radiation.

One aspect of the ambipolar transport is the complicated nature of the internal electrical field, E. That is, the total conduction current density in the semiconductor device is, from Section 1.8 (in one dimension)

$$J = J_n + J_p = e(\mu_n n(x) + \mu_p p(x))\, E(x) + e(D_n n'(x) - D_p p'(x)) \quad (7.37)$$

Invoking the quasineutrality condition[7] in the base region, then $p \cong n + N_A$, where N_A is the dopant concentration in the base. Hence, $n' \cong p'$, and Eq. (7.37) can be rewritten as

$$J \cong e(\mu_n n + \mu_p p)\, E + e(D_n - D_p)\, p' \quad (7.38)$$

Extracting the electric field from this equation, using the Einstein relation, gives

$$E = J/\sigma - p'(kT/e)((\mu_n - \mu_p)/(\mu_n n + \mu_p p)) \quad (7.39)$$

where the conductivity $\sigma = e(\mu_n n + \mu_p p)$. Equation (7.39) gives the appropriate form of the electric field E to use where it appears in the ambipolar transport equations. It is seen that the first term in Eq. (7.39) is the ohmic term, while the second term is the internally generated (Dember) field,[8] as caused by the imbalance of electron and hole diffusion currents. As discussed in Section 1.10, this field arises to speed up the slower particles and slow down the faster particles in their transport in the device, and is one aspect of the definition of ambipolar diffusion. As can be appreciated, insertion of E from Eq. (7.39) into the ambipolar diffusion equations given in Section 1.10 yields a complex system of equations where the carrier densities $n(x)$ and $p(x)$ do not appear explicitly. Hence, full analytic (i.e., closed form) solutions are not obtainable, so that solutions are obtained by numerical integration, using mainframe computer codes. Such numerical solutions are depicted[8] in Fig. 7.9, which are obtained from the PISCES IIB and SPICE type (TRIGSPICE) programs.[8,9] The PISCES IIB program corresponds to a configuration of the exact system of transport equations, including Poisson's equation for the electric field. The figure shows the discontinuities between the linear-low photocurrent regime of Wirth and Rogers and the nonlinear high photocurrent regime. Although Fig. 7.9 cor-

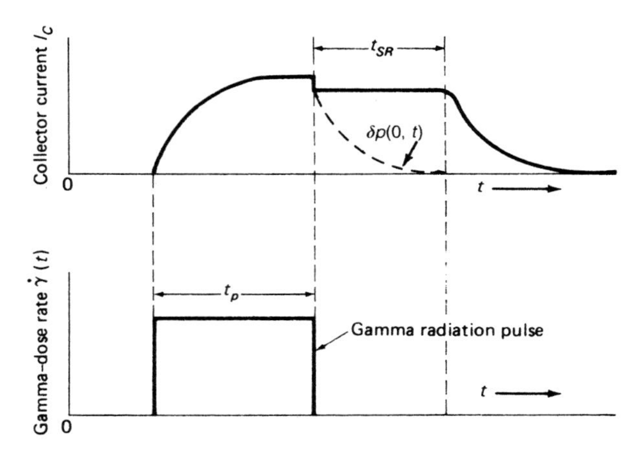

Fig. 7.10 Collector current and dose-rate pulse versus time showing radiation storage time.

responds to infinite medium solutions, extensions for finite devices and ohmic contacts are available.[8]

7.5 RADIATION STORAGE TIME

As seen, operating transistors can be driven into saturation readily by moderate to high dose-rate photon pulses. The time the transistor remains in saturation after termination of the radiation pulse is called radiation storage time, t_{SR}, as illustrated in Fig. 7.10. An equivalent definition will be obtained from the following development, which yields a relationship for radiation storage time in terms of circuit and physical parameters. Let $\delta p(x, t)$ be the excess minority carrier density in the bipolar transistor base satisfying the time-dependent, one-dimensional, diffusion equation, i.e.,

$$D\partial^2\delta p/\partial x^2 - \delta p/\tau + G(t) = \partial\delta p/\partial t \tag{7.40}$$

rewritten from the discussion in Section 1.13. G is the generation rate of excess carriers due to the ionizing gamma pulse. At the end of the pulse duration, the excess carrier density at the collector depletion layer boundary falls, for its source is now absent. When the excess carrier density drops to zero, the collector-base junction ceases to be forward biased, and the collector current also begins to drop, as seen in Fig. 7.10. The radiation storage time is the elapsed time reckoned from the end of the radiation pulse to when the excess carrier density at

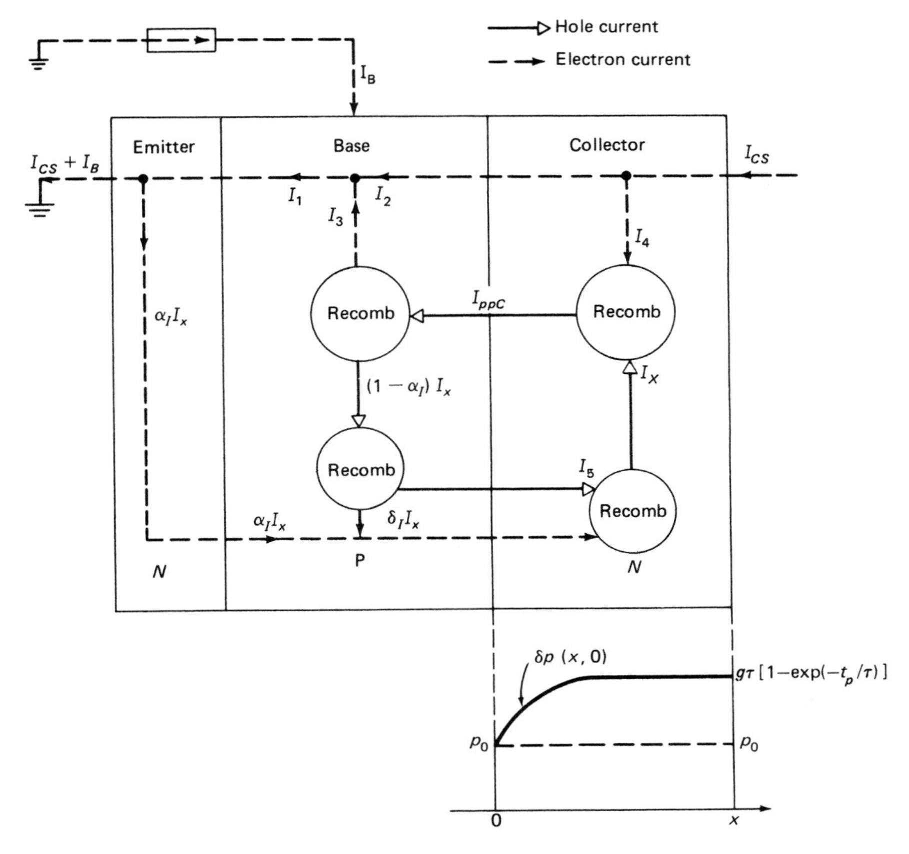

Fig. 7.11 Bipolar transistor radiation storage time equivalent circuit superimposed on its one-dimensional physical model.[10] Figure © 1964 by the IEEE.

Legend:

I_{ppc} is the primary collector photocurrent
I_{CS} is the collector saturation current
I_x is the collector-base junction current
I_g is the base current
h_{FE} is the common emitter current gain; β
h_{FB} is the common base current gain; $\alpha = \beta/(1 + \beta)$
α_J is the inverse α (emitter reverse biased, collector forward biased)
δ_J is the inverse collector efficiency, η_c^{-1}
$I_1 = I_{CS} + I_g + \alpha_J I_X$
$I_2 = \alpha(I_{CS} + I_g + \alpha_J I_X)$
$I_3 = [I_1/(h_{FE} + 1)] - I_g$
$I_4 = (1 - \alpha) I_{CS} - \alpha I_g - \alpha\alpha_J I_X$
$I_5 = (1 - \delta_J) I_X$

the collector depletion layer boundary ($x = 0$), vanishes. Hence, the radiation storage time is available as the root of $\delta p(0, t_{SR}) = 0$, where $\delta p(x, t)$ is the solution of Eq. (7.40). Figure 7.11 depicts a bipolar transistor radiation storage time equivalent circuit, where the transistor is in saturation, super-imposed on a one-dimensional physical model of the transistor.

The collector-base junction current I_x, due to the junction being forward biased by the primary photocurrent I_{ppc}, is obtained from a straightforward nodal analysis of the circuit in Fig. 7.11 as

$$I_X = -\frac{I_{CS} + I_B}{(1+\beta)(1-\alpha\alpha_I)} + \frac{I_B + I_{ppc}}{1-\alpha\alpha_I} \tag{7.41}$$

For reasonably high gain transistors, $\alpha \lessapprox 1$, and $0.02 < \alpha_I < 0.5$. Also, $\alpha_I \cong \delta_I$, so that Eq. (7.41) can be simplified to give

$$-I_x = \frac{I_{CS} + I_B}{1+\beta} - I_B \tag{7.42}$$

Realizing that the above I_x can be replaced by $eDA\partial\delta p/\partial x$, from Fick's law of diffusion, Eq. (7.42), when evaluated at the collector depletion layer boundary ($x = 0$), becomes

$$-eDA\frac{\partial}{\partial x}\delta p(0,t) = \frac{I_{CS} + I_B}{1+\beta} - I_B \tag{7.43}$$

For use below in obtaining the radiation storage time, a needed solution to Eq. (7.40), with boundary conditions given in Eq. (7.43) and evaluated at $t = 0$ is

$$\delta p(x,0) = G\tau[1 - e^{-t_p/\tau}] - \frac{L}{eDA}\left[\frac{I_{CS} + I_B}{1+\beta} - I_B\right]e^{-x/L} \tag{7.44}$$

$\delta p(x, 0)$ is shown in the inset at the bottom of Fig. 7.11. The preceding assumes that the transistor is unsaturated prior to the onset of the radiation pulse and that a constant base current I_{B1} flows during the pulse epoch t_p. After the end of the pulse, the base current attains the value I_{B2}.

The solution to Eq. (7.40) is now obtained using the boundary conditions that in the limit of large x, $\delta p(x, t)$ remains finite, and Eq. (7.43) holds for the initial excess carrier density. The result at $x = 0$ is

$$\delta p(0,t) = \frac{\beta(I_{B2} - I_{B1})}{eA(1+\beta)} \cdot \sqrt{\tau/D}(erf(\sqrt{t/\tau})) - \frac{L}{eDA}\left[\frac{I_{CS} + I_{B1}}{1+\beta} - I_{B1}\right]$$
$$+ G\tau(1 - e^{-t_p/\tau})(1 - e^{-t/\tau}) \quad (7.45)$$

As mentioned, the radiation storage time is the root of $\delta p(0, t_{SR}) = 0$, which, from Eq. (7.45), yields the transcendental equation for t_{SR} as

$$t_{SR} = \ln\left\{\frac{egLA(1 - e^{-t_p/\tau})}{(1+\beta)I_{CS}} - \frac{\beta I_{B1}}{1+\beta} \cdot erfc\sqrt{t_{SR}/\tau} - \frac{\beta I_{B2}}{1+\beta} erf\sqrt{t_{SR}/\tau}\right\} \quad (7.46)$$

Usually $t_{SR} \gg \tau$, so that $erf(t_{SR}/\tau)^{1/2} \approx 1$, and $I_{B1}/I_{B2} \approx 1$; then the above yields to good approximation

$$t_{SR} = \tau \ln\left\{\frac{\Delta I_{CBO}(1 - e^{-t_p/\tau})}{(1+\beta)(I_{CS} + I_{B2})} - I_{B2}\right\} \quad (7.47)$$

where $\Delta I_{CBO} \cong egLA$ is the approximation often used. When the base can be assumed to be open, i.e., $I_{B1} = I_{B2} = 0$, Eq. (7.46), under the previous assumptions regarding t_{SR}, becomes

$$t_{SR} = \tau \ln \dot{\gamma} + \tau \ln[\sqrt{\tau}(1 - e^{-t_p/\tau})] - \ln\left[\frac{I_{CS}}{eGA\sqrt{D}(1+\beta)}\right] \quad (7.48)$$

where t_{SR} and τ are expressed in μs.

When the transistor is driven into saturation by the photocurrent, the well-known diode reverse recovery expression yields a relation between the electrical storage time t_s and the minority carrier lifetime τ, as[11]

$$\tau = t_s[erf^{-1}(1 - I_{CS}/\beta I_B)]^{-2} \quad (7.49)$$

Equation (7.49) has been evaluated for a number of transistor types to yield the empiricism[7] $\tau \cong 0.3\,t_s$. Then

$$t_{SR} = 0.3\,t_s \ln\left\{\frac{(1 - e^{-t_p/0.3\,t_s})I_{ppc}}{(1+\beta)(I_{CS} + I_{B2})} - I_{B2}\right\}$$

Inserting this relationship into Eq. (7.47), where $I_{ppc} = I_{CBO}$, gives

$$t_{SR} = 0.3\,t_s \ln[I_{ppc}\{1 - \exp - (t_p/0.3\,t_s)\}/\{(I_{CS} + I_{B2})/(1+\beta) - I_{B2}\}] \quad (7.50)$$

Figure 7.12(A) depicts experimentally determined t_{SR} as a function of t_s for 180 bipolar transistors for various dose-rate pulses.[12]

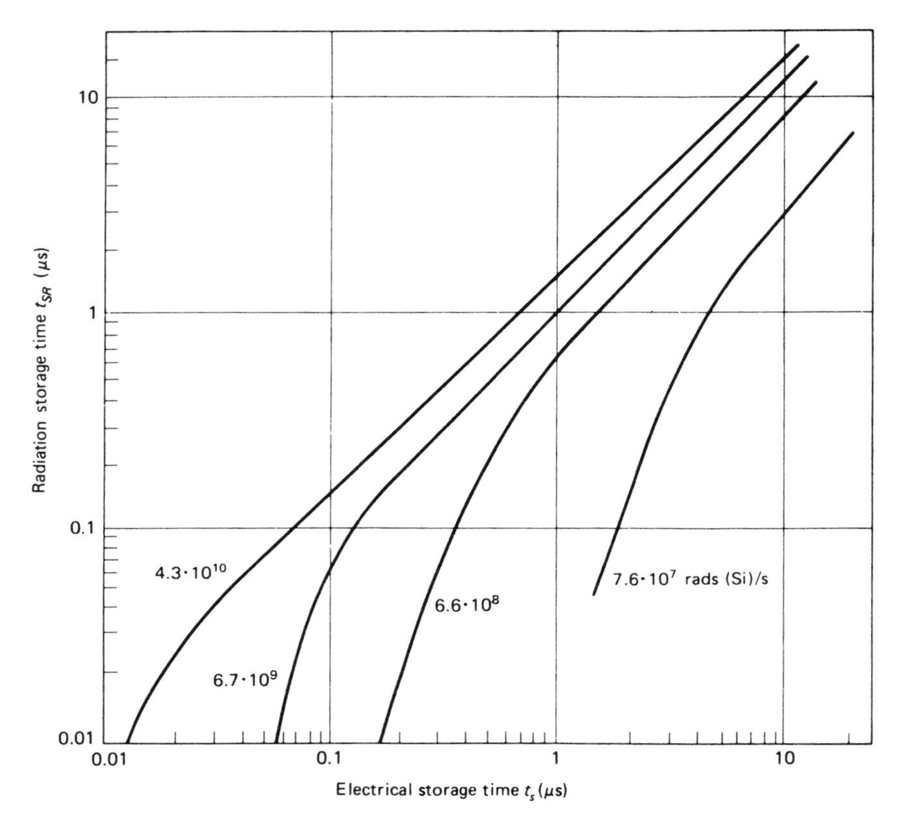

Fig. 7.12(A) Experimentally determined radiation storage time versus electrical storage time for 180 transistors for various dose rate pulses.[12] Figure © 1965 by the IEEE.

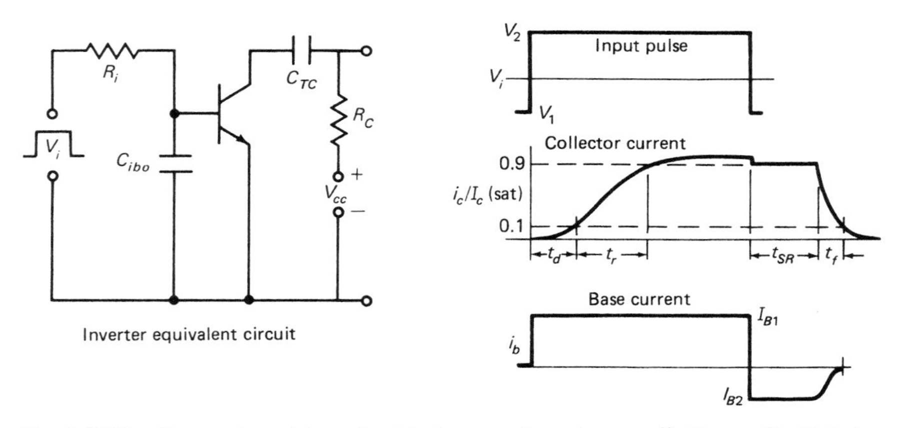

Fig. 7.12(B) Propagation delays for bipolar transistor inverter.[13] Figure © 1962 by McGraw-Hill.

7.6 PROPAGATION DELAY TIME

Storage time discussed in the previous section is one component of the propagation delay time, normally construed as a measure of the time for a state change due to a rectangular input pulse with assumed zero rise and fall time, to propagate through a gate as in a bipolar inverter, depicted in Fig. 7.12(B). As seen in that figure, the propagation delay time consists of: (a) the delay time t_d to bring the inverter from its initial OFF state to the beginning of conduction, i.e., to 10 percent of its saturation current upon onset of the above pulse; (b) the rise time t_r, following t_d, for the collector current to rise from 10 to 90 percent of its saturation value I_c (sat); (c) the storage time t_{SR}, as discussed in the previous section, and (d) the fall time t_f for the collector current to fall back to 10 percent of I_c (sat), as in the figure. These times are given by [13]

$$t_d \cong \tau_D \ln[(1 + V_1/V_2)/(1 - 0.7/V_2)] \tag{7.51}$$

where V_1 and V_2 are the voltage extremes of the input pulse, as in Fig. 7.12(B), and $\tau_D = R_i C_{ib0}$, the input time constant. The rise time is given by

$$t_r \cong \beta\tau_r |\ln(1 - 0.9\ I_{CS}/\beta I_{B1})| \tag{7.52}$$

where I_{B1} is the base current during the positive going phase of the input pulse, $\tau_r = (\omega_T^{-1} + 1.7 R_L C_{TC})$, and $R_L C_{TC}$ is the output time constant. The fall time is given by

$$t_f \cong \beta\tau_r \ln[(1 - I_{B2}/I_{CS})/(0.1 - I_{B2}/I_{CS})] \tag{7.53}$$

and I_{B2} is the base current following the cessation of the input pulse.

The latter two equations for the rise and fall time, respectively, are seen to be functions of the common emitter gain β. Hence, for t_r and t_f, the coupling between the propagation delay times and the damaging effects of radiation is primarily through the effect on β. For example, the effect of neutron fluence Φ_n on β is given by the Messenger-Spratt equation $\Delta(1/\beta) = \Phi_n/\omega_T K$. Incident neutrons or ionizing radiation produces a reduction in β. As seen from the above equations, this causes both rise and fall time to decrease, with radiation storage time decreasing as shown in Fig. 7.12(A).

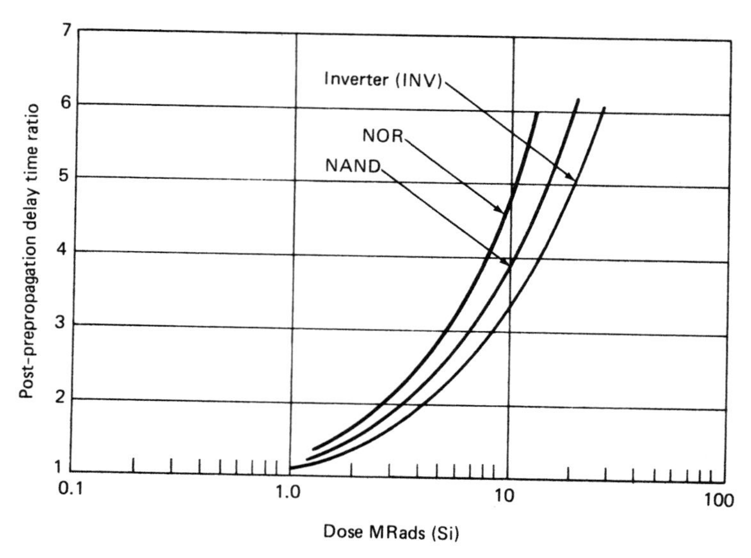

Fig. 7.13 Post- pre- propagation delay time ratio versus ionizing dose for 1.25 μm CMOS/SOS ASIC gate array chips.[14] Figure © 1985 by the IEEE.

Figure 7.13 depicts post- versus pre-irradiation propagation delay time ratios for 1.25 μm feature size CMOS/SOS ASIC (Application Specific Integrated Circuit) gate arrays. The gate types shown are NOR, NAND, and INV (Inverter) chains, each consisting of many gates. The individual pre-irradiation propagation delay times are 0.3 ns per inverter gate and 0.5 ns per NAND or NOR gate. The figure shows that the propagation delay time ratios become excessive only beyond ionizing dose levels of about 1 Mrad (Si). These levels should be construed as those for the chips per se, since any possible dose enhancement factor corrections (Section 10.2) for their metal-layered packages are not included.

Other radiation threshold levels for these gate arrays, fabricated to optimize feature size (minimum lithographic line width) as opposed to radiation hardness, are: (a) upset dose rate of $\sim 10^{11}$ Rads (Si) per sec. (35 ns FXR pulse width), and (b) SEU error cross section of less than $\sim 10^{-7}\,\text{cm}^2$ per bit[14] (Section 8.2).

Practically, gain reduction is manifest to first order in the reduction of fanout, with the driving stage forced out of saturation, thus failing to sink currents of the following gates it drives. This normally overshadows propagation time changes unless substantial fanout margins (Section 12.2) are incorporated into the system of interest.

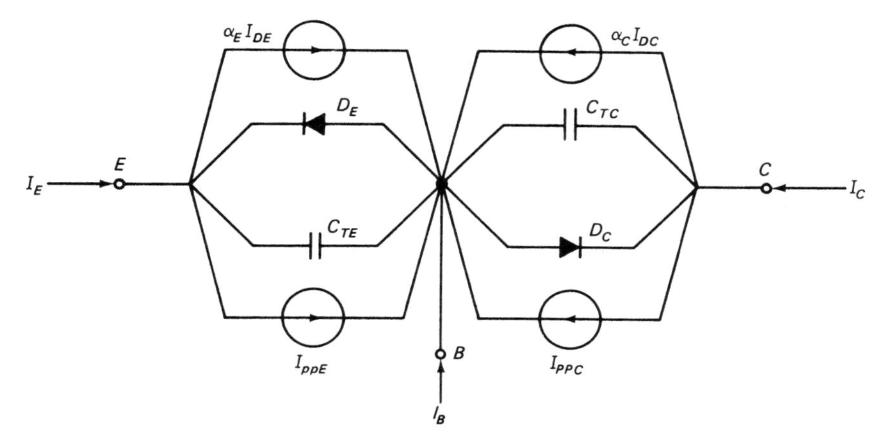

Fig. 7.14 Photocurrent-inclusive Ebers-Moll equivalent circuit for bipolar transistor.

7.7 LATCHUP

The general definition of latchup is that the semiconductor device of interest is transformed (latched up) to an anomolous state that no longer responds to input signals. If this deleterious state does not cause device burnout due to excessive current flow within the device, it can be restored to its previous function by removing and resetting power. Preventive measures must be taken to obviate the possibility of latchup or, at a minimum, to result in a very low latchup probability. At the device level, these measures, including collector gold doping to produce a vanishingly small minority carrier lifetime in the parasitic latchup-inducing system, are discussed in detail in the following. Another measure is the use of dielectrically isolated devices instead of junction-isolated counterparts. At the circuit level, current limiting resistors or inductors can be added in series with the V_{cc} line. Latchup remedies are investigated in detail in Section 12.4.

One of the classic parasitic transistor IC latchup configurations is that of a silicon-controlled rectifier (SCR). Discussion of this switching device provides a basis upon which to examine latchup. The following explanation of four-layer-latchup action is paraphrased from Crowley et al.[15] Applying a positive voltage across the device in Fig. 7.15 as shown, junctions J_1 and J_3 are forward biased, and J_2 is reverse biased. With the gate currents (SCR gate) I_{g1} and I_{g2} both zero, the current I through the system is the same as, for example, through J_2, which is given by

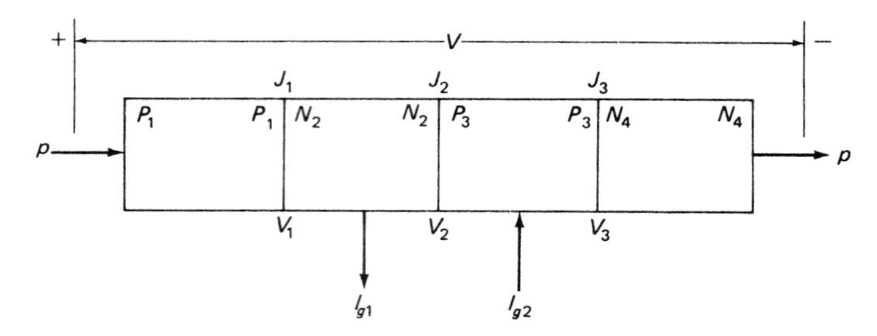

Fig. 7.15 Four-layer *pnpn* diode.[15] Figure © 1976 by the IEEE.

$$I = I(M_h\alpha_h + M_e\alpha_e) + I_{sat} \tag{7.54}$$

where M_h and M_e are the respective avalanche multiplication factors, as defined in Section 3.6, for holes and electrons at junction J_2. α_h and α_e are respective forward current transfer ratios for holes and electrons in regions N_2 and P_3. I_{sat} is the saturation current of the center junction J_2.

In Fig. 7.16, the coordinates of the switching point (V_s, I_s) satisfy

$$M_h(V_s)\alpha_h(I_s) + M_e(V_s)\alpha_e(I_s) = 1 \tag{7.55}$$

When this condition is met, the overall current gain of the device is greater than unity. As this current gain approaches unity, the regenerative action that sustains this structure in the (low impedance) "on" state is initiated. This action can be described using Fig. 7.15 by first following holes emitted by J_1 that are swept across the reverse biased junction J_2 to increase the total majority carrier concentration in region P_3. This latter increase serves to lower the height of the potential barrier at J_3, for the carrier concentration and current are related exponentially to the potential barrier, and so increase the number of electrons emitted at J_3.[15] These electrons are then swept back across J_2, where they contribute to an increase in the majority carrier concentration in region N_2, thereby lowering the potential barrier at J_1. The lowering of the barrier at J_1 increases the injection of holes into N_2, and the regenerative feedback cycle is imminent. When the sum of the current gains exceeds unity, the above potential barrier lowering cycle becomes regenerative, which forces the device into the "on" state. This regenerative action increases the majority carrier concentration in N_2 and P_3 to the point where the

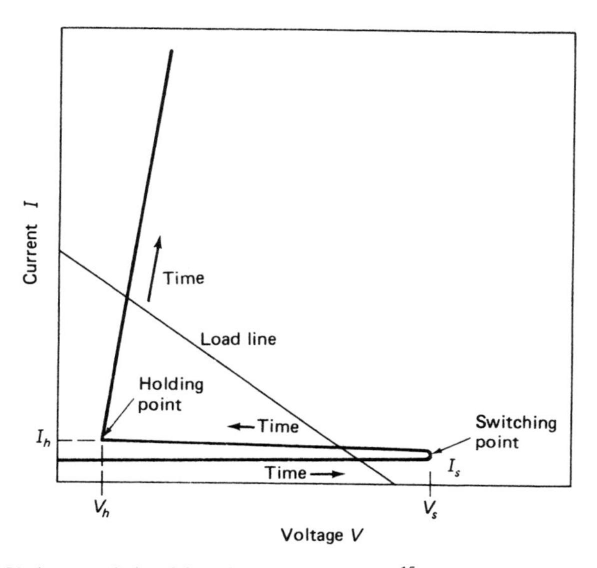

Fig. 7.16 *I-V* characteristic of four-layer *pnpn* device.[15] Figure © 1976 by the IEEE.

barrier at the reverse biased junction J_2 is reduced along with those at J_1 and J_3. This "on" state persists until the voltage across the system is removed or reduced, until the regeneration is quenched, or until the device burns out.

In the low-current "off" state, the current gains are aided by avalanche multiplication at the center junction J_2. As increased majority carrier concentrations in N_2 and P_3 reduce the voltage drop across the reverse biased J_2, the voltage dependent M_e and M_h decrease. The overall current gain continues to decrease because of the current dependent α_e and α_h. At the holding point (V_h, I_h), in Fig. 7.16, there is no contribution to current gain from avalanche multiplication, and the sum of the forward current transfer ratios becomes unity, i.e.

$$\alpha_e(I_h) + \alpha_h(I_h) = 1 \tag{7.56}$$

When gate currents, I_{g1} and I_{g2} are allowed to flow (by being connected to the remainder of the circuitry as part of the larger system), the voltage and current at the switching point satisfy

$$M_h\alpha_h + I + I_{g1}(\partial\alpha_h/\partial I) + M_e\alpha_e + I + I_{g2}(\partial\alpha_e/\partial I) = 1 \tag{7.57}$$

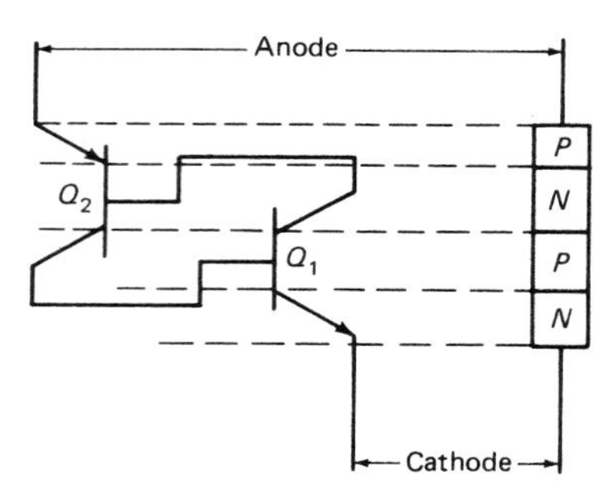

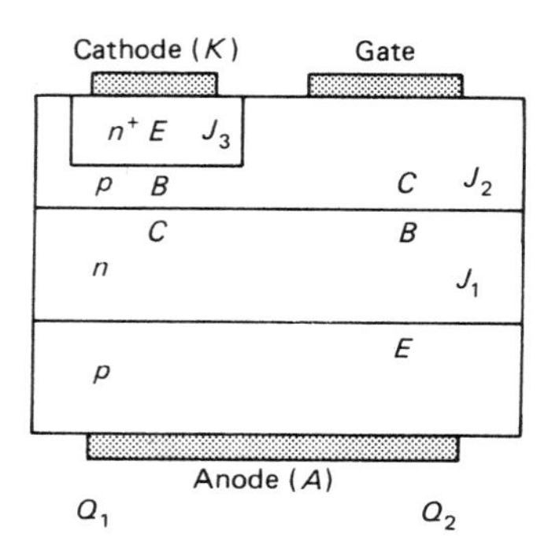

Fig. 7.17 Physical configuration and equivalent circuit of an SCR.[16] Figure © 1968 by John Wiley & Sons.

The gate currents I_{g1} and I_{g2} can come about from external sources, such as induced photocurrents. Even if the radiation-induced photocurrent is small, the derivatives $\partial\alpha/\partial I$ can be very large. Thus a four-layer path may be switched on at a magnitude of induced gate current much less than the holding current I_h.

For an alternative approach to the behavior of the *pnpn* switch, consider the operation of the silicon-controlled rectifier (SCR), which is essentially just such a switch.[15] An SCR is a bistable device whose state can be changed from "off" (high resistance/low current) to "on" (low resistance/high current) by current injection, as discussed below. Figures 7.17 and 7.18 depict the physical configuration, equivalent circuit schematic diagram, and symbol, respectively, of an SCR. In normal operation, the SCR is biased with anode positive and cathode negative. This results in junctions J_1 and J_3 being forward biased and J_2 reverse biased, as in the figure. The three upper layers, which include junctions J_2 and J_3, form an *npn* transistor where the J_3 layer is the emitter, J_2

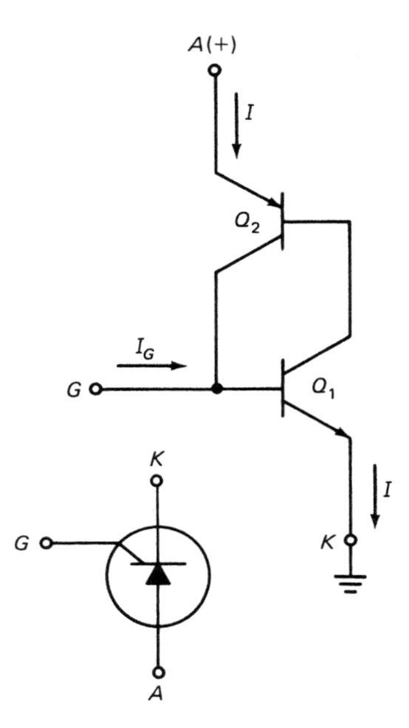

Fig. 7.18 Schematic and SCR Symbol.[16]

layer is the base, and J_1 layer is the collector. Also a *pnp* transistor simultaneously exists where the J_2 layer is the collector, J_1 layer the base, and the anode *p* layer is the emitter. The J_2, or gate, layer is the collector-base junction for both transistors, where "trigger" current is injected to turn the device "on," given that it had been "off." The SCR will now reside in its "on" state until the anode voltage is removed, or until the SCR current falls below its "on" threshold. It then reverts to its "off" state. A typical voltage-current characteristic of an SCR is given in Fig. 7.19. The expression for the anode-cathode voltage V as a function of current I through the SCR is given by[17]

$$V = (kT/e)\ln\left[(I_{s2}/I_{s1}I_{s3})(I_{s1} + a_1 I)(I_{s3} + a_3 I)/(I_{s2} + a_2 I)\right] \quad (7.58)$$

where $I_{s1,2,3}$ are the saturation currents corresponding to the three junctions, and a_1, a_2, a_3 are defined below in terms of the transistor "alphas," viz.

$$a_1 = (1 + \alpha_{1i}\alpha_{2n} - \alpha_{2n}\alpha_{2i} - \alpha_{1i})/(1 - \alpha_{2n}\alpha_{2i} - \alpha_{1n}\alpha_{1i}) \quad (7.59)$$

$$a_2 = (\alpha_{1n} + \alpha_{2n} - 1)/(1 - \alpha_{2n}\alpha_{2i} - \alpha_{1n}\alpha_{1i}) \quad (7.60)$$

$$a_3 = (1 + \alpha_{2i}\alpha_{1n} - \alpha_{1n}\alpha_{1i} - \alpha_{2i})/(1 - \alpha_{2n}\alpha_{2i} - \alpha_{1n}\alpha_{1i}) \tag{7.61}$$

and where

α_{1n} is the transistor "alpha" for J_1 as emitter and J_2 as collector
α_{2n} is the transistor "alpha" for J_3 as emitter and J_2 as collector
α_{1i} is the transistor "alpha" for J_2 as emitter and J_1 as collector
α_{2i} is the transistor "alpha" for J_2 as emitter and J_3 as collector

For the "off" state, $\alpha_{1n} + \alpha_{2n} \equiv f(\alpha) < 1$, which implies that a_2 is negative, while a_1 and a_3 are positive. The conclusion from Eq. (7.58) in this case is that

$$a_2 I + I_{s2} > 0 \tag{7.62}$$

or that $I < I_{s2}/|a_2|$, which is a multiple of the normally small saturation current I_{s2}, so that I is also small.

For the "on" state, $f(\alpha) > 1$, which implies that all "a"s are positive; then I can grow without bound for large but finite V, as seen from Eq. (7.58), so that, with the assumption that I is much larger than the saturation currents, Eq. (7.58) can be rewritten as

$$V = (kT/e) \cdot \ln[(I_{s2}/I_{s1}I_{s3})(I/(I - \alpha_{1n} - \alpha_{2n}))] \tag{7.63}$$

so that in the "on" state, the SCR satisfies the above thinly disguised "diode" equation. That is, the SCR acts as a diode when it is in the "on" state.

Defining corresponding transistor gains β_{1n} and β_{2n}, the above equation yields

$$V = (kT/e) \ln[(I_{s2}/I_{s1}I_{s3})(1 + \beta_{1n})(1 + \beta_{2n})(I/(1 - \beta_{1n}\beta_{2n}))] \tag{7.64}$$

Hence, if the product of the gains approaches unity, then the system will switch to the "on" state, if not already there. For the case of small values of gain, i.e., $\beta_{1n}\beta_{2n} < 1$ with correspondingly small I, the system will revert to the "off" state, if not already there. When in the "off" state, if a gate current is injected, it can cause the *npn* transistor collector current to increase enough to drive the system to the vicinity of $\beta_{1n}\beta_{2n} \simeq 1$, and the system will essentially switch "on." The SCR will remain on even if the gate injection is removed. To turn the device off, the current I must be reduced below the particular threshold current that implies $\beta_{1n}\beta_{2n} \simeq 1$. This is commonly done by removing the anode voltage, although certain

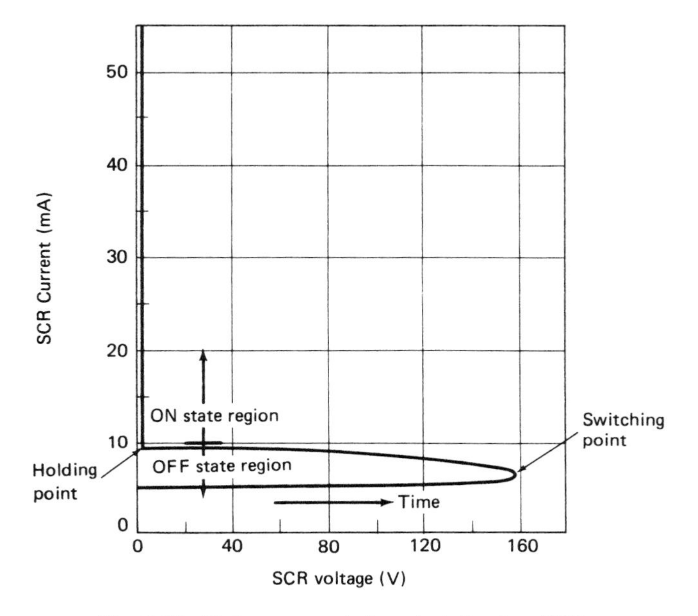

Fig. 7.19 Voltage-current characteristic of an SCR.

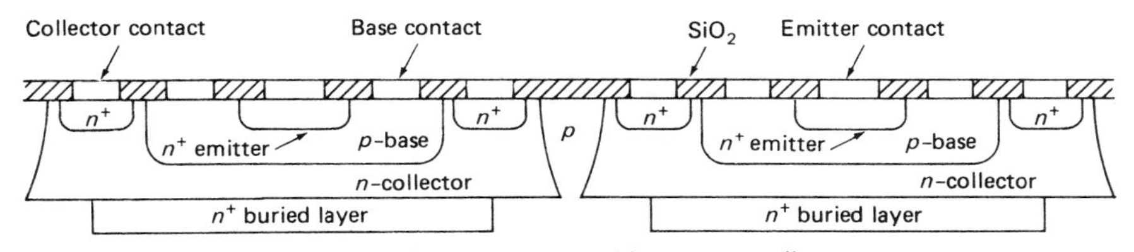

Fig. 7.20 Adjacent junction-isolated *npn* transistors in an integrated circuit.

SCRs have a second gate that can bias the device off. In a multitransistor microcircuit (IC), such as shown in Fig. 7.20, it is seen that the *p* substrate forms part of the parasitic four-layer *pnpn* device. This parasitic SCR can be switched on by any number of spurious manifestations, e.g., a sudden incident ionizing pulse of radiation. This pulse will cause sufficient extraneous current to flow, so that the parasitic SCR will be triggered on. This is one form of latchup. In this case, to prevent the latchup, the parasitic SCR can be denied its "on" state by ensuring that the *p* substrate is so biased that it never becomes forward biased during normal operation. As an aside, one example of a junction-isolated integrated circuit (JIIC), is depicted in Fig. 7.20.

Table 7.1 provides a short list of representative MSI/LSI devices and their dose-rate thresholds for latchup that result in device damage. It is observed that most of the latchup thresholds are greater than 10^8 rads per s. Section 12.4 contains further information on latchup thresholds for various types of semiconductor devices.

In certain CMOS devices, e.g., CMOS multivibrators, latchup is found to occur over only a limited range or set of ranges of dose-rate levels. This phenomenon is termed *latchup window*, and is currently believed to be caused by changes in the forward biasing of the junctions shown in Fig. 7.17 and discussed below. The dose rate generates photocurrents across the equivalent resistances R_S and R_D, representing the substrate and *p*-well, respectively, of the CMOS system, to provide the internal forward biases necessary to turn on the *pnpn* parasitic transistor, as in Fig. 7.21, which is the schematic of Fig. 7.18 redrawn to include R_S and R_D. Moderate photocurrents generated by moderate dose rates would then turn it on. However, higher dose rates (e.g., 10^{11} rads (Si) per s) would generate a proliferation of carriers at sufficiently high injection levels to reduce the above resistances by increasing the material conductivity, (as in conductivity modulation discussed in Section 5.12), so that insufficient forward bias would be available to turn on the parasitic transistor. Hence, a dose-rate window would be apparent in the device operation.[21]

The foregoing argument holds well for a single latchup window for high dose rates. However, in some devices a succession of latchup windows are known to occur at much lower dose rates of about 10^8 rads (Si) per s.[22] It is felt that the multiple latchup window effect is due to the existence of multiple latchup paths that successively wax and wane with increasing dose rate. That is, the corresponding parasitic transistor gain products $\beta_1 \cdot \beta_2$ approach and recede from unity as the dose is increased.[23,24] In such a succession of windows, a dominant latchup window can be about 100 rads wide for a 70-ns pulse, corresponding to an incremental dose rate, Δ rads/pulse width, of about $1.4 \cdot 10^9$ rads (Si) per s.[22] A narrow latchup window would be construed as a few rads in width for the same pulse duration, or an incremental dose rate of about $1.4 \cdot 10^7$ rads (Si) per s.

Another latchup-like phenomenon occurring in NMOS and the NMOS portions of CMOS integrated circuits is called Snap-Back[25] and is somewhat similar to the second breakdown discussed in Section 3.6. It has been detected in commercial CMOS microprocessors and other NMOS devices. The breakdown voltage characteristic of *n*-channel transistors

Table 7.1 Latchup damage thresholds for representative MSI and LSI integrated circuits.[20]

PART NO.	MFR	FUNCTION	TECHNOLOGY	PULSE WIDTH (ns)	THRESHOLD DOSE RATE (10^8 rads per s)
CDP1802CD	RCA	8-bit CPU	Si gate bulk CMOS	18	2–3
CDP1802CD	RCA	8-bit CPU	Si gate bulk CMOS	150	1.2
MMI6701D	MMI	4-bit slice	STTL	4000	3.9
HM1650B	HA	1-K SRAM*	Si gate CMOS	18	0.6
IM650B	ISL	1-K SRAM	Si gate CMOS	18	0.8
54C200	NSC	256-bit SRAM	CMOS	3	100
CD4061	RCA	256-bit SRAM	CMOS	100	32 (Hi rel.)
CD4061	RCA	256-bit SRAM	CMOS	100	65 (Hardened)
82S100	MMI	FPLA**	STTL	20	20
F100-L	Ferranti	16-bit CPU	CDI(Bipolar)***	1000	50

*SRAM stands for *static random access memory*, wherein a bit is representated by the state of a RAM flip-flop. DRAM stands for *dynamic random access memory*, usually an MOS IC, wherein a bit is stored as a charge residing in one node of the RAM. This type of memory needs to be "refreshed" periodically by associated circuitry.

** FPLA stands for Field Programmable Logic Array.

*** CDI stands for Collector Diffusion Isolation.

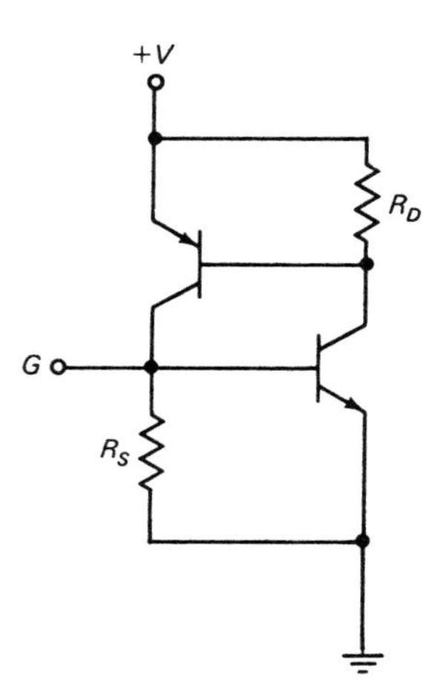

Fig. 7.21 *PNPN* latchup transistor schematic diagram.

exhibits a negative resistance region. This is a necessary condition for regeneration or oscillation, as well as latchup and snap-back. In this case, the negative resistance characteristic of the NMOS device in a microcircuit will provide conditions for a "snap-back" to a spurious stable operation in a reduced drain-to-source voltage state, analogous to the holding state of latchup. This can be induced by extraneous current injection from the drain junction into the substrate, due either to an avalanching drain junction or exposure to an ionizing radiation pulse of sufficient amplitude.

A model and detailed description of snap-back is available.[25] Although also similar to, and able to precipitate latchup, snap-back is not initiated by four layer SCR action, or junction breakdown in a corresponding parasitic lateral bipolar transistor contained within the MOS device. Normal digital logic cycling will terminate the snap-back action, in contrast to invoking a power-down condition to terminate latchup. Snap-back can be minimized by maintaining a large drain current, simultaneous with uninterrupted logic function. Pull-up devices on internal gates are anticipated to preclude snap-back, and associated large output buffers can sustain this phenomenon. Other preventative measures are at hand but result in hampered device performance.[25]

Electrostatic discharge (ESD) can also stimulate snap-back currents in certain NMOS devices.[47] Holes injected into the oxide of a grounded gate MOSFET biased into snap-back are trapped therein. This is almost certainly the reason for a dramatic snap-back induced gate-to-drain dielectric voltage breakdown.[47] Also, snap-back can be shown to cause both a positive and negative threshold voltage shift in MOS devices. In this vein, snap-back also has been shown to result in both hole trapping and interface state generation. When this yields a preponderance of trapped holes, the result is a negative threshold voltage shift. If a preponderance of interface states ensues, the result can be a positive threshold voltage shift. The presence of trapped holes may initiate a degradation of the device hardness due to time-dependent dielectric breakdown.[47] Also, measurements on NMOS and CMOS devices subject to ESD indicate a significant increase in gate hole current during snap-back. The magnitude of this current varies inversely with gate oxide thickness.[47] Snap-back causes oxide damage quite similar to that due to ionizing radiation. It can also precipitate latchup in NMOS/SOI devices.[48] Presently under investigation, it is conjectured as being due to forward biasing of the transistor body-source diode by impact ionization, resulting in a lowering of the threshold voltage. Or, when its V_{GS} is less than its flatband voltage, it behaves like a floating base *npn* bipolar transistor. However, its base hole density is then controlled by the inversion layer under the gate instead of by the usual doping. The latchup occurs when this "bipolar" transistor is activated by the impact ionization current.[48] Also, it is argued that snap-back arises from the same physical mechanism as base current reversal, BV_{CEO}, in advanced *npn* bipolar transistors.[49]

Latchup is a very complex phenomenon, and is potentially catastrophic in unhardened systems. Many device types and technologies have a low probability of latchup, and so one seemingly apparent possibility for latchup hardness assurance is to devise a 100 percent screening test. That is, to only utilize devices in actual systems that pass this screen. This procedure is not recommended, because it is extremely difficult and costly to design a screen, without conflicting requirements, that can adequately encompass all worst-case bias and operating node configurations, worst-case temperatures, and a sufficient number and selection of dose rate pulses to eliminate the possibility of latchup windows. Both common emitter gain and corresponding alpha increase monotonically with temperature, so that the highest operating temperature anticipated should usually be utilized as the worst-case temperature condition.[43] Both β and α also increase with bias voltage, so that the highest expected value of junction bias voltage should also be similarly used.

It should also be noted that bias voltage is often supplied through a monolithically manufactured semiconductor resistor. These resistors can have a relatively large positive temperature coefficient, so that the actual bias on the junction will decrease as a function of temperature for the configuration. The combination of this effect and the temperature coefficients of β and α could produce a worst-case temperature less than the maximum operating temperature, and possibly lead to latchup evidence as a function of temperature. Both β and α usually increase with operating current, so that these currents should be set at their highest anticipated value. Usually, however, bias currents are not independent variables, but are frequently set by worst-case bias voltages, which also result in worst-case bias current.

Finally, the possibility of the multiplicity of latchup windows, as a function of dose rate, will require testing at a number of discrete dose rates up to the expected maximum dose rate. Since the latchup window(s) may be narrow, a large number of dose rate levels may be required. The larger the number of dose rate tests, the smaller the possibility of missing a latchup window. However, using a large number of dose rate probes increases the cumulative ionizing dose and the tendency to damage the device under test. In fact, these tests are usually limited to a few cumulative ionizing dose levels, or the test will be considered destructive.

These difficulties are major, and again latchup screening should not be used for building systems requiring low dose rate failure probabilities with high confidence. Instead, most systems should utilize: (a) devices hardened to eliminate latchup such as CMOS/SOS, dielectrically isolated integrated circuits (DIICs) such as CMOS/SOI, and dielectrically isolated TTL devices; or (b) system level procedures such as power-down sequencing that are used to eliminate latchup during system operation in real time.

7.8 UPSET

The term *upset* has the general meaning of exceeding a tolerance level. As a verb, *to upset* means that a particular agent, such as a transient radiation pulse, has caused a circuit or fragment thereof, to exceed a particular threshold beyond which it no longer functions as intended. For example, if photocurrents cause a memory register to change its state in any fashion not consonant with its normal function, it is said that the photocurrent has upset the memory register. Notice that upset usually pertains to circuit operation, but latchup, discussed in the previous section, normally refers to operation of the particular device in the circuit.

Another example of upset is that of the noise margin upset in an inverter (NOR gate), as part of a digital system that is designed to saturate "on" with an input of at least 5 V and cut "off" with an input of 1 V or less. The corresponding circuit diagram is given in Fig. 7.22. In this circuit, 5 V is the upper acceptance level. This means that if the input

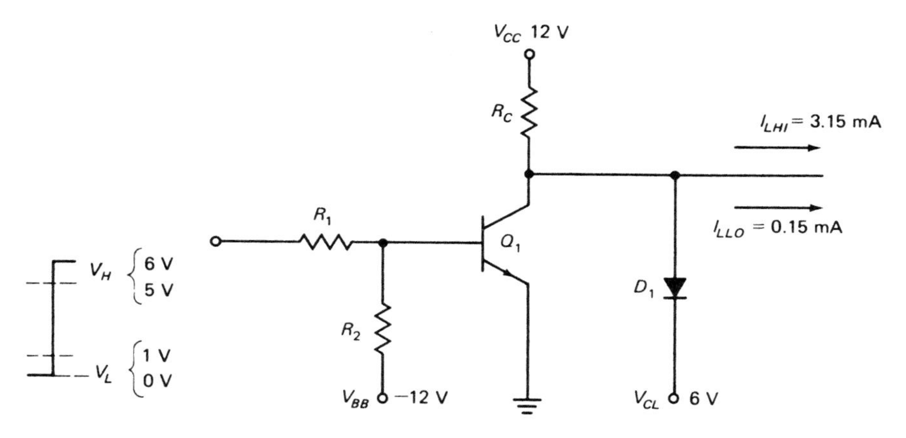

Fig. 7.22 Inverter with clamped collector and source type loads.

Table 7.2 Radiation upset levels of various logic families of integrated circuits.[26]

LOGIC FAMILY	THRESHOLD RANGE (RADS (Si)/s)
Emitter coupled logic (ECL)	4×10^7 to 2×10^8
Integrated Injection Logic (I^2L)	1×10^8 to 1×10^9
Collector Diffusion Isolation (CDI)	3×10^7 to 5×10^7
Transistor transistor logic (TTL)	1×10^8 to 3×10^8
Low power (TTL)	5×10^6 to 4×10^7
Dielectrically isolated DTL	1×10^8 to 8×10^8
Dielectrically isolated TTL	6×10^8 to 5×10^9
Schottky clamped TTL	10^8 to 10^9
CMOS	10^8 to 10^{10}
CMOS/SOS	10^9 to 5×10^{11}
CMOS/SOI	10^9 to 10^{12}

voltage is between 5 and 6 V, the inverter will recognize this input as "high." Similarly, 1 V or less is the lower acceptance level, in that any input between zero and 1 V is interpreted as "low." Hence, the high or 1 level noise margin is 1 V, as is the low or zero level noise margin. For the case of zero input, a photocurrent pulse of such level as to cause a voltage spike sufficient to turn on the inverter (5 V) is said to have upset the device. Actually, conservative upset specifications will assume the inverter will turn on at much less than 5 V, and so a maximum noise upset

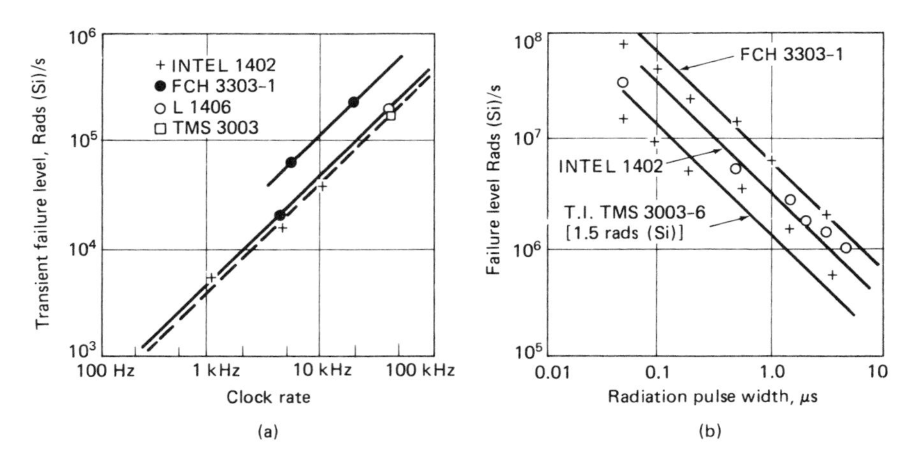

Fig. 7.23 Dynamic MOS shift register dose rate upset failure levels.[27]

specification may be 2 V for this device. Another way of stating the same idea is that the radiation upset threshold of, e.g., a gate, corresponds to that dose rate for which the output exceeds the noise margin, which is 1 V in the above example.

Because the idea of upset is somewhat variegated, no substantive general model of upset exists. Special descriptions have been constructed for special types of upset. Hence, most of the upset data are obtained experimentally. As an example, Table 7.2 depicts transient ionizing radiation upset levels for various digital logic families of integrated circuits. It is seen from the table that the different logic families exhibit transient radiation upset over about four orders of magnitude between about 10^7 to 10^{12} rads (Si) per second. It has been found that the radiation threshold of junction-isolated integrated circuits increases with their propagation delay time. That is, if the device changes its state more slowly, as reflected in a longer propagation time, it can tolerate higher levels of ionizing radiation.

In CMOS systems, the production of substrate currents due to large transient bursts of ionizing radiation can lead to data upset difficulties. Data upset is of major concern in storage devices and systems, such as shift registers and memories.

The sites where data are stored are usually isolated from their associated low-impedance circuitry, so that high-level transient radiation can produce a state reversal of stored binary information. Even though it is usually non-destructive, as is single-event upset discussed in Chapter 8, data upset presents design difficulties for a radiation environment. An example is a static RAM (random access memory), where information is stored in flip-flop circuitry.

For CMOS devices, the predominant dose rate upset mechanism can be rail span collapse.[44,45] Rail span collapse is defined as the temporary reduction in the V_{DD} (drain) − V_{SS} (source) voltage difference across (for instance) a SRAM cell, due to a chip-wide photocurrent-inducing incident pulse of gamma or x-ray radiation. The resulting V_{DD} drop and V_{SS} rise are due to these cell photocurrents all flowing away from the SRAM cells and into the V_{DD} and V_{SS} power supply metallization runs on the chip itself. Since the metallizations possess finite resistance, the photocurrent metallization voltage drops produce, in turn, the SRAM cell voltage drops resulting in impaired function of these devices.

In dynamic memory storage, information is stored as the charged state of a MOS node capacitance. The dynamic memory unit cell requires much less power than the static memory, but the information must be refreshed in the former to compensate for the node capacitance charge leakage. Also a clock is required to step the information through the system. For an increase of pulsed, radiation-induced leakage current, the clock frequency must be increased to refresh the memory more frequently. The required clock frequency increases linearly with the failure level dose rate for MOS shift registers, as seen in Fig. 7.23(a).[27] For dynamic shift registers, the discharge time constant at an MOS storage node must be long compared with the clock ionizing pulse, the upset level is determined by its deposited dose level, for clock periods long compared with the photocurrent pulse width. This can be seen in Fig. 7.23(b), where the critical upset dose rate is plotted as a function of pulse width. Because the hyperbola xy = constant, plots as a straight line with a slope of −1 in "log-log" coordinates, the curves in Fig. 7.23(b) are seen to be hyperbolas of constant dose rate-pulse width product. That is, they are loci of constant ionizing dose.[27] Figure 7.24 depicts photocurrent response for bipolar power transistors, power diodes, and their small signal counterparts.

7.9 TRANSIENT EFFECTS IN CAPACITORS

When capacitors, especially high-capacitance electrolytics, are subject to ionizing radiation, they experience a partial loss of charge. That is, they discharge an amount that depends on the level of ionizing dose that the capacitor has absorbed. The ionizing radiation produces electrons in the dielectric, as well as in the electrodes, that are trapped by energy levels in the forbidden gap of the dielectric material of the capacitor. These

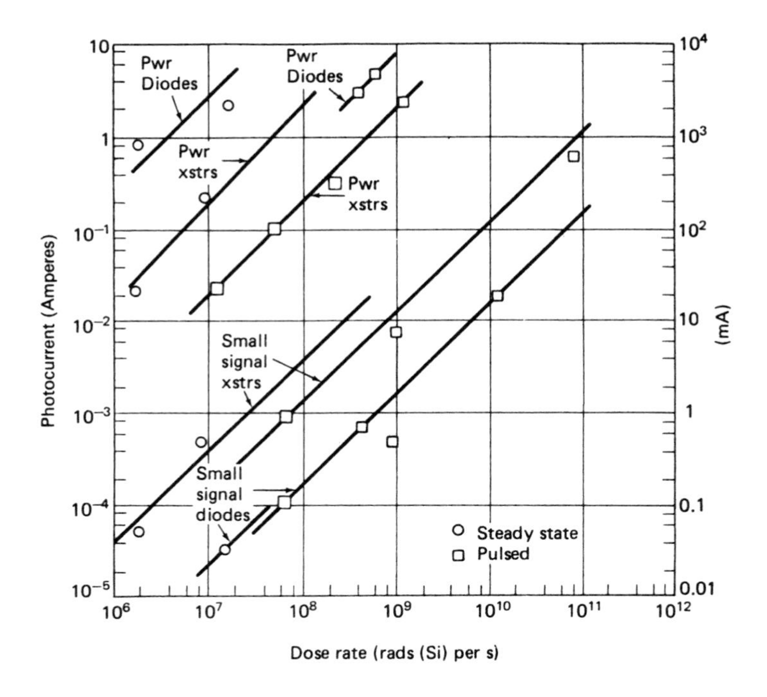

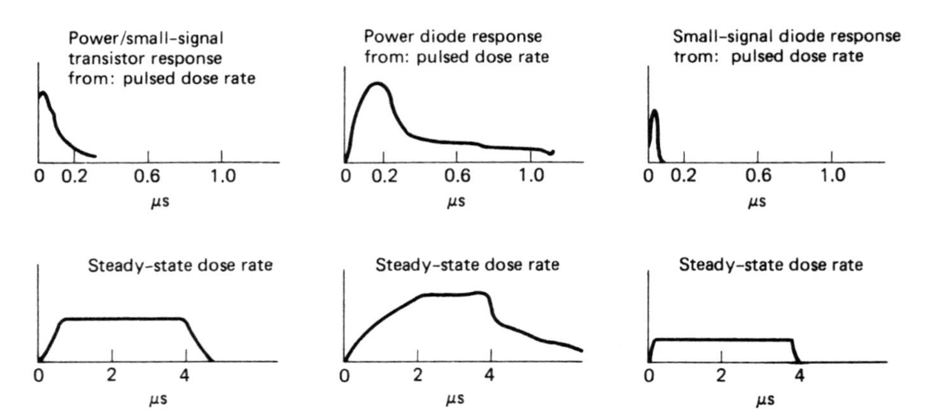

Fig. 7.24 Typical photocurrents in bipolar power and small-signal transistors and power and small-signal diodes, as a function of incident ionizing dose rate.[28]

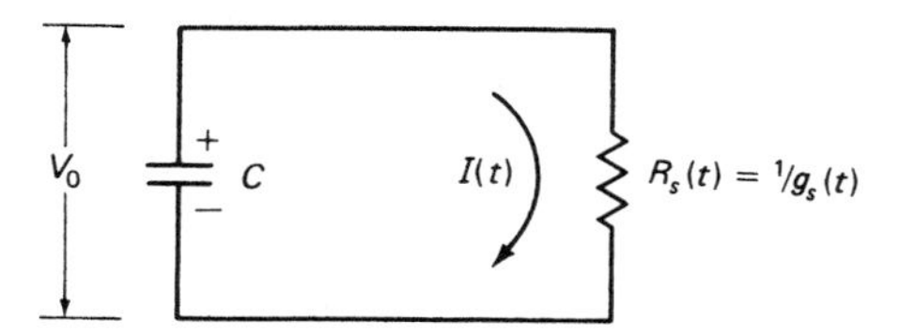

Fig. 7.25 Equivalent circuit for a capacitor subjected to ionizing radiation.

trapped electrons produce an internal electric field within the dielectric, causing conduction currents to flow, as discussed in Section 6.7, which partially discharge the capacitor. If the capacitor is subjected to a pulse of ionizing radiation, the result is reflected in a time-varying change in its shunt conductance.[29] An equivalent circuit for such a capacitor is given in Fig. 7.25. The trapped electrons are subsequently released, mainly through thermal agitation, usually after a time span long with respect to the ionizing radiation pulse. This accounts for the persistence of the increased leakage in the dielectric long after the radiation pulse has subsided.

From Fig. 7.25, the corresponding voltage loop equation is

$$(1/C)\int_0^t I(t')\,dt' + I(t)R_s(t) = V_0 \tag{7.65}$$

where V_0 is the applied voltage across the capacitor prior to the onset of the radiation pulse. Differentiating Eq. (7.65) with respect to time yields

$$\dot{I} + I(\dot{R}_s/R_s + 1/R_sC) = 0; \qquad I(0) = V_0/R_s(0) \tag{7.66}$$

with an integral given by

$$V(t) = IR_s = V_0 \exp - \int_0^t dt'/CR_s(t') \tag{7.67}$$

After the radiation pulse epoch has transpired, the capacitor has been discharged to a level reflected in the following expression, obtained by letting time grow without bound in Eq. (7.67)

$$V_\infty/V_0 = \exp - \int_0^\infty dt'/CR_s(t') = \exp - \int_0^\infty [(\sigma - \sigma_0)/(\varepsilon\varepsilon_0)]\,dt' \tag{7.68}$$

since

$$1/CR_s = g_s/C = (\sigma(t) - \sigma_0)/\varepsilon\varepsilon_0 \tag{7.69}$$

where $\sigma - \sigma_0$ is the conductivity change in the capacitor, ε is its dielectric constant, and g_s is the corresponding conductance. The integral in Eq. (7.68) can be split into two parts. One part is that for times during the pulse epoch, whose duration is T_0, and the second is that following the pulse, For the sake of computational convenience, the pulse is assumed to be rectangular, and constant amplitude of $\dot{\gamma} = \gamma(T_0)/T_0$. From Eq. (7.68) then

$$V_\infty/V_0 = \exp - \left(\int_0^{T_0} [(\sigma - \sigma_0)/\varepsilon\varepsilon_0]\, dt' + \int_{T_0}^{\infty} [(\sigma - \sigma_0)/\varepsilon\varepsilon_0]\, dt' \right) \tag{7.70}$$

A number of studies and experimental results have yielded an empirical relation for each of the above integrands, corresponding to times during and after the pulse.[30] They are: (a) During the pulse, (ε_0 is the permittivity of free space)

$$(\sigma - \sigma_0)/\varepsilon\varepsilon_0 = \dot{\gamma} K_p + \dot{\gamma} \sum_{i=1}^{2} K_{di}\tau_{di}(1 - \exp - t/\tau_{di}) \tag{7.71}$$

where the constants K_p, K_{di}, and τ_{di} are empirically derived, and $\dot{\gamma}$ is the corresponding ionizing pulse dose rate amplitude. Then from integrating Eq. (7.71), and with the above rectangular pulse, one obtains

$$\int_0^{T_0} \frac{\sigma - \sigma_0}{\varepsilon\varepsilon_0}\, dt = \gamma(T_0) \left\{ K_p + \sum_{i=1}^{2} K_{di}\tau_{di}[1 - (\tau_i/T_0)(1 - e^{-T_0/\tau_{di}})] \right\} \tag{7.72}$$

where $\gamma(T_0) = \dot{\gamma} T_0$ is the total dose delivered by the pulse during its epoch. (b) After the pulse, the analogous empiricism is

$$(\sigma - \sigma_0)/\varepsilon\varepsilon_0 = \dot{\gamma}(t)(\exp(T_0/\tau_{di}) - 1) \cdot \sum_{i=1}^{2} \mathrm{K}_{di}\tau_{di} \exp - t/\tau_{di} \tag{7.73}$$

Similarly, from integrating Eq. (7.73)

$$\int_{T_0}^{\infty} [(\sigma - \sigma_0)/\varepsilon\varepsilon_0]\, dt = (\gamma(T_0)/T_0) \sum_{i=1}^{2} K_{di}\tau_{di}^2[1 - \exp - (T_0/\tau_{di})] \tag{7.74}$$

Note that both integrals, Eqs. (7.72) and (7.74), are proportional to the ionizing dose $\gamma(T_0)$. Then Eq. (7.70) can be rewritten as

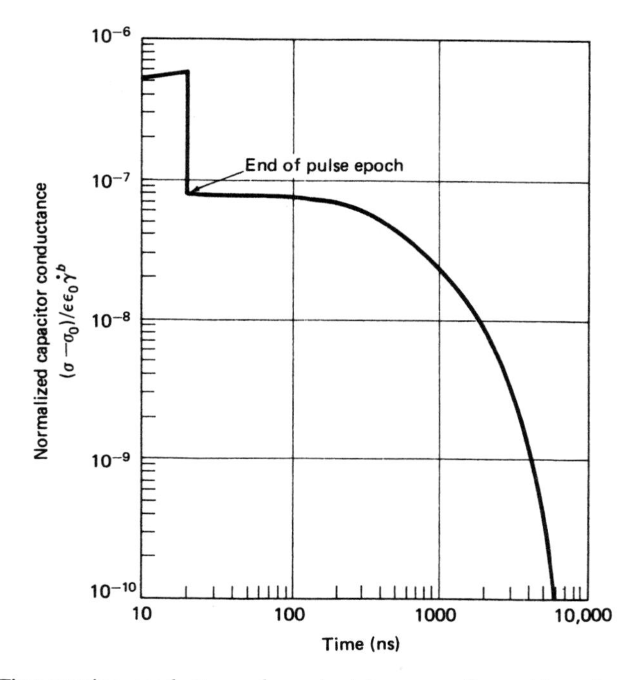

Fig. 7.26 Time-varying conductance for a tantalum capacitor subjected to an ionizing radiation pulse of 20 ns.[31]

$$V_\infty/V_0 = \exp - (\gamma(T_0)/\gamma_c) \tag{7.75}$$

with γ_c as the constant of proportionality. Values of γ_c are tabulated for various types of capacitor dielectrics.[32] A refined expression for $(\sigma - \sigma_0)/\varepsilon\varepsilon_0$, viz,[33]

$$\frac{\sigma - \sigma_0}{\varepsilon\varepsilon_0} = K_p[\dot{\gamma}(t)]^b + \sum_{i=1}^{2} K_{di} \int_{-\infty}^{t} [\gamma'(u)]^{b_i} e^{-(t-u)/\tau_{di}} du \tag{7.76}$$

where the exponents b and b_i are also empirically derived. Equation (7.76) can be numerically integrated for an arbitrary dose rate-time variation and used accordingly in the foregoing. Results are depicted in Fig. 7.26 for a tantalum oxide dielectric capacitor, exposed to a 20-ns rectangular ionizing pulse.[31] Figures 7.27 and 7.28 show the shunt conductance change with time for a 6.8 μF tantalum oxide dielectric capacitor absorbing an ionizing dose of 20 krads (Si).[2,32]

For a rectangular pulse of amplitude $\dot{\gamma}^b$, and width T_0 small compared to

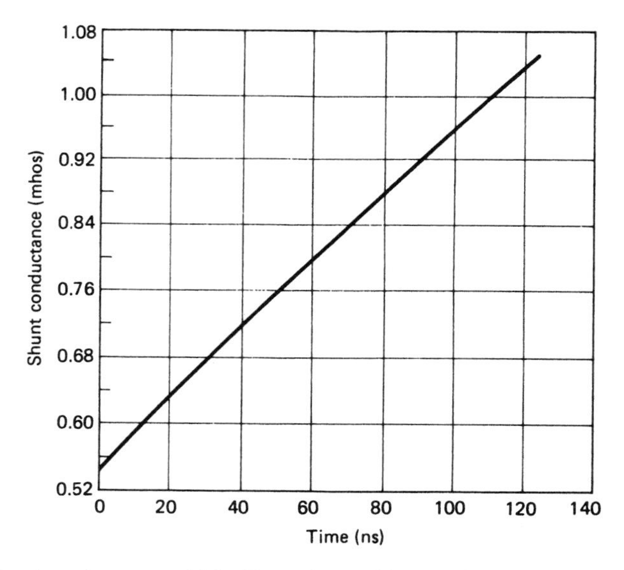

Fig. 7.27 Shunt conductance of 6.8-μF tantalum oxide capacitor during a prompt ionizing radiation pulse that deposited 20 krads (Si) into its dielectric.[30]

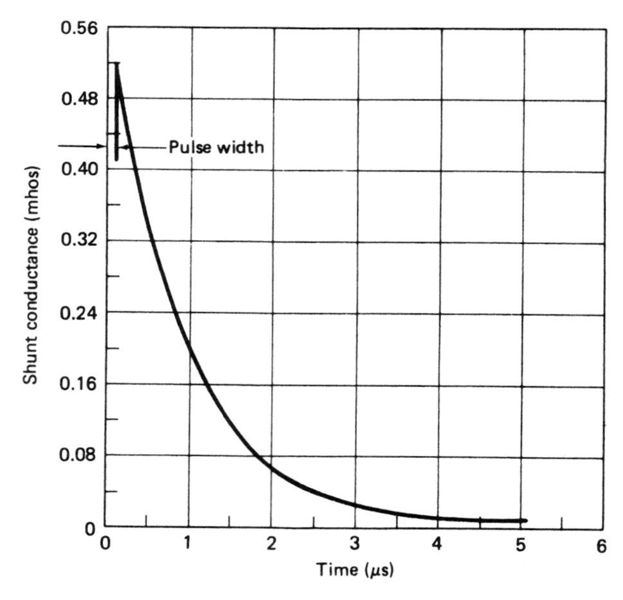

Fig. 7.28 Shunt conductance of 6.8-μF tantalum oxide capacitor after the ionizing radiation pulse.[30]

circuit time constants, i.e., ωT_0, $\omega \tau_{di} \ll 1$ for corresponding frequencies ω, and $b_i \cong b$ for all i, the asymptotic form of Eq. (7.76) ($t \gg T_0$) is given by

$$\int_0^\infty dt/R_s C \cong \ln(V_0/V_\infty) = (K_p + \sum_{i=1}^{2} K_{di}\tau_{di}) \cdot \dot{\gamma}^b T_0 \qquad (7.76\text{A})$$

This provides an expression for capacitor voltage deterioration versus dose rate for various capacitor types using K_p, K_{di}, τ_{di}, or b_i tables.[32] The long T_0 pulse case is approximated by assuming a constant $\dot{\gamma}^b$ in Eq. (7.76) to yield, asymptotically,

$$1/R_s(\infty)C \cong (K_p + \sum_{i=1}^{2} K_{di}\tau_{di}) \cdot \dot{\gamma}^b \qquad (7.76\text{B})$$

The shunt (leakage) resistance $R_s(\infty)$ can then be found for various capacitor values and types for use comparison. Usually, the largest R_s capacitor is the one chosen.

7.10 TRANSIENT EFFECTS IN RESISTORS

Transient radiation effects in resistors are usually small relative to those encountered in semiconductors and capacitors, the latter discussed in the previous section. The transient effects are generally caused by the production of photocurrent due to gamma and x-ray dose rate ionization, primarily from compton processes in the resistor material. This causes conductivity increases which act to form internal leakage paths to shunt the resistor thus lowering its resistance value. This is in contrast to neutron damage discussed in Section 5.13 which results in increased resistance at low to moderate injection levels. It is straight forward to show from Sections 1.8, 1.13, and 7.2 that the conductivity change, $\Delta\sigma$, in the resistive semiconductor material can be written in terms of dose rate as

$$\Delta\sigma = eg(\mu_n + \mu_p)\tau\dot{\gamma} \qquad (7.77)$$

where g_0 is the specific carrier generation rate, (hep per $\text{cm}^3 \cdot \text{rad}$) and τ is the effective carrier lifetime.[34] Equation (7.77) implies that the corresponding conductance G_s is proportional to $\dot{\gamma}$ for the case of low photocurrent injection levels. Considering that G_s^{-1} shunts the given resistor whose value is R_0, the effective resistance R_{eff} is given by

$$R_{eff} = R_0/(1 + R_0 G_s) \tag{7.78}$$

It is seen that the shunting effect is large for large R_0 and vice versa.

For film resistors, measured shunt conductances are obtained empirically as[32]

$$G_s = K_s(\dot{\gamma})^b \qquad \text{(mhos)} \tag{7.79}$$

where the constants K_s and b are determined from measurements of a specific resistor type. K_s is found to vary from 10^{-8} to 10^{-18}, while $0.25 \leqslant b \leqslant 1.0$. For example the shunt conductance expression for a particular solid core film resistor[32] is $G_s = 7.7 \cdot 10^{-13}(\dot{\gamma})^{2/3}$. From Eq. (7.79) in (7.78) for R_0, a one megohm resistor, and $\dot{\gamma} = 10^8$ rad (Si) per sec, $R_{eff} =$ 0.85 megohms.

Diffused resistors are manufactured using the same processing methods as used in modern ICs and are built integrally with them. It was felt that unusually high photocurrents from very high dose rate pulses would severely damage these resistors.[34] This influenced the early heavy use of thin film resistors instead of diffused resistors for use with radiation hard ICs.[34] The damage mechanism was thought to be not only due to high photocurrents directly, but also to conductivity modulation which could cause the diffused resistor to suffer a large drop in its resistance value at these high photocurrent levels, as seen by the conductivity increase given in Eq. (7.77). For such resistors in series with supply lines, the excessive current thereby could cause device burnout. However, the effect of conductivity modulation can be inhibited if the diffused resistors are dielectrically isolated by placing them in dielectric tubs, to minimize their interaction with the surrounding material. The main inhibitory phenomenon is band-to-band recombination under ambipolar conditions at very high current levels.[35] Most diffused resistors are manufactured from relatively low resistivity material, and the high levels of photocurrent required to produce significant conductivity modulation corresponds to ambipolar carrier transport, the latter being discussed in Section 1.10.

To investigate the preceding in detail, Fig. 7.29 depicts two kinds of diffused resistors of the type that are monolithically fabricated within the integrated circuit. In both cases, the resistor per se consists of a p-material diffusion, which rests on an nn^+ substrate. The p-nn^+ junction thus formed provides the reverse-bias junction isolation for the resistor, as explained. That is, this junction isolation maintains circuit currents flowing laterally through the resistor, and not diverted into the substrate and beyond.

The upper part of Fig. 7.29 illustrates the plugged type of diffused resistor. This means that the high potential end of the resistor is con-

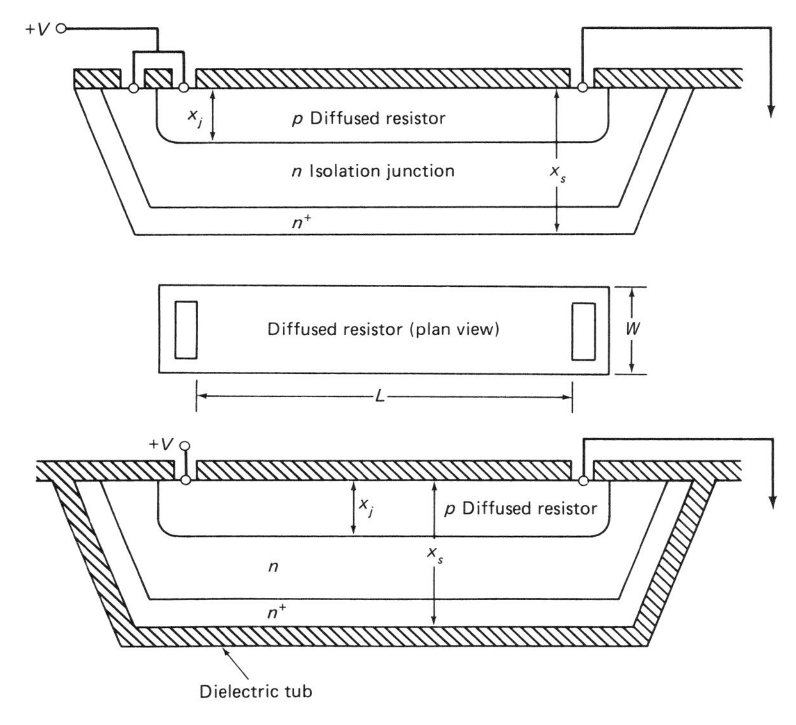

Fig. 7.29 Junction-isolated plugged diffused integrated circuit resistor (upper). Dielectrically isolated unplugged diffused integrated circuit resistor (lower).[36] Figure © 1974 by the IEEE.

nected to the nn^+ substrate, thus shorting the junction at that end. Except at the shorted end of the resistor, junction isolation is provided for it, and the p-nn^+ junction is reverse-biased by virtue of the fact that the nn^+ layer is an equipotential body at $+V$ throughout its length, since no current flows therein, with no ionizing radiation. Hence, the junction is reverse-biased maximally at the low potential end, with the reverse bias along it being given by $+V$, less the corresponding IR drop in the resistor, decreasing toward the plugged end. This is enhanced somewhat by the built-in contact potential across the junction.

The junction reverse bias can be increased further by tapping into suitable potential differences of the proper sign within the integrated circuit.

For low levels of ionizing dose rate, the junction remains reverse biased, as the correspondingly small photocurrent generated within the resistor and nn^+ layer bypasses the junction, mainly by flowing through its shorted end. These systems are sufficiently shallow, in that x_j and x_s are much less than the corresponding diffusion lengths, so that photocurrent can be considered as generated homogeneously throughout their bulk.

The unplugged resistor has no "shorting plug," so that the nn^+ isolation region is left floating (i.e., with no voltage reference). In this case, the p-nn^+ junction is immediately forward biased upon the onset of photocurrent that flows from the p to the nn^+ layer. Actually, small leakage currents would induce a small forward bias with no photocurrent.[36] For both plugged and unplugged resistors, the forward bias occurs mainly near the high-potential end.

In the plugged case, the incident photocurrent skews the potential distribution in both p and nn^+ regions, such that the junction becomes partially forward biased when a threshold incident ionizing dose rate $\dot{\gamma}_c$ is exceeded. When the p-nn^+ junction remains reverse biased, with photocurrent generation sufficiently high to cause conductivity modulation, the resistance $R(\dot{\gamma})$ of the plugged resistor system is[36]

$$R(\dot{\gamma}) = (\rho_{sp}L/W)/(1 + eg_0 x_s \dot{\gamma} \rho_{sp} L^2/2V) \tag{7.80}$$

where ρ_{sp} and ρ_{sn} are the sheet resistances (Section 3.5) of the p-material resistor and nn^+ substrate, respectively. The dimensional and other parameters are illustrated in Fig. 7.29. $R_0 = \rho_{sp}L/W$ is the resistance of the p resistor, as distinct from its attachment to the nn^+ layer.

The threshold dose rate $\dot{\gamma}_c$, at which the junction begins to become forward biased, is[36]

$$\dot{\gamma}_c = 2V/(eg_0 x_s L^2(\rho_{sp} + 2\rho_{sn})) \tag{7.81}$$

The corresponding resistance $R(\dot{\gamma}_c) \equiv R_c$ at the threshold dose rate $\dot{\gamma}_c$, for $\rho_{sn} \ll \rho_{sp}$ (which is the case for the highly doped nn^+ layer), is given by[36]

$$R_c \cong 0.5\rho_{sp}L/W = 0.5R_0 \tag{7.82}$$

This expression can be used as an indication of the radiation hardness of the diffused resistor to ionizing radiation, simply by observing the $\dot{\gamma}_c$ level at which the measured $R_c = 0.5R_0$ is obtained. General results analogous to the foregoing are also obtained for the unplugged resistor.[36]

To investigate important detailed aspects of the preceding, especially at sufficiently high photocurrent-induced injection levels, which imply conductivity modulation, it is evident from the earlier discussions (Section 1.10) that both n and p excess carriers obey the kinetics equation

$$\partial n/\partial t = g_0\dot{\gamma} - U_n + (\nabla \cdot J_n)/e \tag{7.83}$$

where an identical equation holds for the p carriers. $g_0\dot{\gamma}$ is the photocurrent generation rate source, U_n is the carrier current loss due to recombination, and $(\nabla \cdot J_n)/e$ is the carrier current divergence loss due merely to carriers leaving the incremental volume of interest. To develop the recombination term U_n, at equilibrium the thermal generation of hole-electron pairs balances the recombination loss processes, as in

$$Rnp = e_m \tag{7.84}$$

The recombination rate is Rnp, where R is the recombination rate per interaction, as discussed in Section 1.11, and e_m is the thermal hole-electron pair emission rate. For intrinsic material $e_m = Rn_i^2$, and for silicon[37] $e_m = 2 \cdot 10^9\,\text{cm}^{-3}\,\text{sec}^{-1}$. For the nonequilibrium situation at hand, the band-to-band recombination rate, U_{bb}, is

$$U_{bb} = R((n_0 + \delta_n)(p_0 + \delta_p) - n_i^2) \tag{7.85}$$

where $\delta n = n - n_0$, $\delta p = p - p_0$ and n_0 and p_0 are the respective equilibrium values. In the high-injection ambipolar regime, $\delta n \gg n_0$, $\delta p \gg p_0$, so that Eq. (7.85) with Eq. (7.84) yields, for such large $\delta n \simeq \delta p = \delta$,

$$U_{bb} = R\delta^2 = e_m\delta^2/n_i^2 \tag{7.86}$$

Also, the important recombination rate U_R, due to the trapping and recombination states in the energy gap, is given by

$$U_R = \delta/\tau \tag{7.87}$$

where τ is the corresponding minority carrier lifetime. Hence, the total recombination rate U is

$$U = U_R + U_{bb} = \delta/\tau + e_m\delta^2/n_i^2 \tag{7.88}$$

So that, for n-type carriers,

$$U_n = (n - n_0)/\tau_n + e_m(n - n_0)^2/n_i^2 \tag{7.89}$$

and similarly, a total recombination rate for p-type carriers can be written.

For the high injection case under consideration, the recombination loss processes dwarf the losses due to the divergence of the carriers, so that the divergence term in Eq. (7.83) is neglected by comparison. Equation (7.83) is then rewritten[38] using Eq. (7.89), where $\tau_m = n_i^2/e_m = 1.1 \cdot 10^{11}\,\text{cm}^{-3}\,\text{sec}^{-1}$, as

$$\dot{\delta}(t) = g_0\dot{\gamma} - \delta/\tau_n - \delta^2/\tau_m \tag{7.90}$$

with pre- and post-pulse integrals given by, where t_p is the dose rate pulse width,

$$\delta(t) \cong \begin{cases} (2g_0\dot{\gamma}\tau_n)/[\sqrt{(4g_0\dot{\gamma}\tau_n^2/\tau_m) + 1} \cdot \\ \quad \coth\sqrt{(4g_0\dot{\gamma}\tau_n^2/\tau_m) + 1} \cdot (t/2\tau_n) + 1];\, 0 \leq t \leq t_p,\, \delta(0) = 0 & (7.91) \\ [\delta_p \exp - (t - t_p)/\tau_n]/[1 + (\tau_n\delta_p/\tau_m)(1 - \exp - (t - t_p)/\tau_n)]; \\ \quad t > t_p,\, \delta_p \equiv \delta_{\text{pre}}(t_p) & (7.92) \end{cases}$$

For a dose rate pulse $\dot{\gamma}_H$ of magnitude such that high-injection band-to-band recombination rates are greater than those corresponding to the removal of carriers merely through the divergence of the current, the carrier concentration will rapidly rise to its equilibrium value well prior to the end of the pulse duration. Then, only the asymptotic value of $\delta_{\text{pre}}(t)$ need be considered. Therefore, for the high injection ambipolar case, Eq. (7.91) yields (where $4g_0\dot{\gamma}\tau_n^2/\tau_m \gg 1$)

$$\delta_{\text{pre}} \sim (g_0\tau_m\dot{\gamma}_H)^{1/2} = 2.17 \cdot 10^{12}(\dot{\gamma}_H)^{1/2} \tag{7.93}$$

To investigate how the resistivity of the diffused resistor system varies with dose $\tau\dot{\gamma}$, Eq. (7.77) asserts that

$$\Delta\sigma \equiv 1/\rho - 1/\rho_0 = eg_0(\mu_n + \mu_p)\,\tau\dot{\gamma} \tag{7.94}$$

or[38]

$$\rho/\rho_0 = (1 + eg_0(\mu_n + \mu_p)\,\rho_0\tau\dot{\gamma})^{-1} \tag{7.95}$$

τ is the Shockley-Read-Hall effective lifetime[39] given by $\tau = ((n_0 + p_0)\,\tau_0 + 2\tau_\infty\delta)/((n_0 + p_0) + 2\delta)$, where it is seen that $\tau = \tau_0$ for low injection (i.e., small δ) and $\tau = \tau_\infty$ for high injection (i.e., large δ). For p-type carriers, with $\rho_{0p} = (ep_0\mu_p)^{-1}$, as discussed in Section 1.8, and $b = \mu_n/\mu_p$, Eq. (7.95) yields

$$\rho_p/\rho_{0p} = (1 + g_0(1 + b)\,\tau\dot{\gamma}/p_0)^{-1} \tag{7.96}$$

Table 7.3 Dose rate that reduces ρ to one-half in the ambipolar regime.

ρ (OHM CM)	$\dot{\gamma}_c$ (RADS (Si) PER SEC)
0.1	$3 \cdot 10^8$
0.03	$1 \cdot 10^{10}$
0.004	$3 \cdot 10^{12}$

Likewise for n-type carriers, with $\rho_{0n} = (en_0\mu_n)^{-1}$,

$$\rho_n/\rho_{0n} = (1 + g_0(1 + b^{-1})\,\tau\dot{\gamma}/n_o)^{-1} \tag{7.97}$$

for high injection $\dot{\gamma} = \dot{\gamma}_H$, and $\delta_{HI} = (g_0\tau_m\dot{\gamma}_H)^{1/2}$ from Eq. (7.93), for both types of carriers. Inserting the incremental carrier increase due to the incident dose rate (viz., $\delta_{HI} = g_0\dot{\gamma}_H\tau$) into Eq. (7.96) yields

$$(\rho_p/\rho_{0p})_{HI} = (1 + (1 + b)(g_0\tau_m\dot{\gamma}_H)^{1/2}/p_0)^{-1} \qquad ((7.98)$$

Similarly, that for the n-type carrier from Eq. (7.97) gives

$$(\rho_n/\rho_{0n})_{HI} = (1 + (1 + b^{-1})(g_0\tau_m\dot{\gamma}_H)^{1/2}/n_0)^{-1} \tag{7.99}$$

Realizing that the ratio of high-injection excess carrier concentration to the initial carrier concentration δ_{HI}/n_0, δ_{HI}/p_0 is very large under the prevailing ambipolar diffusion conditions, then for both carrier types in Eqs. (7.98) and (7.99) either one becomes, after neglecting unity in their denominators, where $\sigma_i = en_i(\mu_n + \mu_p)$,

$$\rho = \rho_0/F(\dot{\gamma}_H)^{1/2};\; F = \sigma_i(g_0/e_m)^{1/2} = (g_0 n_i^2/e_m)^{1/2} \cdot e(\mu_n + \mu_p)$$
$$= 3.47 \cdot 10^{-7}\,(\mu_n + \mu_p) \tag{7.100}$$

As mentioned earlier, most diffused resistors are manufactured using low resistivity material. Table 7.3, tabulated from computations using Eq. (7.100),[38] illustrates that, for $\rho = 0.004$ ohm cm, it takes $\dot{\gamma}_c = 3.10^{12}$ rads (Si)/sec to reduce its resistivity by 50 percent. Then, as implied by Table 7.3, band-to-band recombination under ambipolar conditions allows such diffused resistors to withstand very high levels of dose rate under operating bias, including conductivity modulation. Experimental data[36] shows that both plugged and unplugged resistors respond similarly. Hence, manufacturers of integrated circuits that include within them monolithically diffused resistors, need have no significant concern regarding their radiation survivability with respect to ionizing radiation. So, to reiterate, for a sufficiently high dose rate, band-to-band recombination removes carriers at a much higher rate than those removed by the physical divergence of the current. As evident from the carrier transport equations,[35] this results in a very rapid rise of the carrier concentration to an equilibrium level, characterized by a balance between dose rate induced carrier generation, and carrier removal by band-to-band recombination. This is followed by a slower decay after the termination of the $\dot{\gamma}$ pulse. This ultimately results in a value of $b = \frac{1}{2}$ in a conductance expression analogous to that in equation (7.79).[35] The preceding reasserts that the combination of low resistivity material of the diffused resistor and its structural geometry as discussed, limit the effect of conductivity modulation, and so limit the high dose rate induced large currents to very much less than corresponding device burnout levels.[40]

PROBLEMS

1. (a) For silicon, show that the bulk photocurrent generation rate has the value 6.5 μamp per cm^3-rad/s.
2. Show that the diffusion component of photocurrent, i_d, for a diode with no bias applied, vanishes.
3. Suppose that the photocurrent producing radiation pulse is so short in duration that it can be represented by an impulse (delta) function of amplitude G, so that $G = g\delta(t)$. Write the corresponding photocurrent pulse from Eq. (7.9).
4. Explain the difference between junction-isolated and dielectrically isolated IC circuitry insofar as their effects on latchup are concerned.
5. How does "sustaining voltage breakdown" result in latchup?
6. In many instances, the collectors of an IC are gold-doped to degrade (decrease) the minority carrier lifetime therein to minimize four-layer latchup. Trace the connection between minority carrier lifetime and Eq. (7.64) to show how gold-doping tends to inhibit latchup. Gold-doping is also used in CMOS ICs to limit parasitic bipolar latchup by reducing the minority carrier lifetime in the substrate.[18]
7. The well-known relation between electrical conductivity and pertinent semiconductor parameters is given by $\sigma = e(n_0\mu_n + p_0\mu_p)$, where n_0 and p_0 are the equilibrium values of the respective carrier densities. Show that the increase in conductivity $\Delta\sigma$ from the incident dose rate is $\Delta\sigma = eg_0(\mu_n + \mu_p)\dot{\gamma}\tau$, where g_0 is the specific carrier generation rate, τ is the effective carrier lifetime, and $\dot{\gamma}$ is the corresponding dose rate.
8. Show that the resistance of a plugged diffused resistor, irradiated by ionizing radiation of rate $\dot{\gamma}$ is $R(\dot{\gamma}) = R_0/(1 + \dot{\gamma}/\dot{\gamma}_c)$. $R_0 = \rho_{sp}L/W$, $\dot{\gamma}_c$ is the threshold

dose rate at which the p-type plugged resistor becomes forward biased with respect to its nn^+ isolation junction, and $\rho_{sn} \ll \rho_{sp}$.

9. In a diffused resistor, show, for the low-injection case, that the asymptotic form of photocurrent-induced excess carriers is $\delta_{\text{pre}} \sim g_0 \tau_n \dot{\gamma}_L$, where $\dot{\gamma}_L$ is the corresponding low injection dose rate.

10. A nuclear reactor solid-state dose rate detector is positioned at a distance from the reactor core such that the incident reactor-produced gamma ray flux corresponds to a dose rate in the detector of 1 krad (Si) per minute. The detector is a hybrid device containing special-purpose silicon diodes with the following single diode equivalent specifications: $A_J = 10^3\,(\mu\text{m})^2$, $W_t = 5\,\mu\text{m}$, and L_p, $L_n \ll W_t$. What is the corresponding detector steady-state photocurrent level?

11. Briefly explain the essential physical aspects of the onset of anomolous nonlinear photocurrent in a bipolar transistor.

12. When a bipolar transistor is driven into the saturation region by photocurrent, the empiricism $\tau = 0.3t_s$ is assumed to hold, where τ is the minority carrier lifetime, and t_s is the electrical storage time. Using Eq. (7.49), show that the ratio of the saturated gain, $\beta_s = I_{CS}/I_B$, to the transistor nominal gain β is $\beta_s/\beta = 0.011$.

13. An x-ray environment simulator (e.g., LINAC) whose pulse width is 50 nsec, dose-scans a silicon device under test (DUT). A latchup window is thereby observed for a total ionizing dose (found by integrating the dose rate over the pulse width) between 500 and 525 rads (Si). The device is latchup-free above and below these levels. What is the corresponding dose rate incident on the device?

14. Certain exacting workers claim that device lots subject to correctly conducted latchup screen testing (100 percent lot test, as opposed to lot sampling tests), using U.S. Mil-STD Handbook 883-C, Methods 1020, 1021.1, are nevertheless damaged. An operational air-to-ground missile uses such devices, in its computers, that are notorious latchers. However, they are strobed in real-time during flight, and those that have latched are momentarily powered down. Discuss whether or not this missile operation proves or disproves the workers' allegations regarding latchup damage.

15. Show that the carrier generation constant, g, for Si is given by $g = 4.05 \cdot 10^{13}$ hep per $\text{cm}^3 \cdot$ rad (Si), as in Eq. (7.1). 3.6 eV are required to create 1 hep by ionization in silicon (Table 1.1).

16. An expression for the transistor-amplified photocurrent $i_c(t)$ in Section 7.3 contains $\beta(t)$, the time-dependent common emitter current gain. The modulus of its fourier transform, $|\overline{\beta}(\omega)|$ is described by the transistor gain-frequency characteristic curve, Fig. 3.26 in Section 3.10. Reference 46 provides an accurate expression for $\overline{\beta}(\omega)$, in that its inverse fourier transform $\beta(t)$ has the proper asymptotic behavior for small and large time. This $\overline{\beta}(\omega)$ is given by $\overline{\beta}(\omega) = ((\alpha_0 - i\omega c_n)/i\omega(1 - \alpha_0 + i\omega c_d)) + (c_n/i\omega c_d)$, where c_n and c_d are defined in the above reference. With $c_n/c_d \cong 1$ and $\beta_0 \gg 1$, show that the corresponding $\beta(t) = \beta_0\,(1 - \exp - (t/\beta_0 c_d))$.

REFERENCES

1. J. L. Wirth and S. C. Rogers. "The Transient Response of Transistors and Diodes to Ionizing Radiation." IEEE Trans. Nucl. Sci. NS-11(5):24–38, Nov. 1964.
2. Wirth and Rogers, op. cit. App. II.
3. J. P. Raymond. IEEE NSRE Conference Short Course, Las Vegas, 1982.
4. D. H. Habing and J. L. Wirth. "Anomolous Photocurrent Generation in Transistor Structures." IEEE Trans. Nucl. Sci. NS-13(6):86–94, Dec. 1966.
5. Habing and Wirth, op. cit.
6. E. E. Enlow and D. R. Alexander. "Photocurrent Modeling of Modern Microcircuit P-N Junction." IEEE Trans. Nucl. Sci. NS-35(6):1467–1474, Dec. 1988.
7. R. M. Warner, Jr. and B. L. Grung. Transistors: Fundamentals For the Integrated Circuit Engineer. J. Wiley Sons, New York, Chap. 8, 1983.
8. A. N. Ishaque, J. W. Howard, M. Becker, and R. C. Block. "Photocurrent Modeling at High Dose Rates." IEEE Trans. Nucl. Sci. NS-36(6):2092–2098, Dec. 1989.
9. A. T. Brown, L. W. Massengill, S. E. Diehl, and J. R. Hauser. "A Model of Transient Radiation Effects in GaAs Static RAM Cells." IEEE Trans. Nucl. Sci. NS-33(6):1519–1522, Dec. 1986.
10. E. A. Carr. "Transient Radiation Effects in Transistors." IEEE Trans. Nucl. Sci. NS-11(5):12–23, Nov. 1964.
11. B. Lax and S. F. Neustadter. "Transient Response of a PN Junction." J. Appl. Phys. 25(9):1148, Sept. 1954.
12. E. A. Carr. "Simplified Techniques For Predicting TREE Responses." IEEE Trans. Nucl. Sci. NS-12(5):30–39, fig. 7, Oct. 1965.
13. A. B. Phillips. Transistor Engineering. McGraw-Hill Book Co., New York, Chap. 16, 1962; M. G. Buehler, B. R. Blaes, and Y. S. Lin. "Radiation Dependence of Inverter Propagation Delay From Timing Sampler Measurements." IEEE Trans. Nucl. Sci. NS-36(6):1981–1989, Dec. 1989.
14. G. J. Brucker, P. R. Measel, P. Oey, K. L. Wahlin, and J. Wert. "SEU, TREE Dose Data For 1.25 μm CMOS/SOS." Hardened Electronics and Research Techn. (HEART) Conf. Monterey, CA, 1985.
15. J. L. Crowley, F. A. Junga, and T. J. Schultz. "Technique For Selection of Transient Radiation Hard Junction Isolated Integrated Circuits." IEEE Trans. Nucl. Sci. NS-23(6):1703–1708, Dec. 1976.
16. F. Larin. Radiation Effects in Semiconductor Devices. Wiley, New York, pp. 138 ff., 226, 1968.
17. J. J. Moll, M. Tannenbaum, J. M. Goldey, and N. Holonyak. "PNPN Transistor Switches." Proc. IEEE 44, Sept. 1956; *also in* F. J. Biondi (ed.). Transistor Technology, Vol. 3. Van Nostrand, Princeton, N.J., pp. 438 ff. 1958.
18. W. R. Dawes, Jr. and G. F. Derbenwick. "Prevention of CMOS Latchup by Gold Doping." IEEE Trans. Nucl. Sci. NS-23(6):2027–2030, Dec. 1979.
19. Larin, op. cit., chap. 10.
20. H. Eisen, K. Pinero, and R. Polimadei. "Nuclear Radiation Effects Data on LSI." Rpt. No. HDL-DS-80-1, Harry Diamond Laboratories, Adelphi, MD, July, 1980.
21. D. Snowden and H. Harrity. DNA Support Services, DNA 5561F, IRT Corp., P.O. Box 80817, San Diego, CA 92138, Jan. 31, 1981.
22. J. Azarewicz. IRT Corp., San Diego, CA (private communication).
23. P. V. Dressendorfer and A. Ochoa, Jr. "An Analysis of the Modes of Operation of Parasitic SCRs." IEEE Trans. Nucl. Sci. NS-28:4288–4291, Dec. 1981.
24. R. L. Pease. "Latchup in Bipolar LSI Devices." Ibid., pp. 4295–4301.

25. A. Ochoa, Jr. et al. "Snap-Back: A Stable Regenerative Breakdown Mode in MOS Devices." IEEE Trans. Nucl. Sci. NS-30:4127–4130, Dec. 1983.
26. N. J. Rudie. Principles and Techniques of Radiation Hardening, vol. 2. Western Periodicals, N. Hollywood, CA pp. 15–17, 1976.
27. J. P. Raymond, D. N. Pocock, and C. W. Perkins. LSI Vulnerability Study. Northrop Research and Technology Center, Hawthorne, CA, Report No. 2865F/NRTC 72-8R, Chap. 3, Oct. 1972.
28. G. C. Messenger. "TREES and Hardness Assurance Technology With Emphasis on Tactical Systems-Transient and Permanent Effects Produced by Dose Rate." Litton Guidance and Control Div., Woodland Hills, CA Seminar notes, March. 1978.
29. J. R. McEwan. "The Calculation of Radiation Induced Charge Loss in Capacitors." Bell Telephone Labs. Memo MEA10205094-001, Aug. 1970.
30. McEwan, op. cit.
31. Rudie, op. cit., pp. 13–14.
32. R. K. Thatcher. Transient Radiation Effects on Electronics. (TREE) Handbook, Battelle Mem. Inst., Columbus OH, Aug. 1969.
33. Thatcher, op. cit.
34. J. G. Fossum et al. "The Effects of Ionizing Radiation on Diffused Resistors." IEEE Trans. Nucl. Sci. NS-21:315–322, Dec. 1974.
35. G. C. Messenger. "Conductivity Modulation Effects in Diffused Resistors at Very High Rate Levels", IEEE Trans. Nucl. Sci. NS-26(6):4725–4729, Dec. 1979.
36. Fossum, et al. loc. cit.
37. G. Bemski. "Recombination in Semiconductors." Proc. IRE, pp. 990–1004, June 1958.
38. Messenger. "Conductivity Modulation Effects in Diffused Resistors at Very High Dose Rate Levels." loc. cit.
39. Phillips, op. cit., p. 85.
40. W. Shedd and J. Cappelli. "The Current Limiting Capability of Diffused Resistors." IEEE Trans. Nucl. Sci. NS-26(6):4720–4724, Dec. 1972.
41. G. S. Brucker, J. Wert, and F. Measel, "Transient Imprint Memory Effect in MOS Memories." IEEE Trans. Nucl. Sci. NS-33(6):1484–1486, Dec. 1986.
42. E. G. Stassinopoulos, G. J. Brucker, O. Van Gunten, and H. S. Kim. "Variation in SEU Sensitivity of Dose-Imprinted CMOS SRAMs." IEEE Trans. Nucl. Sci. NS-36(6): 2330–2338, Dec. 1989.
43. M. S. Cooper, J. P. Retzler, and G. C. Messenger. "Combined Neutron and Thermal Effects on Bipolar Transistor Gain." IEEE Trans. Nucl. Sci. NS-26(6):4758–4762, Dec. 1979.
44. L. W. Massengill and S. E. Kerns. "Transient Radiation Upset Simulations of CMOS Memory Circuits." IEEE Trans. Nucl. Sci. NS-31(6):1337–1343, Dec. 1984.
45. D. G. Mavis, D. R. Alexander, and G. L. Dinger. "A Chip-Level Modeling Approach for Rail Span Collapse and Survivability Analyses." IEEE Trans. Nucl. Sci. NS-36(6): 2239–2246, Dec. 1989.
46. J. Lindmayer and C. Y. Wrigley. Fundamentals of Semiconductor Devices, Section 5.6. D. Van Nostrand Co. Inc., New York, 1965.
47. K. Mistry, D. Krakauer, and B. Doyle. "Impact of Snapback-Induced Hole Injection on Gate Oxide Reliability of N-MOSFETs." IEEE Electron Device Letters, Vol. 11, No. 10, Oct. 1990.
48. J. Gautier and A. J. Auberton-Herve. "A Latch Phenomenon in Buried *N*-Body SOI NMOSFETs." IEEE Elect. Dev. Lett. (EDL) 12(7):372–374, July 1991.
49. J. D. Hayden, D. Burnett, and J. Nagle. "A Comparison of Base Current Reversal and Bipolar Snapback in Advanced *n-p-n* Bipolar Transistors." IEEE Elect. Dev. Lett. 407–409, Aug. 1991.

CHAPTER 8
SINGLE EVENT UPSET

8.1 INTRODUCTION

Single event upsets (SEUs), or soft errors, are mainly logic upset errors that almost always occur in high-density integrated circuits, such as LSI and VLSI devices. They seem to be random with time and position within the IC chip. When a logic cell (e.g., flip-flop) in the chip that has suffered such an upset is later probed, it exhibits no damage or degradation in any of its characteristics. These transitory errors are called single event upsets because they are usually due to a spurious charge produced by the transit of a single ionizing particle through the chip. They are also called soft errors because the corresponding integrated circuit chip is normally not damaged. The single event upset frequency or single event upset bit error rate increases with the packing density of the particular type of integrated circuit. To enhance performance, especially increased speed of operation, the manufacturers of modern-day integrated circuits are tending toward smaller devices, and so increased packing density. Submicron feature size of integrated circuits is essentially extant, hence the SEU-hardening problem is being exacerbated with shrinking device size. Feature size is meant as the smallest dimension of any part of the IC, which is usually the width of a metallic stripe (interconnection).

SEU was predicted in 1962[1] in anticipation of the future advent of high packing density integrated circuits and their susceptibility to cosmic rays. SEU was verified through telemetric measurements in an actual spacecraft in 1975.[2] These errors are due to a number of natural sources, including: (a) galactic cosmic rays; (b) the solar wind flux, mainly protons often included under cosmic rays; (c) heavy particles trapped in the Van Allen belts (energetic protons and small amounts of galactic cosmic rays); (d) alpha particles that are the decay products of naturally occurring radioactive heavy actinides, such as uranium, thorium, and their daughter nuclei. These heavy actinides make up part of the earth's crust, and thus are found in trace amounts within the material that is used to package as well fabricate the IC chip[3]; and (e) fast neutrons (>1 MeV) and protons that are produced in the upper atmosphere by cosmic rays. These last can produce energetic reaction fragment ions, which can cause SEU.[4] They can interact with the silicon of the device, resulting in the production of heavy recoil nuclei ions, such as ^{25}Mg, which can cause SEU. In this regard, SEU has been observed in the computers of commercial aircraft at high altitude due to such recoil nuclei. These fragment ions produce what are called spallation Si(n, α) Mg and Si(n, p) Al reactions.[61] It is also possible for muons, which result from cosmic rays interacting in the atmosphere, to produce SEU in sensitive components, such as charge-coupled devices (CCDs). These effects are important for avionics in the upper atmosphere, and must be addressed for sensitive equipment all the way down to the earth's surface.

The problem of hardening susceptible integrated circuits against SEU is a vexing one. One major reason is that the heavy ion component of galactic cosmic rays is so energetic that it is practically impossible to physically shield systems such as spacecraft. The spacecraft skin and the mass of its internals provide little shielding. SEU hardening emphasizes the circuit and systems approach, as discussed in Sections 8.10 and 12.5. However, for solar wind protons, shielding can be somewhat effective since the range of these protons is negligibly small in materials of interest. This is especially a consideration during an anomolously large solar flare, which occurs very infrequently (about once every few years). The corresponding solar flare proton flux can increase temporarily (~5–100 hours) by three or more orders of magnitude.

Due to the magnetic field of the earth acting as a magnetic spectrometer (except at the poles) a good portion of the incident galactic cosmic rays are diverted away from the earth for low orbit altitudes (~200 miles and down). The major cause of SEU for low earth orbits is due to geographic lacunae in the earth's magnetic field (e.g., South Atlantic Anomaly) where incident galactic cosmic rays are not diverted away from the earth.

For the very high packing densities in LSI and VLSI, each logic cell can be as small as 10 square microns in area. In charge-coupled devices (CCDs), which are mainly used in imaging systems, a single "ON" state could correspond to approximately 50 000 electrons. However, a single alpha particle can ionize enough silicon atoms to produce about 3 million electron hole pairs, creating a spurious charge clearly large enough to upset its state. Therefore, trace amounts (~1 ppm) of the natural actinides lodged within the chip or package material provide sufficient alphas, whose ionization of the device mass can cause SEU in high-packing density chips. Generally, it is found that the SEU error bit rate is a sensitive function of the type and design of the IC, as well as the nature of the incident cosmic radiation.[5] For example, galactic cosmic rays also produce muons in the upper atmosphere. They are members of the family of heavy electrons, or mesons, some with masses up to

Table 8.1 Cosmic-ray-induced SEU for representative LSI and VLSI technologies.

DEVICE TECHNOLOGY	ESTIMATED CRITICAL CHARGE (PICOCOULOMBS)	GEOSYNCHRONOUS DEVICE RATE (CHIP ERRORS PER DAY)	GEOSYNCHRONICS ERROR RATE (CELL ERRORS PER DAY)
54L Series[3] (TTL)	0.70	Depends on cells per chip	2.6×10^{-5}
NMOS RAM[3] (4K)	0.35	2.4×10^{-2}	6×10^{-6}
NMOS RAM[3] (64K)	0.064	2.13	3.3×10^{-5}
SBP 9989[13] (I^2L microprocessor)	0.06	0.02	5×10^{-5}
CCD Memory[3] (64K)	0.006	24.	3.6×10^{-4}
CCD Memory[3] (256K)	0.002	120.	4.7×10^{-4}
54ALS Series[59] (TTL)	0.12	Depends on cells per chip	3.2×10^{-3}
54LS Series[59] (TTL)	0.44	Depends on cells per chip	1.4×10^{-4}
LM 101 Op-Amp[60] (input transistors)	4.96	1.2×10^{-8}	3×10^{-9}
2901A DM[60] (TTL microprocessor)	0.21	3.5×10^{-3}	–
Metal gate 256 × 1 Bulk CMOS RAM[17]	3.5	5.9×10^{-7}	2.3×10^{-9}
2909A DM[60] (microprocessor)	0.006	0.12	–
PMOS Shift register[17] (1024 bits)	0.60	0.017	1.7×10^{-5}
CMM5114 RAM[59] 1K × 4 CMOS/SOS	0.10	8.0×10^{-7}	2.0×10^{-10}
93L422[13] 256 × 4 TTL RAM	0.05	0.74	7×10^{-4}
55182[59] (differential line rcvr)	0.60	1.1×10^{-6}	5.6×10^{-7}

1500 times that of the ordinary electron. Muons usually do not affect MOS RAMs, but are a prime cause of SEU in CCDs, even at ground level, simply because the hypersensitive CCDs are affected by the meager amount of ionization produced by these muons[5] within the chip. An order of magnitude estimate of the SEU bit error rate from the muons is about 10^{-10} per bit per hour for a 64K bit MOS RAM, and about 10^{-8} per bit per hour for a 256K bit CCD memory.[5] Table 8.1 provides cosmic-ray-induced SEU data for representative MOS LSI, VLSI, and bipolar TTL devices. It should be appreciated that SEU is almost wholly a problem peculiar to large-scale ICs, and only to certain families of these, because of the relatively small size of their logic cells, and the circuit and system function of the particular integrated circuit. For a discrete device, such as a transistor or diode, the junction nodal areas are usually so large that their operation remains unaffected by SEU-inducing particles. There are some exceptions, such as *n*-channel power MOSFETs, which are made up of large numbers of very small transistors in parallel.

8.2 CRITICAL CHARGE, LINEAR ENERGY TRANSFER, SEU CROSS SECTION

To examine in quantitative detail how the flux of particles incident on the IC causes SEU, a critical charge Q_c is first defined. It is that quantity of charge (usually measured in pico- or femto coulombs) necessary to change a binary "one" to a zero, or vice-versa, at a particular storage node. SEU errors are construed as the deposition of Q_c at the storage node by electrons supplied by the track of ionized device material produced by the incident particle. It should be realized that the time scale of the SEU-produced charge and its effect on the integrated circuit is usually of the order of a few tens of picoseconds to nanoseconds. This is two or three orders of magnitude shorter than, for example, nuclear weapon-induced dose rate photocurrent production, as discussed in Section 7.2. This short time scale plays a role in the dynamics of SEU with respect to various semiconductor materials and hardening techniques, as will be discussed.

The single event-produced charges initially diffuse radially from the ionization track in the device semiconductor. After sufficient ambipolar diffusion, the electrons and holes thus generated are separated by the large internal electric fields when they are produced in the depletion region of a junction node. The electrons are accelerated to the positive side of the field, while the holes are accelerated to the negative side. Cosmic-ray-induced ionization charge that is generated outside the de-

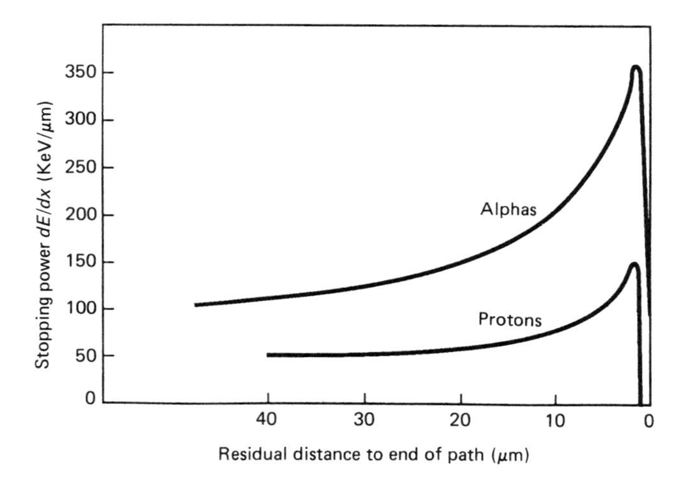

Fig. 8.1A Bragg curves of stopping power versus residual distance to end of path (range) for alpha particles and protons in silicon.

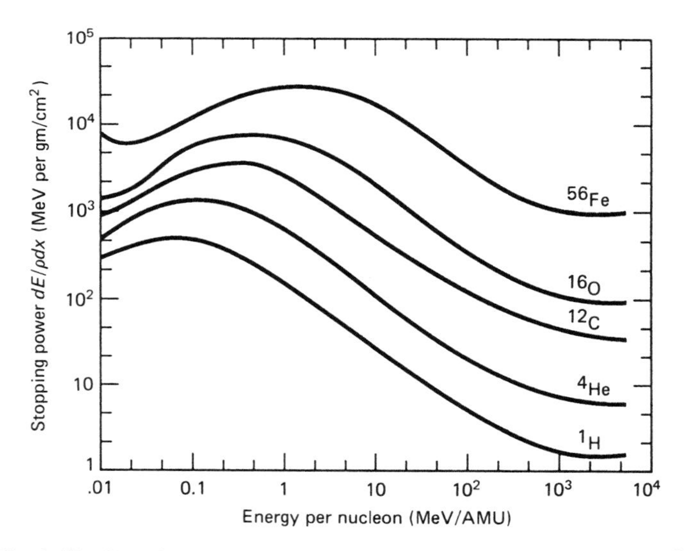

Fig. 8.1B Stopping power in silicon for representative cosmic ray ions.[62]

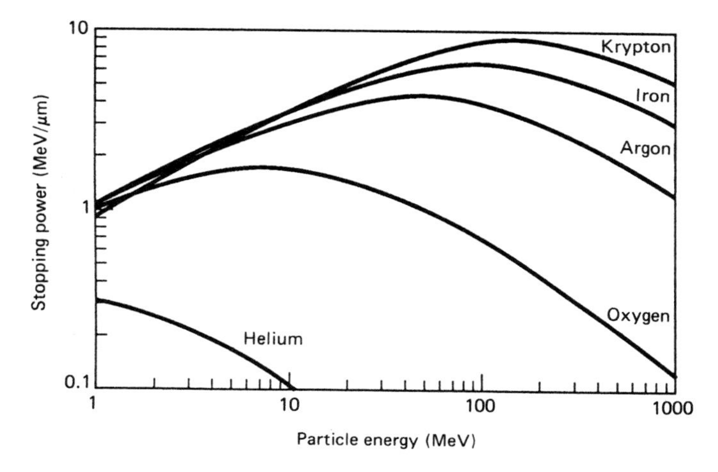

Fig. 8.1C Heavy Ion stopping power in silicon versus particle energy.[62]

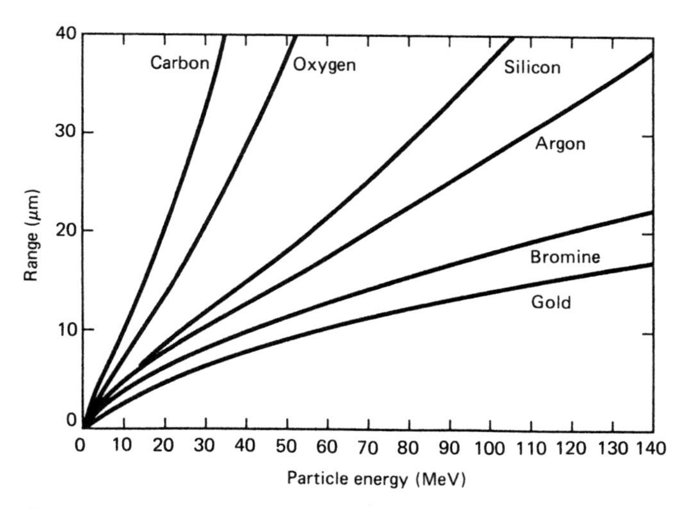

Fig. 8.2 Range-energy curves for heavy ions in silicon.[47] Figure © 1985 by the IEEE.

pletion region neighborhood can diffuse to the edge of the depletion region and be accelerated into the logic storage node, adding to the total effect.

As discussed above, to cause an SEU error, the incident particle must deposit enough ionization energy in the node-sensitive neighborhood of the material to produce the required amount of SEU-inducing charge (i.e., Q_c). This means that the particular incident particle must possess sufficient stopping power in the semiconductor material to deposit the critical charge required for SEU upset; that is, sufficient energy must be deposited by this particle per unit mass of material, or (more frequently) sufficient energy must be deposited per unit track length in the material, to cause SEU. This is expressed as dE/ds, or dE/dx where ds is the incremental length in the material (usually in the direction of the ionization track), dE is the corresponding incremental energy loss, s is a generalized line length, and x usually connotes a specific coordinate dimension.

Figure 8.1A depicts curves of stopping power, called Bragg curves, which are plots of dE/ds or dE/dx for alphas and protons versus the remaining distance to the end of their paths or ranges in silicon. Range is essentially the "crow-flight" penetration distance of the incident particle into the material. Notice that the majority of the particle energy is deposited near the end of the range in the figure, as dE/ds reaches a maximum in that neighborhood. This is a characteristic of such curves, especially for light nuclei. Figures 8.1B and 8.1C depict stopping power in silicon for various cosmic ray ions. Now the range $R(E_0)$, where E_0 is the initial energy of the incident particle, is well approximated for energetic heavy ions by

$$R(E_0) \cong \int_0^{R(E_0)} ds = \int_{E_0}^{0} dE/(dE/ds) = \int_0^{E_0} dE/(-dE/ds) \tag{8.1}$$

where the negative sign merely means that the particle is losing energy to the material. Implied is that the range is not very different from the corresponding "random-walk" distance. dE/ds is the stopping power of the incident ion in the material of interest. Figure 8.2 depicts range-energy curves for heavy ions in silicon.[6] Or, in terms of its Linear Energy Transfer (or LET),

$$|R(E_0)| = \int_0^{E_0} dE/\mathrm{LET} \tag{8.2}$$

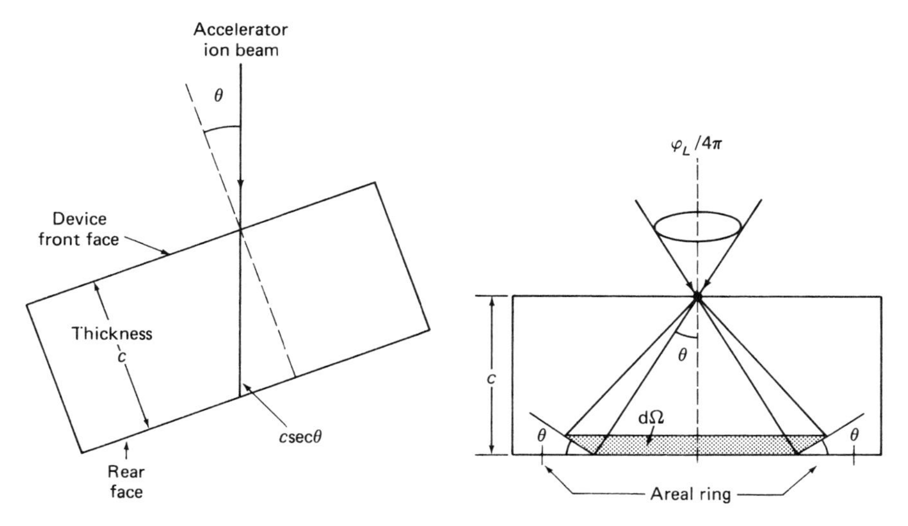

Fig. 8.3 Device rotated in accelerator ion beam to determine seu critical angle.

For the cosmic ray particles and semiconductors of interest, LET is usually an adequate approximation for stopping power.

Another important parameter to be defined and discussed is the critical LET, L_c. First, it must be appreciated that the substantive simulators of cosmic ray particles held in high regard are the very energetic heavy ion beams (e.g., 140 MeV krypton) obtained from large particle accelerators of the cyclotron family, like the Bevelac at UC Berkeley. However, recent experiments have called into question the high-energy heavy-ion beam purity, in terms of the normal assumption of monoenergetic single specie ion beams from high energy accelerators such as the Bevelac.[73] This can have a bearing on the accurate experimental determination of critical LETs of components. The leftmost part of Fig. 8.3 depicts the accelerator beam incident at the face of the assumed thin parallelepiped-shaped sensitive region of the IC device, at an azimuth angle θ with respect to the device face normal. Recall the definition of LET, which is $L = dE/ds \simeq \Delta E/\Delta s$, where ΔE is the increment of energy deposited by the incident particle in its track length segment Δs in the device. Alternatively, the prevalent definition is $L = dE/\rho ds \simeq \Delta E/\rho\Delta s$, which is the energy deposited by the incident particle into the material mass per unit penetrated area $\rho ds \simeq \rho \Delta s$, where ρ is the material density. Common units for L are MeV per mg/cm^2 (i.e., MeV cm^2/mg).

In silicon, 3.6 eV of ionization energy is required to produce one electron-hole pair, where the electron charge is $1.6 \cdot 10^{-19}$ coulombs. For an incident cosmic ray depositing an amount of energy E (in MeV) to ionize the material, the resulting charge generated in silicon in pico coulombs (pC) is $Q = E(\text{MeV})/22.5$. In MOS devices whose memory cells are more than ten square microns in area, the required LET to achieve the deposition of a critical charge is a few MeV per micron of particle track. This implies stopping powers corresponding to very high energy ($\gg$10 MeV) heavy ($Z \gg 2$) cosmic ray ions. There is almost nothing in the terrestrial sphere that can supply such particles naturally and in substantial numbers.

To cause an SEU, a critical amount of energy ΔE_c must be deposited in the device mass to produce the required critical charge Q_c to upset a cell. In Fig. 8.3, assume that the beam is exactly normal to the device initially, so that $\theta = 0$. In this position, the beam ion track length through the device is a minimum given by c, the device thickness. As the device is rotated azimuthally about the beam direction away from its initial normal position, its angular displacement is described by θ. $\theta \leqslant \theta_c$, where θ_c is the critical angle corresponding to a track of sufficient length in the device ($c\sec\theta_c$) for the beam ion to deposit sufficient ionizing energy to provide a critical charge Q_c at a logic node. For $\theta < \theta_c$ in Fig. 8.3, the corresponding LET is given by

$$L \simeq \Delta E/\Delta s = \Delta E/c\sec\theta = L_\perp \cos\theta \tag{8.3}$$

where $L_\perp = \Delta E/c$ is the LET normal to the device face, since c is the device thickness. $L_\perp$ is also referred to as the effective LET.[15]

In the SEU test depicted in the leftmost part of Fig. 8.3, it is assumed that the device has not experienced an SEU at normal incidence using the heavy ion beam available from the accelerator. That is, the available ions cannot deposit sufficient energy along the track length c normal to the beam to produce enough ionization to deposit charge Q_c at the device logic node. However, when the device is rotated to θ_c as described above, the increased track $c\sec\theta_c$ achieves a length such that the associated electronics instrumentation indicates the onset of SEU. This means that a critical amount of energy E_c has been deposited in the device from the beam ion track of length $c\sec\theta_c$. Since $L_{\text{ion}} \equiv L$ is a physical quantity that depends only on the particular ion, the corresponding ion beam particle energy, and the device material properties, it is invariant in the sense that

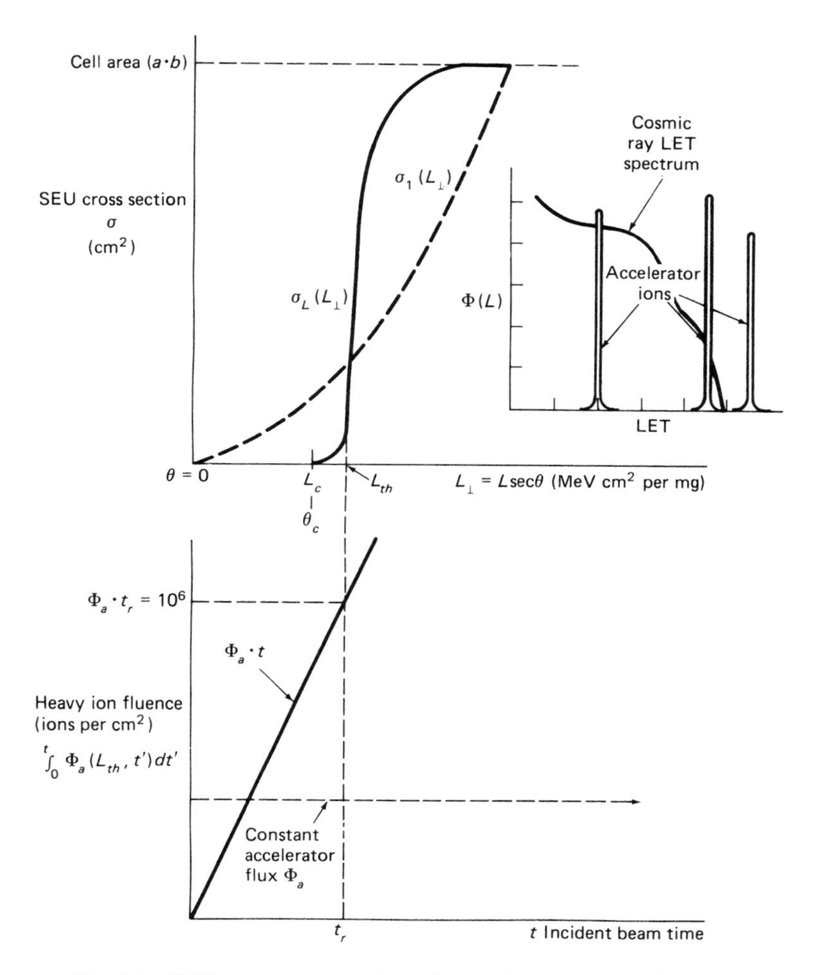

Fig. 8.4 SEU cross sections and accelerator beam heavy ion fluence.

$$L_{\text{ion}} \equiv L = \Delta E/c\sec\theta = \Delta E_c/c\sec\theta_c \tag{8.4}$$

The definition of the critical LET,[7] L_c, is the corresponding $L_\perp$ normal to the device face at the critical angle θ_c (i.e., when $\theta = \theta_c$, $L_\perp = \Delta E_c/c \equiv L_c$). Alternatively, $L_c = Q_c/c$, using suitable conversion factors. Then, from Eq. (8.4),

$$L = L_\perp\cos\theta = L_c\cos\theta_c \tag{8.5}$$

so that $L_c = L\sec\theta_c$ is immediately derived from the accelerator experiment, since θ_c is measured and L is known.

The accelerator test of the SEU susceptibility of a particular integrated circuit consists of measuring the SEU cross section σ_L as a function of $L_\perp = L\sec\theta$, as shown in the upper part of Fig. 8.4. Therefore, at SEU onset, when $\theta = \theta_c$, $L_\perp = L_c$. Unfortunately, SEU onset is frequently not clearly delineated, Fig. 8.4 notwithstanding. Then, for reasons of engineering conservatism, σ_c is sometimes taken at its asymptotic (maximum) value occurring for very large $L_\perp$, as in the figure. More often, σ_c is taken as the value of σ_L at its inflection point.

When the device logic cell sensitive region is assumed to be a thin parallelepiped,[64] the asymptotic SEU cross section usually corresponds to the geometric cross section i.e., the area of the parallelepiped face $a \cdot b$. Once the SEU cross section is known, it can be used in principle to compute the SEU error rate. With the knowledge of the SEU cross section $\sigma_L(L)$ and the incident cosmic ray flux $\Phi(L)$, the SEU error rate, E_r, is given by

$$E_r = \int_{L_{\min}}^{L_{\max}} \sigma_L(L)\,\Phi(L)\,dL \tag{8.6}$$

where $L_{\min}$ and $L_{\max}$ mark the sensible boundaries of the LET flux spectrum $\Phi(L)$, which describes the particular cosmic ray environment. Note that the LET spectrum $\Phi(L) \equiv \Phi(dE/ds)$ is in contrast to its perhaps more familiar cosmic ray energy spectrum $\Phi_E(E)$. The former spectrum or flux dependence is used almost exclusively in this context.

It should be realized that the SEU cross section, like all cross sections, is a semi-empirical parameter usually measured by experiment. However, it can be calculated analytically in certain instances from first principles. Here, it is usually measured in cm^2 as $\sigma_L(L) = \lim_{t\to\infty} E_r(t)/\Phi_a(L, t)$. E_r is the SEU bit error rate, as read from the accelerator instrumentation connected to the device in situ, while Φ_a is the corresponding accelerator beam flux incident on the device as it is undergoing single event upsets. The experiment is repeated for a set of $L_\perp = L\sec\theta$ as the device is displaced through the corresponding set of azimuth angles θ, to yield the upper part of Fig. 8.4.

Another LET parameter, important with respect to SEU accelerator and other heavy ion source tests, is the threshold LET $L_{th} \lesssim L_c$, usually given in SEU tabulations.[16,52,60] L_{th} is defined as the minimum LET for which SEU is assumed to occur when the accelerator beam ion fluence $\int_0^{t_r} \Phi_a(L_{th}, t')\,dt'$ has attained a level of at least 10^6 ions per cm^2.[16] t_r is the corresponding incident ion beam time on the part. An approximate SEU cross section behavior is implied, viz.,

$$\sigma_L \cong \begin{cases} \lim\limits_{t_r \leqslant t \to \infty} E_r(t)/\Phi_a(L_{th}, t); \; t \geqslant t_r; \; \displaystyle\int_0^{t_r} \Phi_a(L_{th}, t')\,dt' = 10^6 \text{ ions per cm}^2 \\ 0 \qquad\qquad ; \; t < t_r; \; \displaystyle\int_0^{t} \Phi_a(L_{th}, t')\,dt' < 10^6 \text{ ions per cm}^2 \end{cases} \tag{8.6A}$$

as seen in the lower part of Fig. 8.4 for an accelerator ion type. For constant flux Φ_a, then $\Phi_a \cdot t_r = 10^6$. L_{th} is the intercept on the $L_\perp$ axis of the extrapolation from the inflection point of σ_L. By its very nature, the single event critical LET L_c is defined in terms of a single incident heavy ion encounter with the device material. The utility of the threshold LET, L_{th}, is that it implies a statistical smoothing of experimentally determined SEU fluctuations from accelerator or other heavy ion source fluxes. This ensues by insuring that statistically sufficient heavy ion SEU interactions occur in the device under test. In the same vein, an alternate definition of L_{th} is the LET corresponding to a measured SEU cross section value that is, for instance, 1 percent of its maximum (asymptotic) value[16] (Fig. 8.4). These two definitions of L_{th} sometimes result in differing values. For detailed information on the LET of a specific device for which SEU data has been tabulated, it is advisable to peruse the raw data if possible, and initiate discussions with the testing entity, and/or with the experimenters themselves. Ultimately, the specific LET will be a range or value judgment, as a result. However, in the context of the electronic use of the part, the precise value of the corresponding LET will usually play a minimal role. It is only whether the LET is low or high, with respect to a nominal reference (e.g., L greater or less than approximately 40 MeV cm^2/mg, a recognized cosmic ray LET spectrum upper bound cutoff), that will be important in the systems analysis or systems design sense.

For SEU testing that utilizes the discrete LETs of a set of the accelerator high-energy heavy-ion family of particles, the fact of their general lack of conformation to an actual galactic cosmic ray flux LET spectrum $\Phi(L)$ in the simulation sense (Fig. 8.4 inset) poses no difficulties. This is mainly because the LET of the particular accelerator ion is effectively varied, simply through varying the device azimuthal angle θ in $L_\perp = L\sec\theta$ between $\theta < \theta < \pi/2$. This LET scan allows the determination of $\sigma_L(L_\perp)$ versus $L_\perp$, as in the upper portion of Fig. 8.4, with a single accelerator ion type, as long as the ion used includes L_c within the corresponding scan. Therefore, spectra $\Phi(L)$ or Φ_a play no direct role, per se, in SEU testing as described.

The use of large accelerators to simulate SEU involves a heavy investment in man-hours, cyclotron beam expertise, and the associated expense. A somewhat limited practical substitute for simulating highly energetic heavy ions are fission fragment ions from the spontaneous fissioning of californium, ^{252}Cf, which is manufactured through successive neutron absorption by the actinide series of nuclides,[8] including uranium, as in a nuclear reactor. The effective half-life of ^{252}Cf for fission is about 2.6 years, producing a wide range of fission fragment masses with the characteristic double-peaked fission fragment yield curve. Using particle time-of-flight methods, efforts are being made to delineate corresponding fission nuclides.[76] Corresponding to one of the peaks is the lighter group with a mean mass number of 102.6 and mean energy of 102.5 MeV. The other, heavy-fragment, peak yields a mean mass number of 142.2 and mean energy of 78.7 MeV. Together, these particles possess a mean LET in silicon of 43 MeV cm^2/ mg. This is about 1.6 times the LET of the 100 MeV ^{56}Fe ions that typify the most intensely ionizing particles of the galactic cosmic ray environment. In fact, the ^{252}Cf fission fragment energies are so large that they are often attenuated to correspond to

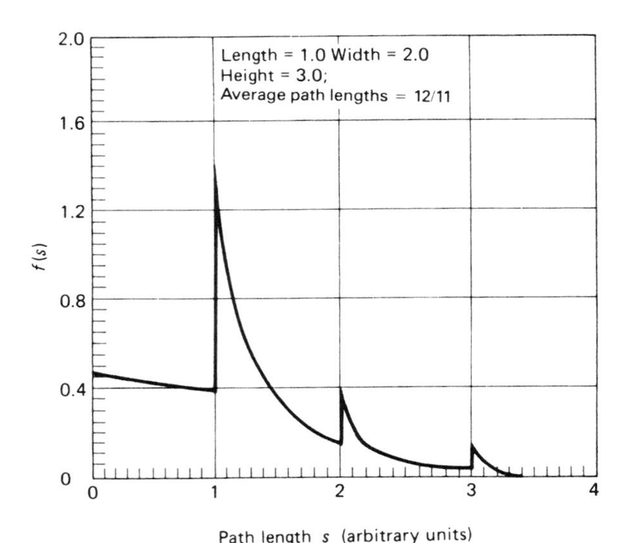

Fig. 8.5 Path length distribution function for a parallelepiped of given dimensions.[11]

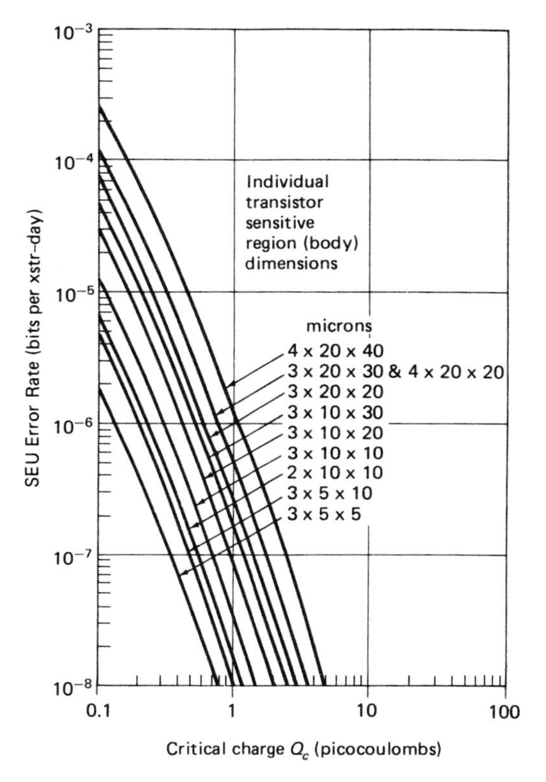

Fig. 8.6 Cosmic ray-induced SEU error rate versus critical charge for various sensitive region transistor dimensions of a static CMOS RAM.[10] Figure © 1980 by the IEEE.

actual cosmic ray environments. Alternatively, this excess particle energy can be used to investigate SEU-induced latchup. However, the range of ^{252}Cf fission fragments in silicon implies a limitation on measuring critical LET,[9] such that $20 \leqslant L_c \leqslant 45\,\mathrm{MeV\,cm^2/mg}$. Often, ^{252}Cf tests are used to identify part sensitivities, with a view to subsequent heavy-ion testing using a high-energy accelerator. ^{252}Cf, with each (spontaneous) fission, produces an average of about 3.85 neutrons, as compared to 2.5 for ^{235}U and 3. for ^{239}Pu. This can represent a substantive source of spurious neutrons, for which appropriate safety measures must be taken. Other isotopes that provide particles of SEU interest include americium (^{241}Am) and curium (^{244}Cm) both of which decay by producing alpha particles. Curium also fissions spontaneously like californium. However, its fission fragment yield per ^{244}Cm nuclear disintegration is only $3 \cdot 10^{-4}$ that of californium.

8.3 SEU ERROR RATE COMPUTATION MODELS

Geosynchronous SEU

To compute the SEU error rate for a particular integrated circuit, it is assumed that the cosmic ray flux is completely isotropic; that is, the device is threaded by cosmic rays from all directions equally. For modeling purposes, it is also usually assumed that the sensitive region of the device is in the form of a thin parallelepiped of dimensions a, b, and c, where c is the thickness (smallest dimension) and ab is the chip face area. Only those projected track lengths in the device long enough to provide a charge Q_c from ionization, and so produce SEU, are of interest. This is taken into account in the error rate computation by first expressing the chip projected area of the parallelepiped as a fraction of the mean projected area, $\bar{A_p}$, which fraction depends on the face aspect angles, with respect to a particular track direction. $\bar{A_p}$ is computed by using the fact that, in the chosen octant of three-dimensional (r, θ, ϕ) space,[10] $A_p = A_x + A_y + A_z$, or $A_p\,(\theta, \phi) = ab\cos\theta + c(b\cos\phi + a\sin\phi)\sin\theta$, where A_x, A_y, and A_z are the projected areas of the faces of the parallelepiped onto the track normal, as seen from that octant (Fig. 8.30), so that

$$\bar{A_p} = \int_0^{\pi/2}\int_0^{\pi/2} A_p(\theta, \phi)\sin\theta\, d\theta d\phi \Big/ \int_0^{\pi/2}\int_0^{\pi/2} \sin\theta\, d\theta d\phi = 1/2(bc + ac + ab) \tag{8.7}$$

Then the SEU bit error rate, E_r, is given by a line integral over all directions within the limits of integration, viz.,

$$E_r = \bar{A_p}\int_{s_{\min}}^{s_{\max}} \Phi(L(s))f(s)\,ds \tag{8.8}$$

where $\Phi(L(s))$ is the integral cosmic ray flux (particles per cm^2-day), as opposed to the differential cosmic ray flux $\varphi_L(L) = \Phi'(L(s))$, which is the flux per unit LET. $s_{\max}$ is the maximum track (chord) length within the parallelepiped sensitive region, which is simply its main diagonal. $s_{\min}$ is the shortest path length that can still result in an SEU (i.e.,

$s_{min} = Q_c/L_{max}$, where L_{max} is the sensibly largest LET of the particular cosmic ray spectrum of interest).

$f(s)$ is the chord distribution function, depicted in Fig. 8.5, which is the probability distribution of all possible track chord lengths in the parallelepiped sensitive region, and is a function of its dimensions.[11,23,44,45] Since it is a parallelepiped, all chord (track) lengths threading it are not identical, as they would be along radii from the center of a sphere. Therefore, certain directions are favored for SEU while others are not, depending on the aspect of the projected area of the parallelepiped, with respect to a particular direction of incident cosmic ray flux. $f(s)$ is normalized so that $\int_0^{s_{max}} f(s)\,ds = 1$. The mean chord length $\bar{s} = \int_0^{s_{max}} sf(s)\,ds = abc/\bar{A}_p$, with a standard deviation of $\bar{s}$. Because $f(s)$ is a cumbersome function of the parallelepiped dimensions, Eq. (8.8) is usually integrated numerically.[10] Some results are given in Fig. 8.6. This figure shows the SEU bit error rate for a static CMOS RAM, as a function of the critical charge Q_c, for various parallelepiped sensitive region dimensions. For a 1024-bit (lK bit) CMOS RAM with the assumptions: (a) $Q_c = 1$ picocoulomb (pC); (b) parallelepiped dimensions $3 \times 10 \times 10$ cubic microns; (c) $\Phi(L(s)) = 0.16$ particles per cm^2 sec; and (d) three transistors per cell sensitive to SEU, Fig. 8.6 shows an expected SEU bit error rate per transistor of $4 \cdot 10^{-8}$ per day. For the cell, the bit error rate is multiplied by 3 to give $1.2 \cdot 10^{-7}$ per cell day, and for a 1024-bit RAM, it is $(1.2 \cdot 10^{-7})(1.024 \cdot 10^3) = 1.2 \cdot 10^{-4}$ per device day.

To develop an expression for ready computation of SEU error rate, the chord line integral, Eq. (8.8), can be transformed (by an integration by parts) into an integral with respect to LET[12], to yield

$$E_r = (S/4) \int_{L_{min}}^{L_{max}} \varphi_L(L)\, C(s(L))\, dL \tag{8.9}$$

where:

$C(s(L))$: integral of the chord length distribution function, i.e., $\int_s^{s_{max}} f(s')\,ds'$

$\varphi_L(L)$: differential cosmic ray spectrum, i.e., $\Phi'(L(s))$, flux/LET

S : surface area of the parallelepiped, i.e., $2(bc + ac + ab)$

L_{min} : minimum LET required to produce SEU, $Q_c/\rho L_{max}$

L_{max} : maximum LET, or LET cutoff for particular LET spectrum used

Since $Q_c = E_c/22.5$ (picocoulombs) for silicon, a corresponding LET

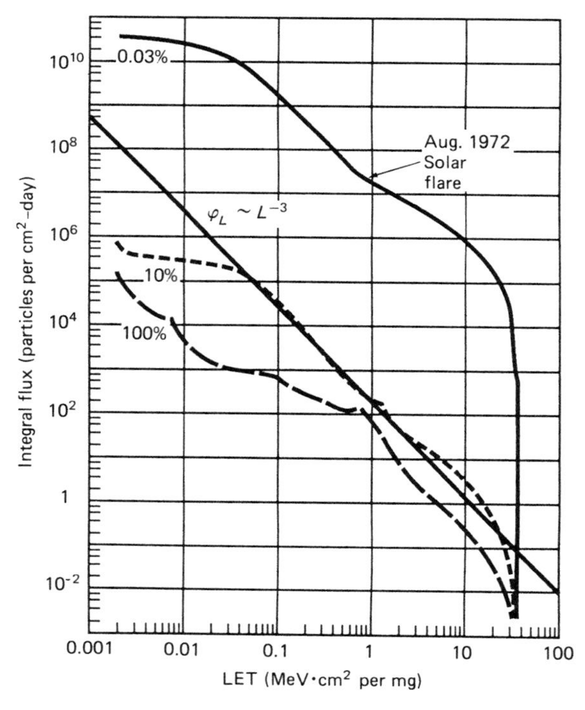

Fig. 8.7 Heinrich curve in silicon of cosmic ray abundance versus let for three different geosynchronous environments.[13] Figure © 1982 by the IEEE.

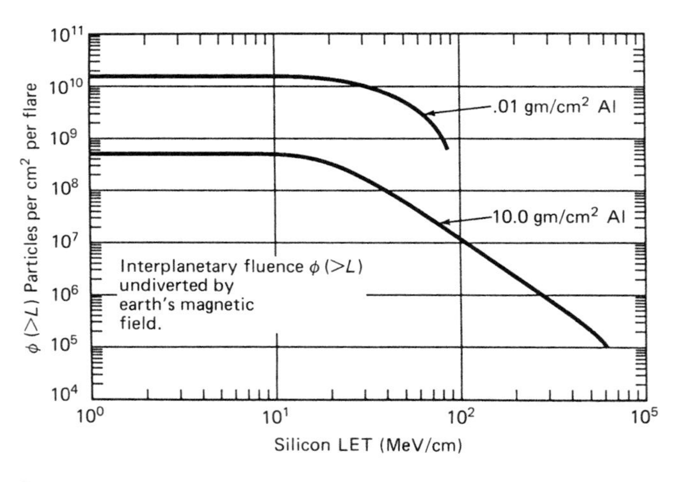

Fig. 8.7A Integral LET spectra for anomolous solar flare protons behind spherical aluminum shields.[27] Figure © 1989 by the IEEE.

of 96.6 MeV cm²/mg yields 1 pC per μm, where E_c has dimensions of MeV. For gallium arsenide, the ionization energy per hep (hole-electron pair) is given in Table 1.1, $Q_c = E_c/30$, and a LET of 56.4 MeV cm²/mg yields 1 pC per μm. Using the thin parallelepiped assumption (i.e., a, $b \gg 3c$), $C(s(L))$ can be approximated by[23]

$$C(s(L)) = \begin{cases} (3/4) \cdot (c/s)^{2.2} \;; & s \geq c \\ 1 - (1/4) \cdot (s/c); & s < c \end{cases} \tag{8.10}$$

As already mentioned, the cosmic ray differential flux $\varphi_L(L)$ is considered a function of the LET L, instead of the more widely used cosmic ray flux particle-energy spectra. Figure 8.7 depicts φ_L in silicon for three different geosynchronous altitude environments.[13] These are called Heinrich Curves of cosmic ray differential flux spectra[14] versus LET. The curve labeled "100%" refers to a solar sunspot maximum condition. This corresponds to the maximum amount of cosmic ray shielding afforded by the interaction between the earth's magnetic field and the solar wind during a sunspot maximum, as discussed in Section 12.4. Hence, the cosmic ray flux environment in the vicinity of the earth is minimal for this case. The curve labeled "10%" corresponds to a solar sunspot minimum condition, where the cosmic ray shielding is much less effective, and so is analogously worse only 10 percent of the time. This is called the "Adams 10 Percent" environment, and is usually taken as the de facto standard cosmic ray geosynchronous environment. The "0.03%" curve is that for the very infrequently occurring anomolously large solar flare condition, where the solar cosmic ray output by far dominates all of the other cosmic ray sources in the vicinity of the earth for a few tens of hours following the onset of the flare. Figure 8.7A depicts a silicon Heinrich Curve for anomolous solar flare protons behind spherical aluminum shielding.[27]

For the galactic cosmic ray environment, as exemplified by the Heinrich Curve for the "Adams 10 percent environment" (as in Fig. 8.7), a useful approximate expression for SEU error rate can be obtained. This is accomplished by first approximating the preceding curve by[7]

$$\varphi_L(L) \cong 5.8 \cdot 10^8 / L^3 \text{ (particles per cm}^2\text{-day LET)} \tag{8.11}$$

Inserting Eqs. (8.10) and (8.11) into Eq. (8.9) yields an estimate of the SEU error rate for the geosynchronous environment

$$E_r \cong 6.5 \cdot 10^{-10} (\sigma_L / L_c^2) \cdot 2 \cdot (1 - 0.9(c/d)^{0.2}) \cong 5 \cdot 10^{-10} \sigma_L / L_c^2$$
$$\cong 5 \cdot 10^{-10} abc^2 / Q_c^2 \tag{8.12}$$

where:

E_r (SEU error rate per cell day) should be construed as an upper bound, and the parameters are dimensioned as follows:

σ_L : SEU cross section (cm^2)
L_c : critical LET (picocoulombs per μm)
a, b, c : parallelepiped sensitive region dimensions (μm)
d : parallelepiped diagonal $(a^2 + b^2 + c^2)^{1/2}$ (μm)
Q_c : critical charge (pC)

The numerical multipliers in Eqs. (8.12) are fitted from SEU model data.[12] To estimate E_r for GaAs devices, replace 5 by 3.5 in the multiplier in Eqs. (8.12).[68]

To use Eqs. (8.12) to compute SEU, the parameters of major interest are a, b, c, σ_L, L_c, and Q_c. σ_L and L_c are usually obtained by experiment, as discussed. Alternatively, $\sigma_L \sim ab$ can be used, as identified in Eqs. (8.12), and similarly $L_c = Q_c/c$. The literature contains tabulations of L_c and L_{th} for literally hundreds of devices, as obtained from accelerator measurements.[15,16] However, for devices that do not appear in these tabulations, these parameters can be estimated. One obvious method is to extrapolate from the tabulations for the device type or family that is apparently most similar to that of interest. This has its pitfalls, however. The dimensions a, b, and c can be obtained from forthcoming manufacturer's process parameters, by arrangement with the specific vendor, or by destructive physical analysis of the part. Q_c can be estimated by using circuit model computer programs (e.g., RADSPICE, SYSCAP), discussed in detail in Section 12.6, or by rough estimation using the "$Q = CV$" method. This method implies that an examination of the pertinent storage cell has already been made to ascertain the salient information state storage nodes, such as those in an integrated circuit memory cell.[17] For bipolar cells, this state is usually stored as charge in the node junction depletion region capacitance. This capacitance, C, can be computed from a knowledge of the manufacturing process physical and electrical parameter design rules, as discussed in Section 3.5. For a known mean voltage swing between the high and low states, ΔV, an estimate of the critical charge can be computed from $Q_c \cong C\Delta V$. This is patently a rough estimate of Q_c, for such a static analysis is an approximation to a more credible transient analysis available from the above circuit model computer codes. Fortuitously, the values of Q_c computed by using either method are reasonably comparable. Hence, for a bipolar base-collector junction assumed to be the information node, ab is the projected area of the base nested in the collector surround, as in Fig. 3.18. c is the thickness of the depletion-layer sensitive region spanning the junction, as discussed in Section 3.4. With the above information, E_r can be computed using the rightmost Eq. (8.12).

For completeness and pedagogical reasons, it is of interest to examine an alternative method for obtaining the computational expressions given in Eq. (8.12) for SEU error rates, which does not use the chord length distribution function $f(s)$ directly.[18] It is based on the range of differential cosmic ray flux incident particles on the device, which can produce SEU, as evinced by their track lengths. Referring to the rightmost part of Fig. 8.3, it is evident that the incident differential cosmic ray flux that has the required energy to deposit a critical charge and incident direction to cause SEU must lie in a solid angle whose half-apex azimuth angle θ satisfies $\theta_c \leqslant \theta \leqslant \pi/2$. This inequality merely stipulates that those cosmic ray particles that cause SEU must have track lengths equal to or greater than $c\sec\theta_c$. This corresponding flux is called $\varphi_e(L)$. If the total differential flux from all 4π solid angle, $\varphi_L(L)$, incident on the thin parallelepiped cell is assumed isotropic, then

$$\varphi_e(L) = 2 \cdot \int_{\theta_c}^{\pi/2} (\varphi_L(L)/4\pi) \cos\theta \, d\Omega = \varphi_L(L) \cdot \int_{\theta_c}^{\pi/2} \cos\theta \, d\Omega/2\pi \tag{8.13}$$

$\varphi_L(L)/4\pi$ is the isotropic cosmic ray flux per unit solid angle. $(\varphi_L(L)/4\pi)\, d\Omega \cos\theta$ is that flux threading an incremental solid angle ring $d\Omega = 2\pi \sin\theta \, d\theta$ incident on the incremental areal ring of the parallelepiped rear face, that is projected onto the particular flux direction, as in the rightmost part of Fig. 8.3. The factor of 2 multiplying the integral in Eq. (8.13) accounts for both front and rear faces of the thin cell. Then, from Eq. (8.13),

$$\varphi_e(L)/\varphi_L(L) = \int_0^{2\pi} d\phi \int_{\theta_c}^{\pi/2} d\theta \cos\theta \sin\theta/2\pi = (1/2)\cos^2\theta_c \tag{8.14}$$

Using Eq. (8.5) for $\cos\theta_c$, Eq. (8.14) then yields

$$\varphi_e(L) = \begin{cases} (1/2)\varphi_L(L) \cdot (L/L_c)^2; & L \leqslant L_c \\ (1/2)\varphi_L(L) \quad ; & L > L_c \end{cases} \tag{8.15}$$

The lower portion of Eq. (8.15) implies that, for LET such that $L > L_c$, all incident particles for which this holds cause SEU. Using Eq. (8.15), the integral cosmic ray flux, $\Phi_e(L)$, which corresponds to all particles that can cause SEU, is given by

$$\Phi_e(L_c) = \int_{L_{\min}}^{L_{\max}} \varphi_e(L)\,dL = \begin{cases} (1/2)\displaystyle\int_{L_{\min}}^{L_c} \varphi_L(L)\cdot(L/L_c)^2\,dL \\ \quad + (1/2)\displaystyle\int_{L_c}^{L_{\max}} \varphi_L(L)\,dL; \; 0 \leqslant L_c \leqslant 40 \\ (1/2)\displaystyle\int_{L_{\min}}^{40} \varphi_L(L)\,dL \qquad L_c > 40 \end{cases} \tag{8.16}$$

where the flux LET spectrum is such that it is assumed negligible for $L_c > 40\,\mathrm{MeV\,cm^2/mg}$. Then, for an average SEU cross section $\bar{\sigma}_L$, the estimated SEU error rate is approximated by the expression

$$E_r \cong \bar{\sigma}_L \Phi_e(L_c) \tag{8.17}$$

since $\bar{\sigma}_L$ is assumed constant. To evaluate the rightmost integrals in Eq. (8.16), a differential LET spectrum fit[18] is used to approximate $\varphi_L(L)$, which is in units of particles per $\mathrm{cm^2}$-day/MeV $\mathrm{cm^2}$ per mg, viz.,

$$\varphi_L(L) = \begin{cases} 65.1/L^2; & 0.07 \leqslant L \leqslant 1 \\ 434/L^{3.8}; & 1 < L \leqslant 40 \end{cases} \tag{8.18}$$

This differential LET spectrum corresponds to the cosmic ray flux maximum that occurs during the epoch of the solar sunspot minimum of the 11-year sunspot cycle. Inserting Eq. (8.18) into Eq. (8.16), with Eq. (8.17), gives the SEU error rate per cell day. It is

$$E_r = \begin{cases} 3.02\cdot 10^{-10}(1 - 0.64L_c^{-0.8})\,\bar{\sigma}_L/L_c^2 \simeq 3\cdot 10^{-10}\,\bar{\sigma}_L/L_c^2; \\ \quad 1 \leqslant L_c \leqslant 40 \\ 2.87\cdot 10^{-10}\,\bar{\sigma}_L/L_c^2 \qquad L_c > 40 \end{cases} \tag{8.19}$$

where the units of the above parameters are the same as for E_r, given in Eq. (8.12).

For epochs other than a sunspot minimum year, the SEU rate is given more accurately by $E_r \cdot h(t)$, where $h(t)$ is a periodic function to reflect the sunspot cycle, viz.,

$$h(t) = 0.71 + 0.29\cos \omega_0(t - t_{pk}) \tag{8.20}$$

where t is in years, t_{pk} is the year of any recent sunspot minimum, and $\omega_0 = 2\pi/11$. For example, for $t_{pk} = 1963.5$, and today $t = 1992$, then $h(1992) = 0.47$. $h(t)$ anticorrelates with the SEU error rate. This mirrors the fact that, for example, during a sunspot maximum (solar wind maximum), the solar wind participates with the earth's magnetosphere in maximally shielding the earth neighborhood from cosmic rays. These maxima correspond to a minimum SEU error rate.

For refinements to include the variation of the SEU cross section with LET, instead of using $\bar{\sigma}_L$ (such as depicted in Fig. 8.4), the SEU error rate can be obtained from

$$E_r = \int_{L_{\min}}^{L_{\max}} \Phi_e(L)\,\sigma_L(L)\,dL \tag{8.21}$$

Using this generic expression for E_r, it is seen that a comparison of the derivation of E_r given by Eq. (8.19) with that in Eq. (8.12) reveals that the chord length distribution function $f(s)$ essentially plays the role of the SEU cross section; therefore, these two expressions for E_r are equivalent. The former is the E_r expression in predominant use currently, mainly because its numerical multiplier is the larger, giving a more conservative SEU error rate.

Proton SEU

Thus far, the cosmic ray environment emphasized has been the galactic component, consisting of very high energy heavy ions, mainly a consideration for spacecraft orbiting at geosynchronous altitudes. Little discussion has thus far focussed on the particles in the Van Allen belts, the South Atlantic Anomaly (SAA), and other phenomena characteristic of medium to low-altitude earth orbits. The Van Allen belts, first discovered about 35 years ago, consist mainly of electrons and protons, whose source is purported to be the decay of neutrons produced by cosmic ray interactions with the very rarified atmosphere at these altitudes. The Van Allen belt constituent concentrations and associated properties are also discussed in detail in Sections 11.5 and 12.5. These neutrons, with a half-life of about 12 minutes, decay to yield protons, electrons, and neutrinos. The chargeless, nearly massless, neutrinos disappear into the cosmos, while the former two particle types are magnetically trapped by the earth's magnetic field as the main constituents that form the Van Allen belts. Some galactic cosmic rays are also trapped in these belts.

With regard to ionizing dose, discussed in Chapter 6, both Van Allen belt electrons and protons have the requisite parameters to produce ionizing radiation damage. However, the electrons do not possess the LET magnitude to cause SEU, whereas the corresponding protons indirectly do. As already mentioned, protons can cause SEU both directly and indirectly. However, their main SEU damage is indirect as wrought by their penetration into the device material to cause nuclear spallation reactions within. The reaction products in silicon include ions of sufficient energy and mass number, such as magnesium and aluminum nuclides, that can cause SEU. Also, with the advent of submicron feature size integrated circuits, protons will have the required LET to cause SEU directly in certain of these chips.

Since protons principally cause SEU indirectly through complex nuclear reactions, a computational expression for proton-induced SEU, e.g., for a spacecraft at Van Allen belt altitudes analogous to Eqs. (8.12) and (8.19), has not yet been derived. However, formidable attempts in this direction are being made.[74,75]

Even though there is no straightforward computational expression for proton-induced SEU (mostly at Van Allen belt altitudes) as there is for the heavy ion-induced upsets at geosynchronous altitudes discussed earlier, there is a general equivalence between the two.[40] That is, there seems to be an approximate first-order empirical correlation between geosynchronous orbit cosmic ray heavy ion-induced SEU rates, and proton-induced SEU rates at intermediate orbits for devices whose upset rates turn out to be less than 10^{-6} per bit day.[40] So that a tentative first-order proton-induced upset rate for a part contained in spacecraft avionics at intermediate altitudes can be gotten from a geosynchronous altitude upset rate computed from the previously discussed corresponding formulas, for the same part.[40] Also, a quantitative method for providing a proton upset rate from geosynchronous heavy ion test data has been introduced.[74]

As mentioned, for present-day device feature sizes, proton-induced SEU is mainly an indirect process. The incident protons induce spallation reactions in the device, which cause SEU, by virtue of the recoil nuclei (e.g., Al, Mg) that are products of the reaction. It is as if these nuclei were introduced ab initio into the device to cause SEU. Insofar as these nuclei simulate SEU produced by the much heavier and energetic cosmic ray ions, further correlations between proton and heavy cosmic ray ion-induced SEU are being explored.[68,70] Strong correlation would be a boon for determining proton-induced SEU, as corresponding methods

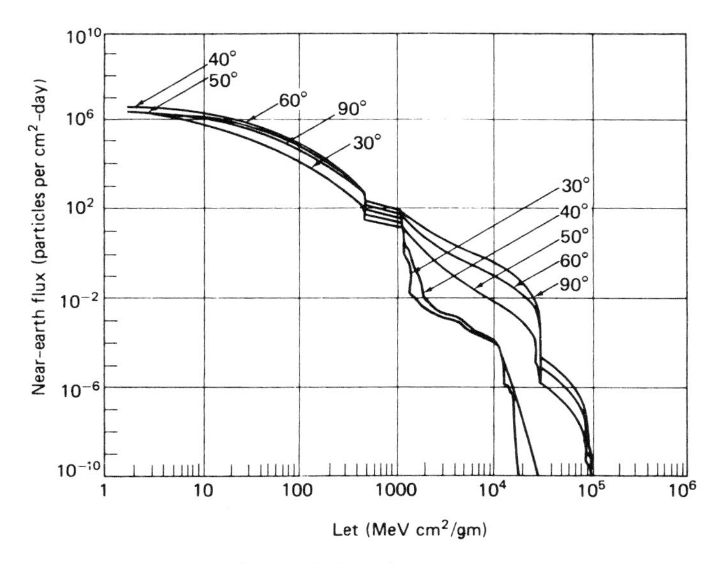

Fig. 8.8 Integral LET spectra (Heinrich Curve) through 25-mil aluminum walls of spacecraft at 460 km altitude for various inclination angles measured from the equator.[67] Figure © 1983 by the IEEE.

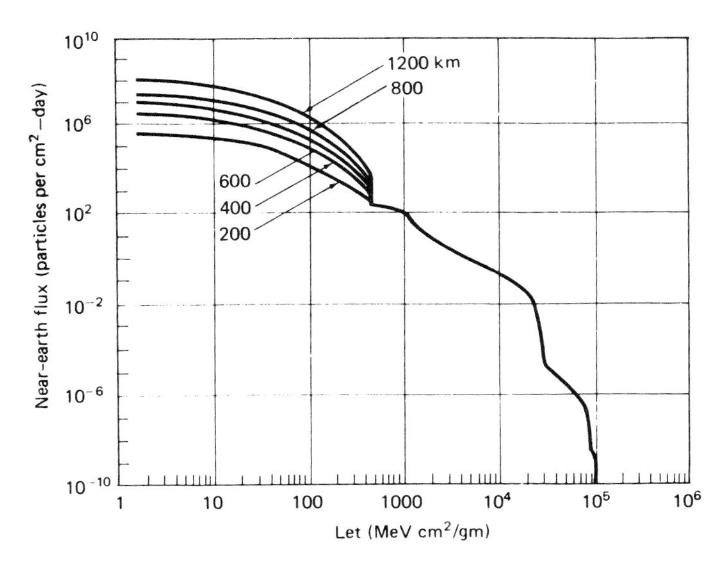

Fig. 8.9 Integral LET spectra (Heinrich Curve) through 25-mil aluminum walls of spacecraft at 60 degrees inclination measured from equator for various altitudes.[67] Figure © 1983 by the IEEE.

for estimating this are not as robust as for cosmic ray heavy ion SEU. As already mentioned, as feature size is further reduced by advancing microcircuit technologies, incident protons will be capable of producing SEU directly. This is because the relatively meager proton-induced SEU ionization charge will nevertheless become sufficient to upset the exceedingly small information nodes in the submicron circuitry of the future.

If the desired Heinrich curves for medium-altitude orbits are available, such as given in Figs. 8.8 and 8.9, knowledge of the threshold LET, L_c, of the device will yield a corresponding flux $\Phi(L_c)$ value. The product of this flux value, with a known average SEU cross section $\bar{\sigma}_L$ (i.e., $\bar{\sigma}_L \Phi(L_c)$), will yield an estimate of the SEU error rate.

However, there is a more accurate computational method for obtaining the proton-induced SEU error rate, which begins with an empirically derived expression for the proton-induced SEU cross section. In units of $10^{-12}\,\text{cm}^2$ per bit per proton/cm^2 it is[19]

$$\sigma_p(A, E) = (24/A)^{14} \cdot (1 - \exp\text{-}0.18(0.18/A)^{1/4}(E - A)^{1/2})^4; \qquad E \geqslant A \tag{8.22}$$

where E (MeV) is the energy of the incident proton, and A (MeV) is an experimentally determined proton SEU energy threshold unique to the device under consideration. That is, $\sigma_p(A, E)$ vanishes for $E < A$, and is depicted in Fig. 8.10. Then, with known proton flux at medium- to low-orbit altitudes,[20,21] $\Phi_p(E)$, which includes the Van Allen belts, and an attenuation factor to account for spacecraft material shielding, the integral

$$E_{pr}(A) = \int_A^\infty \sigma_p(A, E)\, \Phi_p(E)\, dE \tag{8.23}$$

provides the proton-induced SEU error rate, as shown in Fig. 8.11.

$E_{pr}(A)$ can be compared with that due to galactic cosmic rays, as in Fig. 8.12. From this figure, it is seen that the galactic cosmic ray SEU error rate dominates that for protons, except in the central portion of the inner Van Allen belt. This includes the effects of earth magnetic field-solar wind shielding, and the shadowing of the spacecraft afforded by the partial occlusion of the cosmos by the earth itself.

The integral of Eq. (8.23) results in the proton-induced SEU error rate, E_{pr}, which is available in a number of graphs and tabulations.[22] Some of the latter are provided in Section 12.6, where details for obtaining proton-induced SEU error rates for various user devices are

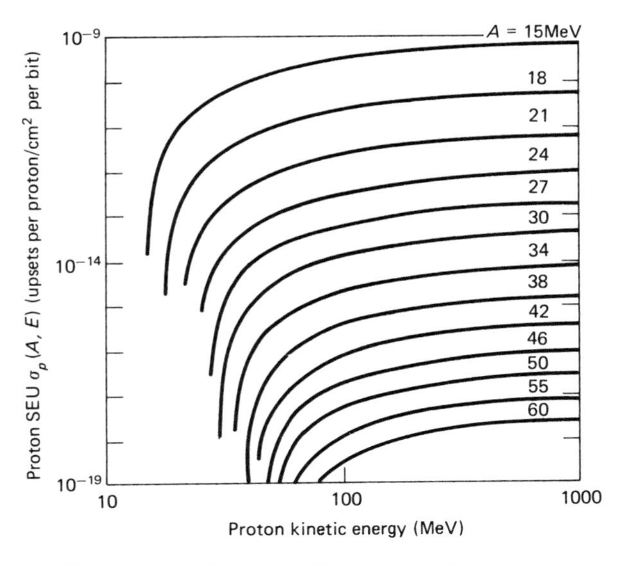

Fig. 8.10 Proton SEU cross section $\sigma_p(A, E)$ versus incident proton energy for various values of the energy threshold parameter A.[40]

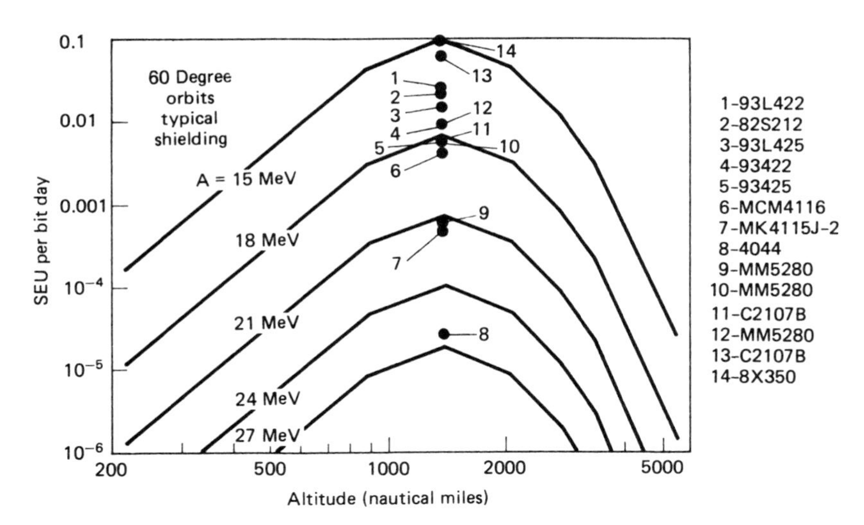

Fig. 8.11 SEU due to protons for certain devices versus altitude and energy threshold parameter A for near-earth orbits inclined at 60 degrees from the equator.[19] Figure © 1983 by the IEEE.

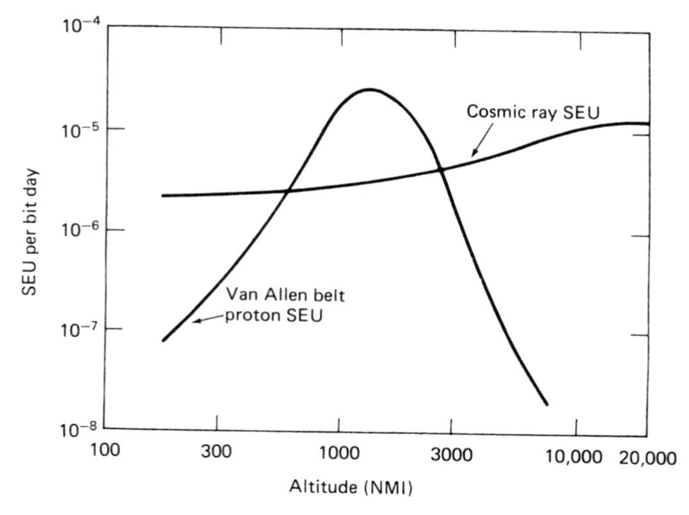

Fig. 8.12 Comparison of SEU error rate versus altitude for 60° circular orbits for moderately sensitive devices ($A = 25$) from Van Allen belt protons and cosmic rays.[40]

given. Figures 8.13 and 8.14 depict E_{pr} versus orbit altitude for various orbit inclination angles, for minimum and maximum solar activity, respectively. It is seen from these figures that, for a given altitude, E_{pr} does not change very much as a function of inclination angle, except for the 30° degree inclination during a solar maximum at low altitudes. Such inclinations are equivalently near-equator orbits, and so are influenced maximally by the earth magnetic field–solar wind interaction. This is especially the case at low altitudes, since the earth's magnetic field strength increases as the orbit drops well below the Van Allen belt proton maximum density altitude. Further, for dense beams of low Z particles (such as protons, deuterons, and tritons), a multiple particle SEU event (not a single-particle multiple event upset) can occur, due to the coherent interaction of two or more such beam particles, which thereby enhance the corresponding SEU cross section[29].

As mentioned earlier, single event upset has been observed in high-flying aircraft. There is evidence that this SEU is due mainly to neutrons and protons at these altitudes interacting with the device silicon or gallium arsenide material to produce heavy particle recoil reactions,[63] as discussed in Section 8.1. These recoils then have sufficient ionization energy to cause SEU. Figure 8.15A depicts SEU error rate versus aircraft avionics altitudes for MOS RAM circuitry with $L_c = 0.5\,\text{MeV cm}^2/\text{mg}$.[63] For these low L_c levels, it is seen that the corresponding SEU error rates are small but perceptible. Figure 8.15F depicts SEU rates versus critical charge at high-flying aircraft altitudes, as well as at sea

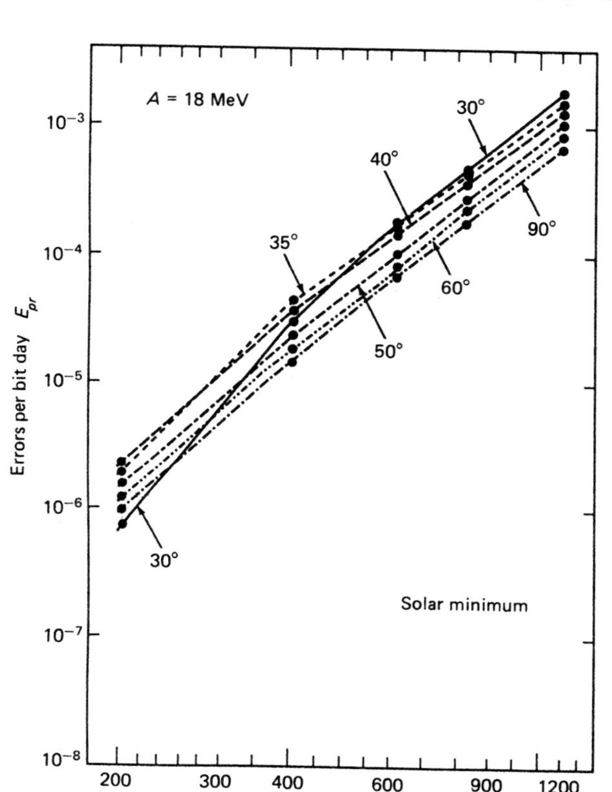

Fig. 8.13 Proton SEU error rates versus orbit altitudes for shielded devices at various orbit inclination angles corresponding to solar activity *minima* for $A = 18$.[22] Figure © 1984 by the IEEE.

level.[69] Specifically, it shows SEU due to (a) low-energy proton interactions and (b) neutron interactions at 55,000 ft, and (c) muon interactions, (d) low-energy proton interactions, and (e) neutron interactions, all at sea level. Hence, it is seen that SEU is relatively small, but not zero, at sea level. SEU can affect sensitive microcircuits at sea level, such as charge coupled devices (CCDs), as already mentioned.

There is an important point to be made at this juncture with respect to the simultaneous competition between proton-induced SEU and the corresponding ionizing dose damage. Recall that the latter is the time integral of dose rate, irrespective of its source. It has been found that the cumulative effect of ionizing dose in the device of interest is such as to increase its sensitivity to SEU, especially for low-proton SEU cross section devices.[40] As already alluded to, this is because the cor-

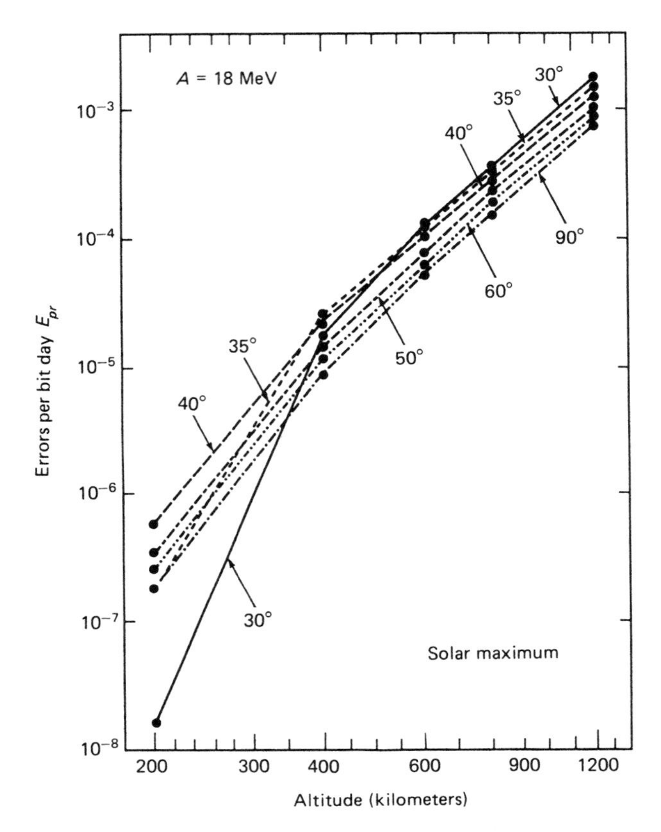

Fig. 8.14 Proton SEU error rates versus orbit altitudes for shielded devices at various orbit inclination angles corresponding to solar activity *maxima* for $A = 18$.[22] Figure © 1984 by the IEEE.

responding ionizing dose damage can mask SEU in the device. This can pose operational difficulties for spacecraft whose orbit passes through the Van Allen belts. As the spacecraft is accumulating ionizing dose from the trapped belt electrons (its ionizing dose main contributor), it is simultaneously being exposed to proton-induced SEU. Also, when measuring device cross sections that are very small (e.g., when using a proton accelerator), the device must be exposed for relatively long periods to collect enough induced charge to obtain a credible SEU cross section measurement, from a statistical standpoint. This can be inimical to the cross section measurement, since the device can fail because its ionizing dose threshold has been reached first. Hence, for certain devices of interest, it may be necessary to permanently degrade a portion of a

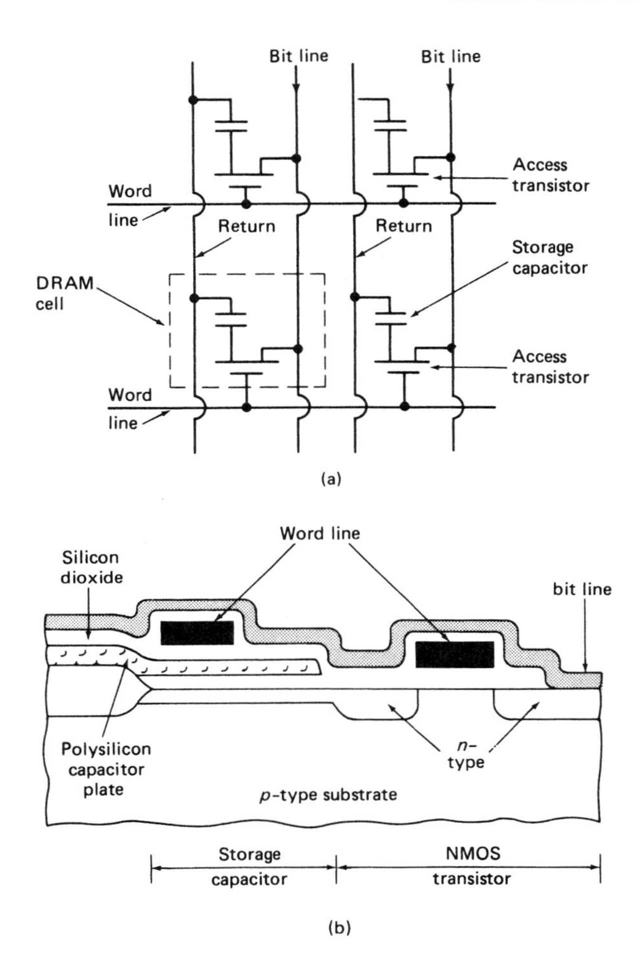

Fig. 8.15 MOSFET *n*-channel DRAM cell: (a) DRAM schematic fragment of four DRAM cells; (b) Vertical section through a single DRAM cell.

part lot in order to measure its correspondingly small SEU cross section. Figure 8.15B depicts the decrease in proton-induced SEU error rate lower bound, as a function of the ionizing dose damage threshold for a family of MOS RAMs, orbiting through the Van Allen belts at 60° with respect to the equator.[40] The figure holds for a single part, in the sense that a test sample number of hard parts ($\sim$1 Mrad (Si)) would be needed for determining SEU error rates less than 10^{-10} per bit day.

In a similar manner, there is an apparent reduction in the proton-induced SEU threshold due to simultaneous competition from dose rate sources. This is shown in Fig. 8.15C, which is a plot of absorbed dose

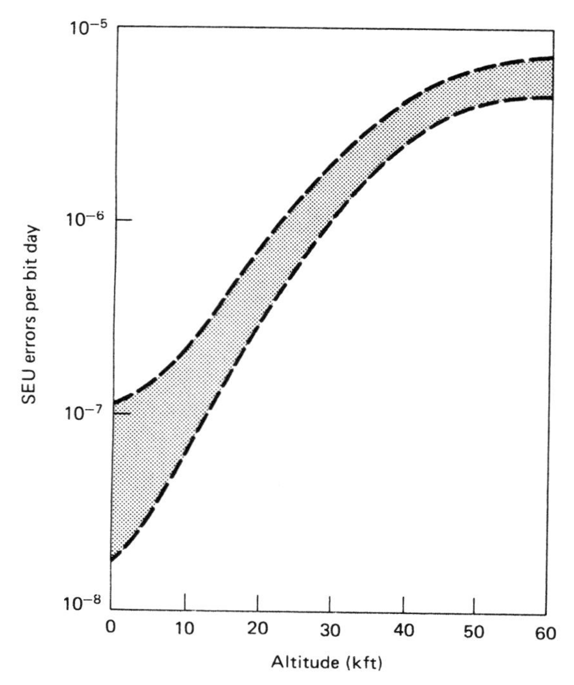

Fig. 8.15A Estimated SEU error rate versus altitude for MOS RAM integrated circuitry whose $L_c = 0.5\,\text{MeV}\,\text{cm}^2/\text{mg}$.[63] Figure © 1989 by the IEEE.

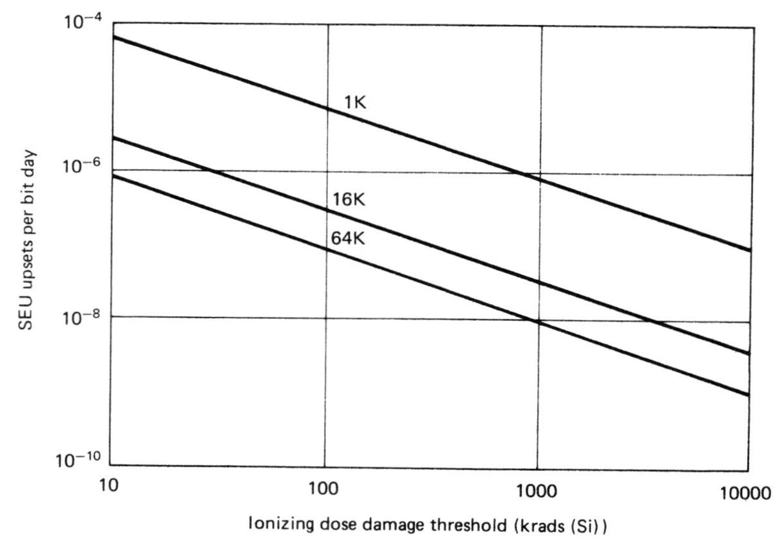

Fig. 8.15B Single part-predicted proton-induced SEU error rate lower bound versus ionizing dose damage threshold for a family of MOS DRAMs in a 60° orbit through the Van Allen belts.[40]

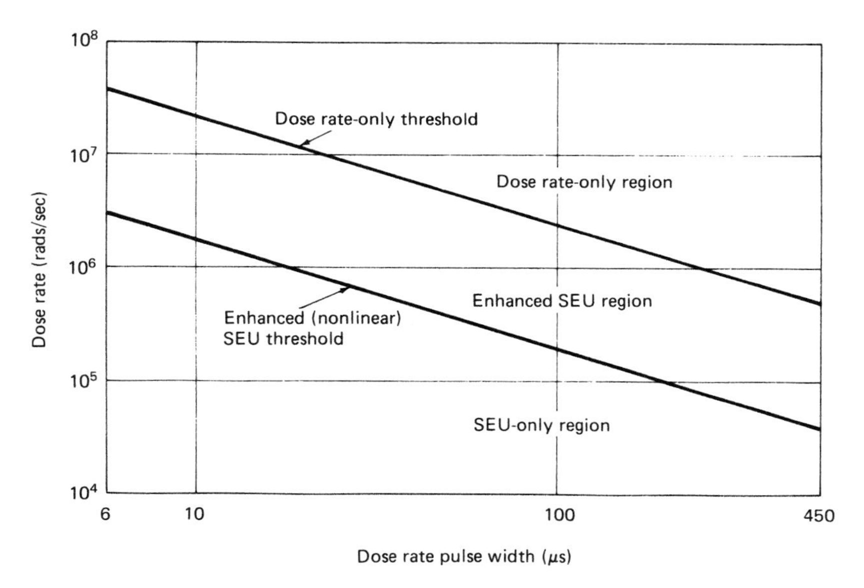

Fig. 8.15C Dose rate versus its pulse width for a CMOS SRAM showing dose rate-only threshold and nonlinear dose rate SEU-enhanced threshold.[65] Figure © 1987 by the IEEE.

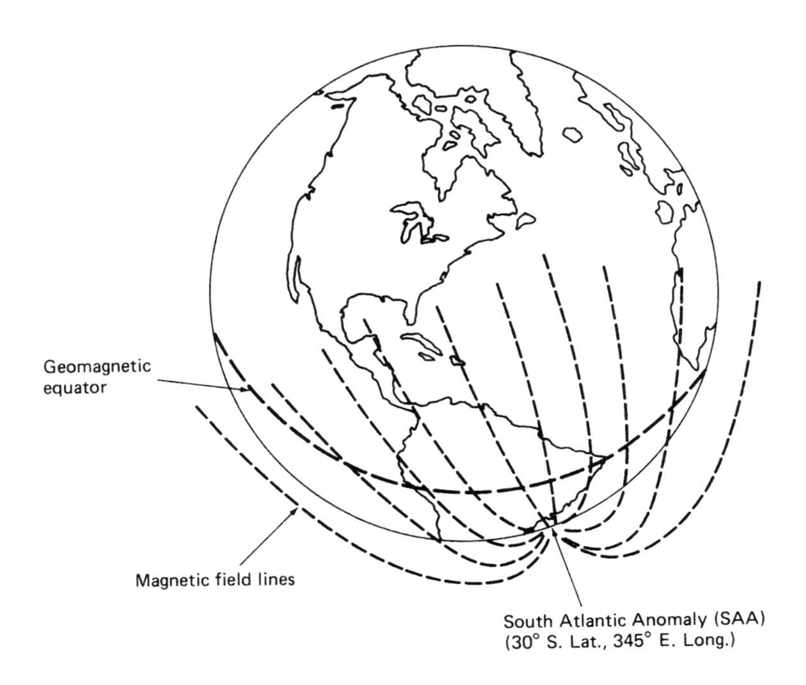

Fig. 8.15D South Atlantic Anomaly.

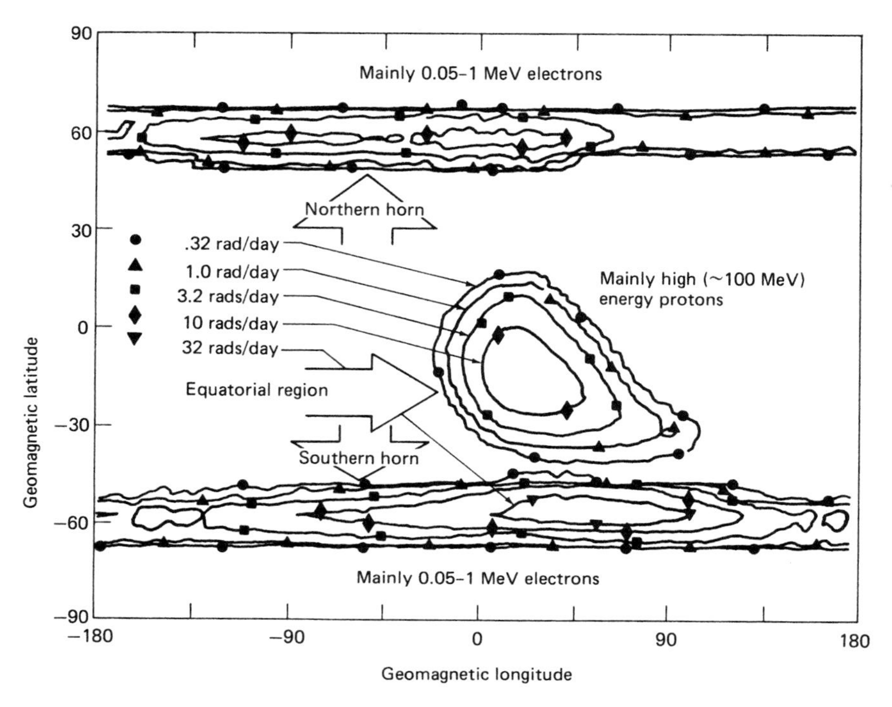

Fig. 8.15E Dose rate fine structure over the South Atlantic Anomaly measured within (840 km, 98.8° Inclination angle) DMSP F-7 spacecraft.[28] Figure © 1984 by the IEEE.

rate versus its corresponding pulse width for a CMOS SRAM.[65] The figure depicts two thresholds. The upper threshold curve is the dose rate upset threshold for this device, now construed as due to spurious photocurrent, as discussed in Section 7.2, plus rail span collapse.[66] Rail span collapse is the transient reduction in the $V_{DD} - V_{SS}$ voltage difference across a particular RAM cell. The V_{DD} drop and V_{SS} rise are due to photocurrent flow away from individual cells into the V_{DD} and V_{SS} power supply metallization runs on the chip, of finite resistance, which thereby produce these voltage drops. This results in a drop in the effective SEU critical charge at a sensitive node in the RAM cell.[65] The lower left region in Fig. 8.15C is where the usual SEU effects predominate. This is exemplified as the region where the number of SEUs per proton accelerator pulse is proportional to the beam fluence. This region is bounded by the lower threshold curve, which is defined as the onset of dose rate for which the SEU rate begins to increase nonlinearly. This is primarily due to the proton-induced nuclear ionizing particle reactions,

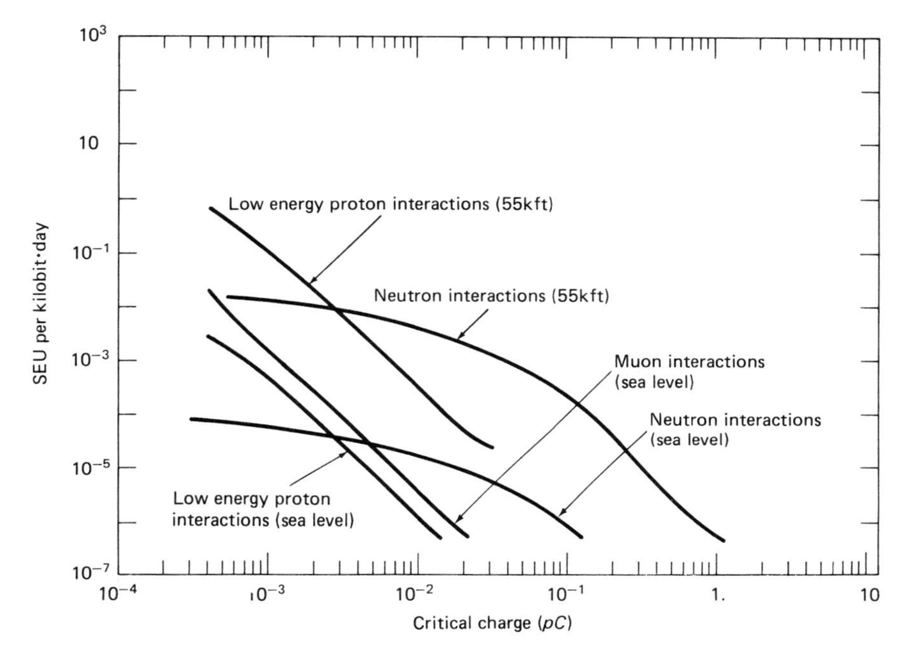

Fig. 8.15F SEU rates at high latitudes versus critical charge, due to protons, neutrons, and muons at high-flying aircraft altitudes and at sea level.[69] Figure © 1988 by the IEEE.

combining with the background dose rate-induced ionization.[65] The upper right region is where dose rate effects due to photocurrents per se predominate over any SEU reactions.

Alpha Particle SEU

Alpha particle-induced SEU errors will also appear in integrated circuits of high packing density and low speed-power product, as mentioned earlier. For example, LSI memories rely on a given amount of charge per cell to represent information, as alluded to in parts (a) and (b) of Fig. 8.15. If the charge level threshold is not maintained by the refresh circuitry, in the case of the dynamic RAM (DRAM), information loss due to alpha particle-induced charge will increase rapidly, as the alpha source is present in the package and chip, as discussed earlier. This holds, of course, for all SEU particle damage. Generally, SEU rates also increase when power supply voltages become meager, especially when the alphas are incident at large angles to the device normal.[24]

Figure 8.15 part (a) depicts a dynamic MOS RAM (DRAM) fragment of four cells. The horizontal lines are called *word* or *select* lines, and the vertical lines are called *bit* or *data* lines. Peripheral memory buffer circuits translate requests to read from or write into specific memory cells signals for the corresponding word and bit lines. A particular DRAM cell at the intersection of a word and bit line consists of one MOS access transistor plus a storage capacitor. The storage capacitor is monolithically built-into the chip and is made up of a thin layer of silicon dioxide as the dielectric. One plate of the storage capacitor comprises one end of the access transistor elongated to extend some distance under the silicon dioxide, as in Fig. 8.15 part (b). The other plate is a layer of conducting polysilicon extending over the silicon dioxide dielectric, as shown in the figure.

To write, the normally off access transistor at the pertinent location is momentarily turned on by a voltage pulse on its gate from the corresponding word line. Almost simultaneously, the data voltage pulse reaches and charges only the appropriate storage capacitor through the corresponding bit line and the on access transistor. After these pulses, the access transistor reverts to its off state.

To read, the access transistor is again turned on as above. This now connects the particular cell by way of the bit line to a sense amplifier, which sends the signal to the input/output circuitry. The latter amplifier refreshes the cell by restoring the bit line to the original voltage (charge) written into the cell.

The alpha-particle-induced electron-hole pair production degrades the information in the cell in a manner similar to cosmic rays. It was determined experimentally for a number of device types that the SEU error rate is linear with the incident alpha particle flux over eight decades of flux.[23] Alpha particles emanating from sources within the device package are the principal causative factor of SEU therein. As mentioned, the sources are the very heavy actinide series of nuclei in the periodic table, such as uranium and thorium. Most of these actinides decay naturally in radioactive chains, of which alpha particles are a by-product, to form stable lead ultimately. For example, each ^{238}U nucleus produces eight alpha particles to decay to ^{208}Pb, and each ^{230}Th nucleus produces six alphas in its decay to ^{208}Pb.

A sensitivity factor, S, can be defined as the number of SEU errors produced per alpha particle. S is given by

$$S = (A_{\text{stor}}/A_{\text{cell}}) \int_0^\infty p(E; Q_c) N(E)\, dE \tag{8.24}$$

where $p(E;\ Q_c)$ is the probability that an alpha particle of energy E causes an SEU. $N(E)$ is the alpha flux energy spectrum, and $A_{\text{stor}}/A_{\text{cell}}$

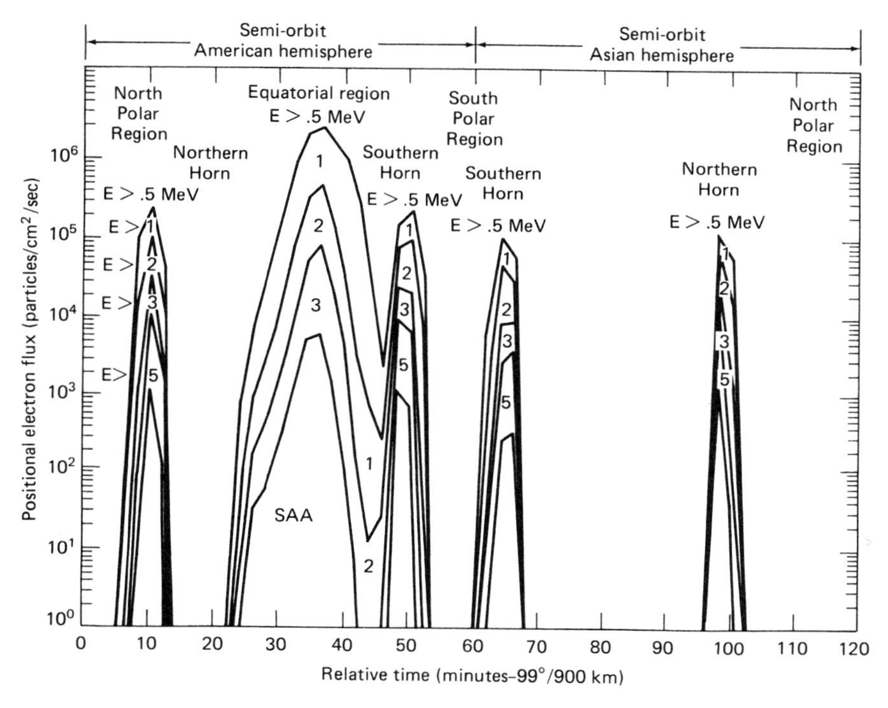

Fig. 8.16 Most severe pass through the South Atlantic Anomaly—instantaneous, integral, omnidirectional, trapped, electrons.[27] Figure © 1988 by the IEEE.

is the fraction of alpha particles that impinge on the sensitive region of the cell. Then the SEU error rate, due to alpha particles E_α, is[26] approximately

$$E_\alpha \cong A_{\text{cell}} \Phi_\alpha S \tag{8.25}$$

where Φ_α is the incident alpha flux of aspect angle to insure the possibility of an SEU. For a particular 4-K bit MOS RAM, $A_{\text{cell}} = 0.027\,\text{cm}^2$, $\Phi_\alpha = 3.8$ per cm^2/h, and a computed $S = 0.015$ SEU per alpha from Eq. (8.24). This particle yields an E_α of about $1.5 \cdot 10^{-3}$ SEU errors per hour.

8.4 SOUTH ATLANTIC ANOMALY, LOW EARTH ORBITS

The magnetic field of the earth has a certain detailed structure that is of concern to the computation of space radiation incident on susceptible electronic systems. In the southern hemisphere, the earth's magnetic field is offset by approxiately 11 degrees from the earth's axis of rotation, and displaced about 500 km toward the western pacific.[27] Hence, there is a polar-like dip in the magnetic field in the vicinity of Brazil, wherein magnetic field lines reenter the earth. This not only produces a singularity in the magnetic field for cosmic rays to intrude, but allows the Van Allen belts to extend down to the atmosphere of the earth. This South Atlantic Anomaly (SAA) is responsible for most of the Van Allen belt radiation received by spacecraft in low and very low earth orbit altitudes, and low inclination angles[27]. On the opposite side of the earth, the Southeast Asian Anomaly evokes stronger field values, but the corresponding radiation is located at higher altitudes[27]. Figures 8.15D, 8.15E, and 8.16, respectively, depict the geographical magnetic field lines in the vicinity of the SAA, its trapped radiation levels, and the associated electron flux, with respect to the relative time of one complete orbit at 900 km altitude, inclined at 99 degrees clockwise with respect to the equator. From Fig. 8.16, it is noticed that there are time periods that are essentially electron flux free, even for Van Allen trapped electron energies down to as low as 0.5 MeV. It can be shown that certain complete orbits during the spacecraft precessional cycle around the earth are electron flux free.[27] This means that with respect to ionizing dose levels (not SEU), extravehicular activity possibly involving astronauts could take place during these flux-free orbits. This does not include extragalactic radiation from gamma ray bursts, whose mean gamma pulse energy flux incident at the top of the earth's atmosphere is about 10^{-2} ergs per cm^2 s, with energies greater than 0.1 MeV.[72] This corresponds to an ionizing dose rate of about 1 krad (Si) per year.

However, the SAA consequences are such that the most intense and penetrating radiation encountered is in the form of protons during the spacecraft SAA span duration. There are a number of models for corresponding low earth orbit proton fluxes.[27] The proton energies at low earth orbits are much larger than those at geosynchronous altitudes. At the latter altitudes, the protons are so "soft" as to be stopped by 50 mils of aluminum, which implies that they are of no significant radiation import. Protons at low earth orbit altitudes impact the spacecraft geographically during its passage through the SAA. The proton fluxes are normally not excessive in that usually only "spot" shielding for certain individual radiation-susceptible parts (Section 13.2) would be required as a worse-case situation.

8.5 FUNNELING AND FUNNELING MODELS

It has been shown that, when a high-energy cosmic ray penetrates a reverse biased semiconductor device through its depletion layers and

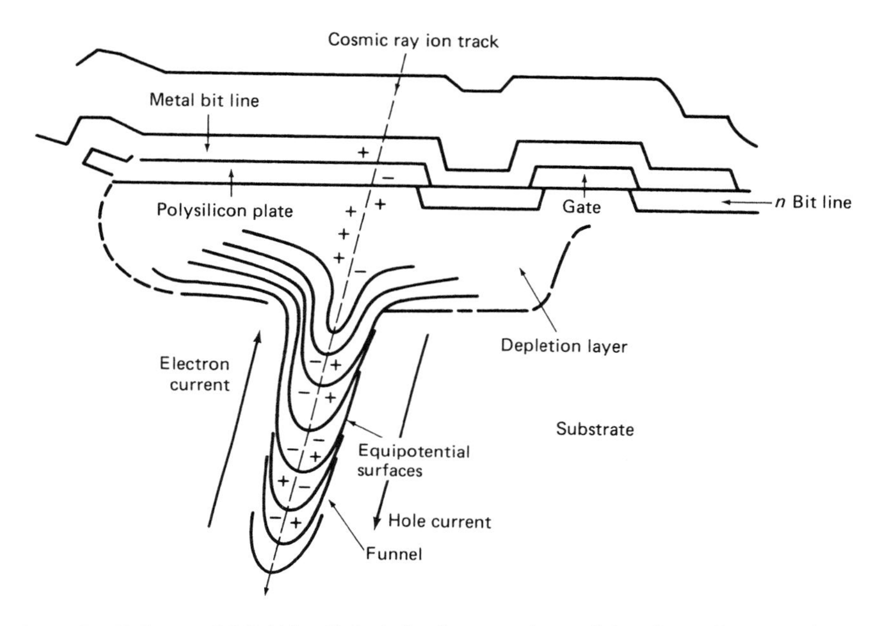

Fig. 8.17 Collapse of DRAM cell depletion layer equipotential surfaces along cosmic ray track to produce funnel in device bulk.

bulk, it produces a track of ions and electrons in its wake due to the ionizing interactions with the material.[30,31] The presence of the track temporarily collapses the depletion layer and so distorts the equipotential surfaces of the depletion layer electric field in the track vicinity. This distortion results in a nesting of funnel-shaped, equipotential surfaces that extend into the substrate bulk along the track, as in Fig. 8.17. This "funneling" of potential surfaces produces a very large gradient, i.e., a large electric field, such that the electrons produced from ionization along the track are propelled back up the funnel into the depletion layer. There is a corresponding hole current directed down into the bulk material. The electrons result in a large increase in charge collected in the depletion layer over that were there no funnel, thus heightening the probability that the critical charge is collected at the device charge storage node. If the particle track impales or passes very near the node most of the charge in the funnel is collected by the node depletion layer in fractions of a nanosecond following the penetration of the cosmic ray particle. If the track lies remote from the node depletion layer, the charge collection time is much longer, reckoned in nano- and microseconds, as its mode of transport is principally by diffusion. After the collection of charge to where its remaining density in the track is comparable to the substrate dopant concentration, the disturbed depletion layer field has relaxed back to its original state. It should be noted that the expressions for SEU error rate, such as Eq. (8.12), do not account for the additional charge collected due to funneling.

A qualitative but detailed description of the funneling phenomenon is based on a number of computer studies and experimental verifications.[30–33] It can be construed as taking place in four phases:

A. The first phase is that of the passage through the material of interest of highly charged energetic heavy cosmic ray ions, which in turn ionize this material to form a track consisting of a wake of ionized particles. This track of electron-hole plasma undergoes energy loss in the material bulk, ultimately coming into thermal equilibrium with it (thermalization), in picoseconds following the strike of the incident heavy ion, within its cylindrical volume of radius, initially less than approximately $0.1\,\mu m$. The track density in this epoch is about $10^{18}-10^{21}$ ionized particles per cm^3, which is orders of magnitude greater than the nominal substrate dopant concentration. This track is considered to traverse the device junction depletion layer, saturating it along the immediate neighborhood of the track. However, this charge saturation is subsequently depleted. As a result, the electric field in the depletion region initially becomes reduced in magnitude as the depletion layer equipotential lines about the track extend down into the substrate, to form a funnel, as depicted in Fig. 8.17.

B. Following thermalization of the track (as described), the cylindrical track immediately expands radially by ambipolar diffusion, since the track charge density is extremely high. Strong electrostatic restoring fields within the track column maintain charge neutrality. During this phase, the junction depletion layer equipotential lines penetrate deep into the substrate, thereby extending the funnel.

C. During this epoch in the track expansion, separation of the electrons and holes occurs mainly on the track periphery, where the track densities are now comparable to the dopant concentrations. The holes are drawn away from the track by the vertical components of the electric field, and down into the substrate. Also, the electrons drift very rapidly up the track and are collected at the sensitive node.

D. As the track density is reduced, the depletion layer begins to revert to its initial configuration in the vicinity of the track, first near the periphery, and then in towards the center. During this phase, charge separation occurs because the track's electrons and

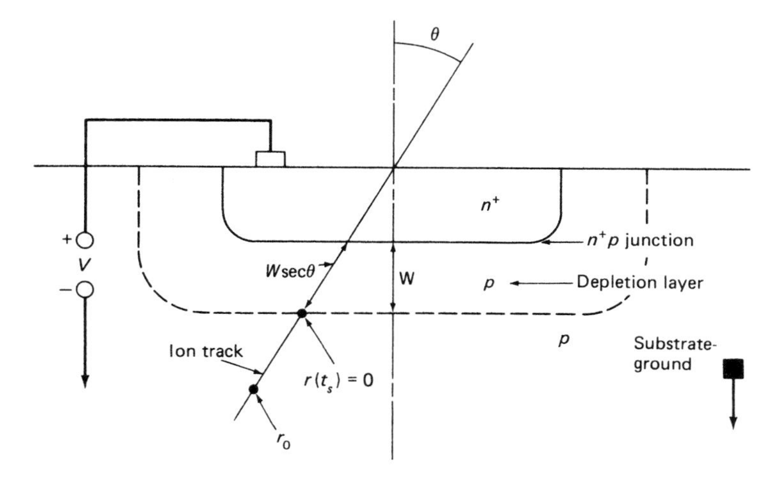

Fig. 8.18 n^+p junction during incident ion strike.

holes are being pulled apart by the screened electric fields. These fields then begin to decrease rapidly, and then, with the dissolution of the track, the junction relaxes back to its original state.

A number of funneling models have been put forward, so that the additional charge they generate (which is often the predominant amount) can be included properly into the SEU error computation. A published model,[34] now discussed, involves the time-dependent funnel length, D_c, following the cosmic ray heavy ion strike through a junction depletion layer, as depicted in Fig. 8.18. The initial (prompt) charge, Q_c, collected due to the electric field within the funnel, as opposed to the delayed charge collected from diffusion outside the funnel, is given by[34]

$$Q_c = e\int_0^{D_c} N_0(z)\,dz = e\bar{N}_0 D_c \tag{8.26}$$

where D_c is the effective funnel length, and $N_0(z)$ is the initial line charge density along the track, composed of hole-electron pairs produced by the incident cosmic ray heavy ions. The track coordinate z is measured from the junction interface, as in Fig. 8.18. $\bar{N}_0$ is an average line charge density taken over the funnel length D_c. Q_c includes both the charge generated in the depletion layer, as well as that from the funnel.

Let D_c be represented by

$$D_c = \bar{v}_d \tau_c + x_d \tag{8.27}$$

where τ_c is the charge collection time, x_d is the initial depth of the depletion layer in the funnel extending into the substrate, and $\bar{v}_d$ is the mean drift velocity of charge upwards toward the junction during τ_c. Assume that the substrate minority carriers are electrons, so that $\bar{v}_d = \mu_n E_L$, where E_L is the average longitudinal electric field given by $E_L = V_0/D_c$. V_0 is the potential between the charge collecting sensitive node and the substrate, and $\mu_n(E_L)$ is the corresponding field-dependent mobility. Inserting these expression into Eq. (8.27) yields a quadratic equation in D_c, whose pertinent root is

$$D_c = (1/2)(x_d + (x_d^2 + 4\mu_n \tau_c V_0)^{1/2}) \tag{8.28}$$

For very small x_d (shallow depletion layer), $D_c \cong (\mu_n \tau_c V_0)^{1/2}$. For very large E_L, the drift velocity becomes asymptotically constant (saturates), so that $\bar{v}_d \cong v_{\text{sat}} \cong 10^7$ cm per sec in silicon. Then Eq. (8.27) becomes

$$D_c = v_{\text{sat}} \tau_c + x_d \tag{8.29}$$

In the initial phases of ion track formation, it is mainly expanding radially by ambipolar diffusion, as discussed. Its radial charge density profile decreases from the central axis of the track to its periphery, where it is of the order of $n \cong p \cong N_A$, the background dopant concentration. It is assumed that the time-dependent ambipolar root mean square diffusion length $r_{rms}(t) = (\overline{r^2(t)})^{1/2} = 2\hat{\beta}(Dt)^{1/2}$, with $D = 25\,\text{cm}^2$ per second in silicon.[34,35] The time-averaged scaling factor, $\hat{\beta}$, to accommodate high-density tracks due to ions heavier than alpha particles, relates the radius of the expanding track column to the ambipolar diffusion length, L_D (viz., $r_{rms} = \hat{\beta} L_D = 2\hat{\beta}(Dt)^{1/2}$). $\hat{\beta}$ is obtained by assuming that the track density radial profile has a gaussian shape, whose level falls to N_A on the track periphery. Therefore, the track charge density is assumed given by $n_D(r, t) \simeq (N_D(0)/4\pi Dt) \exp - (r^2/4Dt)$. At the track periphery, $n_D = N_A$, with a corresponding radius identified by $r_{rms} \cong 2\hat{\beta}(Dt)^{1/2}$, where $\hat{\beta} = ((\ln(N_D(0)/4\pi N_A Dt))^{1/2})_{\text{av}}$, a slowly varying function. The subscript "av" indicates a time average over the duration of the funnel.

In the initial phase, charge neutrality holds strictly. In the second (or charge drift) phase, the charged particles begin to separate to account for track charge motion and the ultimate collection of electrons by the sensitive node. Nevertheless, the electron collection rate equals the hole escape rate into the substrate to still maintain charge neutrality in the track. The radial reduction rate of the track line charge density $\dot{N}_D(t)$ satisfies

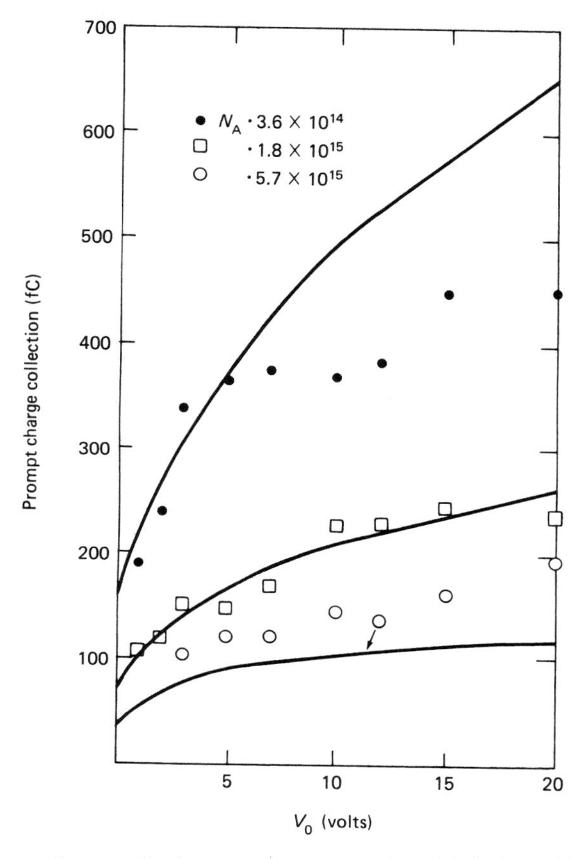

Fig. 8.19 Prompt charge collection measurements and model fit for *p*-Si exposed to Be ions as a function of applied bias.[38] Figure © 1985 by the IEEE.

$$\dot{N}_D(t) = -2\pi r_{rms}(t)J_p(t) = -4\pi N_A\hat{\beta}v_p(Dt)^{1/2} \tag{8.30}$$

since the escaping hole current $J_p = N_A v_p$. The corresponding integral is

$$N_D(t) = N_D(0)(1 - (t/\tau_c)^{3/2}) \tag{8.31}$$

where τ_c is the time necessary to deplete the track charge density near the junction interface, as in Fig. 8.17. It is seen that τ_c can be identified from the above as

$$\tau_c = (3N_D(0)/8\hat{\beta}\pi N_A v_p D^{1/2})^{2/3} \tag{8.32}$$

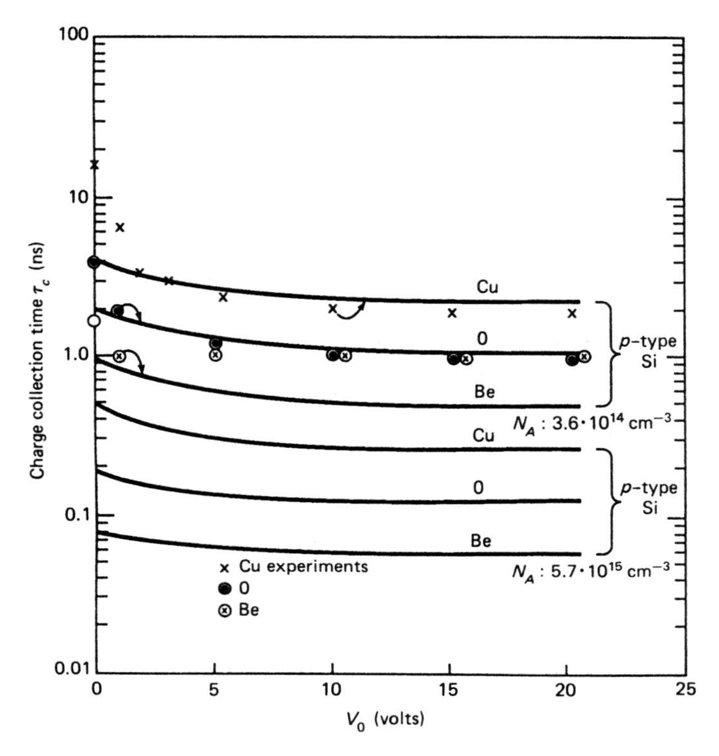

Fig. 8.20 Charge collection time versus bias voltage for Cu-, O-, and Be-incident SEU inducing ions compared with experiments in p-type Si.[38] Figure © 1985 by the IEEE.

Substituting Eqs. (8.32) and (8.28) into Eq. (8.26) yields the prompt charge produced, including both the funnel and the shallow depletion layer, viz.,

$$Q_c = e\bar{N}_0(\mu_n V_0)^{1/2}(3N_D(0)/8\hat{\beta}\pi N_A v_p D^{1/2})^{1/3} \tag{8.33}$$

Note that $\bar{N}_0$ is the mean track line density averaged over the length D_c, whereas $N_D(0)$ is the initial track line density near the junction interface. The difference between $\bar{N}_0$ and $N_D(0)$ is significant only for relatively long track lengths (i.e., large D_c). The preceding refers to the case of positive-applied bias on a p-type substrate. For the opposite case, of a negative-applied bias on an n-type substrate the roles of the electrons and holes are reversed. Also, $Q_c \sim \mu^{1/2}$, so that the charge funnel is longer in a p-type silicon substrate, where electrons are the carriers collected at the sensitive node, by the factor $(\mu_n/\mu_p)^{1/2} \cong 3$ in silicon.

This model is fairly well confirmed in experiments using alpha particles, as well as heavier particles with masses up to oxygen,[35] as seen in Figs. 8.19 and 8.20. In most of these models, the total collected charge is found to be much greater than that known to be from the depletion layer alone. This is strong evidence for the existence of a funnel, certainly in *p-n* junctions. Also experimentally confirmed is the fact that the charge funnel is more pronounced in n^+p junctions with *p*- substrates than the reverse, as predicted by this model. Other models,[32] some closely resembling the preceding,[36] yield similar results. Further modifications of this model involve an ion track plasma screening factor[37] based on skin effect, the use of which asserts that more accurate charge collection values due to funneling can be obtained.

For further insight into the electrodynamics of funneling, another model[36] is described, which approaches funnel charge creation and collection from a different vantage point. An additional sidelight of this model is that it takes account of ion tracks incident at an angle from the normal to the plane of the junction. This model was first applied to ion tracks induced by alpha particles, but is easily generalized to heavier incident SEU ions.

After the n^+p junction (as shown in Fig. 8.18) is struck by an incident ion, the electric field within the depletion layer due to the applied voltage, V_0, drives the ion track hole current, I_p, resulting from the incident particle ionization, down into the *p*-type substrate. This current creates an electric field, $J = \sigma E(r, t)$, in the depletion layer and in the funnel, which satisfies

$$I_p(t) = g_1 e \mu_p E(r, t) \tag{8.34}$$

where the ion track charge generation parameter is g_1 (hole-electron pairs per cm), and r is the ion track position coordinate. As discussed, the columnar ion track charge thermalizes in picoseconds, yielding a corresponding track with a radius of about $0.1\,\mu$m. The hole current, I_p, in the track has a linear charge density of $g_1 = 4 \cdot 10^8$ hep per cm, for SEU-inducing alpha particles in silicon. This is equivalent to a track charge volumetric density of $g_1/\pi r^2 \sim 10^{18}$ hep per cm^3. Due to ambipolar diffusion, this density is reduced to $\sim 10^{16}$ hep per cm^3 after approximately 100 picoseconds. By comparison, the current density exterior to the track is negligible if the nominal background dopant concentrations are much lower than track densities, which is the usual case. Dopant concentrations that are comparable to track densities are also dealt with, as will be seen.

Simultaneously with the preceding, the electric field $E(r, t)$ drives an ion track electron current I_n, resulting from the incident particle ionization, up toward the $r = 0$ coordinate, at the lower boundary of the depletion layer as shown in Fig. 8.18. These ion track electrons possess a velocity $v(r, t)$ given by, with Eq. (8.34),

$$v(r, t) = -\dot{r}(t) = \mu_n E(r, t) = \mu_n I_p(t)/e\mu_p g_1 \tag{8.35}$$

The negative sign in front of $\dot{r}$ signifies that the electron charge motion is in a direction toward smaller r, as they move (drift) under the influence of $E(r, t)$. Consider an electron positioned at an extremity, r_0, of the ion track at zero time, such that it is ultimately collected. Then, integrating Eq. (8.35) gives its position $r(t)$ along the track for $t > 0$,

$$r_0 - r(t) = (\mu_n/eg_1\mu_p) \int_0^t I_p(t')\,dt'; \qquad 0 \leqslant r \leqslant r_0 \tag{8.36}$$

An indicator of the charge collection process ending is when no holes remain in the depletion layer. Let t_s be the time at which this occurs, with the corresponding hole charge $Q_p(t_s) = \int_0^{t_s} I_p(t')\,dt'$, which was contained in the depletion layer, and now collected in the substrate. Also, at this time all of the track electrons originally found between r_0 and $r(t_s) = 0$, the lower boundary of the depletion layer, have also been collected, simply from charge neutrality equilibrium considerations. Then, at t_s, Eq. (8.36) yields

$$r_0 = (\mu_n/e\mu_p g_1)\,Q_p(t_s) = (\mu_n/e\mu_p g_1)(eg_1 W\sec\theta) = (\mu_n/\mu_p)\,W\sec\theta \tag{8.37}$$

since $Q_p(t_s) = eg_1 W\sec\theta$ is the collected hole charge that was contained in the depletion layer segment of the ion track, where W is the effective depletion layer width. It is assumed that the hole charge in the depletion layer on the n^+ side of the n^+p junction is negligible. As for the ion track electrons with positions $r \leqslant r_0$ (i.e., within $r_0 = (\mu_n/\mu_p)\,W\sec\theta$ distance to the lower boundary of the depletion layer), they too were collected by drift, as already discussed. Therefore, the total collection length L is simply r_0 plus the length of the ion track in the depletion layer, $W\sec\theta$. Letting $\mathscr{D}$ be the collection depth in the vertical direction, then

$$\mathscr{L} = (1 + \mu_n/\mu_p)\,W\sec\theta; \qquad \mathscr{D} = (1 + \mu_n/\mu_p)\,W \tag{8.38}$$

Hence, the funnel length is given by $\mathscr{L}$, and the funnel depth by $\mathscr{D} = \mathscr{L}\cos\theta$.

For high substrate dopant concentrations greater than $\sim 10^{16}\,\text{cm}^{-3}$, the currents are poorly contained within the ion track, especially in the middle and later phases of ion track formation. As a result, the current begins to spread isotropically into the substrate, beyond the depletion region. Then, for a track of circular cross section of radius a, with disc spreading resistance[33] $r_{sp} = (4a\sigma)^{-1}$,

$$E = J/\sigma = Ir_{sp} = I/4a\sigma = I/4ae\mu_p N_A \tag{8.39}$$

where $\sigma = N_A e\mu_p$, and the spreading resistance $r_{sp} = (4aN_A e\mu_p)^{-1}$, as discussed in Section 3.13. If Eq. (8.39) is used as an expression for the electric field, and inserted into Eq. (8.35), the analogous derivation results in a corresponding collection depth $\mathscr{D}_r$, given by

$$\mathscr{D}_r = [1 + (3\mu_n g_1/4\mu_p N_A)^{1/3}\, W^{-2/3}]\, W \tag{8.40}$$

If $\mathscr{D}$ and $\mathscr{D}_r$ are evaluated numerically, they yield similar results, and both yield an $N_A{}^{-1/2}$ dependence. This suggests that the models are not sensitive to the assumptions for current flow patterns. For n-type substrates, μ_n/μ_p is replaced by μ_p/μ_n. There is a corresponding funnel on the n^+ side of the n^+p junction, but evaluation of $\mathscr{D}_r$ using Eq. (8.40) gives a negligible length, as alluded to earlier.

After the ion strike, the voltage drop in the quasi-neutral region (one definition of which is the region where $n - p \simeq n_0 - p_0$, which is principally exterior to the track) is given by the product of the carrier current and the resistance represented by the ion track in the depletion layer, plus the spreading resistance from the bottom of the depletion layer, where the track enters the substrate. The current, I_0, corresponds to the applied voltage, V_0, plus the junction diode drop of 0.7 volts across the quasi-neutral region. This results in the temporary collapse of the depletion region in the vicinity of the ion track, as discussed. The current, I_0, is given by

$$I_0 = (V_0 + 0.7)\, G \tag{8.41}$$

where the conductance, G, is obtained from

$$G^{-1} = \mathscr{L}'/eg_1(\mu_n + \mu_p) + 1/4ae\mu_p N_A \tag{8.42}$$

The first term is the resistance of the ion track spanning the depletion layer. $\mathscr{L}'$ is the length of that track segment, and $[eg_1(\mu_n + \mu_p)]^{-1}$ is the corresponding resistance per unit length. The second term is the

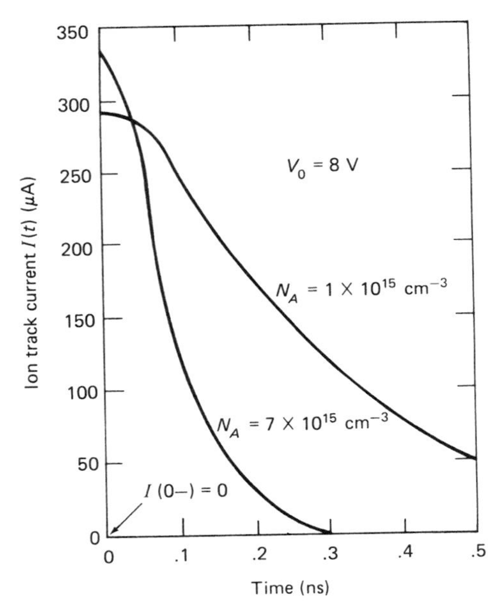

Fig. 8.21 Ion track current I versus time.[36] Figure © 1982 by the IEEE.

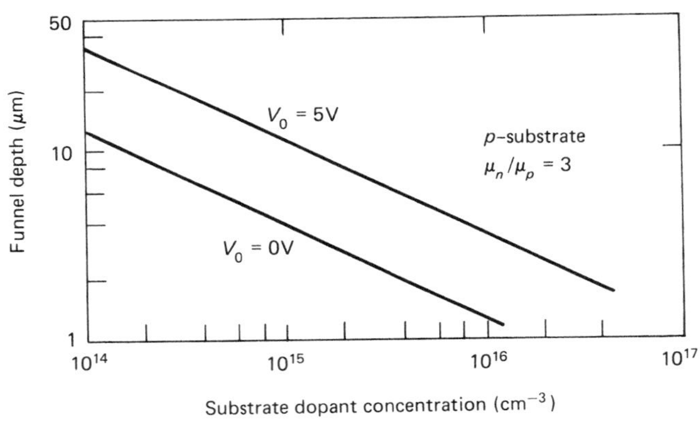

Fig. 8.22 Funnel depth versus substrate dopant density.[36] Figure © 1982 by the IEEE.

spreading resistance reckoned from the depletion layer end of the track column down into the substrate.[33] For cylindrical ion tracks incident at an angle θ to the junction-substrate interface, the spreading resistance[33] $r_{sp} = sn^{-1}\,(1, \sin\theta)/2\pi a e \mu_p N_A \sec\theta$. $sn^{-1}\,(1, \sin\theta)$ is the complete elliptic integral of the first kind, corresponding to the elliptical cross section of the above track column interface. The latter forms the neck of this spreading resistance, like the center of the transverse base current in Fig. 3.29. For normal incidence ($\theta = 0$), $r_{sp} = (4ae\mu_p N_A)^{-1}$, since $sn^{-1}(1,0) = \pi/2$.

The depletion region width $W(t)$ varies during the funneling process. From Eq. (8.37), for $t \leqslant t_s$,

$$W(t)\sec\theta = \int_0^t I_p(t')\,dt'/eg_1 = \mu_p \int_0^t I(t')\,dt'/(\mu_p + \mu_n)\,eg_1 \qquad (8.43)$$

where $I = I_p + I_n$, and $p \simeq n$ in the depletion region. The voltage drop across the depletion region is approximately $eN_A W^2/2\varepsilon\varepsilon_0$, as discussed in Section 3.4. The voltage drop in the quasi-neutral region, $V_{qn}(t)$ is

$$V_{qn}(t) = V_0 + 0.7 - eN_A W^2/2\varepsilon\varepsilon_0 \qquad (8.44)$$

Then,

$$I(t) = V_{qn}(t)\,G \qquad (8.45)$$

assuming that the conductance G is time-independent. Using Eqs. (8.43), (8.44), and (8.45) to eliminate W and V_{qn}, results in a differential equation in $Q(t) = \int_0^t I(t')\,dt'$

$$\dot{Q}(t) = I_0 - GmM^2Q^2; \qquad Q(0) = 0 \qquad (8.46)$$

where $m = eN_A/2\varepsilon\varepsilon_0$ and $M = \mu_p \cos\theta/(\mu_n + \mu_p)\,eg_1$. Its integral, the collected charge, is

$$Q(t) = [(I_0/Gm)^{1/2}/M]\tan h[(I_o Gm)^{1/2}\,Mt] \qquad (8.47)$$

and the total collected charge is $Q(\infty) = (I_0/Gm)^{1/2}/M$. The corresponding current $I = \dot{Q}$ is

$$I(t) = I_0 \mathrm{sech}^2[(I_0 Gm)^{1/2}\,Mt] \qquad (8.48)$$

which is plotted in Fig. 8.21 for two different background dopant concentrations. Note that $I(t)$ in the figure has an initial jump discontinuity, since $I(0-) = 0$ and $I(0+) = I_0$, which is an artifact of the model. Actually $I(t)$ has an initial finite slope, increasing to a maximum value very early in time.[36] The collected charge $Q(t)$, and funnel depth (Figs. 8.22 and

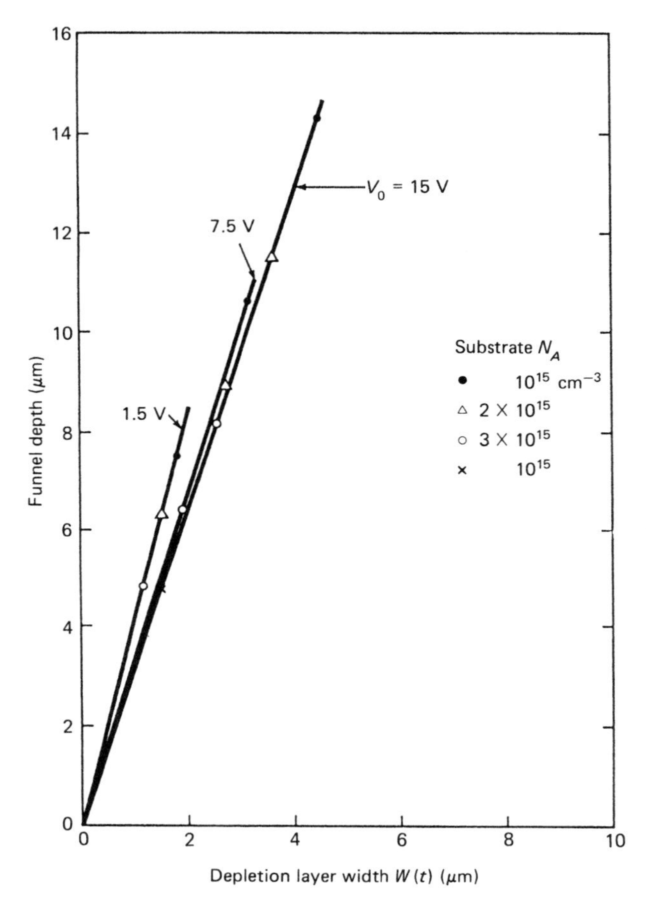

Fig. 8.23 Funnel depth versus depletion layer width compared with experimental data.[36] Figure © 1982 by the IEEE.

8.23) as given above, have been verified experimentally using alpha particles as the incident SEU-inducing ions.[85]

Besides providing detailed descriptions of the electrodynamics of the funneling process, the two major parameters that these and other funneling models provide are the additional prompt charge collected due to the funnel, and the effective funnel length or depth. The additional prompt charge collected, in contrast to the delayed charge collected due to diffusion from the surround to the sensitive node, is of immediate use in the computation of the critical charge for a particular active device or device technology family. The model discussed earlier,[34] and the model discussed immediately above,[36] provide similar results for the additional charge collected due to the funnel. The differences lie in the computation of the equivalent charge collection time. The latter model derives this quantity based on the assumption that the longitudinal (along the ion track) charge separation is operative throughout the creation and demise of the funnel process. The former model derives the charge collection time from charge separation in the radial direction (normal to the ion track). This is because, in the currently established computer simulations,[30,31,33] longitudinal charge separation is not manifest until the late phase of the funneling process, where the ion track densities begin to approach the substrate background dopant concentration. Both models depict reasonable agreement with respect to charge collection for the *p*-type substrate n^+p junction case, except at low values of bias voltage ($0 \leqslant V_0 \leqslant 5$ Volts). For the *n*-type substrate, the latter model seems to underestimate the charge collection, and the funnel length. For *p*-type substrates, the derived funnel lengths of both models are within sensible agreement, which length is about three times the depletion layer width in silicon. A computer model of ion track fundamental kinematics[33] is discussed in quantitative detail in Section 8.7.

8.6 SEU DEVICE SCALING

To gain insight into how SEU depends on device and circuit parameters, certain scaling relations between them have been identified. These provide estimates of SEU susceptibility for microcircuit (IC) device designers and manufacturers. They are useful when planning new process design rules for new device production, as well as for users who are anticipating future device properties to be integrated into their system development.

To appreciate the SEU problem facing the producers and users, Fig. 8.24 illustrates the evolution of microelectronics parameters (i.e., integrated circuit chip sizes and circuit packing densities). It is seen in the figure that the number of bits per chip, and circuits per chip, are increasing almost exponentially with time, while the device feature size is decreasing at a similar rate. This bodes ill in terms of SEU susceptibility of these parts, because of the greatly decreased amount of charge thus required to cause SEU, since cell (individual circuit) dimensions are becoming so minute.

Feature size is usually construed as the smallest dimension on a lithographic processing mask for a particular integrated circuit or device type. It is usually the width of a stripe (circuit connection line or trace), and is much less than the dimension of an integrated circuit device element.

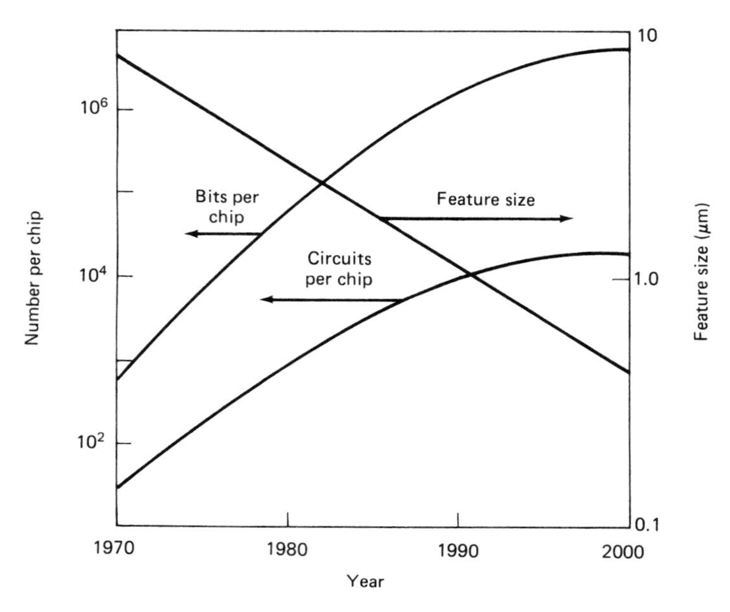

Fig. 8.24 Microcircuit density and dimensions versus production year.[38] Figure © 1985 by the IEEE.

For example, a microcircuit bipolar emitter area dimension might be 10×15 $(\mu m)^2$, while the corresponding stripe width (feature size) could be but $2\,\mu m$. With respect to MOSFET manufacturing, the feature size is usually also the length of the channel between the source and drain. Another parameter that measures microcircuit evolution, in terms of information handling capability, is the functional throughput rate (FTR).[38] For a particular digital part family or type, this is defined as the product of its number of gates by its clock rate (Hz). For constant FTR (iso-FTR), the number of gates and the clock rate are, of course, hyperbolically related, and shown as iso-FTR lines with a common slope in Fig. 8.25. For increasing FTR, the required switching energy per bit per gate decreases to maintain the power rating of the part at levels commensurate with the associated circuitry.

In Section 4.5, Sachs scaling is used to aid in the computation of nuclear weapon parameters versus weapon energy yield. Here, a scaling method for MOSFETs is discussed,[39] which is a forerunner of that to be used for SEU purposes. The scaling rules developed maintain constant electric fields for varying dimensions and other parameters in the device. Specifically, to design a device smaller than a nominal prototype, its

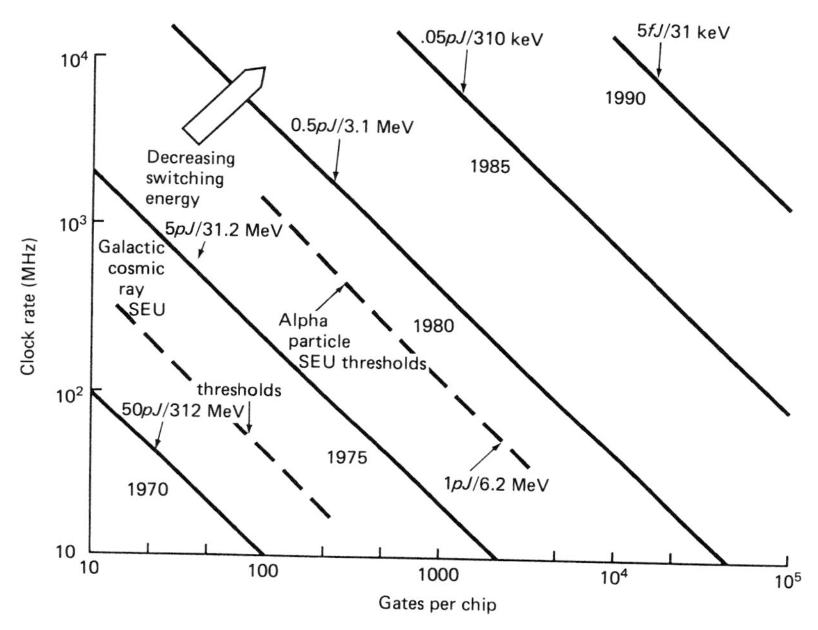

Fig. 8.25 Clock rate versus number of gates per chip showing Iso-FTR lines.[38] Figure © 1988 by the IEEE.

dimensions, voltages, and doping concentrations are scaled by a constant factor to reproduce the same performance. The scaling rules are derived by noting that depletion layer widths are reduced in proportion to the device dimensions, due to the scaled-down voltages and scaled-up dopant concentrations within the device. That is, the depletion layer width, l, can be scaled-down to a smaller width, l', where k is the scale factor, so that from Eq. (3.35) for $N_A \ll N_D$,

$$l' = [2\varepsilon\varepsilon_0 \cdot (\phi_0 + |V_0|/k)/ekN_A]^{1/2} \cong l/k; \qquad \phi_0 \ll |V_0|/k \quad (8.49)$$

The built-in contact potential $\phi_0 \ll |V_0|/k$ is the usual case, so that the above scaling rule is a very good approximation. Similarly, for the decrease in MOSFET threshold voltage, V_T, from Eq. (6.29) et. seq., for n^+ polysilicon gate terminals, the scaled-down threshold voltage V_T' is given by

$$V_T' \cong (t_{\text{ox}}/k\varepsilon_0\varepsilon_{\text{ox}})[-Q_{ss} + (2\varepsilon_0\varepsilon_{\text{si}}kN_A \cdot |V_{\text{sub}}|/k)]^{1/2} \cong V_T/k \quad (8.50)$$

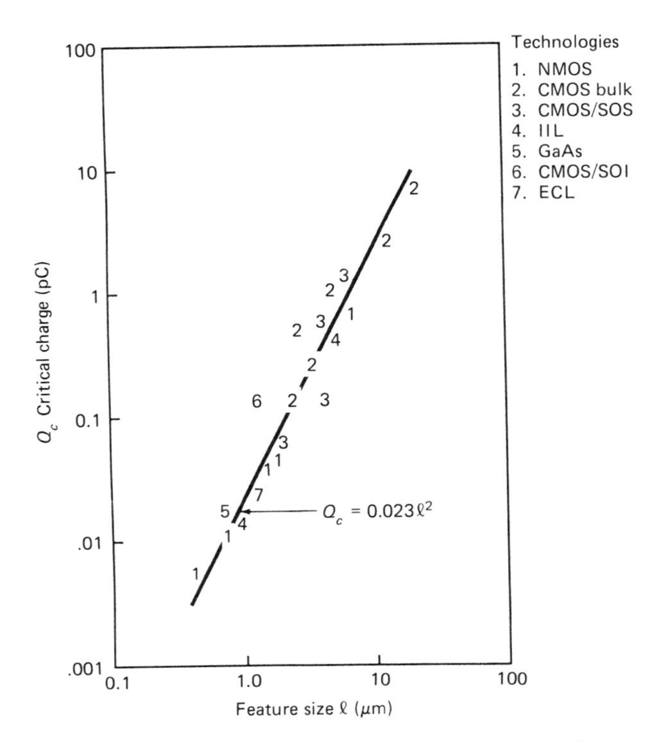

Fig. 8.26 SEU critical charge versus feature size.[40]

where a similar approximation is made (i.e., $|V_{sub}| \ll 2V_f$). t_{ox}/k is the scaled-down gate oxide thickness, because of the proportionately reduced voltages. Since Eqs. (8.49) and (8.50) imply a proportional reduction in all voltages and dimensions, it is seen that the corresponding electric fields remain invariant (i.e., $V'/x' = V/x$).

These scaling rules have implications for the critical charge of a sensitive node in the device. Consider the node-sensitive volume to be a parallelepiped, so that it can be modelled as a parallel plate bit storage capacitor $C = \varepsilon\varepsilon_0 \text{Area}/t$, where t is its thickness, and V is the voltage across C. Then, the preceding scaling rules yield, for the scaled-down critical charge, Q_c',

$$Q_c' = C'V' = [(\varepsilon\varepsilon_0 \text{Area}/k^2)/(t/k)] \cdot (V/k) = Q_c/k^2 \tag{8.51}$$

so that the scaled-down critical charge is inversely proportional to k^2.

In terms of part-type family technologies, the critical charge increases exponentially with feature size, as shown in Fig. 8.26. Interestingly, the critical charge seems independent of semiconductor material (Si or GaAs) or device technology, as seen in the figure.

With respect to the funnel length $\mathscr{L}$, its vertical projection $\mathscr{D}$, and the charge it collects Q_{coll}—as a result of scaling, the Hu[36] funnel model asserts that there are two cases:

(a) Low dopant concentrations ($N_A \ll 10^{16}\,\text{cm}^{-3}$):

$$\mathscr{L}'_{\text{low}} = (1 + \mu_n/\mu_p)(l/k)\sec\theta = \mathscr{L}/\text{k}; \qquad \mathscr{D}'_{\text{low}}(1 + \mu_n/\mu_p)\, l/k = \mathscr{D}/\text{k} \tag{8.52}$$

(b) High dopant concentration ($N_A \gg 10^{16}\,\text{cm}^{-3}$):

$$\mathscr{L}'_{\text{high}} \sim (kN_A)^{-1/2} = \mathscr{L}/\text{k}^{1/2}; \qquad \mathscr{D}'_{\text{high}} = \mathscr{D}_{\text{high}}/k^{1/2}; \qquad Q'_{\text{coll}} = Q_{\text{coll}}/k \tag{8.53}$$

The McLean-Oldham funneling model[34,35] does not discern a dopant concentration distinction, but yields

$$\mathscr{L}' \sim [(V_0/k)\, N_A{}^{-2/3}\, k^{-2/3}]^{1/2} \simeq \mathscr{L}/\text{k}^{5/6}; \qquad Q'_{\text{coll}} = Q_{\text{coll}}/k^{5/6} \tag{8.54}$$

The above Q_{coll} include the charge collected in the depletion layer, as well as that in the funnel proper.

To investigate how scaling affects SEU error rate, E_r, Eq. (8.9) is again used, and the steps of the "approximate" calculation of SEU to obtain E_r, as given in Eq. (8.12), are retraced. However, the scaling factor k is introduced as appropriate in these steps. For example, $s'_{max} = s_{max}/k$ and any $\mathscr{L}' = \mathscr{L}/\text{k}$. The "exact" calculation consists of a numerical integration of Eq. (8.9) using the known differential cosmic ray flux φ_L, the actual integral chord distribution $C(s(L))$, and including the scale factor k as it appears in the development.[13] Figure 8.27 shows a comparison of E_r versus scale factor, for a $15 \times 5 \times 0.5\ (\mu\text{m})^3$ parallelepiped cell of critical charge 1.1 pC, obtained using the approximate, exact, and heuristic calculation discussed below. It is seen in the figure that, for the three cosmic ray environments (viz., solar max., solar min., and 10 percent worst case), E_r is a roughly linear function of the scale factor for the "exact" and "approximate" calculations. This implies that E_r increases with decreasing scale, as expected. However, as also seen in the figure, these E_r approach a mean asymptotically constant behavior for k factors of 10 or more. The E_r corresponding to the heuristic argu-

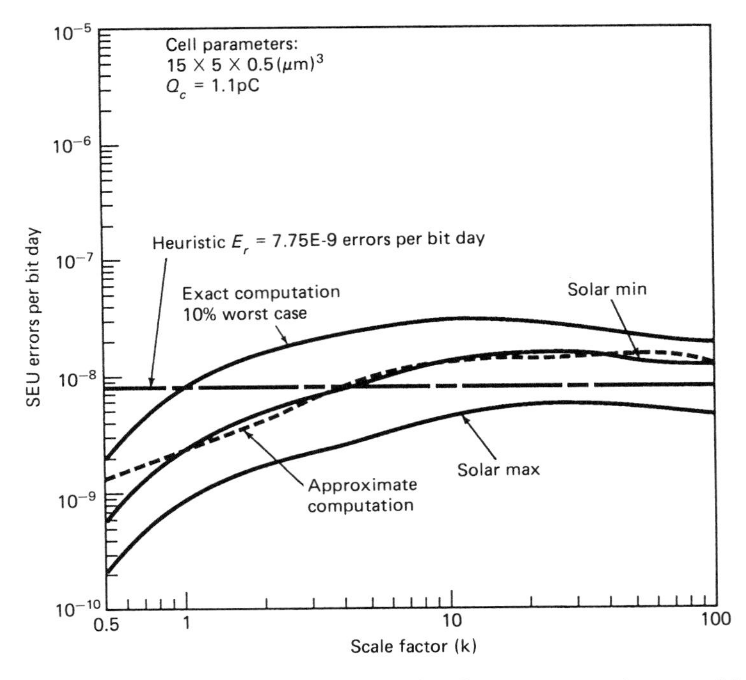

Fig. 8.27 SEU error rate versus scale factor using the exact, approximate, and heuristic formulas for cell parameters given in the inset.[13] Figure © 1982 by the IEEE.

ment is obtained simply by substituting the scale factor k directly into the "approximate" formula for E_r, as given by Eq. (8.12). From Eqs. (8.12) and (8.51), the scaled heuristic SEU error rate E_r' is

$$E_r' \sim a'b'c'^2/Q_c'^2 = (a/k)(b/k)(c/k)^2/(Q_c/k^2)^2 = abc^2/Q_c^2 \quad (8.55)$$

Hence, to this approximation, the SEU error rate is independent of scale change.

For a value of $k \leqslant 10$, the "exact" and "approximate" formulas for E_r show the proper behavior. However, for $k > 10$, all of these E_r are, or have become, roughly constant. This implies that this scaling is beginning to break down for these relatively high values of the scale factor. In actuality, E_r increases with decreasing device parameters, even for $k > 10$, because of other factors related to the actual reduction in device size, but not modelled herein. One major factor is that the SEU charge collection sensitive volume is not merely a parallelepiped, but is a topologically complex volume shape, which is circuit dependent with or without inclusion of the funnel. One extreme that can be appreciated is that, if this phenomenon actually behaved as described by Eq. (8.55); that is, if E_r were independent of scale, the SEU problem would be trivial. Essentially, simple scaling, as developed in the foregoing, degenerates for SEU error rates corresponding to relatively large device parameter changes (i.e., large k). E_r is a complicated function of electronic and material device parameters. Nevertheless, the above scaling rules are valid for derived quantities, such as the critical charge as a function of device dimensions, including feature size.

8.7 MULTIPLE EVENT UPSETS

Besides single event upsets due to energetic heavy ions already described, there are important instances in which multiple event upsets occur.[43] Multiple event upsets are those wherein an incident ion can, for example, cause SEU in a string of memory cell nodes, in an integrated circuit, that happen to lie along the track of the penetrating ion. Multiple event upsets were detected almost as soon as the SEU phenomenon was accepted as a reality.

The role of heavy ion ionization track physical parameters is almost nonexistent with respect to the computation of single event upset. However, such parameters are important in the understanding and computation of multiple event upsets. For example, one multiple upset model used for the calculation of multiple event cross sections depends on the ion track radius, and is discussed below,[42] following a description of ion track transport.

To understand the kinematics of heavy SEU ions within and without the ionization track, it is necessary to investigate their transport. For an ionization track of $4 \cdot 10^{18}$ hole-electron pairs (hep) km assumed cylindrical in cross section, generated by an incident SEU ion, it is generally agreed that the initial track radius is somewhat less than $0.1\,\mu$m.[33,41] The time required to create the initial track of ionized charge is less than 10 ps from the arrival of the incident ion, and possesses an initial charge density of $\sim 10^{18}\,\text{cm}^{-3}$. This very high value of charge density, including relatively high energy "hot" electrons, requires ambipolar transport, discussed in Section 1.11. The corresponding ambipolar transport equation is given by

$$D^*\nabla^2\delta p - \mu^*\bar{E}\cdot\nabla\delta p - \delta p/\tau = \partial\delta p/\partial t \qquad (8.56)$$

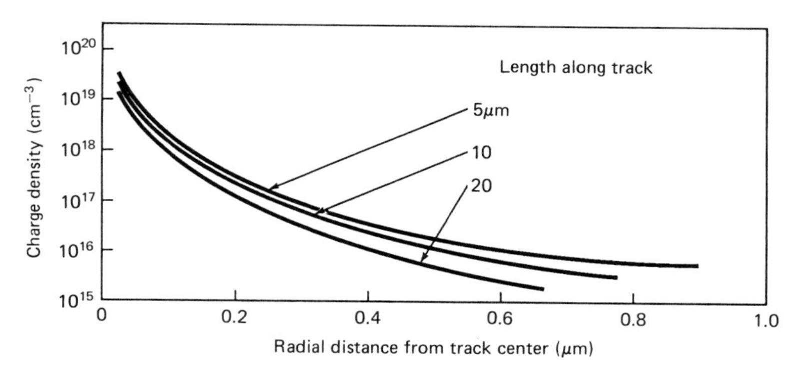

Fig. 8.28 Radial charge density in silicon along 20 GeV iron ion track segment.[42] Figure © 1987 by the IEEE.

where the ambipolar parameters D^* and μ^* are also defined therein. This equation holds for δn as well. The corresponding initial and boundary conditions are:

a. An initial condition on the charge generating function, $\delta p(r, 0)$ is that it is radially symmetric, as for example,[33]

$$\delta p(r, 0) = (N/\pi b^2) \exp - (r/b)^2 \tag{8.57}$$

where the track maximum lineal density, N, lies between 10^7 and 10^{11} hep/cm, and b is an assumed initial ionization track radius. Such track density profiles are depicted in Fig. 8.28, showing monte carlo simulations of their radial density. Implied above is that track radial density gradients are very large, compared to axial ones. For such ambipolar diffusion with negligible electric field drift motion at early times, the ambipolar mobility $\mu^* = 0$.

b. Asymptotically, for relatively long times after track formation, the track charge density becomes greatly reduced, so that D^* and μ^* approach D and μ, respectively.

In rectangular coordinates, the solution to Eq. (8.56) for a constant electric field E_c is, in one dimension,

$$\delta p(x, t) = (N/4\pi Dt) \exp - [\{(x - \mu_i E_c t)^2 + y^2\}/4Dt + t/\tau] \tag{8.58}$$

with a modified mobility, μ_i, defined by[33]

$$\mu_i = n_i\mu_n\mu_p/[n_i\mu_n + (\mu_n + \mu_p)\,\delta p] \tag{8.59}$$

From Eq. (8.58) it can be appreciated that, for early times, the track charge is diffusing radially in ambipolar fashion, with no drift motion due to the electric field. However, for relatively long times (about 500 ps) after track formation, the track begins to manifest drift motion due to the electric field, as the track charge density falls to levels comparable to the dopant concentration in the device cell. Therefore, in the first (early diffusional) phase, the track expands radially.[32] This is followed by a second phase, characterized by charge motion along the track, accelerated by the electric field.

When the heavy SEU-inducing ion penetrates the junction at normal incidence, and traverses the junction depletion layer, the governing equation, Eq. (1.81), can be rewritten as

$$\partial\delta p/\partial t = -\nabla\cdot J_p/e - \delta p/\tau \tag{8.60}$$

Since, in the initial phase, diffusion parallel to the track is negligible, the corresponding diffusion term $-eD_p\partial\delta p/\partial x$ is also negligible. So that, from Eq. (1.85), $J_p/e \cong \mu_p\delta p\,E$, and therefore $\partial J_p/\partial x \cong e\mu_p\delta_p E'(x)$. Using this expression to eliminate J_p in Eq. (8.60) yields, for one spacial dimension,

$$\partial\delta p/\partial t = -(\mu_p E' + 1/\tau)\,\delta p \tag{8.61}$$

with an integral given by

$$\delta p = (N/4\pi Dt)\exp - (\mu_p E' + 1/\tau)\,t \tag{8.62}$$

where the initial condition is $\lim_{t\to 0}(4\pi Dt\delta p) = N$. This implies a time-varying track density decrease of $N/4\pi Dt$. Note that the time-dependent area through which the current flows is $\pi x^2(t)$, where $x^2(t) = 4Dt$, a mean square diffusion distance. Also, $E'(x)$ is readily available from the corresponding Poisson's equation (i.e., $\nabla\cdot E = \rho/\varepsilon\varepsilon_0$, where ρ is the included charge density). That is, $E'(x) = -eN_D/\varepsilon\varepsilon_0$, where ε is the dielectric constant, and ε_0 is the permittivity of free space.

The total δp can be constructed empirically by adding to Eq. (8.62) a term representing the reduction of the high-density track electrons, produced from ionization, to thermal energies. It is characterized by a single time constant $\beta^{-1} = t_0$, where t_0 is the time that marks the transition from the first phase to the second phase. The resulting expression for relatively large time is[33]

$$\delta p = (N/4\pi Dt)(\exp - (\mu_p E' t) - \exp - \beta t) \tag{8.63}$$

Using the time varying area above, the corresponding I_p can be written, using Eq. (1.60) without the diffusion term, to give

$$I_p = 4\pi Dtj_p = 4\pi Dt \cdot e\delta p \mu_p E_0 = -e\mu_p NE_0(\exp - (\mu_p E' t) - \exp - \beta t) \tag{8.64}$$

where, from Eqs. (3.34) and (3.35), the maximum value of the junction depletion layer field $E_{max} \equiv E_0 = (2eN_D(-V_0 + \phi_0)/\varepsilon\varepsilon_0)^{1/2}$. The total electric current, due to both δp and δn, is obtained by summing an expression similar to Eq. (8.63) for δn, with Eq. (8.64) to yield[33]

$$I(t) = -e\bar{\mu}NE_0(\exp - (\bar{\mu}E' t) - \exp - \beta t) \tag{8.65}$$

with $\bar{\mu} = (1/2)(\mu_n + \mu_p)F(E)$. $F(E)$ supplies a mobility correction in its transition from ambipolar mobility to its asymptotic long time value.[33] Hence, the SEU current pulse is given by the difference of two exponential terms. The track decay time constant $1/\mu_p E' = \varepsilon\varepsilon_0/e\mu_p N_D$ is the RC time constant of the junction field region.[33]

The total SEU charge collected, neglecting the diffusion component, is obtained from Eq. (8.65) as

$$\int_0^\infty I(t)\,dt = -eN \cdot \begin{cases} l; \; l < X_j \\ X_j; \; l \geqslant X_j \end{cases} \tag{8.66}$$

where l is the track distance into the junction depletion layer, and X_j is the distance to the junction edge.[33]

As mentioned, the ion-induced charge track parameters are important with respect to multiple event upsets. This also includes the funneling phenomenon. Multiple bit errors pose a difficult problem of detection and correction. This is because even sophisticated error detection and correction (EDAC) methods must resort to elaborate and expensive

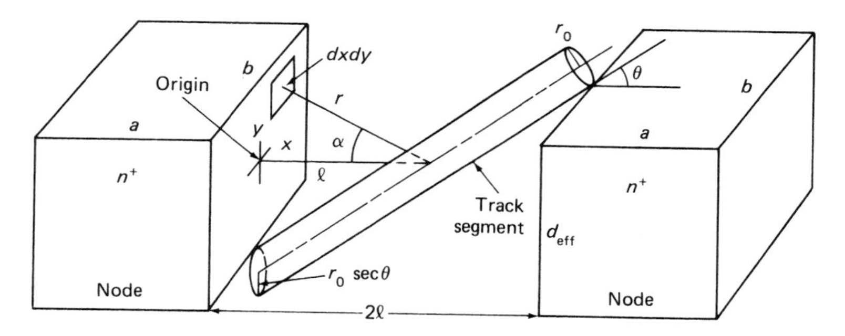

Fig. 8.29 Device showing oblique cosmic ray track segment tangent to two adjacent memory nodes causing double bit SEU.[42] Figure © 1987 by the IEEE.

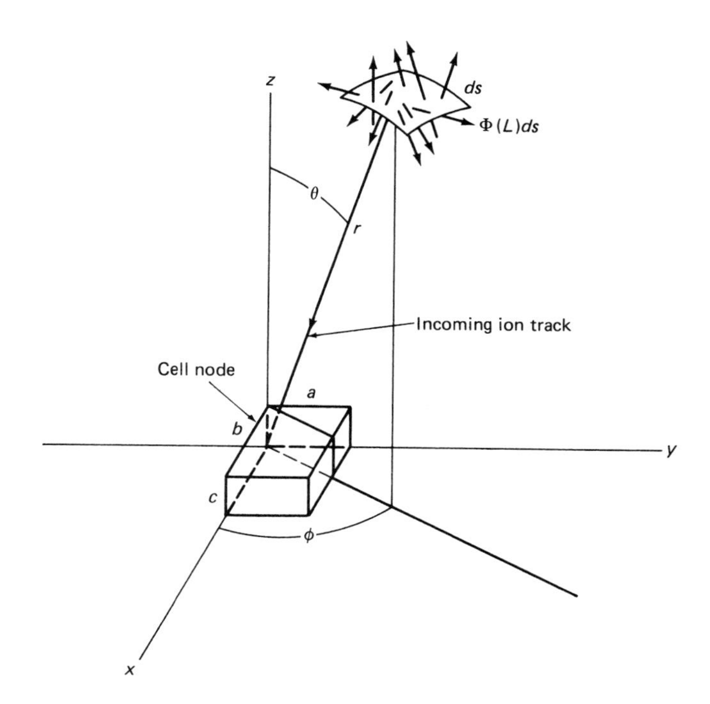

Fig. 8.30 Spherical coordinate system for thin parallelepiped RAM cell node.

redundancy schemes that even then can cope with very few errors per word. High multiple-EDAC methods must use gross redundancies, such as twin or triple central processing units (CPUs) with complex voting protocols.

One criterion for the occurrence of a double-bit SEU error is that of where an obliquely incident ionization track becomes tangent to two adjacent sensitive nodes[42] in for instance, a RAM, as shown in Fig. 8.29. Before investigating the double-bit SEU cross section, it is important to appreciate the computation of the single-bit cross section, σ_1, of a cell node whose sensitive thin parallelepiped area is $ab(\text{cm}^2)$, as in Fig. 8.30.

The SEU error rate, E_r, (errors per bit second) is given by

$$\begin{aligned}
E_r &= \int_{L_{\min}}^{L_{\max}} dL \int_{\theta_c(L)}^{\pi/2} \Phi(L)\, ds(ab\cos\theta/4\pi r^2) \\
&= (2\cdot 2\cdot ab/4\pi) \int_{L_{\min}}^{L_{\max}} dL\Phi(L) \int_{\theta_c(L)}^{\pi/2} d\Omega\cos\theta \\
&= (ab/\pi) \int_{L_{\min}}^{L_{\max}} dL\Phi(L) \int_0^{2\pi} d\phi \int_{\theta_c(L)}^{\pi/2} d\theta\cos\theta\sin\theta
\end{aligned} \tag{8.67}$$

where, in spherical coordinates, $d\Omega = ds/r^2 = \sin\theta\cdot d\theta\cdot d\phi$, with θ and ϕ as the usual polar and azimuthal angles, respectively. r is the distance between the incremental area ds, at an aspect given by θ and ϕ, and the projection of the parallelepiped area, $ab\cos\theta$, in the direction normal to ds, as in Fig. 8.30. In the leftmost integral, $\Phi(L)\,ds$ is the amount of isotropic cosmic ray heavy ion flux, Φ (ions per cm^2 sec), threading ds. $ab\cos\theta/4\pi r^2$ is the fraction of that flux incident on the area ab. Not all of the flux incident on ab causes SEU, but only that portion between a critical angle $\theta_c(L) = \cos^{-1}(L/L_c)$, from Eq. (8.5) (the rightmost portion of Fig. 8.3) and the grazing angle to the parallelepiped (i.e., $\pi/2$). This corresponds to heavy ion flux track lengths sufficient to cause SEU. One of the factors of two accounts for both front and back faces of the parallelepiped by doubling the front face SEU-producing ionizing flux path length contribution. The other factor of two accounts for the edge diffraction effects of the area ab. This accrues to a total equivalent area of twice the geometrical area, in the sense of the extinction cross section contribution in the diffraction theory of optics.

Inserting the above θ_c into Eq. (8.67) yields, after integration over the polar and azimuthal angles,

$$E_r = (ab/L_c^2) \int_{L_{min}}^{L_{max}} \Phi(L)\, L^2 dL$$

$$= (ab)\left[\int_{L_{min}}^{L_c} \Phi(L)(L^2/L_c^2)\, dL + \int_{L_c}^{L_{max}} \Phi(L)\, dL\right] \tag{8.68}$$

Since, in general $E_r = \int_{L_{min}}^{L_{max}} \Phi(L)\,\sigma_1(L)\, dL$, the single-bit SEU cross section $\sigma_u \equiv \sigma_1$ is identified from Eq. (8.68) as

$$\sigma_1(L) = ab \cdot \begin{cases} (L/L_c)^2; & L \leqslant L_c \\ \quad 1 \quad ; & L > L_c \end{cases} \tag{8.69}$$

which is seen to be a quadratic fit in L to the measured SEU cross section, where L above is replaced by $L_\perp$, as in Fig. 8.4 (dashed line).

For the double-bit SEU cross section σ_2, its maximum (worst-case) value is computed. Referring to Fig. 8.29, it is seen that its maximum value occurs at the track segment middle, where the solid angle, Ω_2, subtended by each of the two areas bd_{eff} facing the track middle, is a maximum. By symmetry, each of the areas bd_{eff} subtends the same solid angle Ω_2 where, using Eq. (13.11) to evaluate the integral,

$$\Omega_2 = \int_{4\pi} dA \cos\alpha/r^2 = \int_{-(1/2)b}^{(1/2)b} \int_{-(1/2)(d_{eff}+r_0\sec\theta)}^{(1/2)\,(d_{eff}+r_0\sec\theta)} l dx dy/(x^2 + y^2 + l^2)^{3/2}$$

$$= 4\tan^{-1}[(b/2l)(d_{eff} + r_0\sec\theta)/\sqrt{b^2 + (d_{eff} + r_0\sec\theta)^2 + 4l^2}] \tag{8.70}$$

The double-bit cross section σ_2 is then given in the same form as σ_1,

$$\sigma_2 = (L/L_c)^2\, b(d_{eff} + r_0\sec\theta)(2\Omega_2/4\pi) = (L/L_c)^2\, b[d_{eff} + r_0\sec\theta](\Omega_2/2\pi) \tag{8.71}$$

where Ω_2 is given by Eq. (8.70). The node face vertical dimension d_{eff} is extended by $r_0\sec\theta$, the vertical projection of the track radius, to accomodate the finite track diameter. The additional factor of two in Eq. (8.71) accounts for the areas of the two adjacent cell nodes. It is appreciated that σ_2 is similar in form to σ_1, i.e., $(L/L_c)^2 \cdot$ area. However, the area is reduced, since each bd_{eff} area subtends only a fraction of the total 4π solid angle, as seen from the track middle. The reduction factor is the fraction $\Omega_2/4\pi$ in Eq. (8.71).

Figure 8.31 depicts the ratio σ_2/σ_1 versus $2l$, the cell node separation distance, for various track radii.[42] It is seen from the figure that σ_2/σ_1 becomes significant for thick tracks (those due to very energetic and heavy ions) and small cell node separations, corresponding to micron and submicron features sizes of modern VLSI microcircuits.

Multiple bit upsets can also certainly occur for cosmic ray ions at normal incidence. This is especially so for relatively thick ionization tracks corresponding to high-energy heavy ions like highly ionized iron. Such thick, dense tracks containing high-density "hot" electrons in the high-energy tail of the fermi distribution induce a radially ambipolar diffusion, as discussed earlier. Even without funneling, these tracks can provide sufficient charge to saturate a microcircuit cell neighborhood, to cause multiple bit upsets.[46]

Figure 8.32 shows a detailed multiple-error bit map in a 256K bit DRAM (dynamic RAM) generated by SEU from 168 MeV ^{80}Br ions.[46] It shows that eight memory (cell) bits clustered about the ion track entry puncture were upset. This is corroborated in Fig. 8.33, where the SEU cross section versus LET is plotted. This plot is obtained by using a number of ion types from a tandem Van de Graaf accelerator, spread over various incident energies, culminating in the 168 MeV ^{80}Br ion, which was the most energetic.[46] In this figure, it is seen that the curve proceeds upward with a large slope, instead of becoming asymptotic at the approximate area of a single cell, as in the case of ordinary SEU-like plots. Actually, for the 168 MeV Br ion, the corresponding cross section $\sigma \equiv \sigma_8$, a value about a factor of eight over the single cell area asymptote

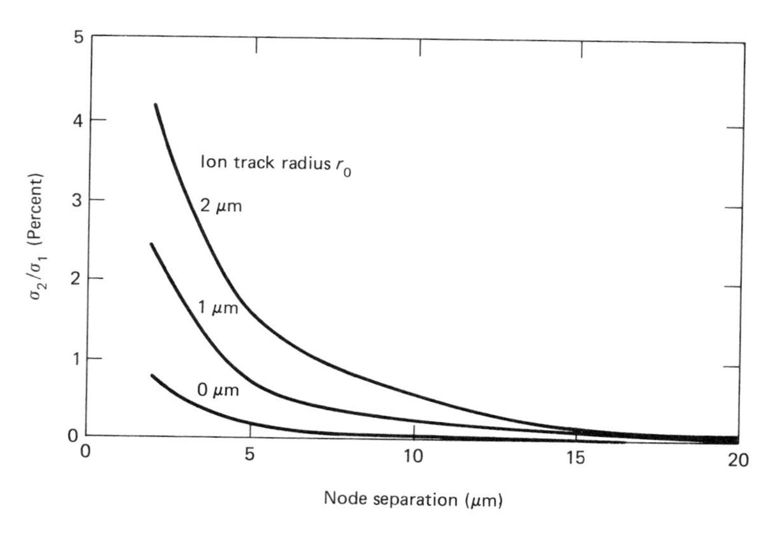

Fig. 8.31 Ratio of the double-bit to single-bit SEU cross section for two adjacent memory nodes for various ion track radii.[42] Figure © 1987 by the IEEE.

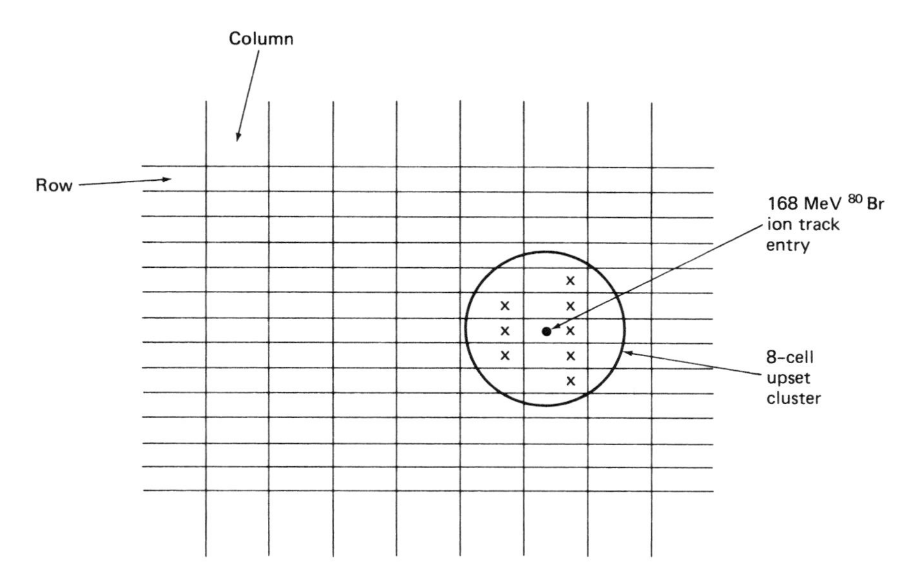

Fig. 8.32 256K DRAM multiple error memory cell bit map caused by single 168 MeV ^{80}Br ionization track.[46] Figure © 1988 by the IEEE.

for an SEU.[46] This corresponds well with the eight-bit charge cluster "hit," as shown in Fig. 8.32.

It is of interest to estimate the amount of charge collected that caused the multiple upsets in the above charge cluster. For this DRAM, the critical charge per memory cell, Q_c, can be computed roughly from $Q_c = C_s \Delta V$. C_s is the cell storage capacitance available from cell dimensions, as discussed in Section 3.5, and earlier in this section. $\Delta V = V_{DD} - V_{IB}$, the voltage swing required to change the state of this cell. V_{DD} is the supply voltage, and in this case V_{IB} is the sense amplifier input bias voltage. For this DRAM, $C_s = 0.05\,\text{pF}$, $V_{DD} = 5\,\text{volts}$, and $V_{IB} = 2.5\,\text{volts}$. This yields $Q_c = 0.125\,\text{pC}$. In Fig. 8.33, it is seen that the LET threshold $L_c = 1\,\text{MeV cm}^2/\text{mg}$, or 0.01 pC per μm for this DRAM. Hence the equivalent charge collection depth was about $0.125\,\text{pC}/(0.01\,\text{pC}/\mu\text{m})$, or $12.5\,\mu$m; so that the ion track segment was at least $12.5\,\mu$m in length.

8.8 ACCELERATOR COUPLED-SPICE CODE CROSS SECTIONS

One very detailed SEU model of an active device entails the coupling of experimentally determined values of actual ion-induced critical charge,

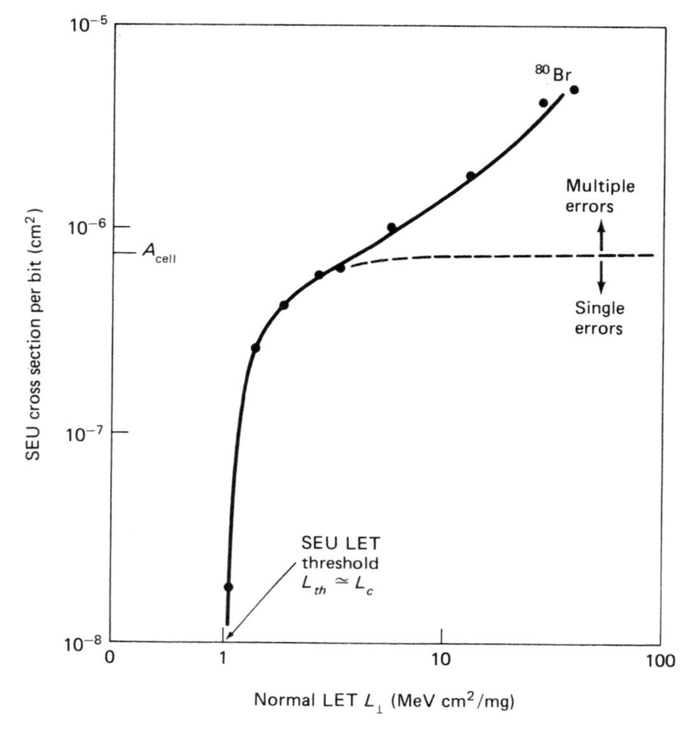

Fig. 8.33 SEU Cross section per bit versus LET for 256K bit DRAM showing multiple event upset from a single ion track at normal incidence.[46] Figure © 1988 by the IEEE.

and SEU cross sections corresponding to its individual elements, to a SPICE-type electrical circuit analytic model. This has been accomplished for a number of microcircuits,[46] including the bipolar LSTTL static RAM (SRAM) portion of a four bit microprocessor.[47] This experiment is discussed with an emphasis on the physical aspects, since the electronics aspects (i.e., the general SPICE modeling) are well known, and can be found in the electrical engineering literature.[71]

The experimental method consists of using an ion beam to cause the production and subsequent injection of SEU charge into various parts of the SRAM cell. This means that, for such an incident ion beam, its crow-flight depth of penetration (range) in silicon, as a function of its energy, can be determined with good accuracy. One reason is that, since the ions are heavy and energetic, they are not deviated from their track very much by scattering, so that their crow-flight distance until stopping is almost the same as their range. However, the main reason is that, for penetrating heavy ions, their stopping power, dE/dx, or LET versus residual range (remaining path distance), increases to a maximum value near the end of their range. That is, most of their energy is deposited into the medium very near their track end. This deposition of a localized amount of energy in position and depth in the device nearly coinciding with the track end provides, through ionization, spurious charge to produce SEU in an individual element of interest in the SRAM cell. Illustrating this property, Fig. 8.1 depicts this energy maximum (Bragg peak) in its plot of stopping power versus residual distance to end of path for alpha particles and protons penetrating silicon. Figure 8.2 shows the corresponding range-energy curves for a number of ions of interest.

These ions, therefore, suitably collimated from accelerators of various types, can be used as precision ionization charge generators to stimulate SEU localized to various parts of the device cell under study. Bromine, ^{80}Br, ions in the energy range from 20–240 MeV corresponding to a LET of 24–39 MeV cm^2/mg have been used in this manner.[46,47] This ion beam is quite narrow, and so can be positioned with micrometer accuracy, to penetrate the device with similar precision to produce ionization where desired by exploiting its Bragg peak. Figure 8.34 depicts bromine ion beams successively penetrating the SRAM cell in its isolation wall, collector, emitter, base, and schottky diode clamp. Dashed lines in the figure connect these ion beam tracks to their corresponding SEU cross section, and deposited critical charge in the lowest portion of the figure. Shown in that figure are the critical charges corresponding to those collected in the various portions of the SRAM. For example, $Q_c = 1.5$ pC is that due to an ion beam track that impaled the collector. Also seen in the lowest portion of the figure is that the cross section curves possess the typical shape, reaching asymptotic values for sufficiently high ion energy and LET.

The two cross section curves for the SRAM are marked "addressed" and "unaddressed." The latter corresponds to the case where the SRAM is loaded with zeroes, and is in a low power state. In this (also low current) state, the SRAM is very sensitive to spurious currents due to SEU disturbances, and so has a relatively large SEU cross section. This is in contrast to the addressed state, which is the fully powered state, being less sensitive to extraneous excitation, and so possessing a smaller cross section, as shown in the figure.

With the knowledge of the critical charge in various elements of the SRAM cell, that data becomes a set of input parameters for various nodes in the SPICE-type simulation of the device. Specifically, they can be modeled as additional transient current generators, similar to the manner by which photocurrent generators are incorporated into SPICE-type programs to make them radiation-inclusive.

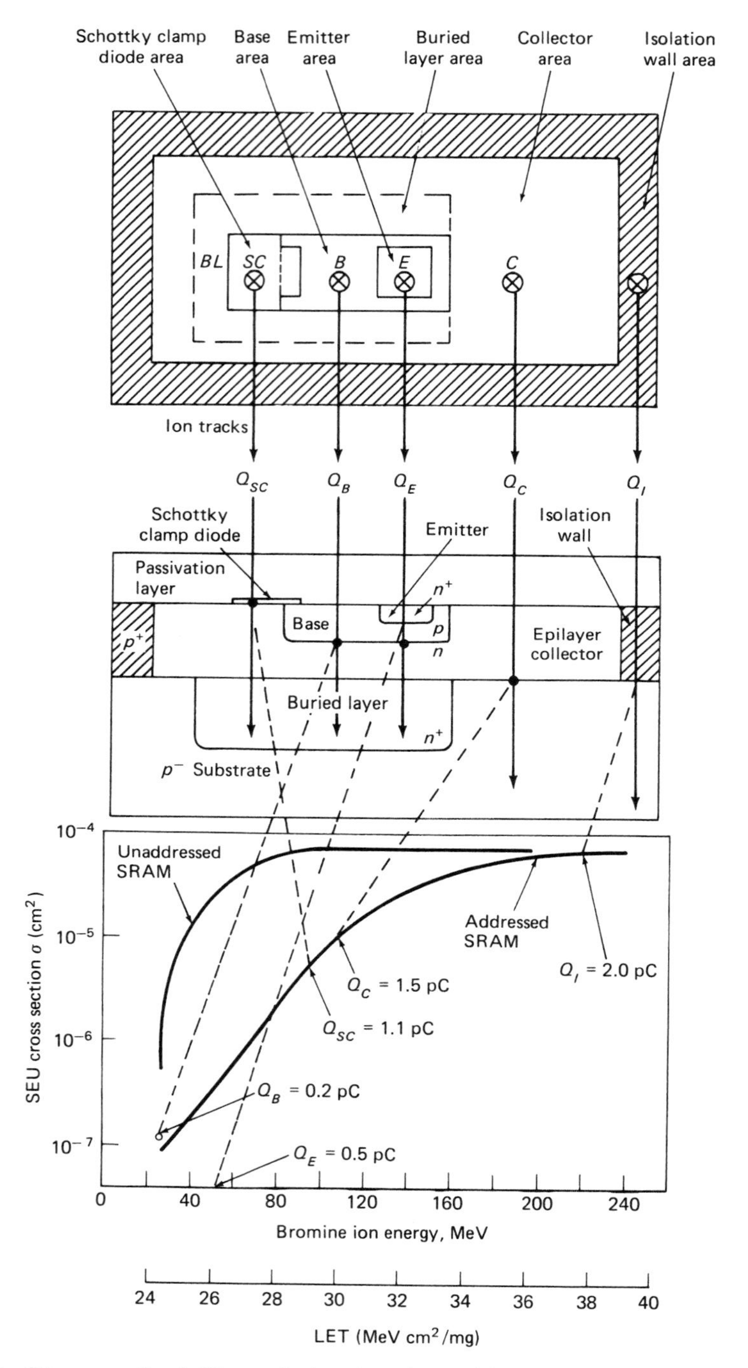

Fig. 8.34 (Upper portions) Plan and elevation views of low-power Schottky (LSTTL) transistor-transistor logic cell of a bipolar static RAM (SRAM) portion of AM2901B four bit slice. Lowest portions shows SEU cross section versus bromine ion energy and LET for individual elements in the SRAM cell in its unaddressed state, and with corresponding critical charge in its addressed state.[46] Figure © 1988 by the IEEE.

8.9 SEU LATCHUP

Single events are well known to include single event latchup[48–52] (SEL) as well as single event upset (SEU). Single event latchup produces "hard" errors, in the sense that the information nodes are not merely overwritten by incoming data, as in the case of SEU. The hard errors may be erased by powering-down that node neighborhood, as in ordinary dose rate-induced latchup, provided node burn-out has not occurred. It could be difficult for a bipolar integrated circuit system to sense a lone single event-induced node latch, if it is a very localized phenomenon,[51] as opposed to a dose rate-induced latch involving a number of nodes, discussed in Section 7.7.

Single event latchup has been observed only in CMOS bulk, and CMOS-Epi devices,[56] to date. Table 12.4 provides a listing of latchup-prone devices. It is well known that CMOS bulk is also very susceptible to dose rate-induced latchup as well, and this is discussed in Section 12.4. Single event latchup has not been observed in bipolar integrated circuits, probably due to their higher latchup energy thresholds, or that the population of such bipolar devices tested for SEL has not yet grown sufficiently large to include a rare SEU-induced latchup. However, single event latchup has been observed in advanced current mode logic (CML) device model situations.[53] CML is a differential, nonsaturating bipolar logic, possessing several internal logic levels, each differing by one *p-n* junction diode drop. These studies depict an SEL rate of about 10^{-7} latchups per cell day.[53] From these model analyses, it has been found that it is the relatively slower diffusion current collection, occurring in microseconds following the onset of the SEU pulse, that is responsible for SEL. This is in contrast to the collection currents from the node depletion layer and funnel, which endure for only a few nanoseconds.

As discussed, ^{252}Cf fission fragments have sufficient energy and LET to cause SEU latchup in integrated circuits. An interesting experiment[54] for determining SEL characteristics in CMOS devices, with ^{252}Cf as the source of heavy ions, uses the device under test (DUT) as part of a monostable RC oscillator circuit. When a ^{252}Cf fission fragment incident on the DUT causes SEL within, the DUT is latched into the usual SCR state, as discussed in Section 7.7. This SCR then triggers the oscillator to produce a corresponding current pulse. Subsequent particles in the fission fragment flux, $^{252}\dot{\Phi}(t)$, repeat the process, and the succession of current pulses (SELs) are counted. The ^{252}Cf source is mounted above the delidded device socket, and the whole placed in an evacuated (10^{-3} torr) chamber. This yields the SEU latchup cross section σ_{SEL} simply as

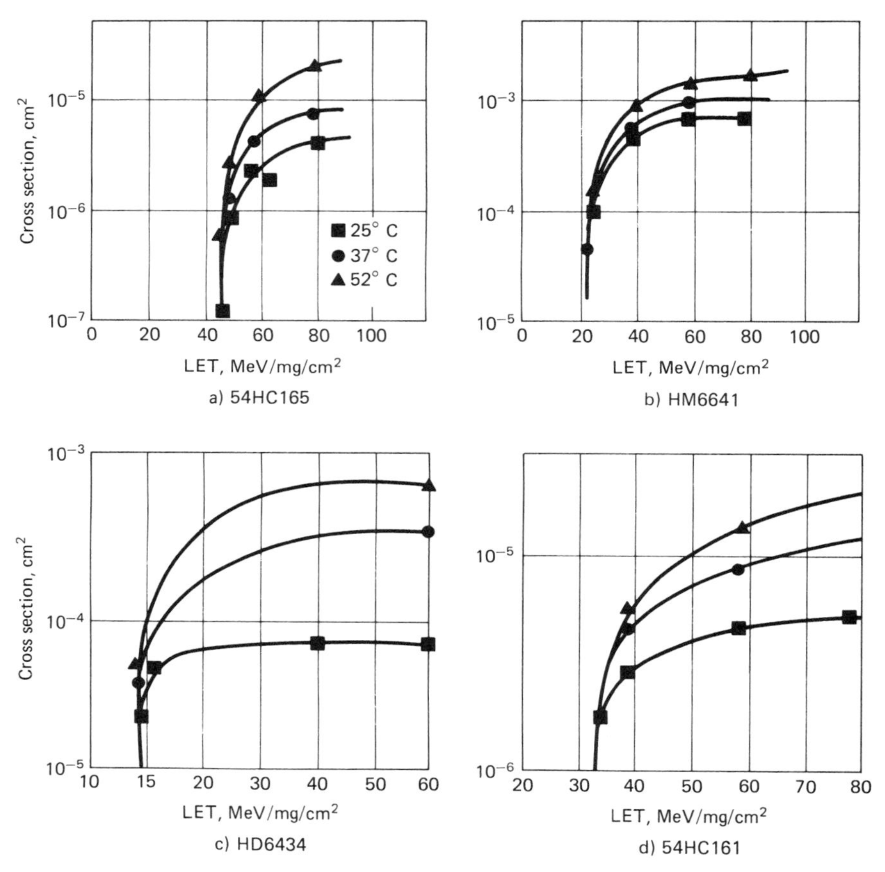

Fig. 8.35 Device latchup cross sections versus LET for various temperatures.[58] Figure © 1986 by the IEEE.

$$\sigma_{\mathrm{SEL}} = \lim_{t \to \infty} [N_{\mathrm{SEL}}(t)/{}^{252}\Phi(t)] \tag{8.72}$$

where $N_{\mathrm{SEL}}(t)$ is the number of SEL at time t, and the fluence ${}^{252}\Phi(t) = {}^{252}\dot{\Phi}(t) \cdot t$, from the calibrated ^{252}Cf source. Devices tested in this manner are tabulated,[54,55] and overlap the devices and SEL parameters values given in Table 12.5.

Recent SEU accelerator measurements[57] have tentatively concluded that the SEL cross section and critical LET, L_c, depend on device ambient temperature. The following development describes the increase in SEL cross section with temperature,[58] but indicates that L_c is essentially independent of temperature. Figure 8.35 depicts the increase in SEL cross section versus LET for four different CMOS devices at three different ambient temperatures.[58] None of these devices show a change in L_c with temperature.

Consider the common CMOS system building block inverter, as depicted in Fig. 8.36. Superposed on this figure is the schematic of a classic SCR latchup path, whether induced by SEL or dose-rate photocurrent. It is seen from the figure that one latchup path consists of a lateral *pnp* parasitic transistor Q_1, directly coupled to a vertical *npn* parasitic transistor Q_2, together comprising an SCR circuit, as discussed in Section 7.7.

Single event latchup can be initiated by an incident heavy ion, whose track is shown in Fig. 8.36, activating parasitic bipolar transistors Q_1 and Q_2 to a high current–low impedance SCR state. From Section 7.7, the latchup criterion is given by $\beta_1 \cdot \beta_2 \cong 1$, where β_1 and β_2 are the respective common emitter current gains of transistors Q_1 and Q_2. From Section 5.10 it is shown that the temperature dependence of the common emitter current gain is $\beta(T) = (\sigma_e L_e/\sigma_b W)(T/T_0)^2$. σ_e and σ_b are the respective conductivities of the parasitic transistor emitter and base that is suffering a latchup, L_e is the diffusion length in its emitter, and W is the corresponding base width. T_0 is the device ambient temperature. The latchup criterion can therefore be rewritten by inserting $\beta(T)$ in the criterion equation to obtain

$$(\sigma_{e1} L_{e1}/\sigma_{b1})(\sigma_{e2} L_{e2}/\sigma_{b2})(T/T_0)^4/W_V W_L \cong 1 \qquad (8.73)$$

where W_V and W_L are the respective base widths of the vertical and lateral transistors. Equation (8.73) asserts that, for device-elevated temperatures over T_0, the parasitic transistor base widths W_V and W_L can be electrically extended from their ambient values, while still allowing latchup to occur, to preserve the above latchup criterion equality. It is seen from Fig. 8.36 that the preceding implies that the electrically expanded effective base widths (e.g., W_L) correspond to increasing the SEL cross section. This is simply because the expanded base width now offers a mainly normal, increased p well-n substrate junction area to the incident SEL heavy ion tracks. Hence, σ_{SEL} increases with temperature, as shown in Fig. 8.34. The detailed electro-physical mechanism by which the base widths extend to increase their effective area is currently under investigation.[57]

Addressing the variation of L_c with temperature,[58] it can be written as

$$L_c \cong Q_c/(W_d + W_f) \qquad (8.74)$$

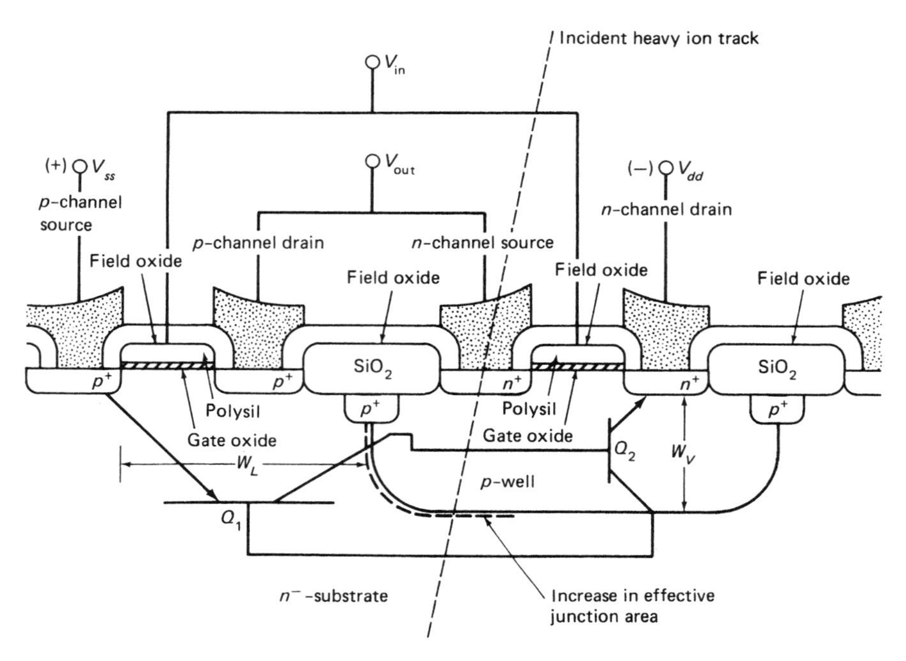

Fig. 8.36 CMOS inverter showing *pnpn* latchup path parasitic transistors and incident ion track.[58] Figure © 1986 by the IEEE.

where W_d is the depletion region width of the p well-n substrate junction, and W_f is the corresponding funnel length, following the penetration of a heavy ion in this neighborhood. Now, $Q_c = C\Delta V_{CB}$, where the depletion layer capacitance of this junction is $C = \varepsilon\varepsilon_0 A/(W_d + W_f)$ and A is the junction area, as discussed in Chapter 3, Problem 2. Then Eq. (8.74) becomes

$$L_c = \varepsilon\varepsilon_0 A\Delta V_{CB}/(W_d + W_f)^2 \tag{8.75}$$

For an abrupt junction, the discussion in section 3.4 yields for the depletion width,

$$W_d = (2\varepsilon\varepsilon_0(\phi_0 + |V_{CB}|)/eN_A)^{1/2} \tag{8.76}$$

ϕ_0 is the built-in (contact) potential, N_A is the dopant concentration of the lightly doped side of the junction, and V_{CB} is the corresponding junction voltage. The funnel length W_f is essentially temperature independent in this context, and L_c in Eq. (8.75) will depend on temperature only through W_d. The temperature dependence of W_d in Eq. (8.76) is evinced only through ϕ_0, since $\phi_0 = (kT/e)\ln(N_A N_D/n_i^2)$, and the intrinsic carrier density $n_i = C_1 T^{3/2} \exp - (E_g/2kT)$, ($C_1$ is a constant), as discussed in Sections 3.3 and 1.5, respectively. With suitable values of the parameters involved,[58] it is readily shown that $\Delta\phi_0/\Delta T \cong 2\,\text{mV/°C}$. Even for very large ambient temperature increase (e.g., $\Delta T = 200°\text{C}$), then $\Delta\phi_0 \cong 400\,\text{mV}$. This implies that any changes in ϕ_0 are dominated by V_{CB}, since it is of the order of 5–10 volts. That $\phi_0 \ll V_{CB}$ is well known, and discussed in Section 3.5. Hence, this argumentation leads to the conclusion that L_c is relatively unaffected by temperature changes.

8.10 SEU HARDNESS APPROACHES

As mentioned before, hardening devices against SEU and SEL is a difficult effort, seldom involving physical shielding methods, but mainly involving system, circuit, and component remedial measures. One of the simplest is judicious part selection, whereby SEU-hard parts are chosen a priori as part of early system planning, in contrast to eleventh-hour retrofitting. Today, there is probably a sufficient number of SEU-hard device tabulations from which to choose corresponding devices for required functions, which provide some diversity for many systems having nominal radiation specifications.

Certain system approaches to SEU hardening are straightforward in the following sense. First, almost no passive or discrete devices (transistors and diodes) are susceptible to SEU. Second, only a few IC function types need be investigated from the system use viewpoint. They include RAMs, logic cells, and certain types of gates, as discussed in detail in Section 12.5. Third, if the IC electronics are analysed from the system viewpoint initially, this rapidly leads to function constraints further limiting part types and technologies to be investigated, from the standpoint of SEU. For example, integrated circuit microcontroller power supply conditioners may be susceptible to SEU, but would not harm system function if temporarily so impacted. The power level thus might experience a transient for a small fraction of a second, but the systems so powered will still function satisfactorily well beyond SEU onset.

Other system approaches attempt to exploit the relatively short time span of the SEU phenomenon and its consequences. The onset of the SEU-inducing ion penetration, to the recovery of the device, generally consumes only a few tens of nanoseconds. If the pertinent systems can be designed to possess low-pass filter characteristics, while maintaining their prescribed function, then such systems would be insensitive to the SEU transient. However, the path of progress in semiconductor microcircuit manufacture is in the direction of increasing the speed of device operation, thereby enhancing device throughput. This is in conflict with the effectively slower system function and operation implied by the superposition of the low-pass filter characteristic onto these integrated circuits. For example, using resistors in the cross-coupled feedback links of IC flip-flops will slow them down to immunize them against SEU. In certain trade-off analyses involving such resistors and other components (e.g., diodes), SEU hardness can be achieved for a corresponding loss in device speed. This approach to SEU hardness and all its ramifications is still applicable, with further details discussed in Section 12.5. Other methods for dealing with SEU include schema from the field of reliability/maintainability, such as redundancy protocols of many types, fault tolerance, and error detection and correction (EDAC) methods.

PROBLEMS

1. In terms of SEU error production in a special-purpose computer from cosmic rays, how would this computer behave in a cruise missile, U-2 reconnaissance plane, or in a NORAD command center under a mountain?
2. Muons that cause SEU errors in CCDs (charge coupled devices) are produced in the upper atmosphere by cosmic rays. Once produced, the muons take on the order of 0.1 ms to traverse the atmosphere to sea level. When the same muons are produced in accelerators, they last only about 1 μs before they decay and disappear. How do the muons produced in the upper atmosphere last long enough to strike earth-bound CCDs to cause SEU errors?
3. Accelerator measurements yield about 10 SEUs for a heavy ion fluence of $1.6 \cdot 10^9$ ions per cm^2 incident on a device. What is the corresponding cross section? Using this cross section, what is the SEU bit rate per device due only to these ions, if their perceived flux is 10^5 ions per cm^2/d? Can this cross section be used to compute SEU rates due to other heavy ions?
4. An incident cosmic ray can enter a chip at any angle, to proceed along a chord therein, producing hole-electron pairs by ionization. The chip-sensitive volume is assumed to be a parallelepiped whose dimensions are a, b, and thickness c. Some workers had construed critical linear energy transfer (LET) as $L_c = Q_c/\bar{s}$, where the mean chord length is $\bar{s} = abc/\bar{A}_p$. (a) Show that $\bar{s} \cong 2c$ for a thin parallelepiped; and (b) Using the above expression for L_c, compute it in units of MeV cm^2/mg, if $Q_c = 10^6$ electron charges and $c =$ 1 micron.
5. An often-used approximate measure of SEU hardness of an active component, such as an integrated circuit, is its experimentally determined

threshold LET (viz., L_c), if known. Is an SEU hard part characterized by a high or low L_c, and why?

6. A particular 1K SRAM (static random access memory integrated circuit) has SEU sensitive memory nodes whose equivalent depletion layer storage capacitances are 0.05 pF. Their corresponding V_{OH} (output voltage–HI) and V_{OL} (output voltage–LO) are 5.5 volts and 2.5 volts, respectively. The nodes are modeled as parallelepipeds of dimensions $3 \times 10 \times 10$ microns. Compute: (a) Their SEU error rate per memory cell day; and (b) The corresponding SEU error rate per device day.
7. Very large (anomolous) solar flares, of duration up to one day, can be anathema to the operation of SEU-sensitive components in spacecraft. The deleterious effects of solar flares can sometimes be put in terms of a factor (e.g., 2000) multiplying the corresponding SEU error rate. This factor is a function of many parameters, so that the above number is only illustrative. Assume that such large flares occur about 1.5 times per 20 years. Assuming that the frequency of these flares is Poisson distributed (rare events). Compute the following: (a) The solar flare probability rate per day; (b) The probability of such a flare occurring only once in the spacecraft mission life of 10 years; (c) The probability of at least one such flare occurring during the above 10-year mission life; and (d) The corresponding equivalent SEU rate, using the above factor of 2000.
8. *N*-channel power MOSFET discrete transistors are an exception to the rule that discrete devices are immune to SEU. These transistors are fabricated of many individual finely structured units in parallel, to provide the needed power-handling capability. Name at least one type of SEU-induced specific process that could cause SEU burnout in these devices.
9. Linear energy transfer implies an essentially constant LET or dE/ds in the device material for the incident particle. The energy lost by these particles is converted mainly into ionization, (i.e., particle–electron interactions of the device material). Therefore, dE/ds is roughly proportional to the electron concentration in the material. That is, $dE/ds \sim NZ$, where N is the number of material atoms per cm^3. Show that this proportionality implies approximately constant LET.
10. For silicon, the relationship between the charge Q generated by an incident particle depositing energy E to ionize the material is Q(picocoulombs) = E(MeV)/22.5. For gallium arsenide, Q(pC) = E(MeV)/30. Derive these relationships, and use Table 1.1. The charge of the electron is $1.6 \cdot 10^{-19}$ coulombs.
11. For the above relations, show that the corresponding LETs L are 96.57 MeV cm^2/mg for silicon, and 56.4 MeV cm^2/mg for GaAs, to both correspond to $L = 1\,pC/\mu m$.
12. SEU is usually associated with spacecraft, since they are exposed to galactic cosmic rays, solar wind flux, etc. Name at least three nonspacecraft systems for which SEU would cause difficulty, and why?
13. Why was the phenomenon of the funnel and funneling postulated?

14. Why does the SEU latchup cross section increase with temperature, as discussed in Section 8.9, since all of the physical dimensions of the latched cell remain essentially constant with temperature?

15. Assume that the thin parallelepiped SEU-sensitive region of volume *abc* comprises the depletion layer of an information node junction. Using Eq. (8.12) for the SEU bit error rate (viz., $E_r = 5 \cdot 10^{-10} abc^2/Q_c^2$), show that E_r is essentially proportional to c^4.

16. The result in Problem 15 shows that the SEU error rate will decrease rapidly for very thin parallelepiped sensitive regions. These usually correspond to very tiny devices. Can the conclusion be drawn that the smaller the device, the smaller the SEU error rate will be?

17. If the orbital altitude of future space-stations is about 150 miles: (a) What fraction of the galactic cosmic rays will the earth itself occlude (shadow) from this spacecraft? (b) Is this a consideration in terms of the space station on-board avionics?

18. If the electronic system designer finds that the computed SEU bit error rates are disturbingly high for the following ICs in the system: voltage regulators; pulse width-modulated supply conditioners; voltage comparators; digital-to-analog converters; and GaAs FET rf amplifiers; enumerate the measures, that the designer should take.

19. In Table 8.1, generally, the smaller the cell critical charge, Q_c, the larger is the SEU error rate per chip day. However, certain devices whose Q_c is smaller than others also possess relatively smaller SEU error rates per chip day. Why?

20. In device dimension-scaling considerations, the Peterson SEU error rate expression, as given in Eq. (8.55), is shown to be independent of such scaling. If it were actually independent of scale, how would this trivialize the SEU problem?

REFERENCES

1. J. T. Wallmark and S. M. Marcus. "Minimum Size and Maximum Packing Density of Non-Redundant Semiconductor Devices." Proc. IRE, pp. 286–298, Mar. 1962.
2. D. Binder, E. C. Smith and A. B. Holman. "Satellite Anomalies From Galactic Cosmic Rays." IEEE Trans. Nucl. Sci. NS-22(6):2675–2680, Dec. 1975.
3. T. C. May and M. H. Woods. "Alpha Particle Induced Soft Errors in Dynamic Memories." IEEE Trans. Elect. Dev. ED-26(1), Jan. 1979.
4. J. S. Browning and D. R. Holtkamp. "Particle LET Spectra From Microelectronics Packaging Materials Subjected to Neutron and Proton Irradiation." IEEE Trans. Nucl. Sci. NS-35(6):1629–1633, Dec. 1988.
5. J. F. Ziegler and W. A. Landford. "Effect of Cosmic Rays on Computer Memories." Science, 206:776–788, Nov. 16, 1979.
6. J. A. Zoutendyk, L. S. Smith and G. A. Soli. "Single Event Upset (SEU) Model Verification and Threshold Determination Using Heavy Ions in a Bipolar Static RAM." IEEE Trans. Nucl. Sci. NS-32(6):4164–4169, Dec. 1985.
7. E. L. Peterson, J. B. Longworthy and S. E. Diehl. "Suggested Single Event Upset Figure of Merit." IEEE Trans. Nucl. Sci. NS-30(6):4533–4539, Dec. 1983.

8. J. H. Stephen et al. "Investigation of Heavy Partiole Induced Latchup Using a ^{252}Cf Source in CMOS RAMs and PROMs." IEEE Trans. Nucl. Sci. NS-31(6):1207–1211, Dec. 1984.
9. J. T. Blandford and J. C. Pickel. "Use of ^{252}Cf to Determine Parameters For SEU Rate Calculation." IEEE Trans. Nucl. Sci. NS-32(6):4282–4286, Dec. 1985.
10. J. C. Pickel and J. T. Blandford. "Cosmic Ray Induced Errors in MOS RAMs." IEEE Trans. Nucl. Sci. NS-27(2):1006–1015, April 1980.
11. M. D. Petroff. Autonetics Division of Rockwell International, Anaheim, CA.
12. P. Shapiro, E. L. Peterson and J. H. Adams, Jr. "Calculation of Cosmic Ray Induced Soft Upsets and Scaling in VLSI Devices." NRL Report. No. 4864, Aug. 26, 1982.
13. E. L. Peterson, P. Shapiro, J. H. Adams, Jr. and E. A. Burke. "Calculation of Cosmic Ray Induced Soft Upsets and Scaling in VLSI Devices." IEEE Trans. Nucl. Sci. NS-29(6):2055–2063, Dec. 1982.
14. W. Heinrich. "Cosmic Ray Abundances Versus Linear Energy Transfer." Radiat. Effects, 34:145–148, 1977; "Propagation of High Energy Ion Beams Through Matter." Radiat. Effects 39:167–171, 1979.
15. D. K. Nichols, W. E. Price, W. A. Kolasinski, R. Koga, J. C. Pickel, J. T. Blandford and A. E. Waskiewicz. "Trends in Parts Susceptibility to Single Event Upset From Heavy Ions." IEEE Trans. Nucl. Sci. NS-32(6):4189–4194, Dec. 1985.
16. D. K. Nichols, L. S. Smith and W. E. Price. "Recent Trends in Parts SEU Susceptibility From Heavy Ions." IEEE Trans. Nucl. Sci. NS-34(6):1332–1337, Dec. 1987.
17. L. L. Sivo, J. C. Peden, M. Brettschneider, W. E. Price and P. Pentecost. "Cosmic Ray Induced Soft Errors in Static Memory Cells." IEEE Trans. Nucl. Sci. NS-26(6): 5042–5047, Dec. 1979.
18. D. Binder. "Analytic SEU Rate Calculation Compared With Space Data." IEEE Trans. Nucl. Sci. NS-35(6):1570–1572, Dec. 1988.
19. W. L. Bendel and E. L. Peterson. "Proton Upset in Orbit." IEEE Trans. Nucl. Sci. NS-30(6):4481–4485, Dec. 1983.
20. E. G. Stassinopoulos and J. M. Barth. "Non-Equatorial Terrestrial Low Altitude Charged Particle Radiation Environment." NASA Goddard Space Flight Center, Report No. X-601-82-9, 1982.
21. E. G. Stassinopoulos. "Orbital Radiation Study For Inclined Circular Trajectories." NASA Goddard Space Flight Center, Report No. X-601-81-28, 1981.
22. W. L. Bendel and E. L. Peterson. "Predicting Single Event Upsets in the Earth's Proton Belts." IEEE Trans. Nucl. Sci. NS-31(60):1201–1206, Dec. 1984.
23. J. N. Bradford. "Cosmic Ray Effects in VLSI in Space Systems and Their Interaction With Earth's Space Environment." Prog. Astro. Aero., 71:549, 1980; "A Distribution Function For Ion Track Length in Rectangular Volumes." J. Appl. Phys, 50, p. 3799, 1979.
24. D. S. Yaney, J. T. Nelson and L. L. Vanskike. "Alpha Particle Tracks in Silicon and Their Effect on Dynamic MOS RAM Reliability." IEEE Trans. Elect. Dev. ED-26, Jan. 1979.
25. T. C. May and M. H. Woods, op. cit.
26. T. C. May and M. H. Woods. "A New Mechanism For Soft Errors in Dynamic Memories." IEEE Reliability Physics (IRPS) Symposium, 1978.
27. E. G. Stassinopoulos and J. P. Raymond. "The Space Radiation Environment For Electronics." Proc. IEEE, 76:1423–1442, Nov. 1988.: J. Bloxham and D. Gubbins. "The Evolution of the Earth's Magnetic Field." Sci. Amer.: 68–75, Dec. 1989.
28. E. C. Mullen, M. S. Gussenhaven and D. A. Hardy. "The Space Radiation Environment at 840 Km." A. F. Geophysics Lab. Hanscom AFB, MA 01731, DNA SEU Conf. Los Angeles, CA, April 1988.

29. G. C. Messenger. "Single Event Upset Considerations For Multiple Particles." Feb. 1985, For Presentation at HEART Conf. Monterey, CA, July 1985.
30. C. M. Hsieh, P. C. Murley and R. R. O'Brien. "A Field Funneling Effect on the Collection of Alpha Particle Generated Carriers. in Silicon Devices." IEEE Trans EDL-2(4), Apr. 1981.
31. Hsieh, Murley and O'Brien. "Dynamics of Charge Collection From Alpha Particle Tracks in Integrated Circuits." IEEE Proc. 19th Ann. Rel. Phys. Conf, Orlando, FL, 1981.
32. H. L. Grubin, J. P. Kreskovsky and B. C. Weinberg. "Numerical Studies of Charge Collection and Funneling in Silicon Devices." IEEE Trans. Nucl. Sci. NS-31(6): 1161–1166, Dec. 1984.
33. G. C. Messenger. "Collection of Charge on Junction Nodes From Ion Tracks." IEEE Trans. Nucl. Sci. NS-29(6):2024–2031, Dec. 1982.
34. F. B. McLean and T. R. Oldham. "Charge Funneling in *N* and *P* Type Silicon Substrates." IEEE Trans. Nucl. Sci. NS-29(6):2018–2023, Dec. 1982.
35. T. R. Oldham and F. B. McLean. "Charge Collection Measurements For Heavy Ions Incident on *N* and *P* Silicon." IEEE Trans. Nucl. Sci. NS-30(6):4493–4500, Dec. 1983.
36. C. Hu. "Alpha Particle Induced Field and Enhanced Collection of Carriers." IEEE Elect. Dev. Lett. EDL-3, No. 2, Feb. 1982.
37. R. M. Gilbert, G. K. Ovrebo and J. Schifano. "Plasma Screening of Funnel Fields." IEEE Trans. Nucl. Sci. NS-32(6):4098–4103, Dec. 1985.
38. E. A. Burke. "The Impact of Component Technology Trends on the Performance and Radiation Vulnerability of Microelectronic Systems." NSRE Conf. Short Course, Sec. 4, Monterey, CA, 1985; R. W. Keyes. "Fundamental Limits in Digital Information Processing." Proc. IEEE, Vol. 69, p. 267, 1981.
39. R. H. Dennard, et al. "Design of Ion-Implanted MOSFETs With Very Small Physical Dimensions." IEEE J. Solid State Circuits, SC-9(5):256–258, Oct. 1974.
40. E. L. Peterson and P. Marshall. "Single Event Phenomena in the Space and SDI Areas." J. Rad. Eff. Res. & Eng., Jan. 1989.
41. J. Bradford. "Nonequilibrium Radiation Effects in VLSI." IEEE Trans. Nucl. Sci. NS-25(5):1144, Oct. 1978.
42. J. S. Cable, N. M. Ghoneim, R. G. Martin and Y. Song. "The Size Effect of an Ion Charge Track on Single Event Multiple-Bit Upset." IEEE Trans. Nucl. Sci. NS-34(6): 1305–1309, Dec. 1987.
43. L. D. Edmonds. "A Distribution Function For Double-Bit Upsets." IEEE Trans. Nucl. Sci. NS-36(2):1344–1346, April 1989.
44. J. N. Bradford. "A Distribution Function For Ion Track Lengths in Rectangular Volumes." J. Appl. Phys. 60(6):3799–3801, June 1979.
45. W. L. Bendel. "Chord Length Distribution Function Through Rectangular Volumes." Nav. Res. Labs., NRL 5369, July 1984.
46. J. A. Zoutendyk, L. S. Smith and G. A. Soli. "Empirical Modeling of Single Event Upset (SEU) in NMOS Depletion Mode Load Static RAM (SRAM) Chips." IEEE Trans. Nucl. Sci. NS-33(6):1581–1585, Dec. 1986; J. A. Zoutendyk, H. R. Schwartz and L. R. Nevill. IEEE Trans. Nucl. Sci. NS-35(6):1644–1647, Dec. 1988.
47. J. A. Zoutendyk, L. S. Smith, and G. A. Soli, op. cit. NS-32(6):4164–4169, Dec. 1985.
48. W. A. Kolasinski, J. B. Blake, J. K. Anthony, W. E. Price and E. C. Smith. "Simulation of Cosmic Ray Induced Soft Errors and Latchup in Integrated Circuit Computer Memories." IEEE Trans. Nucl. Sci. NS-26(6):5087–5091, Dec. 1979.
49. T. Aoki, R. Kasai and M. Tomizawa. "Numerical Analysis of Heavy Ion Particle Induced CMOS Latchup." IEEE Elect. Dev. Lett. EDL-7(5):273–275, May 1986.

50. K. Soliman and D. K. Nichols. "Latchup in CMOS From Heavy Ions." IEEE Trans. Nucl. Sci. NS-30(6):4514–4519, Dec. 1983.
51. J. H. Stephen, T. K. Sanderson, D. Mapper, J. Farren, R. Harboe-Sorensen and L. Adams. "Cosmic Ray Simulation Experiments for the Study of Single Event Upsets and Latchup in CMOS Memories." Trans. Nucl. Sci. NS-30(6):4464–4469, Dec. 1983.
52. R. Koga and W. A. Kolasinski. "Heavy Ion Induced Single Event Upsets of Microcircuits; A Summary of the Aerospace Corp. Test Data." IEEE Trans. Nucl. Sci. NS-31(6):1190–1195, Dec. 1984.
53. J. P. Spratt and D. G. Millward. "Single Event Upset in Bipolar Technologies and Hardness Assurance Support Activities", S.A.I.C., Inc. LA. Jolla, DNA TR-86-126, May 15, 1986.
54. J. H. Stephen, T. K. Sanderson, D. Mapper, M. Hardman, J. Farren, L. Adams and R. Harboe-Sorenson. "Investigation of Heavy Particle Induced Latchup Using a Californium-252 Source in CMOS SRAMs and PROMs." IEEE Trans. Nucl. Sci. NS-31(6):1207–1211, Dec. 1984.
55. M. Reier. "The Use of Cf-252 to Measure Latchup Crossections as a Function of LET." IEEE Trans. Nucl. Sci. NS-33(6):1642–1645, Dec. 1986.
56. D. K. Nichols, W. E. Price, M. A. Shoga, J. Duffey, W. A. Kolasinski and R. Koga. "Discovery of Heavy Ion Induced Latchup in CMOS-Epi Devices." IEEE Trans. Nucl. Sci. NS-33(6):1696, Dec. 1986.
57. W. A. Kolasinski, R. Koga, E. Schnauss and J. Duffey. "The Effects of Elevated Temperature on Latchup and Bit Errors in CMOS Devices." IEEE Trans. Nucl. Sci. NS-33(6):1605–1609, Dec. 1986.
58. M. Shoga and D. Binder. "Theory of Single Event Latchup in Complementary Metal-Oxide Semiconductor Integrated Circuits." IEEE Trans. Nucl. Sci. NS-33(6):1714–1717, Dec. 1986.
59. TRW. OSG, Interoffice Memo, Nov. 15, 1985.
60. D. K. Nichols, W. E. Price, C. J. Malone and T. Smith. "A Summary of JPL Test Data, May 1982–February 1984." Jet Propulsion Laboratories, Pasadena, CA, 1986.
61. R. Silberberg. "Neutron Induced SEU's in the Atmosphere." IEEE Trans. Nucl. Sci. NS-31(6):1183–1185, Dec. 1984.
62. L. C. Northcliffe and R. F. Schilling. Nuclear Data, Vol. A7. Academic Press, 1970.
63. C. S. Dyer, J. Farren, A. J. Sims and J. Stephen. "Measurements of the SEU Environment in the upper Atmosphere." IEEE Trans. Nucl. Sci. NS-36(6):2275–2280, Dec. 1989.
64. J. B. Langworthy. "Depletion Region Geometry Analysis Applied to Single Event Sensitivity." IEEE Trans. Nucl. Sci. NS-36(6):2427–2434, Dec. 1989.
65. M. A. Xapsos et al. "Single Event, Enhanced Single Event, and Dose Rate Effects With Pulsed Proton Beams." IEEE Trans. Nucl. Sci. NS-34(6):1419–1425, Dec. 1987.
66. L. W. Massengill and S. E. Kerns." Transient Radiation Upset Simulations of CMOS Memory Circuits." IEEE Trans. Nucl. Sci. NS-31(6):1337–1343, Dec. 1984.
67. J. H. Adams, Jr. "The Variability of Single Event Upset Rates in the Natural Environment." IEEE Trans. Nucl. Sci. NS-30(6):4475–4480, Dec. 1983.
68. P. J. McNulty. "Predicting Single Event Phenomena in Natural Space Environments." IEEE NSRE Conference Short Course, Chap. 3, Reno, NV, July 1990.
69. R. Silberberg, C. H. Tsao and J. R. Letaw. "Neutron Generated Single Event Upsets in the Atmosphere." IEEE Trans. Nucl. Sci. NS-31(6):1183–1185, Dec. 1984.
70. J. M. Bisgrove, J. E. Lynch, P. J. McNulty, W. G. Abdel-Kader, V. Kletnicks and W. A. Kolasinski. "Comparison of Soft Errors Induced by Heavy Ions and Protons." IEEE Trans. Nucl. Sci. NS-33(6):1571–1576, Dec. 1986.

71. A. R. Newton, D. O. Pederson, A. Sangiovanni-Vincentelli, A. Vladimirescu and K. Zhang. "SPICE Version 2G Users Guide." Dept. Elec. Eng./Comp. Sci., Univ. of Calif., Berkeley, CA 94720, 10 August 1981.
72. J. C. Higdon and R. E. Lingenfelter. "Gamma Ray Bursts." Ann. Rev. Astron. and Astrophys. 28:401–436, 1990; Sci. Amer., Oct. 1976, July 1980, Feb. 1985.
73. R. Koga, N. Katz, S. D. Pinkerton and W. A. Kolasinski. "Bevelac Ion Beam Characterization For Single Event Phenomena." IEEE Trans. Nucl. Sci. NS-37(6):1923–1928, Dec. 1990.
74. J. G. Rollins. "Estimation of Proton Upset Rates From Heavy Ion Test Data." IEEE Trans. Nucl. Sci. NS-37(6):1961–1965, Dec. 1990.
75. W. J. Stapor, J. P. Myers, J. B. Langworthy and E. L. Petersen. "Two Parameter Bendel Model Calculations For Predicting Proton Induced Upset." IEEE Trans. Nucl. Sci. NS-37(6):1966–1973, Dec. 1990.
76. M. Reier. JPL Radiation Effects Section. Progress Reports, Nov. 1991.

CHAPTER 9
ELECTROMAGNETIC PULSE

9.1 INTRODUCTION

One of the four principal emanations from a nuclear detonation, in terms of effects on electronic systems, is the electromagnetic pulse (EMP). Besides being unique, compared with the other three effects, due to neutrons, ionizing dose, and dose rate, EMP encompasses phenomena in the electromagnetic spectrum from very low frequencies of less than a hertz to ultrahigh frequencies (UHF). The phenomenology of EMP is part of the domain of electromagnetic theory, which is reviewed, with a bent toward nuclear EMP, in Chapter 2.

One way to compartmentalize EMP for investigation is to examine EMP output from endoatmospheric detonations and from exoatmospheric detonations. As discussed in Section 4.11, the progenitors of the EMP are mainly the x and gamma rays from the nuclear burst. Through their interactions with the air molecules, they ionize the latter to cause large electric currents flowing radially away from the detonation. If the burst is of moderate yield (0.05–0.50 MT) and is detonated within the atmosphere, the gamma-ray output is confined to a volume on the order of a mile in diameter. The x-ray output immediately interacts with the air to form a fireball whose diameter is also about 1 mile (mi). The diameter of the gamma-ray confinement increases very slowly with yield, so that an increase in yield has very little effect on the volume of this radiation field. Nevertheless, its close-in effects are important, as will be discussed later.

It is when the nuclear detonation occurs exoatmospherically that the EMP has a geographically far-ranging, deleterious effect on electronic systems. Exoatmospheric gamma and x-rays can travel miles before encountering an air molecule, so that there is essentially no fireball. The corresponding photon-source region can be upward of 1000 mi in diameter and about 20 mi thick. If the detonation occurs high above the central United States, this gigantic source of photons can rain down onto the atmosphere below, producing a pancake-shaped region containing prodigious ionization electric currents, as in Fig. 9.6. This results in the production of an electromagnetic field pulse, as discussed in Section 4.11, of considerable strength at sea level. It would essentially span the continental United States, as measured by a line-of-sight tangent cone from

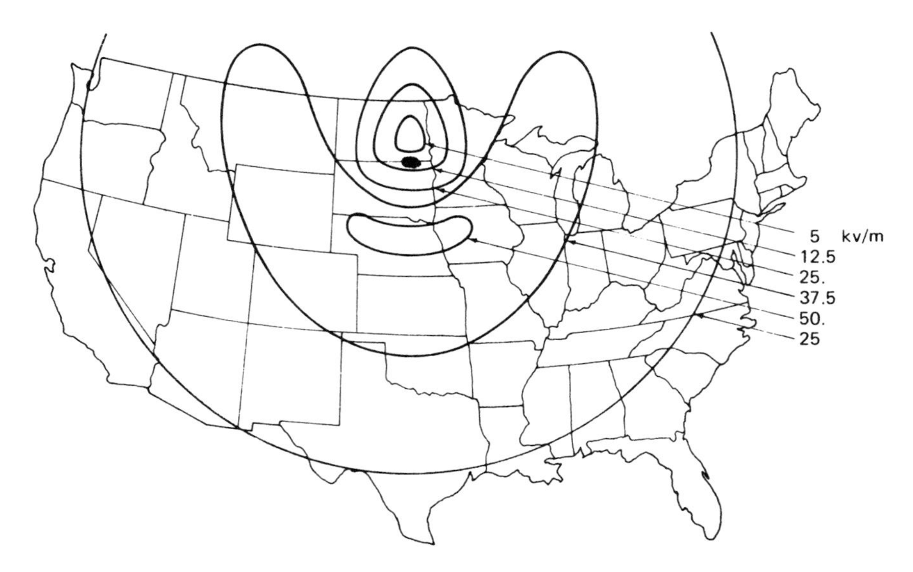

Fig. 9.1 Electric field contours at the earth's surface from a high-altitude (~300 mile) nuclear detonation.[1] Corresponding magnetic field strengths can be up to 200 ampere turns per meter, which is 10 times that of the magnetic field of the earth at sea level. Detonation is above the black disc.

the detonation onto the upper edge of the atmosphere, as shown in Figs. 9.1 and 9.6.

The electric field contours thereon interact with the geomagnetic field according to the magnetic declination at this latitude, which accounts for the asymmetry in Fig. 9.1. These electric fields are only those incident at the earth's surface, in that their ground reflections must be taken into account. That is, at the site of interest, both the direct and reflected fields must be combined to yield the total field strength. The reflected field correction is not trivial, as it is a function of the ground conductivity, which can be as much as 0.02 mhos per meter.

9.2 TIME AND FREQUENCY DESCRIPTION

There are a number of expressions for the time description of the EMP pulse, from simple to very complicated. One such was that used by the Office of Civil Defense. Its studies divide the EMP into endo- and exoatmospheric detonations.[2] For the former, most of the compton electrons attach themselves to the oxygen molecules to produce O_2^- ions. The much lower mobility of these ions corresponds to a much smaller velocity, compared with electrons, so that the recombination time of

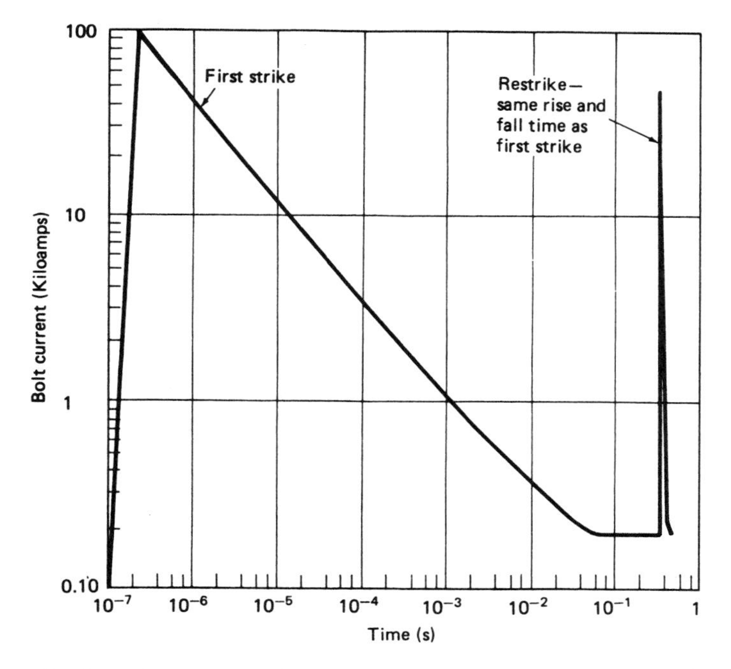

Fig. 9.2 Lightning waveform showing first strike and restrike phase.[3]

these attached electrons with the air molecules from which they were born, is greatly increased over that were there no atmosphere. This results in an increased EMP pulse duration. The ensuing "long" pulse is expressed as

$$E_L(t) = 5.2 \cdot 10^4[\exp - (1.5 \cdot 10^6 t) - \exp - (2.6 \cdot 10^8 t)] \text{ ; volts/meter} \tag{9.1}$$

where time is in seconds.

In the exosphere, no such time stretching of the pulse occurs, because very little attachment and recombination occur in the very rarefied upper atmosphere. This "short" pulse is given by

$$E_s(t) = 6.3 \cdot 10^4[\exp - (1.5 \cdot 10^7 t) - \exp - (2.6 \cdot 10^8 t)] \text{ ; volts/meter} \tag{9.2}$$

Corresponding magnetic field strengths are obtained from $H(t) = E(t)/377$, since the impedance of free space is 377 ohms, as discussed in Section 2.6.

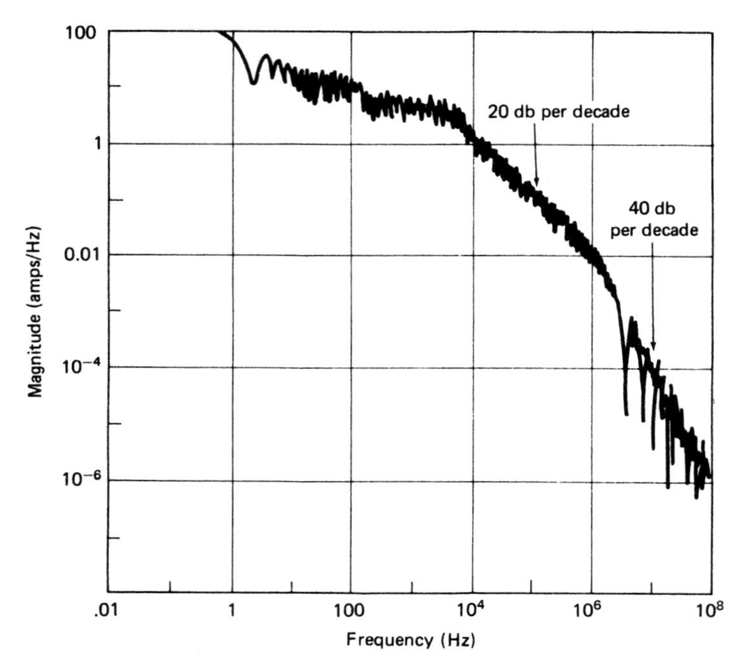

Fig. 9.3 Frequency spectrum of typical lightning waveform.[3]

For a comparison of EMP with lightning, Figs. 9.2 and 9.3 respectively depict a typical lightning waveform in time and the corresponding frequency spectrum. It is interesting to note that lightning striking an aircraft can deliver a factor of 100 larger EMP pulse than an endo-atmospheric nuclear detonation, almost irrespective of yield, at a standoff distance such that the corresponding blast wave does not damage the aircraft fuselage (i.e., ~2 psi overpressure)[40]. The frequency spectrum or frequency signature, of, say, the EMP long pulse is obtained from the modulus of its Fourier transform, or

$$|\bar{E}_L(\omega)| = \left| \int_0^\infty E_L(t) \exp - (i\omega t)\, dt \right| \tag{9.3}$$

where the phase information arg $\bar{E}(\omega)$ is ignored, and $E_L(t)$ in the above integral is obtained from Eq. (9.1). Then

$$|\bar{E}_L(\omega)| = 5.2 \cdot 10^4 \beta / \sqrt{(\omega^2 + \alpha^2)(\omega^2 + \beta^2)} \; ; \; \beta/\alpha \gg 1 \tag{9.4}$$

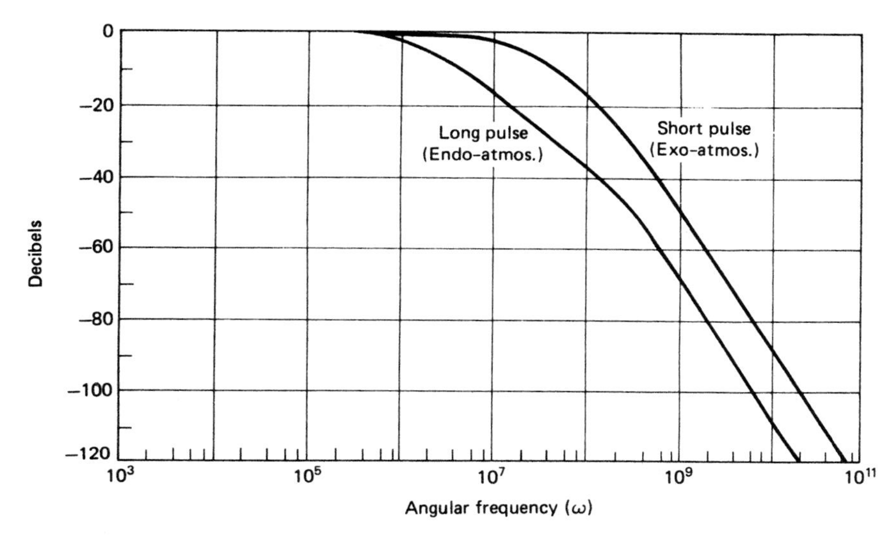

Fig. 9.4 EMP pulse frequency signatures.

where $\alpha = 1.5 \cdot 10^6$, and $\beta = 2.6 \cdot 10^8$. Relative decibel values of these quantities are expressed as, e.g., for the long pulse

$$(DB)_L = 20 \log_{10} |\bar{E}_L(\omega)/\bar{E}_L(0)| \tag{9.5}$$

Plots of DB_L and DB_S, the decibel equivalents of the Fourier transform magnitudes of the long and short pulse, are shown in Fig. 9.4. It is seen there that the EMP pulses extend from very low frequencies to nearly L band frequencies. The lightning stroke pulse, as seen in Figs. 9.2 and 9.3, extends beyond the UHF bands. All the curves exhibit the -40 DB per decade ω^{-2} asymptotic frequency dependence.

9.3 ATMOSPHERIC AND GLOBAL EFFECTS

The EMP of present interest is due mainly to the compton interaction of the x and gamma rays from a high-altitude detonation with the atmosphere molecules to produce recoil electrons, which produce the EMP fields, as discussed in Section 4.11. Besides the above electrons, the radioactive fission debris emits copious quantities of fast electrons. Actually, the prime source of fast electrons of energies greater than 1 MeV are those from the decay of fission products. The above electrons, when carried aloft and injected into the earth's magnetic field at a high

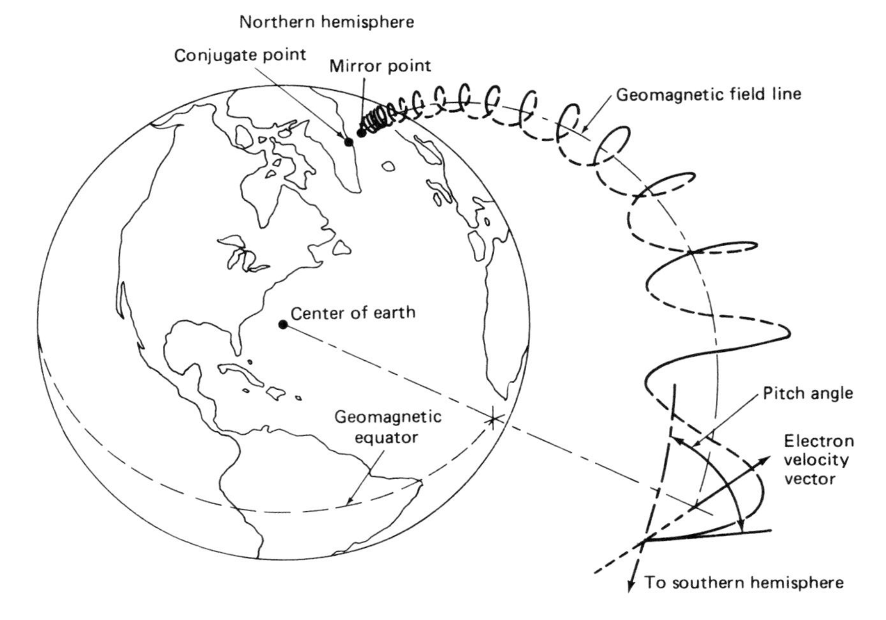

Fig. 9.5 Trapped argus electron motion in the geomagnetic field.

altitude, emit synchrotron radiation that can jam high-frequency communication channels, as discussed below.

For a high-altitude detonation, the hot fireball is accelerated upward at several kilometers per second into the very rarefied upper atmosphere. This fireball carries radioactive fission products upward with it to near ionospheric heights. At these altitudes, the fission product decay electrons can be injected into the geomagnetic field. These are called *Argus electrons*, after the series of experiments that established that injection of such electrons can form a sheath about the earth. The argus electrons become trapped in the geomagnetic field and persist, instead of quickly recombining with or attaching themselves to air ions or molecules, simply because of the tenuous atmosphere at these altitudes. As these electrons have an initial speed at injection, they experience a Lorentz force equal to $(e/c)(\bar{v} \times \bar{H})$, where $\bar{v}$ is the velocity of the electron, and $\bar{H}$ is the geomagnetic field strength. This causes them to spiral along the geomagnetic field lines toward the conjugate point of that particular field line. This is shown in Fig. 9.5.

Employing the construct of electromagnetic field lines, each geomagnetic field line emanates from the magnetic dipole field deep in the

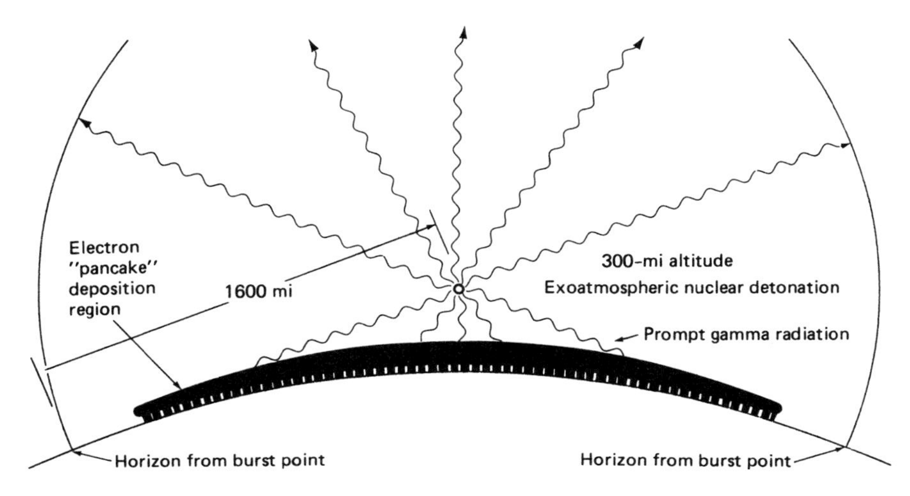

Fig. 9.6 EMP generated by a high-altitude detonation showing pancake deposition region produced by prompt gamma rays.

interior of the earth. The two conjugate points correspond to the two places at which the particular field line pierces the earth surface, as in Fig. 9.5. As the electron spirals along the field line toward the conjugate point, which could be on the other side of the globe from the detonation, the field lines begin to converge toward each polar region, i.e., the magnetic field strength $\bar{H}$ is increasing toward the poles. This results in a component of magnetic force along the magnetic lines in a direction opposite to the general motion of the incoming electron. As the electron them decelerates, its rotational speed begins to increase, since the kinematic energy of the electron is conserved. This causes an increase in the spiral pitch angle of the electron, amounting to a decrease in the forward motion per rotation of the electron. The deceleration increases until the electron spirals into the mirror point, defined to be where the pitch angle is $\pi/2$. The direction of the electron along the field line reverses, and it begins to spiral away from the mirror point, back up the field line toward the opposite mirror point. Hence, the electron is trapped in the geomagnetic field, "bouncing" back and forth between mirror points. The bounce period of these electrons, i.e., the time between successive mirror reflections, is of the order of 0.1–1.0 s, and the gyro period, or time per spiral, is about 1 μs or less. Also, these electrons drift toward the east (positively charged particles drift westward) in their trapped orbits, with a precession period of about 10^4 s, or about 2.8 h once around the earth. Hence, the earth becomes enveloped in a shell

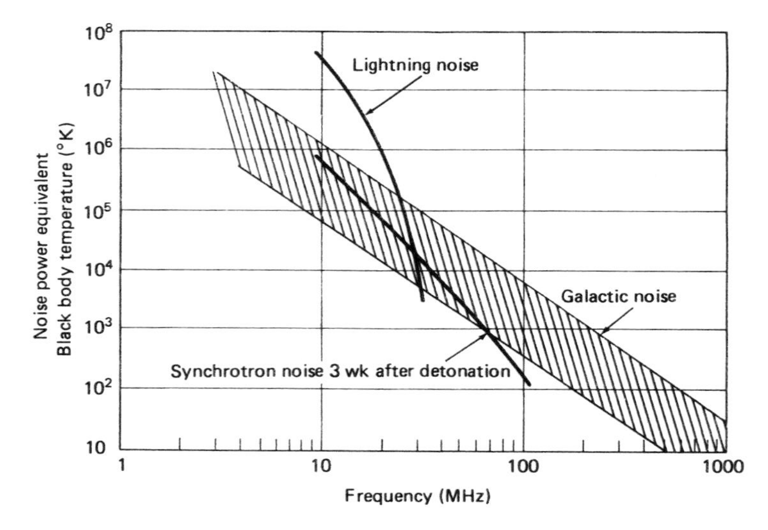

Fig. 9.7 Synchrotron noise from Starfish argus electrons overhead near the geomagnetic equator 3 wk after detonation.[4]

of Argus electrons, producing a temporary artificial ionosphere that ultimately decays through recombination, reattachment, and other dissipative mechanisms. The decay time constant is of the order of days to weeks, depending on the yield, altitude, and many other parameters.

As mentioned, the reason for investigating these electrons is that while spiraling they radiate in a bremsstrahlung sense, because they are accelerated in their circular spiral motion, not unlike particles being accelerated in a synchrotron particle accelerator. This synchrotron radiation can result in sufficient radio frequency noise power to disrupt or jam radio communications channels for hours at a time. A comparison of synchrotron noise 3 wk after the detonation with galactic and lightning noise is depicted in Fig. 9.7. The ordinate of that figure is noise power $S(\omega)$, expressed as the temperature T(°K) of an equivalent black body radiator possessing the same frequency spectrum, and sky solid angle[41] i.e., $S(\omega) = \omega^2 kT/2\pi^2 c^2$.

The experiment used to help verify the existence of Argus electrons was the Starfish event, a 1.4-MT detonation, 400 km above Johnston Island, near the Hawaiian chain, in July 1962.[4] This event disrupted long-haul, high-frequency (HF) communications between Hawaii and the Orient for about 6 h after the detonation. After this period, the synchrotron noise subsided, as seen in Fig. 9.7, when, after 3 wk, it was comparable with galactic noise.

Besides electrons, other charged particles from a high-altitude detonation can also be trapped in the geomagnetic field (i.e., in the lower regions of the Van Allen belts). These include protons whose main source is the decay of neutrons from fission products. The neutron has a half-life of about 12 min, where, in the upper reaches of the stratosphere, it can decay to a proton, electron, and neutrino, before it interacts with any sparse amount of matter there. Other such particles included in the detonation debris are positrons (e^+), deuterons (^2H), tritons (^3H), and alpha particles (He^{++}). High-altitude detonations cause other phenomena, including artificial auroras, etc., where are outside the scope of this discussion.

Electrons trapped in the Van Allen belts produce a fluence incident on a spacecraft within them, whose level strongly depends on the spacecraft orbital parameters. For nearly all orbits of interest, this fluence is less than 10^{15} electrons per cm^2 over a nominal spacecraft lifetime of about 10 years;[42] corresponding proton fluences are less by 3–4 orders of magnitude.[43] This all results in less than 10^4 rads (Si) deposited in semiconductor devices, assumed to be behind 2.5 gm per cm^2 of spacecraft shielding. Minimum shielding is provided by the spacecraft skin, the intervening subsystem housings, the wall thickness of the metal housing the device of interest, and the package thickness of the device itself. Good engineering practice should tend to locate the most ionizing radiation-susceptible devices within the spacecraft in such a manner that additional shielding will be provided by the less susceptible electronics and associated structure.

Extra-terrestrial nuclear detonations in the vicinity of the belts can multiply the trapped electron fluence by as much as two orders of magnitude. These levels correspond to belt electron density saturation, as seen in Fig. 9.7A. However, for most pertinent orbits, this effect will be less than one order of magnitude greater than normal. The corresponding dose deposited behind 2.5 gm per cm^2 of shielding will be less than 100 krads (Si). Thus, required ionizing dose hardness levels for semiconductor devices will be less than 100 krads (Si) for the electronically "pumped-up" belt environment, and less than 10 krads (Si) for the natural belt environment. These generalizations apply to the majority of normally encountered spacecraft environments. However, special cases may occur where, for instance, it is not possible to provide 2.5 gm per cm^2 of shielding for sensitive devices, or it is necessary to utilize an equatorial orbit near the center of the inner Van Allen belt, corresponding to maximum immersion in the belts. For these extreme situations, radiation requirements might exceed 100 krads (Si). It is also customary to add design margins to semiconductor device radiation specifications.

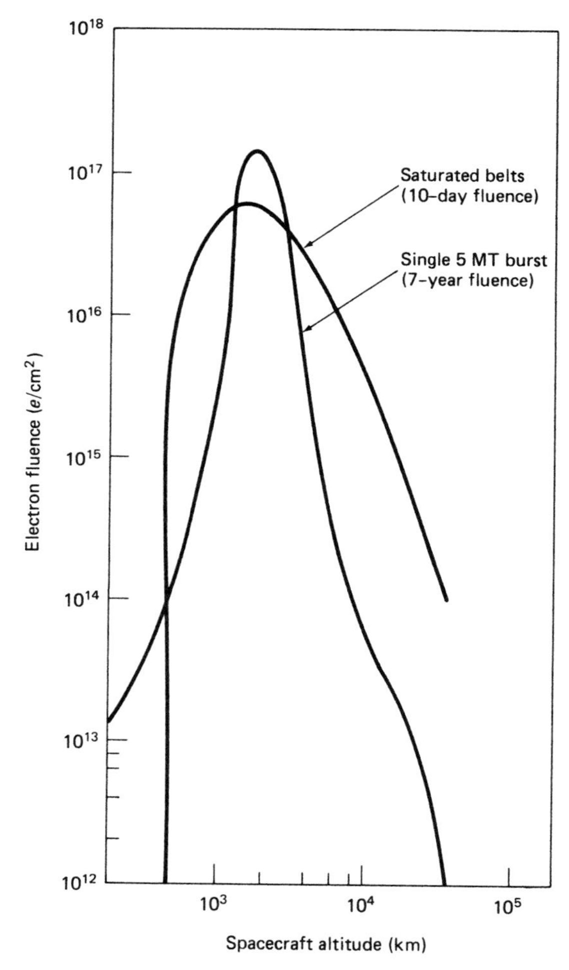

Fig. 9.7A Electron fluences in the Van Allen belt region following extra-terrstrial detonation in that locality.[43]

For example, for ionizing dose, a factor of 10 is often used. In these cases, individual additional "spot" shielding of the sensitive device, as discussed in Section 13.2, should be effective in reducing its deposited dose to tolerable levels.

9.4 SURFACE BURST EMP

As mentioned in Section 9.1, the EMP produced from a surface burst is confined to a relatively small volume about the detonation. This is mainly due to the intimate interaction of the weapon efflux with the sensible atmosphere. At these close-in ranges, such effects as blast overpressure and thermal energy production are major weapon outputs. Nevertheless, there is a surface burst EMP that can cause substantive electromagnetic disruption near ground zero, which bears on military tactical systems. For a burst immediately over the ground or seawater, the compton current production and its ramifications are almost the same as described in Section 4.11. There are certain differences, for the conductivity of the soil or water is usually greater than that of air. Thus the ground or water shunts the electric field and provides a path of low resistance for compton return currents. These currents form a toroidal flow from their production in the air down and back through the ground or water to ground zero, the toroid center. This current looping back through to ground zero produces an azimuthal magnetic field around the toroidal current flow. Simultaneously, the radial electric field, E_r, produces conduction currents in the air, which oppose the compton current to limit the electric field. All this is illustrated in Fig. 9.8. As discussed in Section 4.11, the radial compton current creates a spherically symmetric electric field, resulting in a zero magnetic field. However, the radial components of the compton current density, J, do not vanish, but interact with the radial component of the electric field, as described by[18]

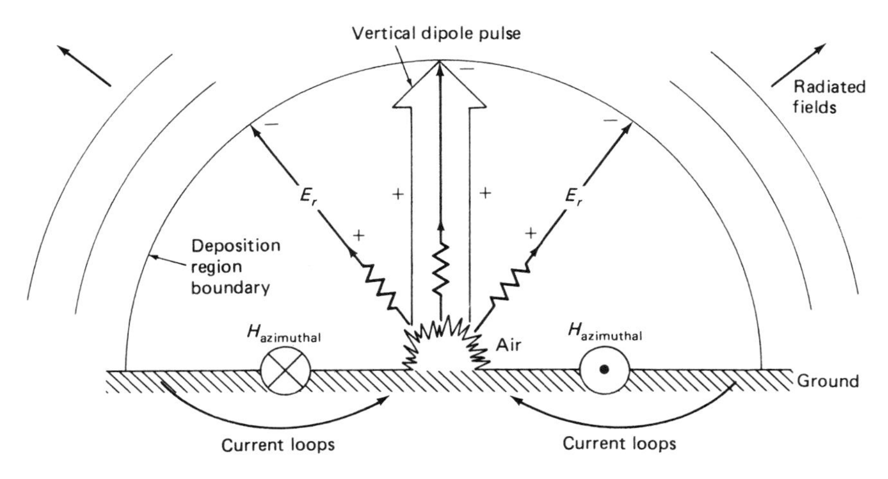

Fig. 9.8 Surface burst EMP showing vertical dipole output and ground return current loop paths.[2]

$$\varepsilon_0 \dot{E}_r + \sigma E_r = -J_r \tag{9.6}$$

where the spatial dependencies of E_r and J_r are neglected temporarily for the sake of discussion. The integral of Eq. (9.6) is given by

$$E_r(t) = -(1/\varepsilon_0) \int_0^t J_r(t')[\exp - \sigma/\varepsilon_0(t - t')]\, dt' \tag{9.7}$$

To examine the early and late time behavior of the above, its Fourier transform is

$$\bar{E}_r(\omega) = -\bar{J}_r(\omega)/\varepsilon_0(i\omega + \sigma/\varepsilon_0) \tag{9.8}$$

Early time corresponds to large ω (high frequencies), so that

$$\bar{E}_r(\omega) \sim -\bar{J}_r(\omega)/i\omega\varepsilon_0 \tag{9.9}$$

Then transforming back into the time domain, for early time, gives

$$E_r(t) \simeq -\int_0^t J(t')\, dt'/\varepsilon_0 = -Q/\varepsilon_0 \tag{9.10}$$

which amounts to the statement that J is charging up the capacitance of the earth. Similarly, for late times, Eq. (9.8) is examined for small ω, which leads to

$$\lim_{t \to \infty} E_r \equiv E_\infty = -J_\infty/\sigma \tag{9.11}$$

Based on estimates of J_∞ and σ, Longmire gives expressions for the above asymptotic fields as[18]

$$\begin{aligned} E_\infty &\approx 2 \cdot 10^{-4}(\alpha + k_1) \text{ volts/meter}, \\ H_\infty &\simeq E_\infty/377 = 5.3 \cdot 10^{-7} \quad (\alpha + k_1) \text{ Amp-turns/meter} \end{aligned} \tag{9.12}$$

where $\alpha \cong 2 \cdot 10^8\,\text{s}^{-1}$. α^{-1} is the fission chain reaction e-folding time of the detonating weapon, and $k_1 \cong 10^8\,\text{s}^{-1}$ is the rate of attachment of electrons onto oxygen atoms in the atmosphere. Attachment is the dominant reaction of the ionic recombination kinetics in this environment. Inserting the requisite numbers into Eq. (9.12) for a surface burst yields $E \cong 60$ kilovolts per meter. For a high-altitude, rarefied atmosphere burst, $\alpha \gg k_1$, so that neglecting the latter yields $E \cong 40$ kilovolts per meter^{-1}.

After the gamma pulse and compton currents reach their maxima, the weapon has essentially disassembled, so that $\alpha \leqslant 0$. Then the corresponding $E \cong 20$ kilovolts per meter, which is maintained until the ion conductivity of the air begins to dominate, causing E to vary as the square root of the electron-ion production rate in the atmosphere.[18]

The temporal development of surface burst EMP can be described as taking place in three phases. The first is the wave phase, in which the conduction current is small compared with the displacement current. Rewriting Eq. (2.2) for surface EMP conditions results in

$$\nabla \times \bar{H} = \frac{4\pi}{c}(\bar{J} + \sigma\bar{E}) + \frac{1}{c}\frac{\partial \bar{E}}{\partial t} \tag{9.13}$$

where $\bar{J}$ is the compton current and σ is the ground conductivity. The conduction current term $4\pi\sigma E \ll \partial E/\partial t$, the displacement term, when $\sigma \ll \alpha/4\pi c = 10^{-3}$ mhos per meter, and so the former is usually neglected. Phase two is termed the diffusion phase, appearing when $\sigma \gg \alpha/4\pi c$ (i.e., when the conduction current dominates over the displacement current). This occurs at electric field saturation; that is, when $E \approx E_\infty$ begins to hold. Now the return conduction current shifts to the ground from the air immediately above, thus forming current loops. These produce an azimuthal magnetic field near the ground, which diffuses up into the air and down into the ground by the well-known skin effect process.[18] The azimuthal magnetic field is also a function of the skin depth in air,[16] which is time-dependent (i.e., $\delta \cong \sqrt{ct_s/4\pi\sigma}$, where t_s is measured from the beginning of the saturation epoch). When this skin-depth height in the atmosphere increases to where it reaches the azimuthal magnetic field vector (tip) at the latter's angular elevation of 60 degrees, as measured from the burst point, the diffusion phase is over.[18] The third, or quasi-static phase occurs when the diffusion has ceased and the transverse (induction) component of the electric field is much less than the longitudinal (electrostatic) component, and the compton and conduction currents begin to cancel each other.

It should be noted that the approximation made earlier with respect to the soil or seawater conductivity being less than that of the atmosphere does not always hold. This can hold near the burst point, when the compton current achieves its maximum, and the local air becomes incandescent, with a corresponding high conductivity. In this case, when the compton return current in the ground surface layers is taken into account, this can result in a reversal of polarity of the azimuthal magnetic field near the ground.[19,20]

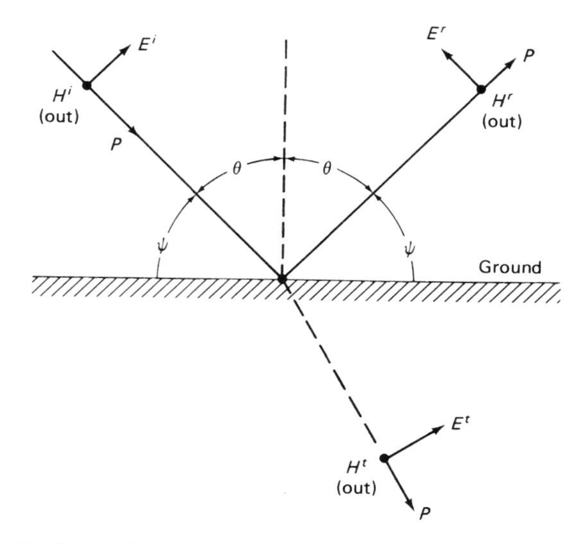

Fig. 9.9 Incident, reflected, and transmitted waves at the air-ground interface.

9.5 SURFACE AND SUBSURFACE REFLECTIONS

As mentioned earlier, the EMP ground-reflected component, combines with the direct (incident) field at the point of interest to yield the total impressed field. The ground has a finite resistivity, so that the ground-reflected field has a substantial amplitude relative to the direct field.

Using the boundary conditions between two dissimilar media for an electromagnetic wave, as discussed in Section 2.7, the equations for plane electric and magnetic field strengths at oblique incidence at the air-ground interface can be developed.[5] Incidentally, this development shows that the incident, reflected, and transmitted (into the ground) electromagnetic waves are all in the same plane. That is, their respective electric vectors, $\bar{E}^i$, $\bar{E}^r$, and $\bar{E}^t$ are all coplanar, and the angle of incidence θ equals the angle of reflection, as seen in Fig. 9.9. The ground-air reflection coefficient, R_v, for the vertical component of the electric field is given by

$$|R_v| \equiv \frac{|\bar{E}^r|}{|\bar{E}^i|} = \frac{(\varepsilon - i\kappa_0)\cos\theta - \sqrt{(\varepsilon - i\kappa_0) - \sin^2\theta}}{(\varepsilon - i\kappa_0)\cos\theta + \sqrt{(\varepsilon - i\kappa_0) - \sin^2\theta}} \tag{9.14}$$

where

$\kappa_0 = \sigma/\omega\varepsilon_0 \simeq 18/f_{MHz}$ for a nominal ground conductivity

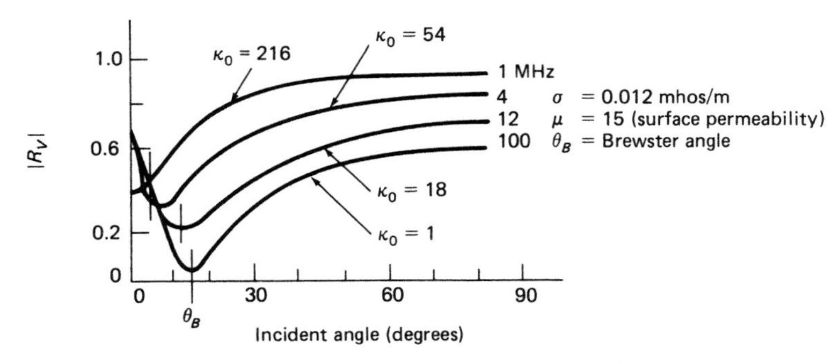

Fig. 9.10 Modulus of the plane wave reflection coefficient $|R_v|$ as a function of incident angle.[6]

σ = ground conductivity
ε_0 = average dielectric constant of the atmosphere
ω = angular frequency of the plane EMP component
ε = ground dielectric constant relative to free space
θ = angle of incidence, where $\psi = \pi/2 - \theta$ is the elevation angle.

$|R_v|$ is plotted versus incident angle, as shown in Fig. 9.10.

Because energy is conserved, $|T_v|^2 + |R_v|^2 = 1$, where the transmission coefficient T_v is given by

$$T_v = |\bar{E}^t|/|\bar{E}^i| = \sqrt{1 - R_v^2} \tag{9.15}$$

It is seen that, from Eq. (9.14), there is an incident angle, θ_B where R_v vanishes. This angle is seen to be a root of $R_v = 0$ and is easily calculated to be $\theta_B = \tan^{-1} \sqrt{\varepsilon}$ for $\varepsilon \gg \kappa_0$, as in the case for a low-loss dielectric. θ_B is recognized to be the well known Brewster angle. Therefore, essentially all of the incident EMP energy will be transmitted into the ground (as none is reflected) for some solid angle of topography under the nuclear detonation whose angular displacement is θ_B radians from the detonation normal to the earth (ground zero).

For normal incidence ($\theta = 0$) of the EMP, the tangential electric and magnetic fields that are transmitted into the ground are given, respectively, by[6]

$$E^T(R) = |\bar{E}^i| \cdot [2Z_2/(Z_0 + Z_2)] \exp - \gamma R \tag{9.16}$$

$$H^T(R) = |\bar{H}^i| \cdot [2Z_0/(Z_0 + Z_2)] \exp - \gamma R \tag{9.17}$$

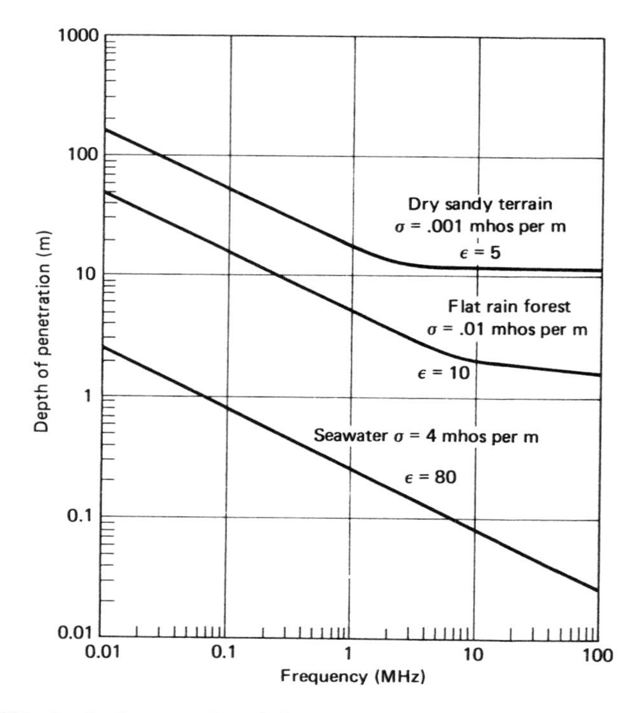

Fig. 9.11 Skin depth of penetration of electromagnetic energy into various topographies.

where the impedance of free space $Z_0 = 377$ ohms, and R is the depth of wave penetration into the ground. Also, $Z_2 = \sqrt{(\mu/\varepsilon)(1 - i\sigma/\omega\varepsilon)}$, the propagation constant is given by $\gamma = i\omega\sqrt{\mu\varepsilon(1 - i\sigma/\omega\varepsilon)}$, and μ is the ground permeability. For a nominal soil, $Z_0 \gg Z_2$, so that from Eq. (9.17) it is seen that the tangential component of the magnetic field strength is doubled over that of the corresponding incident field strength. On the other hand, the tangential electric field, from Eq. (9.16) is but 10 percent that of the incident electric field $\bar{E}^i$.

The propagation constant is defined by $\gamma = \alpha + i\beta$, where α is the plane wave decay constant, and, from Section 2.5, is related to the skin depth of penetration by $\delta = 1/\beta$. From Fig. 9.11, an estimate of the amplitude of the transmitted EMP electric and magnetic fields at various depths for various topographies can be obtained.

9.6 EMP COUPLING TO SHIELDED CABLES

In the previous sections, the origin of the EMP from a nuclear detonation was discussed, with an exoatmospheric EMP wave brought down to the air-ground interface. The wave is then resolved into incident, reflected, and transmitted (into the ground) components. An apt illustration of how EMP influences electronic systems at this point is how the transmitted ground component interacts with shielded electrical buried cables. As cables are a vital part of interconnected electrical and electronic systems, the interaction of the EMP with cables constitutes an important area of investigation, besides the fact that EMP interaction with coaxial cables was one of its first known manifestations, as discussed in Section 4.11. The EMP energy in the wave is incident on the imperfectly shielded cable, and most cables are not perfectly shielded by virtue of their less than completely meshed outer conductor braid or less-than-radio-frequency (RF)-tight cable connectors or even magnetic field coupling directly through solid outer conductor coaxial cable. This energy induces voltages on the internal center conductors of the cable. This generates spurious signal currents to flow in these conductors, which leak into and disrupt or damage the systems connected to each end. Because EMP-derived fields can be of very large amplitude, they can threaten the operation of these systems even to the point of burning out and melting components therein.

First it is necessary to establish the induced electromagnetic field incident on the cable in terms of cable parameters. For a cable at height h above the ground, and an assumed vertically polarized electric field E_{zv} incident on the cable normal to its length, then

$$E_{zv} = E_v^i(1 - R_v \exp - 2ikh) \tag{9.18}$$

R_v is the reflection coefficient for a vertically polarized wave, i.e.

$$R_v = \frac{\varepsilon_r(1 - i\sigma/\omega\varepsilon) - \sqrt{\varepsilon_r(1 - i\sigma/\omega\varepsilon)}}{\varepsilon_r(1 - i\sigma/\omega\varepsilon) + \sqrt{\varepsilon_r(1 - i\sigma/\omega\varepsilon)}} \tag{9.19}$$

where σ, ε, and ε_r are, respectively, the cable conductivity, permittivity, and relative dielectric constant, i.e., $\varepsilon_r = \varepsilon/\varepsilon_0$, and ε_0 is the dielectric permittivity of free space. The propagation number $\kappa = 2\pi/\lambda$, and E_v^i is the EMP-induced incident field amplitude as given in Section 9.1. Note the similarity of the reflection coefficient expressions as given in Eqs. (9.14) and (9.19).

The corresponding transmission line equations for the internal cable current, I, and voltage, V, across the cable, for a variable $E_{zv}(z)$, for sinusoidally varying voltages and currents are, where primes mean the derivative with distance along the transmission line

$$V' + IZ = E_{zv} = Z_T I_0; I' + YV = -Y_T V_0 = -i\omega C_{12} V_0 \qquad (9.20)$$

where

I_0 is the total cable ground return current, including both center conductor and outer conductor braid
Z_T is the transfer impedance between the shield braid exterior and the center conductor
Y_T is the corresponding transfer admittance
V_0 is the shield braid-to-ground voltage
C_{12} is the cable capacitance per unit length
Z and Y are, respectively, the cable impedance and admittance per unit length
$Z_0^2 = Z/Y$ and $\gamma^2 = ZY$ are, respectively, the cable characteristic impedance and cable propagation constant squared.

Equations (9.20) can be solved for V and I for cables of arbitrary length for various cable parameter values, and general results are given in Vance.[8] Here, certain useful results from cable current and voltage computations are described. Note that all solutions of Eqs. (9.20) are Fourier transforms of their actual time dependencies, because of the stipulation of sinusoidal time variation of the important quantities. For example, the corresponding current is actually $I(z, \omega)$, not $I(z, t)$. The latter is available from the inverse Fourier transform of $I(z, \omega)$.

For an electrically short cable ($L \ll \lambda$), terminated in impedance Z_1 at the input end, and Z_2 at the output end, solutions of Eqs. (9.20) are obtained as a special case. The corresponding currents and voltages at each end are

$$I_1 = (I_0 Z_T + V_0 Y_T Z_2)L/(Z_1 + Z_2); \qquad V_1 = -I_1 Z_1 \qquad (9.21)$$

$$I_2 = (I_0 Z_T - V_0 Y_T Z_1)L/(Z_1 + Z_2); \qquad V_2 = I_2 Z_2 \qquad (9.22)$$

The transfer impedance and transfer admittance, respectively, Z_T and Y_T, are the bridges between the currents and voltages on the outside of the cable, due to the EMP, and the sought-after voltages and currents internal to the cable. The goal is the determination of the latter quantities, which affect the electronic devices and systems that are interconnected by these cables.

The transfer impedance Z_T can be defined from the first of Eqs. (9.20) as the cable open circuit voltage per incremental length, V', divided by the total cable return current I_0, i.e.

$$Z_T = V'(z)|_{I=0}/I_0 \equiv V'_{0c}/I_0 \tag{9.23}$$

Similarly, the transfer admittance, Y_T, is the short circuit cable current per incremental length, I', divided by the cable-to-ground voltage V_0, i.e.

$$Y_T = -I'(z)|_{V=0}/V_0 \equiv -I'_{sc}/V_0 \tag{9.24}$$

This discussion will be confined to coaxial cables whose outer shield is woven metal braid, characterized by the parameters:

K = optical braid coverage, $K = 1$ for solid outer braid
α = angle of individual braid plait (bottom of Fig. 9.12)
N = number of wire strands per plait
N_p = number of plaits in the braid
d = diameter of wire strands in plait
a = cable braid shield radius
δ = braid skin depth $(\pi f \mu \sigma)^{-1/2}$

$$e = \begin{cases} \sqrt{1 - \tan^2\alpha}; & \alpha \leqslant \pi/4 \\ \sqrt{1 - \cot^2\alpha}; & \alpha > \pi/4 \end{cases}$$

The general form of the transfer impedance of a braid shield cable is given by[8]

$$Z_T = Z_d + i\omega M_{12} \tag{9.25}$$

where

$$Z_d = \frac{4(1 + i)\,d/\delta}{\pi d^2 N N_p \sigma (\cos\alpha)\sinh(1 + i)\,d/\delta} \tag{9.25a}$$

$$M_{12} = \frac{\pi\mu_0(1 - K)^{3/2}}{6N_p}\begin{cases} \dfrac{e^2}{E(e) - (1 - e^2)K(e)}; & \alpha < \dfrac{\pi}{4} \\ \dfrac{e^2}{\sqrt{1 - e^2}\,[K(e) - E(e)]}; & \alpha \geqslant \dfrac{\pi}{4} \end{cases} \tag{9.25b}$$

The permeability of free space $\mu_0 = 4 \cdot 10^{-7}$. $K(e)$ and $E(e)$ are complete elliptic integrals of the first and second kind, respectively, and are functions of the eccentricity e. Z_d is called the diffusion term, and M_{12} is called the mutual inductance term. Figure 9.13 depicts Z_T normalized to the dc resistance per unit length, R_0, as a function of the frequency of the incident electromagnetic wave, and

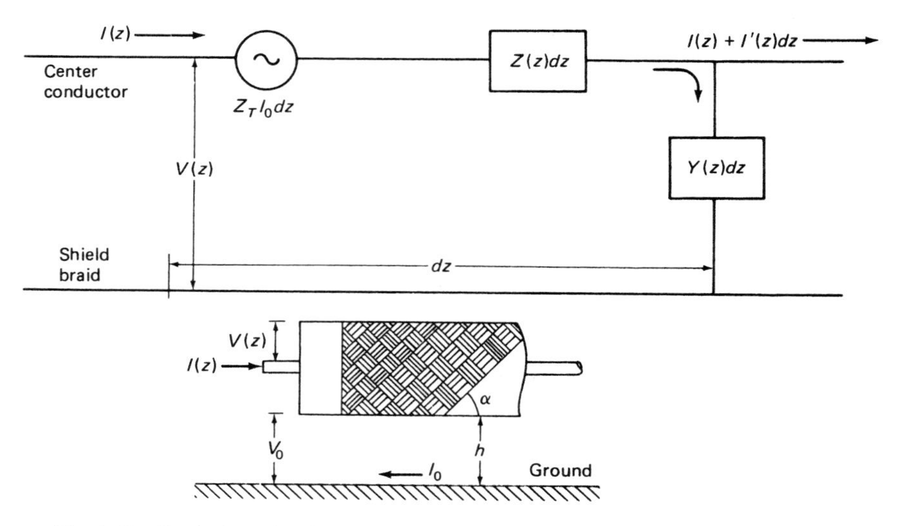

Fig. 9.12 Equivalent circuit for incremental length of braid-shielded coaxial cable.[8,10]

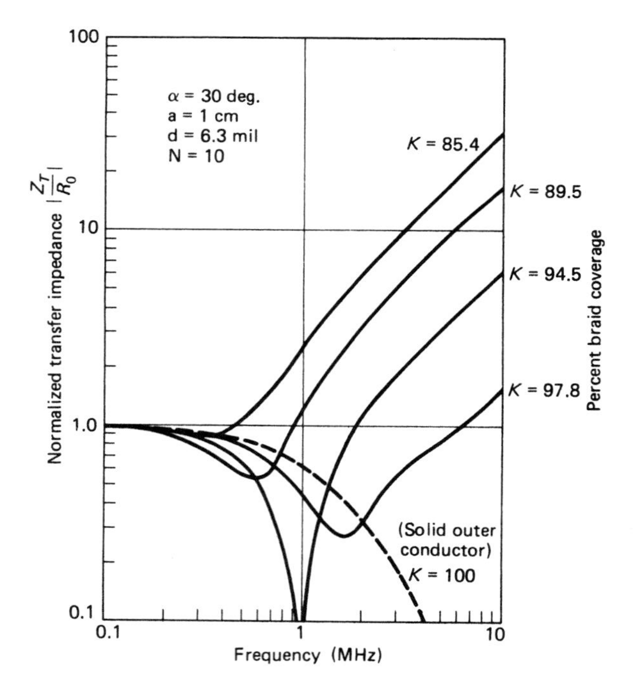

Fig. 9.13 Transfer impedance of braided shield coaxial cable for various values of optical coverage.[7,44] Figure © 1976 by John Wiley & Sons.

$$R_0 = 4/\pi d^2 N N_p \sigma \cos\alpha \tag{9.26}$$

The transfer admittance of this cable is $Y_T = i\omega C_{12}$, where[10,11]

$$C_{12} = \frac{g\pi C_1 C_2 (1 - K)^{3/2}}{6\varepsilon N_p} \begin{cases} [E(e)]^{-1}; & \alpha < \pi/4 \\ \sqrt{1 - e^2}[E(e)]^{-1}; & \alpha \geqslant \pi/4 \end{cases} \tag{9.27}$$

where

C_1 is the capacitance per unit cable length
C_2 is the capacitance per unit length between the cable and outer ground
g is the dielectric correction factor, e.g., for polyethylene dielectric coaxial cable, $g = 1.4$.[8]

For very short and very long cables, certain approximate solutions of the transmission line equations (9.20) yield useful expressions for cable voltage and current.

One measure of the cable shield, which can provide a relationship between the shield current and the center conductor current, is the shielding effectiveness, S, given by

$$S = 20\log_{10}(I_0/I) \qquad \text{(DB)} \tag{9.28}$$

where I_0 is the outer conductor current, and I is the center conductor current. For electrically short ($L \ll \lambda$) cables with very small transfer admittance, and terminated in impedances Z_1 and Z_2 yields

$$I/I_0 = Z_T L/(Z_1 + Z_2); \qquad L \ll \lambda \tag{9.29}$$

When Y_T is not small, and the external shield circuitry is terminated in its characteristic impedance, Z_e, the current ratios are

$$I_1/I_0 = (Z_T + Z_e Z_2 Y_T)L/(Z_1 + Z_2);\ I_2/I_0 = (Z_T - Z_1 Z_e Y_T)L/(Z_1 + Z_2) \tag{9.30}$$

For long cables ($L \gg \lambda$) solutions of Eqs. (9.20) yield approximate ready results. One such is that of a long cable excited by a vertically polarized, normally incident, exponentially decaying EMP pulse, viz., $E_0 \exp - t/\tau$. The resultant cable current, which holds except near the ends, is[7]

$$\bar{I}(z, \omega) = I_0 \tau/((i\omega\tau)^{1/2} + (i\omega\tau)^{3/2}) \tag{9.31}$$

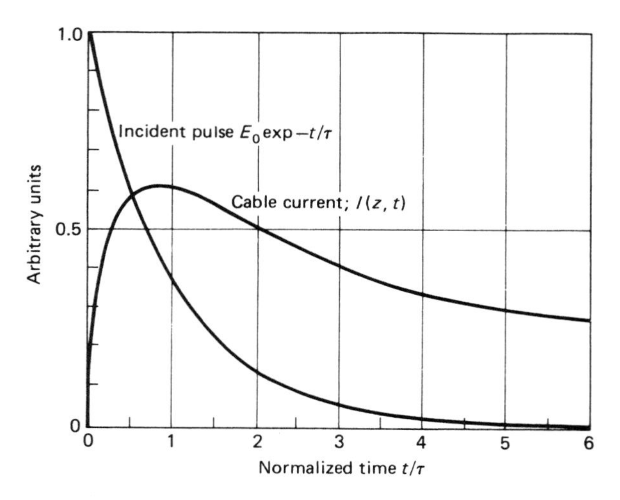

Fig. 9.14 Exponential incident EMP pulse and resulting cable current.[7] Figure © 1976 by John Wiley & Sons.

or, in the time domain, its inverse Fourier transform yields

$$I(z, t) = (2I_0(\exp - t/\tau)/\sqrt{\pi}) \int_0^{(t/\tau)^{1/2}} \exp(u^2)\, du \qquad (9.32)$$

where $I_0 = 10^6 \sqrt{\varepsilon_0 \tau/\sigma} E_0$. This EMP pulse and corresponding cable current is shown in Fig. 9.14. The induced current has its peak value of $i_p \cong 0.61 I_0$ when $t \cong 0.85\tau$.

The induced cable current from a normally incident, vertically polarized, impulse-type wave, whose electric field is given by $E_0\delta(t)$, is

$$I(z, t) = \sqrt{(\varepsilon_0/\sigma\pi t)} \cdot \varepsilon_0(\exp - \tau_d/4t)/L_g \qquad (9.32a)$$

where L_g and τ_d are, respectively, the inductance per unit length of cable, and $\tau_d = \mu_0 \sigma d^2$ with d as the cable burial depth in meters.

For a step function electric field of amplitude E_0, with all else the same as above, an induced cable current results as given by,

$$I(z, t) = \sqrt{\varepsilon_0 t/\sigma\pi}\, E_0(\exp - \tau_d/4t)/L_g \qquad (9.32b)$$

For a short segment of coaxial cable subject to EMP, as shown in Fig. 9.14A, connecting two grounded but elevated metal housings at the same height over a conducting earth, an estimate of the cable center conductor

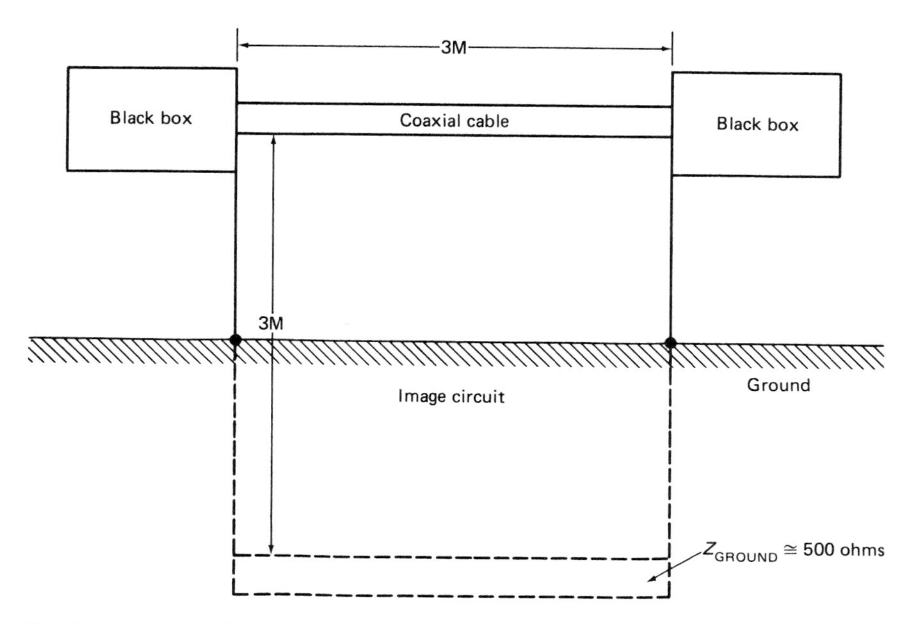

Fig. 9.14A Schematic diagram of short coaxial cable connecting two grounded metallic housings and its image circuit.

current and voltage across the cable can be obtained from Faraday's Law. That is, the EMP-induced voltage V on the cable shield braid, as part of a rectangular loop formed by the cable, the two vertical housing ground straps, and the parallel ground image cable segment is given by

$$V = -10^{-8} A\dot{\phi} \qquad \text{(volts)} \tag{9.33}$$

where ϕ is the EMP-induced magnetic flux threading the loop, and A is the area of the loop. The EMP magnetic flux rate $\dot{\phi}$, from a moderate-yield surface burst, is estimated to be 100 ampere turns per meter (2.5 gauss) delivered in about $2\,\mu$s. Hence, for a $9\,\text{m}^2$ loop area,

$$|V| = 10^{-8} \cdot 9 \cdot 10^4 \cdot 2.5/2 \cdot 10^{-6} = 1.125\,\text{kV} \tag{9.34}$$

on the cable shield braid, with respect to ground. The accompanying cable shield current I_0 is

$$I_0 = 1125/500 = 2.25 \text{ amperes} \tag{9.35}$$

since the ground resistance is estimated at 500 ohms, comprising the major resistance, and image leg, of the rectangular loop. The attenuation through the cable braid and its dielectric is about $-80\,\text{dB}$, corresponding to a magnitude of 10^{-4}. The current in the center conductor of this cable, whose characteristic impedance is 50 ohms, is then $2.25 \cdot 10^{-4} = 0.225\,\text{mA}$. The voltage across the cable from braid to center conductor is $2.25 \cdot 10^{-4} \cdot 50 = 11.25\,\text{mV}$.

9.7 EMP RESPONSE OF ANTENNAS

This section discusses the EMP response of the set of much used antennas associated with aircraft and other mobile and stationary military electronics systems. This discussion does not emphasize inadvertent antennas, such as those made by apertures, cavities, slots, or concomitant openings due to system operational needs, or due to inadequate fastening of the skin of aircraft or missiles. This is covered in the next section on shielding. The former type of antennas are of concern with respect of the EMP susceptibility/vulnerability of the system on which they are mounted. As appreciated from the earlier sections, EMP comprises a broad frequency spectrum from near dc to near GHz frequencies. Hence, the particular antenna must be analyzed in terms of its out-of-band response, as well as its in-band characteristics. An antenna can be completely characterized by its Thevenin equivalent circuit, usually as a two-port network, terminating at the antenna input. A known open circuit voltage, V_{0c}, due to the impressed EMP, and input impedance, Z_{in}, at the antenna input terminals, are usually sufficient to compute the voltage and current in the feedline and further back at the electronics into which the antenna feeds. The same holds for the dual quantity, viz., the short circuit current, I_{sc}, at the antenna input terminals, as well as the input admittance Y_{in}.

Initially consider a cylindrical tubular antenna of height $2h$ and radius a. Assume that the incident EMP electric field vector $\bar{E}_{zv}$ is parallel to the cylinder antenna axis, and that $a \ll \lambda$ at the highest frequency of interest. To obtain the current on this antenna as a function of the incident electric field requires a solution of Maxwell's equations, discussed in Chapter 2, on the antenna surface. This is a classic boundary value problem in electromagnetic theory.[12] One result is that the peak current on a cylindrical antenna of the above type is proportional to its length. Furthermore, an important cylindrical antenna parameter is called Ω, "fatness." It is given by

$$\Omega = 2 \ln (2h/a) \tag{9.36}$$

For example, a missile or aircraft fuselage considered as a cylindrical antenna has a relatively large $h/a = 10$. It is relatively fat with a typical

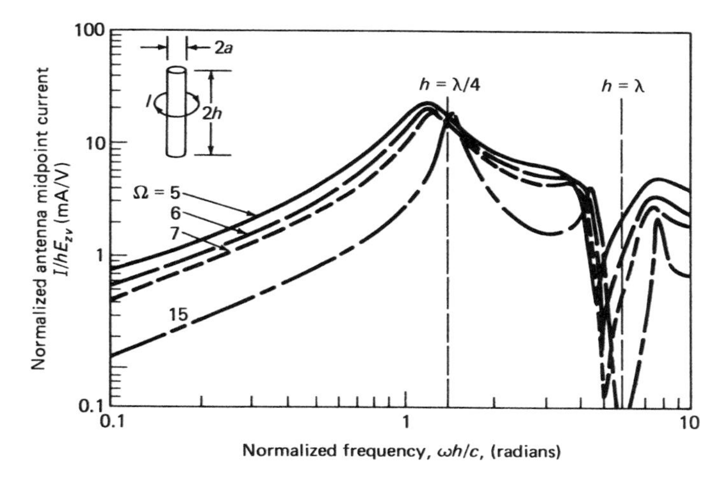

Fig. 9.15 Cylindrical antenna midpoint current for various fatness parameter values versus normalized frequency.[9,13–15] Figure © 1978 by the IEEE.

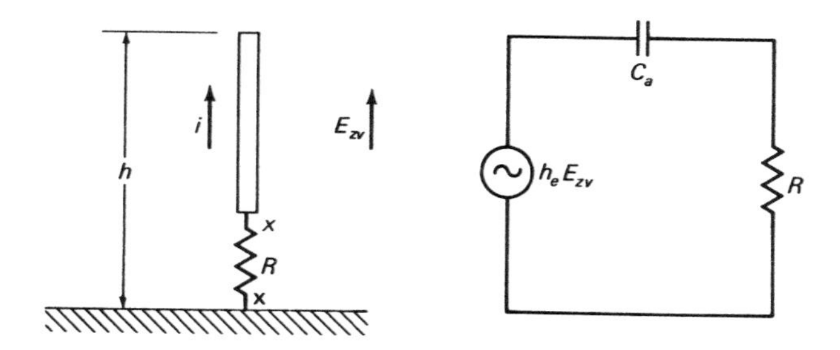

Fig. 9.16 Monopole antenna above perfectly conducting ground and Thevenin equivalent circuit.[7] Figure © 1976 by John Wiley & Sons.

$\Omega \cong 6$. As can be seen from Fig. 9.15, the fat antenna carries more current than the thin antenna (smaller Ω). However, the thin antenna has a higher Q, so that oscillations do not damp out as rapidly as they would for a fat antenna with a lower Q. The normalized frequency abscissa $\omega h/c = hk$, where $k = 2\pi/\lambda$.

Now consider a monopole antenna above a perfectly conducting ground, as in Fig. 9.16, which is a slightly altered version of Fig. 2.4. If the load resistance R were removed, the antenna impedance at "$x - x$" in Fig. 9.16 would appear as a capacitive reactance, with corresponding capacitance C_a given by

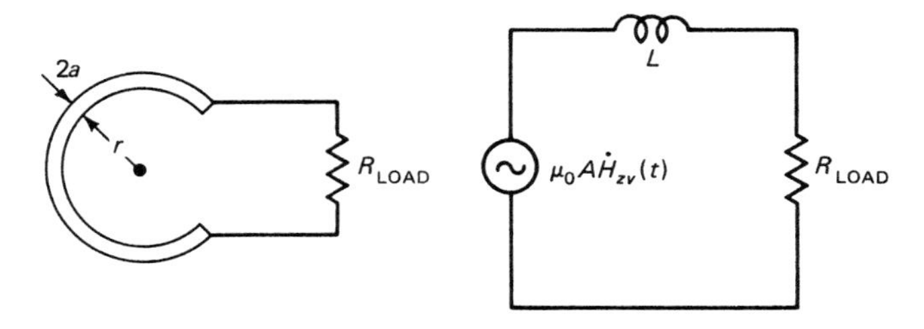

Fig. 9.17 Loop antenna and its equivalent circuit.[7] Figure © 1976 by John Wiley & Sons.

$$C_a = h/60c[\ln(2h/a) - 1] \tag{9.37}$$

with an effective height $h_e = h/2$, and c is the speed of light. A dipole can be represented by a vertical monopole above a perfectly conducting ground, as one arm of the dipole, together with its image extending into the ground, which is the other arm. The dipole capacitance is given by

$$C_a = h/120c[\ln(2h/a) - 1] \tag{9.38}$$

with the dipole effective height $h_e = h$. The EMP-induced voltage at the terminals $x - x$ of the monopole in Fig. 9.16 is given by the product of the EMP electric field E_{zv} by the effective height of the monopole. The current on the monopole, for $(\omega C_a)^{-1} \ll R$ is

$$i(t) = C_a h_e \dot{E}_{zv}(t) \tag{9.39}$$

The energy delivered to the load R is

$$J_R = R(C_a h_e \dot{E}_{zvm})^2 \tag{9.40}$$

where E_{zvm} is the maximum amplitude of the incident EMP electric field, and the overdot signifies the time derivative. When $(\omega C_a)^{-1} \gg R$, little of the energy is dissipated in the load, as most of the energy, J_{st}, is stored in the antenna capacitance C_a, given by

$$J_{st} = \tfrac{1}{2} C_a (h_e E_{zvm})^2 \tag{9.41}$$

The loop antenna behaves in a fashion complementary to the dipole/monopole antenna and is depicted in Fig. 9.17, where:

r is the loop radius
a is the loop half-thickness
R is the load resistance
A is the loop area
H_{zvm} is the maximum amplitude of the incident magnetic field
t_r, t_f are the rise and fall times of H_{zv}, respectively
$L = \mu_0 r(\ln(8r/a) - 2)$, $r \gg a$; loop inductance
μ_0 is the permeability of free space

Where the loop inductive reactance $X_L \ll R$, the loop current is

$$i_L = \mu_0 A\dot{H}_{zv}/R \tag{9.42}$$

The energy dissipated in the loop load R for a triangular magnetic field pulse of rise time t_r and fall time t_f is simply,

$$J_R = \int_0^{t_r+t_f} (\mu_0 A\dot{H}_{zv})^2\, dt/R = (\mu_0 AH_{zvm})^2(1/t_r + 1/t_f)/R \tag{9.43}$$

Where the loop inductive reactance $X_L \gg R$, the loop current is

$$i_L = \mu_0 AH_{zv}/L \tag{9.44}$$

The energy delivered to the loop load in this case is

$$J_R = \int_0^{t_r+t_f} (\mu_0 AH_{zv})^2 R\, dt/L^2 = (\mu_0 AH_{zvm})^2 R(t_r + t_f)/3L^2 \tag{9.45}$$

When the loop is shorted ($R = 0$), the energy thus stored is obtained using Eq. (9.44) and is given by

$$J_L = \tfrac{1}{2}Li_{Lm}^2 = (\mu_0 AH_{zvm})^2/2L \tag{9.46}$$

When computing energy from these antennas that can damage devices connected to them, a conservative approach is to use J_L instead of J_R. For example, all of J_L would be pumped into a transistor load across the loop that is conducting during the field buildup, thus shorting the loop. However, the transistor may be quickly biased off before the pulse subsides. Then J_L would be dissipated across the transistor junction to result in possible damage.

The slot antenna and its complementary dipole are discussed in Section 2.6. As derived from Babinet's principle, the open circuit voltage of the complementary dipole is $V_{0c} = 2h_e E_{zv}$, where $h = h_e$ and E_{zv} is the incident electric field. The short circuit current through the central portion of the slot is $I_{sc} = 2h_e J_z$, where J_z is the surface current due to the perpendicular component of the incident magnetic field. The slot antenna can represent an aperture, close behind where passing cables are exposed to voltages induced on them from the incident fields.[9]

9.8 SHIELDING AND COUPLING

For an initial shielding investigation, consider a simple conducting metal shield in the form of a closed housing for electronic equipment. The latter is to be protected from the effects of incident EMP to a given level of attenuation by this shield. Shielding effectiveness, $S_E(\omega)$, is expressed relative to the nonshielded situation and as a function of the frequency of the EMP wave, i.e.

$$S_E(\omega) = -20 \log_{10} [E_i(\omega)/E_W(\omega)] \qquad \text{(decibels)} \tag{9.47}$$

where E_i is the incident field, E_w is the electric field within the closed housing shield, and ω is the angular frequency of interest. Similarly, for the corresponding incident magnetic field

$$S_H(\omega) = -20 \log_{10} [H_i(\omega)/H_W(\omega)] \tag{9.48}$$

where H_i is the appropriate component of the incident magnetic field, and H_w is the magnetic field in the interior of the shield. The transfer impedance can be defined in this context as

$$Z_T = E_i(\omega)/H_W(\omega) \tag{9.49}$$

The shielding effectiveness of the housing is discussed first in terms of direct current impressed fields. For a direct current (dc) electric field surrounding the housing, Gauss's theorem, from Eq. (2.75), asserts that

$$\oint \bar{E}_i \cdot \bar{n} d^2 \bar{r} = 0 \tag{9.50}$$

where the circle through the integral sign indicates that the integration is about the closed outer surface of the housing. $\bar{E}_i$ is the incident electric field vector, and $\bar{n}$ is the unit normal to the surface of the housing. The right-hand side of Eq. (9.50) asserts that no net charges exist in the interior of the housing. This implies that the internal electronics is sufficiently shielded that no charges exist on its exterior. Equation (9.50) then asserts the fact that $\bar{E}_i$ must vanish in the interior of the housing. That is, the housing effectively shields the interior from the external dc electric field. In other words, the $\bar{E}_i$ field lines all end on charges induced on the outside surface of the housing.

As is well known, there is no analog of the foregoing for a dc magnetic field incident on the housing. Hence, the dc magnetic field can penetrate the housing, with the walls concentrating the magnetic field lines somewhat in the manner of a "magnetic-proof" wristwatch.

If the incident electric field is a function of time, e.g., a sinusoidal variation with time, then the charges induced on the outer surface of the housing move, thus constituting a surface current on the enclosure outer walls. This current, in turn, produces a time-varying magnetic field as described by Maxwell's equations (2.1) and (2.2). This magnetic field penetrates the housing. As seen from the Maxwell equation, (2.2), the corresponding alternating (ac) electric field $\bar{E}_i$, also penetrates the enclosure. For the high-frequency component of the EMP, the increasing inductive effect of the very rapidly expanding and collapsing fields about the metal housing walls attenuates the incident field, as manifest in the skin-effect phenomenon. As an illustration, the ratio of interior and exterior electric fields, for normal incidence, for a spherical enclosure whose wall thickness is much less than the skin depth is given by

$$E_W(\omega)/E_i(\omega) = 3i\omega\varepsilon_0 b/2\sigma t_h; \qquad t_h \ll \delta \tag{9.51}$$

where the skin depth is

$$\delta(\omega) = \sqrt{2/\omega\mu\sigma} \tag{9.52}$$

and t_h, b, σ, and μ are, respectively, the thickness, radius, conductivity, and permeability of the spherical shield enclosure. ε_0 is the permittivity of free space, and ω is one particular frequency of the EMP wave. When the wall thickness is much greater than the skin depth, the corresponding electric field ratio is[16]

$$E_W(\omega)/E_i(\omega) = [3\sqrt{2}i\omega\varepsilon_0 b \exp - R/\delta]/\sigma\delta; \qquad t_h \gg \delta \tag{9.53}$$

where R is the penetration distance into the spherical shell wall.

It is interesting to note that shielding intuition would probably not provide the insight necessary to appreciate that if the shield is embedded in a solid conducting medium, its apparent shielding effectiveness is reduced. That is, for a spherical shell embedded in a conducting ground, the preceding ratio is now given by[16]

$$E_W(\omega)/E_i(\omega) = 3b(\sigma_g + i\omega\varepsilon_g)/2\sigma t_h \equiv (E_W/E_i)_0 \cdot (\varepsilon_g/\varepsilon_0) + 3(b\sigma_g/2\sigma t_h) \tag{9.54}$$

where the subscript g refers to the ground in which the sphere is buried, and the subscript 0 refers to the shell in a vacuum, as implied in Eq. (9.51). Note that for low frequencies, the second term in the rightmost part of Eq. (9.54) decreases, so that the embedding ground degradation first term begins to dominate. This demonstrates the ground degradation of the shielding effectiveness of the sphere.

Apertures (holes) in shields radically degrade shield effectiveness. For a circular aperture, in a semi-infinite conducting sheet of radius R, the radial component of the electric field, E_r, extending into the aperture is[24]

$$E_r = (2E_0 \cos\theta/3\pi)(R/r)^3 \tag{9.55}$$

and the polar component, E_θ, is given by

$$E_\theta = (E_0 \sin\theta/3\pi)(R/r)^3 \tag{9.56}$$

where r is the perpendicular distance of penetration through the aperture into the interior, and θ is the angular displacement from r.

For seams in shields, the process of welding, brazing, or whatever manner of fastening is used to construct the seams sometimes produces a high resistivity discontinuity in the material along the seam. This causes high voltage drops across the seam that, at a minimum, can be a nexus for high fields and attendant heat losses.

Figure 9.18 depicts the currents induced on the housing enclosure from a low-frequency magnetic field. It should be noted, due to the increased inductance of the center portion of the shield, that the current is forced to flow near the edges on all faces. As the frequency of the incident magnetic field is increased, the skin effect begins to dominate and further confines the current flow to the enclosure extremities. For the low-frequency regime, a series LC circuit model is adequate, giving for the ratio of interior to exterior magnetic fields, the following expression for a thin wall sphere whose thickness is very much less than the skin depth[16]

$$H_W(\omega)/H_i(\omega) = R_s/(R_s + i\omega L_s); \qquad t_h \ll \delta \tag{9.57}$$

where the equivalent resistance and inductance are $R_s = (2\pi/3\sigma t_h)$, and $L_s = (2\pi b\mu_0/9)$. For high frequencies, where $t_h \gg \delta$

$$H_W(\omega)/H_i(\omega) = (3\sqrt{2}\cdot\mu\delta \exp - R/\delta)/b \tag{9.58}$$

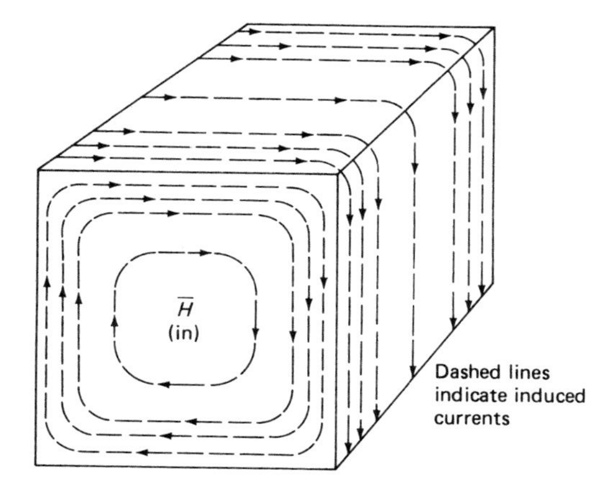

Fig. 9.18 Currents induced on a closed shield from time-varying magnetic fields.[17] Figure © 1968 by the IEEE.

with R being the penetration distance into the shell, μ is the wall permeability, and $\delta = \sqrt{2/\sigma\omega\mu}$.

In a manner similar to the electric field aperture penetration, the corresponding introduced magnetic field varies also in inverse cubed fashion with distance into the shield interior. The magnetic field component within, but near to, the aperature is[16]

$$H_r = (4H_0/3\pi)(R/r)^3 \sin\theta \sin\phi \tag{9.59}$$

where H_r is the radial component of the penetrated magnetic field in a spherical coordinate system whose origin is at the center of the aperture. Also

$$H_\theta = (2H_0/3\pi)(R/r)^3 \cos\theta \sin\phi \tag{9.60}$$

where H_θ is the polar component of the penetrated magnetic field, and the azimuthal component is

$$H_\phi = (2H_0/3\pi)(R/r)^3 \tag{9.61}$$

H_0 is the tangential magnetic field exterior to the aperture.

For an aperture in a conducting plane shield, the penetrating field resembles that of Fresnel diffraction of visible light by an aperture or a knife edge, as can be seen from Fig. 9.19. From the figure, "no aperture" corresponds to the situation where the shielding effectiveness increases with frequency of the incident wave, as the skin-effect absorption comes into play. The various aperture sizes become alternately resonant and antiresonant for UHF and microwave frequency components of the incident magnetic field.

The preceding does not include ferromagnetic materials, such as specially manufactured high-permeability steels. Equation (9.58) shows that the magnetic field absorption by the shield is proportional to $\sqrt{\mu} \exp - \sqrt{\mu}$, by using the definition of skin effect. This implies that a maximum shielding effectiveness occurs for a permeability given by $\mu_{\max} = 2/\omega\sigma R^2$, gotten by equating the derivative of Eq. (9.58) with respect to μ, to zero.

A high μ shield can "saturate through" with heavily increasing magnetization from the incident magnetic field, as the permeability decreases to a very low value. The shielding efficiency suddenly becomes very poor at the onset of the magnetization threshold. That is, the shield suddenly becomes almost transparent to the incident magnetic field. Massive shields avoid this difficulty by making the material sufficiently thick so the above "saturation" never occurs. For sinusoidally varying magnetic fields, the saturation penetration depth, r_s, is[16]

$$r_s = \sqrt{2I_{\max}/B_s\omega\sigma b} \tag{9.62}$$

where B_s is the saturation magnetic flux density, and $I_{\max}$ is the peak circulating current on the shield exterior surface. For an incident magnetic field pulse, the analogous expression for the saturation depth is[16]

$$r_p = \sqrt{\int_0^\infty I(t)\,dt/B_s\pi\sigma b} \tag{9.63}$$

where $\int_0^\infty I(t)dt$ is the total charge collected on the shield outer surface from the pulse.

9.9 EFFECTS ON DEVICES AND COMPONENTS

The major EMP damage that devices and passive components endure is due to voltage and current transients caused by the system response to the EMP. It is vital to be apprised of the failure modes and upset and burnout levels of components of interest, in order to provide some measure of protection for them from the incident EMP. Fortuitously, most components and devices can stand low-duty cycle or transient current, voltage, and power, far in excess of their rated continuous-duty

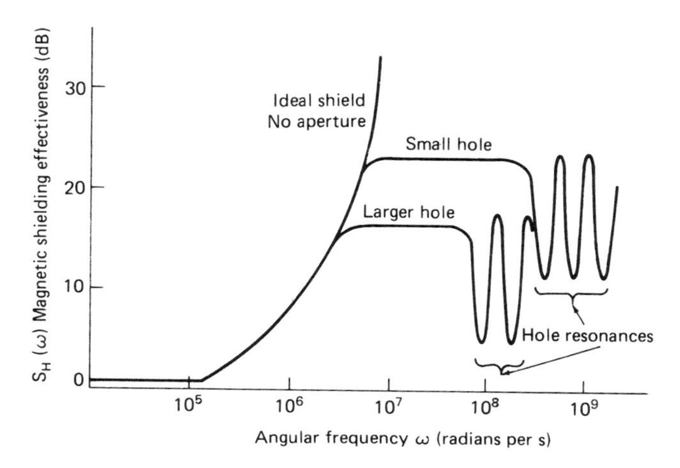

Fig. 9.19 Effects of aperture size in magnetic field shields.[16] Figure © 1976 by Jonn Wiley & Sons.

levels. Also, fortuitously, these thresholds are commensurate with anticipated EMP energy fluence levels, as seen in Problem 2.10.

As discussed in Sections 9.1 and 9.3, the exoatmospheric detonation constitutes a source of the potentially most damaging EMP. This is mainly because the earth line-of-sight tangent cone, with the detonation at the apex, covers a large fraction of a continent. In the case of high-altitude airborne or space-borne systems, there is very little atmosphere intervening from the detonation. Most information on the levels of failure of various components and devices has been predicted on exoatmospheric characteristics of the EMP and the system response in the nanosecond to microsecond regime of the EMP.

Endoatmospheric EMP is much more confined, as discussed, and the frequency spectrum of the associated EMP possesses much more low-frequency content than the exoatmospheric counterpart. Corresponding component EMP data reflect this fact, as their parameters extend into the milli-second region and beyond.

The type of device/component most sensitive to EMP transients is the semiconductor. Keeping in mind that the delivered transients endure on the order of a very few microseconds, corresponding power failure thresholds range from 1–10 W for point contact diodes, 10 W for integrated circuits in general, 20–1000 W for power transistors, 1 kW for zener diodes, and 10 kW for carbon resistors.[21] The relative sensitivity of semiconductors is due mainly to their comparatively small volume, especially their tiny junction areas. Because the thermal time constants of semiconductor device critical volumes are large compared with the impressed EMP derived transients, little generated heat can escape from the vicinity of the junction during the incident EMP pulse duration. Therefore, any thermal model used to described this junction breakdown should contain an adiabatic feature insofar as its thermodynamics are concerned. For example, these thermal properties of a diode junction represent an enhanced sensitivity to the large amount of transient heat impressed on this small area, although when the diode is biased into avalanche breakdown its can dissipate large amounts of heat energy, as does a zener diode. Generally, when such devices are subjected to failure threshold EMP transients, the junction material approaches its melting temperature, whereupon the junction collapses into a short circuit. This is called *thermal second breakdown*, and is discussed in Section 3.6.

Unfortunately, the thermal parameters that describe semiconductor materials are functions of the material temperature.[22] This often thwarts the construction of simple heat-transfer models. Also many electrical parameters of semiconductors change with elevated junction temperature. These temperatures range up to 1700°K, where for example, the melting temperature of silicon is about 1688°K. Examples of temperature-dependent thermal parameters are material density, specific heat or heat capacity, and thermal conductivity. For silicon, the material density is slowly varying with temperature. However, the specific heat increases by about 20 percent for a four fold increase in temperature from about 300–1200°K, while the density varies very little over this range. The thermal conductivity is a sensitive function of temperature, as can be seen in Fig. 9.20, where it varies roughly as $T^{-3/2}$ over the above range. The corresponding specific heats are depicted in Fig. 9.21.

To investigate the analytical predictions of device failure due to semiconductor junctions, a knowledge of the device surge transient threshold failure levels must be available. When a voltage pulse is applied across a semiconductor junction in the reverse direction, the principal voltage drop is directly across the junction per se, in contrast to it being shared by any neighboring material.

An expression for the threshold power absorbed that causes device failure is derived, to provide the means to compute failure threshold data for parts of interest. The threshold power is a function of the incident pulse power and duration, the temperature increase that the junction undergoes from the onset of the pulse, and other thermal parameters. In this model, it is assumed that the junction is semi-infinite, in that its thermal behavior can be described by the one-dimensional form of the heat equation, namely[23]

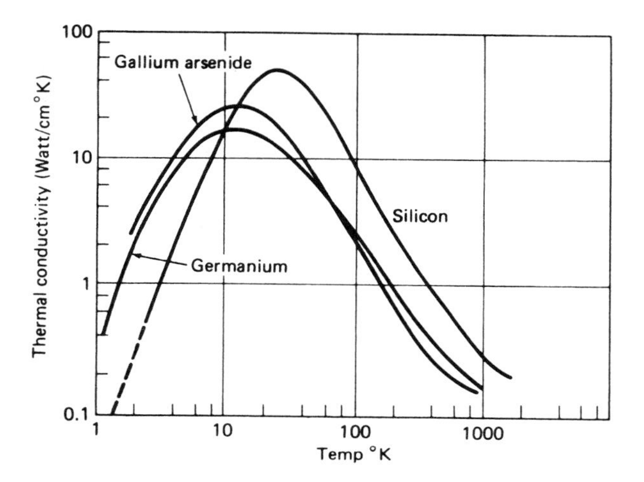

Fig. 9.20 Thermal conductivity of silicon, germanium, and gallium arsenide.[35,36]

$$\frac{\partial}{\partial x}\left(K(T)\frac{\partial T}{\partial x}\right) - \rho(T)c_p(T)\frac{\partial T}{\partial t} = 0 \tag{9.64}$$

where T is the junction temperature (°K), K is the thermal conductivity (watts per cm · °K), ρ is the material density (grams per cm^3), c_p is the specific heat at constant pressure (calories per gram · °K), and $x = 0$ is the position of the junction interface, where it is assumed that the EMP pulse is incident.

For a linear heat conduction description, it is assumed that the above material parameters are constant in Eq. (9.64). It can be then rewritten as

$$\frac{\partial^2}{\partial x^2}T(x,t) - \frac{1}{\alpha}\frac{\partial T}{\partial t}(x,t) = 0; \qquad \alpha = K/\rho c_p \tag{9.65}$$

The boundary conditions are:

a. $$T(x,0) = 0 \tag{9.66}$$

which simply means that the initial temperature of the semi-infinite slab of junction is zero prior to the pulse onset. This implies that any temperature changes can be reckoned from a zero reference. For example, for a temperature change at the junction interface, $\Delta T(0,t) \equiv T(0,t)$.

b. $$T(\infty,t) = 0 \tag{9.67}$$

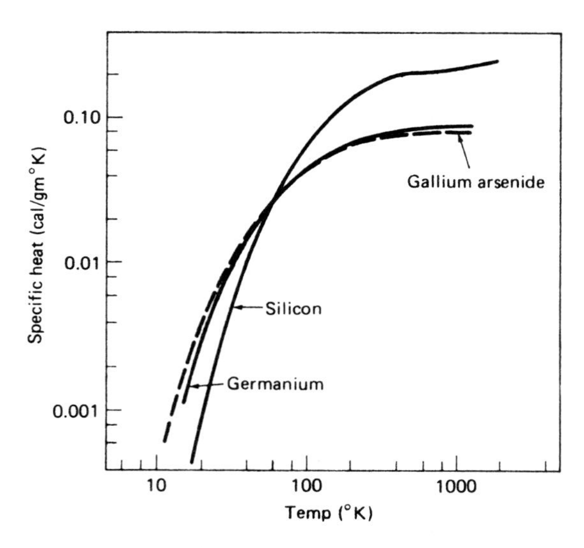

Fig. 9.21 Specific heat of silicon, germanium, and gallium arsenide.[35,36]

implies that deep into the semiconductor, the temperature vanishes. At the junction interface

c.

$$P(t) = -KA\frac{\partial T}{\partial x}(x, t)\bigg|_{x=0} \tag{9.68}$$

which is Newton's law of cooling, where $P(t)$ is the incident EMP heat power, and A is the junction area. Equations (9.65)–(9.68) have a solution given by

$$T(x, t) \equiv \Delta T(x, t) = \frac{1}{A\sqrt{\pi K \rho c_p}} \int_0^t P(t - t')e^{-x^2/4\alpha t'}\, dt'/\sqrt{t'} \tag{9.69}$$

For a rectangular threshold failure power pulse of level P_f and duration t_f, Eq. (9.69) becomes

$$T(x, t_f) \equiv \Delta T(x, t_f) = \frac{P_f}{A\sqrt{\pi K \rho c_p}} \int_0^{t_f} e^{-x^2/4\alpha t'}\, dt'/\sqrt{t'} \tag{9.70}$$

At the junction interface, i.e., at $x = 0$

$$\Delta T_f \equiv \Delta T(0, t_f) = \lim_{x \to 0} T(x, t_f) = \frac{2P_f}{A}\sqrt{t_f/\pi K \rho c_p} \tag{9.71}$$

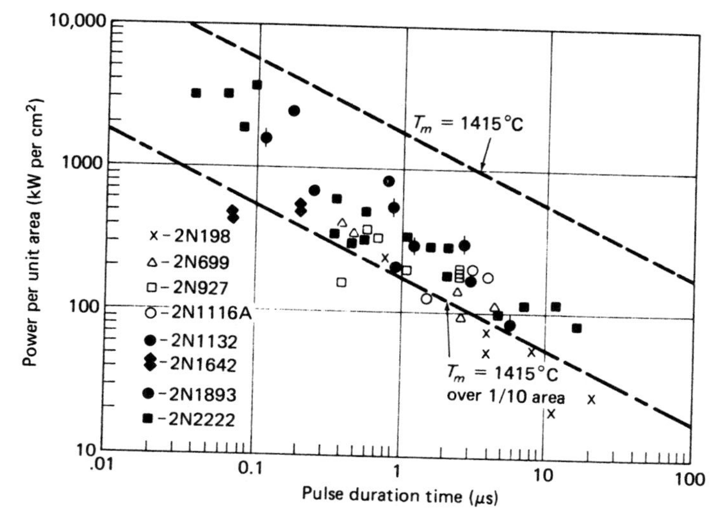

Fig. 9.22 Failure threshold power/pulse duration variations for eight bipolar transistor part types.[23] Figure © 1968 by the IEEE.

From Eq. (9.71), assuming that the junction does not fail beyond the pulse duration time t_f, the sought-after threshold failure power per unit area is given by

$$P_f/A = \frac{\Delta T_f}{2}\sqrt{\pi K \rho c_p / t_f} \tag{9.72}$$

Equation (9.72) is called the Wunsch-Bell model of electrical overstress. It was first used to "best-fit" experimental results from failures of 1200 transistors corresponding to about 80 part types. A subset of those results is shown in Fig. 9.22. Even though the Wunsch-Bell model of device electrical overstress assumes that the semiconductor thermal conductivity, heat capacity, and density are constants, the results as shown in Fig. 9.22 agree well with experiment, when the proper constant K_D is used for the particular device family, as in $P_f/A = K_D t_f^{-1/2}$. Also, as seen from the figure, the fit is better when it is assumed that only 1/10th of the junction area fails, which corresponds to observation,[23] in that the junction hot spot is the focus of failure.

The Wunsch-Bell model holds in a deeper sense, as is shown below, in that it is broadly valid even when the thermal parameters are not constant, but are temperature-dependent.[37] A transformation of the thermal diffusion equation (9.64), where the coefficients depend on temperature, can be made to render it linear. The resulting solution still exhibits the Wunsch-Bell form of Eq. (9.72) for a large class of temperature-dependent thermal parameters. That is, Equation (9.64) is rewritten as

$$(K(T)T_x)_x = ST_t; \qquad S = \rho c_p \tag{9.73}$$

where the subscripts correspond to the particular partial derivatives. Two transformation variables are used.[34] They are (1) for the penetration distance x into the semiconductor, viz.

$$X(x, t) = \int_0^x \sqrt{S(T)/K(T)}\, dx' \tag{9.74}$$

and (2) for the temperature $T(x, t)$

$$Q(X(x, t); t) = \int_0^T \sqrt{K(T')S(T')}\, dT' \tag{9.75}$$

Inserting Eqs. (9.74) and (9.75) into Eq. (9.73), with transformed boundary conditions, finally results in the linear heat equation in the transformed variables

$$\xi_{XX} = \xi_t + CP_0 \xi_X / A \tag{9.76}$$

where $\xi(X, t) = \exp - CQ$. This equation holds if the thermal conductivity $K(T)$ and the density-specific heat product $S(T)$ together satisfy an ordinary differential equation criterion, which is

$$(KS)^{-1/2}\, d/dT(\ln \sqrt{S/K}\,) = C \tag{9.77}$$

C is a constant whose value can be determined, but is not important in this context. It is shown[37] that $K(T)$ and $S(T)$, as taken from the data[35,36] used to plot Figs. (9.20) and(9.21) satisfy the criterion equation, Eq. (9.77), quite well for silicon, gallium arsenide, and germanium from ambient to melting temperatures. For a rectangular failure threshold input power pulse of amplitude P_f and pulse duration t_f, the solution of Eq. (9.76) for K and S pairs that satisfy Eq. (9.77), evaluated at the junction and at the end of the pulse duration t_f, is

$$\xi_f(0, t_f) = 1 - 4 \int_0^{R_f} ierfc(x)\, dx \tag{9.78}$$

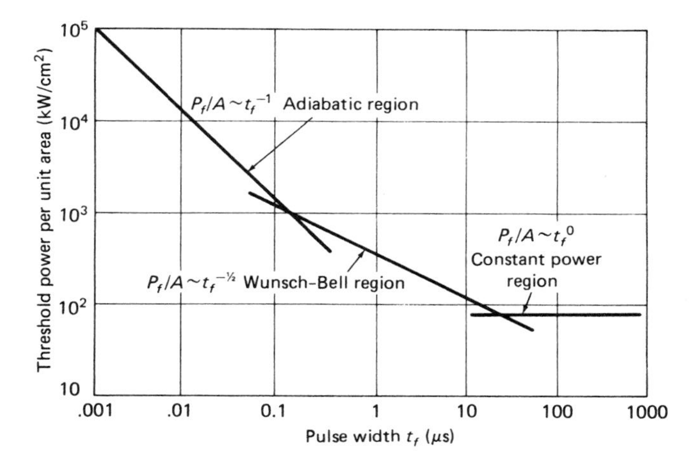

Fig. 9.23 Nominal failure threshold power per unit junction area versus incident pulse width.

where $ierfc(x) = \int_x^\infty erfc(y)dy$, and $R_f = CP_f t_f^{1/2}/2A$. $\xi(X, t)$ is a complicated function of temperature; however at the junction at time t_f, the temperature has reached a constant value T_c, characteristic of its failure mechanism phase change such as melting, and so the transformed temperature $\xi_f(0, t_f)$ is constant. Hence, R_f, being a root of Eq. (9.78), implies that it too is constant, and so thereby it enunciates the Wunsch-Bell relation $P_f/A \sim t_f^{-1/2}$. Therefore, the Wunsch-Bell relation is implied for a large class of physically realizable temperature-varying thermal conductivities, $K(T)$, and density-specific heat products, $S(T)$ that satisfy the criterion equation, Eq. (9.77). Recent findings[45] have shown that the thermal conductivity $K(T)$ of pertinent semiconductors, as thin films is much less than their bulk values. However, the variation of $K(T)$ with temperature is still largely as shown in Fig. 9.20.

It is well known that there are three thermodynamic epochs of failure threshold power variation with incident energy pulse width, as depicted in Fig. 9.23. The first, or adiabatic epoch, corresponds to very short pulse widths, where the thermal inertia of the material is such that device failure occurs before the absorbed heat energy has begun to diffuse into the material beyond the junction. This adiabatic regime implies that $P_f t_f \sim$ constant. The second, or Wunsch-Bell epoch, belongs with $P_f t_f^{1/2} \sim$ constant. The third, or constant-power epoch, where $P_f t_f^0 \sim$ constant, implies thermal equilibrium in the device prior to failure from a suitably long pulse. It can be shown that these three epochs can be

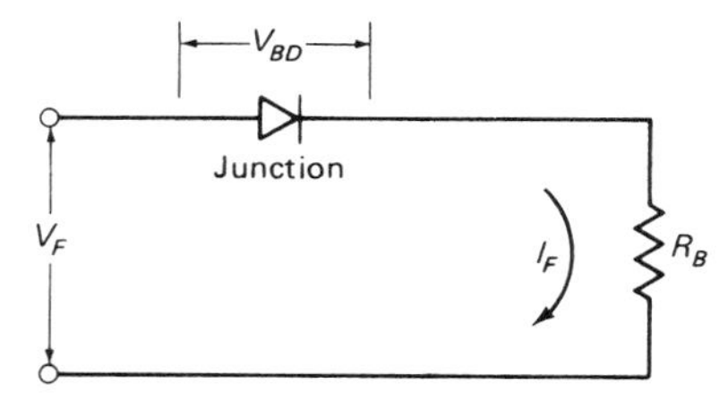

Fig. 9.24 EMP junction failure threshold equivalent circuit.

obtained from the Wunsch-Bell base using a one-dimensional semi-infinite slab model of the junction.[37] Empirically and analytically[37] derived polynomial (three-term) expressions in powers of $t_f^{-1/2}$ have been obtained for the threshold failure power per unit area, each term of which represents one of the above three epochs, as shown in Fig. 9.23. One such is [25]

$$P_f/A = 26A^{1/2}t_f^{-1} + 260A^0t_f^{-1/2} + 690A^{-1/2}t_f^0. \tag{9.79}$$

To compute the damage constant K_D for a particular device in $P_f = K_D t_f^{-1/2}$, certain phenomenological models that use device vendor data are available.[38] One approach features the dopant density as a basis for the empirical determination of the above parameters.[39] Another typical failure junction model uses an equivalent circuit shown in Fig. 9.24.

For the model that uses the equivalent circuit in Fig. 9.24, it is seen that the corresponding loop equations for power and voltage, respectively, are given by

$$K_D t_f^{-1/2} = P_f = I_F V_{BD} + I_F^2 R_B \; ; \; V_F = V_{BD} + R_B I_F \tag{9.80}$$

Eliminating I_F between Eqs. (9.80) yields the desired damage constant as

$$K_D = t_f^{1/2} V_{FP}(V_{FP} - V_{BD})/R_B \tag{9.81}$$

where V_{FP} is the measured junction failure voltage at the end of the pulse time t_f, for an assumed known junction breakdown voltage V_{BD}. As mentioned, there are a number of variations on this method for computing K_D involving elaborations of the above model, and different methods for computing the parameters are also available.[39]

9.10 SYSTEM-GENERATED EMP

System-generated EMP (SGEMP), often referred to as *internal* EMP (IEMP), is produced by electromagnetic pulse energy generated by electric currents, born from ionization due to photon interactions, mainly from x-rays and gamma rays incident on the system structure. This is in contrast to electric currents produced from and by the intervening atmosphere, as in high-altitude EMP (HEMP). SGEMP is manifest as the gammas and x-rays penetrate the system skin to ionize it to yield recoil electrons, as discussed in Section 6.1.[26] The resulting spatial electron fields produce the EMP, which couples to the internals of the system to upset or born out components of many kinds. SGEMP is of prime concern to manufacturers of spacecraft systems, because the high-altitude burst is a potent SGEMP causative agent, as there is virtually no atmospheric attenuation between the nuclear burst radiation and the spacecraft.[27]

The x-ray fluence, the major nuclear radiation output of high-altitude bursts incident on the satellite skin, can be approximated by the appropriate black body distribution

$$\Phi(u) \sim u^3/[\exp(u/kT) - 1] \tag{9.82}$$

where μ is the x-ray energy density of the corresponding fluence $\Phi(u)$, normalized so that $\int_0^\infty \Phi(u)du$ is the total x-ray yield of the burst. T is the temperature of the equivalent black body. As discussed in Section 6.1, photons interact with matter in three different ways. The usual x-ray spectrum is such that the major interactions are photoelectric up to about 100-keV x-rays, in reasonably dense materials, with the compton effect predominating for x rays greater than 100 keV.

After production of electrons by either or both of these processes, the transport of these electrons through the spacecraft outer cover, to include system cavity electron densities, is a complicated process and generally difficult to compute.[26–28] SGEMP can not only be generated in spacecraft, but in all types of air, sea, and land electronic systems. Strictly, SGEMP is construed to refer to a vacuum environment only, i.e., exoatmospheric vehicles, such as spacecraft. Endoatmospheric SGEMP-like effects are termed *low-altitude source region* EMP (SREMP). Spacecraft systems will be used as an example to illustrate the effects of SGEMP in detail, with the understanding that the discussion is applicable to other systems as well.

There are essentially three modes by which SGEMP can be coupled into spacecraft electronics. The first is due to replacement currents. With respect to the spacecraft position, the source exoatmospheric x-ray photons arrive from one broad solid angle sector of space. Such non-isotropic incident x-ray fluence produces the greater portion of electrons on the spacecraft skin most normal to the x-ray fluence direction, resulting in a nonhomogeneous electron surface charge density distribution. This electron charge imbalance on the spacecraft produces induced image charge replacement currents flowing on the outside of the vehicle. These replacement currents can penetrate into the interior through various electrical and electronic apertures, such as telemetry and communication antenna feeds, electro-optical telescopic apertures of various kinds, as well as into and through the solar cell power transmission system.

The second mode by which SGEMP can penetrate the spacecraft is that already mentioned. That is, the incident x rays penetrate the spacecraft skin to produce electrons on the interior walls of the various compartments. The resulting interior electron currents generate cavity electro-magnetic fields that induce voltages on the associated electronics. This produces spurious currents that can cause upset or burnout of these systems.

The third mode is through the x-ray-produced electrons that find their way directly into signal and power cables to cause extraneous cable currents. These currents are also propagated through the spacecraft inter-connection harnesses. Special shielding measures, such as solid outer conductor coaxial cables (hard wire), are employed, which attempt to minimize the first two modes of SGEMP coupling. This, plus the very large number of cables, delineates the fact that the third mode of coupling, i.e., through cable injection currents, constitutes the greatest SGEMP threat to normal operation from among these kinds of coupling.

Straightforward means for minimizing or eliminating SGEMP effects are quite similar to those used in combating EMP. Some of these include the use of transorbs (sec. 12.7) for spurious voltage clipping. Others are decoupling networks consisting of series resistors and shunt zener diodes, and still others consist of series inductors and shunt capacitors. All these are illustrated in Fig. 9.25. Further SGEMP hardening measures include (a) mounting of components close to ground planes to minimize E field coupling, (b) minimizing possible ground loops and routing of cables along ground planes to reduce H field coupling, and (c) using high-density packaging, where allowable, to reduce cavity fields due to electrons threading such cavities or filling the cavities with conducting mesh.

The main avenue of SGEMP-produced spurious current propagation is through hardwire power and signal cabling that interconnects the various subsystems aboard and spacecraft. If the hardwire signal cables in the spacecraft were to be replaced by fiber optic cables, then the SGEMP

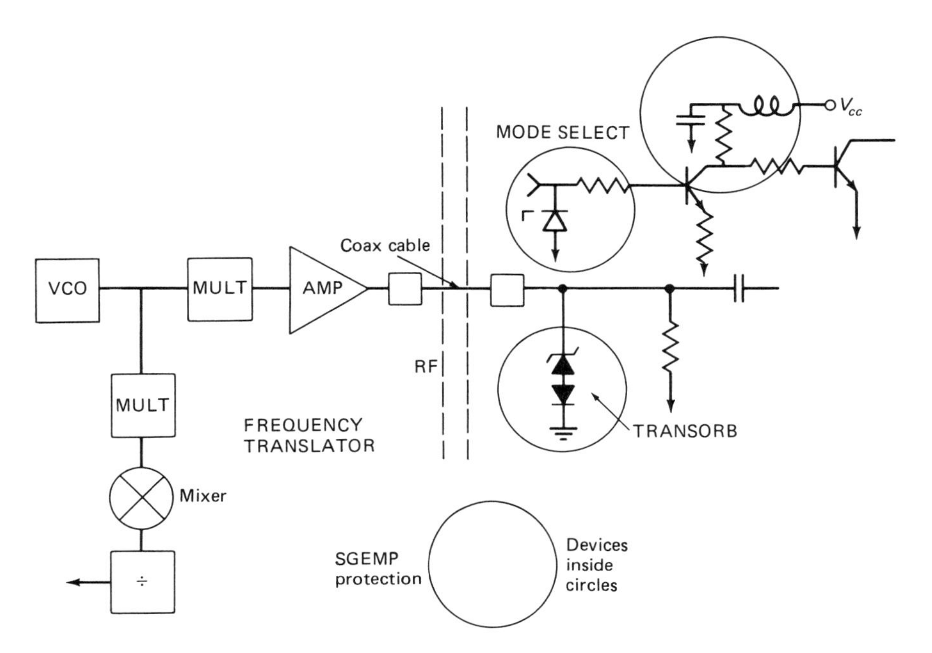

Fig. 9.25 Spacecraft hardwire analog link illustrating circuitry for minimizing SGEMP effects.

susceptibility would be diminished greatly, for fiber optics provide no conductive path for SGEMP-induced transients. The extent of the reduction in susceptibility by such replacement can be appreciated grossly by noting that for a spacecraft, the ratio of interface signal pins to their power lead counterparts is about 10 to 1.

It should be noted that these considerations apply to SGEMP effects only. Transient radiation effects (TREE), such as due to ionizing dose, on the spacecraft electronics and fiber optic cables may be more severe and have been investigated.[30–33]. Even though spacecraft signal links greatly outnumber corresponding power carrying links, the latter still remain to provide a means to couple SGEMP signals into and between subsystems and circuits, simply because the power supplies present common impedances to almost all active system circuitry, thus tying them together with respect to SGEMP propagation.

When the additional interface electronic components required by the fiber optic links are included as part of the spacecraft electronics complement, the weight increase compared to the all-hardwired system must be measured against the increased SGEMP protection. Hence, it is necessary to conduct a refined analysis to decide whether or not the replacement of hardwire signal cables with their fiber optics counterparts will result in increased spacecraft performance.

PROBLEMS

1. For the Office of Civil Defense EMP pulse time variation $E_L(t)$, given by Eq. (9.1), determine from its frequency spectrum $\bar{E}_L(\omega)$, Eq. (9.4), the corner frequency ω_c. ω_c is the angular frequency where both low- and high-frequency asymptotes of $\bar{E}_L(\omega)$ intersect in Fig. 9.4.
2. For an argus electron spiraling around a geomagnetic field line, its Lorentz force is balanced by its centrifugal force to give $Hev/c = mv^2/r$. If the earth's magnetic field is 0.5 gauss, compute the radius of the spiral for a 100-keV electron.
3. The geomagnetic field can also trap protons from a high-altitude burst. These come from the decay of neutrons born of fission debris. Why are such protons never observed in a surface nuclear burst?
4. When a medium-altitude burst detonates, there is an annular swath of earth about ground zero, for which most of the incident EMP will be absorbed into the earth and little will be reflected. If the burst altitude is 3 mi, what is the radius of the swath on dry, sandy terrain?
5. What is the minimum cruising depth for a submarine to avoid any EMP effects, if the low-frequency cutoff of the EMP spectrum is 1 kHz?
6. For the x-ray black body spectrum of Eq. (9.82) show that the spectrum peaks at the energy density $\mu = 2.82\,\text{kT}$. If the energy density $\mu = h\nu$ where ν is the frequency of the x radiation, what is the x-ray temperature T if the radiation peaks at a wavelength of 10 Å (1 Å $= 10^{-8}$ cm)?
7. If a dipole characteristic impedance is 72 ohms, what is the input impedance of the equivalent slot antenna, if the impedance of free space is 377 ohms?
8. If the fuselage of a fighter-bomber resonates as a half-wave antenna at 3 MHz, and that of a heavy bomber similarly resonates at 2.5 MHz, how long is the latter fuselage, if that of the former is 60 ft?
9. Twin structures to be used as storage buildings are to be built and hardened against EMP only. One is placed inside a huge aluminum skin and is situated on a plane earth. The other is embedded into an iron-ore-rich mountain. Which is better protected from EMP?
10. While I am waiting for the traffic light to change in a short tunnel under a reinforced concrete freeway, the AM station on my car radio becomes inaudible. I immediately switch to the FM band and hear those stations fine. Explain in terms of aperture considerations.
11. The original EMP—that from the creation of the universe according to the Big Bang theory—has long since dissipated throughout the cosmos. However, vestiges of this EMP can be picked up by suitable detectors, and they correspond to a noise temperature of about 3°Kelvin. To what frequency and wavelength do they correspond?

REFERENCES

1. J. J. Halpin and J.P. Swirczynski (eds.). Nuclear Weapons Effects on Army Tactical Systems, Vol. 1—Overview, Harry Diamond Laboratory Report no. HDL-TR-1882-I, p. 38, April 1979.
2. U.S. Office of Civil Defense. EMP Threat and Protective Measures. DCPA-TR-61, Aug. 1970.
3. F. N. Holmquist. EMP Environments for the SHF Antenna. TRW IOC 38403 6002-SX-00, June 17, 1981.
4. J. B. Cladis et al. The Trapped Radiation Handbook. Lockheed Missiles and Space Div., Palo Alto, Calif. Report no. LMSC 695953 (DNA 2524H) Fig. 9.11, January 1970.
5. D. H. Towne. Wave Phenomena. Addison-Wesley, Reading, MA. pp. 411 ff, 1967.
6. J. E. Bridges (ed.). DNA EMP Awareness Course Notes. Illinois Institute of Technology Research Inst. (IITRI), Chicage, IL. Report no. DNA 27727, p. 59, August 1971.
7. L. W. Ricketts, J. E. Bridges, and J. Miletta. EMP Radiation and Protective Techniques. Wiley-Interscience, New York, p. 64, 1976.
8. E. F. Vance. Coupling to Shielded Cables. Wiley-Interscience, New York, 1978.
9. Sandia Laboratories. Electromagnetic Pulse Handbook for Missiles and Aircraft in Flight. Doc. no. SC-M-71 0346, Albuquerque, NM, Sept. 1972.
10. H. Kaden. Wirbel und Schirmung in der Nachrichtentechnik. Springer-Verlag, Berlin, 1957.
11. E. F. Vance. "Shielding Effectiveness of Braided Wire Shields," IEEE Trans. Elect. Compat. EMC-17:71–75, May 1975.
12. W. L. Weeks. Electromagnetic Theory For Engineering Applications. Wiley, New York, pp. 330 ff., 1964.
13. K. S. H. Lee, T. K. Liu, and L. Marin. "EMP Response of Aircraft Antennas." IEEE Trans. Elect. Compat. EMC-20(1):94–99, Feb. 1978.
14. K. S. H. Lee. Electrically Small Ellipsoidal Antennas. AFWL Sensor and Simulator Notes, no. 193, Feb. 1974.
15. Boeing Co. Aeronautical System EMP Technology Review. Report no. D224-10004-1, Seattle, pp. 70 ff., April 1972.
16. L. W. Ricketts et al., op. cit. p. 128.
17. D. A. Miller and J. E. Bridges. "Review of the Circuit Approach to Calculate Shielding Effectiveness." IEEE Trans. Elect. Compat. EMC-10(1):52, March 1968.
18. C. L. Longmire. "On the Electromagnetic Pulse Produced by Nuclear Explosions." IEEE Trans. Elect. Compat. EMC-20(1):7–10, Feb. 1978.
19. W. R. Graham, R. L. Schaefer, W. Pine. New Considerations in Close-In EMP. RAND Corp. Santa Monica, CA. Doc. no. RM-6208-PR, Jan. 1970; Air Force Weapon Labs, Albuquerque, NM. Document no. AFWL-TR-70-27, Sept. 1970.
20. C. L. Longmire. Theory of the EMP From Nuclear Surface Bursts. Defense Nuclear Agency (DNA), Washington, D.C. publication, 1972.
21. L. W. Ricketts. Ibid., p. 76, Table 3.1.
22. H. F. Wolf. Semiconductors. Wiley-Interscience, New York, pp. 83–92, 1971.
23. D. C. Wunsch, R. R. Bell. "Determination of Threshold Failure Levels of Semiconductor Diodes and Transistors Due to Pulse Voltages." IEEE Trans. Nucl. Sci. NS-15(6):244–259, Dec. 1968.
24. L. W. Ricketts, Loc. cit. p. 82.
25. D. M. Tasca. Minuteman III Re-entry System Alecs-G Program Piece Parts Support

Test. Final report, General Electric Doc. no. 70SD 401, Contract no. AF04-694-731, Jan. 1970.
26. D. F. Higgins, K. S. H. Lee, and L. Marin. "System Generated EMP." IEEE Trans. Elect. Compat. EMC-20(1) Feb. 1978.
27. K. S. H. Lee and L. Marin. "Charged Particles Moving Near a Perfectly Conducting Sphere." IEEE Trans. Antenn Prop. AP-22, Sept. 1974.
28. L. Marin, K. S. H. Lee, and T. K. Liu. Analytical Calculations on the Photo Electron Induced Currents for the Model of the FLTSATCOM Satellite. Air Force Weapons Lab. (AFWL) Albuquerque, NM. Report no. TN-206, Feb. 1975.
29. M. S. Ash. "Non-Linear Kinetics of Semiconductor Junction Thermal Failure." Proc. 1981 Conf. Electrical Overstress/Electrostatic Discharge, Las Vegas, NV, Sept. 1981.
30. A. H. Johnston and R. S. Caldwell. "Design Techniques For Hardened Fiber Optic Receivers." IEEE Trans. Nucl. Sci. NS-27(6):1425–1431, Dec. 1980.
31. G. H. Sigel, E. J. Friebele, M. E. Gingerich, and L. M. Hayden. "Radiation Response of Large Core Polymer Clad Silica Optical Fibers." IEEE Trans. Nucl. Sci. NS-26(6), Dec. 1979.
32. S. Share and J. Weslik. "Radiation Effects in Doped Silica Optical Waveguides." IEEE Trans. Nucl. Sci. NS-26(6):4802–4807, Dec. 1979.
33. W. H. Hardwicke and H. H. Kalma. "Effects of Low Dose Rate Radiation on Opto-Electronic Components and the Consequences Upon Fiber Optic Data Link Performance." IEEE Trans. Nucl. Sci. NS-26(6):4808–4813, Dec. 1979.
34. M. L. Storm. "Heat Conduction in Simple Metals." J. Appl. Phys. 22(7):940–951, July 1951.
35. C. Y. Ho, R. W. Powell, and P. E. Liley. Thermal Conductivity of the Elements—A Comprehensive Review. Am. Chem. Soc; Am. Inst. Phys. New York, 1975.
36. Y. S. Touloukian et al. Thermophysical Properties of Matter. TPRC Data Series, vol. 1, Thermal Conductivity of Metallic Elements and Alloys. IPI/Plenum Press, New York, 1970.
37. M. S. Ash. "Semiconductor Junction Non-Linear Failure Power Thresholds: Wunsch-Bell Revisited." IEEE Proc. 1983 Electrical Overstress/Electrostatic Discharge Conf., Las Vegas, Nev. Sept. 1983.
38. D. R. Alexander and R. E. Thieme. "Source For Overstress Response Characteristics (SCORCH) User's Manual." AFWL-TR-83-142, Kirtland AFB, Mission Research Corp. Albuquerque, NM, Sept. 1984; C. R. Jenkins and D. L. Durgin. "EMP Susceptibility of Integrated Circuits." IEEE Trans. Nucl. Sci. NS-22(6):2494–2499, Dec. 1975.
39. BDM. Electronic Component Modelling and Testing Program. Final Report no. AFWL-TR-78-62, pt. 1, Albuquerque, NM. March 1980.
40. Proposal: "Atmospheric Electricity Hazards Protection (AEHP) of Advanced Technology Aircraft Electrical/Electronic Subsystems," Lockheed Burbank LR 29940-1, 16 September 1981.
41. J. L. Lawson and G. E. Uhlenbeck. "Threshold Signals", Chap. 5, Vol. 24. MIT Radiation Lab Series. McGraw-Hill, New York, 1950.
42. E. G. Stassinopoulos and J. M. Barth. "Non-Equatorial Low Altitude Charged Particle Radiation Environments." NASA Goddard Space Flight Center, MD, Report No. X-600-86-15, Nov. 1986.
43. D. A. Adams and B. W. Mar. "Predictions and Observations For Space Flight" in Radiation Trapped in the Earth's Magnetic Field, pp. 817–845. C. J. McCormac (ed.). Gordon & Breach, 1966.

44. B. C. Passenheim. How to do Radiation Tests. Fig. 5-14. Ingenuity Ink Publications, San Diego, CA, 1988.
45. J. C. Lambropoulos, M. R. Jolly, C. A. Amsden, S. E. Gilman, M. J. Sinicropi, D. Diakomihalis and S. D. Jacobs. " Thermal Conductivity of Dielectric Thin Films." J. Appl. Phys. 66(9):4230–4242, 1 Nov. 1989.

CHAPTER 10
DOSIMETRY

10.1 INTRODUCTION

The exposure of electronic systems, devices, and components to radiation in radiation facilities plays a central role in the investigation of radiation effects. The actual irradiation of the part or system provides, at least, a simulation of the nuclear and electromagnetic radiation environment it can encounter during its useful life. Hence, attention at a sophisticated and quantitatively accurate level must be given to the many ramifications of both measurement of the exposure and subsequent dose acquired by the system or device under test (DUT) from the incident radiation. Such measurements have a strong statistical aspect that cannot be overemphasized, and its discussion will be given its due in Chapter 14.

Dosimetry is the usual term for the characterization and measurement of the radiation in this context. First, the measurement provides a quantitative description of the amount of radiation from the various sources to which the DUT is exposed. This exposure is incident on and penetrates the DUT. Second, the ensuing interaction of the radiation with the material atoms and/or nuclei results in the material acquiring a measurable dose of radiation. Not all the incident radiation to which the device is subjected results in DUT dose acquisition. For example, some fraction of the incident radiation passes through the device, not interacting at all. This is an important point that will be dealt with in later sections.

10.2 DOSIMETRY FUNDAMENTALS/DOSE ENHANCEMENT

Interactions of radiation with devices, components, and systems deal with radiation flux, fluence, and currents. Such quantities are relatively difficult to measure in the laboratory. They are of primary importance in the analytical and computational aspects of determining the effects of radiation on matter.

The units used to measure absorbed radiation are defined in a manner that seemingly implies that radiation is absorbed through ionization due to x and gamma rays. However, radiation energy released into material can be caused by processes other than ionization, such as displacement damage due to neutrons, as discussed in Section 5.5.

An important but somewhat obsolescent dosimetry unit is the roentgen. This unit has been used for more than 50 years. However, it is only since the early 1960s that the quantity for which the roentgen is a unit has been defined and given the name *exposure*.[1] This event took place at the 1962 International Commission on Radiological Units (ICRU) conference. Simply, one roentgen is the quantity of x- or gamma-rays that, when present in air, produces (through ionization of the air molecules) 1 esu of charge in each cm^3 of dry air at standard temperature (0°C) and pressure (1 atm) (STP); or $2.58 \cdot 10^{-4}$ coulombs (C) per kg of dry air (in SI units). As the value of the charge on the electron is $4.8 \cdot 10^{-10}$ esu ($1.6 \cdot 10^{-19}$ coulombs), the production of 1 esu of charge corresponds to the production of $1/4.8 \cdot 10^{-10}$ or $2.08 \cdot 10^{9}$ ion pairs in STP air. Also, because the amount of energy required to produce one ion pair in air is 32.5 eV, when 1 cm^3 of dry STP air (density of 0.001293 g/cm^3) is exposed to 1 roentgen, the energy released into this air is $(32.5)\ (2.08 \cdot 10^{9})\ (1.6 \cdot 10^{-12})/0.001293 = 84$ ergs per gram of air ($1\ eV = 1.6 \cdot 10^{-12}$ ergs). This is equal to 0.84 rads (air), as shown below.

The most important dosimetry unit measures ionizing dose, or merely dose. This unit is the rad. It is the absorbed dose in amount of 100 ergs per gram in the material of interest. Because the energy absorption amounts are different for different materials exposed to the same incident radiation, the material under study must be specified, such as a dose of one rad (Si) means that 100 ergs of energy are absorbed per gram of silicon. The rad is a CGS unit, while the MKS or SI unit of dose is the gray (Gy), which corresponds to an absorbed dose of 1 joule per kilogram of material, so that 1 Gy = 100 rads. A centigray (cGy) equals 0.01 Gy, or one rad.[53]

As absorbed dose in tissue is often of interest, some biological radiation units are defined in the following. Although all ionizing radiation is capable of producing similar biological effects, the absorbed dose that will produce a particular biological insult may vary from one type of radiation to another. This difference is expressed by a relative biological effectiveness called RBE, defined as the fraction,

$$\text{RBE} = \frac{\text{Absorbed dose of gamma radiation of specified energy}}{\text{Absorbed dose of given radiation with same biological effect}} \quad (10.1)$$

where the gamma radiation (numerator) is usually from a ^{60}Co source. For a particular kind of radiation, the RBE depends on the radiation energy spectrum, dose rate, kind and degree of biological damage, and the organ and/or tissue being irradiated.

The unit of biological dose, called the *RBE dose*, is the roentgen equivalent man (rem). The dose in rems is defined as

$$\textbf{Dose (rems)} = \textbf{dose (rads)} \cdot \textbf{RBE} \tag{10.2}$$

If the incident dose is expressed in grays (Gy), then the corresponding RBE dose unit is the sievert (Sv), defined as[30]

$$\text{Dose (sieverts)} = \text{dose (grays)} \cdot \text{RBE} \tag{10.3}$$

For a dose due to a mixture of radiation types, e.g., alphas, gammas, electrons, neutrons, x-rays, etc., the aggregate absorbed dose in rems is

$$\text{Aggregrate dose (rems)} = \sum_i \text{dose (rads)}_i \cdot (\text{RBE})_i \tag{10.4}$$

where the summation is over the types of radiation, each with its own RBE. A few examples are:[2] The RBE of ^{60}Co gamma rays is, of course, unity. For gamma rays of different energies, the RBE is close to unity, as it is for electrons. However, for alpha particles ingested into the body, the RBE ~10–20 for the production of bone carcinoma. From weapon spectrum neutrons, the acute injury RBE is also close to unity. For the development of cataracts, leukemia, and genetic effects, the RBE is about 4–10. Hence, for specific kinds of biological damage, neutrons and alphas cause much more damage than the same amount of dose from gammas or x-rays. If not directly ingested into the body, alpha particles do potentially negligible damage, as they can be stopped by a few mils of paper.

Other dosimetric units are that of radioactivity, in terms of the emanation rate of particles or quanta of energy from the radioactive material of interest. These units are the curie (Ci), where 1 Ci corresponds to a radioactive substance disintegration rate of $3.7 \cdot 10^{10}$ disintegrations s^{-1}, and the bequerel (Bq) equals 1 disintegration s^{-1}, all irrespective of the particles emanated. So that one curie equals $3.7 \cdot 10^{10}$ bequerels.

A more recently derived unit of dosimetric importance is called *kerma*. Kerma is an acronym for the *k*inetic *e*nergy *r*eleased in *m*aterial. The need for such a quantity arises from the fact that radiation usually deposits its energy in material in essentially two steps. Kerma is an apt descriptor for the first step, which is the transfer or release of incident radiation derived kinetic energy to charged particles of the material, usually through ionization. An example is the release of gamma-ray energy to the relatively loosely bound outer atomic electrons via the compton process, thus ionizing the atoms in the material. Absorbed dose is an apt descriptor for the second step, which is the deposition of the kinetic energy of the charged particles into the material, e.g., through coulomb interactions, such as electron-electron collisions. This is similar to elastic collision energy transfer, as is the case for neutrons. Note that neutrons can produce ionization indirectly, as discussed in Section 5.7.

If the kinematics of the two-step process is such that all the kinetic energy released in the material (kerma) is ultimately absorbed in the material (dose), then the absorbed dose and the kerma are equal. This fortunately is the usual case, with charged particle spacial equilibrium holding throughout the material. Charged particle equilibrium (e.g., in a small mass of material) means that charged particles, such as electrons, carrying an increment of energy out of the corresponding incremental volume, are replaced by charged particles (seldom the same particles) carrying the same amount of energy into this volume. When charged particle equilibrium does not exist, kerma and dose are not equal, as will be seen, and the attendant dosimetry problems are formidable.

A first illustration of the lack of charged particle equilibrium is a simple one of a photon beam incident on a small embedded mass Δm of material, comprising a small part of the surface of a large mass of the same material.[45] Due to any of the three photon-electron interactions discussed in Section 6.1, the resulting electrons produced thereby, and scattered out of this near-surface mass, are certainly greater in number than those entering the mass. Hence, this small mass of material suffers a net loss of electron-conveyed energy, so that charged particle equilibrium does not hold therein. However, succeeding small masses behind Δm lie deeper and deeper within the large mass. At these depths, the electron range $R(E_0)$ is greatly exceeded in all directions by material mass. It can then be appreciated that scattered electron energy out of such a small buried mass can be equal to that brought into this mass by in-scattered electrons, so that charged particle equilibrium can exist within. The minimum thickness of material at which charged particle equilibrium ensues is called the equilibrium thickness. Charged particle equilibrium then holds for this and greater thicknesses.

A second illustration of a lack of charged particle equilibrium is that of a large block of material of high atomic number, Z, with gamma ray photon sources distributed approximately uniformly throughout the material. The block dimensions are large with respect to the range of the (mainly compton) electrons produced therein by the gamma rays, but small with respect to the attenuation length of the resulting photons produced in the block. The atomic number of the material is sufficiently large so that an appreciable amount of the electron energy is lost through its bremsstrahlung photon radiation. Photon energy is therefore lost as this radiation leaks from the block, so that the dose is less than the kerma, at least by the bremsstrahlung loss. Also, charged particle equilibrium cannot exist because of the bremsstrahlung loss, since the electrons will have their energy reduced thereby in amount, depending on their initial energy E_0. However, for electrons of a given energy, E (MeV), the fraction of bremsstrahlung energy, loss compared to their ionization energy loss (electrons can ionize also), is roughly $EZ/(EZ + 700)$. Because $EZ \ll 700$ in most cases, except for very high energy electrons in very high Z materials, the bremsstrahlung radiated fraction is very small. So that kerma and dose are then comparable, and charged particle equilibrium holds approximately.

A third illustration of a lack of charged particle equilibrium is another simple one, that of a thin target of material being irradiated by a source of gamma rays. Here, the range of the electrons produced is much larger than the target dimensions, so that much of the electron energy is lost simply because of electron leakage out of the material. Again, the kerma and dose are not equal.

The consequences of whether or not charged particle equilibrium exists are vitally important in dose measurement. This can be appreciated by realizing that when it holds, the energy deposited, as embodied in the corresponding charged particle equilibrium deposited dose, D_{eq}, is given by[45]

$$D_{eq} = [\Delta E_\gamma + (\Delta E_e = 0)]/\Delta m \qquad \text{(ergs/gm)} \tag{10.5}$$

where

Δm = the small amount of mass under discussion.
ΔE_γ = the difference between the photon-borne energy entering and leaving Δm.
ΔE_e = the difference between the electron-borne energy entering and leaving Δm, which vanishes when charged particle equilibrium holds.

Now, $\Delta m = \rho\Delta x \cdot \Delta A$ for, say, a small parallelepiped mass of area ΔA normal to the incident fluence and thickness Δx. Also, the incident photon energy fluence $\Phi_\gamma = E_{inc}(\gamma)/\Delta A$ (MeV/cm^2).

Then from Section 6.4, it is a straightforward interpretion that

$$E_l(\gamma) = E_{\rm inc}(\gamma)\exp - (\mu/\rho) \cdot (\rho\Delta x) \tag{10.6}$$

where

$E_l(\gamma)$ = photon energy leaving Δm.
$E_{\rm inc}(\gamma)$ = photon energy entering Δm.
μ/ρ = mass absorption coefficient (cm^2/gm).
$\rho\Delta x$ = areal density (gm/cm^2).

The argument in the exponential is usually small, especially for small Δx. Then, retaining only the first two terms in the expansion of the exponential yields, from Eq. (10.6).

$$\Delta E_\gamma \equiv E_{\rm inc} - E_l \equiv E_{\rm inc} \cdot (\mu/\rho)(\rho\Delta x) \tag{10.7}$$

Inserting $E_{\rm inc}$ from $\Phi_\gamma = E_{\rm inc}/\Delta A$ into the right-hand side of the above equation yields

$$D_{\rm eq} \equiv \Delta E_\gamma/\rho\Delta A \cdot \Delta x = \Phi_\gamma\mu/\rho \tag{10.8}$$

which holds only for charged particle equilibrium, as discussed.

Then, for two different materials experiencing the same photon energy fluence Φ_γ, Eq. (10.8) implies that, at this particular energy,

$$D_{\rm eq}(1)/D_{\rm eq}(2) = (\mu_1/\rho_1)/(\mu_2/\rho_2) \tag{10.9}$$

This means, for example, that if charged particle equilibrium holds, measurement of the equilibrium dose $D_{\rm eq}$ in a small amount of embedded detector material (2) can be used to obtain the sought-after dose in the material (1) of interest. If charged particle equilibrium is lacking, little can be ascertained with respect to the deposited dose in the pertinent material in terms of that measured in the detector material.

An important example of lack of charged particle equilibrium is that of dose enhancement. Dose enhancement occurs near the interface of two materials that are dissimilar in terms of their wide disparity of atomic number Z and so necessarily their atomic weights. Dose enhancement is caused by gamma-ray or more importantly x-ray-generated electrons born on the high Z (heavy) side, which are mainly transported across the interface to the low Z (light) side. Due to the relative transport properties of both materials, the low-Z material also scatters electrons back toward the interface. This agglomeration results in a buildup of electron fluence near the interface on the low-Z side and so an enhancement of

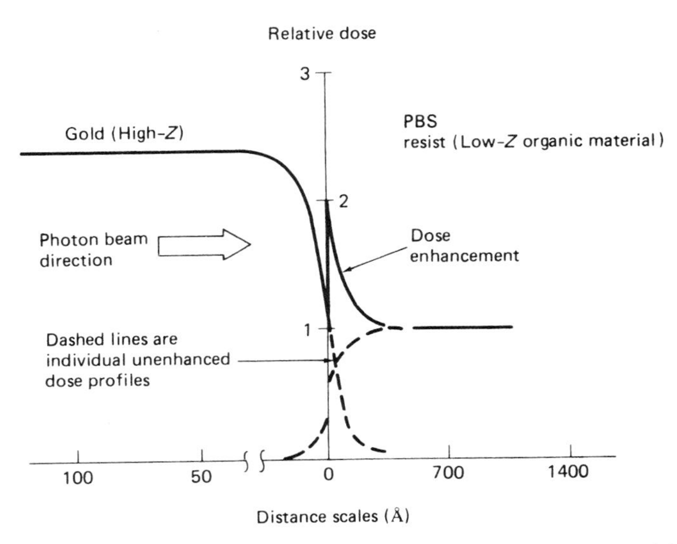

Fig. 10.1 Dose enhancement near a gold-low Z photoresist interface from 1.49-keV x-rays. Resist is a masking coat that prevents chemical action on materials under it, as in photographic and lithographic processes.[34] Figure © 1982 by the IEEE.

deposited dose there. This is not unlike the buildup of thermal neutron flux due to neutron scattering on the reflector side of the core-reflector interface of a thermal nuclear reactor. Levels can be two orders of magnitude over the nonenhanced dose.

X-rays, because of their relatively low energies, interact mainly through the photoelectric effect, which interaction varies like Z^4 yielding an abrupt almost discontinuous dose enhancement peak as depicted in Fig. 10.1. Gamma rays, because they have relatively higher energies, interact mainly through the compton effect, which is relatively independent of Z, so they cause comparatively less dose enhancement, as their effects are nearly equal on both sides of the interface. However, the low-energy tail of the gamma ray spectrum can be important in this context, since these energies are near the x-ray region.

Within an electron range of the interface, electron equilibrium (i.e., charged particle equilibrium) does not exist because of the dose enhancement, as discussed in the preceding first illustration. It can be appreciated that the enhancement process produces a difference between electron energy released and electron energy deposited, so that the kerma and deposited dose are not equal in this relatively thin region.

An important example of dose enhancement can take place within integrated circuits. In an integrated circuit chip, the active region is usually emplaced near its top surface. This is relatively far from the bottom of the chip, which is sometimes fastened to its package by a gold eutectic bond. This bond forms a high-Z (gold) to low-Z (ceramic) interface, potentially subject to dose enhancement in its vicinity. This interface is usually remote from the active region, so that dose enhancement here negligibly affects chip operation. However, two interfaces within the chip active region, forming large Z discontinuities, are (a) the metallization-semiconductor, and (b) the metal package lid – semiconductor, interfaces. Corresponding dose enhancement can be local (a) or chip-wide (b), respectively, depending on the transport parameters of the incident radiation.[5] This must be taken into account with respect to device radiation survivability considerations. Dose enhancement factors for so doing are available for pertinent radiation sources (Table 10.1).[33] Further examples comparing kerma and dose are given in recent reevaluations of the proportion of neutrons and gamma rays in the Hiroshima and Nagasaki nuclear detonations.[35]

Another important example, that of minimizing dose enhancement, occurs when a proper radiation-absorbing filter box is used for testing parts in a ^{60}Co ionizing dose environment. As discussed in Section 6.3, the output of a ^{60}Co cell consists mainly of two gamma ray energies, to simulate the effects of ionizing dose on components under test. In the cell housing, which holds the components being tested, some of the incident ^{60}Co gamma rays undergo compton scattering with the housing walls. The compton effect process results in a reduction in energy of the incident gamma ray emanating from the struck atom, as discussed in Section 6.1. The energy spectrum of the scattered gamma rays has a substantive low energy tail, to provide dose enhancement as discussed above, as they penetrate the parts under test in the cell. To minimize this dose enhancement, a high-Z filter, in the form of a lead liner, can be installed in the ^{60}Co cell. The Pb liner photoelectrically absorbs much of the low energy tail, while allowing most of the primary ^{60}Co gamma rays to penetrate to the components being tested. This photoelectric process in the lead produces fluorescence (mainly due to L-shell Pb electrons), which can, in turn, be absorbed by a thin aluminum liner within the lead liner. This filter technique has been incorporated into an American Society For Testing and Materials standard (viz., ASTM Practice 1249–88). Figure 10.1A depicts the effect of the filter on reducing the soft (low energy) portion of the ^{60}Co source.[50]

In order to calculate the dose enhancement at the interface of two materials of widely disparate atomic number Z, resort is made initially to obtaining solutions from complicated gamma and x-ray photon and electron transport equations.[36] One standard approach is to reduce the

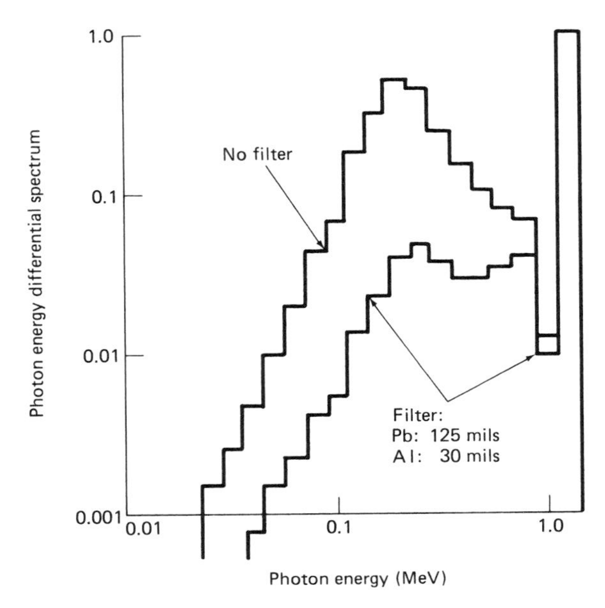

Fig. 10.1A Harry Diamond Laboratories concrete room ^{60}Co source, measured spectra using differential absorption TLD measurements for one source-filter configuration.[50] Figure © 1985 by the IEEE.

transport equations to systems of coupled partial differential, or difference equations.[36] Elaborate monte carlo programs, containing the cross sectional and interaction minutiae of the photoelectric and compton effects, involving the incident photons and resulting electrons, have been used to compute dose enhancement factors for a few dissimilar material interfaces.[37] These monte carlo programs are run on mainframe computers, and yield results of very high accuracy, as corroborated by experimental data.[37]

For more readily available dose enhancement calculational methods, an approach for many and diverse high-Z–low-Z material interfaces is to extrapolate the above monte carlo results to interfaces of interest. This is done by empirically fitting the corresponding dose profiles, derived from monte carlo computations, by easily manipulated functions. The monte carlo results are couched in electron energy currents normalized to the incident monoenergetic photon energy fluence $E_p\phi_p$, where ϕ_p is the monoenergetic equilibrium photon fluence, and $E_p = h\nu_p$ is its energy. Each electron energy current corresponds to photoelectron production from each of the K, L, M, . . . atomic electron shells, as well as from compton-produced electrons. The equilibrium photon fluence ϕ_p corresponds to photons incident on an infinite slab of the high-Z material (i.e., with no dose enhancement from interfaces). The electron currents are assumed to flow in one dimension, as the monte carlo models are also one dimensional, where x is the distance of penetration into the low-Z material side of the interface in gm per cm^2. A prototype normalized electron current is proportional to

$$\begin{aligned} J_e(x) &= \int_{E_0}^{\infty} dE \int_{\Omega} En(x, \Omega)\, v(E)\cos\theta\, d\Omega/E_p\phi_p \\ &= 2\pi \int_{E_0}^{\infty} dE \int_{-1}^{+1} E\varphi(E, x, \mu)\mu d\mu/E_p\phi_p \end{aligned} \tag{10.10}$$

where

$\varphi(E, x, \mu) = n\bar{v}(E)$, the angular electron flux i.e., the number density n of those electrons with velocity vectors $\bar{v}(E)$ in $d\Omega$; having direction cosines $\mu = \cos\theta$, with respect to the x axis; and multiplied by $|\bar{v}(E)|$ ($d\Omega = 2\pi\, |d\mu|$).

E_0 = minimum energy of the photoelectric and compton electrons generated by the incident photons.

Each type of $J_e(x)$ is fitted by a function of the form[38]

$$Y_i(x) = A_i\exp - (B_ix + C_ix^2 + F_ix^3) \tag{10.11}$$

where Y_i is the normalized electron energy current corresponding to the ith type of electron production process in this context. This includes photoelectrons from the K, L, and M electron shells of the material atoms, as well as auger and compton electron contributions, as discussed in Section 6.1.

The electron dose profiles, $D_i(x)$ are obtained from

$$D_i(x) = -Y_i'(x) = A_i(B_i + 2C_ix + 3F_ix^2)\exp - (B_ix + C_ix^2 + F_ix^3) \tag{10.12}$$

where the negative sign multiplying $Y'(x)$ is merely due to the type of function fit chosen in Eq. (10.11) to yield positive $D_i(x)$. That $|D_i(x)| = |Y_i'(x)|$ is the dose is evident, since $|Y_i'(x)|dx$ is the normalized ith type electron energy deposited in an incremental amount of material represented by dx, due to the flow of the electron energy current $Y_i(x)$.

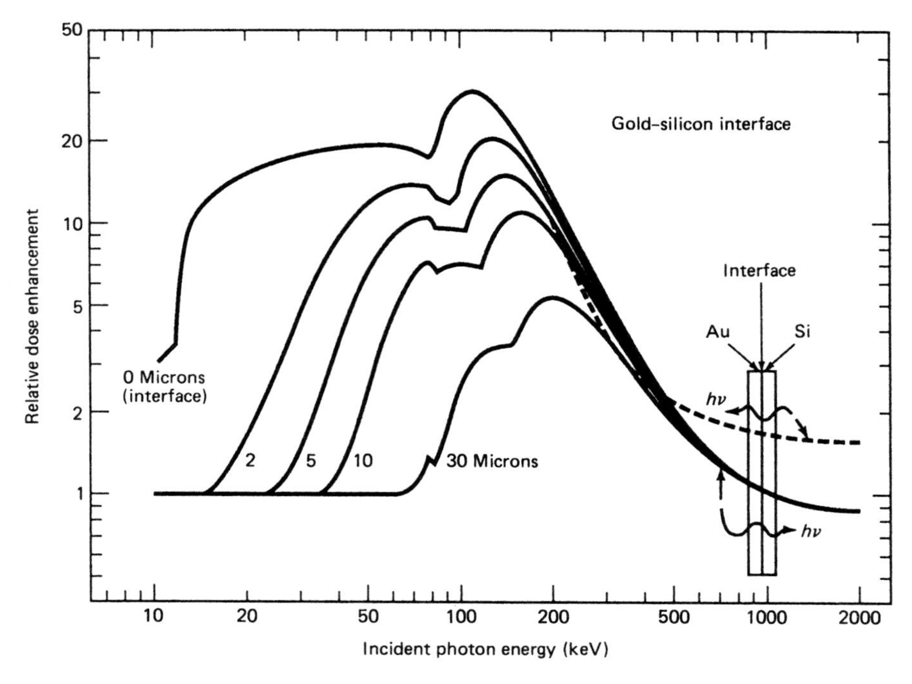

Fig. 10.2 Relative monte carlo dose for Au-Si interface versus energy of incident photons for various distances into silicon from interface.[39] Figure © 1975 by the IEEE.

The relative dose, or dose enhancement factor, $D_r(x)$, is defined as the ratio of the absolute dose $D_{abs}(x)$ to the equilibrium dose D_{eq}. The equilibrium dose is the monoenergetic equilibrium photon dose in an infinite slab of high-Z material, i.e., with no interface, that initiates electron production. $D_{eq} = E_p \mu_{eq} \phi_p$, where μ_{eq} is the photon mass interaction coefficient (cm^2/gm), with the density ρ considered subsumed in μ_{eq}. The absolute dose is the sum

$$D_{abs}(x) = D_{eq} + D_{el}(x) \tag{10.13}$$

where the electron contributed dose $D_{el}(x) = \sum_i -Y_i'(x) \cdot E_p \phi_p = E_p \phi_p \cdot \sum_i D_i(x)$. The relative dose is then given by

$$\begin{aligned} D_r = (D_{eq} + D_{el})/D_{eq} &= 1 + \left(E_p \phi_p \cdot \sum_i D_i(x)/E_p \mu_{eq} \phi_p\right) \\ &= 1 + \sum_i D_i(x)/\mu_{eq} \end{aligned} \tag{10.14}$$

So that,

$$D_r(x) = 1 + (1/\mu_{eq}) \sum_i A_i(B_i + 2C_i x + 3F_i x^2) \exp - (B_i x + C_i x^2 + F_i x^3) \quad (10.15)$$

where the fitting coefficients are A_i, B_i, C_i and F_i. Near the interface, where x is small,

$$D_r(x) \cong 1 + (1/\mu_{eq}) \sum_i A_i B_i \exp - B_i x \quad (10.16)$$

provides a first approximation for the enhancement dose factor.

Figure 10.2 depicts a one-parameter family of relative dose enhancement curves for a gold-silicon interface versus the incident photon energy in the gold, for various penetration distances into the silicon from the interface. From the figure, it is seen that the maximum dose enhancement occurs at the interface for 100 keV incident photons. Also shown is that the incident photon beam direction, whether into the silicon from the gold material or vice versa (dashed line), for very high energy photons (>500 keV), affects the dose enhancement. As seen in the figure, for high-energy photon incidence from Au to Si, the relative dose enhancement $D_r < 1$, indicating dose reduction, while for photon incidence from Si to Au (dashed line), $D_r > 1$, implying dose enhancement. The relative dose reduction is due to the fact that, at these high photon energies, the compton electron current is larger in the silicon than in the gold. In the opposite direction, enhancement occurs because of the strong backscatter from gold of compton electrons produced in the silicon, which had entered the gold. This is also seen in Fig. 10.3, depicting similar behavior for a gold-aluminum interface.[36] From Fig. 10.2, it is appreciated that dose enhancements are large for low incident photon energies, such as those in the photoelectric regime corresponding to nuclear burst photon sources. For high photon energies (e.g., in the 1 MeV region, corresponding to ^{60}Co or similar gamma ray sources), the dose enhancement is appreciably smaller. Nevertheless, this must be taken into account when ascertaining dose levels in silicon active device (integrated circuit) ionizing dose testing. Methods are available for this purpose.[41]

With regard to the various photoelectric and compton electron contributions, Fig. 10.4 depicts them for 100 keV incident photons near a gold-silicon interface. It should be noted that, at small distances from the interface, the K-shell photoelectrons are the major contributor to the dose, as they possess the highest photoelectric cross sections. Since the K-shell electrons within the atom are closest to the nucleus, and so are the energetically most tightly bound, it would seem intuitively correct

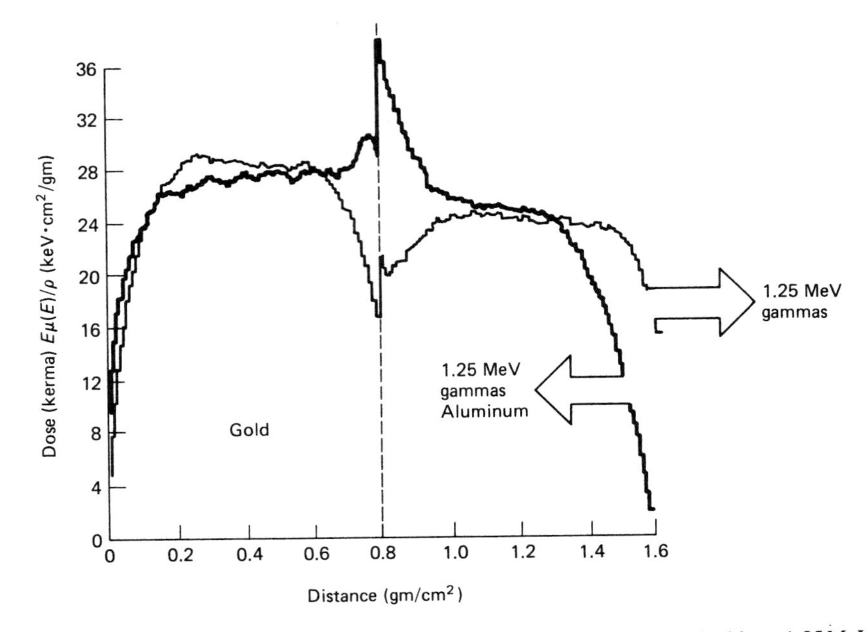

Fig. 10.3 Monte carlo dose profiles for gold-aluminum interface for incident 1.25 MeV gamma rays.[36] Figure © 1981 by the IEEE.

that the more remote electrons in the atom would be more likely to be "photoelectrically" ejected by the incident photons—not the innermost electrons. If the photon energy is much higher than that in the photoelectric regime, then this is the case, which is precisely the compton effect electron production. The K-shell electrons have the highest probability of being photoelectrically ejected by relatively low energy incident photons because, being the mostly tightly bound, they can best couple momentum to the whole atom, which the incident photon brings prior to its annihilation within the atom as part of the photoelectric process, discussed in Section 6.1. Atomic electrons remote from the inner shells are more loosely bound, and so are poorer agents for taking up the photon momentum for the whole atom, which must be conserved together with energy in the photoelectric process. Momentum conservation in the compton process is easily satisfied, since no particle is annihilated as the exiting compton particles (electron and photon) conserve both energy and momentum, with respect to their states prior to their encounter.

So that the more numerous (but lower energy) K-photoelectrons contribute to the steep gradient near the interface, as seen in Fig. 10.4, while the less numerous (but more energetic) L-photoelectrons contribute to the dose farther away from the Au-Si interface. The K-auger electrons

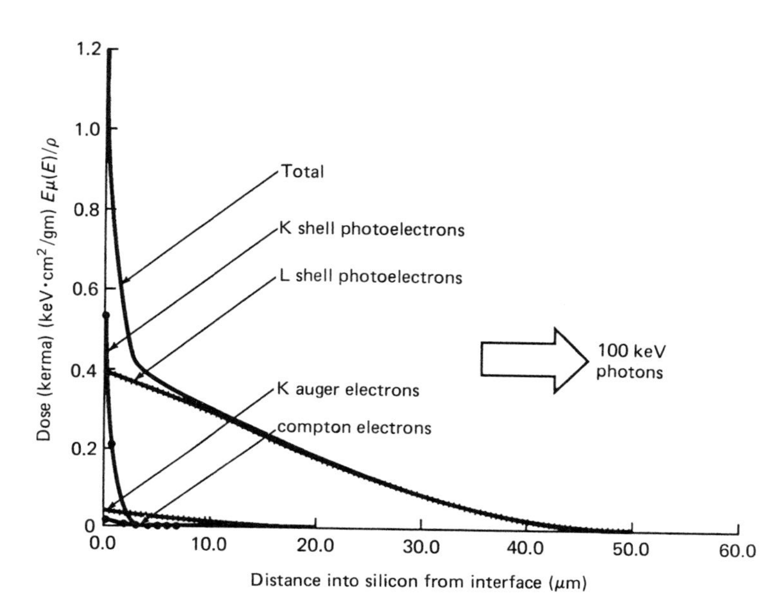

Fig. 10.4 Photoelectric and compton contributions to dose in silicon versus distance in silicon from Au-Si interface.[39] Figure © 1975 by the IEEE.

contribute even less, because the auger process is one of low probability for these parameters. The compton electrons are the minimum contributors, because (for these materials and incident photon energies) the compton effect is not operating at its optimum efficiency, as seen in Fig. 6.2.

To determine values of the fitting coefficients, A_i, B_i, C_i, and F_i, various methods are available.[38] A first-order fit (i.e., the determination of A_i and B_i for the various ith-type processes) is accomplished empirically, as mentioned. Fortuitously, for the Au-Si interface, a single expression for B_i is available to include all of the processes. It is fitted from monte carlo computations, and for all i is given by[38]

$$B = 5.816 \cdot 10^5 E^{-1.678} \qquad (\text{cm}^2/\text{per gm}) \tag{10.17}$$

where E is in eV. For the A_i, there is no single expression for all processes. However, they can be expressed as[38]

$$A_i = (1/4) f_i \mu_i(E_p) R_p(E_i) E_i / E_p \tag{10.18}$$

where

f_i = the probability that another (auger) electron is ejected, instead of a photon in the photoelectric process.

Table 10.1 Average dose enhancement factors* D_r for various device technologies.[51]

IC CHIP/METALLIZATION/TYPE	TYPE OF CHIP PACKAGE LID	5 KeV BLACKBODY X-RAY SPECTRUM INCIDENT THROUGH			15 KeV BLACKBODY X-RAY SPECTRUM INCIDENT THROUGH			GAMMA RAY ENVIRONMENT SPECTRUM**
		20 Mil Al	200 Mil Al	20 Mil Al+ 20 Mil Ta	20 Mil Al	200 Mil Al	20 Mil Al+ 20 Mil Ta	
Al or Polysilicon								
CMOS, NMOS, PMOS, CCD	Ceramic	1.0(1.0)	1.0(1.0)	1.0(1.0)	1.0(1.0)	1.0(1.0)	1.0(1.0)	1.0(1.0)
TTL (54, 54L, 54H, DI)	Kovar	1.7(5.2)	1.9(5.7)	1.9(3.6)	2.9(6.3)	3.4(6.7)	2.5(4.0)	1.4(1.4)
I^2L, ECL, DTL	Kovar	1.7(5.2)	1.9(5.7)	1.9(3.6)	2.9(6.3)	3.4(6.7)	2.5(4.0)	1.4(1.4)
Diodes/Transistors	Gold	4.1(19)	3.9(18)	3.5(11)	8.6(19)	8.0(19)	8.9(18)	2.0(2.0)
Schottky (Pt, W, Ti, Al)								
STTL (54LS, 54S, LSI)	Ceramic	1.9(7.9)	2.1(7.8)	2.9(5.6)	2.8(7.8)	3.1(6.9)	3.0(4.7)	1.5(1.5)
Schottky Diodes	Kovar	3.0(12)	3.0(12)	3.2(9)	6.0(13)	5.8(12)	3.8(7.2)	1.5(1.5)
Schottky Diodes	Gold	4.1(19)	3.9(18)	3.5(11)	8.6(19)	8.0(19)	8.9(18)	2.0(2)
Gold								
TTL DI	All	4.1(19)	3.9(18)	3.5(11)	8.6(19)	8.0(19)	8.9(18)	2.0(2)
RF Devices	All	4.1(19)	3.9(18)	3.5(11)	8.6(19)	8.0(19)	8.9(18)	2.0(2)

* Dose enhancement factor averaged over a 10 μm layer immediately under SiO_2 surface.

** $D_r = 2$ assumed for Au $\geqslant 5\,\mu m$ interface where gammas traverse Si first. For Au first, $D_r \cong 1.4$. For Au $< 5\,\mu m$, D_r assumed proportional to Au thickness. Values in parentheses correspond to max D_r (i.e., at top surface of SiO_2 on Si chip).

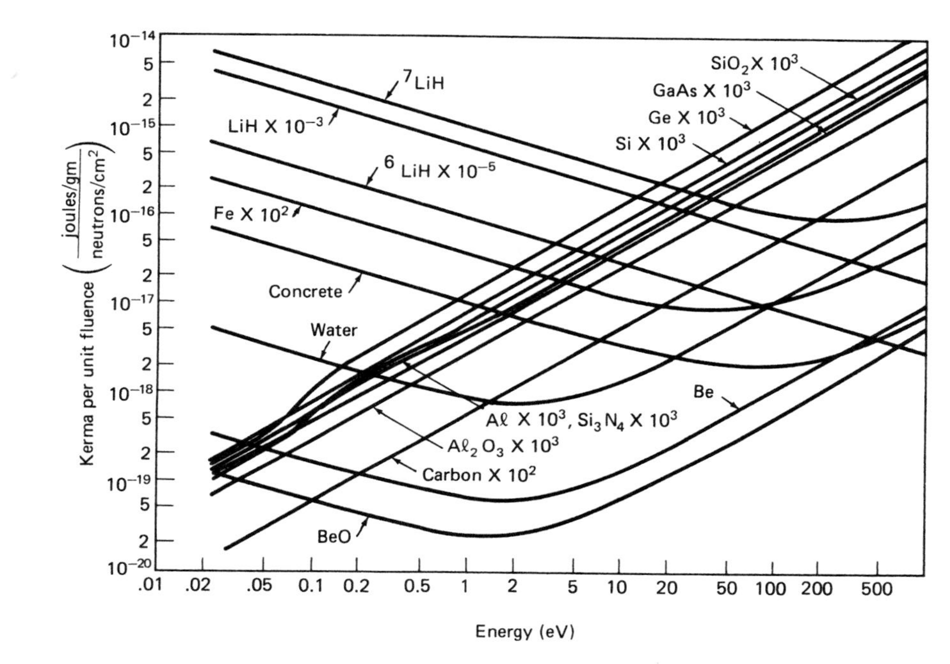

Fig. 10.5 Kerma per unit fluence versus neutron energy for various elements and compounds.[6]

μ_i = the mass interaction coefficient for the ith-type process.
R_p = the photoelectron range.
E_i = the initial energy of the ejected photoelectron.
E_p = $h\nu_p$, the energy of the incident photon.

In gold, the range is obtained empirically for all processes as[38]

$$R_p = 5.745 \cdot 10^{-6} E^{1.586} \qquad (\text{gm per cm}^2) \tag{10.19}$$

Agglomerating the preceding expressions to obtain a first-order relative dose enhancement factor, versus distance into the low Z side of the interface, gives

$$D_r(x) = 1 + (1/4\mu_{\text{eq}} E_p) \sum_i f_i \mu_i(E_p) E_i R_p(E_i) B_i \exp - B_i x \tag{10.20}$$

where it is seen that the maximum dose enhancement occurs at the interface, $D_r(0)$. For Au—Si and Au—CH_2 interfaces, where CH_2 rep-

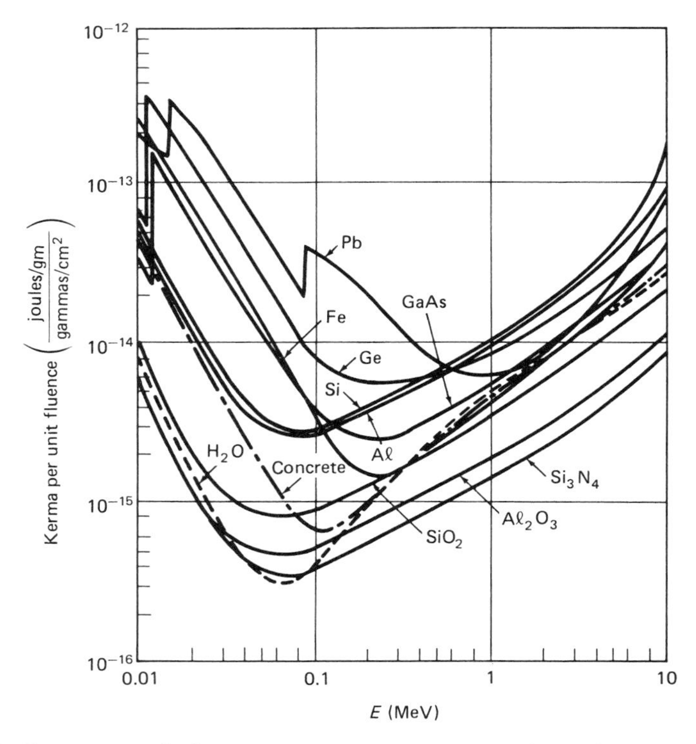

Fig. 10.6 Kerma per unit fluence versus gamma-ray energy for various elements and compounds.[5] Figure © 1979 by John Wiley & Sons.

resents polyethylene[40] values of B_i, R_p, and A_i are available empirically from monte carlo computations. Extrapolation methods to other interface materials can be found.[38,41] Table 10.1 is a tabulation of dose enhancement factors for incident monochromatic x-ray spectra of 5 and 15 keV, plus that for a nominal gamma environment spectrum. These are attenuated through pertinent material thicknesses for various chip metallizations, package lids, and device technologies.[51,52]

Kerma is described quantitatively in the example of incident x-ray or gamma-ray fluence Φ_γ (gammas per cm^2) transferring its energy into a material through photo-electric and compton processes. The amount of energy released into the material, i.e., the kerma, is given by

$$(1.602 \cdot 10^{-6}\,\text{ergs/MeV}) \int_0^\infty N\sigma(E)\Phi_\gamma(E)E\,dE/\rho \qquad (\text{ergs/gram}) \qquad (10.21)$$

where

$\Phi_\gamma(E)$ is the corresponding fluence (cm^{-2})
ρ is the material density (g/cm^3)
σ is the total cross section (cm^2)
N is the number of atoms per cm^3 of material ($\rho N_0/A$)
N_0 is Avogadro's number; $N\sigma/\rho \equiv \mu/\rho$ (cm^2/gm)
A is the mass number of the material (atomic weight)
E is the fluence particle energy (MeV)

Often, the kerma per unit fluence $K(E) = NE\sigma(E)/\rho$ is of interest, so that Eq. (10.21) can be rewritten as

$$(1.602 \cdot 10^{-6}\,\text{ergs/MeV}) \int_0^\infty K(E)\Phi_\gamma(E)\,dE \qquad \text{(ergs/gram)} \tag{10.22}$$

Kerma units can be expressed in a number of ways. The kerma unit is usually ergs per gram, and the kerma per unit fluence can be expressed as joules per gram/fluence unit as in Figs. 10.3–10.6. A "microscopic" kerma can also be defined as the kerma per unit fluence per atom of the material, i.e., $E\sigma(E)$, with units of MeV-millibarn (MeV-mb), as used in Figs. 10.7 and 10.8.

Table 10.2 comprises values of dose or kerma per unit fluence as a function of incident particle energy expressed in rads per particles/cm^2 for aluminum, silicon, and human tissue.

10.3 NUCLEAR ENVIRONMENT CORRELATIONS

As already discussed, dosimetry is the process of quantifying the description of the nuclear radiation dose in a manner that is relevant to the particular radiation type. Consider a radiation fluence spectrum, whether for photons or neutrons, of $\phi(E)$ particles per cm^2-MeV, and a particular radiation effect, $R_{eff}(E)$, such as dose per unit fluence of energy E. Then the total radiation effect, R_{eff} is given by

$$R_{eff} = \int_0^\infty R_{eff}(E)\phi(E)\,dE \tag{10.23}$$

An example of a seemingly trivial, but illustrative, neutron effect is that for $R_{eff}(E) = R_c$, a constant. Then $R_{eff} = R_c\Phi$, where the total fluence $\Phi = \int_0^\infty \phi(E)dE$ particles per cm^2 is constant. Hence, this effect does not depend on the spectral energy details, but is proportional only to the total

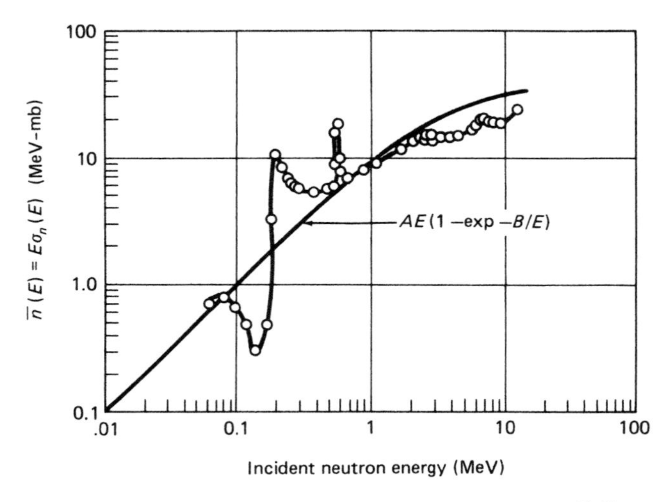

Fig. 10.7 Incident neutron energy lost to atomic processes in silicon.[11-13] Figure © 1975 by the IEEE.

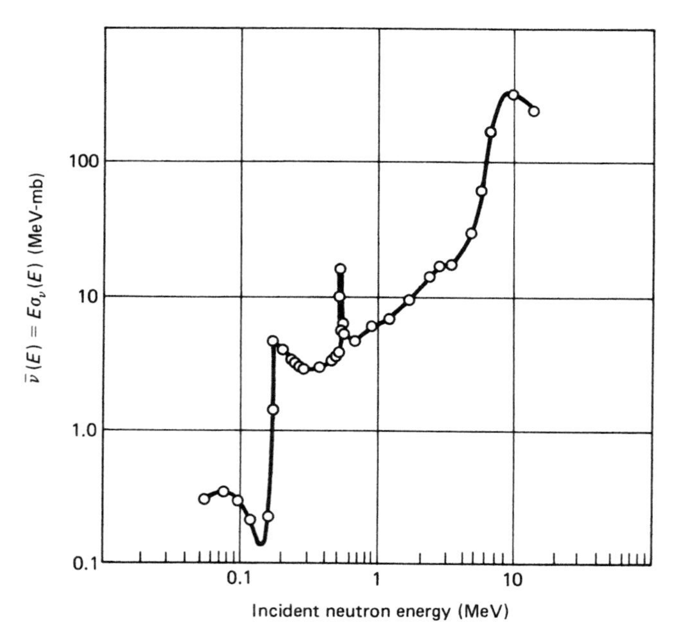

Fig. 10.8 Incident neutron energy lost to electronic processes in silicon.[11-13] Figure © 1975 by the IEEE.

Table 10.2 Conversion factors for electrons, protons, gammas and neutrons.[21–29] (Exponents are suppressed, e.g., $2.5 - 8 \equiv 2.5 \cdot 10^{-8}$)

ENERGY (MeV)	ENERGY DEPOSITED PER UNIT FLUENCE	UNIT CONVERSION	UNIT CONVERSION	UNIT CONVERSION
ELECTRONS				
0.01	2.65−7 rad (Al)/e/cm^2	rad (Al) = 0.979 rad (Si)	rad (Si) = 0.746 roentgen	rad (Si) = 1.35 rad (Tissue)
0.1	5.11−8	= 0.975	= 0.784	= 1.27
1.0	2.36−8	= 0.968	= 0.797	= 1.23
5.0	2.50−8	= 0.965	= 0.768	= 1.26
PROTONS				
2	1.80−6 rad (Al)/p/cm^2	= 0.985	= 0.699	= 0.671
10	5.41−7	= 0.978	= 0.752	= 0.738
30	2.30−7	= 0.976	= 0.775	= 0.768
50	1.54−7	= 0.975	= 0.782	= 0.778
100	9.11−8	= 0.974	= 0.790	= 0.788
GAMMAS				
0.01	4.08−9 rad (Al)/γ/cm^2	= 0.752	= 6.39	= 7.08
0.1	6.18−11	= 0.867	= 1.65	= 1.74
1.25(^{60}Co)	5.14−10	= 0.966	= 0.869	= 0.902
5.	1.43−9	= 0.975	= 0.916	= 0.961
NEUTRONS				
	(Due To Displacements)	(Due To Ionization)		
1.	3.1−11 rad (Si)/n/cm^2	2.6−11 rad (Si)/n/cm^2	3.85−9 rad (Tissue)/n/cm^2	= 67.5
14.	5.3−11	1.2−9	6.67−9	= 5.54

fluence Φ, so that, irrespective of whether the device under consideration is irradiated by the fluence spectrum $\phi(E)$ obtained from a nuclear reactor, neutron producing accelerator, or fixed neutron source, the total effect will be the same if the value of Φ used is the same for all these neutron sources.

However, another effect, such as the displacement of crystalline semiconductor atoms due to incident neutrons, as discussed in Sections 5.5 and 5.6, is very spectral-dependent, so that Eq. (5.5) must be used to obtain the total effect R_{eff}. This amounts to a double integration, over the energies of the recoil atoms as well as the neutron fluence spectrum.

In most instances, $R_{eff}(E)$ is known, for it is essentially the dosimeter or detector response function, which is part of the dosimeter rating specifications. Also, the spectrum, $\phi(E)$, must be measured or known for the individual source facility. Its determination is discussed later.

In the context of transient radiation effects on electronics (TREE), the two major types of radiation to be considered are from photons and neutrons. X-ray and gamma-ray photons deposit energy in materials mainly by ionization, as discussed. It is well known that the amount of bulk ionization effects in silicon depends almost solely on the amount of ionization energy deposited therein. Hence, the appropriate unit for measuring such ionization is the unit of energy deposition, the rad. Therefore, any dosimeter that reads rads directly or whose output can be converted easily and directly to rads, especially in a manner independent of $\phi(E)$, is an optimal instrument for such a measurement.

This is in contrast to neutron-induced displacements in silicon, for example, which manifest a radically different response to the neutron radiation spectrum and for which rads is an inappropriate, though not totally inimical, unit. In this case, a better approach is through the use of the total neutron fluence and spectral shape or instead using an equivalent monoenergetic neutron fluence at a specified energy, which would produce the same concentration of displaced atoms as does the actual neutron spectrum. Two such measures are commonly used. They are (a) the 1-MeV equivalent neutron fluence, and (b) the fission spectrum equivalent neutron fluence. The use of the former is prevalent currently and so is discussed below.

As investigated in Chapter 5, incident neutrons penetrating into silicon cause both displacement and indirect ionization damage. The complexity of these interactions implies that the prudent worker measure both the total neutron fluence and its spectrum. The functional dependence of semiconductor damage on the energy of the incident neutrons has been well established, both theoretically[8] and experimentally.[9] This dependence is directly applicable to the study of both displacement and ionization processes induced by fast neutrons, i.e., those whose energies are above 10 keV. Lindhard's theoretical approach[8] divides the above neutron interactions into atomic (displacement) and electronic (ionization) interactions, and the incident energy E is partitioned between the two processes, such that

$$\bar{n}(E) + \bar{v}(E) = E \qquad (10.24)$$

The function $\bar{v}(E)$ is the energy lost to electronic processes, and $\bar{n}(E)$ is that lost to atomic processes. $\bar{n}(E) = E\sigma_n(E)$ and $\bar{v}(E) = E\sigma_v(E)$ are the respective energies lost per incident neutron per cm^2, and σ_n and σ_v are the corresponding cross sections.

The function $\bar{n}(E)/E$ has been verified experimentally versus incident neutron energy.[9] Smith et al. have compiled these crossections in silicon as a function of energy for these interactions.[11] They have obtained curves of both energy lost to atomic processes, Fig. 10.7 and that lost to electronic processes, Fig. 10.8.

If the assumption is made that the concentration of recombination centers produced in silicon is proportional to the energy lost to atomic processes, which has been confirmed for neutron energies from 0.01–14 MeV, then Fig. 10.7 can be used to determine the neutron damage constant K, as discussed in Section 5.8.

Because it is not often practical to test electronic devices and components in an actual nuclear detonation, the neutron effects of the detonation must be simulated by neutron production facilities, such as nuclear reactors and other neutron sources. The neutron spectra of these sources do not, in general, coincide with the weapon detonation spectra, so that, for the sake of comparing many different neutron sources and spectra, a common 1-MeV monoenergetic equivalent neutron fluence $\phi_{eq}(1\,\text{MeV})$ is defined. It is that a given neutron fluence spectrum can be characterized by ϕ_{eq} (1 MeV), which represents the fluence level of 1 MeV neutrons per cm^2 required to produce the same radiation damage in silicon, as the silicon would endure from the actual neutron fluence energy spectrum. As essentially all the neutron damage is produced by neutron energies above 0.01 MeV, and in keeping with the preceding description, $\phi_{eq}(1\,\text{MeV})$ is defined as[14]

$$\phi_{eq}(1\text{ MeV}) = \int_{.01\text{ Mev}}^{\infty} \phi(E)D(E)\,dE/\bar{D}(1\text{ MeV})\text{ (neutrons per cm}^2) \qquad (10.25)$$

where the damage function, $D(E)$, is the silicon neutron displacement kerma per unit fluence, $K(E)$, discussed earlier and depicted in Fig. 10.7.

$\bar{D}(1\,\text{MeV})$ is called the 1 MeV neutron spectrum average displacement kerma per unit fluence (MeV-mb). A test of the utility of a 1-MeV silicon equivalent neutron fluence was corroborated in experiments performed at several different nuclear reactors.[31] Selected transistors, with activation foils in their immediate neighborhood within the reactors, were irradiated. After withdrawal from the reactors, the foils were used to obtain the neutron fluences and corresponding energy spectra of the reactors, as discussed[31] in Sections 10.4 and 10.7. Because of the rapid variation of $D(E)$ due to strong resonances in the silicon cross section for these processes in the vicinity of 1 Mev, $\bar{D}(1\,\text{MeV})$ is taken as 95 MeV-mb, which is obtained from a weighted average over the entire reactor neutron spectrum.[32]

A semiempirical method to determine $\bar{D}(1\,\text{MeV})$ is to fit the D function in Fig. 10.7 with[32]

$$D(E) = AE(1 - \exp - B/E) \tag{10.26}$$

The constants A and B are determined by a least-squares fit of Eq. (10.10) using sets of experimental data taken from two types of reactors.[32] These results yielded a weighted average 1 MeV displacement kerma per unit fluence for silicon, $\bar{D}(1\,\text{MeV}) = 95 \pm 4$ MeV-mb. For the constants A and B, one reactor data set gave values of $A = 133.3$ and $B = 1.32$, while the other gave values of $A = 177.5$ and $B = 0.75$.[32] The root mean square deviations from the $\bar{D}(E)$ curve of Eq. (10.26) for the two data sets were 29.3 and 24.6, respectively. The recommended resultant value is $\bar{D} = 95$ MeV-mb. This $\bar{D}$ value has the advantage of varying only in a secular fashion with new experimental data or calculations that might result in changes of tabulations of $D(E)$.[32]

10.4 UNFOLDING

Having obtained the damage function $D(E)$, the remaining unknown function in Eq. (10.25) is the neutron fluence (or flux) spectrum $\phi(E)$, for the determination of the 1-MeV equivalent fluence ϕ_{eq} for the particular nuclear source under consideration. It is of interest to discuss how $\phi(E)$ is determined for a particular (reactor) source, using the method of threshold foil spectrometry. This method is used because of its versatility in measuring both pulsed and steady-state reactor spectra, its wide neutron energy range (0.02–14 MeV), wide fluence range (10^{11}–10^{14} neutrons per cm^2), inherent freedom from concomitant gamma-ray interference, excellent reproducibility, and because only peripheral access to the reactor core is required. It is not applicable for foils placed at

Table 10.3 Neutron dosimetry foils for use in the neutron spectrum unfolding method.

FOIL	THRESHOLD ENERGY (MeV)	FLUENCE (n/cm^2)	EMITTED GAMMA ENERGY (MeV)	HALF-LIFE
^{239}Pu(n, f) ^{140}La	0.01	10^{11}–10^{14}	1.593	40.23 h
^{237}Np(n, f) ^{140}La	0.60	10^{12}–10^{14}	1.593	40.23 h
^{238}U(n, f) ^{140}La	1.50	10^{11}–10^{14}	1.593	40.23 h
^{32}S(n, p) ^{32}P	3.00	10^{9}–10^{14}	betas*	14.3 d
^{197}Au(n, γ) ^{198}Au	Thermal	10^{7}–10^{14}	0.412	2.70 d
^{58}Ni(n, p) ^{58}Co	3.00	10^{11}–10^{14}	0.812	71.3 d
^{56}Fe(n, p) ^{56}Mn	4.50	10^{10}–10^{14}	0.835	2.58 h
^{27}Al(n, α) ^{24}Na	7.50	10^{11}–10^{14}	1.369	15.0 h
^{24}Mg(n, p) ^{24}Na	6.30	10^{11}–10^{14}	1.369	15.0 h

* fast electrons

distances greater than 50 cm from the reactor core housed in a low-atomic-number shield, such as lithium hydride or water. Neither can it be used with a neutron-producing linear accelerator (LINAC) source, due to the significant number of gamma-induced fissions, which would produce extraneous neutrons.

The method uses a selected set of foils, each with a different threshold energy of neutrons to initiate the particular nuclear reaction in the foil. The threshold energies, E_T, vary from fractions of an electron volt to about 14 MeV. *Foil* is meant generically and includes powders, liquids, solids, etc., as well as actual thin metallic foils of suitable materials. The foils are covered with cadmium or boron to filter out the thermal or epithermal neutrons and so sense only neutrons greater in energy than 0.01 MeV. The foils are then inserted into the reactor whose neutron flux spectrum $\phi(E)$ is sought. Table 10.3 depicts a typical set of foils for this use. The neutrons from the reactor interact with each foil material to tranform some of its nuclei to a corresponding radio-active isotope. Their exposure time in the reactor is of sufficient duration that an equilibrium level of radioactive isotope concentration ensues, as discussed in the next section. After irradiation and foil removal, the foil activity (radioactivity) is measured. As can be seen from Table 10.2, the emitted activity consists mostly of gamma rays. This measurement gives an indication of the number of radioactive atomic nuclei produced during the irradiation. For the ith foil of a set of n foils

$$N_i = N_{oi} \int_{E_{Ti}}^{\infty} \sigma_i(E)\phi(E)\, dE; \qquad i = 1, 2, 3 \ldots n \tag{10.27}$$

where N_i is the measured concentration of radioactive nuclei from the ith foil, following its irradiation in the reactor. N_{oi} is the original concentration of foil material atoms. $\sigma_i(E)$ is the known cross section for the particular foil neutron activation reaction, and $\phi(E)$ is the desired neutron spectrum. Equations (10.27) are a set of simultaneous integral equations for a common $\phi(E)$. The process of extracting the $\phi(E)$ that is the best approximation for the simultaneous satisfaction of Eqs. (10.27) is called *unfolding*. It can be accomplished using the SAND II unfolding code, which is a well-known computer program used for this task.[15]

10.5 FOIL DOSIMETERS

Of prime importance for the measurement of the fluxes, fluences, and doses already discussed are the detectors of these quantities. As the main type of detection of interest centers on dose determination, the label *dosimeter* is given to these devices. There are various types of dosimeters, often called *foils* in a generic sense. Foils imply all forms of material in which target nuclei are irradiated for the purpose of measurement. For activation measurements, the foils are termed *infinitely dilute*. This means that the concentration of foil nuclei is sufficiently small that such effects as self-absorption or self-shielding, where the outer foil nuclei shield the inner ones from interaction with the incident fluence, are not present.

Generally, foil activation exposures are saturation measurements wherein activated foil nuclei concentrations achieve equilibrium with their radioactive environment. This is enunciated in the foil activation kinetics given in the two coupled equations for a particular foil. The first is the kinetic balance equation for the production and loss of activated nuclei of the ith foil. That is

$$\dot{N}_i = N\sigma_{ip}\varphi(E,t) - N_i\sigma_{ir}\varphi(E,t) - \lambda_i N_i; \qquad N_i(0) = 0 \tag{10.28}$$

The second equation accounts for the "burnup" of the target nuclei of the foil to supply activated nuclei, as given by

$$\dot{N} = -N\sigma_{ip}\varphi(E,t)\ ; \qquad N(0) = N_0 \tag{10.29}$$

where

N_i is the concentration of activated nuclei of the ith foil
N is the concentration of target nuclei of the ith foil
σ_{ip} is the cross section for the production of activated nuclei
σ_{ir} is the removal cross section for loss of activated nuclei

λ_i is the decay constant of the activated nuclei
$\varphi(E, t)$ is the neutron flux of the reactor (neutrons per $cm^2 \cdot sec \cdot MeV$ of energy E) and the corresponding neutron fluence is given by $\phi(E) = \int_0^t \varphi(E, t')dt'$ (neutrons per cm^2 MeV).

In this context, the cross sections are assumed to be spectrum-averaged. This means that, for example

$$\sigma_{ip} = \int_0^{\infty} \sigma_{ip}(E)\varphi(E,t)\,dE \Big/ \int_0^{\infty} \varphi(E,t)\,dE \tag{10.30}$$

For the case of a nuclear reactor at constant power output, i.e., at constant flux $\varphi(E, t) \equiv \varphi_0(E)$ at energy E, so that inserting $\varphi_0(E)$ into Eqs. (10.28) and (10.29) and integrating yields

$$\begin{aligned} N_i(t) &= [N_0\sigma_{ip}\varphi_0/((\sigma_{ir} - \sigma_{ip})\varphi_0 + \lambda_i)] \\ &\quad \times [(\exp - (\sigma_{ip}\varphi_0 t)) - (\exp - (\sigma_{ir}\varphi_0 + \lambda_i t))] \end{aligned} \tag{10.31}$$

In the above, $N_0 \exp - \sigma_{ip}\varphi_0(E)t$, the solution to Eq. (10.29) represents the "burnup" of the foil target nuclei due to their exposure in the reactor. Usually, saturation equilibrium is reached long before anything more than a miniscule percentage of foil nuclei is activated. Hence, the consumption of target nuclei is negligible, so that in Eq. (10.31) the terms $\exp - \sigma_{ip}\varphi_0 t \simeq \exp - \sigma_{ir}\varphi_0 t \simeq 1$. After saturation, since $\sigma_{ip}\varphi_0 \cong \sigma_{ir}\varphi_0$,

$$N_i(\infty) \cong \sigma_{ip} N_0 \varphi_0(E)/\lambda_i \tag{10.32}$$

The saturation activity, A_i, i.e., the decay rate of the activated foil nuclei, is given by

$$A_i = \lambda_i N_i(\infty) = \sigma_{ip} N_0 \varphi_0(E) \tag{10.33}$$

With the spectrum-averaged crossections as exemplified by Eq. (10.30), where $\varphi(E, t) \equiv \varphi_0(E)$, the above results in

$$A_i = N_0 \int_0^{\infty} \sigma_i(E)\varphi_0(E)\,dE; \qquad i = 1, 2, 3, \ldots n \tag{10.34}$$

which are essentially Eqs. (10.27) ready for unfolding, as discussed in Section 10.4.

10.6 THERMOLUMINESCENT DOSIMETERS

Thermoluminescent dosimeters (TLDs) are available in several forms, including pellets, powders, extrusions, and in Teflon or glass capsules. Thermoluminescence is a property of certain insulating materials that, after being irradiated by ionizing radiation, release visible light upon being warmed above ambient temperature. This a form of thermally accelerated phosphorescence. That is, incident radiation produces hole-electron pairs in the TLD material, both of which can be trapped in impurity or displacement defect energy states, as discussed in Section 1.9. When the TLD is warmed to temperatures ranging from the boiling point of water to about 600°F, the electrons return to the valence band emitting fluorescent light.

The light output is characterized by glow peak luminescent resonances, and the integrated light output is a measure of the absorbed dose in the TLD. After readout, they can be annealed by heating to elevated temperatures, then cooled and reused.[16] Materials that are usually used as thermoluminescent dosimeters are lithium fluoride, manganese-activated calcium sulphate, and manganese-activated calcium fluoride. The latter is the most sensitive to gamma rays; however its glow peak resonance is hardly above room temperature, and dissipation of the entrapped energy can occur before readout takes place (fading). CaF_2: Mn is more sensitive than LiF, and its higher mean atomic number of about 16.5 is closer to that of silicon ($Z = 14$) than LiF ($\bar{Z} = 8.14$). The ionization kinetics is a strong function of atomic number, so that the closer the TLD atomic number is to that of silicon, the more like the silicon dose characteristics the TLD absorbed dose will be. Also, the CaF_2: Mn TLD is hardier than the LiF, because it does not saturate at as low a dose. However, below 0.2 MeV incident photon fluence, higher energy filter correction shields are necessary for the CaF_2: Mn, but not for the LiF.[17]

Lower dose measurement limits on TLDs are in the millirad (mr) to microrad (μr) range, as the LiF minimum dose is 10 mr, the $CaSO_4$: Mn is 0.02 mr, and the CaF_2: Mn is about 1 mr. Their respective maximum doses are about 0.7, 10, and 300 krad (Si). TLDs are also sensitive to incident neutrons. They should be calibrated in a radiation field that is measured with a calibrated ionization chamber. Tables 10.4 and 10.5 depict salient properties of TLDs in common use.

10.7 CALORIMETRIC DOSIMETERS

One form of a calorimetric dosimeter is a thin calorimeter that measures the dose by monitoring the temperature rise in a small sample of mate-

Table 10.4 Thermoluminescent dosimeter material properties.[16,17,47] **Table © 1979 by John Wiley & Sons.**

	LiF	$CaSO_4$:Mn	CaF_2:Mn	$Li_2B_4O_7$:Mn	CaF_2(TLD-200)
Minimum dose (rad)	10^{-2}	20×10^{-6}	10^{-3}	.05	10^{-5}
Maximum linear dose (rad)	700	10^4	10^5	—	—
Maximum dose (rad)	10^5	10^5	3×10^5	10^6	10^6
Glow peak (°C)	210	80 to 100	260	200	180
Peak wavelength (Å)	4000	5000	5000	6050	5765
Fading (25°C)	5%/y	~50%/10 h	10%/16 h; then 1%/day	20%/yr	—
Density (g/cm^3)	2.64	2.61	3.18	2.4	3.18
Average Z (photoelectric absorption)	8.14	15.3	16.5	7.4	16.3

Table 10.5 Neutron dose in air per unit neutron fluence.[16,17] Table © 1979 by John Wiley & Sons.

MATERIAL	THERMAL	NEUTRON ENERGY, MeV					
		1	2	3	5.3	8	14.5
CaF_2 : Mn		(Neutron dose units: 10^{-10} rads (air) per n/cm^2)					
Vacuum tube type	1.23	0.59	0.71	0.57	5.4	0.12	1.8
Micro TLD	.88	1.7	1.8	–	7.2	5.9	3.9
LiF							
TLD-100	175	2.0	4.6	5.6	12.3	13.1	20.2
TLD-600	548	7.8	9.6	9.6	19.3	15.8	42.1
TLD-700	None	3.7	7.1	7.3	17.5	14.0	32.4

rial. The thermal characteristics of the material are assumed known, so that simply using the mass law of cooling, that is, $q = mc_v\Delta T$ of the sample material whose mass and specific heat are known, the measured temperature rise ΔT yields the heat energy deposition q. The thinner the calorimeter, the more closely the above cooling law holds, because thicker materials result in a complicated distribution of thermal energy throughout their bulk. Materials that are good thermal conductors are chosen, so that rapid thermal equilibrium is achieved following the incidence of the radiation transient. Beryllium, aluminum, iron, copper, gold, as well as thin lengths of silicon and germanium, are used. The temperature sensor should be such as to minimally perturb the heating capability of the absorber. Thermocouples whose mean atomic number is close to that of the absorber, as discussed in Section 10.6, do not impair its thermal properties very much and so are good sensor candidates. Copper-constantin mounted on a copper foil is an excellent sensor. Sensitivity can be increased by using a small thermistor on both the absorber and the sensor. The thermistor forms one leg of a Wheatstone bridge circuit, and the unbalance due to the incident radiation transient is monitored. In assembling this system, care must be taken to not add any undue material, such as heavy amounts of solder, to avoid the increase in atomic number. Resistance welding is recommended to eliminate this difficulty. If the thermistor thickness is a few mean free paths with respect to the incident radiation, the temperature measurement must be performed for a time sufficiently long to ensure that thermal equilibrium is established in the thermistor. This time is of the order of 100 ms to 1 s.

Thermal isolation is required to measure a small sudden temperature rise in the system. To provide such isolation, fine leads to the temperature sensor should be used. Attempts to insert the calorimeter into a small vacuum chamber can also be made.

For the case of a single pulse from a flash x-ray machine, a method for determining the sensor temperature is to establish a cooling curve for the sensor system, in order to make an extrapolation back to the onset of the pulse. A cooling curve is a plot of decreasing temperature versus time. The characteristics of the calorimeter are functions of the specific heat of the absorber and the temperature behavior of the sensor. If the calorimeter is made of the same material as the device under test (DUT), such as silicon, the absorbed dose in the DUT can be measured directly. If this is not so the calorimeter dose must be converted to the dose of the DUT material of possibly differing Z, so that corrections must be made for differences in stopping power, back-scattering and bremsstrahlung energy losses.

In large complex systems where the thin calorimeter is not used, the absorbed dose can nevertheless be computed from the thermal heating of the system. If it has reasonable physical symmetry, then the transport of the radiation from the source throughout the measuring system can be calculated using well-known analytical techniques, such as the moments method, discrete ordinates (S_n), and monte carlo methods. Once the radiation fluence map is known, the amount of absorbed dose due to the source is immediately obtained.

For measuring heat, the calorimeters above are termed *adiabatic* in the thermodynamic sense. An alternative calorimetric method for large systems is to employ a heated jacket whose temperature is maintained equal to the temperature of the absorber by means of an automatic servo-type regulator. The absorber temperature is held constant, as measured by a substance undergoing a phase change, such as melting ice. Still another method of measuring the amount of radiation-induced heating is to measure a temperature gradient and compute the heating rate, using a known material heattransfer coefficient or one that can be calibrated by comparison with a known electric resistance heater.

10.8 PIN DOSIMETERS

A *p-intrinsic-n* junction (PIN) reverse-biased diode can be used to measure ionizing dose. It is composed of a large central intrinsic region of almost pure silicon, sandwiched between a *p*-region and an *n*-region. The intrinsic region is the relatively large bulk portion of the detector in which charges are produced by the incident ionizing radiation. These charges result in a saturation leakage current, which is a measure of the incident radiation flux. Lithium is often diffused (drifted) into the intrinsic region, as an acceptor, to compensate any donor impurities that are likely to be present from the manufacturing process. This results in a closer approach to a purely intrinsic layer in the PIN diode. An advantage of using semiconductors as radiation detectors is the relatively small amount of energy W required to generate a hole-electron pair, as $W_{Si} = 3.6\,\text{eV}$, and $W_{Ge} = 2.8\,\text{eV}$. If the energy deposited by the radiation-induced ionization is E, then the number of hole-electron pairs produced, N, is simply

$$N = E/W \tag{10.35}$$

The corresponding generated charge $Q_{gen} = eN = eE/W$. For a detector and its associated electronics adjusted so that Q_{gen} can be measured for individual electrons or holes, then their energy can be obtained. Statistical fluctuations in this system proscribe the full width-half maximum (FWHM) number of counts under the energy spectrum peak, with an energy resolution of

$$\Delta E/E = 1.29\sqrt{W/E} = 1.29/\sqrt{N} \tag{10.36}$$

For example, for an 5.5-MeV alpha particle, the energy resolution is about 0.02 MeV.

The PIN charge collection time is very short, measured in nanoseconds, so that it can measure dose rate with good accuracy. It is usually limited to dose rates of up to 10^{10} rads (Si) per s within its linear range. For higher dose rates, its output becomes nonlinear with dose rate, because the internal electric field is perturbed by the correspondingly high current densities. Electron equilibrium in the active volume of the detector must be established before the PIN diode will measure the dose accurately. Many commercial PIN diodes obtain measurable ionization from their high-atomic-number housing material or from the tantalum (high-Z) plate contacts. Thin contacts or contacts attached to degenerate silicon must be used to obviate this difficulty. PIN diodes are also subject to structural degradation due to displacement damage, such as that caused by neutrons. This causes a decrease in the intrinsic silicon minority carrier lifetime, plus a corresponding increase in its resistivity, which leads to spurious leakage currents. Neutron fluences of 10^{12} neutrons per cm^2 or greater with neutron energies of above 10 keV will cause calibration shifts in these instruments. Similarly, they can be damaged by high-energy gamma rays, at large ionizing doses of 10^5–10^6 rads (Si), which will induce displacement damage. Generally, if the minority carrier lifetime can be maintained an order of magnitude larger than the collection

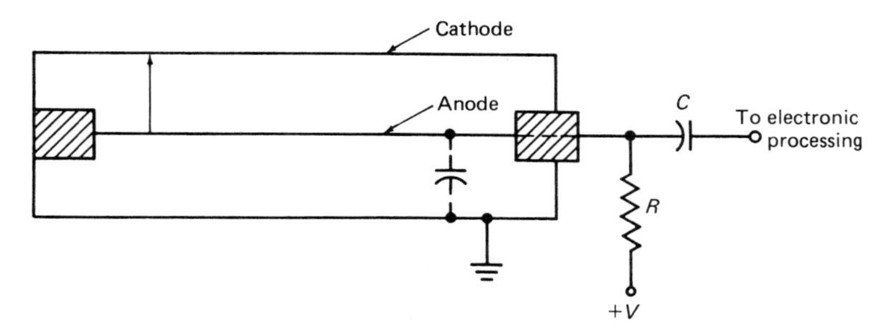

Fig. 10.9 Cylindrical proportional or Geiger-Mueller counter.

time, the PIN diode will function satisfactorily. PIN diodes also can be damaged by high currents from relatively long lasting pulses of electrons or photons or by very repetitive pulses of substantial magnitude. This implies that the PIN diode system and its associated electronics should be checked and calibrated frequently. An example of PIN diode usage is as a radiation detector as part of a circumvention system, discussed in Section 13.9.

10.9 PROPORTIONAL AND GEIGER-MUELLER COUNTERS

The proportional and Geiger-Mueller (GM) counters are usually used to measure electron fluence. They are physically similar, where the differences occur with respect to the various regimes of operation, as characterized by the anode voltage of the counter. These counters consist of two electrodes enclosed in a gas-filled, cylindrical, usually metallic, volume, as depicted in Fig. 10.9. The anode consists of a small-diameter tungsten wire centered on the axis of the cylinder. The cathode is the inside of the thin-walled cylinder housing surrounding the whole. The cathode can be made of an insulated window inserted in the end of the metal cylinder, e.g., graphite-coated glass. To be used as a proportional counter, as explained below, the gas filling is about 90 percent argon and 10 percent methane, at atmospheric pressure or less. In its GM mode of operation, the gas is usually argon at about 0.01 atm plus a small amount of ethanol to quench the discharge initiated by the incident radiation-induced ionization within the tube. Figure 10.10 shows the variation of the total number of electrons collected at the anode as a function of the anode voltage. The two curves correspond to two different energies of incoming electrons. The incident particles simply ionize the gas within, producing a current transient, which is read and processed by associated

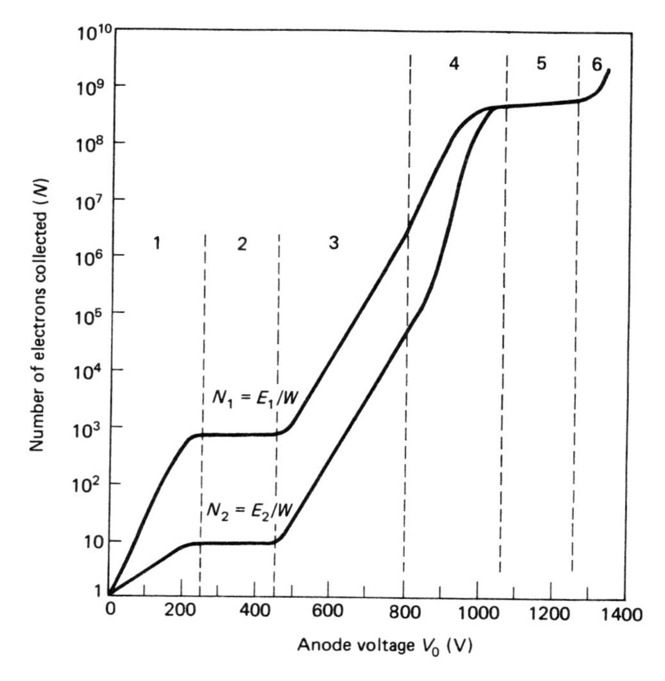

Fig. 10.10 Number of GM/proportional counter electrons collected versus anode voltage.

electronic circuitry. In the first region of Fig. 10.10 the ions and electrons recombine before they are separated by the impressed electric field due to the anode voltage. Therefore, this is not an optimum region to be used for detection. The second, and usable, region is the ionization chamber region, where the counter behaves as an ionization chamber. Here, the number of electrons counted, $N = E/W$, as discussed in the previous section. In region 3, also a usable region, the anode voltage is sufficient for the number of electrons counted to be proportional to the anode voltage, and so this is termed a *proportional counter*. In this region the electrons counted satisfy the proportionality

$$N = ME/W \tag{10.37}$$

where M is the gas multiplication factor, and W is the required ionization energy per ion-pair. The multiplication factor comes about because the incident electrons are accelerated in the relatively high electric field near the tungsten wire anode and thus obtain sufficient energy to produce additional electrons through ionization of the tube gas. In region 4, the

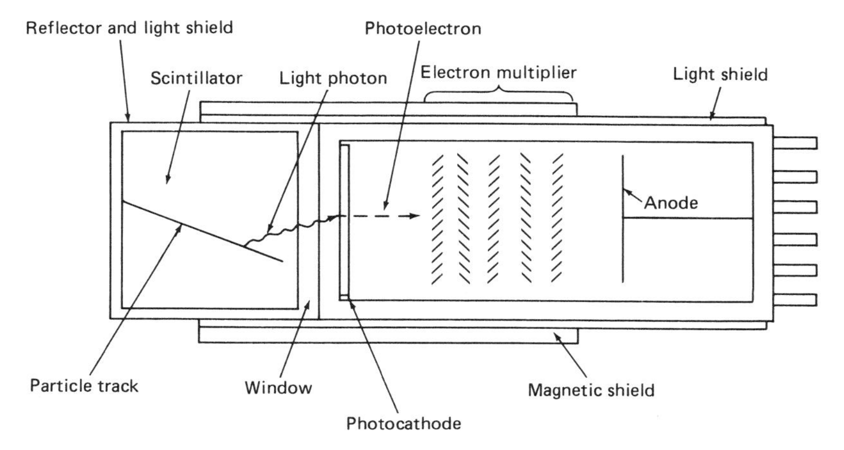

Fig. 10.11 Scintillator counter cross sectional view.[18] Figure © 1979 by John Wiley & Sons.

proportionality breaks down, because the multiplication level ceases to be dependent on the initial ionization. This is exacerbated in region 5, where the spreading discharge is terminated by the loss of electrons to the quenching agent gas. In region 6, the anode voltage is unduly high, causing the counter to discharge continuously, which can cause its ultimate destruction. Generally, the voltage in each region in Fig. 10.10 depends on the gas composition, pressure, temperature, and anode radius.[18]

10.10 SCINTILLATORS

A scintillator detector can be used to count gamma-ray and x-ray photons and moderate energy electrons, to determine the corresponding fluence spectrum. Figure 10.11 shows a cross-sectional diagram of a scintillator detector. An incoming photon incident on the scintillator crystal produces ionization within it, which, with subsequent deexcitation, emits light, similar to hole-electron production and recombination with respect to TLD operation, as discussed in Section 10.6. The scintillator is mounted in a light-tight container to exclude extraneous light from the photomultiplier tube (PM), which is physically closely coupled to the scintillator. The light from the scintillator impinges on the PM cathode, releasing low energy photoelectrons. These electrons are accelerated onto the first "dynode," or secondary emission electrode, due to the dynode potential, similar to the electron-gun portion of a TV picture tube. The

Table 10.6 Properties of scintillator detectors.

TYPE	SCINTILLATORS	REF. INDEX	RELATIVE LIGHT OUTPUT	λ_{max} (Å)
Organic crystals	Anthracene	1.62	1.00	4470
	trans-stilbene	1.63	0.50	4100
Organic liquids	NE213	1.50	0.78	4250
	Toluene base	1.50	0.60	4300
Plastics	NE 102	1.58	0.65	4250
	PVT base	1.58	0.45	4450
Inorganic crystals	Sodium iodide	1.78	2.1	4100
	Cesium iodide	1.78	0.95	4200 to 5700
	Lithium iodide	1.95	0.74	4400
	Zinc sulfide	2.36	3.00	4300
Glass	NE908	1.57	0.20	3990
Gas	Xenon	1.0	1.5	~3300

first dynode releases two or four electrons per incident electron. The secondary electrons are accelerated toward the second dynode, producing further secondary electrons upon impact. This action cascades through the dynode chain, with the resulting current collected at the endplate within the PM tube. The normal PM tube has 12–14 dynodes, providing an electron multiplication of 30–80 dB depending on the PM parameters. PM tube noise can be minimized by cooling the system and using other electronic adjuncts. A low permeability magnetic shield surrounds the whole system to prevent external magnetic fields from distorting the secondary electron orbits through the PM tube. Table 10.6 provides pertinent properties of scintillators used in this system.

10.11 CAVITY IONIZATION CHAMBERS

When it is desired to make an absolute measurement of dose in general or in a system large enough to mount a small cavity within its walls, then the cavity ionization chamber is used.[19] The cavity is filled with gas, and the walls are sufficiently thick for charged particle equilibrium to exist, as discussed in Section 10.2, but not so thick that the primary particles are attenuated within the walls. Figure 10.12 depicts a spherical ionization

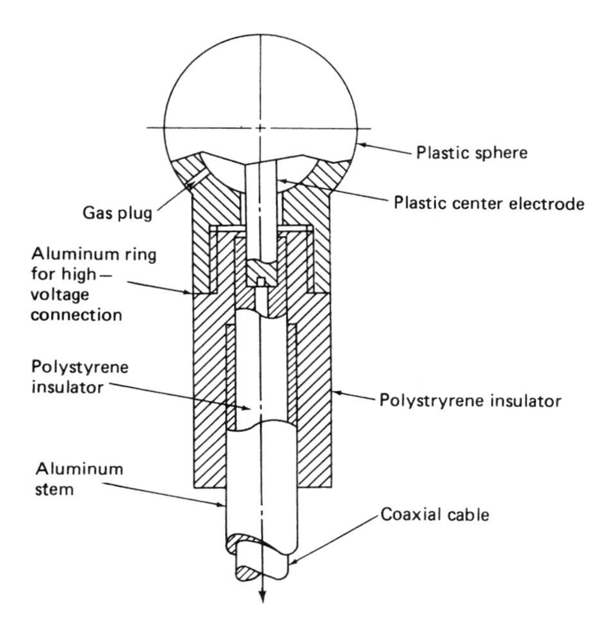

Fig. 10.12 Spherical cavity ionization chamber.

chamber with a 5-mm plastic wall. This device can measure dose from moderate-energy neutrons and gamma rays. The spherical volume is $1\,\text{cm}^3$, with its gas at a pressure of 1 atm and a potential of 250 V. The energy absorbed per gram of cavity gas, D_{cav}, is given by

$$D_{\text{cav}} = WJ \tag{10.39}$$

where W is the energy required to produce an ion-pair in the cavity gas, and J is the measured number of ion-pairs produced per gram of gas. The Bragg-Gray theory asserts, under conditions of charge equilibrium, that[19]

$$D_{\text{wall}}/D_{\text{cav}} = S_{\text{wall}}/S_{\text{cav}} \tag{10.40}$$

and S is the corresponding stopping power, i.e., $S = dE/\rho\, dx$. It is found experimentally that the right-hand side of Eq. (10.40) is independent of energy. If the cavity dimensions are large compared with the particle range, then

$$D_{\text{wall}}/D_{\text{cav}} = (\lambda_{en}\rho)_{\text{cav}}/(\lambda_{en}\rho)_{\text{wall}} \tag{10.41}$$

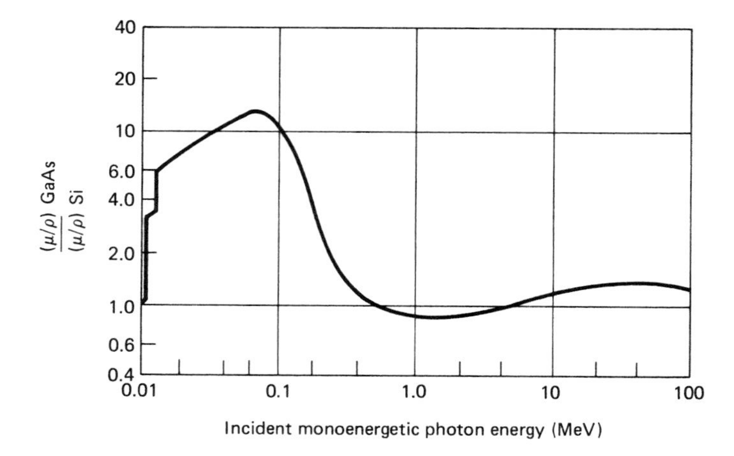

Fig. 10.13 Ratio of mass absorption coefficient of GaAs to silicon as a function of incident photon energy.

where λ_{en} and ρ are the corresponding energy absorption mean free paths and densities. The above expressions also imply that the cavity does not perturb the secondary particle equilibrium in the wall, that almost all such particles originate in the wall, and that the cavity radius is small compared with the particle range at the given gas pressure. If the atomic composition of the wall and the cavity gas can be matched, then the radius constraint no longer applies. Cavity ionization chambers are made with volumes from $1\,cm^3$ to several liters.

10.12 PHOTOGRAPHIC EMULSIONS/RADIOCHROMIC DYES

Photographic emulsion (film) contains grains of silver bromide dispersed in gelatin on a plastic or glass matrix.[20] Emulsions for particle and nuclei track work, such as used in cosmic-ray (balloon) studies, contain almost double the concentration of silver bromide, grains of one-third the size, and up to 20 times the thickness of emulsion compared with commerical-grade photographic emulsions. The incident particle ionizes a few atoms of Ag in each grain to form a latent image, as in the photographic process. After the emulsion is exposed, the "developer" reduces the grains to metallic silver. The inert grains are dissolved in the "hypo" (sodium thiosulphate), and the emulsion is rinsed and dried, again like ordinary film-developing procedures. The emulsion thickness is reduced to about half its former thickness, which must be considered in particle track examination using a microscope. The tracks counted correspond to a fluence measurement for ions, recoil protons, etc. However, for gamma rays and electrons, the absorbed dose is determined from the emulsion blackening, as quantitatively expressed by the optical density $D_0 = \log(I_0/I)$, where I_0 and I are respectively the incident and transmitted light measured through the emulsion. For part of the range, D_0 is proportional to the logarithm of the exposure, but D_0 also depends on the particular batch of emulsion, processing conditions, etc. The optical density must be calibrated as a function of absorbed dose, or exposure, as measured by an absolute dosimeter, such as an ion chamber. For example, $D_0 = 0.3$ can correspond to doses from 1 mrad to 1 rad, depending on the emulsion sensitivity. The photographic emulsion process has been commercially developed to the point where film services are available for equipment and workers, which monitor absorbed dose and can be read daily if desired.

Radiochromic dyes, which are organic dyes suspended in a solid solution in a transparent plastic film host, have been found to be sensitive to ionizing dose, among other radiation types.[48,49] They manifest this sensitivity by a darkening of a e.g., small postage stamp-size square 2 mil sheet of dosimeter, as read in an optical reader.[48] Their stable coloration, as a function of incident radiation dose, can be calibrated to achieve accuracies of better than ± 10 percent for ionizing dose levels between 10^5–10^6 rads (Si).[48] The optical density D_0, corresponding to a particular level of incident radiation, is described above. These dosimeters must be handled and calibrated carefully, as they can be unusually sensitive to their ambient environment.[48]

Other types of radiochromic dosimeters change color for different levels of incident radiation. For these dosimeters, experiments reveal that the ionizing dose level, as read by corresponding readers, shows essentially no dose-rate dependence for dose rate levels between $\sim 10^3$ (^{60}Co) to about 10^{15} (electron pulse) rads (Si) per second.[49]

PROBLEMS

1. For energy (dose) absorbed in silicon, the units are rads (Si), and in air the units are rads (air). Show that the expression that converts rads (Si) to rads (air) is obtained from the relation rads (Si)/rads (air) $= (\mu_{Si}/\rho_{Si})/(\mu_{air}/\rho_{air})$, where μ_{Si}/ρ_{Si} and μ_{air}/ρ_{air} are the respective mass absorption coefficients. That is, for example, $\mu_{Si}/\rho_{Si} = N_0 \sigma_{Si} A_{Si}$ (cm^2/g). Fig. 10.13 depicts $(\mu/\rho)_{GaAs}/(\mu/\rho)_{Si}$ versus incident monoenergetic photon energy.
2. In the determination of $\bar{D}(E)$ discussed in Section 10.3, the empirical fit $D(E)$

$= AE(1 - \exp - B/E)$ is used, where A and B are determined from a least-square fit to the experimental data. For illustration, let $A = 133.3$ and $B = 1.32$. Determine $\bar{D}$ making the required integrations using the Milne integral $E_n(x) = \int_1^\infty (\exp - xy)dy/y^n$. Assume that the above D holds for energies between 0–2.25 MeV, where $E_n(x)$ is a well-tabulated function; yielding $\bar{D} = \int_0^{E_m} D(E')dE'/E_m$ with $E_m = 2.25$ MeV.

3. What is the conversion factor between rads and grays? Also, what is it from dose in air due to the exposure of 1 roentgen, and 1 rad in air?
4. TLDs will not measure the absorbed dose with usable accuracy in materials and configurations where the absorbed dose and the kerma differ widely.[5] Explain this statement.
5. For a particular reactor, the 1-MeV equivalent neutron fluence ϕ_{eq} (1 MeV) often can be gotten from measurements using a sulfur foil only, where N_s is the number of activated sulphur nuclei per cm^3. Show how this computation can be made if the two ratios, ϕ_{eq} (1 MeV)/Φ and N_s/Φ are available at the reactor site, where $\Phi = \int_0^\infty \phi(E)dE$.
6. The total number of tracks per cm^2, counted on a photographic emulsion immediately gives the corresponding value of fluence. Show this by deducing the interpretation that fluence is the total track length per cm^2.
7. A set of foils is placed in a pulsed nuclear reactor that is programmed to undergo a unit impulse flux burst of neutrons, i.e. $\varphi(E, t) = \phi_0(E)\delta(t)$. Show that, within a constant, the saturation activity is given by Eq. (10.34), which is that for a steady-state, constant flux reactor, where the same approximations are used. Hint: Obtain the corresponding solution to Eqs. (10.28) and (10.29) viz., $N_i = N_0\sigma_{ip}\phi_0(E) \exp - (\sigma_{ip}\phi_0(E) + \lambda_i t)$, and $\lambda_i t \ll \sigma_{ip}\phi_0 \ll 1$ is assumed for the required foil exposure duration.
8. Figures 10.7 and 10.8 depict, for silicon, the incident neutron energy (kerma) lost to atomic processes (displacements), and the same for neutron-induced electronic processes (ionization), respectively. For incident energies up to and over 1 MeV, these kerma (viz., $E\sigma_n(E)$ and $E\sigma_\upsilon(E)$) curves look remarkably similar. Why?
9. Why is there no kerma enhancement corresponding to dose enhancement, such as that occurring near the interface of two dissimilar materials?
10. Following neutron testing of semiconductor components, it is necessary to allow them to decay to safe handling levels of radioactivity before taking post-radiation electrical measurements on them. Why is this not necessary after gamma ray (e.g., ^{60}Co) ionizing dose testing of components?

REFERENCES

1. International Commission on Radiation Units and Measurements. Radiation Quantities and Units. ICRU Report no. 11, Sept. 1, 1968. ICRU Publications, P.O. Box 30165, Washington, D.C. 20014.
2. P. N. Stevens and H. C. Claiborne. Weapons Radiation Shielding Handbook. DASA Report No. 1892-5, Washington, D.C., June 1970.

3. F. H. Attix and W. C. Roesch. Radiation Dosimetry, Vol. 1, 2nd ed. Academic Press, New York, 1968.
4. J. A. Auxier. "Kerma Versus First Collision Dose: The Other Side of the Controversy." Health Phys. 17:342, 1969.
5. A. E. Profio. Radiation Shielding and Dosimetry. Wiley-Interscience, New York, p. 302, 1979.
6. H. C. Claiborne, M. Solomito, and J. J. Ritts. "Heat Generation by Neutrons in Some Moderating and Shielding Materials." Nucl. Eng. Design 15:232–236, 1971.
7. R. K. Thatcher and M. L. Green. "TREE Preferred Procedures." Report No. 2028H, 2nd ed., no. 2, pp. 5–11ff, June 1972.
8. O. Lindhard, V. Nielsen, M. Scharff, and P. V. Thompson, Kgl Danske Selskab. Nat. Fys. Medd. 33(10), 1963.
9. A. R. Sattler. "Ionization Produced by Silicon Atoms Within a Silicon Lattice." Phys. Rev. 135(1A), June 1965.
10. G. C. Messenger. "Radiation Effects on Microcircuits." IEEE Trans. Nucl. Sci. NS-13(6), Dec. 1966.
11. E. C. Smith, D. Binder, P. A. Compton, and R. I. Wilbur. "Theoretical and Experimental Determination of Neutron Energy Deposition in Silicon." IEEE Trans. Nucl. Sci. NS-13(6), Dec. 1966.
12. V. C. Rogers, L. Harris, Jr., D. K. Steinman, and D. E. Bryan. "Silicon Ionization and Displacement Kerma For Neutrons From Thermal Energies to 14 Mev." IEEE Trans. Nucl. Sci. NS-22(6), Dec. 1975.
13. R. R. Holmes. In Ballistic Missile Defense Advanced Development Program Weapons Effects Studies, Vol. 2, Radiation Effects in Interceptor Electronics, Suppl. 3. Bell Labs, Western Electric Co., New York.
14. V. V. Verbinski, N. A. Lurie, and V. C. Rogers. "Threshold Foil Measurements of Reactor Neutron Spectra For Radiation Damage Applications." Nucl. Sci. Eng., May 1978.
15. W. N. McElroy, S. Berg, and T. Crockett. "A Computer Automated Iteration Method For Neutron Flux Spectra Determination by Foil Activation," Vol. 1. Air Force Weapons Lab (AFWL), Res. and Tech. Div. AFSC. Kirtland AFB, NM, Report AFWL-TR-76-41, Sept. 1967.
16. A. E. Profio, op. cit. p. 323.
17. R. K. Thatcher and M. L. Green, op. cit. pp. 5–21.
18. A. E. Profio, ibid. p. 268.
19. T. E. Burlin. "Cavity Chamber Theory." *In* Attix and Roesch, "Radiation Dosimetry," 2nd ed. Academic Press, New York, Vol. 1, Chap. 8, 1968.
20. A. E. Profio, ibid. p. 323.
21. M. J. Berger and S. M. Seltzer. Tables of Energy Losses and Range of Electrons and Positrons. NASA SP-3012, 1964.
22. Berger and Seltzer. Additional Stopping Power and Range Tables For Protons, Mesons, and Electrons. NASA SP-3036, 1966.
23. W. K. Barkas and M. J. Berger. Tables of Energy Losses and Ranges of Heavy Charged Particles. NASA NSP-3013, 1964.
24. National Research Council Publications. Studies in Penetration of Charged Particles in Matter. NRCP 1133, 1964.
25. H. E. Johns and J. R. Cunningham. The Physics of Radiology, 3rd ed. C. C. Thomas Publications, Springfield. IL, 1974.
26. G. W. Grodstein. X-Ray Attenuation Coefficients From 10 keV to 100 MeV. National Bureau of Standards, NBS Circular 583, 1957.

27. National Bureau of Standards. Measurement of Absorbed Dose of Neutrons and a Mixture of Neutrons and Gamma Rays. National Bureau of Standards Handbook 75, 1961.
28. W. L. Bendel. "Displacement and Ionization Fractions of Fast Neutron Kerma in TLDs and Silicon." IEEE Trans. Nucl. Sci. NS-24(6), Dec. 1977.
29. G. J. Brucker. Definition of Radiation Units and Conversion Factors For the Basic Types of Nuclear Radiation. RCA Document ST-6822, Aug. 1979.
30. Nuclear News, p. 75, May 1982.
31. V. V. Verbinski, S. P. Cassapakis, R. L. Pease, and H. L. Scott. Simultaneous Neutron Spectrum and Transistor Damage Measurements in Diverse Neutron Fields: Validity of $D(E_n)$. Naval Research Labs., Memorandum Report 3929, Washington D.C., 1979.
32. A. I. Namenson, E. A. Wolicki, and G. C. Messenger. "Average Silicon Neutron Displacement Kerma Factor at 1 MeV." IEEE Trans. Nucl. Sci. NS-29(1):1018–1020, Feb. 1982.
33. D. M. Long, D. G. Millward, and J. Wallace. "Dose Enhancement Effects in Semiconductor Devices." IEEE Trans. Nucl. Sci. NS-29(6), Dec. 1982.
34. J. C. Garth, B. W. Murray, and R. P. Dolan. "Soft X-Ray Induced Energy Deposition in a Three Layered System." IEEE Trans. Nucl. Sci. NS-29(6):1985–1991, Figs. 5a, 5b, Dec. 1982.
35. W. E. Loewe and E. Mendelsohn. "Neutron and Gamma Ray Doses at Hiroshima and Nagasaki," Nucl. Sci. Eng. 81:325–350, 1982.
36. J. C. Garth. "High Energy Extension of the Semi-Empirical Model For Energy Deposition at Interfaces." IEEE Trans. Nucl. Sci. NS-28(6):4145–4151, Dec. 1981.
37. W. L. Chadsey. "POEM—A Fast Monte Carlo Code For the Calculation or X-Ray Photoemission and Transition Zone Dose And Current." Air Force Cambridge Labs. AFCRL-TR-0323 (1975).
38. E. A. Burke and J. C. Garth. "An Algorithm For Energy Deposition at Interfaces." IEEE Trans. Nucl. Sci. NS-23(6):1838–1843, Dec. 1976.
39. J. C. Garth, W. L. Chadsey, and R. L. Shepard, Jr. "Monte Carlo Analysis of Dose Profiles Near Photon Irradiated Material Interfaces." IEEE Trans. Nucl. Sci. NS-22(6):2563–2567, Dec. 1975.
40. W. L. Chadsey. "X-Ray Produced Charge Deposition and Dose in Dielectrics Near Interfaces Including Space Charge Field and Conductivity Effects." IEEE Trans. Nucl. Sci. NS-21(6):235–242, Dec. 1974.
41. T. C. Zietlow. "Dose Enhancement Effects in a Shepard Model 81-22, Cobalt-60 Irradiator." IEEE Trans. Nucl. Sci. NS-34(3):662–666, June 1987.
42. S. M. Seltzer. "Conversion of Depth-Dose Distributions From Slab to Spherical Geometries For Space Shielding Applications." IEEE Trans. Nucl. Sci. NS-33(6): 1292–1297, Dec. 1986.
43. A. B. Chilton, J. K. Shultis, and R. E. Faw. Principles of Radiation Shielding. Prentice-Hall, Englewood Cliffs, NJ, 1984, pp. 35ff.
44. R. L. Pease, A. H. Johnston, and J. L. Azarewicz. "Radiation Testing of Semiconductor Devices For Space Electronics." Proc. IEEE, 76(11):1510–1526, Nov. 1988.
45. K. G. Kerris. "Source Considerations and Testing Techniques." *In* T. P. Ma and P. V. Dressendorfer (eds.). Ionizing Radiation Effects in MOS Devices and Circuits. John Wiley-Interscience, New York, Chap. 8. 1989.
46. G. J. Brucker, R. A. Cliff, V. Danchenko, R. S. Ohanian, M. Sing, and E. G. Stassinopoulos. "Prediction and Measurement of Radiation Damage to CMOS Devices On-Board Spacecraft." IEEE Trans. Nucl. Sci. NS-23(6):1781–1788, Dec. 1976.
47. E. G. Stassinopoulos et al. "Thermoluminescent Detector Performance in Low-Dose

Slow-Rate Environments." Appl. Rad. Isot. 39(4):303–309. Inst. J. Rad. Appl. Instrum. Part A, 1988.

48. V. Danchenko and G. F. Griffin. "Delayed Darkening of Radiation Exposed Radiochromic Dye Dosimeters." IEEE Trans. Nucl. Sci. NS-28(6):4156–4160, Dec. 1981.
49. S. E. Chapell and J. C. Humphreys. "The Dose-Rate Response of a Dye-Polychlorostyrene Film Dosimeter." IEEE Trans. Nucl. Sci. NS-19(6):175–180, Dec. 1972.
50. K. G. Kerris and S. G. Gorbics. "Experimental Determination of the Low-Energy Spectral Component of Cobalt-60 Sources." IEEE Trans. Nucl. Sci. NS-32(6): 4356–4362, Dec. 1985.
51. D. M. Long, D. G. Millward, R. L. Fitzwilson, and W. L. Chadsey. "Handbook For Dose Enhancement Effects in Electronic Devices." Rome Air Development Center, Air Force Sys. Comm. Mar. 1983.
52. J. R. Srour, D. M. Long, R. L. Fitzwilson, D. G. Millward, and W. L. Chadsey. Radiation Effects on and Dose Enhancement of Electronic Materials, Part II. Noyes Publ., Park Ridge, NJ, 1984.
53. U.S. Army Command and General Staff College (USACGSC), Ft. Leavenworth, KS, Subcourse Syllabus, Radiation Effects Section, 1986.

CHAPTER 11
GALLIUM ARSENIDE DEVICES

11.1 INTRODUCTION

In the 1960s decade, germanium discrete devices were rapidly being replaced by silicon counterparts, because of silicon's higher operating temperatures and in spite of the speed advantages inherent in germanium components. Proponents of GaAs discrete devices and integrated circuits assert that the speed and temperature advantage of gallium arsenide over silicon, in addition to its excellent hardness properties, will allow replacement of Si devices by those manufactured from GaAs in many applications.

Gallium arsenide is presently a niche technology under intense development, as evidenced by the large number of articles appearing in technical journals. Discrete gallium arsenide devices, such as varactor diodes and GaAs FET RF amplifiers, have been in use for more than a decade mainly in monolithic microwave integrated circuits (MMICs)[27] as well as in millimeter wave analog signal applications. This is in contrast to GaAs FET digital integrated circuits, memories, and microprocessors.[1,2]

In gallium arsenide, impurity defect surface state densities are $\sim 10^{15}$ per cm^2. The energy levels exist almost wholly in the GaAs forbidden band. This corresponds to about one impurity surface state per GaAs surface atom. Hence, the oxide surface heavily traps charged carriers (holes or electrons) to such an extent that the fabrication of GaAs MOSFET devices of functionally sufficient surface purity is, to date, essentially impossible. Therefore, GaAs FETs imply MESFETs, where the gate consists of a schottky metal-semiconductor junction, in contrast to a silicon MOSFET insulator gate. The macroscopic properties of gallium arsenide are those of a relatively pure material, with a high density of surface states, as mentioned earlier.

The anticipated attraction of the GaAs family of materials is that their microscopic solid-state properties facilitate low- and high-temperature operation and high speed devices. The reason for the high-speed property is that the GaAs electron mobility for the usual electric field levels within the device is about 8000 cm^2/volt-s, more than five times that of silicon, as seen in Table 1.1. GaAs is a direct band gap material, i.e., with opposing

Table 11.1 Comparison of GaAs and silicon MESFETs.[13] © 1970 by the IEEE.

REGION	FUNCTION	GaAs	SILICON
①	Schottky barrier	0.8 eV	0.7 eV
②	Barrier metal	Au/Pt/Ti	Al
③	Ohmic contact	AuGe/Ni	Al
④	Semi-insulating substrate	$>10^7$ ohm · cm	2000 ohm · cm
⑤	Surface dielectric	Si_3N_4	SiO_2
⑥	Interface depletion layer	$N_s > 10^{14}/cm^2$ pins E_{fs}	$N_s \sim 10^{10}/cm^2$ process dependent
⑦	Surface depletion layer	Fixed	Swings from accumulation to inversion
⑧	Junction depletion layer	Long lifetime traps	
	Ionizing dose ΔV_T shift	mV at 100 Mrad (GaAs)	Volts at 1 Mrad (Si)

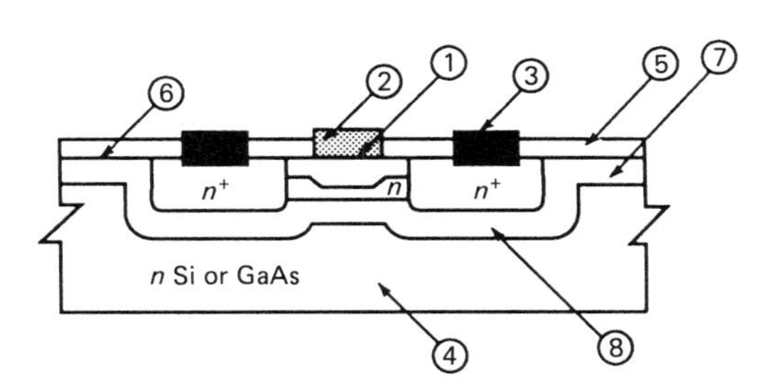

band gap extrema as discussed in Section 1.9 (Table 1.1). The GaAs effective electron mass m_l is relatively small ($\sim 0.06 m_e$).

Since $\mu_n \sim 1/m_l$, this is a principal factor causing the high GaAs mobility. Table 1.1 also gives GaAs hole mobilities, which are only about 5 percent those of electrons. Hence, only n-type GaAs devices are of

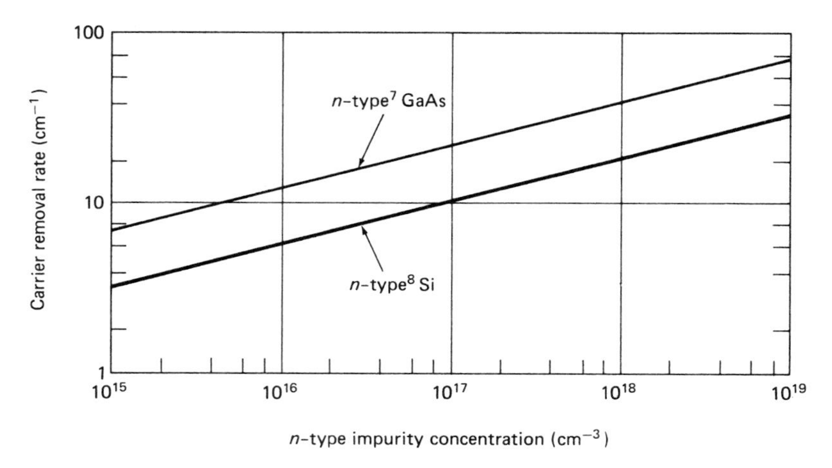

Fig. 11.1 Neutron induced carrier removal rates versus impurity concentration in n-type GaAs and Si materials (assuming $n_0 \cong N_D$ for abscissa).[7,8] © 1970 by the IEEE.

interest for high-speed application. Another important aspect is the plateau representing the operating temperature range in the plot of electron density versus temperature in Fig. 1.14, which is relatively wide for GaAs. This means that GaAs material can be operated stably over a wide temperature range—a feature that is important with respect to military operating temperature specifications for semiconductors. In addition, the intrinsic resistivity of the GaAs substrate is extremely high.

With respect to radiation hardness, it is immediately appreciated that, because FETs are majority carriers, they should suffer little from the vagaries of neutron damage, as discussed in Chapters 5 and 6. However, neutron displacement damage is relatively high in GaAs, as seen by the carrier removal rate (shown in Fig. 11.1). Also, due to the high g_m of GaAs devices, and their small size, f_T of 10–40 GHz is not uncommon. If GaAs bipolar VLSI integrated circuits could be manufactured in volume, they would be resistant to neutrons, as evidenced by the Messenger-Spratt gain relation $\Delta(1/\beta) = \Phi_n/K\omega_T$, as large $\omega_T = 2\pi f_T$ makes for small $\Delta(1/\beta)$, and so minimal gain change with respect to neutron fluence, Φ_n. The major obstacle to the function of discrete GaAs bipolar transistors is their relatively small minority carrier lifetime. As is easily shown, this implies unusably small common emitter current gain, commensurate with the anticipated GaAs element dimensions. The small minority carrier lifetime is one consequence of the fact that gallium arsenide is a direct band gap semiconductor.

However, the GaAs qualities that imply high-speed operation, as seen in GHz discrete GaAs FET stripline amplifiers for example, are partially vitiated in GaAs digital microcircuits. One reason is that the high device packing density of the latter results in the appearance of unwieldy parasitics, such as those in metallization resistances and parasitic capacitances.

Besides the development of semi-insulating GaAs MESFETs and JFETs, heterojunction (merely a junction formed by two different semiconductor materials) injection logic (H-I^2L) and heterojunction bipolar transistor (HBT) circuits are being used to fabricate digital integrated circuits. The heterojunction devices promise even higher speed than GaAs homojunction FETs. All of these FETs are currently in their first developmental phases.

In order to maintain GaAs material semi-insulating, meaning a resistivity sufficiently high (10^7–10^8 ohm cm) for integrated circuit cell isolation, it can be lightly doped with Cr to provide acceptor defect traps. These compensate the inadvertent oxygen donor defect traps introduced by modern (Horizontal Bridgman) fabrication techniques for growing the GaAs crystal. By compensation, it is meant that the desired net dopant concentration, $|N_D - N_A|$, as well as the desired conductivity, be maintained in the GaAs crystal, as described. Similarly, crystals grown by liquid encapsulated Czochralski(LEC) synthesis also possess inadvertent *p*-type dopants, so that a sufficiently high proportion of As must be present during fabrication to maintain a semi-insulating crystalline structure. This high proportion of arsenic ensures the formation of a deep donor (EL2) level defect trap, which plays the same role as chromium above, but here is used to compensate shallow acceptors[4,13] such as carbon.

The following provides a description of the radiation effects in GaAs devices from neutron fluence, ionizing dose, dose rate, and single event upset (SEU), respectively.[5]

11.2 NEUTRON DISPLACEMENT DAMAGE

Neutron-induced displacement damage in GaAs FETs is essentially the same in kind as that in silicon devices. Incident neutrons generate defects in the semiconductor lattice atoms, as discussed in Sections 5.5 and 5.6. The defects introduce defect energy states in the band gap, which can cause minority carrier recombination, thus reducing the minority carrier lifetime, as discussed in Section 5.9. This is inimical to the operation of bipolar transistors, but has a negligible effect on the operation of majority carrier devices, such as MOSFETs, MESFETs, and JFETs. Displacement effects also introduce additional trapping centers, which cause a decrease in the net dopant concentration. In GaAs, it is found that the carrier concentration is reduced in both n- and p-type material,[6] so that it becomes more intrinsic as a result of the neutron damage, as discussed in Section 5.6. Furthermore, both n and p mobilities, μ_n and μ_p, are reduced due to the additional scattering from the presence of the defect states.

As in silicon, two major types of defects are introduced by penetrating neutrons. They are (a) point defects, such as vacancy-interstitial pairs (Frenkel defects), discussed in Sections 5.6 and 5.13; and (b) defect cluster "spikes" in GaAs.[5] Defect damage is evident in a number of ways. One degradation mechanism is the capture of carriers by the defect states, resulting in carrier removal (Section 5.13). Figure 11.1 depicts the carrier removal "rate" in both silicon and gallium arsenide versus impurity concentration in n-type material. It is noted from the figure that the carrier removal rate in Si is roughly half that in GaAs.

In the operation of a MESFET or JFET in the saturation region,[9] as depicted in Fig. 6.4, there are four principal electrical parameters affected by damage due to incident neutrons. They are the pinch-off voltage, V_p; the maximum channel current, $I_{D\max}$; the maximum transconductance, $g_{\max}$; and the cutoff frequency, f_c. Their individual dependencies will be given in turn.

Using the one-dimension portion of the coordinate system given in Fig. 6.4, the one-dimensional Poisson equation for a uniformly doped semiconductor n-region, with depletion layer width h varying along the channel (permittivity ε_s) is given by

$$-V''(x) = E_s' = \rho(x)/\varepsilon_s = eN_D/\varepsilon_s \tag{11.1}$$

From the solutions of Eq. (11.1), the depletion layer widths can be obtained as a function of the pertinent electrical parameters.[9] The depletion layer width h at a distance x from the source, as in Fig. 6.4, is given by the abrupt junction expression derived in Section 3.5 as

$$h(x) = [2\varepsilon_s(V(x) + V_G + V_{bi})/eN_D]^{1/2} \tag{11.2}$$

where:

$V(x)$ is the applied drain voltage at x with respect to the source.
V_{bi} is the built-in (contact) potential of the p^+n junctions.
V_G is the applied gate voltage.

The depletion layer widths at the source and drain ends of the channel are, respectively,

$$h(0) = [2\varepsilon_s(V_G + V_{bi})/eN_D]^{1/2} \qquad (11.3A)$$

$$h(L) = [2\varepsilon_s(V_G + V_{bi} + V_D)/eN_D]^{1/2} \qquad (11.3B)$$

The definition of a pinched-off channel in this context is when $h(L) = a$, that is, the two depletion layers become contiguous at the drain end of the channel. The corresponding voltage is called the pinch-off voltage, V_p, and from Eq. 11.3B satisfies $h(L) = a = (2\varepsilon_s V_p/eN_D)^{1/2}$, where $V_p = V_G + V_{bi} + V_D$, or

$$V_p = eN_D a^2/2\varepsilon_s \qquad (11.4)$$

The current density along the channel is $J_x = \sigma(x)E_x \cong e\mu N_D E_x$, where the field along the channel is $E_x = -V'(x)$. The drain current I_D for the upper half-channel is

$$I_D(x) = e\mu N_D V'(x)(a - h)Z \qquad (11.5)$$

where $(a - h)Z$ is the area normal to current flow at x in Fig. 6.4. Or, with $V'(x) = V'(h)dh/dx$,

$$I_D(h)dx = e\mu N_D(a - h)ZV'(h)dh \qquad (11.6)$$

With $V'(h) = eN_D h/\varepsilon_s$ from Eq. (11.2), integrating Eq. (11.6) yields

$$\int_0^L I_D(h)dx = I_D \cdot L = (e^2\mu N_D^2 Z/\varepsilon_s)\int_{h(0)}^{h(L)} (a - h')h'dh' \qquad (11.7)$$

The maximum channel current, $I_{D\max} = I_{D\text{sat}}$, occurs where the channel is pinched off, i.e., at $h(L) = a$. With a for the upper limit in Eq. (11.7), and integrating using Eqs. (11.3) results in

$$I_{D\max} \equiv I_{D\text{sat}}$$
$$= (e^2\mu N_D^2 Za^3/6\varepsilon_s L)[1 - 3((V_G + V_{bi})/V_p) + 2((V_G + V_{bi})/V_p)^{3/2}] \qquad (11.8)$$

which is also depicted in Fig. 6.4.

The transconductance, $g_m = \partial I_D/\partial V_G$, can be obtained from carrying out the integration in Eq. (11.7), giving

$$g_m = (2Z\mu e N_D/L)[h(L) - h(0)] \tag{11.9}$$

The maximum transconductance, g_{max}, occurs in the linear region in the limit of zero V_D, or maximum slope in Fig. 6.4, which yields, from Eq. (11.9),

$$g_{max} = eN_D a\mu Z/L \tag{11.10}$$

By definition, the cutoff frequency is obtained from

$$f_c \cong g_m/2\pi C_G = \mu V_p/2\pi L^2 \tag{11.11}$$

where the gate capacitance $C_G = 2\varepsilon_s LZ/a$.

The effects of neutron fluence, Φ_n, on V_p (Eq. 11.4), $I_{D\max}$ (Eq. 11.8), g_{max} (Eq. 11.10), and f_c (Eq. 11.11) are neutron-induced changes in carrier removal and mobility within the semiconductor. For carrier removal, reiterating from Section 5.13, expanding the carrier density about $\Phi_n = 0$ to two terms yields

$$n(\Phi_n) \cong n_0 + n'(0)\Phi_n = n_0(1 - \hat{a}\Phi_n) \tag{11.12}$$

where the normalized carrier removal can be defined as $\hat{a} = -n'(0)/n_0$. Empirically,[7] $\hat{a} \cong 7.2 \cdot 10^{-4} N_D^{-0.77}$. Similarly, the neutron-induced mobility

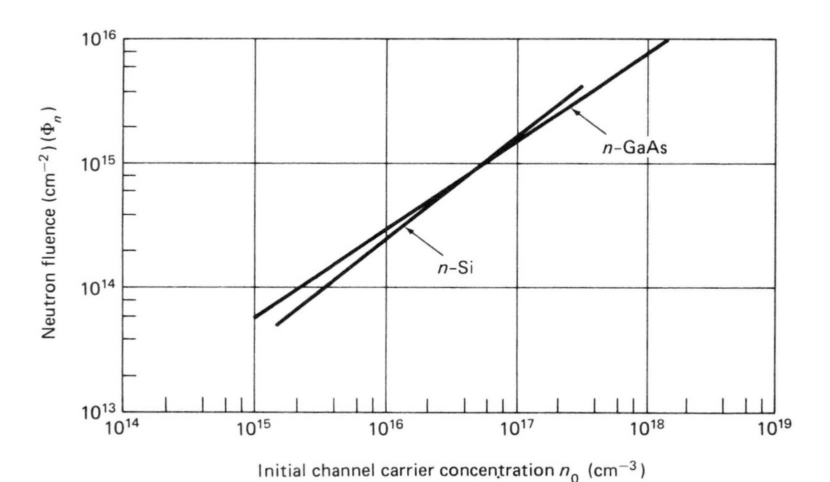

Fig. 11.2 Neutron fluence required to reduce g_m by 20 percent for n-type GaAs and Si material versus preradiation carrier concentration.[10] © 1967 by the IEEE.

change is given by $\mu^{-1} = \mu_0^{-1}(1 + b_\mu \Phi_n)$, where b_μ is an empirically determined mobility damage constant[7] (viz., $b_\mu = 7.8 \cdot 10^{-6} N_D^{-0.64}$). Letting the subscript zero denote preradiation values, the four parameters can now be written in terms of their preradiation levels.[7] Namely, from Eqs. (11.4) and (11.12), and $N_D \simeq n$, $N_{D0} \simeq n_0$,

$$V_p/V_{p0} = N_D/N_{D0} \cong n/n_0 = 1 - \hat{a}\Phi_n \approx \exp - \hat{a}\Phi_n \qquad (11.13)$$

Similarly, from Eqs. (11.8) and (11.12),

$$I_{D\max}/I_{D\max 0} \cong n^2\mu/n_0^2\mu_0 = (1 - \hat{a}\Phi_n)^2/(1 + b_\mu\Phi_n) \approx \exp - (2\hat{a} + b_\mu)\Phi_n \qquad (11.14)$$

From Eqs. (11.10) and (11.12),

$$g_{\max}/g_{\max 0} \cong \mu n/\mu_0 n_0 = (1 - \hat{a}\Phi_n)/(1 + b_\mu\Phi_n) \approx \exp - (\hat{a} + b\mu)\Phi_n \qquad (11.15)$$

and

$$f_c/f_{c0} = \mu V_p/\mu_0 V_{p0} = (1 - \hat{a}\Phi_n)/(1 + b_\mu\Phi_n) \approx \exp - (\hat{a} + b\mu)\Phi_n \qquad (11.16)$$

where the exponentials remain a sizeable fraction in this approximation, which implies that Φ_n is not extremely large. These expressions are also valid for Si with suitable parameter changes.

To obtain an appreciation of the preceding expressions, and to compare them with their silicon counterparts, Fig. 11.2 gives a comparison of $g_{\max}$ for n-type GaAs and Si JFETs. This figure depicts the neutron fluence versus the initial carrier concentration, n_0, required to cause a 20 percent degradation in $g_{\max}$ for JFETs manufactured from both materials.[10] It indicates that, roughly, JFETs of either material fare equally well in this regard. Linear decreases in g_m as a function of neutron fluence Φ_n are found for ion-implanted process JFETs.[13]

Channel hot-electron hardness of MOSFETs has been improved significantly by the implantation of fluorine (e.g., by NF_3 gas) onto the surface of polycrystalline silicon gate electrodes. The improvement can be attributed to the strain relaxation near the Si/SiO_2 interface obtained by the incorporation of fluorine.[29]

Displacement damage in both GaAs and in Si is assumed to be due to incident neutrons as in the preceding. Such damage can also be introduced into the semiconductor by protons and sufficiently energetic electrons, such as those in the Van Allen belts. Another possible source is gamma rays, such as those from a ^{60}Co source.[12]

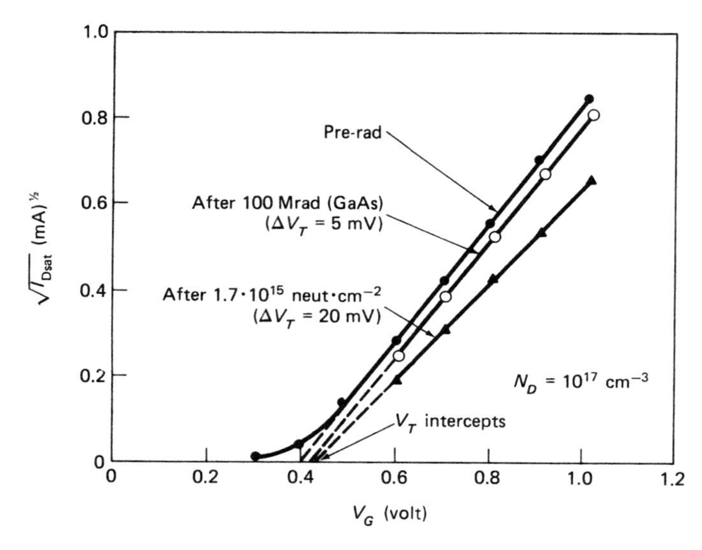

Fig. 11.3 Square root of saturation current versus gate voltage for an enhancement mode GaAs JFET before and after ^{60}Co and 1 MeV equivalent neutron radiation.[13] © 1970 by the IEEE.

11.3 IONIZING RADIATION DAMAGE

GaAs devices as compared to their silicon counterparts are relatively hard to ionizing dose damage. GaAs FETs are MESFETs since the gate-semiconductor interface is a metal schottky barrier junction, (Fig. 11.3A) or JFETs since their gate-semiconductor interface is a *pn* junction. Since there is no gate insulator-oxide in a GaAs FET, little threshold voltage shift, ΔV_T, is encountered from trapped charges. This is as opposed to MOSFETs, where ΔV_T shift is a prime manifestation of ionizing dose damage due to trapped gate oxide charges, as discussed in Section 6.11.

The second reason for GaAs MESFET ionizing dose hardness is that, due to the surface properties of GaAs material today as evidenced by the density of surface defect states, the surface fermi level E_{fs} is pinned at a constant value near the center of the GaAs band gap, as will be discussed.

This imbues GaAs structures with a built-in stability against trapped charge effects produced by ionizing radiation. GaAs devices can withstand ionizing dose levels of more than 10 Mrad (GaAs) with but minor changes in device parameters.[13] This is seen in both Table 11.1, where a comparison with silicon device technology is made,[13] and Fig. 11.3, which depicts very small ΔV_T shifts for both fast (nuclear reactor) neutrons

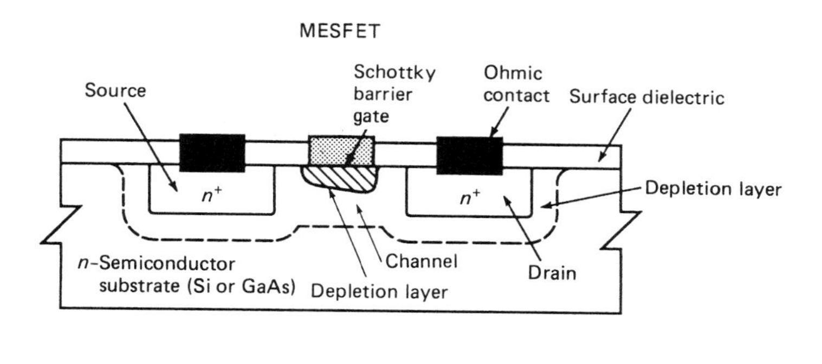

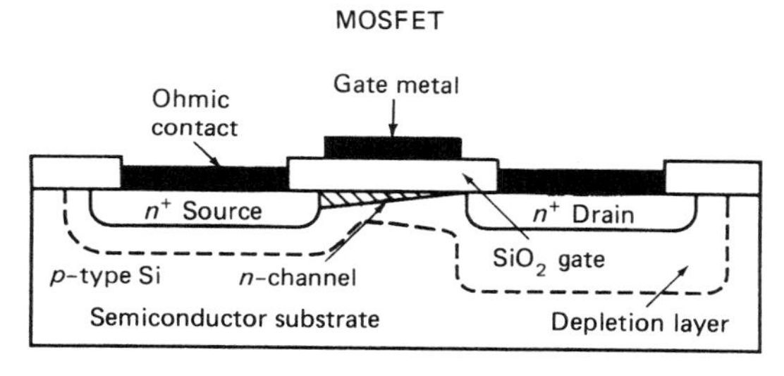

Fig. 11.3A MESFET and MOSFET structures.

and ^{60}Co gamma ray ionizing radiation incident on the GaAs FET.[13] However, above 10 Mrad (GaAs) there is a noticeable degradation of noise figure and radio frequency gain in GaAs FET microwave amplifiers.[14] The deleterious effects of ionizing radiation on these parameters can often be circumvented by sophisticated circuit design and judicious device choice.

In the GaAs MESFET, the electrical stability of the GaAs doped semiconductor is maintained by the very impurity of the material, but these impure surface characteristics augur against manufacture of MOSFET VLSI systems, as discussed above. The impurity stability is manifest by the very high surface and interface defect state densities. This is because the high areal density of defect interface states ($N_{it} = \int_{E_0}^{E_{fs}} D_{it}(E)dE$), forces (pins) the surface fermi level to a constant value, virtually irrespective of external excitation. This, in turn, pins the other electrical parameters such as barrier height and turn-on voltage V_T, thus providing a substantial measure of device hardness. This is seen in the following development.

The potential energy barrier height voltage of the metal-semiconductor system is determined by the metal work function E_m (energy necessary to

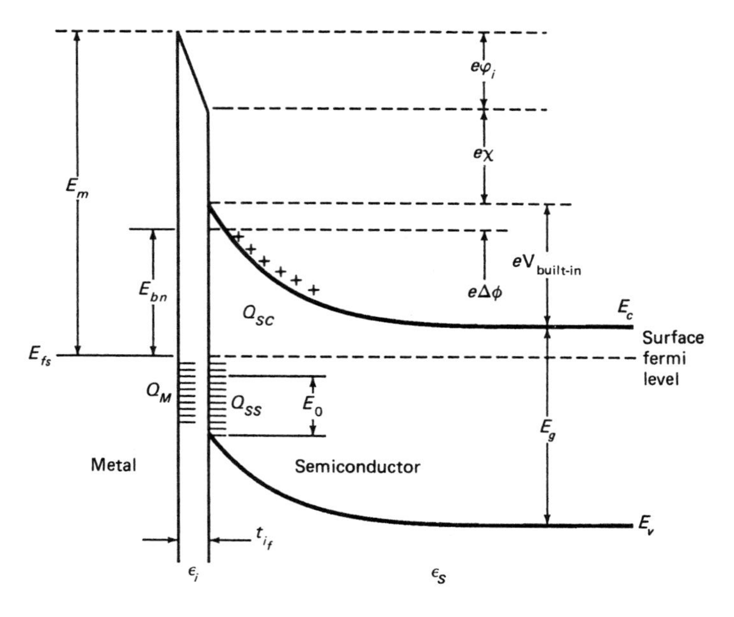

Fig. 11.3B Energy band diagram of metal *n*-type semiconductor schottky barrier junction.[9]

extract an electron through the material surface to the outside) as well as the (per energy) surface state density, D_s, which includes the interface state density per unit energy interval, D_{it}. The expression for the barrier height[34] can be obtained by assuming (a) intimate contact between metal and semiconductor, with at most a few atoms-thick interface layer, which is transparent to electrons, yet can sustain a potential, and (b) D_s depends only on the properties of the semiconductor, and not on the metal.

From Sze,[34] consider a schottky barrier with acceptor state density per energy interval D_s, which possesses an assumed constant energy spectrum spanning from the surface energy E_0 to the fermi level $E_f \geqslant E_0$, as seen in Fig. 11.3B. The corresponding surface charge density, Q_{ss}, is given by parameters depicted in that figure, resulting in

$$Q_{ss} = eD_s(E_g - E_{bn} - e\Delta\phi - E_0) = eD_s(E_{fs} - E_0) \text{ (coulombs/cm}^2\text{)} \tag{11.17}$$

where:

E_{fs} is the fermi energy level at the interface surface (i.e., $E_{fs} = E_g - E_{bn} - e\Delta\phi$).

E_g is the gap width energy.
E_0 is the surface energy; $E_0 = E_{fs}$ before the metal-semiconductor contact was formed.
E_{bn} is the metal-semiconductor barrier height energy.

$e\Delta\phi$ is the lowering of the barrier height energy due to applied bias, and work done against the attraction of the image charge in the metal.[48] Hence $D_s(E_{fs} - E_0)$ is the number of interface and surface states above E_0 that are occupied. $[D_s]$ = the number of surface states per $cm^2 \cdot eV$.

The corresponding space charge in the semiconductor depletion layer, Q_{sc}, is given from Section 3.5 by

$$Q_{sc} = [2eN(E_{bn} + E_{fs} - E_c + e\Delta\phi)]^{1/2} \tag{11.18}$$

Now, Q_M is the equal and opposite induced charge developed on the metal surface

$$Q_M = -(Q_{ss} + Q_{sc}) \tag{11.19}$$

The potential across the interface layer, φ_i, is written simply from Gauss's Law as

$$\varphi_i = -Q_M/(\varepsilon_i/t_{if}) \tag{11.20}$$

where ε_i and t_{if} are the permittivity and thickness of the interface layer, respectively. By studying Fig. 11.3B, it is seen that another expression for φ_i is

$$e\varphi_i = E_m - (e\chi + E_{bn} + e\Delta\Phi) \tag{11.21}$$

By eliminating φ_i between Eq. (11.20) and (11.21), and using Eq. (11.19) to eliminate Q_M, yields

$$\begin{aligned} E_m - e\chi - E_{bn} - e\Delta\phi = {} & [(2e\varepsilon N t_{if}^2/e\varepsilon_i^2)(E_{bn} + e\Delta\phi + E_{fs} - E_c)]^{1/2} \\ & - (e t_{if} D_s/\varepsilon_i)(E_g - E_0 - E_{bn} - e\Delta\phi) \end{aligned} \tag{11.22}$$

The electron affinity, $e\chi$, is defined as the work required to extract an electron from the bottom of the conduction band to "infinity". By inserting numerical values into Eq. (11.22), certain quantities are dwarfed by others. Retaining only the latter quantities allows a reduction in Eq.

(11.22) to yield an expression for the fermi level in the semiconductor surface[34]

$$E_{fs} \cong E_0 - (E_m - e\chi - E_g + E_0)/(1 + e^2 t_{if} D_s/\varepsilon_i) \qquad (11.23)$$

In the limit of very large surface state density D_s, it is seen that the second term vanishes, yielding $E_{fs} \cong E_0$ from Eq. (11.23), so that E_{fs} is spoken of as being "pinned" at the surface energy E_0. For n and p type GaAs, $E_0 \cong 0.75E_g$ and about $0.50E_g$, respectively, where E_g is the band gap width in GaAs.[52]

Pinning the surface fermi level also pins the schottky barrier height, and thus the device electrical characteristics, as already mentioned. The barrier height is a determinant of the threshold "turn-on" voltage V_T for the MESFET. However, the major effect of pinning the fermi level at E_0, because of material surface impurities, is that the maximum value of the fermi distribution, max $f(E_0) \simeq 1.0$ essentially abuts E_0. Similar displacements of the fermi level are discussed in Section 1.6. Since max $f(E_0)$, the occupancy probability of states at the surface energy, is of the order unity, almost all of these states are then occupied by electrons. Upon being subjected to ionizing radiation, the now-filled semiconductor surface or interface states can no longer trap (the radiation-produced) charge, thus rendering this material impervious to damaging effects, as discussed.

It should be noted that occupied surface/interface charge states can act very much like an electrostatic shield.[51] This tends to shield the gate from the remainder of the device to prevent a channel from forming. It is asserted that this shielding property accounted for the very early disappointing results associated with surface transistor experiments.[51]

To discuss ionizing radiation-induced damage in terms of the saturated drain current $I_{Dsat} \equiv I_{Dmax}$, it should be appreciated that GaAs FETs of interest are n-channel enhancement normally-off devices. Normally-off implies that the channel is biased off by only the built-in (contact) potential, V_{bi}, of the gate junction. This is sufficient to totally deplete the channel at $V_G = 0$. For $V_G > V_T > 0$, channel current will begin to flow, and at that point, $V_T = V_{bi} - V_p$, or $V_{bi} = V_T + V_p$. Near the V_T threshold, an expression for the drain current in the saturation region, I_{Dsat}, can be gotten by inserting the above V_{bi} into Eq. (11.8) and expanding the resultant I_{Dsat} about $V_G = V_T$ to three terms. The constant and linear terms in this expansion vanish, leaving a current-voltage square law relationship

$$I_{Dsat} \cong (\mu\varepsilon_s Z/2aL)(V_G - V_T)^2 \qquad (11.24)$$

$I_{Dsat}^{1/2}$ versus V_G is plotted in Fig. 11.3, which reveals a linear dependence for the "current-voltage characteristic," except for small values of saturated drain current. This lends further credence to the expression for I_{Dsat} given in Eq. (11.24).

Photographs of oscilloscope-curve tracer equipment yield[15] voltage-current characteristics of a planar, ion implanted GaAs E-FET (enhancement mode FET), taken before and after its exposure to 10 Mrad (GaAs) ionizing radiation. When the photos are superimposed on each other, no difference is evident. Therefore, any difference must be less than the resolution afforded by the scope trace width, which is equivalent to $\Delta I_{Dsat} \leqslant 50\mu A$.[15,16] From the derivative of I_{Dsat} with respect to V_T in Eq. (11.24) the corresponding ΔV_T is given by

$$\Delta V_T = |\Delta I_{Dsat}|/2(I_p I_{Dsat})^{1/2}; \qquad I_p = \mu \varepsilon_s Z/2aL \text{ amps/(volt)}^2 \tag{11.25}$$

Using parameter values given in Section 10.4, the computed $I_p = 500\,\mu A/V^2$. For a nominal $I_{Dsat} = 500\,\mu A$, and $\Delta I_{Dsat} = 50\,\mu A$, Eq. (11.25) gives $\Delta V_T \cong 50 mV$. This is in approximate agreement with the pre- and post-10^8 rad(GaAs) V_T intercept differences seen in Fig. 11.3. This relatively small ΔV_T (millivolts) can be compared to the relatively large ΔV_T (volts) for an enhancement mode Si MOSFET in Fig. 6.11, and where the ionizing radiation levels are less by one or two orders of magnitude than the GaAs devices.

11.4 IONIZING DOSE RATE DAMAGE

Ionizing radiation dose rate produces a number of damaging effects in GaAs FETs. Very high dose rate levels can result in permanent damage to GaAs material, as well as silicon. Dose rate is usually associated with radiation emanating from pulsed photon (gamma, x ray) sources such as nuclear weapons bursts, although constant dose rate sources such as nuclear reactors are of importance as well. Such photons induce voltages that cause spurious photocurrents to flow in GaAs and other materials, as discussed in Sections 7.2–7.4.

The response of GaAs FET circuits to dose rate is due to the photocurrents generated in the junctions, barriers, and semi-insulating substrates. The principal kinds of such currents are called primary, secondary, and substrate photocurrents, with the latter predominant in GaAs devices. As seen in Chapter 7, photocurrent-induced upset is a function of circuit design parameters, as well as of semiconductor material properties of the associated active devices, whether silicon or gallium arsenide.

It is interesting to note that a relatively long term low-ionizing dose transient photocurrent response exists in some GaAs FETs and in some FET logic.[25] When these devices have been exposed to a dose rate of about 10^{10} rads (Si) per s at room temperature, corresponding to a mere 100 rads (Si) ionizing dose, the measured response recovery times range from 1 to 70 seconds from pulse onset.[25] These transients are attributed to modulation of the FET channel region by charges trapped in deep levels in the GaAs substrate. Deep-level transient spectroscopy (DLTS) techniques have identified an acceptor level (electron) trap, characterized by a threshold activation energy between 0.73 and 0.80 eV, with a capture cross section of about $10^{-14}\,\text{cm}^2$.[25,26] Some advances in mitigating this effect have been made.[30]

The production of photocurrents is discussed in Section 7.2, and the processes are essentially identical, whether in silicon or gallium arsenide. A semi-empirical primary photocurrent expression for GaAs, analogous to Eq. (7.11), is[13]

$$I_{pp}(t) = eg_0\dot{\gamma}A(W + L_n + L_p) = eg_0\dot{\gamma}WL_g(a + L_D) \qquad (11.26)$$

where W is the channel depletion region width, L_n and L_p are the respective diffusion lengths, and $A = WL_g$ is the gate area, where L_g is the gate length. For GaAs structures, $(W + L_n + L_p)$ is replaced by $(a + L_D)$, where a is the channel height and L_D is an empirically derived carrier collection length.[13] For GaAs FETs, $L_D \cong 5\,\mu\text{m}$. The photocurrent generation constant $g_0 = 6.92 \cdot 10^{13}$ hep (hole-electron pairs) per $\text{cm}^3 \cdot$ rad (GaAs).

Also, from the discussion in Section 7.2, the photocurrent due to a rectangular photon pulse of width t_p seconds and magnitude $\dot{\gamma}$ is[13]

$$I_{pp}(t) = eg_0\dot{\gamma}AL_c \cdot \begin{cases} erf(t/\tau_0)^{1/2}; & 0 \leqslant t \leqslant t_p \\ erf(t/\tau_0)^{1/2} - erf[(t - t_p)/\tau_0]^{1/2}; & t > t_p \end{cases} \qquad (11.27)$$

where τ_0 is the minority carrier lifetime in the n^+ region adjacent to the channel, as shown in Fig. 11.5, and L_c is the corresponding carrier collection length. For typical GaAs FETs whose anticipated use is in digital microcircuits, $L_g = 1\,\mu\text{m}$ (short dimension), $W = 10\,\mu\text{m}$ (long dimension), and $a + L_D \cong 5\,\mu\text{m}$. Then, for $\dot{\gamma} = 10^{10}$ rad (GaAs) per second, Eq. (11.26) yields $I_{pp} = 5.3\,\mu\text{A}$.

As discussed in Section 7.3, the secondary photocurrent can be construed as the amplified primary photocurrent in a bipolar transistor. In a FET, an amplification factor μ_{FET}, analogous to the common emitter

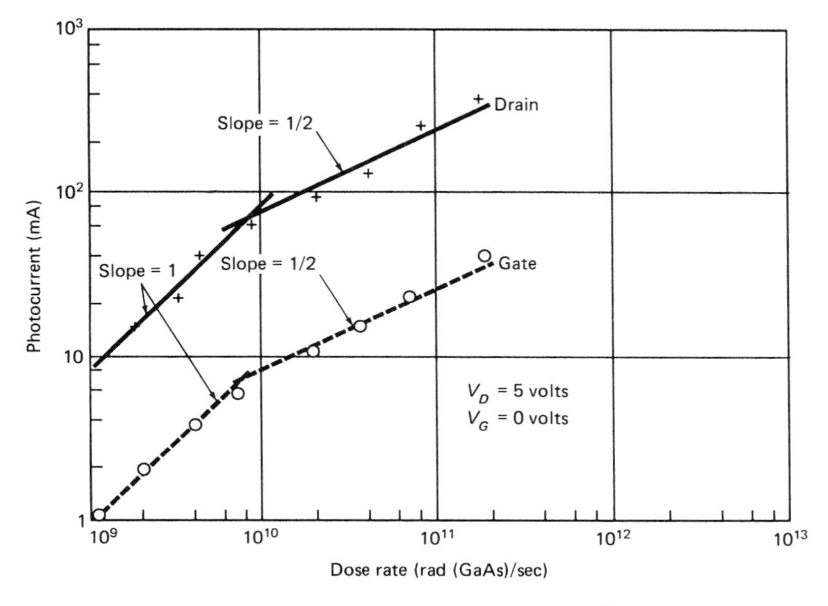

Fig. 11.4 GaAs JFET photocurrent versus incident dose rate.[17] © 1973 by the IEEE.

current gain h_{FE} in a bipolar transistor, is given by $\mu_{FET} = g_m R_G$, where R_G is an equivalent Thevénin output resistance that is reflected back to the gate input. So that the secondary photocurrent I_{sp} can be written as

$$I_{sp} = g_m R_G I_{pp} \tag{11.28}$$

For GaAs FETs, typical values of $g_m = 10^{-3}$ siemens (mhos), and $R_G = 10$ kilo ohms. Using the preceding I_{pp} value yields an $I_{sp} = 10^{-3} \cdot 10^4 \cdot 5.3 \cdot 10^{-6} = 53\,\mu\text{A}$, a factor of ten larger than the primary photocurrent. It should be realized that devices used in digital microcircuits can amplify photocurrents as well.

Experimental results[17,23] involving the $\dot{\gamma}$ pulse-induced photocurrents in GaAs, as a function of the dose rate level, reveal a linear photocurrent dependence for relatively low dose rate, with a transition to an approximate square root dependence for relatively high dose rates. The latter is due to the fact that, for high dose rates, direct recombination is proportional to the square of the excess carrier density, as discussed in Section 1.12.

Figure 11.4 shows plots of experimentally obtained photocurrent data[17] versus dose rate for GaAs JFETs, where this photocurrent dependence is depicted. It is seen in the figure that the drain photocurrent is larger than that of the gate. This is simply because the drain carrier collection volume is the larger. This system can be modeled using a straightforward variant of the simplified kinetics equation for the excess carrier density,[18,23] $\Delta n = \Delta p = \Delta$, resulting from a photocurrent rectangular pulse of width t_p. It is, with $\dot{\gamma}$ constant,

$$\dot{\Delta}(t) = g_0\dot{\gamma} - \Delta/\tau - r[(n_0 + \Delta)(p_0 + \Delta) - n_i^2]; \qquad \Delta(0) = 0 \qquad (11.29)$$

The initial rightmost term in the above equation is the excess carrier production rate. The second term is the loss rate due to all processes other than recombination, and the third is the net loss rate due to recombination. τ is the corresponding carrier lifetime. g_0 is the carrier generation constant, and r is the recombination rate coefficient. For n-type GaAs material, where $p_0 \ll n_0$, then Eq. (11.29) becomes

$$\dot{\Delta}(t) = g_0\dot{\gamma} - \Delta[1/\tau + r(n_0 + \Delta)]; \qquad \Delta(0) = 0 \qquad (11.30)$$

(a) For low injection ($\Delta \ll n_0$), i.e. where $\dot{\gamma}_L$ is relatively small, Eq. (11.30) gives

$$\dot{\Delta}_L + (1/\tau + rn_0)\Delta_L = g_0\dot{\gamma}_L; \qquad \Delta_L(0) = 0 \qquad (11.31)$$

with an integral given by

$$\Delta_L(t) = [\tau g_0\dot{\gamma}_L/(1 + rn_0\tau)]\cdot\begin{cases} 1 - \exp - (1 + rn_0\tau)t/\tau; & 0 \leqslant t \leqslant t_p \\ \{\exp[(1 + rn_0\tau)t_p/\tau] - 1\}\cdot \\ \quad \exp - (1 + rn_0\tau)t/\tau; & t \geqslant t_p \end{cases} \qquad (11.32)$$

(b) For high injection ($\Delta \gg n_0$), where $\dot{\gamma}_H$ is very large, Eq. (11.29) becomes

$$\dot{\Delta}_H + r\Delta_H^2 = g_0\dot{\gamma}_H; \qquad \Delta_H(0) = 0 \qquad (11.33)$$

with a corresponding integral given by

$$\Delta_H(t) = (g_0\dot{\gamma}_H/r)^{1/2}\cdot\begin{cases} \tanh[(g_0\dot{\gamma}_H r)^{1/2}t]; & 0 \leqslant t \leqslant t_p \\ \{\tanh[(g_0\dot{\gamma}_H r)^{1/2}t_p]\}/[1 + (t - t_p)(g_0\dot{\gamma}_H/r)^{1/2}\cdot \\ \{\tanh[(g_0\dot{\gamma}_H r)^{1/2}t_p]\}]; & t \geqslant t_p \end{cases} \qquad (11.34)$$

It is evident from Eq. (11.32) that for low injection, irrespective of its temporal aspects, the excess carrier density Δ_L and so the photocurrent, is proportional to the dose rate $\dot{\gamma}_L$ during the pulse epoch. Similarly, for high injection, from Eq. (11.34), the excess carrier density Δ_H becomes proportional to the square root of the dose rate $\dot{\gamma}_H$. Strictly for high injection, $\dot{\gamma}_H$ is sufficiently large so that $\tanh[(g_0\dot{\gamma}_H r)^{1/2}t] \sim 1$, which holds except for an extremely small initial time period following the onset of the dose rate pulse. This can be seen from the pertinent parameter values,[17] viz., $g_0 = 6.92 \cdot 10^{13}$ hep per $\text{cm}^3 \cdot$ rad (GaAs), $r = 3.3 \cdot 10^{-11}\,\text{cm}^3/\text{s}$, $\tau = 8 \cdot 10^{-7}\,\text{s}$, and $\dot{\gamma}_H = 10^{13}$ rad (GaAs) per s. Hence, this model delineates the dependence of the photocurrent as a function of the dose rate as obtained from experiment, and depicted in Fig. 11.4.

With respect to the experiment, the detailed temporal structure of the radiation pulse shape is essentially reproduced by the measured photocurrent pulse shape to within the experimental resolution time of about 2.5 ns.[17] This indicates that the characteristic reponse time of these GaAs FETs is such that they equilibriate rapidly within the incident dose rate pulse time of the source (flash x-ray accelerator), namely, $t_p = 20$ ns fwhm (full width-half maximum) of the pulse.

That the substrate photocurrent in GaAs FETS is the predominant one has been established,[19] and is shown in the following paragraphs. Substrate photocurrents are much larger than I_{pp} or I_{sp} in virtually all devices fabricated on semi-insulating GaAs substrates.[13] In the following, it is assumed that during the pulse epoch, equilibrium is established rapidly, so that there is a balance between electron-hole pair generation and recombination, that local charge neutrality exists, and that recombination is linear with the local carrier concentration. These assumptions yield a very simple model, such that $\Delta n \cong \Delta p \cong g_0\dot{\gamma}t_p$. The radiation-induced conductivity is then, from Eq. (1.56),

$$\Delta\sigma_{\text{sub}} = e\mu\Delta n = eg_0\dot{\gamma}t_p(\mu_n + \mu_p) = eg_0\gamma_T(\mu_n + \mu_p) \qquad (11.35)$$

For $\dot{\gamma} = 10^{11}$ rad (GaAs)/s, and $t_p = 5$ ns, $\gamma_T = \dot{\gamma}\text{t}_\text{p} = 500$ rad (GaAs) results in $\Delta\sigma_{\text{sub}} = 47.5$ mhos. The preradiation conductivity of the GaAs semi-insulating substrate of $N_D = 10^{17}\,\text{cm}^{-3}$ is given by

$$\sigma_{\text{pre}} = en\mu_n \cong eN_D\mu_n = 1.6 \cdot 10^{-19} \cdot 10^{17} \cdot 9 \cdot 10^3 = 144\,\text{mhos} \qquad (11.36)$$

It is seen that this level of incident radiation increases the conductivity by 33 percent. In otherwords, the incident radiation provides $\Delta n = g_0\dot{\gamma}t_p =$

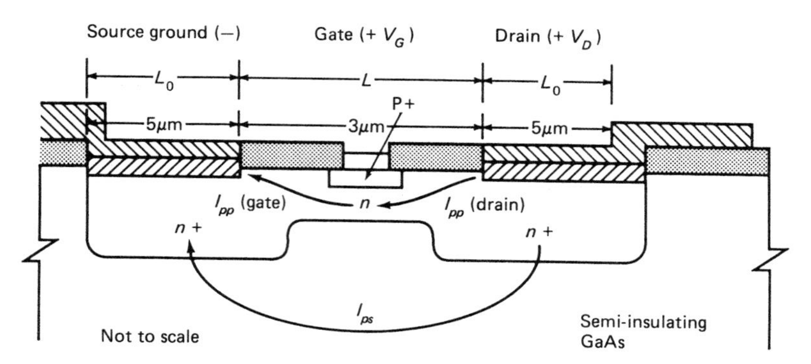

Fig. 11.5 GaAs enhancement mode (E-JFET) cross section with I_{pp} and I_{ps} as shown.[19] © 1983 by the IEEE.

$6.6 \cdot 10^{13} \cdot 10^{11} \cdot 2 \cdot 10^{-9}$ or $3.3 \cdot 10^{16}$ carriers per cm^3, which is an additional one-third of the total preradiation level of $n \cong N_D = 10^{17}\,\text{cm}^{-3}$.

To obtain an expression for the substrate photocurrent I_{ps}, since direct measurement access to an isolated individual device in an integrated circuit is all but impossible, the semi-insulating substrate portion of the GaAs FET is physically modeled.[13] This is done with a fixture made of two relatively elongated thin parallel conducting strips mounted on semi-insulating GaAs material. This represents the source and drain metal separated by the substrate, but elongated to enhance the measurement sensitivity and accuracy. The whole is encapsulated, sealed, and subjected to dose-rate pulsed radiation and measurements taken. Each strip is $W(\mu\text{m})$ long, $L_0(\mu\text{m})$ wide, and separated by $L(\mu\text{m})$, where $W \gg (L + 2L_0)$. A cross section of the fixture is suggested in Fig. 11.5. The resistance of such a configuration, with current in the L direction, is given by[5]

$$R_{\text{sub}} = \pi/\Delta\sigma_{\text{sub}} W \ln(8L_0/L) \tag{11.37}$$

Combining Eqs. (11.35) and (11.37) yields a corresponding I_{ps} as

$$I_{ps} = V_0/R_{\text{sub}} = (e/\pi) g_0 \dot{\gamma} t_p V_0 W(\mu_n + \mu_p) \ln(8L_0/L) \tag{11.38}$$

Since $I_{ps} = V_0/R_{\text{sub}}$ as well, where V_0 is the applied bias voltage across the strips, measurements of I_{ps} for various V_0 and dose rate levels can be made and plotted as shown in Fig. 11.6. To extrapolate these fixture measurements to the actual semi-insulating GaAs FET, an empirical fit to the curves in Fig. 11.6, for $\dot{\gamma} \lesssim 10^{11}$ rad (GaAs)/s is given by

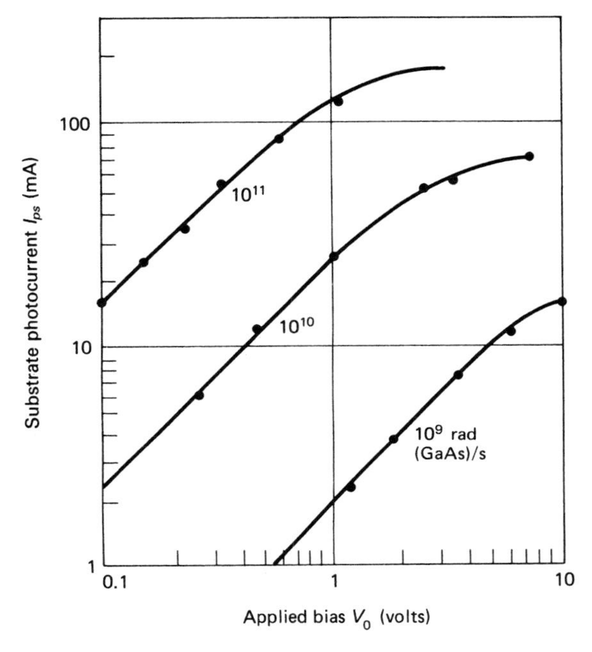

Fig. 11.6 GaAs FET fixture source-drain substrate photocurrent versus ionizing dose rate.[19] © 1983 by the IEEE.

$$I_{ps} = 2.3 \cdot 10^{-11} \dot{\gamma} W V_0 \ln(8L_0/L) \tag{11.39}$$

This I_{ps} can be used in GaAs FET circuit upset modeling of substrate photocurrents in semi-insulating substrates with coplanar contact geometry such that $W/(2L_0 + L) \gg 1$. The deviation from linearity of the curves in Fig. 11.6 is due to high electric field drift velocity saturation because of the field-dependent mobility.[20]

To compute a nominal I_{ps} for a typical GaAs FET, Eq. (11.39) can be used, where $L_0 = 5\,\mu\text{m}$, $L = 3\,\mu\text{m}$, $W = 10\,\mu\text{m}$, $V_0 = 1$ volt, and $\dot{\gamma} = 10^{11}$ rad (GaAs)/s are inserted to give

$$I_{ps} = 2.3 \cdot 10^{-11} \cdot 10^{11} \cdot 10^{-3} \cdot 1 \cdot \ln(8 \cdot 5/3) = 5.96\,\text{mA} \tag{11.40}$$

Hence, it can be seen that the substrate photocurrent $I_{ps} = 5690\,\mu\text{A}$ dwarfs both $I_{pp} = 5.3\,\mu\text{A}$ following Eq. (11.27) and $I_{sp} = 53\,\mu\text{A}$ following Eq. (11.28), for similar devices.

Logic upset levels of GaAs digital microcircuits due to ionizing radiation depend on the associated circuitry as well as the active devices. To model both the device and circuits, a computer program such as SPICE or its variants is usually employed with appropriately placed photocurrent generators within the code.

As a first approximation to an analytic determination of the logic upset levels in GaAs FETs, the device substrate photocurrent I_{ps} can be equated to the GaAs FET drain-source current I_{DS}. This is certainly a sufficient condition to exceed the noise margin, that is, to upset the device. The corresponding drain current is obtained from a variation of Eq. (11.24) viz.,[13]

$$I_{DS} = K'(W/L_G)(V_G - V_T)^2 \qquad (11.41)$$

where L_G is the channel length (short dimension), W is the channel width (long dimension) and K' is the empirically determined parameter appropriate for an enhancement mode JFET (E-JFET) structure. Then, setting Eqs. (11.39) and (11.41) equal to each other results in the upset value of the dose rate according to this criterion,

$$\dot{\gamma}_U = K'(V_G - V_T)^2/2.3 \cdot 10^{-11} V_0 L_G \ln(8L_0/L); \qquad \text{rad (GaAs)/s} \quad (11.42)$$

For E-JFET structures, $L_0/L = 2.5$, $K' = 2 \cdot 10^{-4}\,\text{amps/v}^2$ and $V_G - V_T = V_0 = 1\,\text{volt}$, which results in

$$\dot{\gamma}_U = 0.5 \cdot 10^{11}/L_G\ (\mu\text{m}) \qquad \text{rad (GaAs)/s} \qquad (11.43)$$

Figure 11.7 depicts the upset dose rate $\dot{\gamma}_U$ of the E-JFET versus its channel length L_G.[13]

11.5 SEU ERROR RATE

Single event upset (SEU) phenomena are discussed in Chapter 8. As with the previously discussed radiation damage mechanisms, including neutron fluence, ionizing radiation dose, and its corresponding dose rate, the fundamental SEU processes are essentially identical in GaAs and Si.

Cosmic rays are composed of about 90 percent protons, which mainly emanate from the sun. They are generally of relatively low energy, sufficient to pose only minimal concern with respect to SEU. However, protons in the Van Allen belts can be inimical to spacecraft orbiting within them because of their higher energies. They are mainly a product of the decay of neutrons (whose half-life is about 12 minutes). The latter are produced by cosmic rays interacting with the nuclei of the very rare

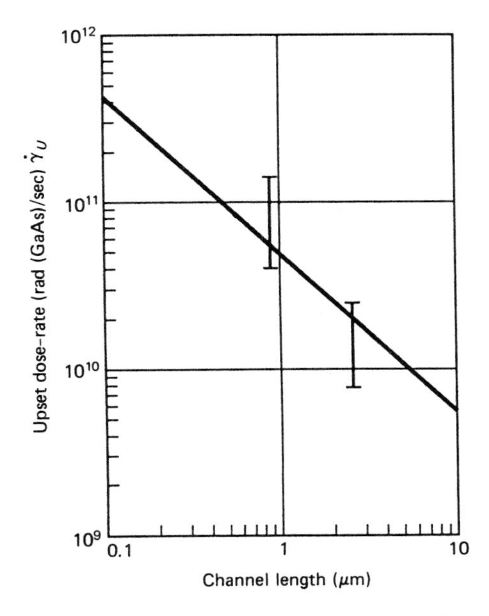

Fig. 11.7 Logic upset dose-rate of enhancement mode FET (E-JFET) as a function of channel length.[13] © 1970 by the IEEE.

particle densities at altitudes of about 600 miles and higher. Electrons are also a product of the decay of these neutron-decay reactions (i.e., $N \rightarrow P + e^- + \nu$(neutrino)), and together with the protons form the Van Allen belts.

Most of the remaining 10 percent of cosmic rays of importance to SEU in spacecraft are the very energetic (up to 2000 MeV), very heavy (through iron), and very highly ionized ions. They are incident on the solar system as debris coming from star-producing enclaves in the cosmos. Being buffeted by galactic magnetic fields, their average travel time is reckoned in millions of years. These heavy ions constitute the major cosmic ray component with respect to SEU in spacecraft at geosynchronous altitude (~23,000 miles). Attempts at coping with these SEU problems can be vexing. One reason is that it is essentially impossible to shield against such energetic heavy particles because of their tremendous material-penetrating capabilities. Since these heavy ions are highly charged (e.g., Fe^{++++++}), the earth's magnetic field can act as a natural spectrometer deflecting them away from the local neighborhood of the earth. However, the deflection efficiency depends on a number of factors, including the cosmic ray incoming direction, spacecraft orbital parameters,

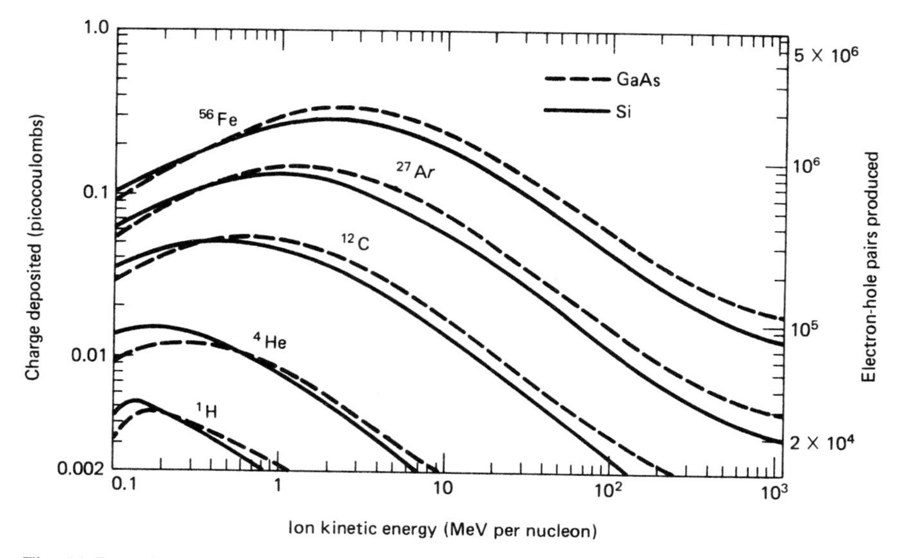

Fig. 11.7A Charge deposited by protons, alphas, and three heavier ions in 1 micron of silicon or gallium arsenide.[24]

and the sunspot cycle. In short, other means must be used to cope effectively with SEU.

A second particle type of concern is the helium ion, He^{++}, or alpha particle. They comprise but a few percent of the total cosmic ray component. Since the earth's crust contains trace amounts of actinide elements such as uranium, thorium, and their daughters, their decay chains ultimately stop at ^{210}Pb, producing alphas at each juncture. Hence, the alphas appear within the device package material itself. Shielding, such as passivation layers, can often be used as a protective measure. However, neutrons can directly enter both the package and chip, to cause nuclear reactions with the GaAs or Si to produce charged particles, as well as alphas within, that cause SEU.

For SEU-produced charge collection within the device, funneling (discussed in Section 8.5) is an interesting phenomenon in semi-insulating GaAs FETs,[21] where the semi-insulating substrate, of almost pure GaAs with or without Cr p^- doping, is used.

One measure of the charge collection depth is the diffusion length (e.g., L_n). Mainly because of the relatively short minority carrier lifetime in GaAs ($\sim 10^{-8}$ s), the diffusion length $L_{\text{GaAs}} \sim (1/200)L_{\text{Si}}$. Therefore, the carrier collection volume in GaAs is generally much smaller than in Si.

Table 11.2 40 MeV proton SEU cross section for various GaAs and Si SRAMs.[13] © 1986 by the IEEE.

SERIAL	VENDOR GENERIC	FUNCTION	PROTON SEU CROSS-SECTION PER BIT (cm^2)
(1)	93L422	1 K Si Bipolar	$5 \cdot 10^{-11}$
(2)	93422	1 K Si Bipolar	$1 \cdot 10^{-11}$
(3)	2147	1 K Si NMOS	$3 \cdot 10^{-12}$
(4)	6504	4 K Si CMOS	$2 \cdot 10^{-14}$
(5)	6147	4 K Si CMOS	$8 \cdot 10^{-15}$
(6)	—	0.256K GaAs JFET (Resistor load)	$8 \cdot 10^{-12}$
(7)	—	0.256K GaAs JFET (Complementary FETs)	$2 \cdot 10^{-14}$
(8)	—	0.256K GaAs MESFET	$2 \cdot 10^{-11}$

In GaAs, the funnel penetration extends into the semi-insulating substrate to produce a correspondingly large amount of charge over that produced in its collection volume, per se. It is known that GaAs and Si are comparable with respect to SEU caused by prompt charge collection.[13] The major difference is due to the differing diffusion lengths, as discussed above. Other differences are due to recombination processes unique to the direct band gap GaAs material, in contrast to the indirect band gap Si material.[13] Actually, GaAs device SEU susceptibility is slightly greater than that for silicon devices for two reasons. The first reason is a bulk effect in that the GaAs device source-to-drain volumes, as well as the gate-to-drain volumes, are sensitive to cosmic ray ions. The gate-to-drain volumes in silicon devices are SEU-insensitive because of the insulator gate, as opposed to the schottky metal-barrier junction-type gate in GaAs devices. The second reason is that the stopping power in GaAs for high-energy ions of interest is slightly greater than that for silicon, hence a given ion will produce more SEU charge in GaAs than in Si. This is seen in the curves in Fig. 11.7A, which are effectively curves of stopping power in silicon and gallium arsenide for ions[24] with mass numbers up to ^{56}Fe.

Another method for comparing the SEU susceptibility of GaAs and Si is through the SEU cross sections of some of their SRAM memory devices. Table 11.2 provides this comparison for 40 MeV protons.[13]

The 93L422 1K Bipolar SRAM is a notoriously SEU susceptible device. Little used nowadays, it possesses the largest SEU cross section in Table 11.2. The cross section of the 256-bit GaAs Complementary JFET (C-JFET), a close circuit relative of CMOS,[5] compares favorably with the

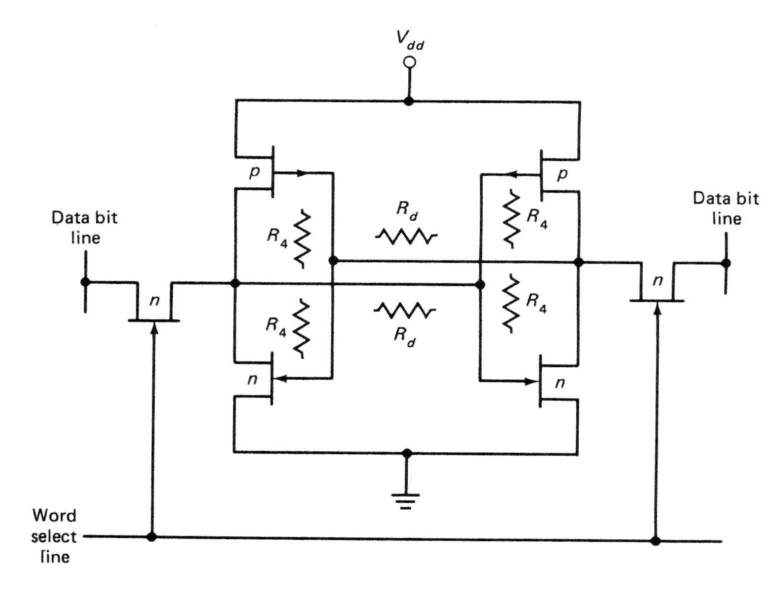

Fig. 11.8 GaAs enhancement mode (E-JFET) complementary SRAM cell.[13] © 1989 by the IEEE.

6504–4K Si CMOS device. The 256-bit GaAs JFET (500 kohm load resistors) compares poorly with its Complementary JFET counterpart, shown in Table 11.2. This is primarily due to an increased noise margin in the C-JFET, and the lower impedance of its cell flip-flop nodes. In the GaAs or Si FET SRAM, it is the HIGH-node at the drain of the OFF-FET that is SEU susceptible from a gate-drain or source-drain particle track impact. In general, as seen from Table 11.2, GaAs SRAMs compare well with the corresponding Si devices, but the hardest SRAMs are made of silicon. Figure 11.8 depicts the circuit of a GaAs FET complementary SRAM cell, or one superposed with decoupling resistors in its feedback paths.

One method for decreasing SEU susceptibility in GaAs or Si SRAMs is to add (monolithically) intra-cell decoupling resistors, as already mentioned. Since the track of the incident cosmic ray strike and subsequent SEU is hardly 100 picoseconds in duration, if the SRAM is caused to respond electrically much slower by using decoupling resistors in its feedback paths, the device can be made essentially immune to SEU. This is currently being explored. However, this suffers from at least three difficulties. The first is that the decoupling resistors and other variations, such as the inclusion of decoupling diodes, lengthen the time response of

Table 11.3 SEU cross sections for 256 bit GaAs SRAMs with two decoupling resistors.[13] © 1989 by the IEEE.

ION TYPE	SEU CROSS SECTION PER BIT (cm^2)
40 MeV proton	$6.27 \cdot 10^{-6}$ (150 kohm resistors)
275 MeV krypton	$1.63 \cdot 10^{-5}$ (150 kohm resistors)
275 MeV krypton	$8.31 \cdot 10^{-5}$ (no resistors)

the device. That is, for the SRAM to be insensitive to SEU, the resistors will slow it down to take on the characteristics of a low-pass filter. This is the antithesis of the device desired performance—a fast time response that translates to high speed. Perhaps there is a sanguine tradeoff between the insertion of passive components in the feedback paths and speed, but this avenue does not seem promising at this juncture. The second difficulty is that, for some military environments, such as those corresponding to high levels of neutron fluence, the corresponding degradation due to displacement damage increases the resistance of these resistors to very high values, as discussed in Section 5.13, thus degrading their effectiveness.

The third difficulty is the practical one of manufacturing SRAMs with these resistors emplaced monolithically, which results in a maintenance of their stability over time in a hardness assurance sense. This has not been established to date. A number of 256-bit GaAs SRAMs with two decoupling resistors (R_d with R_4 shorted in Fig. 11.8) were subjected to various high-energy ion environments to determine their SEU cross sections.[13] The results are shown in Table 11.3.

In the tabulation in Table 11.3, the krypton ion simulates the heavy ion cosmic ray environment. The principal observation is that, for the devices tabulated, the decoupling resistors decrease the sensitivity by a factor of $8.31 \cdot 10^{-5}/1.63 \cdot 10^{-5}$ or 5.1. Figure 11.9 graphically depicts, for complementary GaAs E-JFET SRAMs, data taken from computations modeled using a variant of SPICE.[22] The computational data plotted includes the four decoupling resistor model (R_4 with R_d shorted) only. These plots show the variation of SRAM memory cell write time (a measure of device speed) and SEU errors per bit day, versus the decoupling resistor values. As can be seen from the figure, the SRAM write time increase begins to saturate for large resistor values, as does the bit error rate. However, the latter stays within a factor of about 3.7 between resistor values of

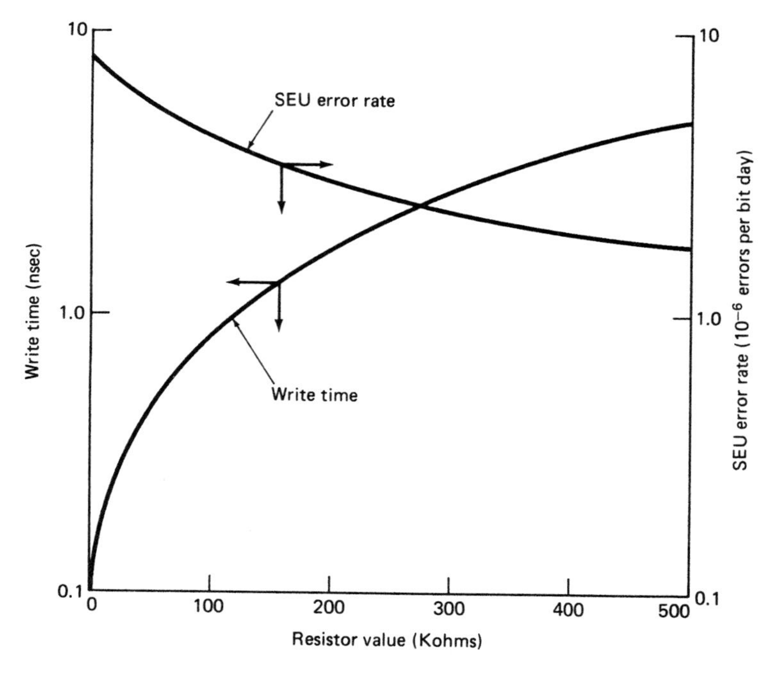

Fig. 11.9 SEU error rate and cell write time versus decoupling resistor values in GaAs complementary-enhancement mode CE-JFETs.[22] © 1986 by the IEEE.

0–500 kilo ohms, while the write time increases by a factor of almost 40 between the same range of resistor values.

As a nominal prototype of the resistive decoupling schema for these devices, the data seem to indicate that this method provides only marginal returns in SEU error rate for an appreciable sacrifice in speed. As mentioned earlier, perhaps other resistive or reactive configurations augur a more favorable trade-off in SEU error rate versus device speed. Active components and circuits in the feedback path show promise without the deleterious effects of resistors, such as diodes or emitter- or collector-followers.[31]

REFERENCES

1. C. M. Vogelsang et al. "Complementary GaAs FET 16 K SRAM." Gallium Arsenide Integrated Circuit Symp. Tech. Digest, pp. 75–78, A-1988.
2. D. L. Harrington, et al. 'RISC Microprocessors." Gallium Arsenide Integrated Circuit Symp. Tech. Digest, pp. 87–90, A-1988.

3. S. M. Sze. Physics of Semiconductor Devices, 2nd ed. Wiley-Interscience, New York, Appendix G, 1981.
4. P. T. Greiling, S. G. Knorr, and C. R. Viswanathan. UCLA Eng. Dept. Short Course, "High Speed Integrated Circuit Technology," Section VII. May 2–6, 1983.
5. R. Zuleeg. "Radiation Effects in GaAs FET Devices." Proc. IEEE 77:389–407, Mar. 1989.
6. L. W. Aukerman *in* R. C. Willardson and A. C. Beer (eds.). Semiconductors and Semimetals, Vol. 4, Chap. 4. Academic Press, New York, 1968.
7. G. H. Marcus and H. P. Breumer. "Radiation Damage in GaAs Gunn Diodes." IEEE Trans. Nucl. Sci. NS-17(6):230, Dec. 1970.
8. R. R. Holmer. "Carrier Removal in Neutron Irradiated Silicon." IEEE Trans. Nucl. Sci. NS-17(6):137–143, Dec. 1970.
9. Sze, op. cit. p. 316 ff.
10. B. Buchanan et al. "Comparison of the Neutron Radiation Tolerance of Bipolar and Junction Field Transistors." Proc. IEEE 55:2188–2197, Dec. 1967.
11. K. Lehovec and R. Zuleeg. "Voltage-Current Characteristics of GaAs JFETs in the Hot Electron Range." Solid State Elect. 13:1415–1426, Nov. 1970.
12. G. E. Brehm and G. L. Pearson. "Gamma radiation damage in epitaxial gallium arsenide." J. Appl. Phys. 43:560, Feb. 1972.
13. R. Zuleeg, op. cit.
14. D. M. Newell et al. "Total Dose Hardness of Microwave GaAs Field Effect Transistors." IEEE Trans. Nucl. Sci. NS-28(6):4403–4410, Dec. 1981.
15. R. Zuleeg and K. Lehovec. "Radiation Effects in GaAs Junction Field Effect Transistors." IEEE Trans. Nucl. Sci. NS-27(5):1343–1354, Oct. 1980.
16. R. Zuleeg. Radiation Effects in GaAs Integrated Circuits" *in* N. G. Einspruch and W. R. Wisseman (eds.). VLSI Electronics Microstructure Science. Academic Press, New York, Chap. 11, 1985.
17. W. S. Ginell, R. Zuleeg, J. L. McNichols, J. K. Notthoff, and K. Lehovec. "Transient Response of Epitaxial GaAs JFET Structures to Ionizing Radiation." IEEE Trans. Nucl. Sci. NS-20(6):171–179, Dec. 1973.
18. A. B. Phillips. Transistor Engineering. McGraw-Hill, New York, p. 80, 1967.
19. R. Zuleeg, J. K. Notthoff and G. L. Troeger. "Channel and Substrate Currents in GaAs FETs Due to Ionizing Radiation." IEEE Trans. Nucl. Sci. NS-30(6):4151–4156, Dec. 1983.
20. Sze, op. cit. p. 324 ff.
21. M. A. Hopkins and J. R. Srour. "Charge Collection Measurements on GaAs Devices Fabricated as Semi-insulating Substrates." IEEE Trans. Nucl. Sci. NS-31(6):1116–1120, Dec. 1984.
22. T. R. Weatherford, J. R. Hauser, and S. E. Diehl. "Comparisons of Single Event Vulnerability of GaAs SRAMs." Trans. Nucl. Sci. NS-33(6):1590–1596 (Table 2), Dec. 1986.
23. G. C. Messenger. "Conductivity Modulation Effects in Diffused Resistors at Very High Dose Rate Levels." IEEE Trans. Nucl. Sci. NS-26(6):4725–4729, Dec. 1979.
24. E. L. Petersen and P. W. Marshall. "Single Event Phenomena in the Space and SDI Arenas." J. Rad. Effects, Res. and Eng., Jan. 1989.
25. M. Simons and E. E. King. "Long Term Radiation Transients in GaAs FETs." IEEE Trans. Nucl. Sci. NS-26(6):5080–5086, Dec. 1978.
26. M. Simons. "Ionizing Radiation Effects in Silicon MOS and Gallium Arsenide FET Structures." Research Triangle Institute Report No. RTI/2019/00-01F, Research Triangle Park, NC, Oct. 1981.

27. W. T. Anderson, M. Simons, A. Christou, and J. Beall. "GaAs MMIC Technology Radiation Effects." IEEE Trans. Nucl. Sci. NS-32(6):4040–4045, Dec. 1985.
28. T. P. Ma and P. V. Dressendorfer (eds.). Ionizing Radiation Effects in MOS Devices And Circuits. Wiley-Interscience, New York, Chap. 7, 1989.
29. Y. Nishioka, K. Ohyu, Y. Ohji, M. Kato, E. F. da Silva, Jr., and T. P. Ma. "Radiation Hardened Micron and Submicron MOSFETs Containing Fluorinated Oxides." IEEE Trans. Nucl. Sci. NS-36(6):2116–2123, Dec. 1989.
30. W. T. Anderson, H. B. Dietrich, E. E. King, R. J. Lambert, and M. Simons. "Reduction of long-term transient radiation response in ion-implanted GaAs FETs." IEEE Trans. Nucl. Sci. NS-29(6):1533–1538, Dec. 1982.
31. G. C. Messenger, O. Dukelow, J. Heller, C. T. Kleiner and M. E. Peacock. "Hardening of bipolar memory cells against SEU." J. Rad. Effects; Res. and Eng. 5(1):12–16, Winter 1986.

CHAPTER 12
COMPONENT AND CIRCUIT HARDNESS DESIGN

12.1 INTRODUCTION

The previous chapters can be construed as a compendium of experimental and theoretical knowledge of nuclear survivability and vulnerability. In those chapters, among other aims, an attempt was made to provide the salient aspects of solid-state theory that lead to device function and circuit operation. Threats to that operation are manifest in various radiation environments. They consist of incident neutrons, gamma- and x-ray photons, plus their time derivatives, the electromagnetic pulse, and cosmic rays. This chapter, and the next, disclose nuclear survivability information about components, circuits, and systems that provide the electronic designer with the wherewithal to do his job. It is presented in narrative form that stresses the "how-to" knowledge, with a handbook flavor instead of the why and wherefore.

12.2 COMPONENT SELECTION: NEUTRON EFFECTS

The immediate result of neutron damage in bipolar transistors is the degradation of the common emitter current gain. This is depicted in Fig. 12.1 and expressed quantitatively in the Messenger-Spratt equation

$$\Delta(1/\beta) \equiv 1/\beta - 1/\beta_0 = \Phi_n/\omega_T K \tag{12.1}$$

where $\omega_T = 2\pi f_T$. f_T is the gain-bandwidth product frequency of the transistor, and K is the damage constant defined in Section 5.9. For silicon, a nominal neutron fluence spectrum, and over a diversity of bipolar transistors, Eq. (12.1) can be written as

$$\Delta(1/\beta) = K_D \Phi_n \tag{12.2}$$

where $K_D = 10^{-7}/f_T$. This implies a universal damage constant $K = 1.6 \cdot 10^6$ neutron-sec per cm^2 under operation at or near the peak (knee) of the transistor gain as a function of collector current, and Φ_n is the

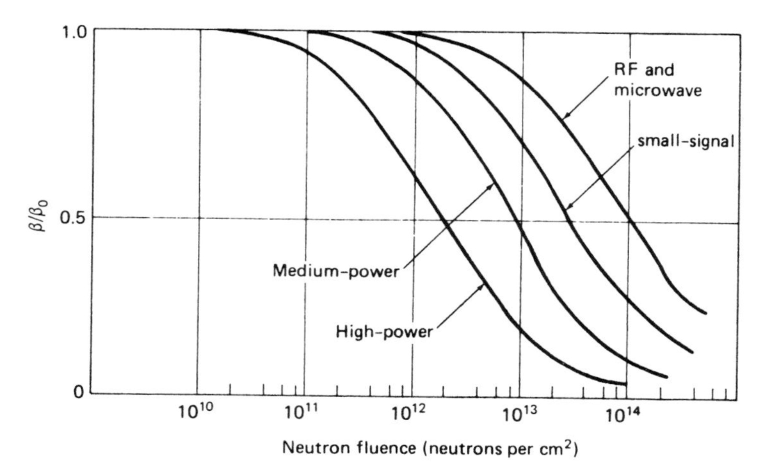

Fig. 12.1 Typical relative common emitter current gain versus incident neutron fluence for various types of bipolar transistors.[1] © 1984 by the IEEE.

neutron fluence. Other things being equal, high-gain, large-f_T transistors should be chosen to minimize the radiation-induced gain loss $\Delta(1/\beta)$. It should be noted that, for small gain changes, $\Delta(1/\beta) \cong -\Delta\beta/\beta^2$. Equation (12.2) can be written as

$$\beta/\beta_0 = 1/(1 + \beta_0 K_D \Phi_n) \tag{12.3}$$

Equations (12.1)–(12.3) are asymptotic gain relations in that they depict gain degradation at relatively long times after exposure to the radiation pulse. The gain degradation immediately after a pulse of incident neutrons can be much greater than that given above. Following an annealing phase, usually of the order of milliseconds, the gain normally reverts to that given by the foregoing expressions. As discussed in Section 5.14, the immediate gain degradation is implicit in the annealing factor, $F(t)$, given by

$$F(t) = \Delta(1/\beta(t))/\Delta(1/\beta_\infty) = (1/\beta(t) - 1/\beta_0)/(1/\beta_\infty - 1/\beta_0) \tag{12.4}$$

or

$$\Delta(1/\beta(t)) = F(t) \cdot \Delta(1/\beta_\infty) \tag{12.5}$$

Annealing factors are usually obtained empirically from experimental data taken for a particular device family. The annealing factor has addi-

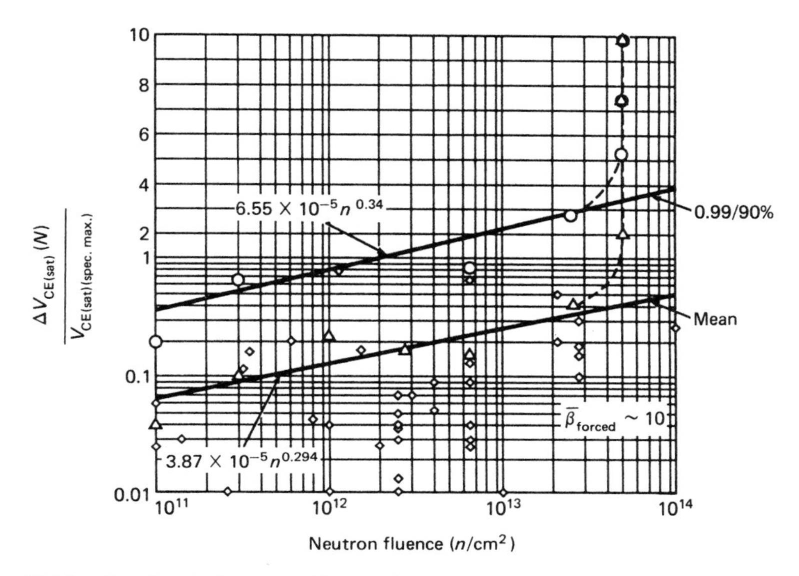

Fig. 12.2A Fractional change in $V_{CE(\text{sat})}$ for general-purpose transistors at 25°C versus neutron fluence. $\Delta V_{CE(\text{sat})}$ is the average change in $V_{CE(\text{sat})}$ for the given test sample.[1,83] © 1984 by the IEEE.

tional properties, e.g., it has a positive jump discontinuity at zero time in that,

$$F(0-) = 0; \quad F(0+) \geqslant 1; \quad F(t) \geqslant 1, t > 0; \quad \lim_{t \to \infty} F(t) = 1 \quad (12.6)$$

Figure 12.2A depicts the fractional change in $V_{CE(\text{sat})}$ versus the incident neutron fluence Φ_n. It shows a changing fractional $V_{CE(\text{sat})}$ for low levels of Φ_n, while almost discontinuously increasing for high Φ_n. This can be appreciated in the following, where the expression for $V_{CE(\text{sat})}$, e.g. in an npn transistor, can be derived easily[82], as

$$V_{CE(\text{sat})} = I_E r_{se} + I_C r_{sc} - (kT/e)\ln[\alpha_R(1 - \beta_f/\beta)/(1 + (1 - \alpha_R)\beta_f)] \quad (12.7)$$

where:

r_{se} and r_{sc} are the emitter and collector saturation resistances, respectively.

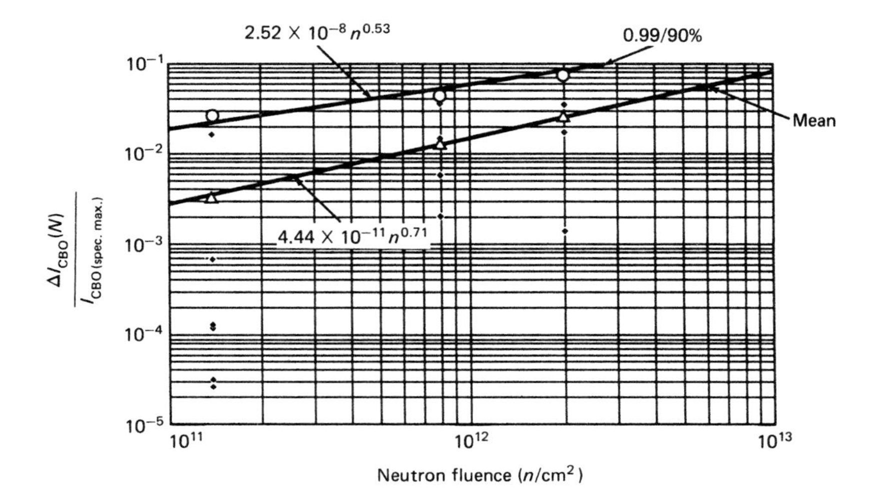

Fig. 12.2B Fractional change in I_{CBO} for general-purpose transistors at 25°C versus neutron fluence. ΔI_{CBO} is the average change in I_{CBO} for the given test sample.[1,83]

$\alpha_R = (1 + R_{CC}/R_{BB})^{-1}$, with R_{CC} and R_{BB} being the effective sheet resistances of the collector and base regions, respectively.

$\beta_f = I_C/I_B$ is the "forced" beta, as fixed by the external circuitry.

Introducing the effect of neutron fluence, Eq. (5.66) gives $\beta^{-1} = \beta_0^{-1} + \Phi_n/K\omega_T$, which, when inserted in Eq. (12.7), results in

$$V_{CE(\text{sat})} = I_E r_{se} + I_C r_{sc} - (kT/e)\ln\{[\alpha_R/(1 + (1 - \alpha_R)\beta_f)] \cdot [1 - \beta_f/\beta_0 - \beta_f \Phi_n/K\omega_T]\} \tag{12.8}$$

From Eq. (12.8), it is seen that, for sufficiently high Φ_n, the logarithm term will change sign, causing the abrupt increase in $V_{CE(\text{sat})}$ in Fig. 12.2A. The phrase "0.99/90%" on the curves means that, in the limit of 99 percent of the data shown, it is valid with a confidence level of 90 percent.

Also, it is to be noted that $\Delta V_{CE(\text{sat})}/V_{CE(\text{sat})}$, versus ionizing dose (as in Fig. 12.12A), depicts a variation similar to that for neutron fluence in Fig. 12.2A. This is reasonable, since it is known that, for ionizing dose damage, $\beta^{-1} \cong \beta_0^{-1} + K_\gamma \cdot \gamma_T$, where γ_T is the ionizing dose level and K_γ is an experimentally determined damage constant for ionizing dose. Hence, its substitution in Eq. (12.8) yields a behavior like that for Φ_n.

The deleterious aspect of increasing $\Delta V_{CE(\text{sat})}/V_{CE(\text{sat})}$ (i.e., increasing $V_{CE(\text{sat})}$ with increasing Φ_n) is that it causes the transistor to come out of saturation for high levels of neutron fluence or ionizing dose. As mentioned, large Φ_n will cause the logarithm term in Eq. (12.8) to change sign, resulting in a rapid increase in the value of $V_{CE(\text{sat})}$, as shown in Figure 12.2A. This forces the transistor into the active region where, with a now decreased β, it sinks very much smaller currents. If it is the output transistor of a logic integrated circuit, its fanout will drop precipitously, thus causing circuit failure.

The behavior of I_{CBO}, the collector leakage current, is shown in Fig. 12.2B. Radiation damage causes changes in surface parameters and consequently I_{CBO} increases.[83]

Other neutron damage effects in bipolar transistors usually include a slight increase in the junction leakage current, I_{CBO}, due to an increase in the recombination rate after irradiation, as seen in Fig. 12.2B. I_{CBO} increases with incident fluence, with typical values of less than 100 nA at Φ_n of 10^{14} neutrons per cm^2. Incident neutrons also increase the junction breakdown voltage BV_{CEO}. This also implies that only small changes occur in BV_{CBO}, even at high fluence levels of 10^{15} neutrons per cm^2. A useful empiricism in this regard is

$$BV_{CEO} \cong BV_{CBO}/\beta^n \tag{12.9}$$

where $n \cong 0.2$ for *npn* and about 0.5 for *pnp* transistors.[2]

Contemporary bipolar digital logic IC families are typified by TTL (transistor-transistor logic) or STTL (Schottky TTL) devices. The use of I^{2}L (current injection logic) integrated circuits puts them in the inventory of relatively radiation-resistant, bipolar logic. Neutron fluence failure thresholds for TTL and STTL are about 10^{14}–10^{15} neutrons per cm^2, as seen in Figs. 12.4 and 12.6, and their ionizing dose failure threshold can exceed 1 Mrad (Si). as in Fig. 12.15.

Whenever possible, attention should be paid to various hardened logic families of devices associated with particular systems. This yields insight into how device engineers attempt to synthesize device families into a system to conform to particular radiation specification constraints.

With the foregoing hardening techniques of part selection, the remaining difficulties and hardening procedures are of somewhat lesser import. For example, certain internal stages in the IC chip can require gains greater than 10 to sink current to satisfy fanout requirements, consistent with potential radiation damage. Fanouts of 10 are the usual vendor specification for gates and flip-flops in integrated circuits. The prudent designer derates such ICs with a view toward radiation hardness require-

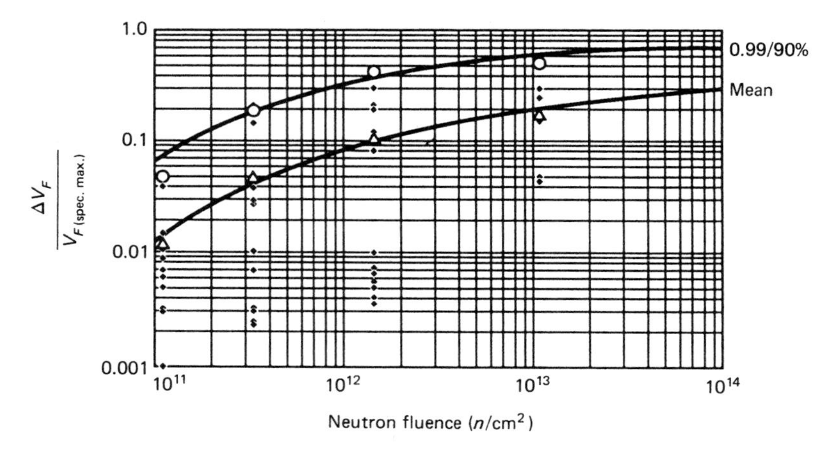

Fig. 12.3A Fractional change in forward voltage V_F, for rectifying diodes at 25°C versus neutron fluence. ΔV_F is the average change in V_F for the given test sample.[1,83] © 1984 by the IEEE.

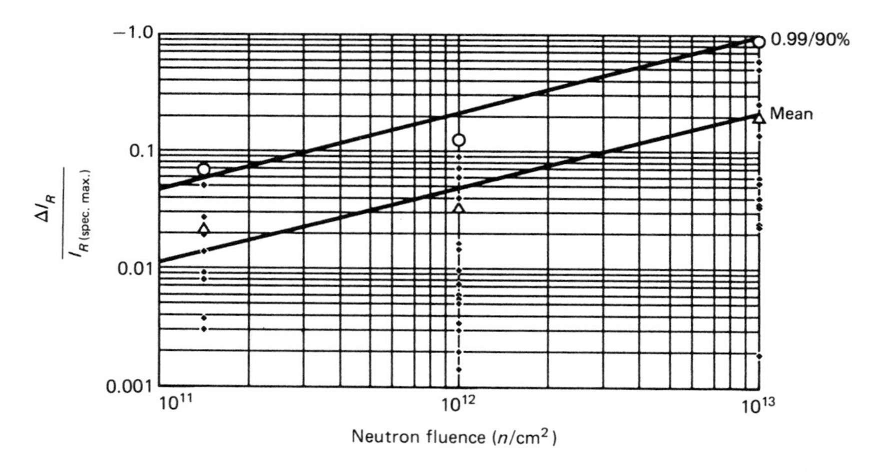

Fig. 12.3B Fractional change in reverse current I_R, for diodes at 25°C versus neutron fluence. ΔI_R is the average change in I_R for the given test sample.[1,83] © 1984 by the IEEE.

ments by assuming a reduction in the fanout capability to, for instance, 8 instead of the usual vendor rating of 10.

In diodes, neutron fluence damage results in two main effects. They are an increase in forward voltage V_F and reverse leakage current I_R, as shown in Figs. 12.3. The leakage current increases because of increased recombination current.[1]

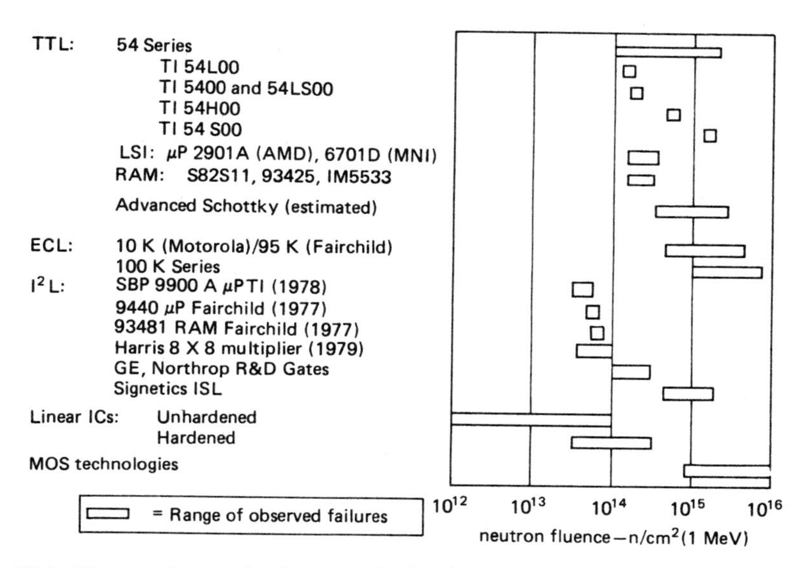

Fig. 12.4 Neutron fluence hardness levels for integrated circuit families circa 1978.[3] © 1980 by the IEEE.

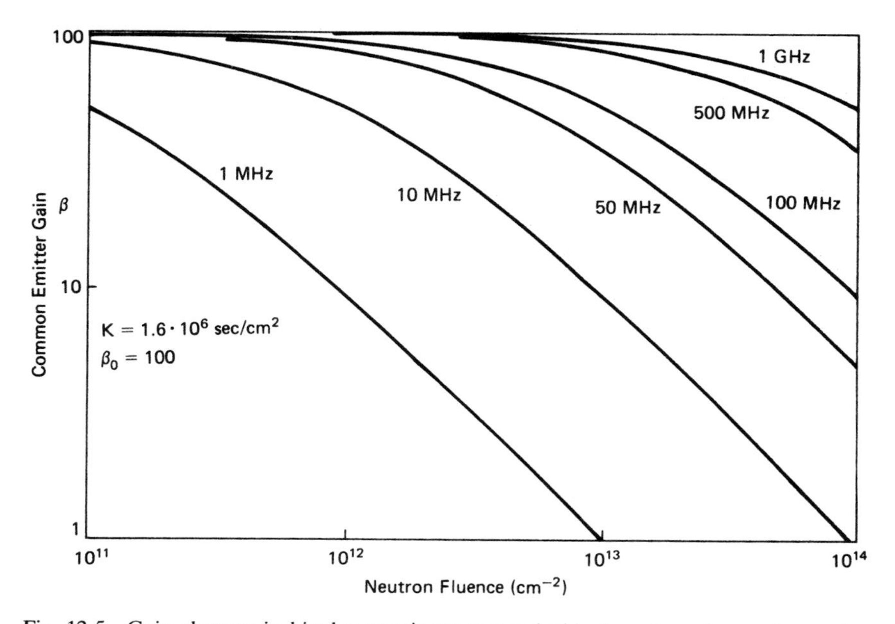

Fig. 12.5 Gain changes in bipolar transistors versus incident neutron fluence for various f_T.[4] © 1985 by the IEEE.

Reference diodes, including zener diodes, are very resistant to neutron fluence damage. However, for precision reference applications, small voltage changes can be important. Voltage reference diode damage depends on the diode breakdown voltage level.[1] Breakdown voltage, V_B, is a function of resistivity and bandgap voltage, which can be written as the following empirical expression[75]

$$V_{\text{Ref}} \equiv V_B \cong 60(E_g/1.1)^{3/2}(N_D/10^{16})^{-3/4} \text{ (volts)} \tag{12.10}$$

The bandgap energy E_g does not vary significantly with neutron fluence, so that combining Eqs. (12.10) and (5.121) gives the principal dependence of V_{Ref} on neutron fluence as

$$V_{\text{Ref}} \cong 60(E_g/1.1)^{3/2}(n_{oi}\exp - 2\Phi_n/k_n)^{-3/4} \tag{12.11}$$

where the remaining parameters are defined in Section 5.13. For small changes in V_{Ref}, corresponding to relatively low levels of Φ_n, it is apparent from Eq. (12.11) that

$$V_{\text{Ref}} \cong V_{\text{Ref0}}[1 + (3/2)\Phi_n/k_n] \tag{12.12}$$

where V_{Ref0} is the zero neutron fluence voltage reference.

For precision applications, temperature-compensated (TC) reference diodes are often used. This system is composed of one or more forward-biased (zener) diodes in series with the reversed-biased reference diode. Because the avalanche ionization rates in zener diodes decrease with increasing temperature,[75] V_B increases with temperature, yielding a positive temperature coefficient. For reversed-biased diodes, evaluating $\partial V/\partial T$ using the diode equations in Section 3.7 reveals a negative temperature coefficient. By appropriate matching, very low composite temperature coefficients can be obtained.[81] However, this diode string is necessarily more susceptible to change from neutron irradiation than a single reference diode, simply from bulk considerations. For a single forward-biased diode in series with a voltage reference diode,[81] the overall reference voltage change is

$$\Delta V_{\text{Ref}}(\Phi_n) \cong (nkT/2e)\ln(1 + \tau_0 K\Phi_n) \tag{12.13}$$

where τ_0 is the preradiation minority carrier lifetime, K is the empirical lifetime damage constant, as in Eq. (5.11), and n is the diode junction ideality factor, all parameters of the forward-biased diode.

Voltage reference diode damage due to neutrons depends on the diode

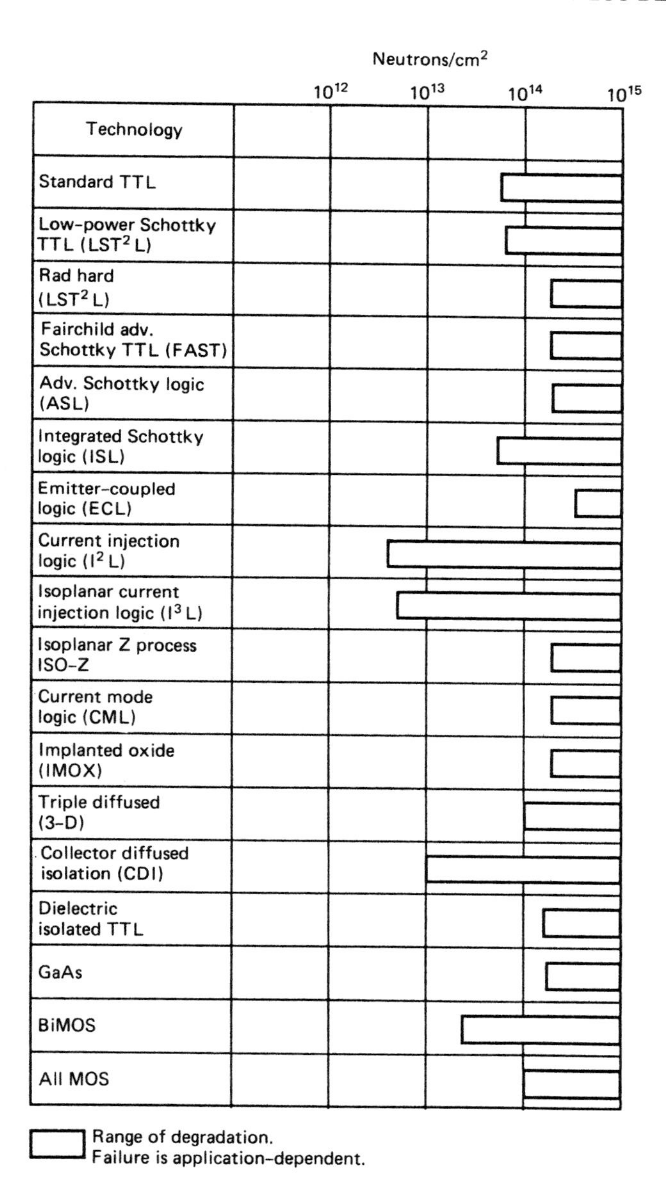

Fig. 12.6 Neutron fluence hardness levels for integrated circuit families circa 1990.[5] © 1990 by the Physitron Corp.

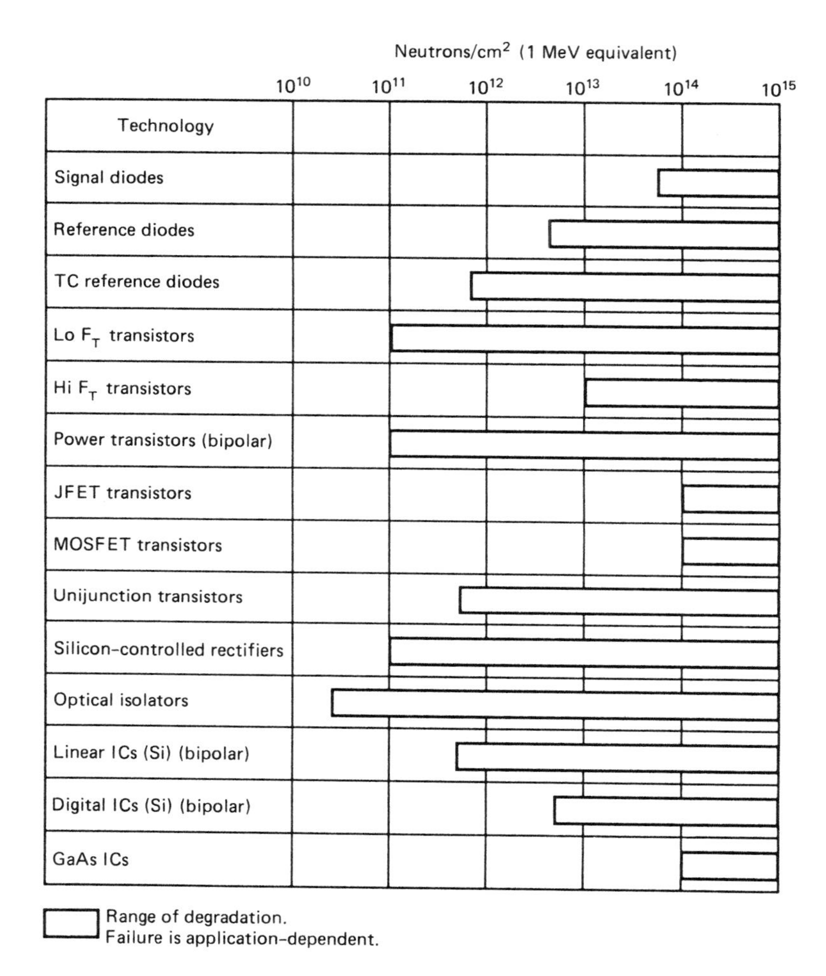

Fig. 12.7 Neutron fluence hardness levels for mainly discrete device families circa 1990.[5] © 1990 by the Physitron Corp.

breakdown voltage level.[1] For general purpose diodes, the reverse bias junction breakdown voltage increases with increasing fluence. This does not cause difficulties for signal diodes, but it does for power and zener diodes, as it shifts the zener breakdown voltage to larger absolute values, which can amount to millivolts at fluences of about 10^{12} neutrons per cm^2. The diode forward voltage drop is not affected significantly by incident neutrons, because most diodes are operated at low-injection levels. However, power diodes are high-injection-level devices, in that large numbers of carriers are injected into the lightly doped base side of the junction, reducing the bulk resistivity of that region. After irradiation, increased recombination occurs in the base region, reducing the diffusion of carriers across this region, which increases its effective resistivity. Most power diodes are constructed on a heavily doped substrate to provide a majority carrier contact with the base, which tends to reduce the base forward voltage drop. Nevertheless, their forward voltage drop can often increase by 3 V at fluence levels of 10^{13} neutrons per cm^2. Diode reverse leakage current is increased, as in transistors, due to incident neutrons. A 50 percent increase in this quantity is typical at fluences of 10^{13} neutrons per cm^2.

Obviously, the neutron-induced degradation of bipolar ICs results from the damage to the individual transistors from which the bipolar IC is built. To include a modicum of hardness in these ICs during their design and construction, collector gold doping, minimum base impedance, high f_T, and short switching times are all used. Most bipolar ICs can tolerate neutron fluences of 10^{13} neutrons per cm^2, but encounter degraded performance at 10^{14} neutrons per cm^2. Neutron irradiation damage levels to components, and families of ICs and other active devices, are depicted in Figs. 12.2–12.6, with advances in technology diversity and trends as seen by comparisons between Figs. 12.4, 12.6 and 12.7.

On occasion, digital logic systems depend on the saturated logic circuit storage time to govern the signal transfer time. From the hardness survivability/vulnerability standpoint, this is poor practice, as it may result in very low neutron-damage thresholds.

Almost all MOS devices are relatively hard to neutron damage from even relatively high fluence levels. This is simply because they are majority carrier devices, and there are no minority carrier lifetime degradation difficulties. However, at very high neutron fluences, of the order of 10^{15} neutrons per cm^2, MOS devices begin to degrade due to carrier removal and mobility degradation.

Other component types, such as SCRs, unijunction transistors, photodiodes, phototransistors, other arcane devices, and solar cells are very sensitive to neutron fluence, to the point that their use should be avoided if at all possible. This is because of their heightened sensitivity to minority carrier lifetime changes in their operation, often due to large junction area, and large effective device volume.

For JFETs, the transconductance, g_m, is the principal parameter sensitive to incident neutron fluence, Φ_n. It is also an important MOSFET parameter, in that for both FET types, the amplification factor $\mu_{\text{FET}} = g_m R_g$, as discussed in Sections 11.2 and 11.4. The neutron degraded transconductance can be written as[87]

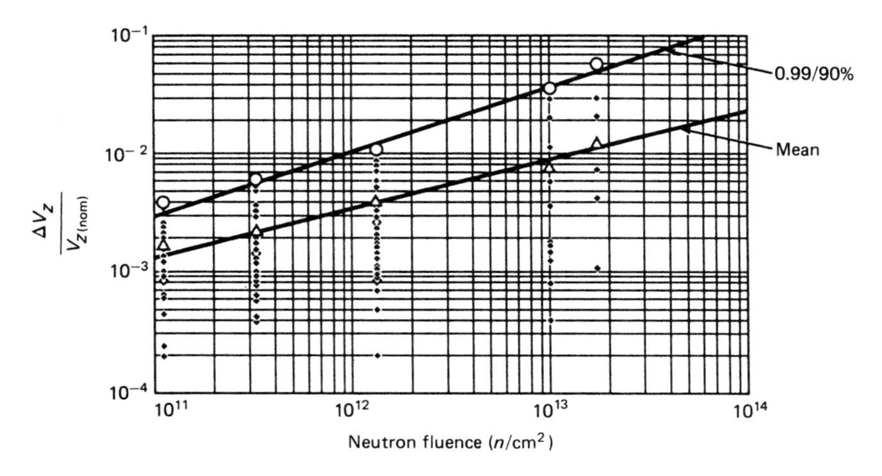

Fig. 12.7A Fractional change in V_Z for zener diodes at 25°C versus neutron fluence. ΔV_Z is the average change in V_Z for the given test sample.[1,83] © 1984 by the IEEE.

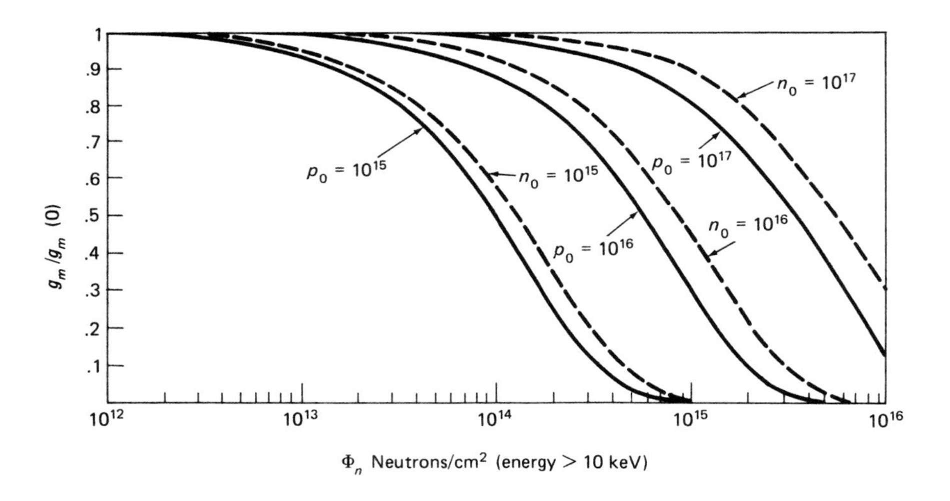

Fig. 12.7B Plots of the theoretical transconductance degradation as a function of incident neutron fluence, with initial channel concentrations n_0, p_0 as parameters.[87] © 1969 by the IEEE.

$$g_m(\Phi_n) = g_m(0)\exp - \Phi_n/K_{gm} \tag{12.14}$$

where $g_m(0)$ is the pre-irradiation transconductance. The transconductance damage constant K_{gm} is given as,

$$K_{gm} = \begin{cases} 398p_0^{0.77}; & p\text{-channel JFET} \\ 93n_0^{0.82}; & n\text{-channel JFET} \end{cases} \tag{12.15}$$

It should be noted that p-channel JFETs are harder to neutrons than their n-channel counterparts. Also, the g_m failure thresholds increase for both p- *and* n-channel devices, with increasing dopant density. Corroborating data indicate that JFETs are not appreciably damaged at neutron levels of $\sim 10^{13}\,\text{cm}^{-2}$, but may be damaged significantly at $\sim 10^{15}\,\text{cm}^{-2}$.[87] This is evident in the g_m plot in Fig. 12.7B.

Incident neutrons also can cause ionization through secondary processes discussed in Section 5.7, and particular aspects are examined in the next section on ionizing dose.

With regard to the combined effects of temperature changes plus those due to neutron fluence, the common emitter current gain change can be written, from Section 5.10 as

$$\Delta(1/\beta) \equiv 1/\beta - 1/\beta_0 = \Phi_n/\omega_T K - 1/\beta_0(1 - 1/(1 + \Delta T/T_0)^2) \tag{12.16}$$

where $\Delta T = T - T_0$, β_0 is the preirradiation gain, and T_0 is the ambient temperature. The three-level defect model discussed in Section 5.9 is used to fit the damage constant K_D as a function of absolute temperature, where $K_D = (\omega_T K)^{-1}$, and K is the universal damage constant. This results in a temperature-dependent K_D[7]

$$K_D(T) = K_D(T_0)(1.1 - 0.1(T/T_0)^4) \tag{12.17}$$

where K_D appears in $\Delta(1/\beta) = K_D\Phi_n$, and $K = 1.6 \cdot 10^6$ neutron seconds per cm^2. Both Eqs. (12.16) and (12.17) are valid for low to moderate values of fluence (10^{11}–10^{13} neutrons per cm^2), base dopant concentration (10^{15}–10^{16} dopant atoms per cm^3), and base current (10^{-6}–10^{-3} amps). For very high fluence levels, K_D becomes only slightly dependent on temperature, and the linear superposition of fluence and temperature in Eq. (12.16) no longer holds. Diffusion parameters such as mobility μ and the diffusion constant D depend on temperature, as seen from the Einstein relationship given in Section 1.8 rewritten here as

$$D/D_0 = (T/T_0) \cdot (\mu/\mu_0) \tag{12.18}$$

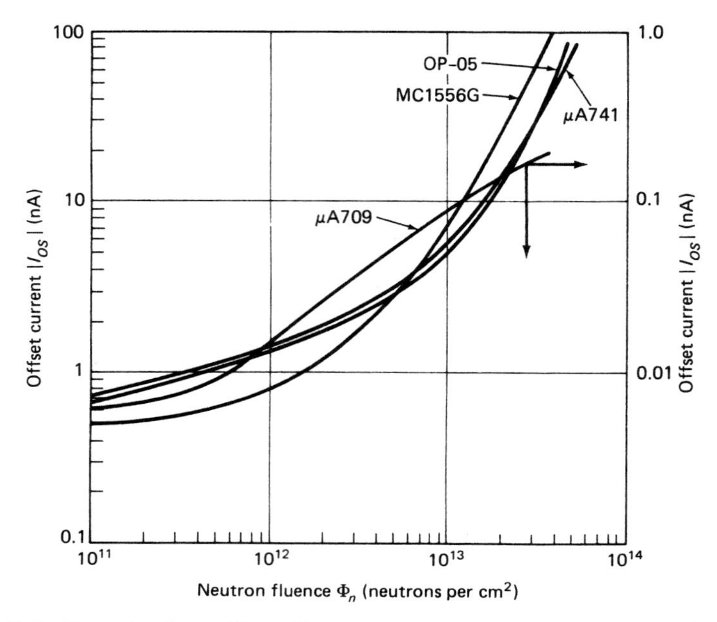

Fig. 12.8 Operational amplifier offset current variation with incident neutron fluence.[1] © 1984 by the IEEE.

The dependence on temperature of the minority carrier lifetime is very slight except for lightly doped bases ($\leqslant 10^{15}\,cm^{-3}$) in high-voltage devices, because the normal recombination occurs by way of deep trap transitions, as discussed in Section 1.9.

Neutron effects in bipolar digital integrated circuits cause permanent changes in their output voltage levels, input current, fanout, and propagation delay time.[1] Most modern TTL ICs experience performance degradation between $5 \cdot 10^{13}$–10^{14} neutrons per cm^2, but, if properly derated on an individual basis, can be used at fluences greater than 10^{14} neutrons per cm^2. Harder digital bipolar ICs, such as advanced Schottky TTL logic (FAST), or emitter coupled logic (ECL), do not degrade for fluences less than 10^{14} and can be used to 10^{15} neutrons per cm^2 when derated. Certain I^2L devices fail at lower fluence levels than TTL or ECL counterparts. Failure mechanisms in the first-generation devices included a loss of gain for the lateral *pnp* transistors and a reduction in gain and/or fanout in the vertical *npn* transistors. These difficulties are minimized in modern I^2L devices, which manifest failure thresholds greater than $5 \cdot 10^{13}$ neutrons per cm^2. Current mode logic (CML) is another latter day form of bipolar logic that has failure thresholds up to 10^{14} neutrons per cm^2.

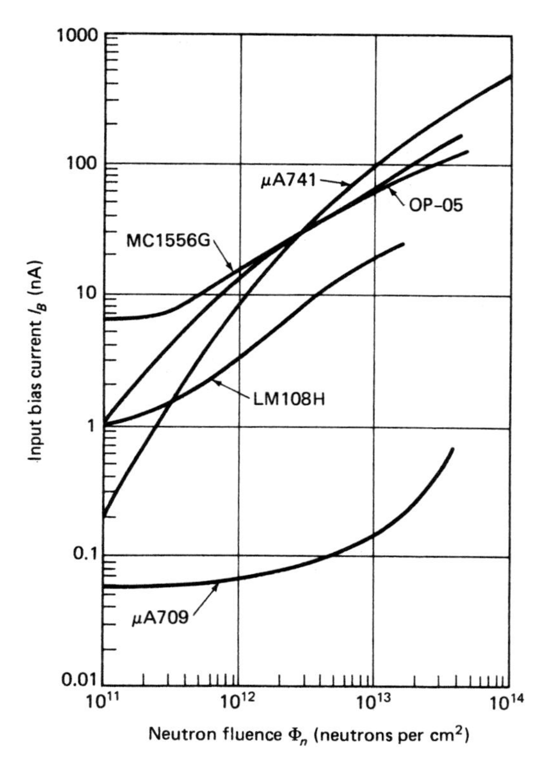

Fig. 12.9 Operational amplifier input bias current variation with incident neutron fluence.[1] © 1984 by the IEEE.

Neutron effects in bipolar linear integrated devices are quite varied because of the great variety of device designs and manufacturing techniques.[1] Changes in the IC parameters are due mainly to neutron-induced gain degradation of the individual transistors. Linear devices containing lateral *pnp* transistors are sensitive to neutron damage, because their f_T is relatively low, as seen from the physical and electrical neutron-dependent gain relationships discussed in Section 5.10.

For operational amplifiers, the important parameters that are affected by neutrons include input bias current, I_B, input offset current, I_{OS}, input offset voltage, V_{OS}, open loop gain, A_V, and slew rate. Figures 12.8–12.11A depict most of these parameter changes as a function of neutron fluence for typical operational amplifiers.[1] Usually, but not always, the first four parameters show significant changes with increasing neutron fluence before the slew rate does.

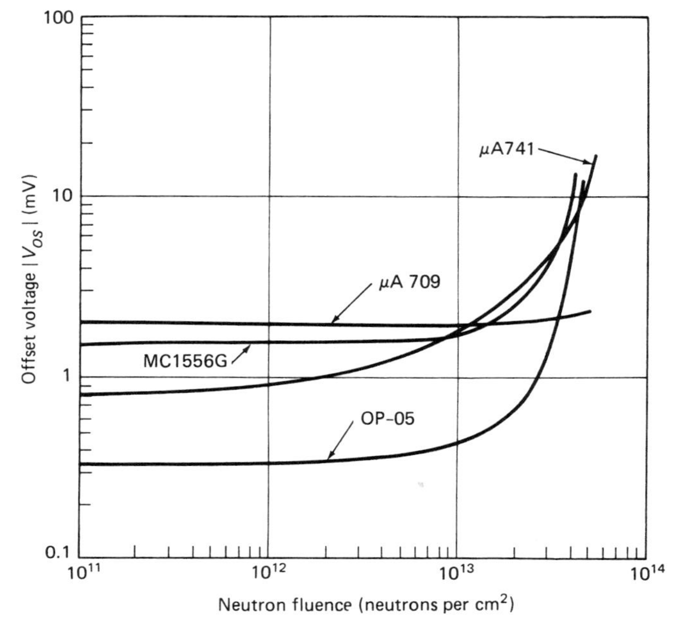

Fig. 12.10 Operational amplifier offset voltage variation with incident neutron fluence.[1] © 1984 by the IEEE.

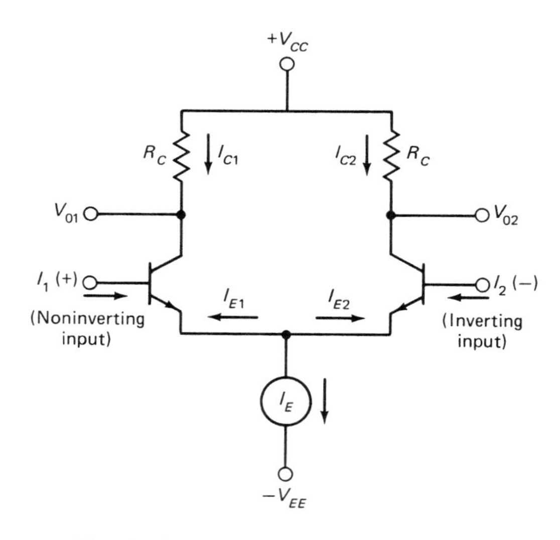

Fig. 12.10A Generalized op-amp input stage.

The input transistors of these ICs operate at very low current levels and therefore tend to have relatively large neutron damage constants, making for a ready gain reduction with neutron fluence. As the neutron fluence increases, thus reducing the transistor gains, the input bias current at each input increases. However, the input transistors are closely matched, so that I_B remains approximately equal in each leg (Fig. 12.10A). Further, I_B will increase before significant changes of V_{OS} and I_{OS} are evident. Most circuits can withstand larger increases of I_B than V_{OS} or I_{OS}, but degradation of any of these parameters can initiate a circuit failure. Even significant degradation of the open loop gain, A_V, will ordinarily not cause a circuit failure.

The slew rate will also be affected, because the gains of the intermediate transistors suffer, thereby reducing the drive currents required to charge internal and external capacitances. This will produce a decrease in the slew rate of most operational amplifiers.

In comparators, I_B, I_{OS}, and gain, all degrade in a manner similar to operational amplifiers. The low-output voltage state, V_{sat}, will also increase with neutron fluence. V_{OS} and V_{sat} are usually the most important from a circuit failure standpoint.

Almost all operational amplifiers (op-amps) and comparators have essentially the same circuit architecture. The input stage, the heart of the op-amp, consists of a differential amplifier (Fig. 12.10A), followed by several stages of amplification with associated functions. Commercial op-amp design uses four basic types of bipolar transistors. They are: supergain vertical *npn* transistors, lateral *pnp*, vertical (substrate) *pnp*, and the garden variety of vertical *npn* transistors.[85] Also, there are about a half dozen op-amp parameters of import that are susceptible to neutron and ionizing dose damage.[84] As mentioned earlier, they include input bias current I_B, input offset current I_{OS}, input offset voltage V_{OS}, open loop gain A_V, slew rate $S_r \equiv \max_{V_{in}} |\dot{V}_{\text{out}}|$ following excitation from a square voltage input pulse V_{in}, and input resistance R_{in}. $\text{Max}_x(\cdot)$ means the maximum of $(\cdot)$ found by choosing from the range of x values. If x is missing, it is implied.

Since bipolar transistors are susceptible to neutron fluence, corresponding changes of the differential amplifier input parameters of the op-amp are mainly due to the degradation of transistor gain. Neutron-sensitive output parameters of comparators are V_{OL} (output voltage-low state) and V_{OH} (output voltage-high state). V_{OL} is actually $V_{CE(\text{sat})}$ of the current sinking transistor in the output stage, as discussed previously. Op-amp parameters susceptible to neutron degradation are linked to degradation of β and leakage currents.[84]

To analyze radiation damage to op-amps, first, the op-amp parameters of interest are expressed in terms of those of the associated transistors. Then the radiation damage aspect is factored in through known dependencies of the above individual transistor parameters for the pertinent radiation types and levels. For example, the change in the input offset voltage ΔV_{OS}, due to incident radiation is given by[84]

$$\Delta V_{OS} \cong (kT/e)\ln[(I_1\alpha_3(1 + \beta_2^{-1}))/(I_1\alpha_3 + I_{CC}/3\beta_2)] \qquad (12.19)$$

where:

- I_1 = bias current of the (noninverting) first stage.
- β_2 = common emitter current gain of the super-gain transistors in the first stage.
- α_3 = common base current gain of the lateral *pnp* transistors in the first stage.
- I_{CC} = power supply current being drawn by the op-amp.

Now, the radiation-induced changes in the individual transistor parameters are incorporated. For β, this is expressed by a variation of Eq. (5.66), i.e. $\Delta(1/\beta_i) = \Phi_n/K_i\omega_{Ti}$ for neutron fluence incident on the ith transistor. If both ionizing radiation and neutron fluence is incident on the op-amp, then

$$\Delta(1/\beta_i) = \Phi_n/K_i\omega_{Ti} + F_i(\gamma_T) \qquad (12.20)$$

where γ_T is the ionizing dose level and an empirically determined function F_i, which increases monotonically. Often $F_i \cong K_{\gamma i}\gamma_T^{\varepsilon_i}$, $(\varepsilon_i > 0)$, where $K\gamma_i$ and ε_i are determined through testing of the particular device of interest. In the event that only ionizing radiation is present, then Φ_n vanishes and $\Delta(1/\beta_i) \cong F_i(\gamma_T)$. In Eq. (12.19), β_2 and I_1 are the more important parameters in terms of ΔV_{OS} susceptibility to neutron-induced degradation. α_3 and I_{CC} usually exert a secondary influence on ΔV_{OS}. It is usually assumed that the same Φ_n and γ_T are incident over the whole op-amp, because of its relatively small size compared to the extent of the radiation fields.

The input bias current is inversely proportional to β of the super gain transistors in the input stage. For matched transistors with equal areas and low offset voltage, the emitter current of each input transistor is $\frac{1}{2}I_1$, independent of β. Then the radiation-induced bias current change ΔI_B for either the noninverting or inverting input, since $I_B = I_1/2\beta_2$, is obtained as[84]

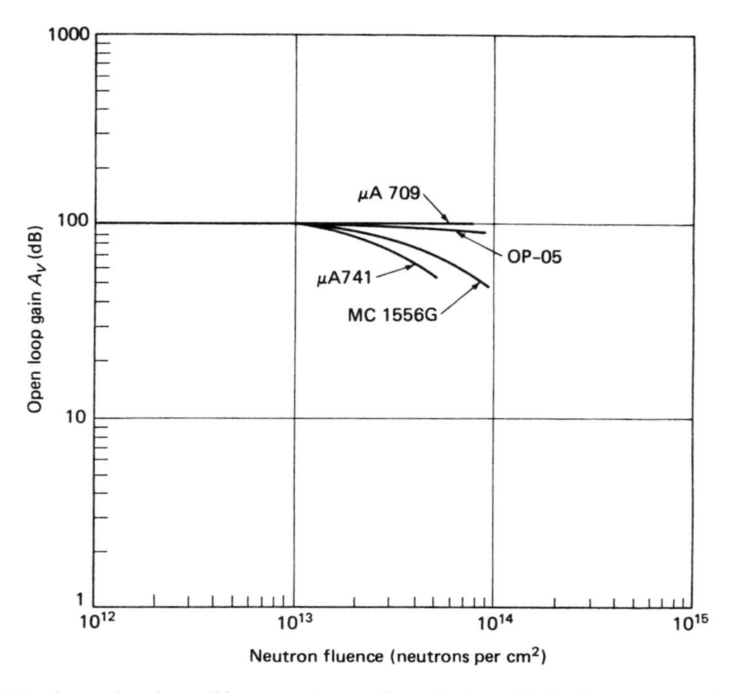

Fig. 12.11 Operational amplifier open loop gain variation with incident neutron fluence.[1] © 1984 by the IEEE.

$$\Delta I_B = (I_1/2)\Delta(1/\beta_2) \tag{12.21}$$

Finally, Eq. (12.20) supplies $\Delta(1/\beta_2)$ in terms of the radiation, which is then inserted into Eq. (12.21). Since the offset current is

$$I_{OS} = I_1 - I_2 = (I_{C1}/\beta_1) - (I_{C2}/\beta_2) \cong (I_E/2)(1/\beta_1 - 1/\beta_2) \tag{12.22}$$

where I_2 is the bias current of the inverting stage, using Eq. (12.20) in

$$\Delta I_{OS} = (I_E/2)[\Delta(1/\beta_1) - \Delta(1/\beta_2)] \tag{12.23}$$

yields the required result in terms of the effects of the incident radiation.

The open loop gain A_V is difficult to measure and compare before and after irradiation. Vendor A_V specifications are often couched in terms unsuitable for the measurement of these devices when irradiated.[84] One recognized approach to this difficulty is to measure and compare the

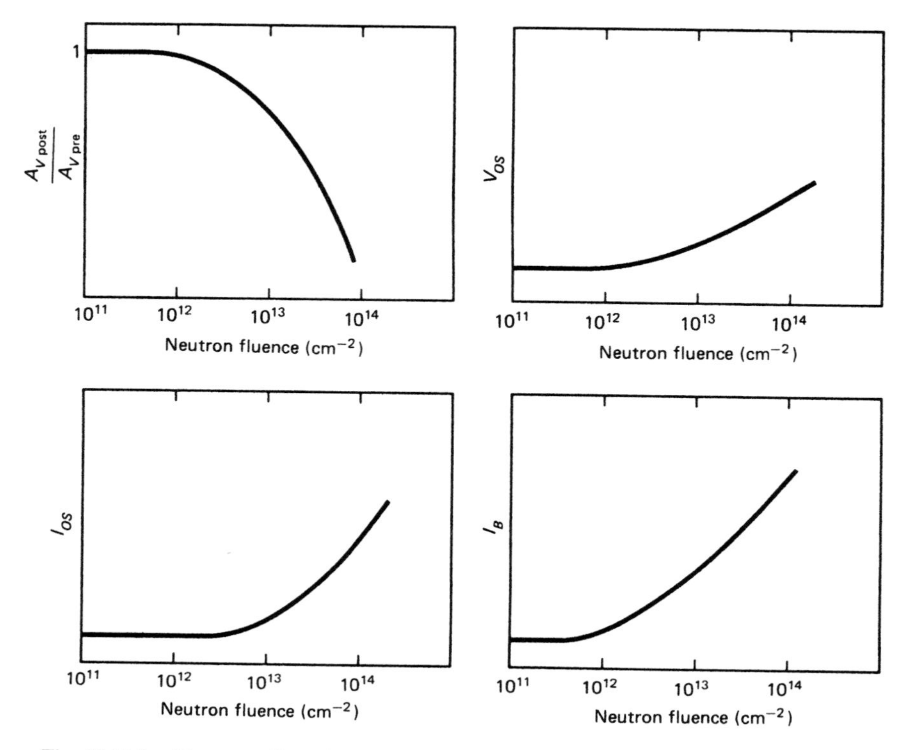

Fig. 12.11A Neutron effects in linear integrated circuits.[1] © 1984 by the IEEE.

device transfer function in its loaded and unloaded state, and define A_V in terms of the maximum input voltage needed to drive the output voltage V_{out} through a given range. Also, there may be large differences between A_V in the positive and negative directions, but they can be discerned from the measured transfer function. Radiation degradation of the output stage drive should be separated from A_V degradation, although it may cause an apparent failure of the A_V measurement.[84]

The slew rate S_r is usually limited by the total current of the first stage, and the size of the external compensation capacitor C_{cp}. So that from its definition given earlier,

$$S_r \equiv \max|\dot{V}_{out}| = \max|I_1/C_{cp}| \qquad (12.24)$$

For all except very large neutron fluence, Eq. (12.24) describes the slew rate degradation, which depends on the degradation of I_1 and α of the lateral *pnp* transistors. The change in slew rate, therefore, is

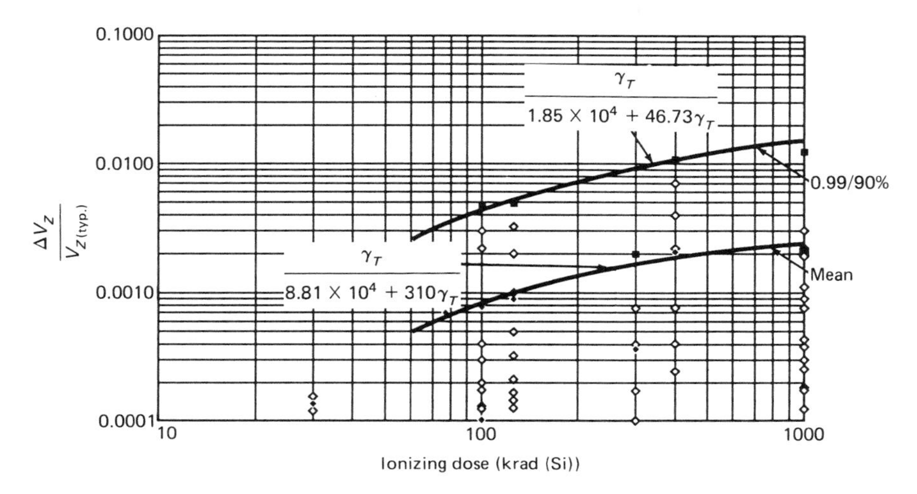

Fig. 12.12 Fractional change in V_Z for zener diodes at 25°C versus ionizing dose. ΔV_Z is the average change in V_Z for the given test sample.[83] © 1986 by Western Periodicals Co.

$$\Delta S_r = \max|\Delta \dot{V}_{\text{out}}| = \max|\Delta I_1/C_{cp}| \tag{12.25}$$

Since I_1 is essentially independent of gain, slew rate degradation is usually negligible, and is mainly due to increased leakage current with increasing radiation.

The input resistance, for multistage op-amps, is generally given by

$$R_{\text{in}} = \partial V_{E1}/\partial I_{B1} + \partial V_{E2}/\partial I_{B2} = (2kT/eI_E)(\beta_1 + \beta_2) \tag{12.26}$$

Since

$$I_B = I_{B1} + I_{B2} = I_{C1}/\beta_1 + I_{C2}/\beta_2 \cong (I_E/2)(1/\beta_1 + 1/\beta_2) \tag{12.27}$$

combining Eq. (12.26) and (12.27) yields, with $\beta_1 \simeq \beta_2$,

$$R_{\text{in}} \cong 4kT/eI_B \tag{12.28}$$

So that $\Delta R_{\text{in}} = (4kT/e)\Delta(1/I_B)$; using Eqs. (12.20) and (12.21) yields the corresponding radiation-induced ΔR_{in}.

A few additional points regarding op-amp radiation degradation can be stated. They are that (a) differential parameters are stable at low to moderate neutron fluences and ionizing dose levels, (b) input impedance decreases and bias current increases with low to moderate levels of

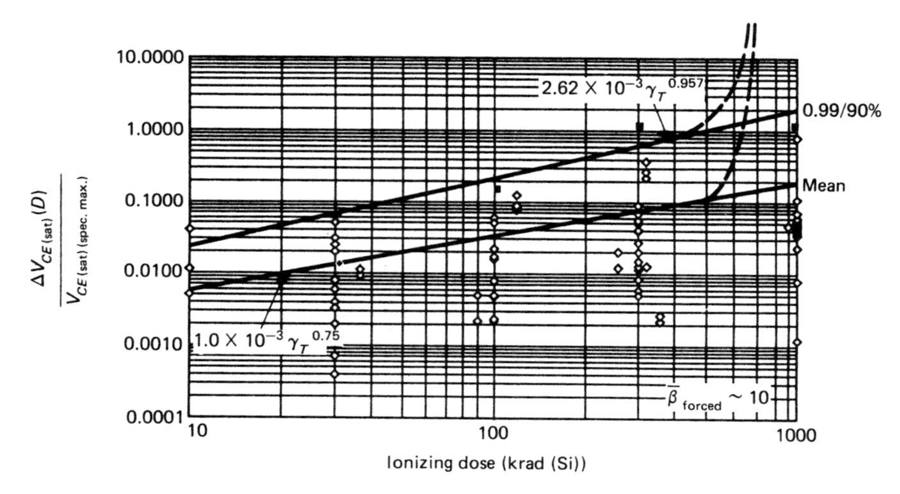

Fig. 12.12A Fractional change in $V_{CE(sat)}$ for general-purpose transistors at 25°C versus ionizing dose. $\Delta V_{CE(sat)}$ is the average change in $V_{CE(sat)}$ for the given test sample.[83] © 1986 by Western Periodicals Co.

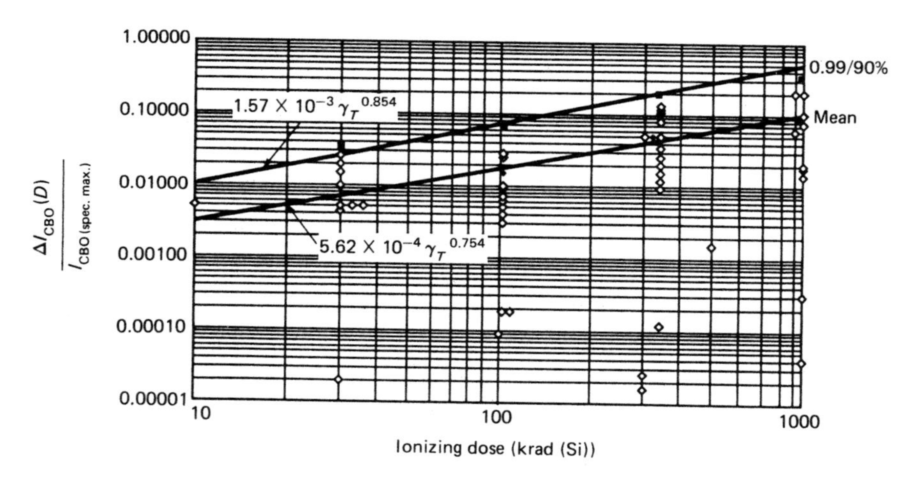

Fig. 12.12B Fractional change in I_{CBO} for general-purpose transistors at 25°C versus ionizing dose. ΔI_{CBO} is the average change in I_{CBO} for the given test sample.[83] © 1986 by Western Periodicals Co.

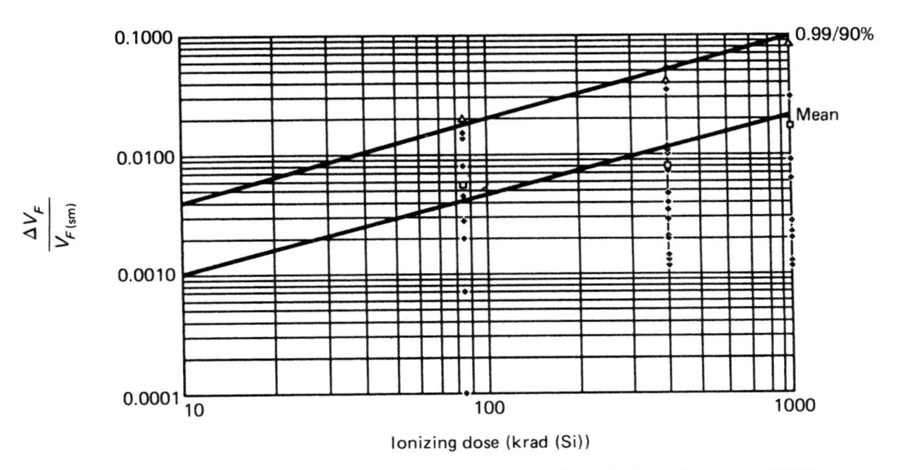

Fig. 12.12C Fractional change in V_F for switching and rectifying diodes at 25°C versus ionizing dose. ΔV_F is the average change in V_F for the given test sample.[83] © 1986 by Western Periodicals Co.

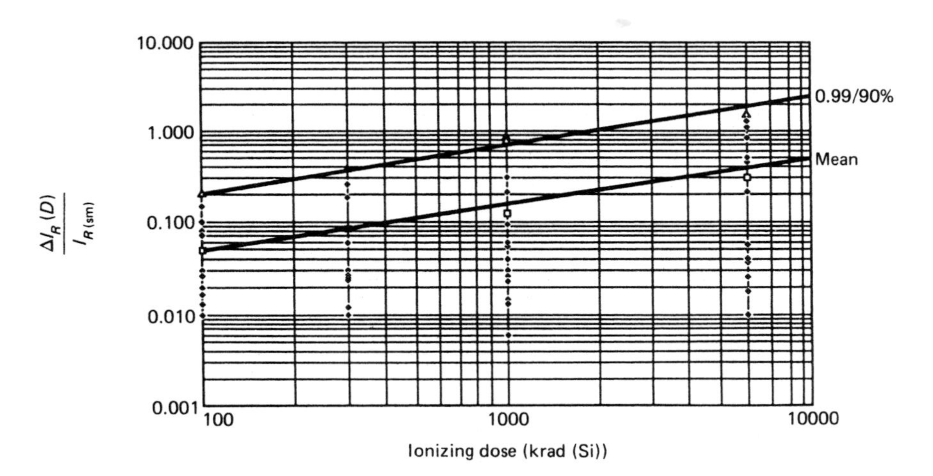

Fig. 12.12D Fractional change in I_R for switching and rectifying diodes at 25°C versus ionizing dose. ΔI_R is the average change in I_R for the given test sample.[83] © 1986 by Western Periodicals Co.

ionizing dose, whereas the corresponding effect of neutron fluence is insignificant except at high fluence levels, (c) changes in input differential parameters are almost always due to ionizing dose or neutron-induced ionization, in contrast to the direct displacement effects of neutron fluence, and (d) input topological balance and routing of power supply metallization remote from input areas is extremely important in op-amp physical design.

For voltage regulators, the line and load regulation parameters usually degrade most rapidly with increasing neutron fluence. However, the output voltage remains remarkably stable in some regulators. Nonetheless, the output voltage of certain regulators drops very rapidly to zero voltage with increasing neutron fluence. The knee of the output voltage-neutron fluence curve is often not consistent for devices of the same type and manufacturer.

12.3 COMPONENT SELECTION: IONIZING DOSE EFFECTS

As discussed in Section 6.5, incident gamma-ray and x-ray photons cause radiation damage to semiconductor components and active devices through the effects of ionization. MOS devices are quite sensitive to ionization dose, in contrast to their relative insensitivity to neutron-induced changes.

In rectifying and switching diodes, ionizing dose damage causes changes in forward voltage, V_F, reverse leakage current, I_R, and breakdown voltage, BV. These changes are barely discernible at ionizing dose levels of 10 krad (Si), but increase at 100 krad (Si) to $\Delta V_F \leqslant 50\,\mathrm{mV}$ and $\Delta I_R \leqslant 10\,\mu\mathrm{A}$.[1]

Voltage reference and zener diode V_Z, or reference voltage changes with ionizing dose, are depicted in Figs. 12.12[1] and 12.16.[5] Insignificant changes in these parameters occur up to 0.1 Mrad (Si), and 0.1–1.0 percent changes are typical between 0.1–1.0 Mrad (Si). The damage mechanism is the creation of partial inversion layers; this process depends on the doping concentration. Heavily doped, low-breakdown-voltage diodes are less susceptible to ionizing dose damage than lightly doped, high-breakdown-voltage diodes.

Microwave diodes of most types, such as PIN, IMPATT, TRAPATT, and Gunn are relatively hard to ionizing dose, as well as neutrons. Their preirradiation parameters are generally unaffected at ionizing dose levels to 1 Mrad (Si) and display only slight changes between 1–10 Mrad (Si).[1]

For field effect transistors (MOSFETs), ionizing dose causes a shift in the gate threshold voltage of n-channel MOS toward negative voltages, and likewise for p-channel MOS. The threshold voltage shifts are due to positive (hole) charges trapped in the gate oxide layer and their migration, as discussed in Sections 6.8 and 6.9. Commercial NMOS devices suffer failure levels of threshold voltage shift for dose as low as 2–10 krad (Si). Further MOS damage effects include changes in channel resistivity and carrier mobility. Changes in MOS drain current leakage are usually larger than predicted by first-order calculations. This implies that some of the leakage current is due to edge leakage possibly caused by surface contaminants near the edge of the junction. Other effects include a decrease in the transconductance and switching speed. Propagation delay time may increase, as well as quiescent supply current, sometimes by orders of magnitude over preirradiation values.

For junction field effect transistors (JFET), the gate-to-source leakage current, I_{GSS}, is one of the most sensitive parameters with respect to ionizing dose, sometimes increasing rapidly beyond 1 Mrad (Si). *P*-channel JFETs are more resistant than *n*-channel JFETs to ionizing dose radiation. Corresponding JFET ionizing dose damage thresholds are manifest between 0.05–0.2 Mrad (Si).[8] Data for seven JFETs from one vendor are given in Table 12.1.[9] Relatively large radiation-induced leakage currents can affect the operation of JFETs, due to their high input impedance. These leakage currents can be computed as diode leakage currents utilizing the ideal diode equation. JFET leakage currents can increase by two orders of magnitude over their preirradiation values after exposure to 10–100 krad (Si) of ionizing radiation.

When an *n*- or *p*-channel depletion or enhancement mode MOSFET is conducting or is cut off, a radiation-inclusive relationship between its drain current, I_D, and threshold voltage, V_T, can be written as

$$I_D = \begin{cases} 0; & \text{cut off } \delta V \leqslant 0 \\ M[(V_{GS} - V_T)V_{DS} - \frac{1}{2}V_{DS}^2]; & \text{unsaturated } \delta V \leqslant V_{DS} \\ M(V_{GS} - V_T)^2/2; & \text{saturated } 0 \leqslant \delta V \leqslant V_{DS} \end{cases} \tag{12.29}$$

where $\delta V = V_{GS} - V_T$. V_{GS} and V_{DS} are the gate-source and drain-source voltages, respectively, and

$$M(\gamma_T) = \bar{\mu}(\gamma_T)\varepsilon W/x_0 L \tag{12.30}$$

W is the channel width, x_0 is the gate oxide thickness, L is the channel length, ε is the oxide permittivity, and $\bar{\mu}$ is the average mobility in the channel. M, V_T, and $\bar{\mu}$ are radiation-dependent, and curves of their dependence on ionizing dose, γ_T, are available.[10]

With regard to MOS devices, NMOS is the least radiation-resistant,

Table 12.1 JFET ionizing dose data from one manufacturer for three radiation levels.[9]

JFET GENERIC NO.	BIAS CONDITION	MEASURED PARAMETERS AT PRE- AND POSTEXPOSURE DOSE IN Mrad (Si)								
		$V_{pinch\ off}$ (V)			I_{GSS} (nA − pA)			G_m (mmhos)		
		PRE	0.1 Mrad	1 Mrad	PRE	0.1 Mrad	1 Mrad	PRE	0.1 Mrad	1 Mrad
2N4416	On	−3.46	−3.47	−3.47	< 20 pa	20 pa	500 pa	5.4	5.4	5.3
	Off	−2.35	−2.35	−2.28	< 20 pa	300 pa	18 pa	6.0	5.8	5.8
2N4856	On	−8.56	−8.56	−8.58	< 20 pa	20 pa	150 pa	8.4	8.4	8.4
	Off	−8.53	−8.53	−8.27	< 20 pa	8.4 pa	400 pa	8.8	8.8	8.8
2N4857	On	−5.03	−5.03	−5.01	< 20 pa	50 pa	110 pa	14.0	13.8	13.8
	Off	−5.08	−5.08	−5.02	< 20 pa	1.5 pa	38 pa	14.0	13.8	13.8
2N4861	On	−1.95	−1.96	−1.95	< 20 pa	80 pa	2.9 pa	15.6	15.4	15.4
	Off	−1.90	−1.88	−1.75	< 20 pa	3.5 pa	340 pa	15.6	15.4	15.4
2N4868	On	−1.83	−1.82	−1.83	< 20 pa	50 pa	600 pa	8.8	8.8	8.8
	Off	−2.41	−2.41	−2.36	< 20 pa	200 pa	12 pa	7.6	7.6	7.6
2N5486	On	−2.07	−2.07	−2.07	< 20 pa	20 pa	400 pa	6.7	6.7	6.7
	Off	−2.04	−2.04	−2.01	< 20 pa	100 pa	8 pa	6.2	6.2	6.2
2N5545	On	−1.08	−1.08	−1.08	< 20 pa	50 pa	600 pa	9.2	9.2	9.0
	Off	−1.10	−1.10	−1.03	< 20 pa	500 pa	15 pa	9.3	9.2	9.2

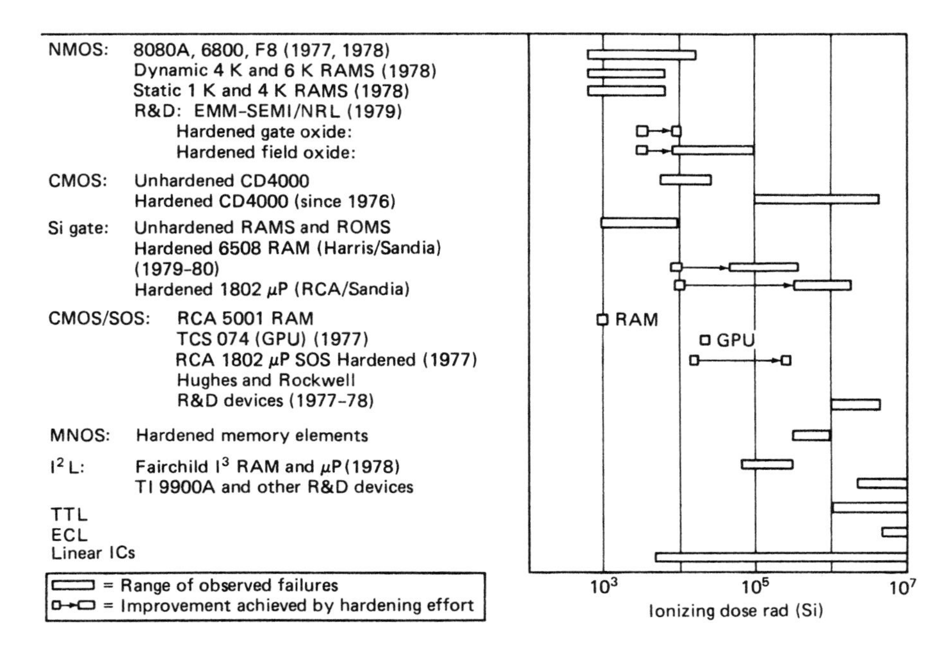

Fig. 12.13 Ionizing dose hardness levels of various MOS and bipolar devices circa 1978.[3] © 1980 by the IEEE.

with failures occurring in the 0.7–7.0 krad (Si) range, for static and dynamic RAMs and microprocessors. Hardness development of such devices is proceeding and shows improvement into the 100 krad–10 Mrad (Si) regime. Aluminum gate terminal devices can be hardened to 10 Mrad (Si). The CD 4000 class of hardened CMOS devices using an aluminum gate terminal is available. The order of increasing ionizing dose radiation hardness is NMOS (least resistant), polysilicon gate CMOS, PMOS, and metal gate CMOS, as shown in Figs. 12.13–12.16.[5]

Other advances in radiation-resistant devices include the 1802 microprocessor hardened to 0.3–2.0 Mrad (Si) and the 6508 dynamic RAM hardened to 0.05–0.4 Mrad (Si). The hardness level of these two devices is strongly dependent on their annealing characteristic. CMOS/SOS and CMOS/SOI devices can be hardened to better than 1 Mrad (Si). MNOS and SNOS nonvolatile memories have been hardened to 0.3 Mrad (Si).

Bipolar digital systems are not strongly affected by ionizing dose radiation and their radiation hardness can easily exceed 1 Mrad (Si) as seen in Figs. 12.15–12.19.

Bipolar construction provides very little contact of any oxide with a

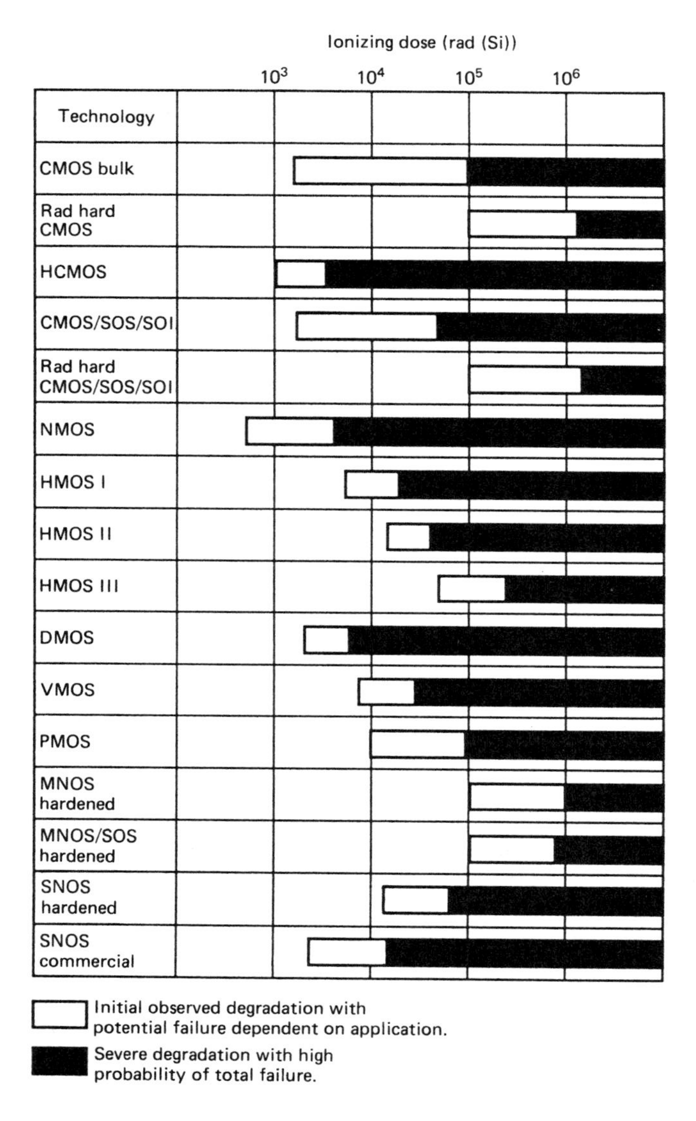

Fig. 12.14 Ionizing dose hardness levels for MOSFET integrated circuit families circa 1990.[5] © 1990 by the Physitron Corp.

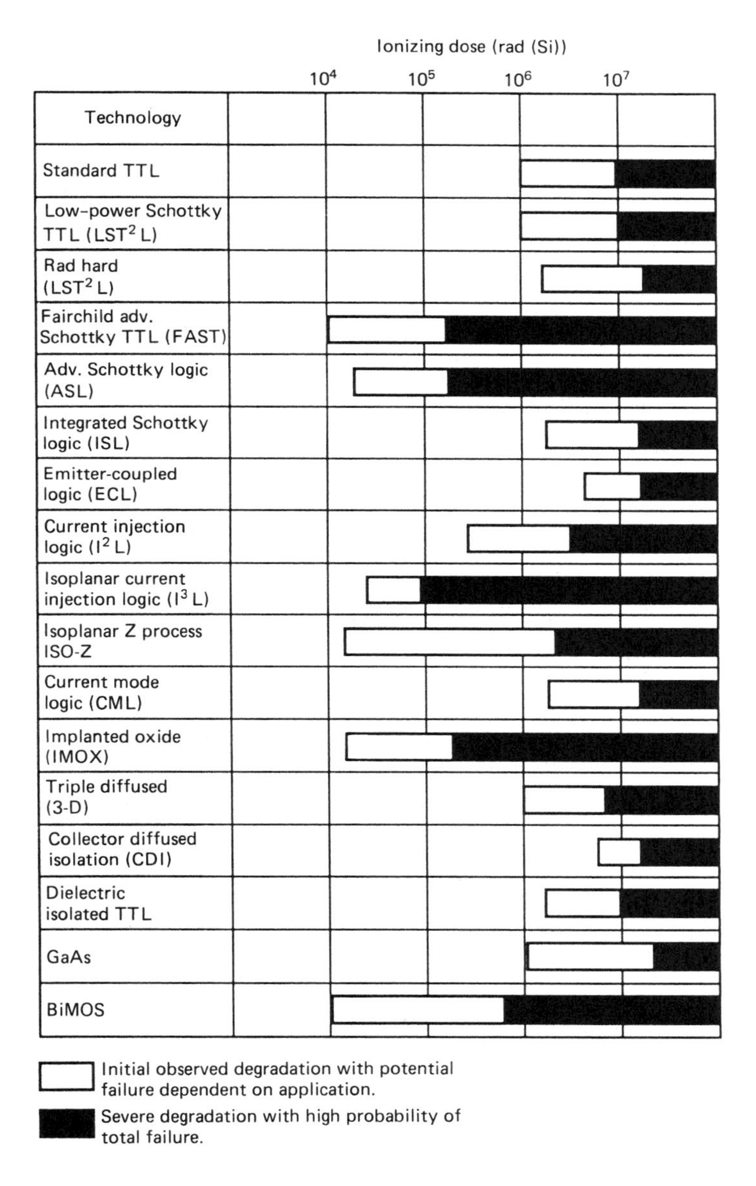

Fig. 12.15 Ionizing dose hardness levels for mainly bipolar integrated circuit families circa 1990. © 1990 by the Physitron Corp.

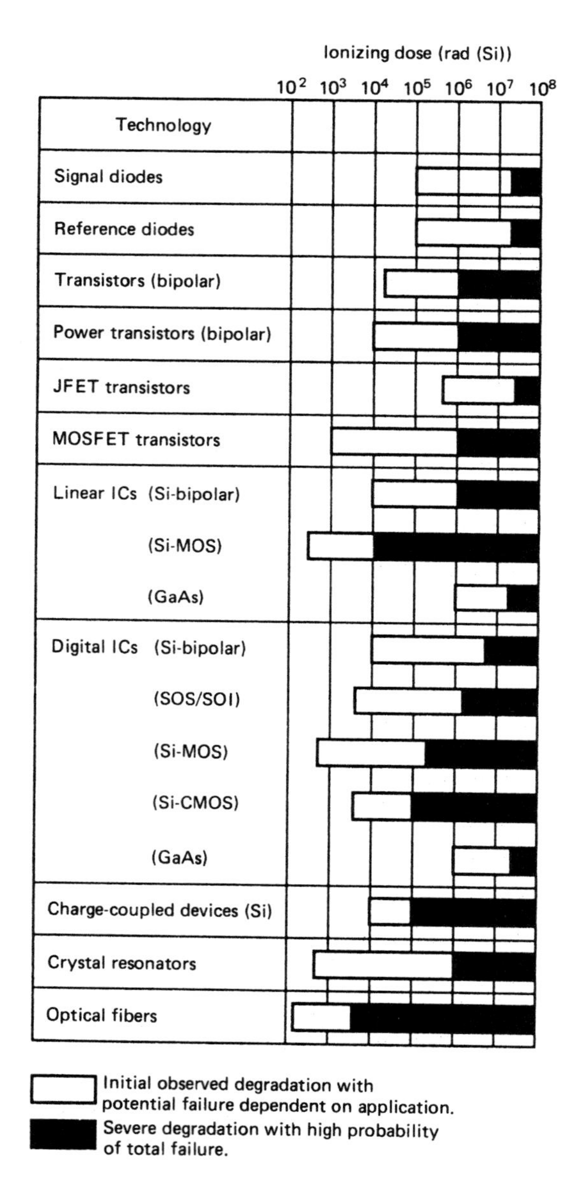

Fig. 12.16 Ionizing dose hardness levels for discrete, linear, and digital device families circa 1990.[5] © 1990 by the Physitron Corp.

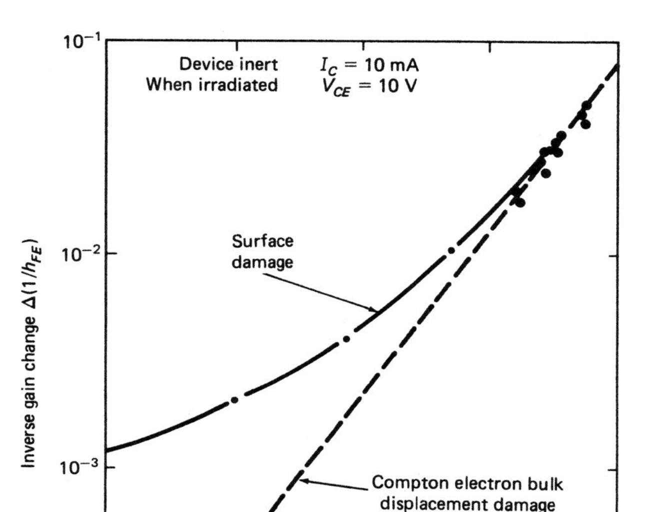

Fig. 12.17 Ionizing dose bipolar gain degradation due to surface charge states and electron bulk displacement damage.[1] © 1984 by the IEEE.

p–n junction. This is in contrast to MOS devices whose gate oxide layer constitutes a relatively large surface in contact with a channel. Ionization damage in this oxide layer degrades the performance of MOS through influencing channel behavior in terms of threshold voltage shifts. However, linear bipolar ICs can be very soft or hard to ionizing dose, depending on the particular device construction and the radiation-resistant processing used if any. For example, relatively low levels of ionizing dose can be devastating to the low current input characteristics of linear bipolar devices, such as the input bias current, and offset voltage of operational amplifiers. Figures 12.13–12.18 depict hardness levels of various MOS and bipolar devices.[2,5,11,12–15,17–30] (sm) in graphs means specification maximum.

To reiterate, bipolar transistors manifest changes in common emitter current gain, h_{FE}, leakage currents, I_{CBO} and I_{BEO}, and saturation voltage, $V_{CE(\text{sat})}$ upon exposure to ionizing dose. Under such radiation,

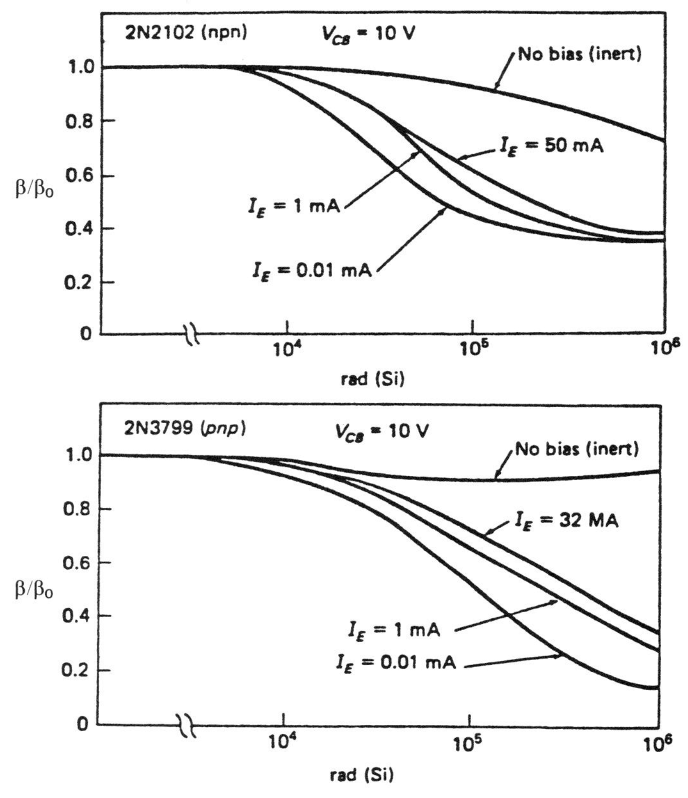

Fig. 12.18 Ionizing dose-induced common emitter gain change of *npn* and *pnp* transistors with in situ bias conditions.[1] © 1984 by the IEEE.

bipolar transistor gain degradation loss is partly due to ionizing dose-induced leakage currents of all types. It is also a function of collector-base bias voltage, as shown in Fig. 12.18. Low current, high gain transistors are unusually sensitive to the effects of ionizing dose. In addition to gain degradation, both the leakage current I_{CBO} and $V_{CE(\text{sat})}$ increase, as mentioned previously. The latter are of lesser importance, amounting to 1 percent changes starting at about 100 krad (Si).

Bipolar digital integrated circuits are relatively hard to the effects of ionizing dose. TTL and I^2L devices are still operable at 100 krad (Si) to 1 Mrad (Si). However, bipolar devices built with oxide isolation, such as advanced Schottky TTL (FAST) parts, and others, are susceptible to failure levels for dose magnitudes greater than 100 krad (Si). This is mainly the result of the ubiquitous leakage currents in such devices. Even when hardened manufacturing processes are implemented to minimize

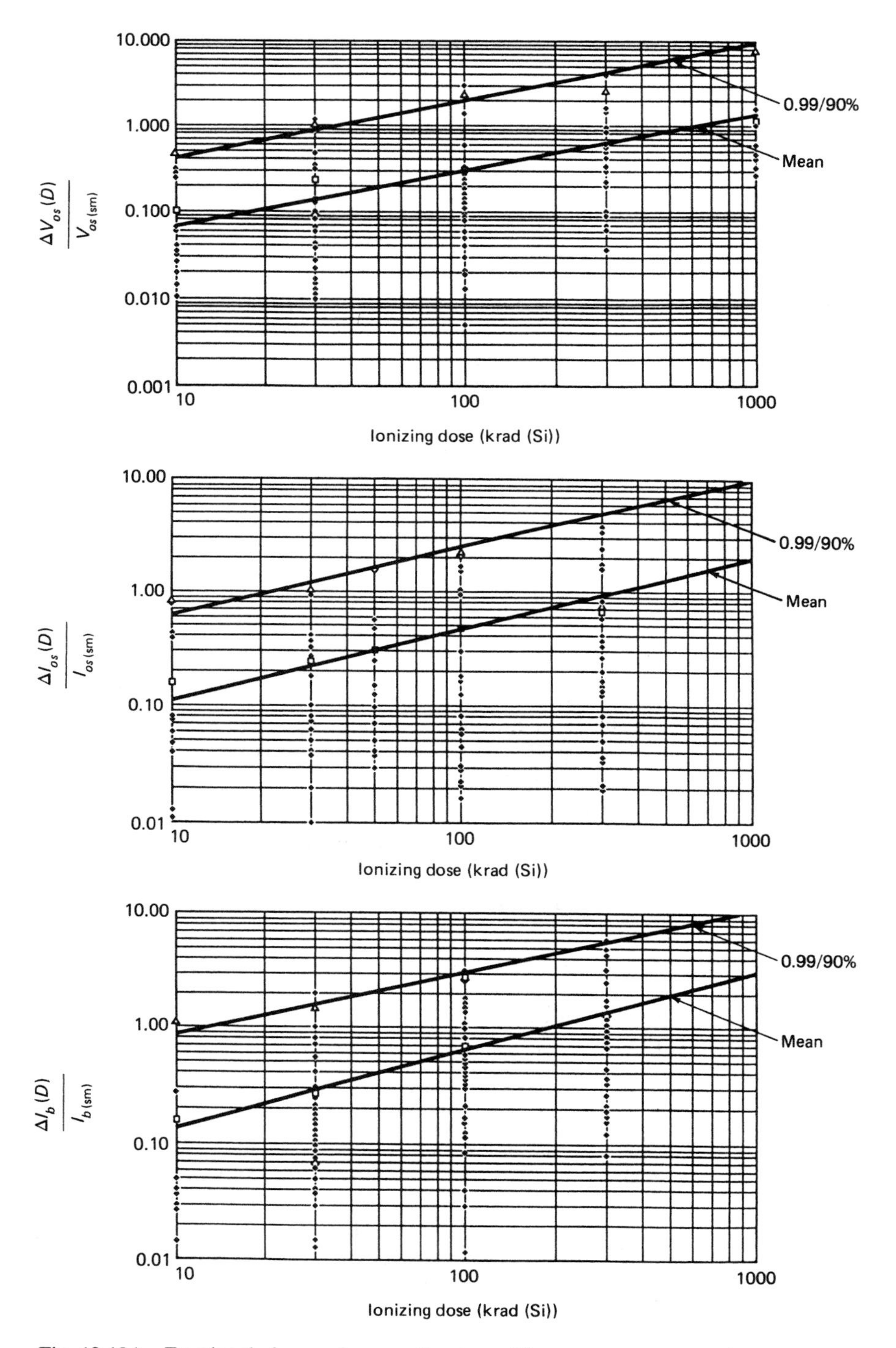

Fig. 12.18A Fractional changes in operational amplifier parameters versus ionizing dose. These data represent the average change for the given test sample at 25°C, divided by the specification maximum value of that parameter.[83] © 1986 by Western Periodicals Co.

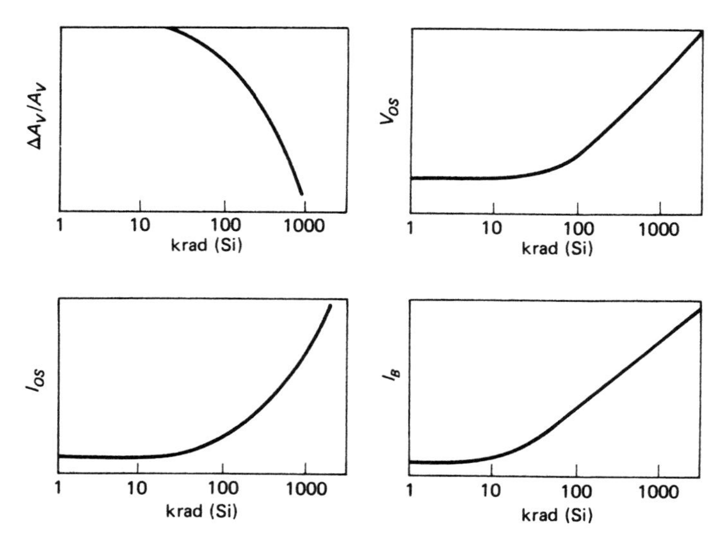

Fig. 12.19 Linear bipolar integrated circuit parameter qualitative variations with ionizing dose.[1] © 1984 by the IEEE.

leakage, these devices may still be susceptible to ionizing dose failure. However, other modern technologies, such as advanced low-power Schottky logic (ALS), isoplanar current-mode logic (CML), and double-diffused (2D) devices, are essentially hard to at least 1 Mrad (Si). Emitter-coupled logic (ECL) remains hard up to 10 Mrad (Si).

Bipolar linear integrated circuits are moderately sensitive to ionizing dose effects. Offset voltage, V_{OS}, offset current, I_{OS}, input bias current, I_B, and open loop gain, A_V, are key parameters of linear ICs that are affected by ionizing dose.

In bipolar linears, the threshold for these parameter changes occurs at about 10 krad (Si). As seen in Fig. 12.19, input bias current changes are more pronounced than those of other parameters. Usually, the mean input bias current change at, for example, 10 krad (Si) is a very small fraction of the maximum I_B. This is not always so, as in the LM-111F voltage comparator/buffer, which has a vendor-specified maximum I_B of 100 nA at 25°C and a measured mean change in I_B of 98.6 nA after exposure to 10 krad (Si).[1] As with bipolar digital integrated circuits, their linear counterparts with oxide isolation suffer significant leakage current damage and subsequent failures at dose levels above 10 krad (Si). Radio frequency (RF) bipolar linear microcircuits are less sensitive than those discussed above, with more than 100 krad (Si) required to produce observable parameter changes.[1]

Table 12.2 Semiempirical photocurrent equations using parameters readily available from vendor data.[31] © 1969 by the IEEE.

POLARITY	P_D (W) @ 25°C	TYPE	PREDICTION EQUATION I_{pp}(mA) =	k_p	
npn	≤0.6	Switching	$k_p \dot{\gamma} C_{cb1} V_{CB1}^{1/3} t_s^{1/2} (6.47 + V_{CB2}^{1/3})$	6×10^{-12}	(12.31)
pnp	≤0.6	"	"	9.5×10^{-12}	
npn	0.8–1.0	"	"	1.7×10^{-11}	
pnp	0.8–1.0	"	"	2.6×10^{-11}	
npn	≥2.0	"	"	4×10^{-11}	
pnp	≥2.0	"	"	6.3×10^{-11}	
npn	All P_D	Linear	$k_p \dot{\gamma} f_T^{-2/5} V_{CB0} (C_{cb1} V_{CB1}^{1/3} + 1.08) \cdot (21.6 + V_{CB2}^{1/3})$	3.24×10^{-13}	(12.32)
pnp	All P_D	"	"	4.8×10^{-13}	

12.4 COMPONENT SELECTION: DOSE RATE EFFECTS

All semiconductor devices and components subject to incident ionizing radiation pulses endure the production of charge carriers and concomitant current flow discussed in Section 7.2. These spurious photocurrents flowing within components and across device junctions can cause functional, and sometimes physical, failure.

In bipolar transistors the base-collector junction area is usually much greater than that of the base-emitter junction. Hence, the base-collector primary photocurrent dominates the transient response. A general expression for the photocurrent due to a pulse of ionizing radiation, as a function of pulse parameters and device dimensions, was derived in Section 7.2. For pulses of duration, t_p, short compared with the minority carrier lifetime, the primary photocurrent, $I_{pp} = eA_c g_0 \dot{\gamma} \sqrt{Dt_p}$.

For particular transistors, whose electrical descriptions are readily obtained from vendor data sheets or catalogs, the following yields two semiempirical generic equations, (12.31) and (12.32), for the corresponding photocurrents. They are contained in Table 12.2, for *npn* and *pnp* switching (digital) and linear operating transistors grouped by their rated power dissipation and other electrical parameters.[31] These electrical parameters include:

V_{CBO}, maximum collector-base voltage (V)
V_{CBI}, reverse bias voltage level at which C_{cb1} is measured (V)

Table 12.3 Bipolar transistor dose-rate-induced primary collector photocurrent (the photocurrent per unit dose rate exponent is lowered, as for example $1.26 \cdot 10^{-7} \equiv 1.26 - 7$).[32]

GENERIC/ MANUFACTURER	I_{ppc}(mA)*	$I_{ppc}/\dot{\gamma}$ amp/rad/s	GENERIC/ MANUFACTURER	I_{ppc}(mA)*	$I_{ppc}/\dot{\gamma}$ amp/rad/s
2N1358/Delco	12600	1.3–7	2N2978/FCH	13.5	1.4–10
2N1722A/TI	3900	3.9–8	2N2484/TI	13.4	1.3–10
2N1908/TI	1900	1.8–8	2N3019/MOT	10.6	1.1–10
2N389/TI	1200	1.2–8	2N1132/FCH	9.6	9.6–11
2N4002/TI	780	7.8–9	2N2984/FCH	6.6	6.6–11
2N4070/SOL	730	7.3–9	2N3499/MOT	6.5	6.5–11
2N1041/TI	550	5.5–9	2N2432A/TI	6.1	6.1–11
2N655/MOT	400	4.0–9	2N1613	5.8	5.8–11
2N5038/RCA	360	3.6–9	2N335B/GE	4.7	4.7–11
2N2851	309	3.1–9	2N2222A/MOT	4.0	4.0–11
2N657/TI	260	2.6–9	2N4430/TRW	3.2	3.2–11
2N4071/SOL	196	2.1–9	2N741A/MOT	3.1	3.1–11
2N3878/RCA	170	1.7–9	2N972/TI	3.0	3.0–11
2N2880/SOL	142	1.4–9	2N914/MOT	2.8	2.8–11
2N3752/SOL	115	1.2–9	2N930/MOT	2.2	2.2–11
2N5004/TI	106	1.1–9	2N3274	2.0	2.0–11
2N3552/TRW	103	1.0–9	2N3253/TI	1.7	1.7–11
2N3792/MOT	100	1.0–9	2N3762/MOT	1.7	1.7–11
2N191/ETC	80	8.0–10	2N2920/MOT	1.6	1.6–11
2N5320/RCA	60	6.0–10	2N2907A/RAY	1.5	1.5–11
2N5005/FCH	45.5	4.5–10	2N4025/FCH	1.4	1.4–11
2N5200/RCA	38.7	3.9–10	2N4022/FCH	1.4	1.4–11
2N5154/FCH	35.9	3.6–10	2N2635/MOT	1.4	1.4–11
2N6248/RCA	35.7	3.6–10	2N338/GE	1.4	1.4–11
2N915/FCH	34.7	3.5–10	2N2219A	1.4	1.4–11
2N188A/ETC	26	2.6–10	2N3251A/RAY	1.4	1.4–11
2N404/RCA	25	2.5–10	2N3467/MOT	1.3	1.3–11
2N696/TI	21	2.1–10	2N4044/DCC	0.76	7.6–12
2N1998/SYL	21	2.1–10	2N708/ITT	0.70	7.0–12
2N1893/GES	20	2.0–10	2N3251A/TI	0.69	6.9–12
2N697/FCH	20	2.0–10	2N3303/MOT	0.56	5.6–12
2N3507/MOT	19	1.9–10	2N4428/TRW	0.40	4.0–12
2N1174/WE	17	1.7–10	2N3014/FCH	0.34	3.4–12
2N910/TI	15	1.5–10	2N3959/MOT	0.28	2.8–12
2N2905A/FCH	14.4	1.4–10	2N4429/TRW	0.19	1.8–12
2N3501/MOT	14.1	1.4–10	2N2412/TI	0.13	1.3–12
2N5153/FCH	13.6	1.4–10	2N2369A/MOT	0.10	1.0–12

* The column labeled I_{ppc}(mA)* corresponds to a photocurrent for a dose rate of 10^8 rad (Si) per s, which is the dose rate output from a nominal 20 KT surface burst at approximately 1 mi.

C_{cb1}, collector bias capacitance at V_{CBI} (pF)
f_T, minimum common emitter gain-bandwidth product (GHz)
$\dot{\gamma}$, dose rate (rads (Si) per s)
t_s, electrical storage time (ns)
V_{CB2}, reverse bias voltage during ionizing pulse (V)
I_{pp}, Primary photocurrent (nA)
P_D, rated power dissipation

Nomograms for the solution of these equations are available.[32] In terms of comparative accuracy between computed and measured photocurrents using the above equations, computed photocurrents of 51 switching transistors are within a factor of 2.5 for 90 percent of those measured. For 74 small and medium signal devices, computed photocurrents are within a factor of 2.5 for 89 percent of those measured.[31] Table 12.3 lists transistor photocurrents, as well as specific photocurrent per dose rate, for a diversity of transistors from high-power to small-signal types. This table can be used as a directory by the reader to match a transistor type whose photocurrent is of interest.

In addition to the primary photocurrent, there is a secondary photocurrent, I_{sp}, which is the sum of the primary photocurrent plus that due to transistor current multiplication of I_{pp}, to yield the augmented collector current discussed in Section 7.3.

At very high dose rates, burnout of the transistor metallization leads can be caused by excessive heating. For this and other excessive transient current difficulties, a small resistor or inductor (ferrite bead) can be inserted in the collector lead to sustain most of the voltage drop, and so limit the current to the affected transistor. Collector-emitter series RC by-passes can also be used.

Power transistors are especially sensitive to ionizing radiation because of their large junction areas, interdigitated construction, and relatively massive bulk. This makes for large numbers of ionized carriers, producing large photocurrents measured in amperes, as seen in Table 12.3. Power transistors can be easily driven into saturation at dose rates of less than 10^8 rad (Si) per s. However, they can at the same time survive dose rates of up to 10^{12} rad (Si) per s. Also, radiation-induced second breakdown can occur in high-voltage power transistors enduring high dose rate levels.

As shown in Fig. 12.26, primary photocurrent can be generated in diodes in the same way as in transistors. Low-power diodes have specific photocurrent generation rates on the order of 10^{-10}–10^{-13} amp/rad (Si) per s. Large power diodes have correspondingly greater photocurrent generation rates measured in 10^{-8}–10^{-9} amp/rad (Si) per s. Ionizing radiation in zener diodes produces photocurrents that modulate the zener

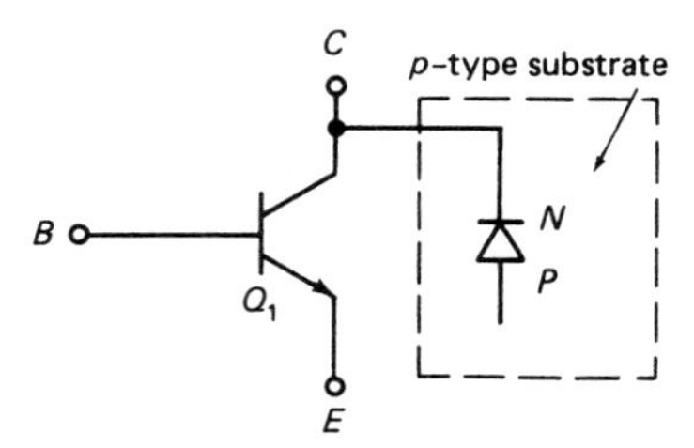

Fig. 12.20 Junction-isolated *npn* transistor equivalent circuit.

voltage. However, this component can be eradicated by suitable filtering. These voltage transients are observed only at high dose rates (large photocurrents), because the conducting zener diode is a low-impedance device.

Bipolar integrated circuits are usually constructed epitaxially, of complicated shapes, with regions of different dopant densities connected by metallization leads, all mounted on a common substrate. Two basic types of ICs, from the fabrication standpoint, are the junction-isolated (JIIC) and the dielectrically isolated (DIIC) types, also discussed in Section 7.7. The junction-isolated transistor equivalent circuit is shown in Fig. 12.20.

The JIIC transistor is so constructed that the *p*-type substrate is in physical contact with the collector, thus forming a junction-like interface. The substrate is provided with a bias to the lowest voltage in the circuit, so that a reverse biased diode junction is produced between the substrate and the collector, as depicted in the figure. This diode electrically isolates the transistor from its neighboring transistors on the substrate chip. However, this substrate diode forms part of a parasitic transistor with the base of the *npn* transistor in Fig. 12.20, which acts as the collector of a parasitic *pnp* transistor. This in turn forms an *npnp* parasitic SCR latchup path, as discussed in Section 7.7.

One method for inhibiting such parasitic SCR action is gold doping of the base and collector of the transistor Q_1 in Fig. 12.20. Gold doping decreases the minority carrier lifetime of the elements of the parasitic transistor path only, in order to strongly degrade their transistor function. As the parasitic transistor gain is a function of the minority carrier lifetime, as discussed in Section 5.9, the thus degraded parasitic transistor gain lessens the probability of it turning on to cause the system to latch. Latchup was the bane of early linear JIIC devices and could occur for dose rates as low as 10^6 rad (Si) per s. A so-called hard latchup locks the device into a large-current, and therefore large-power-dissipating, mode, which often results in catastrophic burnout failure of the device unless it is quickly powered down.

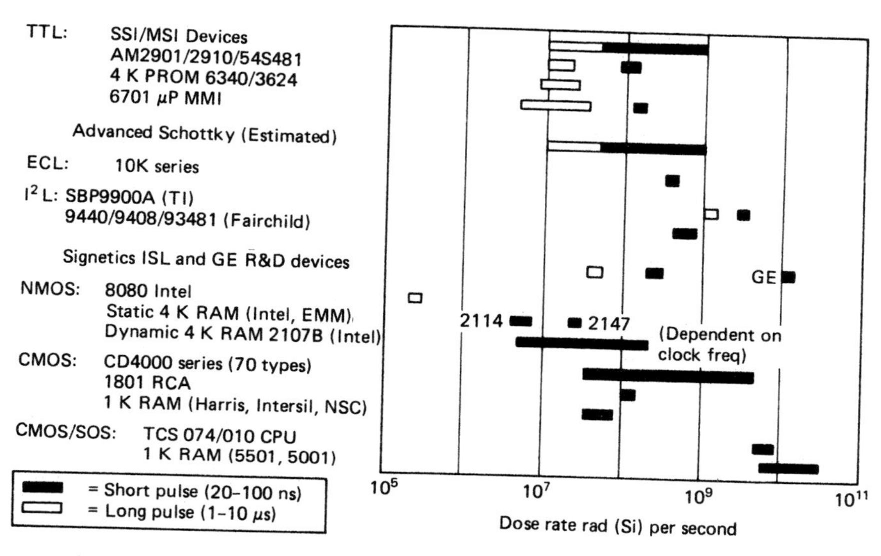

Fig. 12.21 Integrated circuit transient hardness upset threshold levels as a function of dose rate circa 1978.[33] © 1980 by the IEEE.

Dielectrically isolated integrated circuits (DIIC) are constructed so that each transistor is physically isolated from its substrate by using dielectric (SiO_2) layers or tubs between each transistor cell and the common substrate. DIIC devices eliminate the JIIC latchup difficulty. DIICs do not upset at 10^7 rad (Si) per s. Modern versions of these logic ICs can be considered relatively hard to permanent damage from both peak dose rates and high ionizing dose. Specifically, they have survived peak dose rates in excess of 10^{12} rad (Si) per s and can conservatively endure 10^5 rad (Si) ionizing dose.

Latchup is considered to be rare in modern day TTL devices. It has not been observed in emitter-coupled logic (ECL) ICs. Current injection logic (I^2L) integrated circuits do not latch up because their very low required supply voltage is equal to about only one silicon diode forward voltage drop. The latchup propensity may be greater for I^2L systems that operate at 5 V in order to be power-supply-compatible to interface with TTL devices. NMOS and CMOS/SOS/SOI devices are considered to be immune to latchup. However, CMOS bulk devices will readily latch. Destructive or hard latchup, as discussed above, is common in bulk CMOS ICs. Neutron irradiation may be used to effectively harden some CMOS ICs against latchup, through displacement damage that reduces any existing parasitic transistor minority carrier lifetime, as does gold

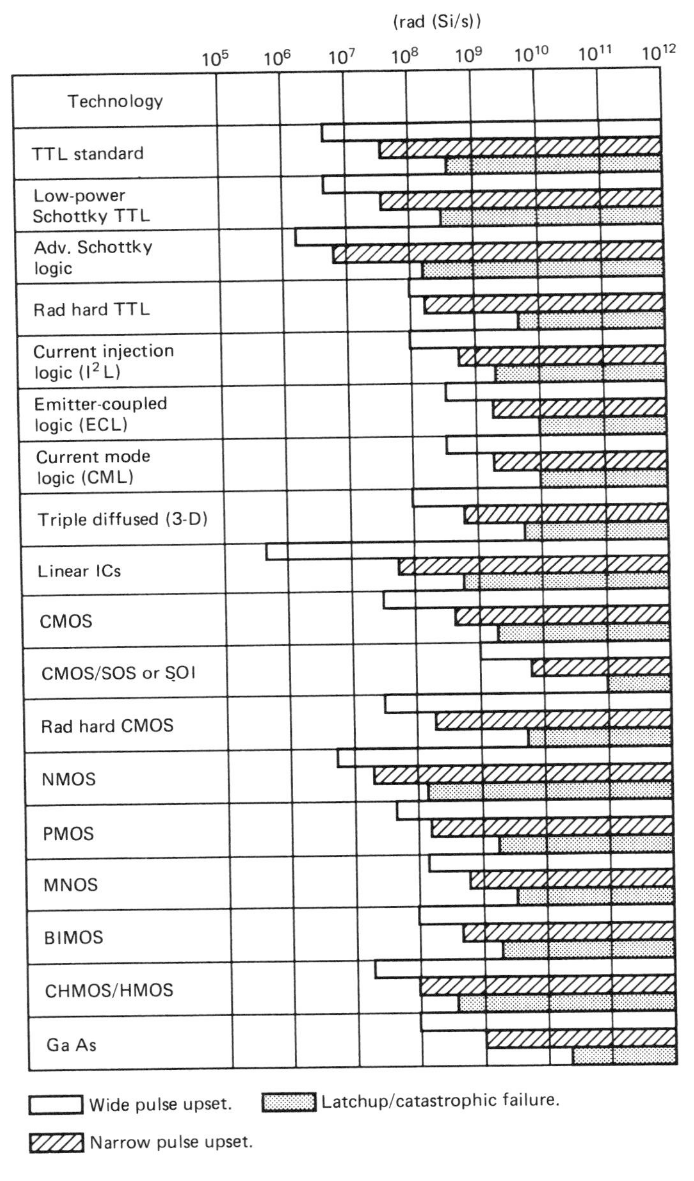

Fig. 12.22 Ionizing dose rate hardness levels for bipolar and MOSFET integrated circuit families circa 1990.[5] © 1990 by the Physitron Corp.

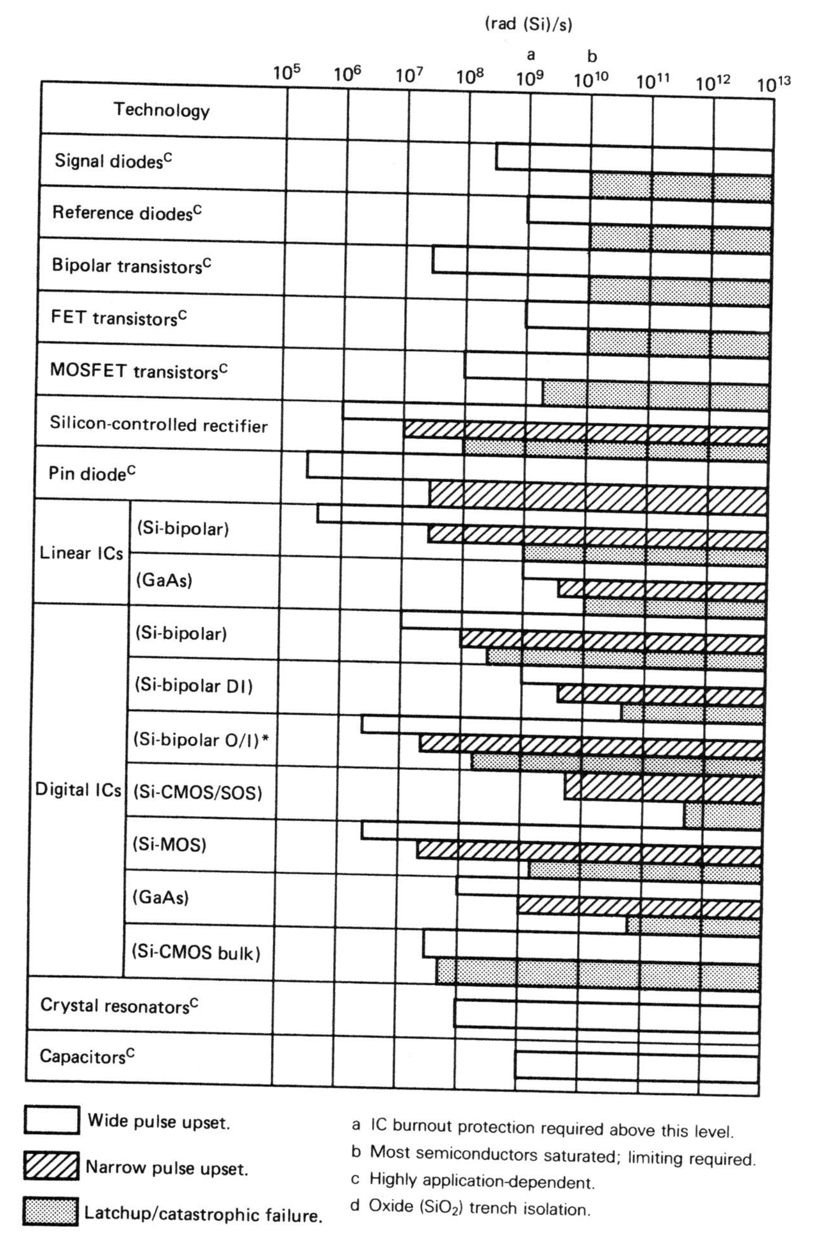

Fig. 12.23 Ionizing dose rate hardness levels for mainly discrete device families circa 1990.[5] © 1990 by the Physitron Corp.

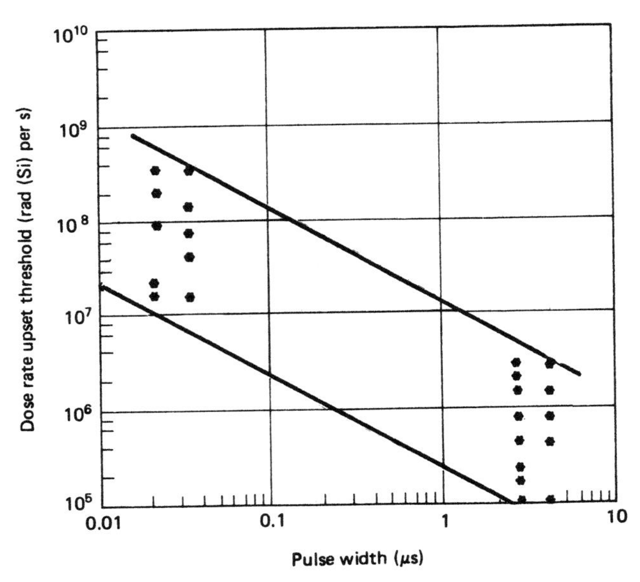

Fig. 12.24 Dose rate upset thresholds versus pulse width for MOS integrated circuits.[1] © 1984 by the IEEE.

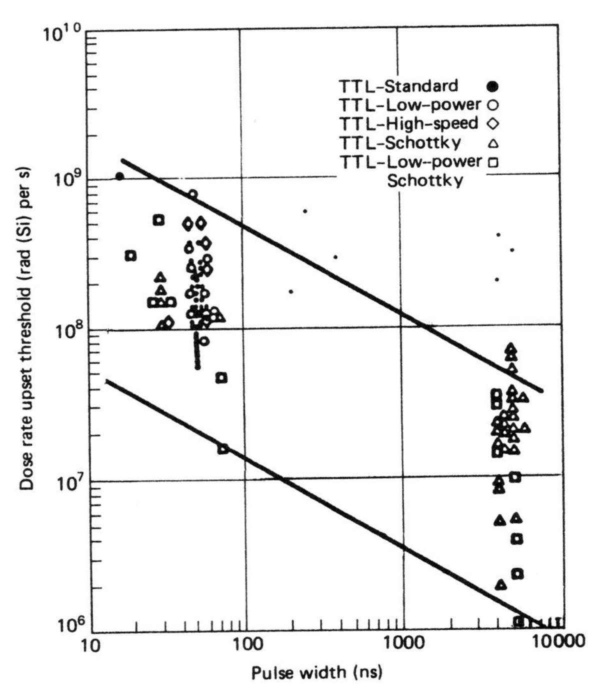

Fig. 12.25 Dose rate upset thresholds versus pulse width for digital bipolar integrated circuits.[1] © 1984 by the IEEE.

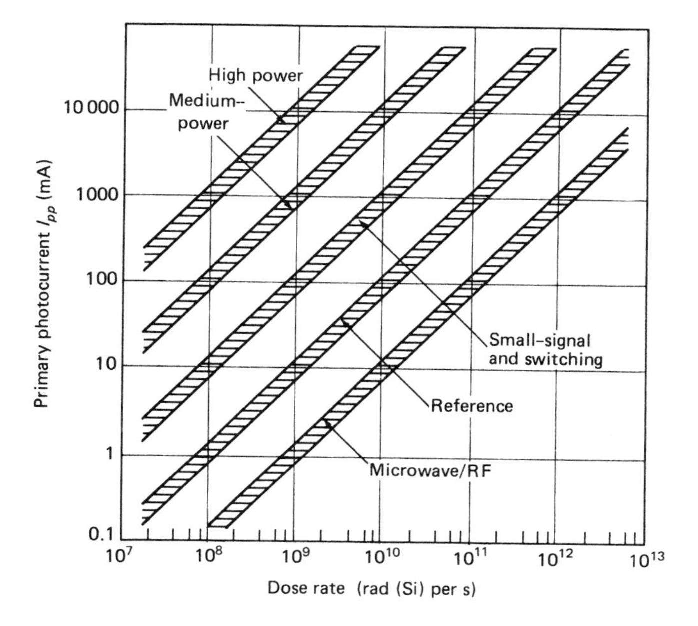

Fig. 12.26 Primary photocurrents for various diodes as a function of dose-rate levels.[1] © 1984 by the IEEE.

doping, previously mentioned. Latchup is far less frequent in bipolar ICs. No latchup was observed in 70 CMOS part types that were "treated" with neutron radiation to reduce their parasitic transistor minority carrier lifetimes, prior to being pulsed to determine latchup. Forty of the 70 part types in the control group (i.e., those not irradiated with neutrons) did latch.[33] The latchup propensity of the above parts in descending order is: bulk CMOS, linear ICs, and schottky digital bipolar ICs. Figure 12.28 depicts latchup thresholds for various devices.

Transient logic upset is usually much less serious than latchup and is discussed in Section 7.8. Transient upset thresholds are shown in Figs. 12.21–12.25. The TTL devices depicted therein include all TTL families, and there is considerable variation in the upset threshold within each family. Limited ECL data reveal a transient upset threshold at about $5 \cdot 10^8$ rad (Si) per s. I^2L devices generally have a higher upset threshold than TTL. NMOS devices are the most sensitive to transient upset, as can be seen from Fig. 12.21. CMOS devices are the next most sensitive, CMOS/SOS and CMOS/SOI are the least sensitive to transient upset.[11–14,15–19,34]

With respect to dose rate-induced burnout, high dose rates can obviously be a causative factor. This can be junction or metallization burnout.

Burnout preventive measures include the use of dielectric isolation, current-limiting resistors, and adequate metallization thickness. Certain part types in CMOS, linear IC, and TTL families are also subject to burnout. Power FETs are subject to burnout because of their high operating voltage. NMOS, I^2L, and ECL are observed to be relatively burnout-free. Again, because of the low supply voltages required, I^2L is the least subject to burnout. It should be noted that burnout can be due to other causes besides ionizing dose rate such as second breakdown and thermal runaway. Finally, high-bias voltages result in a higher burnout susceptibility.

For dose rate effects on diodes, those with high current ratings, such as power rectifiers, and those with high breakdown voltages tend to produce the largest photocurrents, as seen in Table 12.3. Microwave pin diodes, voltage reference diodes, photodiodes, and solar cells, all generate larger photocurrents than conventional signal diodes.[1] Tunnel diodes and gallium arsenide diodes have smaller minority carrier diffusion lengths than signal diodes, so that their photocurrent production is at least an order of magnitude less. Figure 12.25 depicts typical photocurrent levels for several classes of diodes. Increased reverse biasing of diodes increases the junction carrier generation volume and so yields larger photocurrents.

In bipolar transistors, primary photocurrent is generated in both the emitter-base and base-collector junctions. The photocurrent corresponding to the former junction is usually quite small and so is often neglected. Primary photocurrent for the base-collector junction is determined by treating it as a diode junction. This photocurrent will flow in the same direction as I_{CBO}. When the collector-base primary photocurrent divides, the portion that flows in the base is multiplied by $(1 + h_{FE})$ and, if sufficient, turning on the transistor to produce a secondary photocurrent. The magnitude of both primary and secondary photocurrent depends on junction bias, base volume, and circuit impedances.[1] Transistors may also suffer burnout at high dose rates.

In JFETs, primary photocurrent flows between the channel region and both gates. The junction area of the lower gate is usually much larger than that of the upper gate, so that the lower gate photocurrent is dominant. In the JFET, the primary photocurrent flows through the appropriate external resistance and biases it on for sufficiently high currents. This primary photocurrent is amplified as a secondary drain photocurrent. For equal source and drain primary photocurrents, the JFET secondary photocurrent, I_d, is given by

$$I_d = I_{pd}(g_m(2R_g + 2R_{ss} + f_R R_B) - g_d R_L)/(1 + g_m R_B + g_d(2R_B + R_L)) \tag{12.33}$$

where

I_{pd} is the primary drain photocurrent
g_m is the channel transconductance
g_d is the channel conductance
R_{ss} is the substrate bulk resistance
R_B is the tub bulk resistance
R_g is the external gate resistance
R_L is the external load resistance
f_R is the fraction of gate-source photocurrent flowing through R_B.

The JFET normally conducts secondary photocurrent at lower dose rate levels than bipolar transistors, because the JFET gate resistance is much larger than the bipolar transistor base resistance.

MOS transistors also can experience primary and secondary photocurrents, but the latter do not depend on primary photocurrent in the same manner as bipolar or JFET transistors. Here, the primary photocurrent is produced by ionizing dose in the source and drain *p-n* junction regions. Secondary photocurrent is a drain photocurrent produced by a gate voltage shift and may possess both a transient and near-steady-state component. The transient component comes from the gate transient current that flows through gate resistances to shift the gate voltage. Secondary photocurrent is usually much larger than the primary drain photocurrent because of the large value of external gate resistance, and the secondary photocurrent flows in opposition to the primary photocurrent.[1] Dose rate test data on HEXFETs (power FETs of hexagonal configuration) disclose that burnout can occur at dose rates as low as 10^9 rad (Si) per s, appreciably less than that in comparable bipolar power transistors. However, this problem can be mitigated by limiting the photocurrent transient parameters, as is discussed in Sections 12.9 and 13.12.

In bipolar digital integrated circuits, photocurrent response of individual diode and transistor cells is similar to the effects described for their discrete counterparts. Junction-isolated integrated circuits use reverse bias diodes to electrically isolate individual transistor cells from the substrate bulk as discussed in Section 6.15. The corresponding junction areas of collector and substrate are relatively large, so that large photocurrents flow through the substrate and power/ground leads of the device. These photocurrents readily upset the device bias levels when they are sufficiently large. The dose rate upset threshold for digital bipolar ICs depends on the bias voltage employed by the particular IC technology and the incident radiation pulse width. Typical dose-rate-induced upset thresholds for current technologies are given in Table 12.4 and Figs. 12.21–12.25.

Table 12.4 Dose rate upset thresholds for modern digital bipolar technologies.[1] © 1984 by the IEEE.

TECHNOLOGY	DOSE RATE TEST PULSE WIDTH	
	LONG PULSE (1–5 μs) (10^8 rad (Si)/sec)	SHORT PULSE (20–100 ns) (10^8 rad (Si)/s)
TTL	0.1	1.
LSTTL	0.1	1.
I^2L	10.	20.
ECL	5.	50.
CML	0.1	1.
Triple-diffused (3D)	—	3.

Bipolar linear integrated circuits are also susceptible to dose rate upset. The high internal gain of the linears causes larger output transients. Upset threshold values for some linear ICs are shown in Fig. 12.27. It is seen that these upset thresholds are about an order of magnitude lower than for digital bipolar ICs. Also, bipolar linear microcircuits have larger recovery times after the dose rate pulse than bipolar digital ICs. Bipolar linear devices suffer more readily from latchup than bipolar digital ICs. The same holds for dose-rate-induced burnout, because of their higher operating voltages, so that power-supply-limiting measures may be required. Dielectrically isolated devices have higher upset thresholds and shorter recovery times than JIICs.

MOS integrated circuit individual transistors produce dose-rate-induced photocurrent effects similar to those occurring in MOS discrete transistors already discussed. The total photocurrent for large-scale integrated circuits (LSI) can be large enough, at high dose rate levels, to burn them out.

CMOS ICs possess parasitic *pnpn* structures so that dose rate induced latchup occurs in many types. For example, latchup tests conducted on 64 part types[1] of standard CMOS ICs revealed that more than half suffered latchup. They also exhibit latchup windows, discussed in Section 7.7. Junction isolated CMOS bulk on high resistivity substrate is the most latchup prone structure. Latch-up hardening measures for MOS devices include (a) "curing" the parts with reasonably high incident neutron fluence levels, or gold doping, which reduces the substrate minority carrier lifetime of any bipolar parasitic transistors, and (b) current limiting in the power supply voltage leads.

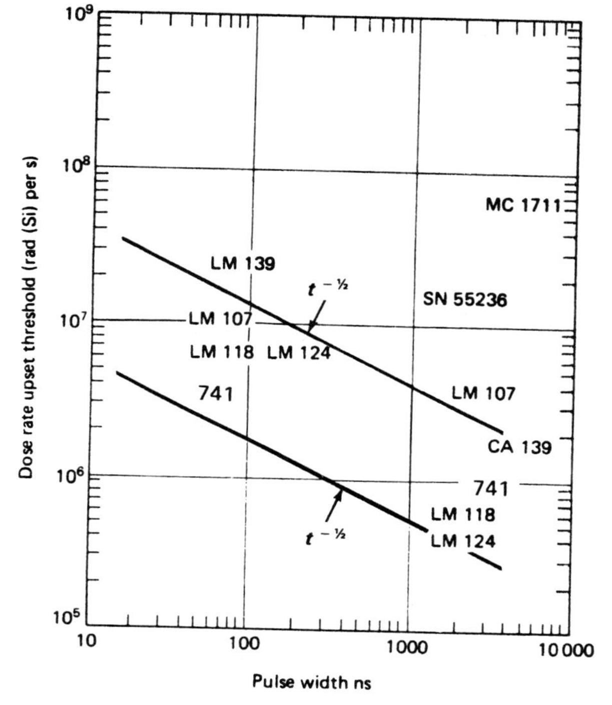

Fig. 12.27 Dose-rate upset thresholds versus pulse width for bipolar linear integrated circuits.[1] © 1984 by the IEEE.

Besides CMOS/SOS and CMOS/SOI devices, CMOS-Epi on bulk, parts are also relatively latchup resistant. This is because they are built on a heavily doped substrate which acts as a shunt for any substrate base bias that might turn-on a parasitic transistor.

CMOS and PMOS ICs upset at dose rate threshold levels corresponding to bipolar technology devices. Figure 12.24 depicts a scatter diagram for CMOS device upset threshold levels for both wide and narrow radiation pulse widths. Upset thresholds for dielectrically isolated CMOS/SOS and CMOS/SOI devices are significantly higher than bulk CMOS by several orders of magnitude. NMOS ICs have low upset thresholds, with some devices being upset by dose rates between $10^7 - 10^8$ rad (Si) per s. Although NMOS upset thresholds are comparatively low, they generally possess no *pnpn* path structure and so are essentially immune to latchup.

Another LSI technology that is immune to dose-rate-induced latchup is that of nonisolated integrated injection logic (I^2L). As mentioned, a

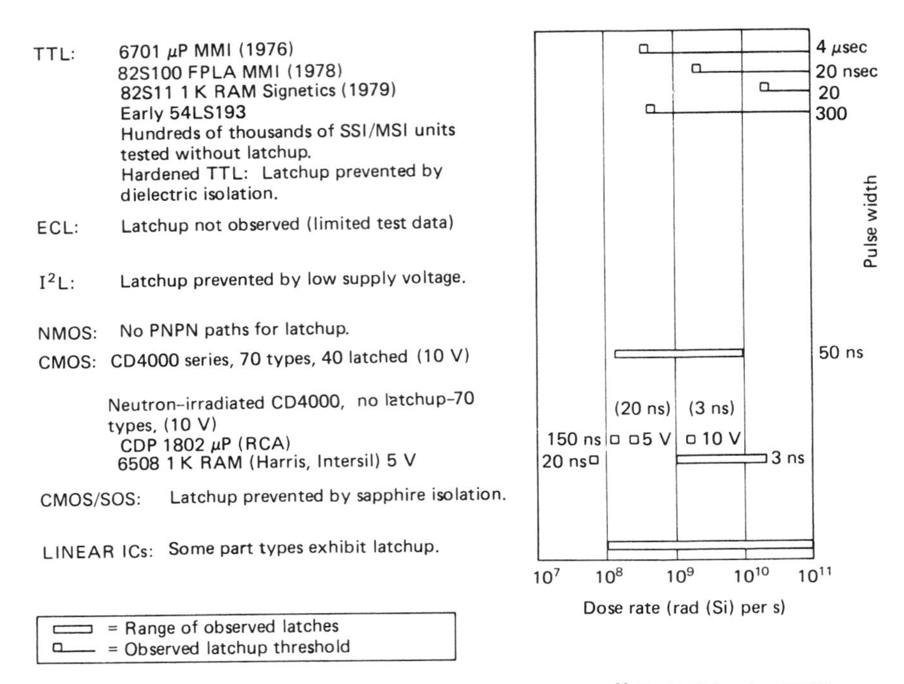

Fig. 12.28 IC latchup threshold levels versus dose rate.[33] © 1980 by the IEEE.

mitigating factor against latchup is its low operating voltage, which is comparable to a single diode drop in silicon. In more detail, it can be shown by circuit analysis, that the I^2L device is essentially latchup free.[33] Schematically, the anode gate of the SCR latchup model circuit, as discussed in Section 7.7, is grounded for I^2L. Any potential latchup-inducing current thence flows from both anode and cathode to ground, instead of from anode to cathode, as in the SCR latchup path. Hence, latchup is virtually impossible in this type of device.[35] Figures 12.22, 12.23, and 12.28 depict latchup thresholds for various IC technologies.[33]

It should be noted that latchup per se is normally a non-destructive phenomenon so that suitably conducted latchup testing in dose rate simulators, such as Flash X-Ray facilities, constitutes a nondestructive test. This is especially the case when the device under test latches and triggers a power-down condition. The U.S. Mil-Std-Hdbk 883C, Method 1020 expressely states that suitably conducted latchup testing shall be non-destructive. A practical application of the non-destructive aspect is the use of bulk CMOS parts in the electronics of certain missile systems. The type of bulk CMOS parts employed therein possess a high susceptibility

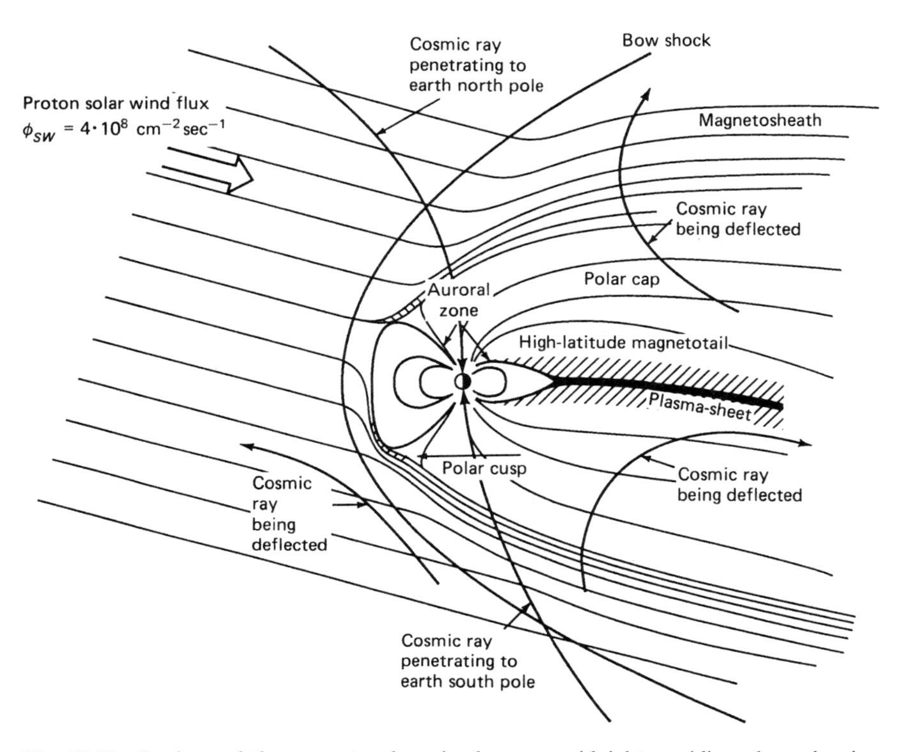

Fig. 12.29 Regions of the magnetosphere in the noon-midnight meridian plane showing cosmic ray trajectories and the solar wind.[36]

to latchup, but are power-down-reset strobed in real time to allow repeated latchup if necessary, recognizing that such latchup does not degrade the part function.

Latchup in CMOS integrated circuits usually draws little current as compared to bipolar microcircuits. Hence, more sensitive monitoring methods must be used for CMOS during testing, or in operational circuitry, to detect latchup.

12.5 COMPONENT SELECTION: SEU EFFECTS

The selection of hard components for the SEU environment is unique in one important aspect, compared to the other radiation environments discussed. Fabricated shielding cannot be effectively utilized. This is because the pertinent cosmic ray ions have super-high energies (MeV-GeV), are highly ionized, and some are very heavy (e.g., Fe^{6+}). This

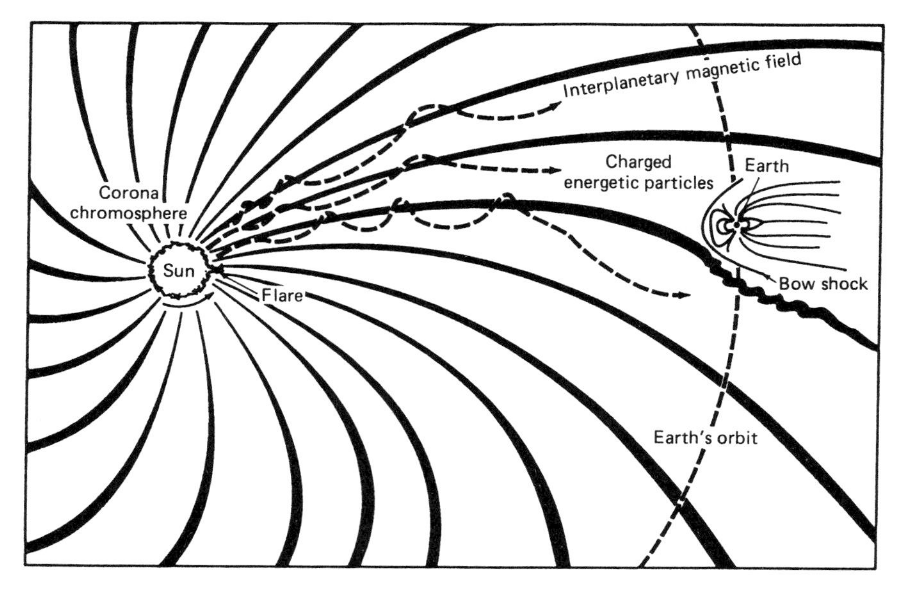

Fig. 12.29A Earth in the solar wind showing the interplanetary magnetic field and emanating solar energetic particles (protons/electrons).[89]

means that attempts at shielding against such particles is not a viable option, so that component selection and system-level hardening approaches are the main available avenues. However, a modicum of natural shielding against galactic cosmic rays is afforded through the combined effects of the solar wind and the earth's magnetic field (except at the poles and the South Atlantic Anomaly), as shown in Figs. 12.29 and 12.29A. The field acts as a gigantic magnetic spectrometer for cosmic ray ions bending their incoming trajectories outward away from the vicinity of the earth. The wind acts as a conductive screen shield, partially enveloping the earth in an ionic bow shock wave, as in the figures. Their effectiveness anticorrelates with the sunspot cycle, in that SEU is a maximum during sunspot cycle (and solar wind) minima, and vice versa. Other natural shields are of a lesser nature, including trapping of cosmic rays in the Van Allen belts and in the Ionosphere, which latter are depicted in Figs. 12.30A and 12.30B. In this regard, certain spacecraft orbits are of sufficiently low altitude that not only are the Van Allen belts not penetrated, but at such low altitudes ($\leqslant$200 km) under the belts, the galactic cosmic ray fluxes are relatively low. In the earth's sensible atmosphere, cosmic rays undergo sufficent attenuating reactions so that, at sea level, the earth is very nearly shielded. However, certain atmospheric reaction

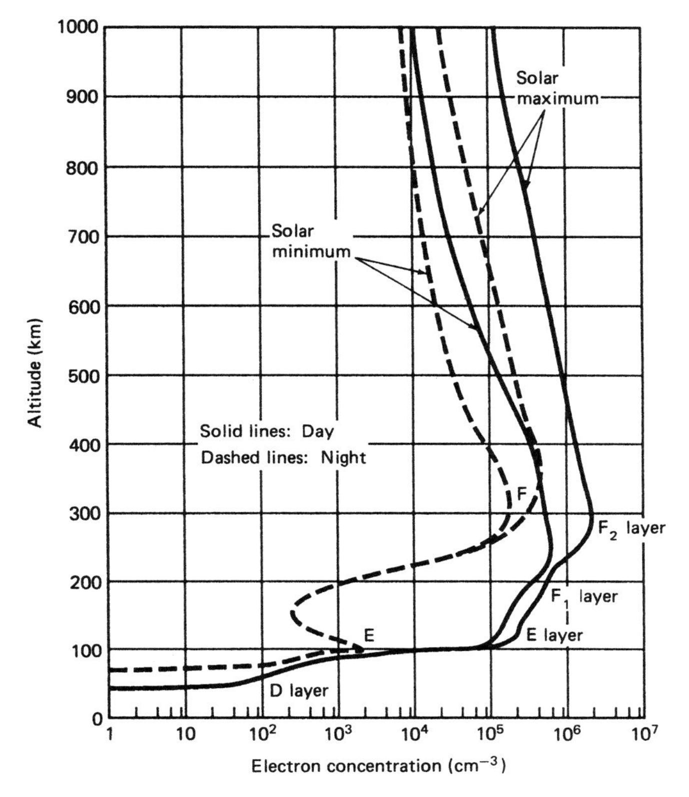

Fig. 12.30A Ionosphere layers of electron concentration merging into the lower region of the inner Van Allen belt.[37] © 1971 by Lockhead Missile Systems Division.

products, such as muons and pions (heavy electrons called mesons) do strike the earth. Being relatively light, they cause little ionization (SEU) except for affecting CCDs and other ultrasensitive components. Since CCDs are used mainly in image-forming systems, the contextual processing by the corresponding electronics minimizes the effect of any SEU induced by these mesons.

The preceding SEU hardening methodology, discussed in Section 8.10 and later in this section, has much in common with hardening against other radiation environments. A good share of the hardening burden rests with judicious part selection at a sophisticated level, and not merely using simple go–no go rules. Toward that end, Fig. 12.31 depicts a bar chart compilation of SEU data. It shows geosynchronous SEU error rate thresholds for a number of modern part technologies, divided mainly into

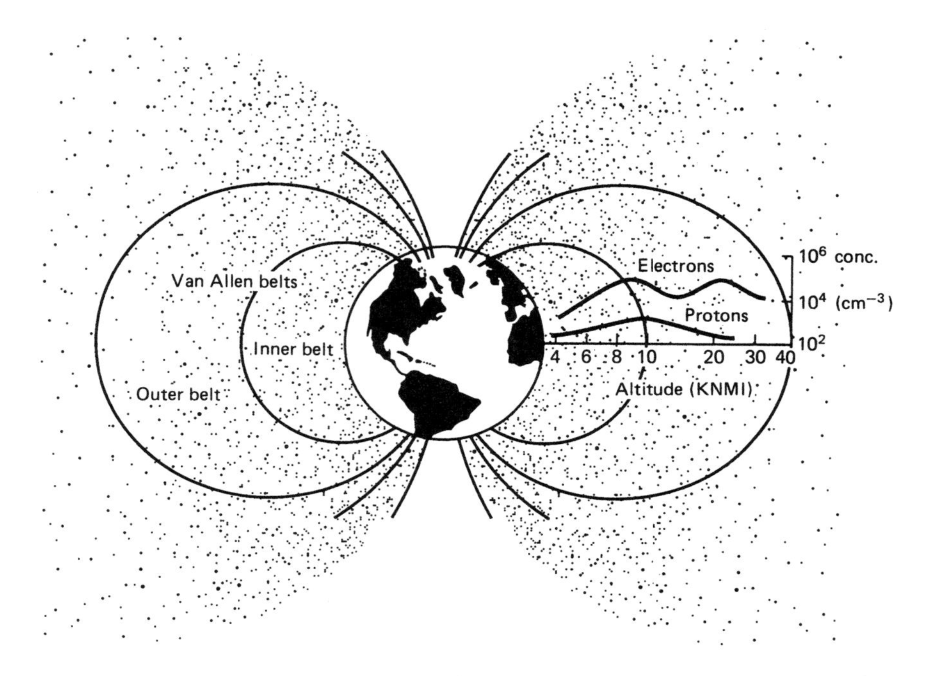

Fig. 12.30B Van Allen belts showing the trapped electrons and protons.[36,37]

MOS and bipolar digital families. It illustrates that, generally, MOS devices are harder to SEU than are bipolar devices.

Besides part selection, certain definite SEU hardening measures have been put forward. One such pertains to hardening flip-flops, latches, and other memory logic integrated circuits. It requires inserting resistances, or diodes, or diode-like devices into the feedback loops of these devices to slow their response, thus making the device insensitive to the sub-nanosecond time scale of the SEU-inducing cosmic ray heavy ion track. As mentioned earlier, there are possibly three further reasons why this method is not completely satisfactory, even though it has been put into production. The first is that slowing the device response to act like a low-pass filter is the antithesis of the modern direction of component manufacture and operation, which is to make such devices achieve even higher speeds. The second is that the feedback path resistors take up valuable chip "real estate," and so reduce the packing density of the integrated circuit. The third is that these resistors can have significant resistivity-temperature coefficients, and thus may not be completely effective over the required temperature range. In this connection, anticipated exo-

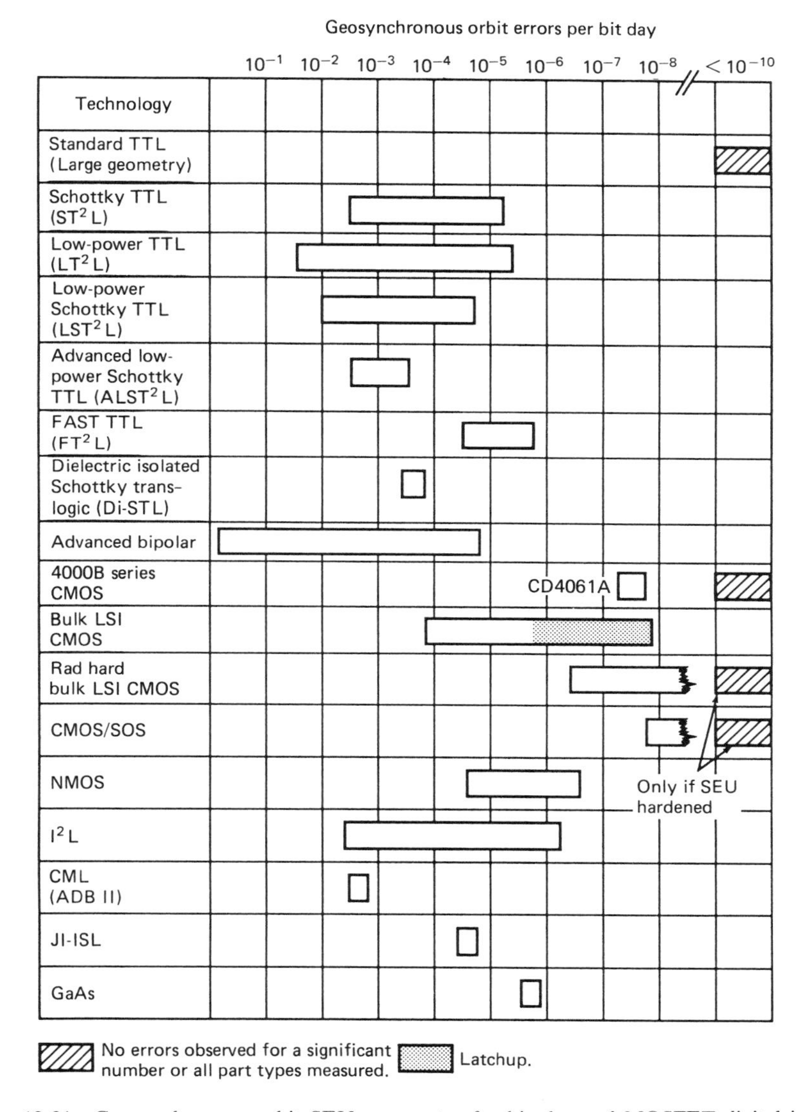

Fig. 12.31 Geosynchronous orbit SEU error rates for bipolar and MOSFET digital integrated circuit families circa 1990.[5,38] © 1990 by the Physitron Corp.

Table 12.5 Devices susceptible to latchup (LU) from heavy ions (ca. 1990).

SER.	DEVICE	VENDOR	TECHNOLOGY	FUNCTION	LATCHUP LET (MeV cm^2/mg)	σ_u(cm^2)	REMARKS/BEAM ION	REFS: IEEE TRANS. NUCL. SCI. NS-XX(Y)
1	SP1016A	ADI	CMOS Bulk	Multiplier	≪37	4E-4	Krypton	38(6) Dec. 1991, p. 1529
2	320C25	TI	CMOS-Epi	DSP	≪37	>1E-2		"
3	SP2100ASG	ADI	CMOS Bulk	DSP	12	1E-3		"
4	MC68882	MOT	CMOS-Epi	Microprocessor	12	2E-3	32 Bit, flt. pt.	"
5	MHS80631	HA	CMOS-Epi	Microprocessor	<25	2E-6	8 Bit	"
6	P1750A	PFS	CMOS-Epi	Microprocessor	<11	7E-3	16 Bit	"
7	68020	MOT	CMOS Bulk	Microprocessor	6			"
8	72016A	IDT	CMOS Bulk	FIFO	13	2E-3	512 × 9	"
9	CS5016	CRY	CMOS Bulk	ADC	≪37		16 Bit	"
10	HM65262	HA	CMOS Bulk	SRAM	≪37	>4E-3	16K × 1	"
11	61CD16	TI	CMOS-Epi	SRAM	<13	6E-3	16K	"

12	28C256	XIC	CMOS Bulk	EEPROM	$\ll 37$		32K × 8	″
13	54AC163	RCA	ACMOS	4 Bit bin. cntr.	37	>5E-6		″
14	54AC163	NSC	ACMOS	4 Bit bin. cntr.	40	1E-5		″
15	ZR34161	ZOR	CMOS-Epi	Sig. processor	17 ± 5	4E-3	12 Micron Epi	″
16	32C016	NSC	CMOS Bulk	Microprocessor	3		Boron	34(6) Dec. 1987, p. 1332
17	27C64	INT	CMOS Bulk	PROM	14			″
18	HM6516	HA	CMOS Bulk	DRAM	<17		Krypton, Argon	″
19	HM6516	HA	CMOS-Epi	DRAM	<29	3E-6	Krypton, Argon	″
20	IN1600	INM	NMOS	DRAM	2	3.1E-5	LU w/mult. SEU	″
21	IN1600	INM	NMOS	DRAM	<3	1.6E-6		″
22	HS6504RRH	HA	CMOS-Epi	DRAM	40		Address latchup only	″
23	54HC165	TI	HCMOS	Shift register	<40	2.5E-5		″
24	80C86	HA	CMOS-Epi	Microprocessor	$\ll 12$	4E-3	Bromine	″
25	3022	RWL	MNOS/SOS	EAROM	<37		Krypton	32(6) Dec. 1985, p. 4189

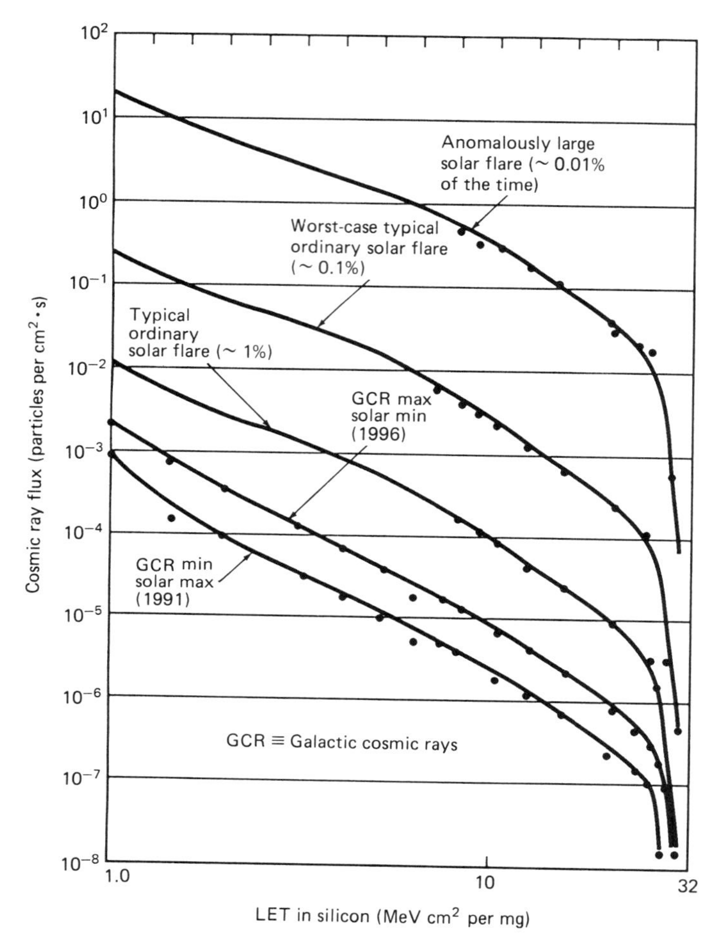

Fig. 12.32 Solar flare and cosmic ray flux at 10700 nmi altitude through 1.23 gm/cm^2 aluminum shielding versus LET in silicon, including no-flare flux during sunspot extrema and flare occurrence frequency.[38] © 1987 by the IEEE.

atmospheric hostile neutron fluence levels are usually too low to cause substantive resistivity changes.

With respect to SEU-induced latchup, the electronic mechanism is similar to dose-rate latchup, discussed in Section 7.7, except that the SEU-induced pulse is one to two orders of magnitude shorter (10–100 picoseconds) than that for dose-rate latchup (10–100 nanoseconds). SEU usually affects only a single component function such as a gate, whereas dose-rate upset can affect a whole circuit or an entire electronic system.

Most SEU-induced latchup occurs in CMOS bulk parts, which are also prone to dose-rate induced latchup. Table 12.5 is a tabulation of SEU-induced latchup susceptibility components, with data obtained from accelerator experiments.

The problem of coping with SEU, and possible SEU burnout, from the infrequently occurring solar flares is a difficult one. As can be seen from Fig. 12.32, solar flare fluxes are from one to three orders of magnitude greater than the no-flare fluxes.

From an unexpected quarter, another part type is susceptible to SEU, and especially to burnout. It is the *n*-channel power MOSFET transistor family, including those manufactured by mature first-line companies. They have a tendency to burn out due to second breakdown, discussed in Section 3.6, when subjected to SEU-type heavy ions.[40] Power MOSFETs are not built like power bipolar transistors, which consist of a relatively large emitter, base, and collector. However, experimentally induced single-event burnout has been reported in bipolar power transistors.[94] Power MOSFETs are comprised of very many parallel cells, each of fine structure, to achieve large operating currents and overall high-power operation, as discussed in Section 13.12. SEU particles can cause a single cell to revert to the second breakdown state, with that cell then heating the neighborhood cells, causing them to participate in this phenomenon as well. This cascades quickly into a tremendous current overload, causing destruction of the part. The hardening procedure is to derate the drain-to-source breakdown voltage (BV_{DSS}), by reducing the peak drain operating voltage. The amount of reduction required depends on the specific part manufacturer, and whether the SEU excitation is due to SEU simulation sources such as particle accelerators, ^{252}Cf, or actual galactic cosmic rays.[41] Several manufacturers also produce power MOSFETs hardened against dose rate burnout.

The following is a list of rules of thumb, suggestions, and methods for SEU hardening, some items of which may be conflicting and require trade-offs. They span the spectrum from mission and system to circuit and part considerations.

1. If possible, use low-altitude equatorial orbits to stay under the Van Allen belts, and away from both poles. The earth's magnetic field necks in at the poles, as the magnetic flux lines enter the earth, exposing the poles to increased incidence from cosmic rays.
2. For short missions, of roughly 1–5 years, attempt to launch during a sunspot cycle maximum. This is when the mean solar wind maximizes, so as to take advantage of the solar wind–earth magnetic field coupling to achieve minimum SEU.

3. Attempt to tolerate SEU error rates, by first examining the system from the viewpoint of subsystem function, and then SEU susceptibility from the piece-part aspect. For example, a particular subsystem may function well even under appreciable levels of SEU error rate.
4. Avionics in spacecraft and very high flying aircraft are SEU susceptible. For atmospheric and ground systems, there is usually a small SEU problem. Also, not all integrated circuits are susceptible to SEU in terms of circuit function. Keep in mind the SEU part function susceptibility hierarchy. It is, from most to least susceptible to SEU:
   (a) DRAMS, SRAMS
   (b) flip-flops, latches
   (c) sequential gates
   (d) combinational gates
   (e) other
5. Where applicable, use error detection and correction (EDAC) circuitry and devices. Memory "scrubbing" schemes should be employed to prevent SEU errors from evolving into double or multiple-bit errors.
6. Fractionate the bits in memory words by physically spreading them over many RAMs (with corresponding circuitry) in the memory array. This minimizes the multiple-SEU error probability.
7. Use SRAMs where feasible, since they have no floating bit lines. Anything "floating" is anathema for SEU hardening.
8. Incorporate heavy I/O decoupling in circuit design to slow response to SEU.
9. Employ large currents in susceptible circuitry, so that possible SEU currents represent only a fraction of normal current.
10. Use "smart" part selection, as discussed in the preceding. However, for recipe purposes, pick parts (if possible) whose $L_c \geqslant 40\,\text{MeV cm}^2/\text{mg}$. This is because the nominal Heinrich cosmic ray spectrum for silicon is essentially cut off for $L_c > 40\,\text{MeV cm}^2/\text{mg}$, implying that no appreciable galactic cosmic ray flux of higher L_c exists in the neighborhood of Earth.
11. Use older integrated circuit part families, with comparatively large feature size, where possible.
12. Minimize impedance of circuits connecting bit lines to sense amplifiers.
13. For SEU-hard memories, insure use of SEU-hard buffering peripheral circuitry, so as not to vitiate the hardness properties of the memories themselves.

14. In IC design, lower part substrate resistivity with a heavily doped buried layer. This tends to shunt the funnel to keep its charge from being collected.
15. If using unhardened *n*-channel power MOSFETs, derate peak operating voltage.
16. Employ integrated circuit chip coatings (passivation layers) to keep alpha particles born in the package from penetrating into the chip. Alpha particles are easily stopped by thin coatings.

12.6 COMPONENT SELECTION: PROTON SEU EFFECTS

Following Peterson et al.,[42] the proton-induced SEU error rate, E_{pr}, is given by an integral, Eq. (8.23), of the product of the proton SEU cross section $\sigma_p(A, E)$, with the corresponding proton flux $\Phi_p(E)$, considered here in the Van Allen belts. The flux, besides being a function of the incident proton energy E, depends on the environment, as expressed in terms of orbit inclination angle with respect to the equatorial line; altitude above the earth; the South Atlantic Anomaly discussed in Section 8.4; and the two extremes of sunspot activity.

The proton SEU cross section depends on the incident proton energy and the proton energy threshold, A. That is, for $E < A$, there is insufficient proton energy to cause the proton-induced nuclear reactions in the chip to produce ions of enough energy and mass number to cause SEU. The E_{pr} is obtained by numerical integration of Eq. (8.23), and the results are presented in suitable form in Table 12.6.

To compute an E_{pr} for a given part of interest in a vehicle whose orbit lies in or near the Van Allen belts, the first step is to obtain its corresponding energy threshold, A. This can be done by a single accurate measurement of the proton SEU cross section for this part, using an energy between about 10–100 MeV. Then A is immediate by using the $\sigma_p(A, E)$ curves to identify A, as in Fig. 8.10, which is a one-parameter family of curves with A as that parameter. Alternatively, A can be obtained from the now-known $\sigma_p(A, E)$, from its formula given in Eq. (8.22). However, A is elusive if the above cross section measurement is unavailable for the particular part of interest. In this case, Table 12.8 can be used for parts that are alike or similar in technology family to those of interest, by interpolating to yield a credible value for A. In this regard, recall that the proton SEU susceptibility decreases (smaller E_{pr}), with increasing values of A. Also, there is a rough correlation between proton SEU and galactic cosmic ray SEU susceptibilities. So that if the latter SEU value exists for this device, it may be used very cautiously to

Table 12.6 Proton SEU error rates in circular earth orbits in Van Allen belts in 1980.[42] © 1984 by the IEEE.

ORBIT INCLINATION	UPSETS PER BIT-DAY FOR $A = 18$ MeV		B (MeV)	
	SOLAR MIN	SOLAR MAX	SOLAR MIN	SOLAR MAX
Altitude: 200 km				
30°	8.20-7	1.58-8	17.5	32.1
35°	2.13-6	1.84-7	12.3	15.4
40°	2.65-6	5.42-7	11.2	12.9
50°	1.68-6	3.17-7	9.3	12.4
60°	1.23-6	2.40-7	10.5	13.5
90°	1.02-6	1.80-7	10.5	12.7
Altitude: 400 km				
30°	3.43-5	1.68-5	22.1	27.5
35°	4.39-5	2.28-5	20.7	23.5
40°	3.97-5	2.13-5	19.0	22.0
50°	2.50-5	1.35-5	18.2	21.4
60°	2.22-5	1.16-5	19.5	22.0
90°	1.66-5	8.58-6	19.6	22.3
Altitude: 600 km				
30°	1.80-4	1.13-4	23.9	25.6
35°	1.81-4	1.15-4	22.3	23.7
40°	1.57-4	1.03-4	21.7	22.7
50°	1.11-4	7.11-5	21.4	22.6
60°	9.13-5	5.73-5	21.8	23.1
90°	7.99-5	5.10-5	22.3	23.4
Altitude: 800 km				
30°	5.04-4	3.59-4	24.6	25.4
35°	4.69-4	3.33-4	23.5	24.5
40°	4.09-4	2.89-4	23.0	23.8
50°	3.03-4	2.13-4	23.1	23.8
60°	2.54-4	1.78-4	23.6	24.5
90°	2.14-4	1.49-4	23.8	24.8
Altitude: 1200 km				
30°	2.08-3	1.62-3	24.1	24.5
35°	1.84-3	1.43-3	23.7	24.1
40°	1.58-3	1.24-3	23.4	23.8
50°	1.23-3	9.56-4	23.6	24.0
60°	1.06-3	8.17-4	23.9	24.3
90°	8.95-4	6.93-4	24.0	24.4

Table 12.7 Proton SEU error rates for selected orbits in high Van Allen proton belts.[42] © 1984 by the IEEE.

ORBIT INCLINATION	ALTITUDE km	UPSETS PER BIT-DAY FOR $A = 18\,\text{MeV}$	B (MeV)
Solar Maximum; Time = 1989.5		$X \cdot 10^{-Y} \equiv X - Y$	
60°	1667	3.03-3	22.6
60°	2593	6.69-3	19.1
60°	3889	3.18-3	11.5
60°	5186	8.19-4	5.0
60°	6389	2.10-4	(0.0)
60°	10371	1.36-6	
Solar Max; Time = 1981.8			
63°	1111	5.69-4	24.4
Solar Min; Time = 1985.8			
63°	1111	7.91-4	23.9
Solar Max; Time = 1989.0			
63°	1111	6.59-4	24.2

extrapolate an appropriate A value. In sum, such interpolations and extrapolations yield only upper or lower bounds for A.

The second step is to obtain an average proton flux spectrum, i.e., $\Phi_p(E)$ for a particular orbit altitude, inclination angle, and solar activity extrema epoch. This is usually available from certain computer orbit program outputs, such as from CREME.[43] The third step is to obtain the distribution of shielding material about the devices of interest in the form of attenuation factors, with which to multiply $\Phi_p(E)$ to get the proton flux incident on the chip itself. The last step is to combine the above factors into Eq. (8.23) to obtain the desired proton SEU error rate, $E_{pr}(A)$. Once the parameter A is obtained, Tables 12.6 and 12.7 can be used to obtain $E_{pr}(A)$ for various orbit parameters. For example, assume an orbit with the following parameters: 800 km altitude, 35° inclination angle, and a solar minimum epoch. The device of interest is assumed to be a MK4116J-2 16K NMOS DRAM. From Table 12.8, it is seen that the corresponding parameter $A = 21.66$. To use Tables 12.6 and 12.7, it is necessary to correct for this A value, since these tabulations correspond to $A = 18$ only. This is done by using the empiricism[42]

$$E_{pr}(A) = E_{pr}(18)(18/A)^{14}(18 + B)/(A + B) \qquad (12.34)$$

Table 12.8 Proton energy threshold *A* for devices from most to least susceptible.[44,45] © 1983 by the IEEE.

SER.	PART NAME	MFR.	FUNCTION	ENERGY (MeV) THRESH "A"	TECHNOLOGY
1	93L422	FSC	1K SRAM	9.88	TTL
2	7164	IDT	64K SRAM	29.85	NMOS
3	93ZL422	FSC	1K SRAM	10.35	TTL
4	2901B	AMD	4 Bit Slice	14.16	TTL
5	8X350	SIG	8 Bit Micro	14.96	LSTTL
6	2909A	AMD	4 Bit Micro	15.60	TTL
7	2909	AMD	4 Bit Micro	16.39	TTL
8	93L422	AMD	1K SRAM	16.42	TTL
9	9407	FSC	9 Bit Reg.	16.58	IIL
10	C2107B	INTEL	4K DRAM	18.27	NMOS
11	93L425	AMD	1K SRAM	17.04	TTL
12	93422	FSC	1K SRAM	17.66	TTL
13	93425A	FSC	1K SRAM	18.00	TTL
14	MCM4116-20	MOTA	16K DRAM	19.80	NMOS
15	MM5280	NSC	4K DRAM	21.16	NMOS
16	9407	FSC	9 Bit Reg.	20.56	TTL
17	54LS395	SIG	4 Bit Sh. Reg.	>21.4	LSTTL
18	54LS374	TI	Octal F/F	>21.4	LSTTL
19	MK4116J-2	MOST	16K DRAM	21.66	NMOS
20	54LS169	AMD	4 Bit Cntr	22.80	LSTTL
21	MSM4044	SSS	4 Bit NAND	26.40	CMOS
22	HM6514	HA	64K DRAM	>50.	CMOS

where B is the correction factor obtained from the last columns in Tables 12.6 and 12.7. From the former table for 800 km altitude, and an inclination angle of 35°, during a solar minimum epoch, $E_{pr}(18) = 4.69 \cdot 10^{-4}$, with a corresponding $B = 23.5$. Inserting this B and A into Eq. (12.34) yields the resultant $E_{pr}(21.66) = 3.23 \cdot 10^{-5}$ errors per bit day. For parameter values other than those listed in the tables, linear interpolation can be used to obtain them, where applicable.

To appreciate proton SEU levels in general, and as compared to geosynchronous SEU, the Combined Release and Radiation Effects (CRRES) spacecraft, recently launched, provides SEU space experimental data,[93] some of which is depicted in Figs. 12.32A and 12.32B. Figure 12.32A shows that, for this orbit, the preponderance of SEU occurs in the Van Allen belt regions ($L \cong R/R_E \cong 1.5$), where R_E is the earth radius at the equator. This contrasts with SEU occurring above the belts in the vicinity of geosynchronous altitudes ($5 \leqslant L \leqslant 7.5$). In Fig.

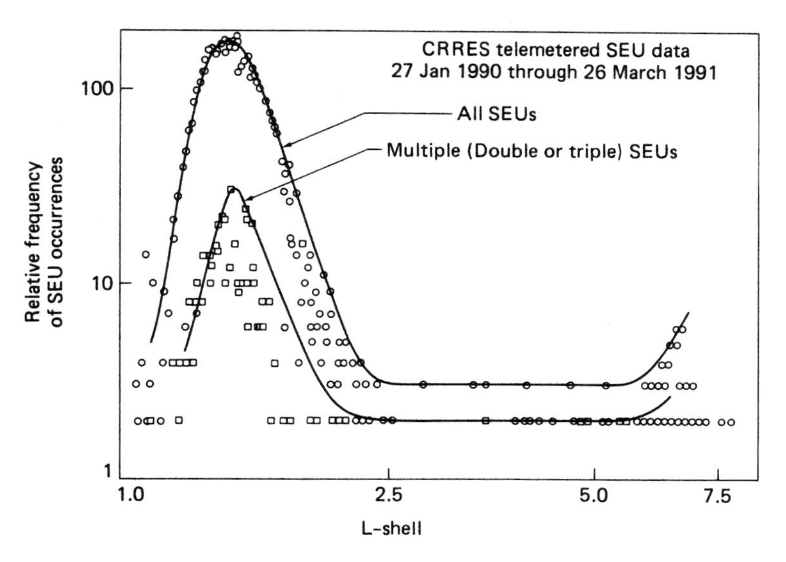

Fig. 12.32A Single and multiple SEUs versus L-shell for seven static RAMs.[93] © 1991 by the IEEE.

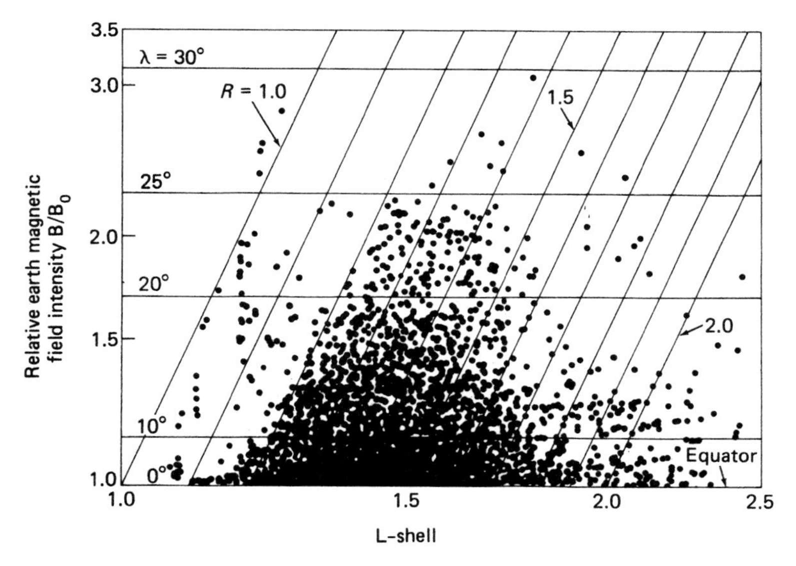

Fig. 12.32B SEUs in the Van Allen belt region shown in B,L coordinates.[93] © 1991 by the IEEE.

12.32A, the curves are "faired-in" over the presented data points. These results corrolate with those sketched in Fig. 8.12, which also compares SEU at geosynchronous and Van Allen belt altitudes. Figure 12.32B details SEU in the Van Allen belts, i.e., mainly proton-induced SEUs.

The relative magnitudes of geosynchronous and Van Allen belt SEU levels, of course, depend on the spacecraft orbital parameters, as well as the part-types on board. The CRRES orbit is highly elliptical, at an inclination angle of 18.2° with respect to the equator. Its perigee is 348 km, the apogee is 33,582 km, and it has a period of 9.87 hrs. Not unlike a Molniya orbit, the spacecraft passes through both the geosynchronous and Van Allen belt altitudes twice during each orbit. The SEU data in the above figures is that telemetered from averages taken over seven representative static RAMs.[93]

For space physics reasons, SEU coordinates in space (spacecraft) are given in terms of corresponding coordinates of the earth's magnetic field magnitude, $B \equiv |\bar{B}|$, and its parameters. In Fig. 12.32B, the B/B_0 ordinate is the earth magnetic field intensity B at a given SEU altitude R, normalized to its field (corresponding flux line) B_0 on the equator at its altitude R_0. For an assumed symmetric earth magnetic dipole field, the Fig. 12.32B abscissa, viz., the McIlwain parameter[37] $L \cong R/R_E$ for L not too large. The connective relation between these quantities is $B/B_0 \cong (R_0/L)^3 \cdot (1 + 3\sin^2\lambda)^{1/2}/\cos^6\lambda$, where λ is the (magnetic) latitude angle (either north or south latitude). Constant values of L (not constant B) correspond to magnetic dipole field intensity surfaces, called L shells.[37]

The above B, L coordinates pertain to a spherically symmetric earth dipole magnetic field (with no dependence on the spherical coordinate azimuth angle ϕ), which is the explanation for the use of the approximation symbols in the preceding expressions. For the actual nonsymmetric earth fields, the B, L coordinate relations are more complex.[37] In the symmetric case, the parametric transformations between the polar r, θ SEU coordinates and the B, L coordinates[37] are: $B/B_0 = (R_0/r)^3 \cdot (4 - 3r/L)^{1/2}$, $r = L\cos^2\lambda = R_0\sin^2\theta$ and $\theta = \pi/2 - \lambda$.

At this juncture, SEUs at L shell values less than 2.0 are arbitrarily attributed to protons, while those with L shell values greater than 2.0 are similarly attributed to heavy cosmic ray ions.[93] When later CRRES data is at hand, it should be much easier to distinguish proton-induced SEUs from those induced by cosmic ray ions. Among other considerations, this should aid in ascertaining the origin of the apparent beginning of an SEU increase in the region of $5 \leqslant L \leqslant 7.5$, shown in Fig. 12.32A.

To compute the critical charge using a radiation-inclusive SPICE-type code,[46] the electronic and physical parameters of the device of interest are formatted as input data. The number of parameters can vary from

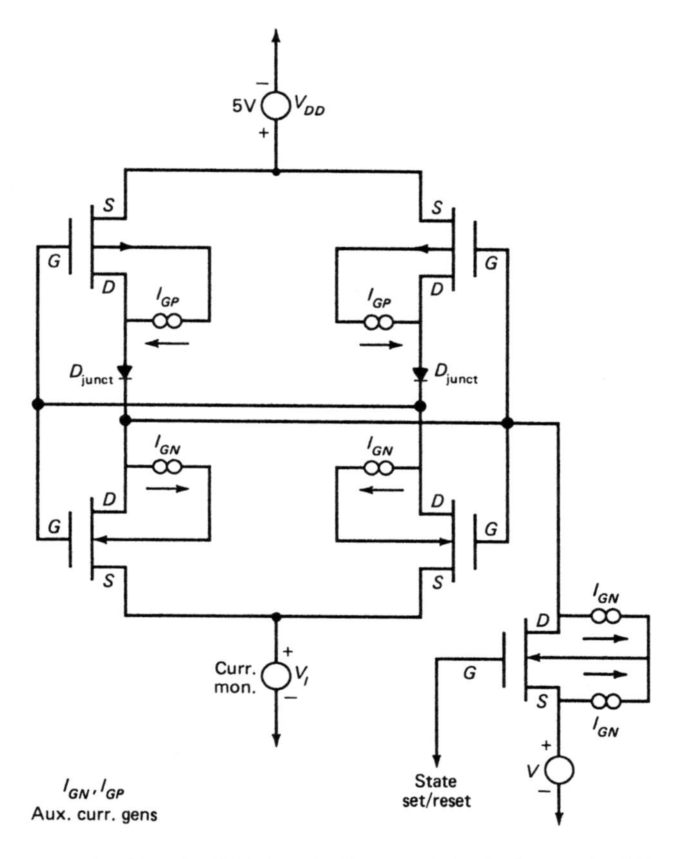

Fig. 12.33 CMM5114/1RZ CMOS/SOS 4K SRAM cell circuit diagram for SEU simulation on SPICE-type code.[46]

10–40 approximately, depending on device circuit complexity and model faithfulness desired. Frequently, only a typical cell of the device that represents the phenomenon to be investigated is modelled, as opposed to the complete integrated circuit. This is done in the first example, which is a typical memory cell, such as in a CMM5114/1RZ 4K CMOS/SOS SRAM. However, in the second example, which is that of a custom multitransistor bipolar *D*-flip/flop, the whole device was modelled. Besides the preceding parameters input to the code, the electrical circuit connections and auxiliary current generator parameters are also entered. The latter provide the SEU current pulse excitation, which suitably mirrors the proper pulse temporal characteristics. Figure 12.33 depicts these

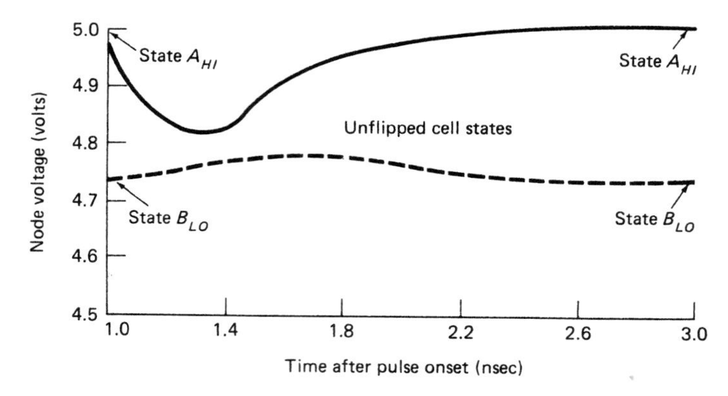

Fig. 12.34 Multitransistor custom bipolar *D*-flip/flop cell excited by SEU pulse insufficient to cause cell change of state. Pulse onset at 1 nsec.

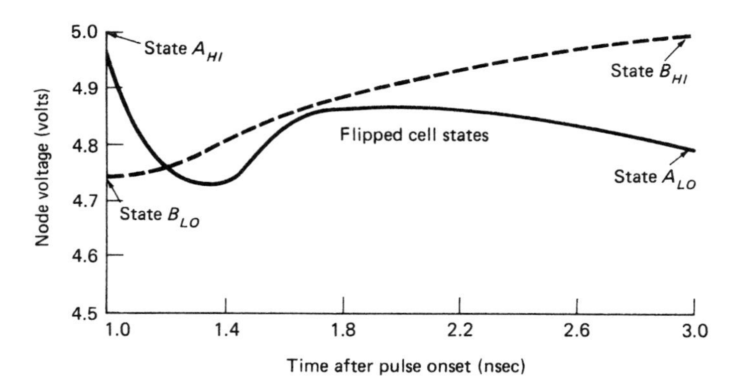

Fig. 12.35 Multitransistor custom bipolar *D*-flip/flop cell excited by SEU pulse sufficient to cause cell change of state. Pulse onset at 1 nsec.

generators (I_{GN}, I_{GD}) in the circuit diagram of the SRAM cell. In that figure, the state SET/RESET transistor and voltage source sets the desired initial state of the cell prior to the SEU pulse onset.

Once the input data is entered, and the corresponding electronic configuration, as represented by the above parameters, is operationally checked in the code, the "production" runs begin. In these examples, trapezoidal SEU pulse shapes of various current amplitudes and FWHM (full-width half-maximum) pulse widths are used. For a given set of pulse parameters, pulse onset is allowed to occur, and corresponding sensitive state node voltages are tracked as a function of time following onset, as

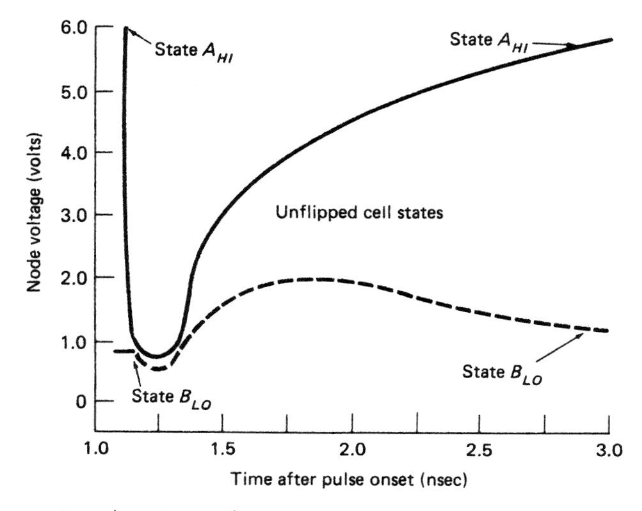

Fig. 12.36 CMM5114/1RZ CMOS/SOS 4K SRAM cell excited by SEU pulse insufficient to cause cell change of state. Pulse onset at 1.1 nsec.

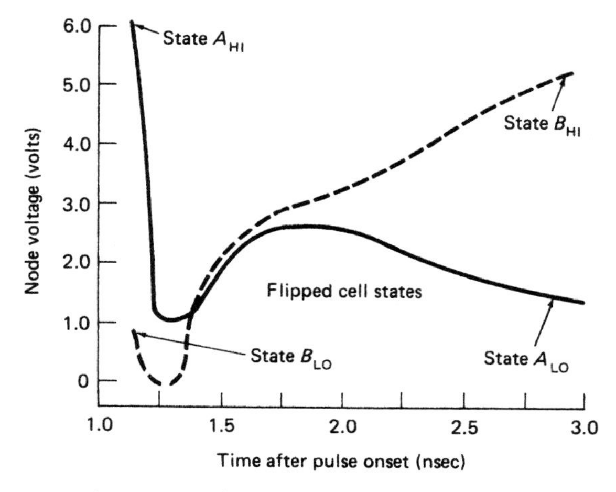

Fig. 12.37 CMM5114/1RZ CMOS/SOS 4K SRAM cell excited by SEU pulse sufficient to cause cell change of state. Pulse onset at 1.1 nsec.

shown in Figs. 12.34–12.37. For the SRAM cell, the SEU pulse current generator corresponding to the node of the "OFF" *p*-channel MOSFET is used as the pulse simulator, as in Fig. 12.33.

Various SEU pulse current amplitudes and durations are run, and the

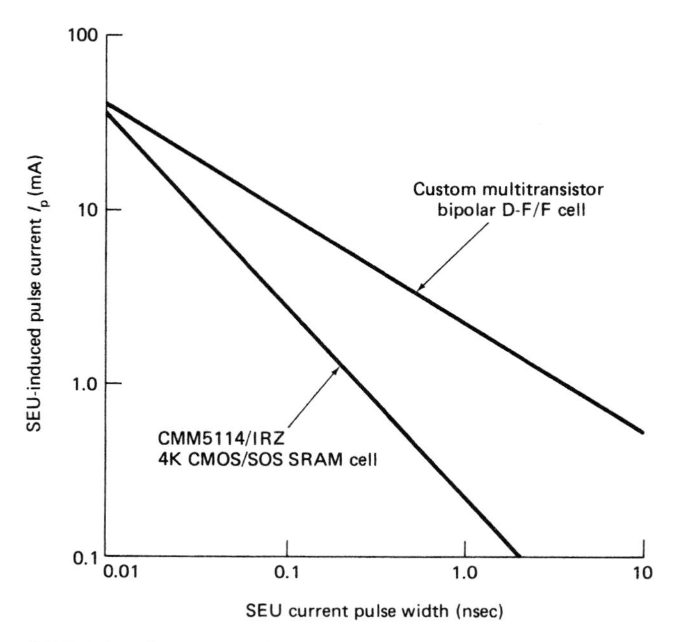

Fig. 12.38 SEU-induced current pulse response for sensitive nodes of two devices versus SEU pulse width as simulated on electronic circuit-device physics SPICE-type program.[46]

results are tracked as mentioned. Initial runs for the *D*-flip/flop and the SRAM cell (as depicted in Figs. 12.34 and 12.36, respectively) are usually obtained. These reveal that the charge collected on the sensitive node, provided by the SEU pulse, is insufficient to cause the device to change its state. By storing and plotting the pulse current amplitudes as a function of the corresponding pulse widths for each run, the approach to the threshold pulse that just causes the device to change its state can be made. Sets of such threshold pulse current amplitude and pulse width pairs are plotted, as shown in Fig. 12.38.

For very short pulse widths, of the order of 0.01–0.10 nsec, which are still representative of the SEU temporal behavior, the desired critical charge Q_c is available from Fig. 12.38. That is, $Q_c = \int_0^{T_p} I_p(t)dt \cong I_pT_p$, where I_p and T_p are the pulse current amplitude and pulse width, respectively. The integral is well approximated by the I_pT_p product for short pulse widths, since I_p is only slowly varying during the pulse duration, as seen by the device time constants implied in Figs. 12.34–12.37.

In Fig. 12.38, computation of the SRAM cell excited by a nominal SEU pulse whose $I_p = 20\,\text{mA}$, and the corresponding $T_p = 10\,\text{psec}$, yields

$Q_c = 0.2\,\text{pC}$ (picocoulombs). Similarly, for the D-flip/flop $I_p = 12\,\text{mA}$ and $T_p = 100\,\text{ps}$, gives $Q_c = 1.2\,\text{pC}$.

12.7 COMPONENT SELECTION: EMP EFFECTS

The electromagnetic pulse generated from nuclear weapons can induce currents of large magnitudes on system structures. As an illustration of the magnitude of such surface currents, the anticipated surface current levels for large military aircraft are of the order of 30,000 amps and of several micro-seconds duration. As investigated in detail in Chapter 9, EMP fields can penetrate any and all types of openings in system housings and induce current flows on electronic circuits, components, and active devices therein. EMP effects are in certain ways similar to electromagnetic interference (EMI) and/or electromagnetic compatibility (EMC) effects. There is a wealth of literature in these areas that is applicable to the EMP problem area.[47,48]

In general, component malfunction can arise from transient upset and burnout due to excessive transient currents. Upset normally occurs at current levels that are less than those required for burnout. Transient upset is essentially the controlling threshold effect in fast digital circuitry. It can cause flip-flops to be triggered, resulting in the malfunction of high-speed computers. Specific items within the computer that can be affected are shift registers, counters, memories, multiplexers and semiconductor switches. Burnout results from excessive currents or voltages that cause overheating or voltage breakdown, resulting in arcing that short-circuits vital part or system nodes.

To achieve a modicum of EMP device or component hardness, certain guidelines include: (a) Devices with large Wunsch-Bell constants should be used, especially at system input/output interfaces; and (b) Additional input and output shunt capacitance should be used wherever possible to slow the circuit response. It is obvious that the aims of (a) and (b) are to give the components and circuits low-pass filter characteristics, in order to shunt the predominantly high-frequency portion that is the bulk of the EMP pulse.

With respect to junction breakdown, those devices that have the largest junction area to withstand the EMP-induced currents and the highest breakdown voltage are best able to sustain EMP energy stresses. Generally, the devices most vulnerable to EMP effects are microwave diodes, transistors, and integrated circuits. Table 12.9 and Fig. 12.39 depict energy required to fail various electronic devices of interest. In Fig. 12.39 and Table 12.9, the failure energy for pulse widths other than the 1-μs

Table 12.9 Estimated energy required for EMP failure of various devices.[47,48] © 1976 by John Wiley & Sons.

DEVICE TYPE	FAILURE ENERGY (μ JOULES)
Point-contact diodes 1N82A–1N69A	0.7–12[a]
Integrated circuits μA709	10[a]
Low-power transistors 2N930–2N1116A	20–1000[a]
High-power transistors 2N1039 (Ge)	1000[a]
Switching diodes 1N914–1N933J	70–100[a]
Zener diodes 1N702A	1000[a]
Rectifiers 1N537	500[a]
Relays (welded contacts)	$2–100 \times 10^3$
Resistors (0.25 W carbon)	10^4

[a] Semiconductor failure energy from 1 μs rectangular pulse.

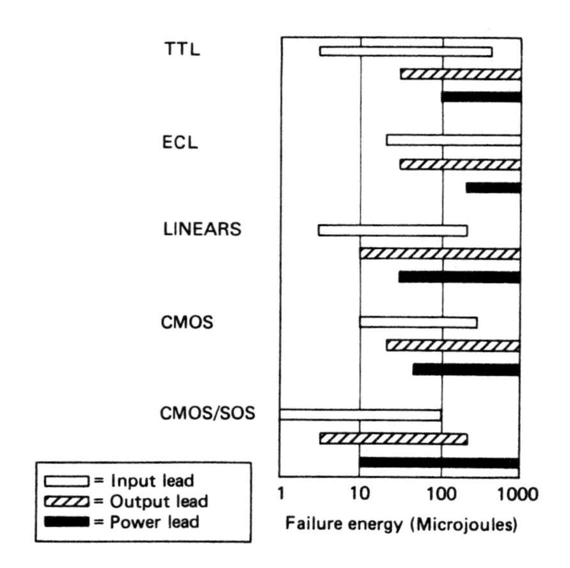

Fig. 12.39 Integrated circuit burnout thresholds versus 1-μs incident EMP pulse failure energy.[33] © 1980 by the IEEE.

pulse shown can be estimated by multiplying the failure energies by the square root of the pulse width in microseconds.

Junctions that are purposely back-biased, such as the substrate in junction-isolated ICs, are vulnerable to a forward biasing EMP spike, which will upset the whole bias voltage structure and may cause junctions to burn out. This can be prevented by judicious installation of current limiters in the bias supply, such as series resistors or inductors.

EMP voltage spikes can damage the front ends of radio frequency receivers. Partial remedies are back-to-back diodes, of ratings that will adequately clip both positive and negative going EMP transients at the receiver input. For strong high-current EMP pulses, spark gaps, varistors, or ultra-fast acting high-power zener-like devices called *transorbs* can be used for general EMP pulse suppression.[49,88]

Semiconductor junction burnout damage thresholds have been investigated extensively, as discussed in Section 9.9. These studies indicate that a junction device, whether a diode or a transistor, can sustain a very large short-duration power pulse surge compared with its continuous-service rating. For an EMP-induced energy damage, the threshold is often orders of magnitude greater than the steady-state specifications might suggest. It turns out that the shorter the EMP pulse duration, the greater is the peak power that the device can withstand. The above studies assume a rectangular pulse, meant to simulate the first (largest) half-cycle of the incident pulse response of the device. The Wunsch-Bell model, discussed in Section 9.9, is a well-used approach in this area. Recall that this relation is given by

$$P_f/A = Kt_f^{-1/2} \tag{12.35}$$

where

P_f is the failure power threshold of the device (kW)
A is the junction area of the device (cm^2)
t_f is the duration of the rectangular EMP power pulse (μs)
K is the damage constant ($kW(\mu s)^{1/2}$ per cm^2)

Equation (12.35) was verified for rectangular pulse widths of $0.1-10\,\mu s$ as applied to 50 different transistor types and 70 different diode types.[5] The various device families have different values of K. Sometimes values of K and A are difficult to obtain without testing the part under investigation. However, extensive experimentation has yielded approximations to values of K/A given by:

a. diodes $K/A = 0.56\,kW\,(\mu s)^{1/2}$ per cm^4; $A \leqslant 0.1\ cm^2$
b. bipolar E-B junctions $K/A = 0.47\,kW\,(\mu s)^{1/2}$ per cm^4; $A \leqslant 0.1\ cm^2$

There are a number of other techniques for obtaining the damage constant K, discussed in the literature.[51–55]

Table 12.10 Semiconductor junction device EMP damage constant guidelines.[56]

TYPE OF SEMICONDUCTOR	DAMAGE CONSTANT K_D (W (s)$^{1/2}$/cm^2)		
	RANGE MINIMUM	RANGE MAXIMUM	RECOMMENDED DAMAGE CONSTANT LOWER LIMIT
Diodes			
Rectifier	5×10^{-1}	2×10^{1}	> 3
Reference	1×10^{-1}	10	> 10
Switching	1×10^{-2}	1	$> 10^{-1}$
Point contact	5×10^{-4}	1×10^{-1}	$> 10^{-2}$
Microwave	3×10^{-4}	3×10^{-2}	$> 3 \times 10^{-3}$
Transistors			
High Power	2×10^{-1}	5×10^{1}	>1
SCR	2×10^{-1}	10	> 1
Germanium	2×10^{-2}	10	$> 2 \times 10^{-1}$
Switching	2×10^{-2}	3×10^{-1}	$> 1 \times 10^{-1}$
Low-power	8×10^{-3}	2	$> 1 \times 10^{-1}$
Integrated Circuits			
Input signal-to-ground	3×10^{-4}	2×10^{-1}	$> 1 \times 10^{-2}$

12.8 CIRCUIT DESIGN MARGINS

Circuit design margins, from the point of view of hardness design, take a variety of forms. In general, their magnitudes depend strictly on the actual predicted circuit failure level and the specification level for the particular radiation environment under consideration, such as neutrons, x rays, gamma rays, and EMP.

Neutron Circuit Margins

For the case of damaging effects of neutrons, the cost-effective approach is to make certain that the vulnerable parts in the radiation critical circuits have adequate design margins.[57,58] For neutrons, one kind of design margin, D_M, can be defined as the ratio of neutron fluence level, Φ_{WC}, at the circuit-specified failure common emitter gain value for the worst case part, h_{FE}(fail), to the specification neutron fluence, Φ_{SP}, i.e.,[59]

$$D_M = \Phi_{WC}/\Phi_{SP} > 1 \quad (12.36)$$

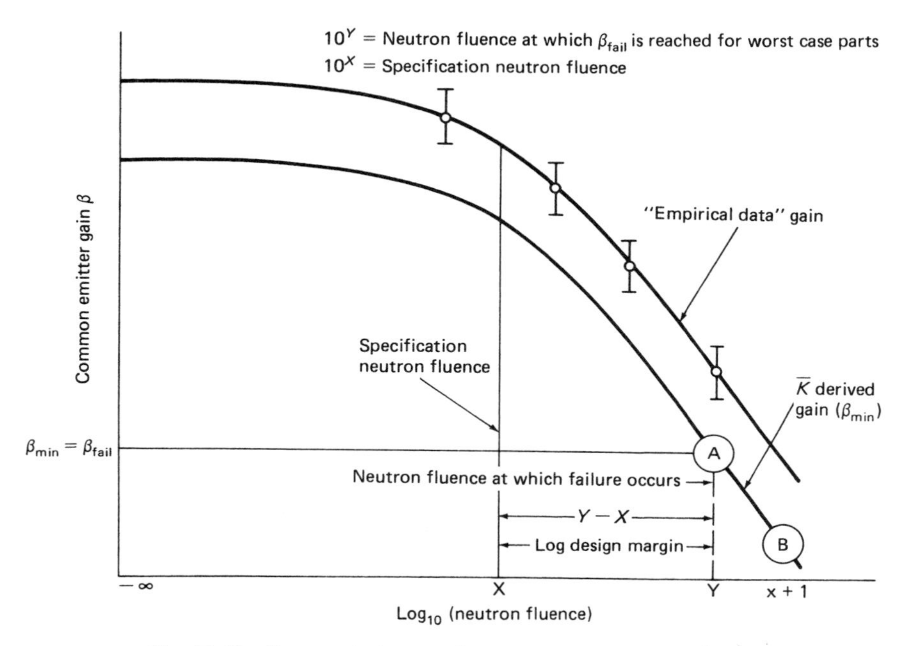

Fig. 12.40 Representative transistor response to neutron fluence.

For example, if it takes 10^{14} neutrons per cm^2 to cause a particular circuit transistor to degrade to its failure gain level, and $5 \cdot 10^{12}$ neutrons per cm^2 is the specification fluence level, then the design margin, $D_M = 10^{14}/5 \cdot 10^{12} = 20$. Statistical aspects are discussed in Section 14.6.

If $D_M > 10$ can be achieved merely by a judicious substitution of parts, then no special controls or procedures are required, in the sense of a hardness-assurance critical-part hierarchy, as discussed in Chapters 14 and 15. For neutron specifications up to about 10^{12} neutrons per cm^2, $D_M > 10$ can be realized for most device types. For circuits that must be potentially subject to higher fluence levels, hardness-assurance controls may be required.

An example of design margin, which can be used for bipolar devices and systems,[60] is depicted in Fig. 12.40. In the figure, the upper curve called "empirical data" gain, is a plot of experimentally obtained transistor gain versus incident neutron fluence. The lower curve is obtained by applying the average damage constant, $\bar{K}$, derived from test data, to the vendor-provided minimum transistor gain, $\beta_{\min}$. That is, the lower curve is plotted from $\beta = \beta_{\min}/(1 - (\beta_{\min}\Phi_n/\omega_T K))$. This lower curve intersects the minimum acceptable value of gain, $\beta_{\min}$, at point A, which

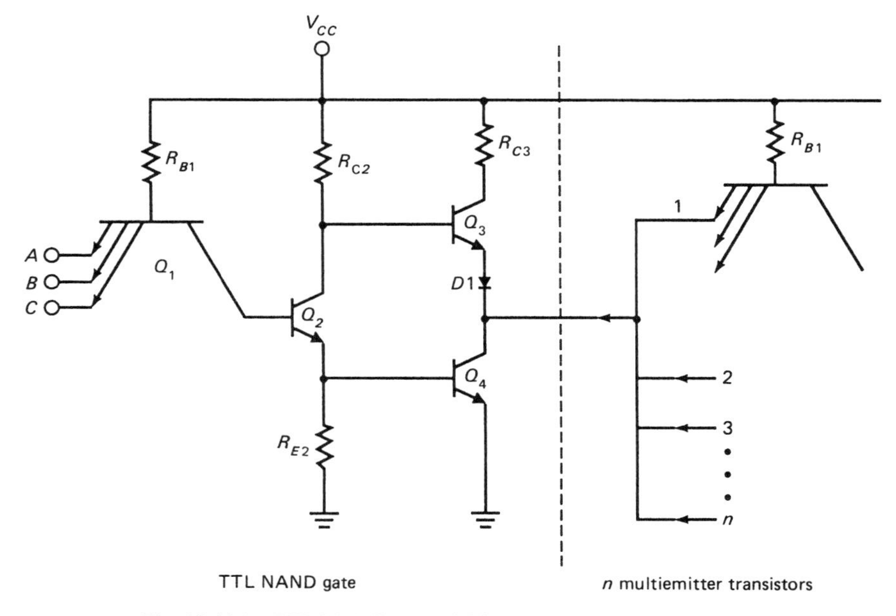

Fig. 12.40A TTL NAND gate sinking n multi-emitter transistors.

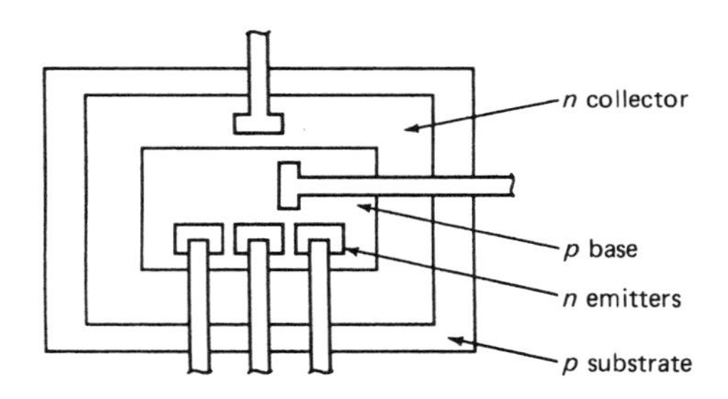

Fig. 12.40B Construction of a multi-emitter bipolar transistor.

point is usually specified in "worst-case" fashion by the circuit designer. This establishes the failure level at 10^Y neutrons per cm^2. If 10^X is the system specification neutron fluence level and $X < Y$, then the design margin is $10^Y/10^X$ or 10^{Y-X}.

In microcircuits, the radiation-induced response comes about through changes in both their dc and switching behavior. Their radiation vulnerability is almost always determined by the tolerance of the circuit design with respect to gain degradation.

One of the most radiation-sensitive electrical parameters associated with digital circuits is the output low-state voltage. Electrical failure occurs when transistors in their ON saturation state are forced out of saturation by radiation-induced gain reduction. This is because the amount of current that the circuit output gate transistor can sink is proportional to its current gain. Thus neutron-induced gain degradation has an impact on its current sinking ability, i.e., its fanout capability.

For the common TTL NAND gate, the threshold neutron fluence for failure, Φ_{th}^{TTL}, can be computed using a circuit-derived failure criterion. That is, the failure criterion can be taken as the state where the TTL gate output voltage change is just greater than the noise margin of the following gates. This is essentially equivalent to a circuit design margin of unity. Then Eq. (5.66) is modified for this case as

$$\Delta(1/\beta_{\min}) \equiv 1/\beta_{\min} - 1/\beta_o \cong \Phi_{th}^{\mathrm{TTL}}/K\omega_T \tag{12.37}$$

where $\beta_{\min}$ is the minimum required gain of the TTL gate output for a fanout of n multi-emitter transistor gates, where (for each) all but one of their emitters are cut off (worst case). A dc circuit analysis (no leakage or load capacitance currents are assumed) of the TTL gate,[58,86] depicted in Figs. 12.40A, gives

$$\beta_{\min}^{-1} = \frac{\begin{array}{c}(1/n)[(V_{CC} - V_{CE(\mathrm{sat})} - V_{BE(\mathrm{sat})}) \cdot (R_{B1}/R_{C2}) \\ + (V_{CC} - V_{BC1} - V_{BE2} - V_{BE4}) - V_{BE4}(R_{B1}/R_{E2})]\end{array}}{(V_{CC} - V_{CE(\mathrm{sat})} - V_{BE(\mathrm{sat})})} \tag{12.38}$$

After inserting the numerical values below, Eqs. (12.37) and (12.38) yield

$$\begin{aligned}\Phi_{th}^{\mathrm{TTL}} = K\omega_T \cdot \{&(1/n)[1 + R_{B1}/R_{C2} \\ &- [1.2 + 0.7(R_{B1}/R_{E2})]/(V_{CC} - 0.9)] - 1/\beta_o\}\end{aligned} \tag{12.39}$$

where

$$\begin{aligned} V_{CE(\mathrm{sat})} &= 0.2\mathrm{V} \\ V_{BE(\mathrm{sat})} &= 0.7\mathrm{V} \\ V_{BC1} &= V_{BE2} = V_{BE4} = 0.7\mathrm{V}. \end{aligned}$$

The damage constant for bipolar ICs to be used at intermediate injection levels is $K = 1.6 \cdot 10^6$ neutron seconds per cm^2, e.g., as in $\Delta(1/\beta) = \Phi_n/K\omega_T$.

Dose Rate Circuit Margins

For dose rate considerations, the notion of circuit design margin is also applicable. In this case, circuits are usually divided into two principal types: digital and analog. Nominally, analog circuits can handle reasonable levels of incident photon dose rates. This is because the resultant dose rate perturbation time of the circuits so exposed is much shorter than the allowed maximum transient time at their output that still ensures adequate circuit performance. Where the relative response time of the analog circuitry and the dose rate perturbation time govern circuit failure criteria, potential failure determination can be based on a circuit design margin, defined as the ratio of excitation time of the driven part or circuit to the duration time of the dose rate pulse. In certain instances, the dose rate perturbation may result in a short inoperative period due to circuit blanking or possible erroneous digital information, as might occur in A/D converters or computer central processing units (CPUs). These should be examined individually and, if the brief disruption causes little more than a miniscule degradation in system performance, then these devices should be considered satisfactory from the point of view of dose rate.

Digital circuitry in which a dose rate transient disruption or any data loss cannot be tolerated must be examined carefully for transient level and response time difficulties.[63] This type of circuitry is generally susceptible to induced logic level perturbation for moderate dose rates. For corresponding types of computers utilizing these circuits or digital circuitry in general, circumvention circuits discussed in Section 13.9 can be used to prevent the loss or scrambling of information therein.

Ionizing Dose Circuit Margins

For ionizing dose specifications that are of moderate level, usually only a few types of semiconductor devices and circuits are susceptible. Potential areas of difficulty lie with MOS circuits, bipolar circuits operating at very low bias currents of the order of a microampere, and possibly operational amplifiers with extremely low input bias currents. For more stringent ionizing dose specifications, proportionately more device types and circuits will be susceptible, so that a more intensive hardening investigation would be necessary. For moderate ionizing dose environments, circuit design margins are usually taken as three to ten.

For bipolar logic gates enduring ionizing dose damage, if $\Delta(1/\beta_{\min}) = F(\gamma_{th})$ can be written where γ_{th} is an ionizing dose threshold level analogous to Φ_{th}^{TTL} in Eq. (12.37), and F is a known function, as in Eq. (12.20), then the threshold ionizing dose for a bipolar TTL NAND gate circuit design margin of unity, $\gamma_{th}^{\mathrm{TTL}}$, can be obtained. It is simply $\gamma_{th}^{\mathrm{TTL}} = F^{-1}(1/\beta_{\min} - 1/\beta_0)$, where $1/\beta_{\min}$ is gotten from Eq. (12.38). For example, following Eq. (12.20), $F(\gamma_T) = K_\gamma \gamma_T^{\varepsilon}$, so that $\gamma_{th}^{\mathrm{TTL}} = F^{-1}[\Delta(1/\beta_{\min})] = [(1/\beta_{\min} - 1/\beta_0)/K_\gamma]^{1/\varepsilon}$. This method can be extended to other gate types and design margins in this context, using corresponding circuit analyses.

EMP Circuit Margins

EMP hardening approaches are related to electromagnetic shielding and electronic circuit hardening. EMP hardness design margins are usually expressed in decibels. For example, an EMP circuit design margin can be defined as

$$\text{EMP hardness margin} = 20\log_{10}(I_{\text{damage}}/I_{\text{spec}}) \qquad \text{(dB)} \qquad (12.40)$$

where I_{damage} is the minimum current that can damage a circuit or device, and I_{spec} is the specified current level impressed at the circuit/device interface connector. For many devices, hardness design margins of 10 dB are deemed satisfactory under normal operating conditions. In the case of device burnout due to EMP, the Wunsch-Bell model can be used to provide a computational method for predicting damage due to burnout. For a given system, the power required to cause a device junction failure is

$$P_f = K_D t_f^{-1/2} \qquad (12.41)$$

where K_D is the appropriate damage constant (Table 12.10), and t_f is the assumed rectangular impressed EMP pulse duration. On the other hand, the usual specifications form of the impressed EMP voltage is often given as a damped sine wave, as in a parallel RLC circuit, i.e.

$$V = V_0(\exp - \omega_0 t/2Q)\sin m\omega_0 t; \qquad m = [1 - (1/2Q)^2)^{1/2} \qquad (12.42)$$

where $V_0, Q = \omega_0 L/R$, $\omega_0 = (LC)^{-1/2}$ are part of the EMP specification; and R, L, and C are the parameters of a laboratory simulator impressed voltage source. This implies that the coupling analysis between the EMP field and the system of interest has been done, with results couched in terms of the above equivalent damped sine wave of voltage impressed on the system input connector pins (direct drive). Hence, it is desired to relate this damped sine wave specification to that of the Wunsch-Bell pulse approximation to establish useful relationships between corresponding parameters. To obtain this relationship, the device burnout failure power, as computed by the Wunsch-Bell relation, is equated to that contained in the damped sine wave of Eq. (12.42). To illustrate, it is initially assumed that the power contained in the first half sine wave period $T/2$ in Eq. (12.42) suffices to provide the burnout failure energy $E(T/2)$ to the device under consideration when indeed it does fail in this manner. That is, from Eqs. (12.41) and (12.42)

$$E(T/2) = (1/R)\int_0^{T/2} V^2(t)\,dt = P_f t_f = K_D t_f^{1/2} \tag{12.43}$$

where R is the corresponding equivalent resistance. Evaluating the above integral using Eq. (12.42), with the assumption that $m \cong 1$, yields

$$E(T/2) = (V_0^2 Q/2\omega R)(1 - \exp - \pi/Q)(4Q^2/(1 + 4Q^2)) = K_D t_f^{1/2} \tag{12.44}$$

For the system $Q \gg 1$, e.g. $Q \simeq 24$, which holds well for EMP impinging on large avionics and ground systems, then Eq. (12.44) becomes, with $\omega_0 = 2\pi f$

$$E(T/2) \cong V_0^2/4fR = K_D t_f^{1/2} \tag{12.45}$$

and, as peak power is implied, $V_0^2/2R = P_f$, and using Eq. (12.41) results in

$$K_D t_f^{-1/2}/2f = K_D t_f^{1/2} \tag{12.46}$$

so that one sought-after relation between the laboratory-impressed damped sine wave and the Wunsch-Bell rectangular pulse from Eq. (12.46) is

$$t_f = 1/2f \tag{12.47}$$

where f is the center frequency of the damped sine wave. If the above is recomputed more accurately, and without the assumption that the device fails in the first half-cycle,[64] the result is then

$$t_f = 1/2.63f \tag{12.48}$$

Even more accurate computations for (a) a rectifying diode dissipating power over one half-cycle only, and (b) for a resistor dissipating EMP power yields $t_f \cong 1/2.25f$.[63,64] From Eq. (12.41), using this last expression for t_f yields

$$P_f = 1.5K_D f^{1/2} \tag{12.49}$$

as the burnout failure power. Corresponding design margins can be obtained from $10\log_{10} P/P_0$, where P_0 is the specification power of the device under consideration.

12.9 PHOTOCURRENT COMPENSATION

Previously discussed methods for limiting the effects of photocurrent on circuits consisted of current-limiting resistors or inductors, usually in series with the collector leads, or bias leads. When the ionizing radiation pulse is incident on many transistors simultaneously, for example as part of an integrated circuit chip, the technique of cancellation or compensation can be used to minimize the effects of the photocurrents produced in the principal transistors. The main idea of compensation is illustrated in Fig. 12.41(a). The diode portions of two other transistors on the same chip, being well-matched physically, and electrically ape the principal transistor Q_1 by ideally removing as much current (charge) from the base of Q_1 as is generated therein by the radiation. That is, the direction of the photocurrent generated in Q_B and essentially bled off to ground is equal and opposite to that generated in Q_1, so that the net circuital effect of that photocurrent vanishes at the base node point; similarly for Q_L and photocurrents leaving the collector node-point.

Without such techniques, Q_1 can be driven easily into saturation by virtue of the photocurrent voltage drop across the base resistor. Variations on this theme are shown in Figs. 12.41(b)–(e). Photocurrent bypass techniques can be used to shunt photocurrent around capacitors and other components.[66] As mentioned, compensation requires careful matching of individual transistors. Also, even though the cancellation is effective over one range of dose rate radiation levels, the combination of Q_B and Q_1 can produce transients at other levels that cannot be compensated easily. This is surely so when using discrete dissimilar devices for compensation, such as a diode-transistor pair. Generally, compensation methods should be used with caution. Additionally, it should be noted that these compensation methods usually provide temperature compensation in instances where the currents vary due to temperature effects.

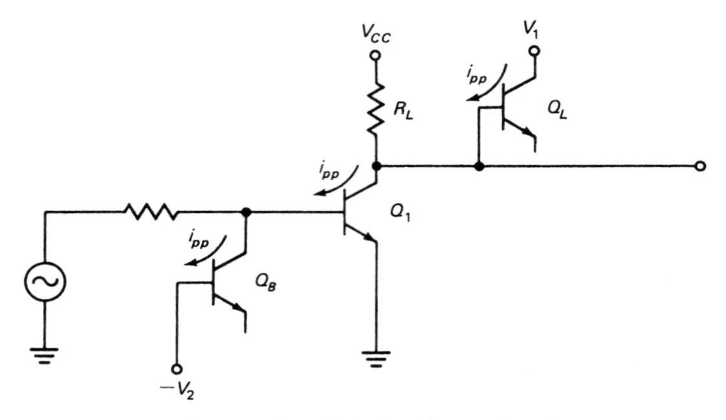

a. Compensation with load and base-emitter shunts

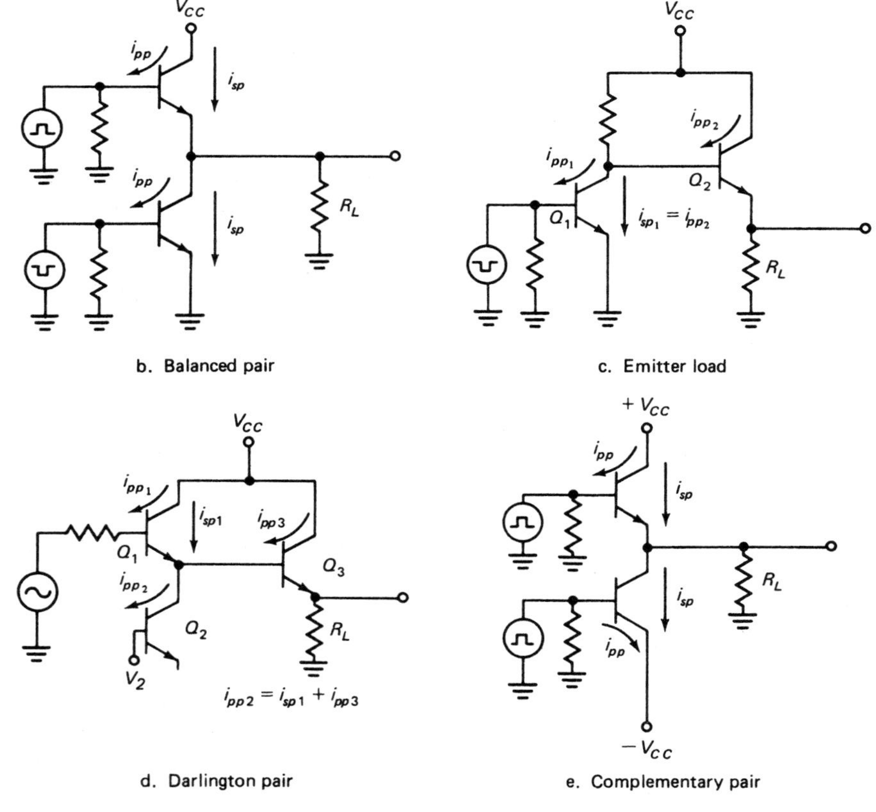

b. Balanced pair

c. Emitter load

d. Darlington pair

e. Complementary pair

Fig. 12.41 Photocurrent compensation techniques for hardening against photocurrent producing ionizing radiation.[67]

12.10 DIFFERENTIAL SYSTEMS

Differential systems include amplifiers, comparators, differential transformers and any device that essentially responds to a comparison signal—usually a difference signal. An example of the differential idea is that of spurious currents induced on a multiconductor cable that tend to be roughly of the same magnitude with the same phase. This is especially so when the individual conductors (pairs) are terminated with the same impedance. If the spurious currents can be maintained as common-mode signals with respect to a given signal line and its return, then the former can be rejected by a differential amplifier or differential transformer. A carefully designed differential amplifier will have both a very high common mode rejection ratio and equal input and output impedance with respect to the signal common (return) or signal ground. Maintainance of equal input and output impedance in the differential system will enhance its common mode rejection capability. Such differential amplifiers have open loop gains in excess of 10^5, so that heavy negative feedback is used with them for electrical stability, which simultaneously makes for eradication of ionizing radiation-produced common mode extraneous currents.[68] Differential amplifiers usually have high input impedances and so are very sensitive to input voltage transients; they also have a tendency to become regenerative at very high frequencies. These negative traits can be overcome largely by careful circuit component choices. The differential amplifier with some minor changes in its circuitry can be converted to a voltage comparator with a detection hysteresis property not unlike a Schmitt trigger. Then this system can be used with two-phased logic in a differential scheme.[68] For example, as used in a digital system, this circuit will maintain a "true" state until the two-phase inputs reverse their state for a given voltage decrement. Then the output would change state to "false." Similarly, when the inputs again reverse, the decrement must be exceeded before the output switches back to the "true" state. Signal level detection would be made as slow as practicable, again to not respond to extraneous signals superimposed on the two-phase logic signals and also to not respond to slight differences in propagation delay. Similarly, differential transformers can be used, e.g., by grounding a center-tapped 1 : 1 primary and secondary to maintain a common primary/secondary impedance from line to ground. Differential transformers also can be employed to convert balanced to unbalanced signals, and vice versa, at various impedance ratios. These usually use ferrite toroid cores with windings of unusual topology, especially at very high frequencies or equivalently very high data rates.[69–71] As is well known, transformers can be construed as low-pass filters, in that while not passing dc, they shunt very-high-frequency signals due to their distributed turns capacitance. Their high-frequency cutoff characteristic can be exploited where the very-high-frequency component of photocurrents or EMP-induced currents are shunted by the transformer. Difficulties with such transformers include turn-to-turn arcing in very-high and ultrahigh-frequency high-power applications, which can cause burnout of the transformer. Finally, the use of these transformers implies a circuit power insertion loss.

12.11 FILTERS

Filters are ubiquitous throughout electronic circuitry. This is also true for circuits subject to radiation, especially EMP. Allusions to filters and their use in radiation protection have been made in many places in this chapter thus far. For EMP and photocurrent pulse applications, low-pass filters are usually employed because of the large amount of high-frequency component appearing, simply because of the short pulse nature of the extraneous signals. Such filters are usually low-pass Pi or T configurations. Low-pass T filters should be used with caution in EMP applications because of the high pulse voltage that can develop across the filter input capacitor, resulting in its degradation or failure, especially where the capacitor resonates with the input signal components. A principal advantage of filters, as opposed to shielding, is that they weigh less and usually do not deteriorate with time, being less subject to corrosion, vibration, and other stresses.

Effective filters have a very low input impedance for frequencies in the stop band. If they are constructed carefully, they have a high dc current capability, low dc resistance, and little cross-coupling between their input and output terminals. Most internal shielding in filters is accomplished by metallic compartmentalization of stages and the use of metallic mesh to fill any cavities within the filter enclosure for SGEMP hardening. Good electrical grounding of the outside of the metal filter housing is essential to good filter performance.

The physical size and weight of filters cover a wide range, being largely dependent on the required current handling capacity. For filtering power bus system interfaces, typical filters include the hybrid dissipative filter (ferrite core plus lossy capacitors to dissipate undesired power), filter pin connectors (filter components embedded in a cylindrical monolithic pin structure), and feed-through capacitor filters (capacitors coaxial with pin terminals and grounded to the chassis or common).[68]

12.12 CURRENT-LIMITING

Current-limiting means are well known to circuit designers as employed for situations of potential transient damage. Perhaps not so well-known is their usefulness in multifarious forms to curb radiation-induced transients, such as dose-rate-induced photocurrents and EMP. For example, under anomolous input conditions, a semiconductor junction can be biased far above safe current levels. The use of a current-limiting resistor will dissipate the excess power instead of the junction. A current-limiting resistor in the input lead of the transistor, together with a separate path to ground, will prevent excessive current from being drawn through the associated base-collector junction. A current-limiting resistor in the emitter lead of a transistor will stabilize it against possible thermal runaway effects produced by excessive transient power dissipated therein. A current-limiting resistor in the collector, or power, lead will reduce the transient power dissipated in the collector junction. A current-limiting resistor in series with a biased substrate lead to its IC superstructure will protect it from excessive currents drawn due to abnormal substrate voltage conditions. For transient pulses, inductors of various types, including ferrite toroidal beads, can often be used in lieu of current-limiting resistors. Ideally, these inductors are purely reactive and so dissipate no power, which is a consideration where power is at a premium. Of course, some small amounts of power are dissipated as inductor core losses.

12.13 CIRCUIT-HARDENING POTPOURRI

This section comprises an assortment of circuit hardness measures. It should be mentioned that components be selected that are least affected by radiation. There are enough anomalies across the diversity of devices and components that such a task is not trivial. Minimum radiation response, and especially minimum ionizing radiation response, should be coupled with fast recovery times and minimum neutron damage in the choice of parts. Of course, all the above must be consistent with electrical circuit requirements.

One method by which electronic circuitry can be made harder to ionizing radiation is the employment of low-circuit impedances wherever practical. Low impedances mitigate the effect of impressed currents to circuit inputs, for example, those induced by photocurrents that can be assumed to act as current generators. The use of clamping methods is another way to prevent large transients from occurring in one stage where effects could be passed on to succeeding stages.

A secondary effect of radiation energy deposited in circuit components is heating. The heat capacity and thermal conductivity of certain components are such that heating can persist for times long after the radiation has ceased. However, as mentioned in Section 9.9, recent research has found that the thermal conductivity of important thin film semiconductors is much less than their corresponding bulk values. In general, modern standard electrical temperature compensation techniques can be used to counteract such temperature excursions. A prudent way of assessing circuit vulnerability to this type of heating requires an estimate of the possible temperature rise and duration in all anticipated critical components and the circuit sensitivity to such temperature induced changes in these component parameters.

Besides circumvention itself, there are some similar but quite simple techniques to accomplish circumventionlike action.[60] For example, relays, magnetic cores, and certain tunnel diodes have two stable states that are relatively radiation-insensitive. They can be prevented from changing state during the onset of the incident radiation pulse, using time-delay or time-multiplexing methods. As an illustration, a specific event sequence can be contrived to occur before a state will change, and which cannot possibly be initiated by radiation-induced saturation of the affected transistors. Triggering can be prevented by time constants long compared with that of the radiation pulse, so that any radiation-induced disturbance will have decayed long before the system is activated.

Physically potting a circuit will sometimes pay for itself in space and weight costs many times over, through preventing air ionization and charge-scattered currents. Potting even low-impedance strip lines or delay lines can significantly reduce the effect of these currents in such systems. Direct-coupling circuits should be used wherever possible, because the recovery of such circuits after a disturbance is a function only of the relaxation time of the associated transistor, not of any reactive coupling components. Often the reactive coupling elements have a very long relaxation time compared with the transistor, thus offering a poor circuit temporal match. Sometimes zener diodes can be used to replace coupling capacitors in these circuits, if the zener photocurrents are small enough.

Digital circuit ON states should correspond to a fully saturated transistor wherever possible. Because dose-rate pulse induced radiation can saturate transistors, it does not make any difference when they are already in their saturated ON state. Further radiation damage mitigation methods and measures are given in the *TREE Handbook*.[71]

12.14 NOISE

Noise is construed here as random fluctuations in current through, or voltage across, suitable nodes within semiconductor devices. When active components are used to measure and amplify small currents and voltages, these signals compete with their noise counterparts.

Noise is usually divided into three types:[75] thermal noise, flicker noise, and shot noise. Thermal noise occurs in any conductor or semiconductor, and is due to thermal motion of the current carriers, whether holes or electrons. Considering a noisy device as having two input and two output terminals, its open circuit output mean squared thermal noise voltage $\overline{E_n^2}$, as measured looking back into its output terminals, is given by

$$\overline{E_n^2} = 4kTR\Delta f \tag{12.50}$$

where k is Boltzmann's constant ($1.38 \cdot 10^{-23}$ J/°K), R is the real part (resistance) of the corresponding output impedance, and Δf is the appropriate noise bandwidth in Hz. To appreciate the magnitude of this noise voltage, assume $R = 1$ kiloohm, and room temperature (293°K), then the root mean squared (rms) thermal noise voltage per root Hz is

$$(\overline{E_n^2}/\Delta f)^{1/2} = E_{rms}/(\Delta f)^{1/2} = 4/\sqrt{\Delta f}\,(nV) \tag{12.51}$$

Flicker noise is the so-called low-frequency $1/f$ or "pink noise," where the noise level is proportional to $1/f^{\alpha}$, $\alpha \approx 1$. It is, of course, most evident at lower frequencies. For most semiconductors, flicker noise is due to surface effects.

Shot noise is the major noise occurring in most semiconductor junctions.[75] At low and intermediate frequencies, its spectrum is usually flat (white noise) with the spectrum amplitude falling off for higher frequencies. The mean squared shot noise current $\overline{i_n^2}$ through a p-n junction is

$$\overline{i_n^2} = 2eI\Delta f \tag{12.52}$$

where I is the device junction current. So that at low injection, and neglecting 1/f noise, the total mean squared noise current $\overline{I_n^2}$ is the sum of the thermal and shot noise, yielding (for reversed-biased junctions)

$$\overline{I_n^2} = (4kTg - 2eI)\Delta f \tag{12.53}$$

where g, the device transconductance is

$$g = \partial I/\partial V = \partial/\partial V[I_s(\exp(eV/kT) - 1)] = (eI_s/kT)\exp(eV/kT) \tag{12.54}$$

Inserting this expression for g into Eq. (12.53) gives

$$\overline{I_n^2} = 2eI_s\Delta f[\exp(eV/kT) + 1] \tag{12.55}$$

where I_s is the diode saturation or reverse current, as given in Eq. (3.89), i.e.

$$I_s = en_i^2 A[(D_n/\tau_n)^{1/2}/N_A + (D_p/\tau_p)^{1/2}/N_D] \tag{12.56}$$

and A is the junction area.

For bipolar transistors, the effect of neutron radiation on component noise current begins with Eq. (5.11), i.e., for times following the cessation of the neutron pulse of width t_p,

$$1/\tau = 1/\tau_0 + \Phi_n/K; \ \Phi_n \equiv \Phi_n(t_p) \tag{12.57}$$

τ_0 is the pre-irradiation minority carrier lifetime, and K is the corresponding damage constant, as discussed in Section 5.9. Assume that the lightly doped side of the junction corresponds to acceptor impurities (i.e., $N_A \ll N_D$). Then the second term in Eq. (12.56) is negligible, compared to the first. Combining Eq. (12.57) with the resultant, Eq. (12.56), yields

$$I_s = (en_i^2 A/N_A)[D_n(1/\tau_0 + \Phi_n/K)]^{1/2} \tag{12.58}$$

Hence, the neutron radiation-altered total mean squared collector noise current from Eq. (12.55) is obtained by inserting I_s from Eq. (12.58), to give

$$I_n^2 = [2e^2 n_i^2 A\Delta f/N_A][D_n(1/\tau_0 + \Phi_n/K)]^{1/2}[\exp(eV/kT) + 1] \tag{12.59}$$

Usually, $\Phi_n/K \gg 1/\tau_0$, so it is seen that the root mean squared noise current $I_{nrms} \equiv (\overline{I_n^2})^{1/2}$ increases only as the fourth root of the incident neutron fluence. If the above inequality is not the case, then I_{nrms} is even less dependent on incident neutrons.

The effects of incident radiation on various active device family noise parameters has been investigated, primarily through experiment, over the past two decades. The pertinent part types, aside from discrete devices, mainly encompass linear technologies, such as operational amplifiers, as opposed to digital device families. This is because the operation of digital circuitry is such that their high and low state levels are usually sufficiently extreme to insure that almost all noise is filtered out by the noise margins incorporated in these levels.

For MOSFETs, 1/f drain noise voltage predominates over other types of noise, seemingly regardless of whether or not the devices are irradiated with ionizing radiation. 1/f drain noise voltage is sensitive to defects at or near the Si–SiO_2 MOS device interface.[90] A well-known model[91] asserts

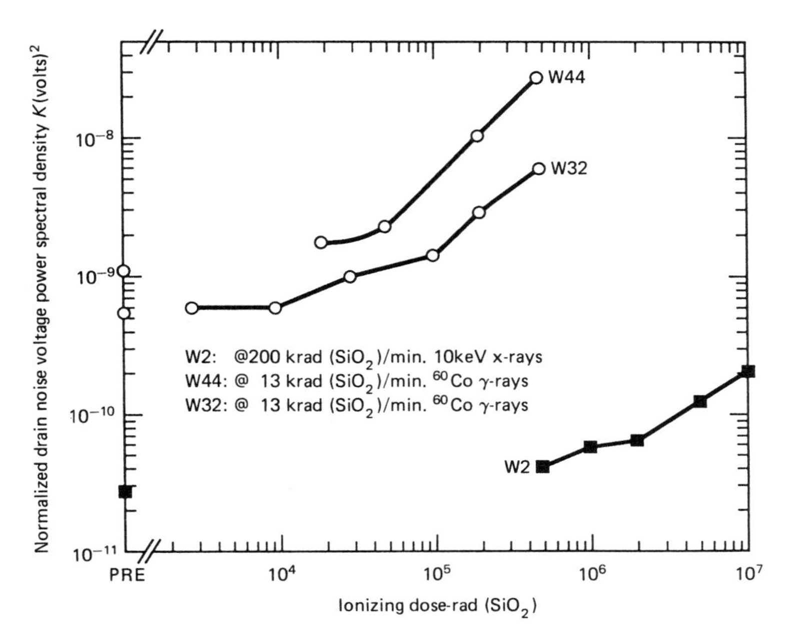

Fig. 12.41A Normalized drain noise voltage power spectral density K versus ionizing dose, at a frequency of 10 Hz, for three MOSFET wafers; W2, W44, W32.[90] © 1990 by the IEEE.

that this 1/f noise is due to fluctuations in the channel current caused by exchange of majority carriers with traps in the above $Si-SiO_2$ region. The corresponding noise power spectral density $S_v(V^2/\text{Hz})$ over a band of frequencies $(S_v \Delta f)$ is proportional to the density of traps whose energy levels are within a few kT of the quasi-fermi energy level.

It has been found[92] that a strong correlation exists between MOSFET pre-irradiation 1/f noise and post-irradiation oxide-trapped charge density, as opposed to post-irradiation interface trapped charge. Good experimental agreement has been obtained with a semi-empirical model, which attributes this noise to tunneling between the channel and the oxide traps. This model represents the drain noise voltage power spectral density S_v as[92]

$$S_v(f) = K \cdot (\bar{V}_d)^2/(V_g - V_T)^2 f^\alpha \text{ (volts)}^2/\text{Hz} \qquad (12.60)$$

where: f = noise frequency (Hz)
$\bar{V}_d$ = mean drain voltage (V)
V_g = gate bias voltage (V)

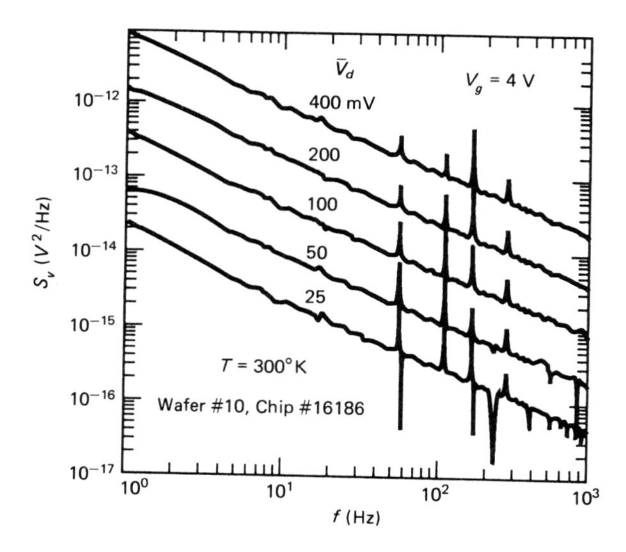

Fig. 12.41B Drain voltage noise power spectral density S_v of excess V_d noise versus frequency. Spurious spikes are 60 Hz pickup.[92] © 1989 by the IEEE.

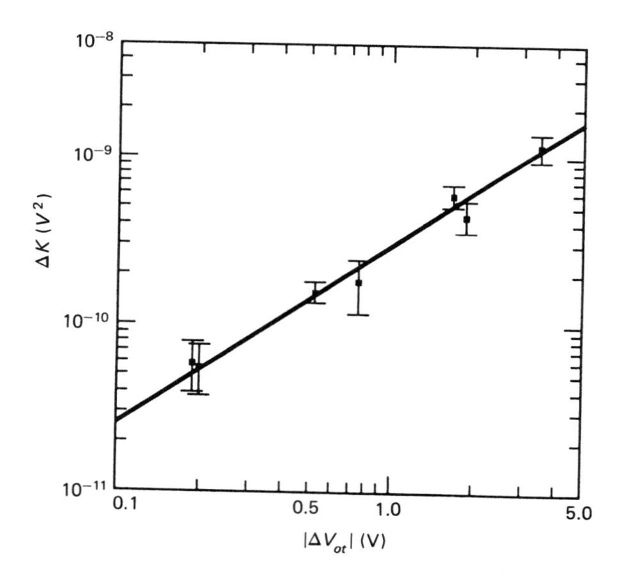

Fig. 12.41C Change in normalized drain voltage noise power spectrol density, ΔK, versus $|\Delta V_{ot}|$.[92] © 1989 by the IEEE.

V_T = gate threshold voltage (V)
α = empirically determined constant $(0.75 \leqslant \alpha \leqslant 1.10)$

K is the normalized drain noise voltage power spectral density in $(\text{volts})^2$. This normalization means that, for $\alpha = 1$ and $\bar{V}_d = V_g - V_T = 1\,\text{volt}$, $K = fS_v$ at room temperature ambient. Figure 12.41A depicts the effect of ionizing dose on K for three representative MOSFET wafers,[90] measured at 10 Hz. Figure 12.41B shows a nominal S_v spectrum for MOSFETs, revealing its $1/f^\alpha$ character.

An analytical expression for K has been obtained[90,92] as

$$K = (e/C_{ox})^2 \cdot (kTD'_{ot})/(LW\ln(t_{\max}/t_{\min})) \tag{12.61}$$

where LW is the channel area, E_g is the SiO_2 band gap energy, $D'_{ot}(E)$ is the density of bulk oxide traps per unit energy interval, $t_{\max}$ and $t_{\min}$ are the respective maximum and minimum tunneling times as part of the noise process.[92] The remaining parameters have their usual definitions, and Eq. (12.61) is assumed to hold for frequencies such that $t_{\max}^{-1} \leqslant f \leqslant t_{\min}^{-1}$.

The observed correlation[92] between the threshold voltage shift component due to oxide traps, ΔV_{ot}, and the normalized drain noise voltage power spectral density K, can be related in simple fashion. ΔV_{ot} and the number density change of ionizing radiation-induced oxide-trapped charge states ΔN_{ot} are connected simply as

$$|\Delta V_{ot}| = e\Delta N_{ot}/C_{ox} \tag{12.62}$$

An assumption is made that, for a given ionizing dose, ΔN_{ot} is proportional to the total number density of oxide traps N_{oT}, with an ionizing dose-dependent constant of proportionality $\lambda(\gamma_T)$, where γ_T is the ionizing dose level. That is

$$\Delta N_{ot} = \lambda N_{oT} = \lambda \int_{E_v}^{E_c} D'_{ot}(E)dE; \qquad 0 \leqslant \lambda \leqslant 1 \tag{12.63}$$

$E_c - E_v = E_g$ in the oxide, and λ increases monotonically with ionizing dose. Also, a further assumption is made that the oxide traps are uniformly distributed in this energy interval, so that D'_{ot} can be factored from the above integral, to give

$$\Delta N_{ot} \cong \lambda E_g D'_{ot} \tag{12.64}$$

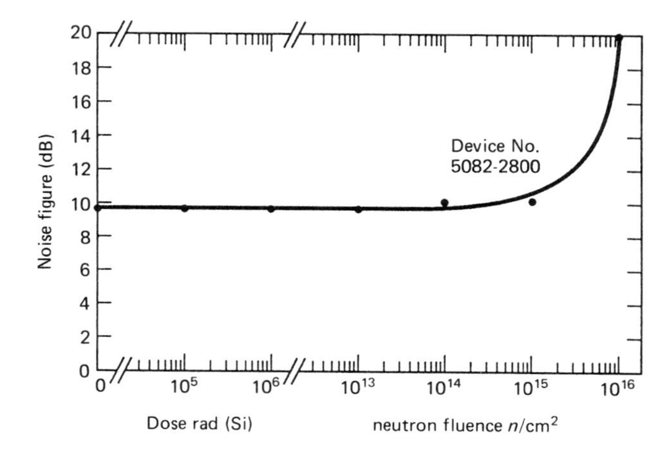

Fig. 12.42 Schottky barrier diode mixer noise figure versus ionizing dose and neutron fluence.[77] © 1973 by John Wiley & Sons.

Substituting Eqs. (12.62) and (12.64) into Eq. (12.61) yields[92]

$$K = (e/C_{ox}) \cdot (kT/\lambda LWE_g \ln(t_{\max}/t_{\min})) \cdot |\Delta V_{ot}| \qquad (12.65)$$

For fixed values of the parameters in Eq. (12.65), it is seen that the normalized noise power K is proportional to $|\Delta V_{ot}|$, and is depicted in Fig. 12.41C.

Another important device type in this context is the schottky barrier diode. These are venerable metal-semiconductor diodes, as opposed to p-n junction diode types, still used as microwave detectors and mixers because of their excellent characteristics at GHz frequencies. The salient noise parameter of microwave devices is the noise figure, closely associated with microwave mixer performance. The noise figure F is defined as $F = (SNR)_{in}/(SNR)_{out}$, where $(SNR)_{in}$ and $(SNR)_{out}$ are the signal to noise ratios at the receiver input and output, respectively. F is usually expressed in dB (i.e., $10 \log_{10}\mathrm{F}$), and the closer F is to unity (zero dB), the better is the receiver sensitivity. Frequently, $NF = 10 \log_{10}\mathrm{F}$ is called the noise figure, while F is called the noise factor. Figure 12.42 depicts the noise figure of a schottky barrier diode microwave mixer.[77] It is seen that the noise figure remains unaffected by reasonably large ionizing dose and neutron fluence levels usually associated with the radiation specifications of many strategic and tactical military systems.

A much-used family of active devices where noise plays a role in their operation is the GaAs MESFET. Again, because of their sterling micro-

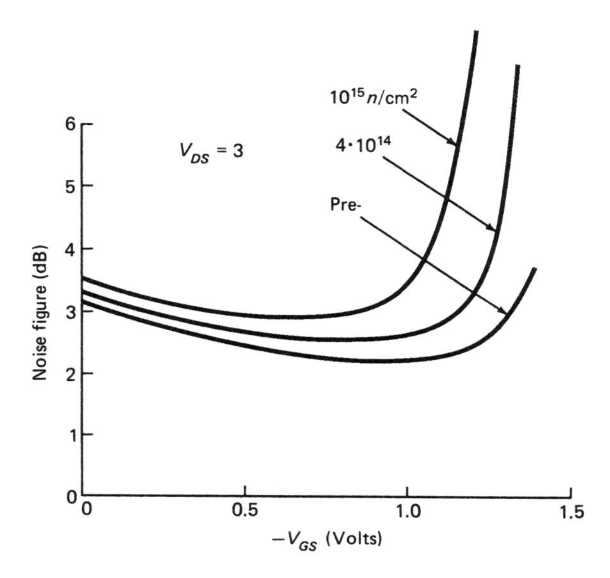

Fig. 12.43 GaAs MESFET (NEC-244-31) noise figure versus gate voltage at 3 GHz for three levels of neutron fluence.[78] © 1980 by the IEEE.

wave frequency characteristics, they are used in the front-end stages of sensitive radar receivers and like systems. Figures 12.43 and 12.44, respectively, illustrate MESFET S-band noise figure and gain versus gate bias for various levels of ionizing dose and neutron fluence.[78,79] It is seen from Fig. 12.43 that noise figure degradation is incipient at a fluence of about 10^{14} neutrons/cm^2, and heavy deterioration is evident at 10^{15} neutrons/cm^2. The increased noise figure and associated gain decrease is related to radiation-induced transconductance degradation, as well as increased parasitic resistance of the device. The low frequency noise can be shown to increase with neutron fluence, but not with ionizing radiation. It also appears to be uncorrelated with the microwave noise component. The ionizing radiation-induced noise figure increase in Fig. 12.44 does not occur until levels of approximately 20 Mrad (Si) are approached.

A final family of active devices where noise is a consideration are low-noise JFETs. Like MESFETs, JFETS are also an essential component in the input stages of low-noise radio frequency (RF) amplifiers. Figure 12.45 depicts thermal drain noise voltage per root frequency, $E_{rms}/f^{1/2}$, for a number of JFETs irradiated to various levels of ionizing dose at discrete measurement frequencies.[80] From the figure it is seen that JFETs

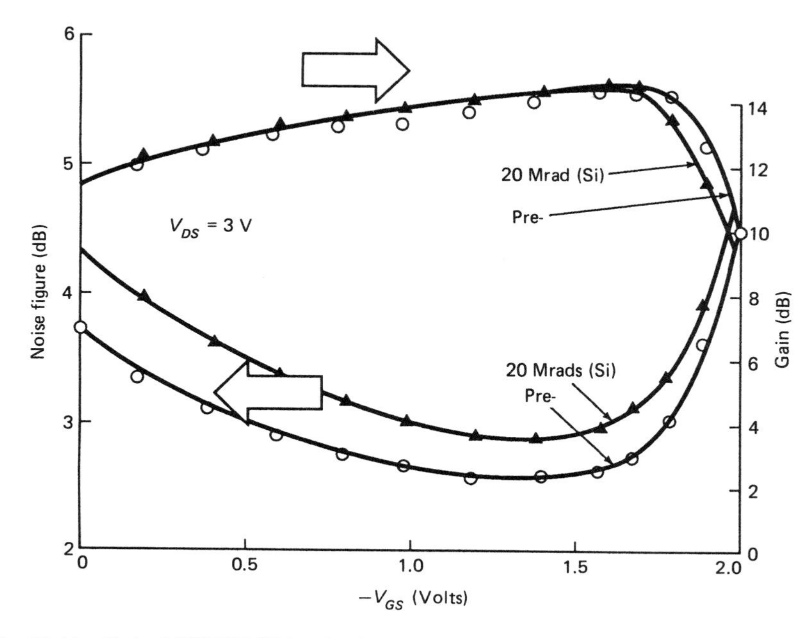

Fig. 12.44 GaAs MESFET (GAT-5-48) noise figure and gain versus gate voltage at 3 GHz for two levels of ionizing dose.[79] © 1979 by the IEEE.

should perform satisfactorily in low-noise amplifiers at ambient temperatures for ionizing dose levels to 100 krad (Si). Above this value, the noise may increase to unacceptable levels, depending on the particular circuitry and application.

As can be appreciated from the foregoing, most of the major noise parameters suffer little degradation from neutron fluence, conservatively up through 10^{14} neutrons/cm^2, and ionizing dose through 1 Mrad (Si). These levels usually fall above the radiation specifications for a good share of military systems. Therefore, the effects of radiation upon noise parameters, in devices where such is a consideration, is often of secondary importance. While noise may be a secondary consideration in the above sense, there remain special systems, such as sensitive radiation or nuclear particle counters, in which radiation-induced noise must be taken into account.

12.15 THERMIONIC INTEGRATED CIRCUITS

Thermionic integrated circuits are a synthesis of integrated circuit-processing techniques and vacuum tube principles.[73,74] These refractory

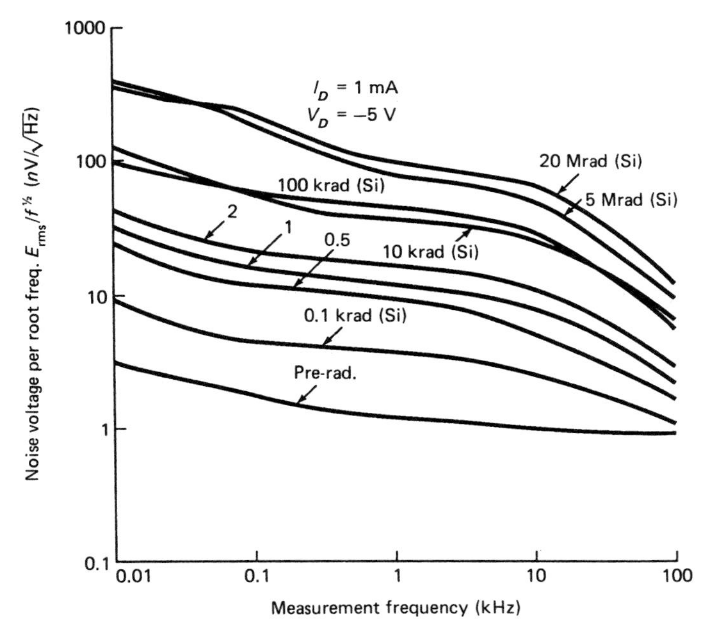

Fig. 12.45 Drain noise voltage for various samples of CM860 JFETs versus measurement frequency for irradiation at different values of ionizing dose.[80] © 1986 by the IEEE.

ceramic packaged, microminiature triode ICs are capable of functioning at both very high temperatures of the order of 1000°C and very high radiation environments. Thermionic emission, as used in vacuum tube technology, provides filament or cathode heater electrons. Instead of vacuum tube-type individual heaters, the heater is common to the thermionic integrated circuit, as the whole die substrate is heated to about 750°C by means of a thin film resistance heater element deposited on the substrate rear. The individual triode cathodes are treated with a coating that readily boils off electrons to provide cathode current.

Thin film deposition and photolithographic techniques of conventional integrated circuit manufacturing processing are used to form the vacuum tube triode microcircuits. The grid and cathode are thus deposited side by side in an interdigitated fashion on a sapphire insulator substrate. The anode is deposited on a contiguous substrate and positioned above the grid and cathode. A gain-bandwidth product of more than 100 MHz can be obtained.

MSI technology in the form of 200 triodes, comprising a 16-bit shift register, with a packing density of 1000 triodes per in.^{2} has been con-

Table 12.11 Thermionic IC radiation and thermal survival levels.[74]

NEUTRON FLUENCE	TOTAL DOSE (GAMMA)	DOSE RATE AND NEUTRON FLUX	PARTICLE TYPE	PULSE WIDTH	TEST TEMPERATURE
10^{17} n/cm^2	250 MRad (Si)	$4.5 \cdot 10^{10}$ rad (Si)/s	photons	20 ns	500°C for 1.5 yr
		$2 \cdot 10^{8}$ rad (Si)/s	protons	4.8 μs	
		$1.4 \cdot 10^{16}$ n/cm^2 s	neutrons	20 ms	

structed. Heater power requirements are but 2.5 mW per triode, using the substrate heater described above.[73,74]

As can be appreciated, thermionic integrated circuits (TICs) possess a high level of thermal ruggedness and nuclear radiation resistance. Table 12.11 depicts the corresponding parameter levels.

In terms of operating efficiency, transient radiation effects can occur in the sapphire (Al_2O_3) substrate and must be taken into account. Incident ionizing radiation (e.g., gammas) on the sapphire produces hole-electron pairs within, leading to a radiation-induced change in its conductivity.[74] This is described by a Shockley-Read-like recombination model.[73,74] The model accounts for both increasing and decreasing induced conductivity. This conductivity change can be reduced by orders of magnitude through the use of chromium doping and neutron irradiation of the sapphire, to provide traps that compensate the ionizing radiation-induced changes in conductivity.[73,74] The conductivity change yields a leakage current in the substrate that results in a replacement current in the electrodes of the individual triodes. This leakage current can be shielded from the TIC proper by the introduction of ground planes between it and the substrate. There are other effects, such as interaction between individual triodes, that can be prevented by suitable shields, repellers, etc.

TIC technology is presently in the research and development phase, and production devices are not available at this juncture. However, because of their extremely high radiation environment tolerance, they may in the future supply certain very difficult hardening niche applications.

REFERENCES

1. J. E. Gover and M. A. Rose. IEEE NSRE Conference Short Course, Colo. Sprgs, CO, Secs. 1, 2, 1984.
2. D. L. Durgin, D. R. Alexander, and R. N. Randall. Hardening Options For Neutron

Effects. Final Report, The BDM Corp., Albuquerque, NM, HDL-CR-75-052-1, app. A, p. 15, Nov. 15, 1976.
3. D. M. Long. Hardness of MOS and Bipolar Technologies. IEEE Annual Conf., Nuclear and Space Radiation Effects. July 1980.
4. W. R. Dawes, Jr. IEEE NSRE Conference Short Course, Monterey, CA, Sec. 2, 1985.
5. M. A. Rose. "Updated Bar Charts of Device Radiation Thresholds, Ca. 1990." Physitron Corp., San Diego, CA, 1990.
6. J. J. Wiczer. "Radiation Hardened Optoelectronic Components: Detectors." Proc. SPIE 616:254–262, 1986.
7. M. S. Cooper, J. P. Retzler, and G. C. Messenger. "Combined Neturon and Thermal Effects on Bipolar Transistor Gain." IEEE Trans. Nucl. Sci. NS-26(6):4758–4762, Dec. 1979.
8. N. J. Rudie. Principles and Techniques of Radiation Hardening, Vol. 2. Western Periodicals Co. N. Hollywood CA, p. 16–4, 1976.
9. Ibid.
10. R. K. Thatcher and J. J. Kalinowski. 1969 TREE Handbook, pp. G-39 ff., Battelle Mem. Inst. Columbus OH.
11. Components Response Information Center (CRIC) of the Harry Diamond Laboratories, Adelphi, MD, Radiation Effects Data Bank HDL-DS-77-1, May 1977.
12. M. G. Knoll. Comparison Study of the Five TTL Families and ECL. Air Force Weapon Labs. Kirtland AFB, NM, AFWL TR-78-5, 1978.
13. CRIC, op. cit. Radiation Effects in LSI Circuits. HDL-DS-80-1, July 1980.
14. M. E. Daniel and F. N. Coppage. "Radiation Hardness of a High Speed ECL Microcircuit." IEEE Trans. Nucl. Sci. NS-22: 2595, Dec. 1975.
15. Mission Research Corp. I^2L Radiation Effects Conference Proceedings, San Diego, CA, Jan. 1975.
16. Rudie, op. cit.
17. D. M. Long. Radiation Effects on Insulator Gate Field Effect (MOS) Integrated Circuits. Martin Marietta Corp. Baltimore, MD, Report DA 28-043-AMC-01520(E), Jan. 1967.
18. E. E. King. "Radiation Hardening of Static NMOS RAMs." IEEE Trans. Nucl. Sci. NS-26(6):5060, Dec. 1979.
19. A. J. Matteucci and M. F. Schneider. "Radiation Testing of CMOS Arrays For Satellites." IEEE Trans. Nucl. Sci. NS-24(6):2285, Dec. 1977.
20. H. Borkan. "Radiation Hardening of CMOS Technologies: An Overview." IEEE Trans. Nucl. Sci. NS-24(6):2043, Dec. 1977.
21. A. London and R. C. Wang. "Radiation Hard Metal Gate CMOS." IEEE Trans. Nucl. Sci. NS-25(6):1172, Dec. 1978.
22. P. R. Measel et al. "Radiation Response of Several Memory Types." IEEE Annual Conf. on Nuclear and Space Radiation Effects. Ithaca, NY, July 1980.
23. T. J. Sanders. "Radiation Hardness of LSI/VLSI Fabrication Processes." IEEE Trans. Nucl. Sci. NS-26(6):5056, Dec. 1979.
24. J. Scarpula et al. "Rapid Annealing Response of the 1102 Bulk CMOS Microprocessor." IEEE Annual Conf. Nuclear and Space Radiation Effects, Cornell Univ., Ithaca, NY, July 1980.
25. E. E. King and R. L. Martin. "Effects of Total Dose on the 1802 CMOS/SOS Microprocessor." IEEE Trans. Nucl. Sci. NS-24(6), Dec. 1977.
26. S. N. Lee et al. "Radiation Hardened Si-Gate CMOS/SOS." IEEE Trans. Nucl. Sci. NS-24(6):2205, Dec. 1977.
27. J. H. Yuan and E. Harrari. "High Performance Radiation Hardened CMOS/SOS

Technology." IEEE Trans. Nucl. Sci. NS-24(6):2199, Dec. 1977.
28. P. Vail. "A Survey of Radiation Hardened Microelectronic Memory Technology." IEEE Trans. Nucl. Sci. NS-24(6):1196, Dec. 1978.
29. Radiation Design Criteria Handbook. Jet Propulsion Labs (JPL). NASA Technical Memorandum 33-763, Aug. 1976.
30. A. H. Johnston and C. A. Lancaster. "A Total Dose Homogeneity Study of the 108A Operational Amplifier." IEEE Trans. Nucl. Sci. NS-26(6):4769, Dec. 1979.
31. J. K. Notthoff. "Technique For Estimating Primary Photocurrents in Silicon Bipolar Transistors." IEEE Trans. Nucl. Sci. NS-16(6):138–143, Dec. 1969.
32. Rudie, op. cit. app. D.
33. Long, op. cit. 1980.
34. J. W. Harrity and P. E. Gammill. "Upset and Latchup Thresholds in CD 4000 Series of CMOS Devices." IEEE Annual Conf. Nuclear and Space Radiation Effects, Cornell Univ., Ithaca NY, July 1980.
35. R. L. Pease and D. R. Alexander. PNPN Latchup in Bipolar Devices. Mission Research Corp., Albuquerque, NM, Report DNA6164F, Jan. 1 1982.
36. B. E. Anspaugh et al. "Solar Cell Radiation Handbook." NASA JPL 82-69, Chap. 5, Nov. 1982.
37. J. B. Cladis. "The Trapped Radiation Handbook." DNA 2524H, Lockheed Missile Systems Division, Palo Alto, CA, Dec. 1971.
38. M. Shoga, J. Gorelick, R. Viswanathan, and E. Smith. "Compilation of Single Event Upset Rates in the Geosynchronous Orbit." Submitted for publication in IEEE Trans. Nucl. Sci.
39. W. K. Roberts. "A New Wide Band Balun." Proc. IRE 45, Dec. 1957.
40. A. E. Waskiewicz, J. W. Groningk, and V. H. Strahan. "Burnout of Power MOS Transistors with Heavy Ions of Cf-252." IEEE Trans. Nucl. Sci. NS-33(6):1710–1718, Dec. 1986.
41. J. H. Hohl and K. P. Galloway. "Analytical Model For Single Event Burnout of Power MOSFETs." IEEE Trans. Nucl. Sci. NS-34(6):1275–1280, Dec. 1987.
42. W. L. Bendel and E. L. Peterson. "predicting Single Event Upsets in the Earth's Proton Belts." IEEE Trans. Nucl. Sci. NS-31(6):1201–1206, Dec. 1984.
43. J. H. Adams, C. H. Tsao et al. "CREME-Cosmic Ray Effects on Microelectronics." Part I, NRL Memo No. 4506, Aug. 1981; Part II, NRL Memo No. 5099, May 1983; Part III, NRL Memo No. 5402, 1984.
44. W. L. Bendel and E. L. Peterson. "Proton Upsets in Orbit." IEEE Trans. Nucl. Sci. NS-30(6):4481–4485, Dec. 1983.
45. D. K. Nichols, et al. "The Single Event Upset (SEU) Response to 590 MeV Protons." IEEE Trans. Nucl. Sci. NS-31(6):1565–1567, Dec. 1984.
46. RADSPICE owned by SAIC, San Diego, and META Software, Sunnyvale, CA; SYSCAP owned by Rockwell International, Autonetics Division, Anaheim, CA.
47. L. W. Ricketts, J. E. Bridges, and J. Mileta. EMP Radiation and Protective Techniques. Wiley, New York, p. 76, 1976.
48. DNA EMP Awareness Course Notes. IIT Research Institute, Chicago, pp. 41 ff. Aug. 1971.
49. Rudie, op. cit. p. 24–14.
50. R. P. Himmel. Thick Film Resistor Adjustment by High Voltage Discharge. Proc. IEEE Electronic Components Conf., 1971.
51. D. C. Wunsch and R. R. Bell. "Determination of Threshold Failure Levels of Semiconductor Diodes and Transistors Due to Pulsed Voltages." IEEE Trans. Nucl. Sci. NS-15(6):244–259, Dec. 1968.

52. D. M. Tasca. Energy-Time Dependence of Second Breakdown in Semiconductors For Submicrosecond Electrical pulses. General Electric Co., Space Div., Valley Forge, PA, No. 67SD7253, Oct. 1967.
53. J. S. Smith. Pulse Power Testing of Microcircuits. Rome Air Development Center (RADC), AF Systems Command, Griffiss AFB, Rome NY, Report RADC-TR-71-59, Oct. 1971.
54. J. B. Singletary and J. A. Hasdal. Methods, Devices, and Circuits For the EMP Hardening of Army Electronics. Report ECOM-0085-F, June 1971.
55. J. B. Singletary and D. C. Wunsch. Final Summary Report on Semiconductor Damage Study, Phase 2. Braddock, Dunn, McDonald, Inc. Report BDM/A 66-70-TR, June 1970.
56. J. P. Retzler. General Hardness Guidelines For Design Engineers. Litton Guidance and Control Systems Div., Woodland Hills, July 1973.
57. R. A. Berger, J. L. Azarewicz, and H. Eisen. "Hardness Assurance Guidelines For Moderate Neutron Environment Effects in Bipolar Transistors and Integrated Circuits." IEEE Trans. Nucl. Sci. NS-25:1555–1560, Dec. 1978.
58. C. W. Perkins, R. W. Marshall, and A. M. Liebschultz. Radiation Effects on Monolithic Microelectronic Circuits. Report No. DA-28-043, AMC-01313(E), FR66-17-278, Nov. 1966.
59. R. Patrick and J. Ferry. Nuclear Hardness Assurance (HA) Guidelines For Systems With Moderate Requirements. Air Force Weapons Labs (AFWL), Kirtland AFB, Report AFWL-TR-76-147, Aug. 1976.
60. J. R. Hauser. Integrated Silicon Device Technology, Vol. 11. Bipolar Transistors. Research Triangle Institute, Durham, NC; ASD-TDR-63-316, Vol. 12, Mar.–Oct. 1965, AF33 (615)-1998, 359 pp.
61. D. Casasent. Electronic Circuits. Quantum Publishers, New York, pp. 88 ff., 1973.
62. B-1 Hardness Assurance Guidelines. ASD, AF Systems Command, Wright Patterson AFB, OH; AFWL, Kirtland AFB NM, March 1976.
63. R. A. Croxall and A. K. Thomas. Subsystem Nuclear Vulnerability Assessment Methodology. TRW Systems Group, Redondo Beach, CA, AFWL TR-77-83 AABNCP Assessment Program, March 31, 1977.
64. EMP Electronic Analysis Handbook. Boeing Aircraft Co., Seattle WA and BDM Corp. Albuquerque, NM, AFWL Kirtland AFB, May 1973.
65. Retzler, op. cit. Fig. 4-3.
66. Thatcher and Kalinowski, op. cit. Fig. K-2, p. K-11.
67. Retzler, ibid.
68. M. S. Ruthroff. "Some Broadband Transformers." Proc. IRE 47:1337, Aug. 1959.
69. Roberts, op. cit. pp. 104, 316.
70. S. Sabaroff. "Longitudinal Isolation Devices For High Frequency Signal Transmission Lines." Patent 2,865,006.
71. Thatcher and Kalinowski, ibid.
72. J. B. McCormick et al. The Development of Integrated Thermionic Circuts For Applications in High Temperature Geothermal Environments. LA-8666-MS, Los Alamos Scientific Labs., Oct. 1980.
73. D. K. Lynn et al. "Thermionic Integrated Circuits: Electronics For Hostile Environments." IEEE Trans. Nucl. Sci. NS-32(6), Dec. 1985.
74. R. W. Klaffky et al. "Radiation Induced Conductivity of Al_2O_3: Experiment and Theory." Phys. Rev. B, 21(8), April 15 1980.
75. S. M. Sze. Physics of Semiconductor Devices. Wiley-Interscience, New York, p. 104 316 ff, 1981.

76. J. E. Gover and T. A. Fischer. "Radiation Hardened Microcircuits For Accelerators." IEEE Trans. Nucl. Sci. NS-35(1):160–165, Feb. 1988.
77. R. J. Chaffin. Microwave Semiconductor Devices: Fundamentals and Radiation Effects. Wiley Interscience, New York, p. 170, Fig. 616, 1973.
78. R. J. Guttman and J. M. Borrego. "Degradation of GaAs MESFETs in Radiation Environments." IEEE Trans. Rel. R-29(3):232–236, Aug. 1980.
79. J. M. Borrego, R. J. Guttman, and S. B. Moghe. "Radiation Effects on Signal and Noise Characteristics of GaAs MESFET Microcircuits." IEEE Trans. Nucl. Sci. NS-26(6):5092–5099, Dec. 1979.
80. J. H. Stephen. "Low Noise Junction Field Effect Transistors Exposed to Intense Ioniztion Radiation." IEEE Trans. Nucl. Sci. NS-33(6):1465–1470, Dec. 1986.
81. D. G. Millward. "Neutron Hardness Assurance Considerations For Temperature Compensated References Diodes." IEEE Trans. Nucl. Sci. NS-25(6):1517–1521, Dec. 1978.
82. A. B. Phillips. Transistor Engineering, McGraw-Hill Book Co., New York NY, p. 223.
83. N. J. Rudie. Principles and Techniques of Radiation Hardening, Vol. III, Third Edition, Sec. 38.2. Western Periodicals Co., N. Hollywood, 1986.
84. A. H. Johnston. "Hand Analysis Techniques For Neutron Degradation of Operational Amplifiers." IEEE Trans. Nucl. Sci. NS-23(6):1709–1714, Dec. 1976.
85. J. V. Wait, L. P. Huelsman, and G. A. Korn. Introduction to Operational Amplifier Theory and Applications, Chap. 1. McGraw-Hill Book Co., New York, 1975.
86. F. A. Ilardi. Computer Circuit Analysis, Chap. 8. Prentice Hall, Englewood Cliffs, NJ, 1976.
87. W. Shedd, B. Buchanan, and R. Dolan. "Radiation Effects on Junction Field Transistors." IEEE Trans. Nucl. Sci. NS-16(6):87–95, Dec. 1969.
88. R. Ghose. EMP Environments and System Hardening Design. Interference Control Technologies, Inc., Gainesville, VA, 1984.
89. D. A. Rosenthal and J. W. Hirman. "A Radio Frequency Users Guide to the Space Environment Services Center Geophysical Alert Broadcasts." Space Environment Lab. NOAA Tech. Memo. ERL SEL-80, Boulder, CO, June 1990.
90. T. L. Meisenheimer and D. M. Fleetwood. "Effect of Radiation-Induced Charge on 1/f Noise in MOS Devices." IEEE Trans. Nucl. Sci. NS-37(6):1696–1702, Dec. 1990.
91. A. L. McWhorter. "1/f Noise And Germanium Surface Properties" *in* Semiconductor Surface Physics. Univ. of Pennsylvania Press, Philadelphia, PA, pp. 207–228, 1957.
92. J. H. Scofield and T. P. Doerr. "Correlation Between Preirradiation 1/f Noise and Postirradiation Oxide Trapped Charge in MOS Transistors." IEEE Trans. Nucl. Sci. NS-36(6):1946–1953, Dec. 1989.
93. A. B. Campbell. "SEU flight data from the CRRES MEP." IEEE Trans. Nucl. Sci. NS-38(6):1849–1858, Dec. 1991.
94. J. L. Titus. "Single event burnout of power bipolar junction transistors." IEEE Trans. Nucl. Sci. NS-38(6):1975–1982, Dec. 1991.

CHAPTER 13
SYSTEM HARDNESS DESIGN

13.1 INTRODUCTION

This chapter attempts to provide the hardness designer with concise statements within security classification limits regarding nuclear survivability and vulnerability of electronic systems. These will include space, strategic avionics, and tactical ground systems. After their portrayal is given from a broad aspect, specific hardening methods will be discussed from the viewpoint of those whose main responsibility is the incorporation of hardness into the overall system design.

13.2 SPACE SYSTEMS

Spacecraft, or synonymously satellite, systems, as well as manned space vehicles, are subjected to a variety of penetrating radiation present in space. Some aspects of that radiation are radically different from the radiation that occurs or could occur within the atmosphere. Examples of the former are cosmic-ray ions. The TREE hardness designer who is part of a space system program must be aware of the specifications and procedures used for determining the dose incident on the spacecraft electronics. He or she must be able to design and recommend appropriate radiation protection measures. The principal objective is to ensure that the design provides sufficient protection so that specified radiation levels do not cause performance degradation over the duration of the mission. This task is twofold. First, the dose due to the radiation incident on each sensitive element within the spacecraft must be computed. Second, if any dose exceeds the specification requirements, the requisite shielding design must be added.

Much of the above is recognized as being common to the task of hardening almost any electronic system. Some unique aspects of space vehicle hardening include the fact that ionizing dose rates are usually very low, and mission duration times are measured in years.

Single-event upset (soft errors), as well as latchup due to heavy cosmic ray ions, can occur and cause difficulties in devices that are part of the on-board electronics. Another unique aspect of space radiation is that there are essentially few near-terrestrial, natural, primary sources of neutrons, x-rays or gamma rays of any consequence. Any such to be found in a spacecraft radiation specification imply a hostile environment.

The prevailing types and sources of penetrating radiation of importance consist of:

a. Solar cosmic rays consisting chiefly of energetic protons, plus a small component of alpha particles, whose energy spectrum extends beyond 100 MeV during solar flares,[1] as well as high Z ions up to and including iron.
b. Magnetically trapped protons with energies up to 40 MeV and electrons ranging from scores of keV to a few MeV that are confined in belts about the earth due to the interaction of the earth's magnetic field with the solar wind. This phenomenon is discussed in Section 9.3. These Van Allen belts correspond to one or more of the naturally occurring charged particle belts with sensible fluences, extending from about 600 to about 12 000 miles above the earth.
c. Galactic cosmic rays whose energies range from about 10 MeV to 1 GeV, which are chiefly protons and heavy ions.

The satellite skin affords little attenuation to high-energy particles that might impinge on a space vehicle at altitudes above 200 km. Hence, radiation protection of the spacecraft electronics is vital for protracted missions at these altitudes and higher. For relatively low-earth orbit (LEO) missions, the trapped radiation, and that due to the South Atlantic Anomaly (Section 8.3), is a principal consideration in terms of incident fluence. For lunar, interplanetary, near polar, and very high-altitude missions, such as for geosynchronous satellites, solar and galactic cosmic ray fluences are the major cause of the production of radiation damage to on-board electronic equipment.

In terms of critical ionizing dose susceptibilities in manned space vehicles, certain types of photographic film, as well as the crew, must be given radiation protection, as depicted in Fig. 13.1. Film, of course, can be stored in suitably shielded cannisters. Fig. 13.1 depicts crew incapacitation probability versus dose, as discussed in Section 13.4. Crew incapacitation dose is different from the mean lethal dose, defined as the 50 percent radiation mortality probability, which in man lies at an ionizing whole-body dose of about 450–550 rad (tissue): 1 rad (Si) = 0.902 rad (tissue) of ^{60}Co gammas.

In unmanned satellites, the electronic systems, including the solar cells, pose the most critical requirements for radiation protection. On-board radiation sources can impose further shielding requirements. These

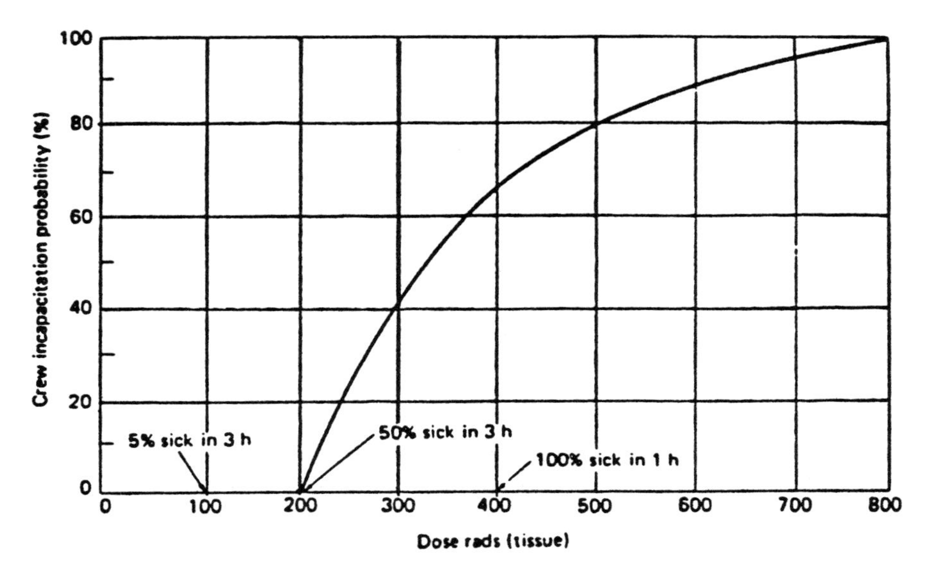

Fig. 13.1 Crew incapacitation versus radiation dose for manned space vehicles.[2]

sources can include small space power nuclear reactors or nuclear radioisotope thermal power generators (RTGs) that supply power to the spacecraft.

Various computer programs (e.g., SOFIP,[90] SOLPRO,[91] ALLMAG,[92] CRUP,[93] and CREME[94]), for both mainframe and desk-top computers, have been written to provide the geomagnetic field, incident particle fluence, flux, and SEU for spacecraft for almost any orbit. The corresponding environments used as input to these codes include the AP (proton 19XX) and AE (electron 19XX) series.[95,96]

The application of radiation hardness techniques to protect satellite systems includes some that are obvious from the common sense standpoint and some that perhaps are a bit unusual. They include attempts to decrease the intrinsic vulnerability of hardness-critical items through the choice of other materials or components, optimization of the radiation protection afforded by the satellite payload masses themselves through radiation shadowing, reconfiguration of its interior to provide maximum areal density (g per cm^2) shielding for the vulnerable subsystems, and simple addition of radiation shielding materials, such as shield housings, for individual pieceparts. Two measures that can be construed as unique

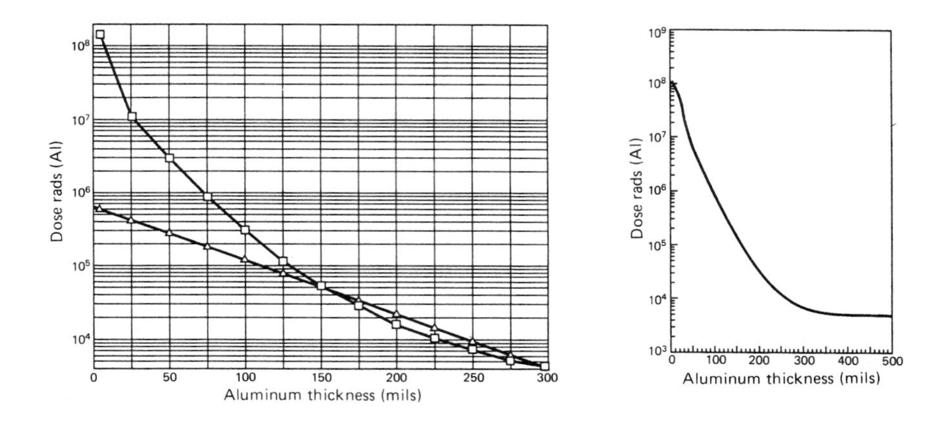

Fig. 13.1A Examples of dose-depth curves.

to spacecraft can be employed. They are the provision for orbital mission abort in the event of an unpredicted, but fatally damaging, solar flare or cosmic-ray burst of particles or to simply change the mission profile, so that the vehicle encounters less but tolerable radiation levels.

Ionizing dose encountered during the mission can be computed and couched in the form of a dose versus depth of spacecraft material plot. Part of the radiation specifications for space systems, including spacecraft of all types, are often couched in dose-depth curves. The dose-depth curve depicts the mission incident ionizing dose as deposited in an incremental volume of material at the center of a solid homogeneous sphere, or slab, versus the sphere radius or slab surface normal depth. Examples of such curves are shown in Fig. 13.1A, 13.2A, and 13.2B. Aluminum is considered the de facto standard material for this purpose. This is because most spacecraft shells, and much of their physical and mechanical internals, are aluminum. Also, the environmental particle interaction scattering and absorption properties of aluminum are well known, and are easily extrapolated to other materials of interest within the spacecraft. The dose-depth curve reflects the nature of the fluences incident on the spacecraft, since it mirrors the particle types and their energy spectra, which constitute the space environment surrounding the spacecraft during its mission life. The dose-depth curves depict the computational results of using particle transport codes to transform the

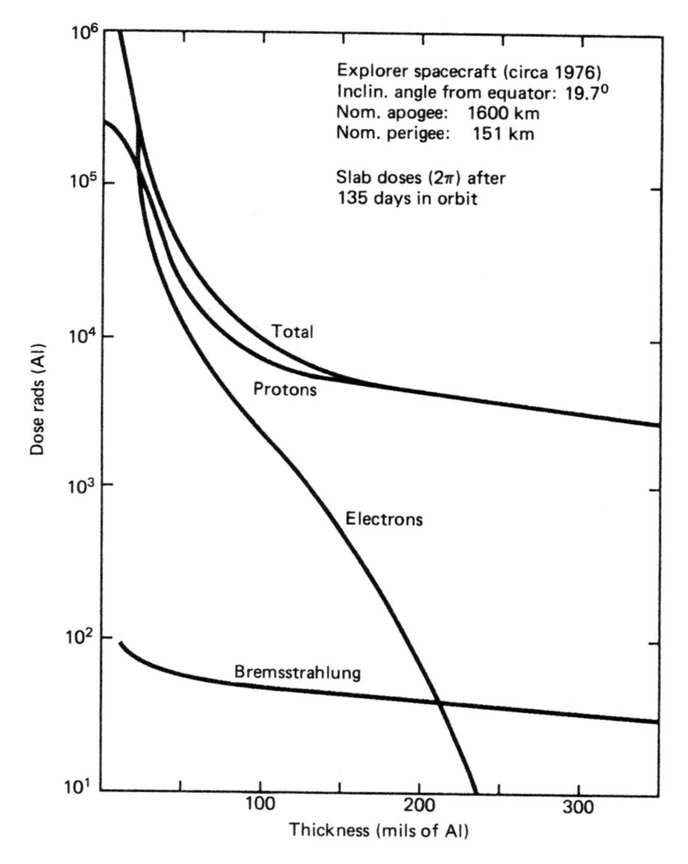

Fig. 13.2A Dose-depth curves computed from standard environment models for Explorer Spacecraft.[46] © 1977 by the IEEE.

pertinent orbit environmental fluences to dose deposited in the aluminum sphere center. These fluences include those due to electrons, protons, x and gamma rays, and other particles peculiar to the particular spacecraft mission. It should be noted that most dose-depth curves fall off with aluminum thickness, and asymptote to a nearly constant low dose level for sufficiently large thickness, as seen in Fig. 13.1A and 13.2A. This asymptotic behavior corresponds to dose deposited from bremsstrahlung generated within the aluminum mass (spacecraft) itself by the incident fluences when they are sufficiently penetrating. This implies that the

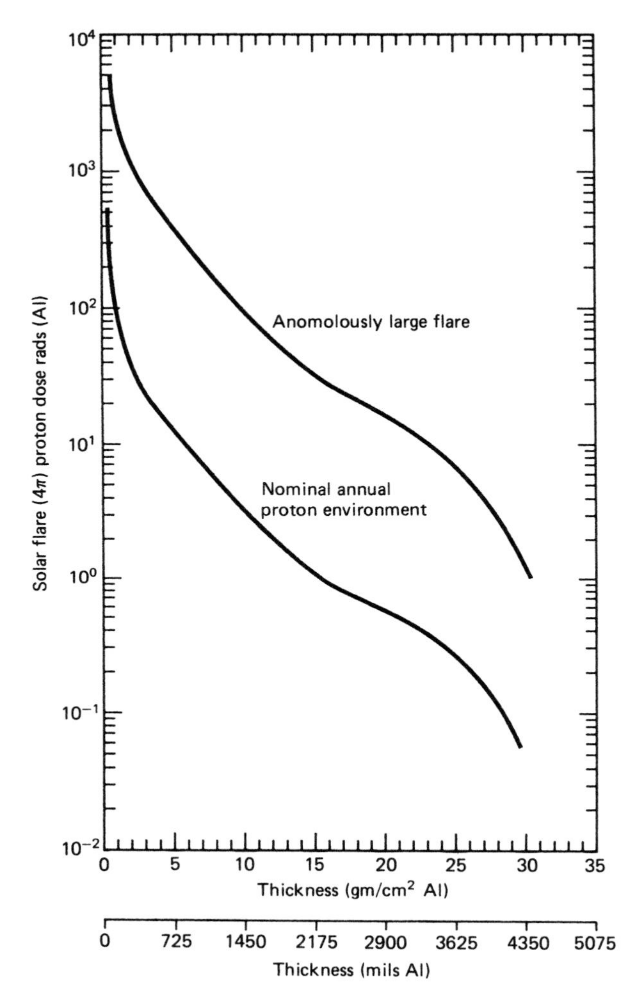

Fig. 13.2B Anomolously large geosynchronous proton solar flare dose-depth curve at 70° W. longitude.[126] © 1990 by the IEEE.

further addition of spacecraft shielding mass extending into the asymptotic region results in only a marginal shielding improvement.

At least two reasons for representing spacecraft radiation specifications in this manner are: First, the somewhat less accessible and often prodigious transport calculational effort that integrates the space orbital environment over the mission life has often already been done. It is then relatively straightforward to translate these results to corresponding dose-depth curves. For this purpose, there are a number of particle transport codes that can provide dose-depth curves from fluences incident on the spacecraft, given by environmental models, such as the NASA AE (electron) and AP (proton) series.[95,96] Second, since the dose-depth curves are essentially integrals of interaction cross sections folded with environmental fluences, they are not unique, in the sense that they can represent a myriad of spacecraft mission environments. This allows certain otherwise classified spacecraft orbits to be masked while their corresponding radiation environment is included in this unclassified fashion.

Infrequently, the dose-depth curve corresponds to the penetration of environmental radiation into a semi-infinite (one-dimensional) aluminum slab, or into both sides of an aluminum slab of finite thickness, instead of a "4π" sphere. The transformation of this radiation environment to a solid sphere is accomplished by the fact that the spherical dose $D_{\text{sphere}}(R)$ is related to that of the infinite thickness slab dose $D_{\text{slab}}(x)$, where x is the penetration distance into the slab,[87] and R is the aluminum sphere radius, viz.,

$$D_{\text{sphere}}(R) = 2[D_{\text{slab}}(R) - RD'_{\text{slab}}(R)] \tag{13.1}$$

As already mentioned, an important application of the dose-depth curve is in the determination of the dose deposited at a particular point in the spacecraft. This usually corresponds to finding the dose deposited in an electronic component at that point. The method to accomplish this task, explained below, is often called the sectoring method. This is done in order to determine whether or not the suspect part, which is usually an active device, can tolerate that dose, given a knowledge of the part's radiation susceptibility. The derivation of how this dose-depth information is used to compute this dose follows.

An expression for the dose $D_a(\bar{r}_0)$ deposited per gram of material in the component shown at vector position $\bar{r}_0$ in the spacecraft in Fig. 13.3 is given by

$$D_a(\bar{r}_0) = c \int_{\Omega} d\Omega \int_0^{\infty} \Phi[E, O_{sp}(R(\bar{\Omega}))]\, E\kappa_{si}(E)\, dE \tag{13.2}$$

where:

c is a conversion factor to express the dose in rads (Si).

$\Phi(E, O_{sp})\, dE$ is the incremental fluence having energy dE about E,

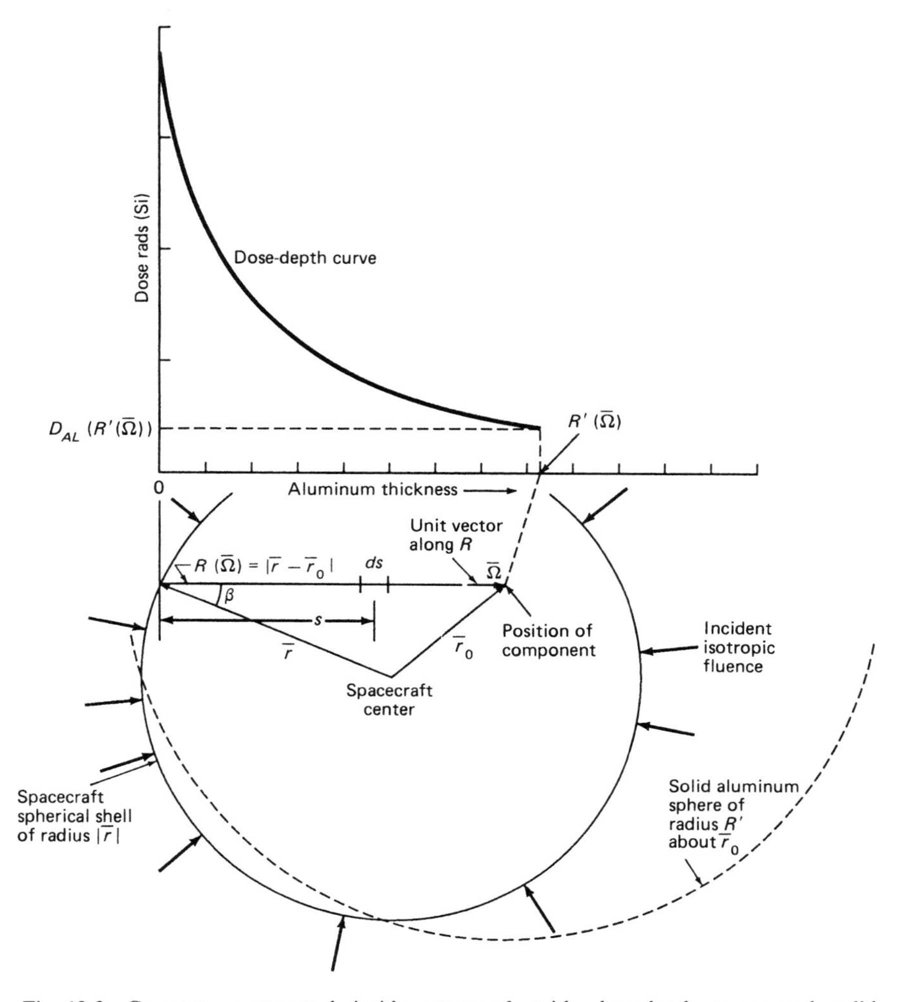

Fig. 13.3 Component mounted inside spacecraft with dose-depth curve and solid aluminum sphere superposed.

incident in the component, that has penetrated a dimensionless distance $O_{sp}(R(\bar{\Omega}))$ into the spacecraft.

$O_{sp}(R(\bar{\Omega}))$ is a dimensionless penetration distance measured in mean free paths along the actual distance $R(\bar{\Omega})$ from the spacecraft shell to the component, as in Fig. 13.3. It is also a function of energy E, but is suppressed in this notation.

$\mu_a = \kappa_a \rho_a$, the linear-attenuation coefficient, where κ_a is the mass-absorption coefficient.
E is the energy of the particle incident on the component.
$\bar{\Omega}$ is a unit vector along $R(\bar{\Omega})$, as in Fig. 13.3.
$d\Omega$ is an incremental solid angle.

O_{sp} is called the optical distance or optical depth, and is defined by the line integral along $R(\bar{\Omega})$, viz.,

$$O_{sp}(R(\bar{\Omega})) = \int_0^{R(\bar{\Omega})} ds/\lambda_a(s, \bar{\Omega}, E) = \int_0^{R(\bar{\Omega})} \mu_a(s, \bar{\Omega}, E)\, ds$$
$$= \int_0^{R(\bar{\Omega})} \kappa_a \rho_a ds \tag{13.3}$$

where $\lambda_a(s, \bar{\Omega}, E) = \mu_a^{-1} = (\kappa_a \rho_a)^{-1}$ is the mean free path for all processes at position ds about s along R, which result in energy deposition. The subscript a represents the generic for all materials. λ_a, μ_a and κ_a are also functions of the energy E of the fluence particles. The corresponding polar and azimuth angles are subsumed in $\bar{\Omega}$, and ρds is the areal density discussed in Section 6.4.

$\kappa_{si}(E) = \mu_{si}(E)/\rho_{si} = N_{si}\sigma_{si}(E)/\rho_{si}$ (cm^2/gm) is the mass absorption coefficient at energy E in the silicon component, where N_{si} is the number of silicon atoms per cm^3, and $\sigma_{si}(E)$ (cm^2) is the microscopic silicon cross section. $E\mu_{si}/\rho_{si}$ is the amount of energy per gram released (kerma) into the component material per unit incident fluence. It is assumed that all of the energy released in the component is absorbed therein. The preceding also holds for the solid aluminum sphere, i.e., $\kappa_{AL}(E) = N_{AL}\sigma_{AL}(E)/\rho_{AL}$.

From the definition of the dose-depth curve, $D_{AL}(R(\bar{\Omega}))$ is the dose deposited in an incremental volume of material at the center of a solid aluminum sphere of radius $R(\bar{\Omega})$, due to the same isotropic fluence as that incident on the spacecraft. It is given by an expression similar to Eq. (13.2), viz.,

$$D_{AL}(R(\bar{\Omega})) = 4\pi c \int_0^{\infty} \Phi(E, O_{AL})\, E\kappa_{AL}(E)\, dE \tag{13.4}$$

where the angular dependence is already integrated to yield 4π for the aluminum sphere. Since the solid aluminum sphere is assumed homogeneous, its λ_{AL} is constant with respect to penetration depth, so that the corresponding optical depth therein is simply $O_{AL} = R(\bar{\Omega})/\lambda_{AL}$. Then, dividing Eq. (13.2) by Eq. (13.4) yields

$$D_a(\bar{r}_0)/D_{AL}(R(\bar{\Omega})) = \left[\int_{\Omega}(d\Omega/4\pi)\int_0^{\infty}\Phi(E, O_{sp})\,E\kappa_{si}(E)\,dE\right] \Big/ \int_0^{\infty}\Phi(E, O_{AL})\,E\kappa_{AL}(E)\,dE \tag{13.5}$$

For most of the spacecraft environments of interest, the compton effect is the dominant interaction with pertinent spacecraft materials. In the incident photon energy regime corresponding to the compton effect, the mass absorption coefficients, μ/ρ, are slowly varying over the majority of spacecraft materials.[43] This implies that $\kappa_a(E) \cong \kappa_{AL}(E)$, or $\rho_a(s, \bar{\Omega})\,\lambda_a \cong \rho_{AL}\lambda_{AL}$. Then $\mu_a = 1/\lambda_a \cong \rho_a(s, \bar{\Omega})/\rho_{AL}\lambda_{AL}$, where $\rho_a(s, \bar{\Omega})$ is the actual density variation along the distance $R(\bar{\Omega})$ in the spacecraft, due to the juxtaposition of generic a materials thereon. So that the optical depth in the spacecraft can be rewritten as

$$O'_{sp}(R'(\bar{\Omega})) = \int_0^R ds/\lambda_a \cong \left(\int_0^R ds\rho_a(s, \bar{\Omega})/\rho_{AL}\right) \Big/ \lambda_{AL} = R'/\lambda_{AL} \tag{13.6}$$

where $R'(\bar{\Omega}) = \int_0^R ds\rho_a(s, \bar{\Omega})/\rho_{AL}$ is called the aluminum equivalent thickness distance from the spacecraft shell to the component, as in the upper portion of Fig. 13.3. Inserting O'_{sp} from Eq. (13.6) into Eq. (13.5) yields

$$D_a(\bar{r}_0) = \left[\int_{\Omega}(d\Omega/4\pi)\,D_{AL}(R)\int_0^{\infty}\Phi(E, O'_{sp})\,E\kappa_{si}(E)\,dE\right] \Big/ \int_0^{\infty}\Phi(E, O_{AL})\,E\kappa_{AL}(E)\,dE \tag{13.7}$$

$D_{AL}(R)$ can be factored into the above angular integration, since it is independent of $\bar{\Omega}$ at this juncture. A functional identity can be obtained by replacing R by R' in Eq. (13.4) to obtain, with $O_{AL}(R') \cong O'_{sp}$,

$$D_{AL}(R)/\int_0^{\infty}\Phi(E, O_{AL})\,E\kappa_{AL}dE \equiv D_{AL}\left(\int_0^R ds\rho_a(s, \bar{\Omega}/\rho_{AL}\right) \Big/ \int_0^{\infty}\Phi(E, O'_{sp})\,E\kappa_{AL}dE \tag{13.8}$$

The substitution of $D_{AL}(R)$ from Eq. (13.8) into Eq. (13.7) yields the sought-after relation between the actual dose deposited in the spacecraft component at $\bar{r}_0$ to that deposited at the center of the solid aluminum sphere (the dose-depth curve) viz.,

$$D_a(\bar{r}_0) \cong \int_{\Omega} (d\Omega/4\pi) \cdot D_{AL}\left(\int_0^{R(\bar{\Omega})} ds \rho_a(s, \bar{\Omega})/\rho_{AL} \right) \qquad (13.9)$$

That is, Eq. (13.9) provides the actual dose deposited in the spacecraft component at $\bar{r}_0$, in terms of the dose-depth information for the aluminum equivalent masses, ρ_a/ρ_{AL}, which add up along all $R(\bar{\Omega})$, from the spacecraft exterior to the component package to the chip within, as in Fig. 13.3.

Consider Fig. 13.4, where it is seen that the component of interest in this example is mounted on a vertical circuit board, which is one of five boards all mounted on a horizontal "mother" board. This whole assembly is encased in an aluminum housing, comprising one of many subsystems cabled together to make up the spacecraft avionics (electronics). For illustrative purposes, symmetry is obtained by mounting the component in the exact center of the middle board (B3). This provides ease in computation, yet elucidates the main ideas involved. This simplifies the calculations for computing the dose deposited in the part to correspond to only one-half the total solid angle, or 2π steradians, about the part. This is then doubled for the total ionizing dose.

From Fig. 13.4, it is also seen that the various solid angles about the part are subtended by rectangular facets, $\Delta\Omega_k$, where k labels the particular facet and solid angle. The facets provide a convenient division for the computation of deposited dose using a personal computer or programmable calculator. For mainframe computer programs of this type, other criteria for dividing (sectoring) the solid angle in a more refined manner, and so into many hundreds of smaller facets, can be employed.

To translate Eq. (13.9) into a usable computational form, its integrals are replaced by sums, to give

$$D_a(\bar{r}_0) \cong \sum_k (\Delta\Omega_k/4\pi) \cdot D_{AL}\left(\sum_j \Delta s_{jk}(\bar{\Omega}_k)\, \rho_{ajk}(\bar{\Omega}_k)/\rho_{AL}\right) \qquad (13.10)$$

where, as mentioned, the index k corresponds to the particular solid angle $\Delta\Omega_k$, i.e., a particular facet such that $\sum_k \Delta\Omega_k/4\pi = 1$. The latter expression is used as a check on the solid angle computation portion of these calculations. The index j labels the position of the line increments Δs_{jk} along the line $R'(\bar{\Omega}_k)$, shown in Fig. 13.3 for a given k facet, where

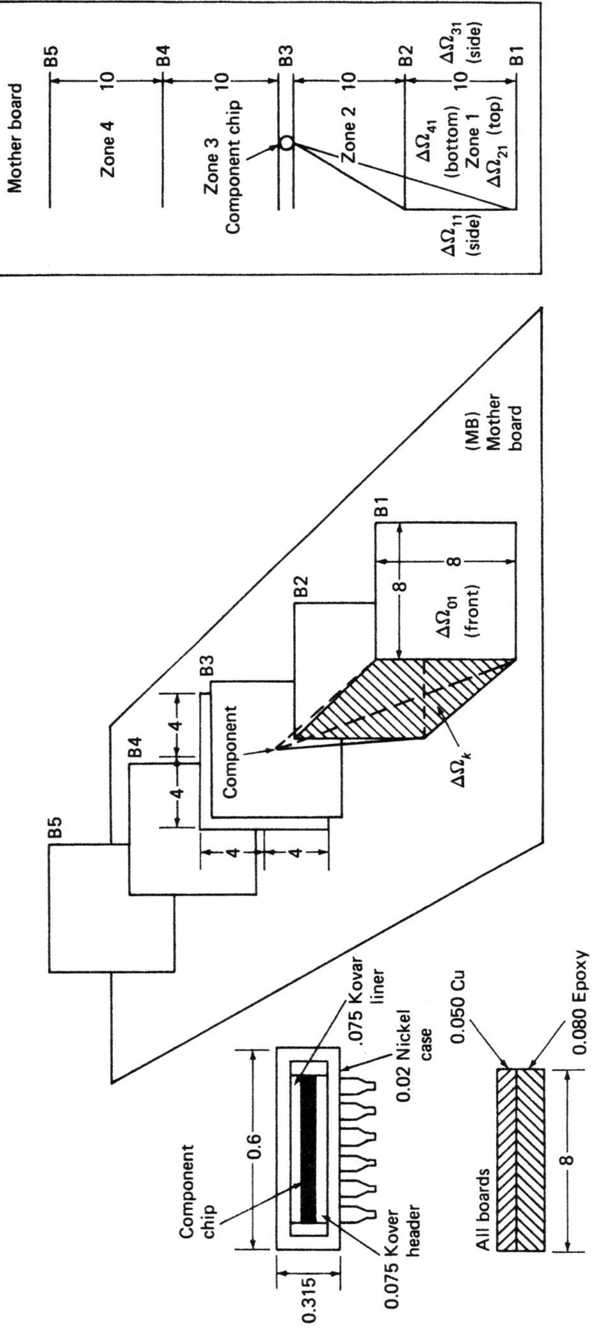

Fig. 13.4 Subassembly layout showing solid angles. (All dimensions in inches)

Table 13.1 2π Dose deposited in part using dose-depth computational method (all dimensions in inches).

(1) ZONE NO.	(2) FACET NO.	(3) BOARD NO.	(3) BOARD LABELS	(4) x_1	(5) x_2	(6) y_1	(7) y_2	(8) z	(9) $\Delta\Omega/4\pi$	(10) ALUM. OUTER SHELL	(11) ALUM. ASSY. HOUS.	(12) BRD. EQUIV. THKNS.	(13) TOTAL* BRD.* EQUIV. THKNS.	(14) PKGE. EQUIV. THKNS.	(15) TOTAL† EQUIV. THK. x_k	(16) DOSE DEPTH $D(x_k)$ Rads (Si)	(17) TOTAL** IONIZING DOSE Rads (Si)
1	$\Delta\Omega_{01}$	3	B1+B2 +B3	4	−4	4	−4	20	0.012246	0.04	0.02	0.2186	0.6558	0.2973	1.0131	32.30	0.396
1	$\Delta\Omega_{11}$	2	B2+B3	10	20	4	−4	4	0.007950	″	″	″	0.4372	″	0.7945	1.217E3	9.675
1	$\Delta\Omega_{21}$	2	″	10	20	4	−4	4	″	″	″	″	″	″	″	″	″
1	$\Delta\Omega_{31}$	2	″	10	20	4	−4	4	″	″	″	″	″	″	″	″	″
1	$\Delta\Omega_{41}$	3	″ +MB	10	20	−4	4	4	″	″	″	″	0.6558	″	1.0131	32.30	0.257
2	$\Delta\Omega_{12}$	1	B3††	0	10	4	−4	4	0.113989	″	″	″	0.2186	″	0.5759	4.582E4	5.223E3
2	$\Delta\Omega_{22}$	1	″	0	10	−4	4	4	″	″	″	″	″	″	″	″	″
2	$\Delta\Omega_{32}$	1	″	0	10	4	−4	4	″	″	″	″	″	″	″	″	″
2	$\Delta\Omega_{42}$	2	″ +MB	0	10	−4	4	4	″	″	″	″	0.4372	″	0.7945	1.217E3	1.387E2
									$\sum_k (\Delta\Omega_k/4\pi)$: 0.500002								$\sum_k$ (2π Dose): 1.584E4 Rads (Si)

* (12) × (No. Brds)
† (10) + (11) + (13) + (14)
** (9) × (16)
†† Part sandwiched between the B3 double board

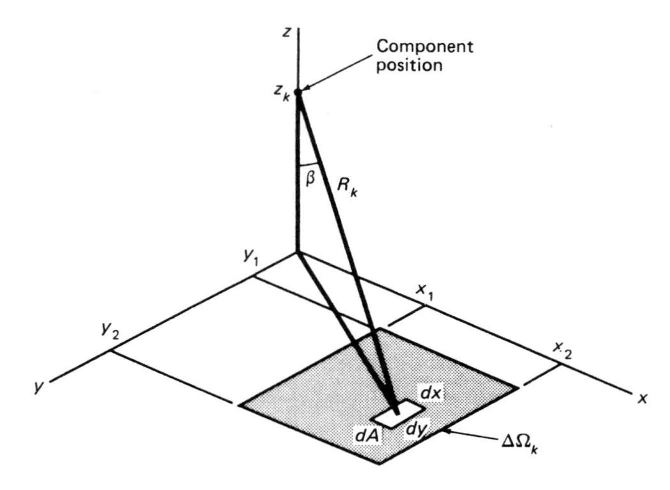

Fig. 13.5 Geometry of computation of solid angles in rectangular coordinates.

Table 13.2 Aluminum mass density equivalent of common spacecraft avionics materials.

MATERIAL	ρ/ρ_{AL}
Circuit board (epoxy)	0.67
Beryllium	0.69
Aluminum oxide (Al_2O_3)	1.3
Stainless steel	2.9
Kovar (29% Ni, 17% Co, 54% Fe)	3.1
Nickel	3.24
Copper	3.3
Solder (63% Sn, 37% Pb)	3.41
Lead	4.21
Tantalum	6.15
Gold	7.16

$\bar{\Omega}_k$ is the unit vector along $R'(\bar{\Omega}_k) = \sum_j \Delta s_{jk}(\bar{\Omega}_k)\,\rho_{ajk}(\bar{\Omega}_k)/\rho_{AL}$, corresponding to the kth facet. $\rho_{ajk}(\bar{\Omega}_k)/\rho_{AL}$ is the aluminum equivalent mass density ratio of the jth line increment along $R'(\bar{\Omega}_k)$. Table 13.2 provides such ratios for common spacecraft avionics materials.

To compute the various solid angles corresponding to the facets $\Delta\Omega_k$, it is seen from Fig. 13.5 that a solid angle facet $\Delta\Omega_k$, reckoned from the component position, is given by

$$\Delta\Omega_k = \int (dA \cos\beta)/R_k^2 = \int_{x_1}^{x_2} \int_{y_1}^{y_2} z_k dx dy/(x^2 + y^2 + z_k^2)^{3/2}$$

$$= \sum_{i=1}^{2} \sum_{j=1}^{2} (-1)^{i+j} \tan^{-1}(x_i y_j / z_k d_{ijk}) \qquad (13.11)$$

where $d_{ijk} = (x_i^2 + y_j^2 + z_k^2)^{1/2}$. Fluence particle paths (rays) that thread a particular $\Delta\Omega_k$ are seen to pierce one or more circuit boards of various materials, as well as the component package shielding material. Each board material and package shield corresponds to an aluminum equivalent density ratio ρ_{ajk}/ρ_{AL}, and thickness Δ_{jk}. In this example, all circuit boards are alike in this respect, having a common aluminum equivalent thickness, $\sum_{j=1}^{2} \Delta_{jk}\rho_{ajk}/\rho_{AL}$ (viz., $0.05 \times 3.3 + 0.08 \times 0.67 = 0.2186$ in. (Fig. 13.4, inset) and using Table 13.2). Then the quantity $R'(\bar{\Omega}_k) \cong x_k = \sum_j \Delta_{jk}\rho_{ajk}/\rho_{AL}$ can be computed for each $\Delta\Omega_k$ facet.

The dose-depth curve can generally be fitted with a sum of exponentials as

$$D_{AL}(x) = \sum_m A_m \exp - \lambda_m x \qquad (13.12)$$

where the number of terms can be made as large as necessary for a given accuracy. For an illustrative dose-depth curve fit used herein, the left portion of Fig. 13.1A single exponential example fitted by $D_{AL}(x_k) = 6.5 \cdot 10^8 \exp - 0.0166x_k$ is used. x_k is usually measured in mils (0.001 in. = 1 mil).

To recapitulate the sectoring algorithm, (1) the peripheral facets bounding the subsystem board assembly, which includes the component of interest, are identified, dimensioned from the component position as reference, and tabulated (Table 13.1, Cols. 1–8); (2) The solid angle (ratio to 4π) that each facet subtends at the position of the component is then computed, tabulated, and checked, so that their total solid angle, divided by 4π, sums to unity (Table 13.1, Col. 9); (3) For each facet, an imagined ray from the component chip is extended out to the facet center, piercing all boards, components, component packages, and housings in between; (4) The actual and aluminum-equivalent thicknesses for each corresponding board, etc. are identified, computed using Table 13.2, and tabulated (Table 13.1, Cols. 10–14). The boards and housings constitute the medium through which the incident radiation must penetrate to reach the component. All else is assumed to be a vacuum, since any gas (such as air) represents negligible attenuation by comparison. For initial conservative computations, all other masses are assumed absent, or are included in an approximate manner discussed later; (5) The equivalent mass thicknesses for each facet are summed to yield the corresponding total equivalent thickness x_k (Table 13.1, Col. 15), and x_k is then inserted into an appropriate representation of the dose-depth curve to yield the dose $D_{AL}(x_k)$ for the kth facet (Table 13.1, Col. 16); (6) Then, the product $D_{AL}(x_k) \cdot \Delta\Omega_k/4\pi$ is computed and tabulated (Table 13.1, Col. 17). This then yields the dose deposited in the component due to radiation that has penetrated the total aluminum-equivalent thickness x_k from the kth facet only, i.e., only through the solid angle $\Delta\Omega_k$; (7) Summing the preceding over all facets yields the sought-after dose deposited in the component (Table 13.1, Col. 17, bottom).

Only half of the example (i.e., for $\sum_k \Delta\Omega_k = 2\pi$ steradians) is completed. The tabular results show that the dose deposited in the component over 4π steradians is then $2 \times 1.584 \cdot 10^4 = 31.7$ krads (Si), utilizing the geometric sysmmetry to double the 2π steradian computed dose. Note that the symmetry is virtually perfect, in that the component is sandwiched between the double board B3. This board is made up of two identical boards back-to-back. Also the kovar header (chip socket) and kovar liner, each of aluminum equivalent thickness $0.075 \times 3.1 = 0.2325$ in. (Fig. 13.4, inset) symmetrically attenuate the radiation incident from the front and rear 2π solid angle sectors, since B3 is the center board.

As can be seen from Column 17 in Table 13.1, each end of this assembly shields the component much better than its girth about board B3, which is relatively much more exposed. This is because of the greater number of boards shielding the end radiation, and with much less exposed solid angle, over that for the broadside radiation incident onto this long narrow subsystem.

If the input data is attended to meticulously, these computations can give results that are comparable in accuracy to those employing mainframe codes of this type. As mentioned, they are readily amenable to personal computer spreadsheet programs, or for hand-held programmable calculator computation. One implementation with which to aid the accuracy of these computations is to fit the dose-depth curve as accurately as feasible, especially the asymptotic bremsstrahlung portion. It should be noted that these computations give deposited doses that are essentially upper or lower bounds. This is because the radiation attenuation of the other subsystems in the spacecraft, between the one of interest and the spacecraft shell, are (a) usually represented by an appended amorphous mass whose attenuation is approximated at the end of the calculation, to yield a lower bound, (b) not included at all, as in this example, which yields an upper bound. The latter corresponds to a radiation-conservative approach.

For components whose radiation susceptibility turns out to be less than the above-computed dose by an a priori ionizing dose margin, there are at least two remedial approaches. The first is obviously to replace the component with a suitable radiation-hard equivalent component. The second is to mount a minishield enclosure about the component in the circuit board, whose physical parameters satisfy the radiation margins and the subsystem physical constraints. Kovar or nickel is usually employed for this purpose. Such a kovar shield is depicted for the component used in this example, besides its nickel package case, in Fig. 13.4, inset.

As alluded to earlier, the foregoing computations can establish a substantively good benchmark deposited dose in a spacecraft component. However, as part of a sensitivity or trade-off study, it is desired to determine, for example, how dose level in the component varies with minishield thickness. Of course, this can be done by using results of multiple runs from the sectoring codes by varying shield thickness, keeping the other parameters constant. Nevertheless, it would be convenient to have an approximate expression relating deposited dose change to shield thickness using the dose-depth curve $D_{\text{sphere}}(R)$. To obtain such an expression, Eq. (13.1) can be construed as a differential equation for the slab dose $D_{\text{slab}}(R)$. Its integral yields, where $D_{\text{sph}} \equiv D_{\text{sphere}}$,

$$2D_{\text{slab}}(R) = R\int_R^\infty D_{\text{sph}}(x)\,dx/x^2 = \int_1^\infty D_{\text{sph}}(Ry)\,dy/y^2 \qquad (13.13)$$

The left-hand side is written as $2D_{\text{slab}}(R)$, which is the dose for a slab irradiated isotropically on both sides, as it would be if it were the cover of the component minishield. This is in contrast to a semi-infinite one-dimensional slab that is irradiated on one side only, which is the basis for the derivation of Eq. (13.1). Successive integration by parts of the above integral gives

$$2D_{\text{slab}}(R) = D_{\text{sph}}(R) - \int_1^\infty (d^2D_{\text{sph}}(Ry)/dy^2)\ln y\,dy \qquad (13.14)$$

$2D_{\text{slab}}(R)$ approaches $D_{\text{sph}}(R)$ for large values of R, where the dose-depth curve becomes almost constant, since large R corresponds to the

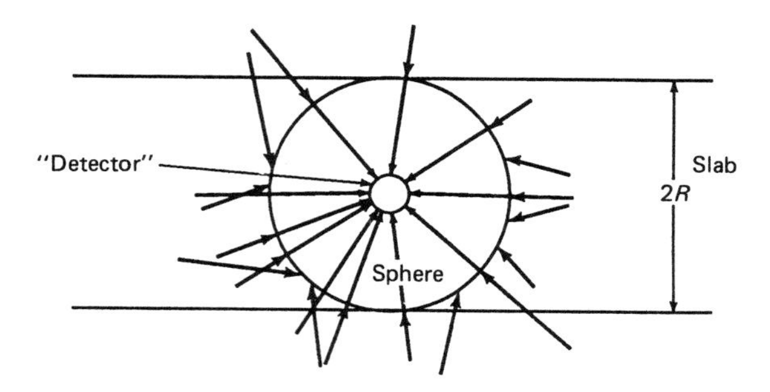

Fig. 13.6 Solid sphere and slab superimposed, showing isotropic fluence particle track segments.

bremsstrahlung region. This can be seen in the dose-depth curves of Fig. 13.1A and 13.2A. Higher-order derivatives of $D_{sph}(R)$ are vanishingly small therein, implying that the second term in Eq. (13.14) can be neglected in comparison to the first for sufficiently large R. Therefore, for increasing values of R,

$$2D_{slab}(R) \lessapprox D_{sph}(R) \tag{13.15}$$

where equality is approached, for sufficiently large R.

Hence, the slab dose $2D_{slab}(R)$ is less than $D_{sph}(R)$, the 4π dose-depth curve. The fact that $2D_{slab}(R) < D_{sph}(R)$ can also be seen from Fig. 13.6. There, it is seen that the set of isotropic fluence particle paths along the solid sphere radii (dotted lines), striking the "detector" common to the center of the slab and sphere, travel shorter paths if they emanate from the sphere circumference than from the slab boundaries. Since the latter paths are longer, those particles suffer more attenuation, so that fewer of them reach the detector. Therefore, the slab dose deposited is less than the dose deposited in the solid sphere central detector (component). Additionally, this difference can be seen in a comparison of fission electrons in an aluminum sphere and slab[3] given in Fig. 13.7.

For changes in dose with depth, Eq. (13.15) is of little use, since the slab dose difference, $\Delta(2D_{slab}(R))$, is a poor approximation to $\Delta D_{sph}(R)$ for almost all but the asymptotic (bremsstrahlung) values of R. However, if R is replaced by $R + \Delta R$ in Eq. (13.13) and the result subtracted therefrom, the dose difference $\Delta(2D_{slab}(R)) = 2D_{slab}(R + \Delta R) - 2D_{slab}(R)$ is obtained in terms of $D_{sph}(R)$. For the usual situation where a mass change corresponding to a fractional change $\Delta R/R \ll 1$, the

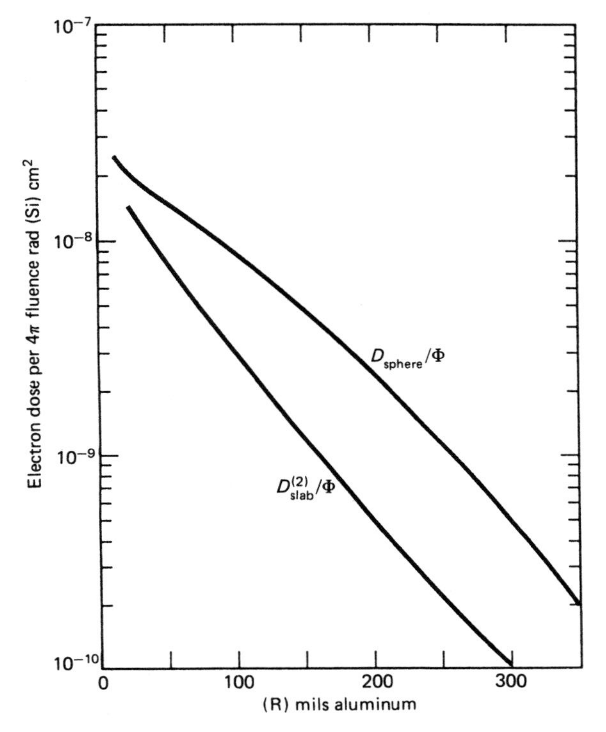

Fig. 13.7 Comparison of deposited dose per fluence at solid sphere center and slab thickness for electrons.[3] © 1979 by the IEEE.

desired approximate expression for slab dose changes corresponding to changes in R on the dose-depth curve is given by, with $2D_{\text{slab}}(R) \equiv D^{(2)}_{\text{slab}}(R)$,

$$D^{(2)}_{\text{slab}}(R + \Delta R) - (1 + \Delta R/R)\, D^{(2)}_{\text{slab}}(R) \cong \Delta D^{(2)}_{\text{slab}}(R) = -(\Delta R/R)\, D_{\text{sph}}(R) \tag{13.16}$$

As in the foregoing, for very large R, in the bremsstrahlung region $\Delta D^{(2)}_{\text{slab}}(R)$ vanishes, as $D_{\text{sph}}(R) \cong D^{(2)}_{\text{slab}}(R)$. This implies that dose changes there are negligible, as is the case. Also implied is that $D^{(2)}_{\text{slab}}(R) \cong D_{\text{sph}}(R) = D_{\text{brem}}(R)$ from Eq. (13.15) for very large R, where $D_{\text{brem}}(R)$ is the bremsstrahlung dose level.

For one albeit crude illustration of the use of Eq. (13.16), assume that the dose-depth curve is given by the right curve in Fig. 13.1A, and that the corresponding spherical spacecraft has a mass of 1000 lbs and a diameter of 3 meters. It then has a homogenized density $\rho = M/V = 4.54 \cdot 10^5/(4/3)\,\pi(150)^3 = 0.032\,\text{gm/cm}^3$. Assume that the position of the subsystem of interest within the spacecraft is 50 cm, or $50 \times 0.032 = 1.6\,\text{g/cm}^2$ in from the shell. That corresponds to $(1.6/2.71)(1000/2.54) = 230$ mils of aluminum, where $\rho_{AL} = 2.71\,\text{gm/cm}^3$. At a distance of 230 mils of aluminum, the above dose-depth curve yields $D_{\text{sph}} \cong 10^4$ Rads (Si). Hence, the corresponding slab dose decrease for a 10 mil aluminum-equivalent mass increase at that position, from Eq. (13.16), is $\Delta D^{(2)}_{\text{slab}} = -(10/230)(10^4) = -435$ Rads (Si). In the case of deposited dose changes for curvilinear mass changes, such as spherical shell thickness, $\Delta D_{\text{sph}}(R)$ from the dose-depth curve can be used to give the corresponding correct dose change directly.

Once the dose-depth curves have been translated into dose absorbed in the internal subsystems, the subsequent hardness considerations are essentially the same as those that provide radiation protection for most other such systems and have already been discussed in earlier chapters. For this reason, the following emphasizes the broad space vehicle dose analysis in making the translation from the known or specified dose to that incident internally.

Protective shielding in this context can be divided into two general types, active and passive. Active shielding defines possible schemes for particle-deflecting electromagnetic fields about the spacecraft that are generated by on-board systems for this purpose. Passive shielding is simply the expeditious positioning of material mass between the incident radiation and the component to be protected. For most practical purposes, the former is yet in its infancy, and the latter is state-of-the-art as seen in the preceding.

For the case of solar cosmic protons and trapped protons in the Van Allen belts, the transport of these particles through a shield of areal density less than 20 g per cm^2 simply corresponds to the first collision loss of the specified primary proton energy incident on the satellite interior through ionization of the shielding material itself. As the incident protons diffuse through the shielding material, they slow down through interactions with these atoms and nuclei. Some of these interactions result in the production of secondary protons and neutrons. For typical protons of the kind under discussion, these secondary particles affect the total absorbed dose by less than 10 percent for shielding thicknesses less than 20 g per cm^2, except for slow protons of 2 MeV or less energy.[4,5] Furthermore, proton-induced secondary particles are usually not an important contributor to ionizing dose. However, secondary gamma radiation is an exception for solar flare protons that are very soft (low energies) and must be taken into account.[6]

From another aspect, being charged particles, proton flux from energetic solar flares, as well as cosmic ray ions generally, are strongly diverted by the earth's magnetic field. This minimizes their penetration to low-earth orbits (~150–400 miles). For example, a 200 km altitude, 30 degree inclination orbit would experience virtually no solar flare flux.[126] The other extreme is a polar orbit, which would be subject to a substantive incoming polar region flux at almost any polar altitude. This is, of course, due to the over-the-pole hole singularities formed where the earth's magnetic field lines converge to enter or diverge to exit the magnetic poles. Normal low-altitude nonpolar orbit exposure is due mainly to the South Atlantic Anomaly, discussed in Section 8.4. Figure 13.7B depicts the (annual) ionizing dose rate for low-earth orbit (LEO) radiation environments versus orbit inclination angles, measured from the equator, for various altitudes from 200–1200 km.[128] It is seen from the figure that the spacecraft LEO ionizing dose environment depends sensitively on its inclination angle, up to about 60 degrees, where upon the polar region lack of magnetic field diversion shielding begins to predominate for larger angles of inclination. Similar SEU considerations ensue for the suborbital trajectories of modern-day long-range ballistic missiles or aircraft paths that fall on routes passing over the polar regions.

For geosynchronous orbits, the geomagnetic shielding is relatively ineffective, because of the earth's weak magnetic field at these altitudes. Nevertheless, the mean annual geosynchronous dose from the nominal solar proton environment is only about 18 rads (Si) behind 300 mils of Al.[126] However, the corresponding dose from a single anomolously large solar flare would be on the order of 600 rads (Si).[126] These cases are depicted in Fig. 13.2B.

For many practical spacecraft shielding design considerations, the resulting shields are usually sufficiently thin that the preponderant particle transport consists of primary particles only. This is in contrast to thick shields with a soft particle energy spectrum, wherein secondary particles represent a large fraction of the resulting total dose. Then complicated transport computational methods, with their attendant uncertainties, especially for very asymmetrically juxtaposed spacecraft mass geometries, must be used. This sometimes calls for techniques, such as elaborate monte carlo programs to establish the incident dose on the electronics accurately.[7–9]

With respect to space shielding calculation accuracy in the prediction of radiation dose, a hierarchy of uncertainties from most to least uncertain is presented. By far, the specification of the spacecraft radiation environment usually corresponds to the major portion of this overall uncertainty. The categorization of maximum tolerable dose thresholds on

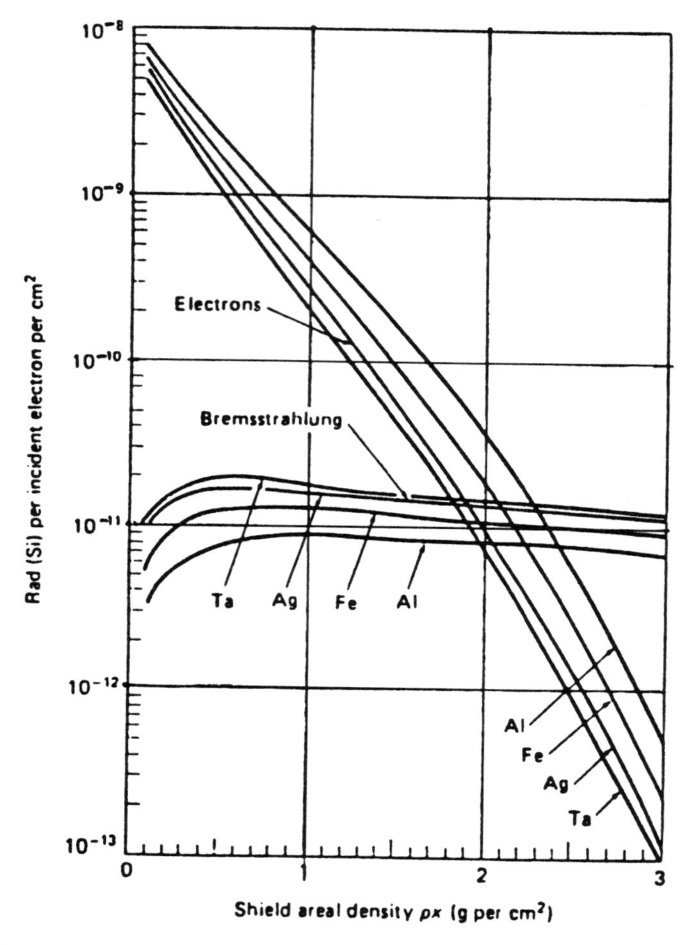

Fig. 13.7A Fission debris electron dose including bremsstrahlung production versus shielding thickness for a family of shielding materials.[14]

systems, and especially the human crew, comprises the second level of uncertainty. For spacecraft crews, acceptable and incurred dose risks are yet ill-defined. The third level is due to uncertainties in the distribution of masses within the spacecraft, especially in large and therefore geometrically complicated systems. This is especially the case for incident radiation environments that are of relatively low energy, which makes the transmission capability of radiation through the shield a sensitive function of its thickness. The last uncertainty level is that of the analy-

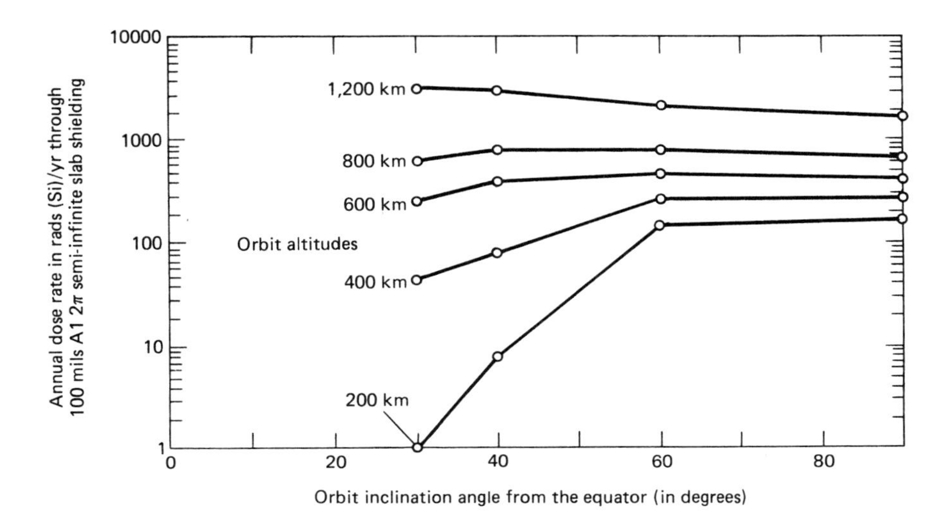

Fig. 13.7B Low-earth orbit (LEO) annual dose rate versus orbit inclination angles from the equator for various orbit altitudes.[128] © 1991 by the IEEE.

tical methods and approximations required to compute the radiation transport through the shield. For simple geometries, the available analytical methods constitute the most accurate element of this hierarchy. Within the analytical methodologies, the greatest source of inaccuracies is due to the geometrical complexity of the satellite mass distribution, which often must be approximated physically in a rudimentary fashion for the sake of a timely and meaningful calculation. Movement of the crew about the spacecraft poses an especially difficult computational problem in the penetration of radiation throughout the satellite interior.

Table 13.2A depicts personnel mean dose rates, computed simply by dividing measured dose by spacecraft mission duration time, for three missions corresponding to various altitudes and orbit inclination angles.[88] Only natural radiation sources, as opposed to anomolously large solar flares and extraterrestrial nuclear bursts, were considered. It may be implied from Table 13.2A that: (a) at relatively low altitudes, essentially under the Van Allen belts (Apollo-Soyuz), the major dose source encountered is during passes through the South Atlantic Anomaly (SAA), discussed earlier and in Section 8.4; (b) Van Allen belt primary electrons did not penetrate into the (Skylab) spacecraft personnel compartments with sufficient flux to be measured on the personnel-carried dosimeters,

Table 13.2A Measured personnel dose rate for three spacecraft missions.[88] © 1976 by the IEEE.

MISSION	MEAN ALTITUDE (km)	INCL. ANGLE (deg)	MEAN DOSE RATE mrad (tissue)/hr.	PARTICLE SOURCE
Apollo–Moon 9–10	Cis-lunar 115 (lunar)	32.5 (lunar)	1.0	Heavy ionized cosmic rays
Skylab 2–4	430	50	3.0	Van Allen belt electrons/protons
Apollo–18 Soyuz–19	200	51.8	0.5	SAA electrons/ protons

and the same was true for the extravehicular activities (EVAs) to 0.3 cm tissue equivalent distance, so that electron dose at tissue depth plus 2 gm/cm^2 of space suit was computed from electron-proton spectrometer data[88]; and (c) For the cis-lunar (between the earth and moon) mission, personnel dose in the Apollo-Moon spacecraft, on the moon, and in the moon lander was mainly due to high-energy cosmic rays and solar particle bursts, once the Van Allen belts were cleared.

Hence, it is appreciated that, with typical spacecraft shielding mass, additional special-purpose radiation shielding, and special personnel garb, astronauts can be subjected to rather lengthy spacecraft missions without suffering undue radiation exposure. Again, this does not include extraterrestrial events such as possible nuclear bursts, galactic gamma bursts[129] (Section 8.4), and large solar flares. Space vehicle "storm cellars" for the latter event within the spacecraft have been considered.[89] Galactic gamma bursts correspond to an ionizing dose rate of about 1 krad (Si) per year in the stratosphere regions.

The dose analysis of heavy charged particles and electrons has interesting aspects, especially with regard to spacecraft electron shielding. Heavy particles, e.g., the high-energy ions from cosmic rays, including alpha particles, can be assumed to be approximately incident isotropically on the spacecraft skin for shields of areal density less than 100 g per cm^2.[4,10] Besides certain estimates of cosmic-ray heavy ion fluences for single-event upset computations, the heavy particle analysis is usually limited to alpha particles from the sun, whose interactions are fundamentally the same as that for protons.[11]

With regard to Van Allen belt trapped electrons, their charge causes them to interact readily with and be stopped by relatively thin shields. Their range in material is roughly 0.5 g per cm^2 per MeV of incident

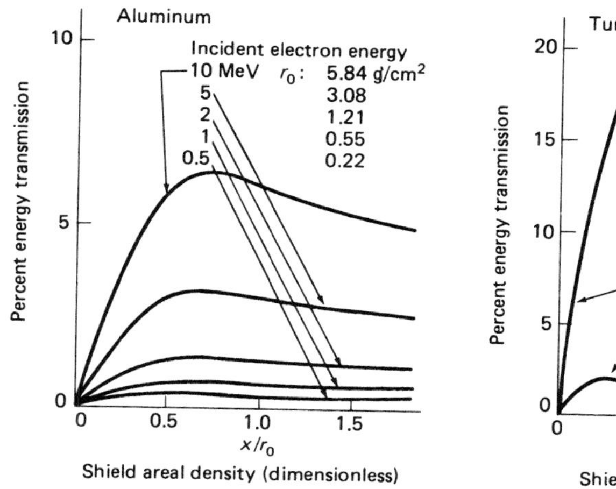

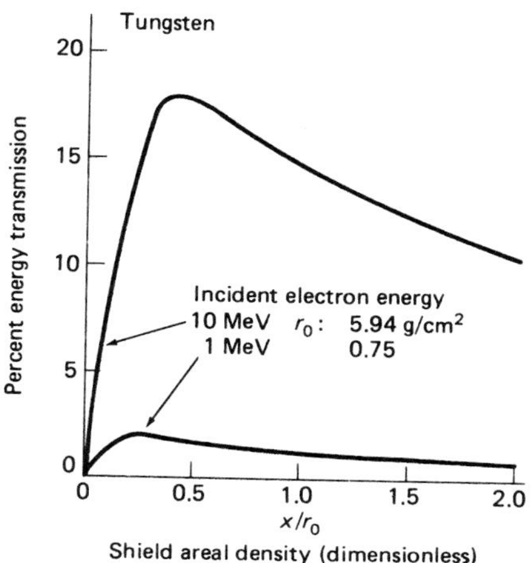

Fig. 13.8 Bremsstrahlung efficiency. Percentage of electron energy converted to bremsstrahlung that penetrates the material shield versus shield areal density for various incident electron energies.[16]

electron energy. However, they can produce bremsstrahlung x rays whose fluence ultimately exceeds that of the electrons for sufficiently thick shields, because of the greater penetrability of the high-energy x rays. An empirical relationship for the bremsstrahlung intensity, I is given by[12]

$$I = C(E)\,ZE^2 \qquad \text{(MeV per incident electron)} \qquad (13.17)$$

where

E is the incident electron energy
Z is the atomic number of the shield material
$C(E)$ is an empirically determined function given by $(0.4 \pm 0.04) \cdot 10^{-3}\,\text{MeV}^{-1}$ for $0.5 \leqslant E \leqslant 3.0\,\text{MeV}$.

Figure 13.7A depicts the electron dose from fission debris, to illustrate the bremsstrahlung presence as a function of shielding thicknesses for various shield materials. The bremsstrahlung can be seen to dominate for large shield thickness, i.e., large areal density, and so provide an ultimate limitation on the effectiveness of the shield. An excellent compendium of bremsstrahlung parameters useful in shield calculations is contained in Berger and Seltzer.[15] Figure 13.8 depicts the fraction of incident electron energy converted to bremsstrahlung that penetrates the slab shield for

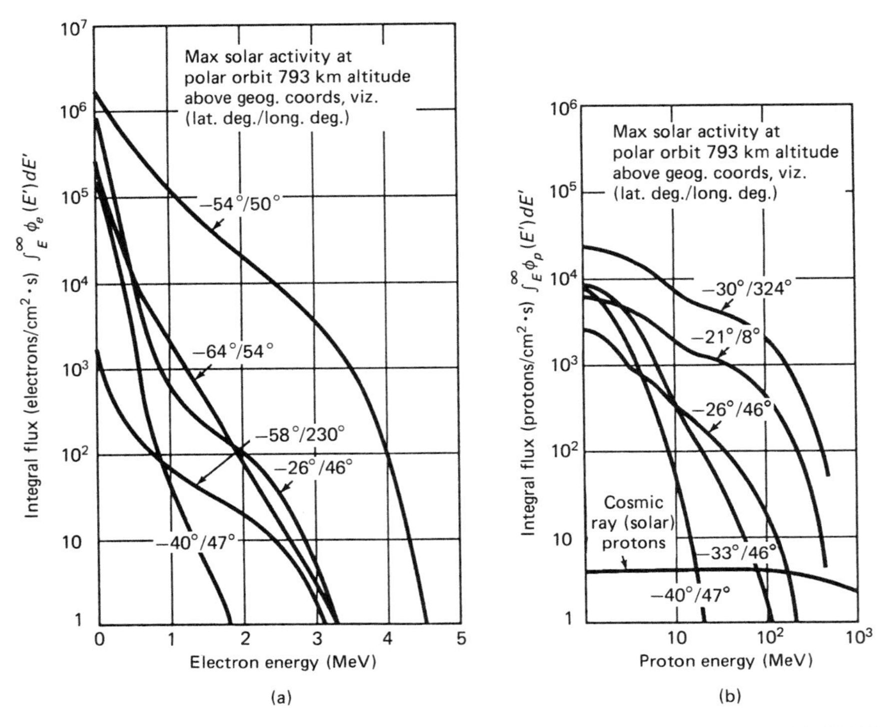

Fig. 13.8A Van Allen belt electron (a) and proton (b) flux spectra for polar orbit (97.5° inclination from equator) at 793 km.[84] © 1976 by the IEEE.

aluminum and tungsten for various areal densities.[16] Figure 13.9 shows the transmission coefficients for electrons incident on aluminum and polyethylene.[17] In terms of weight added to the spacecraft due to shielding mass, this usually amounts to a very small percentage of the total weight.

As can be appreciated at this juncture, the particle radiation that could cause the principal damage to terrestrial-based electronics systems are neutrons, gamma and x rays from nuclear detonations. However, for space systems (including spacecraft) in a benign environment, the principal radiation that they are immersed in (besides cosmic rays) is natural Van Allen radiation consisting of protons and electrons. An additional proton source originating within the solar system is that from solar flares, whose parameters vary roughly in accordance with the 11-year solar sunspot cycle. For a high-altitude spacecraft mean life of a decade, the major amount of natural damaging radiation will be that due to rarely occurring solar flares.

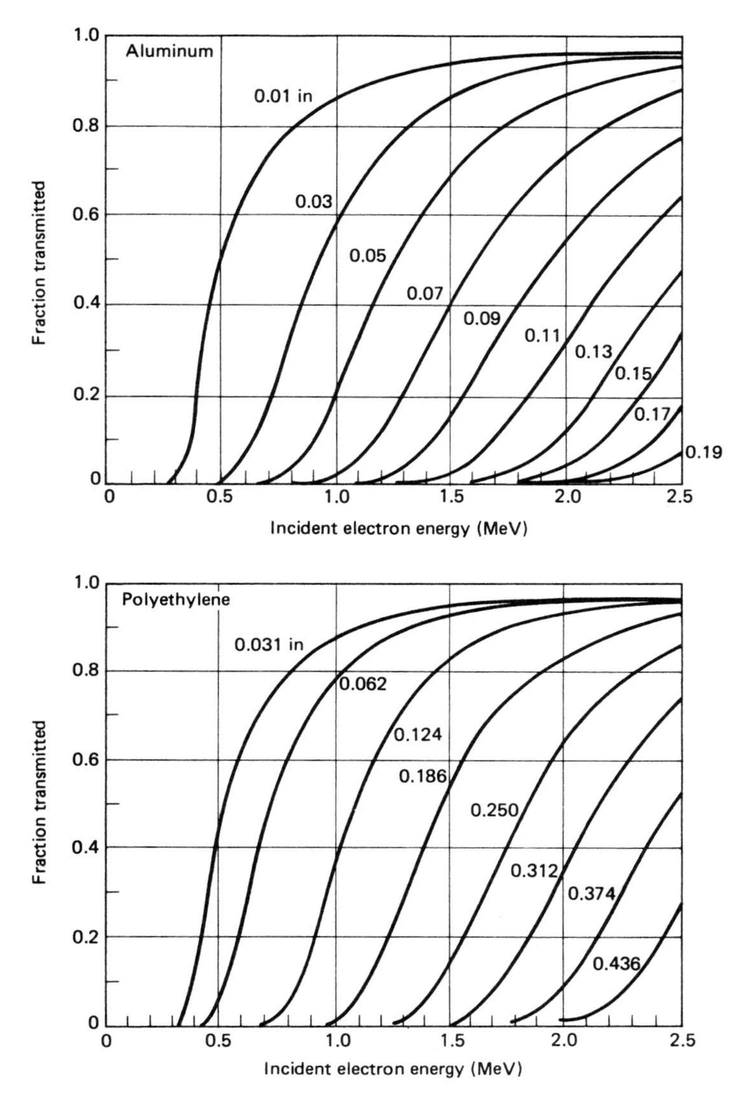

Fig. 13.9 Fraction of electrons transmitted through aluminum and polyethylene of various thicknesses versus incident electron energy.[17]

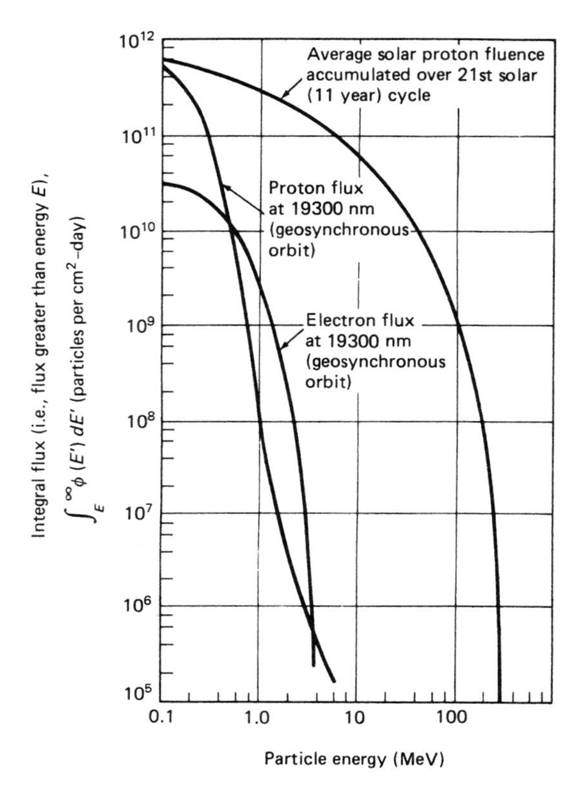

Fig. 13.10 Geosynchronous integral electron and proton flux and solar proton fluence accumulated over the 21st (11-year) solar sunspot cycle versus particle energy.[18]

Radiation from an extraterrestrial nuclear detonation, consisting of neutrons and photons, discussed in Chapters 4–6, is superimposed as a temporary addition to the above secularly changing radiation. Present spacecraft operate mainly in two regions about the earth. The first is the low altitude region between 100–600 nautical miles, and the second is the upper region beyond 10 000 nautical miles altitude, which includes the equatorial geosynchronous satellites. Future satellites may be able to operate wholly within an altitude between 600–6000 nautical miles, which contains the trapped radiation Van Allen belts.

Spacecraft in Molniya orbits experience both types of radiation environments (i.e., electrons and protons) throughout the Van Allen belts at relatively low altitudes, and cosmic ray heavy ions at near-geosynchronous altitudes. A Molniya orbit is highly elliptical, with an inclination angle about 63 ± 6 degrees from the equatorial plane. The orbital period

Table 13.3 Lifetime damage constants for neutrons, electrons, and protons.[19] © 1975 by the IEEE.

SUBSTRATE RESISTIVITY (ohm-cm)	1-MeV NEUTRON DAMAGE CONSTANTS* K_τ(cm²/s) INJECTION LEVEL 10^{-5}	10^{-3}	10^{-1}	10^{0}
n-Type				
1	1×10^{-5}	5×10^{-6}	2×10^{-6}	1.5×10^{-6}
10	6×10^{-6}	3×10^{-6}	1.5×10^{-6}	1×10^{-6}
100	1×10^{-5}	2.5×10^{-6}	5×10^{-7}	3×10^{-7}
p-Type				
1	8×10^{-6}	2×10^{-6}	5×10^{-7}	3×10^{-7}
10	8×10^{-6}	2×10^{-6}	5×10^{-7}	3×10^{-7}
100	2.5×10^{-6}	1.5×10^{-6}	5×10^{-7}	–

	20-MeV PROTON DAMAGE CONSTANTS* K_τ(cm²/s) INJECTION LEVEL 10^{-3}	10^{-1}
n-Type		
1	$(2 - 10) \times 10^{-5}$	$(1 - 5) \times 10^{-5}$
10	–	$\sim 5 \times 10^{-6}$
p-Type		
1	$(1 - 3) \times 10^{-5}$	$\sim 1 \times 10^{-5}$
10	–	$\sim 5 \times 10^{-6}$

	3-MeV ELECTRON DAMAGE CONSTANTS K_τ(cm²/s) INJECTION LEVEL LOW ($\leqslant 10^{-2}$)	HIGH (> 1)
n-Type		
1	$(0.6 - 3) \times 10^{-7}$	$\sim 5 \times 10^{-8}$
10	$(2 - 10) \times 10^{-8}$	$\sim 1 \times 10^{-8}$
p-Type		
1	$(1 - 4) \times 10^{-8}$	$(2 - 8) \times 10^{-9}$
10	$(0.5 - 2) \times 10^{-8}$	$(1 - 4) \times 10^{-9}$
100	$\sim 3 \times 10^{-9}$	$\sim 6 \times 10^{-10}$

* For later comparisons between 1 MeV neutrons and 20 MeV protons, see e.g. references 131, 132.

is extremely long, up to 24 hours (most of which is spent in the apogee lobe) in order to scan selected portions of the earth for relatively long times. With these periods, it possesses apogees (farthest points) up to about 22000 nmi, with corresponding perigees (nearest points) of less than 300 nmi, hence traversing both environments twice every complete orbit.

Figures 13.8A and 13.10 depict typical Van Allen belt trapped-charged particle environments at inner belt and geosynchronous altitudes versus particle energy, respectively. The ordinate is the integral flux $\int_E^\infty \phi(E')\,dE'$, which is the total flux greater than a given energy, E. In both figures, the electron flux spectra are seen to extend to but a few MeV, whereas the (high-energy) proton flux spectra extends to hundreds of MeV. It is also seen that, while the electron flux is larger than the high energy (>30 MeV) proton flux within the heart of the Van Allen belts (Fig. 13.8A), the opposite is the case at the high (geosynchronous) altitude fringes of the belts, (Fig. 13.10). Some proton flux is in the region of cosmic ray energies, and like them can cause single event upset (SEU), as discussed in Sections 8.3B and 12.5.

Because protons are charged particles, their low energy component (less than 1 MeV) is relatively easily attenuated through the process of ionization in the material of interest. In other words, their range is sufficiently small that they can be stopped by the device package material. For the relatively high-energy proton component, the corresponding interaction cross section decreases inversely with energy. These constraints then bracket the effective proton energy, in terms of radiation considerations, to between 1–10 MeV (except for single event upset conditions). Similar factors for electrons limit their effective energy spectrum in this sense to energies around 5 MeV.

Besides being charged particles protons are also heavy with a mass almost the same as a neutron. Hence, reasonably energetic protons can produce neutron-like displacement damage, their charge notwithstanding, as well as ionizing damage. In this regard, sufficiently energetic electrons can also produce displacement as well as ionizing damage. Damage data for these particles is rather sparse compared with that for neutrons. Extrapolation of certain proton and electron experimental data and their comparison to neutron displacement damage in terms of minority carrier lifetime degradation are given in Table 13.2.[19,22] This is done through the minority carrier lifetime damage constant K_τ, defined by Eq. (5.11), rewritten here to give in the post-radiation epoch,

$$\tau^{-1} = \tau_0^{-1} + K_\tau \phi \tag{13.18}$$

where ϕ is the appropriate electron or proton fluence. Table 13.3 gives values of K_τ for 1-MeV neutrons, 3-MeV electrons, and 20-MeV protons for various values of substrate material resistivity and minority carrier injection levels.[18,19] The injection levels therein are the appropriate minority carrier charge densities divided by the corresponding majority carrier densities, such as $\Delta n/p_0$ or $\Delta p/n_0$.

13.3 STRATEGIC SYSTEMS

Hardness design considerations for strategic systems involve many facets of analysis, design, construction, and deployment. Certain aspects of such hardness design involve unique approaches, which will be underscored, though other aspects are common to electronic systems generally. For an illustration of a strategic system, the example of a bomber aircraft will be discussed. There are other strategic systems, e.g., ICBMs launched from ground and submarine launchers and other related programs from which hardening design experience could be used.

BOMBER A/C

A very cost-effective hardness assurance program could be required for the above aircraft. This results in a methodology that would install and maintain a nuclear survivability system capability at a moderate level of confidence. In other words, a basic concept of this effort would include inserting sizable design margins in the system, when this could be available at minimum cost. With this approach, commercially available components could be used for a large part of the avionics. Then hardness controls would be required for the components with inherently small nuclear design margins, relative to anticipated nuclear specifications. The nuclear environment scenarios in which such aircraft are expected to operate are sufficiently moderate that satisfactory hardness design margins could be attained in most cases, so that hardness verification could be largely demonstrated by analysis, as opposed to costly verification testing. In anticipating certain critical areas, analytical methods could be confirmed by testing. Several types of testing could be used, including developmental tests to support design efforts and to provide inputs to the analysis. Also, qualification tests could be performed to verify the analyses and to provide a primary demonstration of compliance.

The aircraft types would have both low- and high-altitude mission profiles, and so could be vulnerable to nuclear weapons effects from both exo- and endoatmospheric detonations. Figure 13.11 depicts nominal mission profiles to which the nuclear effects would apply. It should be noted that x rays are not a factor in the endosphere, for they are absorbed by the atmosphere to form the fireball, as discussed in Section 4.4. However, proton- and neutron-induced single event upset (SEU) discussed in Sections 8.1 and 12.5 could be a factor.

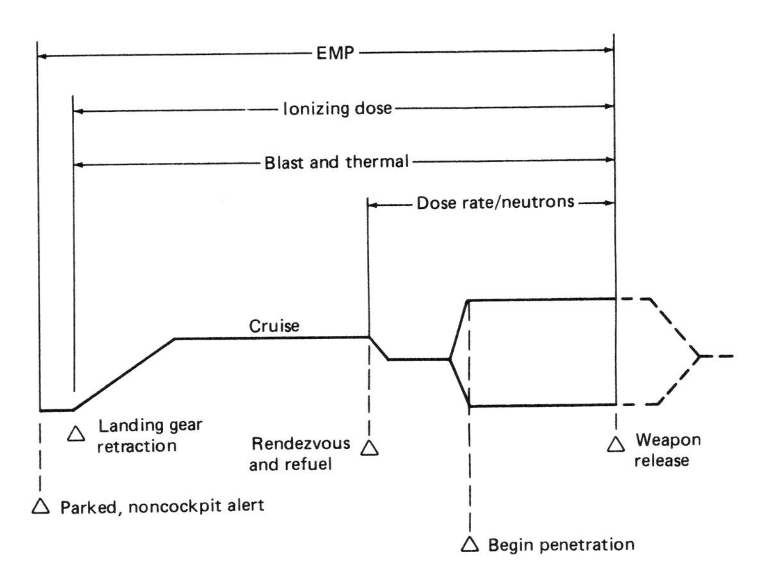

Fig. 13.11 Aircraft mission phases showing nuclear effects.

Blast hardness (overpressure and gusts) would be of concern in terms of structure. The electronic equipment would be designed to meet stringent vibration and acceleration requirements, and in most cases could probably withstand the internal loads resulting from the blast wave. The aircraft would be capable of withstanding the specified overpressure and dynamic pressure environments in atmospheric flight. Verification that the design is in accord with the nuclear blast specifications may be obtained by analytical techniques, supported by static load and dynamic load testing. Large-section static load tests could be employed to verify that the blast specifications are met.

The detonation thermal output hardness requirements for the fireball can usually be satisfied. A major portion of the incident thermal radiation could be handled by reflecting systems and other adjuncts. Any remaining thermal energy could be absorbed by the fuselage skin. Again, thermal hardness design verification would be primarily proved by analysis, including extensive computer-aided methods. This would be closely supported with material, thermodynamic, and structural testing.

The radiation hardness requirement would include specification of ionizing dose, prompt and delayed dose rate, neutron fluence levels, and possible SEU levels. The specification levels could be chosen based on trade-off studies involving a diversity of missions, hardening costs, and the technological ability to achieve system hardness design levels. As a result, the specifications could probably be met with a reasonable capital outlay, assuming that they are within the state of art. The radiation specifications would probably permit mounting a design effort based on a minimum of testing and a maximum of analysis, as discussed above. This approach would also allow the incorporation of actual component radiation-response experimental data into the analysis. In certain cases, computed response based on experimental data of similar devices could be employed. Vendor piece-part testing to support the design could use statistical sampling of each device type. Such sample tests could yield acceptable statistical information. Due to the design margins anticipated within a reasonable design boundary, this approach should be adequate for most component applications.

With regard to TREE hardness, the aircraft in this illustration would be required to survive transients due to incident radiation pulses (dose rate). This could imply a significant hardness design effort, because these systems would rely heavily on on-board computers to coordinate their electrical and electronic functions. However, the extensive use of digital error detecting and correcting (EDAC) techniques could provide a large measure of protection against potential transient upset damage. Hardening the central processing units in these computers might not be necessary in this example. Instead, circumvention methods (Section 13.9) could be employed to clamp memory drivers in the event that incident radiation reached dose-rate levels that say, are an order of magnitude below the dose-rate upset level. At this point, the circumvention system would be activated by a signal from a radiation detection system, which would "freeze" the computer memories before extraneous errors could occur. After the nuclear event, the computer would be restarted from instructions in a hard read-only memory (ROM), initiating an equipment malfunction check and the resumption of normal operation.

The approach to EMP hardening could employ a balance between shielding and electronics hardening. In line with the modus of the maximum use of off-the-shelf components, early studies have indicated that most components can sustain a mean EMP stress. For illustrative purposes, assume that this stress can be expressed as an EMP-induced internal main bus cable current of about 10 amps. A directly related parameter is an assumed worst-case peak EMP fuselage skin current of, say, 30 kiloamps. For the electronics within to endure the effect of such skin current, it is estimated that $20 \log_{10} (3 \cdot 10^4/10) = 70\,\text{dB}$ of attenuation would be required. This could be apportioned between the fuselage skin (20 dB) and installed shielding (50 dB) in the avionics. To meet the latter need, most of the avionics could be located within shielded bays interconnected by high-permeability shielded EMP cables and associated accessories, discussed in Chapter 9. The use of special EMI (electromagnetic interference) protection techniques, which would extend over the life of the system, would also contribute to the EMP hardening requirement. Although the assumed cable current of 10 amps is not particularly stringent for off-the-shelf electronics, modern totally solid-state systems would be carefully designed to achieve this hardness level. Circuit-design hardness methods used might include filtering, surge suppression, and isolation techniques. Because of the conservative measures employed throughout this illustration, it is expected that a considerable design margin would exist in most areas. This would provide high confidence that the above systems would be capable of performing in any reasonable EMP environment that may be specified.

NUCLEAR REACTOR

The second system considered is a nuclear power reactor. After reactor operations personnel, the active semiconductor devices mainly in the reactor instrumentation electronics are the most susceptible to damaging radiation. The later includes such radiation as occurs during normal reactor operation, or during a reactor accident. The instrumentation includes the electronics associated with pressure, temperature, moderator/coolant flow, and nuclear parameter measurement and monitoring. This modern instrumentation is almost wholly comprised of electronic semiconductor devices such as discrete transistors and integrated circuits. The instrumentation in a U.S. nuclear power plant that must survive the aforementioned radiation environment is mainly located within the containment structure. Mainly constructed of heavy steel-lined concrete, this structure houses the reactor core, cooling loops, fuel replacement cranes, and primary and secondary shielding. It is usually the closest reactor surround whose exterior is accessable to personnel.

Under normal operating conditions, the reactor radiation ambient levels are sufficiently low that the personnel and instrumentation alcoves (remote from their respective core probes) in the containment area are not affected. In the U.S., most reactor power is obtained from light-water moderated and cooled reactors (LWRs), as opposed to heavy-water (Canada) or graphite-moderate reactors (USSR). The moderator/coolant water comprises a vital component in the reactor neutron life cycle. The most frequently postulated nuclear accident is a loss of coolant accident (LOCA). The LOCA, including the loss of moderator water, causes the chain reaction in these reactors to stop almost instantaneously, as the reactor becomes subcritical. Under circumstances far less than a LOCA, e.g., if the water moderator temperature merely rises above the operating temperature, or boils, the corresponding water density decrease results in a less effective moderator, which slows down the chain reaction. Thus the LWR is stable in this sense.

However, following the LOCA onset when the reactor is subcritical, there is sufficient heat in the core residual radioactivity to melt core fuel elements, especially when they are exposed above the moderator/coolant water level during the LOCA. This process would be relatively slow compared to a nuclear detonation, for example, but it could cause excessive amounts of radioactive fission products from the damaged core fuel to be dispersed within, and possibly without, the containment volume. This dispersal can be done by the coolant water, in its possibly temporarily uncontrolled flow within the containment structure, until the emergency core cooling system refills the reactor core to again submerge and cool the fuel.

It is vital to determine the amount of radioactive constituents released during the LOCA. The LOCA analysis discussed herein[23,98] has estimated gamma and beta (fast electrons born of fission) dose rates from fission fragment radioactive isotopes released to the ambient air of the containment structure, those isotopes deposited onto containment structure surfaces, and those dissolved in the containment structure sump water. Further LOCA assumptions include partial core meltdown, such that half of the corresponding radioactive iodine isotopes (short-lived gamma emitters) and all of the noble gas isotopes are released. These considerations, along with other assumptions,[23] result in LOCA computations for a 3000 MW (thermal) reactor that yield:

a. Initial peak gamma dose rate following LOCA onset; 5 Mrads (air) per hour
b. 30-day fission product gamma dose; 13.6 Mrads (air)
c. Initial peak beta dose rate following LOCA onset; 24 Mrads (air) per hour
d. 30-day fission product beta dose; 126 Mrads (air)

Figure 13.12 illustrates the time dependence of the preceding computed LOCA "best estimate" of gamma ray dose rate and its integral, the ionizing dose.[23] Figure 13.13 compares two computations of the time-dependent reactor gamma ray energy release rate due to the LOCA. The solid line plot corresponds to the preceding "best estimate" computations performed by SANDIA and IRT corporations.[24] The dashed line is a plot of the results of computations contained in the Nuclear Regulatory Commission (NRC) Regulatory Guide 1.89. The early time disparities in these curves are due to corresponding radioactive sources assumed in the

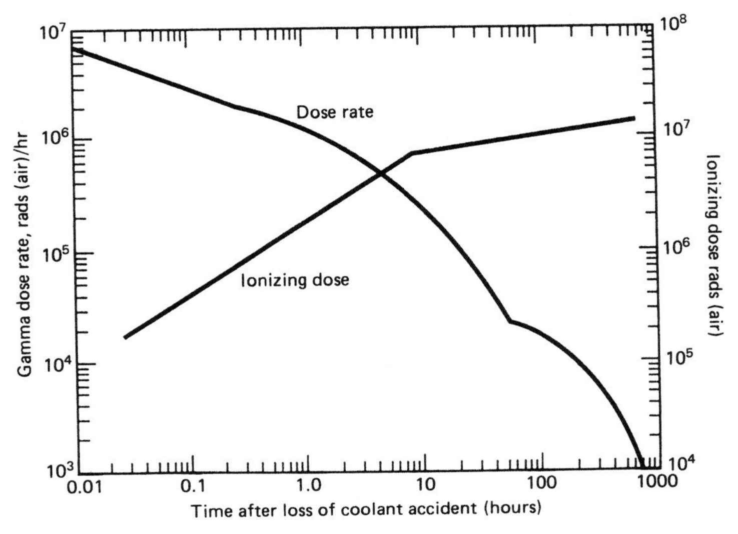

Fig. 13.12 Containment structure gamma dose and dose rate following a loss of coolant accident.[23] © 1980 by the IEEE.

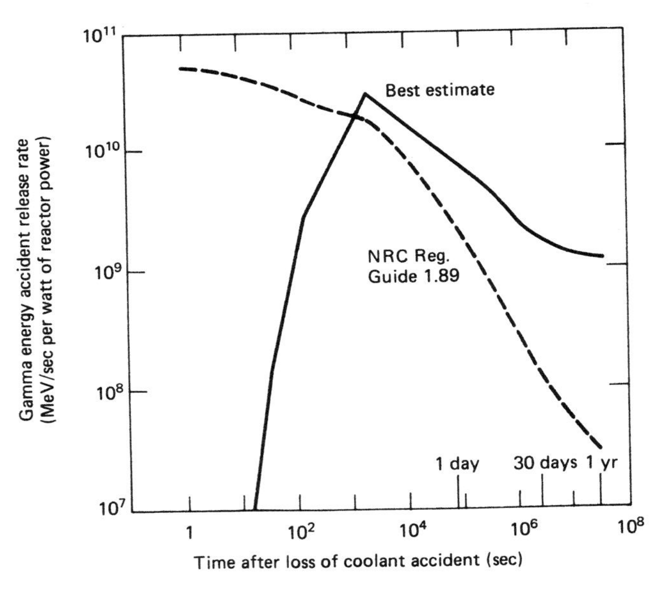

Fig. 13.13 Gamma ray energy release rate following a loss of coolant accident.[23] © 1980 by the IEEE.

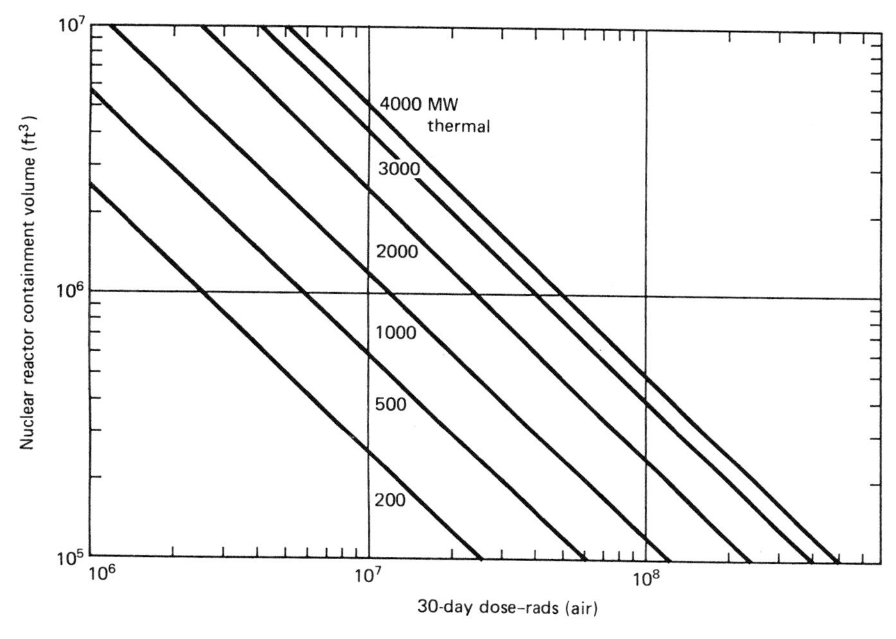

Fig. 13.14 Nuclear reactor containment volume versus 30-day containment dose following a loss of coolant accident for various reactors.[23] © 1980 by the IEEE.

NRC calculations, but not included in the SANDIA/IRT "best estimate" calculations.

The NRC designates certain classes of equipment for safe reactor operation. Class 1E electronic equipment includes all such needed to handle emergency reactor shutdown, containment isolation, reactor core cooling, containment and reactor heat removal, and the prevention of a significant release of radioactive material to the environment. An ionizing gamma dose of 20 Mrad (air) is the maximum allowable for Class 1E equipment located in areas within the containment structure for pressurized LWRs with dry-type containment. If an ionizing dose level less than this value has been specified for sensible technical reasons, an application-specific evaluation must be performed to validate this dose.

Beta dose is often less significant than gamma dose, because beta particles are less penetrating. However, the dielectric insulation of coaxial cables with semi-transparent copper braid outer conductor is considered to be very susceptible to beta dose damage. If, through appropriate shielding measures, the beta dose incident on instrumentation equipment components is attenuated to one-tenth or less of the gamma

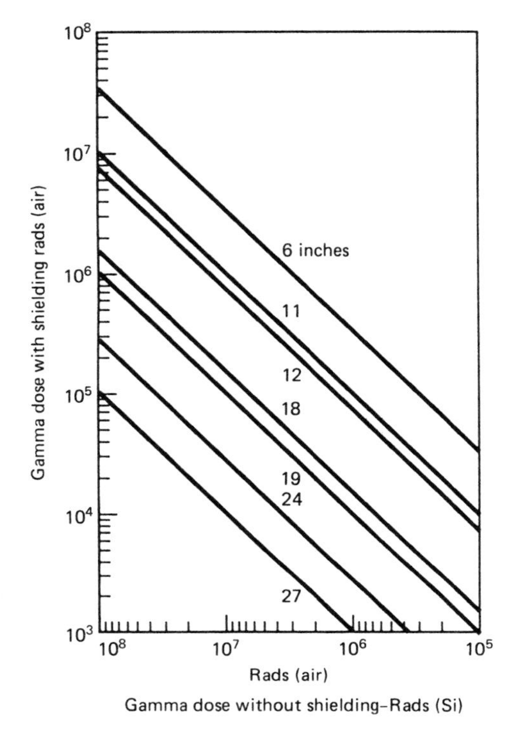

Fig. 13.15 Dose correction factor for concrete shielding.[23]

dose to which this equipment has been qualified, then this equipment may be deemed approved to the sum (beta plus gamma) radiation-dose environment.

The radiation-dose level that can be delivered to the instrumentation electronics in the containment structure depends on reactor power, containment structure volume, amount of instrumentation shielding, and the post-accident duration time through which this equipment must remain functional. The NRC 1E Bulletin 79-01B contains a computational procedure whereby one can estimate radiation hardness requirements for electronic equipment to maintain operation during and after a LOCA. This procedure is illustrated through the following example, which uses Figs. 13.14–13.17 and the following parameters[23]: (i) reactor power level (3000 MW thermal); (ii) instrumentation compartment volume (8000 ft^3); (iii) thickness of instrumentation compartment concrete shield walls (24 inches); (iv) time after LOCA for instrumentation to remain functional (1 hr); and (v) reactor containment structure volume

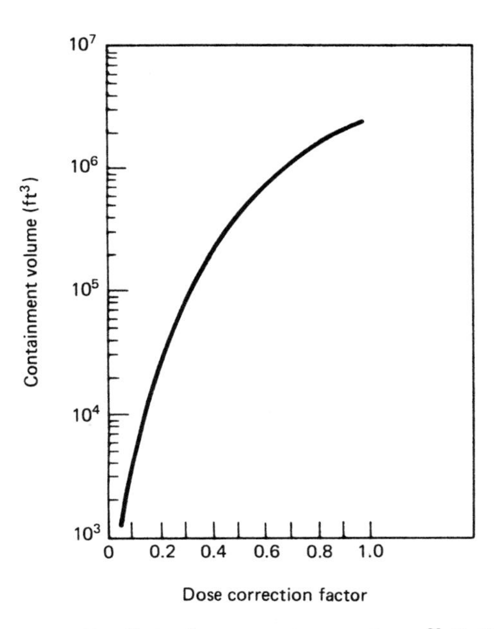

Fig. 13.16 Dose correction factor for compartment volume.[23] © 1980 by the IEEE.

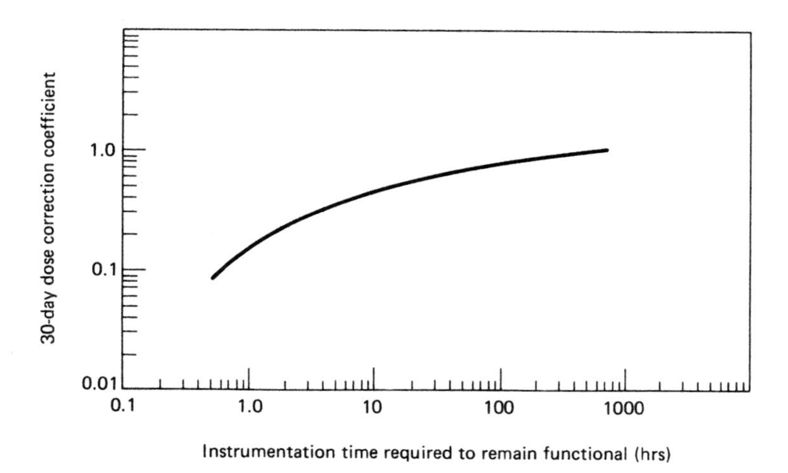

Fig. 13.17 Dose correction for time required to remain functional.[23] © 1980 by the IEEE.

($2.5 \cdot 10^6\, ft^3$). To estimate the corresponding dose that the electronic equipment will receive:

1. From Fig. 13.14, for a 3000 MW (thermal) reactor, and a reactor containment volume of $2.5 \cdot 10^6\, ft^3$, the ionizing dose level in the containment vessel at 30 days following a LOCA is seen to be 16 Mrads (air).
2. To determine the extent to which the 24-inch concrete instrumentation compartment shield attenuates the 30-day integrated dose portion coming from the reactor core, Fig. 13.15 yields this attenuated dose as 45 krads (air).
3. Besides the above dose, there is the major dose portion due to containment ambient radiation released by the LOCA that intrudes directly into the instrumentation compartment. To determine its value, use Fig. 13.16 to obtain a 0.13 correction factor for the $8000\, ft^3$ instrumentation compartment volume. The corresponding instrumentation compartment direct dose is then 0.13×16 Mrads (air) = 2.1 Mrads (air), which is seen to be very much greater than the concrete-attenuated dose.
4. The concrete-attenuated dose and the intrusion dose are summed to yield 45 krads (air) + 2.1 Mrads (air) $\cong$ 2.15 Mrads (air).
5. To obtain the loss of coolant dose at 1 hour after the LOCA, to which the electronics must remain functional, Fig. 13.17 yields a 0.15 correction coefficient to multiply the 30-day dose of 2.15 Mrads (air) to obtain 0.15×2.15 Mrads (air) = 322.5 krads (air). This value at 1 hour after the LOCA dose, to which the instrumentation components must be hardened, is achievable by using Class S space-qualified hardened active parts.

To interpret the above results further, the NRC Regulatory Guide 1.97 specifies that containment area radiation monitors for high-range doses be capable of measuring gamma dose rates up to 10 Mrads (air)/hr for up to 200 days after a LOCA. As mentioned earlier, the peak gamma dose rate almost immediately after the LOCA obtained from the preceding "best estimate" computations (a) and (b) earlier, is 5 Mrads (air)/hr with a 30-day integrated dose of 13.6 Mrads (air). The above NRC dose-rate specification is seen to be a factor of two greater. Therefore, the presumption is made that an ionizing dose requirement for 30 days after the LOCA of, for instance, 20 Mrads (air) (compare to (1) above) can be construed as an acceptable qualification level for this instrumentation.

Also, the preceding indicates that at least some of the instrumentation (high-range dosimeters) in the containment structure must operate for 200 days after the LOCA.[23]

Since the foregoing radiation levels are measured in megarads, some special radiation hardness measures must be used in the electronic design of circuitry, to be capable of withstanding them. This is principally because most commercial active devices that are available are radiation hard to levels approximately an order of magnitude less than the low megarad region. However, electronic circuits and special "rad-hard" parts have been designed to accommodate the latter levels.[25,26] These circuits, though designed for different functions, have many aspects in common. Some of these are enumerated below, with the primary assumption that the damaging radiation is ionizing dose, and secondarily neutron fluence.

1. Bipolar transistors, whether discrete transistors or integrated circuits, are employed almost universally in digital or linear applications, since bipolars are less susceptible to ionizing dose, as discussed in Section 6.15. If FETs are to be used, JFETs are preferable, since they often perform even better than bipolars at the higher ionizing dose regions.
2. For linear devices and circuits, such as operational amplifiers, heavy ac and dc feedback is used, sacrificing gain for stability and reduced radiation susceptibility. This can be made up through the use of multistage circuit approaches.
3. Incorporate large-device gain margins by using high-gain, high-f_T transistors. Hence, even with a large radiation reduced-gain reduction (up to 90 percent), and increased leakage currents, these lower-gain circuits will still function adequately.
4. Minimize the use of large-capacitance bypass capacitors, especially electrolytics, even to the extreme of substituting reversed-biased diode capacitance.[25] The former capacitances are very susceptible to high levels of ionizing radiation, as discussed in Section 7.9.
5. Maintain all interstage dynamic impedances as low as possible. This maximizes the corresponding operating currents.
6. All capacitors should be of the ceramic, silver mica, or solid tantalum types, which provide good temperature performance, as well as good radiation-hardness properties.
7. Carbon-composition resistors have better thermal-structural properties than metal-film resistors.
8. High-quality printed circuit boards should be used, as well as silicone potting compound, with circuitry housed in stainless steel containers.
9. Inductors should be wound with polymide insulated wire.[26]

To place the major reactor accidents in perspective, Table 13.4 provides the amount of radioactive isotopes (radionuclides) released by

Table 13.4 Amount of radioactive nuclides released into the atmosphere (and spilled water within containment) from various nuclear reactor accidents and nuclear weapons tests throughout the world[27,29,32,127] (MCi = Megacuries).

SER.	NUCLIDE RELEASED INTO THE ATMOSPHERE	HIROSHIMA AND NAGASAKI 1945 (MCi)	WINDSCALE 1957 UK (MCi)	THREE MILE ISLAND 1979 U.S.A. (MCi)	CHERNOBYL 1986 USSR (MCi)	WORLD-WIDE ATMOSPHERIC TESTS (MCi)
(1)	^{137}Cs	0.0057	0.0012	No detection in atmosphere	2.4	40.
(2)	^{136}Cs	0.0002	0.000041	"	0.47	150.
(3)	^{134}Cs	0.0002	0.000032	"	1.3	<.04
(4)	^{131}I	0.076	0.016	0.000017	36.	21 000.
(5)	^{133}I	0.021	0.0044	No detection in atmosphere	9.1	49 000.
(6)	^{141}Ce	0.0009	0.00019	"	0.23	12 000.
(7)	^{144}Ce	0.0005	0.00011	"	0.14	1 300.
(8)	^{140}Ba	0.0001	0.000017	"	1.	28 000.
(9)	^{140}La	0.0001	0.000017	"	1.	28 000.
(10)	^{95}Zr	0.0009	0.00020	"	0.23	6 900.
(11)	^{95}Nb	0.0009	0.00020	"	0.23	6 900.

(12)	^{132}Te	0.076	0.016	″	5.3	52 000.
(13)	^{103}Ru	0.0052	0.0011	″	0.76	5 600.
(14)	^{106}Ru	0.0076	0.0016	″	0.16	85.
(15)	^{133}Xe	1.75	0.37	10.	120.	56 000.
(16)	^{89}Sr Released in spilled water	0.0038*	—	0.0012 Ci/liter	—	—
(17)	^{90}Sr ″	—	—	0.0031 ″	—	—
(18)	^{3}H ″	—	—	0.0004 ″	—	—
(19)	^{137}Cs ″	—	—	0.0459 ″	—	—
(20)	^{134}Cs ″	—	—	0.0083 ″	—	—
(21)	Other beta and gamma emitters	—	—	0.0081 ″	—	—
Total in spilled water in containment		—	—	0.228 MCi	—	—
Total released into atmosphere		1.95 MCi	0.411 MCi	10.00 MCi	178.3 MCi	266 975 MCi

* Released in atmosphere.

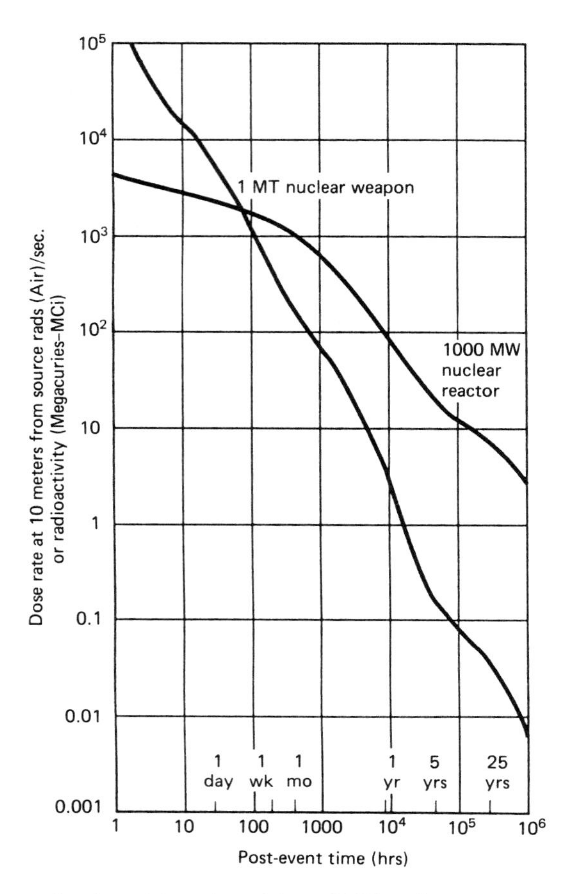

Fig. 13.18 Decay of radioactivity released by a 1 MT nuclear weapon, compared to that from vaporization of a 1000 MW reactor core.[28]

them into the atmosphere, and into the spilled water in the containment structure in the case of the Three Mile Island (TMI) accident. It should be noted that, across the severity spectrum of these events (with the exception of TMI and Hiroshima and Nagaski, but including the nuclear weapon atmospheric tests), there is roughly a factor of one-thousand increase in the amount of released radioactivity into the atmosphere. Also, as seen, the TMI and Chernobyl reactors, respectively, released many times more radioactivity than the combined Hiroshima and Nagaski detonations. For the TMI accident, there was a radioactivity release, consisting of mainly three nuclides (viz., 10 MCi (megacuries) of ^{133}Xe,

17 curies of ^{133}I, and several MCi of ^{85}Kr). However, 0.228 MCi of radioactive nuclides were released into the 3400 m^3 of water with which the containment structure was flooded, to a height of about seven feet, during the accident. Even though the TMI accident was relatively not far-reaching in terms of personnel exposure (maximum dose was 40 mrem whole body to one person offsite), it has cost millions of dollars to decontaminate. Also, not included in the above table was 0.05 MCi of ^{85}Kr that was purposely vented into the atmosphere, after it was ascertained to be safe from the health physics viewpoint. This amount was not considered significant; released nine months after the accident, it amounted to about 0.02 mrem. However, it has been estimated that several MCi of ^{85}Kr was inadvertently vented in the earliest hours of the accident, through breaches in the containment dome.[30]

Figure 13.18 depicts the decay of radioactivity, both in MCi and rads (air) per second released by a 1 MT nuclear weapon, compared to that released from the vaporization of a 1000 MW reactor core.[28] An extreme scenario, for this to occur simultaneously might be the detonation of the nuclear weapon atop the nuclear reactor. Since the nuclear weapon energy release is much more explosive than that in a reactor, the early weapon decay rate is much steeper, as seen in Fig. 13.18. Of course, this figure does not include the blast and thermal energy releases, which are discussed in Sections 4.5 and 4.6. For the reactor, the radioactivity release is relatively long-lived because of its inherently less turbulent kinetics. That their radioactivity levels are seen to be equal at 100 hours after the detonation onset implies nothing more than their relative radioactivity outputs, which depend in good measure on their specifications. Their cumulative radioactivity can be taken as the envelope of the two curves in Fig. 13.18.

13.4 TACTICAL SYSTEMS

Tactical systems are those used in support of a field army. Marine, Navy, and Air Force tactical systems include the fact that they are usually deployed on or near the ground. Further, because of their implied intimate man-machine association, they are generally man-limited in terms of nuclear survivability.[31] The nuclear weapons effects and levels are limited to those anticipated to be encountered in a tactical environment. This usually amounts to an environment in which low-yield weapons are detonated at burst heights near the ground.

A fact that cannot be overemphasized is that the cost of nuclear hardening for tactical systems, as it is for most military systems, is relatively modest when it is incorporated at the inception of the development of such systems. Contrariwise, eleventh-hour add-ons to existing systems in an attempt to harden them, are expensive, unduly extensive, and inherently manifest reduced confidence in their ultimate performance capability.

Again, the weapons effects emphasized herein are those peculiar to the tactical ground situation. The fact that the nuclear weapon outputs possess different delivery speeds, from the speed of light for gammas and x rays to sound speed for blast waves, augurs a synergism deleterious to the defender. For example, if the target is already weakened by a tactically delivered thermal pulse (up to 400 cal per cm^2) traveling at speeds that are orders of magnitude faster than the speed of sound, the structural failure potential of the target to blast shock (~1000 mi/h) is thereby enhanced over that of the structure at ambient temperature. Other like effects include a similar predisposition of materials and structures to greater damage because of multiple nuclear bursts.

Local fallout, especially in inclement weather, also can occur over the battle area from nearby and remote detonations that have taken place previously.

Both surface and high-altitude EMP must be dealt with in the above environment. For surface detonations at or near their ideal height for maximum blast wave overpressure damage from ground zero, the EMP-induced net electron flow will be upward and outward from the earth.[32] For high-altitude EMP, the net electron flow is downward toward the earth. This is due in both cases to the material asymmetries involved.

Effects on electronic equipment that are part of tactical systems, produced by neutrons, ionizing dose, dose rate, and EMP, have been discussed in previous chapters, for much of this type of electronics is common to most military systems. However, shielding for use with tactical systems can be very effective against blast-shock, thermal radiation, and EMP. But neutron and photon shielding requires a great deal of mass, which perhaps rules out shields for all but fixed emplacements. To obtain an attenuation shielding factor of 10 for neutrons and photons requires, respectively, about 10 and 18 in. of concrete, or 15 and 26 in. of earth.[33] As tactical systems are usually mobile systems, alternatives other than massive shielding must be employed.

From the tactical point of view, nuclear survivability is the capability of a system to withstand a nuclear environment without suffering a loss of its ability to accomplish its designated mission. Most tactical systems are man-machine entities in which a human operator or crew is necessary for the system to perform its intended function. For those systems, nuclear survivability requires that the machine portion survive if enough of its crew can also survive long enough to complete the mission. The decision whether to require nuclear survivability for a tactical system is based on its possible use in a nuclear conflict, and the criticality of its mission. If the decision is in the affirmative, then nuclear survivability criteria must be developed and established. A baseline of nuclear effects that the system must survive must be determined. When man is vital to a particular system operation, the corresponding man-machine relationship must be delineated, and the human response to the nuclear environment must be investigated. When crew and equipment are subjected to the same environment, survivability criteria are determined by human vulnerability modified by appropriate shielding factors. If the crew is protected and the equipment is not, then survivability criteria for the equipment are based on the vulnerability of the protected crew. If the crew is exposed and the equipment is protected, survivability criteria are also based on the unprotected crew. As can be seen, human vulnerability is the governing factor. Man is susceptible to all nuclear radiation, except perhaps nuclear EMP. Figure 13.23 depicts the clinical aspects of human acute radiation syndrome. Notice that 2–4 days after initial exposure there is an asymptomatic interlude followed by a recurrence phase that can be lethal depending on the initial levels of radiation.

Casualty and damage mechanisms considered in the development of nuclear survivability for crews include: (a) early transient incapacitation (ETI), which is defined as a temporary period of severe disfunction due to the incident radiation. For example, ETI_{50} is defined as the level at which 50 percent of the crews that are exposed to the corresponding dose levels experience ETI, (b) second-degree skin burns through summer combat uniforms due to absorbed thermal radiation, (c) foxhole collapse due to blast-shockwave overpressures, (d) equipment damage due to blast waves, and (e) casualties of prone crew members flung about by the blast and subsequent hard impact. The effectiveness of the mechanisms (a)–(e) depends on weapon yield, range, level of human protection, and the required percentage of the crew to ensure completion of the tactical mission. The requisite nuclear environments for a particular casualty-producing mechanism are specified by radiation parameters, including ionizing dose, neutron fluence, and peak overpressure. These environments must be augmented by the predetermined percentage of the crew that must survive. Figures 13.19–13.21 depict isodamage or isocasualty curves for the above casualty-producing mechanisms.[34] When all the isocasualty curves are overlaid, and, based on the particular casualty mechanism, the envelope contour of maximum effective range of this combination of nuclear effects can define the nuclear survivability criteria for that particular system. This envelope contour defines the range at which crews will satisfy the operational environment constraints and also

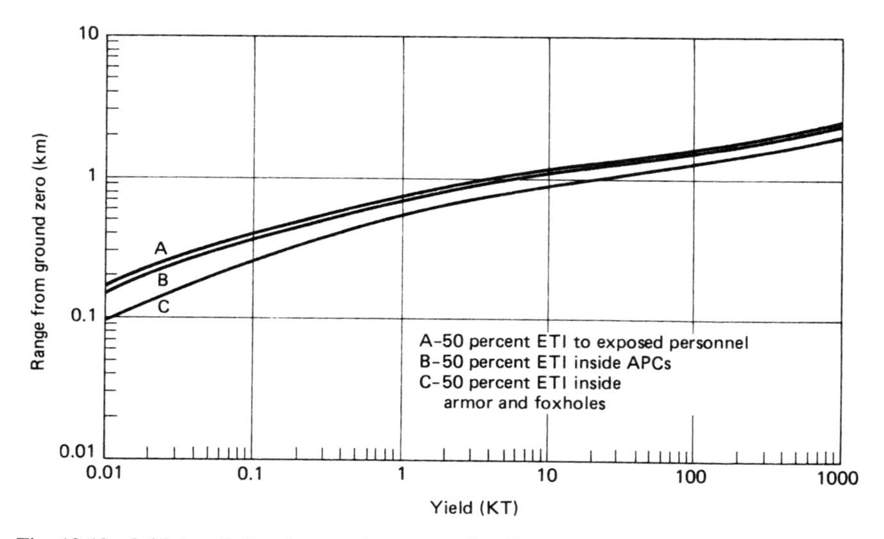

Fig. 13.19 Initial radiation isocasualty curves for 50 percent incidence of early transient incapacitation.[34,36]

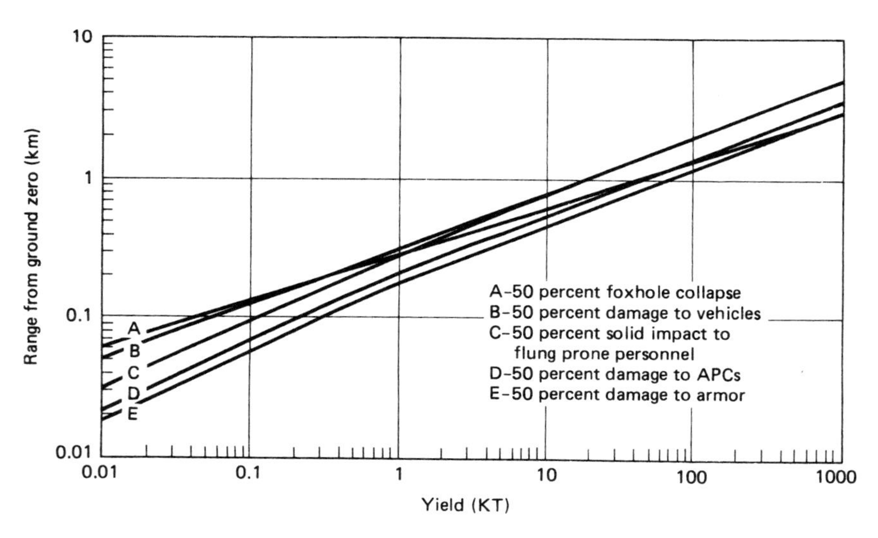

Fig. 13.20 Blast wave isocasualty curves for 50 percent damage incidence.[34,36]

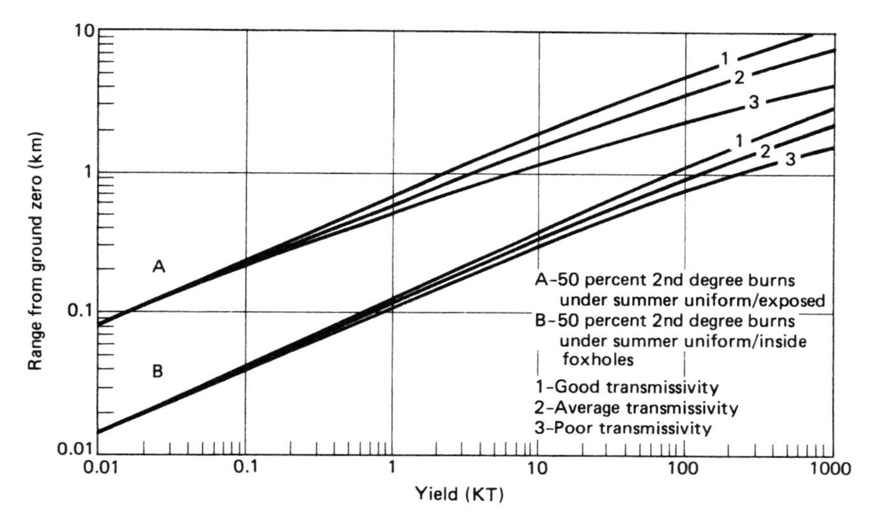

Fig. 13.21 Thermal radiation isocasualty curves for 50 percent damage incidence.[34,36]

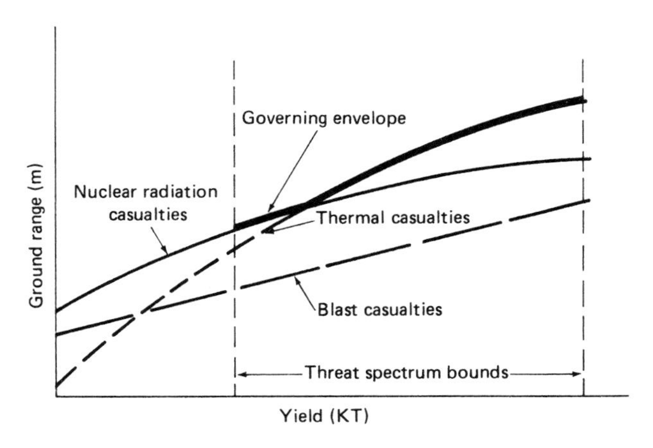

Fig. 13.22 Threat spectrum yield bounds superimposed on the governing envelope.

specifies the range for each weapon yield at which the equipment must survive. Based on battle theatre and tactical deployment considerations, a threat spectrum of weapon-yield bounds is imposed on the above governing envelope. For a tactical situation, the weapon yields employed would extend from subkilotons to tens of kilotons, as opposed to megaton devices, which would usually be employed only in a strategic con-

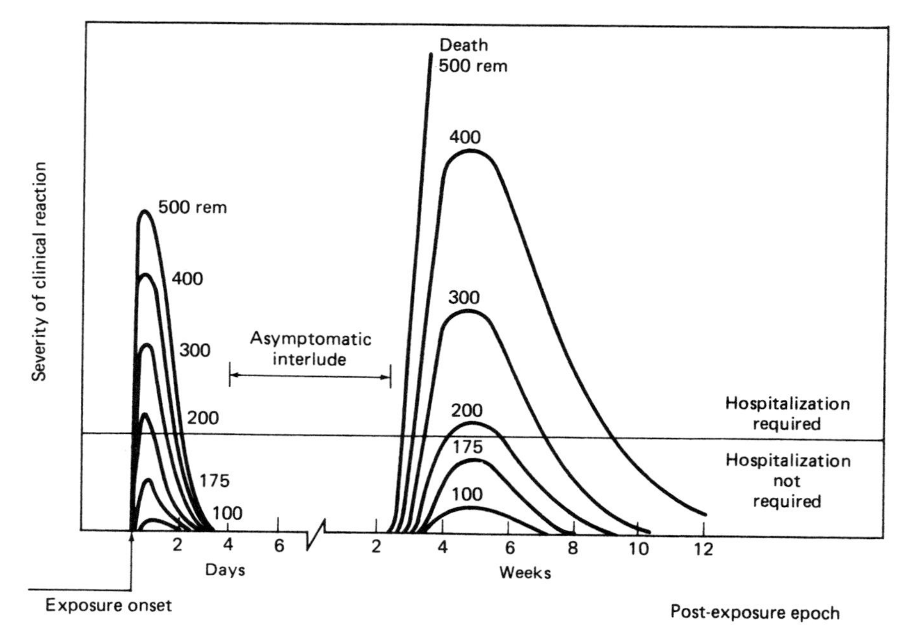

Fig. 13.23 Clinical course of acute radiation syndrome due to various levels of ionizing radiation.[37]

text. Figure 13.22 depicts the superposition of such threat yield-spectrum bounds on the above isodamage curves for which nuclear survivability criteria can be developed for specific equipment. Because maximum values of the salient parameters are probably at the end points corresponding to the threat spectrum limits and at the corners or break points of the governing envelope, then all nuclear weapon effects parameters are computed at the corresponding yield and range paired values. It is assumed that if the equipment can survive at these yield/range paired values, it will survive for all pairs within these limits. These data are now agglomerated to determine the corresponding proper nuclear survivability criteria. The latter do not represent the environments from a single detonation, but are a combination of the most severe environments for each nuclear weapon effect along the governing envelope locus. Thus, material hardened to corresponding specifications can be expected to survive for any yield within the threat spectrum bounds, assuming that a specified percentage of the crew is combat-ready. Ramifications of the foregoing exercise can be repeated for unique situations, such as where the man-machine entity is not paramount. Then other fiduciaries can be used to measure the required survivability. Detailed discussions of such situations are available.[35]

13.5 SYSTEM DESIGN SAFETY MARGINS

A basic concept normally incorporated into the survivability/vulnerability efforts on systems is design safety margins. In man-machine systems, the specified nuclear weapon environment must be relatively moderate. This is because the crew must survive for effective system operation. Hence, it is practical in many instances to incorporate sizable design safety margins with minimum penalties. As can be appreciated at this juncture, design safety margins can often be defined by common sense means. They are usually defined as the ratio between the predicted failure level and the specification level for the particular environment under consideration. Design margin, as used in a statistical context, is discussed in Chapter 14. This approach has particular ramifications for the associated system hardness assurance. In general, when it can be shown that a substantial design margin exists, the degree of control that must be applied at the piecepart level to establish and maintain system hardness is minimal. On the contrary, if the system, device, or component has very little design margin in one or more of its salient parameters, it may be necessary to incorporate stringent procurement/reprocurement controls to ensure that its response distribution is within acceptable limits. Application of special controls could cause significant increase in production costs, lead time, and maintenance costs.

Another aspect of achieving and maintaining system design margins with a measure of confidence, especially for military systems, is to require that pieceparts used be subject to the provisions and requirements of military standards, which are directed toward the procurement of parts with high reliability. The above is usually a contractual requirement for military systems, so that the system maker is enjoined to use such parts. Pertinent governing military standards are embodied in MIL-STD-19500X for discrete devices, and MIL-STD-M-38510X is the corresponding military standard document for integrated circuits.[37] Items listed in these documents are high reliability (HI-REL) and military standard parts, implying that they have been subjected to intensive checks and part screens (100 percent testing of a particular part lot with rejection of failed parts) which minimizes the procurement of marginal devices or components. Most first-line vendors are thoroughly familiar with these military standards. One important result is that the reponse of these part parameters to nuclear weapon radiation effects, including EMP, is usually more uniform and therefore more predictable than those manufactured under conditions different from that implied by the above standards. This uniformity is a key factor in the hardness-assurance programs associated with the systems of interest, as discussed in Chapter 15.

By contrast, it is well to examine system design margins that, in many respects are at the opposite extreme from the manned system and its corresponding moderate radiation specifications. For reentry vehicles (RVs), such as those ejected from an ICBM bus, that in their terminal phase reenter the atmosphere into a hostile nuclear environment, corresponding hardening requirements are extremely difficult to satisfy. One limiting RV vulnerability is neutron-caused melting of the heavy actinide material in its warhead. Another is the melting or detonation of the high-explosive portion of the warhead. Order of magnitude computations put the required incident neutron fluence between 10^{14}–10^{15} neutrons per cm^2. Hence, the total RV hardening specification can emerge by examining the concomitant nuclear radiation effects at the same hostile detonation range that yields the required melting neutron fluence. Such a nuclear hardening specification can be estimated for illustration, as $5 \cdot 10^{14}$ 1 MeV equivalent neutrons per cm^2, 1 Mrad (Si) ionizing dose, and 10^{12} rad (Si) per s for dose rate. Such radiation specifications are so stringent that there are only a few component part technology families that can survive these levels adequately. However, with the incorporation of judicious system design margins, there exist a set of devices from which the electronic system can be fabricated and satisfy the above specifications. Also, there are a number of sophisticated systems built from corresponding part families that approach these radiation specifications. Consolidating this knowledge can probably produce the required RV part set. The following examines the vagaries for meeting these specifications by part type, emphasizing the required system design margins, as synthesized through the design margins incorporated in the component pieceparts.

With regard to various type of diodes, their diversity of characteristics is such that with the use of adequate design margin (derating factors) they can be used in the above environment specifications. This is certainly true for switching diodes. Even though zener diodes evince radiation-induced changes in zener voltage, cautious part choice to include sufficient design margin, plus accommodating circuit design, should provide suitable devices. The forward voltage of rectifier diodes increases with incident radiation, but the use of high speed diodes of higher power dissipation should meet the requirements.

The neutron-induced gain degradation of bipolar transistors can be minimized by choosing bipolars with very high gain-bandwidth product, e.g. with $f_T \geqslant 1$ GHz. Although there a number of work horse signal and switching transistors that satisfy high-level requirements, there are lesser satisfactory hard high-voltage or power transistors available at this juncture. Further, circuit alternatives, such as paralleled transistors, circuit redesign to provide increased gain design margins, or the availability of hardened power transistors must all be considered. In essence, some bipolar hardened transistor development may be in order. Of course, in this high-level nuclear environment, the use of SCRs, unijunction or phototransistors, and most photodevices is out of the question because of their relatively low radiation survivability thresholds.

For digital logic integrated circuits, these RV specifications place the hardness designer between the horns of the following dilemma. The MOS digital ICs can probably survive the neutron fluence level, but would be marginal due to the high ionizing dose requirement. Bipolar digital logic IC families, however, can sustain the ionizing dose specification, but are usually marginal in the face of the neutron fluence level. Ultimately, the direction of the investigation leads to a requirement for a developmental effort that should produce the hardened design and the hardened IC part set, which includes coping with the accompanying upset, latchup, and burnout problems.

With respect to linear ICs, such as operational amplifiers, comparators, and regulators, there are probably several suppliers who produce suitably hardened devices. With sufficient design margin provided by appropriate device choice, some circuit redesign, and possible minor developmental efforts, suitable linear integrated circuits can probably be made available to meet specifications.

13.6 COMBINED RADIATION AND TEMPERATURE EFFECTS

In the derivation of the bipolar common emitter current gain, β, expressed as a function of only physical parameters of the transistor, the equation as given in Section 3.9 does not include very low current effects. These effects are due to (a) recombination in the emitter-base depletion region, and (b) surface recombination effects in regions of the base where the depletion layers have been formed near its surface. When (a) and (b) are incorporated into the above expression, e.g., Eq. (3.133), the result yields two additional terms, which provides a modified gain expression as[38–41]

$$\beta^{-1} = (W^2/2D_b\tau) + (\sigma_b W/\sigma_e L_e) + (SA_s W/A_e D_b)$$
$$+ (WN_b/2D_b n_i)(W_{eb}/\tau_D + SA_s'/A_e)\cdot \exp - (eV_{be}/2kT) \quad (13.19)$$

where

W is the base width
D_b is the base diffusion constant
τ is the base minority carrier lifetime
τ_D is the minority carrier lifetime in the base-emitter depletion region
σ_e, σ_b are, respectively, emitter and base conductivities
L_e is the emitter diffusion length
L_b is the base diffusion length
A_s is the effective area for surface recombination
A_s' is the area of the surface depletion region of the base-emitter junction
A_e is the emitter area
W_{eb} is the width of the emitter-base depletion region
N_b is the base dopant concentration
n_i is the intrinsic carrier concentration
V_{be} is the base-emitter voltage
S is the surface recombination velocity

High-injection (high-current) phenomena, e.g., conductivity modulation and current crowding, are neglected in Eq. (13.19), which is not important for most devices for which design margins have been incorporated for high-reliability applications.

Discarding all but the first term on the right-hand side of Eq. (13.19) can be recognized as being the first step toward the Messenger-Spratt gain degradation expression, as seen from Eq. (5.59) et seq.

For relatively low levels of neutron fluence, moderate gain changes, and moderate temperature changes, Eq. (5.82) adequately represents the combined effects of neutron fluence and temperature. For relatively high values of neutron fluence, the first, or volume recombination, term on the right-hand side of Eq. (13.19) becomes predominant. The temperature dependence of $\Delta(1/\beta)$ becomes nonlinear and simple superposition of fluence and temperature, as in Eq. (5.82), no longer holds. To remedy these fluence and temperature limitations, one can proceed by fitting the temperature variation of the damage constant $K_D = 1/\omega_T K$ in the Messenger-Spratt equation to the three defect level model discussed in Section 5.9. This empirical fit is also discussed in Section 12.2 and is given by[42]

$$K_D(T) = K_D(T_0)[1.1 - 0.1(T/T_0)^4] \qquad (13.20)$$

This is valid for moderate base dopant concentrations (10^{15}–$10^{16}\,\text{cm}^{-3}$) and at low to moderate base current injection levels. For higher dopant concentrations and/or injection levels, K_D experiences only a very slow temperature dependence.

The second term on the right-hand side of Eq. (13.19) is called the *emitter injection efficiency term*. Neutrons have little effect on this term because of the small value of the preirradiation minority carrier lifetime in the emitter. Another reason is that L_e varies only with the inverse square root of the neutron fluence, because $L_e = \sqrt{D\tau_e}$ and $\tau_e \sim 1/\Phi_n$ as discussed in Section 5.9.

When the first term on the right-hand side of Eq. (13.19) is dwarfed by the second term, e.g., when $W^2/2D_b\tau \leq 0.01$, then the current gain is dominated by the emitter injection (second term). This is the situation when no incident neutrons are present. For high-purity silicon, the electron lifetime is about $10\,\mu$s, and the hole lifetime is about $1\,\mu$s. Except for high-voltage transistors, most bipolars are constructed with narrow base widths, typically less than $3\,\mu$m. Therefore, prior to neutron exposure, the initial gain is determined almost entirely by the emitter efficiency, the second term in Eq. (13.19).

The third term in Eq. (13.19) represents surface recombination effects, as discussed in Section 5.9. As mentioned at the outset of this section, the fourth term is the contribution from recombination in the current field region. Now, after exposure to neutrons, the first term of Eq. (13.19) increases as

$$W^2/2D\tau = W^2/2D\tau_0 + K_D\Phi_n \tag{13.21}$$

and similarly for the surface recombination velocity

$$S = S_0 + K_s\Phi_n \tag{13.22}$$

K_s is a surface damage constant that depends on the particular device, including its manufacturing processing. The temperature variation of this term is relatively small. Also,

$$S_0 = \sigma_c v_{th} N_{st} \tag{13.23}$$

where N_{st} is the surface trap density, the thermal velocity $v_{th} \sim T^{1/2}$, and σ_c is the surface trap cross section, so that these surface terms, discussed in Section 1.13, vary little with temperature. In essence, terms containing S as a factor in Eq. (13.19) can be neglected for most devices where surface effects are not important. The temperature-dependence of the field recombination remaining term is also small. This dependence is obtained by using n_i (for silicon) from

$$n_i(T) \sim T^{3/2} \exp - (E_g/2kT) \qquad (13.24)$$

At $T = 300°K$, $n_i \equiv n_0 \cong 1.5 \cdot 10^{10}\,cm^{-3}$, and $E_g = 1.12\,eV$, as discussed in Section 1.5. Using Eqs (5.66), (13.20), (13.21), and (13.24) yields the sought-after combined neutron and temperature effects on the bipolar transistor gain as[45]

$$\beta^{-1} = (\sigma_b W/\sigma_e L_e)(T_0/T)^m + [K_D(T)\Phi_n + (W^2/2D_0\tau)(T/T_0)][1 + A(T_0/T)^r] \qquad (13.25)$$

where $r = ((E_g - eV_{be})/2kT_0) - 3/2$, and $A = (W_{eb}\tau/W\tau_D)(N_b/n_0) \exp - (eV_{be}/2kT_0)$. Values of m range from 1 to about 1.6, and r ranges from 4–8. Eq. (13.25) can be simplified to

$$\beta^{-1} = \beta_0^{-1}(T_0/T)^m + K_D(T)\Phi_n(1 + A(T_0/T)^r) \qquad (13.26)$$

so that for bipolar collector currents below the maximum gain operating point, and for heavy initial neutron exposure, either Eq. (13.25) or (13.26) is valid. As discussed in Section 12.2, for bipolars operating at moderate collector currents and near their maximum gain, Eq. (13.25) can be simplified further to[41]

$$\beta^{-1} \simeq \beta_0^{-1}(T_0/T)^m + K_D(T)\Phi_n[1.1 - 0.1(T/T_0)^4] \qquad (13.27)$$

To use this formula, the room temperature gain β_0 must be known. The $K_D(T_0)$ is obtained from radiation test data or estimated from $K_D = 1/\omega_T K$, where a universal damage constant $K = 1.6 \cdot 10^6$ is employed. The preceding values of m also can be used.

Finally, an approximate conservative expression for the gain degradation is obtained by assuming a temperature dependence for K_D like that for the emitter efficiency term and incorporating low current/surface effects through an empirical constant a, giving

$$\beta^{-1} \simeq \beta_0^{-1}(T_0/T)^m + K_D(T_0)\Phi_n(T_0/T)^a \qquad (13.28)$$

where $a \simeq 1.5$ as obtained from experimental data. A successful experimental program was undertaken to verify these gain formulas.[46]

To recapitulate, some approximate models have been used to extend room temperature bipolar transistor gain behavior to temperature extremes and up to moderate neutron fluence. Transistor gain specifications can be extrapolated without recharacterizing device types for temperature or neutron effects. The latter simpler formulas can be inaccurate at extremely low currents for bipolars where surface effects are significant. However, the underlying model manifests good agreement with experiment over two orders of magnitude of operating current levels and so can be used as a reliable design and analysis tool.

13.7 SYSTEM-HARDENING METHODS: RESET

The next five sections provide typical system-hardening methods and techniques analogous to the component and circuit-hardening techniques discussed in the previous chapter.

The term *reset* implies a scheme for resetting or restarting an electronic system after its temporary malfunction for various reasons. For example, for a special-purpose computer that has experienced a circumvention, the restart of normal operation involves transfer of control to the start of the recovery program. The recovery program performs several functions that include: the recovery of lost time and lost sensor inputs; restarts, ongoing system functions such as certain voltages, discrete signals, etc.; determines a program rollback point; reinitializes program variables and returns program control to a predetermined rollback point. Once this recovery phase is completed, the computer program is restarted. However, the reinitation scheme must provide for lost information contained in the volatile memory. How this is done is discussed in detail in an ensuing section on circumvention.

13.8 SYSTEM-HARDENING METHODS: REDUNDANCY

Even without the impact of the deleterious effects of incident radiation, as the complexities of electronic systems increase, their reliability, can be expected to decrease. One of the techniques for coping with this situation is to use corresponding electronics in redundant configurations. It is well known that for a given noisy information channel capacity, it is more efficient to repeat individual message symbols than to repeat groups of symbols or the whole message. In the same vein, it is a truism that for nonmaintained systems, such as spacecraft, greater payoffs are obtained by employing redundancy at successively lower levels of system design, as illustrated below.[47] However, for standby-maintained systems, the best payoff is gotten from high-level redundancy. Most of the electronics systems considered herein tend toward the former, that is, they are essentially unattended, such as the avionics in an air-to-air missile or satellite electronics.

The general redundancy problem under consideration is that of the allocation of redundancy in electronics systems subject to incident radi-

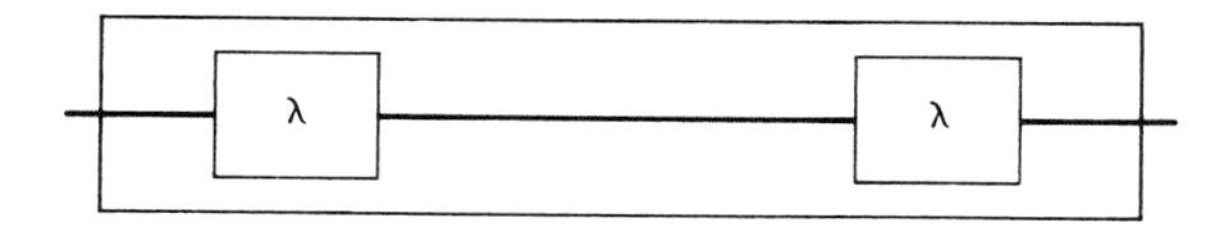

Fig. 13.24 Two units in cascade with failure rates of λ per hour.

ation, with minimal cost. Consider a rudimentary system of two units in cascade (in series) as shown in Fig. 13.24, both with failure rates of λ per hour. For example, let $\lambda = 0.05$ per h, interpreted as a 5 percent probability of failure per hour or with a mean time before failure (MTBF) of λ^{-1} or 20 h.

It is evident that the mean failure probability in either unit in an incremental time Δt is $\lambda\Delta t$. The probability of not failing in Δt hours, i.e., the incremental survival probability is $1 - \lambda\Delta t$. Continuing, the probability of surviving N successive intervals of time, each of Δt hours, such that $N\Delta t = T$ hours, is simply $(1 - \lambda\Delta t)^N$. The limiting survival probability, or reliability R_1, as N becomes very large and Δt becomes very small, with $N\Delta t = T$, gives

$$R_1(\lambda T) = \lim_{N\to\infty} (1 - \lambda T/N)^N = \exp - \lambda T \tag{13.29}$$

For the electronics equipment consisting of the two units shown in Fig. 13.24, the failure rate is 2λ (for cascaded units), and the corresponding reliability $R_2 = R_1^2 = \exp - 0.1t$, since $\lambda = 0.05$ per unit. Now, to improve the system reliability through redundancy, either unit can be duplexed or the whole system can be duplexed in its entirety. If the latter is chosen; then $R_{\mathrm{ent}} = 1 - (1 - R_2)^2$ or

$$R_{\mathrm{ent}} = 2(\exp - (0.1t)) - \exp(-(0.2t)) \tag{13.30}$$

because if the whole system is paralleled with its twin, $(1 - R_2)^2$ is the probability that both units fail. Then the probability that neither fails, R_{ent}, is given by Eq. (13.30). If each unit is separately paralleled however, then

$$\begin{aligned} R_{\mathrm{ea}} = (1 - (1 - R_1)^2)^2 = 4(\exp - (0.1t)) - \\ 4(\exp - (0.15t)) + \exp(-(0.2t)) \end{aligned} \tag{13.31}$$

because $(1 - R_1)^2$ is the probability that either of the two duplexed units fails; the duplexed unit reliability is now $1 - (1 - R_1)^2$ or, for both paralleled units, R_{ea} is given by Eq. (13.31).

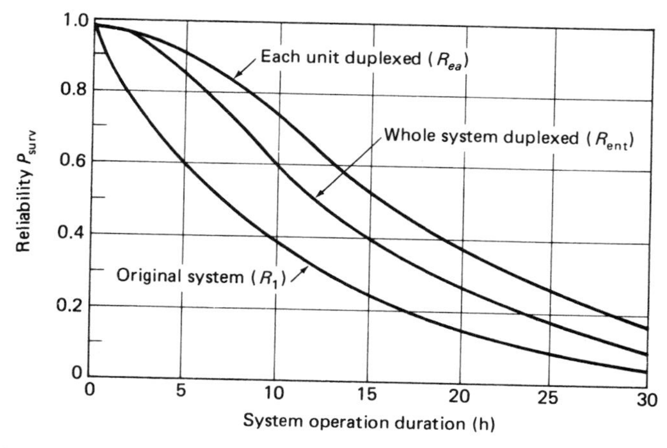

Fig. 13.25 Reliability for high- and low-level redundancy.[48] © 1963 by Prentice-Hall.

Figure 13.25 shows that the case of each unit paralleled consistently yields a higher reliability than duplexing the entire system. Hence, the lower the level of redundancy allocated, the better is the survival probability. It is a straightforward matter to show that the opposite is true for manned equipment. That is, the equipment reliability is higher if the whole system is duplexed when it is manned on a standby basis.[48]

Now consider a very large computer system containing a 1000 IC chip memory. Each IC consists of a 64-K bit RAM, with the whole enduring single-event-upset errors. Assume that the failure rate of each chip is 1 per million hours, or the chip failure rate is $\lambda_c = 10^{-6}\,\mathrm{h}^{-1}$. Assume that the mean failure rate per chip bit is $\lambda_b = 10^{-6}/64000 = 1.6 \cdot 10^{-11}$ (hours)$^{-1}$ or the mean time before failure (MTBF) per bit is λ_b^{-1} or 7.5 million years. For this system then, a failure might occur once every $(1000\lambda_c)^{-1} = 10^3$, or about once every 1000 hours (6 wk), so that the computer failure rate $\lambda_{\text{comp}} = 10^{-3}$ (hours)$^{-1}$.

How should redundancy be allocated within the computer to achieve a given reliability at minimum cost, to abate the single-event upsets within a given maintenance period? Assume that the redundant chips will be used in a parallel arrangement, and that an EDAC system will, when needed, switch from the stricken chip to its redundant twin if there is one. Then the reliability of the computer without redundancy is

$$R = \prod_{i=1}^{n} p_i; \qquad p_i < 1 \tag{13.32}$$

where p_i is the reliability of the ith chip, and the generalized resource allocation cost is given by

$$C = \sum_{i=1}^{n} c_i \tag{13.33}$$

where c_i is the cost of the ith chip. Because $p_i < 1$, it is appreciated that the reliability of each chip must be greater than the overall reliability R, as seen from Eq. (13.32). For a given ith chip, paralleled to achieve an enhanced reliability R_c for each i, the number of paralleled redundant chips n_i required per ith chip is simply gotten from

$$R_c = 1 - (1 - p_i)^{n_i} \tag{13.34}$$

assuming an identical parallel arrangement for all chips, so that

$$n_i = \ln(1 - R_c)/\ln(1 - p_i). \tag{13.35}$$

Assume also that the cost of allocating redundancy is linear with the number of required redundant chips, where the chip price is c_i. Then the cost of adding the redundant chips, $C_k = \sum_{i=1}^{k} C_i n_i$, where k is the total number of redundant chips added. The cost per added chip is then

$$c_k = \sum_{i=1}^{k} c_i n_i / \sum_{i=1}^{k} c_i \tag{13.36}$$

Suppose that in between maintenance periods, T_m, in space, the computer single-event upset error reliability goal is set so that $R_c = \exp - (\lambda_{\text{comp}} T_m) = 0.955$, and that all chips are identical with $p_i = 0.96$ for each. Then from Eq. (13.35) for all i

$$n_i = \ln(1 - 0.955)/\ln(1 - 0.960) = 0.9634 \text{ per chip} \tag{13.37}$$

Hence, for the above 1000-chip memory, about 963 additional chips must be added to achieve the above reliability R_c. This, of course, implies an alternate memory. Of course, the chips in an actual computer are not all identical, so that the redundancy rules would be more varied. The details for determining such rules are well known.[51]

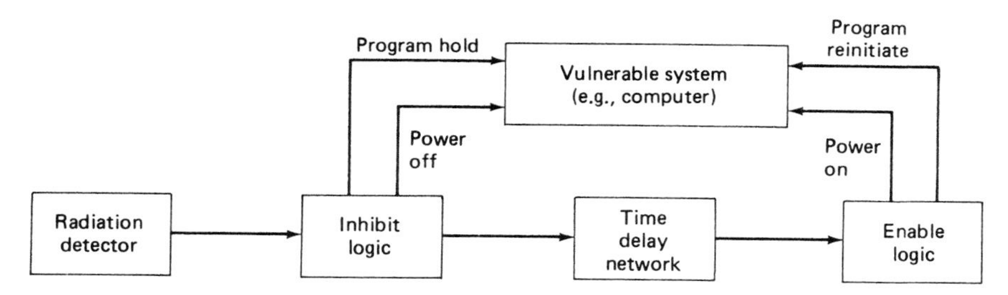

Fig. 13.26 Signal flow diagram of circumvention system.[53] © 1972 by John Wiley & Sons.

13.9 SYSTEM-HARDENING METHODS: CIRCUMVENTION

Circumvention is defined as the electronic process where the system is transferred to a temporary state where all input and output (I/O) ports are clamped to ground at the onset of a radiation pulse. When the pulse has subsided, the combined software and hardware cause the system to revert to its state prior to the pulse event to continue its function. As alluded to earlier, circumvention is used when the incident nuclear pulse amplitude approaches the logic upset level. The circumvention threshold is usually set at a factor of ten lower than the lowest logic upset threshold in the system. This corresponds to a design margin of ten times (10×).

Radiation from an incident nuclear event on an electronic system induces spurious voltages to cause currents to flow that can disrupt proper function. Specifically, these include logic state changes in devices and subcircuits, which alter the contents of registers and memories. To underscore a point elaborated on below, the incident radiation can cause a memory modification of any memory word being accessed by the computer central processing unit expressly at the time that the effects of the nuclear event are incident on the system. In essence then, detection, protection, and recovery measures must be taken to prevent the heart of the electronic system, namely, the computer, from going permanently awry. This combination of protection measures can be taken as another definition of nuclear event circumvention.[52] As implied above, circumvention is useful for system computers that essentially cannot tolerate any data loss, as might occur during nuclear radiation-induced disturbances.

Circumvention is usually accomplished through the use of both software and firmware. The latter consists of additional circuitry to be discussed (ROMs, PROMs, etc.), as well as nuclear event detection system inhibit and recovery programs. Figure 13.26 depicts a flow diagram of a general circumvention system.[53] It consists of a radiation detector that senses the presence of a prompt photon pulse incident on the system. Prompt gamma photons are those directly out of the nuclear detonation during its earliest epoch. The detector then emits a signal of about 0.1–10 ms duration. This detector signal inevitably finds itself in a "race" with, many gamma-ray-pulse-induced signals throughout the system. Those extraneous signals may affect the circuits to be controlled by the circumvention process. The radiation detector output signal must win the race (and it does) to cause the inhibit logic to "freeze" the computer memory store. In particular, destructive readout (DRO) memory current drivers (circuits to be circumvention controlled) would be clamped before the unwanted gamma-ray signals would cause them to randomly overwrite the memory contents. In this description of a circumvention system, it is assumed that the computer memory per se is inherently radiation hard, such as those using magnetic cores. However, only for the particular DRO memory cell that was being read or written on at the time of circumvention, would the possibility of it containing spurious information exist due to the nuclear event. After the time delay of a few milliseconds has passed, and with it any prompt effects due to the nuclear event have subsided, the computer resets and resumes normal operation.

For a DRO core memory, under the assumption that circumvention took place during a read-restore operation, it is desired to restore the affected memory cell to the value it was storing prior to the nuclear event. This can be done using an error detection code method using a check sum. The memory words are divided into blocks, with some bits per block sacrificed to store the check sum. The latter is composed by taking the 2s complement of the arithmetic sum of the words in the block proper. The 2s complement of any binary number is obtained by replacing its 0s with 1s and vice versa and then adding 1 to the result. For example, the 2s complement of the assumed block word binary number 101 is 010 + 001 or 011. If all words in the block are intact, then the arithmetic sum of the block words plus the check sum equals zero. In the above example, 101 + 011 = 000 to three places. Then the remainder of the circumvention recovery cycle ensues. If this sum is not 0, then the particular memory cell addressed at the time of the nuclear event is in error, because all others were and are "safe" in the hard memory. The error is corrected by subtracting the corresponding check sum from it, and inserting the result back into the affected cell. There exist a number of variations on check sum schemes, as well as circumvention methods. However, it is necessary to know which cell, i.e., its address, was being accessed at the onset of the nuclear burst. Because the circumvention cycle freezes the computer's hardened memory address register as well, inspection of this register by the software provides the address of the sought-after cell.

As can be appreciated, this technique is susceptible to a second nuclear event occurring any time during the circumvention recovery phase. Also, the nuclear radiation detector sets a "nuclear" flag to aid the system in determining whether or not an actual nuclear event occurred or whether a nonnuclear disturbance caused the system to initiate a false alarm. The reset or restart of normal computer operation following a circumvention involves the transfer of control to a point in the software sequence that existed prior to the nuclear event. This is discussed in Section 13.7.

13.10 SYSTEM-HARDENING METHODS: COMPUTER MEMORIES

The previous section described the circumvention method for radiation protection of destructive readout (DRO) computer memories. Besides the various types of check sums that can be computed, as discussed in the previous section, there are other methods for protecting memories, including semiconductor volatile memories, which are those that lose their information store upon loss of power. Most of these methods center on the use of a combination of redundancy measures, shielding and hardening methods, and error detection and correction (EDAC) methods.

With respect to the choice of semiconductor device type to use for the memory, recourse must be had to its specific application. For illustration, assume that its use is in a satellite space station where power is at a premium. Under these circumstances, the most attractive memory device technology is CMOS, discussed in Sections 6.9 and 12.3. CMOS power consumption is extremely low, of the order of 1 μW per bit. From the aspect of damaging radiation, the major disadvantage of CMOS is its susceptibility to ionizing dose effects. To deal with this difficulty requires a concerted effort consisting of shielding, choosing specially hardened parts, and possibly using CMOS/SOS or CMOS/SOI. Shielding computations should normally lead to convenient mechanization of sufficient shielding to handle most secularly changing natural space environments without undue space or weight penalties. To obtain the remaining design margins, specially hardened parts are obtained by selection through sample lot and 100 percent screens of electrical and radiation test data. An example of static levels of CMOS drain leakage current as a function of incident ionizing dose is depicted in Fig. 13.27, which provides insight into the spacecraft internal dose levels and leakage currents involved.[54]

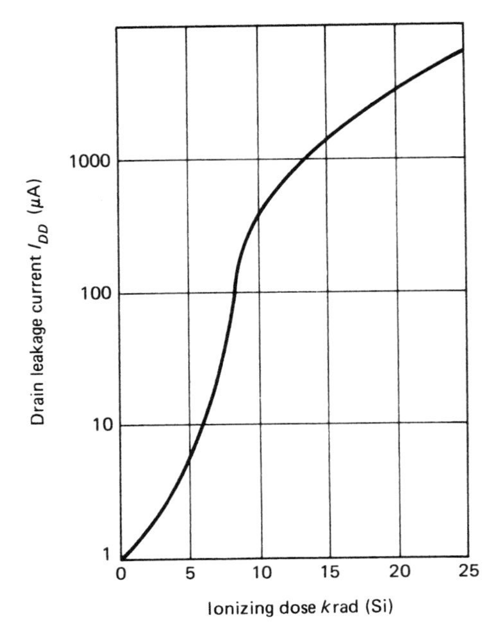

Fig. 13.27 4KX1 bit CMOS RAM leakage current as a function of ionizing dose.[54] © 1978 by the IEEE.

Corresponding satellite "incident dose" levels are one to two orders of magnitude greater than the internally tolerated dose.

A major effect of the incident radiation on the CMOS device in this context is to increase the static drain current, as in Fig. 13.27 and slightly decrease its memory access time. As opposed to CMOS memories, spaceborne CPUs often utilize bipolar logic, e.g., schottky TTL logic because of higher data-processing speed. Bipolar logic is essentially insensitive to reasonable levels of naturally occurring ionizing dose.

With regard to the achievement of increased memory reliability through the use of redundancy measures, both high- and low-level redundancy can be used, even though the system is unattended, as discussed in Section 13.8. High-level redundancy comes about simply by providing a complete standby twin memory, in case of a catastrophic failure of the operating main memory. Also, each memory can be buttressed with EDAC capability with check sum schemes as described in the previous section. This capability allows the memory to continue to operate in the event of a failure of one of the RAM chips in the memory array. If a second RAM fails, immediate transfer to the twin unit, which also possesses the EDAC feature, takes place. These measures produce a marked increase in overall memory and system reliability, compared with using either of these measures alone.

The corresponding quantitative analysis can proceed through the use of the U.S. Department of Defense handbook, *The Reliability of Electronic Equipment*, MIL-HDBK-217, to obtain individual failure rates, λ_i, of the whole spectrum of electronic components and accessories used in the construction of the CMOS memory circuit card. After compartmentalizing the components and devices, each with its particular rates, the gross failure rate can be computed for each, using the convenient formulas in the above handbook. This is finally consolidated into an overall computer reliability probability number, which is a function of all the redundancy and EDAC techniques employed.[55] With only moderate levels of redundancy sophistication, total reliabilities approaching unity can be obtained for computers that use the foregoing memories in applications as discussed.

In the continuing quest for ultra-fast access, radiation-hard, nonvolatile, and NDRO (nondestructive readout) memories for many important system applications, ferroelectric memories have reemerged as a possible candidate.[56] The memory storage potential of ferroelectric materials has been appreciated for almost a half century. However, their initial integration into the monolithic fabrication of semiconductor integrated circuit technology has occurred only within the last decade or so.

Ferroelectric crystalline materials exhibit an electric dipole moment, even in the absence of external field excitation. The aggregate dipole domains within the crystal can be shifted into either of two directional polarization states, depending on the direction of the electric field excitation. The polarization state is read by applying a voltage pulse of given polarity across the ferroelectric capacitor,[57] described in the following paragraph. The corresponding current response to this voltage pulse depends on the previous polarization state of the ferroelectric material. This can be described by a polarization state-field hysteresis (charge/voltage) loop, closely similar to the ferromagnetic hysteresis loop. There is an upper bound on the ferroelectric crystal temperature (curie temperature) beyond which the memory effect is essentially destroyed. The most promising materials are the ionic ferroelectric crystals of the perovskite family, which have the simplest crystalline structure possessing ferroelectricity. These are mainly titanates, e.g., lead zirconate titanate (PZT), barium titanate ($BaTiO_3$), and lead titanate ($PbTiO_3$), the latter two having curie temperatures of 120°C and 490°C, respectively.

A nonvolatile 512-bit MOS DRAM has been built that exploits the electrical polarization states of the ferroelectric storage capacitors to store

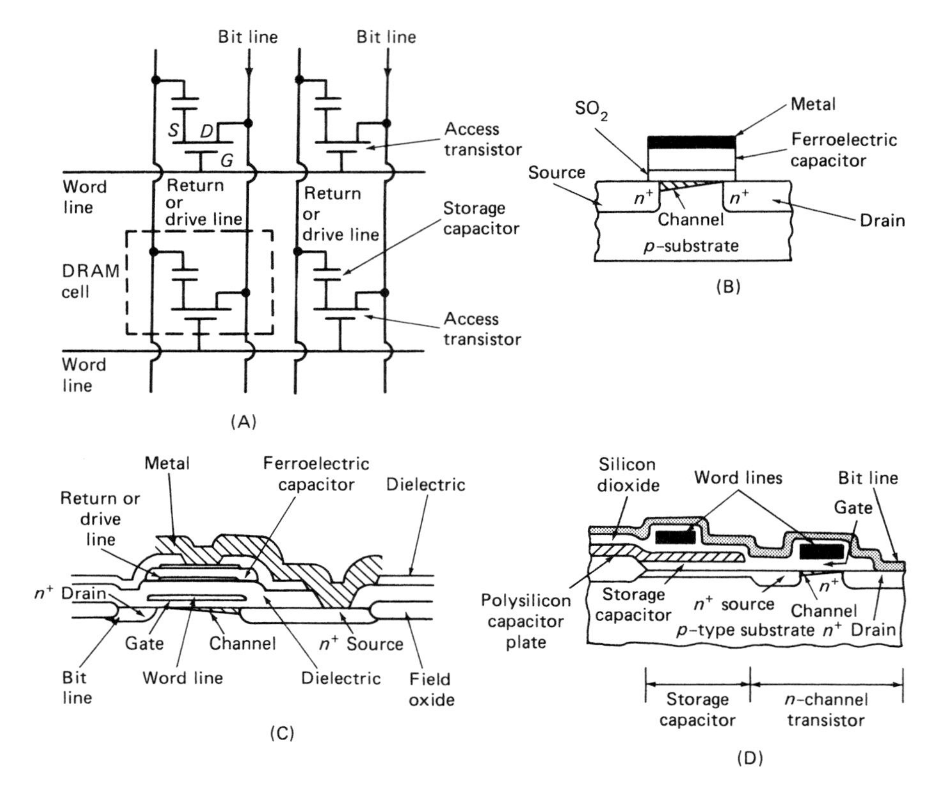

Fig. 13.28 A, Array of four DRAM cells. B, Ferroelectric FET prototype.[57] C, Section through ferroelectric DRAM cell.[57] D, Section through DRAM cell.

information.[57–59] Write speeds between 60–200 ns have been obtained, with the write speed being limited by the intermediate peripheral circuitry.

Figures 13.28C and 13.28D compare the structure of the common MOS DRAM cell with the ferroelectric storage capacitor version.[57] Figure 13.28A depicts a DRAM fragment consisting of four DRAM cells, showing the bit, word, and return or drive lines. Careful scrutiny of these figures reveals that the schematic diagram of Fig. 13.28A and the structures of Figs. 13.28C and 13.28D correspond. Figure 13.28B shows the ferroelectric FET (FEFET) prototype structure.[61] As alluded to above, it is seen from these figures that it is fairly straightforward to modify the design and manufacturing processing of the MOS DRAM cell by insertion of the ferroelectric crystal thin-film material to form the storage capacitor.

The ferroelectric memory cell, consisting of one MOS transistor and one storage capacitor, like the DRAM cell, has a potential for memories of very high packing density. Since the ferroelectric capacitors are non-volatile, no memory refresh charge is necessary. As mentioned, the polarization state is detected by a voltage pulse of given polarity across the capacitor. The amount of capacitor charge perturbed by the pulse depends on its state of polarization. The ferroelectric charge state dissipates very slowly with time. The corresponding memory cells are also relatively insensitive to junction leakage and subthreshold transistor conduction.[57]

Like EEPROMs (electrically erasable programmable read-only memories) ferroelectric memories can be employed in applications where their contents must be written many times. Their read and write times are nearly equal, whereas write times for EEPROMs are typically much longer than read times.

The radiation vulnerability of ferroelectric DRAMs is mainly determined by the ferroelectric capacitor.[60] For example, it is essentially unaffected by SEU because of the relatively large ferroelectric polarization state charge, assuming commensurate ferroelectric capacitor size. Table 13.5 provides calculated and measured threshold levels for ferroelectric DRAMs for the pertinent radiation environments.

In recapitulation, ferroelectric memory cells have been fabricated using processes compatible with semiconductor VLSI manufacturing techniques. These memories can be made NDRO (Non-Destructive Readout) for use in various strategic systems, using several techniques; the most practical is probably a rapid read/restore in combination with EDAC (error detection and correction) software. These memories may replace plated wire, and will have substantial advantages in cost, weight, size, power consumption, and speed. In the future, they may provide a practical cost-competitive solution to the need for nonvolatile RAMs in hardened tactical, avionic, and space systems.

EPROMs (Erasable Programmable Read Only Memories) and EEPROMs (Electrically Erasable PROMs) have enjoyed only limited hardening considerations to date. EPROMs are usually erased by exposing them to UV radiation through their optically transparent package lids. EEPROMs are electronically erased through the use of associated circuitry. Since both of these MOS devices utilize a similar gate structure, their radiation response is similar as well. Their ionizing dose failure threshold can range from merely 5 krads (Si) to 100 krads (Si), the latter attained through the use of special circuits and certain hardening ap-

Table 13.5 Comparison of theoretical and experimental radiation data for ferroelectric DRAMs.[57] © 1987 by the IEDM.

RADIATION STRESS	NEUTRON DAMAGE (Neutrons/cm^2)	IONIZING DOSE-RAD (Si)	IONIZING DOSE RATE-RAD (Si)/s	SEU–^{252}Cf FLUENCE (Fission Fragments/cm^2)
Estimated theoretical threshold	$5 \cdot 10^{16}$	10^8	$2.5 \cdot 10^{13}$	Probably no SEU damage
Experimental data for no observable damage	10^{14}	10^7	$1.2 \cdot 10^{11}$	$6 \cdot 10^6$* ($\sigma_{SEU} < 1.6 \cdot 10^{-7}\,cm^2$)

* Zero SEU measured.

proaches. These devices are most sensitive to ionizing dose, so that major hardening attention should be so focused, subordinating neutron fluence, dose rate, and SEU effects.

One family of these devices is termed electrically floating gate systems, in that the gate includes a polysilicon layer floating gate, insulated from the control gate and the silicon channel. Silicon dioxide layers are used to insulate the polysilicon layer in sandwich fashion. A "one" is stored on the floating gate by increasing the electron charge on the polysilicon, through tunneling or channel hot-electron injection,[130] to establish a "high" threshold voltage for the n-channel transistor. Contrariwise, a "zero" is stored by removing electrons, and also by tunneling or with U-V light[130] and/or adding holes, to establish a negative "low" threshold voltage for the device. Another family, MNOS or SNOS (metal or silicon-nitride-oxide-silicon) will be discussed later.

Incident ionizing radiation tends to oppose the above charge states by reducing or increasing electrons corresponding to the high and low states, respectively. This results in an intermediate charge state, tending to obliterate the stored information. Unhardened floating gate memories can be used in tactical and avionic systems, as long as a substantive hardness assurance program (HAP) is in place, as discussed in Chapter 15. For low to moderate ionizing dose levels in spacecraft environments, they can operate satisfactorily with individual device shielding, as required, in concert with a viable HAP. However, these memories are unsuitable for use in severe radiation environments, such as those corresponding to certain hostile scenarios.

To quantify the preceding device information processing aspects, the threshold voltage, a key device state parameter is given by

$$V_T = V_{T0} + q_s d/\varepsilon \tag{13.38}$$

where

V_{T0} = the initial transistor threshold voltage.
q_s = gate surface charge density.
d = oxide thickness between the control and floating gate.
ε = oxide dielectric constant.

The three processes that contribute to the net loss of charge due to deleterious ionization on the floating gate are: hole injection from the oxide layers on each side of the floating gate, hole trapping in these oxides, and electron emission from the floating gate. For relatively low values of electric field E in floating gate memories, hole injection and

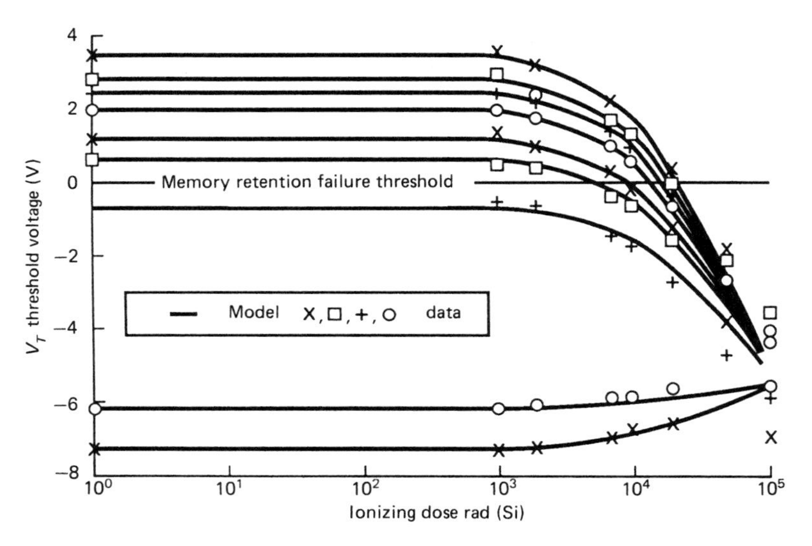

Fig. 13.29 Predicted and measured threshold voltage shift versus ionizing dose for a floating gate memory device.[100] © 1989 by the IEEE.

hole trapping, generally represented by the collection efficiency[100] function $f(E)$, are proportional to the field E in the sense that

$$f(E) = 1 - \exp - kE \cong kE \text{ for } E \leqslant 0.5\,\text{mV/cm} \tag{13.39}$$

For high levels of ionization radiation, the three processes achieve an equilibrium, characterized by an equilibrium threshold voltage V_{TE}.

As discussed earlier, ionizing radiation essentially decreases the threshold voltage span $V_T(D)$ between the "one" and "zero" states. As a function of the ionizing dose level D, in rads (SiO_2),

$$V_T(D) = V_{TE} + (V_{T0} - V_{TE}) \cdot \exp - \alpha D \tag{13.40}$$

where α is a constant characteristic of the specific device manufacturing process, in that it reflects the type and energy level density of the traps in the oxide regions. This straightforward expression yields good agreement with experiments using only the fitting parameter α, as shown in Fig. 13.29. The above asserts that, basically, the charge on the floating gate is well modeled as charge on a parallel plate floating gate capacitor. This charge is proportional to the ionizing dose, and its subsequent change is closely approximated as being proportional to the electric field.

Insufficient knowledge is available at this juncture concerning the detailed fundamental physical nature of the aforementioned oxide traps, in that their effective muting is essential to harden floating gate memory devices.

It should be noted that the UV radiation used to erase the EPROM is tantamount to a heavy dose of ionizing radiation that can result in the creation of an equilibrium threshold voltage V_{TE}, which obliterates the stored information.

With respect to ionizing dose rate damage, the corresponding induced-charge changes on the floating gate occur extremely fast, and so are in phase with the incident radiation pulse. Hence, any change in threshold voltage will depend only on the ionizing dose, and not on the corresponding dose rate. The per-pulse dose is approximately equal to the product of the peak dose rate and the pulse width, (full width-half maximum). In most applications, the associated logic circuitry of these systems will upset at dose rate levels much less than those that might affect these memories per se. Hence, they will be sufficiently hard as compared to the peripheral digital logic.

Normally, reading a floating gate memory is achieved by comparing a read mode voltage with a reference level. For example, above this level, read "one," while below this, read "zero." Schemes to increase the ionizing dose hardness level would include the use of a differential sensing circuit, memories that could be set to a significantly higher "one" threshold by increasing the write cycle drive, and using hardened oxide technology for the floating gate insulating oxide layers. It is certainly conceivably possible to increase their radiation failure thresholds by an order of magnitude using these methods.

The present relatively low ionizing dose failure threshold for floating gate devices makes it unnecessary to evaluate the neutron failure threshold in detail. For any practical radiation scenario, the ionizing dose failure threshold would readily be reached, while the corresponding neutron fluence was below the level which would produce any measurable changes. A semiquantitative neutron failure threshold can be estimated as approximately $10^{14}\,n/cm^2$, at which level the accompanying ionizing dose would be approaching the failure threshold. At $10^{15}\,n/cm^2$, the transconductance and channel resistance would also show significant degradation from this level of neutron fluence.

Floating gate memories are not directly susceptible to single event upset (SEU), since the ionizing charge released by even the highest-energy (LET) cosmic rays in the vicinity of the floating gate will not cause a significant change in its threshold voltage. Again, it must be noted that SEU in the peripheral controlling logic could result in an error.

Floating gate memories have a long-term memory retention problem (Fig. 13.29), in that the charge tends to leak off the floating gate under normal operating conditions. This is exascerbated by incident radiation. Both of these difficulties can be mitigated by using a refresh cycle similar to that used for dynamic memories. In fact, it is probably necessary for space applications, in which the ionizing radiation accumulates over time.

Ionizing radiation effects are well known to be strong functions of the ambient temperature. The radiation analysis and characterization of floating gate devices has primarily assumed room temperature. This is probably adequate for many applications, but it should be kept in mind that elevated temperatures will significantly enhance both normal and radiation-induced leakage currents.

SNOS and MNOS memory function is described in Section 6.11. SNOS is sometimes used as a collective noun to mean both SNOS and MNOS devices, although "NOS" is frequently used as well. Adjustment of the oxide and nitride layer thickness makes the device an effective EPROM. The SNOS device is similar to the floating gate device. However, the charge that controls the memory state consists of a sheet charge at the nitride-oxide interface. SNOS memories have been built that achieve ionizing dose failure thresholds in excess of one megarad (Si).

As in the basic MOSFET, the effect of ionizing radiation is to create hole-electron pairs in the gate oxide and nitride. Here, this charge drifts to the oxide-nitride interface, where it reduces the threshold voltage if the gate is initially in the "one" state, and increases the threshold voltage if the gate is initially in the "zero" state, thus reducing the "window" voltage span between these states. At very large values of ionizing dose, the threshold voltage reaches an equilibrium value and the stored information is lost. Charge storage must occur in a sheet of charge at the interface between two insulators of different dielectric constants in order to satisfy fundamental electrostatics, as discussed in Section 2.7. The dominant radiation effects problem is due to ionizing dose, and so a detailed discussion of dose rate, neutron, and SEU radiation effects is not necessary at this juncture. Write voltages in excess of 20 volts are often used, and burnout has been observed in some devices from single high-energy ions, when write voltage is applied to the gate.

For SNOS devices, a composite expression that combines the effects of normal charge leakage and the effects of radiation-induced leakage on the threshold voltage $V_T \equiv V_T(D, t)$ has been obtained.[97] It is, for $t \geqslant t_0$,

$$V_T(D, t) = (V_{T0} - V_v)(t/t_0)^{-a} \cdot (\exp - \alpha D) - B \int_{t_0}^{t} D'(\tau)(t/\tau)^{-a} \cdot (\exp - \alpha(D(t) - D(\tau))) \cdot d\tau + V_v \tag{13.41}$$

where

a is a tunneling constant.
α^{-1} is a SNOS device time constant
V_v is the threshold voltage before the device has been written, erased, or irradiated.
$V_{T0} \equiv V_T\,(0, t_0)$
t_0 is the time immediately after a write pulse when V_{T0} is measured (typically several seconds).
D is the dose level
$\dot{D}$ is the dose rate.
B is a constant reflecting the normal charge leakage from the gate.

In the absence of radiation, Eq. (13.41) can be written to obtain

$$a = \ln((V_{T0} - V_v)/(V_T(0, t) - V_v))/\ln(t/t_0); \qquad t \geqslant t_0 \quad (13.42)$$

from which the value of a can be determined by measuring $V_T(0, t)$ at two different times following t_0, for example at 5 and 50 seconds. An expression for the threshold voltage of a device irradiated in the "virgin" state (i.e., before it has been written or erased) is[97,99]

$$V_{th}(D, t) = B(\exp(-\alpha D) - 1) + V_v; \qquad t \leqslant t_0 \quad (13.43)$$

where α and B can be obtained by irradiating this device at two radiation levels such as 250 Krads and 1 Mrad. At this point, Eq. (13.41) can be used to model the time and radiation performance of an SNOS memory device. Good agreement between this model and the experimental results are obtained as shown in Fig. 13.30.

With respect to dose rate effects, SNOS memories store such a relatively large amount of information-bearing charge at the oxide-nitride interface that most prompt dose rate pulse-induced ionization does not degrade the memory capability. Therefore, the dose rate upset threshold of the SNOS memory will not be determined by the memory per se, as mentioned above, but by the associated read-write ciruitry, as in floating gate memories. When SNOS devices are properly designed into computer systems, they yield EPROMs with excellent radiation characteristics.

With respect to neutron fluence, SNOS memories have a high degree of inherent hardness to ionizing radiation, and so their response to neutrons usually depends only on displacement damage effects. Since they are majority carrier devices, the only effect of minority carrier lifetime degradation is a small increase in leakage currents. The dis-

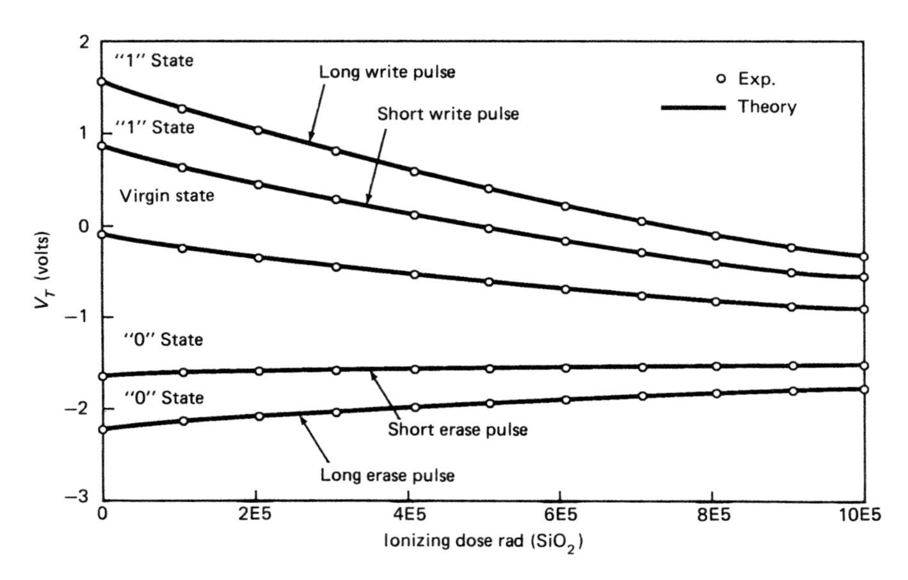

Fig. 13.30 Predicted and measured threshold voltage versus ionizing dose for a SNOS memory device.[99] © 1986 by the IEEE.

placement damage effects of concern are increases in resistivity and decreases in channel mobility. These effects were discussed earlier, and the data provided therein can be used to predict the degradation in transconductance and switching speed, and increase in channel resistance. The neutron failure threshold is usually between 10^{14} and 10^{15} neutrons per cm^2. The response of the transconductance g_m can be obtained as discussed in Sections 11.2 and 12.2, viz., for reasonable values of fluence,

$$g_m = g_{m0}\exp - (K\Phi_n) \cong g_{m0}(1 - K\Phi_n) \tag{13.44}$$

where K is a function of the SNOS channel resistivity. The increase in channel resistance and cutoff frequency $f_c = g_m/2\pi c_g$ is obtained in a similar fashion,

$$f_c \cong f_{c0}(1 - K'\Phi_n) \tag{13.45}$$

where K' is the sum of K and the mobility degradation constant for the appropriate channel resistivity. K and K' are often obtained as fitting constants from radiation experiments, as discussed in Section 5.9.

For cosmic ray environments, a single cosmic ray particle cannot produce enough ionization to deleteriously affect the nitride-oxide interface. However, SNOS devices have been observed to fail catastrophically when exposed to high-energy heavy-cyclotron ions simulating cosmic ray exposure, and with the gate biased in the write mode. This can present a very serious problem in space applications.

The hardening of SNOS memories has proceeded by optimizing the manufacturing process to improve the already very good inherent hardness of the nitride-oxide gate dielectric. Since these devices are used in hardened microprocessors and computers, a large effort has been required to harden the associated microcircuits to a level consistent with the SNOS devices. A circumvention procedure is usually required, for strategic and military space applications as discussed in Section 13.9. For most tactical applications, it is probably not necessary.

Besides ferromagnetic memories, discussed earlier, magnetic memories include ferrite cores, plated wires (magnetic alloy-coated copper wire), and magnetic bubble devices. Due to their high cost, they have received relatively little attention. Even though all of these magneto-inductive memories are inherently hard, with respect to the usual spectrum of radiation environments, they must all be protected from invalid write commands. Further, they cannot be read in appreciable dose rate environments, comparable to the upset levels of the associated circuitry.

A formidable family of radiation-hard nonvolatile memories are state-of-the-art CMOS/SOS and CMOS/SOI RAMs with battery backup. They are, in actuality, volatile, but with battery backup to maintain the memory charge states they can operate-through temporary power supply failures. They possess similar cell density, speed, and power requirements as their associated I/O circuits.

13.11 SYSTEM-HARDENING METHODS: MICROPROCESSORS, COMPUTERS

A microprocessor is usually defined as a whole computer including RAM and ROM memories, central processing unit, and I/O ports, constructed on one (relatively large) chip. Microprocessors of concern here are mainly considered hard or hardenable to sustain substantive levels of incident radiation, as opposed to commercial microprocessors, most of which are quite radiation-sensitive (soft). The most popular commercial microprocessor LSI digital technology today and for the forseeable future is the very radiation-sensitive NMOS. These are used in all the commercial hand calculators and minicomputers. They are used in preference to bipolar technology, because they take much less space per transistor, have much higher input impedance, have better temperature stability and less noise, and are cheaper to produce. NMOS performs better then PMOS because of its higher speed. Details are discussed in Section 6.12.

There seems to be an apparent lag in development between the above LSI and its hardened counterparts, mainly because the commercial markets' demand greatly overshadows hardened IC needs of military systems. This difficulty is being exacerbated with the passage of time. Research and development, both government and commercial, seem to point in the direction of utilizing other types of LSI and VLSI technologies, such as TTL, CMOS, CMOS/SOS and CMOS/SOI, which tend to be radiation-resistant.

That microcomputers, using microprocessors as their central component, provide computing capability in a small package is the overriding reason for their use in almost all commercial and military systems representing the spectrum of hardness specifications. There are more than 85 microprocessors extant that would merit consideration in the context of usage in various radiation environments.[62,64–70] Tables 13.6 and 13.7, respectively, depict the electrical and radiation characteristics of a number of them.

a. 8080—The 8080 is an NMOS device introduced in 1975. It was the first 8-bit microprocessor. By now, it has a broad base of applications that includes a repertoire of various assemblers and compilers, as attested to by their use in some of the popular minicomputers on the commercial market. It is currently available as a Class B device, and is already designed into some military systems. Class B device means that it has withstood a series of burn-in and other tests, as provided by military part test specifications.
b. AM 2901A—The 2901A is not a microprocessor per se. It is called a 4-bit slice and is qualified as a MIL-standard part because of its high performance. The bit slice concept is connected with the fact that the arithmetic logic section of the CPU can be thought of as being segmented into slices, each a certain number of bits in width. The arithmetic logic section of a CPU can be built by linking a number, n, of k-bit slices into a microprocessor whose word length is nk bits. Hence, the word size of the microprocessor can be flexible, because it is determined by the system designer instead of the manufacturer. Bit slices can be used to emulate microprocessors to achieve better performance and radiation hardness, partially through redundancy and partially through the hardness feature of the bit slice.
c. The SA 3227 and SA 3327 are Sandia Corporation CMOS versions

Table 13.6 Electrical properties of representative microprocessors.[63–76]

MICROPROCESSOR	80286	MC68000	CMS6502A	HS80C86	CDP1802	TMS9900	SBP9900	AM2901A
Data word length (bits)	16	16	8	16	8	16	16	16
Max clock freq. (MHz)	8	2	2	5	6.4	3	2	5
Mem. to reg. address min time (μs)/mem acc.	1.5/(2)	1.5/(3)	1.5/(3)	1.5/(2)	2.5/(2)	4.6/(4)	7/(4)	0.6/(3)
Reg. to reg. address min time (μs)/mem acc.	0.5/(1)	1.0/(2)	—	0.6/(1)	—	—	—	0.2/(1)
Load (addr 0–256) min time (μs)/mem acc.	1.5/(2)	1.0/(2)	1.5/(3)	1.5/(2)	2.5/(2)	4.6/(4)	7/(4)	0.6/(3)
Req. mem. cyc. time (μs)	0.15	0.500	0.600	0.11	0.400	0.333	0.500	0.200
Dir. addr. mem. (words)	16 M	64 K	64 K	16 M	64 K	32 K	32 K	64 K
Power rating (W)	0.600	0.500	0.700	0.600	0.150	0.850	0.500	3.2
Technology	HMOS II	HMOS	CMOS	NMOS	CMOS	NMOS	I^2L	STTL

Table 13.7 Ionizing-dose, neutron-fluence, dose-rate, and SEU response of ca, 1990 representative microprocessors.[62–83]

SER.	MICROPROCESSOR	MANUFACTURER	TECHNOLOGY	IONIZING DOSE FAILURE THRESHOLD		NEUTRON FLUENCE FAILURE THRESHOLD (10^{14} n/cm^2)	DOSE RATE THRESHOLD (10^9 Rads/s)	DOSE RATE PULSE WIDTH (ns)	THRESHOLD LET MeV cm^2/mg)	REMARKS/ GEOSYNCH. SEU/Device Day
				Krad (Si)	@ Rads/s					
1.	SA3000/80C85RH	Sandia/HA	CMOS-Epi	1000		1.0	1.9	20–50	120	
2.	AMD 2901A	AMD	LSTTL	5000		4.0	35.	20	7	4E − 3
3.	GP 501	RCA	CMOS/SOS	1000			130.			
4.	80C86	HA	CMOS-Epi	4	39.3		0.1	35	5	3.1E − 6
5.	SA3300/NS32016	Sandia/NSC	CMOS-Epi	5000	27.8		1.0	1000	30/120	no fd*/80 K fd*
6.	SBR 9000	TI	I^2L	2000		0.3	0.07	50	30	
7.	AMD 2901C	AMD	IMOX	80					3.3	
8.	SBP 9989	TI	I^3L	60	150.	0.06			12	5.9E − 4
9.	F9450	FSC	I^3L	300		0.6	0.7		11	
10.	GP001	RCA	CMOS/SOS	1000	112.		45.	20	75	
11.	80286	INT	HMOS II	30	100.					
12.	8080 B	INT	HMOS**	1.5		1.0	1.8	200		
13.	SBP9900 A	TI	I^2L	3000		0.3	2.5	25	20	1.9E − 4
14.	8X300	SIG	LSTTL	1000		0.47			18	
15.	SA2901	Sandia	CMOS	1000					90	5E − 10
16.	Z8002	Zilog	NMOS	6		1.0				

17.	AMD 2909	AMD	LSTTL	5000		0.37	25.	20	2.9	0.12
18.	1750 A	RCA	CMOS/SOS	1000		1.0	1.0	20	120	1E − 7
19.	80386	INT	CMOS-IV Epi	15	1.73				8.5	σ_{bit} = 1E − 6
20.	(ESA) T414	INM***	CMOS-Epi	50	97.				3	σ_{bit} = 2E − 6
21.	F9445	FSC	I^2L	200		0.1	0.3	50	20	1.9E − 4
22.	80C86RH	HA	CMOS-Epi	1000						
23.	SA3470	Sandia	CMOS-Epi	1000						
24.	80C186	INT	CMOS	8	125.					
25.	80186	INT	HMOS II	20	135.					
26.	68020	MOT	CMOS	3	150.					
27.	SA3227	Sandia	CMOS	1000					50	no fd*
28.	SA3327	Sandia	CMOS	1000					100	80 K fd*
29.	F100L	Ferranti	CDI-Bipolar	1000		0.1	0.1			
30.	80C85	HA	CMOS-Epi	100		10.	0.3–0.5		75	
31.	80C31	INT	CMOS-Epi	60				50–1000		
32.	8086 B	INT	HMOS	14			1.0	3	0.2	
33.	GVSC	IBM	CMOS/SOS/SOI	3000	76.8	1.0	1/1E3	30	120	$\dot{\gamma}$ − μP/Mem.
34.	TMP 320C25	TI	CMOS-Epi	52.6	207.		0.26	20–50		
35.	R3000	LSI	CMOS-Epi	100					4	^{60}Co, SRAM Only

* Feedback resistors to mitigate SEU.
** Intel version of NMOS-Epi.
*** INMOS (UK).

of the 2909A bipolar microprocessor, with greatly increased SEU tolerance. It is seen that the SA 3327 has a higher SEU LET than the SA 3227 because of the 80 K ohm feedback resistors. The SA 2901 is a CMOS substitute for the bipolar AMD 2901.

d. The IBM/HWL GVSC device is a variant, with respect to function and architecture, of the 3-chip RCA 1750A microprocessor.

e. The I^2L and I^3L microprocessors, such as the SBR9000, SBP9989, F9445, and F9450, will tend to become obsolete because of power supply incompatibility with other systems and meager noise margins.

ICBM computers must endure severe radiation environments including neutrons, ionizing dose, dose rate transients, and EMP. Even more stringent levels of radiation must be tolerated by the terminal flight phase of reentry vehicles from an ICBM bus, as discussed in Section 13.5. Because of the strategic importance of the ICBM, corresponding guidance and autopilot computer needs have been responsible for the impetus to develop hardened systems designs and hardened semiconductors.

Currently there are two principal ICBM systems being deployed. One is the U.S. Navy *Trident* submarine-launched ballistic missile (SLBM), and the other is the U.S. Air Force Minuteman Missile and its variants. The baseline computer logic technology for both systems is dielectrically isolated bipolar small scale (SSI) and medium scale (MSI) integrated circuits. Table 13.8 depicts the *Trident* autopilot computer and MX flight computer representative part types.[81] The *Trident* C-4 autopilot computer is a noncircumvention, general-register computer, reinitialized by the C-4 guidance computer in the event of logic upset due to transient dose rate pulses. Program storage uses 2048-bit, junction-isolated (JI) bipolar, mask programmable, read-only memories (PROMs). The read/write storage uses a 256-bit JI bipolar random access memory (RAM). All memory chips use an external current-limiting resistor to prevent burnout during transient radiation. The memory power supplies are shorted to ground upon detection of a radiation event to prevent burnout and recover the memories from the possibility of any latchup. All the ICs are low-power Schottky TTL, and no thin film resistors are used in this computer. All other parts, except the ROM and RAM have dielectrically isolated (DI) cells.

The MX flight computer has many characteristics in common with the above *Trident* computer. All the logic ICs are low power Schottky TTL, DI, with aluminum metallization and diffused resistors.

For spacecraft computers, the needs lie in the area of extremely high data throughput for sensor data compression and other purposes. This

Table 13.8 ICBM computers and their part families.[81]

ICBM COMPUTERS		
Computer	Trident C-4	MX Flight
Architecture	Gen Reg Fix Pt	Gen Reg Flt Pt
Data word size	16 bits	32 bits
Logic: Complexity	70 gates	70 gates
Part types	9	10
Technology	DI, LSTTL	DI, LSTTL
Speed	8 ns/gate	8 ns/gate
Power	6 mW/gate	6 mW/gate
Program memory: Size	4096 words	—
Type	JI, LSTTL, 2K ROM	—
Read/write mem: size	256 words	—
Type	JI, LSTTL, 256 RAM	—

ICBM IC PART FAMILIES			
Trident C-4		MX Flight	
Digital ICs		Digital ICs	
(1) Quad2 NAND	DI, LSTTL	(1) Quad2 NAND	DI, LSTTL,
(2) Dual4 NAND	diffused	(2) Dual 4 NAND	diffused
(3) Dual JK FF	resistors	(3) Dual JK FE	resistors,
(4) Dual 4-1 MUX	gold beam	(4) Dual 4-1 MUX	Al metall.
(5) Dual 2-4 DEC	leads	(5) Dual 2-4 DEC	
(6) 4 × 4 Reg File		(6) 4 × 4 Reg File	
(7) 4 Bit ALU		(7) 4 Bit ALU	
(8) 4 Bit Cntr		(8) 4 Bit Cntr	
(9) 4 Bit Shft Reg.		(9) 4 Bit Shft Reg.	
(10) 256 Bit RAM	JI, Al	(10) 1024 Bit ROM	
(11) 2048 Bit ROM	Metall	(11) 1024 Bit PROM	JI, Al metall.
Linear ICs		Linear ICs	
(1) OP AMP	DI, Gold	(1) OP AMP	Thin Flm Nich.
(2) FET Drvr	beam lead	(2) Analog Sw	DI, diffused
	diffused	(3) Quad JFET	resistors,
	resistors,	(4) Video Amp	Al metall.
		(5) VCO	
		(6) Line Rcvr	
		(7) Pl. Wr. Sens Amp	

implies computers with gigabit and larger memories, CPU operations measured in fractions of microseconds, ultrahigh reliability, and greatly increased radiation survivability over long periods, corresponding to a satellite mission life measured in many years. It is anticipated that the increased throughput requirements will be handled with distributed microprocessors operating in parallel, comprising LSI and VLSI devices. Reliability requirements will be addressed with redundant processing and still meet throughput requirements. System-level approaches used for latchup and burnout prevention can be applied to spaceborne processors. Therefore bulk LSI devices can be considered for hard spacecraft computers.

For manned systems, the majority of computer applications will be limited by the human crews with respect to nuclear survivability. These systems continue to use microprocessors, and they will continue to use the latest LSI technologies. Hardening this technology will be considered, but the primary microprocessor criterion probably will be performance, such as speed, word length, instruction length, availability, software support, and reliability.

13.12 SYSTEM HARDENING METHODS: POWER TRANSISTORS

To obtain a suitably radiation-hard power transistor that will survive severe radiation environments is a difficult task. This task is compounded if the power transistor must also support high stand-off circuit voltages.

There are currently two types of power transistors being manufactured. They are the bipolar and MOSFET devices. Until recently, bipolar power transistors were capable of better all-around performance than their MOSFET counterparts. Today, however, MOSFETs, including VMOS (V-groove) and the HEXFET (hexagonal microcell) technology shown in Fig. 13.32, have overtaken the bipolar power transistors, surpassing them in overall performance.

The basic difference in radiation response between bipolar and MOSFET power transistors is the same as for bipolar and MOS devices generally. It is that bipolar power transistors are more sensitive to neutron fluence than to ionizing dose radiation, whereas the opposite is true for MOSFET power transistors. In terms of electrical performance, power MOSFETs lend themselves to use in parallel configurations without current hogging, because of their positive temperature coefficient of resistance which eschews thermal runaway and second breakdown leading to device burnout. This is in contrast to the negative temperature coefficients of resistance in bipolar power transistors. This difference is

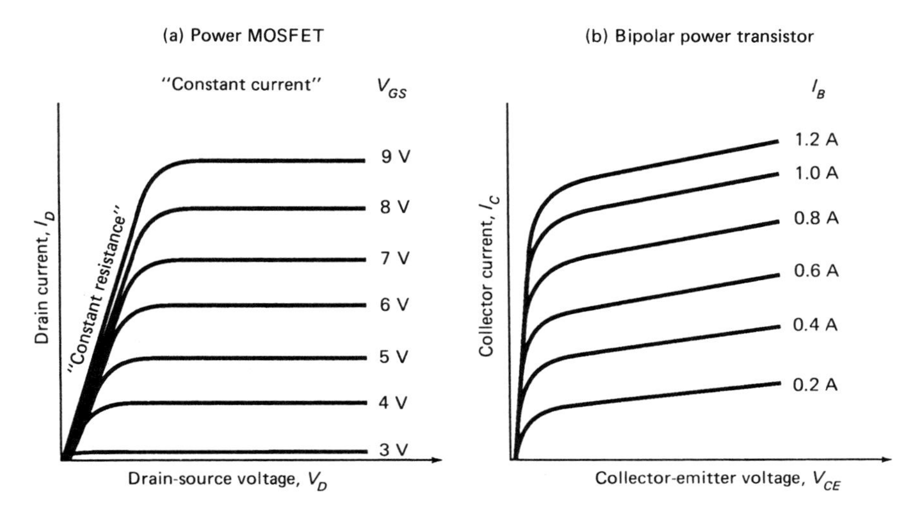

Fig. 13.31 Comparison of idealized output characteristics of power MOSFET and bipolar transistors.

due to the differences in mobility-temperature variation in each transistor. In the relatively high-mobility material of the power MOSFET, the total mobility, μ_T (Section 1.7), decreases with temperature. The corresponding majority carrier velocity decrease is equivalent to a positive temperature coefficient of resistance.

The HEXFET structure consists of about 100,000 hexagonal cells per cm^2 of chip. The cell sources, gates, drains, channels, etc. are all monolithically paralleled, the whole forming a single transistor, as shown in Fig. 13.32. Drain-source current flows up from the common *n* drain-substrate, then horizontally through common channels, then through common sources, and vertically out to the common source terminal. Gate voltage is applied to common gates via the gate terminal. The concomitant MOSFET built-in diodes are shown as the *p* diffusion-*n* drain diodes in the figure. In a bipolar power transistor, the corresponding radiation pulse has a greater tendency to cause second breakdown, usually with accompanying current hogging (Section 3.6). A power MOSFET, when its parasitic bipolars are effectively nullified, is usually harder to radiation-induced turn-on than its bipolar counterpart.

The output (I_C, V_{CE}) characteristic of power bipolars is quite similar to the output characteristic (I_D, V_{DS}) of power MOSFETs, as seen in Fig. 13.31. All MOSFETs have an inherent built-in reversed-biased rectifier diode between the source and drain, with current-carrying ability

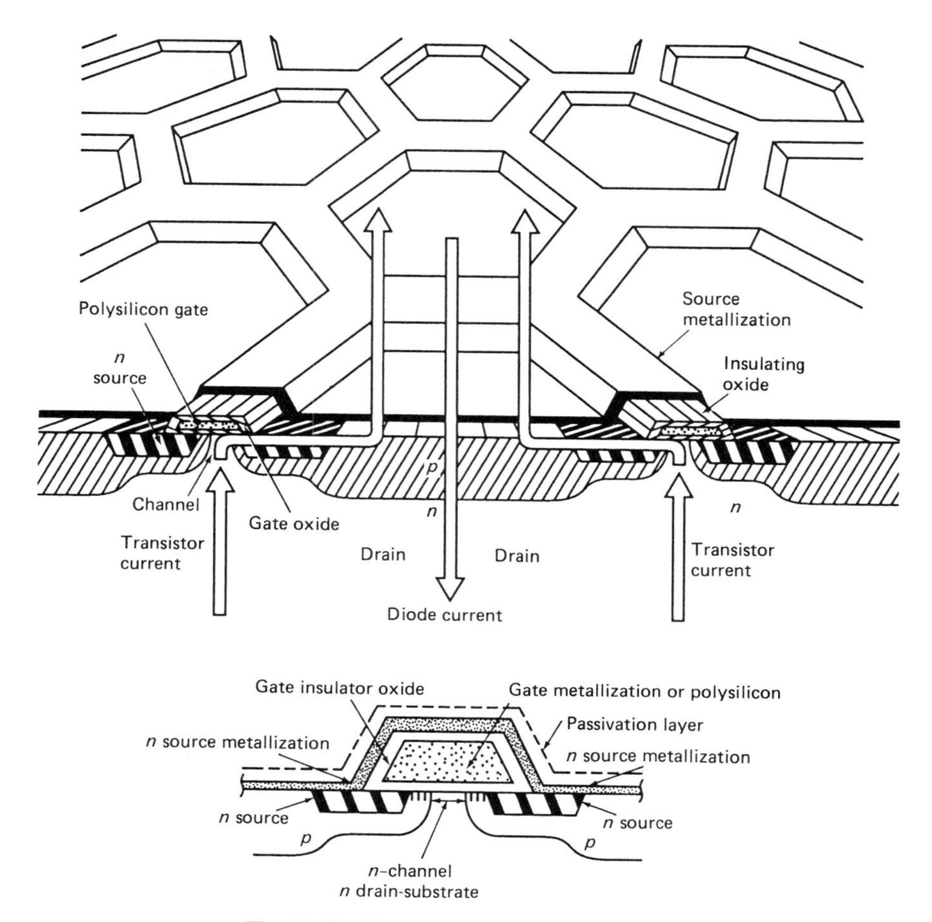

Fig. 13.32 HEXFET structure (not to scale).[101]

comparable to the MOSFET itself. This diode can be extremely useful in certain circuits. For example, it can be used as a clamp diode in an inductive load-switching application. Hence, all circuit design must take this component into consideration. Functionally, MOSFET power transistors are essentially interchangeable with bipolar power transistors. However, the former usually possess a significant performance and cost advantage. Corresponding circuit design is different, of course, reflecting their electrical parameter differences.

As already mentioned, MOSFETs are more sensitive to ionizing dose and less sensitive to neutron fluence than are bipolars. Hardened

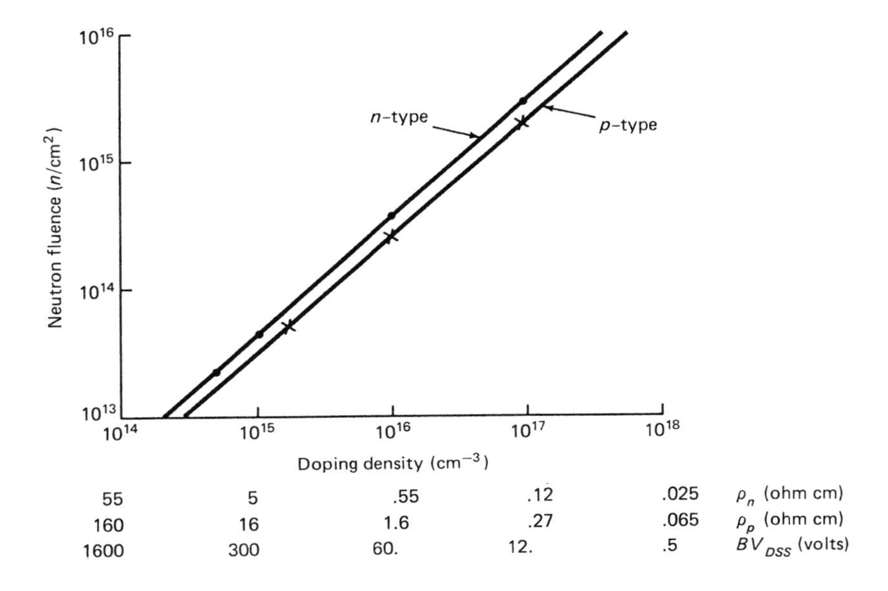

Fig. 13.33 Neutron fluence limits as a function of resistivity, doping density, and breakdown voltage for a MOSFET power transistor.

MOSFET response to dose rate is comparable to that of bipolars. Commercial MOSFETs sometimes show relatively low burnout thresholds in a dose-rate environment. Both MOSFET and bipolar power transistors can suffer second breakdown from sufficiently high dose-rate levels. Their response to EMP-pulsed fields is also comparable, but protection circuits for both can be designed using I/O filter and bypass techniques. With regard to SEU, the MOSFET n-channel power transistor can suffer SEU-induced burnout, but is unlikely in the bipolar power transistor.

The discussion in the following paragraphs will consist first of the response of power MOSFETs to the sequence of accepted radiation effects environments, followed by that for power bipolars to the same environments. The response to neutron fluence of a power MOSFET, as for almost all MOS and bipolars, is much less than that for a power bipolar, simply because MOSFETs are majority carrier devices. Therefore, damage from neutron-induced minority carrier lifetime degradation is foreign to MOSFET operation. In the MOSFET, the major neutron-induced damage effect is an increase in the resistivity of its material at very high ($\geqslant 10^{14}\,cm^{-2}$) fluence levels. Resistivity changes due to incident neutron fluence Φ_n can be expressed through

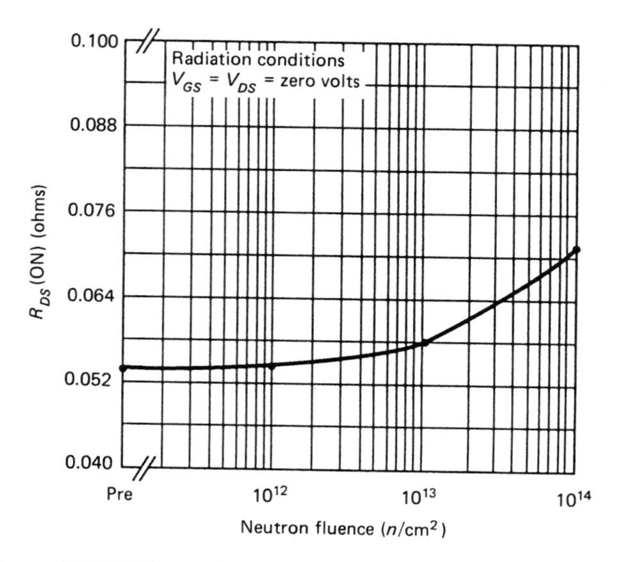

Fig. 13.34 Power MOSFET on resistance, R_{DS}(ON), versus incident neutron fluence.[102]

$$\rho = \rho_0 \exp(\Phi_n/K) \tag{13.46}$$

where ρ_0 is the pre-irradiation resistivity, as discussed in Section 5.13. For n-type material, $K_n = 444n_0^{0.77}$, while for p-type material, $K_p = 387p_0^{0.77}$, where n_0 and p_0 are the pre-irradiation equilibrium majority carrier densities. From Eq. (13.46), for small fractional changes in resistivity.

$$\Delta\rho/\rho \cong \Delta\Phi_n/K \tag{13.47}$$

The graphical result for a limit of $\Delta\Phi_n/K = 0.1$ to permit only small resistivity fractional changes $\Delta\rho/\rho$, is shown in Fig. 13.33. It depicts the corresponding neutron fluence limits for MOSFET device variations in resistivity, doping density, and breakdown voltage. A reasonable limit for devices whose neutron fluence threshold is about $10^{14}\,\text{cm}^{-2}$, is a dopant concentration of about $4 \cdot 10^{15}\,\text{cm}^{-3}$, with a breakdown voltage BV_{DSS} of 120 volts, requiring a resistivity of about 1.5 ohm cm in n-type devices and 3 ohm cm in p-type devices.

Another important parameter in power MOSFETs is the channel ON resistivity, R_{DS}(ON). It is given as[103]

$$R_{DS}(\text{ON}) = [(W/L) \cdot C_{0x} \cdot \mu_e \cdot (V_G - V_T)]^{-1} + F \cdot \rho_0 \exp(\Phi_n/444n_0^{0.77}) \tag{13.48}$$

where

L = channel length.
W = channel width.
C_{0x} = gate oxide capacitance per unit area.
μ_e = channel electron (inversion layer) mobility.
F = a constant depending on the device dimensions.
V_G = gate voltage
V_T = rated threshold voltage

The first term in Eq. (13.48) is the channel resistivity, while the second is that of the drain region. R_{DS}(ON), as a function of neutron fluence for a power MOSFET, is depicted in Fig. 13.34.

Ionizing dose in power MOSFETs results in an increase of positive charge density in the gate oxide, which initially produces a negative shift in threshold voltage, as discussed in Section 6.8. In most of these devices, operation up to one megarad (Si) can be assured by providing sufficient design margin to accommodate a 10-volt negative shift in the threshold voltage. Also, the shift in threshold voltage is a strong function of the applied gate voltage, as shown in Figs. 13.35A and 13.35B, for *n*- and *p*-channel devices, respectively. An approximate expression for the threshold voltage, as a function of ionizing dose for MOSFET power devices, is given by the linear relationship

$$V_T \cong V_{T0} - k_T D \tag{13.49}$$

where V_{T0} is the pre-irradiation threshold voltage, k_T is a measured constant of proportionality, and D is the ionizing dose level. $k_T \cong 10^{-5}$ volts per rad (Si) for a negatively biased power MOSFET gate.

One adverse effect of ionizing dose incident on power MOSFETs is the decrease in breakdown voltage BV_{DSS}. This is due to the ionizing dose-induced distortion of the electric fields, where the device *p-n* junctions terminate at a surface. A simplistic "first-order" explanation is that leakage currents increase with increasing ionizing dose, due to trapping of charges in the oxide covering the field-terminating structures. This causes local changes in the fields, leading to increased leakage current, or conversely, a decrease in breakdown voltage.[104]

Generally, the decrease in breakdown voltage increases with V_{DD} during ionizing radiation. This then requires device selection at a voltage sufficiently high to accommodate the BV_{DSS} drop. However, this will result in a device with an increased R_{DS}(ON), as well as one that is more susceptible to a radiation-induced shift in R_{DS}(ON).[104] This is most

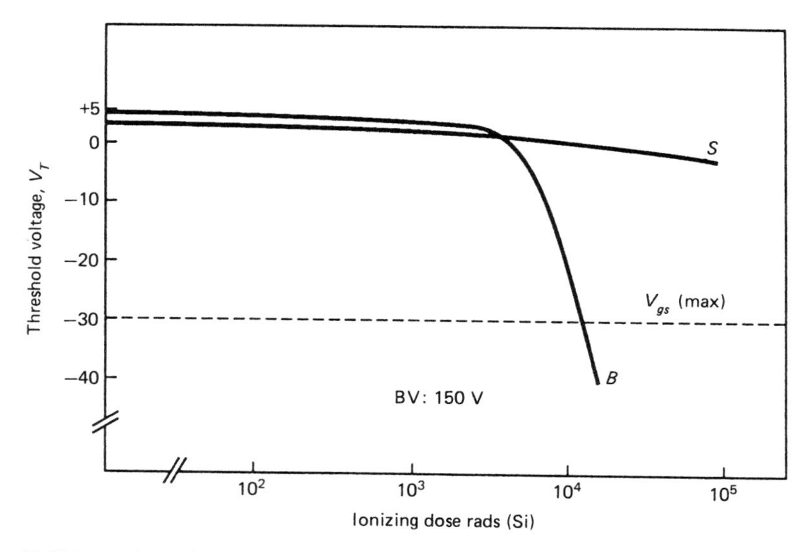

Fig. 13.35A *n*-channel power MOSFET threshold voltage under biased (*B*) and unbiased (*S*) conditions versus ionizing dose.[105,106] © 1983 by the IEEE.

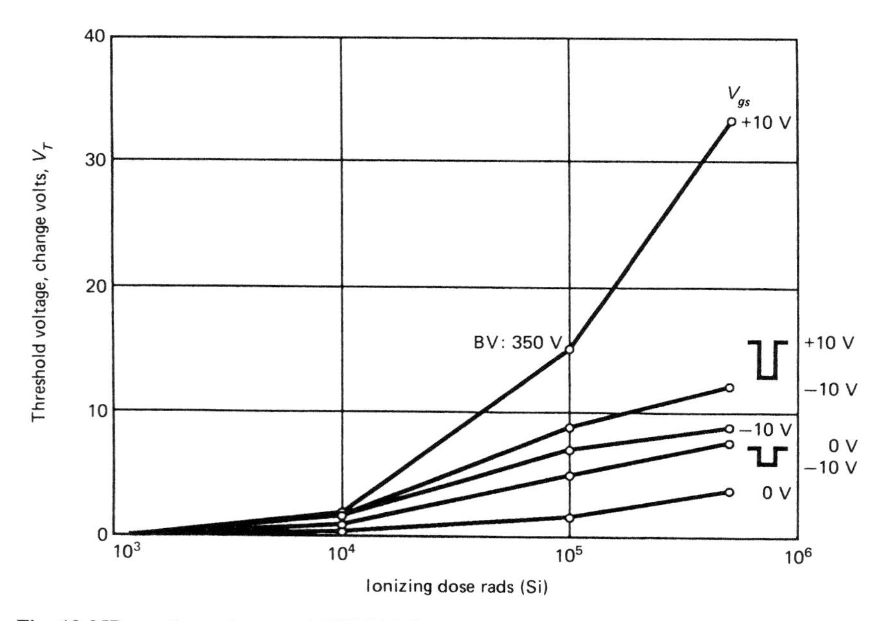

Fig. 13.35B *p*-channel power MOSFET threshold voltage change versus ionizing dose and bias conditions.[105,106] © 1983 by the IEEE.

pronounced in high-BV_{DSS} devices, and is not significant for those rated at 100 volts or less. Figures 13.36A and 13.36B depict breakdown voltage change versus ionizing dose for various values of V_{DS} for 100 and 500 volt breakdown devices, respectively.[105] These figures imply that the voltage breakdown of some higher voltage devices would have to be derated significantly for operation even in moderate ionizing dose environments.

Ionizing dose-rate pulse-induced photocurrent in MOSFET power transistors is very similar to that produced in bipolar counterparts. The steady state junction photocurrent I_{PPS} can be expressed approximately (for silicon) as

$$I_{PPS} \cong eAG(W + L_p + L_n) = (6.48\,\mu\text{amp/cm}^3 \cdot \text{rad (Si) per s}) \cdot AS\dot{\gamma} \tag{13.50}$$

where $S = W + L_p + L_n$, and the remaining symbols are defined in Section 7.2. There is a photocurrent associated with both the drain and source junctions. For a short dose-rate pulse, the junction transient photocurrent $I_{PP}(t)$ is given approximately as

$$I_{PP}(t) \cong I_{PPS} \cdot (t_p/\tau)^{1/2}; \qquad t_p \ll \tau \tag{13.51}$$

where t_p is the dose-rate pulse width and τ is the minority carrier lifetime near the junction of concern.

As mentioned, all power MOSFET transistors contain an inherent source or drain to substrate reversed-biased diode. Also, there is sometimes a superimposed parasitic *pnp* or *npn* structure from the source to the drain. Thus, at very high levels of prompt ionizing-dose pulses, second breakdown in the parasitic transistor could become a precursor for device burnout. High levels of dose rate not only result in reduced breakdown voltage, BV_{DSS}, but in very large drain photocurrents. They can become many times larger than the device vendor current ratings.

Single event upset (SEU) unexpectedly affects certain *n*-channel MOSFET power transistors by causing them to burn out. As discussed earlier, this is due to their parallel cell construction, where a single transistor within the device is initially affected by an SEU. Its incipient burnout triggers the same in adjacent transistors, accompanied by current hogging, with the whole device finally failing. The present circuit design remedy is merely to derate its peak drain operating voltage by at least 30 percent.[107]

Bipolar silicon power transistors have been the venerated choice for an active power device, and are still very much in use in system applications. A typical bipolar device state-of-the-art characteristic is a 10-amp, 90-volt

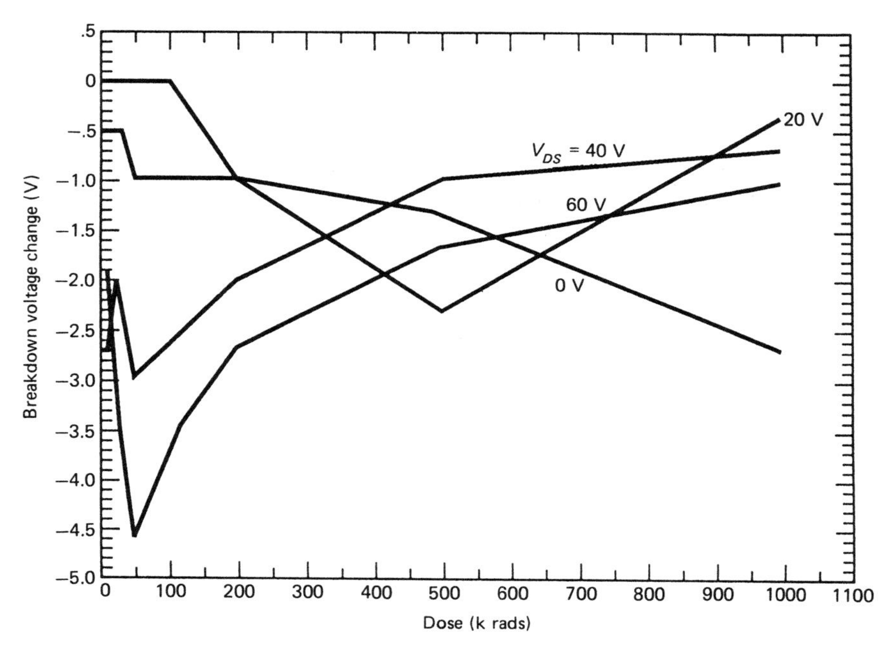

Fig. 13.36A 100 volt MOS power transistor breakdown voltage as a function of ionizing dose for various values of drain to source voltage.[105] © 1983 by the IEEE.

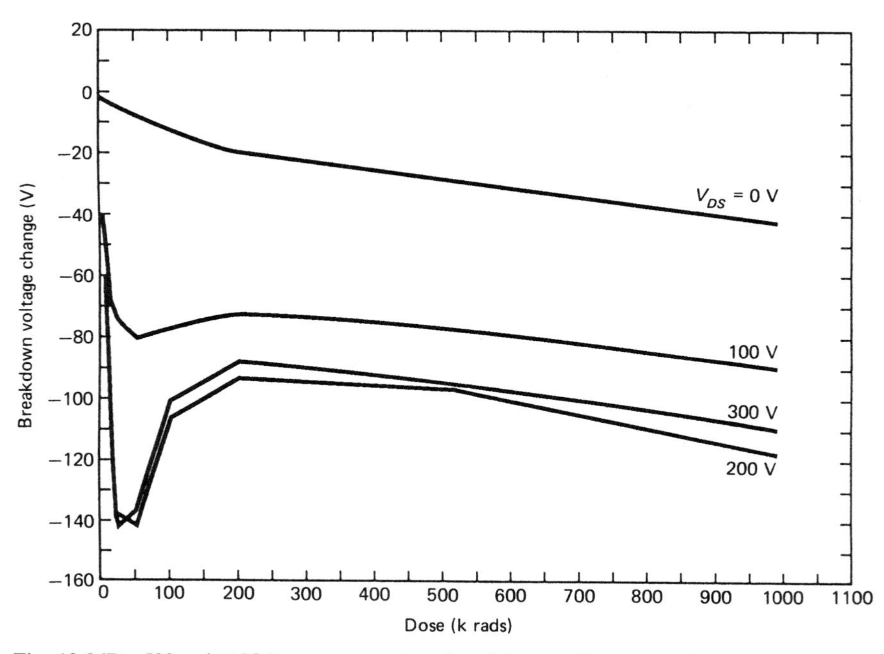

Fig. 13.36B 500 volt MOS power transistor breakdown voltage as a function of ionizing dose for various values of drain to source voltage.[105] © 1983 by the IEEE.

transistor, with a 30 amp collector surge current rating. It possesses a low $V_{CE(\mathrm{sat})}$ of 0.4 volts at 5 amps, with a forced beta of 10.

For x- and gamma radiation, bipolar power transistors show negligible radiation damage effects from ionizing dose, up to approximately one megarad (Si). However, common emitter gain, β, usually begins to degrade at about 10^{12} neutrons per cm^2. Power transistors can withstand current pulses up to about three times the maximum rated current in a prompt ionizing dose-rate pulse environment. However, very large current pulses will result in second breakdown and burnout, as mentioned.

Neutron irradiation of bipolar power transistors can cause major damage by reducing the common emitter gain. As rewritten from Section 5.10, common emitter gain depends on neutron fluence as

$$\Delta(1/\beta) \equiv 1/\beta - 1/\beta_0 = \Phi_n/\omega_T K = K_d \Phi_n \tag{13.52}$$

where the parameters are defined therein. The damage constant K is approximately $1.6 \cdot 10^6$ neutron s/cm^2. K_d, termed the device damage constant, for a typical bipolar power transistor is $5.8 \cdot 10^{-16}\,\mathrm{cm}^2$ per neutron at $I_C = 2$ amps, and $1.3 \cdot 10^{-15}\,\mathrm{cm}^2$ per neutron at $I_C = 18$ amps. How the damage constants vary with collector current is discussed in Section 5.9. K_d achieves a minimum value at a value of I_C for which β is a maximum. Above this peak β, K_d increases rapidly due to emitter crowding. Transistors should not be used in power switching applications above I_C corresponding to this peak β in a neutron environment. This is because the neutron-induced decrease in β results in increases in $V_{CE(\mathrm{sat})}$ and $V_{BE(\mathrm{sat})}$. Unless large design margins are maintained, such operation can result in transistor burnout damage.

It is essential to use the lowest practical value of breakdown voltage and the highest practical value of gain-bandwidth product when designing with power transistors for use in a neutron environment. If it were not for current gain sensitivity, the curves in Fig. 13.33 could also be applied to bipolar power transistors. In any case, these curves can be considered an absolute upper limit for hardened bipolar transistors.

K_d scales as the square of the breakdown voltage, as seen in the following. To increase breakdown voltage, both material resistivity and base width must be increased. Both of these are approximately proportional to the increase in BV_{CEO}, the breakdown voltage. However, increasing the base width reduces the gain-bandwidth product in proportion to the square of the former, so that $K_d \sim (BV_{CEO})^2$. Figure 13.37 depicts the practical neutron fluence limit for hardened bipolar power transistors as a function of breakdown voltage for high reliability systems, with and without hardness assurance control. A comparison of Figs 13.33

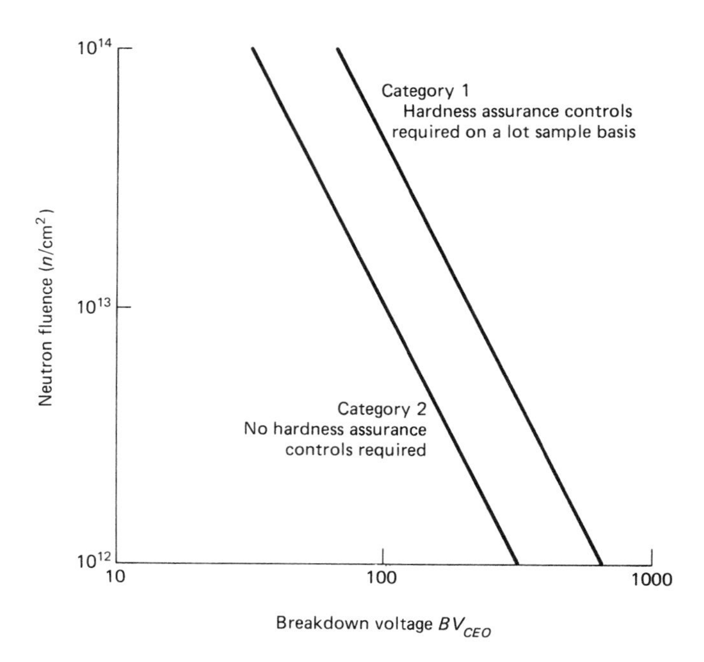

Fig. 13.37 Neutron fluence limit as a function of breakdown voltage, BV_{CEO}, for high reliability bipolar power transistors. The categories in the figure are defined in Chapter 14.

and 13.37 reveals that MOSFET power transistors are about 2–3 times harder to neutron fluence than equivalent bipolar power transistors. Hence, the preferred candidates for power transistors in future hardened electronic systems are MOSFETs, certainly insofar as the neutron environment is concerned.

With respect to ionizing dose, it is already apparent that bipolar transistors would suffer very little in such an environment. This is the case, as typical values of $\Delta(1/\beta) \leqslant 0.01$, for an ionizing dose of about one megarad (Si) produces negligible or no changes in other associated transistor parameters. Hardness assurance programs, as discussed in Chapter 15, typically sum ionizing dose parameter damage with that from neutron fluence, to obtain a total radiation environment parameter change. This is conservative as compared to the alternative of using the "root-sum-squares" approach for computing total parameter radiation damage from disparate sources. On the other hand, the use of bipolar power transistors at low emitter currents is frowned upon, as this practice enhances surface damage effects. Therefore, use of bipolar power transistors at very low currents will render them susceptible to ionizing dose damage. The corresponding circuit designs that would allow low current bias should be avoided for use in radiation environments.

Dose rate considerations for bipolar power transistors are governed by the same relationships for photocurrent generation as those for MOSFET power transistors, as in Eqs. (13.50) and (13.51). A typical bipolar power transistor undergoing an incident dose-rate pulse of, say, 20 ns width could have a collector junction photocurrent of about 4 nA per rad (Si)/s, with a design limit of about 10 nA per rad (Si)/s. For very high prompt dose-rate pulses, transistor burnout can occur, so that circuits should be designed to limit the photocurrent to less than three times the maximum continuous current rating. As alluded to earlier, the photocurrent transient must be within the limits of the SOA of the device.

Single event upset (SEU) considerations do not usually impact bipolar power transistors. This is in contrast to MOSFET power transistors which, as discussed earlier, could suffer SEU pulse-induced burnout. However, SEU-induced burnout in bipolar power transistors has also been reported (Section 12.5). Bipolar power transistors are constructed of essentially one relatively large die. Though its size presents a comparatively large target to incident cosmic ray ions, it has a large critical charge which renders it insensitive to SEU.

13.13 SYSTEM HARDENING METHODS: ANALOG-TO-DIGITAL CONVERTERS (ADC)

An analog to digital converter (ADC) is a system that converts analog signals to digital numbers. It is functionally the opposite of a digital-to-analog converter (DAC). ADCs and DACs usually contain several chips interconnected and assembled in a suitable package, termed a hybrid. Alternatively, the whole is manufactured monolithically on one integrated circuit chip.

During the past decade, there has been a very large increase in the number and types of ADCs with satisfactory electrical properties that meet a host of requirements, from spacecraft to military to commercial applications. However, for an uncomfortably large number of devices, their radiation failure thresholds are relatively low. There are two principal types of ADCs. One is a counter type, and the other is the successive-approximation type, using an input signal-matching programmer.

It is interesting to note that the scanning electron microscope (SEM) can be used as a diagnostic tool to locate radiation-sensitive components in ADC devices.[108] For example, 30 keV electrons being focussed, as part of the electrooptics of the SEM, simultaneously can produce an

Table 13.9 Ionizing dose and neutron fluence response of representative analog-to-digital converters.[108]

SER.	A/D CONVERTER	VENDOR	TECHNOLOGY	BITS	POWER (mW)	SPEED (μs)	IONIZING DOSE FAILURE THRESHOLD Krads (Si)	NEUTRON FLUENCE FAILURE THRESHOLD (10^{13} n/cm^2)	REMARKS
1.	AM 6108	AMD	Bipolar	8	600	0.5	600	>1.0	Tri-state output
2.	AM 6148	AMD	Bipolar	8	600	0.9	600		Like AM 6108
3.	AD 573	ADI	I^2L	10	350	15.	30		Like AD 571
4.	AD 574	ADI	I^2L	12	400	25.	30	<0.3	Two chip Hybrid
5.	AD 574	ADI	Bipolar	12	<400	<25.	>1000		Monolithic
6.	AM 6112	AMD	Bipolar	12	600	3.	600	>1.0	Tri-state output
7.	HI 574A	HA	CMOS/Bipolar	12	500	25.	10		Like AD 574A
8.	HI 674A	HA	CMOS/Bipolar	12	500	12.	10		Faster HI 574A
9.	MN 5211	MicNet	Bipolar	12	915	13.	>600		TTL output
10.	MN 5214	MicNet	Bipolar	12	915	13.	30		TTL output
11.	MN 9022B	MicNet	Bipolar	12	915	13.	>600		Like MN 5214
12.	MP 574A	MicPwr	CMOS/Bipolar	12	600	25.	50		Like AD 574A
13.	MN 5290	MicNet	Bipolar	16	830	40.	>1000		Hybrid, ser/par outputs
14.	TDC 1021J	TRW	Bipolar	4	400	0.4	600		
15.	AM 6148	AMD	Bipolar	8	600	0.9	1000		
16.	TDC 1001J	TRW	Bipolar	8	400	0.4	>2500		
17.	AD 571	ADI	Bipolar	10	400	25.	300		
18.	AD 573	ADI	Bipolar	10	350	15.	75		
19.	AD 7570	ADI	CMOS	10	40	40.	15		
20.	MP 7570	MPS	CMOS	10	40	40.	20		
21.	AD 574A	ADI	Bipolar	12	400	25.	150		Two chip hybrid
22.	ADC 1210	NSC	CMOS	12	135	100.	20		
23.	AM 6112	AMD	Bipolar	12	600	3.	1000	>1.0	
24.	HI 574	HA	CMOS/Bipolar	12	135	25.	10		Hybrid
25.	MN 5216	MicNet	Bipolar	12	600	3.	150		Hybrid
26.	MN 90288	MicNet	Bipolar	12	600	3.	600		Same as MN 5214

ionizing dose rate of 1.5–3 krad (Si) per second within the device being examined. Sufficient SEM exposure can reveal individual chips in an ADC hybrid that have substandard ionizing radiation tolerances.[108] Table 13.9 provides ionizing-dose and neutron fluence radiation data for a representative list of analog-to-digital converters from over half a dozen manufacturers. The devices and corresponding radiation data mainly span the last decade. Essentially, all of these converters are presently available as part of the listed manufacturers' standard product lines. Suggestions were given by the radiation effects community[108] to manufacturers of analog-to-digital converters to enable them to harden their products. In many cases, this knowledge has also been incorporated into their other semiconductor lines.

13.14 SYSTEM HARDENING METHODS: SOLAR CELLS

Solar cell systems are here assumed to principally comprise solar arrays as used on spacecraft. Solar cell radiation damage depends on radiation environments, as well as the cell physical parameters. The main radiation environmental particles are (a) protons and electrons that make up the Van Allen belts through which the spacecraft solar cell arrays may pass in their periodic orbits; (b) neutrons, x, and gamma rays that are produced mainly from a nuclear burst; and (c) heavy particles that occur naturally in small infrequent bursts, such as solar flare protons.

The solar cell is essentially a bipolar junction, heavily doped above the junction (sun side) and lightly doped below. The former is normally termed the emitter, while the latter is called the base. It can also consist of a metal-semiconductor schottky barrier-type of junction, with the metal below. For incident solar photons of energy equal to or greater than the band gap, electron-hole pairs are produced in the cell material. Because of the minority carrier electron-density gradient in the *p*-side material near the junction, these electrons diffuse toward the junction depletion region. Upon reaching it, the depletion layer electric field accelerates them to the opposite side of the junction. Similar carrier kinematics occur for the minority carrier holes generated in the *n*-side material of the junction. Both of these carrier-generated diffusion currents crossing the junction constitute the solar photon-generated useful cell current. This photo-generated current can be represented in a solar cell circuit model, shown in Fig. 13.38. This model is used to determine the current-voltage characteristics of the solar cell, in terms of its physical parameters.[109] How the various radiation environments affect those parameters will provide the link to solar cell electrical

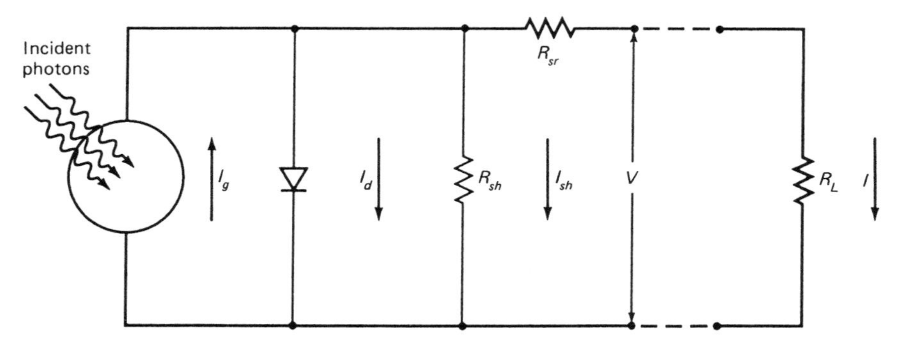

Fig. 13.38 Circuit model of solar cell.[109]

operation as a function of its incident damaging radiation. The model parameters shown in Fig. 13.38 are defined as:

I_g = photo-generated current
I_{sc} = short circuit current
I_d = current through cell as diode
I_0 = cell diode saturation current
I_{sh} = current through cell shunt resistance
R_{sh} = cell shunt resistance ($\sim 10^4$ ohms)
R_{sr} = cell series resistance (fractions of an ohm)
R_L = cell load resistance
V = cell output voltage (hundreds of mV)
V_{0c} = open circuit voltage
I = cell output current (tens of mA)
n = empirical constant (ideality factor ~ 1.25)

Using the diode equation discussed in Section 3.7, and the Kirchoff loop equations in Fig. 13.38, the cell output current I is given implicitly by

$$I = I_g - I_0(\exp(e(V + IR_{sr})/nkT) - 1) - (V + IR_{sr})/R_{sh} \tag{13.53}$$

where the diode saturation current, I_0, is given from Section 3.7 by

$$I_0 = en_i^2 A_J(D_n/p_{n0}L_n + D_p/n_{p0}L_p) \cong en_i^2 A_J((D_n/\tau_n)^{1/2}/N_A + (D_p/\tau_p)^{1/2}/N_D) \tag{13.54}$$

The photo-generated current, I_g, is related to the short circuit current, I_{sc}, of the illuminated cell. The short circuit implies, from Eq. (13.53), with $V = 0$, that

$$I_{sc} \cong I_g - I_0(\exp(eI_{sc}R_{sr}/nkT) - 1) - I_{sc}R_{sr}/R_{sh} \qquad (13.55)$$

With $R_{sr} \cong 0.5$ ohm[110] and $R_{sh} \cong 10^4$ ohms,[109] the exponential in the second term above can be well-approximated to two terms in its series expansion. Equation (13.55) the yields

$$I_{sc} \cong I_g/(1 + (eI_0R_{sr}/nkT) + (R_{sr}/R_{sh})) \cong I_g \qquad (13.56)$$

which, if the latter equality holds, is indicative of a "good" solar cell. As depicted in Fig. 13.44, it has been found experimentally that the short circuit current is very well approximated by

$$I_{sc} = a\ln(L) + b \qquad (13.57)$$

where L is the cell diffusion length, and a, b are empirically determined constants.[111] Now, $L^2 = D\tau$, where D is the diffusion constant, τ is the minority carrier lifetime, and $1/\tau = 1/\tau_0 + K_\tau\Phi$, as discussed in Section 5.9. Then a simple substitution gives (where t_p is the duration of the neutron fluence)

$$1/L^2 = 1/L_0^2 + K_L\Phi(t),\ t \leq t_p; \qquad 1/L^2 = 1/L_0^2 + K_L\Phi(t_p),\ t > t_p \qquad (13.58)$$

Equation (13.58) is sufficiently general in that Φ can represent neutron, proton, or electron fluence, with correponding damage coefficients, K_L, and the zero subscript means preradiation conditions. It should be noted that K_L depends inversely on the injection level,[111] so that as high an injection level as can be tolerated in the cell aids in minimizing diffusion length changes due to incident fluences. Values of the damage coefficients for electrons, protons, and neutrons are discussed in Section 5.9, and tabulated in Section 13.2. Combining Eq. (13.57) and the leftmost Eq. (13.58) yields

$$I_{sc} = I_{sc0} - (a/2)\ln(1 + K_L L_0^2\Phi) \qquad (13.59)$$

Under open circuit conditions, $I = 0$, then Eq. (13.53) yields ($I_{sh} \ll I_g$)

$$V_{0c} \cong (nkT/e)\ln(1 + I_{sc}/I_0) \qquad (13.60)$$

and $I_0(\Phi)$ is available as a function of Φ using Eqs. (13.54) and (13.58). The third parameter that characterizes solar cell operation is P_{max}, given by

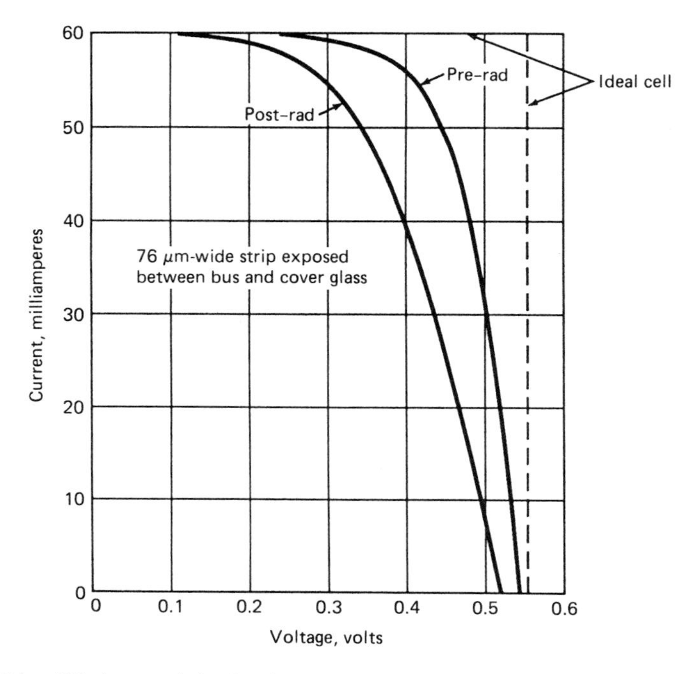

Fig. 13.38A *I-V* characteristic showing damage from 3E13-0.250 MeV proton fluence on partially shielded *n/p* solar cells.[109]

$$P_{\max} = F \cdot I_{sc} V_{0c} \tag{13.61}$$

where F is a solar cell form factor.[109] $P_{\max}$ is the power delivered to the load at a point on the solar cell $I - V$ characteristic where maximum power is being transferred.[112] Figures 13.38A and 13.38B, respectively, depict a partially shielded n/p solar cell $I - V$ and power $I - V$ characteristic, before the after proton irradiation. A fourth parameter is V_{mp}, the cell output voltage during maximum power transfer. Figure 13.39 depicts I_{sc}, V_{0c}, and $P_{\max}$ degradation due to incident neutron fluence for n/p junction silicon solar cells.[113]

Even though neutron fluence damage could be appreciable, its effect is mitigated, since the solar cell annealing factor,[114] $F(t) = (\tau^{-1} - \tau_0^{-1})/(\tau_\infty^{-1} - \tau_0^{-1})$ (Sections 5.14 and 6.15), can have the value $F(0+) \geqslant 10$, but drops precipitously in 10–100 ms to about $F(0.1\text{s}) \cong 2$,[114] as seen in Fig. 13.40. This implies that the solar cell system could probably survive a substantive neutron fluence, by "annealing out" the damage very rapidly, as indicated by the asymptotic $(t > t_p)\, F(t)$. $F(t)$ can be

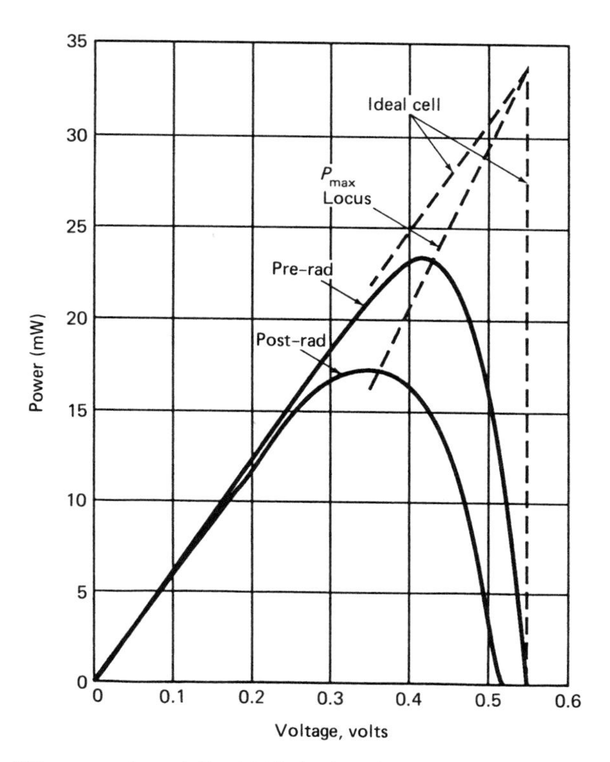

Fig. 13.38B *I-V* power of partially shielded *n/p* solar cell damage from 3E13-0.250 MeV proton fluence.

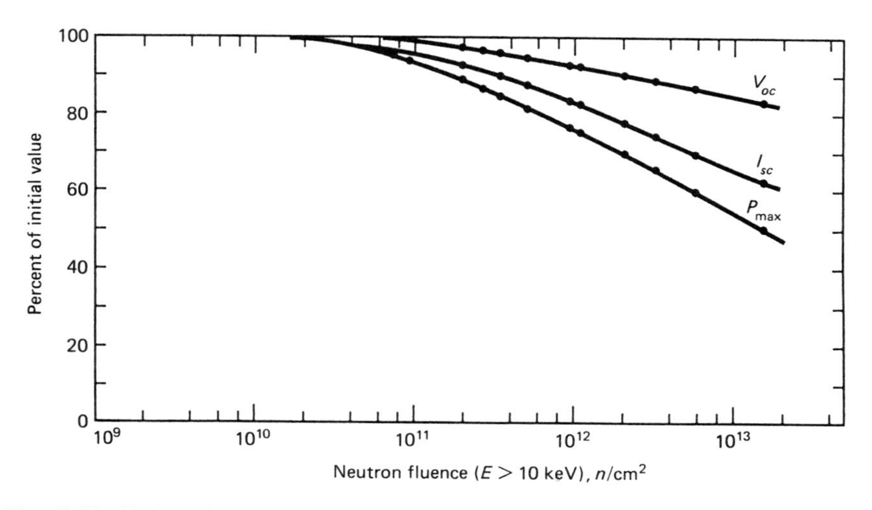

Fig. 13.39 Neutron-induced changes in deep junction 10 ohm-cm *n/p* silicon solar cells.[113] © 1969 by the IEEE.

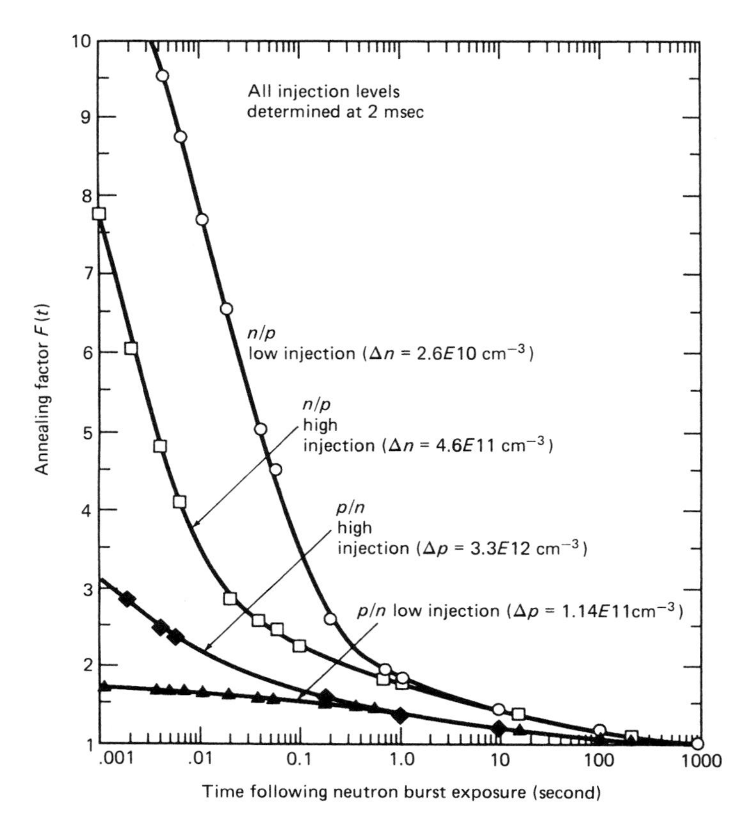

Fig. 13.40 Annealing factor versus time following neutron burst for 10 ohm-cm n/p and 1 ohm-cm p/n cells.[114,124] © 1967 by the IEEE.

computed in terms of $I_{sc}(t)$, I_{sc0}, and $I_{sc\infty}$ by using Eqs. (13.57) and (13.58) to yield

$$F(t) = ((L_0/L_\infty)^{2v} - 1)/((L_0/L_\infty)^2 - 1);\ v = (I_{sc0} - I_{sc})/(I_{sc0} - I_{sc\infty});$$
$$I_{sc\infty} \leqslant I_{sc} \leqslant I_{sc0} \tag{13.62}$$

These parameters are all relatively easy to measure.[109] If it is desired to involve the incident radiation fluence, then $F(t)$ can be rewritten as

$$F(t) = ((1 + K_L L_0^2 \Phi(t_p))^v \cdot (L_p/L_\infty)^{2v} - 1)/((1 + K_L L_0^2 \Phi(t_p)) \cdot (L_p/L_\infty)^2 - 1) \tag{13.63}$$

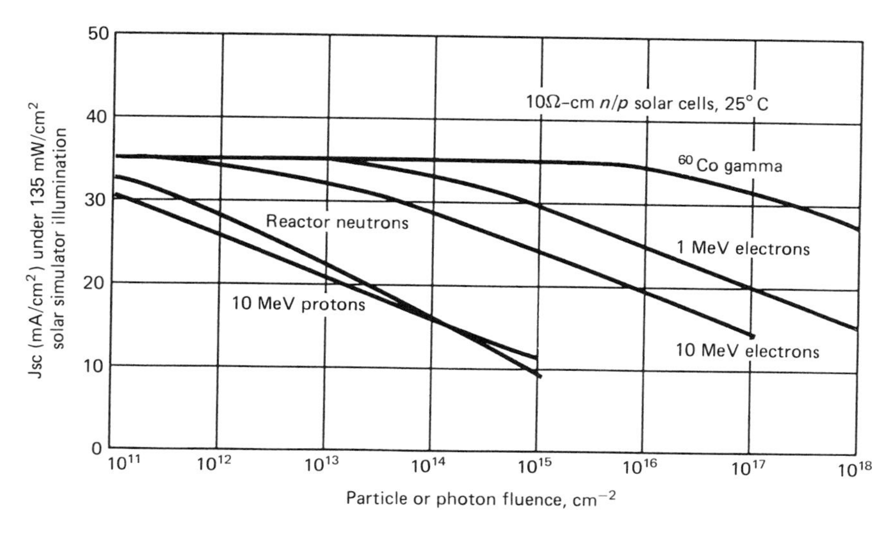

Fig. 13.41 Variation of solar cell short circuit current density with fluence for various radiation types.[115]

where L_p is the diffusion length immediately after the cessation of the radiation, and L_∞ is the (annealed) diffusion length very long after the cessation of the radiation (e.g., very long as compared to the fluence duration time t_p). If L_p/L_∞ can be approximated by unity, $F(t)$ becomes

$$F(t) \cong ((1 + K_L L_0^2 \Phi(t_p))^{\nu} - 1)/K_L L_0^2 \Phi(t_p) \tag{13.64}$$

Ionizing dose in the form of x and gamma rays interact with Si and GaAs solar cells mainly by the production of compton electrons. These electrons can have sufficiently high energies to cause displacement damage in solar cells. ^{60}Co gammas incident on n/p silicon solar cells affect them as shown in Fig. 13.41. The displacement cross section of prompt gammas is very small compared with other types of radiation, and any corresponding damage can usually be neglected.[115]

Ionizing dose rate is the most important aspect of ionizing dose for solar cells, since it can cause spurious photocurrents in the solar cell array, as discussed in Section 7.2. These photocurrents can be troublesome from a high dose-rate source. Their effects can be minimized by current limiting through adjustment of the external load, as well as automatic limiting from the cell series resistance R_{sr}, or with low-value resistors, diodes, and ferrite toroid cores placed in the load lines.

EMP and SEU effects are somewhat alien to solar cell operation by

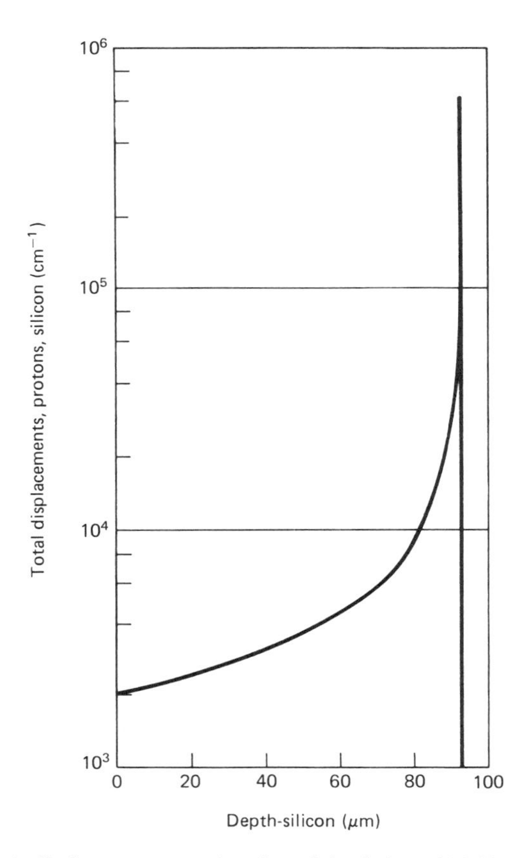

Fig. 13.42 Atomic displacements as a function of depth for a 3 MeV proton in silicon.[117]

virtue of cell construction and their solid-state properties. However, SGEMP could provide difficulties to cell arrays in orbit, again through the creation of spurious currents.

Disregarding damaging radiation, solar cell output parameter dependence on cell temperature and incident illumination intensity is a complicated relationship. Attempts to ascertain these relationships have mainly produced empirical results.[115]

With respect to electron damage, an equivalent 1 MeV electron fluence normalization has been established[115] in much the same manner as that for 1 MeV neutrons discussed in Section 10.3. Variations in *n/p* solar cell base resistivity and cell thickness result in significant variations in response to 1 MeV-equivalent electrons. This is in contrast to other cell parameters, such as cell temperatures (200–370°K) and *p*-type base

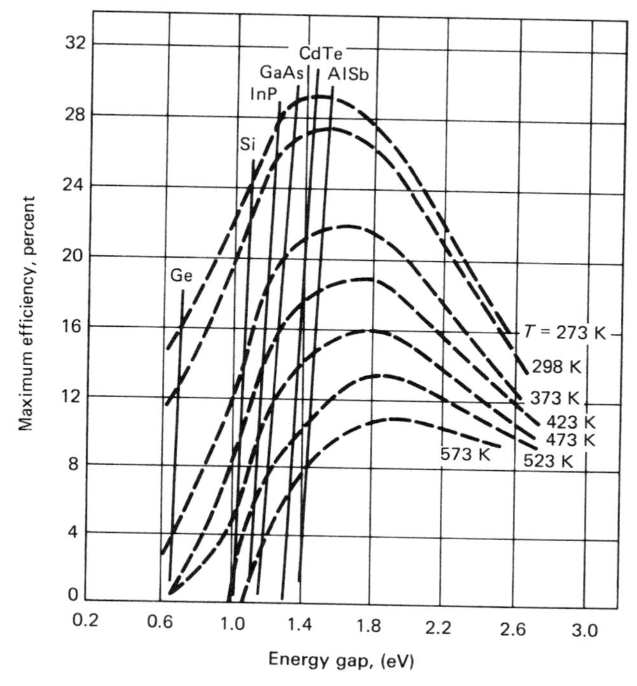

Fig. 13.43 Temperature-dependent maximum efficiency as a function of energy gap for a few solar cell materials.[118] © 1973 by the IEEE.

dopant species, such as boron or aluminum, which have little or no effect on cell response to incident damaging radiation.

Changes in *n*/*p* solar cell response with respect to base resistivity are such that there can be a trade-off between maximum cell power output and cell radiation threshold.[116] This depends on cell thickness. Figure 13.41 depicts $I_{sc}/A_J = J_{sc}$ versus various radiation levels.

Proton damage to *n*/*p* solar cells is somewhat more complex than electron damage. This is because the stopping power of protons less than 5 MeV (Van Allen belt protons) in Si or GaAs is such that their range is less than the normal solar cell thickness. This implies that cell damage is nonuniform with cell thickness. Also, heavy particles like protons lose most of their energy near the end of their range. This energy loss causes mainly atomic displacements, as in the case of neutrons. As these protons are stopped, losing their energy well within the depletion region of the cell junction, they can effectively damage the cell. This holds for protons of energies up to about 1 MeV, as shown in Fig. 13.42. This effect is enhanced for lower energy protons (<1 MeV), as attested to by the

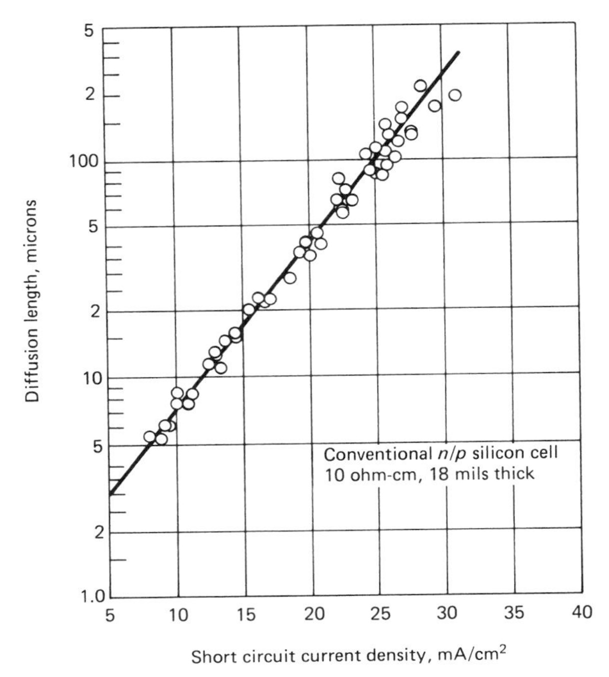

Fig. 13.44 Minority carrier diffusion length versus short circuit current density of conventional n/p silicon solar cells.[109]

unfortunate event of impaired cell operation on board the ATS-1 and Intelsat II-F4 spacecraft.[117] This experience indicated that low-energy proton radiation has a much greater effect on radiation-unprotected cell V_{0c} and P_{max}, compared to electron or high-energy proton fluences.

Radiation damage to even a small portion of the individual solar cell results in an increase in saturation (leakage) current for the whole cell. The cell diode saturation current, I_0, increases with decreased minority carrier lifetime, as discussed in Section 3.7. The minority carrier lifetime decreases due to the displacement damage in the depletion region, caused by the incident protons. The corresponding increased saturation current reduces the cell V_{0c}, as seen in Eq. (13.60). Since a forward-biased cell diode current increases for all diode voltages, the current available to the load decreases, so that P_{max} will also decrease.

The following paragraphs enumerate and provide working explanations of the four principal types of solar cells. Figure 13.43 shows the solar cell theoretical maximum efficiency versus the corresponding cell material band gap for a number of feasible cell materials.[118] As an aside, given that the maximum power output of silicon is less than that of GaAs, as shown in Fig. 13.43, it is anticipated that GaAs material may make a better solar cell than Si, especially at high temperatures, such as produced by a solar cell light concentrator. Concentrators usually consist of individual cell mini-telescopic mirror/lense optics. Not unlike standard telescopes, they focus the incident light onto the cell, making for increased cell efficiency, cell area reduction, and so cell cost.

Since the solar light spectrum is relatively high in the blue and ultraviolet regions, the solar cell spectral response should also be high in these regions, for maximum efficiency. The response of solar cell materials in the above spectral regions indicate that a heavier carrier density will exist near the front surface of the cell, as opposed to the bulk. Hence, by placing the cell junction near the center of this carrier concentration by constructing a shallow junction, relatively more carriers will be collected before recombination than would be the case for a deep junction.

The first type of cell to be described is called the violet cell, and provides for a shallow junction. This junction is less than $0.2\,\mu$m deep, with a decreased series resistance, R_{sr}, of about 0.05 ohms. Tantalum pentoxide antireflection (AR) coatings were developed for this cell, which absorb less blue light and have a high index of refraction.[119] For these cells, $I_{sc}/A_J = 40$ mA per cm^2, compared to 30 mA per cm^2 for conventional cells. Air mass zero (AMO) sunlight efficiencies of about 14 percent are claimed for 2 ohm-cm material violet cells. This cell type is now essentially standard, replacing deep junction cells.

Back Surface Field (BSF) cells have a built-in electric field on their back inside surface, just forward of the cell rear contact. This back surface field is produced by forming an acceptor dopant gradient at the rear surface of the cell, through "alloying-in" an impurity such as aluminum. The field polarity is such as to reverse the direction of carriers in its vicinity to drift them toward the junction. BSF cells can achieve a 20 percent increase in V_{0c} (to about 600 mV) over conventional cells, and are independent of cell thickness. I_{sc} increases of about 10–15 percent, and P_{max} increases of about 13–25 percent are also attained in these cells. However, the effectiveness of the back surface field disappears after only a small amount of damaging radiation (electron fluence) is incident.[120]

The textured surface cell is constructed by etching small, densely packed pyramid shapes into its front surface, to act as multiple reflection light collectors. These cells appear black in plan view because there is no upward reflected light from their surfaces. This cell is also termed the Comsat Non-Reflecting, or CNR, cell. An I_{sc} and P_{max} increase of about 8 percent over the conventional cell is realized with the CNR cell. Another version of the CNR cell is the Vertical Junction (VJ) cell. It has much deeper and more vertically shaped cuts into its front surface than the CNR cell, to catch the light.[120]

Back surface reflector (BSR) cells are similar to textured cells with a reflector on their rear surface. The reflector is used to reflect long wavelength light from the rear surface back through the cell and out through the front, thus again crossing the junction to supply useful current. This amounts to an increase of I_{sc}/A_J and $P_{\max}$ of a few percent. However, this increase is enough to influence all solar cell construction to incorporate this feature for use in space.

Hardening solar cells consists mainly of cell adjuncts, in contrast to primarily hardening them through the use of subtle manufacturing processes, as is done with integrated circuits. For example, adding a transparent cell cover glass of a few mils thickness, plus special tapes to seal all cell edges and inadvertent apertures, hardens cells against most electron and proton fluence, such as that from the Van Allen belts. For n/p cells, the preradiation base resistivity should be low, i.e., high N_A base doping for a low value of saturation current I_0, while maintaining a relatively large diffusion length.

Attempts to artificially heat cells to accelerate their annealing have been considered. One method with some coinage is to temporarily interpose a "black" mirror, with high absorption of sunlight, to heat the cell to raise its temperature.

Studies of both Si and GaAs cells have shown that, for neutrons and protons, maximum radiation hardness is obtained when the diffusion length is large compared to the junction depletion layer thickness.[121] Lithium-doped cells are harder to neutrons and protons than those without Li dopant. However, Li-doped GaAs annealing recovery time is excessively long, especially for military applications.[122] It has been suggested that Li-doped hardness increase is due to Li combining with O_2 single vacancies and divacancies,[123] thus reducing their trapping cross sections for any further recombination processes.[125]

REFERENCES

1. F. B. McDonald. "Review of Galactic and Solar Cosmic Rays," *in* Second Symposium on Protection Against Radiation in Space, NASA Document SP-71, pp. 19–30, Oct. 12, 1964.
2. C. B. Gabbard et al. The Impact of Manned Launches on DOD Space Programs. RDA, Marina Del Rey, CA, Mar. 1977.
3. S. M. Seltzer. "Electron Bremsstrahlung and Proton Depth-Dose Data For Space Shielding Applications." IEEE Trans. Nucl. Sci. NS-26(6):4896–4904, Dec. 1979.
4. W. W. Scott. Estimates of Primary and Secondary Particle Dose Behind Aluminum

and Polyethylene Slabs Due to Solar Flares and Van Allen Belt Protons. Oak Ridge Nat. Labs, ORNL-RSIC-18, July 1967.
5. W. R. Yucker and J. R. Lilley. CHARGE, A Space Radiation Shielding Code. McDonnell Douglas, Astronautics, Huntington Beach, CA, DAC62231, Apr. 1969.
6. R. J. Alexander, Jr. High Energy Nuclear Transport: Protection Against Space Radiation. NASA SP-169, pp. 403–428, 1968.
7. R. J. Alsmiller, Jr., J. W. Wachter, and H. S. Moran. "Calculation of the Neutron and Proton Spectra From Thick Targets Bombarded by 450 MeV Protons and Comparison With Experiment." Nucl. Sci. Eng. 291–294, June 1969.
8. W. S. Snyder, H. A. Wright, J. E. Turner and J. Neufeld. "Calculations of Dose-Depth Curves For High Energy Neutrons and Protons and Their Interpretation For Radiation Protection." Nucl. Appl. 6(4):336–343, April 1969.
9. R. J. Alsmiller, Jr., D. C. Irving, and H. S. Moran. "The Validity of the Straight Ahead Approximation in Space Vehicle Shielding Studies." Nucl. Sci. Eng. 32(1): 56–61, April 1968.
10. H. J. Schaefer. The LET Spectrum of Galactic Radiation Exposure in Space and its Transmission in Shield or Tissue, in Protection Against Space Radiation. NASA Document SP-179, pp. 43–58, 1968.
11. N. J. Rudie. Principles and Techniques of Radiation Hardening, Vol. 3. Western Periodicals Publ., N. Hollywood, CA, app. F, 1980.
12. H. A. Kramers. Phil. Mag. (London) No. 46, p. 836, 1923.
13. R. D. Evans. The Atomic Nucleus. McGraw-Hill, New York, pp. 116–117, 1955.
14. N. J. Rudie. "Principles and Techniques of Radiation Hardening", Vol. 1, Third Edition. Western Periodicals Co., N. Hollywood, CA, pp. 2–47, 1986.
15. M. J. Berger and S. M. Seltzer. "Results of Some Recent Transport Calculations For Electrons and Bremsstrahlung." in Second Symposium on Protection Against Radiation in Space. NASA SP-71, pp. 437–448, Oct. 12, 1964.
16. Berger and Seltzer, op. cit
17. G. D. Magnuson and A. W. McReynolds. Second NASA Symp. NASA SP-71, p. 459.
18. J. H. King. Models of the Trapped Radiation Environment: Low Energy Protons. NASA Report SP-3204, Vol. 4, 1967.
19. V. A. J. Van Lint, G. Gigas, and J. Barengoltz. "Correlation of Displacement Effects Produced by Electrons, Protons, and Neutrons in Silicon." IEEE Trans. Nucl. Sci. NS-22(6):2663–2668, Dec. 1975.
20. E. C. Smith, D. Binder, P. A. Compton, and R. I. Wilbur. "Theoretical and Experimental Determination of Neutron Energy Deposition in Silicon." IEEE Trans. Nucl. Sci. NS-13(6), Dec. 1966.
21. M. E. Wyatt, V. A. J. Van Lint and E. G. Wikner. USAF Materials Report, TR-66-77, April 1966.
22. W. Rosenzweig, F. W. Smith and W. L. Brown. J. Appl. Phys. 35:2707, 1964.
23. J. E. Gover. "Overview of Nuclear and Space Radiation Effects Technology." IEEE NSRE Conference Short Course, Part I, Cornell Univ., Ithaca, NY, July 1980.
24. L. L. Bonzon (Sandia) and N. A. Lurie (IRT). "Best Estimate of LOCA Radiation Signature." NUREG/CR-1237, SAND79-2143, prepared for USNRC, Jan. 1980.
25. D. M. Binkley. "A Radiation Hardened Accelerometer Preamplifier For 100 Mrad (Si) Total Dose." IEEE Trans. Nucl. Sci. NS-29(6):1500–1507, Dec. 1982.
26. D. K. Myers and W. Herzog. "Feasibility Study on the Design and Radiation Testing of Complex Functions For 200 Mrad (Si)." IEEE Trans. Nucl. Sci. NS-29(6):1508–1511, Dec. 1982.
27. P. H. Gudiksen, T. J. Sullivan, and T. F. Harvey. "The Current Status of the Atmos-

pheric Release Advisory Capability (ARAC) and its Application to the Chernobyl Event." IEEE Trans. Nucl. Sci. NS-34(1):584–594, Feb. 1987.
28. S. A. Fetter and K. Tsipis. "Catastrophic Releases of Radioactivity." Sci. Amer. 244(4), Apr. 1981.
29. J. C. Devine. "A Progress Report: Cleaning Up TMI." IEEE Spectrum. table, p. 48, March 1981.
30. EPRI Journal, June 1980, p. 26.
31. J. J. Halpin et al. Guidelines For Nuclear Effects Survivability of Tactical Army Systems. Harry Diamond Labs. HDL-PRL-76-2, Mar. 1976.
32. S. Glasstone (ed.). The Effects of Nuclear Weapons. U.S. Dept. Defense, U.S. AEC. Publ., sect. 3.67, April 1962.
33. Glosstone, ibid.
34. Halpin, op. cit. pp. 164–166.
35. Halpin, ibid.
36. Nuclear Survivability Criteria For Military Equipment: Immediate Transient Incapacitation. U.S. Army Nuclear and Chemical Agency, Fort Belvoir, VA, Document QSTAG-244, 1975.
37. H. B. Gerstner. "Acute Radiation Syndrome in Man." U.S. Armed Forces Med. J. 9(3):313–354, Mar. 1958.
38. A. B. Phillips. Transistor Engineering. McGraw-Hill, New York. pp. 199–203, 1962.
39. A. S. Grove. Physics and Technology of Semiconductor Devices. Wiley, New York, pp. 214–228, 1967.
40. S. K. Ghandi. Semiconductor Power Devices: Physics of Operation and Fabrication Technology. Wiley-Interscience, New York, pp. 152–167, 1977.
41. G. C. Messenger, M. S. Cooper and J. P. Retzler. "Combined Neutron and Thermal Effects on Bipolar Transistor Gain." Trans. Nucl. Sci. NS-26(6):4758–4762, Dec. 1979.
42. Messenger et al., ibid. p. 4758.
43. Ghandi, op. cit.
44. Messenger et al., ibid.
45. Messenger et al., op. cit.
46. E. G. Stassinopolous et al. IEEE Trans. Nucl. Sci. NS-24(6), Dec. 1977.
47. G. H. Sandler. System Reliability Engineering. Prentice-Hall; Englewood Cliffs, NJ, pp. 162ff., 1963.
48. Sandler, ibid.
49. T. C. May and M. H. Woods. IEEE Proc. 1978 International Reliability Physics Symp., IEEE Trans. Electron Dev., ED-26, No. 1, 1979.
50. D. S. Young, J. T. Nelson, and L. L. Vanskike. IEEE Trans. Electron Dev., ed-26, No. 10, 1979.
51. Sandler, op. cit.
52. J. P. Retzler. General Hardness Design Guidelines For Design Engineers. Litton Guidance and Control Div., Woodland Hills, CA, July 1973.
53. L. W. Ricketts. Fundamentals of Nuclear Hardening of Electronic Equipment. Wiley-Interscience, New York, pp. 357ff., 1972.
54. M. Schlenther et al. "In Situ Radiation Tolerance Tests of MOS RAMs." IEEE Trans. Nucl. Sci. NS-25(6), Dec. 1978.
55. U.S. Dept. of Defense Handbook. Reliability Prediction of Electronic Equipment. MIL-HDBK-217C, 9 Apr. 1979, RBRT, RADC, Griffis AFB.
56. D. Bondurant and F. Gnadinger. "Ferroelectronics For Non-volatile RAMS." IEEE Spectrum, pp. 30–33, July 1989.

57. W. I. Kinney, W. Shepard, W. Miller, J. Evans, and R. Womack. "A Non-Volatile Memory Cell Based on Ferroelectric Storage Capacitors." IEDM Tech. Digest. Paper 3.9, pp. 850–851, Dec. 1987.
58. C. T. Sah. "Evolution of the MOS Transistor-From Conception to VLSI." Proc. IEEE, *76*, Oct. 1988.
59. R. Womack, W. Kinney, B. Shepard, W. Miller, and J. Evans. "An Experimental 512-Bit Non-Volatile Memory With Ferroelectric Storage Cell." IEEE Int'l. Solid State Conf. San Francisco, Feb. 12, 1988.
60. G. C. Messenger. "Ferroelectric Memories." DNA-TR-88-94, RDA, Marina Del Rey, CA, June 20, 1988.
61. J. J. Chang. "Non-volatile Semiconductor Memory Devices." Proc IEEE, *64*, pp. 1039–1059, July 1976.
62. J. P. Spratt and D. Z. Harbert. Radiation Hardened Microprocessor Study. Report no. RADC-TR-78-231, USAF Systems Comm. Griffis AFB, Oct. 1978.
63. Spratt and Harbert, op. cit. p. 80.
64. M. E. Baumbaugh, H. A. Eisen, and L. M. Hitt. "Ionizing Radiation Response of the 80286 Microprocessor Under Different Test Conditions." J. Rad. Eff.; Res. and Eng., *5*, 1:38–42, Winter 1986.
65. "Preliminary Radiation Effects Tests of the 80C31." Ltr., Sandia National Labs, CA. 1985.
66. Harris Data Compilation (Adv) on HS-80C85RH. CA. 1984; Briefing Charts CA. 1985.
67. B. G. Blott and A. H. Siedle. "Gamma Ray Effects on an N-Channel MOS Microprocessor (Z8002) In the Zero Bias Condition." IEEE Trans. Nucl. Sci. NS-30(5): 4016–4021, Oct. 1983.
68. A. Gupta and H. D. Toong. "Microprocessors: The First Twelve Years." Proc. IEEE 71(11):1236–1256, Nov. 1983.
69. EDN's (Electronic Defense News Mag.) Eleventh Annual μP/μC Chip Directory, Nov. 15, 1984.
70. K. Karstad. "CMOS-SOS LSI Family Forms High Performance Rad-Hard Processor." RCA CMOS Memory/Microprocessor IC Application Note ICAN 7319, Mar. 1986.
71. P. R. Measel et al. "Development of a Hard Microprocessor." IEEE Trans. Nucl. Sci. NS-28(6), Dec. 1981.
72. N. Wilkin, C. T. Self and H. Eisen. "Ionizing Dose Rate Effects in Microprocessors." IEEE Trans. Nucl. Sci. NS-27(6):1420–1424, Dec. 1980.
73. A. H. Johnston, K. Marks and W. E. Will. "Annealing of Total Dose Damage in the Z8002 Microprocessors." IEEE Trans. Nucl. Sci. pp. 4046–4050, Dec. 1981.
74. E. E. King and G. J. Manzo. "Total Dose Failure Levels of VLSI ICs." IEEE Trans. Nucl. Sci. NS-27(6):1449–1453, Dec. 1980.
75. R. Polimode, H. Eisen and K. Pinero. Nuclear Radiation Effects Data on Large Scale Integrated Circuits. Harry Diamond Labs, HDL-D5-80-1, July 1980.
76. P. G. Coakley, N. J. Colella, J. L. Kaschmitter, C. L. McKnett and D. L. Shaeffer. "Operation of Commercial R3000 Microprocessors in the Low Earth Orbit (LEO) Space Environment." IEEE Trans. Nucl. Sci. NS-38(6):1940–1945, Dec. 1991.
77. M. A. Rose et al. IEEE 1984 Nuclear Science Radiation Effects Conference Short Course. Colo. Springs, CO, Sect. 2.
78. F. W. Sexton et al. "Radiation Testing of the CMOS 8085 Microprocessor Family." IEEE Trans. Nucl. Sci. NS-30(6):4235–4239, Dec. 1983.
79. N. E. Baxter et al. "Evaluation of the Radiation Hardness of the Ferranti F100-L Microprocessor in an Operating System." IEEE Trans. Nucl. Sci. NS-29(6):1737–1739. Dec. 1982.

80. RCA Advanced Technology Labs. Appl. Note, "Generic VHSIC Spaceborne Computer." Oct, 1985.
81. Spratt and Marbert, op. cit. p. 68 ff.
82. V. Nickel and P. Rosenberg. Radiation Hardened Microprocessors. Final Report AFAL Contract F33615-77-C-1001, 31 Dec. 1976.
83. C. L. Axness, S. P. Buchner, G. L. Hash, K. J. Hass, K. H. Hughes, K. Kang, F. W. Sexton and R. K. Treece. "SEU Characterization and Design Dependence of the SA3300 Microprocessor." IEEE Trans. Nucl. Sci. NS-37(6):1861–1868, Dec. 1990.
84. A. B. Holman, E. C. Smith and G. W. Autio. "Verification of Radiation Background Rates in an IR Sensor System." IEEE Trans. Nucl. Sci. NS-23(6):1775–1780, Dec. 1976.
85. S. M. Seltzer. "SHIELDOSE: A Computer Code For Space Shielding Radiation Dose Calculations." NBS Tech. Note 1116, Washington, DC., May 1980.
86. E. G. Stassinopoulos and J. P. Raymond. "The Space Radiation Environment For Electronics." Proc. IEEE 76(11), Nov. 1988.
87. S. M. Seltzer. "Conversion of Depth-Dose Distributions From Slab to Spherical Geometries For Space Shielding Applications." IEEE Trans. Nucl. Sci. NS-33(6): 1292–1297, Dec. 1986.
88. J. V. Bailey. "Dosimetry During Space Missions." IEEE Trans. Nucl. Sci. NS-23(4): 1379–1384, Aug. 1976.
89. M. L. Rossi and M. C. Stauber. "Radiation Protection Design Considerations For Man in Geosynchronous Orbit." IEEE Trans. Nucl. Sci. NS-24(6):2248–2251, Dec. 1977.
90. E. G. Stassinopoulos et al. "SOFIP-A Short Orbital Flux Integration Program." NASA NSSDC/WDC-A-R&S 79-01, GSFC, Greenbelt, MD, Jan. 1979.
91. E. G. Stassinopoulos et al. "SOLPRO-A Computer Code to Calculate Probibilistic Energetic Solar Proton Fluences." NASA NSSDC 75-11, GSFC, Greenbelt, MD, April 1975.
92. E. G. Stassinopoulos et al. "ALLMAG, GDALMG, LINTRA—Computer Programs For Geomagnetic Field And Field-Line Calculations." NASA NSSDC72-12, Feb. 1972.
93. P. Shapiro. "CRUP-Cosmic Ray Upset Program." NRL Memorandum Report 5171, Sept. 14, 1983.
94. J. H. Adams, Jr. "CREME-Cosmic Ray Vs LET Program." NRL Code 6611, Washington, D.C., ca, 1982.
95. D. M. Sawyer and J. I. Vette. "AP8 Trapped Proton Environment For Solar Maximum and Solar Minimum." NASA NSSDC, Rpt. 76-06, GSFC, Greenbelt, MD, Dec. 1976.
96. D. M. Sawyer and J. I. Vette. "AE8 Trapped Electron Environment For Solar Maximum And Solar Minimum." To be published.
97. P. J. McWhorter, S. L. Miller, T. A. Dellin and C. A. Axness. "Retention Characteristics in SNOS Nonvolatile Devices in a Radiation Environment." IEEE Trans. Nucl. Sci. NS-34(6):1652–1657, Dec. 1987.
98. J. E. Gover and J. S. Browning. "Radiation Hardening Systems Considerations: Electrical Systems And Radiation Environments." Report No. SAND 86-1737, Sandia National Laboratories, Albuquerque, NM, Mar. 1987.
99. P. J. McWhorter, S. L. Miller and T. A. Dellin. "Radiation Response of SNOS Nonvolatile Transistors." IEEE Trans. Nucl. Sci. NS-33(6):1414–1419, Dec. 1986.
100. E. S. Snyder, P. J. McWhorter, T. A. Dellin and J. D. Sweetman. "Radiation Response of Floating Gate EEPROM Memory Cells." IEEE Trans. Nucl. Sci. NS-36(6):2131–2139, Dec. 1989.

101. International Rectifier Corp., El Segundo, CA. Reliability Symposium, 1984.
102. P. E. Thibodeau. "Radiation Effects on Power MOSFETs Used in Electronic Power Conversion." International Rectifier Corp., El Segundo, CA. Report, 1987.
103. D. L. Blackburn, T. C. Robbins and K. F. Galloway. "VDMOS Power Transistor Drain-Source Resistance Radiation Dependence." IEEE Trans. Nucl. Sci. NS-28(6): 4354–4359, Dec. 1981.
104. D. A. Grant and J. Gowar. Power MOSFETs: Theory and Applications. J. Wiley Interscience, NY, 1989.
105. D. L. Blackburn, J. M. Benedetto and K. F. Galloway. "The Effect of Ionizing Radiation on the Breakdown Voltage of Power MOSFETs." IEEE Trans. Nucl.Sci. NS-30(6):4116–4121, Dec. 1983.
106. Cf. W. R. Dawes, Jr. "Radiation Effects Hardening Techniques." IEEE NSRE Short Courses, Chaps. 2, July 1985, 1987.
107. A. E. Waskiewicz, J. W. Groninger, V. H. Strahan and D. M. Long. "Burnout of Power MOS Transistors With Heavy Ions of Californium-252." IEEE Trans. Nucl. Sci. NS-34(6):1710–1713, Dec. 1986.
108. M. K. Gauthier and A. R. V. Dantas. "Radiation-Hard Analog-to-Digital Converters For Space And Strategic Applications." Jet Propulsion Lab., Publ. 85-84, Pasadena, CA. Nov. 15, 1985.
109. H. Y. Tada, J. R. Carter, Jr., B. E. Anspaugh and R. G. Downing. "Solar Cell Radiation Handbook." Jet Propulsion Lab. Pub. 82–69, Nov. 1, 1982.
110. M. Wolf and H. Rauschenbach. "Series Resistance Effects on Solar Cell Measurements." Advanced Energy Conversion, Vol. 3, pp. 455–479, 1963.
111. Tada, et al., op. cit. pp. 3–18.
112. W. T. Picciano." Determination of the Solar Cell Equation Parameters Including Series Resistance From Empirical Data." Energy Conversion 9(1), 1969.
113. E. J. Stofel, T. B. Stewart and J. R. Ornelas. "Neutron Damage to Silicon Solar Cells." IEEE Trans. Nucl. Sci. NS-16(5):97–101, Oct. 1969.
114. J. J. Kalinowski and R. K.Thatcher. "Transient Radiation Effects on Electronics (TREES) Handbook." DASA 1420, 2nd Ed., 2nd Rev., p. F-99, Oct. 3, 1969.
115. Tada et al., op. cit. p. 3–40.
116. Tada et al., op. cit. pp. 3–18, 24.
117. Tada et al., op. cit. p. 3–33.
118. J. J. Loferski. "Principles of Photovoltaic Energy Conversion." Proc. 10th IEEE Conference of Photovoltaic Specialists, p. 1, 1973.
119. A. G. Revesz. "Vitreous Oxide Anti-reflection Films in High-Efficiency Solar Cells." Proc. 10th IEEE Conference of Photovoltaic Specialists, p. 180, 1973.
120. Tada et al., op. cit. p. 1–34.
121. C. E. Barnes and J. J. Wiczer. "Neutron Damage Effects in AlGaAs/GaAs Solar Cells." IEEE Trans. Nucl. Sci. NS-31(6):1471–1476, Dec. 1984.
122. Tada et al., op. cit. p. 3–42.
123. I. Weinberg, S. Mehta and C. K. Swartz. "Radiation Damage and Defect Behavior in Ion-Implanted, Lithium Counter-doped Silicon Solar Cells." NASA Tech. Memo. No. 83646, Presented at 17th Photovoltaic Specialists Conf., Kissimee, FL, May 1–4, 1984.
124. B. L. Gregory and H. H. Sander. "Injection Dependence of Transient Annealing in Neutron Irradiated Silicon Devices." IEEE Trans. Nucl. Sci. NS-14(6):116–126, Dec. 1967.
125. J. J. Wysocki. "Role of Lithium in Damage and Recovery of Irradiated Silicon Solar Cells." IEEE Trans. Nucl. Sci. NS-14(6):103–109, Dec. 1967.
126. E. G. Stassinopolous. "Radiation Environment of Space." Chap. 1, Nuclear and

Space Radiation Effects Conference Short Course, Reno, NV, July 1990.
127. Rev. Mod. Phys. 57(3):Part II, July 1985 (whole issue).
128. P. G. Coakley, N. J. Colella, J. L. Kaschmitter, C. L. McKnett and D. L. Shaeffer. "Operation of Commercial R3000 Microprocessors in the Low Earth Orbit (LEO) Space Environment." IEEE Trans. Nucl. Sci. NS-38(6):1940–1945, Dec. 1991.
129. J. C. Higdon and L. E. Lingenfelter. "Gamma Ray Bursts." Ann. Rev. Astron. and Astrophys. Vol. 28, pp. 401–436, 1990; Sci. Amer. Oct. 1976, July 1980, Feb. 1985, and Dec. 1991.
130. IEEE NSRE Conference Short Course Syllabus, Ch. 1. San Drego, CA, July 15, 1991.
131. G. P. Summers, E. A. Wolicki, M. A. Xapsos, P. Marshall, C. J. Dale, M. A. Gehlhausen and R. D. Blice. "Energy Dependence of Proton Displacement Damage Factors for Bipolar Transistors." IEEE Trans. Nucl. Sci. NS-33(6):1282–1286, Dec. 1986.
132. G. P. Summers, E. A. Burke, C. J. Dale, E. A. Wolicki, P. W. Marshall and M. A. Gehlhausen. "Correlation of Particle Induced Displacement Damage in Silicon." IEEE Trans. Nucl. Sci. NS-34(6):1134–1139, Dec. 1987.

CHAPTER 14
STATISTICAL ASPECTS OF HARDNESS DESIGN

14.1 INTRODUCTION

It should be recognized that, in dealing with commercially manufactured semiconductor devices, seldom can the same part type be obtained with identical electrical and physical parameters from different manufacturers, as evidenced by the spread in their characteristics, seen in vendor specification sheets. This comes about through the reality of the various semiconductor batch processes of manufacturing practiced throughout the world today. Even though identical devices are not produced, sufficiently stringent product line standards are usually invoked, so that parameter dispersions are tolerable in that differences can be compensated through normal circuit design. However, not only do the same part types from different vendors differ in their electrical characteristics, but the same holds within an identical part type batch (date code) from the same vendor. This variation of device characteristics also carries over to their relative vulnerability to various types of nuclear radiation, so that, even with their other parameters being equal, some of these parts turn out to be harder than others. For these reasons sample lot testing and selection of hard parts is a valid and often crucial procedure for hardening particular electronic systems.

Hence, the above considerations must be taken into account in systems, circuit, and component hardness design. This asserts that the statistical aspects of the physical and electrical parameters of devices play an important role in their hardness characterization. The influence of these aspects is felt not only during the design phases of system hardening, but, even more importantly, in the hardness-assurance phase, and as the system is deployed in the field. Hardness assurance is discussed in detail in Chapter 15.

14.2 SYSTEM SURVIVAL PROBABILITIES

It is apparent that the sophistication of modern military electronic systems is increasing in such a manner that the imposition of hardening requirements is becoming one of the major influences in their design, construction, and procurement. This can be seen in the increasing scope of their corresponding nuclear survivability specifications.

It goes without saying that one of the important constraints in the hardening of systems is that their nuclear survivability design be accomplished in an efficient manner. One means of initiating a system nuclear survivability investigation is to assign each module of the system with a required survival probability, p_i, consistent with system survival probability, P_s, plus an allocated hardening "cost," C. Assume for illustrative purposes that each module is statistically independent of the others; then[1]

$$P_s = \prod_{i=1}^{n} p_i; \qquad p_i \leqslant 1 \tag{14.1}$$

where n is the number of modules or subsystems in the system. The system manager should strive to maximize P_s consistent with a generalized "cost" C. C is not necessarily a monetary cost, but rather a generalized allocation of all resource costs. It serves to substantively constrain the system to keep it non-trivial, as it would be if $C = 0$. If C vanishes, then for maximum survivability, all of the $p_i \equiv 1$ for an absolutely reliable system at no cost. The cost can be related to the module survival probabilities by

$$C = \sum_{i=1}^{n} k_i p_i \tag{14.2}$$

where the coefficients k_i are known proportionality constants that convert module survival probabilities to module costs. Abstractly, the task is to find the maximum P_s consistent with the two constraints, namely, the cost given by Eq. (14.2) and that $p_i \leqslant 1$ in keeping with the idea of survival probabilities being less than or equal to unity. The solution to the above is well known, (e.g., using the method of Lagrange Multipliers) giving an optimal assignment of module probabilities such that $k_i p_i = C/n$ for all i. That is, the optimal P_s can be attained if the cost, C, is evenly divided among the modules. Each module is allocated a cost C/n, so that its corresponding optimal survival probability $p_i^* = (C/n)/k_i$. The sought-after overall system survival probability is then given by substituting p_i^* into Eq. (14.1). The foregoing provides one interpretation of the meaning of a module probability assignment consistent with an overall system survival probability, and should be construed as metaphor for system reliability in the above sense.

At this juncture, module probabilities that are not mission critical would be removed from Eqs. (14.1) and (14.2). The same holds for any subsystem that consists solely of passive components, because the assumption is made that it is normally inherently hard. System design margins discussed in Section 13.5 must also be established, which would be reflected in derating circuit and device parameters with respect to minimum performance requirements. Then the transition to survival probabilities can be made, as will be discussed.

Once the derating factors have been computed, then circuits that contain devices or components that will operate well within their deratings can also be removed, as above. Those subsystems that are not inherently hard are those that remain to comprise Eqs. (14.1) and (14.2). Then additional analysis must be performed on these circuits, from which the coefficients k_i and circuit survival probabilities p_i are derived. This additional analysis consists of the next level of methodology. It is a detailed function of the incident radiation environments and pertinent device response discussed in the following and in Chapter 15.

14.3 SURVIVAL PROBABILITY DISTRIBUTIONS

As can be appreciated, sample lot testing and corresponding investigations of system piece parts can be described by the language and methods of the statistical inference theory of sampling. Sampling simply consists of removing a small number of parts, without ever replacing them, from a much larger universe of parts and subjecting them to various testing procedures. Statistical inferences are then made about the universe of parts from the sample tested.

For example, the piece part mean value of a particular measured parameter and its standard deviation measured before and after exposure to test radiation are to be construed as estimates of the "true" mean and standard deviation of the whole universe of parts. These statistics are interpreted as the results of a hypothetical computation of the mean and standard deviation, before and after testing, of the total factory output of this part from all manufacturers that have been making it before and will continue making it after the above measurements. Great care must be taken to insure that the sample is actually representative of the future part population. The sample should be selected from several lots representing each manufacturer. Again, the best that can be done is to sample from this universe of parts, measure, compute statistics without part replacement, and draw inferences from these results.

An important probability distribution pertaining to electronic parts and devices affected by radiation is the log-normal distribution. It is simply the well-known normal distribution, where the independent variable is the logarithm of the value of the part parameter of interest, in contrast to merely the value of the part parameter itself. To illustrate, assume that the incremental probability of failure, $dP_f(\Phi_n)$, of a device subjected to a neutron fluence increment $d\Phi_n$ about Φ_n is given by the log-normal distribution, which is

$$dP_f(\Phi_n) = \frac{1}{\sqrt{2\pi}\sigma}\left(e^{-(\ln \Phi_n - \mu)^2/2\sigma^2}\right)\frac{d\Phi_n}{\Phi_n} \tag{14.3}$$

It is understood that Φ_n is normalized to one neutron per cm^2, i.e., the independent variable is Φ_n/Φ_0 where $\Phi_0 = 1$ neutron per cm^2, so that Φ_n/Φ_0 is dimensionless. This kind of normalization to unity is also assumed to hold for other radiation quantities, such as ionizing dose. Also, μ and σ are normalized in the same way.

The cumulative probability distribution to device failure at the fluence level Φ_n is the integral of the above to give

$$P_f(\Phi_n) = \frac{1}{\sqrt{2\pi}\sigma}\int_0^{\Phi_n} e^{-(\ln \Phi'_n - \mu)^2/2\sigma^2}\frac{d\Phi'_n}{\Phi'_n} \tag{14.4}$$

where the parameters $\mu = \overline{\ln \Phi_n}$, and $\sigma^2 = \overline{(\ln \Phi_n)^2} - \mu^2$. Notice that μ and σ^2 depend on $\ln \Phi_n$, not Φ_n itself. The expectation value (not the mean μ), is

$$E(\Phi_n) = \int_0^{+\infty} \Phi_n dP_f(\Phi_n) \Big/ \int_0^{+\infty} dP_f(\Phi_n) = \exp(\mu + \sigma^2/2) \tag{14.5}$$

Also, other statistical parameters are: (a) the median fluence value for the cumulative log-normal distribution, that is, where $P_f(\Phi_{nme}) = \frac{1}{2}$ is the fluence $\Phi_{nme} = \exp \mu$. (b) the mode, which is the fluence corresponding to the most probable value of the log-normal distribution $dP_f(\Phi_n)$ is $\Phi_{nmo} = \exp(\mu - \sigma^2)$; (c) the variance given by

$$\mathrm{var}\,\Phi_n = [\exp(2\mu + \sigma^2)](\exp \sigma^2 - 1) \tag{14.6}$$

with the standard deviation $[\mathrm{var}(\Phi_n)]^{1/2}$. Also, note that, for discrete values of Φ_n computed from a group of measurements, the mean, μ, is obtained from

$$\mu = \overline{\ln \Phi_n} = \frac{1}{N}\sum_{i=1}^{N} \ln \Phi_{ni} \tag{14.7}$$

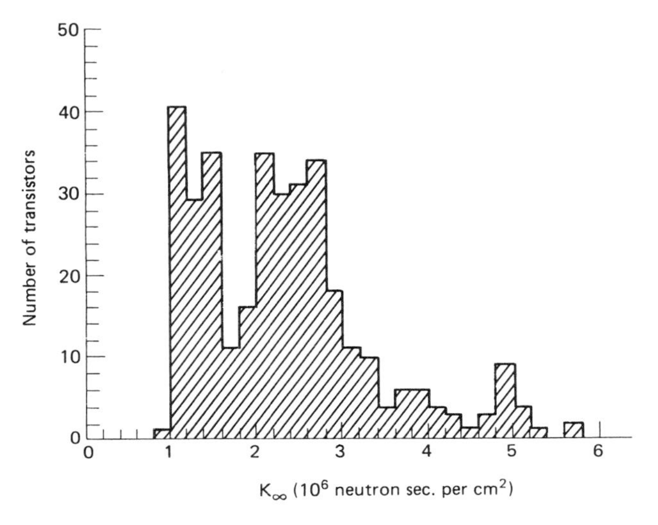

Fig. 14.1 Histogram of measured values of K_∞ for 344 transistors.[2] © 1968 by the IEEE.

where N is the number of fluence measurements made, and μ is related to Φ_{nme} by

$$\Phi_{nme} = \exp \mu = \exp \left(\frac{1}{N} \sum_{i=1}^{N} \ln \Phi_{ni} \right) = \left(\prod_{i=1}^{N} \Phi_{ni} \right)^{1/N} \tag{14.8}$$

That is, the median Φ_{nme} is also the geometric mean of the suitably normalized measurements Φ_{ni}. Of course, all of the above holds for continuous log-normal distributions. For sufficiently small σ, var $\Phi_n \cong (\Phi_{nme}\sigma)^2$, and $\Phi_{nmo} \cong \Phi_{nme}$. There are a number of other probability distributions applicable to statistical use herein, such as the normal, or gaussian, and Weibull distributions.

An initial statistical attempt to correlate radiation parameters from observed experimental data is based on a histogram of 344 bipolar transistors,[2] given in Fig. 14.1, as a function of their lifetime damage constant K, as it appears in the Messenger-Spratt equation (5.66), which for large emitter currents is called K_∞. Figure 14.2 depicts $K' = 1/K$ as a function of the transistor emitter current density J_E as obtained from experimental data.[3] With a knowledge of K_∞, and using Fig. 14.2, the value of K can be determined in a particular application. K_∞ usually corresponds to the value of K at the maximum of current gain versus $I_E \simeq I_C$ (Fig. 3.24) which is about 100 amps per cm^2 for typical bipolar devices. From Fig.

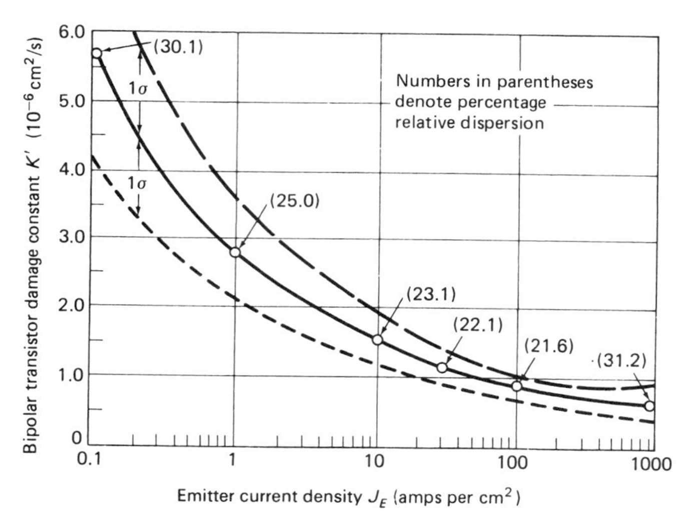

Fig. 14.2 Damage constant K' as a function of emitter current density for *npn* transistors.[5] © 1979 by the IEEE.

14.2 at this value of J_E it is seen that the appropriate $K' \cong 10^{-6}\,\text{cm}^2$ per neutron second. Using Eq. (5.66),

$$\Delta(1/\beta) = \Phi_n/2\pi f_T K = K_D \Phi_n \tag{14.9}$$

where again $K' = 1/K$ and $K_D = 1/2\pi f_T K$. K_D is representative of a given bipolar transistor, f_T is its gain-bandwidth product, and K is to be regarded as a universal damage constant for silicon.

The histogram for K_∞ in Fig. 14.1 is a very useful distribution that can be employed to estimate damage constants in bipolar transistors and as an aid to estimate other transistor parameters as well. A log-normal distribution was fitted to this histogram with $\mu = 14.54$ and $\sigma = 0.423$.[4] From a rigorous statistical point of view, using a log-normal distribution to fit the histogram of Fig. 14.1 is not strictly valid, as a chi-square computation will attest.[5] However, the dispersion of points in Fig. 14.3, which is a normal probability plot of $\ln K_\infty$ can be shown to correspond to a log-normal distribution of K_∞ for a population of 53 transistors, or about one-sixth of the original 344 transistors.[5] So it is in this sense that the log-normal distribution will be retained. For notational convenience, Eq. (14.4) can be rewritten as

$$P_f(\Phi_n) = \frac{1}{\sqrt{2\pi}} \int_{-\infty}^{(\ln \Phi_n - \mu)/\sigma} e^{-y^2/2}\, dy \equiv F_n\left(\frac{\ln \Phi_n - \mu}{\sigma}\right) \tag{14.10}$$

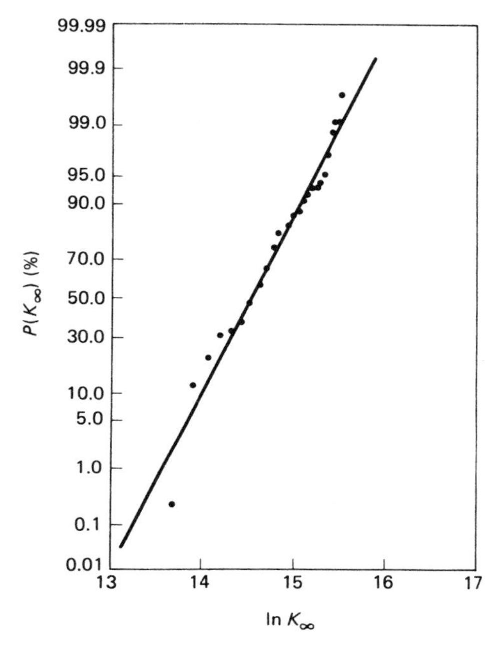

Fig. 14.3 Cumulative plot on normal probability coordinates of $\ln K_\infty$.[5] © 1979 by the IEEE.

where the normal or gaussian cumulative distribution function, F_n, is given by

$$F_n(x) = \frac{1}{\sqrt{2\pi}} \int_{-\infty}^{x} e^{-y^2/2}\, dy; \qquad F_n(x) = 1 - F_n(-x) \tag{14.11}$$

so that the histogram log-normal fit in Fig. 14.3 can be written as

$$P(K_\infty) = F_n\left(\frac{\ln K_\infty - 14.54}{0.423}\right) \tag{14.12}$$

which gives the probability that a particular K as chosen from the histogram is less than or equal to K_∞. The connective relation between the damage constant K and the common emitter current gain β is Eq. (5.66) in the form

$$1/\beta = 1/\beta_0 + K\Phi_n/f_T \tag{14.13}$$

where β_0 is the preirradiation value of the gain. Substituting for K from above into Eq. (14.10) gives, with Eq. (14.11)

$$P_\beta(\beta) = 1 - F_n\left(\frac{\ln K - \mu}{\sigma}\right) = F_n\left(\frac{-\ln(\beta_0/\beta - 1) - \ln(f_T/\beta_0\Phi_n) + \mu}{\sigma}\right) \tag{14.14}$$

For circuit applications, interest lies mainly in the probability that $\beta(\Phi_n)$ is less than a circuit-dependent failure threshold gain value β_T. Then the probability of survival of the ith transistor, $P_{si}(\Phi_n)$, is

$$P_{si}(\Phi_n) = 1 - P_\beta(\beta_T) \tag{14.15}$$

Introducing a circuit tolerance factor $CTF = \beta_{\min}/\beta_T$, where $\beta_{\min}$ is a minimum transistor gain, as given in the vendor specification sheet, allows Eqs. (14.14) and (14.15) to be combined as

$$P_{si}(\Phi_n) = F_n\left(\frac{\ln[(\beta_0/\beta_{\min})(CTF) - 1] + \ln(f_T/\beta_0\Phi_n) - \mu}{\sigma}\right) \tag{14.16}$$

For a conservative approach, assume $\beta_0 = \beta_{\min}$, and if measured values for f_T/β_0 are not available, let $f_T/\beta_0 \cong f_{T\min}/\beta_{\min}$, where $f_{T\min}$ is the minimum specified value of f_T from the vendor specification data.

To recapitulate, for the situation where only vendor data for the transistor under consideration are available, an estimate of the transistor survival probability at an incident fluence level, Φ_n, is then obtained from Eq. (14.16) with the above substitutions giving

$$P_{si}(\Phi_n) = F_n\left(\ln(CTF - 1) + \ln\left[\frac{(f_{T\min}/\beta_{\min}\Phi_n) - \mu}{\sigma}\right]\right) \tag{14.17}$$

A brief digression is appropriate here to explain certain ideas of statistical inference theory: confidence interval, prediction interval, and tolerance interval.

a. A confidence interval is an interval that has a preassigned probability p (e.g., 0.95) of containing the parameter of the population that has been sampled; 0.95 is the preassigned confidence limit or confidence level of the interval.
b. A prediction interval is an interval that has a probability p of containing the very next observation of a series. A prediction interval can also be constructed that has a probability p of containing the

mean of the next r observations. Note that if $r = \infty$, the prediction interval becomes the confidence interval.

c. A tolerance interval has the probability p of containing a proportion P_r or more of the population.

To illustrate, assume that $x_1, x_2, \ldots, x_n$ are a random sample of n independent observations from a normal distribution of the population of x's. As the actual population mean and standard deviation are usually unknown, it turns out that the confidence interval for the mean lies between $\mu_s - k\sigma_s$ and $\mu_s + k\sigma_s$, where μ_s and σ_s are the sample mean and standard deviation.[9] The parameter k is the function $k(p, n)$. For example, for $n = 9$ observations from a normal distribution, and choosing a 95 percent confidence interval, theory gives $k(0.95, 9) = 0.769$.[9] Hence, the 95 percent confidence interval or limit for the population mean $\mu = \mu_s \pm 0.769\sigma_s$. The 95 percent confidence prediction interval for the next observation having a value x_{10}, the tenth observation, is $\mu_s \pm 2.432x_{10}$. The corresponding tolerance interval that has a probability of 0.95 of including the proportion $P_r = 0.95$ (at least 95 percent of the population) is $\mu_s \pm 3.532\sigma_s$; similarly, a 0.99 probability of including 95 percent of the population is $\mu_s \pm 4.265\sigma_s$.

To obtain the survival probability for a 100γ percent confidence interval, $P_{si}(100\gamma\%)$, the following method is used, where $0 \leqslant \gamma \leqslant 1$. A confidence interval or confidence level of γ means that the computed $P_{si}(100\gamma\%)$ contains an estimated mean and standard deviation, m and σ_γ, respectively, that have a probability of $100\gamma\%$ of being their "true" values, and that P_{si} is within the $100\gamma\%$ confidence limit of its true value, as sampled from the universe of part types, as discussed earlier. For the case of the damage constant K in the foregoing histogram, the estimated mean from observed data, m_γ, for $100\gamma\%$ confidence interval is bounded by[6]

$$\mu_c - t_{N-1,1-\gamma} \cdot \frac{\sigma_c}{\sqrt{N}} \leqslant m \leqslant \mu_c + t_{N-1,1-\gamma} \cdot \frac{\sigma_c}{\sqrt{N}} \tag{14.18}$$

and the corresponding estimated standard deviation from observed data, σ_γ, for $100\gamma\%$ confidence interval is given by

$$\sigma_\gamma \leqslant \sqrt{(N-1)/\chi^2_{N-1,\gamma}} \cdot \sigma_c \tag{14.19}$$

where

μ_c is the computed mean from the histogram; $\mu_c = \overline{\ln(K)} = 14.54$

σ_c is the computed standard deviation; $\sigma_c^2 = \overline{(\ln(K))^2} - \mu_c^2 = 0.179$
N is the number of transistors in the histogram
$t_{N-1,1-\gamma}$ is Student's t probability distribution parameter for a 100γ % confidence interval for $N-1$ degrees of freedom, as obtained from a t table.
$\chi^2_{N-1,\gamma}$ is the chi-square probability distribution parameter for a 100γ % confidence interval for $N-1$ degrees of freedom, as obtained from a chi-square table.

For conservative estimates of transistor survival, the lower bound is used for the mean in Eq. (14.18), as well as using the upper bound for σ_c given in Eq. (14.19). For 90 percent confidence, $\gamma = 0.9$, and computations using Eqs. (14.18) and (14.19) result in $m_\gamma = 16.38$, and $\sigma_\gamma = 0.444$, so that from Eq. (14.10)

$$P_K(90\,\%) = F_N\left(\frac{\ln K - 16.38}{0.444}\right) \tag{14.20}$$

Now, retracing the development from Eqs. (14.10)–(14.17) gives the 90 percent confidence transistor survival probability as

$$P_{si,0.9}(\beta) = F_n\left(\frac{\ln[CTF-1] + \ln[(f_{T\,\min}/\beta_{\min}\Phi_n) - 16.38]}{0.444}\right) \tag{14.21}$$

Figure 14.4 is a nomogram that provides a computational means for finding the above 90 percent confidence transistor survival probability as a function of $f_{T\min}$, $\beta_{\min}$, β_T, and the incident fluence Φ_n. The portrayed illustration of its use is for a value of $\beta_{\min}/\beta_T$ of 5, neutron fluence of 10^{12} neutrons per cm^2, $\beta_{\min}$ of 20, and $f_{T\min}$ of 1.1 MHz. This yields from the nomogram a value of $P_{f,0.9}(\beta) = 0.01$, which is actually a failure probability. So that the transistor survival probability $P_{si,0.9}(\beta) = 1 - P_{f,0.9}(\beta) = 0.99$, with 90 percent confidence.

14.4 BINOMIAL STATISTICS OF SAMPLE MEASUREMENTS

The previous section dealt with methods by which the survival probability for a particular confidence level of a transistor, and associated circuit if the fate of the transistor decides the fate of the circuit, could be computed. This transistor survival probability depends only on the vendor gain specifications and the circuit-dependent failure gain threshold. This discussion has to do with the impact of radiation sample lot test results on pertinent binomial probability questions of device behavior.

Consider attribute tests of piecepart device lots, e.g. transistors. That

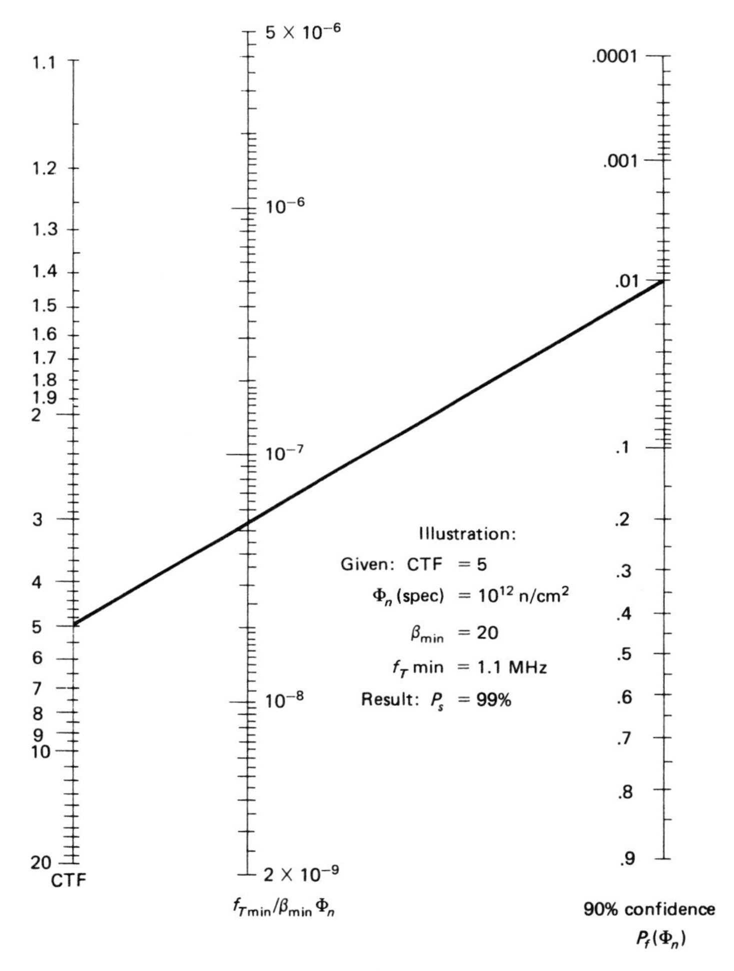

Fig. 14.4 Nomogram for computing the survival probability of a transistor as a function of its incident fluence and vendor data.[1] © 1976 by the IEEE.

is, either the device has the attribute or not, as determined by electrical screening, radiation screening, or merely by inspection, of a particular sample lot. It should be noted that the word *screening* implies testing 100 percent of the samples in the lot under consideration. If the part has the attribute, it has successfully passed the test; if it doesn't it has failed the test. If the probability of success for each part, independently, is p, then $q = 1 - p$ is its probability of failure. Statistically, this is described by the discrete binomial distribution $f(x)$, where $f(x)$ is the probability of the occurrence of x successes out of n trials, with replacement. It is

$$f_n(x) = \binom{n}{x} p^x q^{n-x} \tag{14.22}$$

where the binomial coefficient, viz., $\binom{n}{x} \equiv n!/x!(n - x)!$. With replacement means that, as each part is drawn, it is replaced in the lot after being examined for success or failure, like drawing blind and replacing balls from an urn.

To obtain the cumulative binomial distribution, which is the more useful one, rather than $f(x)$, let a particular parts lot be sampled for the attribute to give a typical sample, $x_1, x_2, x_3, \ldots, x_n$ which consists of a series of 0 and 1, i.e., 0, 1, 1, 0, 0, 1, . . . , where 1 means success and 0 means failure to possess the attribute. An (maximum likelihood) estimation of the probability of success, p, is labeled $\hat{p} = \sum_{i=1}^{n} x_i/n$, which can have the values

$$\hat{p} = 0, 1/n, 2/n, 3/n, \ldots, 1 \tag{14.23}$$

depending on how many successes there are in the particular sample drawn. Statistical sampling theory answers the question: If a particular estimate or success ratio is computed by $\hat{p} = k/n$, where k is the number of successes, what are the corresponding 100γ percent upper and lower confidence limits on the estimate $\hat{p}$ of p? They are:

a. The upper 100γ percent confidence limit is given by the root p_u obtained from the polynomial in p_u given by

$$P_n(x \leqslant k) = \sum_{x=0}^{k} \binom{n}{x} p_u^x q_u^{n-x} = (1 - \gamma)/2 \tag{14.24}$$

b. The lower 100γ percent confidence limit is given by the root p_l, obtained from the polynomial in p_l given by

$$P_n(x \geqslant k) = \sum_{x=k+1}^{n} \binom{n}{x} p_l^x q_l^{n-x} = (1 - \gamma)/2 \tag{14.25}$$

The convention is that if k is 0, the lower confidence limit is taken to be 0, and if k is equal to n, the upper confidence limit is taken to be 1.

One illustration of importance to nuclear survivability is that where all the devices pass the attribute test. Then from Eq. (14.25), the lower (conservative) 100γ percent limit corresponds to $x = k = n$, the single remaining term of the summation, which gives

$$P_n(n = k) = \sum_{x=n}^{n} \binom{n}{x} p_l^x q_l^{n-x} = p_l^n = (1 - \gamma)/2, \qquad (14.26)$$

and to include both tails of the binomial distribution, so that $2(1 - \gamma)/2 = 1 - \gamma$, then

$$n = \ln(1 - \gamma)/\ln p_l \qquad (14.27)$$

Hence, if 90 percent survival probability is desired with 95 percent confidence $n = \ln(0.05)/\ln(0.90) = 29$. Therefore, 0 failures are required out of a sample of 29 devices to achieve "90–95." A simple second example is that a sample of one device corresponds to a 50 percent survival probability with 50 percent confidence or any combination of survival probability and confidence level, such that $p = 1 - \gamma$, as seen from Eq. (14.27).

The computation of the above cumulative binomial probabilities, P_n, is cumbersome when the number of samples becomes large. A method for overcoming this difficulty is through the well-known cumulative incomplete beta function, which is given by

$$F(x, \alpha, \beta) = \frac{(\alpha + \beta + 1)!}{(\alpha!\beta!)} \int_0^x y^\alpha (1 - y)^\beta \, dy \qquad (14.28)$$

If the right-hand side is repeatedly integrated by parts, it easily follows that Eq. (14.24) can be written as

$$P_n(x \leqslant k) = \sum_{x=0}^{k} \binom{n}{x} p^x q^{n-x} = 1 - F(p, k, n - k - 1) \qquad (14.29)$$

and Eq. (14.25) can be written as

$$P_n(x \geqslant k) = \sum_{x=k+1}^{n} \binom{n}{x} p^x q^{n-x} = F(p, k - 1, n - k) \qquad (14.30)$$

Tabulations of the cumulative incomplete beta function $F(x, \alpha, \beta)$ are readily available,[7,8] and they can be used to construct the binomial probability tables used above.

For very large sample sizes, i.e., very large values of n, the binomial distribution approaches the normal distribution, and corresponding confidence intervals are available. One is obtained from the probability that p lies between confidence intervals, with a confidence level of γ, given for very large n by[9]

$$P\left[\hat{p} - d_\gamma\sqrt{\hat{p}(1-\hat{p})/n} < p < \hat{p} + d_\gamma\sqrt{\hat{p}(1-\hat{p})/n}\right] \simeq \gamma \qquad (14.31)$$

where d_γ is given from

$$F_n(d_\gamma) = \gamma/2 \qquad (14.32)$$

where F_n is defined in Eq. (14.11). For example, if $d_\gamma = 1.96$,

$$P[\hat{p} - 1.96\sqrt{\hat{p}(1-\hat{p})/n} < p < \hat{p} + 1.96\sqrt{\hat{p}(1-\hat{p})/n}] \simeq 0.95 \qquad (14.33)$$

gives the approximate 95 percent confidence interval for p for very large samples, n.

Figure 14.5 is a nomogram from which binomially distributed survival probabilities (in terms of failure probabilities) and confidence levels can be derived from attribute test data, such as the foregoing screens discussed, and corresponding to Eqs. (14.24) and (14.25). That is, if any three of the four parameters P_n, n, k, and p are known, the nomogram will yield the fourth. Alternatively, if two pairs of (P_n, p) are known, the nomogram will give the corresponding n and k.

14.5 ONE-SIDED TOLERANCE LIMITS

If a device parameter is known to obey a log-normal distribution, then statistical estimates of this parameter can be made from lot sampling using a relatively small number of samples. For such a parameter, π, of a device and n devices from a lot are sampled for this parameter, it can be shown that the lot should be rejected if a quantity π_L defined by

$$\pi_L = m + K_{TL}(n, \gamma, p)s_l \qquad (14.34)$$

exceeds a given maximum threshold $\pi_{L\max}$. The log-normal estimate of the mean $m = \exp\overline{(\ln\pi)}$, and the estimate of the corresponding variance $S_l^2 = \overline{(\ln\pi)^2} - m^2$. The quantity K_{TL} is called the *one-sided tolerance limit*. It is a well-known statistic of normal distributions and is a function of the sample size n, confidence level γ, and the lot quality proportion called p.[11,12] The latter characterizes the lot, such that if more than the fraction p of the distribution have values of π greater than $\pi_{L\max}$, then

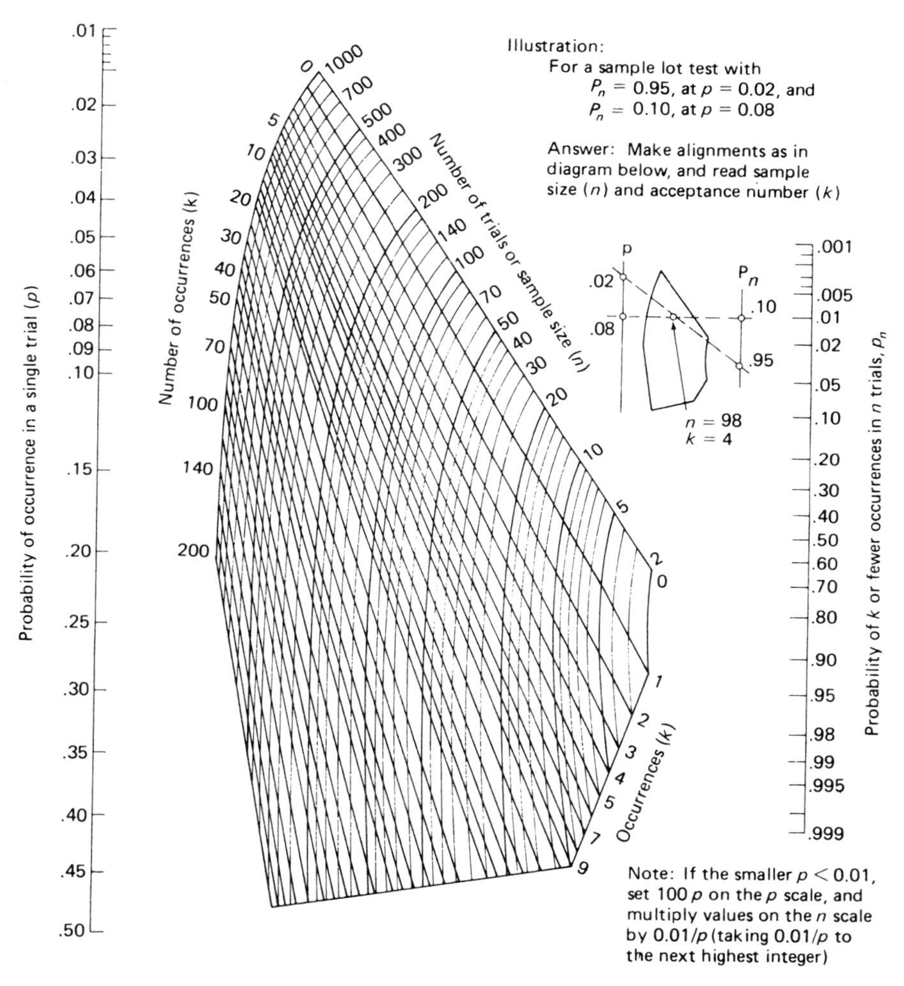

Fig. 14.5 Nomogram for computing binomial survival or failure probabilities with corresponding confidence limits.[10] © 1966 by the IEEE.

the lot will be rejected with probability γ. $\pi_{L\max}$, the upper limit failure threshold, is found through worst-case analysis of the effects of radiation on the circuit that contains the device whose parameter π is of interest. That is, if $\pi_L > \pi_{L\max}$, device failure has occurred. Then again, the device failure threshold may be a lower limit. If so, the lot is rejected in the above sense if $\pi_L < \pi_{L\min}$, where now

$$\pi_L = m - K_{TL}(n, \gamma, p)s_l \tag{14.35}$$

By comparing Eq. (14.18) with Eq. (14.35), it is seen that $K_{TL}(n, \gamma, p) = t_{n-1,1-\gamma}/(N)^{1/2}$, where t is the Student t statistic. For samples larger than 10, K_{TL} can be computed to good approximation using

$$K_{TL}(n, \gamma, p) = \frac{\left[F_n^{-1}(p) + F_n^{-1}(\gamma)\sqrt{\dfrac{\{F_n^{-1}(p)\}^2}{2(n-1)} + \dfrac{1}{n} - \dfrac{\{F_n^{-1}(\gamma)\}^2}{2n(n-1)}}\right]}{[1 - \{F_n^{-1}(\gamma)\}^2/2(n-1)]} \tag{14.36}$$

where $F_n(x)$ is the cumulative normal distribution function. $F_n^{-1}(x)$ is its inverse; that is, given a value of F_n, the corresponding value of x is sought.

As an illustration, it is desired to find the value of $\Delta\pi \equiv \Delta V_{CE}$ of a transistor, where ΔV_{CE} is the change in collector-emitter voltage corresponding to pre- and postradiation conditions, for which it can be predicted with 90 percent confidence that 99 percent of the lot will lie. Hence, $p = 0.99$ and $\gamma = 0.90$. Suppose that the radiation test data from 10 devices yields the (geometric) mean $m = 0.126\,\text{mV}$, with $s = 0.00359\,\text{mV}$. From a K_{TL} table,[12] $K_{TL}(10, 0.90, 0.99) = 3.532$. Then $\Delta V_{CE} \equiv 0.126 + 3.532(0.00359) = 0.1387\,\text{mV}$. Thus the assertion can be made that 99 percent of the transistors will have a ΔV_{CE} value below 0.1387 mV with 90 percent confidence.

Another application using the expression for K_{TL} given in Eq. (14.36) is to the transistor survival probability discussed in Section (14.3).[13] Referring to that section, $\gamma = 0.9$, which gives $F_n^{-1}(0.9) = 1.282$ from tables of the normal distribution. With $n = 53$, inserting in Eq. (14.36) yields.

$$K_{TL}(53, 0.9, p) \cong F_n^{-1}(p)[1 + 0.13\sqrt{1 + 1.93/\{F_n^{-1}(p)\}^2}] \tag{14.37}$$

For practical computations, the required survival probability p will be more than 0.99, so that $F_n^{-1}(p)$ will be very large. Hence, K_{TL} will be very nearly proportional to $F_n^{-1}(p)$, as seen in Eq. (14.37), so that it can be written as

$$K_{TL}(53, 0.9, p) \cong 1.14F_n^{-1}(p) \tag{14.38}$$

Again, Eq. (14.38) is a useful approximation for K_{TL} when F_n^{-1} is large, and when tables of the cumulative normal distribution F_n, and so F_n^{-1}, are more readily available than K_{TL} tables.

Now applying Eq. (14.34) to the case of the log-normal distribution of damage constants discussed in Section (14.3) gives

$$\ln K_{\infty\,\min} = \overline{\ln K_\infty} - K_{TL}(n, \gamma, P_s)s_l \tag{14.39}$$

Also from Section (14.3), $\overline{\ln K_\infty} = 14.54$, $s_l = 0.423$, and

$$\ln K_{\infty\,\min} = -\ln(CTF - 1) - \ln(f_{T\,\min}/\beta_{\min}\Phi_n) - \ln(2\pi) \tag{14.40}$$

where the 2π comes from the fact that $K_D = 1/\omega_T K = 1/2\pi f_T K$ in the development following Eq. (14.9). Combining Eqs. (14.39) and (14.40) yields

$$14.54 - 0.493F_n^{-1}(P_s) = -\ln(CTF - 1) - \ln(f_{T\,\min}/\beta_{\min}\Phi_n) - \ln(2\pi) \tag{14.41}$$

which results in a 90 percent survival probability, or

$$P_s = F_n\left[\frac{\ln(CTF - 1) + \ln(f_{T\,\min}/\beta_{\min}\Phi_n) + 16.38}{0.493}\right] \tag{14.42}$$

It is seen that this is identical to Eq. (14.21) except for the denominator of 0.493 instead of 0.444 therein, reflecting a greater uncertainty in the data. This difference in the denominators can make a significant difference in critical situations where very high survival probabilities are necessary.

14.6 STATISTICS FOR NEUTRON EFFECTS

In the hardening phases of system design, part categorization methods can be used, depending in part on the hardness specification levels.[14] The parts categorization method to be described is referred to as the design-margin breakpoint method, which essentially applies a "breakpoint" category criterion to all parts in the system. It is generally most practical for systems with relatively moderate radiation specifications. Nevertheless, the assumption is implicit that these parts will be properly categorized even under worst-case situations, regarding the most neutron-sensitive portions of the system. An illustration of this method is given by the strategic aircraft hardness design considerations discussed in Section 13.3.

Category I parts are defined as those that will require special scrutiny, from the radiation hardness viewpoint. Category II parts are defined as those that can usually be dismissed from such attention. Category I parts imply the use of a number of statistical tools in their analysis, and they usually appear as some of the parts in systems with more stringent hardness specifications.

The part categories assigned are not time-invariant, but can change as the corresponding parts, circuits, and systems evolve during the life of the program. The application of worst-case assumptions to all parts of the system for the case of high neutron fluences, with simultaneously high survival probabilities, would be far too conservative. Therefore, the aforementioned parts categorization method implies a separate analysis for each hardness-critical part type, whether Category I, II or derivative categories.

The statistical methods employed are based on techniques used in modern industrial quality control.[16] Of course, optimal hardness design in systems utilizes a combination of both the above methods. Following discussion of these methods below, illustrative examples are given. Besides the mere selection of parts from a radiation hardness point of view, circuit hardening can also be an effective way for reducing hardness-assurance costs.

Additionally, circuits should be designed to maximize the use of non-radiation-sensitive parts, consistent with other requirements such as space, weight, and power consumption. They should be designed with an eye for minimizing the radiation sensitivity of certain parameters, such as operating transistors at collector currents at or near their maximum current gain, the output drive current of integrated circuits should have adequate design margin (fanout capability) after irradiation, and generally large design margins should be included to accommodate changes in sensitive circuit parameters.

To evaluate the susceptibility of the system to the radiation environment, a worst-case analysis of each circuit in the system is required, to determine the parameter failure level, π_f, for a particular circuit critical piecepart. Besides the usual knowledge of radiation-sensitive parameter variations, part temperature and aging[24] must be included in the analysis as well. In some systems, the conditions for proper circuit operation will be that the change in a parameter, in contrast to the parameter value itself, must be less than a given failure level, i.e., $\Delta\pi < \Delta\pi_f$, where the failure level $\Delta\pi_f$ is usually determined again by worst-case analysis. An immediate example is the change in the reciprocal gain of a bipolar transistor $\Delta(1/\beta)_f \equiv 1/\beta_f - 1/\beta_{\min}$. The quantity β_f is the gain threshold at which circuit analysis reveals that the circuit will fail for lesser gain levels, while $\beta_{\min}$ is the vendor-specified minimum gain of the transistor. This formulation of gain change introduces a safety factor, for the vendor will nearly always maintain values of β greater than $\beta_{\min}$.

After π_f or $\Delta\pi_f$ for a particular device has been obtained from a worst-case analysis, the corresponding neutron failure fluence should be available. One illustration of this is the well-known relation between neutron fluence and gain change, i.e., Eq. (5.66), which can be written as

$$\Delta(1/\beta)_f \equiv 1/\beta_f - 1/\beta_{\min} = K_D\Phi_f; \qquad K_D = 1/2\pi f_T K \tag{14.43}$$

and can be used to define an appropriate failure fluence Φ_f for bipolar transistors. Usually, for each device type there is only one critical electrical parameter of interest with respect to a given type of radiation. Frequently this parameter is often the one that is also electrically critical in the circuit. The failure fluence for bipolar transistors also can be obtained through a circuit design margin $M = \beta_{\min}/\beta_f$.[17] Equation (5.66) also can be written in terms of failure fluence conditions as

$$1/\beta_f - 1/\beta_i = K_D\Phi_f \tag{14.44}$$

where β_i is the preirradiation value of gain. Multiplying the above by $\beta_{\min}$, where $\beta_{\min}/\beta_i \ll M$ is assumed, gives

$$K_D\beta_{\min}\Phi_f \simeq M \tag{14.45}$$

which is then used to define the failure fluence Φ_f.

Once the worst-case analysis provides the parameter failure threshold value π_f or $\Delta\pi_f$, a lot sample of corresponding devices can be subjected to increasing fluence levels to measure their individual failure fluence levels, Φ_{fi}. With these data, plots of $\pi(\Phi)$ or $\Delta\pi(\Phi)$ versus incident fluence Φ can be made for the samples, extrapolating if necessary to intersect the π_f, or $\Delta\pi_f$ line to yield the respective Φ_{fi} for each device, as shown in Fig. 14.6. With this information, two important statistics can be computed that characterize the sample lot. They are:

a. the geometric mean failure fluence Φ_{mf}, because it is assumed that log-normal statistics are in order, i.e.

$$\Phi_{mf} = \exp(\overline{\ln\Phi_f}) = \left(\prod_{i=1}^{n}\Phi_{fi}\right)^{1/n} \tag{14.46}$$

where $\overline{\ln\Phi_f} = \sum_{i=1}^{n} \ln\Phi_{fi}/n$, and

b. the corresponding variance is given by

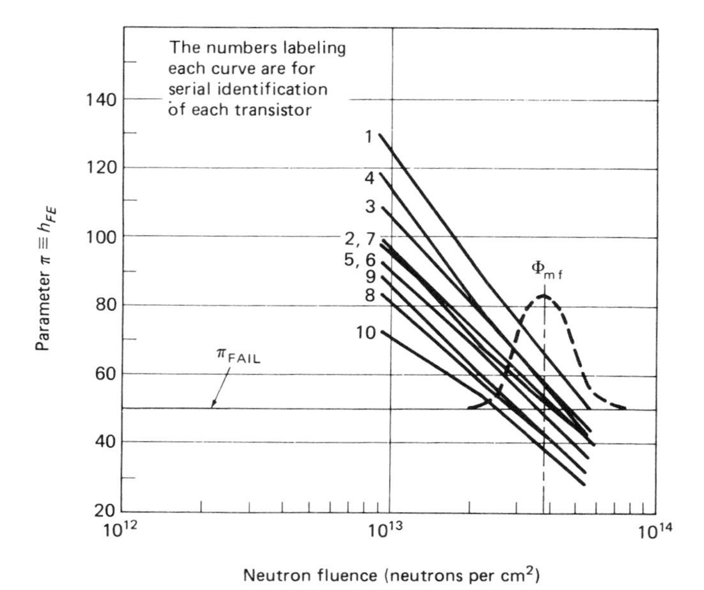

Fig. 14.6 Fluence to failure measurements for bipolar transistors.

$$s_l^2 = \frac{1}{n-1} \sum_{i=1}^{n} (\ln \Phi_{fi} - \overline{\ln \Phi_f})^2 = \frac{n}{n-1} [\overline{(\ln \Phi_{fi})^2} - (\overline{\ln \Phi_f})^2] \quad (14.47)$$

The preceding characterization measurements can often be dispensed with if data already exist on the part under consideration. The measurement of the damage constant discussed in Section 14.3 is essentially a characterization of all silicon bipolar transistors, and an important result is the definition of a universal damage constant K.

The design margin is a system parameter that accounts for unknown factors that could cause failure. It is an attempt to balance the consequences of failure against the cost involved in reducing risk, as perfectly trouble-free systems can never be guaranteed. The design margin for neutrons can be defined as Φ_{mf}/Φ_{spec}, where Φ_{spec} is the specified fluence. This definition is based on failure fluence considerations, but is still useful when some parameter is the cause of failure, as will be seen. Φ_{mf}, the geometric mean failure fluence, is used, for it is well approximated by a log-normal distribution for most device populations.

As already indicated, the design margin is used to categorize devices into categories I and II. These are basically divisions of devices that require further hardening attention (category I) and those that do not (category II). Design margins other than the above can be defined for any parameter $\pi(\Phi)$, such as $\pi_f/\overline{\pi(\Phi_{\text{spec}})}$. This implies that $\pi_f/\overline{\pi(\Phi_{\text{spec}})} > 1$, when $\pi(\Phi)$ increases with fluence. The reciprocal, $\overline{\pi(\Phi_{\text{spec}})}/\pi_f$ would be used for a design margin when $\pi(\Phi)$ decreases with fluence. The overbar denotes the geometric mean.

In hardening electronic systems, in contrast to circuits, design margins are compared with defined system category criteria. For moderate radiation specifications, all parts of the system can be covered by a single category criterion. System category criteria are well delineated. For example, if the design margin is 10 or more, the corresponding system is designated as category II, and no significant further attention is given it in this context. However, if the design margin is less than 10, it is designated as category I. The latter category is divided into subcategories, as discussed in Section 14.7 and Chapter 15.

For stringent radiation specification levels, each part type must be considered separately, and different category criteria must be developed for each part type, as discussed. For the hardness-assurance program, discussed in Sections 15.1 and 15.2, all Category I parts must be analyzed statistically, as mentioned. If the design margin is greater than 100, the part is designated as hardness noncritical, so that no further attention is given it in the hardness assurance sense. Usually, there are a modest number of critical part types that are likely to cause system failure due to incident radiation. For these, a quantitative estimate must be made of their survival probability. These estimates can be expressed as a 100γ percent confidence level that a certain part will have a specified neutron fluence survival probability P_s.

If it is assumed that individual piecepart failure values of fluence or parameter failure threshold values are distributed log-normally, then the logarithm of the part categorization criterion (PCC) is defined by

$$\ln(PCC) = K_{TL}(n, \gamma, p) \cdot s_l \tag{14.48}$$

where s_l can be obtained from Eq. (14.47) in the case of failure fluences or using the same form for s_l with appropriate changes for the case of parameter threshold values. K_{TL} is the one-sided tolerance factor for a normal distribution as discussed in the previous section. Equation (14.48) is usually written as

$$PCC = \exp(K_{TL}(n, \gamma, p) s_l) \tag{14.49}$$

PCC gives a quantitative indication of the dispersion of the parameters of interest. To illustrate, consider a specified fluence of $2.5 \cdot 10^{13}$ neutrons per cm^2 for a particular sample lot of 20 bipolar transistors whose $\overline{\ln\Phi_f} =$ 31.92 as obtained from observed data, and $s_l = 0.40$. Their required survival probability is 0.9999 at the 90 percent ($\gamma = 0.9$) confidence level. Using Eq. (14.46), $\Phi_{mf} = 7.3 \cdot 10^{13}$ neutrons per cm^2. Hence the design margin is $\Phi_{mf}/\Phi_{\text{spec}} = 7.3 \cdot 10^{13}/2.5 \cdot 10^{13} = 2.92$. Now, $K_{TL}(20, 0.9, 0.9999) = 4.892$, so that the part categorization criterion is

$$PCC = \exp(4.892)(0.40) = 6.83 \tag{14.50}$$

From Section 14.5, it is seen that PCC measures dispersion about a mean. As the *PCC* of 6.83 exceeds the design margin of 2.92, this part is acceptable for systems use, but remains classified as category I. This is because its dispersion exceeds the design margin with high probability. Therefore, further hardness assurance attention is given to it, and its associated circuitry, to assure that future lots remain acceptable.

Now, instead of using fluence-to-failure data, the part categorization criterion is computed for the preceding sample lot of transistors using a device parameter, namely, their damage constants K_{Di}. As discussed in Section 14.3, it was shown that the damage constant population can be assumed to be log-normally distributed. As before, $\Phi_{\text{spec}} = 2.5 \cdot 10^{13}$ neutrons per cm^2, and, as part of the worst-case analysis, the transistor is considered to fail if its current gain β is less than 40. The device average damage constant $\bar{K}_D$ is computed from

$$\bar{K}_D = \exp\left(\sum_{i=1}^{n} \ln K_{Di}/n\right) = \left(\prod_{i=1}^{n} K_{Di}\right)^{1/n} \tag{14.51}$$

where the individual damage constant of the ith transistor is obtained from Eq. (5.66) in the form

$$K_{Di} = (1/\beta_{i\text{post}} - 1/\beta_{i\text{pre}})/\Phi \tag{14.52}$$

Φ is the incident fluence level used to irradiate the transistors, and the subscripts *pre* and *post* refer to their pre- and postirradiation states. The failure limit K_{Df}, i.e., the K_D corresponding to the failure gain β_f and specified fluence is obtained from

$$K_{Df} = (1/\beta_f - 1/\beta_{\min})/\Phi_{\text{spec}} \tag{14.53}$$

and $\beta_{\min}$ is the vendor-specified minimum current gain.

An illustration consists of $\beta_{\min} = 75$, $\beta_f = 40$, and $\Phi_{\text{spec}} = 2.5 \cdot 10^{13}$ neutrons per cm^2; $K_{Df} = 4.667 \cdot 10^{-16}$. From the computations, $\bar{K}_D = \exp(-35.9) = 2.551 \cdot 10^{-16}$, and, because β_f decreases with fluence, the appropriate design margin equation to use is $K_{Df}/\bar{K}_D = 4.667 \cdot 10^{-16}/2.551 \cdot 10^{-16} = 1.83$. $K_{TL} = 4.892$ as before, and the corresponding K_D statistical computations give $s_l(K_D) = 0.35$, so that $PCC = \exp(4.892)(0.35) = 5.54$. Again, this PCC is greater than the computed design margin of 1.83, so that this part is acceptable for systems use, but remains Category I from these statistics. Other aspects are discussed with respect to hardness assurance-considerations in Chapter 15.[17]

14.7 STATISTICS FOR IONIZING DOSE EFFECTS

Many aspects of the structure and statistics for ionizing dose effects are identical to those for neutron effects, discussed in the preceding section. The main differences will be highlighted in the following paragraphs. As in neutron effects above, the design breakpoint categorization method is used, depending on the radiation levels involved. For moderate radiation environments, the breakpoint method is widely used. For severe radiation environments, and where part survival probabilities are relatively lower, many devices become Category I and statistical methods must be employed. This usually requires radiation characterization testing, and the calculation of the ionizing dose design margin. Category I parts require that the statistical method use the computation of the PCC for each hardness-critical part corresponding to the characterization data.

In an ionizing dose environment, or in a neutron environment, radiation test data are necessary to accomplish parts characterization. If an existing data base for the part or parts of interest is available, tests may not be necessary. The radiation characterization consists of exposing parts to increasing ionizing dose levels, until the parameter of interest, π, exceeds a specified failure level, π_f, as determined from worst-case circuit analysis. For the ith part of the sample lot, the corresponding parameters, π_i, are all plotted on the same graph as a function of the ionizing dose level. Each π_i curve will normally intersect the common horizontal line representing π_f, as in Fig. 14.7. The projection of these intersections onto the ionizing dose level abscissa yields a set of ionizing dose failure levels $\{D_{fi}\}$. These then provide the data from which pertinent statistics are computed and used to categorize the part/circuit application and provide the piecepart hardness-assurance criteria.

For each radiation-sensitive part type, there may be more than one electrical parameter that is of interest, in terms of how it is affected by radiation. These parameters for each part type make up the bulk of device radiation data found in radiation data banks. Also, there can be more than one value of π_f for a particular parameter, depending on the circuit application. Analogous to the neutron fluence statistics, the

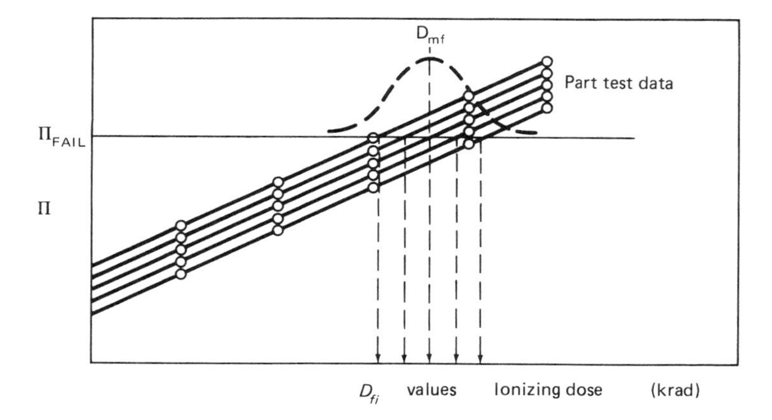

Fig. 14.7 Relationship of parameters used to determine D_{fail}.

ionizing dose mean failure level, D_{mf}, D_f, and D_{fi} are all normalized to one rad $(\cdot)$, with

$$D_{mf} = \exp(\overline{\ln D_f}) = \exp \sum_{i=1}^{n} \ln D_{fi}/n = \left(\prod_{i=1}^{n} D_{fi}\right)^{1/n} \qquad (14.54)$$

and the ionizing dose design margin (TDM) is

$$TDM = D_{mf}/D_{\text{spec}} \qquad (14.55)$$

where D_{spec} is the specification level for ionizing dose in this application. In space systems, D_{spec} is normally less than the system ionizing dose specification, as the spacecraft housing plus the surrounding masses partially shield the parts under consideration. In such cases D_{spec} incident on the part can be computed as described in Section 13.2.

Part characterization is used to determine whether hardness-assurance lot acceptance testing is required for a particular part or circuit application, as alluded to earlier. As in the case of neutrons, there are two major categories of part classification, with respect to ionizing dose considerations, depending on the ionizing dose design margin. The latter is then compared to a PCC level, which is dependent on part parameter values.

Similar to the preceding section on neutrons, the ionizing dose design margin (TDM) criterion, as obtained from lot acceptance test data, is given by

$$TDM = \pi_f/\exp(\overline{\ln \pi}) \qquad (14.56)$$

for increasing parameter values and

$$TDM = [\exp(\overline{\ln \pi})]/\pi_f \tag{14.57}$$

for decreasing parameter values, where $\overline{\ln \pi} = \sum_{i=1}^{n} \ln \pi_i / n$, with $s_{ln}^2(\pi) = n[\overline{(\ln \pi)^2} - (\overline{\ln \pi})^2]/(n - 1)$, and K_{TL} is defined previously. Resort to context must be made to avoid ambiguity in Eqs. (14.56) and (14.57), where the symbol TDM is used in both instances to define ionizing dose design margin.

For the case of sample lot failure ionizing dose testing,

$$TDM = (\exp \overline{\ln D_f})/D_{\text{spec}} \geqslant \exp K_{TL} s_{ln}(D_f) \tag{14.58}$$

where

$$\overline{\ln D_f} = \sum_{i=1}^{n} \ln D_{fi}/n \text{ and } s_{ln}^2(D_f) = n[\overline{\ln D_f)^2} - (\overline{\ln D_f})^2]/(n - 1) \tag{14.59}$$

As a first illustration of the foregoing,[19] consider the categorization of an operational amplifier using input bias current I_{IB} as the radiation-susceptible parameter for a sample of five parts, and using D_f values. Assume an ionizing dose specification $D_{\text{spec}} = 150$ krad (Si), a survival probability of 0.999 at a 90 percent confidence level ($\gamma = 0.9$). Worst-case circuit analysis yields $\Delta I_{IBf} = 10$ nA, and plotting ΔI_{IB} versus ionizing dose for all five parts on the same graph, on which the ΔI_{IBf} line is superimposed, yields a set $\{D_{fi}\}$ as discussed. With this set, D_{mf} is computed using Eq. (14.54) to get

$$D_{mf} = \exp(\overline{\ln D_f}) = 520 \text{ krad (Si)} \tag{14.60}$$

so that $TDM = D_{mf}/D_{\text{spec}} = 0.52/0.15 = 3.5$, and from Eq. (14.59) a corresponding $s_{ln}(D_f) = 0.331$. For $n = 5$, $\gamma = 0.90$, and $p_s = 0.999$, the $K_{TL} = 6.112$. Note that in Eq. (14.60) and Eq. (14.54), $\exp(\overline{\ln D_f}) = \left(\prod_{i=1}^{n} D_{fi}\right)^{1/n}$ still retains the dimensions of krad (Si), assuming that each D_{fi} has the same dimension. Then

$$PCC = \exp K_{TL} s_{ln} = \exp(6.112)(0.331) = 7.58 \tag{14.61}$$

Since PCC is greater than TDM for this operational amplifier, it is still deemed usable in the system, but maintained as a Category IM part.

Category IM means that the part is so labelled because of design margin considerations. The companion Category IH corresponds to parts or subcircuits incorporated solely for radiation hardening, i.e., they are hardness-dedicated. One example would be a PIN diode radiation detector in a circumvention circuit.

Example 2 is an interesting one for the categorization of a bipolar transistor, using as a basis the gain degradation due to ionizing dose environment.[20] Here, $D_{spec} = 150\,\text{krad (Si)}$, $n = 5$, $\gamma = 0.95$, and $p_s = 0.999$. From the vendor part type sheet, $\beta_{min} = 50$, and worst-case analysis of the circuit where the part is used yields $\beta_f = 39$. Now

$$\Delta(1/\beta)_f \equiv 1/\beta_f - 1/\beta_{min} = 1/39 - 1/50 = 0.0055 \tag{14.62}$$

It should be noted that $\Delta(1/\beta)_f$ as computed uses β_{min}, because this is the worst-case pretest acceptance criterion for future lots to be hardness-assurance-tested. For the case where parts are selected for higher values, $\Delta(1/\beta)_f$ should be computed using the preirradiation value of β.

Continuing, the above five part samples are irradiated at increasing dose levels until they all have sustained at least $\Delta(1/\beta)_f = 0.0055$. Again, a plot of $\Delta(1/\beta)$ versus ionizing dose level is made, with the $\Delta(1/\beta)_f$ line superimposed. The intersections of $\Delta(1/\beta)_i$ with this line yield the corresponding set $\{D_{fi}\}$. In this case, assume that the $\{D_{fi}\}$ results in a

$$D_{mf} = \exp(\overline{\ln D_f}) = 418\ \text{krad (Si)} \tag{14.63}$$

with a corresponding $s_{ln}(D_f) = 0.165$. Now $TDM = D_{mf}/D_{spec} = 418/150 = 2.8$. For $n = 5$, $\gamma = 0.95$, and $p_s = 0.999$, $K_{TL} = 7.501$. Then

$$PCC = \exp(7.501)(0.165) = 3.46 \tag{14.64}$$

Here, also, PCC is greater than TDM, so that the part is acceptable for system use, but continued as Category IM, which also means that it requires continued sample lot testing, as discussed.

Example 3 describes the categorization of a bipolar transistor using as a basis collector leakage current I_{CBO}.[21] Here $D_{spec} = 30\,\text{krad (Si)}$, $n = 5$, $\gamma = 0.95$, and $p_s = 0.999$. A worst-case circuit analysis yields $I_{CBOf} \equiv \pi_f = 30\,\text{nA}$. The part samples are to be irradiated at increasing dose levels until all have reached $\pi_f = 30\,\text{nA}$. It is assumed in this case that none of the part curves of I_{CBO} versus ionizing dose intersect the π_f horizontal line, so that in this situation, the alternate approach is to use the π_i values corresponding to D_{spec} and compute the TDM from them. Then Eq. (14.56) gives the appropriate TDM as

$$TDM = \pi_f/\exp(\overline{\ln \pi}) \tag{14.65}$$

Then with $\exp(\overline{\ln \pi}) = 3.61\,\text{nA}$, and $s_{ln}(\pi) = 0.186$

$$TDM = 30/3.61 = 8.31 \tag{14.66}$$

For $n = 5$, $\gamma = 0.95$, and $p_s = 0.999$, $K_{TL} = 7.501$. Hence

$$PCC = \exp(7.501)(0.186) = 4.04 \tag{14.67}$$

Since PCC is less than TDM, and TDM is close to 10, this part might possibly be recategorized as Category II, but in the interim will still require sample lot acceptance testing.

The next two examples illustrate lot acceptance testing for (a) single-dose level lot acceptance testing using π_i values, and (b) multiple ionizing dose lot acceptance testing using D_f values.

a.[22] One transistor type at a single ionizing dose level: $D_{spec} = 150\,\text{krad (Si)}$, $n = 5$, $\gamma = 0.95$, $p_s = 0.999$, $V_{CB} = 5\,\text{V}$, $\pi_f(I_{CBO}) = 15\,\text{nA}$. The sample lot of five parts is irradiated at the above D_{spec} and voltage conditions. The results yield the parameters $\{\pi_i(I_{CBO})\}$ from which $\exp(\overline{\ln \pi}) = 0.618$, and $s_{ln}(\pi) = 0.169$. For increasing parameter values, use the criterion equation (14.56), i.e.

$$TDM = \pi_f/\exp(\overline{\ln \pi}) = 15\text{ nA}/0.618\text{ nA} = 24.3 \tag{14.68}$$

The corresponding $K_{TL} = 7.501$, so that

$$PCC = \exp(7.501)(0.169) = 3.6 \tag{14.69}$$

As $TDM > PCC$, and greater than 10, this part lot is acceptable as category II.

b.[23] Consider an operational amplifier using bias current ΔI_{IB} as a basis, $D_{spec} = 150\,\text{krad (Si)}$, $n = 6$, $\gamma = 0.95$, $p_s = 0.999$, $\Delta\pi_f = 90\,\text{nA}$. Irradiating the six samples until all have attained $\Delta\pi_f = 90\,\text{nA}$, results in plots of ΔI_{IB} versus ionizing dose, with the $\Delta\pi_f$ line superimposed, to yield a set of values $\{D_{fi}\}$. Computations give $\exp(\overline{\ln D_f}) = 1380\,\text{krad}$ (Si), $s_{ln}(D_f) = 0.250$. From Eq. (14.57)

$$TDM = [\exp(\overline{\ln D_f})]/D_{spec} = 1380/150 = 9.2 \tag{14.70}$$

For $n = 6$, $\gamma = 0.95$, $p_s = 0.999$, $K_{TL} = 6.612$, so that

$$PCC = \exp(6.612)(0.25) = 5.2 \tag{14.71}$$

As TDM > PCC, the lot is acceptable for system use and might be recategorized to Category II.

To conclude, parts categorization can be compartmentalized using the Parts Categorization Criterion (PCC) in the hardness-assurance process. For moderate radiation requirements, the radiation sample tests compare the mean failure level with the design breakpoint levels, based on the system specifications. In an overarching sense, with respect to the preceding, if their ratio (design margin) is greater than 100×, (100 times), the part is assumed hardness noncritical. If the ratio lies between 100× and 10×, the part is designated Category II. If the ratio lies between 10× and 3×, the part is designated Category IM or IH. If it is less than 3×, the part is rejected.

Category I is construed to require PCC lot testing of said parts. As a result of these tests, the part may indicate a 99 percent survival probability, P_s, with 90 percent confidence at the system specification level. If this is the case, these parts may be recategorized to Category II, providing the data comprises an adequate statistical sample of the parts being processed. Also, the user may agree to slight variations of P_s and confidence level after suitable negotiations with the supplier.

REFERENCES

1. D. R. Alexander, D. L. Durgin, R. N. Randall, and J. J. Halpin. "Hardening Options For Neutron Effects: An Approach to Tactical Systems." IEEE Trans. Nucl. Sci. NS-23(6):1691–1696, Dec. 1976.
2. G. C. Messenger and E. L. Steele. "Statistical Modelling of Semiconductor Devices For the TREE Environment." IEEE Trans. Nucl. Sci. NS-15(6):133–139, Dec. 1968.
3. F. Larin. Radiation Effects in Semiconductor Devices. J. Wiley, New York, p. 163, 1968.
4. Alexander et al., op. cit.
5. A. I. Namenson. "Statistical Treatment of Damage Factors For Semiconductor Devices." IEEE Trans. Nucl. Sci. NS-26, pp. 4691–4694, Oct. 1979.
6. D. L. Durgin, D. R. Alexander, and R. N. Randall. "Hardening Options For Neutron Effects." HDL Final Report HDL-CR-74-052-1, Nov. 1976.
7. M. Abramowitz and I. Stegun. Handbook of Mathematical Functions. Dover, New York, Tables 26.7, 26.9, 1968.
8. E. Pearson and H. Hartley (eds.). Biometrika Tables for Statisticians, Vol. 1. Cambridge Univ. Press, 1954.
9. A. Mood. Introduction to the Theory of Statistics, McGraw-Hill, New York, 1951.
10. IEEE Spectrum. "Western Electric Plots Short Cuts to Probability Computation." p. 72, Dec. 1966.
11. M. Natrella. Experimental Statistics. National Bureau of Standards Handbook 91, pp. 2–14 ff., 1963.
12. Natrella, ibid. Table T-14.
13. Namenson, op. cit.

14. R. Berger, H. Eisen, A. Namenson, E. Wolicki, and R. Scace. "Piece Part Neutron Hardness Assurance Guidelines For Semiconductor Devices." DNA 5910F, Naval Research Labs, Oct. 6 1981.
15. R. P. Patrick and J. Ferry. Nuclear Hardness Assurance Guidelines For Systems With Moderate Requirements. AFWL-TR-76-147, Sept. 1976.
16. J. Juras, F. Gryna, Jr., and R. Bingham (eds.). Quality Control Handbook. McGraw-Hill, New York, 1974.
17. G. C. Messenger. "Hardness Assurance Considerations For the Neutron Environment." Trans. Nucl. Sci. NS-22(6):2308–2313, Dec. 1975.
18–23. K. Martin, W. Price, and M. Gauthier. Hardness Assurance Guidelines For Total Dose Radiation Effects. DNA5909F JPL, Pasadena, CA, Feb. 1 1982.
24. M. S. Ash and H. C. Gorton. "A Practical End-of-Life Model For Semiconductor Devices." IEEE Trans. Reliability 38(4):485–493, Oct. 1989.

CHAPTER 15
HARDNESS ASSURANCE

15.1 INTRODUCTION

This chapter draws heavily on the authors' knowledge of specific systems that mainly utilize bipolar parts. In that regard, the content is oriented toward bipolar discrete devices and integrated circuits. Nevertheless, the hardness assurance technology contained herein has been, and can be directly extended to systems using MOS integrated circuits.

Because most radiation hardness difficulties lie in the electronics portions of modern systems, the major emphasis of a hardness-assurance program is electronics assurance. The prime features of electronics hardness assurance are related to electronic circuit design that is based on radiation hard parts. These are parts or part types that either through sample lot part selection or intrinsic design, such as dielectrically isolated CMOS/SOS RAMs, are deemed hard relative to their soft counterparts. These parts are required to remain functional during and after neutron degradation of their electrical operating characteristics, to tolerate transient pulsed ionization radiation, single event upset, total ionization dose, EMP, and to be insensitive to thermomechanical effects. Hardness-assurance activities are defined to ensure that only controlled, hard part types are used and that higher levels of assembly (system level) are also hard. The scope of the electronics hardening effort extends from semiconductor device process controls at the vendor plant to simulated nuclear weapon effects testing of the entire system. Exposure of parts to nuclear reactor, cyclotron, and flash x-ray radiation are three examples of simulation testing.

At the part production level, process controls and tolerance levels are established for such items as purity and resistivity of basic silicon chips, diffusion furnace temperature, dopant concentrations, diffusion times, lead metallization cross sections, and lead bonding procedures. For example, control of transistor base width, diffusion processing, and corresponding surface quality are three examples of importance in maintaining the required common emitter transistor current gain, with respect to its resistance to degradation due to incident neutron radiation. Proper control of transistor minority carrier lifetime, diffusion parameters, and mask deterioration is important in order to limit photocurrents under ionizing radiation from gamma or x rays. Control of the deposition of lead metallization and its cross section minimizes the change of metallization burnout due to excessive currents and also maintains the required surface conditions to minimize the effects of ionization currents. The use of certified bonding machinery operated by approved bonders, followed by periodic bond testing, ensures the control of device bond strengths.

During, and/or subsequent to the parts production process, hardness-related electrical parameters, such as the gain-bandwidth product, f_T, common emitter current gain, β, and leakage currents are measured. Also, device samples are tested for current carrying capacity, and static bond pull tests are made to ensure bond integrity. Such data are to be reviewed on a closely controlled, continuing basis and compared with previous data to determine process trends, both by vendors and by the customer. This phase of the hardness-assurance effort is called *supplier data monitoring*, and modern industrial quality control statistical measures, such as those discussed in the previous chapter, are brought to bear. In this regard, the supplier data related to hardness parameters are monitored and analyzed to track vendor performance to hardness requirements. Vendors are required to provide acceptance data, screen data, and from time to time, other production data as deemed necessary. Such a monitoring effort provides the earliest possible warning of potential hardness degradation.

The hardness-assurance program includes radiation testing of various levels of assembly to verify part and assembly hardness. The purpose of the hardness-assurance verification testing is to ascertain that the controls and screens do actually result in hard parts, and that the latter have been used in their hardened circuits and assemblies. These tests provide an experimental confirmation of part hardness that cannot be achieved completely by electrical or mechanical testing of merely the parts themselves. For example, during production, assembly hardness could be adversely affected by variations in manufacturing procedures and other difficulties due to the vagaries of production. However, direct radiation-testing of these assemblies will ensure that radiation hardness has not been compromised during production. The significance of the information obtained in verification testing at the assembly level is that the test results contain all influences on assembly hardness, both known and unknown. Furthermore, testing at the system level will provide confidence with respect to system hardness by direct measurement, which is a paramount objective of the hardness-assurance program.

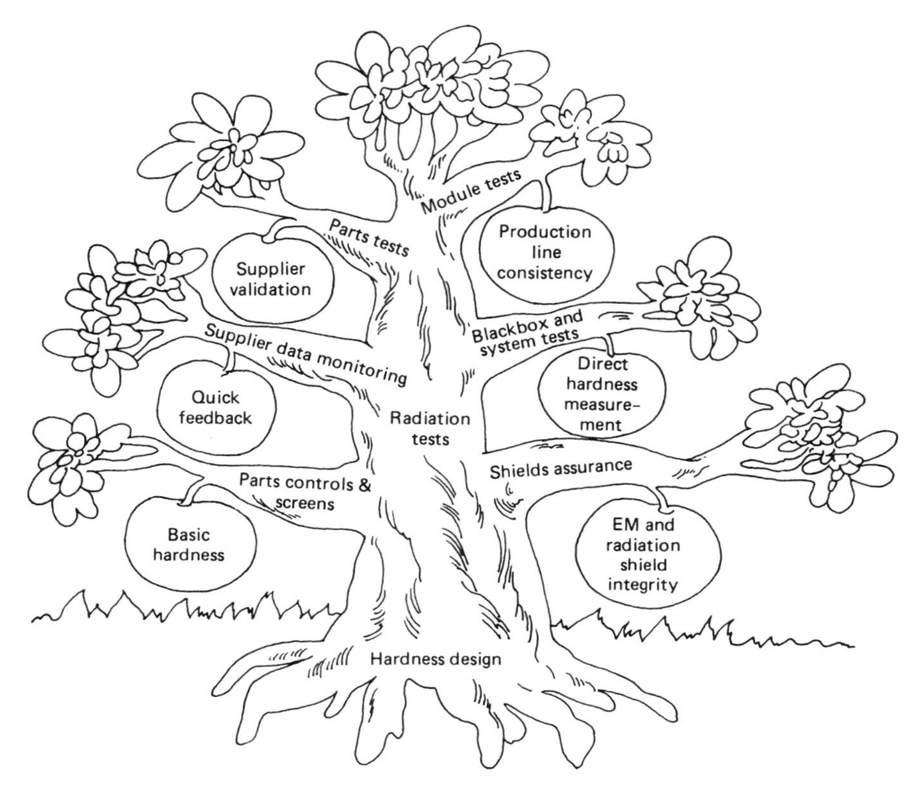

Fig. 15.1 Hardness assurance tree.

15.2 HARDNESS-ASSURANCE TREE

The hardness-assurance program can be conceptualized as consisting of interrelated efforts in three areas; electronics, radiation shielding, and electromagnetic shielding. The electronics portion can be subdivided into semiconductor vendor screens and controls, vendor data-monitoring analysis, and flash x-ray, linear accelerator, and nuclear reactor radiation testing at the semiconductor, module, and system assembly levels. The radiation shielding portion addresses the adequacy and completeness of ionizing radiation shields involving fabrication control, radiographic, and other inspections, including configuration assurance inspection. The electromagnetic shielding portion also includes fabrication control, integrity, assembly, and configuration inspection, plus radio frequency testing of cables, subassemblies, and systems.

The major parts of the system hardness-assurance program can be illustrated using the paradigm of the hardness assurance tree, firmly rooted in a hard system design. The many branches represent the major activities of hardness assurance at various levels of assembly, in that the production hardware retains hardness levels equivalent to that of the baseline design. The fruits of the tree represent the specific benefits resulting from the various hardness-assurance activities, as illustrated in Fig. 15.1.[1] The fundamental hardness of the systems coming off the production line can be directly related to the semiconductor manufacturing processes as monitored by part controls and screens. Assurance that a supplier's screens and controls have been implemented properly and are maintained continuously is provided by the vendor data monitoring activity (with quick feedback for problem areas), an effort in which appropriate data summaries and comparisons can be supplied with dispatch by computer. Continuing verification of parts hardness is provided by radiation-testing of samples from selected deliveries of lots of each part type. Radiation tests conducted on samples from three levels of assembly, viz., modules, "black boxes" (subsystems), and the system, provide assurance that the fabrication and assembly processes are under control and that production line consistency is being maintained. Tests at the higher levels of assembly also furnish direct measurement of system hardness. Radiation and electromagnetic shielding integrity is provided by specific assurance activities related to these shields.

15.3 DESIGN FEATURES REQUIRING HARDNESS ASSURANCE

The hardness-assurance program is also oriented toward the specific hardness design features of the system under consideration. For neutron displacement effects, the design features requiring hardness assurance lie within the semiconductor electronic parts and assemblies (circuits). For ionizing dose effects, the design features requiring hardness assurance are similiar to those for neutrons, e.g., the semiconductor devices and circuits and in some cases a radiation-initiated circumvention system designed to eliminate system malfunction due to radiation transients, as discussed in Section 13.9.

A summary of those features requiring hardness assurance is presented in Table 15.1. Several of the design features requiring hardness assurance are influenced by the several nuclear environments and their effects. For example, semiconductor parts design is listed for three different effects, namely, ionization damage, displacement damage, and thermomechanical damage. All three effects exert an influence on hardened part parameters

Table 15.1 Design features requiring hardness assurance.

DESIGN FEATURES REQUIRING ASSURANCE	RADIATION EFFECTS	ENVIRONMENTS
Semiconductor parts design	Ionization	Gamma rays, x rays, and neutrons
	Displacement damage	Neutrons
	Thermomechanical damage	X rays, blast, and thermal
Electronic assemblies design	Ionization	Gamma rays, x rays, and neutrons
	Displacement damage	Neutrons
Circumvention initiation	Ionization	Gamma rays and x rays
Radiation shields	Ionization and thermomechanical damage	X rays
EMP shields	Induced currents	EMP

necessitating hardness assurance in the design, fabrication, and testing of semiconductor devices. However, the critical areas differ. For example, the die (chip) processing is critical from a neutron hardness standpoint requiring strict control of diffusion process parameters. Photocurrent limitation and metallization quality is critical for ionization rate effects. The hardened manufacturing processes utilizing lower temperatures for the high-temperature processing steps such as oxidation and diffusion, and the ultraclean processing precautions, are critical from an ionizing dose standpoint. The quality of header (package) chip bonds and lead wire metallization bonds is important with respect to thermomechanical effects. Another important design feature requiring hardness assurance is the achievement and maintenance of shield integrity, in order to attenuate incident environments to innocuous levels within the system.

15.4 PRODUCTION-ASSURANCE RELATIONSHIPS

Figure 15.2 depicts the relationships between the various portions of the electronics hardness-assurance efforts toward hardware production flow. Assurance that the vendor screens and controls have been implemented properly and are continuously maintained is obtained from the supplier (vendor) data monitoring activity. Verification that the pertinent activities are resulting in hard parts is provided by radiation-testing of Category 1M samples from each delivery of each part type. This is called hardness assurance verification testing (HAVT). Radiation tests that are conducted

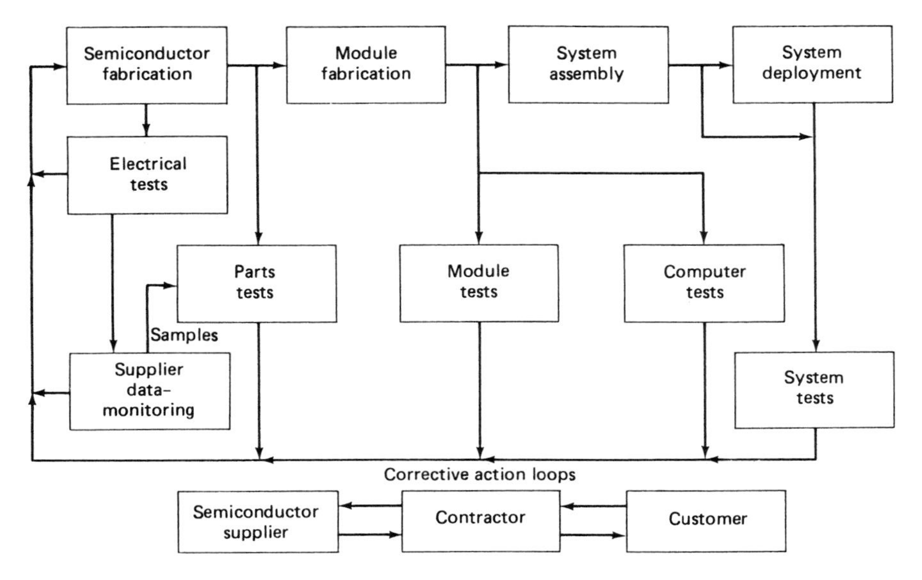

Fig. 15.2 Production-assurance relationships.

on assemblies (assembly HAVT) to obtain assurance that the fabrication and assembly processes are under control include three levels of assembly, viz., modules, major subsystems components (black boxes), and the final system assembly. The significant aspects of the program are test, analysis, and corrective action. The program is based on electrical and mechanical testing by the supplier and electrical/radiation testing throughout the production cycle by the contractor user. The testing program provides an extensive and continuing data base characterizing the hardness of the system, and the testing operation is followed by detailed analyses. As with other types of quality assurance programs that have extremely low failure rates, the hardness-assurance program utilizes overstress testing and analysis to obtain quantitative estimates of hardness capability.

Three types of output are obtained from the analyses. They are: (1) for immediate problems-pass/fail quality assurance analysis, (2) for potential problems-trend tracking of electrical/mechanical parameters from vendor and contractor testing, and (3) for hardness-assurance estimation.

The final phase of the program is corrective action. As difficulties and potential problems arise, they are evaluated, and appropriate corrective action is identified. Changes in processing, testing, and/or analysis as warranted are implemented, or, in the case of semiconductor device difficulties, negotiations with the supplier would eliminate such problems.

15.5 SEMICONDUCTOR DEVICE CONTROLS AND SCREENS

Semiconductor devices are the active electronic elements of major military systems and, as such, provide the radiation hardness foundation for the system under consideration. For this reason much effort has been expended in the design and fabrication of radiation-resistant semiconductor devices. Two separate but closely allied procedures should be provided to ensure that procured parts are hard. They are hardness controls and hardness screens.

The manufacturing processes critical to radiation-sensitive parts parameters are identified, and associated controls on these parts are "baselined" in appropriate procurement documentation. *Baselining* is a term used for the contractual fixing of semiconductor processing details. The vendor manufacturing process and the control and test specifications in effect at the time of the purchase order award constitute the baseline with which the vendor is contractually bound. All critical operations are documented in the baseline, and no deviations may be made without prior contractor concurrence.

Process control is the activity of performing the various steps in the process, as prescribed in the baseline documentation. Process control is, of course, verified by inspecting the quality of the output. Inspection may be performed on all the parts or on a sample lot.

Specific values of each radiation-sensitive parameter are established as acceptance criteria. Acceptance tests, or screens, are performed on semiconductor devices to ensure that they meet corresponding specifications. A screen is defined as a 100 percent test, i.e., a test in which each semiconductor device is tested and either accepted or discarded. This is as differentiated from sample tests, where far fewer than 100 percent of parts are tested. Sample acceptance tests are used when (a) the test is destructive, (b) their use is indicated by economic considerations, or (c) when a high degree of confidence is evidenced in the underlying process controls. Details of particular hardness controls and screens and the actual acquisition of hardness data are discussed below with respect to the corresponding radiation environment.

For the neutron environment, the major effect of neutron radiation incident on a bipolar transistor is a decrease in the base minority carrier lifetime, which manifests itself in a decrease in transistor current gain, as discussed in Section 5.10. It is borne out by massive test data that control of the neutron-induced gain degradation can be achieved through process control of f_T. Maintenance of the f_T value is dependent upon careful monitoring of the diffusion process, control of impurity density of the source semiconductor boule, temperature and duration of the boule in the furnace, depth of base diffusion, general cleanliness of the process, and other factors. Because of the close relationship between the physical base width and the gain-bandwidth product, close control of the former during the diffusion process is essential. Neutron hardness of all transistors and certain integrated circuits can be assured at the vendor facility by an f_T screen of all delivered parts. There are, however, certain integrated circuits for which an f_T control terminal is not accessible through the IC case. For these, either special control transistors diffused on the basic wafer, or selected integrated circuits can be scribed and bonded with leads, thus allowing them to be packaged with direct access to the control transistors. In these cases, the f_T acceptance test is performed on a sampled basis in contrast to a screen.

Another parameter whose radiation response influences the degradation of transistor gain is the *p-n* junction leakage current. An increase in certain base current parameters will generally result in a transistor gain decrease. One of these parameters is the leakage current in MOSFET transistors, whose magnitude is a function of surface imperfections and contamination at Si-SiO_2 interfaces, in the SiO_2 per se, and on the SiO_2 surface. Contamination control measures are therefore required during the prediffusion cleaning and passivation processes. Junction leakage currents are measured as a 100 percent acceptance test, i.e., a screen test of the finished device. Measurement of leakage currents in discrete semiconductor devices is a simple procedure, whereas measurement for every junction in an IC is not practical because of their sheer numbers and inaccessibility. Therefore, the acceptance tests are performed on a typical accessible element of an IC that is representative of the condition of the surface.

For ionization radiation rate effects, such as those from gamma and x rays, an important endeavor is the prevention of photocurrent-induced metallization burnout. Prevention of metallization burnout in a hostile environment requires that the photocurrents produced by the ionizing radiation and multiplied by transistor action to produce secondary photocurrents be minimized, and that the metallization cross section be sufficient to withstand the maximum value of photocurrent. Photocurrent production, being a function of minority carrier lifetime and active device volume, is in effect controlled as discussed under semiconductor device vendor controls and screens for neutron effects. With respect to metallization cross section, it has been determined that a minimum cross sectional area can be established to handle safely radiation-induced photocurrents in dielectrically isolated microcircuits. The production processes involved in metallization cross section control are: metallization deposit times, evaporation rates, and mask inspection. A screen test cannot be applied to the metallization acceptance test, because available test methods are destructive; hence resort must be made to sample testing. One example of an acceptance test requires the passage of a 1-μs, 5-Amp square wave pulse through a metallization stripe of the smallest cross section on the chip and is performed on samples from each wafer. Alternatively, the vendor can make direct physical measurements of stripe thickness of sample devices.

For thermomechanical effects, an immediate consideration is that of lead bonds and bonding. A device parameter to be considered is low-strength bonds that may not withstand thermomechanical damage due to incident radiation. Also, the susceptibility of device bonds to shock-stress-induced damage is minimized by assurance of strength and quality of bonds to specified standards based on appropriate computational and test results. As mentioned already, bond strength during production is controlled by initial certification and periodic recertification of bond machines and operators. Substantial amounts of testing have established the force that may be exerted for testing bonds without damage or degradation of good bonds. One set of vendor-required pull values for 1-mil wires are: 2 grams per bond for gold to gold connections, and 0.8 grams per bond for aluminum to aluminum. A sampling program can be conducted by vendors on their production lines. They test bonds by using a bond-pulling machine in which a hook engages the bond wire and exerts the requisite force. Another method is the use of compressed air jets to apply this force. A sample may be taken of the output of each bond machine periodically. If any unsatisfactory bonds are found, the entire output of that machine for that period may be rejected.

15.6 DATA-MONITORING

As seen in Fig. 15.2, the supplier data-monitoring program (SDM) plays a key role in the successful function of production assurance. The purpose of the SDM is to acquire, store, and analyze vendor data that relate to device hardness characteristics for diodes, transistors, and integrated circuits. The basis of the program is that variations in supplier processes that affect radiation hardness are revealed by the analysis of device test data. Additionally, the SDM program complements the hardness-testing activity by delineating the device types most in need of testing and by enhancing the interpretation of test results through comparison of the tested sample to the large underlying part population distribution. Conversely, SDM benefits from the testing activity by the availability of experimental results to reinforce data analysis findings.

Acceptance tests are the principal source of data regarding the characteristics and performance of semiconductor devices prior to their actual use. Analyses of such data provide the ability to distinguish whether or not deterioration or improvement has occurred. The historical information on part behavior over a period permits a realistic and effective correlation of acceptance test values with the ultimate parts hardness characteristics.

Parameter data from suppliers of devices are acquired from measurements performed at the supplier facility in the course of his producing the qualifying parts for production. Of prime importance are those parameters directly related to hardness. These include leakage currents, common emitter gain, β, gain-bandwidth product, f_T, IC metallization thickness, and bond strength. Additional parameter data, including other acceptance data, are obtained for their potential aid in evaluating the primary hardness parameter data. The latter are usually stored in peripheral storage media, such as disk or tape, for ready access by the contractor's computers. Data summaries, distributions, and histograms are available for comparison with preceding lot data. The monitoring of these data displays provides immediate feedback of potential difficulties before they can become critical.

15.7 DATA-ANALYSIS METHODOLOGY

Data analysis can be broken down into two main types. The first is essentially routine interrogation of data yielding relatively fixed form outputs. It is a type of data reduction that isolates potential problem areas on the basis of preset variability limits. The second type involves analysis performed on an individual part basis to follow up the routine preliminary results. A prime example of a data-reduction method is the automated comparison of different lot means and standard deviations specified to predetermined limits. A statistical distribution histogram, as shown in Fig. 15.3, depicts the frequency of occurrence of gain versus the gain value. Acceptance limits are included to define the margin of safety. The combination of an approximately normal distribution for the gain of the transistors with a *three-sigma* limit very close to the specification limit represents a situation worth continued monitoring. For example, a small shift in the distribution could produce an unacceptably high rate of rejection of parts, as seen in the figure.

The tracking of parameter trends is monitored for each device type where measured data of hardness control parameters are available. An

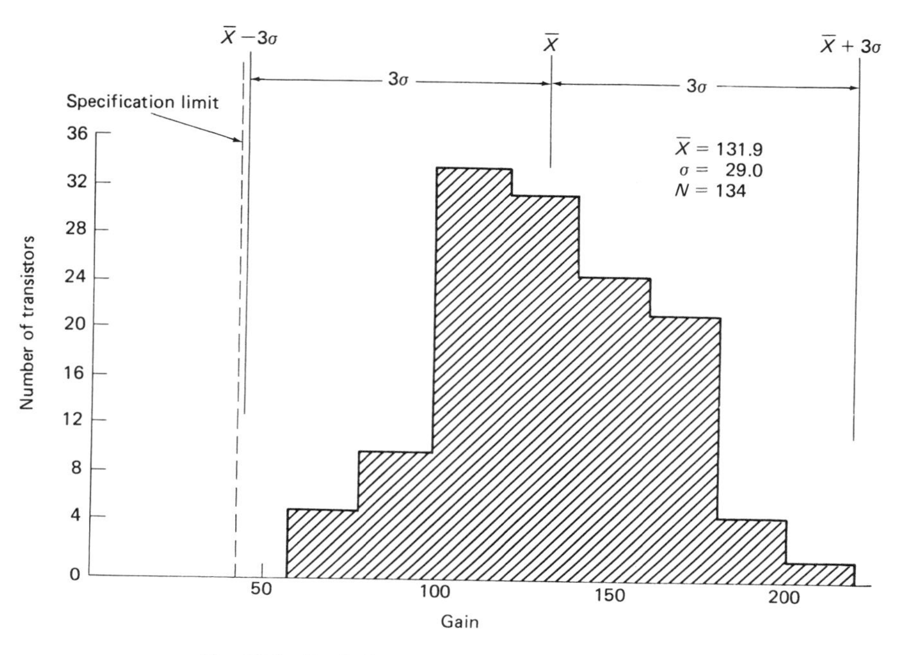

Fig. 15.3 Typical power transistor gain distribution.

"*x-bar, sigma*" chart with plots of the mean and standard deviation of the set of data indicates the typical behavior of the devices and their variation in this behavior, such as Fig. 15.3. A *state-of-control* condition exists when all points fall within the required control limits. This means that the result represented by each point is consistent with the estimated process average and associated process variation. An *out-of-control* condition exists when a point exceeds the upper or lower monitoring limits. This means that the result represented by that point is significantly different from the estimated process average and associated process variation. Such a condition requires an engineering review.

In addition to the above charts, certain parameter trends are tracked by plotting the parameter average, maxima, and minima as a function of the corresponding part fabrication chronology, as shown in Figs. 15.4 and 15.5. These figures show typical data for gain-bandwidth product being monitored as a parameter related to neutron hardness. The data in Fig. 15.4 depict a high-volume, mature, integrated circuit manufacturing process showing good control from lot to lot. This describes a situation that would allow a significant reduction in sampling. Fewer lots and smaller samples, could be used without compromising the program. The

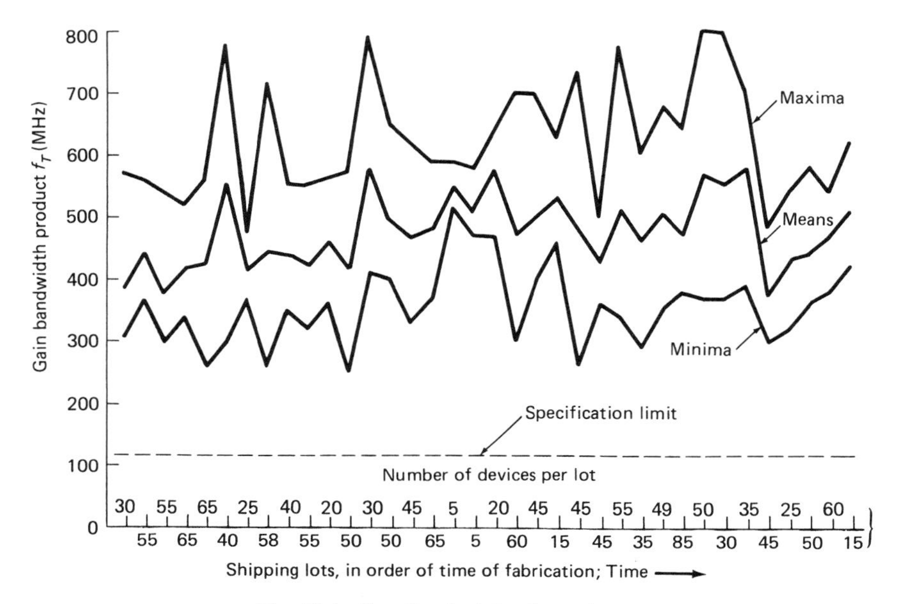

Fig. 15.4 Supplier f_T data of a mature process.

data shown in Fig. 15.5, however, illustrate a broad continuous variation of the mean, indicative of a low-volume, immature process, and also suggest a controlled adjustment of diffusion parameters by the vendor in an attempt to increase his yield. As it turns out, the data from the corresponding wafer lot control group of transistors show areas of broad variability and definite trends with time. Evaluation of the distribution histograms of various lots signal that the variability seen in the trend charts is a strong effect. It is not due to a very few maverick devices, for example. Continued close attention to this device is clearly warranted in the context of the hardness-assurance program.

Another valuable hardness-assurance tool is provided by the example of the statistical distribution plots of incremental burn-in* leakage current change of a bistable flip-flop. The latter, ΔI_{leak}, is the leakage current difference prior to and after burn in. As seen from Fig. 15.6, the device population is fragmented into two distinct lots: 87.5 percent of the parts of lot A show a ΔI_{leak} ranging from 0.2–1.6 nA, but only 4.2 percent of the parts of lot B fall into that range, whereas 95.8 percent of lot B are

* Burn-in refers to the standard operational testing of an item from the production line, with all aspects of its function stressed at least to their specification levels for a protracted period.

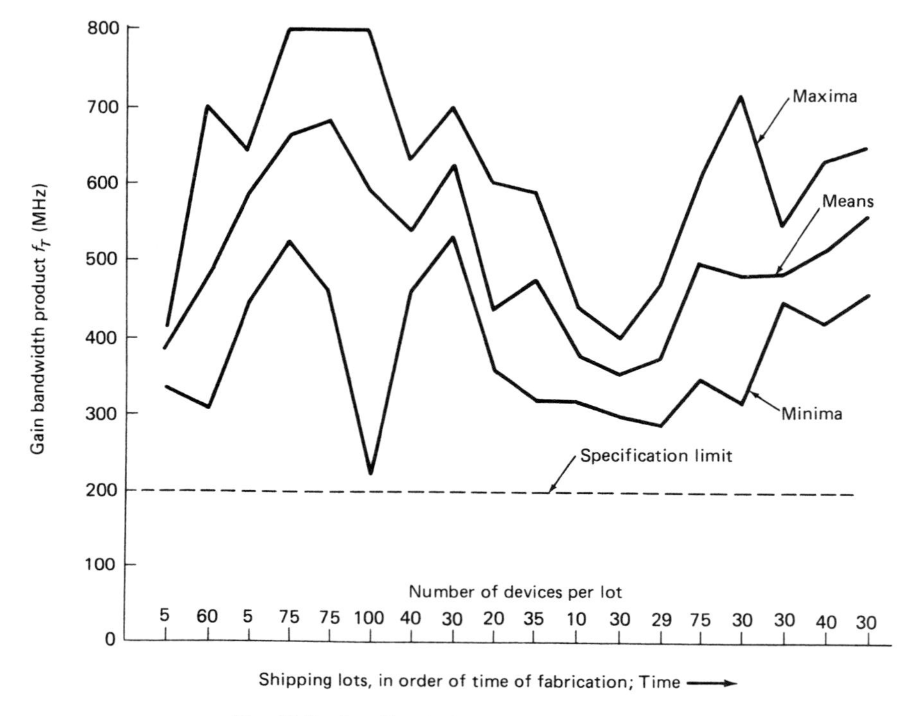

Fig. 15.5 Supplier f_T data of an immature process.

widely dispersed outside a normal distribution. This clearly demonstrates a significant difference between the two lots. Supplier corrective action is clearly required. Furthermore, lot B could possibly be restricted to non-critical usage. Generally, a number of factors may influence process consistency that are detectable by the analyses. These include tolerance variations in go-no go test equipment, changes in the dominant process or process variable, new personnel, new equipment or fixtures, new batches of raw materials or new sources of materials, and simply pure mistakes.

The use of scatter diagrams, as in Fig. 15.7, can also reveal lot-to-lot variations. Here, the data presented in Fig. 15.6 as an ΔI_{leak} distribution are given as a scatter diagram of post burn in leakage current versus pre-burn in leakage current. The degree of scatter in lot B is clearly shown to be significantly different from lot A.

Another example of a change in monitored supplier data that would trigger analysis is shown in Fig. 15.8, as a plot of the output impedance of a low-level switch versus its shipping lot chronology. The data therein

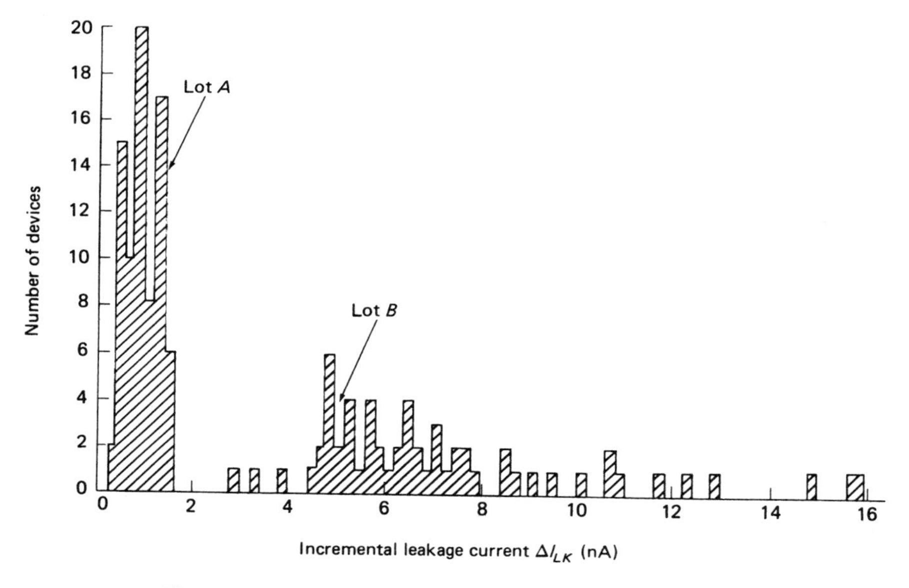

Fig. 15.6 Distribution of incremental burn in leakage current.

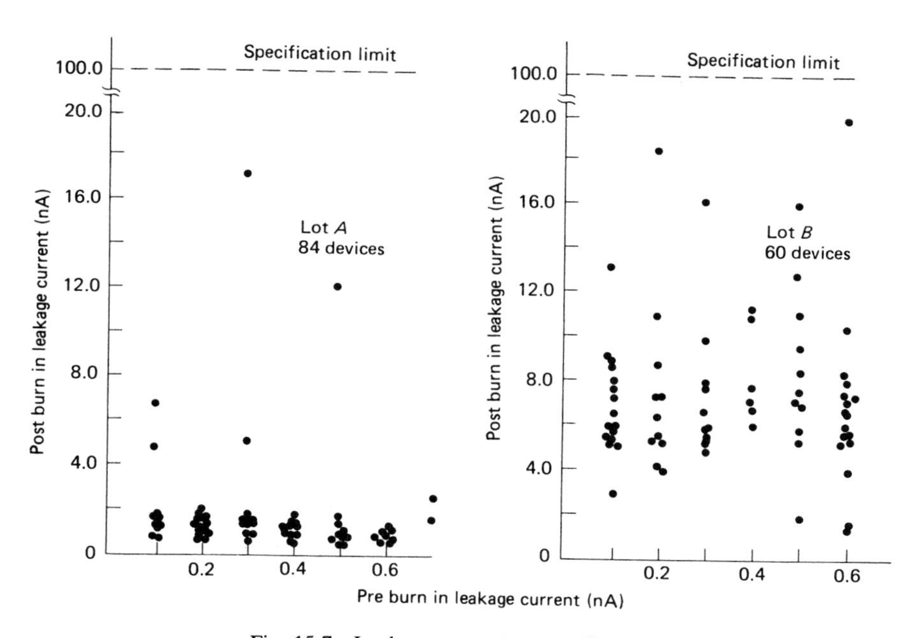

Fig. 15.7 Leakage current scatter diagrams.

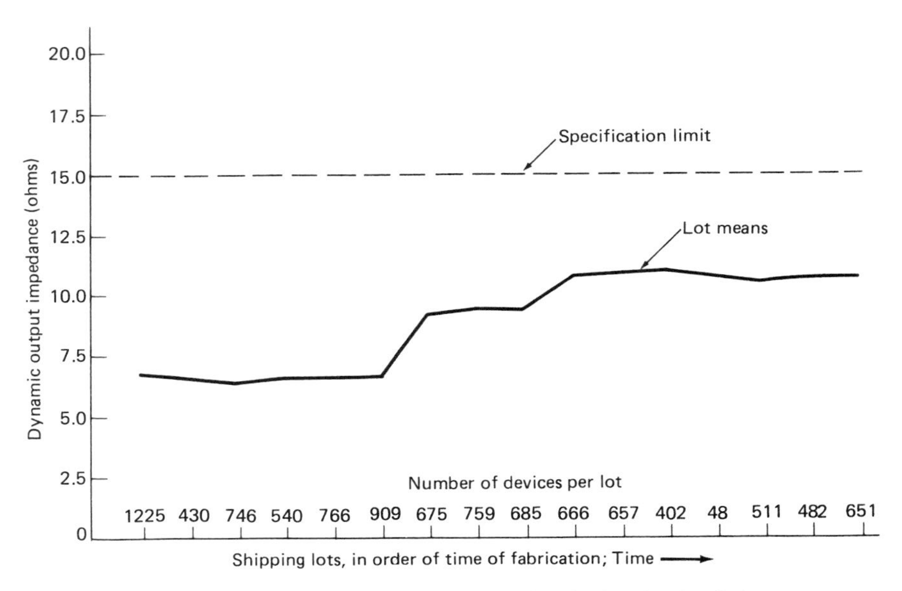

Fig. 15.8 Supplier output impedance data for low-level switch.

show a significant shift in output impedance in the direction of decreasing safety margin. In response to the observed shift, samples were drawn from representative lots, one prior to the shift and one after.

The dynamic output impedance of the above switch was measured as a function of reciprocal base current. The inverse gain of the device, β_i, is calculated from[2]

$$r_d = r_T + kT/eI_B\beta_i \tag{15.1}$$

where r_d and r_T are the dynamic and static output impedances, respectively, and I_B is the base current. As indicated in Fig. 15.9, both forward and inverse gain have shifted in the same direction, indicating a change in emitter and collector efficiency, rather than a change in collector, emitter, or load series resistance. Further investigation revealed that the supplier had indeed made a process change from an epitaxial buried layer to a diffused buried layer structure, corresponding in lot time to the detected shift in dynamic output impedance. Isolation of the cause of the shift effectively resolved the problem.

Another mode of representation is through the use of cumulative probability plots. As an example, a plot of the measured cumulative failure probability of integrated circuit bonds versus their breaking

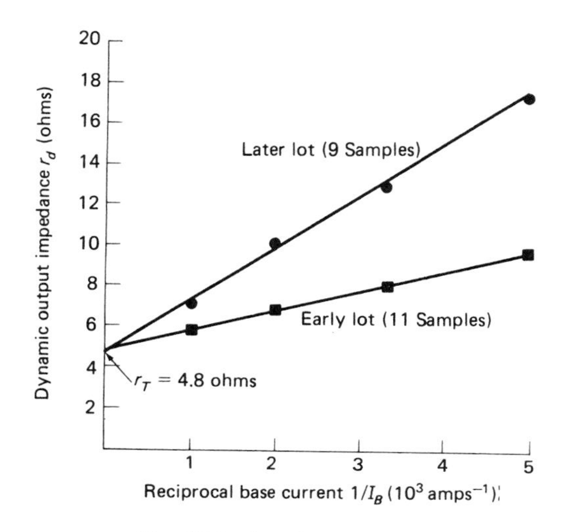

Comparison of lots

Parameter	Early lot	Later lot
f_T	260 MHz	210 MHz
r_T	4.8 Ω	4.8 Ω
β	270 ± 50	97 ± 6
β_i	27	10

Fig. 15.9 Supplier output impedance data analysis for low-level switch.

strength is depicted in Fig. 15.10. That the curves are straight lines implies that the cumulative bond failure probability distribution with breaking strength is normally distributed. That there are two curves means that the distribution is bimodal (has two peaks). The low failure rate tail of the distribution, which is the flatter line in Fig. 15.10 is determined primarily by poor bonds, i.e., low separation strength. Failures at higher breaking strengths are typically caused by failure of the bond lead wires themselves. It should be noted from the figure that there is a very small but finite bond failure probability at zero stress. That is, there is a non-zero probability that the bond can break spontaneously without the application of any significant force.

15.8 HARDNESS-ASSURANCE VERIFICATION: RADIATION TESTS

The hardness-assurance verification testing (HAVT) effort is the radiation test verification portion of the hardness-assurance program. The main objective of HAVT is to provide direct experimental verification that procedural methods, including screens and controls used during the production processes, are adequate to assure that previously established design standards of hardness are maintained. Experimental measurements of hardware response made during and after exposure to simulated radiation environments constitute the heart of HAVT. The latter represents a supplement to routine inspection, electrical testing, and process

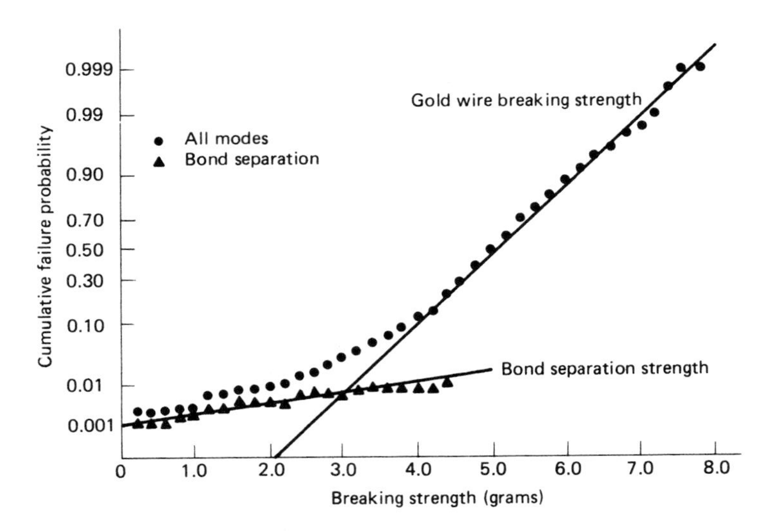

Fig. 15.10 Cumulative failure probability versus lead breaking strength.

control by providing a direct measure of part test hardness that can be related to system hardness capability.

Various levels of assembly are also tested, and the selection of the appropriate level is based on many factors including cost, timeliness, destructive or nondestructive test, and, of course, the determination of the optimal level. It is the level that provides the optimum data from a system radiation effects impact standpoint. For example, parts, modules, and systems are tested for neutron and high-level ionizing radiation effects. Parts radiation data provide the fastest feedback at lowest cost, and module measurements provide the earliest available data on hardness of parts integrated into the system. Systems data provide a direct measure of system hardness capability.

For each radiation effect that has an impact on system performance, there is a specific level of assembly to be tested and type of measurement to be made before, during, and after irradiation. Table 15.2 shows that for the permanent types of effects, such as gain degradation, radiation-induced burnout, thermomechanical damage, and ionizing dose damage, functional electrical tests before and after irradiation are adequate to measure the impact. However, operational tests with measurements made during exposure are required for all transient effects, including those due to neutron ionization as well as those due to gamma and x rays.

Threshold measurements at several levels of assembly are required to discern logic upset. To assure system protection for low-level ionization,

Table 15.2 Radiation tests.

EFFECT	MEASUREMENT	LEVEL OF ASSEMBLY
Neutron gain degradation	Functional test before and after exposure	Parts, modules, and systems
Neutron ionization (transients)	Operational test	Modules
Logic upset (transient effects)	Threshold screen threshold test	Detector Selected parts and modules and computer enabled and disabled
Radiation–induced burnout Transient malfunction	Operational test and functional test before and after exposure	Parts, Modules, and systems
Thermomechanical effects	Functional test before and after exposure	Parts and computers
Ionizing radiation dose	Functional test after exposures	Parts and modules

circumvention as discussed in Section 13.9, must be initiated properly. In this connection, threshold tests are conducted on parts, modules, and on-board computers. These are nondestructive tests permitting the use of the very hardware that is ultimately fielded. Radiation detectors are screened to select the proper threshold response, and certain other parts, modules, and computer-like subsystems can be tested on a sample basis. Tests are conducted at the module level for the effects of neutron ionization. The latter produces a longer duration ionizing environment than that induced by gamma and x rays. Neutron ionization influences system transient performance, thus requiring testing at least at the circuit (module) level. Testing at the parts level is usually not necessary, for part testing in the high-intensity long-pulse ionizing radiation environment provides data only on equilibrium photocurrents that are applicable directly to neutron ionization effects. For thermomechanical effects, the major concern lies in the area of dissimilar metallic bonding. Because this represents a problem area at the part level, parts are tested. The Nevada Test Site (NTS) can provide a means for obtaining data on a multitude of identical bonds exposed simultaneously to nuclear radiation, such as the Dining Car Event experiment.[9]

15.9 HARDNESS-ASSURANCE VERIFICATION: PARTS

There are four major types of radiation damage for which parts testing is performed. They are transistor gain degradation, logic upset, latchup/

burnout, and thermomechanical effects. Nuclear weapon environments are: neutron fluence, low-level ionizing radiation, high-level ionizing radiation, and pulsed x-ray radiation. Assuming that bond survivability is measured by testing statistically significant numbers of bonds to demonstrate hardness capability, followed by strict process control and verified by physical rather than by radiation testing, the prime effort of HAVT is in the neutron and ionizing radiation area. These environments are simulated readily in existing facilities. Because all part types are not equally sensitive to the effects of each environment, each part type is exposed only to those environments for which the electrical response of the part could impact system performance.

There are a number of factors that influence the choice of parts to be tested. Prime among these are the part type hardness safety margin in its most radiation-sensitive circuit application and the system/mission criticality of the parts in which each part type appears. Secondary factors are the number of parts per system by part type and the vendor history of the production of radiation-resistant parts.

Test samples are selected from production part lots at relatively equal time intervals throughout the part production run. This provides the maximum assurance that the vendors maintain uniform part hardness throughout the production program and permits the determination of any degrading trends, so that appropriate corrective action can be taken. Lots or individual samples within lots may be selected and modified according to results obtained from the monitoring of supplier electrical test data. If the supplier data-monitoring activity indicates a significant change from the norm in electrical data for a particular production lot, then this lot may be sampled for test instead of or in addition to lots previously scheduled.

Neutron testing for hardness-assurance purposes is usually performed on unpowered parts at multiple fluence levels. Irradiation of unpowered parts, at a level significantly above the specification level, represents a worst-case test. Satisfactory response under these conditions provides confidence in parts hardness at lower cost than would be achieved by irradiating powered parts.

Evaluation of neutron data at the specification fluence consists of a comparison of the radiation-induced degradation increments and the values of selective device electrical parameters with preestablished limits. These limits are defined on the basis of worst-case radiation circuit analysis and a device hardness qualification test program. Failure under these circumstances will initiate an immediate investigation and recommendation for corrective action.

For discrete transistors, the minority carrier lifetime damage constant

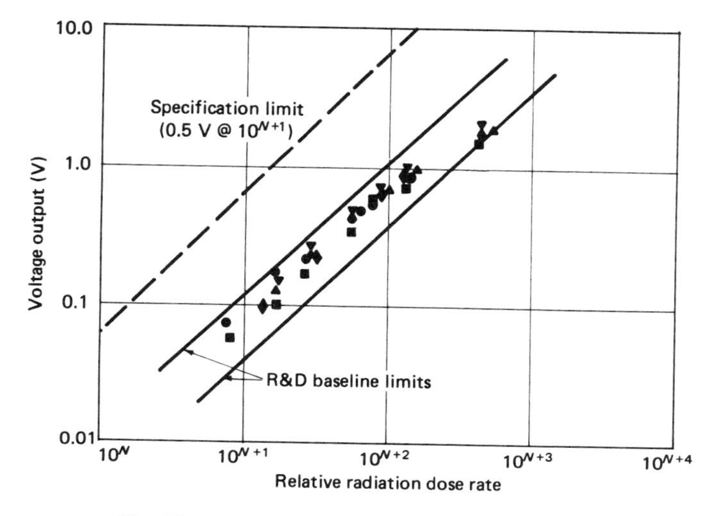

Fig. 15.11 Read preamplifier threshold test data.

and device h_{FE} degradation as a function of neutron fluence would be currently tracked. For integrated circuits, those electrical parameters that are application-critical and neutron-sensitive are chosen for hardness tracking. Typical parameters are device gain, output sink current, output transistor saturation voltage, device dynamic output impedance, and dc output voltage offset, depending on the type of integrated circuit under test.

Photocurrent sensitivity determination is achieved by testing the appropriate device types, properly powered and operating under worst-case loading conditions, in a simulated ionizing radiation environment. Satisfactory hardness in a high-level gamma-ray pulse environment is demonstrated by the proper response to a pre- and post-irradiation electrical input pulse, in addition to the recovery of the radiation-induced output transient within a specified time. Threshold testing consists of monitoring oscilloscope traces of the output of a semiconductor device, during a series of increasing dose/dose rate, low-level gamma radiation, and plotting the data to determine the ionizing dose-rate threshold for each device.

In ionizing radiation threshold testing, the threshold of each part is determined graphically from the plot of output data versus the exposure dose rate. The threshold dose rate, which represents an interpolation of the test data, is that dose rate that would produce a specified threshold output. Typical ionizing radiation threshold data are shown for a read preamplifier IC in Fig. 15.11. Proper functioning of the associated computer requires a minimum read preamplifier threshold output at a specific dose rate. From the figure, consistency of test results is seen to be good, with all data points lying between the R & D baseline limits and with all samples showing a factor of 10 safety margin for threshold dose rate.

In high-level ionizing radiation testing, experience indicates that this high-level test leaves the semiconductor device essentially unaffected or results in device latchup or burnout, both of which are tantamount to catastrophic failure. Because of the go-no go nature of this type of test, the primary data record consists of an oscilloscope photo on which are recorded the pre- and post-radiation response to an electrical stimulus and the corresponding radiation-induced output transient. When comparison of the two electrically induced responses demonstrate that the electrical integrity of the part has not been compromised, then examination of the radiation-induced transient recovery time should verify the absence of an actual or incipient latchup. Data for tracking trends of hardness versus device chronology are obtained in the form of recovery times for the specific irradiation. When such data exhibit internal consistency and good correlation with the R&D baseline data, a satisfactory safety margin is usually implied, so that the corresponding parts will perform satisfactorily.

15.10 HARDNESS-ASSURANCE VERIFICATION: MODULES

The hardness-assurance program includes the radiation testing of various levels of assembly. These radiation tests provide confidence in part and assembly hardness that cannot be achieved by electrical or physical tests. A module is the smallest assembly of electronics produced as a unit in the manufacturing process. During production, the hardness of assemblies could be affected adversely by variations in manufacturing procedures and problems due to personnel-related factors in production operations. Direct radiation testing of assemblies is performed because the resulting data contain all influences impacting the assembly hardness, including those known and unknown. Testing at the assembly level, in addition, provides data more readily applicable to system hardness determination.

Modules are not all equally sensitive to the effects of radiation environments. Consequently, each module is tested only in those environments for which its electrical response could influence system performance. For example, some modules containing digital circuitry are relatively insensitive to neutron degradation. Certain analog-type modules, however, are relatively unaffected by the transients caused by ionizing radiation.

Because of these considerations, the data obtained by module testing may often be separated conveniently for discussion by radiation type, viz., neutrons and ionizing radiation. Test success or failure is based on module parameter changes during and after exposure. Pass or fail criteria take the form of allowable limits on voltage and current signals at critical points within the circuitry. Besides the voltage and current limitations, a secondary pass/fail criterion is the circuit or module transient recovery time. All module circuits must revert to a condition of stability within a specified time after the cessation of the radiation.

The two types of data that are of interest in parts testing are also relevant to module testing. These are (1) the success or failure of a test module to meet required limits of performance at a predetermined radiation level, and (2) the identification of "softening" trends, which are manifest by substantive parameter drifts as a function of successive tests. The former type of data are used to accept or reject the particular test sample and the population it represents. The latter type of data are used to identify changes in hardness, to prevent potential future failures. The pass/fail decision is made by comparing test data with test requirements. Trend data are more involved, often requiring higher levels of exposure than the hardware must meet, to obtain measurable quantities of degradation or deviation from required specifications.

The effects of neutron irradiation are to degrade the performance of semi-conductor parts and hence degrade the performance of the module. The basic information from neutron testing is obtained from module functional (electrical) tests before and after radiation. In the case of test data of a power amplifier for example, its output impedance, as measured by output potential drop, is degraded with increasing neutron fluence. However, when only a slight degradation at higher than specification levels occurs, together with a negligible degradation at the specification level, the module can be considered robustly hard.

Ionizing dose-rate effects are limited to transient response and catastrophic damage that may be caused by transients. These two types of information are obtained by real-time data acquisition during exposure for transient response determination and by functional testing after irradiation for permanent damage evaluation. The timing requirements for module operation are often related to the time duration of circumvention, i.e., circuit incapacitation during the existence of the effects from the radiation environment. This transient malfunction is acceptable for a finite time, of the order of milliseconds or less after the onset of the incident environment. Hence, the information required from an ionizing radiation test is primarily that the hardware output has returned to its preirradiation state, within specified limits and within a specified time.

15.11 CATEGORIZING HARDNESS-CRITICAL ITEMS

Certain system radiation specifications are moderate, in the sense that the application of worst-case assumptions to all parts of the system, such as would be the case for stringent high neutron and gamma fluence levels, is far too conservative. Instead, the following procedural method applies a categorization to each hardness-critical item in such a system. Examples of these types of systems include the strategic systems discussed in Section 13.3.

Hardness-critical items (HCI) are those whose proper function in a system can be affected by incident nuclear-weapon-produced or natural radiation. Mostly, these are active devices such as powered microcircuits (ICs), diodes, and discrete transistors. Items that are patently not hardness-critical from an electrical standpoint include chassis, relays, cable connectors, etc. Given a sufficiently high level of radiation, all these could become hardness-critical. However, the system would in that case be destroyed by the concomitant nonnuclear effects, such as the shock wave overpressure and thermomechanical effects that would be present co-radially with such high levels of radiation. Even without the non-nuclear aspect, the assumption at the outset is that the scenario chosen results in a radiation specification that is meaningful in the above sense.

A basic idea that is part of the hardness-critical-item concept is that of design margins, as discussed in Chapters 13 and 14. The hardness-critical-item categorization is built on the structure of the design margin. It is important to realize that this categorization of parts and subassemblies secularly evolves over the life of the project, as the individual subsystem designs are refined. That is, for the system specifications under discussion there are two main hardness-critical-item categories, namely, category I and category II. The former category is often subdivided into two sub-categories, IH and IM, to delineate the difference between a hardness dedicated category I, and a design margin category I part, respectively. The definitions of hardness-critical categories I and II are generally applied at the parts level, so that parts control will result in the maintenance of the design hardness, in accord with the hardness-assurance program concepts described earlier. Although parts criticality is usually very closely tied to the response of higher tier elements in the design, e.g., piecepart design margin as related to circuit usage, designation of a part as a category I hardness-critical item will not generally necessitate a category I designation of the higher tier element into which such a part is incorporated. For example, the initial achievement of hardness to neutron fluence is circuit-dependent. However, after the circuit design is firm and subject to configuration controls, the hardness level can be maintained through piecepart controls, i.e., stringent controls for category I pieceparts and less stringent controls for category II parts. Therefore, designating circuits as category I is not necessary, except for circuits expressly required for the achievement of hardness, called *hardness-dedicated* Category IH *circuits*.[3] An example is the radiation circumvention/clamp circuits discussed in Section 13.9, which are always classified as category IH. Similarly, in some instances a circuit may be hardness-dedicated, hence category IH hardness-critical, though some or all of the pieceparts therein will be category II, because they have adequate design margins. Examples include feed-through capacitors and surge-arresting devices employed for EMP protection, as well as photocurrent clamping circuits used to minimize the effects of gamma dose-rate transients.

In the following, the criteria to be used in the determination of hardness-critical items (HCI) including parts, components, and design elements, will be addressed for each of the nuclear weapon specification environments. These are blast, thermal, and transient radiation effects on electronics (TREES) that are induced by neutrons, gamma and x rays, and the electromagnetic pulse (EMP). The nuclear blast environment is usually specified in terms of shock-wave overpressure and accompanying hypervelocity wind gusts. Such overpressures generate crushing effects on system structures, as discussed in Chapter 4. HCI category I designation of components and equipment is applicable to those items that must be design-hardened to withstand the blast environment.

As an illustration, the majority of aircraft structural designs are driven by the normal worst-case flight loads, i.e., the nuclear blast environment usually imposes lower structural loads than the designed flight loads on the airframe surfaces. Because the nuclear blast requirements do not drive the design of those items in this case, they are designated as category II.

Examples of category I items can be weapon or bomb bay doors and associated hardware that is sensitive to blast overpressure, or horizontal stabilizers, which are sensitive to hypervelocity wind gusts. The division point between category I and category II for blast is the design point for the normal worst-case flight loads, still using the airframe example. This point is usually fixed for the airframe structure, so detailed definitions of design margins are not necessary to define this division. However, if worst-case flight loads shift, reevaluation is necessary. The thermal environment is normally specified in terms of a pulse shape of the thermal (heat) flux and the corresponding cumulative thermal (heat) fluence. This thermal pulse shape is dependent on weapon yield and detonation altitude, among other factors. Because thermal requirements can be severe, considerable effort must be expended on appropriate design hardness. Examples for light structures, such as aircraft, include composite structural components, honeycomb panels, and radomes. If successive thermal pulse exposures are part of the specification, then complications can ensue. To illustrate, a surface coating may provide protection to the underlying structure for an initial thermal exposure; however as a result of that exposure, its reflection characteristics could be degraded so that protection is inadequate for subsequent expsosures. The HCI designation with regard to thermal environment is similar to that utilized for blast, as discussed in the foregoing. Any component whose design is driven by the thermal requirement as influenced by the system thermal specification is designated category I. All other components are designated category II. Examples of category I hardness-critical items with respect to thermal effects may be aircraft composite fairings, radomes, honeycomb structures, optical sensors, and surface coatings.

Neutron fluence impacts the function of semiconductor devices in almost every system. The transistor electrical parameter most affected by neutron damage and generally most important to circuit design is the common emitter current gain, h_{FE} or β. Hence, design margins are usually defined with reference to this parameter. As linear integrated circuits, such as operational amplifiers, frequently reflect a composite gain related to internal transistors, this approach is also applicable to these devices. This is in contrast to IC logic circuitry, which is not usually significantly affected by neutrons at the normal environment neutron specification level. In a few specific circuits, such device parameters as breakdown voltage and propagation delay times may be of greater relevance than gain. In these situations, neutron-induced changes and related design margins should be developed in a manner similar to the gain design margin.

For neutron effects, the design point that usually marks the boundary between category I and II is a neutron fluence design margin of one order of magnitude. This means that if the device functions effectively, even when subjected to an order of magnitude or more neutron fluence level over that as given by its neutron specification, it is designated as category II. This is in accord with the usual definition of design margin. The latter is defined as the ratio of the neutron fluence level at which failure/unacceptable response occurs, to the specification fluence level. Failure/unacceptable response is based on circuit level operational requirements and is taken to be the point at which the circuit operation exceeds the design tolerance limits. This is usually determined through circuit analysis utilizing piecepart response data. Thus, each semiconductor piecepart degradation is related to circuit operational requirements.

As mentioned, if specific circuits or subcircuits are incorporated into the design solely for the purpose of radiation hardening, they are considered hardness-dedicated and are designated as HCI category IH. For a non-hardness-dedicated circuit designed without including hardness-dedicated parts and/or subcircuits, the circuit hardness criticality designation is category II. Pieceparts within both hardness dedicated and other circuits will be either category I or II, depending on design margin considerations.

An example that illustrates the boundaries between categories I and II is given in Fig. 12.40 in Section 12.8. The uppermost curve therein, labeled empirical data, is a representative plot of transistor gain as a function of neutron fluence. This curve is drawn through the medians of the distribution of sample data points, and the error extremities are indicated by the error bars. The lower curve is obtained by applying the average damage constant, as derived from test data, to the (minimum) transistor gain, as given by Eq. 5.66, and published as vendor data. This curve passes through the minimum acceptable value of gain, β_{min}, as specified by the circuit designer (point A) which establishes the failure fluence level 10^Y neutrons per cm^2. The design margin is then given by the ratio $10^Y/10^X$, where 10^X is the specification level. Point B is the category I/category II breakpoint and is located at a neutron fluence of 10^{X+1}, i.e., an order of magnitude above the specification level. The designation of the piecepart under consideration is determined by whether point A is to the left (category I) or right (category II) of point B. The piecepart in this example is category I.

Ionizing dose rate is the environment that most directly accounts for the production of photocurrents in the junctions of semiconductor devices and that results in electrical transients in the affected circuits. In this context, both prompt and delayed dose rates must be considered. The majority of analog circuitry and included parts in most systems are designated category II for gamma-ray-dose-rate effects, because in most cases, the circuit perturbation time is much shorter than the tolerable transient response time at the circuit output. However, in cases where relative response times are the governing failure factors, the "order of magnitude" design margin for category I and II determination may be based directly on the ratio of excitation time of the driven part to the duration of the radiation-induced transients, rather than an order of magnitude in dose rate. Usually, if it can be established that the response time of the controlled analog circuitry or device is greater than 1 ms, the driving circuitry and parts can be designated as category II.

Analog circuits and pieceparts that are determined to be category I on the above basis, as well as analog electronics not responsive analytically

Table 15.3 Typical endoatmospheric threat, hardness-critical-item design margins.[4]

RADIATION TYPE	CATEGORY	DESIGN MARGIN*	LEGEND
Neutrons	II	$\Phi_F > 10\Phi_{spec}$	Φ_F: Incident fluence
	I	$3\Phi_{spec} < \Phi_F < 10\Phi_{spec}$	Φ_{spec}: Spec. fluence
Ionizing dose	II	$D_F > 10D_{spec}$	D_F: Incident dose
	I	$3D_{spec} < D_F < 10D_{spec}$	D_{spec}: Spec. dose
Dose rate	II	$\dot{\gamma}_{ups} > 10\dot{\gamma}_{circ}$	$\dot{\gamma}_{circ}$: Incident dose rate
		$\dot{\gamma}_{BO} > 10\dot{\gamma}_{spec}$	$\dot{\gamma}_{BO}$: Burnout dose rate
		$t_{crit} > 5t_{prop}$	t_{crit}: Prop delay threshold
	I	$3\dot{\gamma}_{CIRC} < \dot{\gamma}_{UPS} < 10\dot{\gamma}_{CIRC}$	$\dot{\gamma}_{spec}$: Spec. dose rate
		$3t_{PROP} < t_{CRIT} < 5t_{PROP}$	$\dot{\gamma}_{ups}$: Upset threshold
			t_{prop}: Prop. delay
EMP	II	$20\log_{10} I_{ups}/I_{spec} > 10\text{DB}$	I_{ups}: Incident upset curr.
		$20\log_{10} V_{ups}/V_{spec} > 10\text{DB}$	I_{spec}: Spec. curr.
		$W_{dam}/W_{spec} > 3$	W_{dam}: Incident damage pwr.
			W_{spec}: Spec. rated power
	I	$3 < I_{UPS}/I_{SPEC}, V_{UPS}/V_{SPEC} < 10$	V_{ups}: Incident upset volt.
		$2W_{SPEC} < W_{DAM} < 3W_{SPEC}$	V_{spec}: Spec. volt.

* Parts whose hardness design margin is less than 3X are categorically rejected for systems use.

to the described perturbation time approach to categorization, should be considered from a functional standpoint. In some cases the perturbation may cause a short inoperative period, circuit blanking, or possible erroneous digital information, as in analog-to-digital converters. The resultant effects on subsystem operation should be examined for significance with respect to mission-completion capability. If the only result of an analog circuit response to the dose-rate pulse is a brief disruption that does not degrade the system performance, the circuit will be category II for gamma dose rate.

Ionizing dose deleteriously affects a few types of semiconductor devices for moderate radiation specifications implied here. Due to the resulting ionization caused by incident gamma rays, as discussed in Chapter 6, semiconductor devices are potentially susceptible. These include MOSFETs, as they can undergo gate threshold voltage shifts; bipolar transistor circuits operated at very low bias currents, e.g., less than $1\,\mu$A; and possibly operational amplifiers with extremely low offset currents.

For the ionizing dose environment, the breakpoint between categories I and II is also taken at $10\times$ the specification level. In other words, the breakpoint is associated with a design margin of 10 referred to the specification level. The median of the response distribution will be used as the basis for the determination of the design margin.

This is shown in Table 15.3, which depicts a sample endoatmospheric threat, hardness-critical-item design margin and category divisions. EMP hardening approaches to minimize magnetic coupling and to provide shielding to attenuate electric field effects usually employ controlled-wiring, cable-routing and electrical grounding schema similar to electromagnetic compatibility (EMC) measures. Each component with a significant role in the EMP hardening approach and/or whose design is impacted by the EMP requirement is usually defined to be category I hardness-critical. Examples of category I components/designs are shielded bay doors, shielded bay structures, conduits, conduit connectors, and terminals. Usually no quantitative division point is presented for EMP shielding effectiveness associated with the foregoing. Because the shielding effectiveness of a given configuration is strongly dependent upon the assembly, maintenance practices, aging, corrosion, etc., difficulties could be encountered in practice regardless of the potential shielding effectiveness obtained in an ideal configuration.

Electronic interface pieceparts may be either voltage- or current-sensitive. However, the sensitivity can be related generally to the impressed current at the connector pins by use of appropriate circuit parameters.The design hardness margin for EMP is defined as the logarithmic relationship of the current level that will produce device damage, $I_{w\mathrm{dam}}$, to the determining EMP current level impressed on the device at the interface connector, $I_{w\mathrm{spec}}$. That is

$$\text{Hardness margin (dB)} = 20\log_{10}(I_{w\,\mathrm{dam}}/I_{w\,\mathrm{spec}}) \tag{15.2}$$

The breakpoint between hardness critical categories I and II for EMP is sometimes taken at the hardness margin of 10 dB. Other aspects of hardness margins, both for EMP and the preceding environments, are discussed in Sections 12.6 and 13.5.

15.12 HARDNESS-ASSURANCE COSTS: SUMMARY

The question of hardness-assurance costs and general hardness costs is vexing. With respect to the latter, modern semiconductor systems are sufficiently complicated and their technology is changing so rapidly, that it is difficult to extrapolate from earlier experience. It would be a superlative achievement if a plot of hardness costs as a percentage of system cost versus required hardness levels could be made available. Especially a one-parameter family of such curves, where the parameter is designated as the epoch in which the hardness effort was introduced into a particular program, e.g., ranging from hardness at program inception to an eleventh-

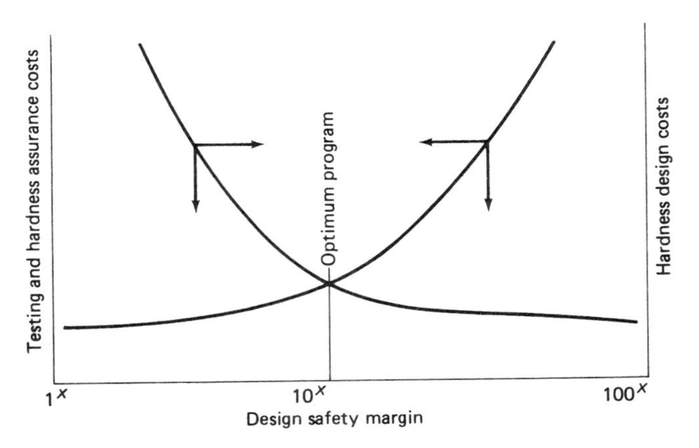

Fig. 15.12 Hardening costs.

hour effort to a hardness retrofit for a completed system. Unfortunately, for security and other reasons, such a set of curves would be probably impossible to obtain. At this juncture, the best that can be stated in this regard is that the later a hardness program is introduced into a system in its development, the costlier the hardness will be, with the costs increasing approximately exponentially with the introduction epoch.

Insofar as hardness-assurance costs are concerned, somewhat more qualitative information is forthcoming. If hardness-testing plus hardness-assurance costs are plotted as a function of design safety margin, the resulting curve is monotone increasing, as in Fig. 15.12. But, if, on the same curve, hardness-design costs are plotted versus the same abscissa, they are monotone decreasing. This is because once the system "core" hardness is in place and paid for, marginally increasing hardness-design efforts will produce rapidly increasing design safety margins.

However, to achieve greater design safety margins, hardness-testing plus hardness-assurance costs merely keep climbing at a superlinear rate. Nevertheless, there exists an optimum program, as seen in Fig. 15.12, that for a simultaneously minimum testing plus hardness-assurance cost and minimum hardness-design costs, an optimum design safety margin will ensue. The intersection of these two curves, in the figure, corresponds to an optimum design safety margin.

Figure 15.13 shows both hardening costs and equivalent hardening *benefit dollars* as a function of relative hardness level, assuming that the system is a priori hard to EMP or it has been made so. The vertical dashed line depicts a possible optimum program, where most of the hardening benefit has been accrued and the corresponding hardness costs

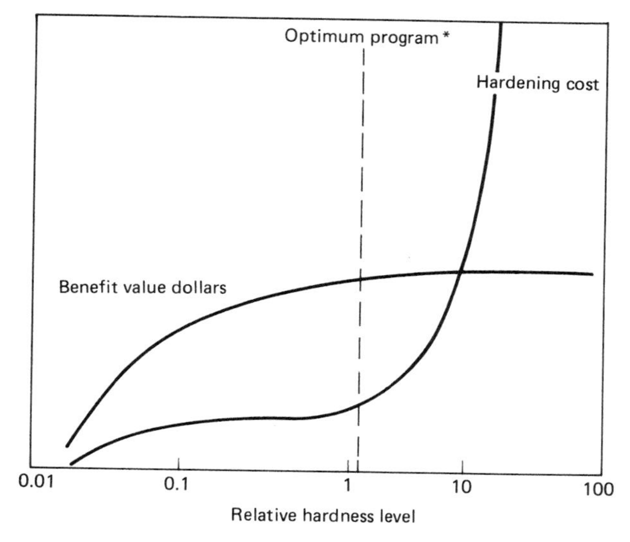

Fig. 15.13 Radiation effects costs and benefits.

have not become excessive. It is seen that its dependence on relative hardness level in this range is sufficiently weak that the optimum line can be displaced to the left by almost an order of magnitude in hardness level abscissa and achieve roughly comparable hardness benefit at roughly comparable cost. This optimum corresponds to a moderate hardness level system requirement. However, for very stringent, high hardness levels, there is no option but to endure rapidly increasing hardness costs with but marginally increasing hardness benefit. Again, these curves are qualitative in their attempt to represent the distillation of many man-years of system experience in hardness survivability/vulnerability for electronic systems.

Systems that must operate during (operate-through) and after nuclear radiation exposure are hardened by careful design. System hardness is generally verified during a research and development program. A hardness-assurance program becomes necessary as the status of the system program progresses from the development phase to the production phase. Satisfactory performance of production systems predicated upon the premise that hardness safety margins achieved during the development program are not allowed to degrade significantly during system production must be proved.

The hardness-assurance program that has been described provides the required assurance that the safety margins are being maintained. The program ensures the continuing performance of production systems both during and after radiation exposure. A summary of hardness activities and the resulting benefits has been presented above. Hardness-related electrical and mechanical semiconductor part parameters have been identified for part-vendor control and screen. Continuous monitoring and analysis of vendor data provide the early identification of certain types of parts softening, resulting in timely vendor corrective action. Radiation testing of part sample lots confirms the conclusions drawn from vendor monitoring and verifies part radiation response to simulated environments. Finally, higher assembly radiation testing ensures that radiation hardness is maintained in the delivered production systems.

15.13 HARDNESS MAINTENANCE

The focus of this exposition of nuclear hardness maintenance is somewhat centered on Army material requirements. However, its content is readily seen to be applicable across the services. Hardness maintenance can be defined[5] as: That combination of inspection, test, repair, and logistic activities accomplished during the operational phase of a nuclear-hardened system, to insure that the hardness built into the system is retained throughout the system lifetime. Collateral with this definition is that of hardness surveillance,[5] defined as: Those long-term inspection and test activities conducted during and beyond hardness maintenance to assure that systems are, in fact, being properly maintained with the desired nuclear hardness.

Military equipment that has been fielded will suffer difficulties even under normal environmental conditions during its deployment, which could adversely affect its nuclear survivability. These problems are usually associated with repair, redesign, reprocurement, or operations. For example, if maintenance or repair is performed poorly, especially with little or no regard for the hardness-critical features of the equipment, it is obvious that its nuclear survivability could be compromised. From the point of view of maintenance and repair personnel unschooled in nuclear survivability, hardness design incorporated into the equipment is often transparent, in that it seems to serve no apparent form, fit, or function. Hence, these aspects are easily ignored, removed, or bypassed in restoring equipment to its operational capability. To prevent this situation, not only must proper instructions be included in the equipment technical manuals, but corresponding personnel must be schooled in order to be made acutely aware of the nuclear survivability aspect embedded in the equipment. Unfortunately, examples are many where provisioners have supplied, and repair personnel have installed, circuitry, components, and parts that look and perform like their nuclear-survivability qualified counterparts, but would not perform adequately in a nuclear environment. This can be exascerbated when equipment is redesigned, and/or where the original equipment manufacturer source no longer exists. It is vital that a paper trail exist to trace this equipment's hardness baseline through corresponding drawings and design documentation. As already alluded to, an item's nuclear survivability can be degraded through continued normal operating conditions, corrosion, and age. For example, corrosion could reduce the effectiveness of special EMP grounding and shielding procedures by degrading wiring ohmic contacts, and metal-to-metal electronic shielding enclosure integrity. To assure that detailed nuclear survivability procedures are implemented throughout all phases of the acquisition cycle, the Army has established a Life Cycle Nuclear Survivability Program, under Army Regulation 70–60.[5]

A paramount aspect of military deterrent forces is the ability of their personnel and equipment to survive in the nuclear battle arena. As nuclear survival requirements are now part and parcel of the various mission areas of the services, its preservation, especially during deployment, is mandatory. Life Cycle Nuclear Survivability Programs have been established to provide logisticians, maintenance personnel, trainers, and military personnel operators with the wherwithal necessary to maintain nuclear survivability throughout the deployment phases of new military equipment. The overall thrust of this program is the development of nuclear survivability maintenance approaches and methods that are consistent with the repair, replacement, redesign, and reprocurement procedures used by the services.

To ensure survival of tactical and strategic military equipment in the hostile nuclear theater, it must be protected against nuclear damage effects. Certain equipment may only be required to survive high-altitude EMP (HEMP), while others may be required to survive both HEMP and SREMP from a tactical nuclear threat. Aspects of these requirements, vis á vis military equipment construed as man-machine entities, are discussed in Section 13.4.

An example of a tactical nuclear threat would be that produced by a low-yield close-in burst at or near the surface, resulting in a shock/blast wave, concomitant thermal radiation, prompt nuclear radiation (neutrons and gammas), and surface EMP (SREMP). As discussed in previous chapters, these nuclear burst-produced environments have wide-ranging effects on military equipment. For example, the blast and thermal outputs result in bending, crushing, charring, or burning portions or all of a particular type of equipment. The EMP and nuclear-radiation environments can produce electrical transients that can both upset and burn out components, circuits, and subassemblies. To a lesser extent, nuclear radiation can permanently alter the electrical properties of semiconductor devices and associated materials, such as dielectrics and insulators. For large metal housing installations, such as communications shelters, the metal housing can be struck by gamma photons, producing compton electrons. These may create spurious SGEMP and corresponding currents therein. System-generated EMP (SGEMP) is discussed in Section 9.10.

It is imperative that nuclear survivability-unique requirements be incorporated early into the directives, statements of work, regulations, and other documentation that support system acquisition. This is in keeping with the proviso that highly efficient, moderate-cost system hardening, with high confidence, is obtained when nuclear-hardening aspects and specifications are asserted, and inserted, at the inception of military equipment procurement, design, and construction. This is also discussed in Section 13.3, as applied to strategic systems.

As described in detail in Section 15.11, military equipment items are partitioned, in terms of their nuclear survivability, into hardness-critical item categories. Such hardness-critical items (HCIs) are clearly identified on corresponding documentation and accompanying drawings. They are also labelled in the narrative portion of task analysis documents, as well as in piece-part lists. They are identified through a coded lexicon in the Army Logistics Support Analysis Record.[6,7,8] Failure to designate parts with correct HCI categories can result in untoward consequences when operating equipment in nuclear environments. A semingly trivial example is that of negligently substituting a plastic knob for a metal knob on a communications transceiver. The plastic knob can melt or deform sufficiently in a thermal radiation environment so that its graduations become undecipherable. The HCI-correct metal knob is required because it will remain intact in this environment, as well as not becoming harmfully radioactive. The military hardness survivability discipline is replete with similar examples.

With respect to the whole area of the procurement of hardened military systems from industry, it is apparent that the latter frequently construe the task of hardening as one more exasperating burden in the fabrication of their system according to military specifications. Nuclear hardening is often an intangible in the spectrum of technical requirements of system designers. If sufficient attention to system hardness is not brought to bear assertively and sufficiently early, as reckoned from program inception, eleventh-hour fixes may be required. This results in expensive, time-consuming equipment retrofitting to include hardening.

Often, this yields only a low to medium confidence level for the hardness-modified equipment. For example, hardening a large, complex system should normally consume approximately 0.1–1.0 percent of the total system costs, excluding deployment costs. Unfortunately, certain systems extant have run up hardening costs a factor of 5–10 larger, due to insufficient attention paid to hardness requirements early enough in the system procurement cycle. Approximately every decade, the then-current U.S. Department of Defense issues a directive enjoining corresponding industries to pay sufficient attention to the matter of hardness for systems procured by them.

REFERENCES

1. G. C. Messenger. Hardness Assurance. Rockwell International, Report BR 71-899/201, Oct. 1971.
2. D. Casasent. Electronic Circuits. Quantum Publishers, New York, Section 3.9, 1973.
3. B-1 Hardness Assurance Guidelines. Aeronautical Systems Div. Air Force Systems Command, Wright-Patterson AFB, OH, Mar. 1976.
4. G. C. Messenger. Hardness Assurance. Radiation Hardness Electronics Technology Conference, Melbourne, FL, Oct. 21–23, 1980.
5. F. M. Connell, D. B. Stewart, and R. K. White. "Logistics Support of Nuclear-Survivable Equipment." HDL-CR-89-013-2, US Army Lab. Command., June 1989.
6. F. M. Connell, D. B. Stewart, R. K. White, and J. J. Halpin. "Provisioning Hardness Critical Items For Army Equipment." HDL-CR-89-013-1, Feb. 1989.
7. R. Winklareth. "Integrated Logistic Support For Maintaining The Nuclear Survivability of Army Tactical Systems." HDL-CR-86-071-1, May 1987.
8. R. K. White and F. M. Connell. "Logistics Support of Nuclear-Survivable Equipment." HDL-CR-87-069-1, April 1987.
9. J. E. Lohkamp and J. M. McKenzie. "Measurement of the Energy Dependence of Neutron Damage in Silicon Devices. (Dining Car Event)." IEEE Trans. Nucl. Sci. NS-22(6):2319–2325, Dec. 1975.

APPENDIX A ANSWERS TO PROBLEMS FOR CHAPTERS 1–10

CHAPTER 1 ANSWERS

1.(a). Bohr stationary state energy levels for atoms of atomic number Z are obtained in vacuum from:

$$(1.1)\ mv^2/r = Ze^2/r^2; \qquad (1.2)\ mvr = n\hbar;$$
$$(1.3)\ E = \tfrac{1}{2}mv^2 - Ze^2/r.$$

1.(b). For $n = 1$, $Z = 1$, $E_1 = -me^4/2\hbar^2 = (9 \cdot 10^{-28}) \times (4.8 \cdot 10^{-10})^4/2(6.6 \cdot 10^{-27}/6.28)^2 = (2.16 \cdot 10^{-11}\,\text{ergs})/(1.6 \cdot 10^{-12}\,\text{ergs/eV}) = -13.6\,\text{eV}$.

1.(c). In unrationalized CGS units in a medium of dielectric constant ε,

$$(1.1)\ mv^2/r = Ze^2/\varepsilon r^2; \qquad (1.2)\ \text{same}; \qquad (1.3)\ E = \tfrac{1}{2}mv^2 - Ze^2/\varepsilon r.$$

Eliminating v and r from these equations, after some algebraic manipulation, $E_n = -mZ^2e^4/2\varepsilon^2\hbar^2n^2$. So that for $n = 1$, $Z = 1$, $E_1 = -13.6\,\text{eV}/\varepsilon^2 = -13.6/(11.7)^2 = -0.1\,\text{eV}$. This is an approximation, and as a reference check[30] reveals that the effective mass $m^* = m_n/m = 0.33$ should be used. So that now $|E_1| = 13.6m^*/\varepsilon^2 = 13.6 \times 0.33/(11.7)^2 = 0.033\,\text{eV}$. Hence, donor electrons are easily available in the lattice, with respect to energy, since room temperature thermal energy is about $0.026\,\text{eV}$.

1.(d). For ground state electrons, $Z = 1$, $n = 1$ in 1(a) above. Eliminating v between these equations yields the ground state Bohr radius, $r_1 = \hbar^2/me^2 = (6.6 \cdot 10^{-27}/2\pi)^2/(9 \cdot 10^{-28})(4.8 \cdot 10^{-10})^2 = 0.53 \cdot 10^{-8}\,\text{cm} = 0.53\,\text{Å}$.

2. Proper normalization means that the Fermi distribution $f(E)$ must be multiplied by a constant, N_f, so that $\int_0^\infty N_f f(E)dE = 1$, in keeping with the idea of a probability function. Hence, $N_f \int_0^\infty dE/(1 + \exp(E - E_f)/kT) = 1$. Integrating (easy), $N_f = [kT\ln(1 + \exp(E_f/kT)]^{-1} \cong E_f^{-1}$, since $\exp(E_f/kT) \gg 1$ at room temperature.

3. Since $m^* = \hbar^2/(d^2E/dk^2)$, and for a free electron $E = p^2/2m = (\hbar k)^2/2m$, where m is the free electron mass; differentiating to get $d^2E/dk^2 = \hbar^2/m$, and substituting yields $m^* = m$.

4. This is a problem in integrating $n_0 = \int_{E_c}^{\infty} N_e(E)f(E)dE$. Substituting N_e and f into the integrand, with $x = (E - E_c)^{1/2}$ and $(1 + \exp(E - E_f)/kT)^{-1} \cong \exp - (E - E_f)/kT$, since $E_c - E_f \gg 3kT$, and using $\int_0^{\infty} x^{1/2}e^{-x}dx = \Gamma(3/2) = \sqrt{\pi}/2$, where $\Gamma(x)$ is the gamma function, and the number of minima in the reciprocal lattice space $M = 1$, yields n_0 in Eq. (1.27).

5. Since $n_0 = p_0$ for intrinsic material, equating Eqs. (1.27) and (1.29) yields $N_c \exp(E_f - E_c)/kT = N_v \exp(E_v - E_f)/kT$. Solving for E_f gives Eq. (1.33). The resulting second term for silicon intrinsic at 300°K is about 0.01 eV, so that $E_f \cong \frac{1}{2}(E_c + E_v)$.

6. The number density of silicon atoms $N_{si} = \rho_{si}N_0/A_{si}$, where $\rho_{si} = 2.33\,\text{gm/cm}^3$, Avogadro's number $N_0 = 0.602 \cdot 10^{24}$ atoms per mole, and A_{si} is 28 gm/mole. So $N_{si} = 2.33 \times 0.602 \cdot 10^{24}/28 = 5 \cdot 10^{22}$ Si atoms per cm^3.

7. Now $np = n_i^2$, $p \cong n_i^2/N_D$, and $n \cong n_i^2/N_A$. For $N_D = 10^{14}/\text{cm}^3$, $n_i = 1.45 \cdot 10^{10}$, then $p \cong (1.45 \cdot 10^{10})^2/10^{14} = 2.1 \cdot 10^6$ holes/cm^3; $n = n_i^2/p = (1.45 \cdot 10^{10})^2/2.1 \cdot 10^6 = 10^4$ electrons/cm^3 and $N_A \cong n_i^2/n = (1.45 \cdot 10^{10})^2/10^{14} = 2.1 \cdot 10^6$ acceptor atoms per cm^3.

8. The collision rate is the reciprocal of the mean free time between collisions, so that $(10^{-6}/10^7)^{-1} = 10^{13}$ collisions per sec.

9. No, because Fick's Law asserts that the diffusion component of the *particle* current is proportional to the *negative* gradient of the carrier density without regard to charge. The negative electron charge and the hole positive charge combine with the negative gradient to establish the correct sign of the corresponding *electric* current.

10. From Eqs. (1.66) and (1.67), for the recombination rate of traps, U, it is seen that $\min(E_t - E_i)$ makes for max U. Hence the statement in question (10) follows.

11.(1). From Problem 6, $N_{si} = 5.10^{22}$ Si atoms per cm^3. Since $n_o p_0 = n_i^2 = (1.45 \cdot 10^{10})^2$, and for intrinsic silicon $n_0 = p_0$, there are $1.45 \cdot 10^{10}$ hole-electron pairs per cm^3. Hence, $1.45 \cdot 10^{10}/5 \cdot 10^{22} = 2.9 \cdot 10^{-13}$ or 1 hep (hole-electron pair) per $3.4 \cdot 10^{12}$ Si atoms.

11.(2). $\sigma_{\text{isi}} = (n_0\mu_n + p_0\mu_p)e = 1.45 \cdot 10^{10}(1350 + 480) \cdot 1.6 \cdot 10^{-19}$ (coulombs/electron) $= 4.25\,\mu$mhos (μsiemens) per cm.

$$\rho_{\text{isi}} = 1/\sigma_{\text{isi}} = 1/4.25 \cdot 10^{-6} = 235 \text{ kilohm cm.}$$

11.(3a). $N_D(\text{Sb}) = 5 \cdot 10^{22}/10^7 = 5 \cdot 10^{15}$ donor atoms per cm^3.

11.(3b). $n = 5 \cdot 10^{15} + 1.45 \cdot 10^{10}$(intrinsic electron density) $= 5 \cdot 10^{15}$ electrons per cm^3. $p = n_i^2/n \cong n_i^2/N_D = (1.45 \cdot 10^{10})^2/5 \cdot 10^{15} = 4.21 \cdot 10^4$ holes per cm^3.

11.(3c). So, $1.45 \cdot 10^{10}/4.21 \cdot 10^4 = 3.45 \cdot 10^5$.

11.(3d). $\sigma = (n\mu_n + p\mu_p)e \cong N_D\mu_n e = 5 \cdot 10^{15} \times 1350 \times 1.6 \cdot 10^{-19} =$ 1.08 mhos(siemens)/cm.

$\rho = 1/\sigma = 0.92$ ohm cm.

12.(a). At very low temperatures, the ambient thermal energy is insufficient to ionize the dopant atoms. Hence, they play no role in providing carriers over those available in the intrinsic state.

12.(b). At very high temperatures, the fermi level approaches the center of the energy gap, which is the intrinsic state of the semiconductor, where $np = n_i^2$. However, n_i is very large at these high temperatures, implying that the main source of carriers is their thermal generation from the valence band to the conductance band. This source of carriers completely dominates those provided by the dopants, thus reducing the latter to negligible importance.

13. (e) None of the above.

14.(a). From Problem 1.1, the donor and acceptor energy levels are given by $E_{D,A} = -m^*_{D,A}e^4Z^2/2\varepsilon^2\hbar^2n^2$ (ergs) $\varepsilon_{Si} = 11.7$, $1\,\text{eV} = 1.6 \cdot 10^{-12}$ ergs. Since the donor and acceptor atoms are hydrogen-like, their $Z = 1$ above, but their effective masses are $m^*_D = 0.33m_0$, $m^*_A = 0.55m_0$, as discussed in Section 1.5. For the ground state $n = 1$. Then, doing the computation on your calculator gives $E_D = 0.033\,\text{eV}$, $E_A = 0.055\,\text{eV}$. Since 0.026 eV corresponds to room temperature ($kT = 8.6 \cdot 10^{-5} \cdot 300°\text{K} = 0.026\,\text{eV}$), it is appreciated from their E_D and E_A values that these donors and acceptors are easily ionized within the silicon lattice, to perform their doping function.

14.(b). The ratio of E_D and E_A in germanium, with respect to silicon, is seen to be given by $(\varepsilon_{Si}/\varepsilon_{Ge})^2 = (11.7/16.3)^2 = 0.51$.

15. Inserting Eq. (1.77A) into the expression for n_p in the problem statement, with $p \cong N_A$, results in $n_p = n_i^2 \cdot N_A{}^{((0.018/kT)-1)} \cdot 10^{-0.306/kT}$.

Now, $0.018/kT - 1$ is, respectively -0.31 at room temperature (298°K), 1.71 at liquid N_2 temperature (77°K), and zero at 210°K.[35] Hence, at the latter temperature, n_p is essentially independent of the dopant density, implying that nominal transistor function is virtually nonexistant in this temperature region, as discussed in Section 1.10.

CHAPTER 2 ANSWERS

1. To yield Eq. (2.14), it is seen from Eq. (2.11) that $\nabla \cdot \bar{A} + \partial\phi/c\partial t = 0$ must hold. From $\bar{B} = \nabla x\bar{A}$, it is clear that the gradient of a scalar potential, ∇V, can be added to $\bar{A}$, since $\nabla x(\bar{A} + \nabla V) = \nabla x\bar{A}$, because $\nabla x\nabla V \equiv 0$. Hence, $\bar{A}$ can be replaced by $\bar{A} + \nabla V$. Also, to keep $\bar{E}$ invariant in Eq. (2.9), viz., $\bar{E} = -\nabla\phi - \partial\bar{A}/c\partial t = 0$, ϕ can be replaced by $\phi - \partial V/c\partial t$. Inserting these two extensions, viz., $\bar{A} + \nabla V$ and $\phi - \partial V/c\partial t$ into $\nabla \cdot \bar{A} + \partial\phi/c\partial t = 0$ yields $\nabla^2 V - \partial^2 V/c^2\partial t^2 = 0$, a wave equation.

2. The Biot and Savart Law, Eq. (2.15), can be rewritten in circular coordinates as $dB = Idl \sin\theta/cr^2$.

2.(a). Magnetic field at the center of a circular loop of wire carrying current I. Here, $\theta = \pi/2$ always; $B = \mu H = (I/c)\oint dl/r^2 = (I/c)\int_0^{2\pi} rd\varphi/r^2$. Or, $H = 2\pi I/\mu cr$.

2.(b). Magnetic field at distance r_0 from an infinitely long wire carrying a current I. $B = \mu H = (2I/c)\int_0^\infty dl \sin\theta/r^2 = (2Ir_0/c)\int_0^\infty dl/(r_0^2 + l^2)^{3/2}$. Or $H = 2I/\mu cr_0$.

3. By p material, it is meant that the current is primarily due to holes. So that the Hall electric field $E_y = v_x B_0/c$, $v_x = I_0/pe$ for holes. Eliminating v_x, as was done for the negative electron current in Eq. (2.23), the result here is still given by $E_y = RI_0B_0$, but the Hall coefficient now is $R = 1/pec > 0$.

4. For plane electromagnetic waves, the equation of a stationary wave front is $\bar{k} \cdot \bar{r} =$ constant, as it asserts that the locus of all points $\bar{r}$ are projected onto that plane wave front. $\bar{k} = 2\pi\hat{n}/\lambda$ is the propagation vector, where $\hat{n}$ is the unit vector normal to the wave front (i.e., it points in the direction of the now-moving wave front. $\bar{r}$ is the vector to a given point on the plane wave front from the origin. Without any loss in generality, consider a one dimensional plane wave, so that $kx = \omega t$, or $2\pi x/\lambda = 2\pi\nu t$. Then $x = vt$ is the plane wave front coordinate, traveling at speed $v = \nu\lambda$.

5. The one-dimensional wave equation for the E_x, or H_x, field in vacuum is $\partial^2 E_x/\partial x^2 = \partial^2 E_x/c^2\partial t^2$. E_x (or H_x) $= f(x - ct) + g(x + ct)$ satisfies this wave equation, as seen by substitution. $f(x - ct)$ is a wave of shape given by the function f, traveling in the $+x$ direction; with speed c. $g(x + ct)$ is a wave of shape g, traveling in the $-x$ direction, with speed c.

6. For the short stub antenna, the computation proceeds as given in Section 2.6 for the dipole antenna, but the current in the stub is triangular, with an effective antenna length. This amounts to the stub-effective length being half of its physical length l. Since the dipole radiation resistance $R_r = 80\pi^2(l_e/\lambda)^2$ following Eq. (2.62), the corresponding short stub radiation resistance $R_s = 20\pi^2(l/\lambda)^2$.

7. The equation of continuity is a statement of charge conservation. That is, $\nabla \cdot \bar{j} = -\partial\rho/\partial t$ merely states that the charge density is decreasing in a given volume by virtue of the current density $|\bar{j}| = |\rho e\bar{v}|$ diverging $(\nabla \cdot \bar{j})$ out of the volume.

8. Assuming $\mu = \varepsilon = 1$ in the free space, where $\sigma = \bar{J} = 0$, following the development given in Section (2.3) yields an $\bar{E} = -\nabla\phi_h - (1/c^2)\partial^2\bar{Z}/\partial t^2$. ϕ_h is the scalar potential for the Hertz potential $\bar{Z}$ (i.e., where $\bar{H} = (1/c)\partial(\nabla X\bar{Z})/\partial t)$. With these two expressions, $\bar{E}$ and $\bar{H}$ can be eliminated in Eqs. (2.2) and (2.3). This yields $\nabla^2\phi_h + (1/c^2)\partial^2(\nabla \cdot \bar{Z})/\partial^2 t = 0$, analogous to Eq. (2.10), as well as $\partial/c\partial t[(\nabla^2\bar{Z} - (1/c^2)\partial^2\bar{Z}/\partial t^2] - \nabla(\phi_h + \nabla \cdot \bar{Z})) = 0$, analogous to Eq. (2.11). To uncouple these equations, let $\nabla \cdot \bar{Z} + \phi_h = 0$. Then both $\bar{Z}$ and ϕ_h satisfy wave equations, as seen by substituting for $\nabla \cdot \bar{Z}$ in the above equations, to yield $\nabla^2\bar{Z} - (1/c^2)\partial^2\bar{Z}/\partial t^2 = 0$ and $\nabla^2\phi_h - (1/c^2)\partial^2\phi_h/\partial t^2 = 0$, respectively.

9.(a). In spherical coordinates, assume $\bar{Z}(r,t) = \bar{Z}_0(r)\exp i\omega t$, which then yields (from its wave equation) $\bar{Z}_{0rr} + (2/r)\bar{Z}_{0r} = (1/c^2)\bar{Z}_{0tt}$, where the r and t subscripts denote partial derivatives. From the solution to this equation, it follows that $\bar{Z} = (\bar{p}_0/r)\exp - i(\bar{k} \cdot \bar{r} - \omega t) \equiv \bar{p}(t)(\exp - i\bar{k} \cdot \bar{r})/r$, where $\bar{k} = (2\pi/\lambda)\bar{n}$, and $\bar{n}$ is a unit vector pointing outward in the direction of, and normal to, the wave front. $\bar{r}$ is the ray vector from the center of the dipole to a point on the spherical wave front. $\bar{p}(t) = \bar{p}_0 \exp i\omega t$ is the dipole moment.

9.(b). Since the vector potential $\bar{Z}$ is now known from 9.(a), the magnetic field is then $\bar{H} = (1/c)\partial(\nabla X\bar{Z})/\partial t = (1/c)\nabla X\bar{Z}_t = ik\nabla X\bar{Z}$ where $k = 2\pi/\lambda$. This yields $\bar{H} = (ik(1/r^2 + ik/r)\exp - i(\bar{k} \cdot \bar{r} - \omega t))(\bar{p}_0 X\bar{r}_0)$, where $|\bar{p}_0|$ is the magnitude of the dipole moment, and $\bar{r}_0$ is a unit vector along the ray vector $\bar{r}$ (i.e., $\bar{r} = \bar{r}_0 r$). The near field is the close-in region where $r \ll \lambda$, and the far field is the region where $r \gg \lambda$. So that, $\bar{H}_{\text{near}} = [(k/r^2)\exp i(\omega t - \bar{k}\cdot\bar{r} + \pi/2)]\cdot(\bar{p}_0 X \bar{r}_0)$, while $\bar{H}_{\text{far}} = (k^2/r)\exp i(\omega t - \bar{k}\cdot\bar{r} + \pi))(\bar{p}_0 X \bar{r}_0)$.

9.(c). Retarded potential means that the above expressions for $\bar{Z}$ hold only for $t \geqslant r/c$, which is the time it takes a wave to emanate from the dipole out to $\bar{r}$. For $t < r/c$, $\bar{Z} = 0$, which means that insufficient time has elapsed for the generated field to get out to $\bar{r}$. Alternatively, the field wave front at $\bar{r}$ and t, now, is due to what emanated from the dipole r/c seconds ago.

10.(a). The Poynting vector $\bar{S} = \bar{E} \times \bar{H} = \bar{n}EH\sin\theta = \bar{n}EH$ (MKS or SI units), where unbarred quantities are the corresponding scalars, $\bar{n}$ is a unit vector normal to the $\bar{E} \times \bar{H}$ plane, and $\bar{E}$ and $\bar{H}$ are at right angles to each other, so that $\theta = \pi/2$. From Section 9.2:

In the endo-atmosphere:

$$E = H = E_{en}(t) = 52[\exp - (1.5\cdot 10^6 t) - \exp - (2.6\cdot 10^8 t)]$$

In the exo-atmosphere:

$$E = H = E_{ex}(t) = 63[\exp - (1.5\cdot 10^7 t) - \exp - (2.5\cdot 10^8 t)]$$

where E is in kilovolts per meter, $H = E/377$ (Ohm's Law), where 377 ohms is the impedance of free space, as well as in the atmosphere to good approximation, and the time is in seconds. Therefore, $S = E^2/377$ watts per m^2.

The time, $t_{\max}$, at which the EMP maximum occurs is obtained as the root of $\dot{S}(t_{\max}) = 0$. Since the temporal portions of E and H are of the form (exp-at) $-$ (exp-bt), the derivative at $\dot{S} = 0$ gives $t_{\max} = (\ln(b/a))/(b - a)$. So that:

Endo-atmosphere:

$$t_{\max} = -[\ln(1.5\cdot 10^6/2.6\cdot 10^8)]/(2.6\cdot 10^8 - 1.5\cdot 10^6) = 19.9\,\text{ns}$$

Exo-atmosphere:

$$t_{\max} = -[\ln(1.5\cdot 10^7/2.6\cdot 10^8)]/(2.6\cdot 10^8 - 1.5\cdot 10^7) = 11.6\,\text{ns}$$

$$S_{\max\,\text{endo}} = E_{en}^2(t_{\max}^{en})/377 = 6.68 \text{ megawatts per m}^2$$

$$S_{\max\,\text{exo}} = E_{ex}^2(t_{\max}^{ex})/377 = 6.59 \text{ megawatts per m}^2$$

10.(b).
$$W_{fen} = \int_0^\infty S(t)_{en} dt$$
$$= [(5.2 \cdot 10^4)^2/377] \int_0^\infty [(\exp - 1.5 \cdot 10^6 t)$$
$$- (\exp - 2.6 \cdot 10^8 t)]^2 dt$$
$$= 2.35 \text{ joules/m}^2$$
$$W_{fex} = \int_0^\infty S(t)_{ex} dt$$
$$= [(6.3 \cdot 10^4)^2/377] \int_0^\infty [(\exp - 1.5 \cdot 10^7 t)$$
$$- (\exp - 2.6 \cdot 10^8 t)]^2 dt$$
$$= 0.44 \text{ joules/m}^2$$

CHAPTER 3 ANSWERS

1. For $e\phi/kT$ small (i.e., the potential energy $e\phi \ll kT$), so that $\sinh(e\phi/kT) \cong e\phi/kT$. The two equations corresponding to Poisson's equations, Eq. (3.22) are then: $\phi''_> - \kappa^2\phi_> = -4\pi e N_D/\varepsilon$ for $x > 0$; $\phi''_< - \kappa^2\phi_< = 4\pi e N_A/\varepsilon$ for $x < 0$, where $\kappa^2 = 8\pi e^2 n_i/\varepsilon kT$. Using the same boundary conditions from Eqs. (3.27) and (3.28), and also multiplying ε by $4\pi\varepsilon_0$, the solutions are: $\phi_> = (eN_D/\varepsilon\varepsilon_0\kappa^2)[1 - (\cosh\kappa(l_+ - x)/\cosh\kappa l_+)]$ and $\phi_< = (eN_A/\varepsilon\varepsilon_0\kappa^2)[(\cosh\kappa(l_- + x)/\cosh\kappa l_-) - 1]$.

To determine l_+ and l_-, two relations are needed. The first is obtained by using $\phi'_>(0) = \phi'_<(0) = -E_0$ from Eq. (3.26), to give $N_D \tanh\kappa l_+ = N_A \tanh\kappa l_-$. The second is obtained from Eq. (3.31) viz., $\phi_0 - V_0 = \phi_>(l_+) - \phi_<(-l_-)$, which yields

$$\phi_0 - V_0 = (e/\varepsilon\varepsilon_0\kappa^2)[N_D(\cosh\kappa l_+ - 1)/\cosh\kappa l_+ + N_A(\cosh\kappa l_- - 1)/\cosh\kappa l_-]$$

For vanishing κ, it is seen that the above $\phi_>$ and $\phi_<$ reduce to Eqs. (3.29) and (3.30). Similarly, the above two relationships for l_+ and l_- reduce to Eqs. (3.32) and (3.33). Since κ^2 depends inversely on temperature, vanishing κ removes such dependencies from the other parameters in the preceding.

2.(a). $(-V_0)^{1/2}$ is positive simply because the corresponding development holds only for reversed-biased junctions, hence-V_0 is therefore positive.

2.(b). From Eq. (3.35) e.g., for $N_A \gg N_D$ and $|V_0| \gg \phi_0$, the total junction depletion layer width $l = (2\varepsilon\varepsilon_0 V_0/Ne)^{1/2}$, where $N \equiv N_D$, the lightly doped side. Inserting this into Eq. (3.38) yields $C/A_j = \varepsilon\varepsilon_0/l$. This is the capacity per unit area of a parallel plate capacitor of width l, and material of dielectric constant ε, between the two plates.

3. For a linearly graded junction, the parallel plate capacitance per unit junction area analog is still valid, i.e., $C/A_J = \varepsilon\varepsilon_0/l_m$, where now l_m is the total depletion layer width. Assuming a constant linear grade, m, and that $|V_0| \gg \phi_0$, Eq. (3.58) gives the appropriate l_m. When this l_m is inserted in the above expression for the capacitance per unit junction area, the result yields Eq. (3.59) viz., $C/A_J = (me(\varepsilon\varepsilon_0)^2/12(-V_0))^{1/3}$. Here also, the junction is assumed reversed-biased, so that $(-V_0)$ is positive.

4. Zener diode operation is constant equilibrium avalanche operation throughout the junction. From Eq. (3.70), $I_p' = 0$ yields $I_p = \alpha_n I/(\alpha_n - \alpha_p)$. Since $I_p + I_n = I$ (constant), $I_n = -\alpha_p I/(\alpha_n - \alpha_p)$.

5. Punch-through occurs when the depletion layers reach the junction end terminals, which act like regions of very high S_n or S_p. Junction breakdown is where the junction acts like a short circuit, maintaining none of its physical or electrical behavior.

6. $\omega\beta = \omega_T$, or $2\pi f\beta = 2\pi f_T$, yielding $\beta f \cong 1/T_{tr}$, where T_{tr} is the transit time.

7. The space charge regions bordering the junction produce an internal electrostatic field by which the internal or contact potential exists. The voltmeter would need current flowing through it to generate a reading. No current flows if there is no forward bias on the junction. If bias is applied to make current flow, the internal electrostatic field collapses to allow current flow, with a corresponding vanishing contact potential.

8. The sheet resistance of the base is given by $\rho_s = \rho/t$, where t is the base thickness of 1 mil. So that $\rho_s = 0.92/2.54 \cdot 10^{-3} = 362$ ohms per square.

9. The depletion layer capacitance expression, $C_d = A_J[e\varepsilon\varepsilon_0/2(N_A^{-1} + N_D^{-1})(\phi_0 - V_0)]^{1/2}$, can be used as long as $(\phi_0 - V_0) > 0$. So that,

a. $C_d(0.5 \text{ volts})/C_d(-1 \text{ volt}) = C_d/5\,\text{pF} = [(0.90 - (-1))/(0.90 - 0.5)]^{1/2}$, then $C_d = 5 \cdot (1.9/0.4)^{1/2} = 11\,\text{pF}$.

b. For $\eta = 1$, the diffusion capacitance is $C_v = (\tau/kT/e)I = (10^{-6}/0.026)(10^{-2}) = 0.39\,\mu F$.
c. $C_v/C_d = (0.39 \cdot 10^{-6})/(11 \cdot 10^{-12}) = 35{,}500$

10. Two diodes in series ensure that the current is the same through each diode, where one diode is forward biased and the other is reversed biased. From the diode equation in the paragraph following Eq. (3.78), the voltage across a diode is given by $I = I_0(\exp(eV_0/\eta kT) - 1)$. Here, $I = I_0$, the value of the reversed bias diode current, since it is the latter that limits the current through this series circuit. Inserting I_0 for I, with the given numbers, yields $V_0 = 0.036$ volts across the forward-biased diode. So that the voltage across the reversed-biased diode is $5 - 0.036 = 4.964$ volts, which is seen to be the major portion of the applied voltage across this circuit.

11. $L/W < 1$ implies that $L < W$, so that the ratio W/L counts the number of squares in parallel to the current flow in the L direction. Hence, the bulk resistance $R = \rho_s/(W/L)$, or the sheet resistance of each square divided by W/L, which is still identical to the original expression for R.

12. Since $h_{FE} = I_C/I_B$, $h_{FE} \cdot I_B = I_C$. Taking partial derivatives with respect to I_B yields $I_B(\partial h_{FE}/\partial I_B) + h_{FE} = h_{fe}$, using the definition $h_{fe} = \partial I_C/\partial I_B$. Rewriting, with the appropriate chain rule, gives $I_B(\partial h_{FE}/\partial I_C) \cdot (\partial I_C/\partial I_B) + h_{FE} = h_{fe}$, or $h_{fe} = h_{FE} + I_B(\partial h_{FE}/\partial I_C)h_{fe}$, yielding $h_{fe} = h_{FE}/[1 - (I_C/h_{FE})(\partial h_{FE}/\partial I_C)]$, the sought-after relation between the two gains.

CHAPTER 4 ANSWERS

1. First, for the variation in air density with altitude r, $\rho(r)$, where the origin is taken at the burst altitude R_0, the density decreases by a factor of ten for each 10 mile ($5.28 \cdot 10^4$ ft) increase in altitude. This asserts that $\rho(r - h) = \rho(r)/10$, where r is measured down from the burst altitude $R_0 = 1.5 \cdot 10^5$ ft, and $h = 10$ miles $= 5.28 \cdot 10^4 \times 30.5 = 1.61 \cdot 10^6$ cm. Inserting $\rho(r) = Ac^r$, where A and c are constants to be determined, and $\rho(R_0) = \rho_0$ (sea-level air density), into the above difference equation yields $\rho(r) = \rho_0 \exp - m(R_0 - r)$, where r is in cm, $m = 1.43 \cdot 10^{-6}\,\text{cm}^{-1}$, $R_0 = 1.5 \cdot 10^5 \times 30.5 = 4.58 \cdot 10^6$ cm, and $\rho_0 = 1.1$ gm/l.

For neutrons, the physical attenuation factor (in Eq. 4.55) between the burst altitude and ground is $\exp - (\int_0^{R_0} \rho dr/2.38 \cdot 10^4)$. Inserting $\rho(r)$ from the preceding, and integrating, yields exp $-$ $[\rho_0(1 - \exp - mR_0)/2.38 \cdot 10^4\,\text{m}]$. For a 0.1 MT yield, Eq. (4.55) gives $\Phi_n = 2 \cdot 10^{25} \times 0.1 \times e^{-32.34}/(4.58 \cdot 10^6)^2 = 8.6 \cdot 10^{-4}$ neutrons/cm^2 $= 2 \cdot 10^{-12}$ rad (tissue), which is negligible.

For gamma rays, the corresponding physical attenuation factor (in Eq. 4.56) is exp $-$ $(\int_0^\infty \rho dr/(1070 + 1.5Y^2)$ between the burst altitude and ground. Here, $m = 4.36 \cdot 10^{-5}\,\text{ft}^{-1}$. Then with the same $\rho(r)$, the attenuation factor becomes exp $-$ $(9.33 \cdot 10^{-4}\rho_0(1 - e^{-mR_0})/m)$ or exp $-$ $(9.33 \cdot 10^{-4} \times 1.1(1 - \exp - [4.36 \cdot 10^{-5} \times 1.5 \cdot 10^5])/4.36 \cdot 10^{-5}) =$ exp $-$ 23.54. So, letting $f = 1/2$ and $\alpha(Y) \equiv 1$ in Eq. (4.56) gives the gamma dose as $D_\gamma = 7 \cdot 10^{13} \times .1 \times \exp - 23.54/(1.5 \cdot 10^5)^2 = 1.86 \cdot 10^{-8}$ rad (Si) $= 1.86 \cdot 10^{-8} \times 0.902 = 1.68 \cdot 10^{-8}$ rads (tissue), which is negligible.

For thermal energy radiation, Eq. (4.41) gives $Q_0 = (Y\exp - \alpha R)/R^2$ (calories/cm^2), where R is the sea-level slant range in miles and Y is the yield in KT. The sea-level slant range, R, equivalent to altitude R_0 is gotten by equating the horizontal and vertical air masses per unit area. For ρ_0, the sea-level density, $\int_0^R \rho_0 dr = \int_0^{R_0} \rho_0 e^{-mh} dh$, yielding $R = (1/m)(1 - e^{-mR_0}) \cong 1/m = 4.34$ miles. From Fig. 4.14, the transmittance $T_r \equiv e^{-\alpha R} \cong 0.75$ for 4.34 miles of very clear air. Hence $Q_0 = 100 \times 0.75/(28.4)^2 = 0.09\,\text{cal/cm}^2$.

The solar constant is about 2 cal/cm^2 · min. at sea-level normal incidence. Hence, the foregoing types of radiation would cause little harm to a ground observer. Thought question: For the thermal energy case, for example, why is it not necessary to convert R^2 in the denominator of Q_0 as well, to equivalent slant range? Answer: $1/R^2$ represents the geometrical attenuation factor which, unlike the material attenuation in the Q_0 numerator, is independent of the transport medium. The above formulas are approximate in the sense that they hold for endo-atmospheric bursts, in the main. The energy output composition of an exo-atmospheric burst is different from that of an endo-atmospheric burst.

2.(a). From Eq. (4.55), $\Phi_n(1\,\text{mi}) = 2 \cdot 10^{25} \times (\exp - (1.1 \times 5.28 \cdot 10^3 \times 30.5/2.38 \cdot 10^4))/(5.28 \cdot 10^3 \times 30.5)^2 = 4.52 \cdot 10^{11}$ neutrons/cm^2.

2.(b). From Eq. (4.55), $\lim_{S \to 0} \int \Phi_n dS = \lim_{r \to 0} 2 \cdot 10^{25} Y \int_0^{2\pi} \int_0^{\pi} [(\exp - \rho r/2.38 \cdot 10^4)/r^2](r^2 \sin\theta d\theta d\phi) = 2 \cdot 10^{25} \cdot 4\pi Y/0.602 \cdot 10^{24} = 8\pi \times 1 \times 10^{25}/0.602 \cdot 10^{24} = 417.6$ moles.

3. The ambient radiation from fallout, $R_r \sim t^{-1.2}$ (t in hours). In two weeks, $R(336\,\text{hrs}) \sim (336)^{-1.2} = 9.3 \cdot 10^{-4}$. So $R_r(336)/R_r(1) = 9.3 \cdot 10^{-4} \cong 0.1$ percent.

4. For a purely spherically symmetric burst, $\bar{E} = \bar{E}(r)$ only. In spherical coordinates, $\nabla X\bar{E} = (\bar{a}_r/r\sin\theta)(\partial/\partial\theta(E_\theta\sin\theta) - \partial E_\theta/\partial\phi) + (\bar{a}_\theta/r)((1/\sin\theta)\partial E_r/\partial\phi - \partial/\partial r(rE_\phi)) + (\bar{a}_\phi/r)(\partial/\partial r(rE_\theta) - \partial E_r/\partial\theta)$, where $\bar{a}_r$, $\bar{a}_\theta$, $\bar{a}_0$ are the corresponding unit vectors. Since $\bar{E}$ is a function of r only, $E_\theta = E_\phi = 0$ and $\partial E_r/\partial\phi = \partial E_r/\partial\theta = 0$. Then, as seen from the expression for the curl, $\nabla X\bar{E}(r) = 0$. Then, emerging from Maxwell's equations, $\bar{H}$ = constant, no wave equations of radiation can exist, and so no EMP can ensue.

5.(a). This is worked out in detail in Section 10.2.
5.(b). Discussion is still going on. However, the roentgen is usually construed as a unit of dose nowadays.
5.(c). There are $2.08 \cdot 10^9$ ion pairs produced per cm^3 of STP air (one from each air molecule) per esu. However, this amount of air contains $1.293 \cdot 10^{-3} \times 0.602 \cdot 10^{24}/28.8 = 2.7 \cdot 10^{19}$ molecules per cm^3. Therefore, only $2.08 \cdot 10^9/2.7 \cdot 10^{19} = 7.7 \cdot 10^{-11}$ is the fraction of air ionized.

6.(a). The double-pulse behavior is a result of the burst radiation output (x ray) interacting with the atmosphere to form a fireball, as discussed in detail in Section 4.6. In an exo-atmospheric burst, there is no atmosphere with which the burst can interact. This then results in only one burst maxima, which is simply descriptive of the initial radiation efflux.
6.(b). The subtraction of Eq. (4.35) from Eq. (4.37) yields $3.2 \cdot 10^{-2}Y^{1/2} - 1.47 \cdot 10^{-4}Y^{1/3} = \Delta t = 1$ sec. Its root is the yield, which is 979 kT.

7.(a). It is seen that Eq. (4.19) for ΔP, is quadratic in $(Y/R^3)^{1/2}$, whose positive root for $\Delta p = 2$ psi gives $Y = 8.08 \cdot 10^{-14}R^3$, where the range R is now in feet.
7.(b). $R = (1\,\text{MT}/8.08 \cdot 10^{-14})^{1/3} = 2.31 \cdot 10^4$ ft = 4.38 mi.
7.(c). $\Phi_n = (2.10^{25} \cdot 1/(2.31 \cdot 10^4 \times 30.5)^2)\exp - (1.1 \times 2.31 \cdot 10^4 \times 30.5/2.38 \cdot 10^4) = 0.29$ neutrons/cm^2.
7.(d). $D_\gamma = (7.10^{13}Y/R^2)\exp - \rho R/\lambda = (7.10^{13} \times 1/(2.31 \cdot 10^4)^2)\exp - (1.1 \times 2.31 \cdot 10^4/(1070 + 1.5)) = 6.59 \cdot 10^{-6}$ rads (Si).
7.(e). $Q = 1000 \times 0.55/(4.38)^2 = 28.67$ cal/cm^2.
7.(f). Nothing required for 7.(c) and 7.(d). For 7.(e), 20 mil aluminum aircraft skin with 2 mils of standard aircraft paint will blister at 30 calories/cm^2.[11] Use readily available heat-resistant paint.
7.(g). For neutron fluence at sea level:

$$\Phi_n = 5.7 \cdot 10^7 r\exp - (4.62 \cdot 10^{-5}r); \; (r \text{ in cm}).$$

For gamma ray dose at sea level:
$D_\gamma = 5.66R\exp - (1.03 \cdot 10^{-3}R)$; ($R$ in ft).
For thermal energy at sea level:
$Q = 2.25 \cdot 10^{-3}R\exp - (2.65 \cdot 10^{-5}R)$; ($R$ in ft; $T_r = 0.55$).

Φ_n, D_γ, and Q are now of the form $x\exp - mx$, which has a maximum value of $(m\exp 1.0)^{-1}$ occurring at $x = m^{-1}$. This results in $\Phi_{n\max} = 4.54 \cdot 10^{11}$ neutrons/cm^2 at $2.16 \cdot 10^4$ cm (710 ft), $D_{\gamma\max} = 2.03$ krads (Si) at 973 feet, and $Q_{\max} = 31.3$ calories at $3.77 \cdot 10^4$ feet (7.15 mi).

7.(h). For all of the above types of radiation, it is seen that, for small R and Y, the allowable levels of fluence are proportional to the range, up to a maximum range. Beyond this maximum, the allowable fluences decrease for increasing range. This is because the yield versus range (for 2 psi) increases as R^3, which for large yields is stronger than the diminution now afforded mainly only by the exponentially decreasing attenuation.

8. The expression for the shock wave p_r, for both large and small p, results follows from $p_r = 2p(1 + 3p/(7p_0 + p))$. For small p, $p_r \simeq 2p$, while for large p, $p_r \simeq 8p$.

9. $(1\,\text{MT}/1\,\text{KT})^{1/3} = 10$; the answer is 400 feet.

10. Superimposed plots of exclusion radii, or keepout slant ranges from ground zero for ionizing radiation (gamma ray) plus neutron dose, and static overpressure, can be made. Keepout range is that distance corresponding to a given level of a radiation or blast threshold. For less than this slant range, the threshold is exceeded. For gamma ray dose in rads (Si), Eq. (4.56) is used where the fission factor $f = 1/2$ and $\alpha(Y) \cong 1$ for yields between 0.1–100 KT. Gamma ray enhancement, as discussed in Section 4.10, is negligible for these yields. Rewriting Eq. (4.56) where the slant range R is now in yards, Y is in KT, and the ionizing-dose rads (Si) $\cong$ rads (tissue), gives with sea-level density of 1.1 grams per liter,

$$D_\gamma \text{ (rads tissue)} = (3.89 \cdot 10^9 Y/R^2)\exp - (R/324)$$

Similarly, for neutrons, using Eq. (4.55) and the fact that $4.4 \cdot 10^8\, n/\text{cm}^2 = 1$ rad (tissue), results in

$$D_n \text{ (rads tissue)} = (5.05 \cdot 10^9 Y/R^2)\exp - (R/236)$$

Hence, summing the previous equations gives

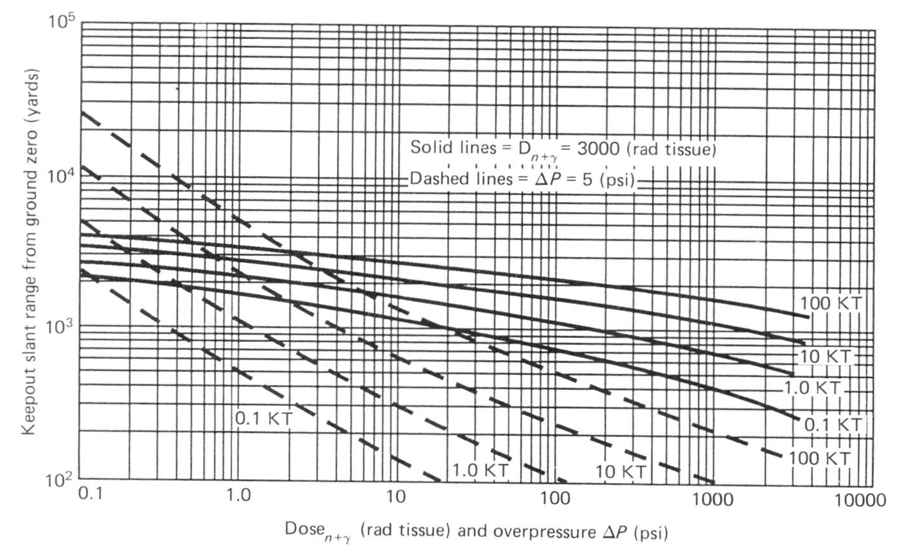

Fig. A.1 Keepout slant range from ground zero versus ionizing radiation plus neutron dose, and overpressure, for various endoatmospheric burst yields from 0.1–100 KT.

$$D_{n+\gamma} = (10^9 Y/R^2)(3.89 \exp - (R/324) + 5.05 \exp - (R/236))$$

For the combined dose of ionizing radiation and neutrons, $D_{n+\gamma}$ versus slant range for yields from 0.1–100 kT are plotted in Fig. A.1. In the same way, for static peak overpressure, ΔP, Eq. (4.19), is used to give

$$\Delta P(\text{psi}) = 1.22 \cdot 10^8 (Y/R^3) + 3.7 \cdot 10^4 (Y/R^3)^{1/2}$$

Peak overpressure versus keepout slant range for the same yields as above are plotted superimposed on the radiation curves in Fig. A.1. From this figure, it is seen that the radiation dose specifications are generally controlling for low-yield bursts, whereas blast overpressure specifications are controlling for high-yield bursts. The truck (and the communications equipment) will be destroyed at the given 5 psi level. From the figure it is appreciated that, for $\Delta P = 5$ psi, the corresponding dose $D_{n+\gamma} = 5$ rads (tissue), which is far less than the 3000 rads (tissue) threshold for the immediate incapacitation of personnel. Hence, it is seen that overpressure dominates the keepout range for 100 KT, which is approximately 2000 yards. For the personnel, immediate incapacitation at

3000 rads (tissue) (from the figure) occurs at approximately 300 yards for an 0.1 KT burst.

The answers can be couched within the following specification examples, viz.,

(a) For a 0.1 KT burst at 300 yds
2.5 psi
*$3 \cdot 10^{12}$ n/cm^2
*$1 \cdot 10^{10}$ rads/sec
*2600 rads (tissue)
*2060 rads (Si)

(b) For a 100 KT burst at 2000 yds
*5 psi
$2 \cdot 10^{11}$ n/cm^2
$1 \cdot 10^{8}$ rads/sec
200 rads (tissue)
210 rads (Si)

where the asterisk implies controlling specifications when combining (a) and (b). Thus, the approximate specification would be 5 psi, $3 \cdot 10^{12}$ neutrons/cm^2, $1 \cdot 10^{10}$ rads per s, 2600 rad (tissue), and 2060 rads (Si).

11. The function of the neutron bomb is to supply copious amounts of neutrons, with a minimum of other damaging component outputs such as blast and thermal energy. Hence, its yield must be small to minimize the collateral outputs. The size limitation is probably less than 1 KT, e.g., between 0.1–1 KT.

CHAPTER 5 ANSWERS

1. From Eqs. (5.59) and (5.65), $|\alpha_T| \cong |1 + i\omega/i\omega_T| = 1 - \beta^{-1}$. For $|\alpha_T| = (1 + (\omega/\omega_T)^2)^{1/2}$, the high-frequency asymptote is ω/ω_T, while the low-frequency asymptote is $|\alpha_T| \cong 1$. At the corner frequency ω_c, coincidence between the high- and low-frequency asymptotes occurs. So that $\omega_c/\omega_T = 1$, or $\omega_c = \omega_T$.

2. In Eqs. (5.85) and (5.86), $\psi_{BE} = [\exp(eV_{BE}/kT)] - 1$, and $\bar{\psi}_{BE}(x,s)$ is the corresponding Laplace transform with respect to time. For small V_{BE}, the first two terms in the expansion of the exponential results in $\psi_{BE} \simeq eV_{BE}/kT$. Then $\bar{\psi}_{BE}(x,s) \cong e\bar{V}_{BE}(x,s)/kT$, yielding the linearized Eqs. (5.87) and (5.88), where $\bar{V}_{BE}(x,s)$ is the Laplace transform of $V_{BE}(x,t)$, with respect to the time variable.

3. In Eqs. (5.89) and (5.90), the quantity σ contains the lifetime τ_n, and the diffusion length L_n. To yield the neutron fluence explicitly in the above equations, $\tau_n \cong K/\Phi_n$ where $\tau_0^{-1} \ll \Phi_n/K$, and so $L_n = (D_n K/\Phi_n)^{1/2}$ can be inserted therein. This can also be done for the Ebers-Moll equations, Eqs. (5.91)–(5.98).

4. The diode equation, Eq. (3.78), contains the saturation or leakage current, which is $I_0 = e(n_{p0}D_n/L_n + p_{n0}D_p/L_p)$ as given by Eq. (3.88). To include the effects of neutrons, from the preceding problem $L_n = (D_nK/\Phi_n)^{1/2}$, so that $D_n/L_n = (D_n\Phi_n/K)^{1/2}$ and $D_p/L_p = (D_p\Phi_n/K)^{1/2}$ inserted into I_0 yields $I_0 = e(D_n^{1/2}n_{p0} + D_p^{1/2}P_{n0})(\Phi_n/K)^{1/2}$.

5. The gain change due to neutron fluence, as well as temperature, is given from Eq. (5.82) as $\Delta(1/\beta) = (\Phi_n/\omega_TK) - \beta_0^{-1}[1 - 1/(1 + \Delta T/T_0)^2]$, where $\Delta T = T - T_0$. This can be rewritten as $1/\beta - 1/\beta_0 = \Phi_n/\omega_TK - 1/\beta_0 + T_0^2/T^2\beta_0$, so that $1/\beta = \Phi_n/\omega_TK + (1/\beta_0)(T_0/T)^2$.

6. From the system of equations (5.125)–(5.132), substituting for N_{DV}, N_{OV}, N_{VV}, and N_{IV} in Eq. (5.130) yields the fifth-degree algebraic equation in N_V, viz., $N_V = K_V\Phi_n - (K_{DV}N_{Di}N_V)/(1 + K_{DV}N_V) - (K_{OV}N_{Oi}N_V)/(1 + K_{OV}N_V) - K_{VV}N_V^2 - (K_VK_{IV}N_V\Phi_n)/(1 + K_{IV}N_V)$ as can be seen by multiplying through by the common denominator. It can be reduced to a cubic or quadratic equation by ignoring the presence of certain impurities. This is done in Section 5.13, where only donor and oxygen vacancies are included, which approximates the situation for low neutron fluence levels.

7. The derivation of Eq. (5.165) is given in the latter part of Section 5.15. The apparent first-order dependence on Φ_n would come from $D_n = L^2/\tau$. However, $\tau^{-1} \sim \Phi_n/K_n$ and $1/L^2 \sim \Phi_n/K_L$ (Eq. (5.117)). Thus $D_n = K_L/K_n$ with no first-order dependence on Φ_n. That is, there are no terms dependent on minority carrier lifetime, and hence no first-order degradation due to neutron fluence.

8. The two-transistor Darlington circuit's main feature is that the first transistor directly drives the second, giving an overall gain $\beta_D = \beta^2$. β is the post-radiation gain of each transistor. For each transistor, $\Delta(1/\beta) = K_1\Phi_n$, where $K_1 = (\omega_TK)^{-1}$. Then $\Delta(1/\beta_D) = 1/\beta^2 - 1/\beta_0^2 = (1/\beta - 1/\beta_0)(1/\beta + 1/\beta_0) = \Delta(1/\beta)(1/\beta + 1/\beta_0)$. Using $\Delta(1/\beta) = 1/\beta - 1/\beta_0 = K_1\Phi_n$ to eliminate $1/\beta$ from the previous equation yields $\Delta(1/\beta_D) = (K_1\Phi_n)(2/\beta_0 + K_1\Phi_n)$. For very high Φ_n, $\Delta(1/\beta_D) \cong (K_1\Phi_n)^2$.

9. The cascade recoil is proportional to the neutron fluence, as given by Eq. (5.7). A cascade of 1000 silicon recoil atoms per $10^{-16}\,\text{cm}^3$ results from one fast neutron, and implies that $10^{16}\,\text{n/cm}^2$ is needed to maintain the same cascade production per cm^3. Since $\rho = 2.33\,\text{gm/cm}^3$, $\Phi_n = 10^{16}/2.33 = 4.3 \cdot 10^{15}$ neutrons/cm^2 per gm of Si. It is seen that the total number of recoil atoms produced is $4.3 \cdot 10^{15} \times 10^3 = 4.3 \cdot 10^{18}$ per gm of Si.

10. In an appreciable neutron fluence environment, the silicon resistivity increases, as discussed in Section 5.13. This can thus increase the resistance of the monolithically built-in feedback resistors in the flip-flop ICs, thereby increasing their feedback response, and so making the circuit slower.

11. $N\sigma = \lambda^{-1} = \frac{1}{4}$ of a primary available per cm^3 per incident fast neutron, or the probability of producing a primary per cm^3 per incident fast neutron is 0.25. Hence, the product of the number of secondaries per primary and the number of available primaries per cm^3, $\bar{n}_s/\lambda = 4000 \cdot 0.25 = 1000$, which reconciles with the total number of recoils.

12.(a). From Section 3.3, the expression for the potential drop in the base is given by $\phi_0 = (kT/e)\ln(p(x)/n_i) \cong (kT/e)\ln[(N_D(x) - N_A(x))/n_i]$. Then, the electrostatic field across the base $E_0 = |\phi_0(X_{jE}) - \phi_0(X_{jC})|/|X_{jE} - X_{jC}|$, which is seen to be constant for an exponentially graded base. So, substituting for the potential drop in the above yields $E_0 = [(kT/e)/(X_{jE} - X_{jC})] \cdot \ln[|N_D - N_A|_{\mathrm{jE}}/|\mathrm{N_D} - \mathrm{N_A}|_{\mathrm{jC}}]$. Inserting the parameter values gives $E_0 = ((0.025\,\text{volts})/(2 \cdot 10^{-4}\,\text{cm})) \cdot \ln(2 \cdot 10^{18}/5 \cdot 10^{15}) = 750\,\text{V/cm}$.

12.(b). By definition, $\eta = \ln(|N_D - N_A|_{jE}/|N_D - N_A|_{jC}) = \ln(2 \cdot 10^{18}/5 \cdot 10^{15}) = 6$.

13. Two points define a line, three a plane, and four a volume (cluster). Thus, four defects are required for even the smallest possible cluster.

14.(a). From Eq. (5.111), $f_{TN}(\eta) = (1/2\pi)((W_b^2/D_n)(\eta - 1 + \exp - \eta)/\eta^2)^{-1}$. For the reverse-direction transistor, the collector and emitter roles are reversed. Such transistors are used today in chopper circuits and bilateral switches. The doping profiles are reversed in this type of transistor in that $f_{TI}(\eta) = f_{TN}(-\eta)$. Hence, $f_{TN}(\eta)/f_{TI}(\eta) = (-\eta - 1 + \exp\eta)/(\eta - 1 + \exp - \eta) = 100$, with a root $\eta = 6.28$.

14.(b). As noted in Problem 5.12, a descending, exponentially graded base implies a constant electric field E_0. Integrating Eq. (5.101) through the base of width W_b yields $E_0 = (kT/e)(\eta/W_b)$. To obtain W_b, use Eq. (5.111), i.e.,

$$\begin{aligned} W_b &= ((D_n\eta^2/2\pi f_{TN})/(\eta - 1 + \exp - \eta))^{1/2} \\ &= ((40/2\pi \cdot 3 \cdot 10^8)((6.28)^2/(6.28 - 1 + \exp - 6.28)))^{1/2} \\ &\cong 4\,\mu\text{m} \end{aligned}$$

Then $E_0 = (0.026)(6.28/4 \cdot 10^{-4}) = 410\,\text{V/cm}$.

15. Inserting E_0 into Eq. (5.102) yields $p'(x) - (\eta/W_b)p(x) = -J_p/eD_p$; $p(W_b) = 0$, whose integral is $p(x) = \int_x^{W_b}(J_p eD_p)\exp[(\eta/W_b)\cdot(x - x')]dx'$. Continuing with the above steps gives $\tau_b = (1/D_p)\int_0^{W_b}dx \int_x^{W_b}dx' \exp((\eta/W_b)(x - x'))$, which, after integration, yields the result given in Eq. (5.109). Again, the result holds for both n- and p-type doped bases, so the subscripts n and p are used interchangeably to match the context.

CHAPTER 6 ANSWERS

1. From $E^2 = p^2c^2 + m_0^2c^4$, which is a general relationship for particle energies, there are two cases, (i) uninteresting in this context are particles of non-zero rest mass at rest, implying $p = 0$, so $E = m_0c^2$ for them, and (ii) pertinent here are zero-rest mass particles, mainly photons like x and gamma rays, so that $E = pc$ for them. For such particles $E = h\nu$ also, and so $p = E/c = h\nu/c = h/\lambda$. For a one-gram transistor chip that absorbs 100 Mrad $= 10^8$ rads $= 10^{10}$ ergs in 1 μsec, the corresponding momentum or impulse imparted to the chip, $p = 10^{10}/3 \cdot 10^{10} = (1/3)$ tap (CGS unit of momentum). MIL-STD-883C requires that such transistors endure an impulse, I, corresponding to an acceleration of 1500 g for 0.5 msec. So, $I = 1.5 \cdot 10^3 \times 0.98 \cdot 10^3 \times 0.5 \cdot 10^{-3} = 735$ taps. Hence, the device could probably withstand the impulsive stress. However, the thermal damage is given by the temperature rise $\Delta T = Q/mC_h = 10^3$ joules/1 $\times$ 0.7 (joules/gm°C) $=$ 1400°C for a silicon device. Silicon melts at 1415°C, so the chip will not only melt, but will probably vaporize, as the energy input occurs in 1 μsec, corresponding to a dose rate of about 10^{14} rads per sec. If the chip doesn't completely vaporize, the ablation of its exterior would induce an imploding shock wave sufficient to destroy its interior.

2. For ^{60}Co, $N_{\text{Co}} = N_0\exp - t/\tau_{\text{mean}} = N_0\exp - (t\ln 2/\tau_{1/2}) = N_0\exp - 0.693t/5.3$. So that it will take $t = -5.3 \cdot \ln(0.016)/0.693 = 31.6$ years for the source strength to drop to 1.6 percent of its initial activity.

3.(a). The intensity difference from the front to the back of the slab is given by the limit as $\Delta x \to 0$ of $I(x) - I(x + \Delta x) = N\sigma\Delta x \cdot I(x)$, where the right-hand side is the amount of intensity removed from the incident beam. Or, $I(x) - (I(x) + I'(x)\Delta x + \ldots) = N\sigma\Delta x \cdot \text{I}(x)$. As $\Delta x \to 0$, $I'(x) = -N\sigma I(x)$, with an integral $I(x) = I_0\exp - N\sigma x$.

3.(b). For $I = I_0\exp - \mu x$, where $\mu = N\sigma$, the mean free path $\lambda_0 =$

$\bar{x} = \int_0^\infty xI(x)dx / \int_0^\infty I(x)dx = \int_0^\infty xe^{-\mu x}dx / \int_0^\infty e^{-\mu x}dx = 1/\mu$, hence $\lambda_0 = (N\sigma)^{-1}$.

4. Since the interactions of interest for gamma rays with materials are mainly compton processes, the interaction probabilities are roughly proportional to the material electron concentration. For example, the energy loss with depth of penetration or stopping power $dE/dx \sim -NZ$, the electron concentration, where the material particles per cm^3, $N = \rho N_0/A$, and N_0 is Avogadro's number. Hence, $dE/\rho dx \sim -N_0 Z/A$. For most pertinent materials $Z/A \sim 1/2$, then $dE/\rho dx$ is roughly constant for most materials for the above interactions. So that the differential areal density $d\xi \equiv \rho dx$ (grams per cm^2) is a sensible unit of penetration. Also, and perhaps more important, is that ξ is the penetrated mass per unit area, or loosely the "mass penetrated."

5. In pair production, a high-energy photon (gamma ray) in the vicinity of a nucleus can be annihilated, thereby producing an electron-positron pair. Both the electron and positron have the same mass (different charge sign), so that to conserve energy m_0c^2 for an electron is 0.511 Mev, and for pair production the initiating photon energy must be at least 2×0.511 Mev = 1.022 Mev of energy.

6. MOSFETs are majority carrier devices. Hence, they are not affected to first-order by the vagaries of the neutron degradation of minority carrier lifetime.

7. Hole trapping in the gate oxide near the Si/SiO_2 interface in the channel, and the generation of fast interface states thereon.

8. The flatband voltage, V_{FB}, is that gate voltage in MOS devices required to oppose the work function energy differences between the SiO_2 gate insulator and the silicon bulk. From an energy viewpoint, V_{FB} restores the energy bands spanning the silicon bulk, gate insulator, and aluminum gate lead to their flatband condition. This is discussed in Section 6.9, and the flatband energies are depicted in Fig. 6.20. V_{FB} changes due to ionizing radiation are similar to those that cause changes in the threshold voltage V_T. They are sometimes not as drastic as the latter because there are compensating effects, as discussed in Section 6.9. V_T, the MOS gate threshold or 'turn-on' voltage is initially shifted to more negative values in both PMOS and NMOS devices. The additional negative voltage is required to compensate the trapping of ionizing radiation-induced holes (positively charged) in the gate oxide.

9. Figure 6.21 reveals that the slope corresponding to the exponent n in $V_{FB} \sim t_{ox}^n$, and $V_T \sim t_{ox}^n$ is $n = 0.36$.

10.(a). In the SNOS and MNOS devices, the "nitride-oxide" gate insulator is harder than merely an SiO_2 gate insulator, in that the amount of trapped holes per rad of deposited incident radiation is much lower.

10.(b). PMOS and NMOS hardening differences are discussed in detail in Section 6.12. Also note that Fig. 6.21 infers that PMOS are harder than NMOS, since the flatband voltage shift is much less with negative bias (PMOS normal bias polarity) than with positive gate bias (NMOS normal bias polarity).

11. Discussed in detail in Section 6.13. A prime difference is that CMOS/SOS (and CMOS/SOI) is not susceptible to latchup, in contrast to bulk CMOS. Also, they generate about two orders of magnitude less photocurrent than bulk CMOS.

12. Electrons at the gate oxide-channel interface.

13. Consider the transistor in Figure 6.35, whose digits are assumed to be of average length b and width a. Then the digit area that "sees" only the collector is $N(a - 2w)b$, and the total underlying area is Nab, where N is the number of digits on the chip. Hence $\alpha_{ss} = N(a - 2w)b/Nab$, so that $\beta_{ss} = \alpha_{ss}/(1 - \alpha_{ss}) = (a - 2w)/2w$. With an increasing number of digits confined to a fixed chip substrate area, $\beta_{ss} = \lim_{a \to 2w} (a - 2w)/2w$, which is seen to yield a very small β_{ss} in the limit of large numbers of ever-thinner interdigits.

14. Discussion confined to the areas specified in the problem statement.

15. $((1\,\text{MeV} \cdot \text{mb})/(\text{neut}))(10^{-27}\,\text{cm}^2/\text{mb})((5 \cdot 10^{22}/2.33)$ Si atoms/gm) $\cdot$ $(1.6 \cdot 10^{-6}$ ergs/MeV) divided by 100 ergs/gm rad (Si)) $= 3.43 \cdot 10^{-13}$ rads (Si) per neutron per cm^2.

CHAPTER 7 ANSWERS

1. From Eq. (7.1), $g = 4.05 \cdot 10^{13}$ (hole-electron pairs/cm^3 rad) $\times$ $1.6 \cdot 10^{-19}$ (coulombs/hep) $= 6.48 \cdot 10^{-6}$ coulombs/cm^3 rad $= 6.48\,\mu$amps/cm^3 (rad/sec).

2. From equation (7.6) with no bias applied, $\partial p_n(0+, t)/\partial x = 0$, and $\partial n_p(0-, t)/\partial x = 0$ as seen from equations (7.7) with $V_0 = 0$, so that $i_d = 0$.

3. From Eq. (7.9), for $G(t) = g\delta(t)$ yields

$$I_{pp}(t) = eA\{W_t g\delta(t) + (\pi t)^{-1/2}[(D_n)^{1/2}\exp - t/\tau_n + (D_p)^{1/2}\exp - t/\tau_p]\}$$

4. Dielectric isolation (DI) implies that adjacent component isolation is obtained by embedding each component in a dielectric tub. This precludes the existence of intercomponent electrical latchup paths. Junction isolation (JI) implies that component isolation is obtained by assuring that the substrate has the lowest voltage with respect to any other element in the component (cf. Section 12.4). This electrical configuration can be upset by an incident radiation pulse. This can induce a latchup situation.

5. Sustaining voltage breakdown is synonymous with BV_{CEO} breakdown. In a bipolar, $BV_{CEO} \simeq BV_{CBO}/\beta^{1/n}$ (Eq. 12.9). BV_{CEO} breakdown can induce latchup by uniting two latchup path fragments, through the thus shorted junction, to form a four layer latchup path. This a very infrequent occurrence, and can be associated with snapback induced latchup (Section 7.7).

6. Gold doping inhibits latchup by providing additional defect energy levels within the band gap. This serves to reduce the β of the parasitic transistors. Gold doping in the collector leaves the operation of the intentional transistor unaffected. This then curbs the propensity of the device to latch by reducing the parasitic transistor gains.

7. The incident pulse of ionizing radiation provides a carrier photocurrent in the form of hole-electron pairs. Hence, the electrical conductivity increases thereby as $\sigma + \Delta\sigma = e((n_0 + \Delta n)\mu_n + (p_0 + \Delta p)\mu_p)$. For such currents, the corresponding carrier density increments, $\Delta n = \Delta p = \Delta_c$. So that $\Delta\sigma = e\Delta_c(\mu_n + \mu_p)$. The hole-electron pairs constitute a charge density $e\Delta_c = eg_0\dot{\gamma}\tau$, where τ is the effective carrier lifetime. Hence, $\Delta_c = g_0\dot{\gamma}\tau$, and inserting in the above expression for conductivity yields $\Delta\sigma = eg_0(\mu_n + \mu_p)\dot{\gamma}\tau$.

8. Use Eq. (7.81), the expression for $\dot{\gamma}_c$, to eliminate the applied voltage V in Eq. (7.80). Then, with $R_0 = \rho_{sp}L/W$ and $\rho_{sn} \ll \rho_{sp}$ for the highly doped nn^+ layer, the relationship in the problem statement for $R(\dot{\gamma})$ ensues. Actually,[32] $R(\dot{\gamma}) = R_0/(1 + \dot{\gamma}/\dot{\gamma}_c)$ for $\dot{\gamma} \ll \dot{\gamma}_c$; $R(\dot{\gamma}) = R_0(\dot{\gamma}_c/4\dot{\gamma})^{1/2}$ for $\dot{\gamma} > \dot{\gamma}_c$.

9. For low injection, $4g_0\dot{\gamma}\tau_n^2/\tau_m \ll 1$, so that, from Eq. (7.91), the expression for δ_{pre} is obtained.

10. Using Eq. (7.12), $I_{ss} \cong eA_J g\dot{\gamma}_{ss}W_t = (1.6 \cdot 10^{-19})(10^{-5})(4.1 \cdot 10^{13})(10^3/60)(5 \cdot 10^{-4}) = 0.55\,\text{pA}$.

11. For high photocurrents, the base-emitter potential drop can exceed the breakdown voltage BV_{EBO}. At this instant of time, the emitter and base are connected through this low-impedance (high-current) breakdown path. The corresponding current pulse is then amplified to produce the very large anomolous collector current discontinuity. Or, the photocurrent in the base circuit can turn the transistor on, causing it to amplify, as in the preceding.

12. Equation (7.49), viz., $\tau = t_s[erf^{-1}(1 - I_{CS}/\beta I_B)]^{-2}$ can be inverted to yield

$$I_{CS}/\beta_{IB} = \beta_s/\beta = 1 - erf(t_s/\tau)^{1/2} = 1 - erf(3.33)^{1/2} = 0.011.$$

13. $(525 - 500)/50 \cdot 10^{-9} = 5 \cdot 10^8$ rads (Si) per second.

14. Discussion in terms of how the detailed definition of latchup impinges on latchup testing limitations to prevent burnout.

15. The number of hole-electron pairs created per cm^3 per rad (Si) is[16]: 1 rad (Si) $= (100\,\text{ergs/gm}) \cdot (2.33\,\text{gm/cm}^3) \cdot ((1/1.6 \cdot 10^{-12})\,\text{eV/erg})(1/3.6\,\text{eV per hep}) = 4.05 \cdot 10^{13}$ hep per cm^3.

16. The inverse fourier transform $\beta(t) = (1/2\pi)\int_{-\infty}^{+\infty}\bar{\beta}(\omega)\exp(i\omega t)d\omega$, where $\bar{\beta}(\omega)$ is the expression given in the problem statement. $\bar{\beta}(\omega)$ can be rewritten in this case as the more familiar Laplace transform by letting $i\omega = s$. The result is $\bar{\beta}(s) = ((\alpha_0 - sc_n)/s(1 - \alpha_0 + sc_d)) + (c_n/sc_d)$. Then, with $c_n/c_d \cong 1$ and $\beta_0 \gg 1$, the corresponding $\beta(t) \cong \beta_0(1 - \exp - (c_d/\beta_0)t)$.

CHAPTER 8 ANSWERS

1. In the case of both the cruise missile (very low altitude mission profile) and the NORAD computer in the mountain, the atmosphere provides shielding, except possibly for CCD devices, so that very low SEU error rates will occur. In the U-2 case, the mission profile is very high altitude, so that the atmosphere provides only partial shielding, with higher SEU rates expected.

2. The muons produced in the upper atmosphere by cosmic rays are traveling at ultrahigh speed toward the earth. Their speed is sufficiently

large so that time slows down for them, relative to a ground observer, so that they can traverse the thickness of the atmosphere in about 10^{-4} s to cause SEU in CCDs. This is as compared to their natural decay time of 1 μsec in their relativistically reduced track length through the atmosphere. This is a classic example of special relativity.

3. SEU error rate = $\sigma\phi_a$, where ϕ_a is the accelerator beam flux, and the device cross section is $\sigma(cm^2)$. So that $\sigma = 10/1.6 \cdot 10^9 = 6.25 \cdot 10^{-8}\,\text{cm}^2$ per device. SEU (error rate per device) = $6.25 \cdot 10^{-8} \times 10^5$ (ions per cm^2 day) = $6.25 \cdot 10^{-3}$ SEU errors per device day. To determine the SEU bit error rate for a device containing 10^3 bits, e.g., a 1K-bit (1024 bits) RAM, then the SEU (error rate per bit) = $6.25 \cdot 10^{-8} \times 10^5/10^3 = 6.25 \cdot 10^{-6}$ errors per bit day. These results hold for this device only, since it is assumed to possess this unique SEU cross section.

If this cross section was measured for this device, and corresponds to a given type of accelerator beam ion (e.g., krypton), then strictly speaking, these measurements are unique to this device and ion type. However, engineering extrapolations can be made to encompass other ions, as well as the whole cosmic ray ion spectra. Extrapolations across device types are much more difficult to make.

4.(a). $\bar{s} = abc/\bar{A}_p = abc/\tfrac{1}{2}(ab + bc + ca) \cong 2c$, since ac, ab, $\ll ab$ for a thin parallelepiped.

4.(b). $Q_c = 10^6 \times 1.6 \cdot 10^{-19} = 0.16\ pC$, $L_c = Q_c/2c\rho_{\text{si}} = 0.16\text{pC} \times 22.5\,\text{MeV}/\ pC/2 \times 10^{-4} \times 2330\,\text{mg/cm}^3 = 7.73\,\text{MeV cm}^2/\text{mg}$.

5. An SEU-hard device has a high critical LET. For high L_c, it takes much more ionizing radiation produced-deposited charge per unit penetration distance, or deposited charge in a penetrated mass per unit area, to upset the device than for a low L_c component.

6.(a). $Q_c \cong C_{\text{stor}} \cdot \Delta V = 0.05 \cdot 10^{-12}(5.5 - 2.5) = 0.15\,\text{pC}$ (picocoulombs). The SEU is given by $5 \cdot 10^{-10} abc^2/Q_c^2 = 5 \cdot 10^{-10} \times 10 \times 10 \times (3)^2/(0.15)^2 = 2 \cdot 10^{-5}$ SEU errors per memory cell per day.

6.(b). $2 \cdot 10^{-5} \times (1024)$(1K SRAM) $= 2.05 \cdot 10^{-2}$ SEU errors per device day.

7. For a Poisson distribution, $P_r(k \text{ flares}) = (p_r t)^k \exp - (p_r t)/k!$ in time t (days), where the individual solar flare probability rate per day, p_r, is given by:

a. $p_r = 1.5/20 \times 365 = 2.05 \cdot 10^{-4}$/day (The day is the appropriate time unit since the flare duration is approximately 1 day).

b. For a 10-year mission life, $p_r t = 2.05 \cdot 10^{-4} \times 3650 = 0.75.$; $P_r(1 \text{ flare}) = 0.75 \exp - 0.75 = 0.354$.
c. P_r (at least 1 flare, i.e., 1 or more flares) $= 1 - P_r(\text{zero flares}) = 1 - \exp - 0.75 = 0.528$.
d. Equivalent SEU error rate per cell day $= 2.05 \cdot 10^{-4}/2000 = 1.03 \cdot 10^{-7}$, or $0.354/2000 = 1.77 \cdot 10^{-4}$ SEU error rate per device day.

8. SEU-induced second breakdown will cause an individual finely structured transistor to burn out, thus provoking current-hogging processes within to induce cumulative damage, ultimagely to ruin the device.

9. $dE/ds \sim NZ = \rho N_0 Z/A$, where N_0 is Avogadro's number, ρ is the material density, and Z/A is the ratio of atomic number to mass number of the device material. Then the material LET is approximately constant when the incremental areal density $d\xi = \rho ds$ is used to write LET as $dE/d\xi$.

10. From Table 1.1, the ionization energy required to produce an electron (hep) is 3.6 eV for Si and 4.8 eV for GaAs. For Si, $3.6\,\text{eV}/1.6 \cdot 10^{-19} = 3.6/0.16 = 22.5\,\text{MeV/pC}$. For GaAs, $4.8\,\text{eV}/1.6 \cdot 10^{-19} = 30\,\text{MeV/pC}$.

11. For Si, $L = E(MeV)/\rho c = 22.5 \times 1\,\text{pC}/(2.33 \cdot 10^3\,\text{mg/cm}^3) \cdot 10^{-4} = 96.57\,\text{MeV cm}^2/\text{mg}$. For GaAs, $L = E(\text{MeV})/\rho c = 30 \times 1\,\text{pC}/(5.32 \cdot 10^3\,\text{mg/cm}^3) \cdot 10^{-4} = 56.4\,\text{MeV cm}^2/\text{mg}$.

12. (1) An ICBM on an over-the-pole trajectory, which would be exposed to SEU, because the magnetic field lines at the north pole singularity would allow cosmic rays to intrude. (2) High-flying aircraft, which can be exposed to cosmic rays and their reaction products existing at the upper reaches of the atmosphere. (3) Ground equipment on an atmosphereless satellite like the moon. With no atmosphere with which to interact, cosmic rays are incident down to the corresponding ground level.

13. Because the charge produced by the known heavy ion track in the device, without considering a funnel, falls far short of that actually collected at a given information node.

14. Even though the physical dimensions remain constant, there is an electrically effective increase in the parasitic SCR current-carrying area of

the p well -n substrate junction as in Fig. 8.36. This is due to the positive temperature dependence of the transient βs.

15. The node depletion layer (flat plate) capacitance $C = \varepsilon\varepsilon_0 ab/c$, as discussed in Section 3.5. Also, $Q_c \cong CV$, where V is the voltage swing across the junction. Inserting these expressions yields $E_r = 5 \cdot 10^{-10} c^4 / ab(\varepsilon\varepsilon_0 V)^2$.

16. No, as shown in Fig. 8.26, $Q_c \simeq 0.023 l^2$, where l is the feature size, then $E_r = 5 \cdot 10^{-10} abc^2/(0.023)^2 \cdot l^4$. Letting c be comparable to l, it is seen then that $E_r \sim l^{-2}$, so that the smaller the device (feature size), the greater will be the SEU error rate.

17.(a) This is a straightforward solid angle computation, with h/R being a small fraction, where $h = 150$ miles altitude, and $R = 4000$ mile radius of the earth. Show that this fraction $\Omega/4\pi = (1/2)(1 - (2h/R)^{1/2}) = 0.36$. (b) No, except for the polar regions, solar flares and the South Atlantic Anomaly (SAA), the space station is partially shielded from galactic cosmic ray-induced SEU by the earth's magnetic field. However, it is a consideration with respect to ionizing dose and dose rate from the depressed altitude of the Van Allen belt fluxes in the vicinity of the SAA, as shown in Figures 8.15D and 8.16.

18. None, since for all of these device types their function following SEU onset will not substantially affect system operation.

19. Because, there are many more cells per chip in the smaller cell case. That is, the chip SEU error rate is the product of the cell SEU error rate by the number of cells in the chip, since the individual cell probabilities are assumed independent and equal.

20. If E_r were independent of scale, the device size would make no difference, and then submicron devices would suffer the same SEU as their larger counterparts; so there would be no SEU bar to tiny devices. In reality, these models break down for large scaling ratios, hence the preceding does not hold, as discussed in Section 8.6.

CHAPTER 9 ANSWERS

1. From Eq. (9.4), $\bar{E}_L(\omega) = 5.2 \cdot 10^4 \beta((\omega^2 + \alpha^2)(\omega^2 + \beta^2))^{-1/2}$. The low-frequency asymptote $\bar{E}_{\text{low}} = 5.2 \cdot 10^4/\alpha$, while the high-frequency asymptote $\bar{E}_{\text{hi}} \simeq 5.2 \cdot 10^4 \beta/\omega^2$. Then, when $\bar{E}_{\text{low}} = \bar{E}_{\text{hi}}$ at $\omega = \omega_c$, the result yields $\omega_c = (\alpha\beta)^{1/2}$.

2. $Hev/c = mv^2/r$ yields $r = mvc/eH$. Since $v = (2E/m)^{1/2}$, $r = (2mE)^{1/2}c/eH$, or $r = (2 \cdot 10^{-27} \cdot 10^5 \times 1.6 \cdot 10^{-12})^{1/2} \cdot 3 \times 10^{10}/4.8 \cdot 10^{-10} \times 0.5 = 22.4$ meters.

3. The high-altitude burst produces neutrons in the rarified very high atmosphere. The mean life of the neutron is about 12 minutes, which can then decay in that sparse environment to produce an electron, a proton, and a neutrino, simply because of the lack of matter with which the neutron can interact. At the surface, the sea-level air density and the material environment is substantive enough for the neutron to be absorbed quickly with high probability, well before its mean life transpires. Further, it is generally asserted that the decay of neutrons produced by cosmic ray interaction with the sparse atmosphere at high altitudes provides the protons and electrons that are the main constituents of the Van Allen belts.

4. At the Brewster angle, $\sqrt{\varepsilon} = \tan \theta_B = r_{\text{swath}}/HOB$ (height of burst). So that $r_{\text{swath}} = \sqrt{\varepsilon} \cdot HOB = \sqrt{5} \cdot 3 = 6.71$ miles.

5. The skin depth at 1 kHz is obtained from $\delta = 1/(\pi f \mu \sigma)^{1/2}$. For $f = 10^3$ Hz, $\sigma = 4$ siemens (mhos)/meter, $\mu \cong 4\pi \cdot 10^{-7}$ henries/meter; $\delta = (\pi \cdot 10^{-3} \cdot 4\pi \cdot 10^{-7} \cdot 4)^{-1/2} = 7.96$ meters. 95 percent attenuation corresponds to e^{-3} instead of e^{-1} in depth, yielding $3 \times 7.96 = 23.9$ meters.

6.(a). From Eq. (9.82), $\Phi(u) \sim u^3/(\exp(u/kT) - 1)$. At the spectrum peak, $\Phi'(u_m) = 0$ yields the transcendental equation $3 - x_m = 3 \exp - x_m$, where $x_m = u_m/kT$. The root of this equation is $x_m = 2.82$, so $u_m = 2.82kT_m$ holds at the peak, and T_m is called the spectrum (peak) temperature.
6.(b). $h\nu = hc/\lambda = 2.82kT_m$, from which $T_m = 6.6 \cdot 10^{-27} \times 3 \cdot 10^{10}/ 2.82 \times 1.38 \cdot 10^{-16} \times 10^{-7} = 5.1 \cdot 10^6$ °K.

7. For a slot and a metallic antenna counterpart, $Z_{sl} \cdot Z_m = Z_0^2/4$. For a dipole antenna $Z_m = 72$ ohms, so that $Z_{sl} = Z_0^2/4Z_m = (377)^2/4 \times 72 = 494$ ohms.

8. Since $L_{\text{bomber}}/L_{\text{fighter}} = 3\,\text{MHz}/2.5\,\text{MHz}$, $L_{\text{bomber}} = (3/2.5)60 = 72$ feet.

9. If the iron ore mountain is sufficiently rich in iron so that its electromagnetic attenuation (equivalent skin depth is less) is greater than the aluminum building, then the mountain is a better EMP shield.

10. The reinforced concrete freeway electromagnetically represents an embedded array of metal reinforcing bars (rebar). The array aperture dimensions are probably small with respect to the wavelength of the AM station (~300 meters), but large with respect to the FM wavelength (~3 meters). Hence, the FM signal is little attenuated by the rebar array, while the opposite is true for the AM signal.

11. $E = kT = h\nu$; $\nu = kT/h = (1.38 \cdot 10^{-16})3/6.6 \cdot 10^{-27}$ yields (a) 62.7 GHz, and (b) $\lambda = 3 \cdot 10^{10}/62.7 \cdot 10^9 = 0.48$ cm.

CHAPTER 10 ANSWERS

1. The dose in silicon is:

$D_{si} = (1.602 \cdot 10^{-6}$ ergs/MeV)N_{si} (Si atoms/cm^3)$\sigma_{si}(E)$(absorption cross section/Si atom)$E\phi_\gamma(E)/\rho_{si}$ where $\phi_\gamma(E)$ is the incident photon fluence at energy E, and $\rho_{si} = 2.33$ gm/cm^3. For air, the dose is: $D_{air} = 1.602 \cdot 10^{-6}(N_{air}\sigma_{air}(E)/\rho_{air}(E))\, E\phi_\gamma(E)$. Therefore, $D_{si}/D_{air} = (\mu_{si}/\rho_{si})/(\mu_{air}/\rho_{air})$, where $\mu = N\sigma$.

2. $D(E) = AE(1 - \exp - B/E)$, $\bar{D}(E) = \int_0^{E_m} D(E)dE/E_m$, where $E_m = 2.25$ MeV, So $\bar{D}(E) = (A/E_m)\int_0^{E_m}(1 - e^{-B/E})EdE$. Furthermore, $\bar{D}(E) = (A/E_m)(E_m^2/2 - \int_0^{E_m}(\exp - B/E)EdE = AE_m(\frac{1}{2} - E_3(B/E_m))$, where the Milne integral $E_n(x) = \int_1^\infty (\exp - xy)dy/y^n$. So that $\bar{D}(E) = AE_m(\frac{1}{2} - E_3(1.32/2.25)) = 133.3 \times 2.25(\frac{1}{2} - 0.19) = 93\,\text{MeV} \cdot mb$.

3.(a). 1 rad is the absorption of 100 ergs/gm (CGS unit of dose), while 1 gray (Gy) is the absorption of 1 joule/kg (MKS unit of dose), so that 1 gray (Gy) $= 10^7$ ergs/10^3 gm $= 10^4$ ergs/gm $= 100$ rads. A recently introduced SI unit is 1 centigray (cGy) $\equiv 0.01$ Gy $= 1$ rad (Sec. 10.2).
3.(b). The "exposure" of 1 roentgen results in a dose of 85 ergs (Section 10.2) in STP air. So that this particular exposure level is equivalent to the energy depostion of 0.85 rads (air).

4. If the absorbed dose and the kerma differ widely, this means that not all of the energy released in a material is absorbed as dose. TLDs measure only the dose absorbed, which could be only a fraction of the kerma, and hence will give a false reading of energy released. This is especially the case where particle equilibrium does not exist.

5. $\phi_{eq}(1\,\text{MeV}) = (\phi_{eq}(1\,\text{MeV})/\Phi)/(N_s/\Phi) \cdot N_s$; $\Phi = \int_0^\infty \phi(E)dE$.

6. Fluence, ϕ, is defined as the number of particles threading a 1 cm^2 area summed over all directions. Each such threading particle represents

a path or track of varying length depending on its speed. $\sigma\phi$ is the total number of interactions summed over all such tracks, or $N\sigma\phi$ is the number of interactions per cm^3, where σ (cm^2) is the per target interaction probability and N is the number of target atoms per cm^3. The number of interactions per cm^3 is still given by $N\sigma\phi$, if ϕ is construed as the total track length per cm^2.

7. Rewriting Eq. (10.29) for a unit burst of neutron fluence gives $\dot{N} = -N\sigma_{ip}\phi_0(E)\delta(t)$; $N(0) = N_0$. Integrating yields $N = N_0 \exp - \sigma_{ip}\phi_0(E)$, and inserting N into Eq. (10.28) gives

$$\dot{N}_i + (\lambda_i + \sigma_{ir}\phi_0(E)\delta(t))N_i = N_0\sigma_{ip}\phi_0(E)\delta(t) \exp - \sigma_{ip}\phi_0(E).$$

The integral of this equation is $N_i = N_0\sigma_{ip}\phi_0(E) \exp - (\sigma_{ip}\phi_0(E) + \lambda_i t) \approx N_0\sigma_{ip}\phi_0(E)$, since $\lambda_i t \ll \sigma_{ip}\phi_0 \ll 1$ is the case for the required foil exposure time. This is to be compared with $A_i = \lambda_i N_i(\infty) = N_0\sigma_{ip}\phi_0(E)$ from Eq. (10.33).

8. To first order, and for neutrons in the vicinity of 1 MeV in energy, the atomic processes that cause displacements also cause ionization damage, due to the ionization created by the primary recoil silicon ions, as discussed in Section (5.7). Since the displacement cross section $\sigma_n(E)$ then roughly mirrors both processes in this energy range, these curves should look similar to each other.

9. The definition of kerma implies, in this case, the release of gamma or x-ray photon energy into the material by means of their interaction with the material electrons. The transport, scattering, and subsequent redistribution of these electrons (secondary particles) results in dose enhancement. Hence, the primary photon radiation kerma plays no direct part in dose enhancement. The variation of kerma with penetration distance is thus monotonic at most, with no enhancement hump at interfaces.

10. Incident neutrons, besides causing displacement damage, can activate the material nuclei into which they penetrate, principally through (n, γ) reactions. These now radioactive nuclei decay with characteristic half-lives, and yield gamma rays in the process. Enough time following neutron irradiation must elapse so that when electrically testing components after irradiation, their gamma-ray emission levels are not harmful. Gamma ray testing does not generally activate material nuclei of interest, such as silicon, so that the corresponding components are not made radioactive. Gamma rays damage these materials principally by ionization, as discussed in Sections 6.2–6.5.

GLOSSARY

A	mass number
	energy constant
	cable cross section area
	antenna loop area
	junction area
$\bar{A}$	vector potential
A_1, A_2, A_3	empirical constants
A_J	emitter-base junction area
ACT	Actel Corp.
A_c	collector area
	collector deep-storage transport factor
ADI	Analog Devices Inc.
A_e	emitter area
A_E	emitter deep-storage transport factor
A_i	saturation activity
A_{cell}	cell area
APC	armored personnel carrier
A_s	recombination surface area
	effective surface recombination area
ASL, ALS	advanced schottky logic
ASC	Allied Signal Corp.
A_{DV}, A_{OV}, A_{IV}	resistivity coefficients
ALT	Alpha Industries
A_{stor}	sensitive region of cell area
A_s'	base-emitter depletion region surface area
AMD	Advanced Micro Devices Corp.
$ASIC$	application specific integrated circuit
ATM	Atmel Corp.
A_z	z component of vector potential
$\bar{A}_p$	average projected area
a	impact parameter
	radius of dipole
	radiation constant
	plate radius
	antenna element radius
	coaxial cable shield wire radius
	antenna loop half-thickness
	energy constant
$\bar{a}$	mean acceleration

$\bar{a}_1, \bar{a}_2, \bar{a}_3$	basis vectors
a_{ij}	Ebers-Moll coefficient
$\bar{a}_{ij}(x, s)$	Laplace transform with respect to time of a_{ij}
a^2/V	gas condensation pressure
B	empirical constant
	structure width
$\bar{B}$	magnetic induction
B_n	normal component of magnetic vector
B_0	static magnetic field
BOX	buried oxide
BPTEOS	boron phosphorous TEOS
B_s	saturation magnetic flux density
BV_{CEO}	breakdown collector-emitter voltage (open base)
BV_{CBO}	breakdown collector-base voltage (open emitter)
b	spherical shield radius
	volume of gas atoms
b_1, b_2, b_3	basis vectors of reciprocal lattice
bar(s)	synonym for die (dice)
C	allocated cost
	conductance
	capacitance
C_a	antenna capacitance
	capacitance of short monopole
C_c	total system cost
CCD	charge coupled device
CDI	collector diffused isolation
C_{TC}	collector capacitance associated with Ebers-Moll bipolar equivalent circuit
C_i	cost of *i*th chip
C_{TE}	emitter capacitance associated with Ebers-Moll bipolar equivalent circuit
C-JFET	complementary J-FET (GaAs)
CML	current mode logic
CMOS	complementary MOS
CMOS bulk	garden variety CMOS
CMOS epi-on bulk	CMOS on epi-layer on bulk substrate
CMOS/SOI	CMOS on insulator (substrate)
CMOS/SOS	CMOS on sapphire (Al_2O_3)
CPU	(computer) central processing unit
CTF	circuit tolerance factor
CVD	chemical vapor deposition
CRY	Crystal Semiconductor Corp.
C_{ox}	capacitance of unit area of oxide
CYP	Cypress Corp.

C_{nk}	electron capture probability of *k*th energy level
$C(E)$	empirical function of energy
C_{cb1}	collector bias capacitance
C_n	electron recombination capture probability
c	speed of light
	dopant concentration
c_c	collector capacitance
	collector depletion capacitance
c_e	emitter capacitance
c_p	specific heat at constant pressure
	hole recombination capture probability
c_{11}	elastic constant
c_d	drag coefficient
c_0	sea level sound speed
c_g	channel capacitance
c_D	diode capacitance
c_{pk}	hole capture probability of the *k*th energy level
c_1	coaxial cable capacitance per unit length
c_2	coaxial cable capacitance to ground
c_v	specific heat at constant volume
$3D$	triple diffused
$D(t, \tau)$	defect density
D^*	ambipolar diffusion coefficient
$\bar{D}$	electric displacement
D_B	diffusion coefficient of minority carriers in base
D_E	diffusion coefficient of minority carriers in emitter
D_C	diffusion coefficient of minority carriers in collector
D_f	ambient dose level
D_1	annealed (permanent) defect density
D_n	electron diffusion coefficient
D_p	hole diffusion coefficient
D_i^*	intrinsic ambipolar diffusion coefficient
D_{it}	number density of interface traps/cm$^2 \cdot$ eV
DDC	Data Device Corp
$D(E)$	neutron displacement kerma per unit fluence
$\bar{D}(1\,\text{MeV})$	1 MeV mean neutron displacement kerma per unit fluence
D_{cav}	energy absorbed per gram of gas
D_{wall}	energy absorbed per gram of cavity wall
D_0	optical density
	initial magnitude of diffusion coefficient
D_{ot}	number density of oxide traps/cm$^3 \cdot$ eV
D_M	design margin
D_b	base diffusion constant
D_{fi}	failure dose of *i*th transistor
D_{mf}	ionizing dose mean failure level

DML	ionizing dose design margin for lot test data
$DMOS$	power FET structural variant
D_γ	ionizing dose
D_{spec}	specification level of ionizing dose
$DIIC$	dielectrically isolated integrated circuit
DI	dielectrically isolated
$DITTL$	dielectrically isolated TTL
$DLTS$	deep level transient spectroscopy
DTL	diode transistor logic
d	coaxial cable burial depth
	diameter of wire strands in plait
$dP_f(\phi)$	incremental log-normal failure probability
dT/dx	stopping power (incremental energy loss per unit length)
	derivative of temperature with respect to distance
dE/dx	stopping power (incremental energy loss per unit length)
$dE/\rho dx$	stopping power (incremental energy loss per unit areal density)
dN/dE	differential neutron flux
die (dice)	synonyms for chip (chips)
$d\bar{l}$	infinitesimal directed length vector
$d^3\bar{r}$	differential volume integration element
$d^2\bar{r}$	differential area integration element
dQ/dx	charge stopping power
E	energy
	electric field magnitude
	energy density
	x-ray energy
	energy of carrier incident on junction potential barrier
	radiation-induced ionization energy
	burnout failure energy
	electric field intensity magnitude
ECL	emitter coupled logic
$E(e)$	elliptic integral of the second kind
E_0	geometric mean static electric field
	electric field strength at the junction
	maximum internally generated electric field
E_c	energy of conduction band bottom
E_f	Fermi energy level
$E_L(\omega)$	Fourier transform of $E_L(t)$
E_r	radial electric field
E_m	maximum electric field strength
E^T	tangential electric field
E_s	short EMP electric field pulse
	subsurface electric field
E_L	long EMP electric field pulse

E_d	head-on collision energy transferred to recoil atom
E_n	incident neutron energy
E_R	recombination energy level
E_i	internal electric field
	intrinsic Fermi energy level
E-JFET	enhancement mode JFET (GaAs)
E_r	single event, (soft error), upset bit rate
E_v	energy of valence band top
E_g	energy of forbidden gap width
E_t	trap energy level
	recombination center energy level
$\bar{E}_a$	applied electric field intensity
$\bar{E}_i$	internally generated electric field
$E(\lambda)$	emitted energy spectral density
$\bar{E}_i$	incident electric field
$\bar{E}_t$	transmitted component of electric field
$\bar{E}_r$	reflected component of electric field
E_∞	asymptotic electric field intensity
$\bar{E}_i(\omega)$	incident field transform
$\bar{E}_W(\omega)$	electric field within shield
E_{ci}	energy of conduction band bottom for intrinsic material
E_{vi}	energy of valence band top for intrinsic material
E_{cn}	*n*-side of junction bottom of conduction band
E_{cp}	*p*-side of junction bottom of conduction band
E_{vn}	*n*-side of junction top of valence band
E_{vp}	*p*-side of junction top of valence band
E_{fn}	equilibrium Fermi energy on *n*-side of junction
E_{fp}	equilibrium Fermi energy on *p*-side of junction
E_{th}	threshold energy
E_{ox}	electric field in gate oxide
ECL	emitter coupled logic
EPR	electron paramagnetic resonance
E_{fL}	failure energy for 1-μs pulse
E_{Ti}	threshold interaction energy of the *i*th species
E_{zv}	vertically polarized electric field component
E_{iv}	vertically polarized incident electric field
E_{zvm}	maximum of E_{zv}
ESD	electrostatic discharge
ESR	electron spin resonance
ETI	early transient incapacitation
$E_{i\max}$	maximum emergy received by recoil atom
$E_{\max}$	maximum available incident neutron energy
E_{DV}	donor vacancy trap energy level
E_{OV}	oxygen vacancy trap energy level
E_{VV}	divacancy trap energy level
e^-	electron symbol

e^+	positron symbol
e	charge of electron
	eccentricity
$erf(x)$	error function
$erfc(x)$	complementary function
F	noise figure
$F(x)$	impurity grading function
$F(t)$	annealing factor
F_x	x component of force
$F(x, \alpha, \beta)$	beta function
$FAMOS$	floating gate avalanche-injection MOS memory
F_n	cumulative normal distribution
F_n^{-1}	inverse of F_n
$FAST(TTL)$	Fairchild advanced schottky TTL
$FIPOS$	fully isolated porous silicon
FSC	Fairchild Semiconductor Corp.
FUJ	Fujitsu Ltd.
f	gamma ray fission factor
	frequency
	alpha gain cutoff frequency
$f(E)$	Fermi distribution
f_j	fraction of incident energy consumed in atomic processes
f_T	unity gain transistor gain-bandwidth frequency
$f(s)$	path length distribution
G	ionization pulse amplitude
$GaAs\ FET$	gallium arsenide FET
g	carrier generation constant
	coaxial cable dielectric correction factor
	generation rate of hole-electron pairs
	acceleration of gravity
$g_s(t)$	shunt conductance
g_n	electron generation rate
g_n'	excess electron carrier generation rate
g_m	transconductance
g_{m0}	transconductance initial value
g_{0n}	thermal equilibrium electron generation rate
g_{0p}	thermal equilibrium hole generation rate
g_p	hole generation rate
g_p'	excess hole carrier generation rate
H	height of structure
$\bar{H}$	magnetic field intensity vector
$H(t)$	Heaviside step function
H_0	nominal value of magnetic field

$H(t)$	EMP magnetic field pulse
$\bar{H}_i(\omega)$	incident magnetic field Fourier transform
$H_W(\omega)$	magnetic field within cable shield
H_r	radial component of magnetic field
H_θ	polar component of magnetic field
H_ϕ	azimuthal component of magnetic field
H_∞	asymptotic magnetic field intensity
H_{zvm}	maximum incident EMP magnetic field
^2H	deuteron (hydrogen isotope)
^3H	triton (hydrogen isotope)
H_e^{++}	alpha particle (helium ion)
H^T	tangential magnetic field
H^i	incident magnetic field
HA, *HAR*	Harris Corp.
HBT	heterojunction bipolar transistor
HCMOS	CMOS structural variant
HIT	Hitachi Ltd.
HMOS	MOS variant
HEMT	high electron mobility transistor
HEXFET	hexagonally configured power FET
HON, *HWL*	Honeywell Inc.
h	Planck's constant
	antenna element height
	junction height
	plate thickness
	pillbox height
$\hbar$	$h/2\pi$
h_{FE}	steady-state (dc) common emitter gain
hep	hole-electron pair
h_{fe}	dynamic common emitter gain
$h_{FE}(t)$	time-varying common emitter current gain
h_c	antenna effective height
I^2L	current injection logic
I^3L	isoplanar current injection logic
IMOX	implanted oxide
INM	Inmos corp. (UK)
INT	Intel Corp.
ISO-Z	isoplanar *Z* process
ISL	integrated schottky logic
I	impulse
	radiation intensity
	transmitted light intensity
	current in center conductor of coaxial cable
	electric current
$I(z, t)$	coaxial cable induced current

I'	coaxial cable short circuit current per unit length
I_D	drain current
I_{DC}	base-collector junction breakdown current
I_{DE}	base-emitter junction breakdown current
I_{DD}	drain current
IDT	Integrated Device Technologies Inc.
I_E	emitter current
I_{damage}	device damage current
I_0	coaxial cable ground return current
	incident light intensity
	external cable shield current
	current on outside of coaxial cable shield
	steady-state current
	diode reverse current
	saturation current density
I_B	base current
I_1	coaxial cable input current
	current flow in emitter resistance
$I_{\max}$	peak circulation current
I_{leak}	leakage current
I_c	collector current
I_p	hole current
I_E	emitter current
I_r	radial current
I_s	diode reverse saturation current
	switching point current
I_h	holding current
I_g	gate current
$I_{w\text{dam}}$	EMP-induced damage current
$I_{w\text{spec}}$	EMP-specification current
I_{upset}	upset damage current
I_2	coaxial cable output current
I_{sc}	dipole short circuit current
ΔI_B	base current increase
I_{si}	saturation current of *i*th junction
I_{B1}	base current during pulse
I_{B2}	base current following pulse
I_L	leakage current
I_{sE}	base-emitter reverse saturation current
I_{pB}	hole current in transistor base
I_{nE}	electron current into base from emitter
I_{sR}	surface recombination hole current
I_{Ep}	emitter hole current
I_s	source current generator
I_{IB}	input bias current
$\overline{I}_E(x, s)$	Laplace transform with respect to time of emitter current

$\bar{I}_C(x,s)$	Laplace transform with respect to time of collector current
I_{sat}	saturation current
IBM	International Business Machines Co.
I_{DSS}	saturated drain-source current
I_{CBO}	collector-base (leakage) current, emitter open
I_{IBf}	failure input bias current
I_{ppc}	primary collector photocurrent
I_{ppE}	primary emitter photocurrent
I^2L	Integrated injectioin logic
ITI	immediate transient incapacitation
I_{cs}	collector saturation current
I_{pp}	primary collector photocurrent
i	$\sqrt{-1}$
i_d	delayed photocurrent
i_L	loop current
i_R	equivalent recombination current
i_B	base current
i_5	emitter diffusion current
i_p	coaxial cable peak induced current
	prompt photocurrent
i_2	collector junction depletion current
i_1	collector diffusion current
i_3	emitter junction depletion region current
i_4	carrier generation base current
i_{JI}	parasitic element current
i_{VR}	volume recombination current
i_{SR}	surface recombination current
i_{ppD}	diode primary photocurrent
$\langle i,j,k\rangle$	Miller indices
J	Joule energy
	number of ion pairs produced per gram of gas
	current density
$\bar{J}$	total electric current density
J_c	channel current density
J_E	emitter current density
J_{EO}	emitter saturation current density
J_L	shorted loop energy
J_R	energy delivered to load R
J_r	radial current density
$J_n^{(0)}, J_n^{(1)}$	zero- and first-order electron current density
J_n	electron current density
J_p	hole current density
J_s	surface current density
J_t	tunnel current density
J_{cm}	compton electron current density

J_s	reverse current density
J_{ng}	generated electron current density
J_{nr}	electron recombination current density
J_{pr}	hole recombination current density
J_{pg}	generated hole current density
J_{st}	energy stored in antenna capacitance
JFET	junction FET
JIIC	junction isolated integrated circuit
JI	junction isolated integrated circuit
JITTL	junction isolated TTL
$\bar{j}_n$	electron particle current; current density
$\bar{j}_p$	hole particle current; current density
j_{pB}	hole current density in transistor base
K	phonon wave number
	neutron damage constant
	coaxial cable braid coverage factor
	damage constant
	thermal conductivity
	thermal failure damage constant
	normalized noise power spectral density
K_1	empirical constant
K_0	empirical constant
K_L	neutron damage constant
$K_0(x)$	modified Bessel function
K_D	thermal damage constant
K'	inverse damage constant
$\bar{K}_D$	average damage constant
$K(E)$	kerma per unit fluence
$K_D(T)$	temperature-dependent damage constant
K_τ	minority carrier lifetime damage constant
K_{TL}	one-sided measure of sample standard deviation
K_{Di}	damage constant of *i*th transistor
K_{Df}	damage constant corresponding to failure limit
K_∞	large emitter-current neutron damage constant
$\bar{K}_j$	reciprocal lattice vector
K_{ln}	neutron damage constant at low injection for *n*-type material
K_{lp}	neutron damage constant at low injection for *p*-type material
K_n, K_p	neutron damage constant for *n* and *p* material, respectively
K_{ln}^*	neutron damage constant at low injection for *n*-type material low impurity limit
K_{lp}^*	neutron damage constant at low injectioin for *p*-type material low impurity limit
K_V	vacancy coefficient
K_{DV}	donor vacancy coefficient
K_{OV}	oxygen vacancy coefficient

K_{VV}	divacancy coefficient
K_{IV}	interstitial coefficient
$K_n(E)$	neutron damage constant as a function of energy
K_p, K_{di}	empirical constants
$K(e)$	complete elliptic integral of first kind
kT	kiloton (weapon yield)
k	propagation number
	wave number
	Boltzmann constant
$\bar{k}$	wave vector
k_n	electron wave vector
k_p	hole wave vector
k_n, k_p	empirical constants
k_1	electron attachment rate
k_i	proportionality constant
$LOCOS$	local oxidation of silicon
$LPCVD$	low-pressure chemical vapor deposition
LSI	large scale integrated circuits
	LSI Logic Corp.
$LSTTL$	low power schottky TTL
LDD	lightly doped drain
L	channel length
	diffusion length
	scalar angular momentum
	linear energy transfer (stopping power)
	coaxial cable length
	loop inductance
	resistor length
$\bar{L}$	angular momentum vector
L_c	diffusion length of minority carriers in collector
	critical linear energy transfer
L_g	coaxial cable inductance per unit length
L_b	diffusion length of minority carriers in base
L_e	diffusion length of minority carriers in emitter
L_n	diffusion length for minority electron carriers
L_p	diffusion length for minority hole carriers
L_s	equivalent inductance of coaxial cable shield
L_{p0}	preirradiation diffusion length
L_{Di}	intrinsic Debye length
L_{th}	threshold linear energy transfer
LET	linear energy transfer
$L_\perp$	normal linear energy transfer
l	length
	wall thickness
l_e	effective length of dipole

l_+	width of depletion layer on *n*-said of junction
l_-	width of depletion layer on *p*-side of junction
MBE	molecular beam epitaxy
MESFET	metal oxide gate FET
MNOS	metal nitride oxide gate silicon FET
MNOS/SOS	MNOS on sapphire (Al_2O_3)
MSI	medium scale integration
M	gas multiplication factor
	number of conduction band minima
	bomb plasma mass
	circuit design margin
M_D	diode constant
M_n	electron multiplication
M_p	hole multiplication
M_s	figure of merit for thick walls
M_h	avalanche multiplication factor for holes
M_e	avalanche multiplication factor for electrons
MED	Marconi Electronic Devices Corp.
M_l	lofted mass
MIC	Micron Technologies Inc.
M_{AC}	base-collector breakdown current multipliers
MIT	Mitsubishi Corp.
M_{AE}	base-emitter breakdown current multipliers
MMI	Monolithic Memories Inc.
MMIC	monolithic microwave integrated circuit
MT	megaton (weapon yield)
MODFET	modulation doped FET
MOT	Motorola Corp.
M_{12}	coaxial cable mutual coupling coefficient
MPS	Micro Power Systems Inc.
MTA	Matra Harris Semiconductor Inc.
m	log-normal estimate of mean value
	mass of electron
m_a	plate areal density
m_o	rest mass
m_γ	standard mean value for $100\gamma\%$ confidence interval
m_s	figure of merit for thin walls
m^*	effective mass
m^*_{de}	effective mass for electron density of states
$m^*_\parallel$	effective mass component
$m^*_\perp$	effective mass component
m^*_{dh}	effective mass for hole density of states
m^*_{ce}	electron conductivity effective mass
m^*_{ch}	hole conductivity effective mass
m^*_e	electron effective mass

m_h^*	hole effective mass
m_+	slope of dopant grade on n-side of junction
m_-	slope of dopant grade on p-side of junction
N	number density of impurity ions
	number of samples
	number density of lattice atoms
	number density of silicon atoms
	number of target atoms per cm^3
	number of coaxial cable braid strands per plait
	number of electron-hole pairs
N_A, N_a	acceptor dopant atom density
N_{AO}, N_{AW}	acceptor dopant concentration at base extremities
N_B	dopant concentration in base
N_C	dopant concentration in collector
N_d	number density of displaced atoms
N_D	donor concentration
N_c	number density of recombination centers
	effective density of states in conduction band
NEC	Nippon Electric Corp.
N_E	dopant concentration in emitter
N_i	concentration of radioactive nuclei of the ith species
$NIEL$	non-ionizing energy loss
NSC	National Semiconductor Corp.
N_V	vacancy concentration
N_v	effective density of states in the valence band
N_0	oxygen atom impurity concentration
	areal density of holes
	Avogadro's number
N_{VV}	divacancy concentration
N_I	interstitial concentration
$N_e(E)$	number density of available electron states
$N_h(E)$	number density of available hole states
N_m	number density of target particles
$N(E)$	alpha particle flux energy spectrum
N_s	surface recombination density
N_t, N_T	number density of traps
n_i	intrinsic carrier density
N_{it}	number density of interface-trapped charges/cm^2
N_{ot}	number density of oxide-trapped charges/cm^3
$N_s(E_i)$	number density of displaced atoms of energy E_i
N_i	number of moles of ith species
N_s	ionized dopant concentration
N_{BC}	net base impurity concentration immediate to collector
N_{BE}	net base impurity concentration immediate to emitter
N_{st}	surface areal trap density

N_D, N_d	donor dopant atom density
N_{DV}	donor-vacancy complex concentration
N_{Di}	initial donor concentration
N_{Oi}	initial oxygen impurity
N_{OV}	oxygen vacancy complex concentration
N_{IV}	interstitial-vacancy concentration
N_{ss}	surface state density
$NMOS$	n channel MOS
n	steady-state majority carrier concentration in n-type material
	fanout number
	integer
	electron density
	empirically determined exponent
	number of sides of many sided volume
	number of modules or circuits
	number of trials
$n(E)$	number density of incident neutrons of energy E
$\bar{n}$	unit vector
n_1	free electron concentration
	initial excess electron density
n_i	intrinsic carrier density
n^-	lightly doped n-type region
n^+	heavily doped n-type region
n^{++}	very heavily doped n-type region
n_∞	asymptotic electron density
n_k	free electron concentration of the kth energy level
$\bar{n}(E)$	energy lost to atomic processes
n_{pb}	base minority carrier density
n_s	surface electron concentration
n	electron density
$n'(\phi)$	carrier removal rate
n_{BE}	base-emitter empirical constant
n_n	electron concentration in n-type region
n_p	electron concentration in p-type region
n_{p0}	electron concentration at immediate p-side of junction
n_{n0}	electron concentration at immediate n-side of junction
n_0	equilibrium density of electrons
	number density of incident particles
n_{0i}	pre-radiation majority carrier density
OWI	Omni-Wave Inc.
P	linear momentum
p	hole concentration or density
	average power

	incident thermal power
	pressure
	thermal pulse power density
ΔP	overpressure
PFS	Performance Semiconductor Corp.
P_{β}	survival probability corresponding to given gain β
P_k	cumulative probability distribution
P_r	population fraction
$P_{si, 0.9}$	survival probability for 90 percent confidence interval
P_{si}	survival probability for the ith transistor
P_t	probability of carrier tunnelling through potential barrier
P_{∞}	asymptotic hole concentration
P_0	reference pressure
P_s	system survival probability
$P_n(x \leq k)$	binomial probability of k successes at most
P	probability of success in binomial trials
PIN	p-intrinsic-n diode
PLA	programmable logic array
$PMOS$	p channel MOS
PZT	lead zirconate titanate ($PbZr_xTi_{1-x}O_3$)
p	dimensionless pressure variable
p	hole concentration near surface
p_0	equilibrium hole concentration
$\bar{p}$	electric dipole moment
p_1	initial excess hole concentration
p_s	stagnation pressure
p_k	peak overpressure
p_e	hole concentration at emitter junction
p_s	surface hole concentration
p_i	reliability of ith chip
p^+	heavily doped p-type region
p^{++}	very heavily doped p-type region
$p(0)$	initial hole concentration
p_p	hole concentration in p-type region
p_n	hole concentration in n-type region
$\hat{p}$	maximum likelihood estimate of probability of success
p_{av}	average pressure
p_{p0}	hole concentration at the immediate p-side of the junction
p_{n0}	hole concentration at the immediate n-side of the junction
Q	thermal energy density
	resonant circuit figure of merit
	electric charge
$Q(t)$	excess base minority carrier charge

Q_A	positive surface charge density
Q_D	positive donor space charge density
Q_G	gate charge
Q_c	channel charge corresponding to collector current
Q_c	critical charge
Q_{ss}	gate insulator space charge density
Q_{gen}	charge generated per electron-hole pair
QML	qualified manufacturers list
q	charge of electron
	dynamic pressure
	binomial failure probability
q_e	charge per unit length
$\bar{q}_B$	mean surface charge on base area
R	depth of EMP penetration into ground
	antenna resistance to ground
	loop load resistance
	radius of circular aperture
	equivalent load resistance
	recombination rate
	Hall effect coefficient
	bomb radius
	universal gas constant
	distance from detonation point
	slant range
R_a	radius of cylindrical collector
R_B	base resistance
R_C	collector resistance
	computer reliability
RCA	Radio Corporation of America
R_D	coaxial cable dc resistance per unit length
	dynamic drain resistance
R_e	emitter resistance
R_c	a constant
R_i	base resistance connected to positive supply voltage
R_1	reliability (survival probability)
R_n	recombination rate for electrons
R_p	recombination rate for holes
R_k	rate of introduction of recombination centers for the kth energy level
R_s	dynamic source resistance
R_r	reflection coefficient
R_s, R_D	equivalent resistances
R_s	resistance of shield
$R_s(t)$	shunt resistance as a function of time
R_r	radiation resistance

R_L	load resistance
R_{ent}	reliability of entire system
R_{each}	reliability of each subsystem
R_{eft}	total radiation effect
$R_{eff}(E)$	radiation effect per unit energy
$R(E_0)$	range of particle of energy E_0
R_{DL}	diode leakage resistance
R_{gd}	drain surface leakage resistance
R_{gs}	source surface leakage resistance
R_{CL}	base-collector junction leakage resistance
R_{EL}	base-emitter junction leakage resistance
R_s	drain or source resistance
RIE	reactive ion etch
ROX	recessed field oxide
RTL	resistor-transistor logic
RCTL	resistor capacitor-transistor logic
R_s	fireball radius
RWL	Rockwell Int'l Corp.
r	scalar radius
	radius of Bohr orbit
	dimensionless radius variable
	roentgen
	rad
$\bar{r}$	radius vector
	ray vector
r_B	base resistance
r'_B	base spreading resistance
r_c	collector resistance
r_e	emitter resistance
r_d	dynamic output impedance
$\bar{r}_i$	ith ray vector
S	sensitivity factor
	density-specific heat product
	surface recombination velocity
	shield effectiveness
SAN, SNL	Sandia Corp.
SEI	Seiko Ltd.
$\bar{S}$	Poynting vector
SEL	single event latchup
SEQ	SEEQ Inc.
SIMS	secondary mass spectroscopy
SIMOX	separation by ion implantation of oxygen
SOA	safe operating area
SOI	silicon on insulator (SiO_2)
SOS	silicon on sapphire (Si on Al_2O_3)

SIPOS	semi-insulating polycrystalline oxygen doped silicon
SIG	Signetics Corp.
SNY	Sony Inc.
$S(\omega)$	noise power
SOR	Sorep Inc.
S_{cav}	mean stopping power of cavity
S_{wall}	mean stopping power of cell wall
$\overline{S}_E(\omega)$	shielding effectiveness from electric field
$S_H(\omega)$	shielding effectiveness from magnetic field
S_V	noise/power spectral density
$S_{p\max}$	maximum surface recombination velocity
SNOS	polysilicon nitride oxide silicon
SOI	silicon on insulator
SSI	small scale integration
s	Laplace transform variable
s_l	log-normal estimate of standard deviation
T	absolute temperature
	kinetic energy
T_0	initial temperature
	reference absolute temperature
TEL	Teledyne Inc.
T_{tr}	transistor base transit time
T_r	transmittance
T_D	time constant
TI, TIX	Texas International Corp.
T_r	transmission coefficient
ΔT	temperature difference
TDM	total (ionizing) dose design margin
TEOS	Ti-etched oxide silicide layer
TFT	thin film transistor
THO	Thomson Mil/Space Inc.
T_I, T_N	storage time constant
T_m	maintenance period
TOS	Toshiba Ltd.
TTL	transistor-transistor logic
TRW	Thompson Ramo Wooldridge Corp.
t_0	pulse width
t_{0x}	gate oxide thickness
t_s	electrical storage time
t_{sR}	radiation storage time
t_s	shock wave arrival time
t_{PROP}	propagation delay time
t_{crit}	critical propagation delay time
t_r, t_f	rise and fall time, respectively

U	single trap level recombination rate
U_0	potential barrier height
u_1, u_2	base transport factor expansion coefficients
$\bar{u}$	mean particle velocity
	shock wave speed
V	voltage
	volume
	scalar function
V'	coaxial cable open circuit voltage per unit length
V_1	cable input voltage
V_2	cable output voltage
V_0	applied junction voltage
	dc forward-bias voltage
V_s	switching point voltage
V_h	holding point voltage
V_{GS}	gate-source voltage
V_{DS}	drain-source voltage
V_{FB}	flatband voltage
V_{cb}	collector-base voltage
V_{mg}	midgap voltage
V_p	pinch-off voltage
ΔV_T	gate threshold voltage shift
V_{GT}	gate threshold voltage
V_T	gate threshold voltage
V_B	breakdown voltage
V_u	wind speed
V_D	diode voltage
V_A	applied voltage
V_H	Hall voltage
	barrier voltage
V_{0c}	open circuit voltage
V_{sc}	short circuit voltage
V_{CE}	collector-emitter voltage
V_{BE}	base-emitter voltage
V_{CC}	applied dc supply voltage to bipolar circuits
$\bar{V}_{BC}(x, s)$	Laplace transform with respect to time of base-collector voltage
$\bar{V}_{BE}(x, s)$	Laplace transform with respect to time of base-emitter voltage
V_{BC}	base-collector voltage
V_{sub}	substrate voltage
$V_{CE}(\text{sat})$, $V_{CE(\text{sat})}$	saturation value of collector-emitter voltage
V_{upset}	upset damage voltage
VHSIC	very high speed integrated circuits

VLSIC	very large scale integrated circuits
VMOS	power FET structural variant
VUV	vacuum ultraviolet
v	velocity
	incident carrier velocity
	neutron average speed
$\bar{v}$	vector velocity
	average velocity
v_g	group velocity
$v(k)$	velocity of individual packet waves
v_n	electron carrier velocity
v_p	hole carrier velocity
$\bar{v}_s$	mean phonon velocity
v_{th}	thermal velocity
var	variance
W	potential barrier thickness
	radiated power density
	Lambertian flux
	channel width
	resistor width
W_B	base width
W_E	emitter width
W_{EB}	width of emitter-base depletion region
W_m	modified black body radiated power
W_n	*n*-side junction depletion layer thickness
W_p	*p*-side junction depletion layer thickness
W_t	depletion region width
W_{sc}	thickness of collector junction depletion region
W_{se}	thickness of emitter junction depletion region
W_{damage}	thermal damage power
W_{spec}	specification power
w	depletion layer width
X	distance variable
$\bar{X}$	mean value of X
XIC	Xicor Inc.
XIL	Xilinx Corp.
X_L	loop inductive reactance
XPS	x-ray photoelectron spectroscopy
x	x-coordinate
x_{jc}	base-collector dopant interface
x_{je}	base-emitter dopant interface
x_0	oxide thickness
x_1	base-emitter junction coordinate
x_2	base-collector junction coordinate

Y	admittance
	bomb energy yield
	Young's modulus
Y_T	transfer admittance
Z	atomic number
	impedance
Z_0	coaxial cable characteristic impedance
	impedance of free space
Z_i	atomic number of *i*th specie of atom
ZIL	Zilog Inc.
Z_T	transfer impedance
Z_{dip}	input impedance of dipole antenna
Z_{slot}	input impedance of slot antenna
ZMR	zone melt recrystallization
ZOR	Zoran Inc.
α	alpha particle
	damped sine wave decrement
	attenuation coefficient
	thermal diffusivity
	common base current gain, h_{FB}
	fission chain reaction *e*-folding time
	plane wave decay constant
α_h	forward current transfer ratio for holes
α_e	forward current transfer ratio for electrons
α_f	inverse common base gain
α_I	current transfer coefficient
a_n	electron ionization rate
α_p	hole ionization rate
α_N	normal common base current gain
α_T	base transport factor
α_{ik}	transistor "alpha" for J_i as "emitting" and J_k as "collecting"
α_I	inverse common base gain
β	common emitter gain
β_f	failure gain level
β_i	inverse common emitter gain
β_D	Darlington transistor gain
β_N	common emitter gain midfrequency
β_0	low-frequency
	preirradiation
β_{min}	minimum value of gain
β_{ik}	transistor gain for J_i as "emitting" and J_k as "collecting"
$\beta_{i\text{post}}$	postirradiation common emitter gain
$\beta_{i\text{pre}}$	preirradiation common emitter gain

Γ	gamma-ray flux
γ	propagation constant
	fractional confidence level
γ_T	ionizing dose
$\dot{\gamma}_T$	ionizing-dose rate
$\dot{\gamma}_{\text{spec}}$	specification dose rate
$\dot{\gamma}(t)$	dose rate (gamma dot)
$\dot{\gamma}_{B0}$	device burnout dose rate
δ	deflection
	skin depth
	excess carrier concentration
δ'	excess bulk carrier generation rate
$\delta(t)$	unit impulse (delta) function
δ_n	excess electron carrier concentration
δ_p	excess hole carrier concentration
$(\delta n)_0$	initial electron excess carrier concentration
$(\delta p)_0$	initial hole excess carrier concentration
δ_{ij}	Kronecker delta
∇	gradient operator
$\nabla \cdot$	divergence operator
$\nabla \times$	curl operator
ε	dielectric constant
	oxide permittivity
	emissivity
$\overline{\varepsilon}_1, \overline{\varepsilon}_2$	unit vectors
ε_f	fermi energy level
ε_p	thermal hole production rate
ε_0	permittivity (dielectric constant of free space)
ε_{si}	dielectric constant of silicon
ε_{0x}	dielectric constant of gate oxide
$\varepsilon_{\text{SiO}_2}$	dielectric constant of silicon dioxide (gate)
ζ	transformed temperature variable
η	dimensionless velocity variable
	density ratio
	emitter efficiency
	dopant grading factor
η_{EE}	emitter proportionality constant
θ	critical angle
	mean bomb temperature
	kT/e

θ_B	Brewster angle
θ_i	angle of incidence
κ	thermal conductivity
	mass absorption coefficient (μ/ρ)
	thermal diffusivity
λ	mean free path
	wavelength
	failure rate per hour
λ_e	electron mean free path in air
λ_c	chip failure rate
λ_b	bit failure rate
λ_i	individual failure rate
	decay constant of the *i*th type of activated nucleus
λ_n	electron mean free path
λ_p	hole mean free path
λ_T	total mean free path
λ_γ	gamma-ray mean free path in air
λ_{comp}	computer failure rate
λ_{en}	energy absorption mean free path
μ	linear attenuation coefficient
	ground permeability
	spherical shield permeability
	statistical mean
μ^*	ambipolar mobility
$\bar{\mu}$	average mobility
	mean plasma atomic weight
μ	magnetic permeability
μ_n	electron mobility
μ_i	impurity scattering mobility
μ_i^*	intrinsic ambipolar mobility
μ_p	hole mobility
μ_T	total mobility
μ_0	magnetic permeability of free space
μ_s	sample mean
ν	frequency
	Poisson's ratio
$\nu(E)$	energy lost due to electrical processes
Ξ_c	deformation potential constant in conduction band
Ξ_v	deformation potential constant in valence band

ξ	dimensionless density variable
ξ	areal density
π	device parameter value
	hep escape probability
π_f	parameter failure value of the *i*th transistor
ρ	average plasma density
	charge density
	electrical resistivity
	ambient air density
	shock (blast) wave density
	density
ρ_B	electrical resistivity in base
ρ_i	preirradiation resistivity
$\rho_{n,p}$	resistivity due to electrons, holes
ρ_0	sea level air density
ρdx	incremental areal density
σ	electrical conductivity
	radiation constant
	statistical standard deviation
	cross section
σ_c	surface trap cross section
	compton cross section
σ_g	spherical shield ground conductivity
σ_e	emitter electrical conductivity
σ_m	cross section for minority carrier absorption
σ	cross section for production of recoil atoms
σ_n	electron capture cross section
σ_p	hole capture cross section
$\sigma(\theta)$	scattering cross section
$\sigma_i(\varepsilon)$	interaction cross section for *i*th nuclear species
σ_γ	standard deviation for $100\gamma\%$ confidence interval
σ_s	sample standard deviation
	surface charge density
$\langle\sigma_c\rangle$	fluence averaged cross-section for production of recombination atoms
$\sigma_v(E)$	cross section for electrical processes
$\sigma_b;\sigma_B$	electrical conductivity in base
σ_T	total charge density
σ_{ir}	removal cross section of activated nuclei
σ_{ck}	cross section for production of recoil atoms for *k*th energy level
τ	mean free time
	minority carrier lifetime
	decay constant

	minority carrier lifetime in base
$\overline{\tau^2}$	mean square average free time
τ_c	minority carrier lifetime in collector
τ_D	minority carrier lifetime in depletion region
τ_i	preirradiation minority carrier lifetime
τ_n	electron minority carrier lifetime
τ_p	hole minority carrier lifetime
τ_1, τ_2	lifetimes in two level recombination model
ϕ	scalar potential
	magnetic flux
ϕ_F	Fermi potential
ϕ_+, ϕ_n	potential on n-side of junction
ϕ_-, ϕ_f	potential on p-side of junction
ϕ	flux
$\phi(E)$	neutron fluence spectrum
$\mathbf{\Phi}_n$	neutron fluence
$\mathbf{\Phi}$	total fluence
ϕ_{eff}	effective fluence
$\varphi(E)$	spectral fluence
ϕ_0	contact potential
ϕ_s	surface potential
$\phi_{eq}(1\,\text{MeV})$	1-MeV equivalent neutron fluence
ϕ_{Fn}	Fermi potential on n-side of junction
ϕ_{Fp}	Fermi potential on p-side of junction
$\mathbf{\Phi}_\alpha$	alpha flux
$\mathbf{\Phi}_{th}$	threshold neutron fluence
$\mathbf{\Phi}_{WC}$	worst-case damage neutron fluence
$\mathbf{\Phi}_{\text{spec}}, \mathbf{\Phi}_{sp}$	specification neutron fluence
χ	electron affinity
ψ	intrinsic potential
	elevation angle
	total flux
ψ_F	Fermi potential
$\psi(V_0)$	$[\exp(eV_0/kT) - 1]$
ψ_{BE}	$[\exp(eV_{BE}/kT) - 1]$
ψ_s	surface potential
Ω	solid angle
	mho, conductance unit
	phonon angular frequency
$\mho$	fatness parameter
ω	angular frequency
ω_c	cutoff angular frequency
	upper corner transistor angular break frequency
ω_T	angular frequency corresponding to gain-bandwidth product

INDEX

INDEX